Felix Spindler
(personal copy)

Transition Metals for Organic Synthesis
Volume 1

Edited by
M. Beller and C. Bolm

Further Reading from Wiley-VCH:

R. Mahrwald (Ed.)

Modern Aldol Reactions
2 Vols.

2004, ISBN 3-527-30714-1

de Meijere, A., Diederich, F. (Eds.)

Metal-Catalyzed Cross-Coupling Reactions
2nd Ed., 2 Vols.

2004, ISBN 3-527-30518-1

Krause, N., Hashmi, A.S.K. (Eds.)

Modern Allene Chemistry
2 Vols.

2004, ISBN 3-527-30671-4

Cornils, B., Herrmann, W.A. (Eds.)

Aqueous-Phase Organometallic Catalysis
2nd Ed.

2004, ISBN 3-527-30712-5

Transition Metals for Organic Synthesis

Building Blocks and Fine Chemicals

Second Revised and Enlarged Edition

Volume 1

Edited by
M. Beller and C. Bolm

WILEY-VCH

WILEY-VCH Verlag GmbH & Co. KGaA

Edited by

Professor Dr. Matthias Beller
Leibniz-Institute for Organic Catalysis
University of Rostock
Buchbinderstraße 5–6
18055 Rostock
Germany

Professor Dr. Carsten Bolm
Department of Chemistry
RWTH Aachen
Professor-Pirlet-Straße 1
52056 Aachen
Germany

■ All books published by Wiley-VCH are carefully produced. Nevertheless, authors, editors, and publisher do not warrant the information contained in these books, including this book, to be free of errors. Readers are advised to keep in mind that statements, data, illustrations, procedural details or other items may inadvertently be inaccurate.

Library of Congress Card No.: applied for

British Library Cataloguing-in-Publication Data
A catalogue record for this book is available from the British Library.

**Bibliographic information published
by Die Deutsche Bibliothek**
Die Deutsche Bibliothek lists this publication in the Deutsche Nationalbibliografie; detailed bibliographic data is available in the Internet at <http://dnb.ddb.de>

© 2004 WILEY-VCH Verlag GmbH & Co. KGaA, Weinheim

All rights reserved (including those of translation in other languages). No part of this book may be reproduced in any form – by photoprinting, microfilm, or any other means – nor transmitted or translated into machine language without written permission from the publishers. Registered names, trademarks, etc. used in this book, even when not specifically marked as such, are not to be considered unprotected by law.

Printed in the Federal Republic of Germany
Printed on acid-free paper

Composition   K+V Fotosatz GmbH, Beerfelden
Printing   Strauss GmbH, Mörlenbach
Bookbinding   Litges & Dopf Buchbinderei GmbH, Heppenheim

ISBN   3-527-30613-7

Preface to the Second Edition

Is there really a need for a second edition of a two-volume book on the use of Transition Metals in Organic Synthesis after only 6 years? How will the community react? Are there going to be enough interested colleagues, who will appreciate the effort (and spend their valuable money in times of shortened budgets)? Do we, the editors, really want to invest into a project, which, for sure, will be most time-consuming? All of these questions were asked about three years ago, and together with Wiley/VCH we finally answered them positively. *Yes*, there has been enough progress in the field. *Yes*, the community will react positively, and *yes*, it is worth spending time and effort in this project, which once more will show und underline the strength of modern transition metal chemistry in organic synthesis.

The Nobel Prize in Chemistry 2001, which was awarded to K. Barry Sharpless, Ryoji Noyori (who both are authors in this book), and William S. Knowles for their contributions in asymmetric catalysis, nicely highlighted the area and demonstrated once more the high synthetic value of the use of transition metals for both small-scale laboratory experiments and large-scale industrial production.

During the past six years the field has matured and at the same time expanded into areas, which were rather unexplored before. Taking this development into account we decided to pursue the following concept: On the one hand the authors of the first edition were asked to up-date their original chapters, and most of them kindly responded positively. In a few cases the contributions of the first edition were reused and most often up-dated by an additional chapter written by another author. Some fields are now covered by other authors, which proved most interesting, since the same topic is now presented from a different perspective. New research areas have been summarized by younger active colleagues and leading experts.

It should be clearly stated that the use of transition metals in organic synthesis can not be fully covered even in a two-volume set. Instead, the present book presents a personal selection of the topics which we believe are the most interesting and actual ones. In general, the focus of the different contributions is on recent research developments since 1998. Literature up to mid – sometimes end – of 2003 has been taken into account. Hence, we believe the new book complements nicely the first more general edition of this book.

Transition Metals for Organic Synthesis, Vol. 1, 2nd Edition.
Edited by M. Beller and C. Bolm
Copyright © 2004 WILEY-VCH Verlag GmbH & Co. KGaA, Weinheim
ISBN: 3-527-30613-7

Most importantly, as editors we thank all contributors for their participation in this project and, in some case, for their patience, when it took longer than expected. We also acknowledge the continuous stimulus by Elke Maase from Wiley/VCH, who did not push but challenged. It remains our hope that the readers will enjoy reading the new edition and discover aspects, which will stimulate their own chemistry and create ideas for further discoveries in this most timely and exciting area of research and science.

Aachen, June 2004 *Carsten Bolm*
Rostock, June 2004 *Matthias Beller*

Contents

Preface V

1	**General** *1*	
1.1	**Basic Aspects of Organic Synthesis with Transition Metals** *3*	
	Barry M. Trost	
1.1.1	Chemoselectivity *4*	
1.1.2	Regioselectivity *6*	
1.1.3	Diastereoselectivity *8*	
1.1.4	Enantioselectivity *9*	
1.1.5	Atom Economy *11*	
1.1.6	Conclusion *12*	
1.1.7	References *13*	
1.2	**Concepts for the Use of Transition Metals in Industrial Fine Chemical Synthesis** *15*	
	Wilhelm Keim	
1.2.1	General Concepts *15*	
1.2.2	Use of Transition Metals in Fine Chemical Synthesis *17*	
1.2.3	Catalyst Preparation and Application *23*	
1.2.4	The Future *24*	
1.2.5	References *25*	
2	**Transition Metal-Catalyzed Reactions** *27*	
2.1	**Hydroformylation: Applications in the Synthesis of Pharmaceuticals and Fine Chemicals** *29*	
	Matthias Beller and Kamal Kumar	
2.1.1	Introduction *29*	
2.1.2	Hydroformylation: Applications for Pharmaceuticals and Natural Products *29*	
2.1.3	Hydroformylation: Synthesis of Agrochemicals and their Intermediates *41*	

Transition Metals for Organic Synthesis, Vol. 1, 2nd Edition.
Edited by M. Beller and C. Bolm
Copyright © 2004 WILEY-VCH Verlag GmbH & Co. KGaA, Weinheim
ISBN: 3-527-30613-7

2.1.4	Hydroformylation: Examples of Fine Chemical Synthesis	43
2.1.5	Conclusions and Outlook	50
2.1.6	References	51

2.2 New Synthetic Applications of Tandem Reactions under Hydroformylation Conditions 57
Peter Eilbracht and Axel M. Schmidt

2.2.1	Introduction	57
2.2.2	Hydroformylation and Isomerization	58
2.2.3	Hydroformylation and Reduction of the Oxo Aldehydes	59
2.2.4	Hydroformylation and Additional Carbon-Heteroatom Bond Formations	60
2.2.4.1	Hydroformylation in the Presence of Oxygen Nucleophiles	61
2.2.4.2	Hydroformylation in the Presence of Nitrogen Nucleophiles	64
2.2.4.2.1	Synthesis of O,N- and N,N-Acetals	65
2.2.4.2.2	Synthesis of Imines and Enamines	67
2.2.4.2.3	Hydroformylation in the Presence of Other N-Nucleophiles	71
2.2.4.2.4	Amines via Hydroformylation/Reductive Amination (Hydroaminomethylation)	71
2.2.4.3	Hydroformylation in the Presence of Other Heteroatom Nucleophiles	81
2.2.5	Concluding Remarks	82
2.2.6	References	82

2.3 Multiple Carbon-Carbon Bond Formations under Hydroformylation Conditions 87
Peter Eilbracht and Axel M. Schmidt

2.3.1	Introduction	87
2.3.2	Hydroformylation in the Presence of Carbon Nucleophiles	88
2.3.2.1	Hydroformylation in the Presence of Stable Wittig Reagents	89
2.3.2.2	Hydroformylation in the Presence of Allyl Silanes and Allyl Boranes	89
2.3.2.3	Hydroformylation in the Presence of Nucleophilic Hetarenes	92
2.3.3	Hydroformylation and Subsequent Mixed Aldol Reactions	93
2.3.4	Hydroformylation and Other C-C Bond-forming Reactions	99
2.3.4.1	Hydroformylation/Amidocarbonylation Sequences	100
2.3.4.2	Fischer Indole Synthesis with Oxo-Aldehydes	103
2.3.4.3	Hydroformylation and Carbonyl Ene Reactions	104
2.4.4	Hydrocarbonylation/Insertion Sequences Leading to Ketones	105
2.3.5	Concluding Remarks	109
2.3.6	References	109

2.4	**Hydrocarboxylation and Hydroesterification Reactions Catalyzed by Transition Metal Complexes** *113*	
	Bassam El Ali and Howard Alper	
2.4.1	Introduction *113*	
2.4.2	Intermolecular Hydrocarboxylation and Hydroesterification of Unsaturated Substrates *113*	
2.4.2.1	Hydrocarboxylation of Alkenes *113*	
2.4.2.2	Hydroesterification of Alkenes *117*	
2.4.2.3	Hydrocarboxylation and Hydroesterification of Allenes and Dienes *120*	
2.4.2.4	Hydrocarboxylation and Hydroesterification of Simple and Hydroxyalkynes *122*	
2.4.3	Intramolecular Cyclocarbonylation of Unsaturated Compounds *126*	
2.4.4	Conclusion *130*	
2.4.5	References *130*	
2.5	**The Amidocarbonylation of Aldehydes** *133*	
	Axel Jacobi von Wangelin, Helfried Neumann, Dirk Gördes, and Matthias Beller	
2.5.1	Introduction *133*	
2.5.2	The Cobalt-Catalyzed Amidocarbonylation *134*	
2.5.3	The Palladium-Catalyzed Amidocarbonylation *141*	
2.5.4	Outlook *146*	
2.5.5	References *146*	
2.6	**Transition Metal-catalyzed Alkene and Alkyne Hydrocyanations** *149*	
	Albert L. Casalnuovo and T. V. Rajan Babu	
2.6.1	Introduction *149*	
2.6.2	Alkene Hydrocyanation *149*	
2.6.3	Alkyne Hydrocyanation *151*	
2.6.3.1	Nickel Phosphite-catalyzed Reactions *151*	
2.6.3.2	$Ni(CN)_4^{2-}$-catalyzed Reactions *151*	
2.6.3.3	Addition of R_3SiCN *152*	
2.6.4	New Directions in Nickel-catalyzed Alkene Hydrocyanation *153*	
2.6.4.1	New Ligands *153*	
2.6.4.2	Catalytic Asymmetric Hydrocyanation *153*	
2.6.5	Conclusions *155*	
2.6.6	References *156*	
2.7	**Cyclopropanation** *157*	
	Andreas Pfaltz	
2.7.1	Introduction *157*	
2.7.2	Metal-catalyzed Decomposition of Diazo Compounds *157*	
2.7.3	Enantioselective Cyclopropanation with Copper Catalysts *158*	
2.7.4	Dinuclear Rhodium Catalysts *163*	

2.7.5	Simmons–Smith Reaction *167*
2.7.6	Kulinkovich Hydroxycyclopropane Synthesis *167*
2.7.7	References *168*
2.8	Cyclomerization of Alkynes *171*
	H. Bönnemann and W. Brijoux
2.8.1	Introduction *171*
2.8.2	Transition Metal-Catalyzed Syntheses of 6-Membered Carbocycles *173*
2.8.2.1	Benzenes and Cyclohexadienes *174*
2.8.2.2	Quinones *177*
2.8.2.3	Phenylenes *178*
2.8.2.4	Naphthalenes and Phenanthrenes *178*
2.8.3	Transition Metal-Catalyzed Syntheses of 6-Membered Heterocycles *179*
2.8.3.1	Pyranes, Pyrones, Pyridones, and Sulfur-Containing Heterocycles *179*
2.8.3.2	Pyridines *182*
2.8.3.2.1	Pyridine *185*
2.8.3.2.2	Alkyl-, Alkenyl-, and Arylpyridines *185*
2.8.3.2.3	2-Amino- and 2-Alkylthiopyridines *188*
2.8.3.3	Bipyridyls *189*
2.8.3.4	Isoquinolines *190*
2.8.3.5	Miscellaneous *191*
2.8.4	List of Abbreviations *193*
2.8.5	References *193*
2.9	**Isomerization of Olefin and the Related Reactions** *199*
	Sei Otsuka and Kazuhide Tani
2.9.1	Introduction *199*
2.9.2	Allylamines *199*
2.9.2.1	Characteristics of the Catalysis *200*
2.9.2.2	Mechanisms *201*
2.9.2.3	Synthetic Applications *201*
2.9.3	Allyl Alcohols *203*
2.9.4	Allyl Ethers *205*
2.9.5	Unfunctionalized Olefins *206*
2.9.6	Asymmetric Skeletal Rearrangements *207*
2.9.6.1	Epoxides *207*
2.9.6.2	Aziridines *208*
2.9.7	References *208*
2.10	**Coupling of Aryl and Alkyl Halides with Organoboron Reagents (Suzuki Reaction)** *211*
	Alexander Zapf
2.10.1	Introduction *211*
2.10.2	Mechanism *212*
2.10.3	Coupling of Aryl Halides *213*

2.10.3.1	Phosphine Ligands	213
2.10.3.2	Carbene Ligands	220
2.10.3.3	Other Ligands	222
2.10.4	Coupling of Alkyl Halides and Tosylates	223
2.10.6	Summary and Outlook	225
2.10.7	References	226

2.11	**Transition Metal-Catalyzed Arylation of Amines and Alcohols**	**231**
	Alexander Zapf, Matthias Beller, and Thomas H. Riermeier	
2.11.1	Introduction	231
2.11.2	Catalytic Amination Reactions	231
2.11.2.1	Palladium-Catalyzed Arylation of Aromatic and Aliphatic Amines	231
2.11.2.2	Palladium-Catalyzed Synthesis of Primary Anilines	243
2.11.2.3	Nickel-Catalyzed Arylation of Primary and Secondary Amines	244
2.11.2.4	Copper-Catalyzed Arylation of Primary and Secondary Amines	244
2.11.3	C–O Coupling Reactions	246
2.11.4	References	253

2.12	**Catalytic Enantioselective Alkylation of Alkenes by Chiral Metallocenes**	**257**
	Amir H. Hoveyda	
2.12.1	Introduction	257
2.12.2	Zr-Catalyzed Enantioselective Carbomagnesation Reactions	257
2.12.2.1	Catalytic Enantioselective Addition Reactions	257
2.12.2.2	Zr-Catalyzed Kinetic Resolution of Unsaturated Heterocycles	263
2.12.2.3	Zr-Catalyzed Kinetic Resolution of Cyclic Allylic Ethers	266
2.12.2.4	Other Related Catalytic Enatioselective Olefin Alkylations	267
2.12.3	Summary and Outlook	268
2.12.4	References	268

2.13	**Palladium-Catalyzed Olefinations of Aryl Halides (Heck Reaction) and Related Transformations**	**271**
	Matthias Beller, Alexander Zapf, and Thomas H. Riermeier	
2.13.1	Introduction	271
2.13.2	Mechanism	272
2.13.3	Catalysts	274
2.13.4	Asymmetric Heck Reactions using Chiral Palladium Catalysts	281
2.13.4.1	Mechanistic Features of Asymmetric Heck Reactions	283
2.13.4.2	New Catalyst Systems for Asymmetric Heck Reactions	285
2.13.5	Recent Applications of Heck Reactions for the Synthesis of Natural Products, Complex Organic Building Blocks and Pharmaceuticals	288
2.13.6	Miscellaneous	299
2.13.7	Concluding Remarks	300
2.13.8	References	300

2.14	**Palladium-Catalyzed Allylic Substitutions** *307*
	Andreas Heumann
2.14.1	Introductory Remarks and Historical Background [1] *307*
2.14.2	Reactions of π-Allyl Palladium Complexes [15] *308*
2.14.3	Catalytic Introduction of Nucleophiles *309*
2.14.4	Mechanism – Stereochemistry *310*
2.14.5	Allylic Reductions – Hydrogenolysis – Eliminations *311*
2.14.6	Protective Groups *311*
2.14.7	Trimethylenemethane (TMM) Cycloadditions *312*
2.14.8	Allylic Rearrangements *312*
2.14.9	Enantioselective Reactions *312*
2.14.10	Preparative Glossary *315*
2.14.11	References and Notes *315*

2.15	**Alkene and Alkyne Metathesis in Organic Synthesis** *321*
	Oliver R. Thiel
2.15.1	Introduction *321*
2.15.2	Alkene Metathesis *321*
2.15.3	Enyne Metathesis *328*
2.15.4	Alkyne Metathesis *330*
2.15.5	Outlook *331*
2.15.6	References *332*

2.16	**Homometallic Lanthanoids in Synthesis: Lanthanide Triflate-catalyzed Synthetic Reactions** *335*
	Shū Kobayashi
2.16.1	Introduction *335*
2.16.2	Lewis Acid Catalysis in Aqueous Media *335*
2.16.2.1	Aldol Reactions *336*
2.16.2.2	Allylation Reactions *337*
2.16.2.3	Diels–Alder Reactions *338*
2.16.2.4	Micellar Systems *338*
2.16.2.5	Recovery and Reuse of the Catalyst *340*
2.16.3	Activation of Nitrogen-containing Compounds *340*
2.16.3.1	Mannich-type Reaction *341*
2.16.3.2	Aza Diels–Alder Reactions *344*
2.16.3.3	1,3-Dipolar Cycloaddition *347*
2.16.3.4	Reactions of Imines with Alkynyl Sulfides *348*
2.16.4	Asymmetric Catalysis *348*
2.16.4.1	Asymmetric Diels–Alder Reaction *348*
2.16.4.2	Asymmetric [2+2]-Cycloaddition *353*
2.16.4.3	Asymmetric Aza Diels–Alder Reaction *354*
2.16.4.4	Asymmetric 1,3-Dipolar Cycloaddition *355*
2.16.5	Miscellaneous *356*
2.16.6	References *358*

2.17	**Lanthanide Complexes in Asymmetric Two-Center Catalysis** 363	
	Masakatsu Shibasaki, Hiroaki Sasai, and Naoki Yoshikawa	
2.17.1	Heterobimetallic Lanthanide Complexes in Asymmetric Two-Center Catalysis 363	
2.17.1.1	Introduction 363	
2.17.1.2	Catalytic Enantioselective Nitroaldol Reactions Promoted by LnLB Catalysts 364	
2.17.1.3	Second-Generation LLB Catalyst 365	
2.17.1.4	Catalytic Asymmetric Conjugate Additions by LnSB 366	
2.17.1.5	Catalytic Enantioselective Hydrophosphonylations 367	
2.17.1.6	Enantioselective Direct Aldol Reactions 369	
2.17.2	Alkali Metal-Free Lanthanide Complexes in Asymmetric Two-Center Catalysis 371	
2.17.2.1	Catalytic Enantioselective Epoxidations 371	
2.17.3	La-Linked-BINOL Complex 374	
2.17.4	Enantioselective Cyanosilylation of Aldehydes Catalyzed by Ln-Ln Homobimetallic Complexes 375	
2.17.5	Conclusions 377	
2.17.6	References 377	
2.18	**Bismuth Reagents and Catalysts in Organic Synthesis** 379	
	Axel Jacobi von Wangelin	
2.18.1	Introduction 379	
2.18.2	Carbon-Carbon Bond-Forming Reactions 380	
2.18.2.1	Bismuth(0) 380	
2.18.2.2	Bismuth(III) 381	
2.18.2.3	Organobismuth Compounds 385	
2.18.3	Carbon-Heteroatom Bond-Forming Reactions 388	
2.18.4	Outlook 392	
2.18.5	References 392	
3	**Transition Metal-Mediated Reactions** 395	
3.1	**Fischer-Type Carbene Complexes** 397	
	Karl Heinz Dötz and Ana Minatti	
3.1.1	Synthesis and Reactivity 397	
3.1.2	Carbene-Ligand Centered Reactions 398	
3.1.2.1	Carbon-Carbon Bond Formation *via* Metal Carbene Anions 398	
3.1.2.2	Carbon Nucleophile Addition to α,β-Unsaturated Carbene Complexes 400	
3.1.3	Metal-Centered Reactions 402	
3.1.3.1	[3+2+1] Benzannulation 402	
3.1.3.2	Cyclopropanation Reactions 409	
3.1.3.3	Photoinduced Reactions of Carbene Complexes 412	
3.1.4	Synthesis of Five-Membered Carbocycles 414	

3.1.5	Group 6 Metal Carbenes in Catalytic Carbene Transfer Reactions 418
3.1.6	References 421

3.2	**Titanium–Carbene Mediated Reactions** 427
	Nicos A. Petasis
3.2.1	Introduction 427
3.2.1.1	Precursors to Titanium Carbenes 427
3.2.1.2	Geminal Bimetallic Derivatives 429
3.2.2	Carbonyl Olefinations 430
3.2.2.1	Carbonyl Methylenations with the Tebbe Reagent 431
3.2.2.2	Carbonyl Olefinations with Dimethyl Titanocene and Related Derivatives 433
3.2.2.3	Carbonyl Methylenations with CH_2Br_2–Zn–$TiCl_4$ and Related Systems 437
3.2.2.4	Carbonyl Alienations 439
3.2.3	Alkyne Reactions 440
3.2.4	Nitrile Reactions 440
3.2.5	Olefin Metathesis Reactions 442
3.2.6	Ring-opening Metathesis Polymerizations (ROMP) 443
3.2.7	References 444

3.3	**The McMurry Reaction and Related Transformations** 449
	Alois Fürstner
3.3.1	Introduction 449
3.3.2	Some Lessons from Inorganic Chemistry: The Family of McMurry Reagents 450
3.3.3	Recommended Procedures 452
3.3.3.1	Titanium–Graphite and Other Supported Titanium Reagents 452
3.3.3.2	The $TiCl_3$/Zn Reagent Combinations 453
3.3.3.3	Activation of Commercial Titanium 455
3.3.4	McMurry Coupling Reactions in Natural Product Synthesis 456
3.3.5	Nonnatural Products 458
3.3.6	Titanium-induced Cross-Coupling Reactions 461
3.3.6.1	Mixed Couplings of Aldehydes and Ketones 461
3.3.6.2	Keto-Ester Cyclizations 462
3.3.6.3	Synthesis of Aromatic Heterocycles 463
3.3.7	References 466

3.4	**Chromium(II)-Mediated and -Catalyzed C-C Coupling Reactions** 469
	David M. Hodgson and Paul J. Comina
3.4.1	Introduction 469
3.4.2	Allylic Halides 470
3.4.3	1,1-Di- and 1,1,1-Trihalides 472
3.4.4	Alkenyl and Aryl Halides (and Enol Triflates) 474
3.4.5	Alkynyl Halides 476

3.4.6	Alkyl Halides	476
3.4.7	Transformations Involving C=O and C=C Reduction	477
3.4.8	References	479

3.5 Manganese(III)-Based Oxidative Free-Radical Cyclizations 483
Barry B. Snider

3.5.1	Introduction	483
3.5.2	Oxidizable Functionality	484
3.5.3	Oxidants and Solvents	485
3.5.4	Common Side Reactions	486
3.5.5	Cyclization Substrates	487
3.5.6	References	489

3.6 Titanium-Mediated Reactions 491
Rudolf O. Duthaler, Frank Bienewald, and Andreas Hafner

3.6.1	Introduction – Preparation of Titanium Reagents	491
3.6.2	Addition of Allyl Nucleophiles to Aldehydes and "Ene" Reactions	493
3.6.3	Aldol-Type Addition of Enolates to Aldehydes	499
3.6.4	Addition of Alkyl-Nucleophiles to Aldehydes	503
3.6.5	Cycloadditions and Miscellaneous Reactions	508
3.6.6	References	513

3.7 Zinc-Mediated Reactions 519
Axel Jacobi von Wangelin and Mathias U. Frederiksen

3.7.1	Introduction	519
3.7.2	Preparations and Coupling Reactions	519
3.7.2.1	Zinc Insertion into C-X Bonds	520
3.7.2.2	Transmetalations	521
3.7.3	Cross-Coupling Reactions	525
3.7.3.1	Unsaturated Coupling Partners	525
3.7.3.2	Saturated Coupling Partners	528
3.7.3.3	Carbometalations	529
3.7.4	Organozinc Additions to C=X	530
3.7.4.1	Alkylzinc Nucleophiles	530
3.7.4.2	Arylzinc and Vinylzinc Nucleophiles	531
3.7.4.3	Alkynylzinc Nucleophiles	533
3.7.5	Asymmetric Conjugate Additions	536
3.7.5.1	Copper Catalysis	536
3.7.5.2	Nickel Catalysis	538
3.7.5.3	Oxa- and Phospha-Conjugate Additions	538
3.7.6	Aldol Reactions	539
3.7.7	Cyclopropanation	541
3.7.8	Reactions of Zinc Enolates	543
3.7.8.1	Reformatsky-type Reactions	543
3.7.8.2	Amino Acid Syntheses	544

3.7.8.3	Palladium-Catalyzed Reactions	545
3.7.8.4	Miscellaneous Reactions	546
3.7.9	Summary and Outlook	547
3.7.10	References	547

3.8 The Conjugate Addition Reaction 553
A. Alexakis

3.8.1	Introduction	553
3.8.2	General Aspects of Reactivity	553
3.8.3	Enantioselectivity	556
3.8.4	References and Notes	560

3.9 Carbometalation Reactions of Zinc Enolate Derivatives 563
Daniella Banon-Tenne and Ilan Marek

3.9.1	Introduction	563
3.9.2	Intramolecular Carbometalation	563
3.9.3	Intermolecular Carbometalation	569
3.9.4	Conclusions	571
3.9.5	Acknowledgements	572
3.9.6	References	572

3.10 Iron Acyl Complexes 575
Karola Rück-Braun

3.10.1	Introduction	575
3.10.2	Acyl Complexes Derived from Pentacarbonyl Iron	575
3.10.3	Phosphine-Substituted Chiral-at-Iron Derivatives and Analogs	576
3.10.4	Diiron Enoyl Acyl Complexes	578
3.10.5	Iron-Substituted Enones and Enals	580
3.10.6	References	582

3.11 Iron–Diene Complexes 585
Hans-Joachim Knölker

3.11.1	Introduction	585
3.11.2	Preparation of Iron–Diene Complexes	585
3.11.3	Iron-Mediated Synthesis of Cyclopentadienones	588
3.11.4	Synthetic Applications of Iron–Butadiene Complexes	591
3.11.5	Synthetic Applications of Iron–Cyclohexadiene Complexes	594
3.11.5.1	Iron-Mediated Total Synthesis of Carbazole Alkaloids	595
3.11.5.2	Iron-Mediated Diastereoselective Spiroannulations	596
3.11.6	References	598

3.12 Chromium-Arene Complexes 601
Hans-Günther Schmalz and Florian Dehmel

3.12.1	Introduction	601
3.12.2	Preparation	602

3.12.3	Nucleophilic Addition to the Arene Ring *602*
3.12.4	Ring Lithiation *604*
3.12.5	General Aspects of Side Chain Activation *605*
3.12.6	Side Chain Activation via Stabilization of Negative Charge *606*
3.12.7	Side Chain Activation via Stabilization of Positive Charge *607*
3.12.8	Stabilization of Radicals in the Benzylic Position *608*
3.12.9	Additions to Complexed Benzaldehydes and Related Substrates *609*
3.12.10	Cross-Coupling Reactions *610*
3.12.11	Solid Phase Chemistry *611*
3.12.12	Arene-Cr(CO)$_3$ Complexes as Catalysts *612*
3.12.13	References *613*

3.13	**Pauson-Khand Reactions** *619*
	D. Strübing and M. Beller
3.13.1	Introduction *619*
3.13.2	Stoichiometric Pauson-Khand Reactions *620*
3.13.3	Catalytic Pauson-Khand Reactions *622*
3.13.4	Stereoselective Pauson-Khand Reactions *624*
3.13.5	Synthetic Applications *627*
3.13.6	Transfer Carbonylations in Pauson-Khand Reactions *629*
3.13.7	Conclusions and Outlook *630*
3.13.8	References *631*

Subject Index *633*

List of Contributors

Elisabetta Alberico
CNR –
Instituto di Chimica Biomolecolare
sezinone di Sassari, Regione Baldinca
07040 Li Punti (Sassari)
Italy

Alexandre Alexakis
Dépt. de Chimie Organique
Univ. de Genève
30, quai Ernest Ansermet
1211 Genève 4
Switzerland

Howard Alper
Department of Chemistry
University of Ottawa
10 Marie Curie
Ottawa, Ontario
K1N 6N5
Canada

Isabel W. C. E. Arends
Laboratory of Organic Chemistry
and Catalysis
Delft University of Technology
Julianalaan 136
2628 BL Delft
The Netherlands

Terry T.-L. Au-Yeung
Department of Applied Biology
and Chemical Technology
The Hong Kong Polytechnic University
Hong Kong
China

Jan-E. Bäckvall
Department of Organic Chemistry
Arrhenius Laboratory
Stockholm University
106 91 Stockholm
Sweden

J.-M. Basset
Laboratoire COMS CPE Lyon –
UMR CNRS 9986
43 bd du 11 novembre 1918
69616 Villeurbanne Cedex
France

Oliver Beckmann
Institut für Organische Chemie
der RWTH Aachen
Professor-Pirlet-Str. 1
52056 Aachen
Germany

Matthias Beller
Institut für Organische
Katalyseforschung
an der Universität Rostock e.V.
Buchbinderstraße 5–6
18055 Rostock
Germany

Frank Bienewald
Ciba Spezialitätenchemie AG
Postfach
Klybeckstrasse 141
4002 Basel
Switzerland

Transition Metals for Organic Synthesis, Vol. 1, 2nd Edition.
Edited by M. Beller and C. Bolm
Copyright © 2004 WILEY-VCH Verlag GmbH & Co. KGaA, Weinheim
ISBN: 3-527-30613-7

List of Contributors

Hans-Ulrich Blaser
Solvias AG
Catalysis Research, WRO-1055.628
P.O. Box
4002 Basel
Switzerland

Carsten Bolm
Institut für Organische Chemie
der RWTH Aachen
Professor-Pirlet-Str. 1
52056 Aachen
Germany

Helmut Bönnemann
MPI für Kohlenforschung
Kaiser-Wilhelm-Platz 1
45470 Mülheim an der Ruhr
Germany

Armin Börner
Institut für Organische
Katalyseforschung
an der Universität Rostock e.V.
Buchbinderstr. 5–6
18055 Rostock
Germany

Stephen M. Brown
Zeneca Process Technology
Development
Huddersfield Works
P.O. Box A38, Leeds Road
Huddersfield HD2 1FF
United Kingdom

Werner Brijoux
MPI für Kohlenforschung
Kaiser-Wilhelm-Platz 1
45470 Mülheim an der Ruhr
Germany

Albert L. Casalnuovo
The Dupont Company
Central Research and Development
Department
P.O. Box 80328
Experimental Station
Wilmington, Delaware 19880-0328
USA

Albert S.C. Chan
Department of Applied Biology and
Chemical Technology
The Hong Kong Polytechnic University
Hong Kong
China

Shu-Sun Chan
Department of Applied Biology
and Chemical Technology
The Hong Kong Polytechnic University
Hong Kong
China

Isabelle Chellé-Regnault
Laboratoire de Chimie Organique
Place Louis Pasteur, 1
1348 Louvain-la-Neuve
Belgium

Pedro Cintas
Departamento de Quimica Organica
Facultad de Ciencias-UEX
06071 Badajoz
Spain

Paul J. Comina
School of Chemistry
University of Reading
PO BOX 224, Whiteknights,
Reading RG6 6AD
UK

Rosenildo Correa da Costa
Institut für Organische Chemie
Universität Erlangen-Nürnberg
Henkestraße 42
91054 Erlangen
Germany

Florian Dehmel
Institut für Organische Chemie
Universität Köln
Greinstraße 4
50939 Köln
Germany

Kanae Doda
Laboratoire de Chimie Organique
Place Louis Pasteur, 1
1348 Louvain-la-Neuve
Belgium

Karl Heinz Dötz
Institut für Organische Chemie
und Biochemie
Universität Bonn
Gerhard-Domagk-Str. 1
53121 Bonn
Germany

Raphal Dumeunier
Laboratoire de Chimie Organique
Place Louis Pasteur, 1
1348 Louvain-la-Neuve
Belgium

Rudolf O. Duthaler
Novartis Pharma AG
WSJ-507.109
Postfach
4002 Basel
Switzerland

Peter Eilbracht
FB Chemie, Org. Chemie I
der Universität Dortmund
Otto-Hahn-Straße 6
44227 Dortmund
Germany

Bassam El Ali
Chemistry Department
King Fahd University of
Petroleum & Minerals
QF 31261 Dhahran
Saudi-Arabia

B. L. Feringa
University of Groningen
Department of Chemistry
Nijenborgh 4
9747 AG Groningen
The Netherlands

Richard W. Fischer
Anorganisch-Chemisches Institut
TU München
Lichtenbergstr. 4
85747 München
Germany

Giancarlo Franciò
Institut für Technische Chemie und
Makromolekulare Chemie
Bereich Technische Chemie
Worringerweg 1
52074 Aachen
Germany

Mathias U. Frederiksen
Chemistry Department
Stanford University
Stanford, CA 94305-5080
USA

Gregory C. Fu
MIT Department of Chemistry
77 Massachusetts Avenue
Room 18-411
Cambridge, MA 02139
USA

Alois Fürstner
MPI für Kohlenforschung
Kaiser-Wilhelm-Platz 1
45470 Mülheim an der Ruhr
Germany

Arnaud Gautier
Laboratoire de Chimie Organique
Place Louis Pasteur, 1
1348 Louvain-la-Neuve
Belgium

Paul R. Giles
Laboratoire de Chimie Organique
Place Louis Pasteur, 1
1348 Louvain-la-Neuve
Belgium

Serafino Gladiali
Dipartimento di Chimica
Università di Sassari
Via Vienna 2
07100 SASSARI
Italy

John A. Gladysz
Institut für Organische Chemie
Universität Erlangen-Nürnberg
Henkestraße 42
91054 Erlangen
Germany

DIRK GÖRDES
Institut für Organische
Katalyseforschung
an der Universität Rostock e.V.
Buchbinderstraße 5–6
18055 Rostock
Germany

HELENA GRENNBERG
Department of Organic Chemistry
Arrhenius Laboratory
Stockholm University
106 91 Stockholm
Sweden

ANDREAS HAFNER
Ciba Specialties Chemical Ltd.
Additive Research
P.O. Box 64
1723 Marly 1
Switzerland

TAMIO HAYASHI
Kyoto University
Department of Chemistry
Faculty of Science
Sakyo, Kyoto 606-01
Japan

HORST HENNIG
Universität Leipzig
Fakultät für Chemie und Mineralogie
Institut für Anorganische Chemie
Johannisallee 29
04103 Leipzig
Germany

WOLFGANG A. HERRMANN
Anorganisch-Chemisches Institut
TU München
Lichtenbergstr. 4
85747 München
Germany

ANDREAS HEUMANN
Faculté de St.-Jérome
ENSSPICAM, UMR-CNRS 6516
Université d'Aix-Marseille
13397 Marseille Cedex 20
France

DENNIS J. HLASTA
Johnson+Johnson
Pharmaceutical Research
& Development
Welsh and McKean Roads
PO Box 776
Spring House, PA 19477
USA

LUKAS HINTERMANN
Institut für Organische Chemie
der RWTH Aachen
Professor-Pirlet-Str. 1
52056 Aachen
Germany

DAVID M. HODGSON
The Dyson Perrins Laboratory
University of Oxford
South Parks Road
Oxford OX1 3QY
UK

JENS HOLZ
Institut für Organische
Katalyseforschung
an der Universität Rostock e.V.
Buchbinderstr. 5–6
18055 Rostock
Germany

AMIR HOVEYDA
Department of Chemistry
Merkert Chemistry Center
Boston College
Chestnut Hill, Ma 02067
USA

YASUSHI IMADA
Department of Chemistry
Graduate School of Engineering
Science
Osaka University
1–3, Machikaneyama, Toyonaka
Osaka 560-8531
Japan

HENRI B. KAGAN
Universite Paris-Sud
Institut de Chimie Moleculaire d'Orsay
Laboratoire des Reactions Organiques
Selectives
91405 Orsay Cedex
France

TSUTOMU KATSUKI
Department of Molecular Chemistry
Graduate School of Science
Kyushu University
Hakozaki, Higashi-ku
Fukuoka 812-8581
Japan

WILHELM KEIM
Institut für Technische Chemie und
Petrolchemie der Technischen
Hochschule Aachen
Templergraben 55
52074 Aachen
Germany

HANS-JOACHIM KNÖLKER
Institut für Organische Chemie
Technische Universität Dresden
Bergstraße 66
01069 Dresden
Germany

SHŪ KOBAYASHI
Grad. School of Pharmaceutical
Sciences, University of Tokyo
Hongo, Bunkyo-ku
Tokyo 113-0033
Japan

HARTMUTH KOLB
Department of Chemistry, SP-227
The Scripps Research Institute
10550 N. Torrey Pines Rd
La Jolla, CA 92037
USA

FRITZ E. KÜHN
Anorganisch-Chemisches Institut
TU München
Lichtenbergstr. 4
85747 München
Germany

KAMAL KUMAR
Institut für Organische
Katalyseforschung
an der Universität Rostock e.V.
Buchbinderstraße 5–6
18055 Rostock
Germany

JUNG LEE
Johnson+Johnson
Pharmaceutical Research
& Development
Welsh and McKean Roads
PO Box 776
Spring House, PA 19477
USA

WALTER LEITNER
Institut für Technische Chemie
und Makromolekulare Chemie
Bereich Technische Chemie
Worringerweg 1
52074 Aachen
Germany

JACQUES LE PAIH
Institut für Organische Chemie
der RWTH Aachen
Professor-Pirlet-Str. 1
52056 Aachen
Germany

T. O. LUUKAS
L'Oréal
Advanced Research
Aulnay sous Bois
France

ILAN MAREK
Department. of Chemistry
Technion – Israel Institute of
Technology, Technion City
Haifa 32000
Israel

ISTVÁN E. MARKÓ
Laboratoire de Chimie Organique
Place Louis Pasteur, 1
1348 Louvain-la-Neuve
Belgium

ANA MINATTI
Institut für Organische Chemie
und Biochemie
Universität Bonn
Gerhard-Domagk-Str. 1
53121 Bonn
Germany

CHRISTIAN MÖSSNER
Institut für Organische Chemie
der RWTH Aachen
Professor-Pirlet-Str. 1
52056 Aachen
Germany

KILIAN MUÑIZ-FERNANDEZ
c/o K. H. Dötz
Kekule-Institut für Organische Chemie
und Biochemie
Gerhard-Domagk-Str. 1
53121 Bonn
Germany

SHUN-ICHI MURAHASHI
Department of Applied Chemistry
Okayama University of Science
Ridai-cho 1-1, Okayama 700-0005
Japan

JEAN-LUC MUTONKOLE
Laboratoire de Chimie Organique
Place Louis Pasteur, 1
1348 Louvain-la-Neuve
Belgium

HELFRIED NEUMANN
Institut für Organische
Katalyseforschung
an der Universität Rostock e.V.
Buchbinderstraße 5–6
18055 Rostock
Germany

RONNY NEUMANN
Department of Organic Chemistry
Weizmann Institute of Science
Rehovot
Israel 76100

HISAO NISHIYAMA
School of Material Science
Toyohashi University of Technology
Tempaku-cho
Toyohashi 441
Japan

RYOJI NOYORI
Research Center for Materials Science
Nagoya University
Chikusa, Nagoya 464-8602
Japan

TAKESHI OHKUMA
Research Center for Materials Science
Nagoya University
Chikusa, Nagoya 464-8602
Japan

SEI OTSUKA
Tokushima Research Institute
Otsuka Pharmaceutical Co. Ltd.
Kawauchi-cho
Tokushima 771-01
Japan

CHIARA PALAZZI
Institut für Organische Chemie
der RWTH Aachen
Professor-Pirlet-Str. 1
52056 Aachen
Germany

NICOS A. PETASIS
Department of Chemistry
University of Southern California
Los Angeles, CA 90089-1661
USA

ANDREAS PFALTZ
Institut für Organische Chemie
der Universität Basel
St.-Johanns-Ring 19
4056 Basel
Switzerland

FREDDI PHILIPPART
Laboratoire de Chimie Organique
Place Louis Pasteur, 1
1348 Louvain-la-Neuve
Belgium

T. V. RAJANBABU
Department of Chemistry
The Ohio State University
100 W. 18th Avenue
Columbus, OH 43210
USA

OLIVER REISER
Institut für Organische Chemie
Universität Regensburg
Universitätsstr. 31
93053 Regensburg
Germany

THOMAS H. RIERMEIER
Degussa AG
Projekthaus Katalyse
Industriepark Höchst, G 830
65926 Frankfurt am Main
Germany

KAROLA RÜCK-BRAUN
Institut für Organische Chemie
TU Berlin
Straße des 17. Juni 135
10623 Berlin
Germany

HIROAKI SASAI
ISIR
Osaka University
8-1 Mihogaoka, Ibaraki-Shi
Osaka 567
Japan

GUNTHER SCHLINGLOFF
Ciba Specialities
Köchlinstr. 1
79630 Grenzach-Wyhlen
Germany

HANS-GÜNTHER SCHMALZ
Institut für Organische Chemie
Universität Köln
Greinstraße 4
50939 Köln
Germany

AXEL M. SCHMIDT
FB Chemie, Org. Chemie I
der Universität Dortmund
Otto-Hahn-Straße 6
44227 Dortmund
Germany

JAYSREE SEAYAD
Institut für Organische
Katalyseforschung
an der Universität Rostock e.V.
Buchbinderstraße 5–6
18055 Rostock
Germany

K. BARRY SHARPLESS
Department of Chemistry
The Scripps Research Insitute
10666 North Torrey Pines Road
La Jolla, CA 92037
USA

ROGER A. SHELDON
Laboratory of Organic Chemistry
and Catalysis
Delft University of Technology
Julianalaan 136
2628 BL Delft
The Netherlands

M. SHIBASAKI
Graduate School
of Pharmaceutical Sciences
University of Tokyo
7-3-1 Hongo, Bunkyo-ku
Tokyo 113-0033
Japan

GEORGIY B. SHUL'PIN
Semenov Institute
of Chemical Physics
Russian Academy of Sciences
ul. Kosygina, dom 4,
Moscow 119991
Russia

DENIS SINOU
Universite Claude-Bernard Lyon 1
Lab. de Synthese Asymmetrique
U.M.R.
U.C.B.L. CNRS 5622
C.P.E. Lyon – Batiment 1918
69622 Villeurbanne Cedex
France

BARRY B. SNIDER
Department of Chemistry
Brandeis University
415 South Street, MS 015
Waltham, MA 02454-9110
USA

FELIX SPINDLER
Solvias AG
Catalysis Research, WRO-1055.6
P.O. Box
4002 Basel
Switzerland

HEINZ STEINER
Solvias AG
Catalysis Research, WRO-1055.628
P.O. Box
4002 Basel
Switzerland

DIRK STRÜBING
Institut für Organische
Katalyseforschung
an der Universität Rostock e.V.
Buchbinderstraße 5–6
18055 Rostock
Germany

MARTIN STUDER
Solvias AG
Catalysis Research, WRO-1055.6.17
P.O. Box
4002 Basel
Switzerland

KENNETH S. SUSLICK
School of Chemical Sciences
University of Illinois
at Urbana-Champaign
Chemical & Life Sciences Laboratory
600 S. Mathews Av.
Urbana, Illinois 61801
USA

KAZUHIDE TANI
Department of Chemistry
Osaka University
Toyonaka
Osaka 560-8531
Japan

DANIELLA BANON-TENNE
Department of Chemistry Technion
Israel Institute of
Technology, Technion City
Haifa 32000
Israel

OLIVER R. THIEL
Amgen Inc.
Small Molecule Process Research
One Amgen Center Drive, Thousand
Oaks California 91320
USA

WERNER R. THIEL
Institut für Anorganische Chemie
Technische Universität Chemnitz
Straße der Nationen 62
09107 Chemnitz
Germany

ANNEGRET TILLACK
Institut für Organische
Katalyseforschung
an der Universität Rostock e.V.
Buchbinderstraße 5–6
18055 Rostock
Germany

MASAO TSUKAZAKI
Laboratoire de Chimie Organique
Place Louis Pasteur, 1
1348 Louvain-la-Neuve
Belgium

BARRY M. TROST
Chemistry Department
Stanford University
Stanford, CA 94305-5080
USA

CHRISTOPHER J. URCH
Zeneca Agrochemicals
Jealott's Hill Research Station
Bracknell
Berkshire RG42 6ET
United Kingdom

AXEL JACOBI VON WANGELIN
Institut für Organische
Katalyseforschung
an der Universität Rostock e.V.
Buchbinderstraße 5–6
18055 Rostock
Germany

PETER WASSERSCHEID
Institut für Technische Chemie und
Makromolekulare Chemie
Bereich Technische Chemie
Worringerweg 1
52074 Aachen
Germany

THOMAS WESKAMP
Anorganisch-Chemisches Institut
TU München
Lichtenbergstr. 4
85747 München
Germany

KEIJI YAMAMOTO
Kyoto University
Department of Chemistry
Faculty of Science
Sakyo, Kyoto 606-01
Japan

NAOKI YOSHIKAWA
Graduate School
of Pharmaceutical Sciences
University of Tokyo
7-3-1 Hongo, Bunkyo-ku
Tokyo 113-0033
Japan

ALEXANDER ZAPF
Institut für Organische
Katalyseforschung
an der Universität Rostock e.V.
Buchbinderstraße 5–6
18055 Rostock
Germany

1
General

1.1
Basic Aspects of Organic Synthesis with Transition Metals

Barry M. Trost

Chemistry is described as a central science – one in which phenomena are defined at a molecular level. Understanding functions ranging from material science to biology occurs increasingly at the molecular level. At the heart of such an exercise is synthesis. Designing structure for function requires the greatest flexibility in putting together the molecular edifice. A key in synthesis is efficiency, which may be defined as the ability to convert readily available building blocks into the target molecule in relatively few synthetic operations that require minimal quantities of raw materials and produce minimal waste.

Synthetic efficiency may be divided into two major sub-categories – selectivity [1] and atom economy [2]. Four types of selectivity categorize reactions. First, differentiation among bond types is termed chemoselectivity. Such selectivity can be rather simple, such as selective additions to a carbon–carbon double bond in the presence of a carbon–oxygen double bond or vice versa. Alternatively, such differentiation can be quite subtle, such as differentiating among several carbonyl groups in the same molecule. Second, orienting reactants with respect to each other is termed regioselectivity. Markovnikov vs. anti-Markovnikov additions to a carbon–carbon double bond are classical illustrations. The regioselectivity of the additions of the equivalent of allyl anions to carbonyl groups represents a continuing contemporary challenge [3].

The remaining selectivity issues revolve around stereochemistry. Controlling relative stereochemistry, termed diastereoselectivity, is generally simpler than controlling absolute stereochemistry, termed enantioselectivity. Because of this fact, a frequent approach to the latter problem is to convert it into one of controlling relative stereochemistry. In reactant design, such a strategy has given birth to the concept of chiral auxiliaries [4]. While such an approach is useful and practical, its requirement of a stoichiometric amount of the auxiliary clearly defines it as a less desirable one. Controlling absolute stereochemistry in which the chiral inducing agent is needed only catalytically is clearly the penultimate goal.

Selectivity helps assure that reactions proceed with minimal byproducts that must be separated and disposed of. However, it does not tell the whole story. An additional issue relates to the question of how much of what one puts into a pot ends up in a product. Too often, this issue, which may be called atom economy [2], is sacrificed to resolve problems of selectivity. Consider the Wittig olefination

Transition Metals for Organic Synthesis, Vol. 1, 2nd Edition.
Edited by M. Beller and C. Bolm
Copyright © 2004 WILEY-VCH Verlag GmbH & Co. KGaA, Weinheim
ISBN: 3-527-30613-7

[5] which was introduced to solve the problem of regioselectivity of introduction of a double bond. For the synthesis of a methylenecycloalkane, a reagent whose mass is 357 transfers a unit of mass 14. The remainder of the reagent which constitutes >95% of its original mass becomes byproducts that must be disposed of. Nevertheless, the uniqueness of the process makes it important. It also indicates that an opportunity exists for invention of a more atom economical one.

The development of reactions and reagents that achieve both selectivity and atom economy must be a prime goal of synthetic chemistry. Furthermore, creating new types of bond forming reactions that also address the twin issues of selectivity and atom economy enhance opportunities for simplification of synthetic strategies. The ability of transition metal complexes to catalyze organic reactions constitutes one of the most powerful strategies to address these fundamental issues. Choice of the transition metal combined with the design of the ligand environment provide opportunities for electronic and steric tuning of reactivity to a high degree.

In drawing upon examples from my laboratories to illustrate the principles, I am giving a personal account of my conversion to the world of transition metals to solve problems of selectivity and atom economy. The many chapters that follow provide the readers with the vast scope of the effort throughout the world and the rewards to date. Allylic alkylation becomes a good starting point [6] since the initial question is why bother to use transition metal catalyzed reactions when such alkylations proceed in the absence of catalysts with suitably reactive leaving groups. The answer to the question is embodied in the concept of selectivity – the transition metal catalyst provides an avenue for controlling chemo-, regio-, diastereo-, and enantioselectivity not possible in its absence. It also allows use of more easily handled, generally more readily available, and less noxious substrates. Other examples will illustrate reactions that are not possible in the absence of a transition metal. As these reactions are discussed, it will become apparent that a key phenomenon that underlies much of the ability of transition metals to function as they do is initial coordination. This prerequisite combined with the issues of selectivity leads me to compare transition metal complexes to the active sites of enzymes and to dub them the 'chemists' enzymes'.

1.1.1
Chemoselectivity

Consider the reactivity of the bromoester 1 towards nucleophiles (Eq. (1)) [7]. There is little question that simple treatment with a nucleophile in a solvent that promotes S_N2 reactions like DMF leads exclusively to substitution of bromide (Eq. (1), path a). On the other hand, employing a solvent like THF in which such direct substitutions are slower, addition of a Pd(0) complex completely changes the course of the reaction to substitution of the allylic ester (Eq. (1), path b).

1.1.1 Chemoselectivity

$$\text{(1)}$$

The phenomenon responsible for the reversal of chemo-selectivity is coordination, a prerequisite for the Pd(0) complex to effectively achieve ionization. The double bond proximal to the ester provides initial coordination to Pd(0), which sets the stage for the palladium to promote an intramolecular ionization of even relatively poor leaving groups like carboxylates to a π-allylpalladium intermediate as depicted in Scheme 1. In the absence of such precoordination, even relatively reactive leaving groups like bromide or iodide remain unreactive towards palladium under the above conditions.

Whereas main group organometallics preferentially add across polarized unsaturation, notably carbonyl groups; carbametallation when the metal is a transition metal normally involves addition across a relatively nonpolarized carbon–carbon unsaturation. This selectivity stems from the preferential coordination of the latter to the metal. Carbapalladation, a key step in the Heck arylation and vinylation [8], occurs across carbon–carbon double bonds selectively even in the presence of carbon–oxygen double bonds. The fact that unactivated carbon–carbon triple bonds are better ligands than unactivated carbon–carbon double bonds even allows discrimination between these two types of carbon–carbon unsaturation – a key issue in an approach to vitamin D and its analogues as illustrated in Eq. (2) [9].

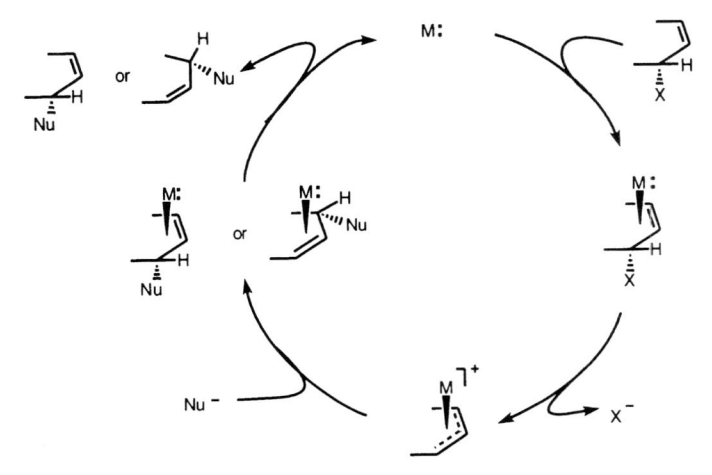

Scheme 1 A metal-catalyzed allylic alkylation.

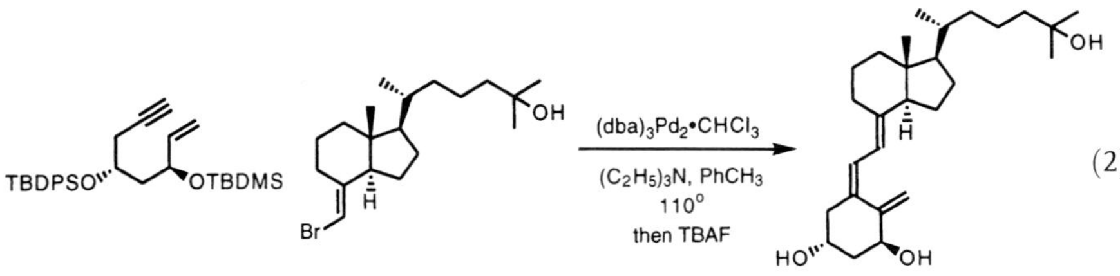

1.1.2
Regioselectivity

Transition metal complexes may control the orientation of chemical reactions. Allylic alkylations illustrate this phenomenon in an intermolecular case. Whereas, palladium-catalyzed reactions normally are dominated by steric effects leading to attack on an unsymmetrical π-allylpalladium intermediate at the less substituted carbon (Eq. (1)) [10], more electropositive metals like molybdenum [11] or tungsten [12] promote reaction at the more electron deficient allyl terminus, which will be the more substituted one. As shown in Eq. (3), the molybdenum-catalyzed reaction was employed to make a product which is a fragment of the saponaceolides [13].

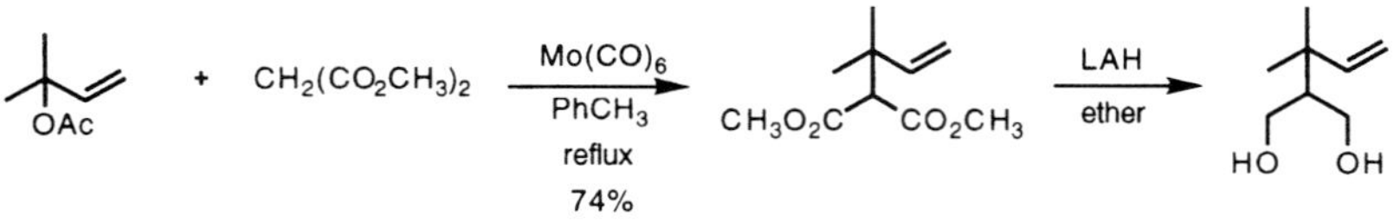

Directing reactions along regiochemical pathways not possible in a nonmetal-catalyzed reaction constitutes another aspect of transition metal chemistry. Consider the Alder ene reaction [14] as shown in Eq. (4), path a. By virtue of the mechanism of the concerted thermal reaction, the resultant product between an alkene and an alkyne is a 1,4-diene regioselectively. On the other hand, migration of a vinyl hydrogen H_b can lead to the synthetically particularly useful 1,3-diene, Eq. (4), path b, which participates in Diels–Alder and other cycloadditions. In this case, the role of a catalyst would be not only to increase the rate of the reaction and permit it to proceed at temperatures significantly below its thermal version, which may permit reactions to proceed which otherwise might fail, but also to redirect the regioselectivity when desired.

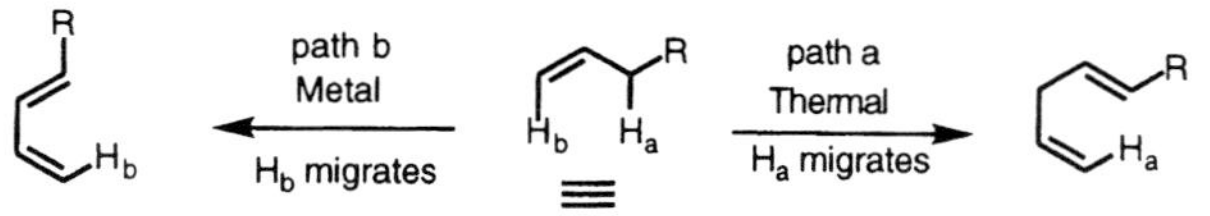

Using a carbametallation as shown in Eq. (5) as the key C–C bond forming reaction produces an intermediate **2** for which β-elimination of hydrogen can involve either H_a or H_b. The weaker allylic H_b bond might be anticipated to eliminate more facilely thereby producing the 1,3-diene. As shown in Eq. (6), this process can indeed be realized [15] and served as a key step in the synthesis of the isolactaranes stereopolide [16] and merulidial [17] in which the subsequent regio- and diastereoselective Diels–Alder reaction of 2-bromomethylmaleic anhydride introduces all of the remaining carbon atoms in both total syntheses.

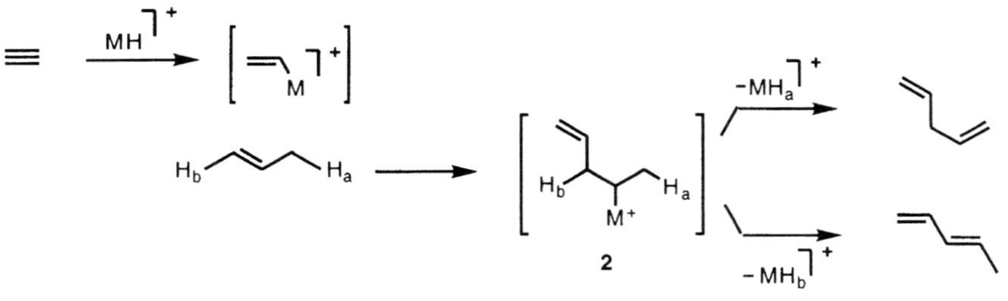

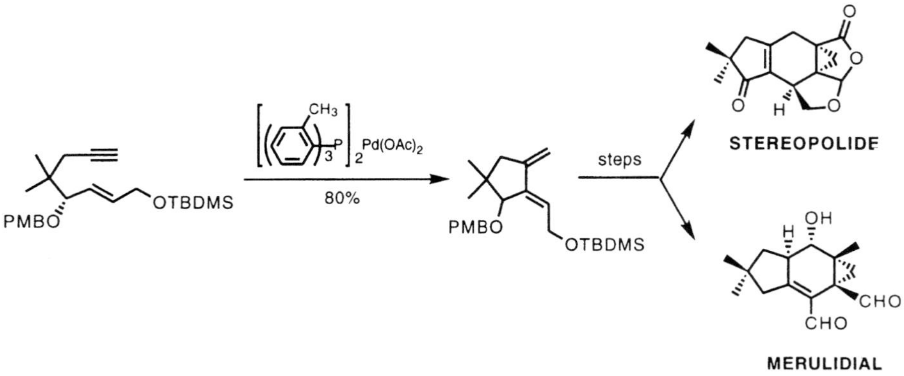

1.1.3
Diastereoselectivity

Changing the 'rules' of reactivity is an exciting prospect offered by transition metal-catalyzed reactions. Allylic alkylations nicely illustrate this phenomenon. The stereochemical rule for S_N2 reactions is substitution with inversion of configuration. Examination of Scheme 1 indicates that a metal-catalyzed reaction effects substitution with net retention of configuration regardless of the regioselectivity, i.e. the nucleophile approaches the same face of the allyl fragment from which the leaving group departed. This result stems from either a double inversion mechanism (as depicted) [18] or a double retention mechanism (not depicted) [19]. Equation (7) illustrates employment of this phenomenon for direct substitution in the elaboration of the steroid side chains of the ecdysones, insect molting hormones [20]. Another example of this principle is shown in Eq. (8). The diastereoselectivity of the Diels–Alder reaction arising from an *endo* transition state generates the all *cis* isomer such as 3. Cyclization to the quinuclidine system 4, common in alkaloids represented by ibogamine [21] and catharanthine [22], then requires an S_N2' substitution with retention of configuration which is equally accessible by the transition-metal catalyzed chemistry. The final cyclization to form ibogamine required a new type of reactivity that derived from the Heck arylation but initiated by an electrophilic aromatic substitution by a palladium(+2) salt and terminated by reductive cleavage of a σ-C–Pd bond. Thus, the availability of two palladium mediated reactions as illustrated in Eq. (8) created a four-step synthesis of ibogamine from the Diels–Alder partners.

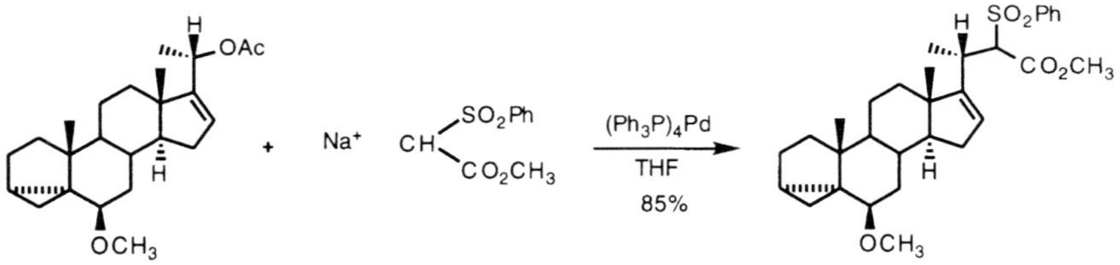

(7)

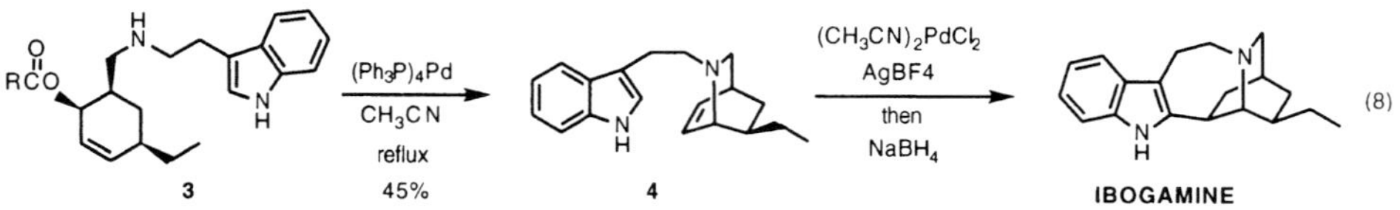

(8)

Controlling relative stereochemistry also applies to sp^2 carbon in terms of alkene geometry. The geometry of trisubstituted alkenes is difficult to control by most olefination protocols; whereas, 1,2-disubstituted alkenes are readily available by such methods in either the *E* or *Z* configuration [5]. If the disubstituted alkene can be specifically converted to a trisubstituted alkene by stereospecific replacement of a vinyl C–H bond by a C–C bond, a new way to create trisubstituted alkenes of defined geometry will be available. Activation of C–H bonds by transition metals is one of their fundamental reactions. Using pre-coordination to direct regioselectivity, a new strategy for stereocontrolled construction of trisubstituted alkenes emerges as illustrated in Eq. (9) using a Ru catalyst [23, 24].

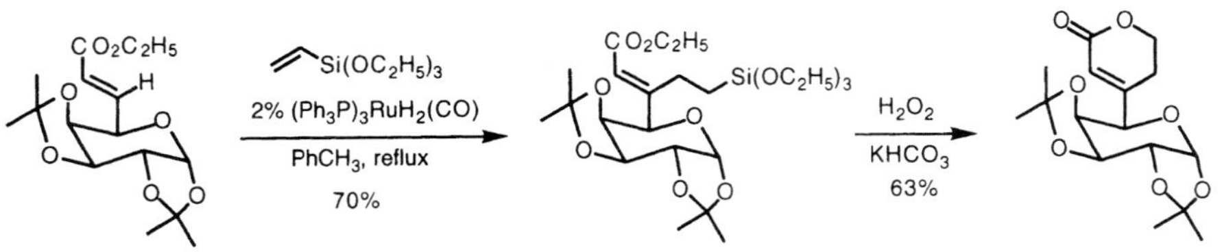

$$\tag{9}$$

1.1.4 Enantioselectivity

Controlling absolute stereochemistry certainly must be classified as one of the major challenges of contemporary organic synthesis. Doing so wherein the asymmetric inducing entity is used only catalytically is the most effective approach [25]. The pioneering studies of Knowles [26], Kagan [27], and others [25] on asymmetric catalytic hydrogenation proved the principle that transition metal complexes can indeed achieve excellent enantioselectivity. How far can this concept be pushed? In all cases of successful enantioselectivity, the bond forming event introducing stereochemistry occurs within the coordination sphere of the metal. Can reactions in which the enantiodiscrimination occurs outside the coordination sphere of the metal also proceed with synthetically useful ee's? Using the chemists' enzymes concept, can chiral space analogous to an active site of an enzyme be created to influence the absolute stereochemistry? Fig. 1 illustrates the concept in the case of a complex for asymmetric allylic alkylation [28, 29]. This catalyst system efficiently induces asymmetry via a number of different mechanisms. Equation (10) illustrates an example of inducing stereochemistry in an ionization event leading to the synthesis of nucleosides [30]. Since the starting dibenzoate derives from furan in one step, a six step synthesis of the complex of the polyoxin-nikkomycin complex results. Equation (11) illustrates inducing enantioselectivity in the nucleophilic addition step wherein an enantioconvergence occurs since both enantiomers of the racemic butadiene monoepoxide produce the same enantiomeric product [31]. This simple synthesis of vinylglycinol in a protected form from cheap, commercially available starting materials makes it an excellent building block. For example, vigabatrin, an anti-epileptic in the *S* form, is available in either enantiomeric form in only four steps [32].

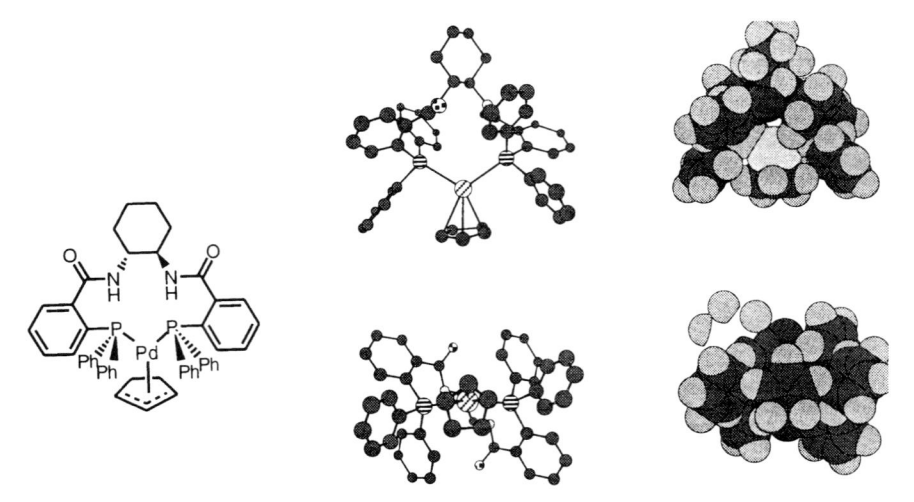

Fig. 1 A chiral pocket for asymmetric induction.

(10)

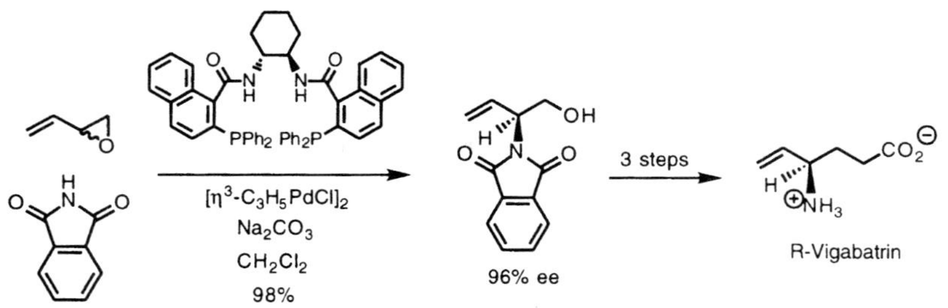

(11)

1.1.5
Atom Economy

While most attention has focused on solving problems of selectivity, that is not sufficient for synthetic efficiency. Consideration of maximal use of raw materials and minimal generation of waste calls for solutions to be atom economical – i.e. as many as possible of the atoms of the reactants should end up in the product with the ideal being the product as simply the sum of the reactants, i.e. the reaction involves only additions with anything else being required catalytically. The serendipitous discovery that a ruthenium complex catalyzed a novel cycloaddition of 1,5-cyclooctadiene provides easy entry to the energy rich tricycle **5** nearly quantitatively [33]. The starting materials are also made by addition reactions – cyclooctadiene by the dimerization of butadiene catalyzed by a nickel complex [34] and the diol from acetylene and 2 equivalents of acetone [35]. Thus, a fairly complicated bridged bicycle arises by a series of three additions from butadiene, acetylene, and acetone.

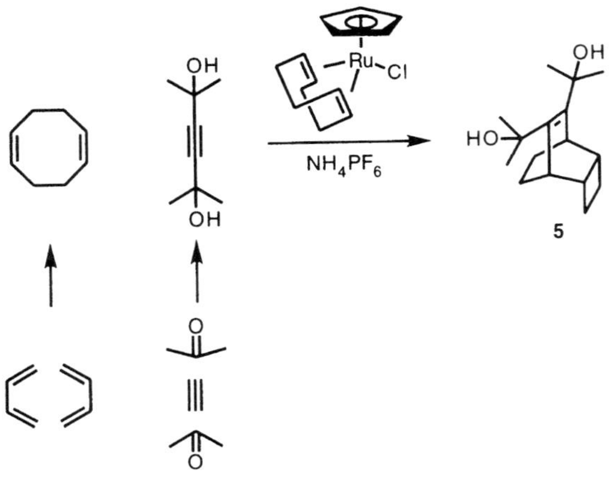

(12)

Activation of C–H bonds provides prime opportunity to rationally invent new addition reactions. Equation (9) illustrates one such example. Activation of the C–H bond of terminal alkynes allows its addition across 'activated' alkynes [36]. Hydrogen shuffling by certain palladium complexes allows isomerization of alkynes into π-allylpalladium intermediates that also leads to additions [37]. Combining these ideas led to a structurally complex macrodiolide in which the final three steps were only addition reactions as illustrated in Eq. (13) [37].

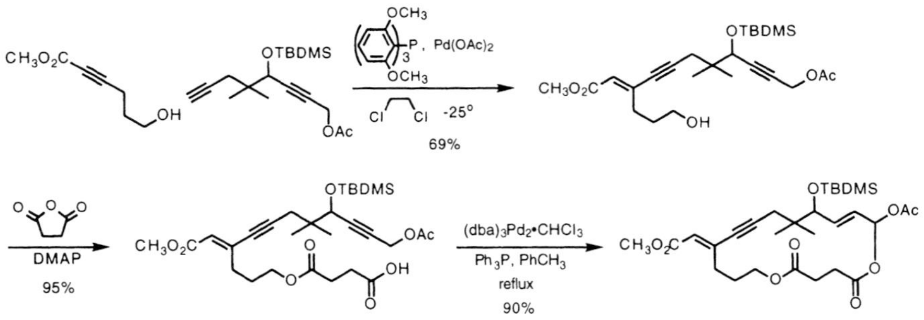

$$\tag{13}$$

1.1.6
Conclusion

Opportunities to invent new reactions catalyzed by transition metals to solve problems of selectivity and to do so with as much atom economy as possible appear infinite. The range of transition metals and their sensitive response to their ligand environment assure the truth of that statement. In the ideal, we can achieve both objectives, which is clearly what we must strive for. On the other hand, problems must be solved to meet the needs of society today. Thus compromises must also be made. Clearly, the practice of organic chemistry today with respect to accessing sophisticated structures for various practical end uses arose from the availability of new paradigms for molecular transformations derived from organometallic chemistry. As our understanding of the underlying reactivity principles increases, our ability to rationally invent new synthetic reactions that move us toward the ideal will undoubtedly increase. At this point, it is already clear that catalysts have gone far beyond their traditional function as simple rate enhancers to become the ultimate arbiter of which path a reaction will take [38].

1.1.7
References

1. B. M. Trost, *Science* **1983**, *279*, 245.
2. B. M. Trost, *Science* **1991**, *254*, 1471; B. M. Trost, *Angew. Chem., Int. Ed. Engl.* **1995**, *34*, 259.
3. For a few interesting recent developments, see: B. S. Guo, W. Doubleday, T. Cohen, *J. Am. Chem. Soc.* **1987**, *109*, 4710; A. Yanagisawa, S. Habane, K. Yasue, H. Yamamoto, *J. Am. Chem. Soc.* **1994**, *116*, 6131.
4. For an early example of the concept of a chiral auxiliary, see: A. I. Meyers, G. Knaus, K. Kamata, M. E. Ford, *J. Am. Chem. Soc.* **1976**, *98*, 567. Chiral oxazolidinones have proven to be exceptionally useful, see: D. A. Evans, M. D. Ennis, D. J. Mathre, *J. Am. Chem. Soc.* **1982**, *104*, 1737. Pseudoephedrine appears to be a very practical chiral auxiliary for enolate alkylations, see: A. G Myers, J. L. Gleason, T. Yoon, D. W. Kung, *J. Am. Chem. Soc.* **1997**, *119*, 656. For an overview, see: J. Seyden-Penn, *Chiral Auxiliaries and Ligands in Asymmetric Synthesis*, Wiley, New York, **1995**.
5. For a review, see: S. E. Kelly in *Comprehensive Organic Synthesis* (Eds.: B. M. Trost, I. Fleming, S. L. Schreiber), Pergamon Press, Oxford, **1991**, Vol. 1, pp. 729–817.
6. For a revew, see: S. A. Godleski in *Comprehensive Organic Synthesis* (Eds.: B. M. Trost, I. Fleming, M. F. Semelhack), Pergamon Press, Oxford, **1991**, Vol. 4, pp. 585–662.
7. B. M. Trost, T. R. Verhoeven, *J. Am. Chem. Soc.* **1980**, *102*, 4730; B. M. Trost, M. Lautens, *J. Am. Chem. Soc.* **1987**, *109*, 1469; B. M. Trost, V. J. Gerusz, *J. Am. Chem. Soc.* **1995**, *117*, 5156.
8. R. F. Heck in *Comprehensive Organic Synthesis* (Eds.: B. M. Trost, I. Fleming, M. F. Semmelhack), Pergamon Press, Oxford, **1991**, Vol. 4, pp. 833–864; A. de Meijere, F. E. Meyer, *Angew. Chem., Int. Ed. Engl.* **1994**, *33*, 2379.
9. B. M. Trost, J. Dumas, M. Villa, *J. Am. Chem. Soc.* **1992**, *114*, 9836.
10. B. M. Trost, M.-H. Hung, *J. Am. Chem. Soc.* **1984**, *106*, 6837; B. Akermark, S. Hansson, B. Krakenberger, A. Vitagliano, K. Zetterberg, *Organomet.* **1984**, *3*, 679; B. M. Trost, C. J. Urch, M.-H. Hung, *Tetrahedron Lett.* **1986**, *27*, 4949; for reviews, see: Ref. [6] and B. M. Trost, *Acc. Chem. Res.* **1980**, *13*, 385; J. Tsuji, *Organic Synthesis with Palladium Compounds*, Springer-Verlag, Berlin, **1980**.
11. B. M. Trost, M. Lautens, *Tetrahedron* **1987**, *43*, 4817; B. M. Trost, C. A. Merlic, *J. Am. Chem. Soc.* **1990**, *112*, 9590.
12. B. M. Trost, M.-H. Hung, *J. Am. Chem. Soc.* **1983**, *105*, 7757; B. M. Trost, G. B. Tometzki, M.-H. Hung, *J. Am. Chem. Soc.* **1987**, *109*, 2176; G. C. Lloyd-Jones, A. Pfaltz, *Angew. Chem., Int. Ed. Engl.* **1995**, *34*, 462. For a recent report of similar regioselectivity with an Ir catalyst, see: R. Takeuchi, M. Kashio, *Angew. Chem., Int. Ed. Engl.* **1997**, *36*, 263.
13. J. R. Corte, unpublished results.
14. B. B. Snider, *Comprehensive Organic Synthesis* (Eds.: B. M. Trost, I. Fleming, L. A. Paquette), Pergamon Press, Oxford, **1991**, Vol. 5, pp. 1–28.
15. B. M. Trost, G. J. Tanoury, M. Lautens, C. Chan, D. T. MacPherson, *J. Am. Chem. Soc.* **1994**, *116*, 4255; B. M. Trost, D. L. Romero, F. Rise, *J. Am. Chem. Soc.* **1994**, *116*, 4268.
16. B. M. Trost, J. Y. L Chung, *J. Am. Chem. Soc.* **1985**, *107*, 4586. Also see: B. M. Trost, P. A. Hipskind, J. Y. L. Chung, C. Chan, *Angew. Chem., Int. Ed. Engl.* **1989**, *28*, 1502.
17. B. M. Trost, P. A. Hipskind, *Tetrahedron Lett.* **1992**, *33*, 4541.
18. B. M. Trost, L. Weber, *J. Am. Chem. Soc.* **1975**, *97*, 1611; B. M. Trost, T. R. Verhoeven, *J. Am. Chem. Soc.* **1980**, *702*, 4730; T. Hayashi, A. Yamamoto, T. Hagihara, *J. Org. Chem.* **1986**, *51*, 723.
19. J. W. Fuller, D. Linebarrier, *Organomet.* **1988**, *7*, 1670; D. Drvorak, I. Stary, P. Kocovsky, *J. Am. Chem. Soc.* **1995**, *117*, 6130.
20. B. M. Trost, Y. Matsumura, *J. Org. Chem.* **1977**, *42*, 2036.

21 B. M. Trost, S. A. Godleski, J.-P. Gent, *J. Am. Chem. Soc.* **1978**, *100*, 3930.
22 B. M. Trost, S. A. Godleski, J. L. Belletire, *J. Org. Chem.* **1979**, *44*, 2052.
23 B. M. Trost, K. Imi, I. W. Davies, *J. Am. Chem. Soc.* **1995**, *117*, 5371.
24 For updated regioselective aromatic substitution, see: F. Kakuichi, S. Sekine, Y. Tanaka, A. Kamatani, M. Sonoda, N. Chatini, S. Murai, *Bull. Chem. Soc. Jpn.* **1995**, *68*, 62. For elaboration of trisubstituted acylcycloalkenes, see: F. Kakuichi, Y. Tanaka, T. Sato, N. Chatani, S. Murai, *Chem. Lett.* **1995**, 679.
25 R. Noyori, *Asymmetric Catalysis in Organic Synthesis*, John Wiley, New York, **1994**; I. Ojima, *Catalytic Asymmetric Synthesis*, VCH, New York, **1993**.
26 W. S. Knowles, M. J. Sabacky, *J. Chem. Soc., Chem. Commun.* **1968**, 1445; B. D. Vineyard, W. S. Knowles, M. J. Sabacky, G. L. Bachman, D. J. Weinkauff, *J. Am. Chem. Soc.* **1977**, *99*, 5946.
27 T.-P. Dang, H. B. Kagan, *J. Chem. Soc., Chem. Commun.* **1971**, 481; W. Dumont, J.-C. Poulin, T.-P. Dang, H. B. Kagan, *J. Am. Chem. Soc.* **1973**, *95*, 8295.
28 B. M. Trost, D. L. Van Vranken, C. Bingel, *J. Am. Chem. Soc.* **1992**, *114*, 9327; B. M. Trost, D. L. Van Vranken, *Angew. Chem., Int. Ed. Engl.* **1992**, *31*, 228.
29 For reviews, see: B. M. Trost, *Acc. Chem. Res.* **1996**, *29*, 355; B. M. Trost, D. L. Van Vranken, *Chem. Rev.* **1996**, *96*, 395.
30 B. M. Trost, Z. Shi, *J. Am. Chem. Soc.* **1996**, *118*, 3039.
31 B. M. Trost, R. C. Bunt, *Angew. Chem., Int. Ed. Engl.* **1996**, *35*, 99.
32 B. M. Trost, R. C. Lemoine, *Tetrahedron Lett.* **1996**, *37*, 9161.
33 B. M. Trost, K. Imi, A. F. Indolese, *J. Am. Chem. Soc.* **1993**, *115*, 8831.
34 P. W. Jolly, G. Wilke, *The Organic Chemistry of Nickel*, Academic Press, New York, **1975**, Vol. II.
35 W. Ziegenbein, *Chemistry of Acetylenes* (Ed.: H. G. Viehe), Marcel Dekker, New York, **1969**, pp. 169–256.
36 B. M. Trost, M. Sorum, C. Chan, A. E. Harms, G. Rühter, *J. Am. Chem. Soc.* **1997**, *119*, 698; B. M. Trost, M. C. McIntosh, *J. Am. Chem. Soc.* **1995**, *117*, 7255.
37 B. M. Trost, W. Brieden, K. H. Baringhaus, *Angew. Chem., Int. Ed. Engl.* **1992**, *31*, 1335.
38 M. E. Davis, S. L. Scrib, *Selectivity in Catalysis*, American Chemical Society, Washington, DC, **1993**.

1.2
Concepts for the Use of Transition Metals in Industrial Fine Chemical Synthesis

Wilhelm Keim

1.2.1
General Concepts

In general, the chemical industry is undergoing a dramatic change, as indicated in Fig. 1.

Currently, the fine chemical industry is establishing a strong position for itself. Of course, the borderlines between fine chemicals, basic chemicals (commodities), specialities, and life science chemicals are fluid. A commonly used grouping embraces

- Standard intermediates with a range of applications: these products will undergo further transformation and can be sold to any customer
- Chemicals ready to be formulated to give an end product
- Custom-made chemicals for a single customer.

To understand the fine chemicals market one has to look at its main characteristics as outlined in Tab. 1.

The pharmaceuticals industry provides the most important clients. This implies that good manufacturing practice (GMP) often is a must and process changes can be undesirable.

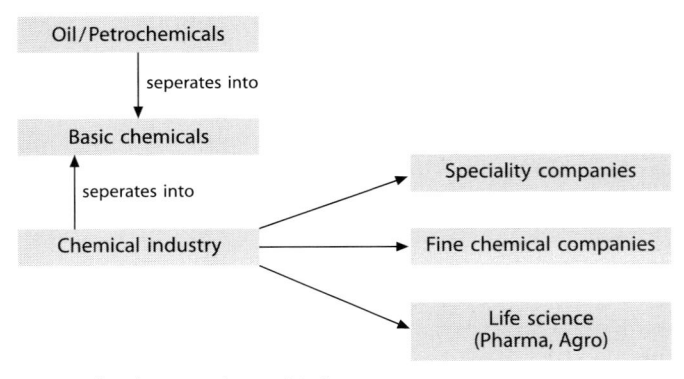

Fig. 1 The changing chemical industry.

Transition Metals for Organic Synthesis, Vol. 1, 2nd Edition.
Edited by M. Beller and C. Bolm
Copyright © 2004 WILEY-VCH Verlag GmbH & Co. KGaA, Weinheim
ISBN: 3-527-30613-7

Tab. 1 Main characteristics of the fine chemical market

Characteristic	Description
Products	Many thousands, highly fragmented, produced by a large number of companies, of which the big players are often multinational chemical companies with captive consumption
	Complex molecules – several functionalities
Product life cycle	Around 30 years, i.e., a short life cycle
	Time to market is important
Product volume	<10 000 t/a
Product differentiation	Very low, produced to specifications, high quality, uniformity, purity, close link with customer
Price	>$ 10/kg – producer aims at dominant supplier position
Capital intensity	Moderate
Processing	Batch, multi-step, often specialized technique, small plant size
Equipment	Simple, usually multi-purpose
Raw material	Often captive source, often unique raw material position
Scheduling	Short term, high flexibility
R&D focus	Process development
Environment	By-products cause difficulties, regulatory compliance

Very important characteristics are time for manufacture and time to market. The development cycle of life science products is becoming increasingly shorter, and great pressure is applied by life science companies, which outsource more and more of their intermediates. Nowadays, the producers are also under increased competitive pressure from China, India, and other countries.

To fully understand the importance of transition metals for fine chemical synthesis, one always must bear in mind the main driving forces in this industry, some of which are outlined in Fig. 2.

Fig. 2 Driving forces for fine chemicals.

1.2.2
Use of Transition Metals in Fine Chemical Synthesis

Historically, non-catalytic routes were used for the syntheses of fine chemicals. The pressure on production cost (cutting of processing steps, batch versus continuous), the need for waste minimization, safety aspects, changes in raw materials and many other things have led fine chemical producers to look at transition metal-based processes (catalysis).

The choice of the industrial manufacturing process for a chemical product is determined primarily by economic and environmental considerations.

Economic Considerations

Low price, high quality (e.g., purity), and secure availability are important for the buyer of a fine chemical. In order to meet these requirements, the producer of a fine chemical might have a unique source of raw material or a prime position in a certain technology. A classical example of the advantage of raw material availability is BASF's fine chemicals business based on acetylene.

An illustration of how one feedstock can be used for a variety of useful chemical reactions is given in Fig. 3.

In the future, we can also expect that potential advantages in raw materials will emerge out of biotechnology. Regarding technology (see also Tab. 1), transition metals may lead to new reaction pathways that are not found in conventional or-

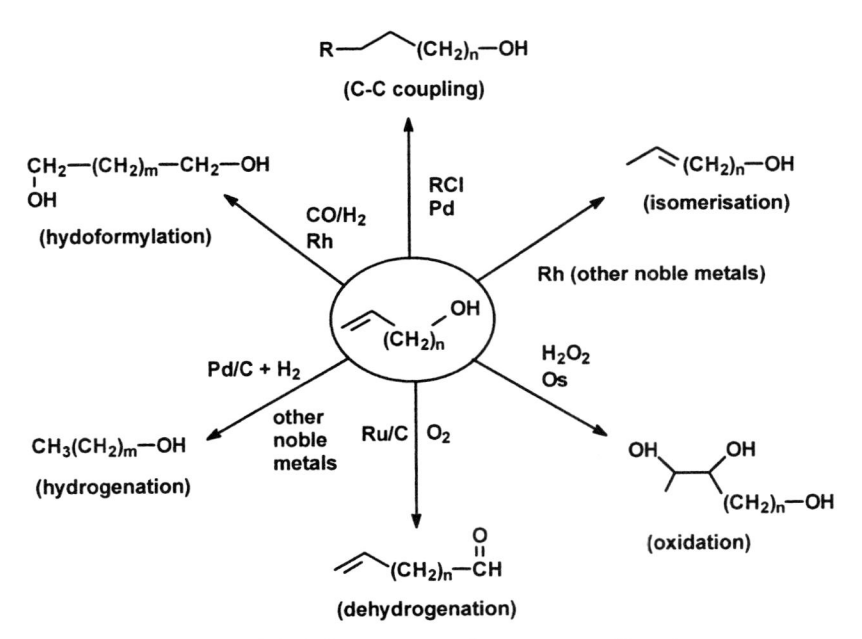

Fig. 3 Use of unsaturated alcohols in a variety of transition metal-based reactions.

ganic chemistry, thus leading to a reduction in the number of reaction steps (change in reactivity). Of course, they may also provide routes to novel products. In addition, the use of transition metals can reduce the severity of the reaction conditions, which can lead to substantial cost savings (energy efficiency). One of the greatest virtues of applying transition metals rests in the improvement of all kinds of selectivity: chemoselectivity, regioselectivity, and stereoselectivity. In this regard, asymmetric synthesis provides an excellent example of the importance of selectivity. However, one must be aware that even if the enantioselective catalytic step yields an *ee* as high as 95%, this may not be good enough for pharmaceuticals unless economic processes are available to purify to values required.

Despite the technological advantages of transition metal-catalyzed routes to fine chemicals, some producers will shy away from using them because of lack of know-how in handling complexes/catalysts (e.g., air sensitivity), but this situation will certainly change in the future.

Environmental Concerns
Concerns about our environment are already forcing us to modify many older technologies in order to reduce atmospheric pollution or to eliminate hazardous waste. Government legislation will force changes in the way companies produce chemicals. The process with the highest yield may not always be the appropriate choice. Instead, there will be a balance between the selectivity of the products and the yield. Even small quantities of by-products in fine chemical synthesis will no longer be acceptable if they create unacceptable environmental hazards. It may be more desirable to suffer a slight yield loss than to have to dispose of a highly toxic by-product associated with a higher-yield process. In this respect, Sheldon has reported that in fine chemicals manufacture 5–50 kg by-products per kg products (for pharmaceuticals 25–100 kg) can be expected. But one must be aware that these figures are based on complete waste disposal and do not include the use of by-products within the whole product strategy of a company. Also, in fine chemicals synthesis, the goals of "green chemistry" and sustainable chemistry must be aimed at. Transition metals are used in the synthesis of fine chemicals either catalytically or stoichiometrically, as illustrated in Fig. 4.

As a general rule in industry, catalytic routes are preferred over stoichiometric ones whenever possible. However, this often does not hold for fine chemicals, which are manufactured predominantly by multi-step syntheses, and the added value can be very high.

A great variety of chemical reactions can be carried out utilizing transition metals, for example:

- Hydrogenation
- Isomerization
- Dehydrogenation
- Asymmetric synthesis
- Oxidation
- Hydrosilylation
- C–C coupling.

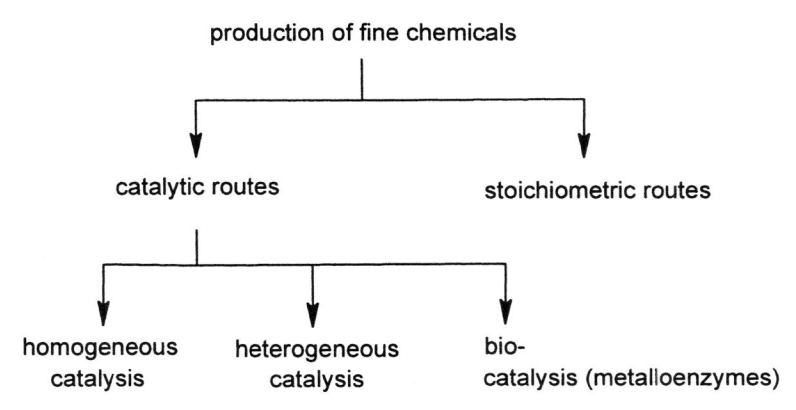

```
                    production of fine chemicals
                              │
              ┌───────────────┴───────────────┐
              ▼                               ▼
       catalytic routes              stoichiometric routes
              │
     ┌────────┼────────┐
     ▼        ▼        ▼
homogeneous heterogeneous  bio-
 catalysis   catalysis   catalysis (metalloenzymes)
```

Fig. 4 Transition metals in fine chemicals synthesis.

Hydrogenation
Hydrogenation reactions are very useful "work-horses" for the fine chemical industry. Product yields are often above 95%, and many catalysts are industrially available. They can be heterogeneously operated in fixed bed or slurry (Pd on carbon, Pt on carbon). But also organometallic compounds are finding broad applications. A common homogeneous catalyst applied in hydrogenations is the Wilkinson complex $(Ph_3P)_3RhCl$. The main advantages of homogeneous catalysts compared to their heterogeneous counterparts are:

mild operating conditions; chemoselective reduction of one organic function in the presence of other reducible functions, e.g., selective reduction of a carbon-carbon double bond in the presence of –CHO, –NO_2 or –CN; regioselective hydrogenation of just one of several similar reducible groups; enantiomeric reductions made possible by using chiral ligands (see below).

As an example of the importance of selectivity in fine chemical hydrogenation processes, the selective reduction of a substituted alkyne to yield the fragrance linalool is shown in Eq. (1). This process is performed by BASF on a commercial scale.

$$\text{alkyne substrate} \xrightarrow[\text{Catalysts}]{H_2} \text{Linalool} \qquad (1)$$

Hydrogenation by transition metals can be applied, for instance, for: C-C multiple bonds, aromatic rings, carbonyl compounds, nitro and nitroso compounds, halonitroaromatics, imines, nitriles, oximes, reductive alkylations, reductive aminations, hydrogenolysis, and transfer hydrogenation.

Dehydrogenation

Dehydrogenation is an endothermic process operating at higher temperatures. At the temperatures needed, transition metal complexes often do not survive, and heterogeneous metal catalysts are preferred (Pd > Pt > Rh, also Fe and others).

Oxidation

Oxidation reactions are extensively used in the synthesis of fine chemicals. The majority of the processes employed industrially involve transition metal complexes in one- or two-electron step reactions. They can be stoichiometric or catalytic. In the past, many stoichiometric oxidations were carried out with traditional oxidants like permanganate and dichromate, producing vast amounts of inorganic effluents which are difficult to dispose of. The search for alternative routes has generated a broad spectrum of catalytic oxidation agents which embrace homogeneous liquid-phase, heterogeneous gas-phase and biochemical enzymatic oxidations. As oxidants, molecular oxygen, hydrogen peroxide, and alkyl hydroperoxides are used preferentially. The myth that stoichiometric reactions are inherently inferior to catalytic ones does not hold as a rule. This is demonstrated in the Wacker process, which is a stoichiometric oxidation made catalytic by ingenious chemistry.

Compared to the "two-electron" Wacker process, transition metal-catalyzed free-radical processes are more important for the synthesis of fine chemicals. As an example, the oxidation of p-tert-butyltoluene in the presence of a Co^{II}/Br^- catalyst system is shown in Eq. (2).

$$\text{p-tert-butyltoluene} \xrightarrow{O_2, [Co/Br]} \text{p-tert-butylbenzaldehyde} \longrightarrow \longrightarrow \text{lilial (fragrance)} \quad (2)$$

Very often, selective oxidations of primary and secondary alcohols to the corresponding aldehyde and ketone can be performed (geraniol to citral via a Pt/C catalyst).

Equations. (3) and (4) show two commercial reactions, emphasizing the broad approach possible in oxidation reactions:

$$\text{oleic acid} + O_2 \xrightarrow{RuO_4} \text{pelargonic acid} + \text{azelaic acid} \quad (3)$$

$$\text{butadiene} + CH_3COOH/O_2 \xrightarrow{Pd} \longrightarrow \text{1,4-butanediol} \quad (4)$$

C–C Coupling

C–C coupling embraces a wide spectrum of reactions, and is one of the most important reactions in fine chemical synthesis. It is impossible to give full credit to the myriad reactions used or of potential use based on transition metals. Clearly,

Tab. 2 C–C coupling reaction

Hydroformylation

olefin + CO/H$_2$ $\xrightarrow{\text{Co or Rh}}$ aldehyde $\longrightarrow$ alcohol and acid

Carbonylation

RX + CO + H$_2$O $\xrightarrow{\text{Rh or Pd}}$ R–C(=O)–OH

The reverse reaction is called decarbonylation

Cross-coupling

Organic halide + organometallic reagent e.g.

R$_1$X + R$_2$MgX $\xrightarrow{\text{Pd}}$ R$_1$–R$_2$ + MgX$_2$

Heck reaction, Suzuki reaction, Buchwald-Hartwig reaction, Sonogashira reaction and many more

Cyclopropanations

C=C + N$_2$CHCOOEt $\xrightarrow{\text{Rh}}$ cyclopropane-COOEt

Hydrocyanation

RCH=CH$_2$ + HCN $\xrightarrow{\text{Ni}}$ R–CH$_2$–CH$_2$–CN

Oligomerization

n-olefins $\xrightarrow{\text{Ni, Rh, Pd}}$ dimers, oligomers, polymers

Telomerization

1,3-diene + HY $\xrightarrow{\text{Pd}}$ C$_8$H$_9$Y (for butadiene)

Metathesis

R^{1}CH=CHR^1
R^{2}CH=CHR^2 $\xrightleftharpoons{\text{Ru}}$ R^1–CH=CH–R^1 + R^2–CH=CH–R^2

---

the formation of new carbon-carbon bonds is the most important process in organic synthesis, and consequently its realization and application using heterogeneous and homogeneous transition metals has been widely explored. Some examples, which emphasize the potential of transition metals in fine chemical synthesis, include hydroformylation, carbonylation/decarbonylation, cross-coupling, cyclopropanation, hydrocyanation, oligomerization, telomerization, metathesis, and many more.

Tab. 2 lists further selected examples of C–C coupling.

Tab. 3 Examples of asymmetric synthesis applying transition metals

Isomerization

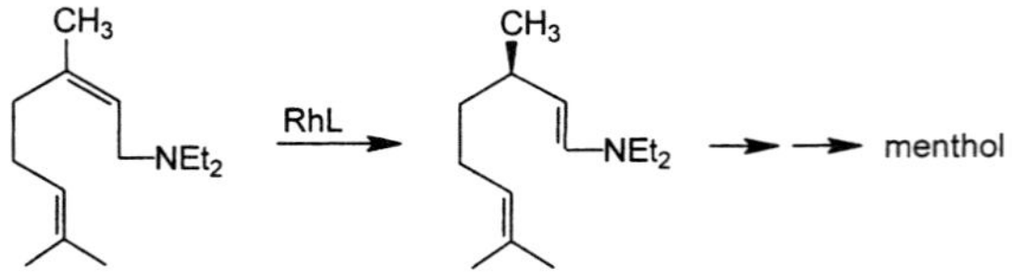

$\xrightarrow{\text{RhL}}$ $\longrightarrow$ menthol

Tagasago menthol process

C–C bond formation

a) $R_2C=CH-CH=C(Me)_2$ + $N_2CHCOOEt$ $\longrightarrow$ cyclopropane-COOEt

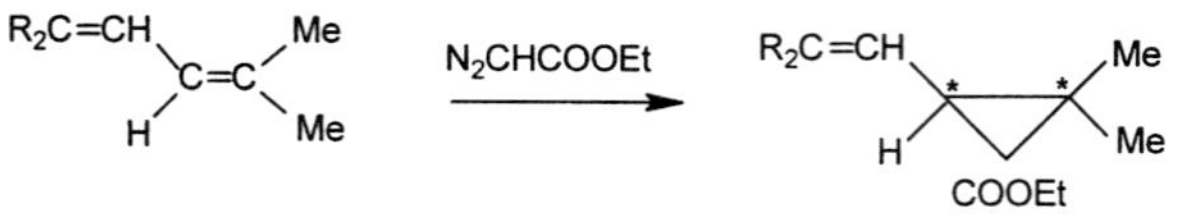

Pyrethroid insecticides

b) R_1R_2HCMgX $\xrightarrow[\text{PdL*}]{R_3X}$ $R_1R_2HCR_3$ + MgX_2

Oxidation

a) cis-alkenol-(CH$_2$)$_8$CH$_3$ $\xrightarrow{\text{TiL*}}$ (7R, 8S) Dispalur

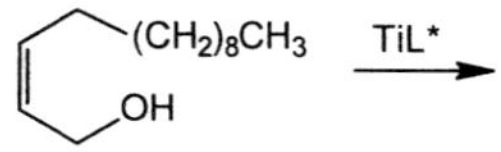

b) Glycidol synthesis (Sharpless)

c) Jacobsen epoxidation

Hydroxylation

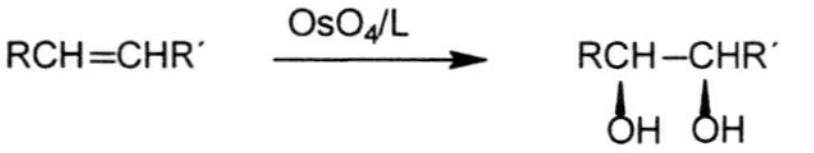

$RCH=CHR'$ $\xrightarrow{OsO_4/L}$ $RCH(OH)-CHR'(OH)$

Hydrogenation

a) imine $\xrightarrow[\text{IrL}]{H_2}$ amine $\longrightarrow$ Metachlor (Herbicide)

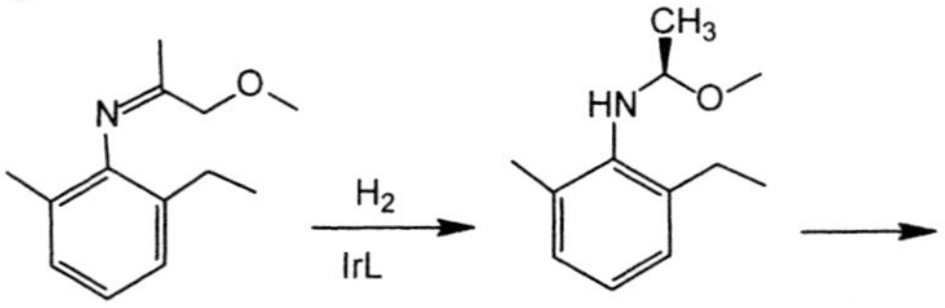

b) L-dopa synthesis

c) L-phenylalanine synthesis (Aspartame)

Isomerization

Isomerization reactions of double bonds are easily catalyzed by transition metals. $Rh(PPh_3)_3Cl$ has been used in the synthesis of several natural products. Two commercial processes are presented in Eqs. (5) and (6).

$$\text{dehydrolinalool} \xrightarrow{V} \text{citral} \qquad (5)$$
(BASF)

$$\text{N,N-diethylgeranylamine} \xrightarrow{Rh\ (chiral)} \text{diethylcitranellalenamine} \qquad (6)$$
(Tagasago menthol process)

Hydrosilylation

Hydrosilylation is the Pt-catalyzed addition of an H–Si group to a carbon-carbon double bond, and is used in the "curing/hardening" of silicone polymers.

Asymmetric Synthesis

Within a remarkably short time, chirality has become a key issue facing the fine chemical industry. New standards with regard to technology and quality as well as regulatory criteria require the application of single isomers. Thus, a growing demand for enantiomerically pure compounds and commercially viable production technologies can be foreseen. Tab. 3 shows a brief overview of a variety of asymmetric syntheses already used in the production of pharmaceuticals, agrochemicals, flavors, fragrances, and many other compounds.

The range of chiral catalysts used is only limited by the ingenuity of the synthetic chemist. Chiral bidentate ligands are used widely: BINAP, DIPAMP, BPE, PENNPHOS, CHIRAPHOS, DUPHOS, DIOP, DEGUPHOS.

Mainly noble metals are used. Critical issues for chiral catalysts are catalyst performance with respect to selectivity, turnover number, turnover frequency, cost of catalyst, availability of starting materials, development time, and purity requirements for substrates and reagents. One should also remember that pharmaceuticals demand higher purity than agro chemicals. Asymmetric chemistry will be the best route to compounds with single chiral centers; compounds with multiple chiral centers may favor biocatalytic routes. One can also speculate that biocatalysis will be combined with classical transition metal chemistry. J. E. Bäckvall, for example, has used an enzyme for the kinetic resolution of a racemic alcohol and a ruthenium catalyst to racemize the unwanted enantiomer back to the racemate.

For a deeper insight into the use of transition metals in fine chemicals the reader is referred to some very good books [1] and journals [2] in this field.

1.2.3
Catalyst Preparation and Application

For the synthesis of fine chemicals, it is preferable to use commercially available catalysts, and a variety of catalyst manufacturers provide heterogeneous catalysts off the shelf.

In homogeneous catalysis the picture is much more complex. One can buy transition metal compounds, which are already active or are active after special treatment techniques. An example of a precursor complex for metathesis is shown in the chemical formula (**1**) below.

$$\mathbf{1} \quad \begin{array}{c} \text{Cl} \quad \text{PCy}_3 \\ \diagdown| \\ \text{Ru}=\text{CH--Ph} \\ \diagup| \\ \text{Cl} \quad \text{PCy}_3 \end{array} \qquad \text{(Grubbs)}$$

The complex *1* can be used directly in metathesis reactions. Also, for cross-coupling reactions, $(Ph_3P)_4Pd$ compounds are available from the supplier. One often prepares the catalyst *in situ* starting from a transition metal salt in the presence of a ligand. The basic approaches available for catalyst preparation are shown in Fig. 5 for a catalyst amenable for Shell's SHOP-process.

One can start directly using complex *2* as a precursor complex. If the latter is very expensive, one can apply the reduction route or an oxidation route, as shown in Fig. 5.

Great attention must also be paid to the proper choice of the reactor and the chemical engineering aspects. Chemists should not hesitate to ask for help from an experienced engineer.

Great consideration must also be given to the mode of operation: continuous or batch. For small volume production – as a rule of thumb < 1000 tonnes – multi-purpose batch reaction systems will be used. For companies not involved in continuous production, a cultural problem is often faced.

1.2.4
The Future

The use of transition metals/organometallic complexes in organic reactions in industry is still in its infancy. Some of the apparent reasons are lack of knowledge and the perception that organometallics are difficult to handle and to come by.

Ni^{2+} + Ph_2PCH_2COOH

↓ reduction

catalyst ← [complex with Ph, Ph, P, Ni, O, O, R] **2**

↑ oxidation

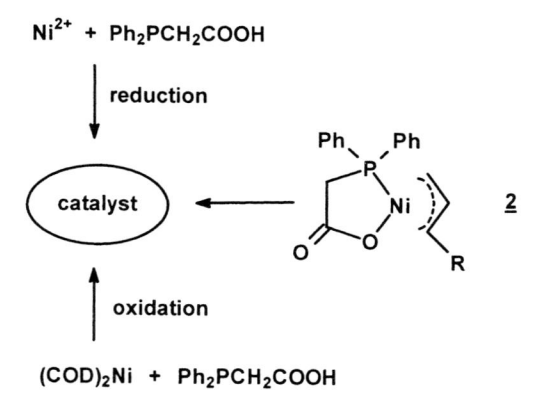

$(COD)_2Ni$ + Ph_2PCH_2COOH

Fig. 5 Preparation of a homogeneous catalyst.

Many of the reactions carried out have been developed only within the last two decades, and a study of the relevant literature tempts one to conclude that organic chemists have finally discovered the use of transition metals/organometallics in syntheses, most players in the field of organometallic synthesis having hitherto been inorganic chemists by training and thinking and therefore oriented more toward structure and less toward reactions. Great impetus for the future will come via technology transfer from academia to industry.

For fast development of new/improved processes, it is highly desirable to find a catalyst in a short time. Massive use of parallel and combinatorial screening and testing methods can be foreseen.

The demand for asymmetric synthesis will increase as legislation requires the use of enantiomerically pure pharmaceuticals and agrochemicals.

Regarding homogeneous catalysts, there is a great need to heterogenize/immobilize them to make commercial operation easier and less costly.

The author is firmly of the opinion that the use of transition metals in fine chemical synthesis has a bright future.

1.2.5
References

Books

1 RASE, H. F., *Commercial Heterogeneous Catalysis;* CRC Press LLC, Boca Raton, 2000, 488 pages, ISBN 0-8493-9417-1
2 TSUJI, J., *Transition Metal Reagents and Catalysts;* Wiley-VCH, Weinheim Germany, 2000, ISBN 0-471-63498-0
3 ROBERTS, S. M., *Catalysts for Fine Chemicals Synthesis;* Wiley-VCH, Weinheim, Germany, 2000, ISBN 0-471-98123-0
4 BHADURI, S., *Homogeneous Catalysis;* Wiley-VCH, Weinheim, Germany, 2000, ISBN 0471-37221-8
5 SCHLOSSER, M., *Organometallics in Synthesis;* Wiley-VCH, Weinheim, Germany, 2001, ISBN 0471-98416-7
6 NISHIMURA, S., *Heterogeneous Catalytic Hydrogenation;* John Wiley & Sons, New York, 2001, ISBN 0-471-39698-2
7 CYBULSKI, A., MOULIJN, J. A., SHARMA, M. M., SHELDON, R. A., *Fine Chemicals Manufacture – Technology and Engineering;* Elsevier Publishing Company, The Netherlands 2001, ISBN 0-444-82202
8 SHELDON, R. A., VAN BEKKUM, H., *The Wide Scope of Catalysis,* Wiley-VCH, Weinheim Germany, 2000, ISBN 3-527-29951-3
9 OMAE, I., *Application of Organometallic Compounds;* John Wiley & Sons, New York, 1998, ISBN 0-471-97604-0

Journals

Catalysts and Catalytic Reactions
Advances in Catalysis
Catalysis Reviews
Journal of Catalysis
Applied Catalysis
Catalysis Letters
Topics in Catalysis
Catalysis Today
Journal of Molecular Catalysis
Reaction Kinetics and Catalysis Letters

2
Transition Metal-Catalyzed Reactions

2.1
Hydroformylation: Applications in the Synthesis of Pharmaceuticals and Fine Chemicals

Matthias Beller and Kamal Kumar

2.1.1
Introduction

The synthetic routes to aldehydes are probably more diverse and numerous than the routes to any other class of carbonyl compounds, reflecting not only the intrinsic value of aldehydes as synthetic intermediates but also the scarcity of truly general routes to aldehydes themselves. However, exceptionally, the transition metal-catalyzed reaction of carbon monoxide and hydrogen with an alkene, *the hydroformylation reaction*, a one-step transformation of an olefin to an aldehyde having one carbon more, is one of the world's largest homogeneously catalyzed processes in industry [1], and is also an important tool for organic synthesis [2]. The technical significance of this reaction, traditionally known as the "Oxo synthesis", is based on the facts that a new carbon-carbon bond is formed and the aldehydes produced are easily converted into a multitude of industrially important secondary products [3]. As a straightforward addition reaction of inexpensive starting materials, it is a clean and practical method. Although the reaction was discovered back in 1938 by O. Roelen, significant academic and industrial activity still exists in this field, which is of great industrial potential.

Applications of hydroformylation and related reactions in industrial and organic synthesis have greatly increased in the last few years. Naturally, some excellent reviews on this topic have appeared before [4]. Therefore in this chapter we have made an attempt to summarize only the new and latest trends being used and applied by organic chemists in this field.

2.1.2
Hydroformylation: Applications for Pharmaceuticals and Natural Products

Among the different catalysts in recent years, rhodium carbonyl complexes, which ensure higher chemo- and regioselectivity with respect to other metal derivatives, have been especially used for the synthesis of molecules of biological importance [2c, 4a]. For example, the synthesis of 2-arylpropanoic acids [5], a commercially important class of anti-inflammatory and analgesic agents [6], using hydroformyla-

Transition Metals for Organic Synthesis, Vol. 1, 2nd Edition.
Edited by M. Beller and C. Bolm
Copyright © 2004 WILEY-VCH Verlag GmbH & Co. KGaA, Weinheim
ISBN: 3-527-30613-7

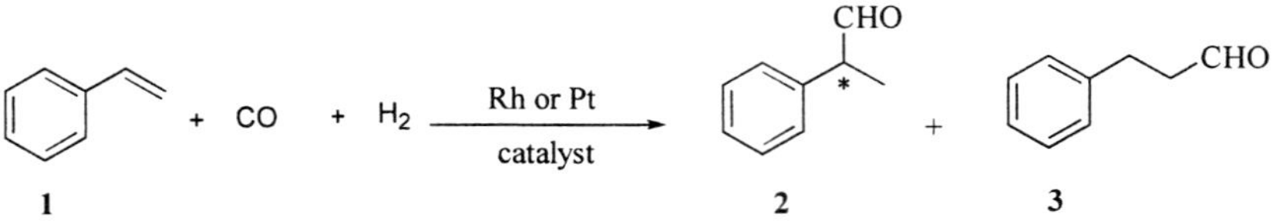

Scheme 1 Hydroformylation of styrene.

tion of styrene and related vinylaromatics (Scheme 1) [7], has gained increasing popularity. Moreover, styrene represents a useful model compound to prove the catalyst performance of hydroformylation catalysts with respect to regio- and enantioselectivity. In the early 1990s (R,R)-[(bicyclo[2.2.2]-octane-2,3-diyl)bis(methylene)]-bis(methylene)]bis[diphenylphosphine]$PtCl_2$/$SnCl_2$ (92% regioselectivity to chiral aldehyde and 85% *ee*) [8] and $Rh(CO)_2$(acac)/(hydroxy-*iso*-butoxy-P)$_2$-2R,4R-pentanediol (98% regioselectivity and 90% *ee*) [9] were reported as interesting chiral catalyst systems. Shortly afterwards, rhodium complexes formed *in situ* from $Rh(CO)_2$(acac) and (R,S)-BINAPHOS and (S,R)-BIPHEMPHOS were developed, which are capable of hydroformylating styrenes giving hydratropaldehyde with high regioselectivity (88–90%) and optical yields (94%) [10].

In the last few years, new chiral ligands have been evaluated with regard to the induction of high enantioselectivity in the course of the hydroformylation [11, 12]. But still the most difficult problem in enantioselective hydroformylation is the simultaneous control of both regio- and enantioselectivity, which limits the structural variety of suitable alkenes for enantioselective hydroformylation significantly [4g]. For example, catalysts based on platinum are known to promote fairly high enantioselection; the regioselectivity, however, toward the production of branched aldehyde, is too low to excite interest [13]. Significant results obtained in asymmetric hydroformylation toward commercial drugs are listed in Tab. 1. More recently, Claver and coworkers demonstrated elegantly that chiral diphosphite ligands derived from D-glucose show promising results in the asymmetric hydroformylation of styrene derivatives with regioselectivities up to 98% and *ee*'s up to 91% [14]. Interestingly, using a perfluoroalkyl-substituted BINAPHOS-ligand $((R,S)$-3-H^2F^6-BINAPHOS) in the rhodium-catalyzed asymmetric hydroformylation of substituted styrenes, high regio- (up to 96%) and enantioselectivities (up to 94%) were obtained recently [15].

Apart from the syntheses of profene analogs, hydroformylation has been reported to be the key step in the synthesis of Pheniramine and several structurally related compounds. For example, N,N-dimethylcinnamylamines produce 2-aryl-4-(dimethylamino)butanals smoothly by $HRh(CO)(PPh_3)_3$-catalyzed hydroformylation. The resulting amino aldehydes can be further converted in two steps into Pheniramines, a well-known first generation family of H_1 antihistaminic agents. The overall yields of the synthesis range between 60 and 70% [16]. Comparably good results have been obtained more recently in the preparation of Pheniramines starting from acetals **5** [17] (Scheme 2).

In principle, the best method to produce the intermediate 3-aryl-3-(2-pyridyl)-aldehydes consists in the regioselective hydroformylation of easily accessible 1-aryl-

Tab. 1 Asymmetric hydroformylation of vinylaromatics to aldehyde precursors of anti-inflammatory agents

Substrate	Catalyst precursor	P (atm) $CO/H_2 = 1$	Temp. (°C)	2-Aryl-propanal or -butanal			Antiin-flammatory agent	Ref.
				Yield (%)	ee (%)	Conf.		
4-Iso-butylstyrene	Rh(CO)$_2$(acac)/(iso-BHA-P)$_2$- 2R,4R-pentanediol[a]	14[b]	r.t.	98	82	(S)	Ibuprofen	9
4-Iso-butylstyrene	Rh(CO)$_2$(acac)/(R,S)-BINAPHOS	100[c]	60	88	92	(S)	Ibuprofen	10
(E)-Propenylbenzene	Rh(CO)$_2$(acac)/(R,S)-BINAPHOS	100[c]	60	10	92	(R)	Indobufen	10
6-Methoxy-2-vinyl-naphthalene	Rh(CO)$_2$(acac)/(iso-BHA-P)$_2$- 2R,4R-pentanediol[a]	14[b]	r.t.	99	85	(S)	Naproxen	9
6-Methoxy-2-vinyl-naphthalene	[Rh(COD)$_2$]BF$_4$/L[d]	110[c]	r.t.	68	39	(S)	Naproxen	12

a) iso-BHA = hydroxy-iso-butyric alcohol
b) CO/H$_2$ ≅ 2
c) CO/H$_2$ ≅ 4
d) L = Carbohydrate-phosphinite A

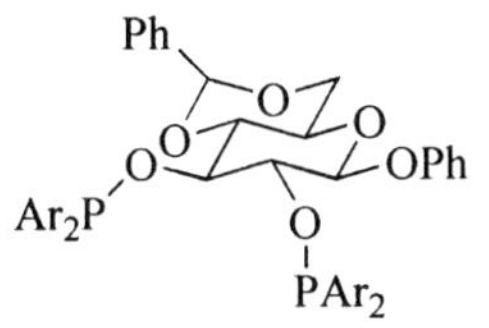

Ar = 3,5-(CF$_3$)$_2$C$_6$H$_3$

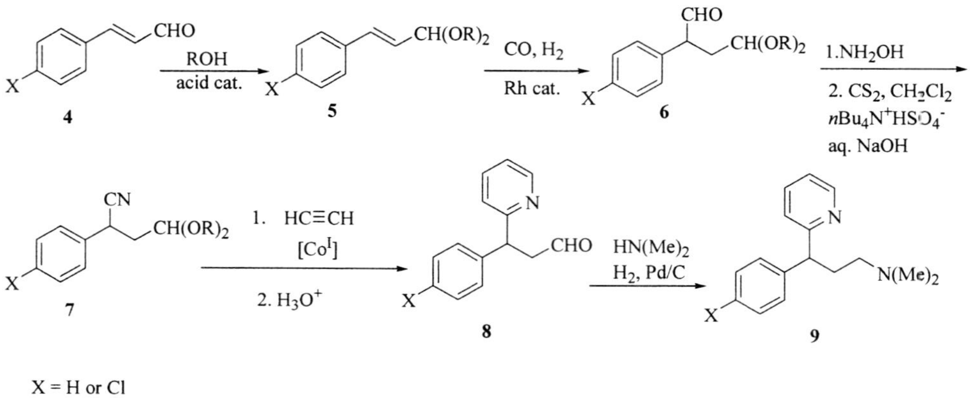

Scheme 2 Preparation of Pheniramines.

1-(2-pyridyl)ethenes (Scheme 3). However, these substrates are almost regiospecifically transformed into isomeric 2-aryl-2-(2-pyridyl)propanals **12** [16, 18] (Fig. 1). It was suggested that the pyridinic nitrogen represents in this type of olefin the preferred site of coordination for the catalytically active rhodium hydride carbonyl complex, so promoting the formation of the branched intermediate α-alkyl complex and hence 2-aryl-2-(2-pyridyl)propanal **12** [18].

As a matter of fact, 1,1-diarylethenes and 1-phenyl-1-(4-pyridyl)ethene are regiospecifically formylated to 3,3-diarylpropanals in high yields [19]. These results are in contrast to those obtained with vinyl aromatic and heteroaromatic olefins, which are characterized by high or complete α-regioselectivity in the presence of rhodium carbonyl complexes. The prevalence of the branched aldehyde is generally attributed to the strongly favored formation of the benzylic α-alkyl-rhodium intermediate stabilized by the aromatic group on the carbon bonded to the metal [7]. For the same reason, the rhodium-catalyzed hydroformylation of 1,1-diphenylethene leads to a large amount of the branched σ-alkyl-rhodium intermediate, as demonstrated by deuterioformylation experiments [20]. However, the more branched alkyl species does not undergo CO migratory insertion, and only β-hydride elimination to the starting olefin takes place. On the contrary the less branched σ-alkyl species, in spite of its low concentration in the reaction solution, is converted into the corresponding σ-acyl-complex and hence into 3,3-diphenylpropanal [20].

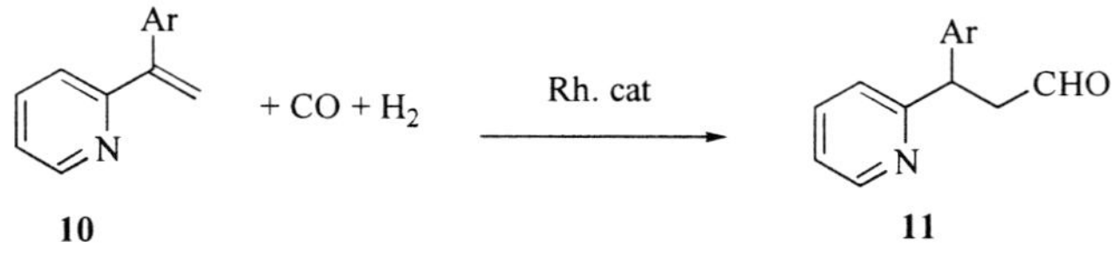

Scheme 3 Hydroformylation of 1-aryl-1-(2-pyridyl)ethenes.

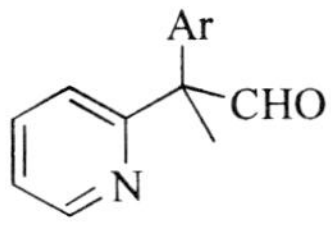

Fig. 1 2-Aryl-2-(2-pyridyl)-propanal.

Platinum-catalyzed reductive amination of the above aldehydes using different amines gives rise to interesting pharmaceuticals having broad spectrum therapeutic activity, such as Phenpiprane, Diisopromine, Tolpropamine. Here the product yields range between 70 and 85% [19b]. Alternatively, the oxo-aldehyde can be converted into the enamine, which affords the final drug by $NaBH_4$ reduction in methanol.

Interestingly, one-step conversion of 1,1-diarylethenes into the corresponding 1-(N,N-dialkylamino)-3,3-diarylpropanes is also possible, because rhodium-carbonyl complexes have been shown to be efficient catalysts for the reductive amination of aldehydes under hydroformylation conditions [21]. This transformation is viable in high yield only if the preformed 3,3-diphenylpropanal is allowed to react with excess of the appropriate amine in the same oxo-reactor [19b]. Unfortunately, the above outlined reaction is not suitable for Phenpyramine (Milverine), an antispasmodic agent [22], from 3,3-diphenylpropanal and 4-aminopyridine. This latter compound promotes the reduction ability of rhodium complexes toward alcohols under hydroformylation conditions [23]. Hence, the corresponding alcohol 3,3-diphenylpropan-1-ol is almost exclusively formed. Therefore, an alternative pathway based on oxidation, amide formation, and reduction was developed to yield Milverine [24] in 35% overall yield (Scheme 4).

Also 4,4-diarylbutanals, which are precursors of a number of valuable pharmaceuticals, are accessible via hydroformylation reactions. In particular, the synthesis of 4,4-bis(4-fluorophenyl)butanal and of the corresponding alcohol appeared attractive: Fluspirilene [25], Penfluridol [26], Lidoflazine [27] (Fig. 2) and other therapeutically active compounds embody in their structure the 4,4-bis(4-fluorophenyl)butyl group.

Obviously the regioselective hydroformylation of 3,3-bis(4-fluorophenyl)propene should provide a straightforward route to 4,4-bis(4-fluorophenyl)butanal (Scheme 5). Unfortunately, in this reaction the starting olefin tends to isomerize under

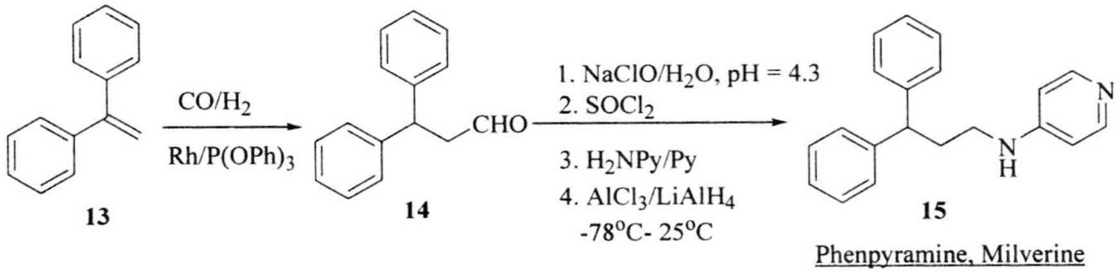

Scheme 4 Synthesis of Milverine starting from 1,1-diphenylethene.

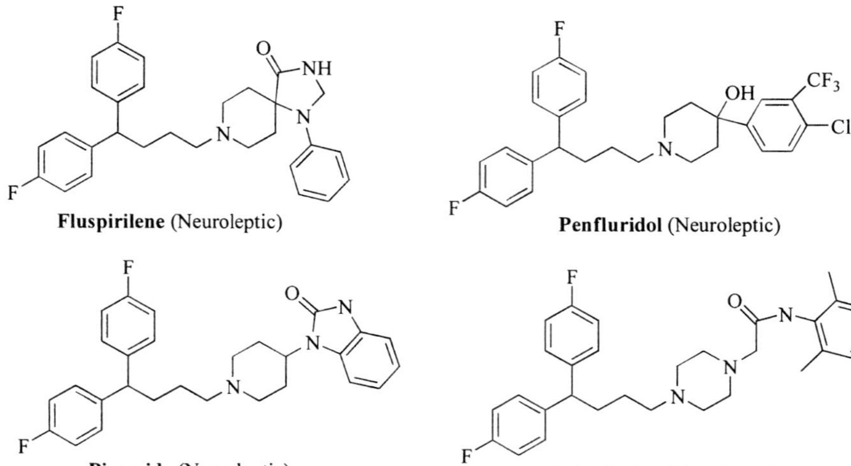

Fluspirilene (Neuroleptic)

Penfluridol (Neuroleptic)

Pimozide (Neuroleptic)

Lidoflazine (Vasodilator)

Fig. 2 Pharmaceuticals derived from 4,4-bis(4-fluorophenyl)butanal.

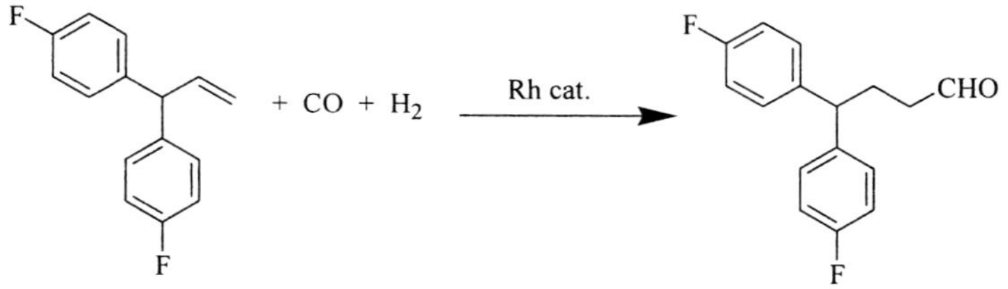

Scheme 5 Regioselective hydroformylation of 3,3-bis(4-fluorophenyl)propene to 4,4-bis(4-fluorophenyl)butanal.

usual oxo-conditions even in the presence of rhodium-phosphine catalysts: for example, at 100 atm ($CO/H_2=1$) and 90 °C in the presence of $HRh(CO)(PPh_3)_3$/PPh_3 (1:2 molar ratio), a high yield of oxo-product was obtained (~99%), but the linear aldehyde (53%) was accompanied by a significant amount (46%) of the undesired branched isomer. To solve this problem, two different synthetic oxo-methodologies have been developed starting from commercially available 4,4′-difluorobenzophenone. Schemes 6 and 7 depict the general features of the processes.

The first pathway involves the hydroformylation of 1,1-bis(4-fluorophenyl)ethene to 3,3-bis(4-fluorophenyl)propanal. As expected, this aldehyde was formed with high yield (85%) at 100 atm ($CO/H_2=1$) and 100 °C in the presence of $[Rh(CO)_2Cl]_2$/$P(OPh)_3$ (Rh/P molar ratio 1:1). Homologation of the product, accomplished via reaction with the instant ylide $CH_3OCH_2\text{-}O\text{-}^+PPh_3$ Br^-/$NaNH_2$ in ethyl ether [28] afforded the desired aldehyde in about 70% overall yield (Scheme 6) [29].

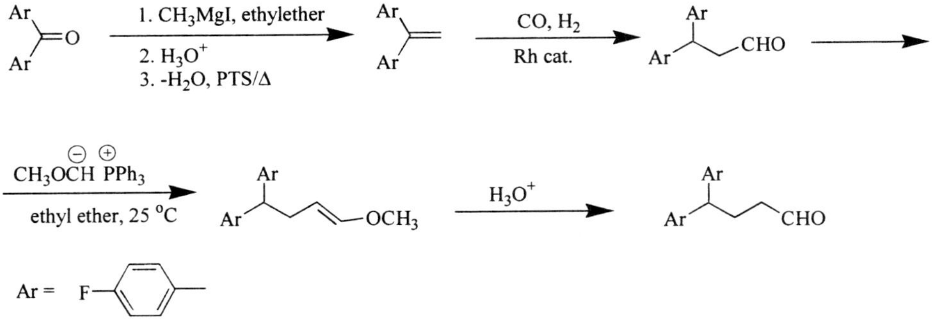

Scheme 6 Preparation of 4,4'-bis(4-fluorophenyl)butanal starting from 4,4'-difluorobenzophenone.

Alternatively, 4,4'-difluorobenzophenone is transformed into 1,1-bis(4-fluorophenyl)-2-propene-1-ol, which is smoothly hydroformylated to the corresponding hydroxyfuran **18**. **18** is further transformed in two steps into 4,4'-bis(4-fluorophenyl)butan-1-ol (Scheme 7).

The yield of the final alcohol **20** is about 60% based on 4,4'-difluorobenzophenone. The corresponding chloride or bromide can be successfully employed for the selective alkylation to form Fluspirilene [30] or Penfluridol [26] respectively.

Among other molecules of medicinal importance, 2-chromanol is an interesting precursor of various pharmaceuticals [31]. This structurally simple intermediate is prepared in good yields by several methods: however, none of them seems to be fully suitable for a semi-industrial scale production. A strategy involving hydroformylation of 2-hydroxystyrene derivatives leading to 2-chromanol is quite successful in getting the important intermediate in satisfactory yields (Scheme 8). Encouraging results (~85% yield of linear aldehyde) were obtained in the hydroformylation of 2-benzyloxystyrene in water/toluene biphasic system catalyzed by Rh(CO)$_2$acac modified with a water-soluble Xantphos ligand [32].

In another report, the rhodium-catalyzed domino hydroformylation of 1,5-dienes has been exploited to synthesize hexahydro-4H-chromenes [33] (Scheme 9),

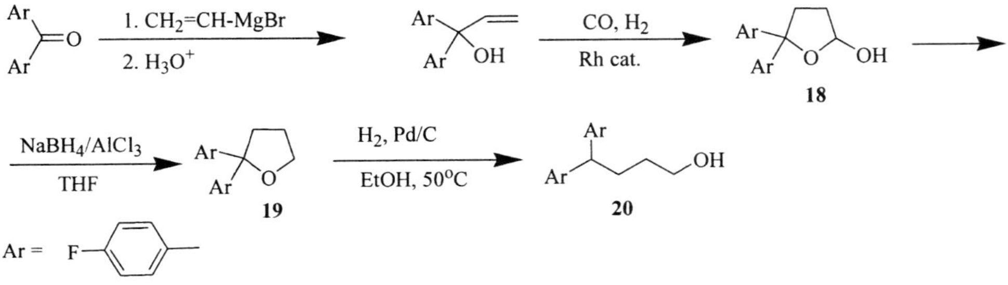

Scheme 7 Preparation of 4,4'-bis(4-fluorophenyl)butan-1-ol starting from 4,4'-difluorobenzophenone.

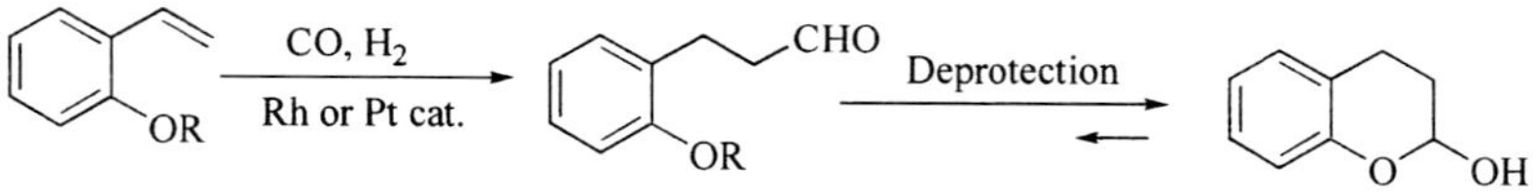

Scheme 8 Hydroformylation of 2-hydroxystyrene derivatives leading to 2-chromanol.

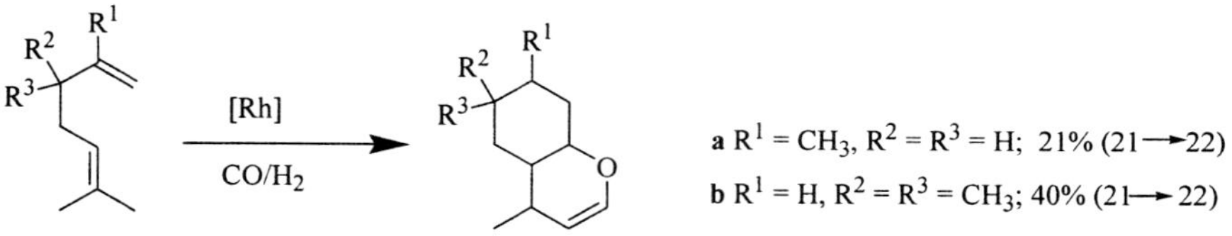

Scheme 9 Domino synthesis of hexahydro-4H-chromenes using hydroformylation reactions.

a $R^1 = CH_3$, $R^2 = R^3 = H$; 21% (21 → 22)
b $R^1 = H$, $R^2 = R^3 = CH_3$; 40% (21 → 22)

albeit in low yields. This multi-step procedure starts with a hydroformylation of one double bond followed by carbonyl ene reaction, a second hydroformylation, and an enol ether-forming cyclization reaction.

In recent years the hydroformylation reaction has been used in multi-step organic syntheses of medicinally and biologically important molecules more often than before [34]. For example, carba-D-fructofuranose **24a** has been prepared in eight steps from the highly oxygenated cyclopentene **26** in an overall yield of 32% (Scheme 10) [35]. Carba-D-fructofuranose **24a** mimics fructose 2,6-biphosphate in polarity and shape after phosphorylation (**24b**), and thus may act as a stable modulator of the isozyme of 6-phosphofructo-2-kinase activity with potentially useful therapeutic effects on cancer cell metabolism (Fig. 3).

The inseparable mixture of aldehydes **27** and **28** was reduced using NaBH$_4$ in MeOH. Finally, compound **29** was hydrogenated (Pd/H$_2$, EtOH, 98%) to form carba-D-fructofuranose **24a**. As components of antiviral carbocyclic nucleosides and as substrates for glycosyl transferases [36], these pseudo-sugars have been shown to enhance metabolic stability of potential drugs.

Sheldon and co-workers demonstrated that bioactive melatonin (N-acetyl-5-methoxytryptamine) can be synthesized elegantly from N-allylacetamide in a novel one-pot reaction (Scheme 11) [38]. Here, the use of an inverted aqueous two-phase catalytic system increases the selectivity of the Rh-phosphine-catalyzed hydroformylation step and simplifies the catalyst/product separation.

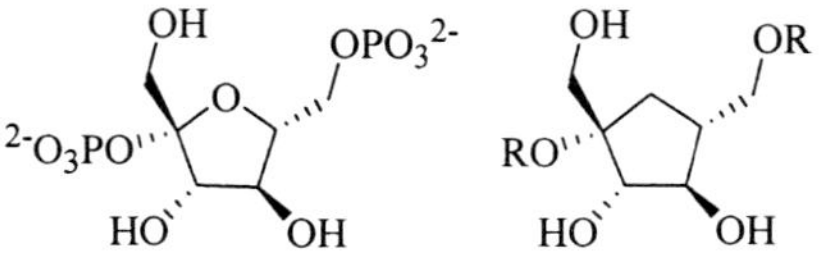

24a, R = H
24b, R = PO$_3^{2-}$

Fig. 3 Structures of fructose 2,6-biphosphate and carbafructofuranose.

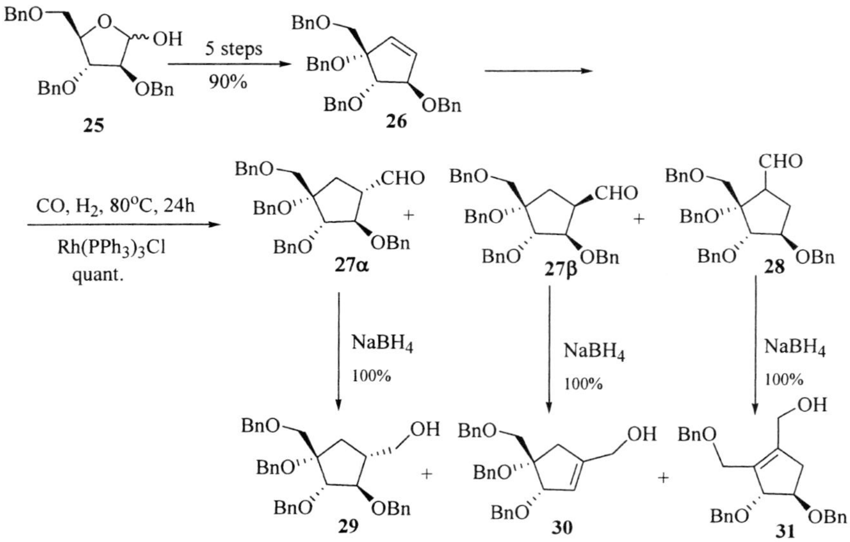

25 26 27α 27β 28 29 30 31

Scheme 10 Synthesis of potentially active carba-sugar analogs using hydroformylation of oxygenated cyclopentene.

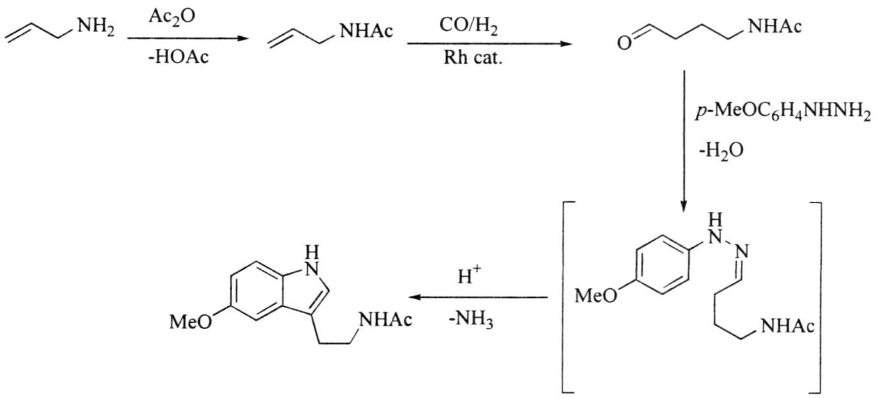

Scheme 11 A short synthesis of *N*-acyl-5-methoxytryptamine.

Cyclic amino acids are of increasing biological importance because of their relationship to active natural products, e.g., alkaloids [39] (piperidines), kainic acid analogs (pyrrolidines) [40], which are used as peptidomimetics [41]. A highly enantioselective synthesis of cyclic α-amino acids involving a one-pot, domino hydrogenation-hydroformylation sequence using a single rhodium catalyst was reported recently (Scheme 12) [42]. In the presence of both Rh-DuPHOS and Rh-BIPHEPHOS (Fig. 4), hydrogenation of the unsaturated amino acid ester **32 b** (rt, 30 psi of H_2, 3 h) followed by hydroformylation (80 °C, 80 psi of CO/H_2, 3 h) gave **34 b** and **35 b**

Scheme 12 Tandem hydrogenation-hydroformylation for synthesis of cyclic α-amino acids.

in a 2:1 ratio and 63% isolated yield. Importantly, the enantiomeric excess of **34b** and **35b** was found to be >95% *ee*. A similar reaction sequence involving **32a** with Rh-DuPHOS and Rh-PPh$_3$ gave **34a** and **35a** in ca. 1:1 ratio in 81% isolated yield. The *ee* of **34a** and **35a** was also shown to be ≥95%. Even using a single catalyst Rh(I)DuPHOS for both hydrogenation and hydroformylation, the reaction worked well under high pressure of CO/H$_2$ (400 psi) to yield the desired products.

Enantiopure non-proteinogenic amino acids continue to attract considerable interest, not only as precursors of unusual peptides and compounds possessing useful biological properties, but also as building blocks in organic synthesis [43]. For example, (3*R*)- and (3*S*)-piperazic acid moieties (Fig. 5) are embodied in Azinothricins, a family of anti-tumor antibiotics. Several synthetic approaches to these het-

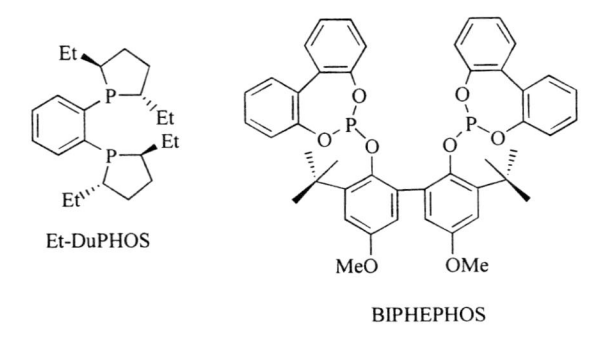

Fig. 4 Ligands for tandem hydrogenation-hydroformylation reactions.

Fig. 5 (3*S*)-Piperazic acid.

erocyclic compounds have been reported [44]. Here, the hydroformylation of the appropriate olefinic substrates offers a useful way to prepare large quantities of these intermediates.

As an example, chiral 3-piperazic acid is accessible following the strategy outlined in Scheme 13. The starting olefin 1,2-dicarbethoxy-1,2,3,4-tetrahydropyridazine is easily obtained from fairly cheap 1,3-butadiene and diethylazodicarboxylate in two steps [45]. The hydroformylation reaction carried out at 80–100 °C and 100 atm total pressure gave chemoselectivities up to 75% using a cobalt catalyst and almost quantitative yield in the presence of rhodium catalysts [46]. Since the known asymmetric synthesis of (R)- or (S)-3-piperazic acid is rather laborious [44], an enantioselective hydroformylation of 1,2-dicarbethoxy-1,2,3,4-tetrahydropyridazine represents a promising route to the above non-proteinogenic amino acid derivatives [47].

An interesting example of an enantiotopic differentiation in the hydroformylation was recently reported by Ojima and co-workers [48]: 4-tosylamino-1,6-heptadienes react in the presence of $Rh(CO)_2(acac)$ and the ligands BIPHEPHOS (Fig. 4) and SB-P-1011 (Scheme 14) to afford chiral heterocyclic products in high yield and selectivity (Scheme 14). It has been shown that these compounds represent useful precursors of alkaloid derivatives [49].

As unnatural antibiotics, 1-methylcarbapenems are among the most extensively investigated β-lactams in the last two decades. 1-Methylcarbapenems have a wide range of positive biological properties including strong antibacterial activity, resistance to β-lactamase, and metabolic stability [50]. Of particular note is the resistance to renal dehydropeptidase [51]. The β-lactam ring exhibits a rather surprising stability under hydroformylation conditions [52]. This fact allowed the synthesis of 1-β-methylcarbapenem via hydroformylation. 4-Vinyl-β-lactam (44) was subjected to a diastereoselective hydroformylation catalyzed by Rh(I) complexes with various chiral chelating phosphorus ligands (Scheme 15) [52]. The chemoselectivity of the reaction is generally high when using phosphine-phosphite ligands structurally related

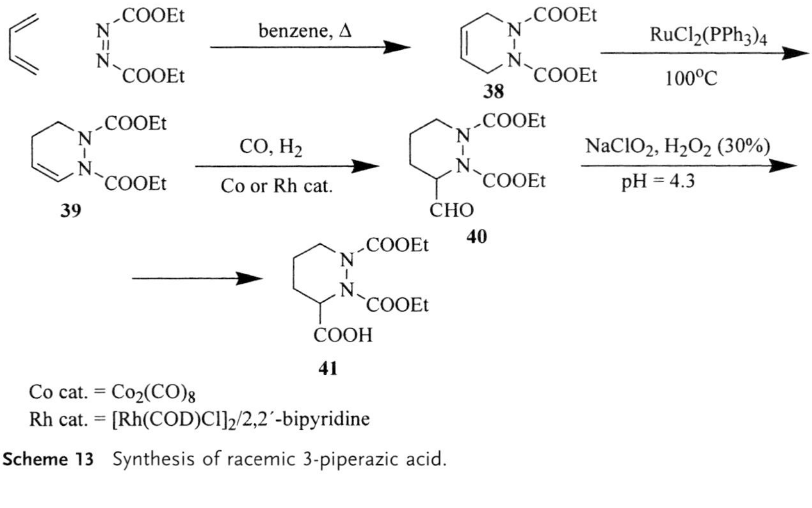

Co cat. = $Co_2(CO)_8$
Rh cat. = $[Rh(COD)Cl]_2$/2,2'-bipyridine

Scheme 13 Synthesis of racemic 3-piperazic acid.

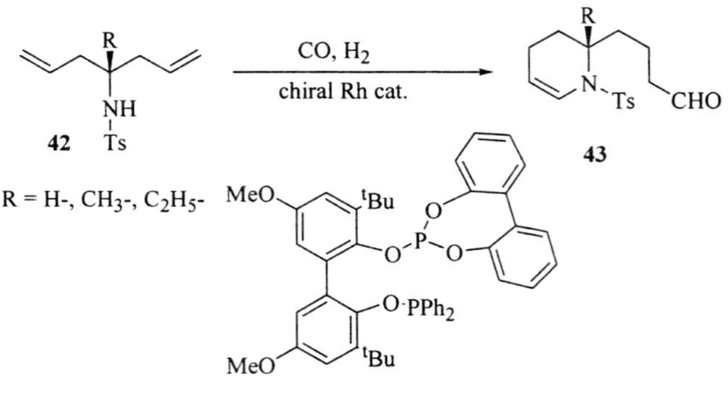

Scheme 14 Hydroformylation of 4-tosylamino-1,6-heptadienes to chiral heterocyclic precursors of alkaloid derivatives.

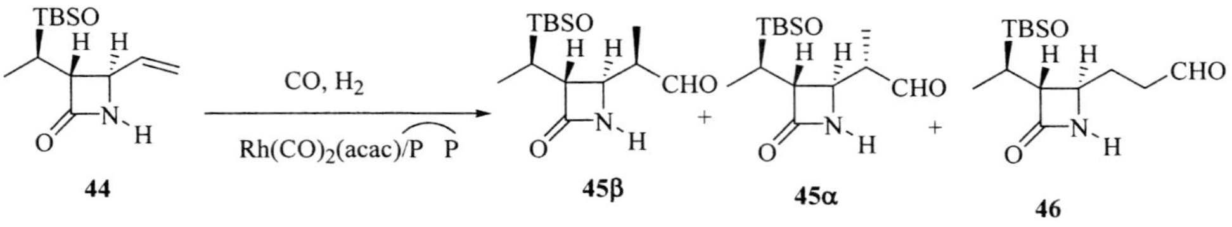

Scheme 15 Diastereoselective hydroformylation of 4-vinyl-β-lactams.

to BINAPHOS. The best results have been achieved carrying out the hydroformylation in the presence of the new ligand (R)-2-Nap-BIPNITE-p-F (I, Fig. 5) at 60 °C and 50 atm ($CO/H_2 = 1:1$). Here, a chemoselectivity of 95% and a regioselectivity of 74% toward the branched aldehyde were obtained, whereas the diastereoselectivity toward the formation of the desired α-epimer reached 96%.

In a recent report, the asymmetric hydroformylation of 4-vinyl-β-lactam was also catalyzed by a rhodium catalyst in the presence of chiral phosphine ligands. Here, the catalyst system consists of a zwitterionic rhodium catalyst, $(NBD)Rh^+$

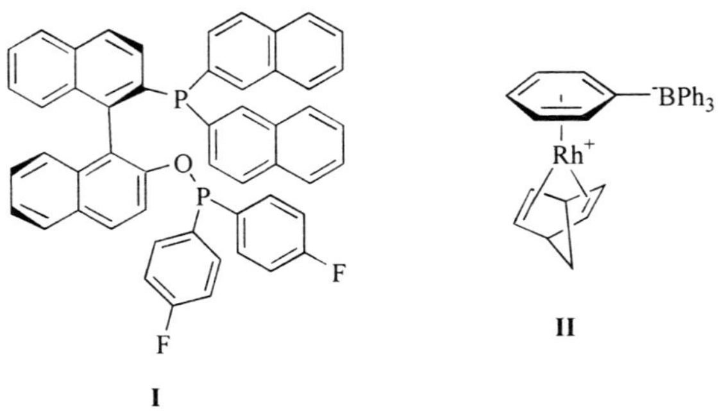

Fig. 6 (R)-2-Nap-BIPNITE-p-F and zwitterionic rhodium catalyst for diastereoselective hydroformylations.

($C_6H_5B^-PPh_3$) (II, Fig. 6) and (S,S)-2,4-bis(diphenylphosphino)pentane [(S,S)-BDPP] and gives branched aldehydes in high regio- and stereoselectivity [53].

2.1.3
Hydroformylation: Synthesis of Agrochemicals and their Intermediates

Both homogeneous and heterogeneous transition metal catalysis-based synthetic methodologies are nowadays commonly being explored for the production of agrochemicals [54]. Therefore it is not surprising that apart from its applications in the synthesis of pharmaceuticals, hydroformylation reactions are also used for the synthesis of agrochemicals. Especially the rhodium-catalyzed hydroformylation reaction can contribute to the solution of the manifold synthetic problems connected with molecules displaying herbicide, insecticide, or fungicide activity. For example, 2-aryloxypropanals derived from the hydroformylation of aryl vinyl ethers can easily be oxidized to the corresponding 2-aryloxypropanoic acids (Scheme 16) [4c].

Among these acids, several compounds are powerful herbicides, and their market is steadily growing [55]. In general, aryl vinyl ethers are rather reactive substrates toward hydroformylation, the regioselectivity to the more branched aldehyde being ≥90% in the presence of rhodium catalysts [4c]. 2,4-Dichlorophenyl, 2-methyl-4-chlorophenyl, and 1-naphthyl vinyl ethers have been transformed into the racemic 2-aryloxypropanoic acids in 60–85% yield applying $Rh_6(CO)_{16}$ as precatalyst [56]. Because the desired biological activity generally resides in only one enantiomer [55], asymmetric synthesis of these molecules is highly desired. Although first trials of catalytic asymmetric hydroformylation were disappointing [57], the newer generations of chiral ligands such as bulky phosphino-phosphites, e.g. BINAPHOS [10a], should allow for progress in this area.

Oxo-reactions can play an important role also in the preparation of another class of important chiral agrochemicals, namely Fenvalerate and its structural analogs. Fenvalerate [58] is a powerful broad-spectrum pyrethroid insecticide (Fig. 7).

A key part of Fenvalerate has been prepared by regioselective hydroformylation of 2-methyl-1-(4-chlorophenyl)propene (Scheme 17) at high pressure (100 atm)

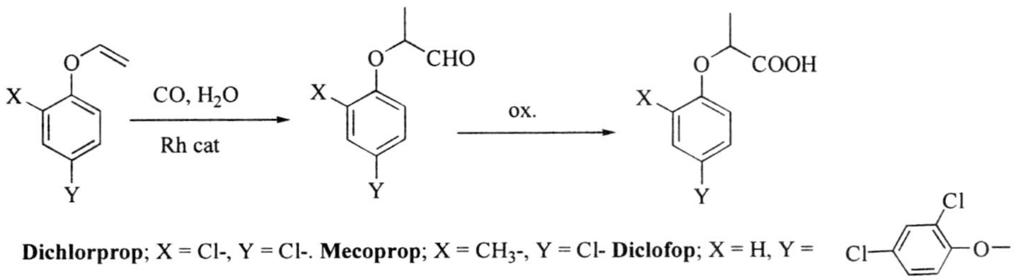

Dichlorprop; X = Cl-, Y = Cl-. **Mecoprop**; X = CH_3-, Y = Cl- **Diclofop**; X = H, Y =

Scheme 16 Hydroformylation of aryl vinyl ethers toward herbicide precursors.

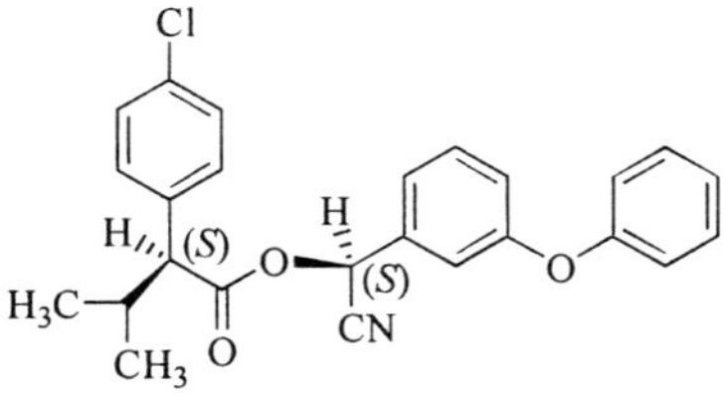

Fig. 7 Fenvalerate.

(S,S)-Fenvalerate

and 90–120 °C in the presence of common rhodium carbonyl complexes as catalyst precursors. As far as the regioselectivity is concerned, the desired branched aldehyde is the main product; however, it is accompanied by a variable amount (up to 26%) of the linear aldehyde 3-methyl-4-(4-chlorophenyl)butanal derived from double-bond isomerization of the starting olefin [59]. This isomerization reaction is significantly inhibited when a small amount of PPh$_3$ is present in the reaction medium. In fact, the highest yield of the desired aldehyde (88%) was reached by carrying out the reaction at 90 °C and 100 atm (CO/H$_2$=1) using Rh$_6$(CO)$_{16}$/PPh$_3$ (molar ratio=1:2) as the catalytic system. The obtained aldehyde was subsequently oxidized to the corresponding acid in almost quantitative yield. It has to be pointed out that Fenvalerate has two stereogenic centers. The diastereomer having both stereogenic centers of (S)-configuration among the four possible ones displays the highest insecticide activity. Thus, the enantioselective hydroformylation of 2-methyl-1-(4-chlorophenyl)propene could represent a promising approach to (S)-3-methyl-2-(4-chlorophenyl)butanoic acid, the key intermediate for the synthesis of Fenvalerate.

Obviously, other active agrochemicals related to Fenvalerate, as for example RP-40 (Fig. 8), can be conveniently prepared following the synthetic procedure outlined in Scheme 17.

In addition to the examples shown, hydroformylation reactions also serve as an interesting tool to synthesize intermediates for new agrochemicals like arylpropionaldehydes [60], 4-hydroxybutanals [61], and [(1-phenylethenyl)phenyl]propionaldehyde [62].

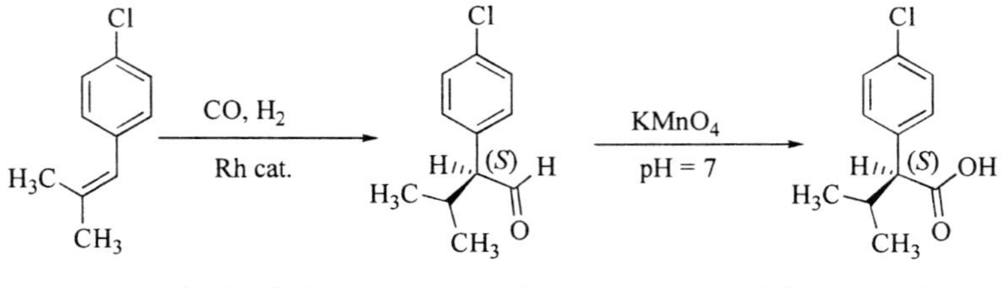

Scheme 17 ynthesis of a key intermediate for the preparation of the insecticide Fenvalerate.

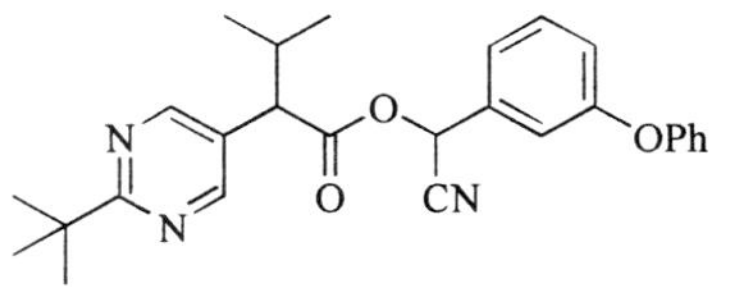

Fig. 8 Acaricide RP-40.

2.1.4
Hydroformylation: Examples of Fine Chemical Synthesis

Aldehydes which are produced via hydroformylation usually serve as building blocks for a variety of other products. Because of the versatile chemistry of the aldehyde group [63], they can be further converted via reduction, oxidation, or other reactions to give alcohols, amines, carboxylic acid derivatives, aldol condensation products, and many others. Thus, hydroformylation reactions offer several possibilities for the synthesis of fine chemicals intermediates. Clearly, because of the potential formation of different isomers and double-bond isomerization reactions, a careful optimization of reaction conditions and the actual catalyst is often required.

An example of the synthesis of an interesting fine-chemical building block is provided by the asymmetric hydroformylation of vinyl acetate. This reaction allows in principle for the conversion of an inexpensive bulk chemical in one step into a useful three-carbon building block [4a]. In Tab. 2, representative results obtained on the enantioselective hydroformylation of vinyl acetate are summarized.

From the data reported in Tab. 2 it is evident that the introduction of atropisomeric phosphino-phosphite ligands in 1993 enabled for the first time excellent regio- and enantioselectivities to be achieved. However, because of the limited catalyst stability and the high price of the ligand system, commercial applications of these systems seem to be difficult even in the fine chemicals industry.

Homobimetallic rhodium complexes with the binucleating racemic tetraphosphine ligand $(Et_2PCH_2CH_2)(Ph)PCH_2P(Ph)(CH_2CH_2PEt_2)$ are active and selective hydroformylation catalysts for a variety of terminal olefins [64a]. The striking activity of this homobimetallic rhodium complex was demonstrated to be due to a cooperative effect of both metal centers [64a, 65]. The racemic tetraphosphine ligand has been resolved into the enantiomerically pure (R,R)- and (S,S)-form, and the corresponding chiral bimetallic rhodium complexes have been prepared [66]. These complexes have been shown to be effective catalysts for the asymmetric hydroformylation of vinyl esters [67]. For instance, vinyl acetate was converted into (S)-2-acetoxypropanal with 80% regio- and 85% enantioselectivity. More recently, it was shown that hydroformylation of vinyl acetate in the presence of 0.5% of a rhodium(I) complex of the chiral diazaphospholidine ligand ESPHOS [68] gave ex-

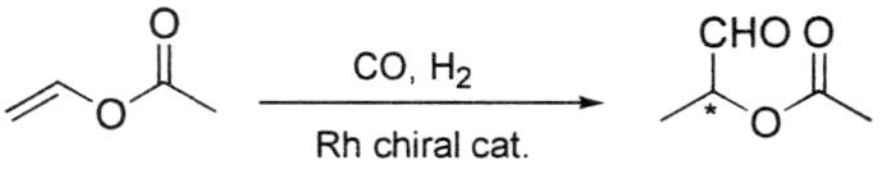

Scheme 18 Hydroformylation of vinyl acetate.

Tab. 2 Enantioselective hydroformylation of vinyl acetate (Scheme 18)

Catalyst precursor	P (atm) CO/H_2=1	Temp. (°C)	2-Acetoxypropanal			Ref.
			Yield (%)	ee (%)	Conf.	
[Rh(CO)$_2$Cl]$_2$/ (R)-BINAP	70	80	5[a]	47	(S)	64[b]
Rh(CO)$_2$(acac)/L[b]	100	20	89	41	(R)	64[b]
Rh(CO)$_2$(acac)/(iso-BHA-P)$_2$-2R,4 R-pentandiol	90	50	99	50	(S)	9
Rh(CO)$_2$(acac)/(R,S)-BINAPHOS	100	60	86	92	(S)	10[a]
Rh(CO)$_2$(acac)/(S,R)-BIPHENPHOS	100	60	55	90	(R)	10[b]
Rh$_2$(allyl)$_2$(Et-Ph-P$_4$)/HBF$_2$[c]	6	85	n.d.	85	n.d.	65
[Rh(acac)(CO)$_2$]/ESPHOS	8	60	90	89	(S)	69

a) Only 6% conversion of the substrate after 40 h was obtained.
b) L = B
c) Et-Ph-P4 = (Et$_2$PCH$_2$CH$_2$)PhPCH$_2$PPh(CH$_2$CH$_2$PEt$_2$)

L = B

ESPHOS

cellent results [69]. At 60 °C and 8 bar of synthesis gas pressure, a mixture of branched and linear isomers was obtained in 90% yield (n:iso=5:95), and the major product had 89% ee.

The asymmetric hydroformylation of allyl acetate was recently reported [70]. Here, rhodium complexes prepared by mixing Rh(CO)$_2$(acac) and chiral phosphino-phosphites derived from the reaction of enantiomerically pure cis- or trans-3-diphenylphosphinotetrahydrofuran-4-ol with atropoisomeric chlorophosphites were used as the catalytic precursors (Scheme 19). Typically, allyl acetate was converted into (+)-2-methyl-3-acetoxypropanal in 64% yield at 80 °C and 40 atm (CO/H_2=1) in the presence of the rhodium phosphino-phosphite complex (see Scheme 19). The enantiomeric excess, however, was rather low (up to 44%).

Another interesting class of alkenes for the development of a branched-selective hydroformylation are allylic ethers. The resulting products of such a reaction would be propionate aldols, a structural sub-unit contained in numerous bio-active natural products. One of the advantages of such a process is that the products (protected β-hydroxyaldehydes) are ready for further chain extension without the need for further protecting group manipulations. Directing groups have been shown to be an effective method for controlling the course of such reactions [71, 72]. As a recent example, Breit reported a general and effective method for the diastereoselective hydroformylation of methallyl esters [73]. Here, dibenzophosphol-5-ylmethyl ethers (47) were

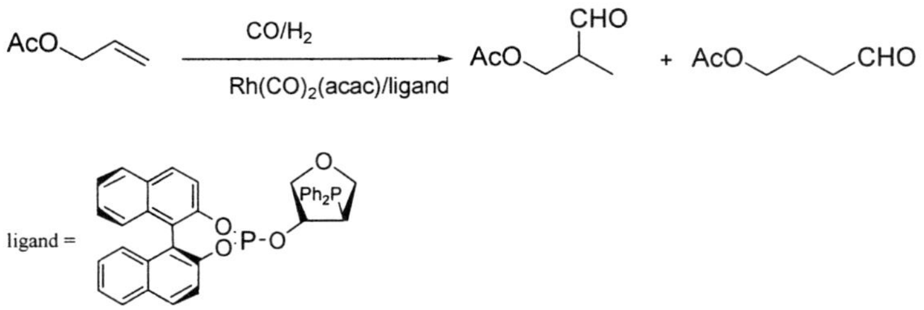

Scheme 19 Asymmetric hydroformylation of allyl acetate.

prepared in 77% yield [74] and subsequently subjected to hydroformylation reactions. As shown in Tab. 3, various aldehydes are produced, mainly as a single (>98:2) regioisomer [75, 76]. The reaction of the 1,1-disubstituted olefin (Tab. 3, entry 6) is especially remarkably because the branched (quaternary) aldehyde is produced with 92:8 regioselectivity. This is probably the first example of a highly branched-selective hydroformylation of a 1,1-dialkyl-substituted olefin.

Many methods have been developed for the stereoselective synthesis of 1,3-polyol chains, a recurring motif in several important natural products [77]. With a notable exception in the work of Rychnovsky [78], many of these strategies have

Tab. 3 Directed hydroformylation of dibenzophosphol-5-ylmethyl ethers of allylic alcohols

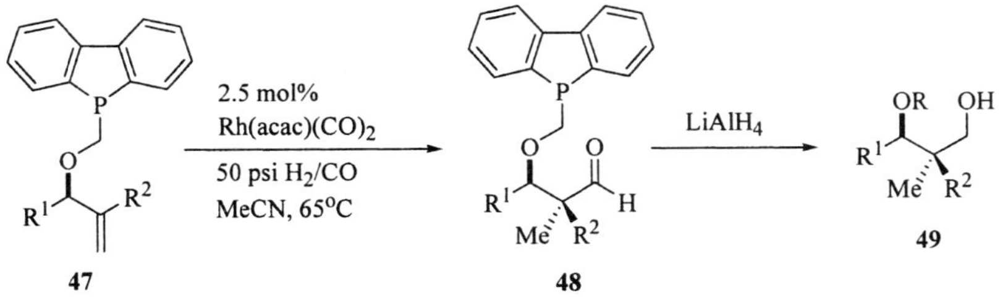

Entry No.	R^1	R^2	rs	ds	Yield 48 (%)
1	Me	H	>98:2	81:19	92
2	Ph	H	>98:2	86:14	96
3	i-Pr	H	>98:2	94:6	94
4	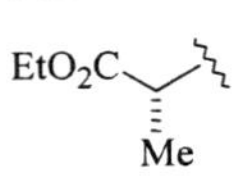 EtO$_2$C/Me	H	>98:2	90:10	98
5	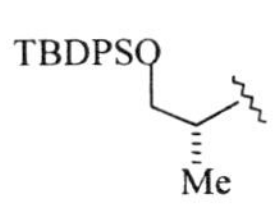 TBDPSO/Me	H	>98:2	93:7	87
6	H	Me	92:8	–	88

focused on the construction of β-hydroxy carbonyls or their equivalents by way of aldol addition or allylation reactions or related processes. However, Leighton et al. [79] demonstrated elegantly that the rhodium-catalyzed hydroformylation of defined enol ethers is an effective method for the diastereoselective synthesis of β-hydroxy aldehydes [80].

It should be noted that, despite the vast body of literature on hydroformylation reactions, examples using acyclic enol ethers as substrates are relatively rare. Tab. 4 outlines the results for the hydroformylation of several enol ethers. It was observed that the *tert*-butyl group in the acetal position is not necessary for high

Tab. 4 Rhodium-catalyzed hydroformylation of enol ethers

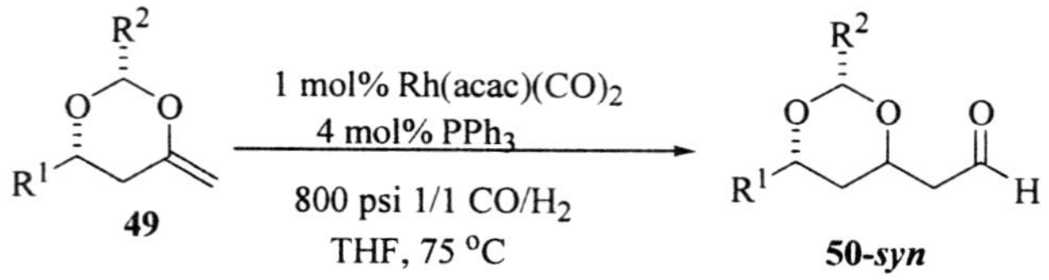

Entry	R^1	R^2	Major product	Yield (%)	Regiosel.
1.	H	t-Bu	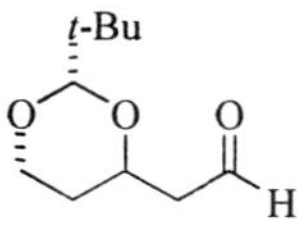	81	12:1
2.	Me	t-Bu	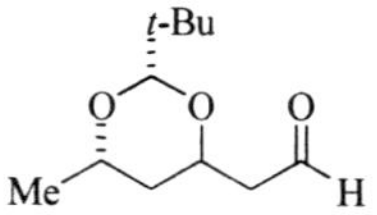	72	13:1
3.	Me	Me	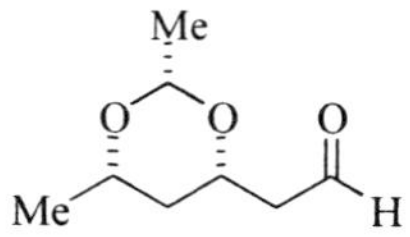	75	13:1
4.	BnO(CH$_2$)$_2$	Me	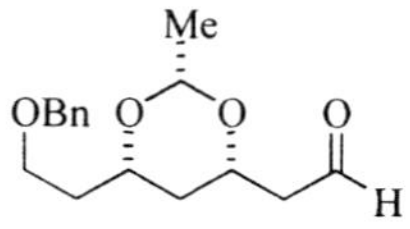	71	9:1
5.	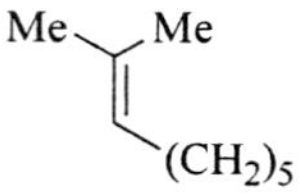 Me$_2$C=C(CH$_2$)$_5$—	Me	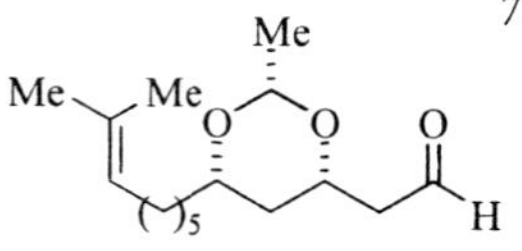	71	11:1

selectivity. Indeed, no difference was observed by using the corresponding acetaldehyde-derived acetal (Tab. 4, entry 2 versus entry 3).

The hydroformylation of cyclic enol ethers like 2,3-dihydrofuran in the presence of rhodium catalysts was carefully investigated a few years ago [81]. The binuclear complex [Rh$_2${μ-S(CH$_2$)$_3$NMe$_2$}(COD)$_2$] in conjunction with suitable ligands was used as the catalytic system under mild conditions (40–80 °C; 5 atm CO/H$_2$=1), generally achieving very high conversions and chemoselectivities (Scheme 20). Usually, tetrahydrofuran-2-carbaldehyde is the prevailing regioisomer in this reaction. For example, a 75% yield was reached using tris(o-tert-butylphenyl)phosphite as ligand. On the other hand, tetrahydrofuran-3-carbaldehyde is selectively formed when 2,5-dihydrofuran is hydroformylated at 80 °C and 5 atm using the above dinuclear rhodium complex modified with tris(trimethyl)phosphite as the catalytic precursor. In another selective hydroformylation of 2,5-dihydrofuran using HRh(CO)(PPh$_3$)$_3$, tris(2-tert-butylphenyl)phosphite along with 1,4-bis(diphenyl-phosphino)-butane (dppb) and ethanolamine in iso-propanol under 40 atm pressure of syngas at 55 °C for 2 h, for 99% raw material conversion a selectivity of 87% was observed [82].

The preparation of enantiomerically pure tetrahydrofuran-2-carboxylic acid starting form 2,3-dihydrofuran, an easily available material only recently commercialized by Eastmann-Kodak at about US $10 per kg, appears attractive. The chiral acid is a valuable building block for therapeutically active molecules such as Terazosin and structurally related compounds [83].

Among other substrates for hydroformylation, vinyl fluoride appears to be interesting because α-fluoroacrylic acid esters represent valuable monomers to produce polymeric materials with interesting properties such as optical glasses, high performance optical fibers, membranes, coatings etc. [84]. Moreover, α-fluoroacrylic acid derivatives are employed as starting compounds for the synthesis of more complex organofluorine compounds [85]. It is known that vinyl fluoride, an easily available and important monomer used in the polymer industry to produce high-performance homo- and co-polymers, can be hydroformylated under standard conditions in the presence of rhodium carbonyl complexes to 2-fluoropropanal in good yields (up to 81%) [86]. Careful oxidation to 2-fluoroalkanoic acid proceeded in about 70% yield (Scheme 21) [87]. Dehydrogenation of 2-fluoroalkanoic acid to α-fluoroacrylic acid was accomplished by a chlorination-dehydrochlorination reaction in more than 60% yield.

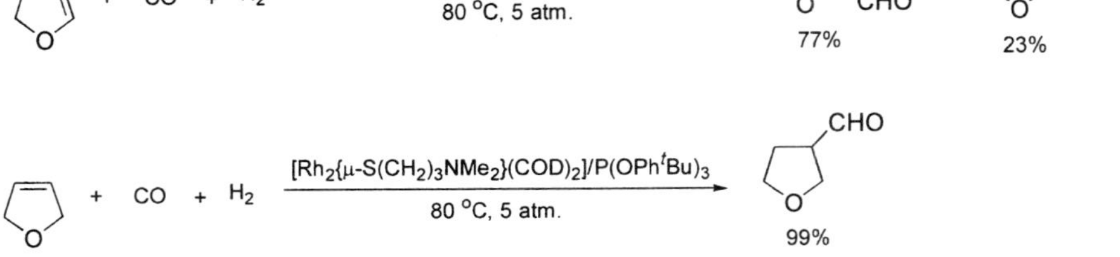

Scheme 20 Hydroformylation of cyclic enol ethers.

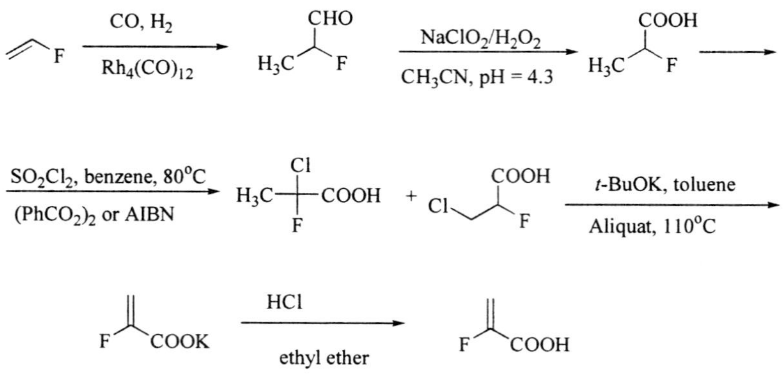

Scheme 21 Hydroformylation of vinyl fluoride followed by oxidation to α-fluoroacrylic acid.

A straightforward preparation of 2-(trifluoromethyl)acrylic acid, a useful intermediate for the preparation of several fluorinated therapeutically active molecules [88], was accomplished starting from 3,3,3-trifluoropropene. This olefin presents the peculiarity of undergoing hydroformylation giving a very high yield (90–95%) of the branched aldehyde in the presence of rhodium catalysts and of the linear isomer if cobalt catalysts are employed (Scheme 22) [89]. The resulting fluoroaldehydes were *in situ* α-selenenylated and oxidized with 30% H_2O_2 to produce the desired unsaturated acids in 73–75% overall yield [90].

From an academic point of view, silylaldehydes are interesting bifunctional reagents, with the trialkylsilyl group behaving as a latent carbanion [91]. Such aldehydes are difficult to prepare by conventional organic synthetic methods. However, hydroformylation of commercially available trialkylvinylsilanes under mild conditions offers an efficient route to silylaldehydes (Scheme 23). The first example of a regioselective formation of α- or β-trialkylsilylaldehydes by hydroformylation of vinylsilanes was reported in the mid 1990s [91]. Here, the cationic rhodium complex $Rh(COD)_2BPh_4$ catalyzed the reaction, giving the branched isomer as the major product (60–70% of the total aldehyde produced) at 100 °C and 14

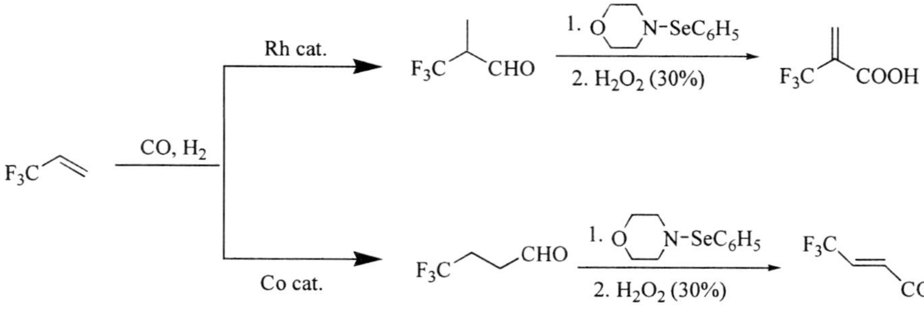

Scheme 22 Hydroformylation of 3,3,3-trifluoropropene and subsequent functionalization.

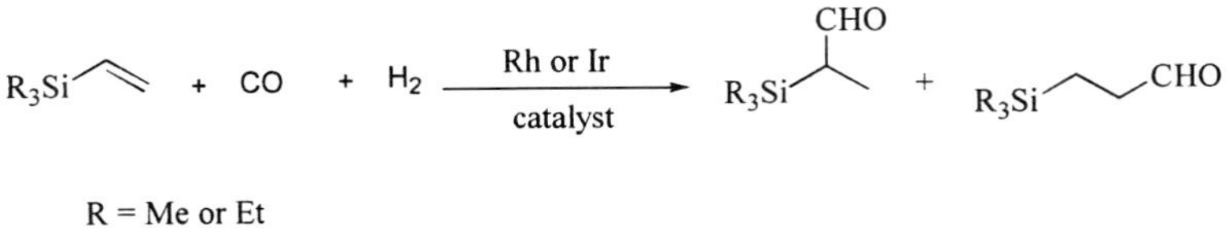

2.1.4 Hydroformylation: Examples of Fine Chemical Synthesis | 49

Scheme 23 Hydroformylation of trialkylvinylsilanes.

atm (CO/H$_2$ = 1:2). The addition of 2 equivalents of PPh$_3$ caused a complete shift in the selectivity, affording the β-trialkylsilylaldehyde as the major product using triethylvinylsilane as the substrate (branched to linear isomer ratio = 7:93). Interestingly, iridium complexes such as hydrated IrCl$_3$ or [Ir(COD)$_2$]+BF$_4^-$ showed opposite regioselectivity, producing the linear silylaldehyde without the addition of phosphine. The isolated yields range from 30% for the branched silylaldehyde to 91% for the linear one.

In recent years, apart from synthetic applications, considerable attention has been paid to the attachment of homogeneous catalysts to insoluble supports in an

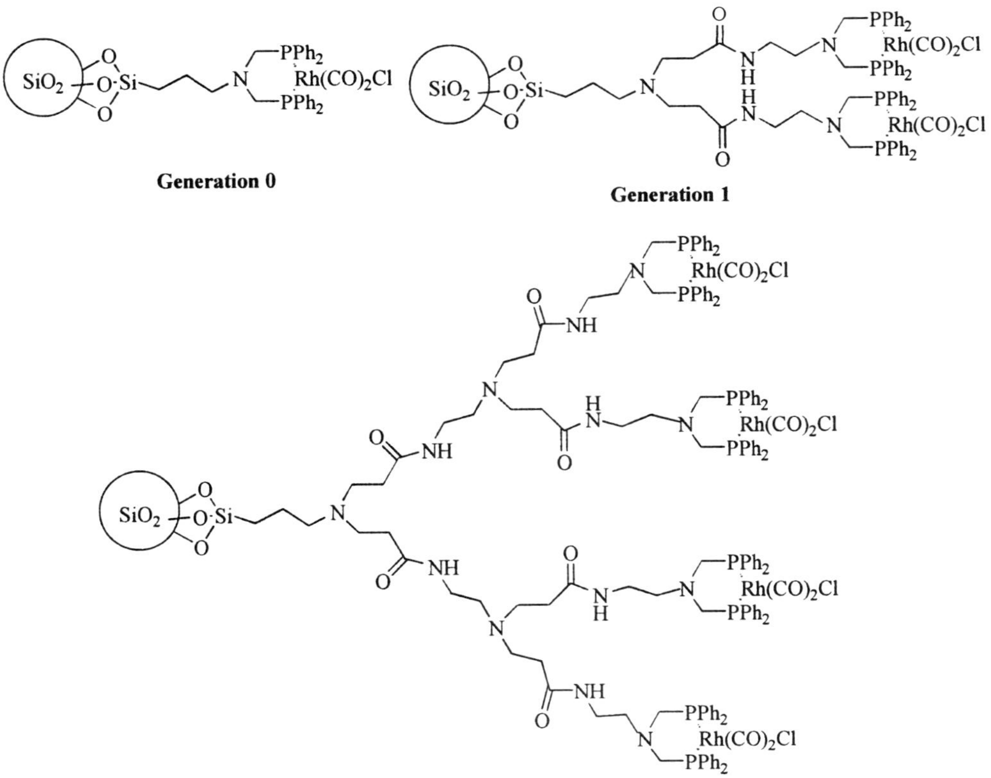

Fig. 9 Proposed structures of different generations of Rh-PPh$_2$-PAMAM-SiO$_2$ complexes.

attempt to combine the practical advantages of heterogeneous catalysis with the efficiency of homogeneous catalysis [92, 93]. Alper et al. reported that a silica-supported polyamidoamine dendrimer system (Rh-PPh$_2$-PAMAM-SiO$_2$) is an active and regioselective catalyst for hydroformylation of aryl olefins and vinyl esters. Polyaminoamido dendrimers, up to the fourth generation, were constructed on the surface of silica gel particles (Fig. 9) [94].

Excellent selectivities, favoring the branched aldehydes obtained from aryl olefins and vinyl esters, were observed by using the corresponding Rh(I) complex. The heterogeneous Rh(I) catalyst has been recycled and reused (up to four cycles) without significant loss of selectivity or activity.

Another recent trend is the combination of hydroformylation reactions with other transformations to new one-pot or domino processes. For example, hydroformylation of alkenes followed by reductive amination, the so-called hydroaminomethylation reaction, can be controlled to yield selectively linear amines from internal olefines [95]. Also, this method constitutes an economically attractive and environmentally favorable synthesis of linear aliphatic amines [96]. Similarly, a number of domino reactions involving sequential, hydroformylation-Mukaiyama aldol, hydroformylation-allylboration, hydroformylation-Wittig olefination and hydroformylation-hydrogenation combinations find increasing popularity [97].

2.1.5
Conclusions and Outlook

Active research in the field of hydroformylation over more than six decades has led to a number of important bulk chemical processes. However, comparably few industrial applications of this method for the synthesis of more complicated organic products are known. Many organic chemists still doubt whether this catalytic reaction can actually be exploited to produce valuable intermediates and specialty chemicals at economically viable costs, hence replacing old preparative methods based on conventional organic chemistry. This indifference toward the reaction has obviously grown because of the difficulty in controlling selectivity throughout the course of the hydroformylation. This situation has significantly improved during the past decade. Thus, control of regioselectivity of terminal alkenes in favor of the linear aldehydes is possible today. However, there is still no practical solution to the problem of achieving a general *iso*-selective hydroformylation of simpler aliphatic terminal alkenes.

At present, different research groups, both academic and industrial, are engaged in developing more active and selective hydroformylation reaction systems by (a) varying the structures of the ligands in transition metal complexes [3, 98] and/or introducing in the process bi- or polymetallic complexes as catalysts precursors [3, 99], (b) employing different olefins in hydroformylation to synthesize organic intermediates of industrial/medicinal importance [2, 3], (c) using newer techniques involving heterogeneous catalysis on a solid support [100] and biphasic hydroformylation using water-soluble rhodium complexes with hydrophilic ligands [101], and (d) the use of new chiral ligands to achieve higher enantioselectivities in asymmetric hydroformylation of a variety of olefins [102]. Despite significant improvements in the past decade, the simultaneous control of regio- and stereoselectivity is still a distant goal. The use of rhodium complexes with bulky phosphites, which ensure good chemo- and regioselectivities in the hydroformylation of functionalized olefines [103], is worthy of deeper study and further experimentation.

Hydroformylation reactions today provide a number of aldehydes bearing additional functionalities, which are more difficult to prepare by conventional organic synthetic routes. Oxo reaction of olefines containing different oxygen, nitrogen, halogen, sulfur, and silicon functional groups provides various di- or trifunctional synthons for the preparation of rather complex molecules like natural products, antibiotics, and other bio-active molecules.

Nevertheless, more endeavors are required to address selectivity, activity, and recycling problems before organic chemists both in industry and in the academic area will apply this reaction with more confidence in their synthetic strategies.

2.1.6
References

1 L.A. VAN DER VEEN, P.C.J. KAMER, P.W.N.M. VAN LEEUWEN, *Angew. Chem. Int. Ed.* **1999**, *38*, 336.
2 (a) B. CORNILS in *New Synthesis with Carbon Monoxide* (Ed.: J. FALBE), Springer, Berlin, **1980**, pp. 1–225. (b) P. PINO, *J. Organomet. Chem.* **1980**, *40*, 223. (c) C. BOTTEGHI, R. GANZERLA, M. LENARDA, G. MORETTI, *J. Mol. Catal.* **1987**, *40*, 129. (d) G.W. PARSHALL, S.D. ITTEL, *Homogeneous Catalysis*, John Wiley, New York, **1992**, 2nd edn.
3 M. BELLER, B. CORNILS, C.D. FROHNING, C.W. KOHLPAINTNER, *J. Mol. Catal.* **1995**, *104*, 17.
4 (a) C. BOTTEGHI, S. PAGANELLI, A. SCHIONATO, M. MARCHETTI, *Chirality* **1991**, *3*, 355. (b) G. CONSIGLIO in *Catalysis in Asymmetric Synthesis* (Ed.: I. OJIMA), VCH, New York, **1993**, p. 273. (c) C. BOTTEGHI, M. MARCHETTI, S. PAGANELLI, *Trends in Organometallic Chemistry*, Council of Scientific Information, Trivandrum, India, **1994**, pp. 433–463. (d) S. GLADIALI, J.C. BAYON, C. CLAVER, *Tetrahedron: Asymmetry* **1995**, *6*, 1453. (e) C. Botteghi, M. Marchetti, G. del Ponte, *Quimica Nova* **1997**, *20*, 30. (f) B. BREIT, W. SEICHE, *Synthesis* **2001**, 1. (g) H.-W. BOHNEN, B. CORNILS, *Adv. Catal.* **2002**, *47*, 1. (h) B. VISHWANATHAN, S. SIVASANKER, A.V. RAMASWAMY, *Catalysis* **2002**, 311. (i) J.F. HARTWIG, *Science* **2002**, *297*, 1653. (j) L.A. VAN DER VEEN, C.J. PAUL, P.W.N.M. VAN LEEUWEN, *CATTECH* **2002**, *6*, 116. (k) B. BREIT, *Acc. Chem. Res.* **2003**, *36*, 264.
5 S. KOTHA, *Tetrahedron* **1994**, *50*, 3653.
6 H.R. SONAWANE, N.S. BELLUR, J.R. AHUJA, D.G. KULKARNI, *Tetrahedron: Asymmetry* **1992**, *3*, 163.
7 A. VAN ROOY, E.N. ORIJ, P.C.J. KRAMER, P.W.N.M. VAN LEEUVEN, *Organometallics* **1995**, *14*, 34.
8 G. CONSIGLIO, S.C.A. NEFKENS, *Tetrahedron: Asymmetry* **1990**, *1*, 417.
9 J.E. BABIN, G.T. WHITEKER, *PCT Int. Appl. WO 93/03839* (Union Carbide) **1993**; *Chem. Abstr.* **1993**, *119*, 159872.
10 (a) N. SAKAI, S. MANO, K. NOZAKI, H. TAYAKA, *J. Am. Chem. Soc.* **1993**, *115*, 7033. (b) T. HIGASHIZAMA, N. SAKAI, K. NOZAKI, H. TAYAKA, *Tetrahedron Lett.* **1994**, *35*, 2023. (c) N. SAKAI, K. NOZAKI, H. TAKAYA, *J. Chem. Soc., Chem. Commun.* **1994**, 395.

11 F. Agbossou, J. F. Carpentier, A. Mortreux, Chem. Rev. **1995**, 95, 2485.
12 (a) T. V. Rajanbabu, T. V. Ayers, Tetrahedron Lett. **1994**, 35, 4295. (b) S. Gladiali, D. Fabbri, L. Kollar, J. Organomet. Chem. **1995**, 491, 91. (c) G. J. H. Buisman, M. E. Martin, E. J. Vos, A. Klootwijk, P. C. J. Kramer, P. W. N. M. van Leeuwen, Tetrahedron: Asymmetry **1995**, 6, 719. (d) A. M. Masdes-Bult, S. Orejin, A. Castillon, C. Claver, Tetrahedron: Asymmetry **1995**, 6, 1885. (e) A. Orejin, A. Castellanos, A. M. Masdeu, C. Claver, S. Castillon, Proceedings of 10th International Symposium on Homogeneous Catalysis, Princeton, August **1996**, p. A4.
13 (a) S. Naili, J.-F. Carpentier, F. Agbossou, A. Montreux, Organometallics **1995**, 14, 401. (b) A. Scrivanti, V. Beghetto, A. Bastianini, U. Matteoli, G. Menchi, Organometallics **1996**, 15, 4687.
14 M. Dieguez, O. Pamies, A. Ruiz, S. Castillon, C. Claver, Chem. Eur. J. **2001**, 7, 3086.
15 G. Francio, W. Leitner, Chem. Commun. **1999**, 1663.
16 C. Botteghi, G. Chelucci, G. Del Ponte, M. Marchetti, S. Paganelli, J. Org. Chem. **1994**, 59, 7125.
17 C. Botteghi, S. Paganelli, M. Marchetti, G. Del Ponte, H. A. Stefani, U. Azzena, B. Sechi, An. Quim. Int. Ed. **1998**, 94, 210.
18 C. Botteghi, S. Paganelli, L. Bigini, M. Marchetti, J. Mol. Catal. **1994**, 93, 279.
19 (a) C. Botteghi, M. Marchetti, S. Paganelli, B. Sechi, J. Mol. Catal. **1997**, 118, 173. (b) C. Botteghi, L. Cazzolato, M. Marchetti, S. Paganelli, J. Org. Chem. **1995**, 60, 6612.
20 R. Lazzaroni, G. Uccello-Barretta, S. Scamuzzi, R. Settambolo, A. Caiazzo, Organometallics **1996**, 15, 4657.
21 T. Baig, J. Molinier, P. Kalck, J. Organomet. Chem. **1993**, 455, 219 and references therein.
22 (a) P. Masi, A. Monopoli, A. D. Saravalle, C. Zio, Ger. Offen. 3 002 909 (Italiana Schoum S. p. A.), **1980**; Chem. Abstr. **1980**, 93, 239250. (b) Pharma Project Structures, Acc. No. 7291, P. J. B. Publishers Ltd., Richmond, Surrey, UK, **1996**.
23 T. Mizoroki, M. Kioka, M. Suzuki, S. Sakatani, A. Okumura, K. Maruya, Bull. Chem. Soc. Jpn. **1984**, 57, 577.
24 E. Dalcanale, F. Montanari, J. Org. Chem. **1986**, 51, 567.
25 Pharma Project Structures, Acc. No. 949, P. J. B. Publishers Ltd., Richmond, Surrey, UK, **1996**.
26 (a) K. Sindelar, M. Rajsner, I. Cervena, V. Valenta, J. O. Jilek, B. Kakac, J. Holubek, E. Svatek, F. Miksik, M. Protiva, Collect. Czech. Commun. **1973**, 38, 3879. (b) Pharma Project Structures, Acc. No. 544, P. J. B. Publishers Ltd., Richmond, Surrey, UK, **1996**.
27 Pharma Project Structures, Acc. No. 945, P. J. B. Publishers Ltd., Richmond, Surrey, UK, **1996**.
28 M. Schlosser, B. Schaub, Chimia **1982**, 36, 396.
29 C. Botteghi, S. Paganelli, M. Marchetti, P. Pannocchia, Proceedings of 10th International Symposium on Homogeneous Catalysis, Princeton, August **1996**, p. B26.
30 J. Elks, G. R. Ganellin, Dictionary of Drugs, Chapman and Hall, London, **1990**.
31 (a) G. Bartoszyk, R. Devant, H. Boettcher, H. Greiner, J.-J. Berthelon, M. Brunet, M. Noblet, J.-J. Zeiller, Merck, EP 0707007A, **1996**. (b) B. Junge, R. Schohe, P.-R. Seidel, T. Glaser, J. Traber, U. Benz, T. Schuurman, J.-M. De Vry, Bayer AG, US Patent 5 137 902, **1992**. (c) G. Yiannikouros, P. Kalaritis, J. G. Streenrod, R. Scarborough, Cor Therapeutics Inc., USA, WO 0187871, **2001**. (d) D. A. Clark, S. W. Goldstein, R. A. Volkmann, J.-F. Eggler, G. F. Holland, B. Hulin, R. W. Stevenson, D. K. Kreutter, M. E. Gibbs, et al., J. Med. Chem. **1991**, 34, 319.
32 C. Botteghi, S. Paganelli, F. Moratti, M. Marchetti, R. Lazzaroni, R. Settambolo, O. Piccolo, J. Mol. Cat. **2003**, 200, 147.
33 R. Roggenbuck, P. Eilbracht, Tetrahedron Lett. **1999**, 40, 7455.
34 For applications of hydroformylation in heterocyclic synthesis, see: (a) C. L. Kranemann, B. E. Kitos-Rzychon, P. Eil-

BRACHT, *Tetrahedron* **1999**, *55*, 4721. (b) D.J. BERGMANN, E.M. CAMPI, W.R. JACKSON, A.F. PATTI, *Aust. J. Chem.* **1999**, *52*, 1131. (c) D.J. BERGMANN, E.M. CAMPI, W.R. JACKSON, A.F. PATTI, *Chem. Commun.* **1999**, 1279. (d) R. LAZZARONI, R. SETTAMBOLO, M. MARIANI, A. CAIAZZO, *J. Organomet. Chem.* **1999**, *592*, 69. (e) B. SCHMIDT, B. COSTISELLA, R. ROGGENBUCK, M. WESTHUS, H. WILDEMANN, P. EILBRACHT, *J. Org. Chem.* **2001**, *66*, 7658. (f) R. SETTAMBOLO, A. CAIAZZO, R. LAZZARONI, *Tetrahedron Lett.* **2001**, *42*, 4045. (g) R. ROGGENBUCK, A. SCHMIDT, P. EILBRACHT, *Org. Lett.* **2002**, *4*, 289.

35 (a) M. SEEPERSAUD, M. KETTUNEN, A.S. ABU-SURRAH, T. REPO, W. VOELTER, Y. AL-ABED, *Tetrahedron Lett.* **2002**, *43*, 1793. (b) M. SEEPERSAUD, M. KETTUNEN, A.S. ABU-SARRAH, W. VOELTER, Y. AL-ABED, *Tetrahedron Lett.* **2002**, *43*, 8607.

37 (a) S. OGAWA, N. MATSUNAGA, H. LI, M.M. PALCIC, *Eur. J. Org. Chem.* **1999**, *3*, 631. (b) S. OGAWA, T. FURAYA, H. TSUNODA, O. HINDSGAUL, K. STABGIER, M.M. PALCIC, *Carbohydr. Res.* **1995**, *271*, 197.

38 G. VERSPUI, G. ELBERTSE, F.A. SHELDON, M.A.P.J. HACKING, R.A. SHELDON, *Chem. Commun.* **2000**, 1363.

39 H. TAKAHARA, H. BANDOH, T. MOMOSE, *Tetrahedron* **1993**, *49*, 11205.

40 A.F. PARSONS, *Tetrahedron* **1996**, *52*, 4149.

41 S. HANESSIAN, G. MCNAUGHTON-SMITH, H.-G. LOMBART, W.D. LUBELL, *Tetrahedron* **1997**, *53*, 12789.

42 (a) E. TEOH, E.M. CAMPI, W. ROY, A.J. ROBINSON, *Chem. Commun.* **2002**, 978. (b) E. TEOH, E.M. CAMPI, W.R. JACKSON, A.J. ROBINSON, *New J. Chem.* **2003**, *27*, 387.

43 I.H. ASPINALL, P.M. COWLEY, G. MITCHELL, R.J. STOODLEY, *Chem. Commun.* **1993**, 1179 and references therein.

44 K.J. HALE, V.M. DELISSER, S. MANAVIAZAR, *Tetrahedron Lett.* **1992**, *33*, 7613 and references therein.

45 G. MENCHI, U. MATTEOLI, A. SCRIVANTI, S. PAGANELLI, C. BOTTEGHI, *J. Organomet. Chem.* **1988**, *354*, 215.

46 G. MENCHI, S. PAGANELLI, U. MATTEOLI, A. SCRIVANTI, C. BOTTEGHI, *J. Organomet. Chem.* **1993**, *450*, 229.

47 G. MENCHI, S. PAGANELLI, U. MATTEOLI, A. SCRIVANTI, C. BOTTEGHI, unpublished results.

48 Z. LI, M. TZAMARIOUDAKI, D.M. IULA, I. OJIMA, *Proceedings of 10th International Symposium on Homogeneous Catalysis*, Princeton, August **1996**, p. B46.

49 R.B. HERBERT in *Comprehensive Organic Chemistry* (Ed.: E. HASLAM), Pergamon Press, Oxford, **1979**, Vol. 5, pp. 1045–1119.

50 (a) K. ODA, A. YOSHIDA, *Tetrahedron Lett.* **1997**, *38*, 5687. (b) A.H. BERKS, *Tetrahedron* **1996**, *52*, 331 and references therein.

51 D.H. SHIH, L. CAMA, B.G. CHRISTENSEN, *Tetrahedron Lett.* **1985**, *26*, 587.

52 K. NOSAKI, W. LI, T. HORIUCHI, H. TAYAKA, *J. Org. Chem.* **1996**, *61*, 7658.

53 H.S. PARK, E. ALBERICO, H. ALPER, *J. Am. Chem. Soc.* **1999**, *121*, 11697.

54 J. CROSBY, *Pestic. Sci.* **1996**, *46*, 11.

55 A. WILLIAMS, *Pestic. Sci.* **1996**, *46*, 4.

56 C. BOTTEGHI, G. DELOGU, M. MARCHETTI, S. PAGANELLI, B. SECHI, *J. Mol. Catal.* **1999**, *143*, 311.

57 M. MARCHETTI, G. DELOGU, C. BASOLI, *9th International Symposium on Homogeneous Catalysis*, Jerusalem, August 21–26, **1994**, pp. 104–105.

58 H.P.M. VIJVERBERG, M. OORTGIESEN in *Stereoselectivity of Pesticides* (Eds. E.J. ARIENS, J.J.S. VAN RENSEN, W. WELLING), Elsevier, Amsterdam, **1988**, p. 151 and references therein.

59 C. BOTTEGHI, D. DALLA BONA, S. PAGANELLI, M. MARCHETTI, B. SECHI, *An. Quim. Int. Ed.* **1996**, *92*, 101.

60 (a) P. HEN, Y. FUJITA, H. ONE, *Jpn. Kokai Tokkyo Koho* **1989**, JKXXAF JP 01203346 A2 19890816. (b) H. ONO, T. KASUGA, S. KYONO, *Jpn. Kokai Tokkyo Koho* **1989**, JKXXAF JP 01121233 A2 19890512. (c) I. SHIMIZU, H. NOMURA, K. UCHIDA, Y. MATSUMURA, Y. ARAI, *Jpn. Kokai Tokkyo Koho* **1989**, JKXXXAF JP 01013047 A2 19890117.

61 T. SUZUKI, H. UCHIDA, K. MARUMO, *Jpn. Kokai Tokkyo Koho* **1989**, JKXXAF JP 01121234 A2 19890512.

62 I. SHIMIZU, H. NOMURA, K. UCHIDA, Y. MATSUMURA, Y. ARAI, *Jpn. Kokai Tokkyo*

Koho **1989**, JKXXXAF JP 01013048 A2 19890117.

63 (a) S. Patai (ed.): *The Chemistry of the Carbonyl Group*, Wiley Interscience, New York, **1966**, 1970. (b) J. Falbe (ed.): *Methoden der Organischen Chemie* (Houben-Weyl), Thieme: Stuttgart, Germany, **1983**, Vol. E3 (Aldehyde).

64 (a) M.E. Broussard, B. Juma, S.G. Train, W.J. Peng, S.A. Laneman, G.G. Stanley, *Science* **1993**, *260*, 1784. (b) N. Sakai, K. Nosaki, K. Mashima, H. Takaya, *Tetrahedron: Asymmetry* **1992**, *3*, 583.

65 (a) W.-J. Peng, S.G. Train, D.K. Howell, F.R. Fronczek, G.G. Stanley, *J. Chem. Soc. Chem. Commun.* **1996**, 2607. (b) R.C. Matthews, D.K. Howell, W.-J. Peng, S.G. Train. W.D. Treleaven, G.G. Stanley, *Angew. Chem. Int. Ed.* **1996**, *35*, 2253.

66 G.G. Stanley, P. Albuquerque, D.K. Howell, W.J. Peng, F.R. Fronczek, *Proceedings of 10th International Symposium on Homogeneous Catalysis*, Princeton, August **1996**, p. CL–12.

67 G.G. Stanley, *Catalysis of Organic Reactions*, Marcel Dekker, New York, **1995**, pp. 363–372.

68 S.W. Breeden, M. Wills, *J. Org. Chem.* **1999**, *64*, 9735.

69 S. Breeden, D.J. Cole-Hamilton, D.F. Foster, G.J. Schwarz, M. Wills, *Angew. Chem. Int. Ed.* **2000**, *39*, 4106.

70 A. Kless, J. Holz, D. Heller, R. Kadyrov, R. Selke, C. Fischer, A. Börner, *Tetrahedron: Asymmetry* **1996**, *7*, 33.

71 A.H. Hoveyda, D.A. Evans, G.C. Fu, *Chem. Rev.* **1993**, *93*, 1307.

72 (a) W.R. Jackson, P. Perlmutter, E.E. Tasdelen, *Chem. Commun.* **1990**, 763. (b) Z. Zhang, I. Ojima, *J. Organomet. Chem.* **1993**, *454*, 281. (c) L. Ren, C.M. Crudden, *J. Org. Chem.* **2002**, *67*, 1746 and references therein.

73 (a) B. Breit, *Angew. Chem. Int. Ed.* **1996**, *35*, 2835. (b) B. Breit, *Chem. Commun.* **1997**, 591. (c) B. Breit, *Eur. J. Org. Chem.* **1998**, 1123. (d) B. Breit, M. Dauber, K. Harms, *Chem. Eur. J.* **1999**, *5*, 2819. (e) B. Breit, *Chem. Eur. J.* **2000**, *6*, 1519.

74 Procedure adapted from a report on the transformation of MOM ethers to O,S-acetals. See: H.E. Morton, Y. Guindon, *J. Org. Chem.* **1985**, *50*, 5379.

75 I.J. Krauss, C.C.-Y. Wang, J.L. Leighton, *J. Am. Chem. Soc.* **2001**, *123*, 11514.

76 For other diastereoselective hydroformylations of similar cyclic and acyclic systems see: (a) B. Breit, S.K. Zahn, *Tetrahedron Lett.* **1998**, *39*, 1901. (b) B. Breit, S.K. Zahn, *J. Org. Chem.* **2001**, *66*, 4870. (c) B. Breit, G. Heckmann, S.K. Zahn, *Chem. Eur. J.* **2003**, *9*, 425.

77 S.D. Rychnovsky, *Chem. Rev.* **1995**, *95*, 2021.

78 S.D. Rychnovsky, U.R. Khire, G. Yang, *J. Am. Chem. Soc.* **1997**, *119*, 2058.

79 (a) I.J. Krauss, C.C.-Y. Wang, J.L. Leighton, *J. Am. Chem. Soc.* **2001**, *123*, 11514. (b) J.L. Leighton, D.N. O'Neil, *J. Am. Chem. Soc.* **1997**, *119*, 11118.

80 For synthesis of β-hydroxy aldehyde and therefrom 1,3-diol from hydroformylation of epoxides, see: R. Weber, U. Englert, B. Ganter, W. Keim, M. Möthrath, *Chem. Commun.* **2000**, 1419.

81 (a) A. Polo, J. Real, C. Claver, S. Castillon, J.C. Bayon, *Chem. Commun.* **1990**, 600. (b) A. Polo, C. Claver, S. Castillon, A. Ruiz, J.C. Bayon, J. Real, C. Mealli, D. Maisi, *Organometallics* **1992**, *11*, 3525.

82 K. Kinoshita, K. Odaka, *Jpn. Kokai Tokkyo Koho* **1996**, JKXXAF JP 08295683 A2 19961112.

83 J. Elks, G.R. Ganellin, *Dictionary of Drugs*, Chapman and Hall, London, **1990**, p. 1159.

84 (a) L.S. Boguslavskaya, I.Y. Panteleeva, T.V. Morozova, A.V. Chuvatkin, N.N. Kartashov, *Russian Chem. Rev.* **1990**, *59*, 906 and references therein. (b) K.-R. Gassen, D. Bielefeldt, A. Marhold, P. Andres, *J. Fluor. Chem.* **1991**, *55*, 149 and references therein. (c) C. Botteghi, U. Matteoli, S. Paganelli, V. Fassina, V. Castelvetro, M. Aglietto, in: *Proceedings of the 5th International Symposium on the Conservation of Monuments in the Mediterranean Basin*, Seville, Spain, 5–8 April **2000**, p. 222.

85 (a) C. Wakselmann, *Macromol. Symp.* **1994**, *82*, 77, and references therein.

(b) K. W. Laue, G. Haufe, *Synthesis* **1998**, 1453.
86 I. Ojima, K. Kato, M. Okabe, T. Fuchikami, *J. Am. Chem. Soc.* **1987**, *109*, 7714.
87 C. Botteghi, S. Paganelli, B. Vicentini, C. Zarantonello, *J. Fluor. Chem.* **2001**, *107*, 113.
88 C. Botteghi, G. Del Ponte, M. Marchetti, S. Paganelli, *J. Mol. Catal.* **1994**, *93*, 1.
89 I. Ojima, K. Kato, M. Okabe, T. Fuchikami, *J. Am. Chem. Soc.* **1987**, *109*, 7714.
90 F. Outurquin, C. Paulmier, *Synthesis* **1989**, 690.
91 C. M. Ruden, H. Alper, *J. Org. Chem.* **1994**, *59*, 3091.
92 (a) A. N. Ajjou, H. Alper, *J. Am. Chem. Soc.* **1998**, *120*, 1466. (b) D. A. Annis, E. N. Jacobsen, *J. Am. Chem. Soc.* **1999**, *121*, 707 and references therein.
93 (a) A. J. Sandee, V. F. Slagt, J. N. H. Reek, P. C. J. Kamer, P. W. N. M. van Leeuwen, *Chem. Commun.* **1999**, 1633. (b) B. Li, X. Li, K. Asami, K. Fujimoto, *Chem. Lett.* **2003**, *32*, 378. (c) C. P. Mehnert, R. A. Cook, N. C. Dispenziere, M. Afeworki, *J. Am. Chem. Soc.* **2002**, *124*, 12932.
94 (a) S. C. Bourque, F. Maltais, W.-J. Xiao, O. Tardif, H. Alper, P. Arya, L. E. Manzer, *J. Am. Chem. Soc.* **1999**, *121*, 3035. (b) S. C. Bourque, H. Alper, L. E. Manzer, P. Arya, *J. Am. Chem. Soc.* **2000**, *122*, 956.
95 A. M. Seayad, M. Ahmed, H. Klein, R. Jackstell, T. Gross, M. Beller, *Science*, **2002**, *297*, 1676.
96 (a) M. Ahmed, A. M. Seayad, R. Jackstell, M. Beller, *J. Am. Chem. Soc.* **2003**, *125*, 10311. (b) A. M. Seayad, K. Selvakumar, A. Moballigh, M. Beller, *Tetrahedron Lett.* **2003**, *44*, 1679.
97 P. Eilbracht, L. Bärfacker, C. Buss, C. Hollmann, B. E. Kitsos-Rzychon, C. L. Kranemann, T. Rische, R. Roggenbuck, A. Schmidt, *Chem. Rev.* **1999**, *99*, 3329.
98 F. P. Pruchnik, *Organometallic Chemistry of Transition Elements*, Plenum Press, New York, **1990**, p. 691.
99 (a) N. L. Lewis, *Chem. Rev.* **1993**, *93*, 2693. (b) S. A. Laneman, F. R. Fronczek, G. G. Stanley, *J. Am. Chem. Soc.* **1988**, *110*, 5585.
100 (a) M. Lenarda, L. Storaro, R. Ganzerla, *J. Mol. Catal.* **1996**, *111*, 203. (b) M. E. Davis, *CHEMTECH* **1992**, 498.
101 (a) W. A. Hermann, C. W. Kohlpaintner, *Angew. Chem. Int. Ed. Engl.* **1993**, *32*, 1524. (b) B. Cornils, E. G. Kuntz, *J. Organomet. Chem.* **1995**, *502*, 177. (c) S. R. Waldvogel, M. T. Reetz, *Proceedings of 10th International Symposium on Homogeneous Catalysis*, Princeton, August **1996**, p. A10. (d) M. S. Goedheijt, B. E. Hanson, J. N. H. Reek, P. C. J. Kamer, P. W. N. M. Leeuwen, *J. Am. Chem. Soc.* **2000**, *122*, 1650. (e) E. Paetzold, G. Oehme, C. Fischer, M. Frank, *J. Mol. Catal.* **2003**, 95 and references therein.
102 P. Eilbracht, *Stereoselective Synthesis* (Houben-Weyl), **1996**, Vol. 4, p. 2503.
103 G. D. Cuny, S. L. Buchwald, *J. Am. Chem. Soc.* **1993**, *115*, 2066.

2.2
New Synthetic Applications of Tandem Reactions under Hydroformylation Conditions

Peter Eilbracht and Axel M. Schmidt

2.2.1
Introduction

"Tandem reactions", also known as "domino reactions", "reaction cascades" or "sequential transformations", combine several synthetic steps in a single operation [1]. These reaction sequences are usually considered to be related to "biomimetic" procedures, the multi-step squalene cyclization to lanosterol being a prominent example [2]. Tandem reactions without change of reaction conditions and without addition of further reagents require only one single setup of starting materials, reagents, and solvents, and no isolation of intermediates is necessary. Therefore they save time and materials and furthermore avoid waste of chemicals and solvents. At present, considerable efforts are concentrating on the development of new procedures, even including total synthesis of more complex target molecules [3].

Among many ways of combining polar, radical, pericyclic, metal-mediated and other reactions to give tandem procedures, those reaction steps which are homogeneously catalyzed by transition metal complexes are especially valuable tools in organic synthesis. These usually provide effective and selective transformations under mild reaction conditions. Numerous transition metal-catalyzed reactions as addition reactions are highly atom economic, among them the hydroformylation reaction ("oxo reaction") of alkenes with hydrogen and carbon monoxide leading to synthetically useful aldehydes ("oxo aldehydes"). Hydroformylation is established as an important industrial tool for the production of aldehydes and products derived therefrom [4]. Hydroformylation, however, is also applied in the synthesis of more complex target molecules [5], including stereoselective and asymmetric syntheses [6].

Because of the versatile chemistry of the aldehyde group [7], the products of hydroformylation are easily further converted via reduction, oxidation, nucleophilic attack at the carbonyl group, or electrophilic attack in the acidic α-position to give alcohols, amines, carboxylic acid derivatives, aldol condensation products and many others (Scheme 1). Therefore, following the growing interest in tandem procedures, hydroformylation has frequently been integrated in reaction sequences of this type [8]. These also include conversions of the intermediate metal acyl sys-

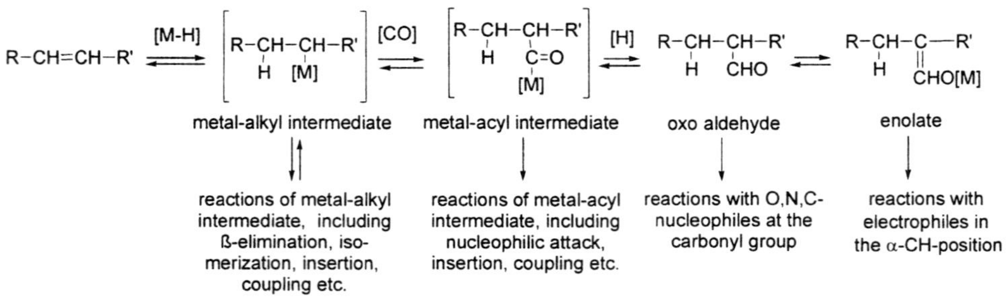

Scheme 1

$$R-CH=CH-R' \xrightleftharpoons{[M-H]} \left[\begin{array}{c} R-CH-CH-R' \\ | \quad \quad | \\ H \quad [M] \end{array} \right] \xrightleftharpoons{[CO]} \left[\begin{array}{c} R-CH-CH-R' \\ | \quad \quad | \\ H \quad C=O \\ \quad \quad | \\ \quad \quad [M] \end{array} \right] \xrightleftharpoons{[H]} \begin{array}{c} R-CH-CH-R' \\ | \quad \quad | \\ H \quad CHO \end{array} \xrightleftharpoons{} \begin{array}{c} R-CH-C-R' \\ | \quad \quad \| \\ H \quad CHO[M] \end{array}$$

metal-alkyl intermediate | metal-acyl intermediate | oxo aldehyde | enolate

↓↑ | ↓ | ↓ | ↓

reactions of metal-alkyl intermediate, including β-elimination, isomerization, insertion, coupling etc. | reactions of metal-acyl intermediate, including nucleophilic attack, insertion, coupling etc. | reactions with O,N,C-nucleophiles at the carbonyl group | reactions with electrophiles in the α-CH-position

tems which are formed under the hydroformylation conditions and can be trapped by addition of nucleophiles or insertion of unsaturated units (Scheme 1).

In this brief survey, the scope of tandem hydroformylation sequences is described, and, in addition to a more comprehensive review covering the literature up to 1998/99 [8], some more recent synthetic applications are added. The material is ordered according to the type of the additional bond forming reaction, e.g. C-H (isomerization and reduction), C-O (acetals, enol ethers), or C-N (aminals, enamines). C-C bond formations, such as coupling, aldol addition, olefin insertion with ketone formation, or other types of conversion are described in the next chapter of this volume ("Multiple C-C Bond Formations under Hydroformylation Conditions").

2.2.2
Hydroformylation and Isomerization

Hydroformylation of internal olefins usually proceeds with low regioselectivity and is slower than isomerization and hydroformylation of terminal alkenes (Scheme 2). Therefore mixtures of regioisomeric aldehyde products are obtained [4].

Selective synthesis of linear aldehydes from internal olefins, e.g., *n*-pentanal from 2-butene (Scheme 2), is considered as one of the major challenges in indus-

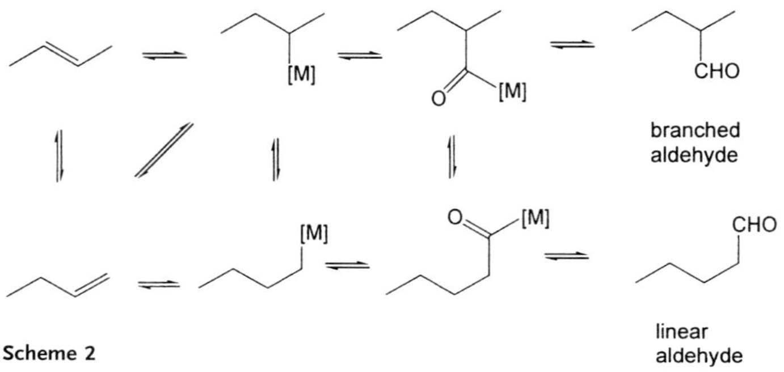

Scheme 2

trial hydroformylation chemistry, since with this method inexpensive feedstocks of internal olefin mixtures such as butenes, hexenes, octenes, or unsaturated fatty acid derivatives could be used for terminally functionalized linear products. In order to achieve this goal, catalysts with a high isomerization activity and a high selectivity to form the linear metal alkyl and metal acyl intermediates are required. Although the formation of terminal alkenes from internal olefins is thermodynamically unfavorable, the former are hydroformylated much more rapidly than the internal systems. Earlier efforts toward this goal are compiled in several reviews [9] and have revealed that strongly electron-withdrawing monophosphite ligands and bulky diphosphites with wide bite angles such as BIPHEPHOS lead to high amounts of linear aldehydes from internal olefins. Since diphosphite ligands, however, suffer from low stabilities, various XANTPHOS-type ligands with large bite angles have been developed. These show high activities and *n*-selectivities in the hydroformylation of 2-octene, 4-octene, and functionalized alkenes to linear aldehydes [10, 11]. More recently, various other diphosphite and diphosphine ligands have been synthesized and applied to the isomerization/hydroformylation procedure [12–14]. Some modified binaphthyl diphosphines as ligands in rhodium-catalyzed isomerization/hydroformylation of 2-pentene, 2-butene, 2-octene and 4-octene give *n/iso* ratios of up to 95:5 [15].

As an alternative concept, dual homogeneous catalytic Ru/Rh systems have been developed for consecutive isomerization/hydroformylation reactions to convert internal olefins to linear *n*-aldehydes with up to 99% selectivity [16]. In another approach, hydroformylation of short-chain internal olefins like 2-pentene to linear aldehydes was performed in an aqueous two-phase system with a recyclable cobalt-TPPTS catalyst [17]. The biphasic methodology, however, is restricted to lower olefins. Hydroformylation of water-insoluble higher internal olefins such as 4-octene has been achieved in biphasic systems with use of rhodium complexes of reusable, water-soluble calix[4]arene-phosphines, albeit with low selectivity toward linear aldehydes [18].

The concept of hydroformylation of internal olefins to linear aldehydes was also applied to the hydroaminomethylation of internal olefins to linear amines. This method combines hydroformylation with reductive amination and is described below [19].

2.2.3
Hydroformylation and Reduction of the Oxo Aldehydes

Reduction of oxo aldehydes to alcohols under hydroformylation conditions is one of the commonest parallel and consecutive reactions lowering yields or selectivities and leading to undesired side products. Thus, alcohol formation via reduction of oxo aldehydes may also lead to formic acid esters (with CO) and acetals (with oxo aldehydes).

Under controlled conditions, reduction of the oxo aldehydes is used as a method for the direct synthesis of primary alcohols from alkenes with homologation of

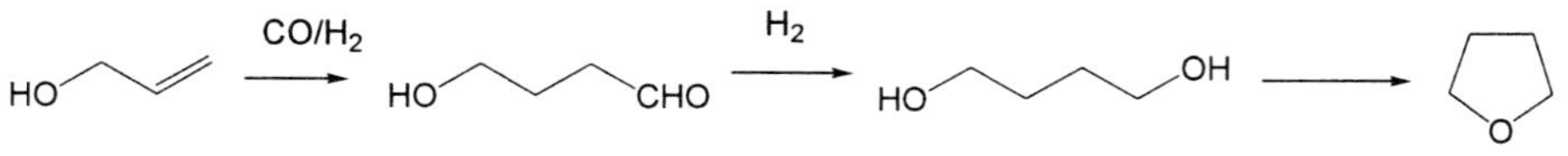

Scheme 3

$$HO\diagup\!\!\!\diagdown\!\!=\quad\xrightarrow{CO/H_2}\quad HO\diagup\!\!\!\diagdown\!\!\diagdown CHO\quad\xrightarrow{H_2}\quad HO\diagup\!\!\!\diagdown\!\!\diagdown\!\!\diagdown OH\quad\longrightarrow\quad\text{(tetrahydrofuran)}$$

the carbon chain. In industrial processes, these products are important as solvents, plasticizers, or detergents. Up to now, however, all commercial processes are still performed in stepwise procedures, since no process has been developed allowing this conversion under mild conditions. Considerable efforts have therefore been made to optimize the direct process. Earlier results and mechanistic considerations are compiled in a recent review [8].

Usually alcohols are the major products if hydroformylation is carried out with cobalt or rhodium catalysts under forcing conditions with higher hydrogen pressure in the presence of phosphines. Furthermore, the addition of amines or other N-donors or alcohols promotes direct alcohol formation [20]. Methodical and mechanistic investigations of rhodium-catalyzed hydroformylation/reduction showed that aldehydes need not necessarily be the intermediates in direct alcohol formation, and therefore an alternative mechanism via protonation of the metal acyl intermediate and hydrogen transfer to a hydroxycarbene intermediate was proposed [21–23].

More recently, van Leeuwen et al. [24] developed a switchable homogeneous hydroformylation catalyst covalently tethered to a polysilicate. The immobilized recyclable rhodium complex [Rh(A)CO]+ [A = N-(3-trimethoxysilane-n-propyl)-4,5-bis(diphenyl-phosphino)phenoxazine] can be switched reversibly between the hydroformylation mode, the hydroformylation/hydrogenation/reduction mode (by addition of 1-propanol), and the hydrogenation/reduction mode (by changing the atmosphere from CO/H_2 to H_2).

Hydroformylation/reduction of functionalized alkenes leads to the corresponding alcohols [25]. A useful application of this type is the direct conversion of allylic alcohol to 1,4-butanediol or tetrahydrofuran (Scheme 3).

Various other combinations of hydroformylation/reduction sequences are feasible. Thus the sequences might be combined with alcohol homologation and the introduction of one or several carbon monoxide units, resembling the Monsanto/Cativa acetic acid process or Fischer-Tropsch-type procedures.

2.2.4
Hydroformylation and Additional Carbon-Heteroatom Bond Formations

If hydroformylation is performed in the presence of hetero nucleophiles such as alcohols or amines, these can attack the *in situ*-formed aldehyde group to give aldehyde addition and condensation products such as acetals, enol ethers, imines, or enamines. With unsaturated alcohols or amines the corresponding cyclic products are expected. These nucleophiles, however, can also attack the metal acyl intermediate of the hydroformylation process, leading to carboxylic acid derivatives. By choosing the appropriate reaction conditions (e.g., higher hydrogen pressure) and catalyst systems, these problems can mostly be overcome.

2.2.4.1
Hydroformylation in the Presence of Oxygen Nucleophiles

Hydroformylation of alkenes in the presence of alcohols, depending on the reaction conditions, leads to various addition and condensation products of the initial aldehyde system (Scheme 4). One option is hemiacetal/enol ether and acetal formation. On the other hand, the metal acyl intermediate may be trapped to form esters in an overall hydroesterification. This latter reaction prevails at lower hydrogen partial pressures or with palladium catalysts in the absence of hydrogen. The presence of alcohols under hydroformylation conditions may also enhance the reduction of oxo aldehydes to alcohols (see above).

Various examples of intermolecular hemiacetal and acetal formations under hydroformylation conditions in the presence of alcohols or diols are described [4, 8]. In addition, acetalization can also be achieved via transacetalization by use of acetals, like 2,2-dimethoxy-propane, or with ortho esters as the source of alcohols in the presence of an acidic cocatalyst.

Thus, direct acetalization of oxo aldehydes is used to protect sensitive aldehyde products, especially in asymmetric hydroformylation with formation-sensitive stereogenic centers in the α-position of the aldehyde group [6, 8]. The acetal formation prevents racemization of the aldehyde product. A representative example of this method is the asymmetric hydroformylation of 2-ethenyl-6-methoxynaphthalene in the presence of triethyl orthoformate (Scheme 5) [6]. The acetal of the branched hydroformylation product is an intermediate in the synthesis of naproxen.

Of synthetical interest is also the hydroformylation of unsaturated alcohols leading to cyclic hemiacetals (lactols) with ring sizes depending on the distance between the hydroxy function and the double bond. Oxo aldehydes bearing a remote alcohol function spontaneously cyclizes, especially if five- or six-membered rings can be formed. By variation of the reaction conditions, subsequent conversions of the hemiacetals can be integrated in the one-pot sequence. Thus, these hemiace-

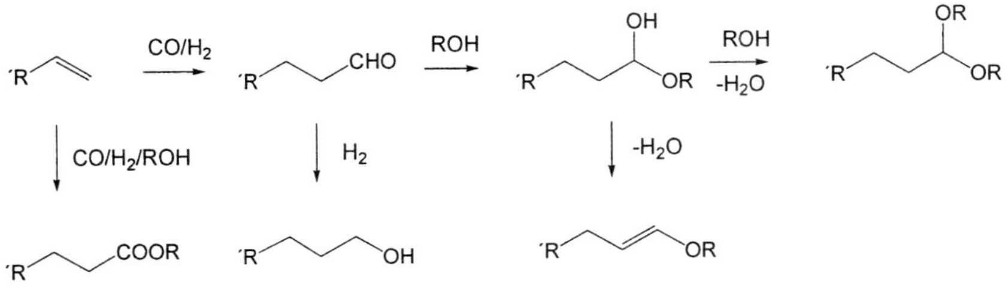

Scheme 4

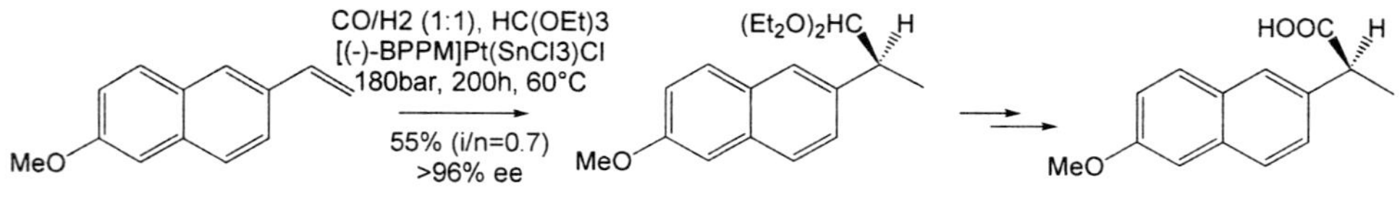

tals give the corresponding enol ethers after elimination, or various acetals if other alcohols are added (Scheme 6) [26].

The products obtained by this type of tandem hydroformylation offer access to a wide range of further transformations. Reduction of the aldehydes, hemiacetals, or enol ethers leads to diols; oxidation to lactones and other reactions are performed at the double bond of the enol ether (e.g., epoxidation, dihydroxylation, allylic substitution). These reactions enable the synthesis of interesting acyclic or heterocyclic compounds, including important precursors, e.g., as industrial intermediates for ethers and resins or subunits of naturally occurring products with biological and pharmacological activities.

Numerous examples of rhodium-catalyzed ring-forming hydroformylations using allylic, homoallylic, and other unsaturated alcohols are reported and compiled in an earlier review [8]. Thus, allylic alcohols under hydroformylation conditions lead to furan-type lactols, which are easily oxidized to butyrolactones. As an example, rhodium-catalyzed enantioselective lactol synthesis via tandem hydroformylation/acetalization of cinnamyl alcohol is achieved with chiral ligands (Scheme 7) [27].

Di- or tetrahydropyrans with vinyl side chains obtainable by diastereoselective ring-closing metathesis or by addition of vinylmagnesium chloride to appropriately functionalized tetrahydropyranones are converted to spirocyclic hemiacetals under hydroformylation conditions (Scheme 8) [28]. Oxidation yields the corre-

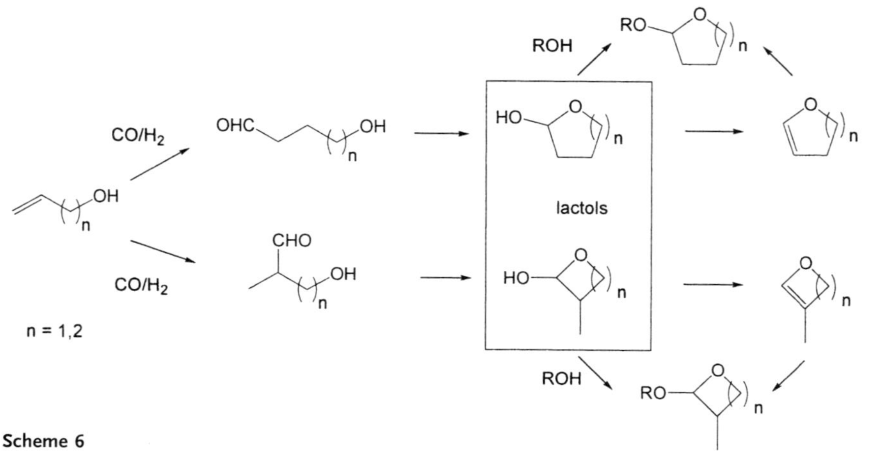

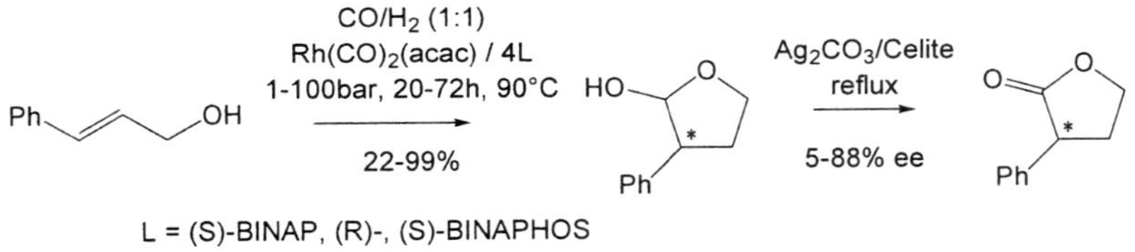

2.2.4 Hydroformylation and Additional Carbon-Heteroatom Bond Formations

Scheme 7

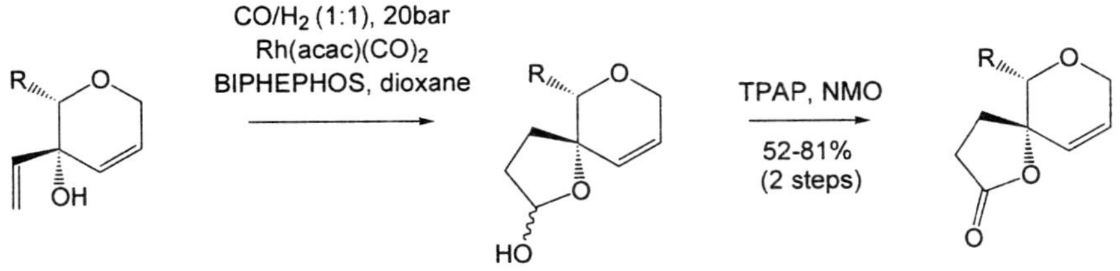

L = (S)-BINAP, (R)-, (S)-BINAPHOS

Scheme 8

R = Me, Ph

sponding lactones. Spirocyclic γ-butyrolactones of this type are widespread in nature and play a key role as synthetic intermediates.

δ-Lactols and -lactones derived therefrom are obtainable from the corresponding homoallylic alcohols. With dehydration the corresponding dihydropyrans are prepared. This procedure can also be applied to the synthesis of monocyclic pyrans and spiropyrans as potential precursors and building blocks for natural products such as pheromones or antibiotics. A representative example is the synthesis of the pyranone subunit of the Prelog-Djerassi lactone. For this purpose various 1,2-disubstituted homoallylic alcohols were used (Scheme 9) [29].

Similarly, hydroformylation of chiral homoallylic alcohols to give the corresponding δ-lactols has been used in the total synthesis of leucascandrolide A [30]. Hydroformylation of homoallylic alcohols with quaternary centers in the allylic and/or the homoallylic position(s) selectively leads to the n-hydroformylated products forming spiropyran derivatives under these reaction conditions. The cyclic

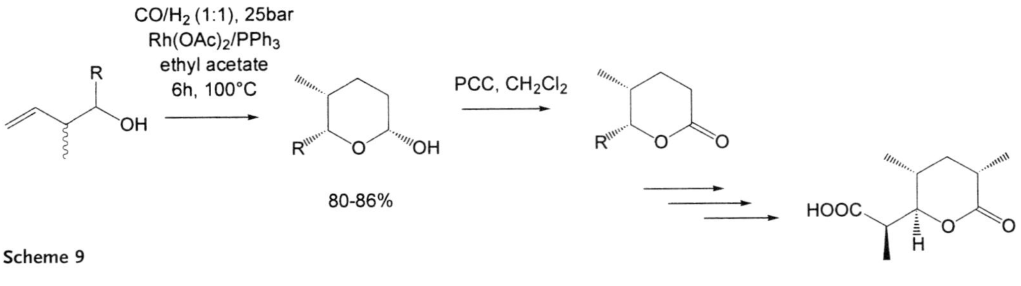

Scheme 9

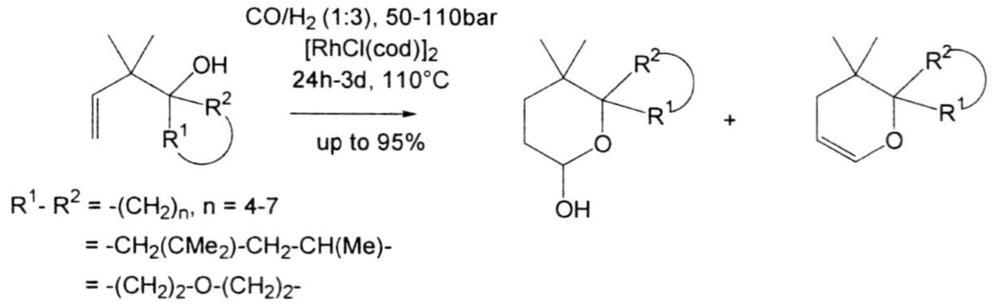
CO/H2 (1:3), 50-110bar
[RhCl(cod)]2
24h-3d, 110°C
up to 95%
OH
R2
R1
+
O
R1- R2 = -(CH2)n, n = 4-7
= -CH2(CMe2)-CH2-CH(Me)-
= -(CH2)2-O-(CH2)2-

Scheme 10

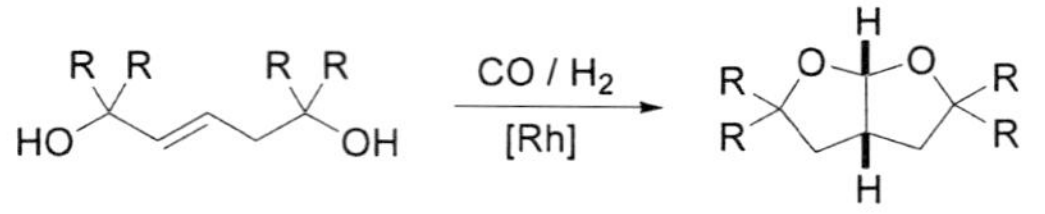
R R
HO
OH
CO / H2
[Rh]
H
O
R

Scheme 11

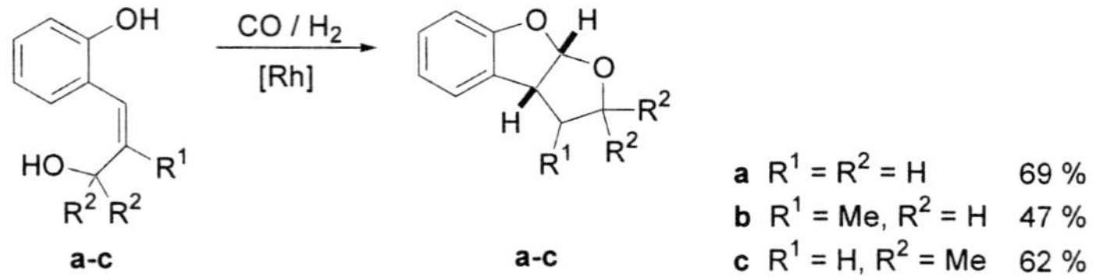
OH
HO
R1
R2 R2
a-c
CO / H2
[Rh]
O
H
R2
a R1 = R2 = H 69 %
b R1 = Me, R2 = H 47 %
c R1 = H, R2 = Me 62 %

Scheme 12

system in the homoallylic position can vary in the ring size. Substituents or heteroatoms in the rings are also tolerated (Scheme 10) [31].

Similar tandem hydroformylation/cyclizations have been applied to various unsaturated terpene alcohols such as *iso*-pulegol, myrtenol, or pinocarveol [8, 32]. Propargylic alcohols lead to saturated or unsaturated lactols or directly to lactones, depending on the reaction conditions and the catalyst [8, 33]. Benzofurans, benzopyrans, or benzooxepins are formed from phenols or benzyl alcohols with unsaturated side chains [8]. Rhodium-catalyzed tandem hydroformylation/acetalization of α,ω-alkenediols gives facile access to perhydro-furo[2,3*b*]furans and perhydro-furo [2,3*b*]pyrans in good yields (Scheme 11) [34].

Similarly, benzoannelated tetrahydrofuro[2,3*b*]-furans are obtained by hydroformylation of *ortho*-hydroxy cinnamyl alcohols. Compounds of this type are substructures of aflatoxins and related natural products (Scheme 12) [34].

2.2.4.2
Hydroformylation in the Presence of Nitrogen Nucleophiles

Hydroformylation in the presence of nitrogen nucleophiles such as amines leads to a variety of nitrogen products (Scheme 13, only *n*-products shown). After addi-

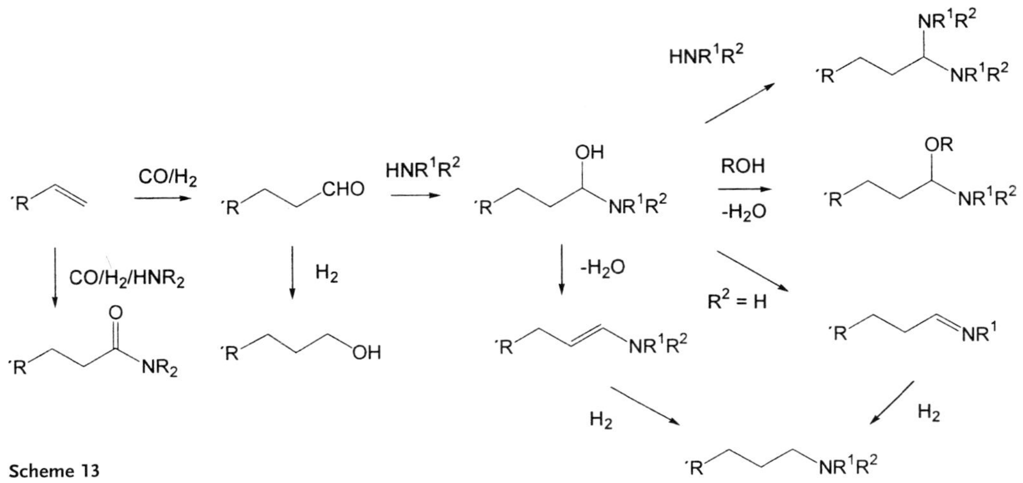

Scheme 13

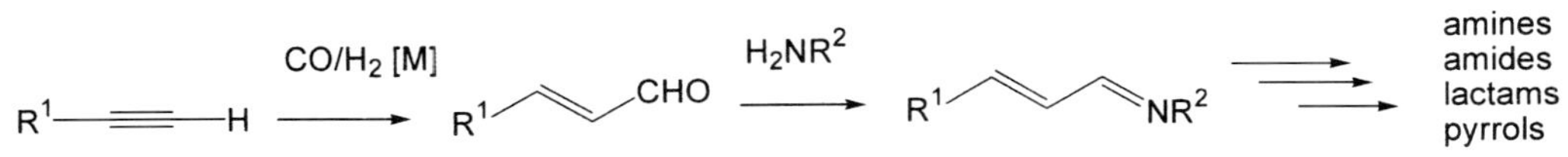

Scheme 14

tion of the amine to the oxo aldehyde imines, enamines or N,N-acetals can be formed, depending on the reaction conditions and catalysts. These can be reduced to saturated amines or react with alcohols to form O,N-acetals. Instead of aldehyde formation, carboxylic acid amides can also be formed by nucleophilic attack of the metal acyl intermediate.

All reactions of this type can also proceed in intramolecular versions if unsaturated amines or alcohols are used. Similar reaction sequences can be applied to alkynes, leading to saturated or unsaturated amines, amides, lactams, or pyrroles (Scheme 14).

With sufficient selectivities, these tandem hydroformylation procedures are synthetically extremely versatile and have been applied to the synthesis of various target molecules [35]. This chapter concentrates on more recent examples, since the older literature is covered in a recent review [8].

2.2.4.2.1 Synthesis of O,N- and N,N-Acetals

Transition metal-catalyzed transformations of unsaturated amines offer a convenient synthetic access to cyclic O,N-hemiacetals. If performed in the presence of alcohols or orthoesters O,N-acetals are formed, and with additional N-nucleophiles N,N-acetals are obtained. These compounds are synthetically attractive building blocks and were therefore used as key step in the synthesis of various natural

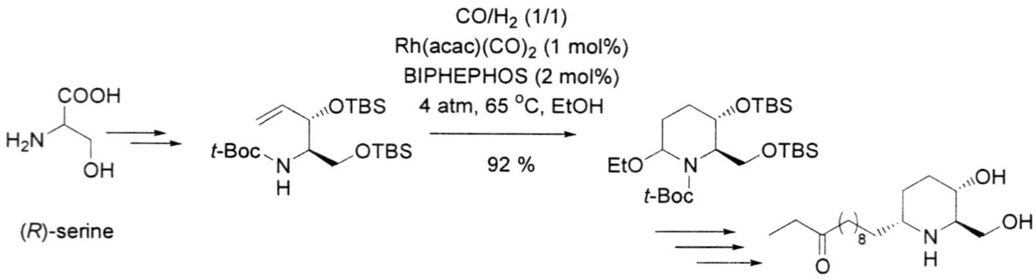

Scheme 15

(R)-serine → t-Boc protected intermediate → [CO/H₂ (1/1), Rh(acac)(CO)₂ (1 mol%), BIPHEPHOS (2 mol%), 4 atm, 65 °C, EtOH, 92%] → cyclized intermediate → (+)-prosopinine

products [8, 35]. Thus, the synthesis of (+)-prosopinine starting from enantiopure D-serine requires the construction of a cyclic O,N-acetal functionality, which is effectively achieved via regioselective Rh-catalyzed cyclohydrocarbonylation, providing the required functionality for the attachment of the side chain (Scheme 15) [36].

Similarly, deoxoprosphylline [36], pipecolic acids, izidines [37], and the bicyclic alkaloids (±)-isoretronecanol, (±)-trachelanthamidine [38] and 6-epi-poranthellidine [39] were synthesized via tandem hydroformylation/cyclization. More recently, highly efficient syntheses of azabi-cyclo[4.4.0]alkane amino acids were achieved by Rh-catalyzed cyclohydrocarbonylation of dipeptides bearing a terminal olefin moiety and a heteroatom nucleophile [40]. Here, the amine function as well as a second O-, S-, or N-Boc function are present in the acyclic starting material, to form the bicyclic system with N,O-, N,N- or N,S-subunits in one step (Scheme 16).

This method was also used to prepare diazabicycloalkanes and oxazabicycloalkanes containing medium and large rings from N-alkenylpropane-1,3-diamines and 2-(alkenylamino)ethanols in excellent yields without the need for high dilution. Selective ring opening of these compounds leads to large heterocyclic rings (Scheme 17) [41].

Rhodium-catalyzed reactions of *ortho*-alkenylaminobenzylamines or -benzamides with syngas in excellent yields give quinazolines and quinazolinones containing a fused alicyclic ring (Scheme 18) [42, 43].

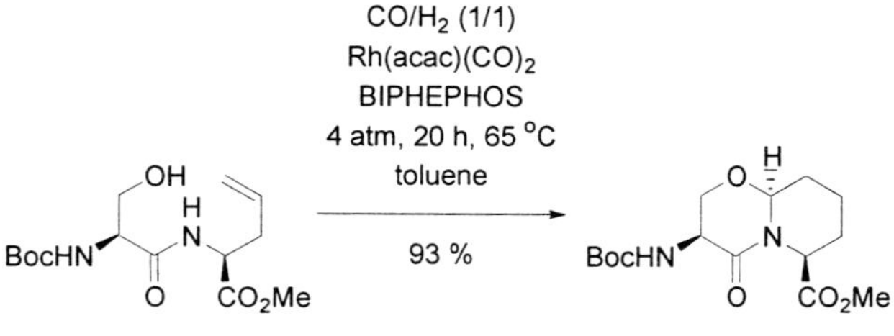

Scheme 16

BocNH-CH(CO₂Me)-C(=O)-N(H)-CH(CH=CH₂)-CH₂OH → [CO/H₂ (1/1), Rh(acac)(CO)₂, BIPHEPHOS, 4 atm, 20 h, 65 °C, toluene, 93%] → bicyclic product

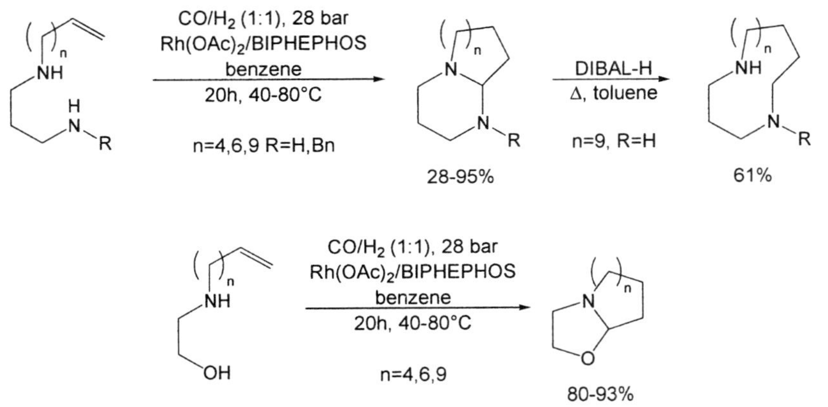

2.2.4 Hydroformylation and Additional Carbon-Heteroatom Bond Formations | 67

Scheme 17

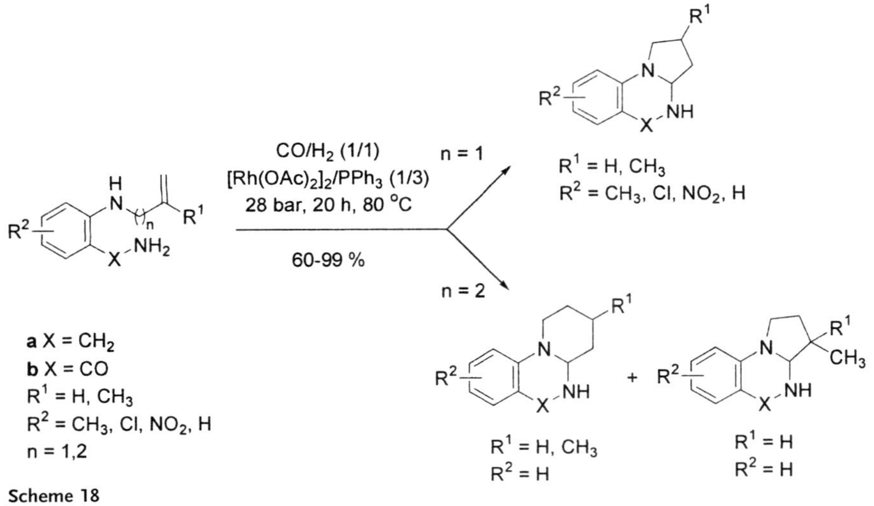

Scheme 18

The quinazolines and quinazolinones thus obtained have a wide range of biological activities and are currently marketed as pharmaceutical agents. Similar other examples of allylic and homoallylic diamines are reported and compiled in a recent review [8].

2.2.4.2.2 Synthesis of Imines and Enamines

The hydroformylation of olefins in presence of primary and secondary amines offers access to imines and enamines [8]. These intermediates often undergo hydrogenation in an overall hydroaminomethylation (see below) or further carbonylation, e.g., in amidocarbonylations starting from olefins (see Chapter 2.3). With allylic or homoallylic amines, the corresponding dihydropyrroles and tetrahydropyridines with imine or enamine substructures are obtained. While intermolecular imine and enamine formation from primary amines without further reduction is rare [8], hydroformylation of unsaturated secondary amines or acylated amines with hydroformylation and intramolecular enamine formation has become a valuable tool in heterocyclic ring synthesis.

Thus *ortho*-vinylanilines obtained via Heck reaction of *ortho*-haloanilines cyclize to indoles in a hydroformylation/condensation sequence, sometimes in good yields, to give substituted tryptamines and tryptopholes (Scheme 19) [44]. These products are common structural units in indole alkaloids with important biological and pharmacological properties.

A similar formation of indoles has been achieved starting from *ortho*-nitrostyrenes after reduction of the nitro groups under hydroformylation conditions [45].

The cyclohydrocarbonylation of 2-(alkoxycarbamoyl)-4-pentenoate selectively leads to the corresponding pipecolic acid in quantitative yield (Scheme 20) [36, 37]. It is noteworthy that the reactions under these conditions yield O,N-acetals instead of enamines if alcohols are used as solvent (see above).

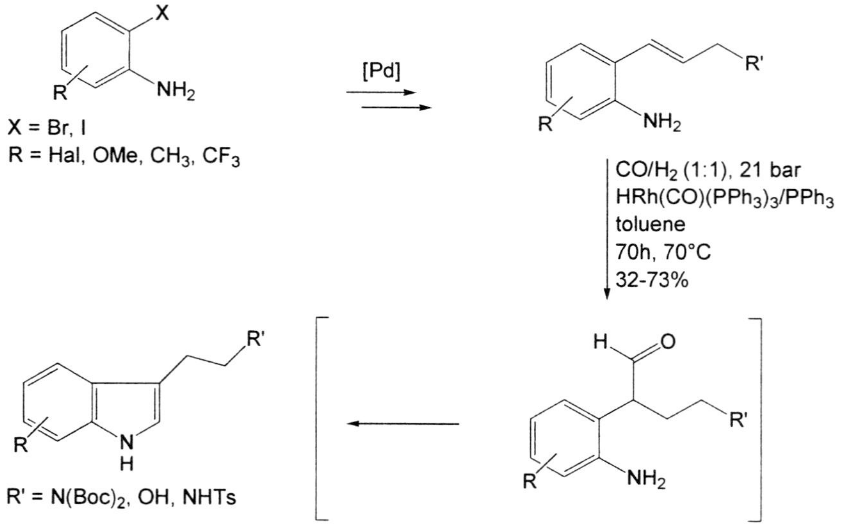

Scheme 19

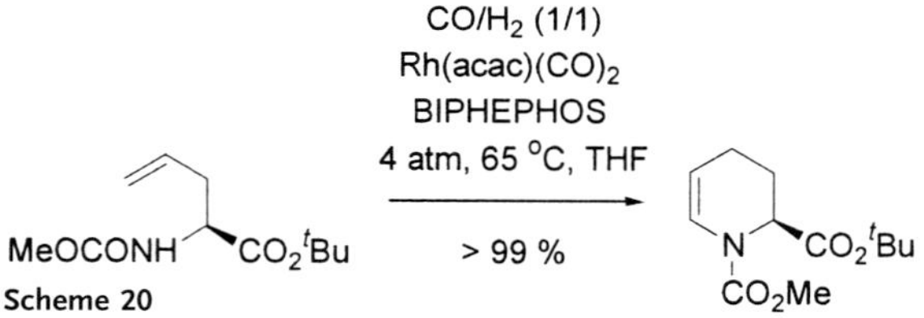

Scheme 20

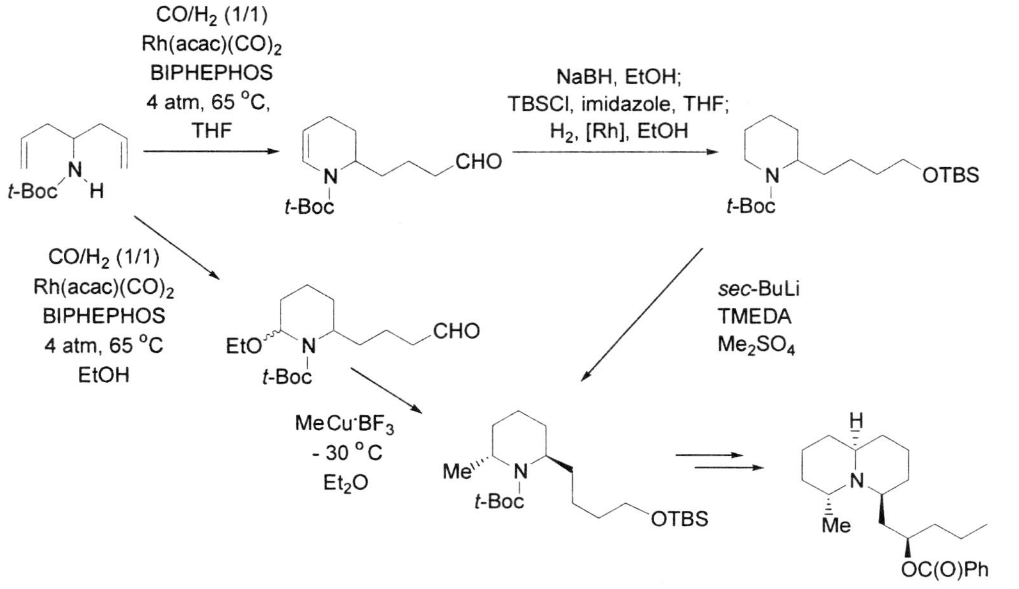

Scheme 21

6-*epi*-porantherilidine

Similarly, depending on the reaction time and solvent, protected diolefinic amines and similar systems undergo hydroformylation/condensation reaction sequences leading to O,N-hemiacetals, enamines, or O,N-ketene-acetals, respectively (Scheme 21) [46]. This procedure has been applied to the synthesis of 6-epi-porantherilidine [39].

More recently, this procedure has been performed in the presence of the chiral ligand BINAPHOS with high stereoselectivities (Scheme 22) [35, 47].

Enantioselective hydrogenation followed by a hydroformylation/cyclization sequence can be achieved in a single-pot version to form five- and six-membered cyclic α-amino acids with good yields and *ee*s >95%. Rh(I)-DuPHOS is used as the sole catalyst system to perform both stereoselective hydrogenation of the prochiral dienamide esters and hydroformylation of the resulting homoallylic amine [48].

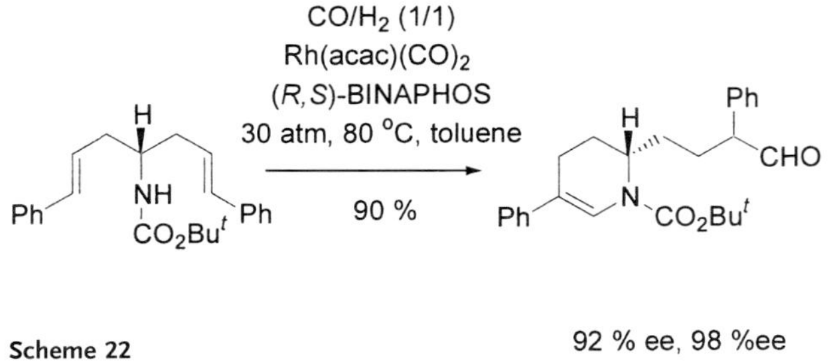

Scheme 22

92 % ee, 98 %ee

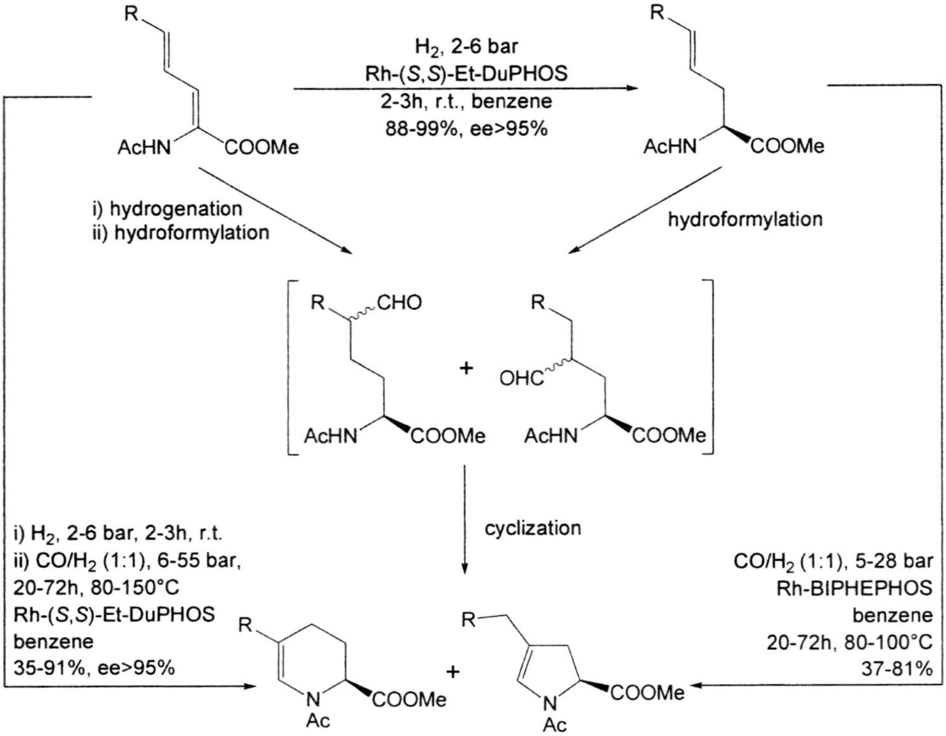

Scheme 23

A one-pot hydroaminovinylation reaction has been achieved for the synthesis of sulfonated and phosphonated enamines using the zwitterionic rhodium complex [Rh$^+$(cod)(η^6-PhBPh$_3$)$^-$] together with a chelating phosphine ligand as the catalyst (Scheme 24). The regio- and stereoselectivities of the reactions are mainly excellent [49].

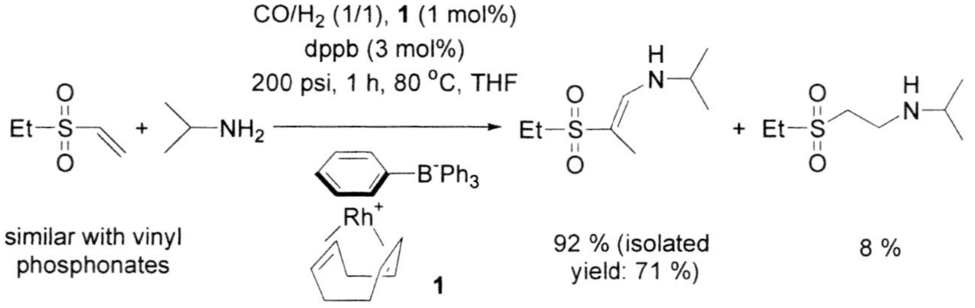

Scheme 24

2.2.4.2.3 Hydroformylation in the Presence of Other N-Nucleophiles

In principle, various other N-functions can be used as nucleophiles in tandem hydroformylation reactions. Amides usually are less nucleophilic, but in the absence of stronger nucleophiles they also add to *in situ*-generated oxo aldehydes. Various examples of the use of amides as N-nucleophiles have been mentioned above and are described in an earlier review [8]. Hydrazine and substituted hydrazines offer an interesting version of hydroformylation in the presence of N-nucleophiles, readily forming hydrazones with good regioselectivities (Scheme 25) [50]. Further reduction of the hydrazones is observed only under harsh conditions [51].

These hydrazones can be used for various types of further conversions, e.g., in Fischer indole syntheses if hydrazones formed from aryl hydrazines are used [52]. This version was recently applied to one-pot and tandem procedures of indole syntheses directly starting from alkenes and aryl hydrazines under hydroformylation conditions (see next chapter) [53].

2.2.4.2.4 Amines via Hydroformylation/Reductive Amination (Hydroaminomethylation)

Hydroformylation of alkenes in the presence of amines leads to an overall hydroaminomethylation if the initial hydroformylation of the alkene is followed by a condensation of the intermediate aldehyde with a primary or secondary amine to form an enamine or imine, respectively, and finally by a hydrogenation to give a saturated secondary or tertiary amine (Scheme 26). This reductive amination of *in situ*-generated oxo aldehydes (or, looking from the amine side, this amine alkylation) has been extensively reviewed [8]; therefore only some selected and some more recent examples are described here.

Efficient catalysts for hydroaminomethylation are rhodium complexes such as [RhCl(cod)]$_2$ or Rh(acac)(CO)$_2$ [8, 54]. More recently, various new catalysts have been applied in the hydroaminomethylation of alkenes, e.g., the zwitterionic rhodium complex [Rh$^+$(η^6-Ph-BPh$_3$)$^-$] [55], the cationic [Rh(L-L)(cod)]$^+$ BF$_4^-$ bearing a

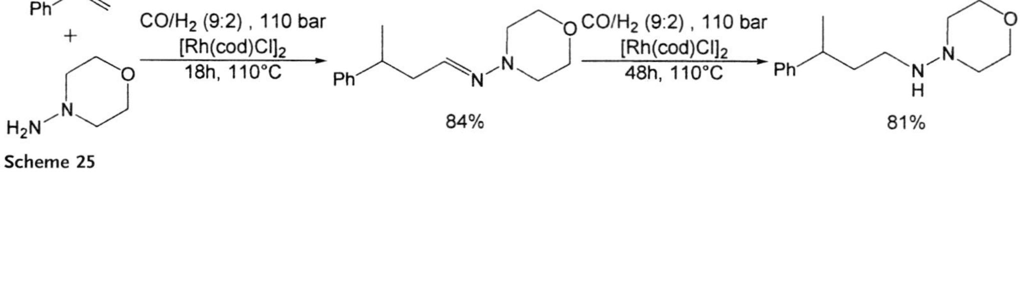

Scheme 25

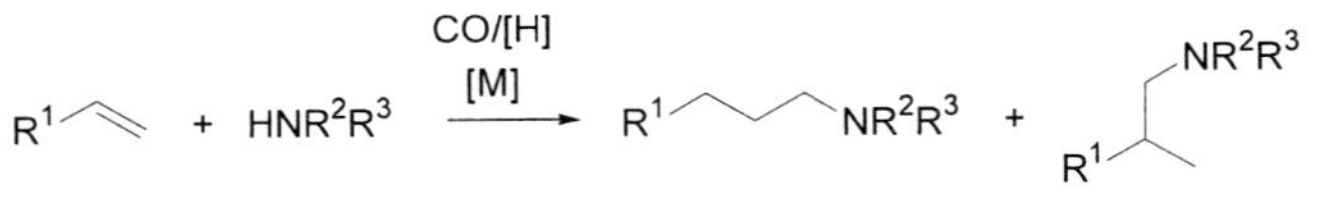

Scheme 26

R¹–CH=CH₂ + HNR²R³ →[CO/[H], [M]] R¹–CH₂CH₂–NR²R³ (n-product) + R¹–CH(NR²R³)–CH₃ (iso-product)

chelating P-N-ligand [56], various rhodium and ruthenium complexes, and mixtures thereof [57]. Efficient hydroaminomethylation of terminal as well as internal aliphatic and aromatic olefins with various amines is described using the rhodium carbene catalyst [Rh(cod)(Imes)Cl] [58].

Intensive studies toward improvements of the regioselectivity are reported [59, 60]. Thus, XANTPHOS as a ligand in rhodium-catalyzed hydroaminomethylation proved an effective ligand to control the *n*-selectivity [60]. Hydroaminomethylation can be performed in supercritical CO_2 [61] or in ionic liquids [62] and is successfully applied to solid-phase synthesis of saturated amines from immobilized olefins [63].

Almost every type of primary and secondary amine is tolerated, leading selectively to the corresponding secondary or tertiary amines. Primary amines are selectively mono-alkylated, and long chain 1-alkenes can also be used as substrates in the hydroaminomethylation procedure. The resulting products are useful for a variety of applications such as intermediates for surfactants, emulsifiers, rust inhibitors, fabric softeners, finishing agents, insecticides, and bactericides. The scope of the hydroaminomethylation reaction sequence has also been extended to nitro compounds acting as precursors for primary amines [64].

Various amine compounds of pharmaceutical interest are also easily available via hydroaminomethylation using 1,1-diphenylethylenes or allylated phenothiazines, iminodibenzyls, carbazoles, and pyrazoles as olefinic compounds in good yields [65–67]. More recently, tolterodine, an important urological drug, has been prepared with good yields starting from 1-[2-hydroxy-5-methyl)phenyl]-1-phenylethylene via stepwise hydroformylation and reductive amination (Scheme 27) [68]. This synthesis can also be performed in good yields in a rhodium-catalyzed one-pot tandem procedure starting from the diaryl ethene precursor [69]. Attempts at enantioselective hydroformylation of the 1,1-diaryl alkene catalyzed by $Rh(CO)_2(acac)$ in the presence of (S,R)-BINAPHOS and other enantiopure ferrocenyl-diphosphines afforded only low yields of the expected chiral aldehyde with enantiomeric excesses not exceeding 8% [68, 69].

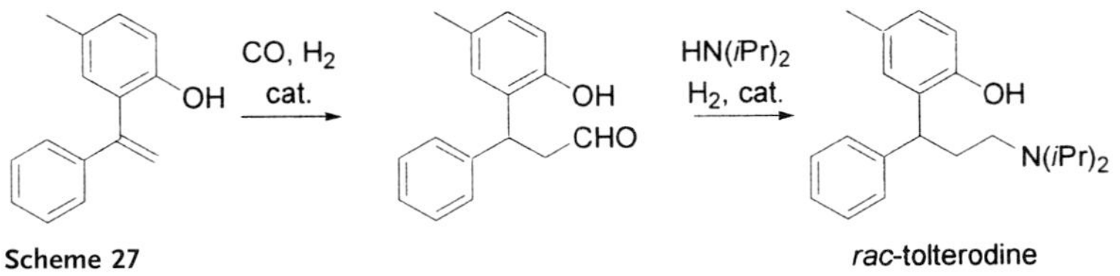

Scheme 27

[Diaryl ethene] →[CO, H₂, cat.] [aldehyde intermediate with CHO] →[HN(iPr)₂, H₂, cat.] rac-tolterodine [N(iPr)₂ product]

Hydroaminomethylation in the presence of ammonia usually leads to mixtures of secondary and tertiary amines with only small amounts of primary amines, depending on the reaction conditions and catalysts [8, 70]. More recently, selective formation of secondary amines from olefins and ammonia has been reported [71]. Selective formation of primary amines via hydroformylation/reductive amination in the presence of ammonia was achieved with a bimetallic Rh/Ir catalyst system under two-phase conditions [72].

If the hydroaminomethylation protocol starting from diolefins is used, a,ω-diamines can also easily be generated (Scheme 28) [73, 74]. Diamines with long aliphatic chains separating the two amino functions thus obtained are of importance as synthetic and biological surfactants, or in the production of membrane or bioactive compounds. Heterofunctionalized a,ω-diolefins are similarly transformed to the corresponding diamines. Representative examples are spermine or spermidine analogs [75].

Hydroaminomethylation is also used in the construction of dendrimers [76]. Here, divergent and convergent strategies with wide variabilities can be used. A selected example is shown in Scheme 29.

In a similar manner, polymers with unsaturated chains or side chains can be converted to polyamines [77–79]. Conjugated diolefins usually undergo hydroformylation with low selectivities [81]. Mostly, hydrogenation of at least one double bond occurs, and mixtures of various saturated and unsaturated amines and diamines are obtained [74]. Similarly to alkenes, alkynes may also serve as unsaturated compounds in hydroaminomethylation reaction sequences. Although synthetically attractive, few investigations toward hydroformylation and hydroaminomethylation of alkynes in the presence of N-nucleophiles are known. Usually a preferred transformation to furanonic derivatives is observed under hydroformylation conditions [8].

One-pot methods with simultaneous and selective introduction of two identical or two different alkyl substituents into a primary amine appear to be extremely difficult, if not impossible. For this problem, the hydroaminomethylation protocol offers a straightforward solution [8], since selective bisalkylation of primary amines and ammonia is achieved with two equivalents of the olefin [71]. Even unsymmetrically substituted tertiary amines can be generated if two olefins with clearly different hydroformylation reaction rates (e.g., styrene and cyclohexene) are used (Scheme 30) [71].

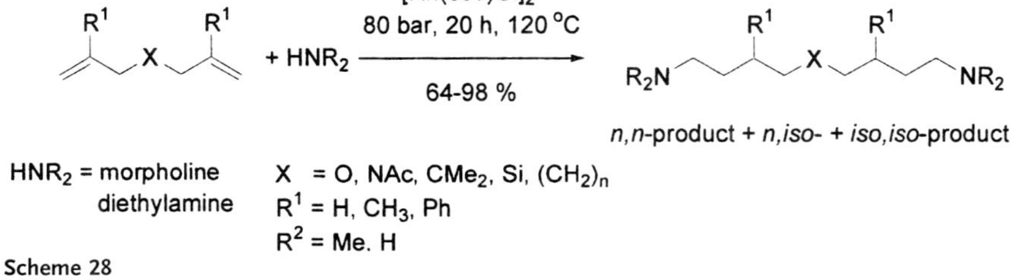

HNR_2 = morpholine, diethylamine
X = O, NAc, CMe_2, Si, $(CH_2)_n$
R^1 = H, CH_3, Ph
R^2 = Me, H

[Rh(cod)Cl]$_2$, 80 bar, 20 h, 120 °C, 64–98 %

n,n-product + n,iso- + iso,iso-product

Scheme 28

Scheme 29

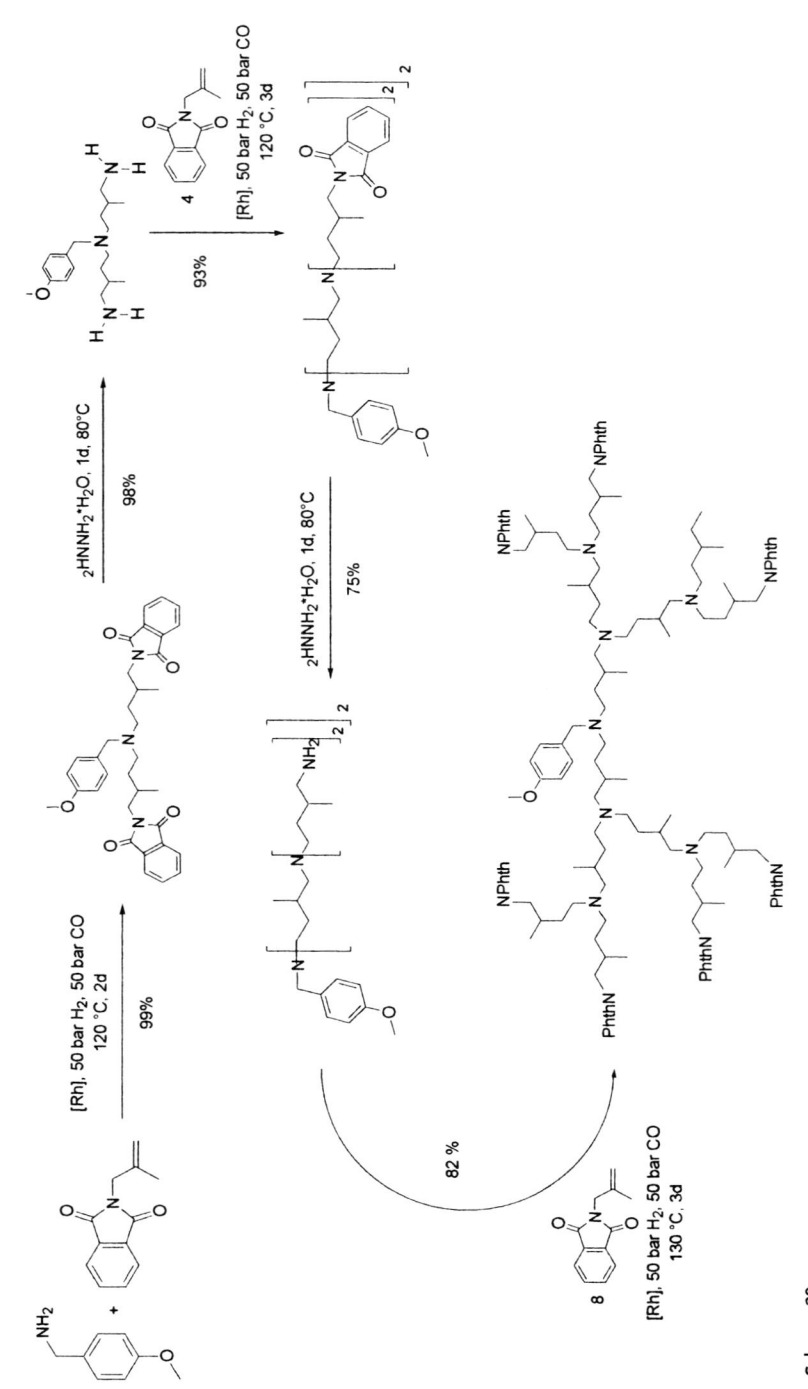

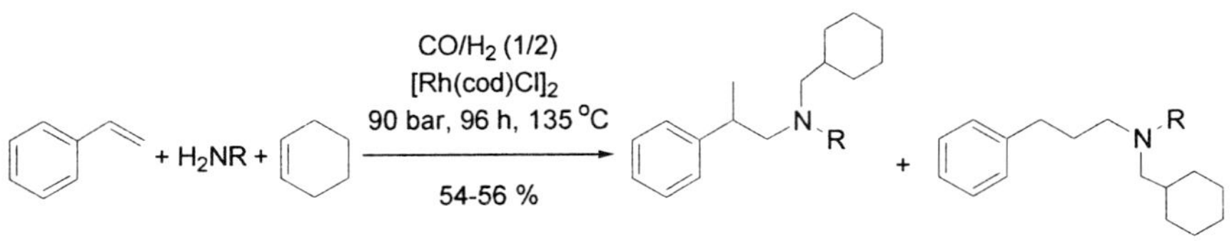

2.2.4 Hydroformylation and Additional Carbon-Heteroatom Bond Formations

R = (CH$_2$)$_3$CH$_3$, CEt$_2$Ph

Scheme 30

This method, however, is limited to a few cases. More widely applicable is an alternative protocol where a preformed aldehyde is added to the hydroaminomethylation reaction mixture. Here, the primary amine first reacts with the preformed aldehyde in a rapid condensation. Hydrogenation follows to give a secondary amine, which only then can undergo hydroaminomethylation with the olefin in a final step leading to various types of unsymmetrical tertiary amines. By the use of ketones instead of aldehydes the synthetic potential can be broadened (Scheme 31) [71]. The use of unsaturated ketones leads to azacyclic ring systems (see below).

Hydroaminomethylation of disubstituted internal alkenes usually occurs with low regioselectivities and additional isomerizations. Thus hydroaminomethylation of unsaturated fatty acids such as oleic acid esters or oleic alcohol with secondary amines gives mixtures of regioisomers (Scheme 32) [82]. If primary amines are

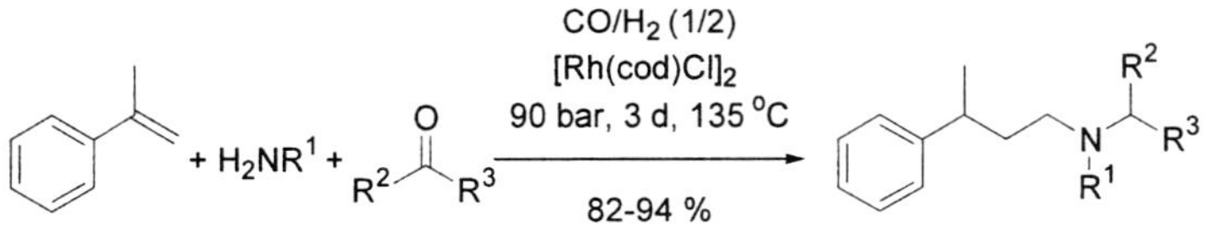

Scheme 31

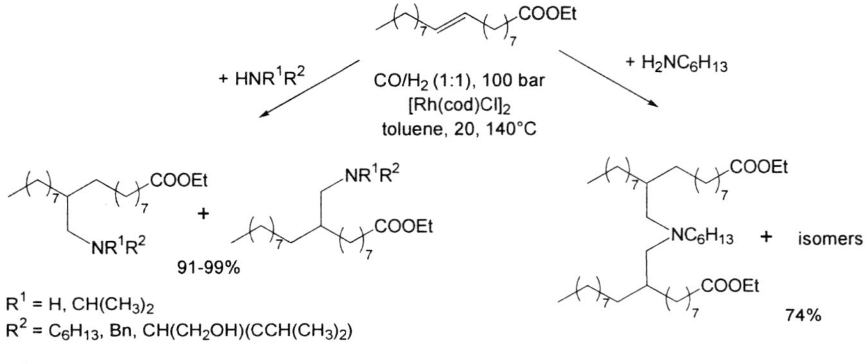

R^1 = H, CH(CH$_3$)$_2$
R^2 = C$_6$H$_{13}$, Bn, CH(CH$_2$OH)(CCH(CH$_3$)$_2$)

Scheme 32

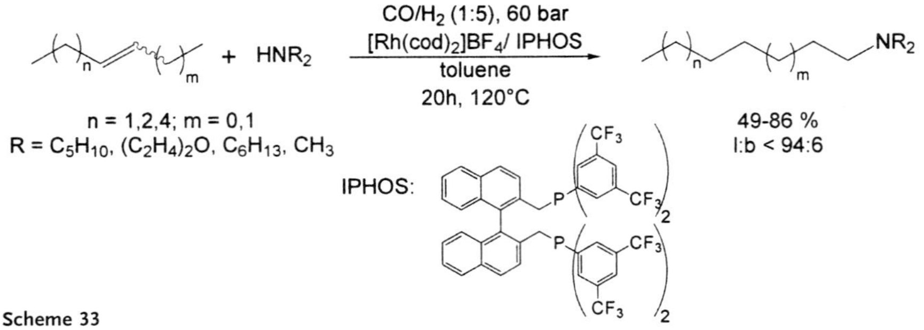

Scheme 33

used in a one-pot procedure, two fatty acid chains can be linked to the amine, giving highly functionalized branched fatty acid derivatives.

Interestingly, the selective synthesis of linear aliphatic amines from internal olefins or olefin mixtures was achieved through a catalytic one-pot reaction consisting of an initial olefin isomerization followed by hydroformylation and reductive amination. Specially designed phosphine ligands were used in the presence of rhodium catalysts (Scheme 33) [83]. This formation of a single product from a mixture of reactants has been designated a "perfect reaction" [84].

Hydroaminomethylation tolerates various functional groups in the olefinic species, such as ethers, tertiary amines, ketones, esters, amides, silylgroups, and many others, depending on the type of amine used [85–87]. Thus, numerous polyfunctional amine products are obtainable with high versatility. On the other hand, acidic and nucleophilic OH-, NH-functions or halides may react with the *in situ*-generated aldehyde, leading to other products including ring formations. Thus 1,4-diamines are prepared if starting from methallyl chloride and secondary amines under hydroaminomethylation conditions via a one-pot nucleophilic substitution/hydroformylation/reductive amination sequence [88].

Interestingly, hydroaminomethylation of the unsaturated steroids (including steroid allylic alcohols) with various secondary amines, because of the allylic alcohol functionality, proceeds diastereoselectively, in moderate to good yields, to the corresponding amino alcohols (Scheme 34) [89].

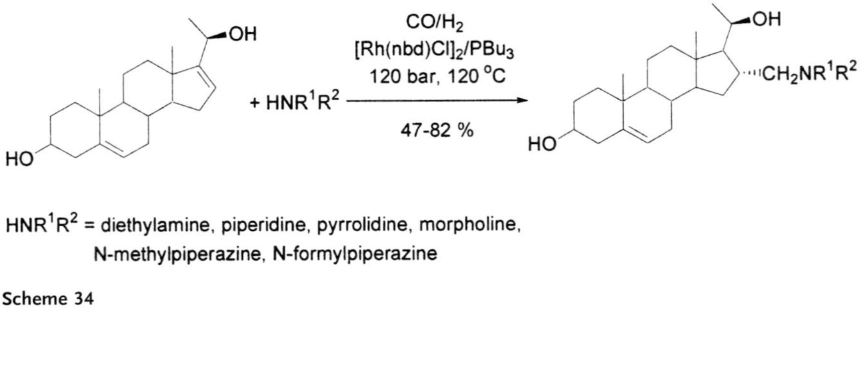

HNR^1R^2 = diethylamine, piperidine, pyrrolidine, morpholine, N-methylpiperazine, N-formylpiperazine

Scheme 34

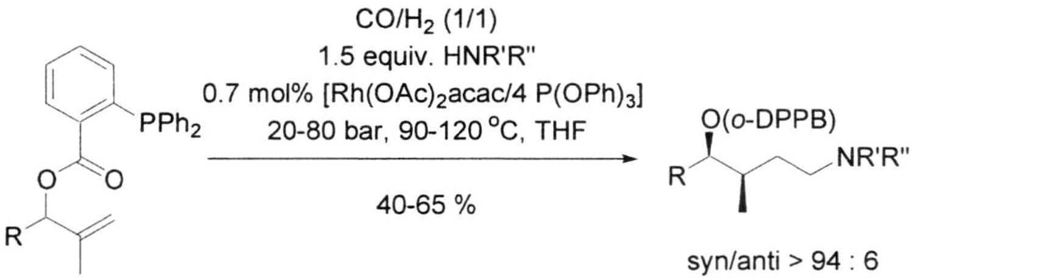

2.2.4 Hydroformylation and Additional Carbon-Heteroatom Bond Formations

Scheme 35

Reaction conditions: CO/H$_2$ (1/1), 1.5 equiv. HNR'R", 0.7 mol% [Rh(OAc)$_2$acac/4 P(OPh)$_3$], 20-80 bar, 90-120 °C, THF, 40-65 %, syn/anti > 94 : 6

o-DPPB = *ortho*-diphenylphosphorylbenzoate

Similarly, *ortho*-DPPB functionalized methallylic alcohols undergo diastereoselective hydroaminomethylation (>88% *de*), leading to the corresponding secondary or tertiary amines in good yields (Scheme 35). This diastereoselectivity is induced by a precoordinating effect of the stereodirecting phosphine group in the *ortho*-DPPB moiety [90].

Unsaturated amines or amides under hydroformylation conditions undergo intramolecular ring closure. The catalytic cycle of the hydroformylation reaction offers two reaction pathways of the metal acyl intermediate, leading to lactams on one hand or cyclic amines on the other (Scheme 36) [8].

The generation of lactams proceeds, via nucleophilic attack of the nitrogen function, to the rhodium-acyl species, which presumably is precoordinated to the metal. Cyclic amines are obtained from unsaturated amines if the intermediate undergoes hydrogenolysis to the aldehyde followed by intramolecular reductive amination with the amino group. The chemoselectivity of these reactions is controllable by the ratio of hydrogen and carbon monoxide and/or the chosen ligand. Numerous examples of both lactam and cyclic amine formation are known and described in an earlier review [8]. Here, only more recent examples of intramolecular hydroaminomethylations are discussed.

Thus, intramolecular hydroaminomethylation of arylmethallylamines [91], alkenamines [92], or methallylbenzamide [93] directly leads to five-, six- or seven-membered cyclic amines (Scheme 37). Here, NaBH$_4$ is used as an additional reducing agent.

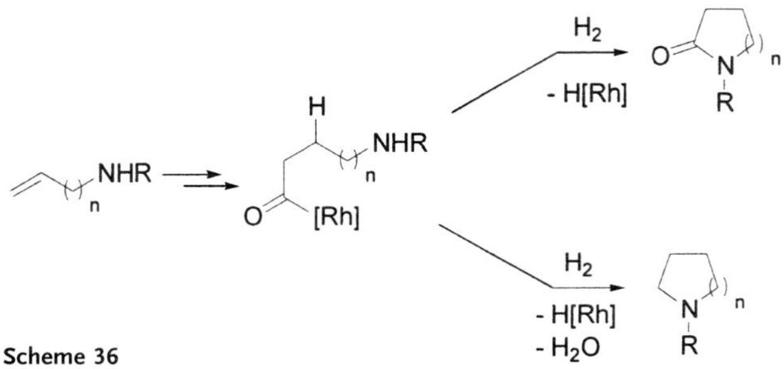

Scheme 36

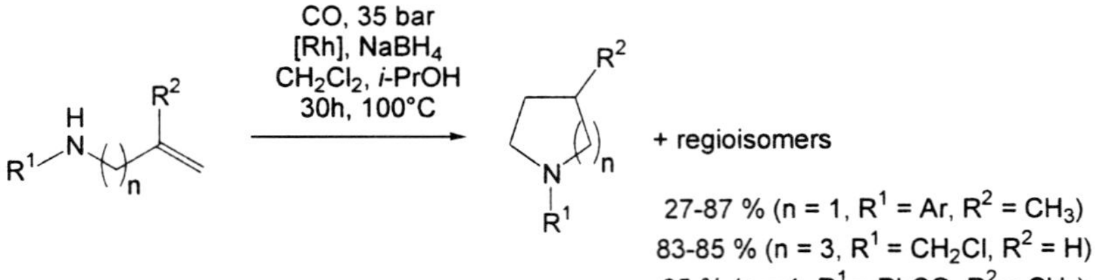

Scheme 37

Reaction conditions: CO, 35 bar, [Rh], NaBH$_4$, CH$_2$Cl$_2$, i-PrOH, 30h, 100°C

27-87 % (n = 1, R^1 = Ar, R^2 = CH$_3$)
83-85 % (n = 3, R^1 = CH$_2$Cl, R^2 = H)
85 % (n = 1, R^1 = PhCO, R^2 = CH$_3$)

Cyclic amines instead of lactams are formed as the major products of hydroaminomethylation in supercritical carbon dioxide (scCO$_2$), whereas the cyclic amide is formed preferentially in conventional solvents (Scheme 38) [61]. This selectivity switch is interpreted as an effect of reversible formation of the carbamic acid in the solvent CO$_2$, which reduces the tendency for intramolecular ring closure at the Rh-acyl intermediate. Thus, scCO$_2$ simultaneously acts as a solvent and as a temporary protecting group during homogeneously rhodium-catalyzed hydroaminomethylation of ethyl methallylic amine.

In this investigation, an interesting dimer formation was observed in up to 52% yield by changing the reaction conditions [61]. This product is interpreted to be formed by a double cyclocondensation of two hydroformylated allylamines followed by a transannular aldol-type reaction (Scheme 39).

The cyclization of various unsaturated amines via rhodium-catalyzed intramolecular hydroaminomethylation without the need for high dilution was extended to cyclic amines of medium and large ring sizes in yields of up to 85% (Scheme 40) [94]. High regioselectivities for non-branched products can be obtained when BIPHEPHOS is used as a ligand in the hydroformylation reaction.

In principle, α,β-unsaturated aldehydes generated from alkynes under hydroformylation conditions may undergo various subsequent reactions in the presence of amines. On one hand nucleophiles can react in a Michael-type 1,4-addition to form an aminoaldehyde [95]. On the other hand the aldehyde can undergo condensation with a primary amine to give an unsaturated imine (Scheme 41). After

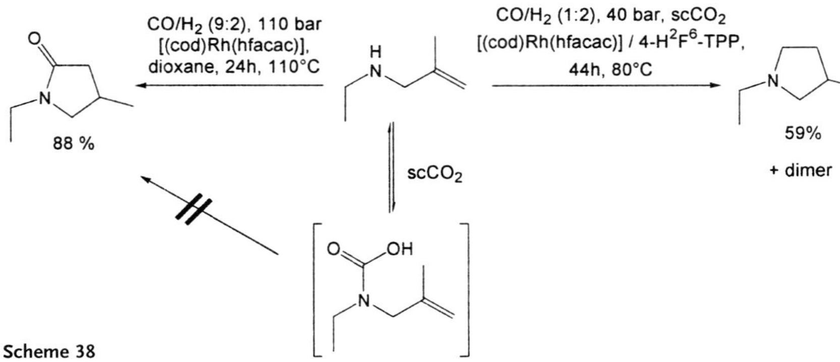

Scheme 38

CO/H$_2$ (9:2), 110 bar, [(cod)Rh(hfacac)], dioxane, 24h, 110°C — 88 %

CO/H$_2$ (1:2), 40 bar, scCO$_2$, [(cod)Rh(hfacac)] / 4-H^{2}F^6-TPP, 44h, 80°C — 59% + dimer

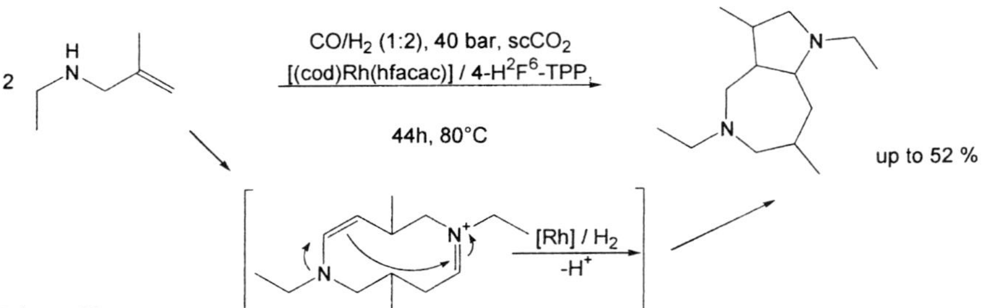

Scheme 39

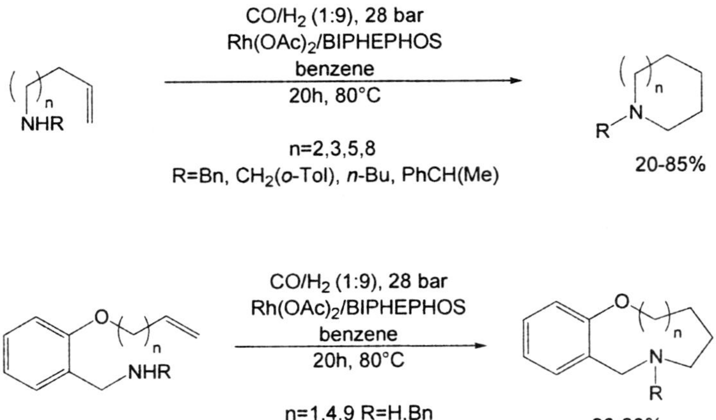

Scheme 40

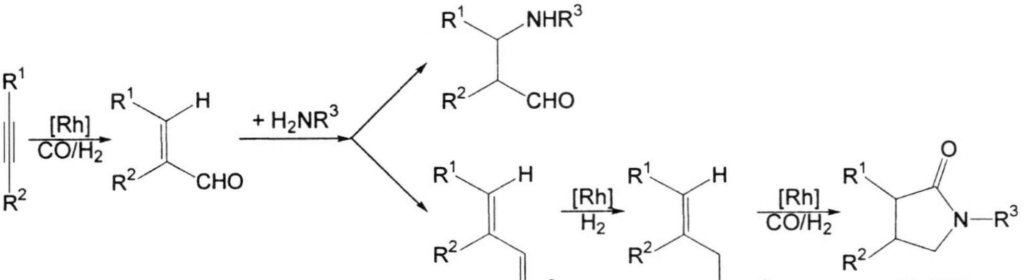

Scheme 41

selective hydrogenation of the imine unit, the resulting allyl amine can undergo further carbonylation and intramolecular lactam formation. In this way, various lactams are obtained, albeit in moderate yields [96]. Major by-products in this multistep sequence are the corresponding furanone and mono- or bisamides resulting from an insertion of the amine into the metal-acyl species during the hydroformylation cycle.

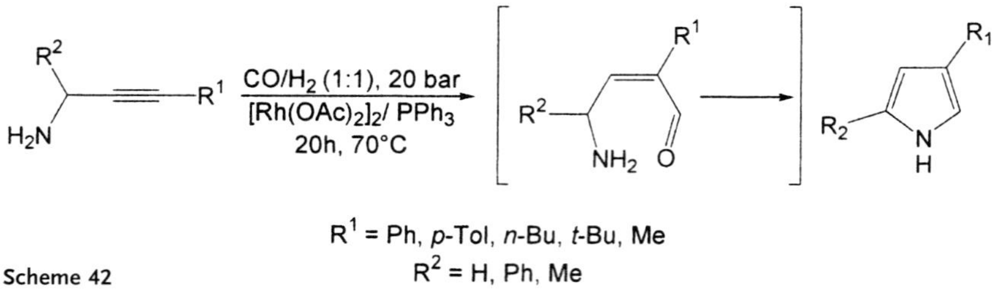

Scheme 42

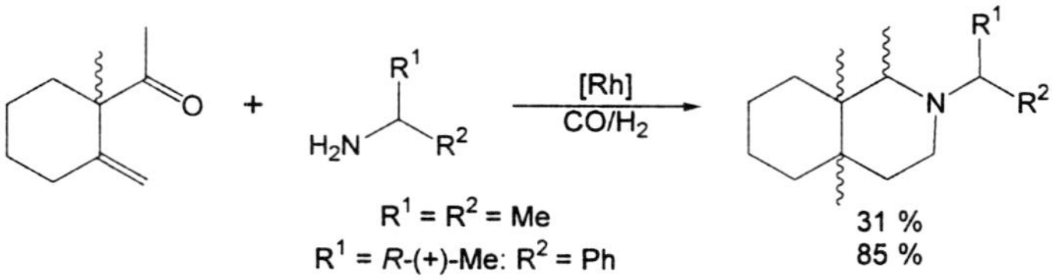

R^1 = Ph, p-Tol, n-Bu, t-Bu, Me
R^2 = H, Ph, Me

Scheme 43

R^1 = R^2 = Me — 31 %
R^1 = R-(+)-Me; R^2 = Ph — 85 %

Intramolecular versions of these rhodium-catalyzed reactions are more successful. Thus, aryl- and alkylpropargylamines with syngas give 3-substituted or 2,4-disubstituted pyrroles in good to excellent yields. This one-pot synthesis includes hydroformylation of the alkyne to form unsaturated aldehydes and intramolecular condensation with the amine group. Alkylpropargylamines lead to the corresponding pyrroles in lower yields than those obtained with aryl propargyl amines (Scheme 42) [97, 98].

Simultaneous reductive amination of a keto group and an *in situ*-generated aldehyde from an unsaturated ketone can be applied to intramolecular cyclizations, yielding N-heterocycles, if substrates with carbonyl and olefin functionality are used (Scheme 43) [71].

Rhodium-catalyzed hydroaminomethylation of α,ω-diolefins in the presence of primary amines or secondary α,ω-diamines has been applied to macroheterocyclic

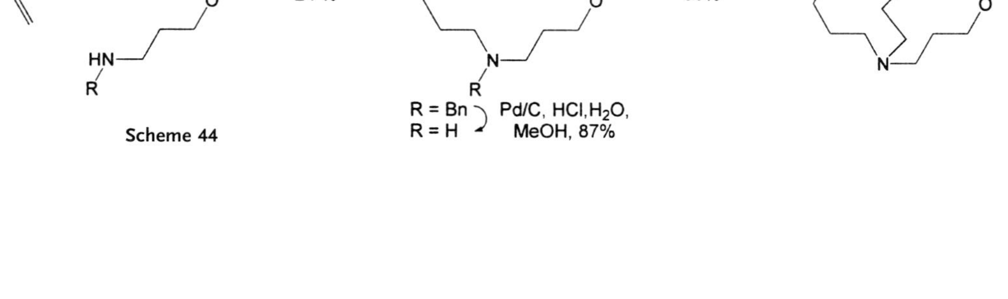

Scheme 44

R = Bn ⎫ Pd/C, HCl, H$_2$O,
R = H ⎭ MeOH, 87%

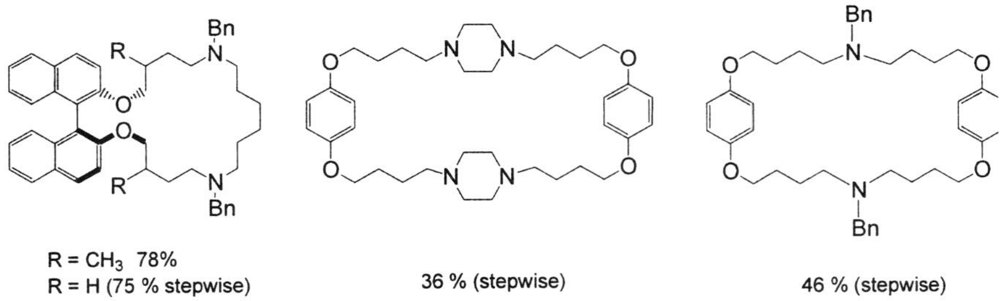

R = CH₃ 78%
R = H (75 % stepwise)

36 % (stepwise)

46 % (stepwise)

Scheme 45

ring synthesis [99, 100]. Compared to common strategies, this methodology offers a very efficient synthetic route to substituted macrocyclic polyamines with high variability. Starting from (hetero)diallylic systems, 12- to 36-membered polyheterocycles have been readily obtained in up to 78% yield. In addition, such macrocyclic systems can be debenzylated, and the resulting macrocyclic diamines undergo a second ring-closing bis(hydroamino-methylation) to give cryptand systems (Scheme 44) [99].

This procedure allows versatile construction of numerous macrocyclic systems with varying building blocks. More recently, rigid and flexible aromatic and chiral binaphthyl systems have been integrated into macrocyclic systems of this type (Scheme 45) [100].

2.2.4.3
Hydroformylation in the Presence of Other Heteroatom Nucleophiles

Heteroatoms other than oxygen and nitrogen up to now have rarely been used in tandem hydroformylation procedures, although they offer high potential in the construction of functionalized acyclic and heterocyclic systems. Thus, Ojima et al. reported highly efficient syntheses of azabicyclo[4.4.0]alkane amino acids by Rh-catalyzed cyclohydrocarbonylation of dipeptides bearing a terminal olefin moiety and an S-heteroatom nucleophile to form the bicyclic system with N,S-subunit in one step (Scheme 46) [40].

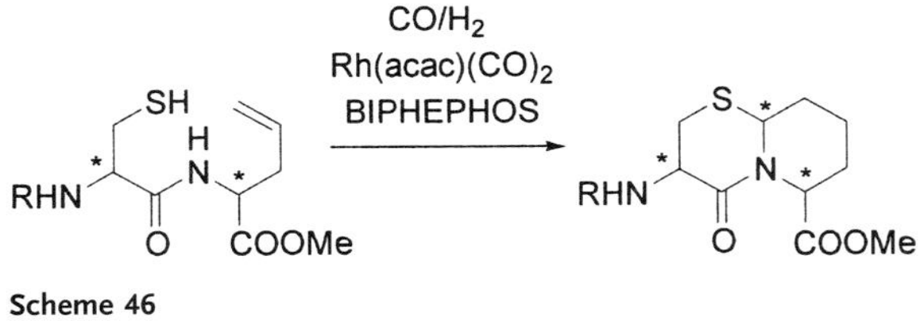

Scheme 46

2.2.5
Concluding Remarks

In conclusion, transition metal-catalyzed hydroformylation sequences starting from easily accessible functionalized or non-functionalized unsaturated compounds followed by reduction, isomerization, or C-O- or C-N-bond-forming steps offer convenient and versatile synthetic applications in the construction of homologous, functionalized carbon skeletons and heterocyclic systems. It can be expected that more interesting examples and applications will be presented in the future. An important expansion of these procedures is offered by hydroformylation sequences with additional C-C- bond formations, as described in a recent review [8] and in the next chapter.

2.2.6
References

1 TIETZE, L. F. *Chem. Rev.* **1996**, *96*, 115–136.
2 WENDT, K. U.; SCHULZ, G. E.; COREY, E. J.; LIU, D. R. *Angew. Chem.* **2000**, *112*, 2930–2952. *Angew. Chem. Int. Ed.* **2000**, *39*, 2812–2833.
3 (a) PARSONS, P. J.; PENKETT, C. S.; SHELL, A. J. *Chem. Rev.* **1996**, *96*, 195–206. (b) NICOLAOU, K. C.; MONTAGNON, T.; SNYDER, S. A. *Chem. Commun.* **2003**, 551–564.
4 (a) BOHNEN, H. W.; CORNILS, B. *Adv. Catal.* **2002**, *47*, 1–64. (b) CORNILS, B.: *J. Mol. Catal. A-Chem.* **1999**, *143*, 1–10. (c) BELLER, M.; CORNILS, B.; FROHNING, C. D.; KOHLPAINTNER, C. W. *J. Mol. Catal. A-Chem.* **1995**, *104*, 17–85. (d) CORNILS, B.; HERRMANN, W. A.; KOHLPAINTNER, C. W. *Angew. Chem.* **1994**, *106*, 2219–2238; *Angew. Chem. Int. Ed. Engl.* **1994**, *33*, 2144–2163. (e) CORNILS, B.; HERRMANN, W. A. *Applied Homogeneous Catalysis with Organometallic Compounds*, VCH, Weinheim, 1996. (f) VAN LEEUWEN, P. W. N. M.; CLAVER, C. (eds.) *Rhodium Catalyzed Hydroformylation*, Kluwer Academic Publishers, Dordrecht, 2000.
5 (a) BURKE, S. D.; COBB, J. E.; TAKEUCHI, K. *J. Org. Chem.* **1990**, *55*, 2138–2151. (b) BREIT, B.; ZAHN, S. K. *Tetrahedron Lett.* **1998**, *39*, 1901–1904. (c) HORNBERGER, K. R.; HAMBLETT, C. L.; LEIGHTON, J. L. *J. Am. Chem. Soc.* **2000**, *122*, 12894– 12895. (d) BREIT, B. *Chem. Eur. J.* **2000**, *6*, 1519–1524. (e) DREHER, S. D.; LEIGHTON, J. L. *J. Am. Chem. Soc.* **2001**, *123*, 341–342. (f) LIU, P.; JACOBSEN, E. N. *J. Am. Chem. Soc.* **2001**, *123*, 10772–10773. (g) BREIT, B.; ZAHN, S. K. *J. Org. Chem.* **2001**, *66*, 4870–4877. (h) SUN, P.; SUN, C.; WEINREB, S. M. *Org. Lett.* **2001**, *3*, 3507–3510. (i) SUN, P.; SUN, C. X.; WEINREB, S. M. *J. Org. Chem.* **2002**, *67*, 4337–4345. (k) SEEPERSAUD, M.; KETTUNEN, M.; ABU-SURRAH, A. S.; REPO, T.: VOELTER, W. AL-ABED, Y. *Tetrahedron Lett.* **2002**, *43*, 1793–1795. (l) WEINREB, S. M. *Acc. Chem. Res.* **2003**, *36*, 590–560.
6 (a) GLADIALI, S.; BAYON, J. C.; CLAVER, C. *Tetrahedron: Asymmetry* **1995**, *6*, 1453–1474. (b) EILBRACHT, P. in *Methoden der Organischen Chemie* (Houben-Weyl) Vol. E 21c: *Stereoselective Synthesis* (Eds.: HELMCHEN, G.; HOFFMANN, R. W.; MULZER, J.; SCHAUMANN), Thieme: Stuttgart, 1995, 2488–2733. (c) AGBOSSOU, F.; CARPENTIER, J.-F.; MORTREUX, A. *Chem. Rev.* **1995**, *95*, 2485–2506. (d) BREIT, B.; SEICHE, W.; *Synthesis* **2001**, 1–36. (e) BREIT, B.; *Acc. Chem. Res.* **2003**, *36*, 264–275.
7 (a) PATAI, S. (Ed.) *The Chemistry of the Carbonyl Group*, Wiley-Interscience, New York, 1966, 1970. (b) FALBE, J. (Ed.) *Methoden der Organischen Chemie* (Houben-

Weyl); Thieme: Stuttgart, Vol. **E3** (Aldehyde), 1983.
8 EILBRACHT, P.; BÄRFACKER, L.; BUSS, C.; HOLLMANN, C.; KITSOS-RZYCHON, B.; KRANEMANN, C. L.; RISCHE, T.; ROGGENBUCK, R.; SCHMIDT, A. *Chem. Rev.* **1999**, *99*, 3329–3365.
9 (a) KRAMER, P.C.J.; REEK, J.N.H.; VAN LEEUWEN, P.W.N.M. in VAN LEEUWEN, P.W.N.M.; CLAVER, C. (eds.) *Rhodium-catalyzed Hydro-formylation*), Kluwer Academic Publishers, Dordrecht, **2000**, p. 35–62. (b) FROHNING, C.D.; KOHLPAINTNER, C.W. in CORNILS, B.; HERRMANN, W.A., (eds.) *Applied Homogeneous Catalysis with Organometallic Compounds* Wiley-VCH, Weinheim, **1996**, Vol. 1, p. 3–25.
10 MEESSEN, P.; VOGT, D.; KEIM, W. *J. Organomet. Chem.* **1998**, *551*, 165–170.
11 (a) VAN DER VEEN, L.A.; KAMER, P.C.J.; VAN LEEUWEN, P.W.N.M. *Angew. Chem.* **1999**, *111*, 349–351, *Angew. Chem. Int. Ed.* **1999**, *3*, 336–338. (b) VAN DER VEEN, L.A.; KAMER, P.C.J.; VAN LEEUWEN, P.W.N.M. *Organometallics* **1999**, *18*, 4765–4777.
12 SELENT, D.; WIESE, K.-D.; RÖTTGER, D.; BÖRNER, A. *Angew. Chem.* **2000**, *112*, 1694–1696; *Angew. Chem. Int. Ed.* **2000**, *39*, 1639–1641.
13 JACKSTELL, R.; KLEIN, H.; BELLER, M.; WIESE, K.-D.; RÖTTGER, D. *Eur. J. Org. Chem.* **2001**, *20*, 3871–3877.
14 SELENT, D.; HESS, D.; WIESE, K.D., et al.; *Angew. Chem.* **2001**, *113*, 11739–1741; *Angew. Chem. Int. Ed.* **2001**, *40*, 1696–1698.
15 KLEIN, H.; JACKSTELL, R.; WIESE, K.D.; BORGMANN, C.; BELLER, M. *Angew. Chem.* **2001**, *113*, 3505–3508; *Angew. Chem. Int. Ed.* **2001**, *40*, 3408–3411.
16 BELLER, M.; ZIMMERMANN, B.; GEISSLER, H. *Chem. Eur. J.* **1999**, *5*, 1301–1305.
17 BELLER, M.; KRAUTER, J.G.E. *J. Mol. Catal. A-Chem.* **1999**, *143*, 31–39.
18 SHIRAKAWA, S.; SHIMIZU, S.; SASAKI, Y. *New J. Chem.* **2001**, *25*, 777–779.
19 (a) SEAYAD, A.M.; AHMED, M.; KLEIN, H.; JACKSTELL, R.; GROSS, T.; BELLER, M. *Science* **2002**, *297*, 1676–1678. (b) SEAYAD, A.M.; SELVAKUMAR, K.; AHMED, M. BELLER, M. *Tetrahedron Lett.* **2003**, *44*, 1679–1683.
20 (a) KANEDA, K.; YASUMURA, M.; HIRAKI, M.; IMANAKA, T.; TERANSIHI, S. *Chem. Lett.* **1981**, 1763. (b) KANEDA, K.; IMANAKA, T.; TERANISHI, S. *Chem. Lett.* **1983**, 1465.
21 MACDOUGALL, J.K.; SIMPSON, M.C.; GREEN, M.J.; COLE-HAMILTON, D.J. *J. Chem. Soc., Dalton Trans.* **1996**, 1161–1172 and earlier references cited.
22 SIMPSON, M.C.; COLE-HAMILTON, D.J. *Coord. Chem. Rev.* **1996**, *155*, 163–207.
23 SIMPSON, M.C.; CURRIE, A.W.S.; ANDERSEN, J.-A.M.; COLE-HAMILTON, D.J.; GREEN, M.J. *J. Chem. Soc., Dalton Trans.* **1996**, 1793–1800.
24 SANDEE, A.J.; REEK, J.N.H.; KAMER, P.C.J.; VAN LEEUWEN P.W.N.M. J. *Am. Chem. Soc.* **2001**, *123*, 8468–8476.
25 BOTTEGHI, C.; GANZERLA, R.; LENARDA, M.; MORETTI, G. *J. Mol. Catal.* **1987**, *40*, 129.
26 CHALK, A.J. in RYLANDER, P.N.; GREENFIELD, H.; AUGUSTINE, R.L. (eds.), *Catalysis of Organic Reactions,* Marcel Dekker, New York, **1988**, p. 43.
27 NOZAKI, K.; LI, W.; HORIUCHI, T.; TAKAYA, H. *Tetrahedron Lett.* **1997**, *38*, 4611–4614.
28 SCHMIDT, B.; COSTISELLA, B.; ROGGENBUCK, R.; WESTHUS, M.; WILDEMANN, H.; EILBRACHT, P. *J. Org. Chem.* **2001**, *66*, 7658–7665.
29 WUTS, P.G.M.; OBRZUT, M.L.; THOMPSON, P.A. *Tetrahedron Lett.* **1984**, *25*, 4051–4054.
30 HORNBERGER, K.R.; HAMBLETT, C.L.; LEIGHTON, J.L. *J. Am. Chem. Soc.* **2000**, *122*, 12894–12895.
31 KITSOS-RZYCHON, B.; EILBRACHT, P. *Tetrahedron* **1998**, *54*, 10721–10732.
32 (a) SIROL, S; KALCK, PH. *New J. Chem.* **1997**, *21*, 1129–1137. (b) SIROL, S.; GORRICHON, J.P.; KALCK, P.; NIETO, P.M.; COMMENGES, G. *Magn. Reson. Chem.* **1999**, *37*, 127–132.
33 FUKUTA, Y.; MATSUDA, I.; ITOH, K. *Tetrahedron Lett.* **2001**, *42*, 1301–1304.
34 ROGGENBUCK, R.; EILBRACHT, P. *Org. Lett.* **2002**, *4*, 289–291.
35 OJIMA, I.; MORALEE, A.C.; VASSAR, V.C. *Top. Catal.* **2002**, *19*, 89–99.
36 OJIMA, I.; VIDAL, E.S. *J. Org. Chem.* **1998**, *63*, 7999–8003.

37 Ojima, I.; Tzamarioudaki, M.; Eguchi, M. *J. Org. Chem.* **1995**, *60*, 7078–7079.
38 Eguchi, M.; Zeng, Q.; Korda, A.; Ojima, I. *Tetrahedron Lett.* **1993**, *34*, 915–918.
39 Ojima, I.; Iula, D.M. in: S.W. Pelletier (ed.) *The Alkaloids*, Vol. 13, Pergamon, London, **1998**, p. 371.
40 Mizutani, N.; Chiou, W.-H.; Ojima, I. *Org. Lett.* **2002**, *4*, 4575–4578.
41 (a) Bergmann, D.J.; Campi, E.M.; Jackson, W.R.; Patti, A.F. Bergmann, D.J. *Chem. Commun.* **1999**, 1279–1280. (b) Bergmann, D.J.; Campi, E.M.; Jackson, W.R.; Patti, A.F.; *Aust. J. Chem.* **1999**, *52*, 1131–1138. (c) Bergmann, D.J.; Campi, E.M.; Jackson, W.R.; Patti, A.F. Saylik, D. *Tetrahedron Lett.* **1999**, *40*, 5597–5600.
42 Campi, E.N.; Jackson, W.R.; McCubbin, Q.J.; Trnacek, A.E. *Aust. J. Chem.* **1994**, *47*, 1061–1070.
43 Campi, E.N.; Jackson, W.R.; Trnacek, A.E. *Aust. J. Chem.* **1997**, *50*, 1031–1034.
44 Dong, Y.; Busacca, C.A. *J. Org. Chem.* **1997**, *62*, 6464–6465.
45 Ucciani, E.; Bonfand, A. *J. Chem. Soc., Chem. Commun.* **1981**, 82.
46 Ojima, I.; Iula, D.M.; Tzamarioudaki, M. *Tetrahedron Lett.* **1998**, *39*, 4599–4602.
47 (a) Hua, Z.; Mizutani, N.; Zhang, P.Y.; Ojima, I. in: 21st Am. Chem. Soc. National Meeting, Abstracts (**2001**) ORGN394. (b) Zhang, P.Y.; Lee, S.Y.; Ojima, I. in: 220th Am. Chem. Soc. National Meeting, Abstracts (**2000**) ORGN107.
48 (a) Teoh, E.; Campi, E.M.; Jackson, W.R.; *Chem. Commun.* **2002**, 978–979. (b) Teoh, E.; Campi, E.M.; Jackson, W.R.N.; Robinson, A.J. *New J. Chem.* **2003**, *27*, 387–394.
49 Lin, Y.-S.; Ali, B.E.; Alper, H. *J. Am. Chem. Soc.* **2001**, *123*, 7719–7720.
50 Beller, M.; Jackstell, R.; N.N.; publication in preparation.
51 Rische, T.; Eilbracht, P. unpublished.
52 Verspui, G.; Elbertse, G.; Sheldon, F.A.; Hacking, M.A.P.J.; Sheldon, R.A.; *Chem. Commun.* **2000**, 1363–1364.
53 Köhling, P.; Schmidt, A.M.; Eilbracht, P. *Org. Lett.* **2003**, *5*, in press.
54 (a) Rische, T.; Eilbracht, P. *Synthesis* **1997**, 1331–1337. (b) Rische, T.; Eilbracht, P. *Tetrahedron* **1999**, *55*, 7841–7846.
55 Lin, Y.S.; El Ali, B.; Alper, H. *Tetrahedron Lett.* **2001**, *42*, 2423–2425.
56 Kostas, I.D.; Screttas, C.G. *J. Organomet. Chem.* **1999**, *585*, 1–6.
57 Schulte, M.M.; Herwig, J.; Fischer, R.W.; Kohlpaintner, C.W. *J. Mol. Catal. A* **1999**, *150*, 147–153
58 Seayad, A.M.; Selvakumar, K.; Ahmed, M.; Beller, M. *Tetrahedron Lett.* **2003**, *44*, 1679–1683.
59 Schaffrath, H.; Keim, W. *J. Mol. Catal. A.* **1999**, *140*, 107–113.
60 Ahmed, M.; Seayad, A.M.; Jackstell, R.; Beller, M. *J. Am. Chem. Soc.* in press.
61 Wittmann, K.; Wisniewski, W.; Mynott, R.; Leitner, W.; Kranemann, C.L.; Rische, T.; Eilbracht, P.; Kluwer, S.; Ernsting, J.M.; Elsevier, C.L. *Chem. Eur. J.* **2001**, *7*, 4584–4589.
62 Eilbracht, P.; Greiving, H.; Mersch, C. DE 10010046-A1, **2001** (19. 02. 2001).
63 Dessole, G.; Marchetti, M.; Taddei, M. *J. Comb. Chem.* **2003**, *5*, 198–200.
64 Rische, T.; Eilbracht, P. *Tetrahedron* **1998**, *54*, 8441–8450.
65 Botteghi, C.; Cazzolato, L.; Marchetti, M.; Paganelli, S. *J. Org. Chem.* **1995**, *60*, 6612–6615.
66 Rische, T.; Eilbracht, P. *Tetrahedron* **1999**, *55*, 1915–1920.
67 Botteghi, C.; Corrias, T.; Marchetti, M.; Paganelli, S.; Piccolo, O. *Org. Proc. Res. Develop.* **2002**, *6*, 379–383.
68 Botteghi, C.; Corrias, T.; Marchetti, M.; Paganelli, S.; Piccolo, O. *Org. Proc. Res. Develop.* **2002**, *6*, 379–383.
69 Donsbach, M.; Eilbracht, P.; Buss, C.; Mersch, C. Ger. Pat. DE 10033016 A1 (07. 07. 2000).
70 Knifton, J.F.; Lin, J.J. *J. Mol. Catal.* **1993**, *81*, 27–36.
71 Rische, T.; Kitsos-Rzychon, B.; Eilbracht, P. *Tetrahedron* **1998**, *54*, 2723–2742.
72 Zimmermann, B.; Herwig, J.; Beller, M. *Angew. Chem.* **1999**, *111*, 2515–2518; *Angew. Chem. Int. Ed.* **1999**, *16*, 2372–2375.

73 Drent, E., Breed, A. J. M. (Shell Int. Research Maatschappij B. V.) EU Patent 457 386, **1992**, *Chem. Abstr.* **1992**, *116*, 83212h.
74 Kranemann, C. L.; Eilbracht, P. *Synthesis* **1998**, 71–77.
75 Eilbracht, P.; Kranemann, C. L.; Bärfacker, L. *Eur. J. Org. Chem.* **1999**, 1907–1914.
76 Koc, F.; Eilbracht, P. manuscript in preparation.
77 Jachimowicz, F.; Hansson, A. in: Augustine, R. L. (ed.) *Catalysis of Organic Reactions*, Dekker, New York, **1985**, 381–390.
78 Jachimowicz, F.; Hansson, A. (Grace, W. R. & Co.) Can. Patent 1 231 199, 1984; *Chem. Abstr.* **1988**, *109*, 38485u.
79 Jachimowicz, F. (Grace, W. R. & Co.) GE 3 106 139, 1981.
80 Sunder, A.; Turk, H.; Haag, R.; Frey, H. *Macromolecules* **2000**, *33*, 7682–7692.
81 Cornils, B. in: Falbe, J. (ed.) *New Syntheses with Carbon Monoxide*; Springer, Berlin, **1980**, 1–225.
82 Behr, A.; Fiene, M.; Buss, C.; Eilbracht, P. *Eur. J. Lipid. Sci. Technol.* **2000**, *102*, 467–471.
83 Seayad, A.; Ahmed, M.; Klein, H.; Jackstell, R.; Gross, T.; Beller, M. *Science* **2002**, *297*, 1676–1678.
84 Hartwig, J. F. *Science* **2002**, *297*, 1653–1654.
85 Hartwig, J. F. *Science*; **2002**, *297*, 1653–1654.
86 Eilbracht, P.; Kranemann, C. L.; Bärfacker, L. *Eur. J. Org. Chem.* **1999**, 1907–1914.
87 Bärfacker, L.; Rische, T.; Eilbracht, P. *Tetrahedron* **1999**, *55*, 7177–7190

88 Rische, T.; Eilbracht, P. *Tetrahedron* **1999**, *55*, 3917–3922.
89 Nagy, E.; Benedek, C.; Heil, B.; Törös, S. *Appl. Organomet. Chem.* **2002**, *16*, 628–634.
90 Breit, B. *Tetrahedron Lett.* **1998**, *39*, 5163–5166.
91 Zhou, J.-Q.; Alper, H. *J. Org. Chem.* **1992**, *57*, 3328–3331.
92 (a) Zhang, Z.; Ojima, I. *J. Organomet. Chem.* **1993**, *454*, 281. (b) Gomes da Rosa, R.; Ribeiro de Campos, J. D.; Buffon, R. *J. Mol. Catal.* **1999**, *137*, 297–301.
93 Ojima, I.; Zhang, Z. *J. Org. Chem.* **1988**, *53*, 4422–4425
94 (a) Bergmann, D. J.; Campi, E. M.; Jackson, W. R.; Patti, A. F.; Saylik, D. *Tetrahedron Lett.* **1999**, *40*, 5597–5600. (b) Bergmann, D. J.; Campi, E. M.; Jackson, W. R.; Piatti, A. F.; Saylik, D. *Aust. J. Chem.* **2000**, *53*, 835–844.
95 Jones, M. D. *J. Organomet. Chem.* **1989**, *366*, 403–408
96 Bärfacker, L.; Hollmann, C.; Eilbracht, P. *Tetrahedron* **1998**, *54*, 4493–4506.
97 Campi, E. M.; Fallon, G. D.; Jackson, W. R.; Nilsson, Y. *Aust. J. Chem.* **1992**, *45*, 1167–1178.
98 Campi, E. M.; Jackson, W. R.; Nilsson, Y. *Tetrahedron Lett.* **1991**, *32*, 1093–1094.
99 (a) Kranemann, C. L.; Costisella, B.; Eilbracht, P. *Tetrahedron Lett.* **1999**, *40*, 7773–7776. (b) Kranemann, C. L.; Eilbracht, P. *Eur. J. Org. Chem.* **2000**, 2367–2377.
100 Angelovski, G.; Eilbracht, P. *Tetrahedron* **2003**, *59*, 8265–8274.

2.3
Multiple Carbon-Carbon Bond Formations under Hydroformylation Conditions

Peter Eilbracht and Axel M. Schmidt

2.3.1
Introduction

"Tandem reactions", also known as "domino reactions", "reaction cascades" or "sequential transformations", combine several synthetic steps to a single operation without change of reaction conditions and without addition of further reagents, thus requiring only one single setup of starting materials, reagents, and solvents and no isolation of intermediates [1]. Considerable efforts are being concentrated on the development of new procedures, including the total synthesis of more complex target molecules [2]. Multiple carbon-carbon bond-forming reaction sequences are of especially high value for the rapid construction of new carbon skeletons from easily available starting materials, and in this regard the hydroformylation reaction ("oxo reaction") of alkenes is an interesting method to be included in multiple carbon-carbon bond forming sequences, since hydroformylation not only forms itself a new carbon-carbon bond but also leads to aldehydes ("oxo aldehydes") as synthetically useful starting materials for numerous further carbon-carbon bond forming conversions [3], e.g., via carbon nucleophile attack at the aldehyde carbonyl group or carbon electrophile attack in the acidic α-position (Scheme 1).

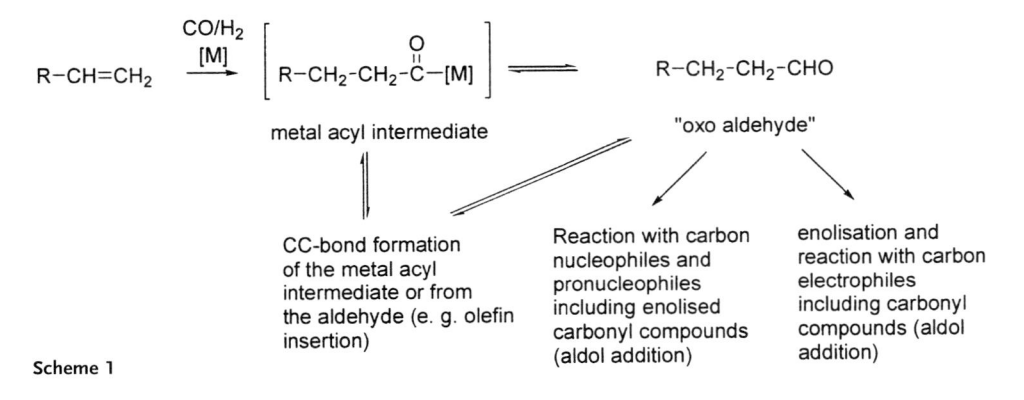

$$R-CH=CH_2 \xrightarrow{CO/H_2, [M]} \left[R-CH_2-CH_2-\overset{O}{\underset{\|}{C}}-[M] \right] \rightleftharpoons R-CH_2-CH_2-CHO$$

metal acyl intermediate → CC-bond formation of the metal acyl intermediate or from the aldehyde (e. g. olefin insertion)

"oxo aldehyde" → Reaction with carbon nucleophiles and pronucleophiles including enolised carbonyl compounds (aldol addition)

→ enolisation and reaction with carbon electrophiles including carbonyl compounds (aldol addition)

Scheme 1

The homogeneously transition-metal-catalyzed hydroformylation is not only established as an important industrial tool for the production of aldehydes and products derived therefrom [4], but it is also applied in the synthesis of more complex target molecules [5], including stereoselective and asymmetric syntheses [6]. Therefore, following the growing interest in tandem procedures, hydroformylation has frequently been integrated in reaction sequences of this type with various options of additional conversions of the final aldehyde product or the intermediates in the catalytic cycle of hydroformylation [7].

These tandem hydroformylation sequences also include multiple carbon-carbon bond formations via further conversions of the aldehyde products as described above and, especially under the hydroformylation conditions, via conversions of the intermediate metal alkyl or metal acyl systems formed in the catalytic cycle, e.g., by the addition of nucleophiles or insertion of unsaturated units.

In this survey, tandem hydroformylation sequences with additional carbon-carbon bond formations are described, and, in addition to a more comprehensive review covering the literature up to 1998/99 [7], some more recent synthetic applications are added. Other tandem sequences of hydroformylation accompanied or followed by isomerization, reduction, and additional C-O or C-N bond-forming reactions are described in the previous chapter of this volume ("New synthetic applications of tandem reactions under hydroformylation conditions").

2.3.2
Hydroformylation in the Presence of Carbon Nucleophiles

As discussed above, under hydroformylation conditions, various types of additional C-C bond formations can occur, via reactions of either the oxo aldehyde or its enolized carbon pronucleophiles as well as via reactions of the metal acyl intermediate, e.g., through olefin insertion (Scheme 1). In this section, reactions of oxo aldehydes with C-nucleophiles except enolates are described (Scheme 2), whereas enolate addition and other hydroformylations with additional C-C bond formations will be discussed in the following Sections 3 and 4.

Various hydroformylation sequences with additional C-C bond formation via attack of carbon nucleophiles to the *in situ*-formed oxo aldehydes are reported. These nucleophiles or pronucleophiles cannot be the reactive carbanion reagents, e.g., with electropositive alkaline or alkaline earth metals as counterions, because of unwanted side reactions with the catalyst. However, stable Wittig reagents, allylsilanes, allylboranes, electron-rich arenes or heteroarenes, enolates, or stabilized

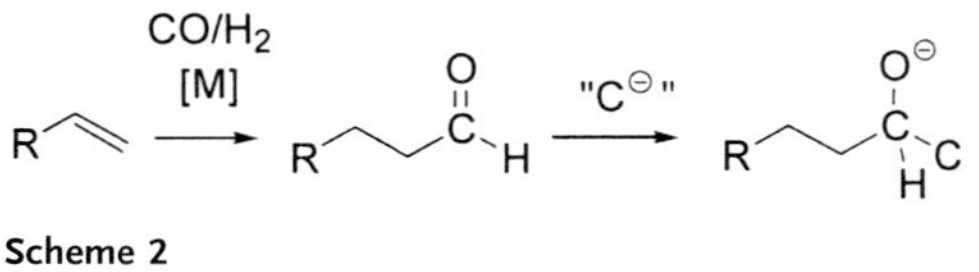

Scheme 2

2.3.2 Hydroformylation in the Presence of Carbon Nucleophiles

enolate derivatives are potential carbon pronucleophiles for tandem hydroformylation reactions with additional C-C bond formations.

2.3.2.1
Hydroformylation in the Presence of Stable Wittig Reagents

Hydroformylation in the presence of stable phosphorus ylides leads to a tandem hydroformylation/Wittig olefination procedure with direct olefin formation from the oxo aldehyde. This procedure, with or without a consecutive hydrogenation of the resulting olefin, was used in a diastereoselective version starting from an o-DPPB-modified methallylic alcohol to give the saturated product (Scheme 3) [8].

Similarly, the rhodium(I)-catalyzed sequential silylformylation/Wittig olefination of terminal alkynes with hydrosilanes and carbon monoxide in the presence of stabilized P-ylides leads to substituted 2,4-dienoic esters in a one-pot procedure (Scheme 4). The $\alpha,\beta,\gamma,\delta$-unsaturated esters are generated with high (2E,4Z) stereoselectivity in good to excellent yields. No further hydrogenation occurs. Conversions of the products in cycloaddition reactions are presented [9].

2.3.2.2
Hydroformylation in the Presence of Allyl Silanes and Allyl Boranes

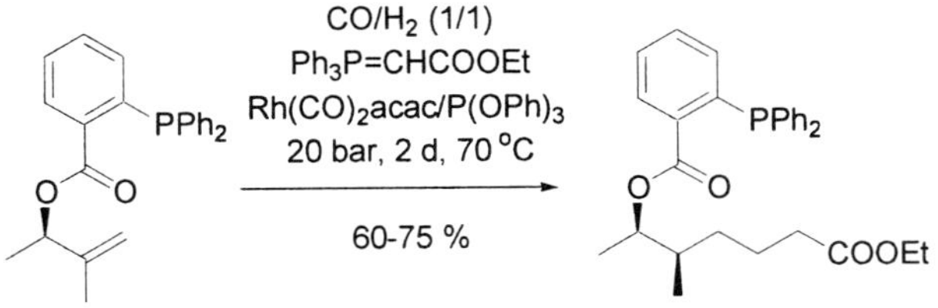

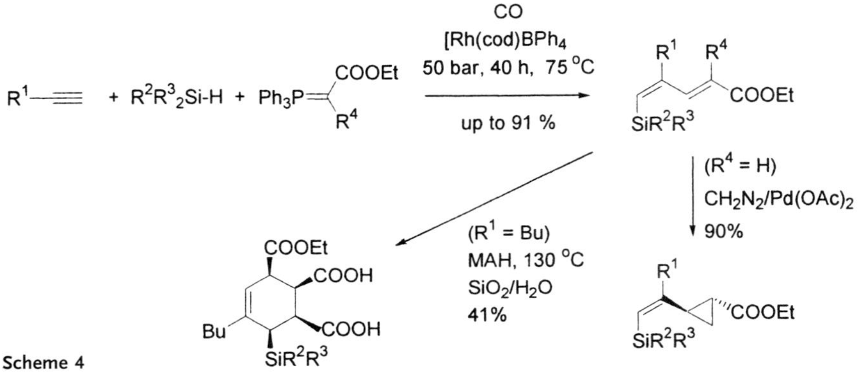

Allylsilanes and allylboranes are potential allyl anion equivalents, stable enough under hydroformylation conditions to be included in subsequent allyl addition reactions to the *in situ*-formed oxo aldehydes. Thus bisallylsilanes react under hydroformylation conditions with carbonylative coupling of the two allyl groups (Scheme 5) [10]. The reaction is interpreted to proceed via hydroformylation of one of the allyl groups followed by intramolecular Sakurai reaction and double-bond migration to form an enol and the keto group observed in the final isomeric products.

An interesting alternative is the use of silylformylation instead of hydroformylation in a tandem intramolecular silylformylation/allyl silylation of alkenes [11] (Scheme 6) and alkynes [12], allowing a rapid synthesis of polyol fragments for polyketide/macrolide synthesis.

This tandem procedure has been successfully applied in a formal total synthesis of Mycoticin A [13]. The scope and utility of these reactions was expanded to (*Z*)- and (*E*)-crotyl groups, leading to the stereospecific incorporation of both *anti* and *syn* propionate units into the growing polyol chain (Scheme 6) [14].

With repetitive application of the same procedure, the power of this methodology for the rapid assemblage of polyketide-like structures is demonstrated (Scheme 7) [14].

An appealing entry to condensed 1,5-oxazadecalin systems is achieved via tandem hydroformylation/allylboration/hydroformylation sequences starting from an *N*-al-

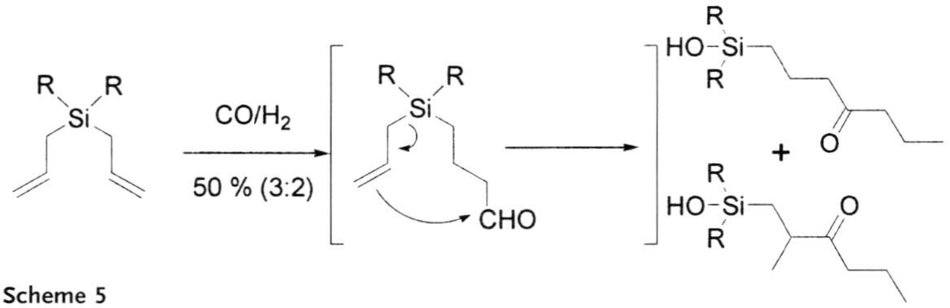

Scheme 5

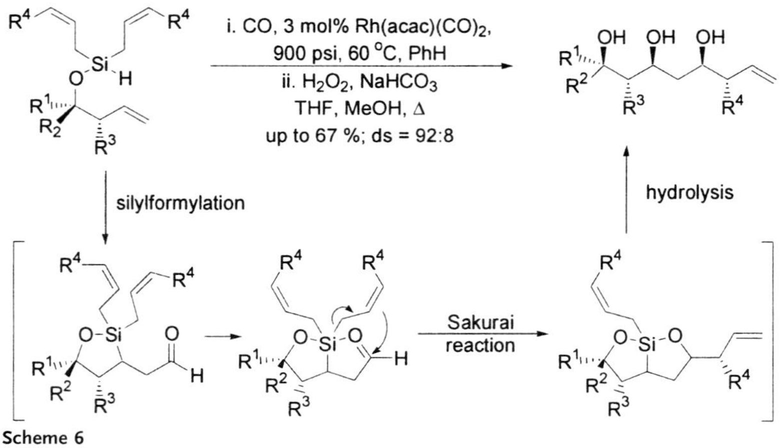

Scheme 6

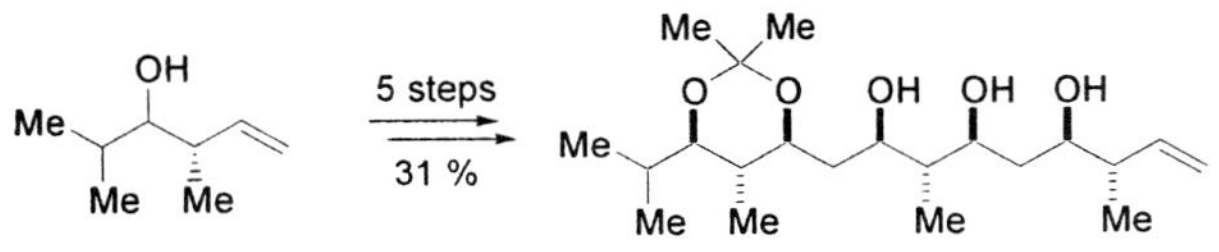

lyl-γ-amidoallylboronate, readily prepared from a boronate aldehyde (Scheme 8) [15]. Hence, regioselective hydroformylation generates an aldehyde which undergoes diastereoselective intramolecular allylboration to give a vinyl derivative. The reaction does not stop at this stage, since this alkene moiety again undergoes *n*-selective hydroformylation to give an equilibrium mixture of lactols and an open-chain δ-hydroxy aldehyde. Reductive removal of the Cbz group furnished the indolizidine in a further domino type process consisting of hydrogenation, cyclization/enamine formation, and hydrogenation.

Similarly (E)-alkoxyallylboronates were used as the starting point for intramolecular allylboration reactions leading to the *trans*-disubstituted hydrooxepans as a mixture of anomers (Scheme 9) [16].

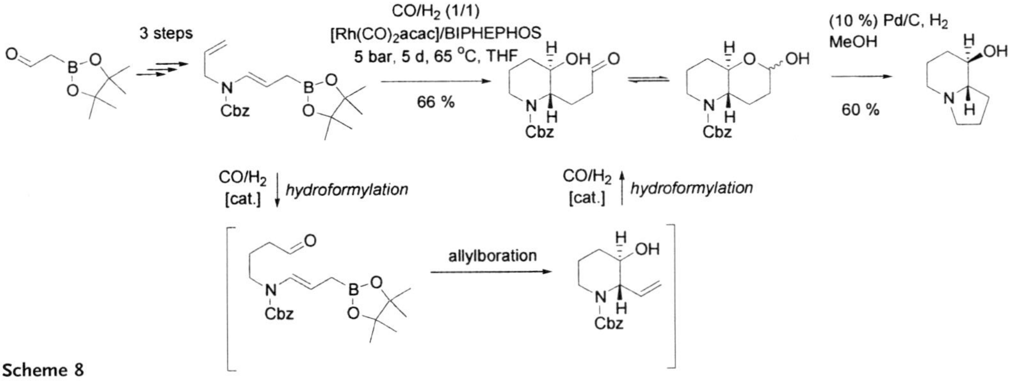

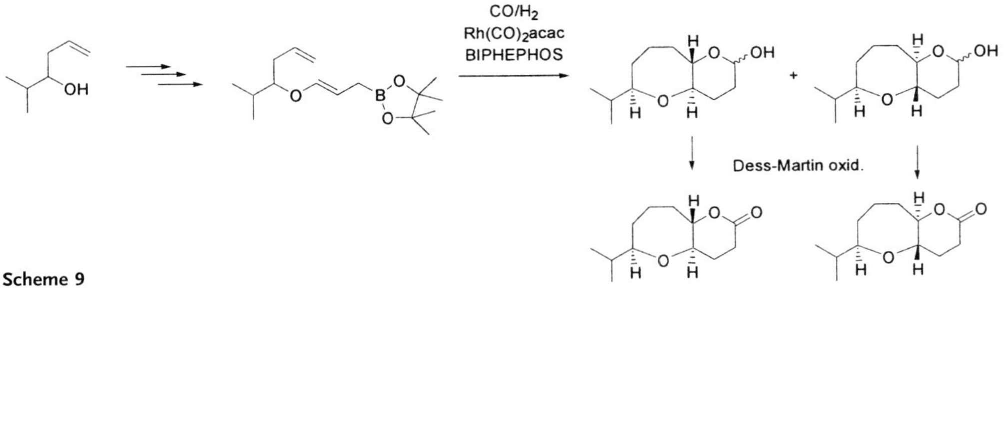

Allylboronates attached to 2-vinyl-tetrahydropyrane, if subjected to the hydroformylation protocol described above, in an intramolecular allylboration form hydrooxepane rings stereounselectively as a mixture of diastereoisomers. Here, dehydration occurred to give enol ether units, apparently during the longer reaction period [16].

2.3.2.3
Hydroformylation in the Presence of Nucleophilic Hetarenes

Electron-rich hetarenes can also act as pronucleophiles in tandem hydroformylation procedures. Thus, hydroformylation of N-allyl-pyrroles leads to 5,6-dihydroindolizines via a one-pot hydroformylation/cyclization/dehydration process (Scheme 10) [17]. The cyclization step represents an intramolecular electrophilic aromatic substitution in the α-position of the pyrrole ring. This procedure was expanded to various substrates bearing substituents in the allyl and in the pyrrole unit.

A similar Pictet-Spengler type intramolecular electrophilic aromatic substitution in the α-position of an indole unit was observed in hydroformylation of indole-substituted terminal alkenes supported on a solid phase [18].

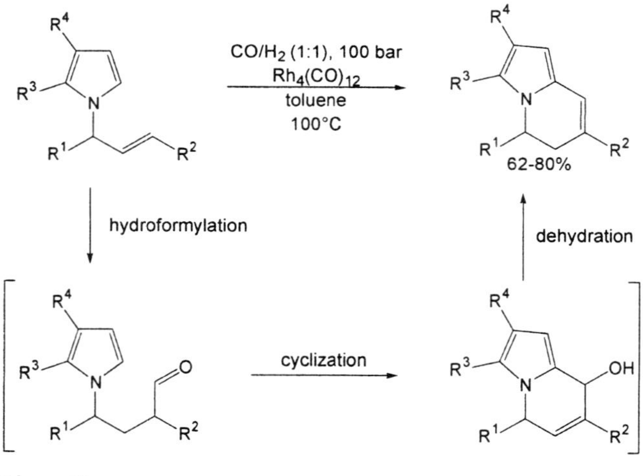

Scheme 10

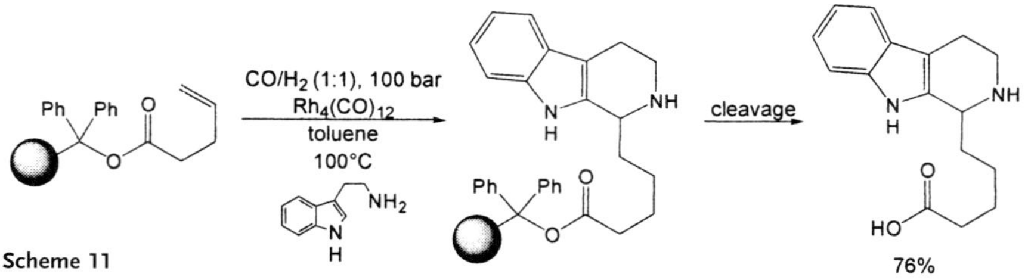

Scheme 11

2.3.3
Hydroformylation and Subsequent Mixed Aldol Reactions

Aldol addition of aldehydes represents one of the most important reactions in synthetic organic chemistry [19]. Self-condensation of oxo aldehydes is observed as an unwanted side reaction under hydroformylation conditions [4]. On the other hand, self-condensation of oxo aldehydes is one of the most important transformations, leading to functionalized new carbon skeletons as β-hydroxy aldehydes, a,β-unsaturated aldehydes, or hydrogenation products thereof [4]. Some of these self-condensation products lead to important industrial compounds such as 2-ethyl-hexanol, obtained via propene hydroformylation followed by aldol addition and dehydration and reduction (Scheme 12).

Product selectivity is a major problem in tandem hydroformylation/homo aldol reaction sequences, since under hydroformylation conditions dehydration of the initial aldol adducts easily occurs, followed by hydrogenation of the reactive a,β-unsaturated aldehydes. Therefore product mixtures are often obtained. Although considerable efforts have been made to overcome these problems [4, 7], none of the commercial processes (e.g. 2-ethyl-1-hexanol) seem to be run as a tandem hydroformylation procedure without change of the reaction conditions for the different steps. Intramolecular aldol reactions of dialdehydes following hydroformylation of conjugated (e.g. butadiene) and nonconjugated (e.g. 1,5-cyclooctadiene) dienes lead to cyclization products of different ring size, mostly, however with low yields [7].

Mixed intermolecular aldol condensations following hydroformylation of alkenes in the presence of preformed carbonyl compounds are rare, because of the inherent problems of chemoselectivity. However, diastereoselective Knoevenagel condensation of *in situ*-generated oxo aldehydes with stabilized enols or enolates are known in inter- and intramolecular versions. Thus o-DPPB-modified allylic and methallylic alcohols undergo regio- and stereoselective intermolecular Knoeve-

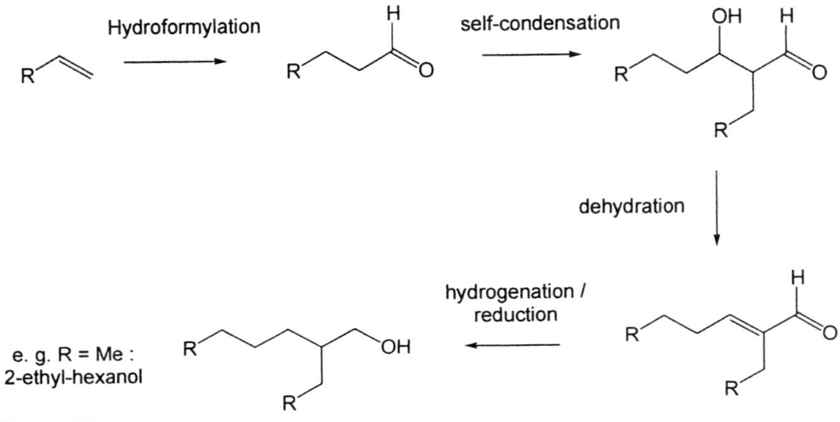

Scheme 12

nagel condensation if rhodium-catalyzed hydroformylation is performed in the presence of malonic or acetoacetic esters, piperidine, and BIPHEPHOS [20]. With malonic mono esters only, under the reaction conditions, an additional decarboxylation follows (Scheme 13) [20].

Various intramolecular versions of mixed aldol reactions followed by hydroformylation of unsaturated aldehydes or ketones are reported. Here the problems of chemo- and regioselectivity are overcome by choice of the substrate or by use of preformed enolate functionalities. Thus rhodium-catalyzed hydroformylation of 2-formyl-*N*-allyl-pyrrole unselectively leads to the *n*- and the *iso*-product (approx. 1:1) with complete cyclization of the former after prolonged reaction times to give 7-formyl-5,6-indolizine in up to 46% yield (Scheme 14) [21]. Here only one of the aldehyde groups can act as the enolate nucleophile.

Comparable hydroformylation and aldol cyclization of an β,γ-unsaturated ketone bearing a quaternary center in the presence of PTSA chemo- and regioselectively leads to a single cyclization product. Here, similarly to the stepwise version, the oxo aldehyde preferentially reacts as the electrophilic carbonyl component and the ketone as the nucleophilic enol to form the five-membered ring product. Dehydration and hydrogenation of the resulting enone readily occurs under the reductive reaction conditions used (Scheme 15) [22].

While the saturated ketone can be obtained in nearly quantitative yield, this synthetically unfavorable loss of functionality can be overcome by use of the corresponding unsaturated silyl enol ethers, which undergo selective hydroformylation at the monosubstituted double bond followed by a Mukaiyama type aldol addition (Scheme 16) [22, 23]. Using this method in aprotic solvents the silyl enol ether moiety reacts as the C-nucleophile with complete transfer of the silyl function to the aldol hydroxy group.

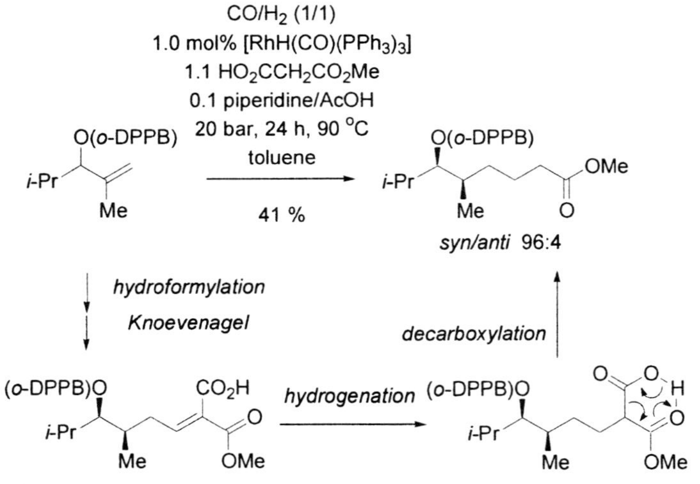

Scheme 13

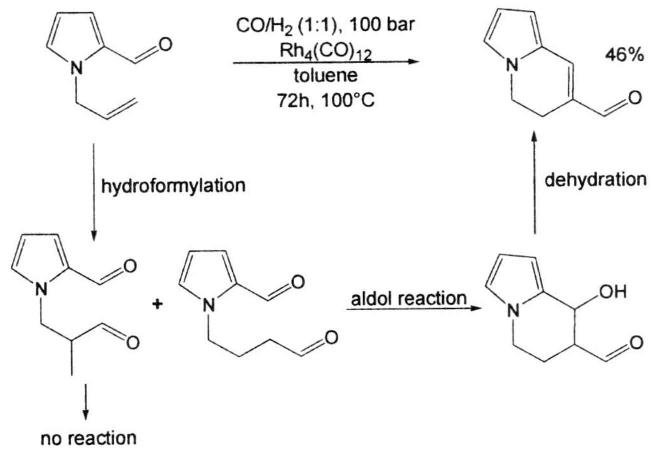

2.3.3 Hydroformylation and Subsequent Mixed Aldol Reactions

Scheme 14

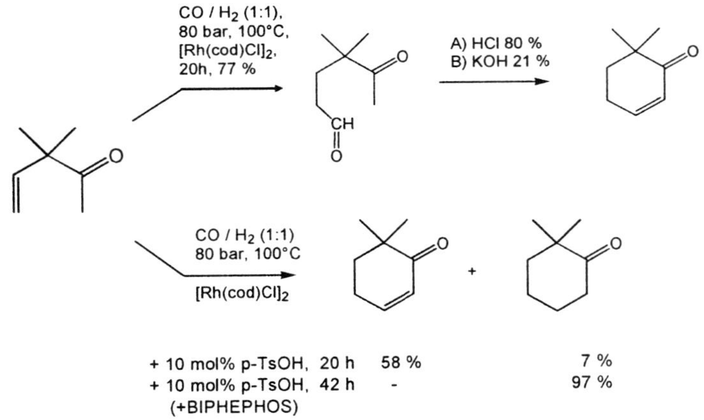

Scheme 15

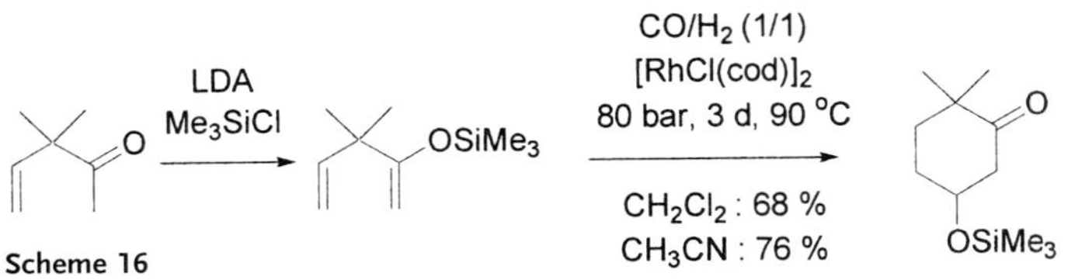

Scheme 16

This method can also be applied to the silyl enol ethers of other unsaturated ketones, aldehydes, or esters [23, 24]. Thus with γ,δ-unsaturated ketones, depending on the method used, either acylated cyclopentenes or silylated cyclopentanols are obtained (Scheme 17). With these substrates, regiocontrol toward seven-mem-

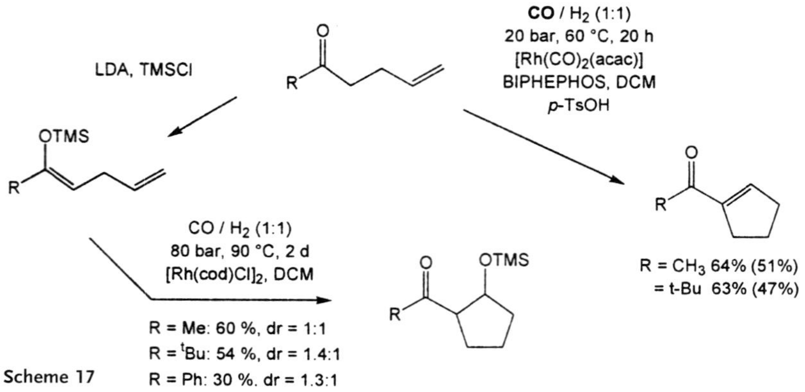

Scheme 17

bered rings, under the hydroformylation conditions, has so far failed with the corresponding regioisomeric silyl enol ethers (because of isomerizations), but is successful with additional regiodirecting ester groups [23, 24].

Unmodified unsaturated esters under tandem hydroformylation/aldol reaction conditions without cyclization only give the corresponding aldehydes, whereas the silylated ester enolates smoothly cyclize in an intramolecular Mukaiyama type condensation (Scheme 18) [23, 24].

Interestingly, unsaturated aldehydes after conversion to the corresponding silyl enol ethers under hydroformylation conditions selectively give the aldol cyclization product, whereas the unmodified aldehydes unselectively lead to product mixtures (Scheme 19) [23, 24].

These examples clearly demonstrate that, starting from unsaturated carbonyl compounds as precursors for dicarbonyl systems, the preformed carbonyl unit can conveniently be activated to serve as the enolate unit in aldol reactions following the hydroformylation step *in situ*, generating the second (aldehyde) carbonyl group. Thus the vinyl groups acts as an uncreative precursor of aldehydes. In the presence of this unit, silyl enol ethers and other enolate equivalents (such as bo-

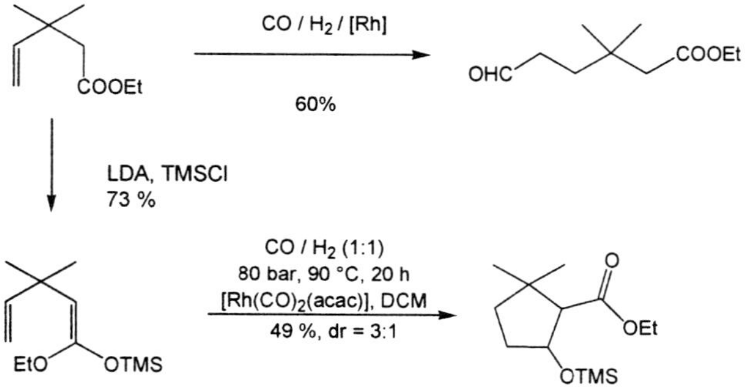

Scheme 18

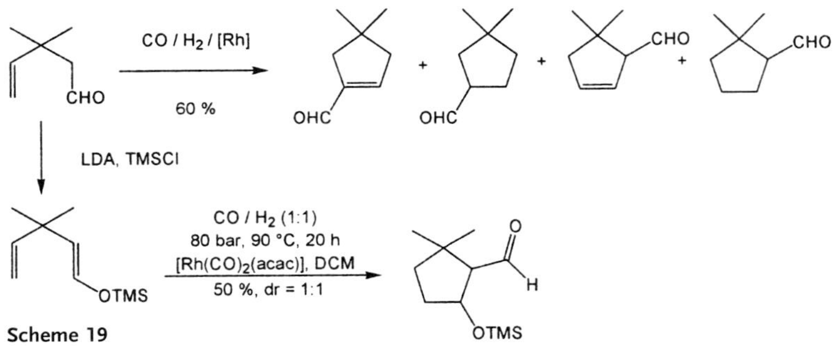

Scheme 19

ron enolates) of the preformed carbonyl unit are easily obtained and allow an effective chemo-, regio-, and stereocontrol of the cyclization immediately following the hydroformylation step [24].

Similarly, tandem hydroformylation/aldol sequences can be applied to the formation of bicyclic and spirocyclic compounds. Thus, silyl enol ethers of to 3-vinyl and 3-allyl cycloalkanones give ring-anellated products (Scheme 20) [24].

Using the same methods for 2-allylcycloalkanones can in principle lead to three different aldol product types from the aldehydes obtainable via hydroformylation (Scheme 21). Usually the spiro compounds are preferred, as already demonstrated [23], but, according to more recent results, various methods, such as regiodirecting and/or reversibly blocking ester groups or the above-described use of enolate equivalents, are available to achieve selective control of the reaction outcome [24].

An interesting expansion of these hydroformylation/aldol reaction sequences is achieved if the conversions are performed in the presence of amines. Depending on the amine type (primary or secondary), steric effects, and reaction conditions, different product types are obtainable. Conversions of 3,3-dimethyl-4-penten-2-one under hydroformylation conditions in the presence of amines give aldol products as described above with either hydroaminomethylation/reductive amination products or aldol products with additional 1,4-addition of the amine to the enone or re-

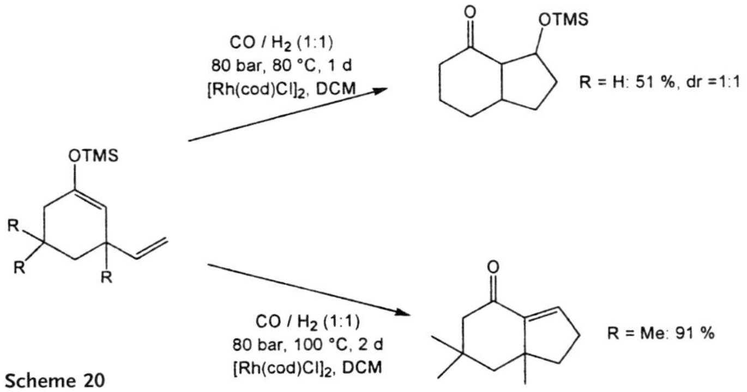

Scheme 20

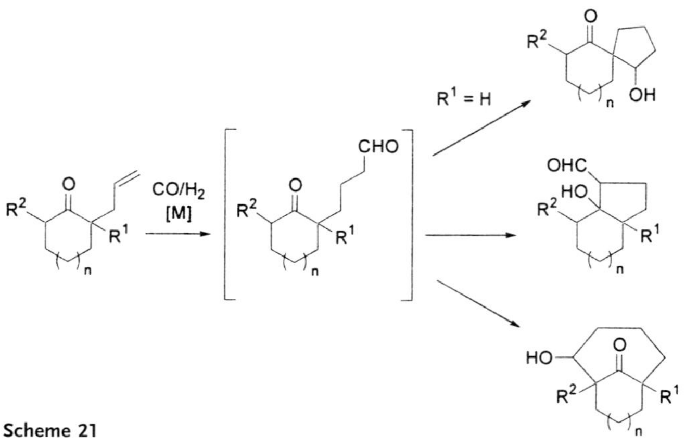
CO/H2
[M]
R1 = H
CHO
OHC
HO
OH

Scheme 21

ductive amination of the ketone function of the saturated product (Scheme 22) [23].

Tandem hydroformylation/aldol reactions in the presence of amines have also been applied to other mono- and diolefinic substrates [7]. Thus, divinylsilanes under hydroformylation conditions in the presence of secondary amines give silacyclohexane derivatives. In this reaction, two equivalents of carbon monoxide are incorporated, leading to the formation of dialdehydes, which cyclize to form functionalized six-membered silacyclic rings including imine/enamine formation, elimination, and hydrogenation steps (Scheme 24) [25]. The reaction proceeds with up to quantitative yields and 1,4-*trans*-selectivity.

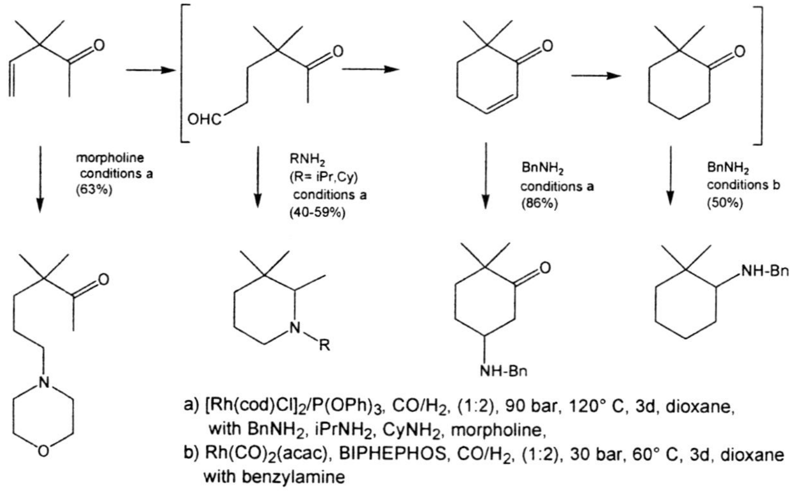
morpholine
conditions a
(63%)
RNH2
(R= iPr,Cy)
conditions a
(40-59%)
BnNH2
conditions a
(86%)
BnNH2
conditions b
(50%)
NH-Bn
a) [Rh(cod)Cl]2/P(OPh)3, CO/H2, (1:2), 90 bar, 120° C, 3d, dioxane, with BnNH2, iPrNH2, CyNH2, morpholine,
b) Rh(CO)2(acac), BIPHEPHOS, CO/H2, (1:2), 30 bar, 60° C, 3d, dioxane with benzylamine

Scheme 22

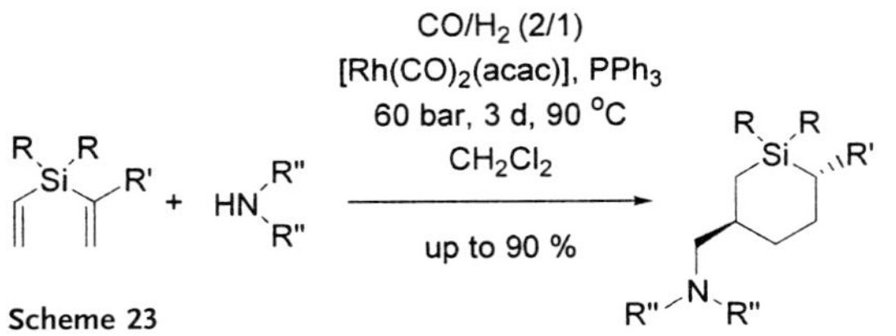

Scheme 23

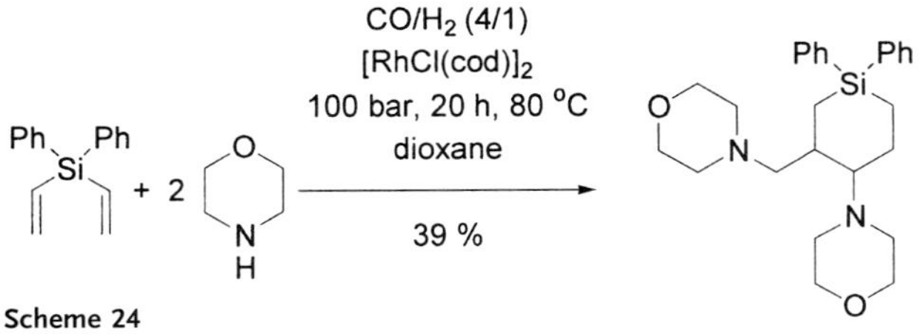

Scheme 24

By varying the hydroformylation conditions, the same substrates are converted via an alternative pathway; with incorporation of two equivalents of morpholine, a silaheterocyclic diamine is formed (Scheme 24) [25].

Even more complicated is the self-condensation of enamine/imine functionalities obtained with allylic arylamines under hydroformylation conditions (Scheme 25) [26]. Here, a final intermolecular electrophilic aromatic substitution of the resulting iminium species completes the sequence to form a polycyclic system.

2.3.4
Hydroformylation and Other C-C Bond-forming Reactions

The tandem hydroformylation reactions described above result from an attack of a carbon nucleophile to the *in situ*-generated oxo aldehyde or via enolization of the oxo aldehydes to react with electrophiles. Other types of tandem hydroformylations with additional C-C bond-forming steps are described in this section.

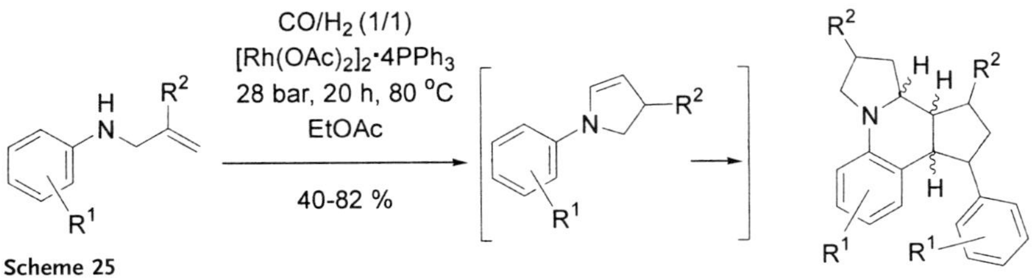

Scheme 25

2.3.4.1
Hydroformylation/Amidocarbonylation Sequences

Enamines generated *in situ* from aldehydes and the NH function of amides in the presence of cobalt or palladium catalysts can be hydrocarboxylated in an overall amidocarbonylation to form N-acylated α-amino acids [27]. This reaction was found by Wakamatsu while studying the oxo process of acrylonitrile with $Co_2(CO)_8$ as catalyst (Wakamatsu reaction [28]) and is now in a palladium-catalyzed version considered as a powerful new tool in the synthesis of natural and non-natural amino acids and derivatives used as nutrient additives, sweeteners, polyamides, surfactants, agrochemicals, and pharmaceuticals.

By combining this method with hydroformylation to provide the required aldehyde, an attractive access to amino acids directly from olefin feedstocks could be achieved, and various applications using simple alkenes [29] and functionalized alkenes [30] are described with use of cobalt or rhodium catalysts or bimetallic catalysts of both metals [27]. A tandem hydroformylation/amidocarbonylation sequence first described by Stern et al. [29] uses dodecene and acetamide to give 2-(acetylamino)tetradecanoic (Scheme 27).

Lin and Knifton [31] demonstrated the positive influence of chelating phosphine ligands (e.g. dppe) in tandem hydroformylation/amidocarbonylation of tetradecene. If functionalized olefins are used, tandem hydroformylation/amidocatrbonylation leads to interesting N-acetyl amino acid derivatives (Scheme 28)

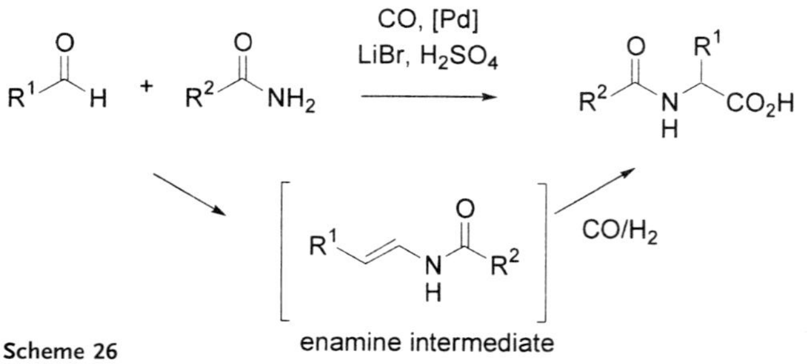

Scheme 26

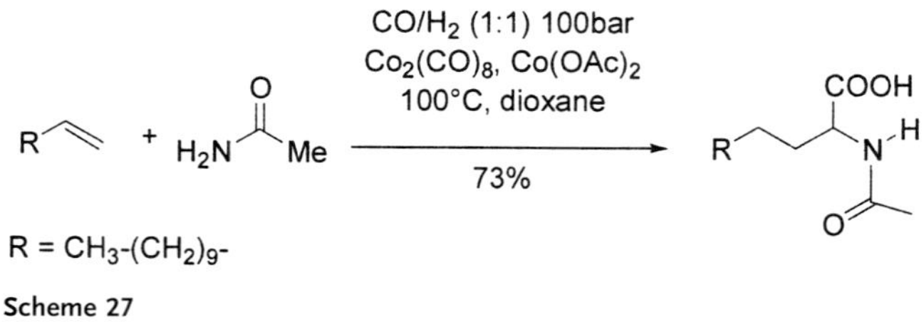

Scheme 27

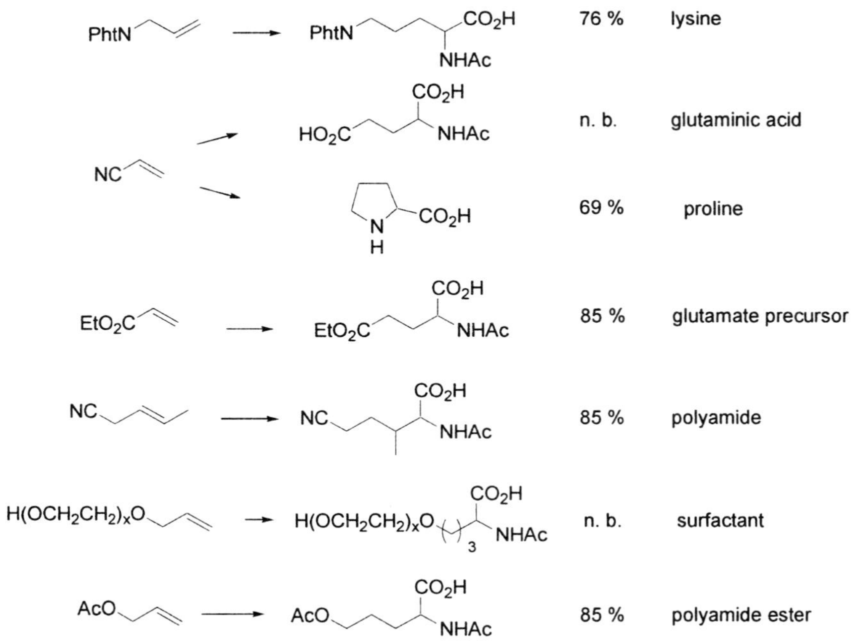

2.3.4 Hydroformylation and Other C-C Bond-forming Reactions | 101

Scheme 28

such as lysine, glutaminic acid, and proline, as well as cyano-, polyoxyethylene- and O-acetyl-functionalized derivatives [31].

Ojima et al. [32] obtained bicyclic N-α-ethoxyamide via rhodium-catalyzed hydroformylation in the presence of triethyl *ortho*-formic acid, which was carbonylated to α-amido esters in varying yields by using octacarbonyl dicobalt as catalyst

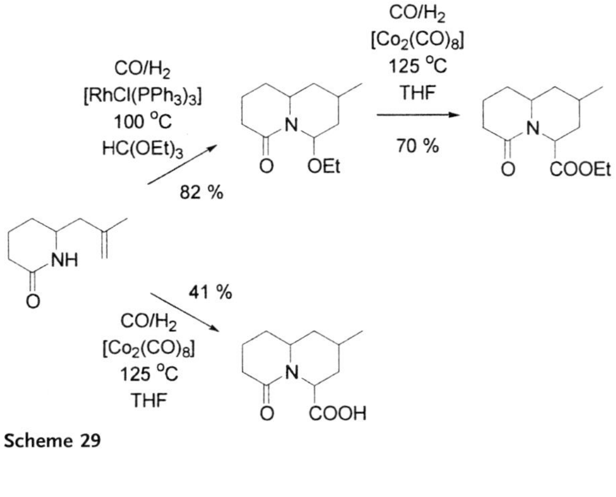

Scheme 29

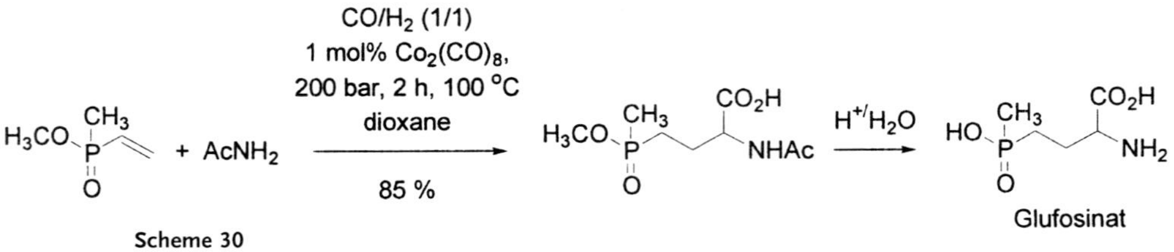

Scheme 30

$H_3CO-P(=O)(CH_3)-CH=CH_2$ + AcNH$_2$ →[CO/H$_2$ (1/1), 1 mol% Co$_2$(CO)$_8$, 200 bar, 2 h, 100 °C, dioxane, 85%] $H_3CO-P(=O)(CH_3)-CH_2-CH_2-CH(NHAc)-CO_2H$ →[H$^+$/H$_2$O] HO-P(=O)(CH$_3$)-CH$_2$-CH$_2$-CH(NH$_2$)-CO$_2$H

Glufosinat

(Scheme 29). This two-step conversion can also be performed in a tandem procedure directly from the olefin to give the corresponding acids.

Glufosinate is a naturally occurring hydroxyphosphoryl amino acid showing herbicide and antibiotic activities. This compound is commercially available as the nonselective herbicide BASTA™. Glufosinate is obtainable by various methods, including amidocarbonylation if starting from the corresponding aldehydes or acetals of the methyl phosphinates [33]. Direct tandem hydroformylation/amidocarbonylation followed by hydrolysis is achieved if starting from the easily available methyl vinyl phosphinate (Scheme 30) [34].

More recently, various stibine ligands have been used in hydroformylation/amidocarbonylation of cyclohexene and l-pentene catalyzed by Co$_2$(CO)$_8$. These ligands not only enhance the activity of the catalyst, they also increase the selectivity in comparison to classical phosphine ligands. All reactions are carried out at low syngas pressures (25 bar) [35].

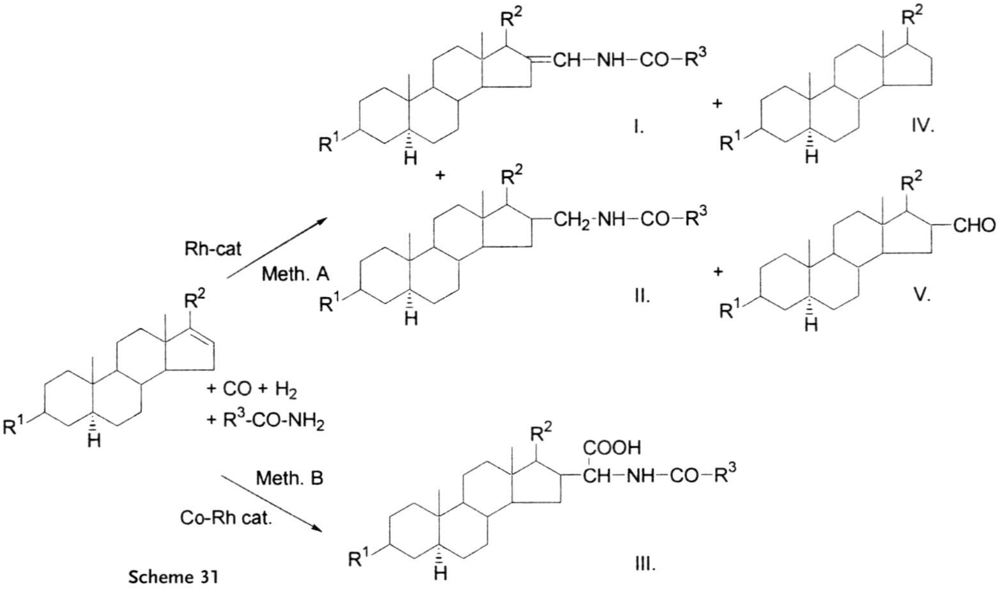

Scheme 31

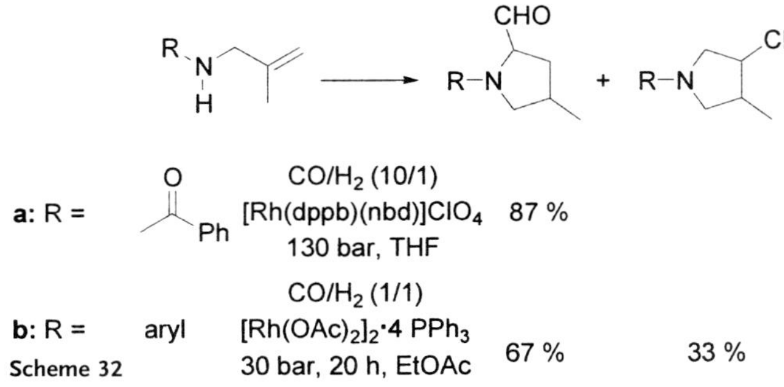
R
N
H
CHO
R–N
+
R–N
CHO
a: R =
O
Ph
CO/H2 (10/1)
[Rh(dppb)(nbd)]ClO4
130 bar, THF
87 %
b: R =
aryl
CO/H2 (1/1)
[Rh(OAc)2]2·4 PPh3
30 bar, 20 h, EtOAc
67 %
33 %

Scheme 32

Androstene and pregnene derivatives were functionalized by amides under hydroformylation conditions with rhodium or binary rhodium-cobalt catalysts. Whereas the reaction catalyzed by $[Rh(nbd)Cl]_2$ modified with PPh_3 results in the unsaturated amidomethylidene derivatives, the rapid hydrogenation of these compounds takes place in the presence of a basic phosphine ligand. If a binary rhodium-cobalt system of $[Rh(nbd)Cl]_2$ and $Co_2(CO)_8$ is used, amidocarbonylation of the steroids occurs with high chemo- and regioselectivity (Scheme 31) [36].

A similar reaction sequence starting from *N*-methallylamides or *N*-methallylanilines, respectively, is reported to proceed via hydroformylation of the olefinic double bond and consecutive intramolecular enamine condensation, followed by a further hydroformylation of an enamine double bond and resulting in 2- and 3-formylpyrrolidines (Scheme 32) [37–39].

2.3.4.2 Fischer Indole Synthesis with Oxo-Aldehydes

In various cases, hydroformylation has been used to generate the aldehydes required for the Fischer indole synthesis. Furthermore, Sheldon et al. reported a one-pot synthesis of melatonin starting from *N*-allylacetamide via regioselective hydroformylation and Fischer indole synthesis, although in this procedure reaction vessels were changed as well as the reaction conditions (Scheme 33) [40].

Since hydrazones are not easily hydrogenated under hydroformylation conditions (see above), both steps could be combined to a single tandem procedure directly starting from olefins and arylhydrazines without changing the reaction conditions [41]. Under hydroformylation conditions, the *in situ*-generated oxo alde-

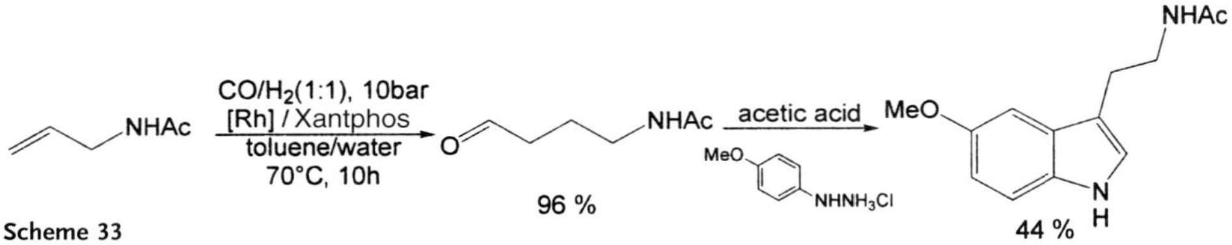
NHAc
CO/H2(1:1), 10bar
[Rh] / Xantphos
toluene/water
70°C, 10h
O
NHAc
96 %
acetic acid
MeO
NHNH3Cl
MeO
NHAc
N
H
44 %

Scheme 33

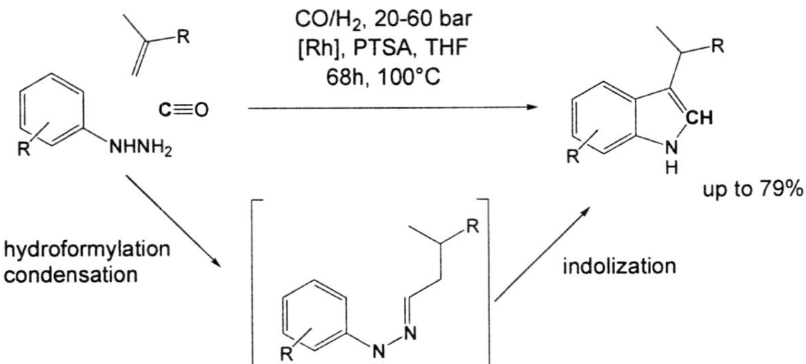

Scheme 34

[Scheme 34: Hydroformylation/Fischer indole synthesis. An arylhydrazine (Ar-NHNH$_2$) and an alkene under CO/H$_2$, 20–60 bar, [Rh], PTSA, THF, 68 h, 100 °C proceed via hydroformylation/condensation to a hydrazone intermediate, then indolization to give a 3-substituted indole, up to 79%.]

hydes are trapped by the hydrazine to give hydrazones. In the presence of a Brønsted acid, this hydrazone is directly converted to the indole without isolation. This procedure leads to 3-substituted indoles if unsubstituted phenylhydrazine is used and to 3,5- or 3,7-disubstituted indoles if substituted arylhydrazines are used. Classical Fischer indolizations of aldehydes may suffer from side reactions, like the aldol processes. Therefore, protected aldehydes are often used, e.g., acetals or aminals. In contrast to this, in the tandem hydroformylation/Fischer indole protocol the oxo aldehyde exists only in low stationary concentrations, thus preventing undesired side reactions. The selectivity of this tandem approach can be increased by using benzhydrylidene-protected aryl hydrazines. Yields of up to 79% for tryptamine derivatives can be conveniently achieved (Scheme 34).

2.3.4.3
Hydroformylation and Carbonyl Ene Reactions

Hydroformylation can also be combined with a carbonyl ene reaction. This reaction sequence is observed in rare cases if nonconjugated diolefins are selectively hydroformylated at one of the double bonds and the resulting aldehyde reacts with the remote second double bond. Thus, conversion of limonene in a one-pot procedure forms two diastereoisomers of an alcohol if PtCl$_2$(PPh$_3$)$_2$/SnCl$_2$/PPh$_3$ or

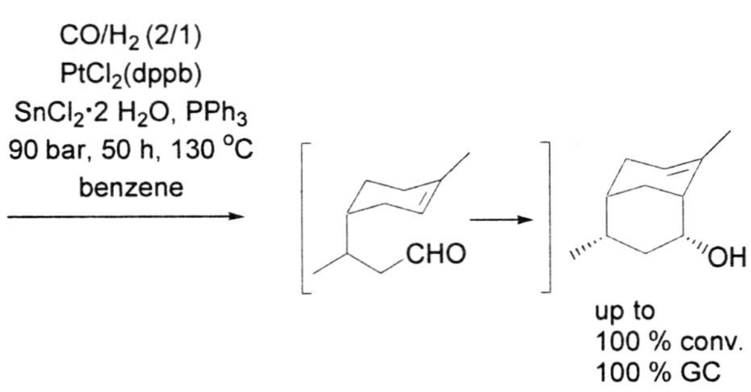

Scheme 35

[Scheme 35: Limonene under CO/H$_2$ (2/1), PtCl$_2$(dppb), SnCl$_2$·2 H$_2$O, PPh$_3$, 90 bar, 50 h, 130 °C, benzene, gives an aldehyde intermediate (CHO) which undergoes carbonyl ene cyclization to a bicyclic alcohol; up to 100% conv., 100% GC.]

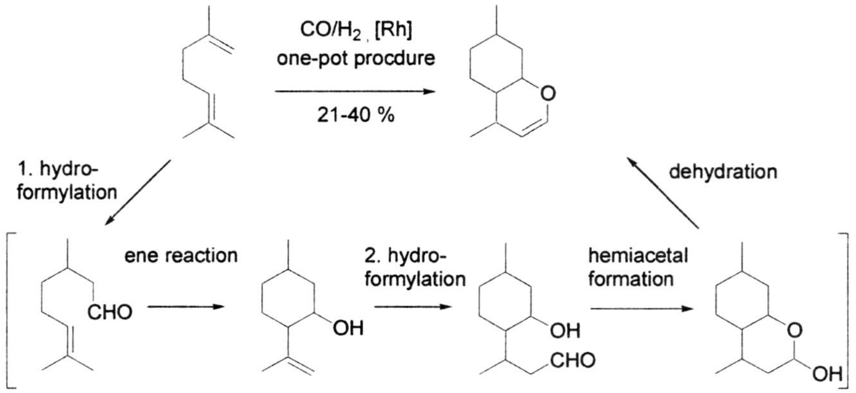

Scheme 36

PtCl$_2$(diphosphine)/SnCl$_2$/PPh$_3$ catalyst systems are used (Scheme 35). Best results are achieved with the PtCl$_2$(dppb) complex. The mechanism of the final intramolecular cyclization step resembles an acid-catalyzed carbonyl ene reaction [42].

Wilkinson's catalyst allows the synthesis of bicyclic hydrogenated chromane derivatives directly from acyclic 1,5-dienes [43]. This one-pot sequence merges four steps starting with a hydroformylation of the terminal double bond, followed by a metal-induced carbonyl-ene reaction and hydroformylation/acetalization/dehydration of the cyclic alcohol (Scheme 36).

2.4.4
Hydrocarbonylation/Insertion Sequences Leading to Ketones

Under hydroformylation conditions, the formation of ketones from alkenes is occasionally observed [4]. This reaction type (hydrocarbonylation) has been optimized toward ketone formation in several cases. Thus, starting from ethylene, diethyl ketone formation is a well-established process (Scheme 37) [4, 7].

The same reaction type has been applied to various higher olefins such as propene, butenes, cycloalkenes, or styrenes, as well as alkynes and dienes, mostly giving mixtures of regioisomers. Homogeneous and heterogeneous catalysts of mo-

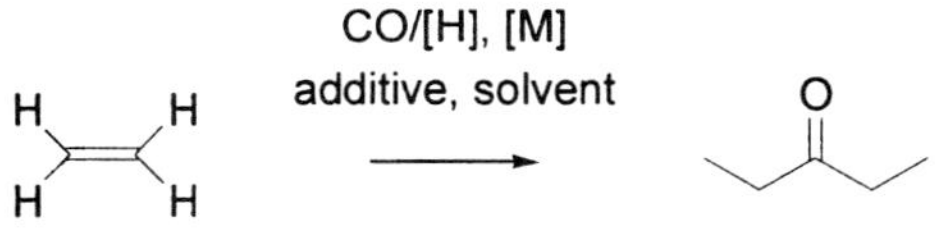

[M]: e.g.: CO$_2$(CO)$_8$, Rh$_2$O$_3$, Rh(H)(CO)(PPh$_3$)$_3$

Scheme 37

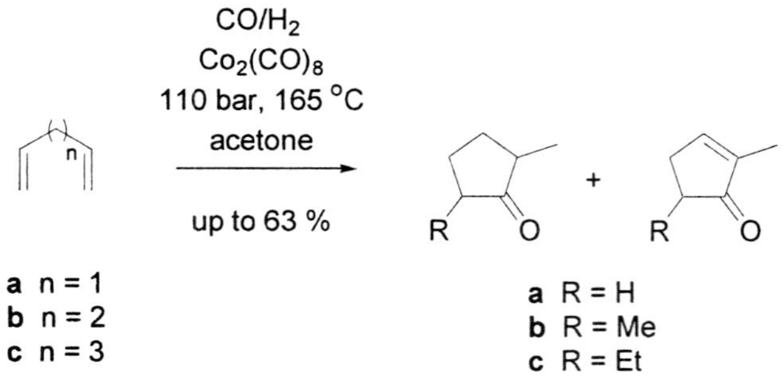

Scheme 38

CO/H$_2$
Co$_2$(CO)$_8$
110 bar, 165 °C
acetone

up to 63 %

a n = 1
b n = 2
c n = 3

a R = H
b R = Me
c R = Et

Scheme 38

lybdenum, cobalt, nickel, palladium, rhodium, and iridium have been used. Mechanistically, the initial steps of the reaction pathway follows the classical hydroformylation mechanism [4, 7]. The olefin undergoes hydrometalation and carbon monoxide insertion to form the metal acyl complex. If, instead of hydrogenolysis, a second olefin insertion takes place followed by a terminating reductive elimination, saturated ketones are formed. Formation of unsaturated ketones via β-H-elimination can also occur (Scheme 38). The regioselectivity is determined in the different insertion steps involved.

Reactions of this type are extensively discussed in a recent review [7], so only selected examples are given here. An interesting synthetic application was first described by Klemchuk [44], reporting that the formation of cyclopentanones from α,ω-dienes is achieved under hydroformylation conditions. Independently of the chain length between the two olefin units, cyclopentanones are the preferred products (Scheme 38). This conversion is also carried out under cobalt or rhodium catalysis and water gas shift conditions in a CO atmosphere, leading to a mixture of saturated and unsaturated cyclopentanones (Scheme 39) [7, 45, 46].

According to accepted mechanistic considerations the reaction is initiated by a hydrometalation of one of the two double bonds (Scheme 39). The metal alkyl complex then undergoes CO insertion to give a metal acyl complex. A further olefin insertion leads to the kinetically favored five-membered ring with a metal alkyl moiety at the exocyclic methyl group. Final hydrogenolysis or β-elimination liberates the product.

Conversion of 3,3-disubstituted 1,4-pentadienes leads to cyclopentanones with quaternary centers [47]. Cobalt and rhodium complexes can be used as catalyst precursors. Hydrogen or water is used as the hydrogen source. The quaternary center blocks isomerization and enhances the cyclization through a Thorpe-Ingold effect [48]. This method is applied in the synthesis of the aromatic sesquiterpene (±)-α-cuparenone (Scheme 40) [49] with moderate diastereoselectivity of 3:1 (trans:cis) [50].

Further applications of this method are demonstrated with 1,4-dienes and 1,5-dienes bearing various substitution patterns to give substituted cyclopentanones [51] and spiro-cyclopentanone if starting from 1,1-divinylcycloalkanes (Scheme 41)

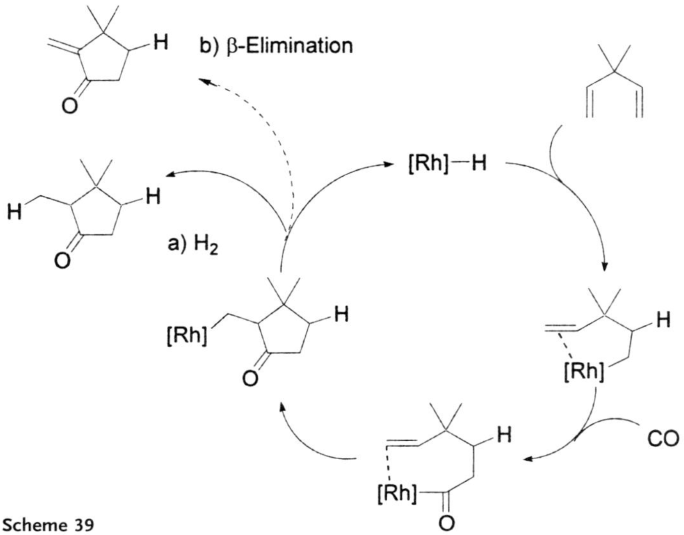

2.4.4 Hydrocarbonylation/Insertion Sequences Leading to Ketones

Scheme 39

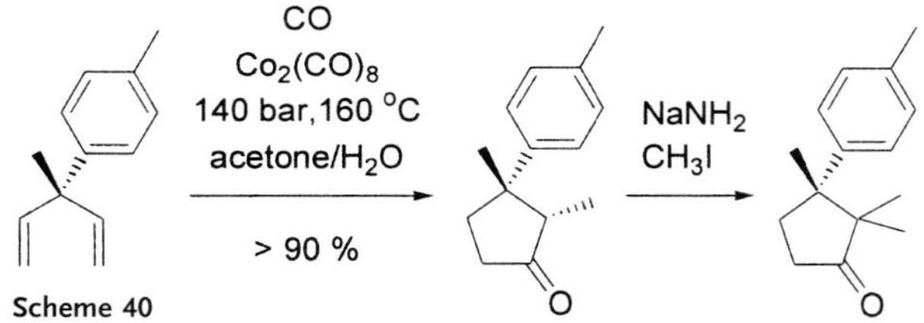

Scheme 40

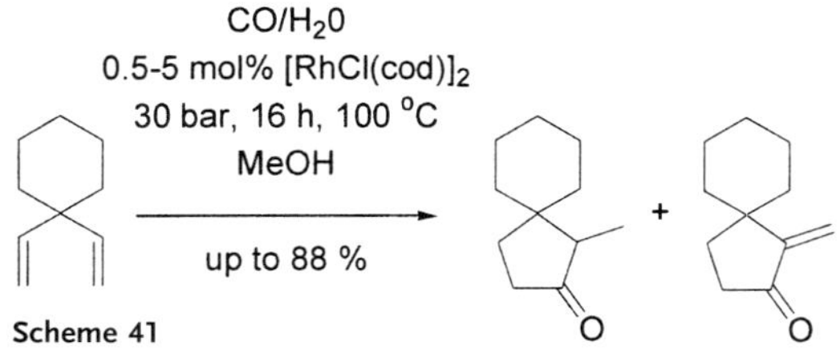

Scheme 41

[52]. In contrast to homogeneous rhodium catalysis with [RhCl(cod)]$_2$, the use of a polymer-attached CpRh(cod) complex provides the α,β-unsaturated cyclopentenones in higher yields [53].

A variety of functional groups in the C3-side chain of 3,3-disubstituted 1,4-pentadienes is tolerable under the reaction conditions of rhodium- or cobalt-catalyzed hydrocarbonylative cyclization [54, 55].

The system PdCl$_2$(PPhl$_3$)$_2$/SnCl$_2$/PPhl$_3$ proved to be an efficient catalyst precursor for the hydrocarbonylative cyclization of terpenoid 1,4-dienes such as *trans*-iso-

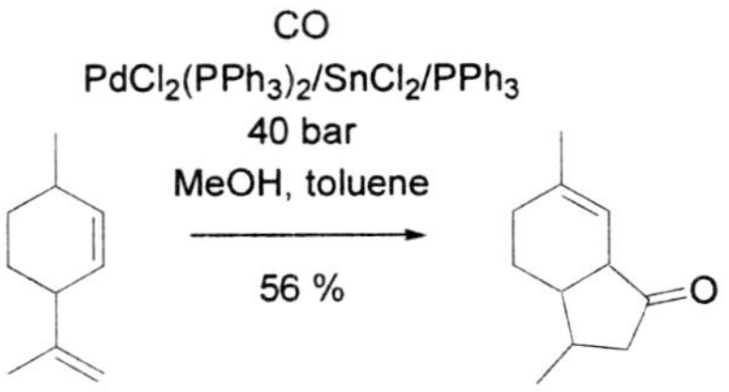

Scheme 42

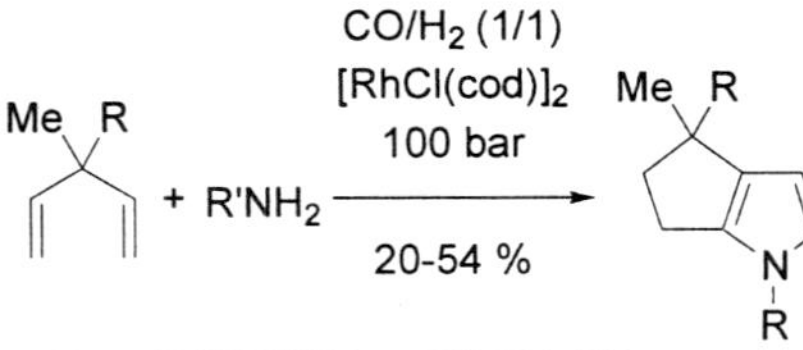

Scheme 43

R = -Me, -CH(OH)CH₃, -CH₂CH₂OH
R' = i-Pr, Bu, Bn, (R)-(+)-phenylethyl, p-methoxyphenol

limonene to form an unsaturated hydroindanone derivative as a mixture of diastereoisomers (Scheme 42) [56, 57]. The double bond generated results from a β-H-elimination of the alkyl metal intermediate in the final step.

An interesting extension of carbonylative cyclization of 1,4-dienes is achieved if the reaction is performed under hydroformylation conditions in the presence of primary amines to give bicyclic pyrroles [58]. Here, a 1,4-dicarbonyl intermediate formed from the cyclized rhodium alkyl intermediate (Scheme 39) reacts to form a pyrrole in the final step (Scheme 43).

Cyclocarbonylation is also observed with a bisallyl carbamate reacting with HCo(CO)₄ to form a ketone. A cobalt acyl olefin complex has been postulated to be an intermediate which then undergoes a consecutive conversion to a five-membered cyclic ketone (Scheme 44) [59].

The alternating copolymerization of carbon monoxide and ethene leading to polyketones is a reaction of remarkable fundamental and industrial relevance. This reaction is catalyzed by Pd(II) compounds stabilized by chelating diphosphine ligands (Scheme 45) [60].

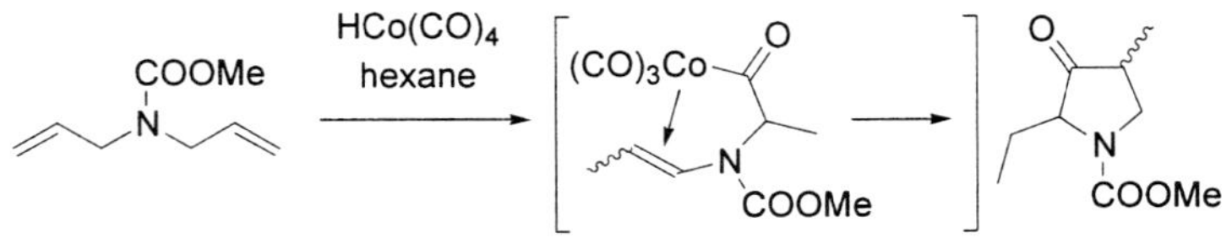

Scheme 44

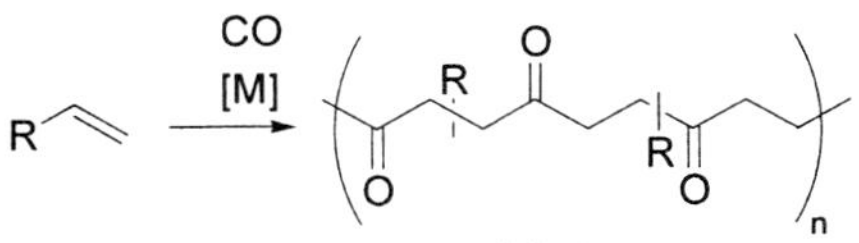

Scheme 45

polyketone

The chain lengths can be controlled by varying the reaction conditions and the catalyst precursors.

2.3.5
Concluding Remarks

Transition metal-catalyzed hydroformylation sequences with various additional C-C bond-forming steps seem to be a powerful tool in the construction of new functionalized carbon skeletons and heterocyclic systems starting from easily accessible functionalized or non-functionalized unsaturated compounds. This is an important expansion of the tandem hydroformylation procedures combined with isomerizations, reductions, or C-O or C-N bond-forming steps, as described in a recent review [7] and in the preceding chapter in this volume. It can be expected that more interesting examples and applications will come in the near future.

2.3.6
References

1 TIETZE, L. F. *Chem. Rev.* **1996**, *96*, 115–136.
2 (a) PARSONS, P. J.; PENKETT, C. S.; SHELL, A. J. *Chem. Rev.* **1996**, *96*, 195–206. (b) NICOLAOU, K. C.; MONTAGNON, T.; SNYDER, S. A.; *Chem. Commun.* **2003**, 551–564.
3 (a) PATAI, S. (ED.) *The Chemistry of the Carbonyl Group*, Wiley-Interscience: New York, 1966 and 1970. (b) FALBE, J. (ed.) *Methoden der Organischen Chemie* (Houben-Weyl); Thieme: Stuttgart, Vol. **E3** (Aldehyde), 1983.
4 (a) BOHNEN, H. W.; CORNILS, B.; *Adv. Catal.* **2002**, *47*, 1–6. (b) Cornils, B., *J. Mol. Catal. A-Chem.* **1999**, *143*, 1–10. (c) BELLER, M.; CORNILS, B.; FROHNING, C. D.; KOHLPAINTNER, C. W. *J. Mol. Catal. A-Chem.* **1995**, *104*, 17–85. (d) CORNILS, B.; HERRMANN, W. A.; KOHLPAINTNER, C. W. *Angew. Chem.* **1994**, *106*, 2219–2238; *Angew. Chem. Int. Ed. Engl.* **1994**, *33*, 2144–2163. (e) CORNILS, B.; HERRMANN, W. A. *Applied Homogeneous Catalysis with Organometallic Compounds*, VCH: Weinheim **1996**. (f) VAN LEEUWEN, P. W. N. M.; CLAVER, C. (Eds.) *Rhodium-catalyzed Hydroformylation*, Kluwer Academic Publishers, Dordrecht, **2000**.
5 (a) BURKE, S. D.; COBB, J. E.; TAKEUCHI, K. *J. Org. Chem.* **1990**, *55*, 2138–2151. (b) BREIT, B.; ZAHN, S. K. *Tetrahedron Lett.* **1998**, *39*, 1901–1904. (c) HORNBERGER, K. R.; HAMBLETT, C. L.; LEIGHTON, J. L. *J. Am. Chem. Soc.* **2000**, *122*, 12894–12895. (d) Breit, B. *Chem. Eur. J.* **2000**, *6*, 1519–1524. (e) DREHER, S. D.; LEIGHTON, J. L. *J. Am. Chem. Soc.* **2001**, *123*, 341–342. (f) LIU, P.; JACOBSEN, E. N. *J. Am. Chem. Soc.* **2001**, *123*, 10772–10773.

(g) Breit, B.; Zahn, S.K. *J. Org. Chem.* **2001**, *66*, 4870–4877. (h) Sun, P.; Sun, C.; Weinreb, S.M. *Org. Lett.* **2001**, *3*, 3507–3510. (i) Sun, P.; Sun, C.X.; Weinreb, S.M. *J. Org. Chem.* **2002**, *67*, 4337–4345. (k) Seepersaud, M.; Kettunen, M.; Abu-Surrah, A.S., Repo, T., Voelter, W. Al-Abed, Y. *Tetrahedron Lett.* **2002**, *43*, 1793–1795. (l) Weinreb, S.M. *Acc. Chem. Res.* **2003**, *36*, 590–650.

6 (a) Gladiali, S.; Bayon, J.C.; Claver, C. *Tetrahedron: Asymmetry,* **1995**, *6*, 1453–1474. (b) Eilbracht, P. in *Methoden der Organischen Chemie* (Houben-Weyl) Vol. E 21c: *Stereoselective Synthesis* (Eds.: Helmchen, G.; Hoffmann, R.W.; Mulzer, J.; Schaumann), Thieme, Stuttgart, **1995**, 2488–2733. (c) Agbossou, F.; Carpentier, J.-F.; Mortreux, A. *Chem. Rev.* **1995**, *95*, 2485–2506. (d) Breit, B.; Seiche, W. *Synthesis* **2001**, 1–36. (e) Breit, B.; *Acc. Chem. Res.* **2003**, *36*, 264–275.

7 Eilbracht, P.; Bärfacker, L.; Buss, C.; Hollmann, C.; Kitsos-Rzychon, B.; Kranemann, C.L.; Rische, T.; Roggenbuck, R.; Schmidt, A. *Chem. Rev.* **1999**, *99*, 3329–3365.

8 Breit, B.; Zahn, S.K. *Angew. Chem.* **1999**, *111*, 1022–1024; *Angew. Chem., Int. Ed. Engl.* **1999**, *38*, 969–971.

9 Eilbracht, P.; Hollmann, C.; Schmidt, A.M.; Bärfacker, L. *Eur. J. Org. Chem.* **2000**, 1131–1135.

10 a) Bärfacker, L. Thesis, Dortmund University, **1999**. b) Bärfacker, L.; Eilbracht P. unpublished results.

11 Zacuto, M.J.; Leighton, J.L. *J. Am. Chem. Soc.* **2000**, *122*, 8587–8588.

12 O'Malley, S.J.; Leighton, J.L. *Angew. Chem., Int. Ed.* **2001**, *40*, 2915–2917.

13 Dreher, S.D.; Leighton, J.L. *J. Am. Chem. Soc.* **2001**, *123*, 341–342.

14 Zacuto, M.J., O'Malley, S.J., Leighton, J.L. *J. Am. Chem. Soc.* **2002**, *124*, 7890–7891.

15 (a) Hoffmann, R.W.; Brückner, D.; Gerusz, V.J. *Heterocycles* **2000**, *52*, 121–124. (b) Hoffmann, R.W.; Brückner, D. *New J. Chem.* **2001**, 369–373.

16 Hoffmann, R.W.; Krüger, J.; Brückner, D. *New J. Chem.* **2001**, *25*, 102–107.

17 (a) Settambolo, R.; Savi, S.; Caiazzo, A.; Lazzaroni, R. *J. Organomet. Chem.* **2000**, *601*, 320–323. (b) Settambolo, R.; Caiazzo, A.; Lazzaroni, R. *Tetrahedron Lett.* **2001**, *42*, 4045–4048.

18 Dessole, G.; Marchetti, M.; Taddei, M. *J. Comb. Chem.* **2003**, *5*, 198–200.

19 Trost, B.M.; Fleming, I.; Heathcock, C.H. (Eds.), *Comprehensive Organic Synthesis*, Pergamon, Oxford, Vol. 2, **1991**.

20 Breit, B.; Zahn, S.K. *Angew. Chem.* **2001**, *113*, 1964–1967. *Angew. Chem. Int. Ed.* **2001**, *40*, 1910–1913.

21 Settambolo, R.; Savi, S.; Caiazzo, A.; Lazzaroni, R. *J. Organomet. Chem.* **2001**, *619*, 241–244.

22 Hollmann, C.; Eilbracht, P. *Tetrahedron Lett.* **1999**, *40*, 4313–4316.

23 Hollmann, C.; Eilbracht, P. *Tetrahedron* **2000**, *56*, 1685–1692.

24 (a) Hollmann, C., Thesis Dortmund University, **2000**. (b) Keränen, M.D.; Hollmann, C.; Fresu, S., Scognamillo, S.; Eilbracht, P. manuscripts in preparation.

25 Bärfacker, L.; El Tom, D.; Eilbracht, P. *Tetrahedron Lett.* **1999**, *40*, 4031–4034.

26 Anastasiou, D.; Campi, E.M.; Chaouk, H.; Fallon, G.D.; Jackson, W.R.; McCubbin, Q.J.; Trnacek, A.E. *Aust. J. Chem.* **1994**, *47*, 1043.

27 (a) Beller, M.; Eckert, M. *Angew. Chem.* **2000**, *112*, 1027–1044; *Angew. Chem. Int. Ed.* **2000**, *39*, 1010–1027. (b) Knifton, F. in: *Applied Homogeneous Catalysis with Metal Complexes* (Eds.: Herrmann, W.A.; Cornils, B.), VCH, Weinheim, **1996**, p. 159.

28 Wakamatsu, H.; Uda, J.; Yamakami, N. *J. Chem. Soc. Chem. Commun.* **1971**, 1540–1541.

29 Stern, R.; Hirschauer, A.; Commereuc, D.; Chauvin, Y. GB-B 2.000.132 A, **1978** [*Chem. Abstr.* **1979**, *91*, 192 831].

30 (a) Ojima, I.; Hirai, K.; Fujita, M.; Fuchikami, T. *J. Organomet. Chem.* **1985**, *279*, 203. (b) Ojima, I.; Okabe, M.; Kato, K.; Kwon, H.B.; Horvath, I.T. *J. Am. Chem. Soc.* **1988**, *110*, 150.

31 (a) Lin, J.J.; Knifton, J.F. *J. Organomet. Chem.* **1991**, *417*, 99. (b) Knifton, J.F.; Lin, J.J.; Storm, D.A.; Wong, S.F. *Catal. Today* **1993**, *18*, 355; (c) Lin, J.J.;

Knifton, J. F.; Yeakey, E. L. (Texaco Inc.), US-A 4.918.222, 1987 [*Chem. Abstr.* **1990**, *113*, 115 869]; (d) Lin, J. J.; Knifton, J. F. *Adv. Chem. Ser.* **1992**, *230*, 235–247; (e) Lin, J. J.; Knifton, J. F. *CHEMTECH* **1992**, *22*, 248.

32 Ojima, I.; Zhang, Z. *Organometallics* **1990**, *9*, 3122.

33 (a) Jägers, E.; Böhshar, M.; Kleiner, H.-J.; Koll, H.-P. (Hoechst AG), DEB 3.913.891, **1990** [*Chem. Abstr.* **1990**, *113*, 41329]; (b) Jägers, E.; Erpenbach, H.; Koll, H.-P. (Hoechst AG), DE-B 3.823.885, **1990** [*Chem. Abstr.* **1990**, *113*, 41325]; (c) Jägers, E.; Erpenbach, H.; Bylsma, F. (Hoechst AG), DE-B 3.823.886, **1990** [*Chem. Abstr.* **1990**, *113*, 41326].

34 (a) Takigawa, S.; Shinke, S.; Tanaka, M. *Chem. Lett.* **1990**, 1415; (b) Sakakura, T.; Huang, X.-Y.; Tanaka, M. *Bull. Chem. Soc. Jpn.* **1991**, *64*, 1707.

35 Gomez, R. M.; Sharma, P.; Arias, J. L.; Perez-Flores, J.; Velasco, L.; Cabrera, A. *J. Mol. Catal. A* **2001**, *170*, 271–274.

36 Nagy, E.; Benedek, C.; Heil, B.; Török, S. *Appl. Organomet. Chem.* **2002**, *16*, 628–634.

37 Ojima, I.; Zhang, Z. *J. Org. Chem.* **1988**, *53*, 4422.

38 Anastasiou, D.; Campi, E. M.; Chaouk, H.; Fallon, G. D.; Jackson, W. R.; McCubbin, Q. J.; Trnacek, A. E. *Aust. J. Chem.* **1994**, *47*, 1043.

39 Anastasiou, D.; Campi, E. M.; Chaouk, H.; Jackson, W. R.; McCubbin, Q. J. *Tetrahedron Lett.* **1992**, *33*, 2211.

40 Verspui, G.; Elbertse, G.; Sheldon, F. A.; Hacking, M. A. P. J.; Sheldon, R. A. *Chem. Commun.* **2000**, 1363–1364.

41 Köhling, P.; Schmidt, A. M.; Eilbracht, P. *Org. Lett.* **2003**, *5*, in print.

42 Diaz, A. de O.; Augusti, R.; dos Santos, E. N.; Gusevskaya, E. V. *Tetrahedron Lett.* **1997**, *38*, 41–44.

43 Roggenbuck, R.; Eilbracht, P. *Tetrahedon Lett.* **1999**, *40*, 7455–7456.

44 Klemchuk, P. P. US Patent 2 995 607, **1959**; *Chem. Abstr.* **1962**, *56*, 1363e.

45 Kobori, Y.; Takesono, T. JP 61 277 644, **1986**; *Chem. Abstr.* **1987**, *107*, 23010v.

46 Keil, T.; Gull, R. GE 3837452 A1, **1990**; *Chem. Abstr.* **1990**, *113*, 171548r.

47 Eilbracht, P.; Acker, M.; Totzauer, W. *Chem. Ber.* **1983**, *116*, 238–242.

48 Eliel, E. L.; Wilen, S. H. *Stereochemistry of Organic Compounds;* Wiley, New York, **1994**, p. 682.

49 Eilbracht, P.; Balss, E.; Acker, M. *Tetrahedron Lett.* **1984**, *25*, 1131–1132.

50 Eilbracht, P.; Balss, E.; Acker, M. *Chem. Ber.* **1985**, *118*, 825–839.

51 Eilbracht, P.; Acker, M.; Rosenstock, B. *Chem. Ber.* **1989**, *122*, 151–158.

52 Eilbracht, P.; Acker, M.; Hädrich, I. *Chem. Ber.* **1988**, *121*, 519–524.

53 Dygutsch, D. P.; Eilbracht, P. *Tetrahedron* **1996**, *52*, 5461–5468.

54 Eilbracht, P.; Hüttmann, G.-E. *Chem. Ber.* **1990**, *123*, 1053–1061.

55 Eilbracht, P.; Hüttmann, G.-E.; Deussen, R. *Chem. Ber.* **1990**, *123*, 1063–1070.

56 (a) Naigre, R.; Chenal, T.; Ciprès, I.; Kalck, P.; Daran, J.-C.; Vaissermann, J. *J. Organomet. Chem.* **1994**, *480*, 91. (b) Chenal, T.; Naigre, R.; Ciprès, I.; Kalck, P.; Daran, J.-C.; Vaissermann, J. *J. Chem. Soc., Chem. Commun.* **1993**, 747.

57 Lacaze-Dufaure, C.; Lenoble, G.; Urutigoity, M.; Gorrichon, J.-P.; Mijoule, C.; Kalck, P. *Tetrahedron Asymmetry* **2001**, *12*, 185–187.

58 Kranemann, C. L.; Kitsos-Rzychon, B.; Eilbracht, P. *Tetrahedron* **1999**, *55*, 4721–4732.

59 Garst, M. E.; Lukton, D. *J. Org. Chem.* **1981**, *46*, 4433–4438.

60 (a) Sen, A. *Acc. Chem. Res.* **1993**, *26*, 303. (b) Drent, E.; Budzelaar, P. H. M. *Chem. Rev.* **1996**, *96*, 663–681. (c) Drent, E.; van Broekhoven, J. A. M.; Budzelaar, P. H. M. in *Applied Homogeneous Catalysis with Organometallic Compounds* (Cornils, B., Herrmann, W. A., Eds.) VCH, Weinheim, **1996**, Vol. 1, p. 333. (d) Sommazzi, A.; Garbassi, G. *Prog. Polym. Sci.* **1997**, *22*, 1547–1605. (e) Nozaki, K.; Hijama, T. *J. Organomet. Chem.* **1999**, *576*, 248–253. (f) Bianchini, C.; Meli, A. *Coord. Chem. Rev.* **2002**, *225*, 35. (g) van Leeuwen, P. W. N. M.; Zuideveld, M. A.; Swennenhuis, B. H. G.; Freixa, Z.; Kamer, P. C. J.; Goubitz, K.; Fraanje, J.; Lutz, M.; Spek, A. L. *J. Am. Chem. Soc.* **2003**, *125*, 5523–5539.

2.4
Hydrocarboxylation and Hydroesterification Reactions Catalyzed by Transition Metal Complexes

Bassam El Ali and Howard Alper

2.4.1
Introduction

The hydrocarboxylation and hydroesterification of olefins, alkynes, and other unsaturated substrates are reactions of industrial potential or demonstrated value [1–6]. The utilization of transition metal complexes as catalysts in these carbonylation reactions has increased significantly. The transformation of alkenes, alkynes, and other related substrates in the presence of group VIII metals and carbon monoxide affords carboxylic acids or carboxylic acid derivatives depending on the source of proton used [1, 3, 5]. Cobalt, nickel, and iron carbonyl, as well as palladium complexes, are the most frequently used catalysts [1–3, 7]. Several hydrocarboxylation methods have appeared in the last 20 years [6–10]. However, the scope of existing methodologies are usually limited by the need for high pressures of carbon monoxide or for concentrated inorganic acids. While many catalysts have been successfully employed in these reactions, they often lead to mixtures of products [11–13].

This chapter focuses on recent developments on the catalytic and regioselective hydrocarboxylation and hydroesterification reactions.

2.4.2
Intermolecular Hydrocarboxylation and Hydroesterification of Unsaturated Substrates

2.4.2.1
Hydrocarboxylation of Alkenes

The synthesis of carboxylic acids from olefins, carbon monoxide, and water is one of the first examples of the use of metal complexes in hydrocarboxylation reactions. A number of patents and publications which have appeared on this subject reveal that drastic conditions (high pressures and high temperatures) are often required to effect this transformation, and that the reaction generally affords mixtures of straight chain and branched chain acids [11–13]. The regioselective hydrocarboxylation of olefins to branched-chain acids was achieved under mild condi-

Transition Metals for Organic Synthesis, Vol. 1, 2nd Edition.
Edited by M. Beller and C. Bolm
Copyright © 2004 WILEY-VCH Verlag GmbH & Co. KGaA, Weinheim
ISBN: 3-527-30613-7

tions to form acids in high yields by the use of carbon monoxide and oxygen in acidic media (HCl), and catalytic amounts of palladium chloride and copper(II) chloride at room temperature [14]. The concentration of HCl and the presence of oxygen have a significant influence on the rate and on the yield of the reaction. Of particular interest is the application of this reaction to the preparation of non-steroidal anti-inflammatory agents including ibuprofen and naproxen. Repetition of the hydrocarboxylation reaction of p-isobutylstyrene, in the presence of (R)-(–)- or (S)-(+)-binaphthyl-2,2-diyl hydrogenphosphate (BNPPA), gave optically active ibuprofen (Eq. 1) [15].

$$\text{p-isobutylstyrene} + CO + O_2 \xrightarrow[\text{HCl, H}_2\text{O, THF} \atop \text{r.t., 1 atm}]{\text{PdCl}_2, \text{CuCl}_2, L} \text{ibuprofen} \quad (1)$$

Recently, a new catalytic process was reported for the selective synthesis of linear carboxylic acids, in high yields and good selectivities (80–100%), from alkenes and formic acid catalyzed by palladium acetate in the presence of 1,4-bis(diphenylphosphino)butane (dppb) at 6.8 atm of carbon monoxide and 150 °C (Eq. 2) [16].

$$R^1\text{-C}\!=\!\text{CH}_2 + \text{HCOOH} + \text{CO} \xrightarrow[\text{DME, 6.8 atm} \atop \text{150°C, 3-6 h}]{\text{Pd(OAc)}_2,\ \text{dppb}} R^1\text{-CH-CH}_2\text{-CO}_2\text{H} + R^1\text{-C(CH}_3\text{)-CO}_2\text{H} \quad (2)$$
$$\text{(R}^2\text{)} \qquad\qquad\qquad\qquad\qquad\qquad\qquad\ (R^2) \qquad\qquad (R^2)$$

R1- alkyl, aryl, silyl, pyrrolidone, ...etc
R2- H, CH3, C2H5, ...etc

Monosubstituted and 1,1-disubstituted olefins behaved in an analogous fashion. Useful bifunctional products were obtained from this reaction, including keto-acids, diacids, cyanoacids, and others. Some results of the application of the Pd(OAc)$_2$–dppb–HCO$_2$H–CO catalytic system for the hydrocarboxylation of various alkenes are given in Tab. 1.

A possible mechanism for this reaction is outlined in Scheme 1. Palladium acetate may react with CO and dppb to form complex **A**. Reaction of **A** with formic acid can produce the palladium formate complex **B** as the active catalytic species, which is converted to **C** on treatment with olefin. CO insertion into the Pd-alkyl bond gives **D** which, on reductive elimination in the presence of formic acid and carbon monoxide, would afford the anhydride **E**, and regenerate **B**. As anhydrides are thermally unstable, decarbonylation can occur to form the carboxylic acid. Use

Tab. 1 Hydrocarboxylation of alkenes by Pd(OAc)$_2$–dppb–HCO$_2$H–CO [a].

Alkene	Time [h]	Yield [%]	l[b] [%]	b[b] [%]
o-CH$_3$C$_6$H$_4$–CH=CH$_2$	3	92	90	10
2,4,6-(CH$_3$)$_3$C$_6$H$_2$–CH=CH$_2$	3	98	100	0
(CH$_3$)$_3$C–CH=CH$_2$	3	94	100	0
NC–CH$_2$–CH=CH$_2$	16	74	86	14
OHC–C(CH$_3$)$_2$–CH$_2$–CH=CH$_2$	16	97	100	0
HOOC–(CH$_2$)$_2$–CH=CH$_2$	16	94	86	14
Ph–C(CH$_3$)=CH$_2$	6	82	100	0

a) Reaction conditions: Pd(OAc)$_2$ (0.02 mmol); Dppb (0.04 mmol); HCOOH (10.0 mmol); alkene (5.0 mmol); DME (5.0 ml); 150 °C; 6.8 atm.
b) l=linear, b=branched.

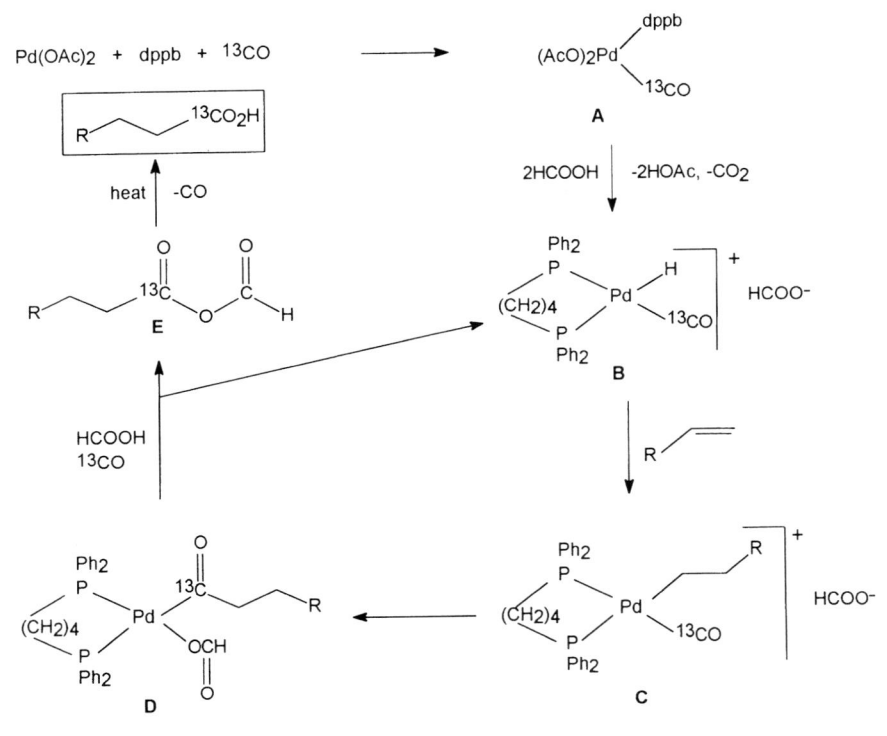

Scheme 1

of ^{13}C-labeled carbon monoxide in the reaction affords the carboxylic acid with the label at the acid carbon.

High yields and excellent selectivities for cycloalkylacetic acids, several of which have anti-inflammatory properties or are important intermediates in medicinal chemistry [17], were realized by direct hydrocarboxylation of the corresponding methylenecycloalkanes using the catalytic system Pd(OAc)$_2$–dppb–HCO$_2$H–CO [18].

Carboxylic acid groups can be attached at the terminal olefin sites to form polycarboxylic acids via catalytic hydrocarboxylation of polybutadienes. The production of such polymers with carboxylated backbones is of particular interest due to potential application of these polymers in films and surface coatings [19, 20]. Thus, the hydrocarboxylation of polybutadienes was studied, with different selectivities obtained subject to the nature of the catalytic system [21]. Specifically the use of PdCl$_2$(PPh$_3$)$_2$–SnCl$_2$ led to the placement of carboxylic acid units at 1,2-positions (Eq. 3). Both 1,4- and 1,2-carboxylate polymers are formed when the reaction was effected with PdCl$_2$(PPh$_3$)$_2$–PPh$_3$ at 170 °C in benzene, or with PdCl$_2$–CuCl$_2$ in THF [21].

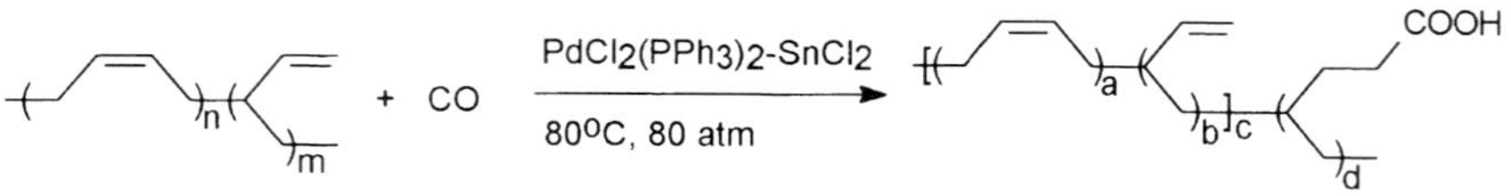

$$\text{(3)}$$

Recently, the hydrocarboxylation of 1,2-polybutadiene was achieved with full conversion of pendant double bonds by using the Pd(OAc)$_2$–HCOOH catalytic system (CO, dppb, 150 °C, 6.8 atm) (Eq. 4) [22].

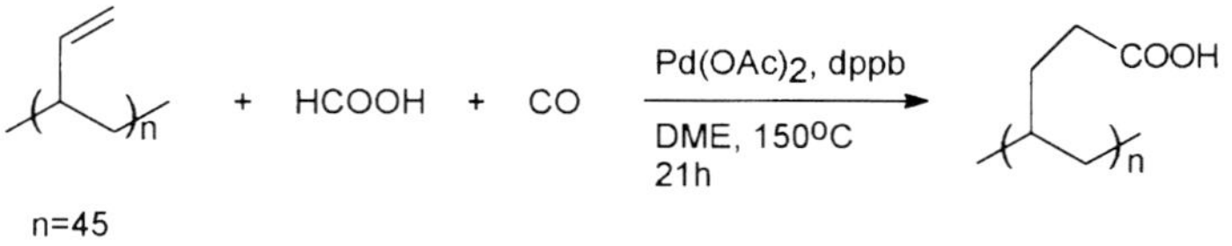

$$\text{(4)}$$

The IR, ^{13}C and ^{1}H NMR spectra, and iodine value determination of the product shows total saturation of vinyl groups, with complete selectivity for the hydrocarboxylation of straight chain acid units and no cross linking or chain scission [22]. The hydrocarboxylation of 1,2-polybutadiene also occurred under oxidative carbonylation conditions [14], with 44% of the vinyl groups reacted [22].

The hydrocarboxylation of fluorinated olefins [3,3,3-trifluoropropene (TFP) and pentafluorostyrene (PFS)] leads to the formation of fluorinated acids useful in organic synthesis. The palladium complex PdCl$_2$(dppf) [dppf = 1,1-bis(di-phenylphosphino)ferrocene] in the presence of 10 equivalents of SnCl$_2$ at 125 °C and at 10 atm CO, showed the highest catalytic activity with TFP (yield = 93%, selectivity = 99% in linear acid), and the catalyst PdCl$_2$(dppb) afforded PFS hydrocarboxylation products in excellent yield and selectivity (Eq. 5) [23].

$$R_f\text{-CH}=\text{CH}_2 + \text{CO/H}_2 \xrightarrow[\text{AcOH - H}_2\text{O}]{\text{[Pd]}} R_f\text{-CH}_2\text{CH}_2\text{-COOH} \quad (5)$$

$R_f = CF_3, C_6F_5$.

The regioselective hydrocarboxylation of alkenes to linear carboxylic acids was also achieved by the use of oxalic acid as the source of hydrogen and carbon monoxide, with the catalytic system $Pd(OAc)_2$-dppb-PPh_3 at 150 °C and 20 atm (Eq. 6) [18].

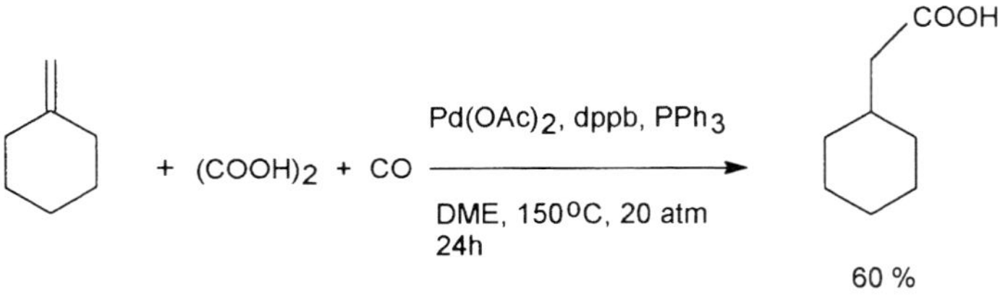

$$\text{methylenecyclohexane} + (\text{COOH})_2 + \text{CO} \xrightarrow[\text{24h}]{\text{Pd(OAc)}_2\text{, dppb, PPh}_3 \atop \text{DME, 150°C, 20 atm}} \text{cyclohexylmethyl-COOH} \quad (6)$$

60 %

Recently, a study was made of the kinetics of 1-heptene hydrocarboxylation catalyzed by $PdCl_2(PPh_3)_2$ in dioxane at 110 °C [24]. From a mechanistic viewpoint it is interesting that the authors claim that three types of hydride complexes $HPdCl_{2-m}(CO)_m$ ($m = 0-2$) are involved in this process.

2.4.2.2
Hydroesterification of Alkenes

The hydroesterification of alkenes is, like hydrocarboxylation, an industrially important reaction and is of interest from a synthetic point of view [6, 25, 26]. Palladium chloride in combination with copper(II) chloride catalyze the hydroesterification of olefins in acidic alcohol in the presence of carbon monoxide and oxygen (1 atm). Branched chain esters were obtained as the principal and, in some cases, as the only reaction product [27]. The use of diols in place of monoalcohols resulted in the regioselective monohydroesterification of alkenes [28]. However, a ratio of 10:1 of diol:olefin has to be used (Eq. 7).

$$C_8H_{17}\text{-CH}=CH_2 + HO(CH_2)_2OH \xrightarrow[\text{CO, O}_2\text{, HCl} \atop \text{1.0 atm, r.t.}]{\text{PdCl}_2\text{, CuCl}_2} C_8H_{17}\text{-CH(CH}_3)\text{-CO}_2(CH_2)_2OH + C_9H_{19}CO_2(CH_2)_2OH$$

70 % 20 %

(7)

The cationic hydridoaquopalladium(II) complex, $trans$-$[(Cy_3P)_2HPd(H_2O)]^+BF_4^-$, in conjunction with dppb and p-toluenesulfonic acid (p-TsOH), has been more recently used as a catalyst for the hydroesterification of various olefinic substrates (Eq. 8) [29].

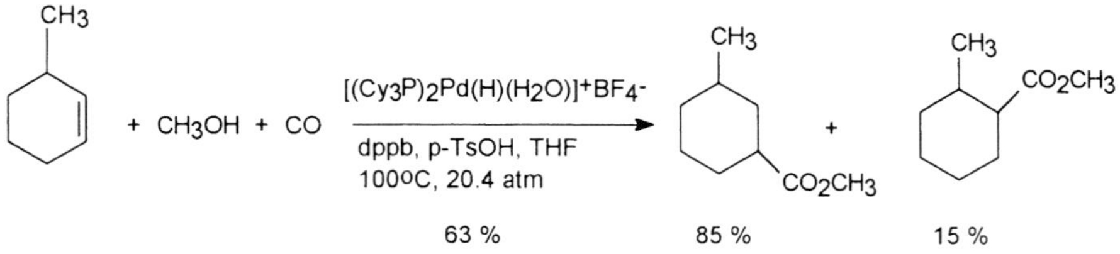

[Reaction scheme showing methylcyclohexene + CH3OH + CO with [(Cy3P)2Pd(H)(H2O)]+BF4−, dppb, p-TsOH, THF, 100 °C, 20.4 atm, giving methylcyclohexane (63%), and two ester products: linear CO2CH3 substituted methylcyclohexane (85%) and branched isomer (15%)]

(8)

Dppb as an added ligand (equimolar with respect to the palladium complex), favors the formation of straight chain carboxylic esters as the major products. The use of other alcohols (ethanol, isopropanol) with different alkenes also give high yields and selectivity for linear carboxylic esters [29].

Palladium acetate immobilized on montmorillonite is another effective catalyst for the hydroesterification of olefins with carbon monoxide and methanol, in the presence of PPh$_3$ and an acid promoter, with branched chain esters formed as the major products (Eq. 9) [30].

$$\text{Ph-CH}=\text{CH}_2 + \text{CH}_3\text{OH} + \text{CO} \xrightarrow[\text{PPh}_3, \text{C}_6\text{H}_6]{\text{Pd-Clay, HCl}} \text{Ph-CH(CH}_3\text{)-CO}_2\text{CH}_3$$
600 psi, 125 °C

(9)

Another heterogeneous catalyst is palladium/graphite in combination with copper(II) chloride and lithium chloride which results in the oxidative dicarbonylation of alkenes using a 20:1 ratio of CO/O$_2$ (e.g. preparation of alkyl succinates) (Eq. 10) [31]. The ratio of diester and dimethyl carbonate formed is sensitive to the nature of the palladium catalyst precursor, e.g. Pd/graphite or PdCl$_2$.

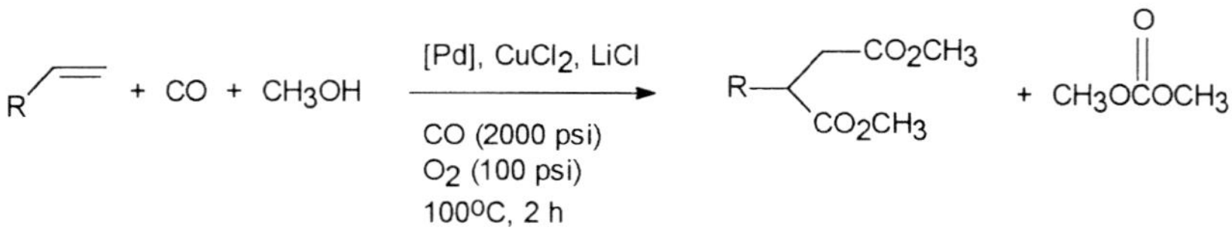

$$\text{R-CH=CH}_2 + \text{CO} + \text{CH}_3\text{OH} \xrightarrow[\substack{\text{CO (2000 psi)}\\\text{O}_2\text{ (100 psi)}\\100°\text{C, 2 h}}]{[\text{Pd}], \text{CuCl}_2, \text{LiCl}} \text{R-CH(CO}_2\text{CH}_3\text{)-CH}_2\text{-CO}_2\text{CH}_3 + \text{CH}_3\text{OCOCH}_3$$

(10)

The asymmetric hydroesterification of methyl methacrylate to a 1,4-diester was realized with [(R,R)-DIOP]PdCl$_2$ as the catalyst precursor (Eq. 11) [32]. Enantioselectivities of about 40% have been obtained. Similarly, dimethyl succinate was the major product using methyl acrylate as the reactant, although the isomeric 1,1-diester was also formed in substantial quantities [33].

2.4.2 Intermolecular Hydrocarboxylation and Hydroesterification of Unsaturated Substrates

$$CH_3O_2C-C(CH_3)=CH_2 + CO + CH_3OH \xrightarrow[C_6H_6,\ 120°C,\ 216\ atm]{PdCl_2[(R,R)\text{-DIOP}]} CH_3O_2C-CH(CH_3)-CH_2-CO_2CH_3 \quad (11)$$

The hydroesterification of vinylsilanes catalyzed by a palladium complex intercalated into montmorillonite, in the presence of an appropriate amount of PPh_3 and p-TsOH, afforded β-silylesters in high yield and excellent regioselectivity (Eq. 12) [34a]. The use of $Co_2(CO)_8$ as a catalyst also resulted in high regioselectivity and excellent yields of β-silylesters [34b]. The hydroesterification of 1,2-polybutadiene to the corresponding polyester has been carried out using $Pd(OAc)_2$ and dppb as catalysts in the presence of p-TsOH, resulting in full conversion of the pendant vinyl groups [22].

$$(CH_3)_3Si\text{-}CH=CH_2 + CH_3OH + CO \xrightarrow[\substack{CH_2Cl_2,\ p\text{-TsOH} \\ 100°C,\ 500\ psi}]{\text{Pd-Clay, } PPh_3} (CH_3)_3Si\text{-}CH_2CH_2\text{-}CO_2CH_3 \quad (12)$$

Formate esters have attracted considerable interest as possible C_1 intermediates [26]. In this respect it is interesting to note that $RuCl_2(PPh_3)_3$ catalyzes the hydroesterification of ethylene with methyl formate, but the catalytic activity was rather low, and the type of olefins were very limited [35]. However, $Ru_3(CO)_{12}$ does catalyze the same process in quite an effective manner. The reaction of ethene at 230°C and 90 atm of N_2 yields up to 92% of propionic acid ester. The advantage of HCO_2CH_3 vs. CH_3OH/CO is that additional CO is unnecessary; indeed the addition of CO diminishes the ester yield, indicating the importance of the *in situ* decomposition of methyl formate into CH_3OH/CO during the catalytic cycle. The use of higher olefin substrates leads to a mixture of linear and branched isomers in low yields [36]. $Ru_3(CO)_{12}$ in conjunction with $(CH_3)_3NO \cdot 2H_2O$ provides another catalytic system for the hydroesterification of less reactive olefins, such as cyclohexene, with alkylformate in the presence of carbon monoxide (Eq. 13). The best yield (68%) was obtained with benzyl formate [37].

$$\text{C}_6\text{H}_{10} + HCO_2R + CO \xrightarrow[200°C,\ 20\ atm]{Ru_3(CO)_{12},\ (CH_3)_3NO\cdot 2H_2O} \text{C}_6\text{H}_{11}\text{-}CO_2R \quad (13)$$

In addition, the hydroesterification of ethene with methyl formate was catalyzed by the complex $[PdH(Cl)(PBu_3)_2]$, generated *in situ* by addition of 1 equivalent of $NaBH_4$ to $PdCl_2(PBu_3)_2$. Extra carbon monoxide is not required for the reaction and methyl propanoate was formed with high selectivity using methanol as the solvent (Eq. 14) [38].

$$CH_2=CH_2 + HCO_2CH_3 \xrightarrow[MeOH, 130]{PdCl_2(PBu_3)_2, NaBH_4} CH_3CH_2CO_2CH_3 \quad (14)$$

The simple, mild, palladium(II)–copper(II)–CO–O_2–HCl system catalyzes the oxidative carbonylation of formate esters and olefins. Higher yields were obtained by using an excess rather than an equimolar quantity of formate ester to substrate (Eq. 15) [39a].

$$R\text{-}CH=CH_2 + HCO_2R' \xrightarrow[\substack{HCl, CO, O_2 \\ \text{Dioxane, 1 atm, r. t.}}]{PdCl_2, CuCl_2} R\text{-}CH_2CH_2\text{-}CO_2R' + R\text{-}\underset{\underset{CH_3}{|}}{CH}\text{-}CO_2R' \quad (15)$$

Palladium(0) complexes [Pd(PPh$_3$)$_4$ or Pd(dba)$_2$ (dba = dibenzylideneacetone)] in the presence of dppb, catalyze the reaction of alkenes and formate esters to form linear carboxylic esters as the major products, while the branched chain isomer was obtained by the use of a palladium(II) complex PdCl$_2$(PPh$_3$)$_2$ (Eqs. 16, 17) [39b].

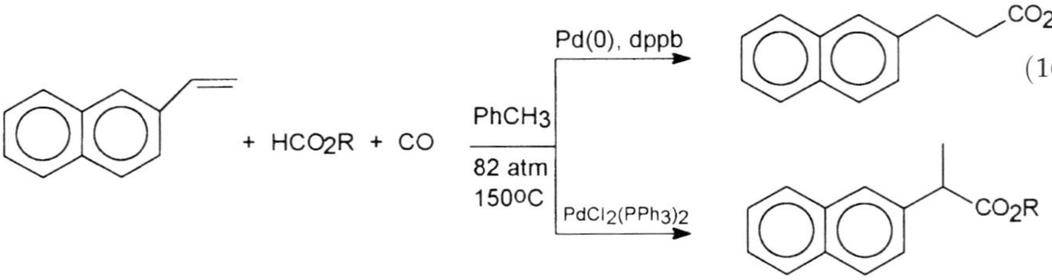

(16)

(17)

2.4.2.3
Hydrocarboxylation and Hydroesterification of Allenes and Dienes

The hydroesterification and hydrocarboxylation of allenes gives different results depending on the metal catalyst used [40]. When the catalytic system PdCl$_2$–CuCl$_2$–CO–O$_2$–HCl (1 atm and room temperature) described above was applied to allene, methyl-2-methoxymethylacrylate was formed in 85% yield (Eq. 18) [41a].

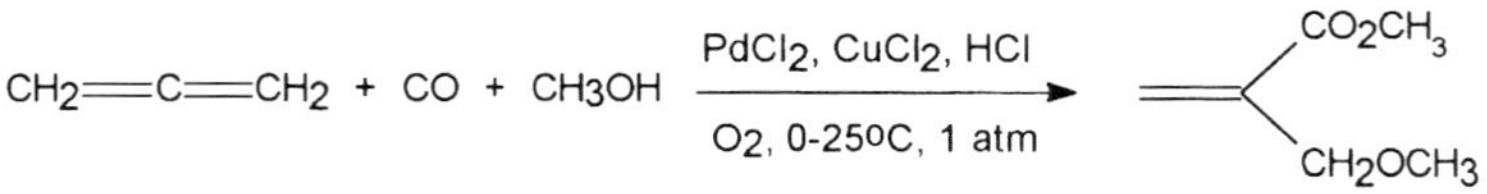

$$CH_2=C=CH_2 + CO + CH_3OH \xrightarrow[O_2, 0-25°C, 1\,atm]{PdCl_2,\,CuCl_2,\,HCl} \underset{CH_2OCH_3}{\overset{CO_2CH_3}{\diagup\!\diagdown}} \quad (18)$$

The product yield was low for the corresponding methoxyester obtained from vinylidene cyclohexane [41a]. The regioselective hydrocarboxylation of allenes to β,γ-

unsaturated carboxylic acids was attained using phase-transfer catalysis (PTC). Here allenes react with carbon monoxide in an aqueous base–toluene two-phase system, with cetyltrimethylammonium bromide (CTAB) as phase-transfer agent, and nickel cyanide as the catalyst, to give β,γ-unsaturated carboxylic acids in 48–66% yields (Eq. 19) [41b].

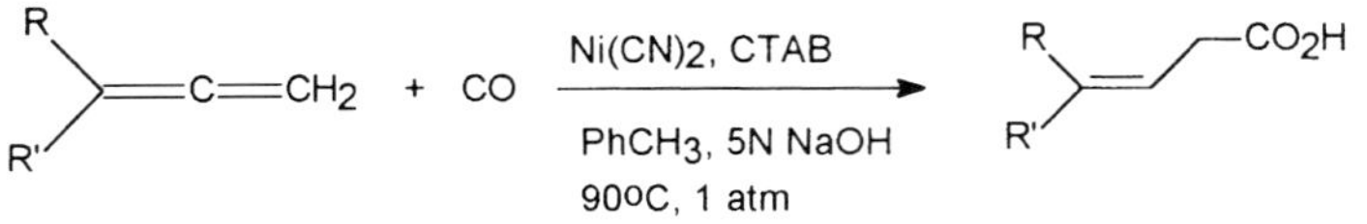

$$\underset{R'}{\overset{R}{>}}=C=CH_2 \;+\; CO \;\xrightarrow[\substack{PhCH_3,\;5N\;NaOH \\ 90^\circ C,\;1\;atm}]{Ni(CN)_2,\;CTAB}\; \underset{R'}{\overset{R}{>}}=\!\!\diagup\!\!\diagdown_{CO_2H} \tag{19}$$

α-Vinylacrylic acids were obtained by the carbonylation of α-allenic alcohols using trans-$[(Cy_3P)_2Pd(H)(H_2O)]^+BF_4^-$ and p-TsOH as catalysts at 20 atm of CO and 100 °C in THF (Eq. 20) [42].

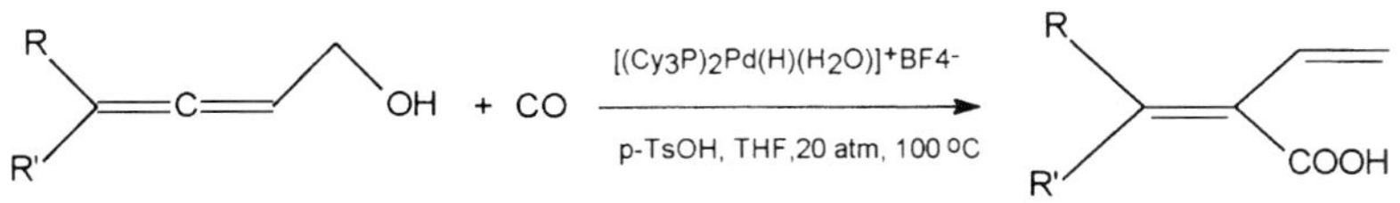

$$\underset{R'}{\overset{R}{>}}=C=\!\!\diagup\!\!\diagdown_{OH} \;+\; CO \;\xrightarrow[p\text{-TsOH, THF, 20 atm, 100 °C}]{[(Cy_3P)_2Pd(H)(H_2O)]^+BF_4^-}\; \underset{R'}{\overset{R}{>}}\!\!=\!\!\overset{\diagup\!\!=}{\underset{COOH}{\diagdown}}$$

R, R' ≠ H (20)

The reaction of 1,3-butadiene with methyl formate catalyzed by $PdCl_2$ in the presence of HCl afforded methyl esters of pentenoic acid in low yields apart from C_9 telomers and butadiene dimers (Eqs. 21, 22) [36].

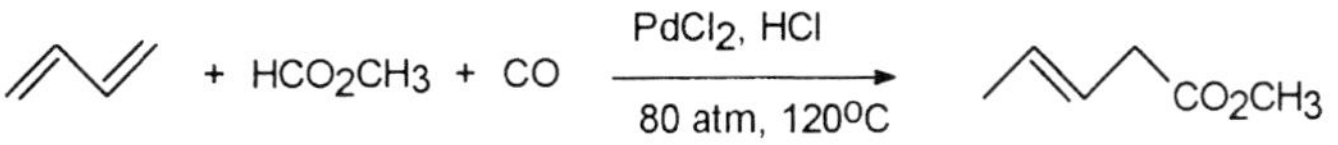

$$\diagup\!\!\diagdown\!\!\diagup\!\!\diagdown \;+\; HCO_2CH_3 \;+\; CO \;\xrightarrow[80\;atm,\;120^\circ C]{PdCl_2,\;HCl}\; \diagup\!\!\diagdown\!\!\diagup\!\!\diagdown_{CO_2CH_3} \tag{21}$$

$$2\;\diagup\!\!\diagdown\!\!\diagup\!\!\diagdown \;+\; HCO_2CH_3 \;+\; CO \;\xrightarrow[80\;atm,\;120^\circ C]{PdCl_2,\;HCl}\; \diagup\!\!\diagdown\!\!\diagup\!\!\diagdown\!\!\diagup\!\!\diagdown\!\!\diagup\!\!\diagdown_{CO_2CH_3} \tag{22}$$

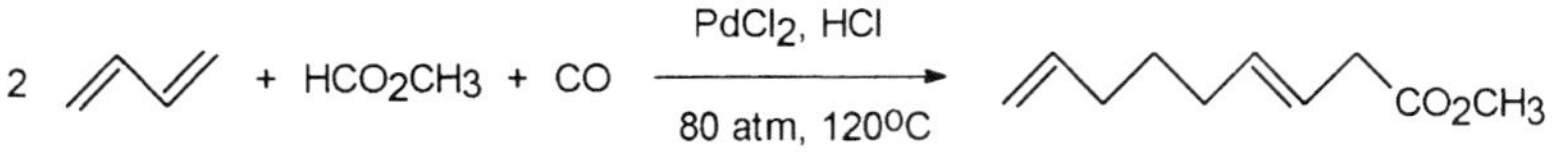

The oxidative dicarbonylation of 1,3-butadiene to generate dimethyl hex-3-ene-1,6-dioate – a precursor of adipic acid – resulted using Pd/graphite, in combination with $CuCl_2$ and LiCl (Eq. 23) [31].

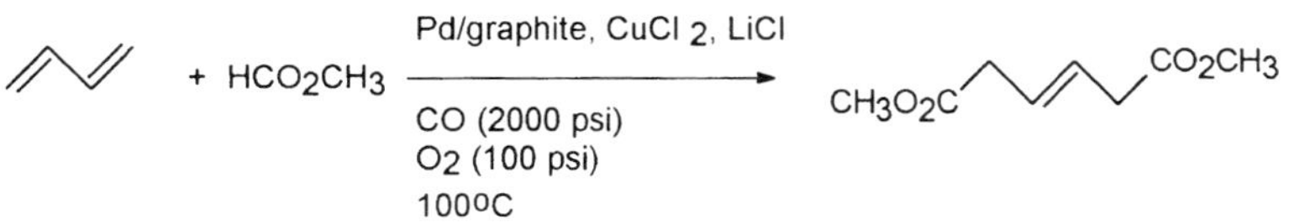

$$\diagup\!\!\diagdown\!\!\diagup\!\!\diagdown \;+\; HCO_2CH_3 \;\xrightarrow[\substack{Pd/graphite,\;CuCl_2,\;LiCl \\ CO\;(2000\;psi) \\ O_2\;(100\;psi) \\ 100^\circ C}]{}\; CH_3O_2C\diagup\!\!\diagdown\!\!\diagup\!\!\diagdown_{CO_2CH_3} \tag{23}$$

Other conjugated dienes were converted to functionalized β,γ-unsaturated esters using $PdCl_2$–$CuCl_2$–CO–O_2–HCl–CH_3OH. In order to get good yields it was essential to dry the mixture of PdCl$_2$, methanol, and conc. HCl over molecular sieves prior to addition of the diene and cupric chloride. It is also important to add a quaternary ammonium salt such as Aliquat-336 to prevent polymerization. Thus, *trans*-methyl-5-methoxy-3-nonenoate was obtained in 60% yield [43]. Di- and trisubstituted 1,3-dienes were converted to γ,δ-unsaturated acids by use of formic acid, carbon monoxide, and catalytic quantities of Pd–C/PPh$_3$/dppb in 1,2-dimethoxyethane. The hydrocarboxylation of isoprene, for example, occurs under 6.2 atm of CO and at 110 °C to form the corresponding β,γ-unsaturated acid in 52% yield (Eq. 24) [43b].

$$\text{isoprene} + HCO_2H + CO \xrightarrow[\substack{DME,\ 6.2\ atm \\ 110\ ^\circ C}]{Pd\text{-}C,\ dppb,\ PPh_3} \text{product} \qquad (24)$$

2.4.2.4
Hydrocarboxylation and Hydroesterification of Simple and Hydroxyalkynes

The metal complex catalyzed hydrocarboxylation and hydroesterification of alkynes and alkynols is of value for the synthesis of a,β-unsaturated acids and their derivatives [1, 6, 8]. Phase-transfer catalysis is a beneficial milieu for the reaction of alkynes with carbon monoxide, aqueous base, toluene as the organic phase, and catalytic amounts of Ni(CN)$_2$ and cetyltrimethylammonium bromide (CTAB) as the phase-transfer agent to form unsaturated acids in 30–62% yields. The application of the same catalytic system to diynes gives unsaturated diacids in good yields (Eq. 25) [44].

$$HC{\equiv}C(CH_2)_nC{\equiv}CH + CO \xrightarrow[\substack{PhCH_3,\ NaOH \\ 90\ ^\circ C,\ 1\ atm}]{Ni(CN)_2,\ CTAB} \underset{65\text{-}68\ \%}{CH_2{=}\underset{COOH}{\overset{COOH}{C}}(CH_2)_n\underset{COOH}{\overset{}{C}}{=}CH_2} \qquad (25)$$

The direct regioselective hydrocarboxylation of alkynes to saturated carboxylic acids can be achieved using cobalt chloride, potassium cyanide, and nickel cyanide, under phase transfer conditions. Polyethylene glycol (PEG-400) was used as the phase-transfer agent, with the branched saturated acids formed as the major products (43–65%) (Eq. 26) [45].

$$Ph{-}C{\equiv}CH + CO \xrightarrow[\substack{Ni(CN)_2,\ PhCH_3,\ KOH \\ 90\ ^\circ C,\ 1\ atm}]{CoCl_2,\ KCN,\ PEG\text{-}400} \underset{56\ \%}{Ph\text{-}\overset{COOH}{CH}\text{-}CH_3} + \underset{2\ \%}{Ph\text{-}CH_2CH_2\text{-}COOH} \qquad (26)$$

Moreover, alkynes are hydrocarboxylated with formic acid in the presence of a catalytic amount of Pd(OAc)$_2$, dppb, and PPh$_3$ at 120 psi of CO and 100–110 °C, to

2.4.2 Intermolecular Hydrocarboxylation and Hydroesterification of Unsaturated Substrates

produce the corresponding α,β-unsaturated carboxylic acids. The use of a mixture of the two ligands, dppb and PPh$_3$, significantly improves the yields (Eq. 27) [46].

$$R-C\equiv C-R' + HCOOH + CO \xrightarrow[\substack{DME,\ 100-110\ ^\circ C \\ 120\ psi}]{Pd(OAc)_2,\ dppb,\ PPh_3} \underset{1}{\overset{R\diagup\hspace{-2pt}=\hspace{-2pt}\diagdown R'}{\underset{HOOC\hspace{12pt}H}{}}} + \underset{2}{\overset{R\diagup\hspace{-2pt}=\hspace{-2pt}\diagdown R'}{\underset{H\hspace{12pt}COOH}{}}} \quad (27)$$

The distribution of the products (**1** and **2**) is approximately 9:1 in favor of **1** when R is phenyl or a straight chain alkyl group; **2** is favored when R is *t*-Bu, and is the exclusive product when R is SiMe$_3$. Internal alkynes (R$_5$R' H) also undergo catalytic hydrocarboxylation with formic acid, but the regioselectivity is not as high as for terminal alkynes [46]. The result of experiments using 1-deuterio-4-phenyl-1-butyne with HCO$_2$H (Eq. 28), or 4-phenyl-l-butyne with

$$R-C\equiv C-D \xrightarrow{HCOOH} \underset{56\%}{\overset{R\diagup\hspace{-2pt}=\hspace{-2pt}\diagdown D}{\underset{HOOC\hspace{12pt}H}{}}} + \underset{44\%}{\overset{H\diagup\hspace{-2pt}=\hspace{-2pt}\diagdown R}{\underset{D\hspace{12pt}COOH}{}}} \quad (28)$$

R = PhCH$_2$CH$_2$

HCOOD and DCOOH (Eqs. 29, 30) indicate that:

$$R-C\equiv C-H \begin{cases} \xrightarrow{HCOOD} \underset{43\%}{\overset{R\diagup\hspace{-2pt}=\hspace{-2pt}\diagdown D}{\underset{HOOC\hspace{12pt}H}{}}} + \underset{40\%}{\overset{R\diagup\hspace{-2pt}=\hspace{-2pt}\diagdown H}{\underset{HOOC\hspace{12pt}D}{}}} + \underset{17\%}{\overset{R\diagup\hspace{-2pt}=\hspace{-2pt}\diagdown H}{\underset{D\hspace{12pt}COOH}{}}} \quad (29) \\[10pt] \xrightarrow{DCOOH} \underset{91\%}{\overset{R\diagup\hspace{-2pt}=\hspace{-2pt}\diagdown H}{\underset{HOOC\hspace{12pt}H}{}}} + \underset{9\%}{\overset{R\diagup\hspace{-2pt}=\hspace{-2pt}\diagdown H}{\underset{H\hspace{12pt}COOH}{}}} \quad (30) \end{cases}$$

R = PhCH$_2$CH$_2$

1. Terminal alkynes do not undergo oxidative addition of the *sp* C–H(D) bond.
2. Since none of the above experiments lead to a product in which the incoming 'H' or 'D' and 'COOH' units are *gem* to each other, a metal vinylidene intermediate is unlikely here. The most probable mode of bonding for alkynes is that in which the alkyne is coordinated to the metal in a π-fashion.
3. Formic acid protonates the metal center since deuterium incorporation occurs into a vinyl position in the product with HCOOD but not with DCOOH.
4. The 1,1-disubstituted product, **1**, results from a combination of *cis* and *trans* addition of the 'H' and 'COOH' moieties, whereas the 1,2-disubstituted product, **2**, arises from 'cis' addtion to give the (*E*) stereoisomer.

On the basis of these deuterium labeling studies and other experimental results, a reaction mechanism has been proposed (Scheme 2) involving electron-rich Pd(0)

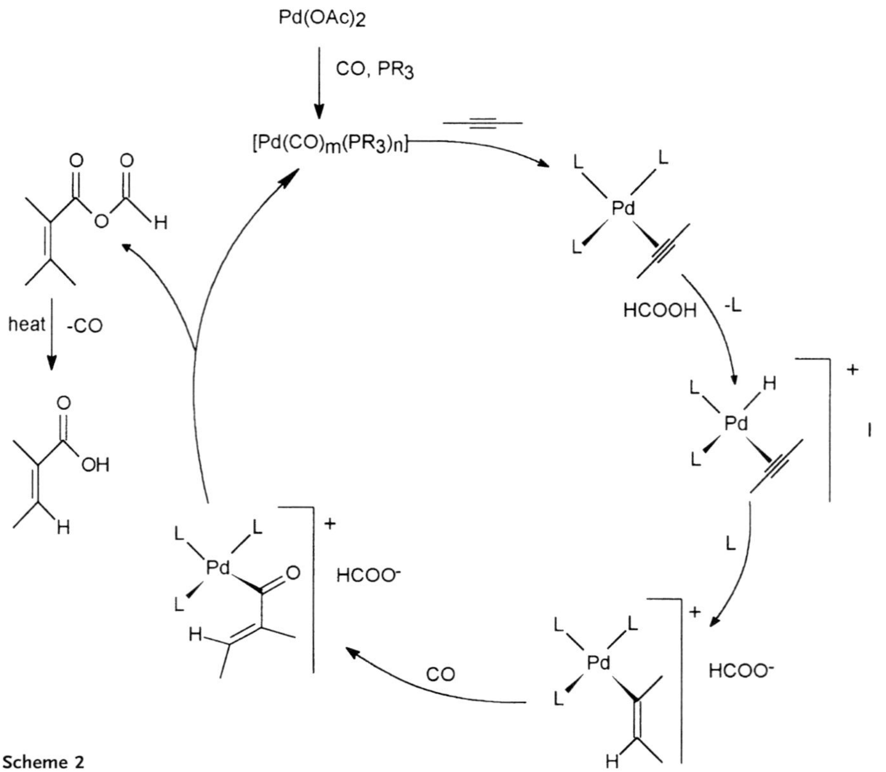

Scheme 2

species which are known [47] to form Pd–H bonds in the presence of strong acids.

The hydrocarboxylation of alkynes with formic acid or oxalic acid, [(CO$_2$H)$_2$], was also achieved using the heterogeneous catalyst Pd/C, in the presence of dppb and PPh$_3$ at 40 atm of CO and 110 °C. The catalytic activity of the heterogeneous and homogeneous systems were similar. The yield of α,β-unsaturated acids are good (61–78%) and the reaction is regioselective [48].

Formic acid reacts with terminal alkynes, PdCl$_2$, CuCl$_2$, CO, and O$_2$ at room temperature affording monosubstituted maleic anhydrides and the corresponding maleic and fumaric acids in 30–75% total yield. The regioselectivity of the reaction depends on the type of alkyne used (Eq. 31) [49].

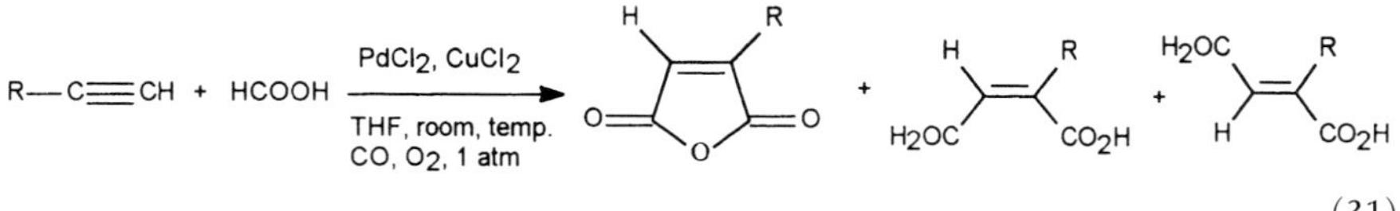

(31)

2.4.2 Intermolecular Hydrocarboxylation and Hydroesterification of Unsaturated Substrates

Terminal alkynes undergo regioselective hydroesterification to unsaturated *cis*-diesters using $PdCl_2$, $CuCl_2$, HCl, alcohol, carbon monoxide, and oxygen. These reactions are complete within 2 h at room temperature and atmospheric pressure (Eq. 32) [49b].

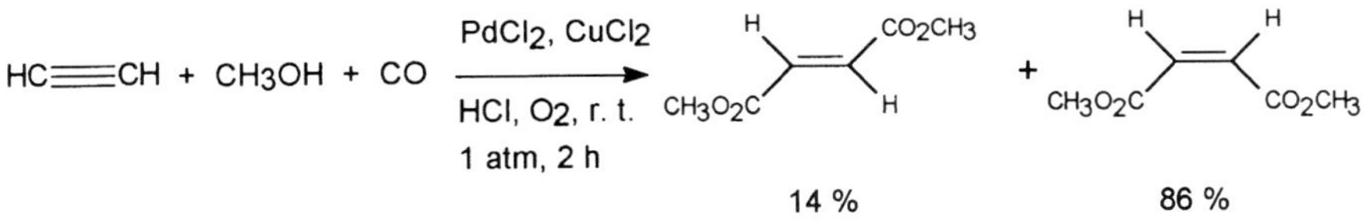

$$HC{\equiv}CH + CH_3OH + CO \xrightarrow[\text{1 atm, 2 h}]{\substack{PdCl_2,\ CuCl_2 \\ HCl,\ O_2,\ r.\ t.}} \underset{14\%}{\overset{H}{\underset{CH_3O_2C}{>}}{=}\underset{H}{\overset{CO_2CH_3}{<}}} + \underset{86\%}{\overset{H}{\underset{CH_3O_2C}{>}}{=}\underset{CO_2CH_3}{\overset{H}{<}}}$$

(32)

A highly efficient cationic palladium catalyst has been developed for the methoxycarbonylation of alkynes. An interesting application is the selective production of methyl methacrylate from propyne. The active palladium complex is formed by the combination of a ligand containing a 2-pyridylphosphine moiety with a palladium(II) species and a proton source containing weakly coordinating anions. A high turnover number (50 000 mol product per mol Pd per hour) and excellent selectivity to methyl methacrylate of up to 99.9% can be realized under mild conditions (Eq. 33) [50].

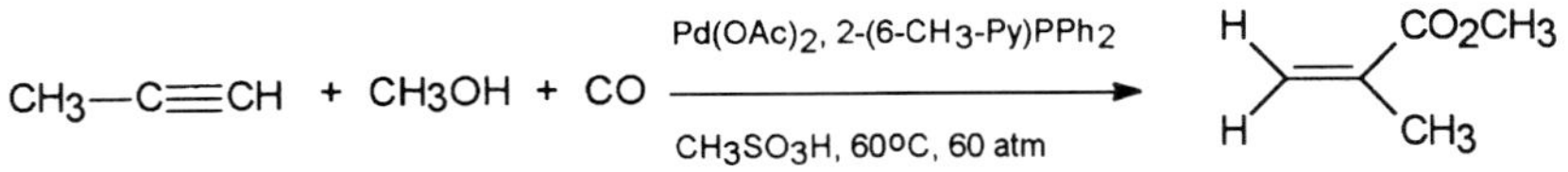

$$CH_3-C{\equiv}CH + CH_3OH + CO \xrightarrow[\text{CH}_3SO_3H,\ 60°C,\ 60\ atm]{Pd(OAc)_2,\ 2\text{-}(6\text{-}CH_3\text{-}Py)PPh_2} \overset{H}{\underset{H}{>}}{=}\underset{CH_3}{\overset{CO_2CH_3}{<}}$$

(33)

The conversion of alkynes into *tert*-alkyl esters was realized using $Pd(OAc)_2$ and dppb at 150 °C and 80 atm of CO. The use of primary and secondary alcohols gave low yields of esters (Eq. 34) [51a].

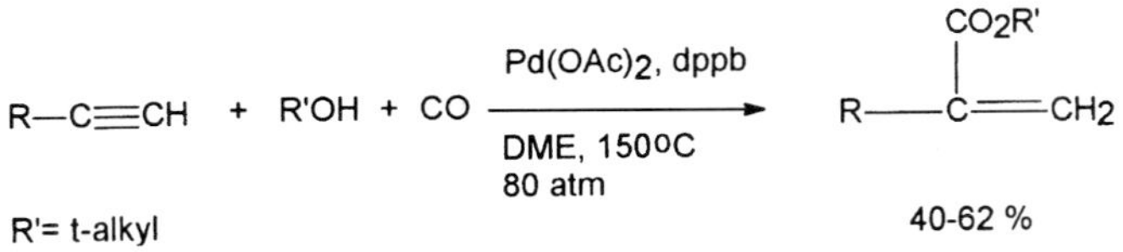

$$R-C{\equiv}CH + R'OH + CO \xrightarrow[\substack{DME,\ 150°C \\ 80\ atm}]{Pd(OAc)_2,\ dppb} R-\underset{\underset{40\text{-}62\%}{}}{\overset{CO_2R'}{C}}{=}CH_2$$

R' = t-alkyl

(34)

However, the regioselective hydroesterification of alkynes and alkynols was achieved by the use of formate esters and the catalytic system, $Pd(OAc)_2$–dppb–PPh_3 in the presence of *p*-TsOH at 20 atm of CO and 100 °C (Eq. 35) [51b].

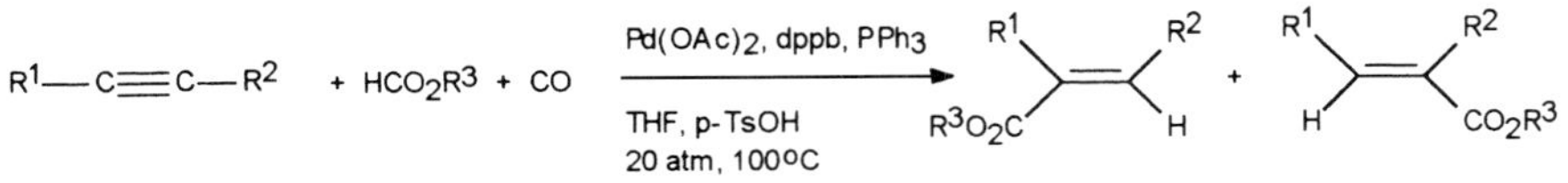

$$R^1-C{\equiv}C-R^2 + HCO_2R^3 + CO \xrightarrow[\substack{THF,\ p\text{-}TsOH \\ 20\ atm,\ 100°C}]{Pd(OAc)_2,\ dppb,\ PPh_3} \overset{R^1}{\underset{R^3O_2C}{>}}{=}\underset{H}{\overset{R^2}{<}} + \overset{R^1}{\underset{H}{>}}{=}\underset{CO_2R^3}{\overset{R^2}{<}}$$

(35)

Alkynols also react under these conditions in a regioselective manner, e.g. treatment of 1-ethynyl-1-cyclohexanol with *n*-butyl formate gave the *trans*-unsaturated ester in 50% yield (Eq. 36) [51b].

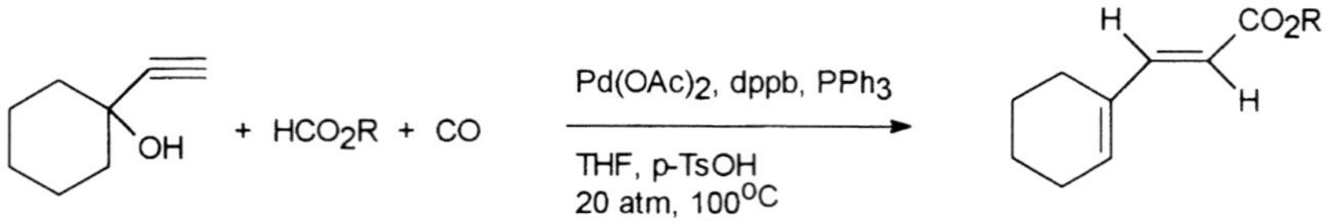

$$\text{1-ethynyl-1-cyclohexanol} + HCO_2R + CO \xrightarrow[\substack{\text{THF, p-TsOH} \\ \text{20 atm, 100°C}}]{\text{Pd(OAc)}_2,\ \text{dppb, PPh}_3} \text{trans-unsaturated ester} \quad (36)$$

Unsaturated diacids were obtained from the nickel cyanide catalyzed carbonylation of alkynols under phase-transfer conditions. The stereochemistry of the reaction is sensitive to the nature of the quaternary ammonium salt, Q^+X^-, used as the phase-transfer agent (Eq. 37) [52].

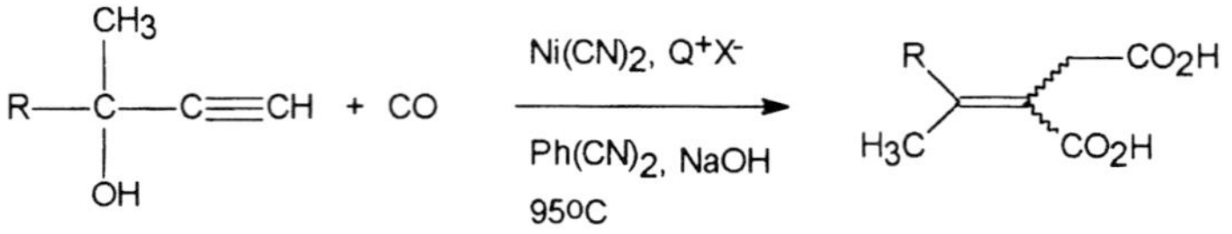

$$R-\underset{\underset{OH}{|}}{\overset{\overset{CH_3}{|}}{C}}-C\equiv CH + CO \xrightarrow[\substack{\text{Ph(CN)}_2,\ \text{NaOH} \\ \text{95°C}}]{\text{Ni(CN)}_2,\ Q^+X^-} \underset{H_3C}{\overset{R}{\diagup}}\!\!=\!\!\underset{CO_2H}{\overset{CO_2H}{\diagdown}} \quad (37)$$

The cationic hydridopalladium complex, *trans*-[(Cy$_3$P)$_2$Pd(H)(H$_2$O)]$^+$BF$_4^-$, catalyzes the hydrocarboxylation or the hydroesterification of alkynols and alkynediols to dienoic acids and esters, and to cross-conjugated diesters (Eq. 38) [53].

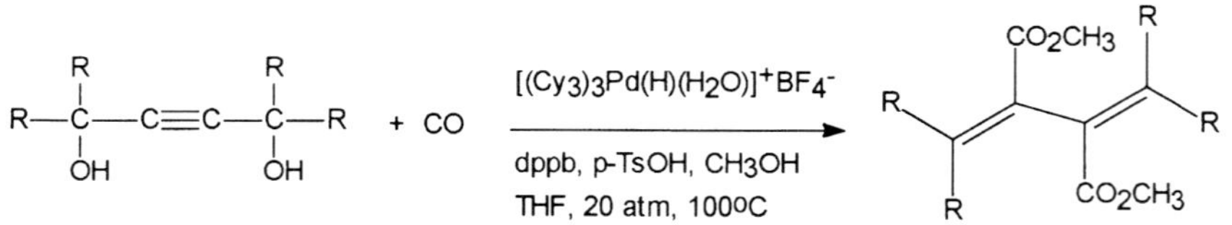

$$R-\underset{\underset{OH}{|}}{\overset{\overset{R}{|}}{C}}-C\equiv C-\underset{\underset{OH}{|}}{\overset{\overset{R}{|}}{C}}-R + CO \xrightarrow[\substack{\text{dppb, p-TsOH, CH}_3\text{OH} \\ \text{THF, 20 atm, 100°C}}]{[(Cy_3)_3Pd(H)(H_2O)]^+BF_4^-} \text{cross-conjugated diester} \quad (38)$$

2.4.3
Intramolecular Cyclocarbonylation of Unsaturated Compounds

Appropriate unsaturated alcohols can undergo the intramolecular version of the hydroesterification reaction affording lactones [54]. The γ,δ- and δ,ε-unsaturated alcohols were converted to five- and six-membered ring lactones by the use of PdCl$_2$–CuCl$_2$–HCl–CO–O$_2$ in THF. The reaction is regioselective but not stereoselective and the yields of lactones are reasonably good (35–80%) [55, 56a]. The addition of poly-L-leucine as the chiral ligand results in the synthesis of optically active lactones in up to 61% enantiometric excess (Eq. 39) [56b].

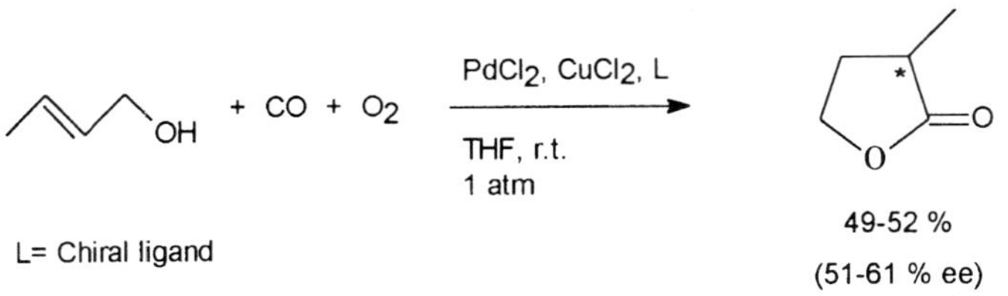

2.4.3 Intramolecular Cyclocarbonylation of Unsaturated Compounds

$$\text{allyl alcohol} + CO + O_2 \xrightarrow[\text{THF, r.t.} \atop \text{1 atm}]{\text{PdCl}_2, \text{CuCl}_2, L} \gamma\text{-lactone} \qquad (39)$$

49-52 %
(51-61 % ee)

L= Chiral ligand

A novel palladium-catalyzed asymmetric cyclocarbonylation of allylic alcohols to γ-butyrolactones was described very recently [56c]. Hence, treatment of 2-methyl-3-phenyl-3-buten-2-ol with a mixture of 1:1 CO and H_2, together with $Pd_2(dba)_3$, $CHCl_3$, and (–)-bppm as the chiral ligand (CH_2Cl_2, 100 °C), affords the γ-lactone in excellent chemical yield and high enantiomeric excess (Eq. 40). The best enantioselectivity (89% ee) was obtained with aryl groups as substituents [56c].

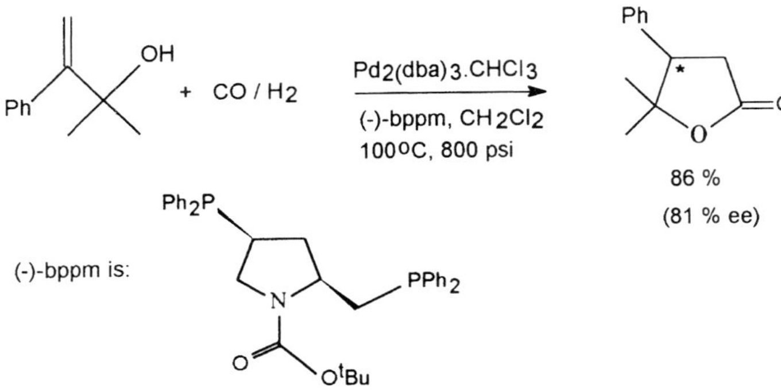

$$\text{substrate} + CO/H_2 \xrightarrow[\text{(-)-bppm, CH}_2\text{Cl}_2 \atop 100°C, 800 \text{ psi}]{Pd_2(dba)_3 \cdot CHCl_3} \text{product} \qquad (40)$$

86 %
(81 % ee)

(-)-bppm is: [pyrrolidine with Ph₂P, PPh₂, and O=C-O^tBu substituents]

Negishi and co-workers described a method for cyclocarbonylation of o-allylbenzyl halides to produce benzo-annulated enol lactones. The reaction is catalyzed by $PdCl_2(PPh_3)_2$ with triethylamine (NEt_3) used as a base (Eq. 41) [57].

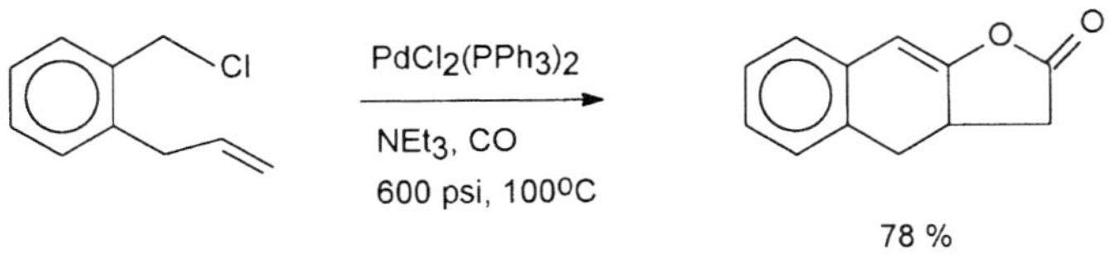

$$\text{o-allylbenzyl chloride} \xrightarrow[\text{NEt}_3, \text{CO} \atop 600 \text{ psi}, 100°C]{PdCl_2(PPh_3)_2} \text{benzo-annulated enol lactone} \qquad (41)$$

78 %

The Pd(0) complex, $Pd(dba)_2$, with added dppb, catalyzes the lactonization of unsaturated alcohols under neutral conditions. Secondary and tertiary allylic alcohols were cyclocarbonylated at 40 atm of CO and 190 °C affording γ-butyrolactones in 45–92% yield (Eq. 42). In addition 2-(5H)-furanones were prepared by this method when alkynols were employed as substrates (Eq. 43) [58].

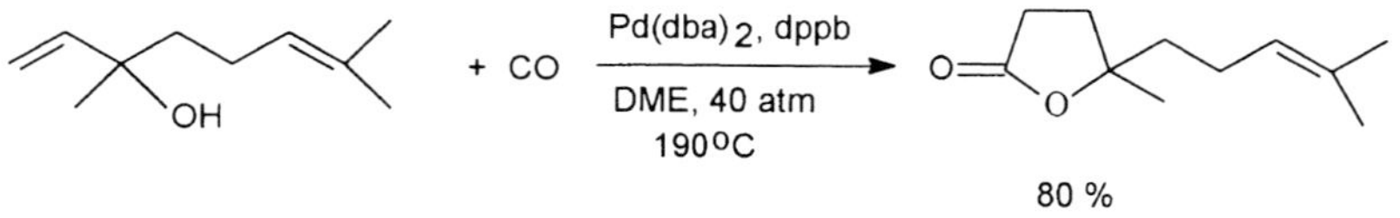

$$\text{(42)}$$

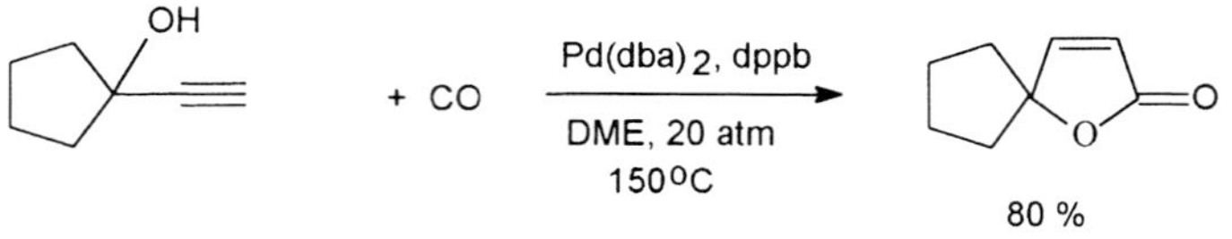

$$\text{(43)}$$

The synthesis of 3(2H)-furanones was realized from the reaction of alkynols and haloarenes under CO or CO_2 atmosphere. Different transition metals were used in this reaction (Eq. 44) [59].

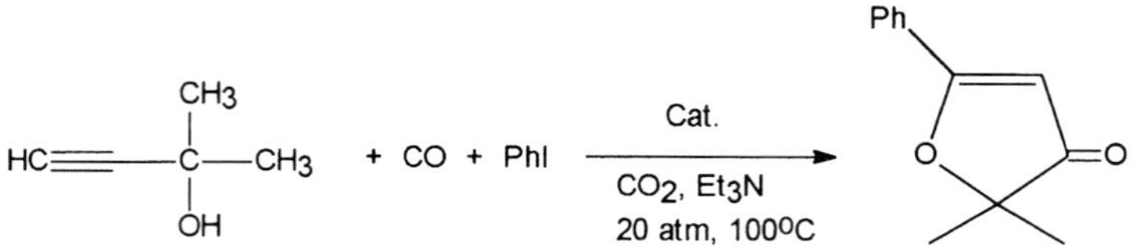

$$\text{(44)}$$

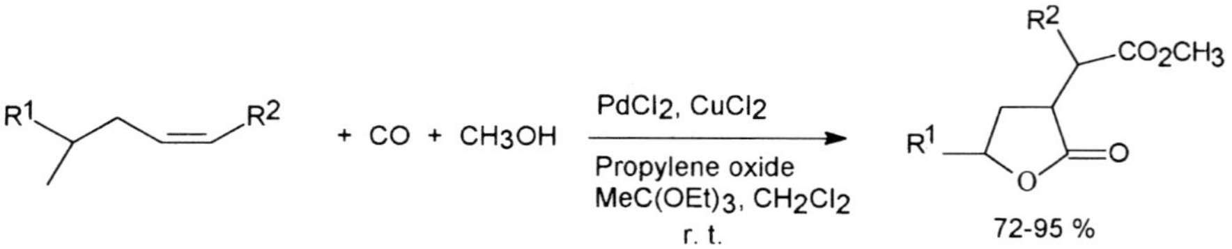

$$\text{(45)}$$

Palladium(II) complexes catalyze the hydroesterification and the cyclocarbonylation of 3-buten-1-ols in the presence of propylene oxide and ethyl orthoacetate (room temperature, 1 atm), affording lactones with an ester side chain in good yields (72–95%). The process occurs via stereospecific *cis*-addition (Eq. 45). Under the same reaction conditions, 4-(trimethylsilyl)-3-butyn-1-ols afford α-methylene-γ-butyrolactones in high yields (81–83%) (Eq. 46). The role of propylene oxide may be to quench the formed hydrogen chloride thus maintaining neutral reaction conditions [60].

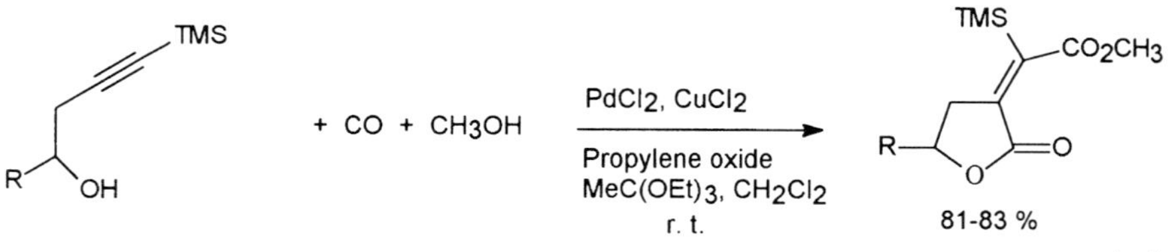

$$\text{(46)}$$

The intramolecular oxycarbonylation of unsaturated polyols was realized in the presence of $PdCl_2$ and $CuCl_2$. Using a variety of carbohydrate-derived substrates with up to five free hydroxy groups afforded bicyclic lactones in high chemo-, regio-, and diastereoselectivity (Eq. 47) [61].

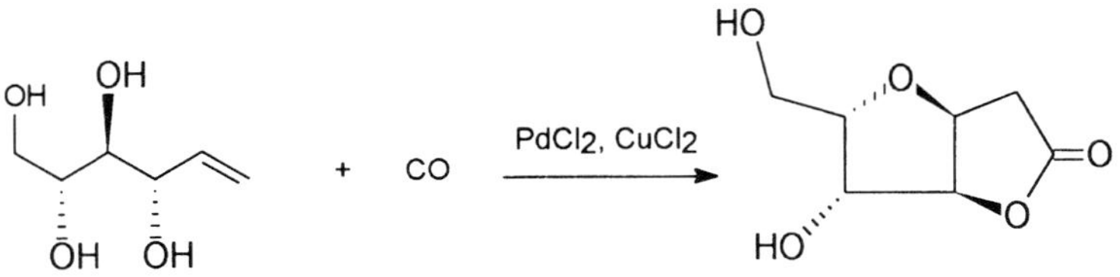

$$\text{(47)}$$

Recently, the synthesis of five- and seven-membered ring lactones, and five-, six-, and seven-membered ring lactams, via intramolecular cyclocarbonylation of allylphenols, 2-allylanilines, and 2-aminostyrenes has been described using palladium based catalyst systems. The regiochemical control of these reactions depend on the relative pressures of the gases, the choice of solvent, the nature of the metal catalyst and added ligand. For example, treatment of 2-allylphenol ($R^1 = R^2 = H$, Eq. 48) with a 1:1 mixture of CO and H_2, a catalytic amount of $[Pd(PCy_3)(H)(H_2O)]^+BF_4^-$, and dppb at 120 °C in CH_2Cl_2 resulted in the formation of the five-membered ring lactone in 76% yield while the seven-membered ring lactone was obtained as a principal product (92–95%) by the use of the cationic palladium complex or $Pd(OAc)_2$, but in *toluene*. 2-Allyl-phenols containing a methyl group as a substituent on the allyl chain (R^1 or $R^2 = CH_3$, Eq. 48) were converted into seven-membered ring lactones (89–95%) only in toluene [62].

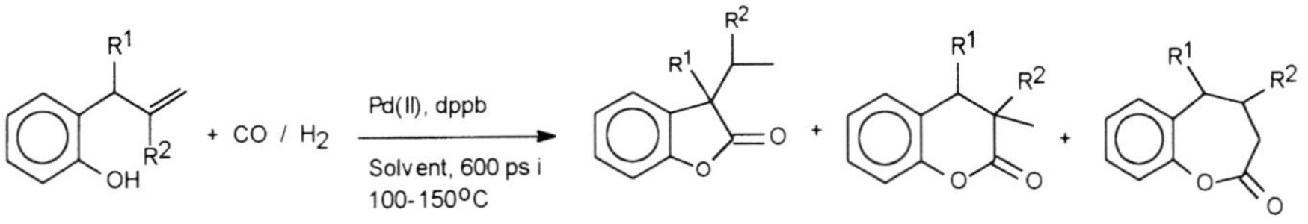

$$\text{(48)}$$

A similar intramolecular carbonylation of 2-allylanilines resulted in the formation of six- and seven-membered ring lactams as major products. The six-membered ring lactams were obtained in high yields using $Pd(OAc)_2$ and PPh_3 as the catalyst in CH_2Cl_2 at 80 °C, with 1:1 CO/H_2 (600 psi). However, the seven-membered ring lactams were formed by the use of dppb instead of PPh_3 at 100 °C. Five- and six-membered ring lactams were prepared by the cyclocarbonylation of 2-aminostyrenes in the presence of $Pd(OAc)_2$ as the catalyst and tricyclohexylphosphine (PCy_3) or dppb as the ligand (Eq. 49) [62].

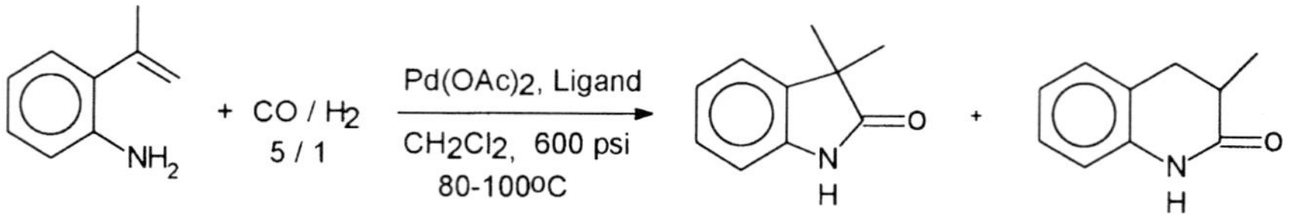

Ligand	Yield, %	Selectivity, %	
PCy3	95	100	0
dppb	75	0	100

(49)

2.4.4
Conclusion

Recent advances in hydroesterification, hydrocarboxylation, and lactonization reactions catalyzed by transition metals and their complexes demonstrate the versatility of these processes in terms of utility in synthetic organic chemistry. Palladium catalysts are particularly useful catalysts for these reactions. Considering the regio- and stereochemical control of many of these processes, and the prochiral nature of a significant proportion of the reactants, it is anticipated that the asymmetric synthesis of acids, esters, and lactones will be an important area of substantial development in the next five years.

2.4.5
References

1. I. TKATCHENKO, *Comprehensive Organometallic Chemistry*, Pergamon Press, New York, **1982**, Vol. 8.
2. F. R. HARTLEY, S. PATAI, *The Chemistry of Metal-Carbon*, Wiley, New York, **1982**, Vol. 3.
3. G. W. PARSHALL, *Homogeneous Catalysis*, Wiley, New York, **1980**.
4. A. L. WADDAMS, *Chemicals from Petroleum*, Gulf, Houston, Texas, **1978**.
5. C. MASTERS, *Homogeneous Transitionmetal Catalysis*, Wiley-Interscience, New York, **1993**.
6. G. W. PARSHALL, S. D. ITTEL, *Homogeneous Catalysis*, Wiley-Interscience, New York, **1993**.
7. A. MULLEN IN *New Synthesis with Carbon Monoxide* (Ed.: J. FALBE), Springer-Verlag, Berlin, **1980**.
8. J. P. COLLMAN, L. S. HEGEDUS, J. R. NORTON, R. G. FINKE, *Principles and Applications of Organotransition Metal Chemistry*, University Science Books, Mill Valley, **1987**.
9. D. FOSTER, A. HERSHAMN, D. E. MORRIS, *Catal. Rev. Sci. Eng.* **1981**, *23*, 89.
10. Y. SOUMA, H. SANO, J. IYODA, *J. Org. Chem.* **1973**, *38*, 2016.
11. J. TSUJI, *Palladium Reagents and Catalysts*, Wiley, Chichester, **1995**.
12. G. GAVINATOR, L. TONIOLO, *J. Mol. Catal.* **1981**, *10*, 161.
13. J. F. KNIFTON, *J. Org. Chem.* **1976**, *41*, 2885.

14 H. Alper, B. Woell, B. Despeyroux, J. H. Smith, *J. Chem. Soc., Chem. Commun.* **1983**, 1270.
15 H. Alper, N. Hamel, *J. Am. Chem. Soc.* **1990**, *112*, 2803.
16 B. El Ali, H. Alper, *J. Mol. Catal.* **1992**, *77*, 7.
17 M. Kuchor, B. Brunova, J. Crimova, V. Rejholec, *Eur. J. Med. Chem.* **1978**, *13*, 263.
18 B. El Ali, H. Alper, *J. Org. Chem.* **1993**, *58*, 3595.
19 P. Molyneux, *Water Soluble Synthetic Polymers*, CRC Press, Boca Raton, Florida, **1984**, Vols. 1–2.
20 B. G. Clubley, *Chemical Inhibitors for Corrosion Control*, Royal Society of Chemistry, Cambridge, **1990**.
21 P. Narayanan, B. G. Clubley, D. J. Cole-Hamilton, *J. Chem. Soc., Chem. Commun.* **1991**, 1628.
22 A. Nait-Ajjou, H. Alper, *Macromolecules* **1996**, *29*, 1784.
23 C. Botteghi, G. Del Ponte, M. Marchetti, S. Paganelli, *J. Mol. Catal.* **1994**, *93*, 1.
24 T. E. Krön, Yu. G. Noskov, M. I. Terekhova, E. S. Petrov, *Zh. Fiz. Khim.* **1996**, *70*, 82; *Chem. Abstr.* **1996**, *124*, 288513.
25 R. A. Sheldon, *Chemicals from Synthesis Gas*, D. Reidel, Dordrecht, **1983**, Chap. 5, p. 104.
26 G. A. Olah, A. Molnar, *Hydrocarbon Chemistry*, John Wiley, New York, **1995**, p. 276.
27 B. Despeyroux, H. Alper, *Ann. N. Y. Acad. Sci.* **1983**, *415*, 118.
28 S. B. Ferguson, H. Alper, *J. Chem. Soc., Chem. Commun.* **1984**, 1349.
29 K. T. Huh, H. Alper, *Bull Korean Chem. Soc.* **1994**, *15*, 304.
30 C. W. Lee, H. Alper, *J. Org. Chem.* **1995**, *60*, 250.
31 J. J. Lin, J. F. Knifton, *Catal. Lett.* **1996**, *37*, 199.
32 G. Consiglio, L. Kollar, R. Kolliker, *J. Organomet. Chem.* **1990**, *396*, 375.
33 G. Consiglio, S. C. A. Nefkens, C. Pisano, F. Wenzinger, *Helv. Chim. Acta* **1991**, *74*, 323
34 (a) B. Lee, H. Alper, *J. Mol. Catal.* **1996**, *111*, L3. (b) R. Takenchi, N. Ishii, M. Sugiura, N. Sato, *J. Org. Chem.* **1992**, *57*, 4189.
35 P. Isnard, B. Denise, R. P. A. Sneeden, J. M. Cognion, P. Durual, *J. Organomet. Chem.* **1983**, *256*, 135.
36 W. Keim, J. Becker, *J. Mol. Catal.* **1989**, *54*, 95.
37 T. Kondo, S. Yoshii, Y. Watanable, *J. Mol. Catal.* **1989**, *50*, 31.
38 J. Grevin, P. Kalck, *J. Organomet. Chem.* **1994**, *476*, C23.
39 (a) M. Mlekuz, F. Joo, H. Alper, *Organometallics* **1987**, *6*, 1991. (b) I. J. B. Lin, H. Alper, *J. Chem. Soc., Chem. Commun.* **1989**, 248.
40 T. L. Jacobs, *The Chemistry of the Allenes* (Ed.: S. R. Landor), Academic Press, New York, **1982**.
41 (a) H. Alper, F. W. Hartstock, B. Despeyroux, *J. Chem. Soc., Chem. Commun.* **1984**, 905. (b) N. Satyanarayana, H. Alper, I. Amer, *Organometallics* **1990**, *9*, 284.
42 M. E. Piotti, H. Alper, *J. Org. Chem.* **1994**, *59*, 1956.
43 (a) S. T. Fergusson, H. Alper, *Mol. Catal.* **1986**, *34*, 381. (b) G. Vasapollo, A. Somasunderam, B. El Ali, H. Alper, *Tetrahedron Lett.* **1994**, *35*, 6203.
44 I. Amer, H. Alper, *J. Organomet. Chem.* **1990**, *383*, 573.
45 J. T. Lee, H. Alper, *Tetrahedron Lett.* **1991**, *32*, 1769.
46 D. Zargarian, H. Alper, *Organometallics* **1993**, *12*, 712.
47 J. K. Stille, D. E. James, *J. Am. Chem. Soc.* **1975**, *97*, 674.
48 B. El Ali, G. Vasapollo, H. Alper, *J. Org. Chem.* **1993**, *58*, 4739.
49 (a) D. Zargarian, H. Alper, *Organometallics* **1991**, *10*, 2914. (b) H. Alper, B. Despeyroux, J. B. Woell, *Tetrahedron Lett.* **1983**, *24*, 5691.
50 E. Drent, P. Arnoldy, P. H. M. Budzelaar, *J. Organomet. Chem.* **1994**, *475*, 57.
51 (a) B. El Ali, H. Alper, *J. Mol. Catal.* **1991**, *67*, 29. (b) B. El Ali, H. Alper, *J. Mol. Catal.* **1995**, *96*, 197.
52 N. Satyanarayana, H. Alper, *Organometallics* **1991**, *58*, 6956.
53 K. T. Huh, A. Orita, H. Alper, *J. Org. Chem.* **1993**, *58*, 6956.

54 H. M. Colquhoun, D. J. Thompson, M. V. Twigg, *Carbonylation*, Plenum Press, New York, **1991**.
55 H. Alper, D. Leonard, *J. Chem. Soc., Chem. Commun.* **1985**, 511.
56 (a) H. Alper, D. Leonard, *Tetrahedron Lett.* **1985**, *26*, 5639. (b) H. Alper, N. Hamel, *J. Chem. Soc., Chem. Commun.* **1990**, 135. (c) W. Y. Yu, C. Bensimon, H. Alper, *Chem. Eur. J.* **1997**, *3*, 417.
57 G. Wu, I. Shimoyama, E. Negishi, *J. Org. Chem.* **1991**, *56*, 6506.
58 B. El Ali, H. Alper, *J. Org. Chem.* **1991**, *56*, 5357.
59 Y. Inoue, K. Ohuchi, I. F. Yen, S. Imaizumi, *Bull. Chem. Soc. Jpn.* **1989**, *62*, 3518.
60 Y. Tamaru, M. Hojo, Z. Yoshida, *J. Org. Chem.* **1991**, *56*, 1099.
61 T. Gracza, T. Hasenohrl, U. Stahl, V. Jäger, *Synthesis* **1991**, 1108.
62 B. El Ali, K. Okuro, G. Vasapollo, H. Alper, *J. Am. Chem. Soc.* **1996**, *118*, 4264.

2.5
The Amidocarbonylation of Aldehydes

Axel Jacobi von Wangelin, Helfried Neumann, Dirk Gördes, and Matthias Beller

2.5.1
Introduction

The amidocarbonylation of aldehydes provides a highly efficient access to N-acyl-α-amino acid derivatives by the reaction of an aldehyde, amide, and carbon monoxide under transition metal catalysis. Ajinomoto chemist H. Wakamatsu serendipitously discovered this reaction when observing the formation of amino acid derivatives as by-products in the cobalt-catalyzed oxo reaction of acrylonitrile [1]. The reaction was further elaborated to an efficient one-step synthesis of racemic N-acyl-α-amino acids (Scheme 1) [2, 3].

Since the N-acyl-amino acid structural motif is central to a large number of compounds, several syntheses of interesting targets were realized by this procedure. Important applications of N-acyl-amino acids range from the direct marketing as pharmaceuticals, agrochemicals, and surfactants to their use as building blocks for further chemical manipulations [4]. Examples (Scheme 2) include the pharmaceuticals captopril and N-acetylcysteine, the herbicide Flamprop-isopropyl, anionic sarcosinate tensides, and simple dipeptides such as the sweetener aspartame. The presence of the acyl fragment in the molecule imparts hydrophobic properties that enable the use of sarcosinates (N-acyl-N-methylglycines) with fatty acid chains as environmentally friendly detergents. Besides the range of direct applications, racemic N-acetyl-α-amino acids are important intermediates in the synthesis of enantiomerically pure α-amino acids via enzymatic hydrolysis, which is still the method of choice for the large-scale preparation of enantiomerically pure amino acids (e.g. N-acetylvaline or N-acetylmethionine) [5].

The efficiency and feasibility of the three-component amidocarbonylation reaction particularly springs from the atom-efficient utilization of the ubiquitous and

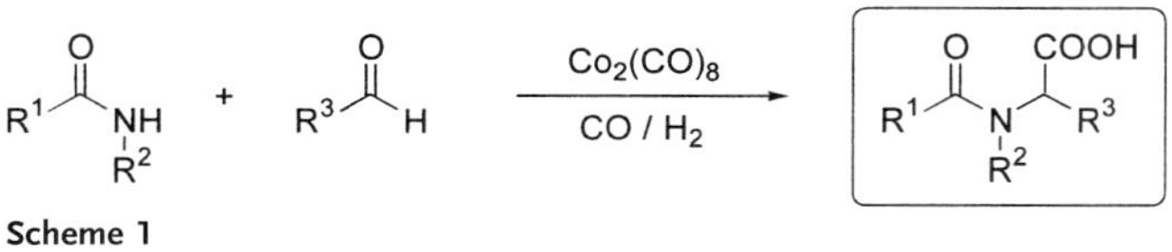

Scheme 1

Transition Metals for Organic Synthesis, Vol. 1, 2nd Edition.
Edited by M. Beller and C. Bolm
Copyright © 2004 WILEY-VCH Verlag GmbH & Co. KGaA, Weinheim
ISBN: 3-527-30613-7

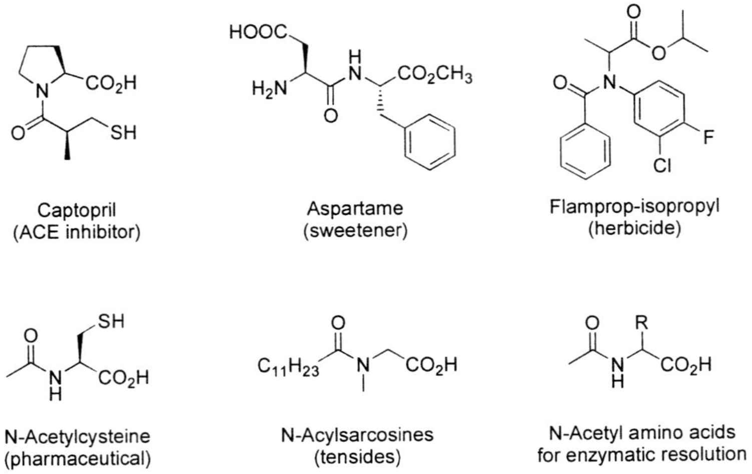

2.5 The Amidocarbonylation of Aldehydes

Scheme 2

- Captopril (ACE inhibitor)
- Aspartame (sweetener)
- Flamprop-isopropyl (herbicide)
- N-Acetylcysteine (pharmaceutical)
- N-Acylsarcosines (tensides)
- N-Acetyl amino acids for enzymatic resolution

cheap starting materials aldehyde, amide, and CO [6]. In the early years, cobalt catalysts were exclusively employed in this reaction. Later on, palladium catalyst systems were also shown to be highly active. Major developments and applications of both systems, which exhibit strong distinctions in their general applicability and scope as well as the underlying mechanistic details, are discussed below.

2.5.2
The Cobalt-Catalyzed Amidocarbonylation

The active catalyst in cobalt-catalyzed amidocarbonylation is [HCo(CO)$_4$]/ [Co(CO)$_4$]$^-$, which is generated *in situ* from Co$_2$(CO)$_8$ (typically 1–5 mol%) in the presence of CO/H$_2$ [7]. The reaction is carried out at 70–160 °C and syngas pressures of 50–200 bar (CO/H$_2$=1/1 to 4/1). Most procedures use 0.1–3 M solutions of the reactants in solvents like dioxane, THF, DME, ethyl acetate, acetone, or benzene.

As shown in Scheme 3, the reaction commences with the condensation of the employed amide and aldehyde to give several equilibrating species including the α-hydroxyalkylamide, 1,1-bisamide, N-acylimine, and N-acylenamine [2a, 8]. Subsequent addition to the cobalt catalyst generates alkyl complex **II**. However, the specific nature of the active catalyst species is still not clear [7, 9]. In principle, the formation of **II** can proceed via nucleophilic substitution of hydroxyalkylamide **I** with [Co(CO)$_4$]$^-$ or insertion of the N-acylenamine into [HCo(CO)$_4$]. Cobalt-centered carbonylation of alkyl complex **II** affords acyl complex **III**, which can release the desired N-acyl-α-amino acid via hydrolysis or gives an oxazolonium salt via intramolecular ring closure in the presence of a water-trapping reagent. The reaction proceeds with 100% atom utilization, as one equivalent of water that is gener-

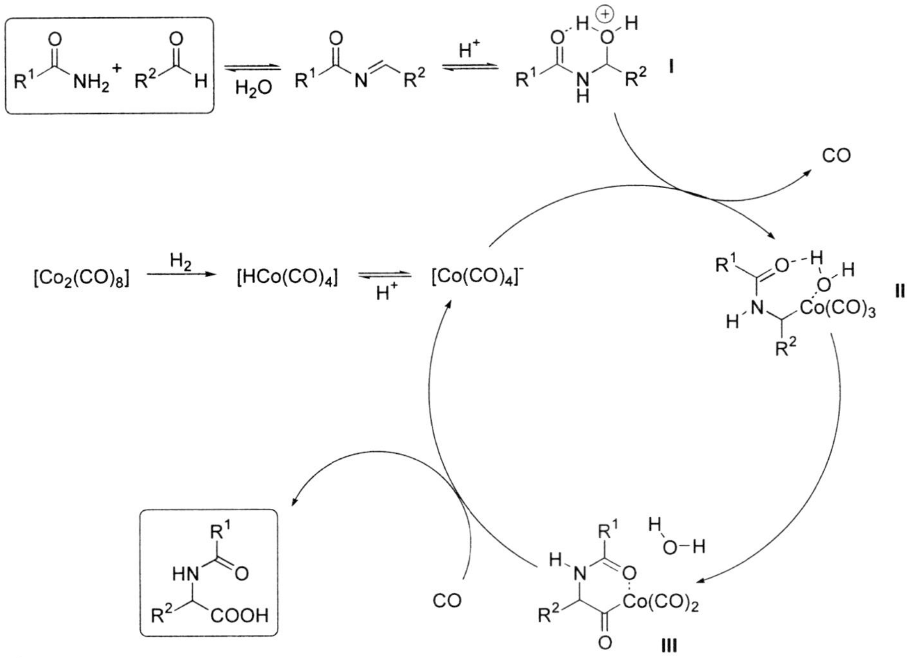

Scheme 3

ated in the initial amide-aldehyde condensation step is consumed in the final quenching of acyl complex **III**.

Generally, emphasis is laid on the key metal-catalyzed CO insertion step, and thus the amidocarbonylation of aldehydes is classified as a carbonylation [10] of an alkyl-X bond. On the other hand, the reaction can also be viewed as an amidoalkylation of weakly nucleophilic carbon monoxide with the intermediate N-acyliminium ion as the electrophilic species [3]. The immediate mechanistic vicinity of a wide range of powerful α-aminoalkylations [11] in combination with the inherent transition metal catalysis clearly accounts for the continued interest that revolves around the amidocarbonylation reaction. The dual nature of the underlying mechanism is illustrated in Scheme 4.

Numerous aldehydes have been used successfully in amidocarbonylation for the synthesis of natural and unnatural amino acid derivatives, though the cobalt-catalyzed process is limited to α-hydrogen bearing aldehydes and formaldehyde. Diamidocarbonylation with a primary amide and two equivalents of formaldehyde affords N-acylamino diacetic acids, which are of potential use as glufosinate intermediates [12]. Cyclic and secondary amides, such as 2-pyrrolidinone, can only be amidocarbonylated with formaldehyde [13]. Research activities have also been devoted to the optimization of the general procedure, including the use of ligated cobalt catalysts, the addition of co-catalysts, and the development of a milder two-step process. Lin reported on the catalytic performance of various cobalt/ligand

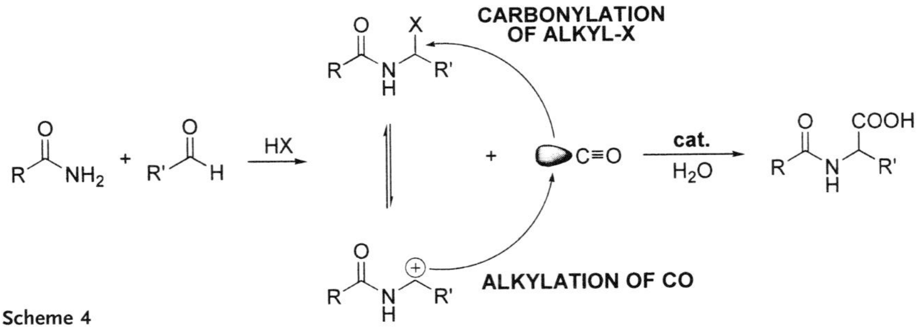

2.5 The Amidocarbonylation of Aldehydes

Scheme 4

systems in the synthesis of N-acetylglycine. Basic phosphines, such as PBu$_3$, were shown to allow low-pressure conditions (55 bar). The addition of Ph$_2$SO or succinonitrile resulted in improved selectivity and facilitated the catalyst recovery [14]. The addition of acid co-catalysts (pK_a < 3, e.g., trifluoromethanesulfonic acid) allowed for low-temperature conditions with higher attendant reaction rates. Moreover, no addition of hydrogen is needed under these conditions (Scheme 5) [15]. A two-step process for the reaction of paraformaldehyde with N-methyl-amides on a > 250 kg scale was developed at Hoechst AG [16]. The N-methyl-α-hydroxymethyl-amide is generated in the presence of acid at 80 °C and subsequently carbonylated (50–70 °C, 10–50 bar CO) under cobalt catalysis to afford the glycine derivative in high yields. The key to this reaction lies in the higher activation energy for the methylol formation and the higher stability and selectivity of the cobalt catalyst at lower temperatures (see sarcosinate synthesis in Scheme 11).

The preparation of ^{13}C-labeled N-acyl-amino acids [17] by amidocarbonylation has also been demonstrated. Intramolecular amidocarbonylations (e.g., to N-benzoyl-pipecolinic acid) significantly broadened the scope of the reaction. In the presence of water-trapping reagents (molecular sieves etc.), oxazolones can be accessed in straightforward manner [9].

Considerable efforts were also directed toward the extension of the procedure to other starting materials that undergo *in situ* transformation to aldehydes. A broad range of olefins, acetals, epoxides, alcohols, and chlorides were demonstrated to be effective alternative starting materials in cobalt-catalyzed amidocarbonylation reactions.

As olefins are ubiquitously available and inexpensive feedstock, the domino hydroformylation-amidocarbonylation provides an interesting direct route to amino acid derivatives [1, 18]. Here, cobalt and rhodium carbonyls and bimetallic complexes were shown to effectively catalyze the hydroformylation step [19]. Stern ob-

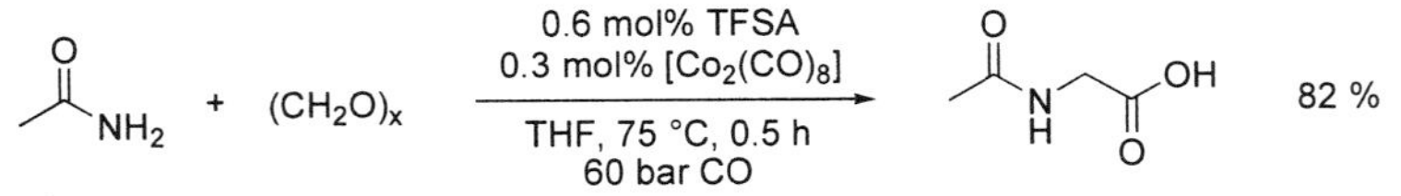

Scheme 5

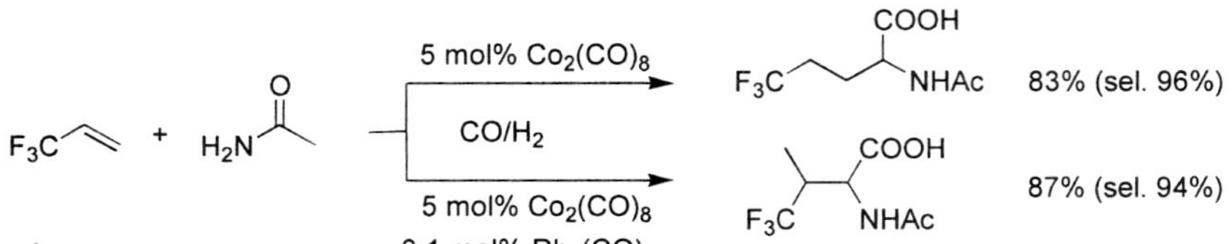

2.5.2 The Cobalt-Catalyzed Amidocarbonylation

Scheme 6

tained 2-(acetylamino)tetradecanoic acid from dodecene and acetamide in 73% yield on a >100 g scale. Aspects of n/iso selectivity with trifluoropropene (5 mol% $Co_2(CO)_8$) were investigated in a detailed manner by Ojima (Scheme 6) [19]. The addition of 0.1 mol% $Rh_6(CO)_{16}$ to the cobalt catalyst gave branched N-acetyltrifluorovaline, which indicated that the hydroformylation step governs the regioselectivity of the domino process.

The positive effect of chelating phosphine ligands in domino hydroformylation-amidocarbonylation was demonstrated by Lin and Knifton in reactions with tetradecene. The addition of 1 mol% of 1,3-bis(diphenylphosphinyl)propane (dppp) to 2 mol% of [$Co_2(CO)_8$] enhanced the yield of N-acetyl-α-aminohexadecanoic acid by roughly 30% [14]. Interesting cyano, polyoxyethylene, and O-acetyl functionalized N-acylamino acid derivatives can be obtained from functionalized olefins (Table 1).

Tab. 1 Functionalized α-amino acids from olefins

Olefin	Amidocarbonylation product	Yield (%)	Application
 PhtN~~	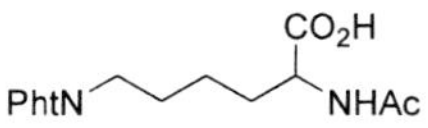 PhtN~~~(CO2H)NHAc	76	Lysine [20]
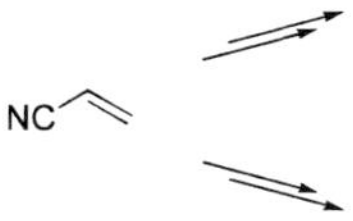	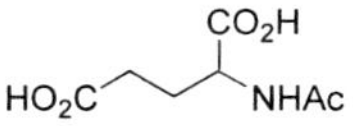 HO2C~(CO2H)NHAc	n.d.	Glutaminic acid [9]
NC~	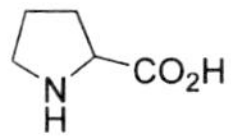 pyrrolidine-CO2H	69	Proline [9]
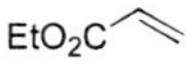 EtO2C~	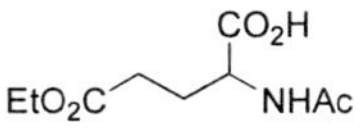 EtO2C~(CO2H)NHAc	85	Glutamate [21]
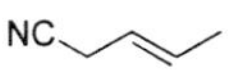 NC~~	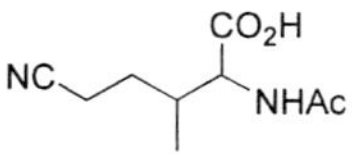 NC~(CO2H)NHAc	85	Polyamide [14a]
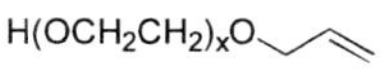 H(OCH2CH2)xO~~	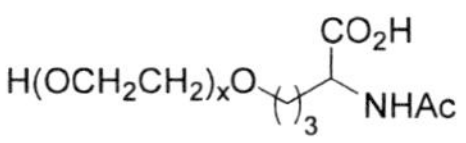 H(OCH2CH2)xO-(}3-(CO2H)NHAc	n.d.	Surfactant [14a]
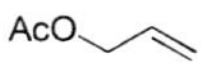 AcO~~	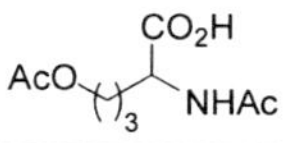 AcO-(}3-(CO2H)NHAc	85	Polyamide ester [14a]

Diamidocarbonylation products may also be synthesized in moderate yields from terminal diolefins [14].

Employment of acetals, as masked aldehyde equivalents, cleanly affords the amino acids or, under exclusion of water, the corresponding esters [1b]. The use of acetals is of particular advantage if other functional groups in the starting material require protection, for example in intramolecular amidocarbonylations (Scheme 7).

The *in situ* generation of the aldehyde can be achieved by a preceding rearrangement of epoxides and allyl alcohols. Ojima demonstrated the use of styrene oxide and propene oxide in the presence of [Ti(O*i*Pr)$_4$] or [Al(O*i*Pr)$_3$] as co-catalysts [19a]. This variant of amidocarbonylation proceeds via N-acyl-α-alkoxyamines. The transition metal-mediated isomerization of allylic alcohols [HRh(CO)(PPh$_3$)$_3$, Fe$_2$(CO)$_9$, RuCl$_2$(PPh$_3$)$_3$, PdCl$_2$(PPh$_3$)$_2$] was also shown to be compatible with amidocarbonylation conditions (Scheme 8) [22]. Alcohols that form stable carbonium ions can be hydrocarbonylated to the corresponding aldehyde under oxo conditions. Yukawa synthesized N-acetyl-O-methyltyrosine from 4-methoxybenzylalcohol in 50% yield. Cyclopropylmethanol was converted to N-acetyl-3-cyclopropylalanine in similar manner [23].

N-α-Alkoxyalkylamides, intermediates in the amidocarbonylation of acetals, have been carbonylated to interesting N-acyl-amino acids. Ojima prepared bicyclic N-α-ethoxyamides by rhodium-catalyzed hydroformylation in the presence of triethyl orthoformate, which were carbonylated to the corresponding ethyl esters in good or moderate yields. The corresponding N-acylamino carboxylic acids could also be obtained directly in a cobalt-catalyzed domino hydroformylation-amidocarbonylation reaction (Scheme 9, top). Izawa prepared N-acylamino acid esters in good yields by a two-stage synthesis involving anodic oxidation of cyclic amides and subsequent cobalt-catalyzed carbonylation of the resultant α-methoxyamides. Remarkably high diastereoselectivities were observed in the synthesis of teneraic acid from enantiomerically pure pipecolinic acid (Scheme 9, bottom) [24].

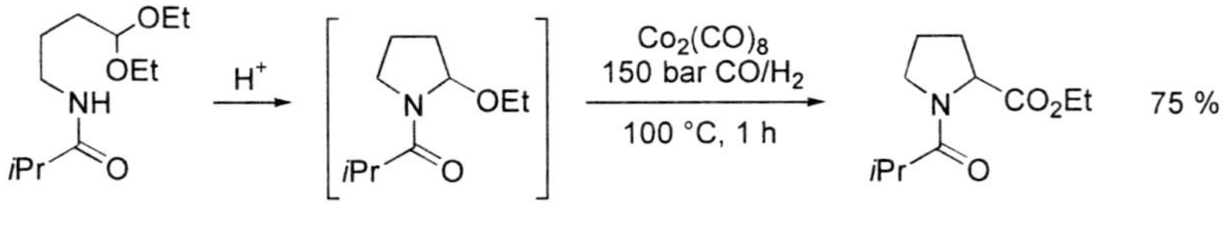

Scheme 7

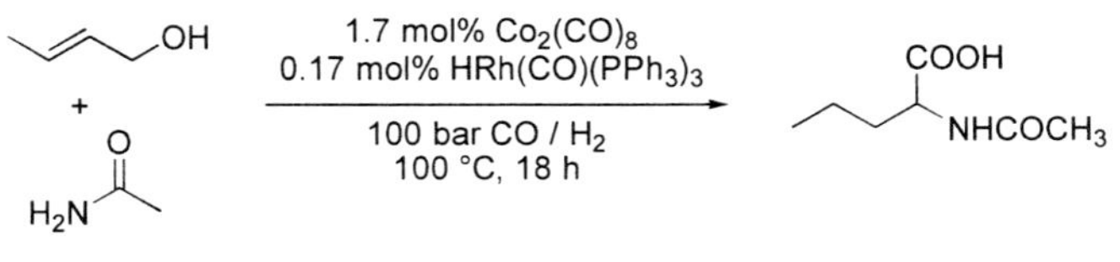

Scheme 8

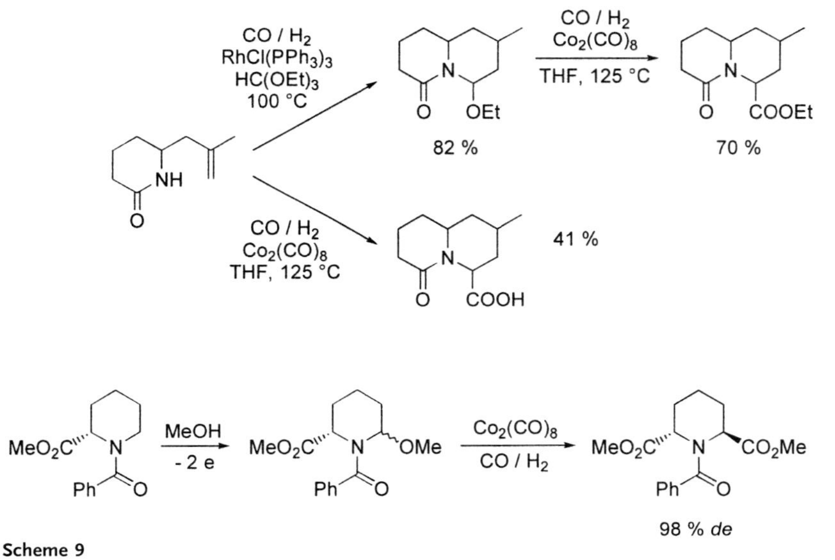

2.5.2 The Cobalt-Catalyzed Amidocarbonylation

Scheme 9

Industrially important applications of the cobalt-catalyzed amidocarbonylation include the preparation of N-acetyl-phenylalanine, sarconisates, and glufosinate. N-Acetyl-D,L-phenylalanine is a key intermediate for the synthesis of aspartame (methyl ester of L-aspartyl-L-phenylalanine). Amidocarbonylation routes for N-acetyl-phenylalanine can start either from phenylacetaldehyde, styrene oxide, or benzyl chloride (Scheme 10). Lewis acid-catalyzed rearrangement of styrene oxide to phenylacetaldehyde followed by amidocarbonylation gives the product in 92% yield [25]. Benzyl chloride is another suitable starting material for the synthesis of N-acetyl-phenylalanine. By the clever selection of the individual reaction parameters, de Vries achieved good yields (up to 79%) of N-acylarylalanines, although high catalyst concentrations (12.5 mol% [$Co_2(CO)_8$]) and pressures (275 bar) were required. Furthermore, the liberated chloride ions deactivate the catalyst [26].

The disadvantages of both processes are currently the low catalyst productivity and space time yields. Currently, L-phenylalanine is produced by tyrosine fermentation. However, an enzymatic aspartame process developed by the Holland Sweetener Company uses D,L-phenylalanine as starting material [24].

Sarcosinates of fatty acids are useful anionic tensides with low hardness sensitivity and good dermatological digestibility. N-Acylsarcosines are manufactured on an annual >10 000 ton scale by Schotten-Baumann reaction of fatty acid chlorides with sarcosines [27]. Although secondary amides generally give lower yields in the amidocarbonylation, N-methyl fatty acid amides smoothly react with paraformaldehyde. Lin obtained N-methyl lauroyl glycine at 200 bar CO/H_2 (3/1), 120 °C in the presence of 3 mol% $Co_2(CO)_8$ in 87% yield [14e]. The two-step pilot plant process developed by Hoechst affords yields in excess of 98% under very mild conditions

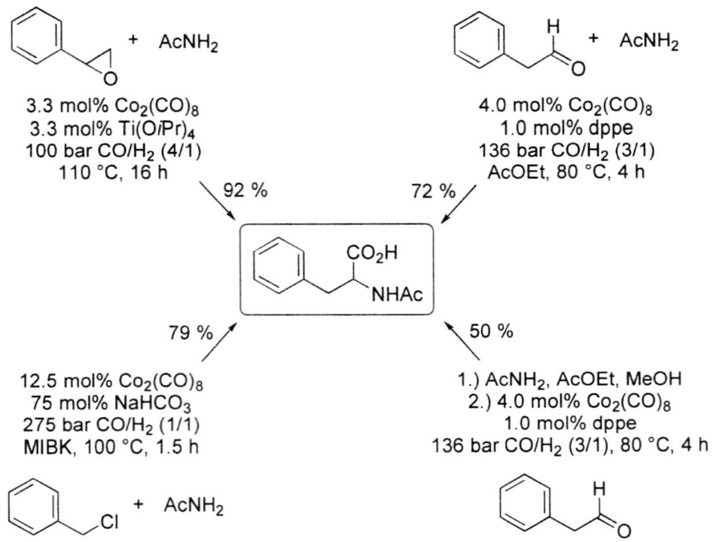

Scheme 10

[Scheme 10 depicts four routes converging on N-acetyl phenylalanine (PhCH2CH(NHAc)CO2H):
- Styrene oxide + AcNH2, 3.3 mol% Co2(CO)8, 3.3 mol% Ti(OiPr)4, 100 bar CO/H2 (4/1), 110 °C, 16 h → 92%
- PhCH2CHO + AcNH2, 4.0 mol% Co2(CO)8, 1.0 mol% dppe, 136 bar CO/H2 (3/1), AcOEt, 80 °C, 4 h → 72%
- PhCH2Cl + AcNH2, 12.5 mol% Co2(CO)8, 75 mol% NaHCO3, 275 bar CO/H2 (1/1), MIBK, 100 °C, 1.5 h → 79%
- PhCH2CHO: 1) AcNH2, AcOEt, MeOH; 2) 4.0 mol% Co2(CO)8, 1.0 mol% dppe, 136 bar CO/H2 (3/1), 80 °C, 4 h → 50%]

(Scheme 11) [16]. The cobalt carbonyl catalyst was regenerated by sequential precipitation (as $Co(OH)_x$), melting, and treatment with synthesis gas (120 bar).

Glufosinate (phosphinotricine), a natural phosphinyl amino acid, exhibits herbicide and antibiotic activity and is marketed as a non-selective herbicide under the trade name BASTA©. Jägers effected conversion of 3-[butoxy(methyl)phosphoryl]-propylaldehyde diethylacetal into the N-acetylamino acid in 79% yield in 1 h. Among the various routes that can be adopted for the synthesis of phosphinotricine via amidocarbonylation [28], the domino hydroformylation-amidocarbonylation of methylvinylphosphinate constitutes a particularly efficient method with 80% overall yield (Scheme 12) [29].

To the best of our knowledge at present, none of the cobalt-catalyzed amidocarbonylation procedures is applied on an industrial scale. Although the production of sarcosinates on the tonne scale demonstrated the industrial viability of the method. However, often unsatisfactory catalytic activity of the catalyst (TON < 100) and relatively harsh conditions (T > 100 °C; p(CO/H2) > 100 bar) limit the possibilities of the application of the method. Hence, there exists interest in the development of more efficient catalysts and conditions.

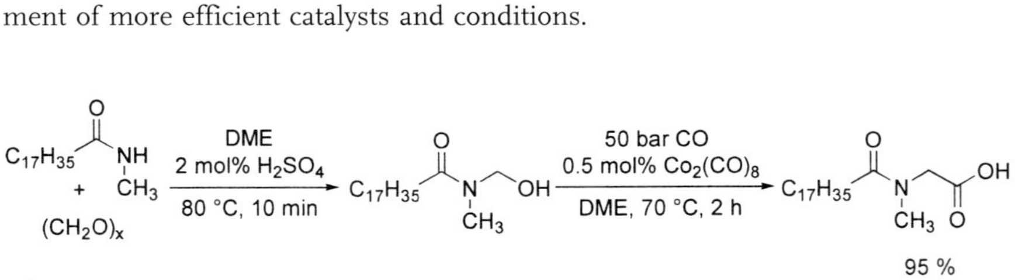

Scheme 11

[Scheme 11: $C_{17}H_{35}C(O)NHCH_3$ + $(CH_2O)_x$ → (DME, 2 mol% H_2SO_4, 80 °C, 10 min) → $C_{17}H_{35}C(O)N(CH_3)CH_2OH$ → (50 bar CO, 0.5 mol% $Co_2(CO)_8$, DME, 70 °C, 2 h) → $C_{17}H_{35}C(O)N(CH_3)CH_2CO_2H$, 95%]

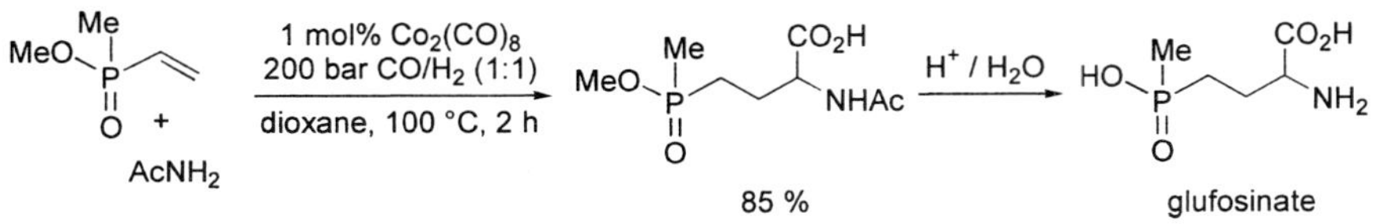

Scheme 12

2.5.3
The Palladium-Catalyzed Amidocarbonylation

In 1987, researchers at Hoechst found that palladium complexes are capable catalysts for the amidocarbonylation of aldehydes [30]. Later on, extensive screening efforts implicated an optimized procedure for the palladium-catalyzed version of this reaction. The presence of a strong acid (1 mol% H_2SO_4) and halide anions (10–35 mol% of LiBr or nBu_4NBr) were proven highly beneficial. The reaction was typically run at 80–120 °C and 30–60 bar CO. Under optimized conditions, the desired N-acyl-α-amino acids could be afforded with TONs of up to 60 000 (TOF >1000 h^{-1}). Both palladium(0) [e.g., $Pd_2(dba)_3$, $Pd(PPh_3)_4$] and palladium(II) complexes [e.g., $PdBr_2$, $Pd(OAc)_2$] were successfully employed as catalyst precursors. Among various dipolar aprotic solvents (DMF, DMAc, MeCN, etc.) that can be used in the palladium-catalyzed amidocarbonylation, high-boiling NMP was shown to provide optimal selectivities (Scheme 13) [31, 32].

As with the cobalt-catalyzed process, the reaction is assumed to commence with the simple condensation of amide and aldehyde (Scheme 14). In recent mechanistic investigations, Kozlowski identified N-acylimine and N-acylenamine species as key intermediates in the palladium-catalyzed amidocarbonylation [33]. In the presence of HX, the resulting equilibrating imine-enamine species are converted to an intermediate α-halo imine (**I**). Oxidative addition of the C(sp^3)-X bond to palladium(0) gives rise to the formation of palladium(II) alkyl complex **II**, which rapidly inserts CO. However, β-hydride elimination of the corresponding enamine from alkyl complex **II** is not competitive under the reaction conditions. Irreversible cleavage of the resulting acyl complex **III** might involve either intramolecular ring closure release to give oxazolones (followed by hydrolysis) or direct intermolecular attack of water onto **III**.

Unlike the reaction under cobalt catalysis, the palladium-catalyzed reaction is not limited to aldehydes containing α-hydrogens. Furthermore, the two metal-catalyzed reactions differ in the presence of halide ions, which is essential under palladium catalysis in order to facilitate the oxidative addition of **I** to palladium(0). However, the palladium catalysts are more tolerant to various functional groups.

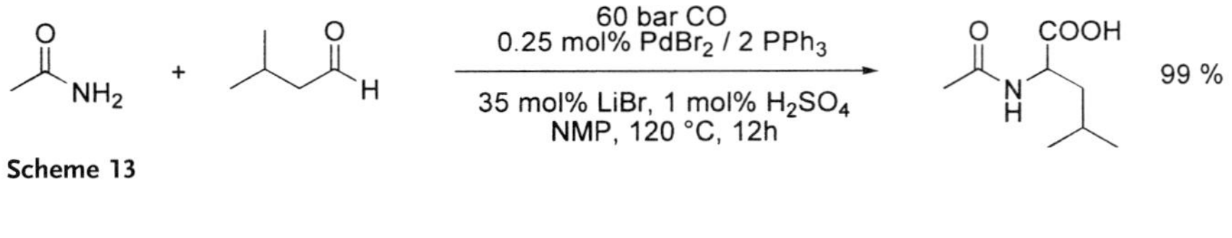

Scheme 13

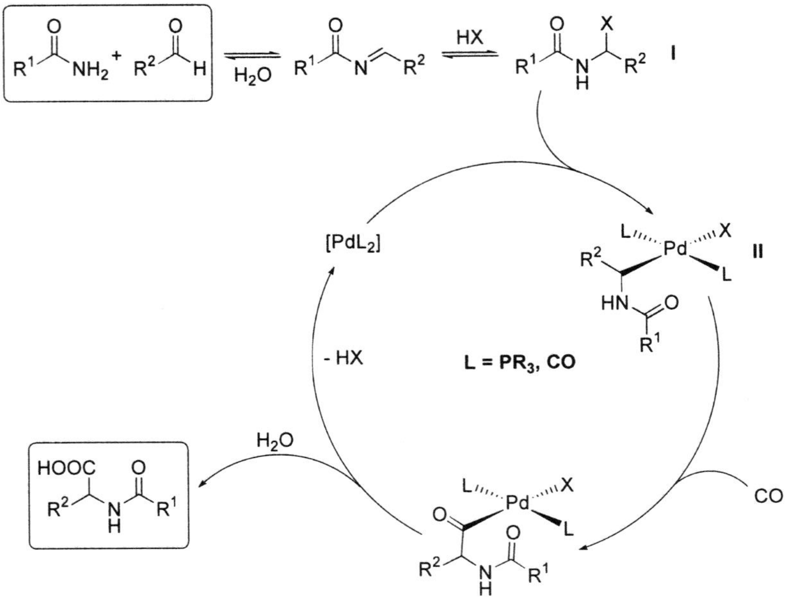

2.5 The Amidocarbonylation of Aldehydes

Scheme 14

By comparison, the palladium-catalyzed amidocarbonylation outperforms the cobalt-catalyzed version by a factor of 10–100 in terms of catalyst activity. Under optimal conditions, catalyst turnover numbers of 60 000 (TOF >1000 h^{-1}) can be achieved in the synthesis of *N*-acetylleucine [32c].

Among the numerous synthetic applications of palladium-catalyzed amidocarbonylation reactions, those that are not amenable to cobalt catalysis or have not otherwise been described will be especially highlighted. Important applications of the palladium-catalyzed variant include the synthesis of hydantoins from substituted ureas as amide equivalents [34], the amidocarbonylation of commercially more attractive nitriles [35], and the preparation of arylglycines [36].

Hydantoins possess high significance in amino acid production as well as in pharmaceutical research as low molecular weight *N*-heterocycles [37]. Complementing the classic Bucherer-Bergs multicomponent reaction [38], the palladium-catalyzed ureidocarbonylation constitutes a three-component (urea, aldehyde, CO) synthesis of substituted hydantoins (Table 2). Unlike the Bucherer-Bergs reaction, the ureidocarbonylation reaction provides access to hydantoins containing diverse substituents in the 1-, 3-, and 5-positions with good selectivities [34]. With monosubstituted ureas, 3-substituted hydantoins are obtained.

With regard to raw material costs, the amide components are the cost-determining factor in most cases. Commercially more attractive nitriles, which can be viewed as amide equivalents, thus constitute an interesting alternative. The combination of the amidocarbonylation protocol with the *in situ* transformation of ni-

Tab. 2 Ureidocarbonylation of aldehydes

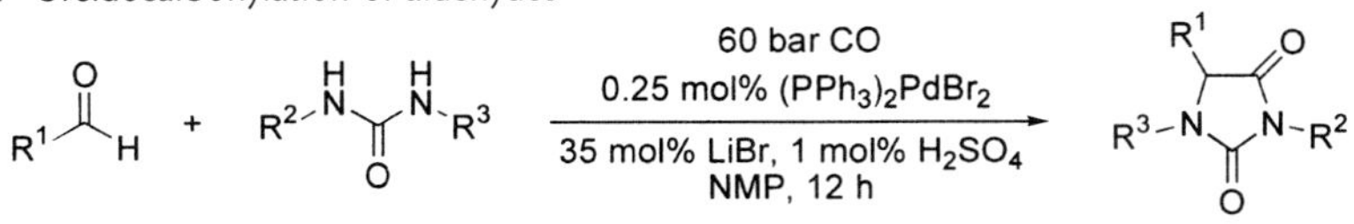

$$R^1CHO + R^2NH-C(O)-NHR^3 \xrightarrow[\text{35 mol\% LiBr, 1 mol\% H}_2\text{SO}_4]{\text{60 bar CO, 0.25 mol\% (PPh}_3\text{)}_2\text{PdBr}_2} \text{hydantoin}$$

	R^1	R^2	R^3	T (°C)	Yield (%)	TON
1	Ph	Me	H	80	75	300
2	Cy	Et	H	100	51	204
3	Cy	Ph	H	100	64	256
4	Ph	Bn	H	100	50	200
5	Cy	Et	Et	100	89	356
6	i-Bu	Me	Me	120	61	244
7c	H	Ph	Ph	130	93	372
8	H	Me	Me	100	73	292
9	Ph	Me	Me	100	85	340
10	m-ClC$_6$H$_4$	Me	Me	100	79	316

triles to amides [35] significantly extends the scope of the reaction. The one-pot amidocarbonylation of nitriles can be performed via preceding nitrile hydrolysis in conc. sulfuric acid [39] or by passing a stream of HCl through a nitrile/formic acid solution [40] and was shown to afford the desired N-acyl-amino acids in good yields (Scheme 15).

Highly potent antibiotics such as the cyclic glycopeptides vancomycin, β-avoparcin, and chloropeptin contain an N-acyl-α-arylglycine motif [41], and thus make this class of compounds an attractive synthetic target. An atom-efficient synthesis by transition metal-catalyzed amidocarbonylation has until now proved elusive, since the classical cobalt catalysts require the presence of α-hydrogen atoms. Systematic studies have shown that amidocarbonylation of benzaldehydes in the presence of palladium catalysts allows for the synthesis of functionalized, racemic N-acetyl-α-arylglycines (Table 3) [36]. Generally, electron-rich benzaldehydes react faster than those with electron-withdrawing substituents, though useful yields could be achieved in all cases.

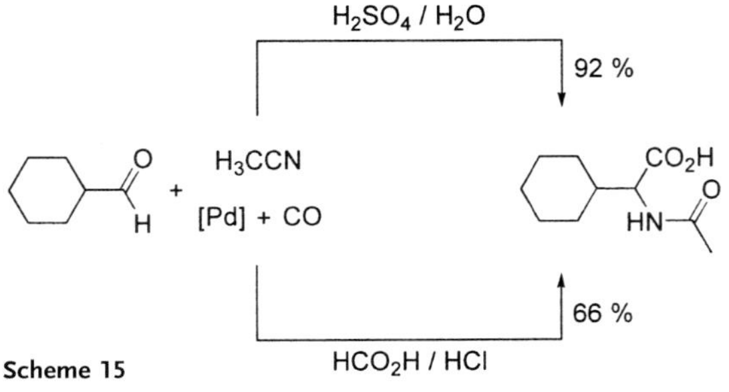

Scheme 15

Tab. 3 N-Acetyl-α-arylglycines from palladium-catalyzed amidocarbonylation

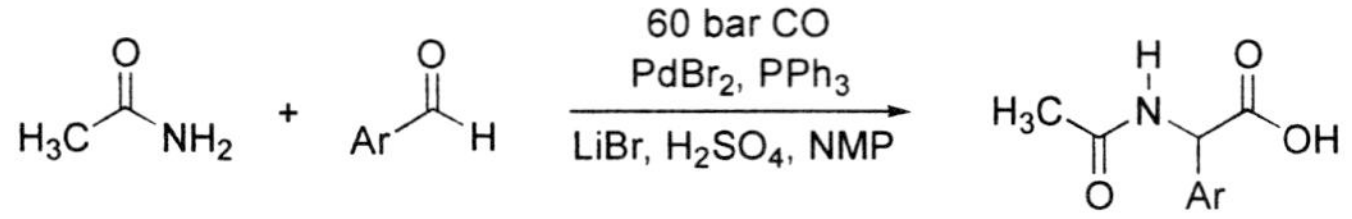

$$H_3C-C(O)-NH_2 + Ar-CHO \xrightarrow[\text{LiBr, H}_2\text{SO}_4\text{, NMP}]{\substack{\text{60 bar CO} \\ \text{PdBr}_2\text{, PPh}_3}} H_3C-C(O)-NH-CH(Ar)-COOH$$

	Ar	T (°C)	t (h)	Yield (%)	TON
1	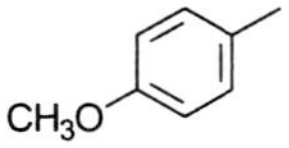 4-CH$_3$O-C$_6$H$_4$	100	15	75	300
2	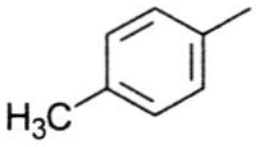 4-H$_3$C-C$_6$H$_4$	120	12	95	380
3	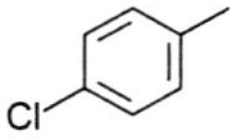 4-Cl-C$_6$H$_4$	100	15	65	260
4	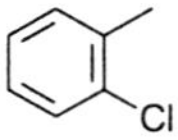 2-Cl-C$_6$H$_4$	100	15	56	224
5	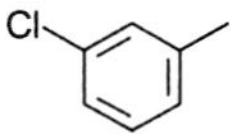 3-Cl-C$_6$H$_4$	100	15	63	252
6	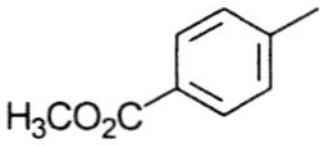 4-H$_3$CO$_2$C-C$_6$H$_4$	120	12	89	356
7	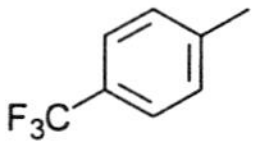 4-F$_3$C-C$_6$H$_4$	120	12	82	328
8	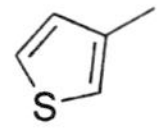 2-thienyl	125	60	42	168

The relevance of palladium-catalyzed amidocarbonylation for natural product synthesis was demonstrated by the multi-gram scale preparation of the central amino acid of chloropeptin I [(S)-3,5-dichloro-4-hydroxyphenylglycine] via the combination of amidocarbonylation and enzymatic hydrolysis [42]. This synthesis is three steps shorter than literature procedures for similar derivatives, and documents the advantage of the two-stage combination of palladium-catalyzed amidocarbonylation and enantioselective, enzymatic hydrolysis with acylases (Table 4) [43].

Amidocarbonylations with secondary amides are, in general, more difficult. However, N-alkyl- and N-aryl-N-acyl-amino acids can be prepared in moderate yields with Pd/C catalysts. The reaction of Paracetamol© with paraformaldehyde gave N-acetyl-N-(4-hydroxyphenyl)glycine in 70% yield [44]. N-Substituted glycines constitute important building blocks of peptoids [45].

Recent screening experiments by our group implicated an optimized set of conditions for palladium-catalyzed amidocarbonylation. For the first time it was dem-

Tab. 4 Domino amidocarbonylation – enzymatic hydrolysis

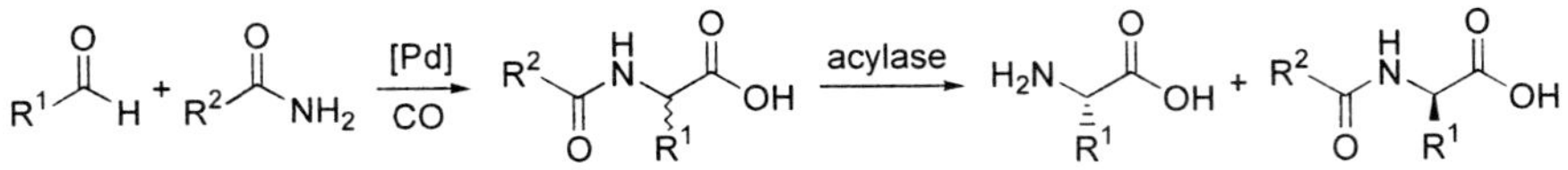

Amidocarbonylation				Enzymatic hydrolysis		
(R,S)-N-Acyl-amino acid		Yield (%)	TON	Enzyme	(S)-Amino acid Yield (ee) (%)	(R)-N-Acyl-aminoacid Yield (ee) (%)
R^1	R^2					
p-ClBn	Me	75	300	AA	40 (>99)	46 (94)
MeSC$_2$H$_4$	Me	75	300	PKA	32 (>99)	40 (86)
Cy	Bn	83	332	PA	38 (>99)	49 (94)
Cy	MeOCH$_2$	83	332	PKA	44 (99)	47 (97)

AA: Aspergillus spp. Acylase; PKA: pig liver acylase.

onstrated that palladium-catalyzed amidocarbonylations of aldehydes can be run with significantly lower halide concentrations (<20 mol%) without a major yield decrease. While phosphine-free catalyst systems give best yields at low CO pressure, phosphine-ligated palladium catalysts lead to better yields at higher CO pressure. At low palladium loadings (<0.1 mol%), unwanted condensation reactions of aldehydes and amides become increasingly competitive (Scheme 16) [46].

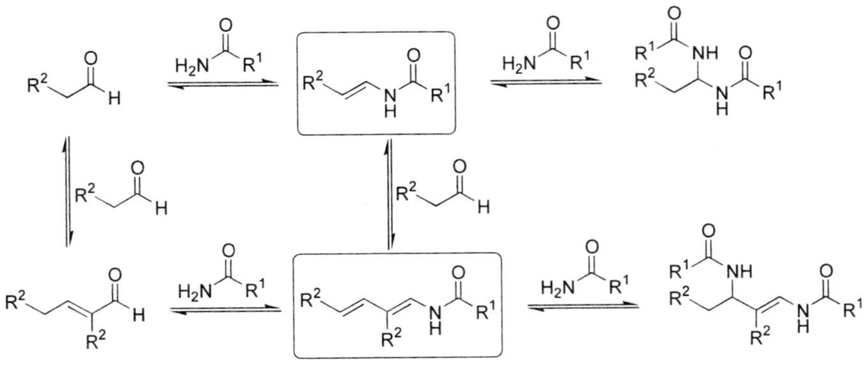

Scheme 16

2.5.4
Outlook

Despite significant progress in the design of more efficient catalysts and conditions for the transition metal-catalyzed amidocarbonylation of aldehydes, the classical Strecker reaction is still the benchmark reaction when it comes to industrial applications. Nevertheless, the amidocarbonylation route has potential advantages over the Strecker method, which makes it interesting for further elaboration. Amidocarbonylation benefits from cheap carbon monoxide as starting material. The resulting N-acyl-amino acids can be subjected to efficient enzymatic racemic resolution strategies to obtain the corresponding deacylated amino acids in enantiomerically pure form. Such strategy saves two reaction steps compared to the Strecker route. In our opinion, the combined chemo- and biocatalytic two-step sequence of amidocarbonylation-enzymatic resolution constitutes the most direct and efficient approach to enantiomerically pure amino acids on >100 g scales. Clearly, future challenges for amidocarbonylations include optimization of the reaction conditions (reduction of catalyst amounts, low-pressure-low-temperature conditions), extension to other reactants (higher functional group tolerance), the use of amines and ammonia instead of amides, and the development of an efficient asymmetric procedure by using chiral ligands.

2.5.5
References

1 (a) WAKAMATSU, H.; UDA, J.; YAMAKAMI, N.; *Chem. Commun.* **1971**, 1540. (b) WAKAMATSU, H.; UDA, J.; YAMAKAMI, N. DE-B 2115 985, **1971**. (c) WAKAMATSU, H. *Kagaku* **1989**, 44, 448.
2 (a) PINO, P.; PARNAUD, J.-J.; CAMPARI, G.; *J. Mol. Catal.* **1979**, 6, 341. For reviews, see: (b) KNIFTON, J.F. in *Applied Homogeneous Catalysis with Organometallic Compounds* (Eds.: CORNILS, B.; HERRMANN, W.A.), VCH, Weinheim, **1996**, 159. (c) OJIMA, I. *Chem. Rev.* **1988**, 88, 1011. (d) OJIMA, I. *J. Mol. Catal.* **1986**, 37, 25.
3 A related metal-free amidocarbonylation protocol (Koch carbonylation) using sulfuric acid was concurrently developed but never found further applications because of the severe reaction conditions and its limitation to glycine derivatives: WITTE, H.; SEELIGER, W., *Liebigs Ann. Chem.* **1972**, 755, 163.
4 SZMANT, H.H. *Organic Building Blocks for Chemical Industry*, Wiley, New York, **1989**.
5 DRAUZ, K.; WALDMANN, H. *Enzyme Catalysis in Organic Synthesis*, VCH, Weinheim, **1995**.
6 DYKER, G. in *Organic Synthesis Highlights IV* (Ed.: SCHMALZ, H.-G.), Wiley-VCH, Weinheim, **2000**, 53.
7 OJIMA, I.; ZHANG, Z. *Organometallics* **1990**, 9, 3122.
8 (a) MAGNUS, P.; SLATER, M. *Tetrahedron Lett.* **1987**, 28, 2829. (b) ZAUGG, H.E. *Synthesis* **1970**, 49.
9 IZAWA, K. *Yuki Gosei Kagaku Kyokaishi* **1988**, 46, 218.
10 COLQUHOUN, H.M.; THOMPSON, D.J.; TWIGG, M.V. *Carbonylation*, Plenum, New York, **1991**.
11 HIEMSTRA, H.; SPECKAMP, W.N. in *Comprehensive Organic Synthesis*, Vol. 2 (Eds.: TROST, B.M.; FLEMING, I.), Pergamon, Oxford, **1991**, 1007. (b) KARSTENS, W.F.J.; KLOMP, D.; RUTJES, F.P.J.T.; HIEMSTRA, H. *Tetrahedron* **2001**, 57, 5123.

12 STERN, R.; REFFET, D.; HIRSCHAUER, A.; COMMEREUC, D.; CHAUVIN, Y. *Synth. Commun.* **1982**, *12*, 1111.

13 LIN, J.J. (Texaco Inc.), US-A 4 620 949, **1986**.

14 (a) LIN, J.J., KNIFTON, J.F. *J. Organomet. Chem.* **1991**, *417*, 99. (b) KNIFTON, J.F.; LIN, J.J.; STORM, D.A.; WONG, S.F. *Catalysis Today* **1993**, *18*, 355. (c) LIN, J.J.; KNIFTON, J.F.; YEAKEY, E.L. (Texaco Inc.), US 4918222, **1987**.

15 DRENT, E.; KRAGTWIJK, E. (Shell Int. Research), GB 2252770, **1991**.

16 (a) BELLER, M.; FISCHER, H.; GROSS, P.; GERDAU, T.; GEISSLER, H.; BOGDANOVIC. S. (Hoechst AG), DE-B 4415712, **1995**. (b) BOGDANOVIC, S.; GEISSLER, H.; BELLER, M.; FISCHER, H.; RAAB K. (Hoechst AG), DE-B 19545641 A1, **1995**.

17 YUAN, S.S.; AJAMI, A.M. *J. Labelled Compd. Radiopharm.* **1985**, *22*, 1309.

18 (a) WAKAMATSU, H. *Sekiyu Gakkaishi* **1974**, *17*, 105. (b) STERN, R.; HIRSCHAUER, A.; COMMEREUC, D.; CHAUVIN, Y. (Institut Francaise du Petrole), US 4264515, **1981**.

19 (a) OJIMA, I.; HIRAI, K.; FUJITA, M.; FUCHIKAMI, T. *J. Organomet. Chem.* **1985**, *279*, 203. (b) OJIMA, I.; OKABE, M.; KATO, K.; KWON, H.B.; HORVATH, I.T. *J. Am. Chem. Soc.* **1988**, *110*, 150.

20 AMINO, Y.; IZAWA, K. *Bull. Chem. Soc. Jpn.* **1991**, *64*, 613.

21 LIN, J.J. (Texaco Inc.), US 4720573, **1988**.

22 HIRAI, K.; TAKAHASHI, Y.; OJIMA, I. *Tetrahedron Lett.* **1982**, *23*, 2491.

23 (a) YUKAWA, T.; YAMAKAMI, N.; HOMMA, M.; KOMACHIYA, Y.; WAKAMATSU, H. (Ajinomoto Co., Inc.), JP 4985011, **1974**. (b) AMINO, Y.; IZAWA, K. *Bull. Chem. Soc. Jpn.* **1991**, *64*, 1040.

24 (a) SHONO, T.; MATSUMURA, Y.; TSUBATA, K. *J. Am. Chem. Soc.* **1981**, *103*, 1172. (b) SHONO, T.; MATSUMURA, Y.; TSUBATA, K. *Tetrahedron Lett.* **1981**, *22*, 2411.

25 LIN, J.J.; KNIFTON, J.F. *Catal. Lett.* **1997**, *45*, 139.

26 DE VRIES, J.G.; DE BOER, R.P.; HOGEWEG, M.; GIELENS, E.E.C.G. *J. Org. Chem.* **1996**, *61*, 1842.

27 (a) MIKHALKIN, A.P. *Russ. Chem. Rev.* **1995**, *64*, 259. (b) WIELAND, T. in *Methoden der Organischen Chemie (Houben-Weyl)* Vol. XI/2, 4th edn., **1958**, 305. (c) GREENSTEIN, J.P.; WINITZ, M. *Chemistry of the Amino Acids*, Krieger, Malabar, **1961**, 1831, 2375.

28 (a) JÄGERS, E.; BÖSHAR, M.; KLEINER, H.-J.; KOLL, H.-P. (Hoechst AG), DE 3913891, **1990**. (b) JÄGERS, E.; ERPENBACH, H.; KOLL, H.-P. (Hoechst AG), DE 3823885, **1990**. (c) JÄGERS, E.; ERPENBACH, H.; BYLSMA, F. (Hoechst AG), DE 3823886, **1990**.

29 (a) TAKIGAWA, S.; SHINKE, S.; TANAKA, M. *Chem. Lett.* **1990**, 1415. (b) SAKAKURA, T.; HUANG, X.-Y.; TANAKA, M. *Bull. Chem. Soc. Jpn.* **1991**, *64*, 1707.

30 JÄGERS, E.; KOLL, H.-P. (Hoechst AG), EP-B 0.338.330 B1, **1989**.

31 BELLER, M.; ECKERT, M. *Angew. Chem. Int. Ed.* **2000**, *39*, 1010.

32 (a) BELLER, M.; ECKERT, M.; VOLLMÜLLER, F.; GEISSLER, H.; BOGDANOVIC, S. (Hoechst AG), DE-B 19627717, **1996**. (b) BELLER, M.; ECKERT, M.; VOLLMÜLLER, F.; BOGDANOVIC, S.; GEISSLER, H. *Angew. Chem. Int. Ed.* **1997**, *36*, 1494. (c) BELLER, M.; ECKERT, M.; VOLLMÜLLER, F. *J. Mol. Catal.* **1998**, *135*, 23.

33 FREED, D.A.; KOZLOWSKI, M.C. *Tetrahedron Lett.* **2001**, *42*, 3403.

34 BELLER, M.; ECKERT, M.; MORADI, W.; NEUMANN, H. *Angew. Chem. Int. Ed.* **1999**, *38*, 1454.

35 BELLER, M.; ECKERT, M.; MORADI, W.A. *Synlett.* **1999**, 108.

36 BELLER, M.; ECKERT, M.; HOLLA, E.W. *J. Org. Chem.* **1998**, *63*, 5658.

37 (a) SYLDATK, C.; MÜLLER, R.; SIEMANN, M.; KROHN, K.; WAGNER, F. in *Biocatalytic Production of Amino Acids and Derivatives* (Eds.: ROZZELL, J.D.; WAGNER, F.), Hanser, Munich, **1992**, 75. (b) SYLDATK, C.; MÜLLER, R.; PIETZSCH, M.; WAGNER, F. in *Biocatalytic Production of Amino Acids and Derivatives* (Eds.: ROZZELL, J.D.; WAGNER, F.), Hanser, Munich, **1992**, 129.

38 (a) WARE, E. *Chem. Rev.* **1950**, *46*, 403. (b) BUCHERER, H.T.; STEINER, W. *J. Prakt. Chem.* **1934**, *140*, 291.

39 (a) HENECKA, H.; KURTZ, P. in *Methoden der Organischen Chemie (Houben-Weyl)*, Vol. 8, 4th ed., **1952**, 654. (b) SEELIGER, W.; HESSE, K.-D. (Hüls AG), US-A 3846419, **1974**.

40 (a) Becke, F.; Gnad, J. *Justus Liebigs Ann. Chem.* **1968**, *713*, 212. (b) Becke, F.; Fleig, H.; Pässler, P. *Justus Liebigs Ann. Chem.* **1971**, *749*, 198.

41 (a) Williams, D. H.; Searle, M. S.; Westwell, M. S.; Mackay, J. P.; Groves, P.; Beauregard, D. A. *Chemtracts-Organic Chemistry* **1994**, *7*, 133. (b) Rao, A. V. R.; Gurjar, M. K.; Reddy, K. L.; Rao, A. S. *Chem. Rev.* **1995**, *95*, 2135. (c) Matsuzaki, K.; Ikeda, H.; Ogino, T.; Matsumoto, A.; Woodruff, H. B.; Tanaka, H.; Omura, S. *J. Antibiotics* **1994**, *47*, 1173. (d) Roussi, G.; Zamora, E. G.; Carbonnelle, A.-C.; Beugelmans, R. *Tetrahedron Lett.* **1997**, *38*, 4401.

42 Beller, M.; Eckert, M.; Geissler, H.; Napierski, B.; Rebenstock, H.-P.; Holla, E.-W. *Chem. Eur. J.* **1998**, *4*, 935.

43 (a) Chenault, H. K.; Dahmer, J.; Whitesides, G. M. *J. Am. Chem. Soc.* **1989**, *111*, 6354. (b) Verkhovskaja, M. A.; Yamskov, I. A. *Russ. Chem. Rev.* **1991**, *60*, 1163.

44 Beller, M.; Moradi, W. A.; Eckert, M. *Tetrahedron Lett.* **1999**, *40*, 4523.

45 Kessler, H. *Angew. Chem. Int. Ed.* **1993**, *32*, 543.

46 Gördes, D.; Neumann, H.; Jacobi von Wangelin, A.; Fischer, C.; Drauz, K.; Krimmer, H.-P.; Beller, M. *Adv. Synth. Catal.* **2003**, *345*, 510.

2.6
Transition Metal-catalyzed Alkene and Alkyne Hydrocyanations

Albert L. Casalnuovo and T. V. RajanBabu

2.6.1
Introduction

Organonitriles are key intermediates for a variety of businesses including polymers, fibers, agrochemicals, cosmetics, and pharmaceuticals. In principle, the transition metal-catalyzed hydrocyanation of alkenes or alkynes offers a direct and economical way to produce such organonitriles (Eq. 1). DuPont's production of the nylon-6,6 precursor ADN (adiponitrile) from 1,3-butadiene, for example, represents a very successful commercial application of this reaction. As shown in Eqs. (2) and (3), a triarylphosphite nickel complex hydrocyanates butadiene in a two-step process to give an overall anti-Markovnikov addition of HCN. This process, which is used to produce over 1 billion pounds of ADN annually, was made possible during the 1960s when Drinkard and co-workers discovered that zero-valent nickel phosphite complexes catalyze the hydrocyanation of unactivated alkenes [1–3]. Although this remains the only commercial application of this catalyst system, this discovery sparked several pivotal mechanistic studies on transition metal-catalyzed reactions [4–12] and continues to offer exciting opportunities for ligand/catalyst design.

$$RCH=CH_2 + HCN \longrightarrow RCH_2CH_2CN \quad (1)$$

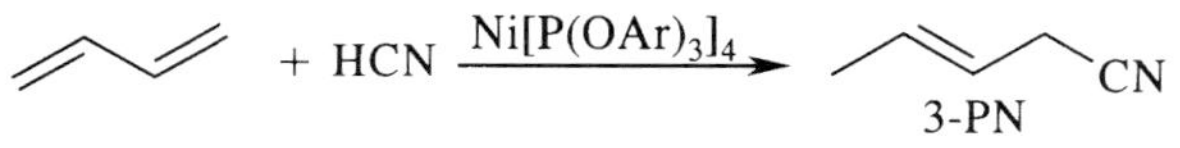

$$\text{butadiene} + HCN \xrightarrow{Ni[P(OAr)_3]_4} \text{3-PN} \quad (2)$$

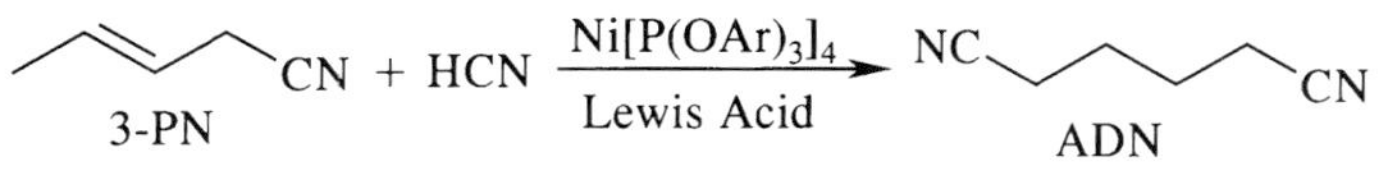

$$\text{3-PN} + HCN \xrightarrow[\text{Lewis Acid}]{Ni[P(OAr)_3]_4} \text{ADN} \quad (3)$$

2.6.2
Alkene Hydrocyanation

Some of the earliest work on alkene hydrocyanation reported the use of catalysts such as dicobalt octacarbonyl [13] or copper(I) salts [14]. For example, catalysis

Transition Metals for Organic Synthesis, Vol. 1, 2nd Edition.
Edited by M. Beller and C. Bolm
Copyright © 2004 WILEY-VCH Verlag GmbH & Co. KGaA, Weinheim
ISBN: 3-527-30613-7

with dicobalt octacarbonyl effected the hydrocyanation of simple terminal and internal alkenes, styrene, conjugated dienes, and cyclopentadiene Diels–Alder adducts. Unlike the nickel-catalyzed process (*vide infra*), the major products corresponded to an overall Markovnikov addition of HCN. Unfortunately, relatively high reaction temperatures were needed (130 °C) and relatively low catalyst turnover was obtained (1–8 mol nitrile per mol Co). Since the discovery of the nickel-catalyzed process, little further work on these catalysts has been reported.

The scope and mechanism of the nickel-catalyzed hydrocyanation of alkenes has been extensively described [5, 15]. Although a comprehensive review will not be included here, some relevant details are provided as background. The typical hydrocyanation catalysts are zero-valent triarylphosphite nickel complexes, $Ni(P(OAr)_3)_x$ ($x=3, 4$; $Ar=Ph$ or 2-, 3-, or 4-tolyl). These catalysts tolerate a fairly broad range of substrates and functional groups. However, alkenes conjugated with electron-withdrawing groups (e.g. nitriles, esters, aldehydes) are usually poor substrates and can poison the catalyst. Fortunately, these activated alkenes can usually be hydrocyanated by traditional base-catalyzed methods.

Conjugated dienes are readily hydrocyanated and the resulting allylic nitriles are produced by reductive elimination from stable and often detectable η^3-allyl nickel cyanides (e.g. **1**). In the case of the ADN process, this reductive elimination step is reversible and the nickel-catalyzed equilibration of the regioisomeric allylic nitriles is achieved at higher temperatures (Eq. 4). Isolated dienes (e.g. 1,5-cyclooctadiene) give products typical of conjugated dienes when alkene isomerization is available through the classic insertion/β-hydride elimination mechanism. Vinylarenes, similar to conjugated dienes, give rise to η^3-benzyl nickel cyanides. After reductive elimination of NiL_2 as a result, HCN tends to add to vinylarenes in an overall Markovnikov addition to yield a preponderance of the branched nitrile [5, 16].

$$\diagup\!\!\!\diagup \xrightarrow[\text{HCN}]{NiP_4} \underset{\mathbf{1}}{NiP_2CN} \;\rightleftharpoons\; \diagup\!\!\!\diagup\!\!\!\diagdown_{CN} + \underset{NC}{\diagup\!\!\!\!\diagdown}\diagup\!\!\!\diagup \quad (4)$$

Hydrocyanation of isolated alkenes usually requires the addition of catalytic amounts of Lewis acids (e.g. $ZnCl_2$, $AlCl_3$, BR_3) and tends to give an overall anti-Markovnikov addition. Internal alkenes hydrocyanate much slower than terminal alkenes and often give a preponderance of linear nitriles when alkene isomerization is possible. The Lewis acids (L.A.) promote alkene isomerization by increasing the concentration of cationic nickel hydride isomerization catalysts (Eq. 5).

$$NiP_xH(CN) + L.A. \rightleftharpoons [NiP_xH]^+ + [NC-L.A.]^- \quad (5)$$

Loss of catalytic activity is observed if the zero-valent nickel hydrocyanation catalysts undergo irreversible oxidation by a second equivalent HCN. This oxidation is enhanced by Lewis acids (Eq. 6). The overall oxidation is believed to be second order in HCN concentration, thus maintaining a low HCN concentration is crucial for high catalyst turnover.

$$NiP_x + 2HCN \rightarrow Ni(CN)_2 + H-2 + P \tag{6}$$

2.6.3
Alkyne Hydrocyanation

2.6.3.1
Nickel Phosphite-catalyzed Reactions

Zero-valent nickel arylphosphite complexes and, to a lesser extent, zero-valent palladium phosphine or phosphite complexes will catalyze the addition of hydrogen cyanide to alkynes [17–21]. For example, nickel-catalyzed alkyne hydrocyanation has been used in the preparation of β- and γ-amino acid derivatives (Eq. 7) [18] and as a route to α-alkylidene γ-lactones [17].

$$R-\!\!\!\equiv\!\!\!-\!\!\!\diagdown_{Ft} \xrightarrow[\text{Ni[P(OPh)}_3]_4]{\text{HCN}} RCH=C(CN)CH_2Ft \longrightarrow \longrightarrow RCH_2CH(CO_2H)CH_2NH_2$$

Ft = phthalimido
(7)

The mechanism of the reaction is thought to be similar to alkene hydrocyanation, including a stereospecific *cis* addition of HCN to the alkyne [21]. Studies by Jackson and Perlmutter have shown that the overall regioselectivity of the addition is strongly dependent on the steric properties of the alkyne substituents (other than H) and whether the substrate is a terminal alkyne. Terminal alkynes generally favor the formation of branched nitriles, even when relatively bulky R_3Si alkyne substituents are present. However, both *tert*-butyl- and phenylacetylene have been reported to give an excess of the terminal nitrile. The hydrocyanation of internal alkynes favors branched nitriles in which the nitrile group is attached to the least hindered carbon.

2.6.3.2
$Ni(CN)_4^{2-}$-catalyzed Reactions

Funabiki *et al.* have reported an intriguing catalyst for alkyne hydrocyanation that does not utilize organophosphorus stabilized nickel or palladium complexes [22]. Based on earlier work involving the stoichiometric hydrocyanation of alkynes with $[Co(CN)_5H]^{3-}$, Funabiki *et al.* developed the $Ni(CN)_4^{2-}$-catalyzed hydrocyanation of alkynes without the use of hydrogen cyanide. As shown in Eq. (8), treatment of the alkyne with KCN, a reducing agent such as Zn or $NaBH_4$, and a catalytic amount of Ni(CN)$_4^{2-}$ effects alkyne hydrocyanation and subsequent hydrogenation of the unsaturated nitrile.

$$R^1\!\!\equiv\!\!R^2 + KCN_{(aq)} \xrightarrow[\text{NaBH}_4 \text{ or Zn}]{\text{Ni(CN)}_4^{2-}} R^1(NC)C\!=\!CHR^2 \longrightarrow R^1CH(CN)CH_2R^2 \qquad (8)$$

The hydrocyanation/hydrogenation is applicable to both terminal and internal alkynes although symmetrical alkynes were surprisingly unreactive. Branched nitriles are strongly favored by this catalyst system but, unlike the nickel phosphite catalyzed process, the regioselectivity is relatively insensitive to substituent steric effects. Deuterium labeling studies showed that borohydride was the hydrogen atom source for the hydrocyanation whereas the solvent provided the hydrogen source for the alkene hydrogenation. Other attractive features of this system include the use of an air- and moisture-stable catalyst precursor, the use of the cyanide ion as both a reagent and a ligand for the catalyst, and the ease of separating an aqueous catalyst solution from the product and substrate. One apparent drawback of this system is the low catalyst turnover (≤8 mol RCN per mol Ni).

2.6.3.3
Addition of R$_3$SiCN

An interesting and potentially powerful variation of the hydrocyanation reaction is the addition of silyl cyanides to unsaturated carbon–carbon bonds. Chatani et al. have reported the palladium- and nickel-catalyzed addition of trimethylsilyl cyanide to terminal alkynes [23]. Terminal aryl alkynes were the most reactive substrates and gave exclusively branched nitriles in the presence of catalytic amounts of PdCl$_2$ and pyridine (Eq. 9). Highly selective *syn* additions (~95% Z isomer) were observed for *para*- or *meta*-substituted aryl alkynes, whereas less stereoselectivity (~80% Z isomer) was observed with *ortho*-substituted alkynes. Surprisingly, when the reactions were carried out without solvent in the presence of excess trimethylsilyl cyanide and a zero-valent nickel catalyst, 5-amino-pyrroles were obtained in 55–65% yield (Eq. 10) [23a, 24]. The mechanism of this latter reaction is not well understood but results from the reaction of 3 equivalents of TMSCN with the alkyne. Terminal aliphatic alkynes generally behaved similarly to the aryl alkynes but the yields and stereoselectivity were somewhat lower.

$$Ar\!\!\equiv\!\!H + TMSCN \xrightarrow[\text{Toluene}]{\text{PdCl}_2/\text{Py}} \underset{NC\quad TMS}{\overset{Ar\quad H}{\diagup\!\!=\!\!\diagdown}} \qquad (9)$$

$$Ar\!\!\equiv\!\!H + 3\ TMSCN \xrightarrow{\text{NiCl}_2/i\text{-Bu}_2\text{AlH}} \underset{NC\ \ N\ \ N(TMS)_2}{\overset{Ar}{\text{pyrrole}}} \qquad (10)$$

Allenes also proved to be reactive in the presence of the PdCl$_2$/pyridine catalyst and afforded vinylsilanes in which silicon was bound to the central carbon of the allene (Eq. 11) [25]. To our knowledge similar additions of silyl cyanides to isolated alkenes have not been reported.

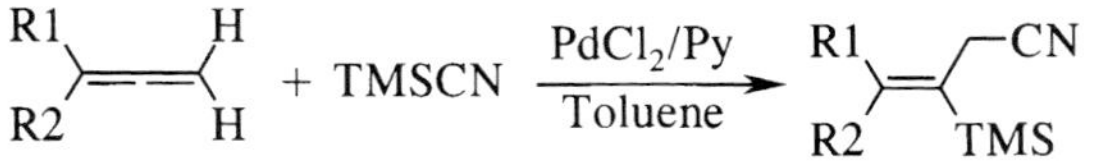

$$\underset{R2}{\overset{R1}{\diagup}}=\underset{H}{\overset{H}{\diagdown}} + \text{TMSCN} \xrightarrow[\text{Toluene}]{\text{PdCl}_2/\text{Py}} \underset{R2}{\overset{R1}{\diagup}}=\underset{\text{TMS}}{\overset{\text{CN}}{\diagdown}} \quad (11)$$

2.6.4
New Directions in Nickel-catalyzed Alkene Hydrocyanation

2.6.4.1
New Ligands

Most of the new developments in this field have centered on the design of new ligands to improve upon the reaction characteristics of the nickel triarylphosphite catalysts or to induce asymmetry (i.e. asymmetric catalysis) in the product nitriles (Sect. 2.6.4.2). For example, Pringle and co-workers reported a nickel catalyst containing the chelating phosphite ligand **2** derived from 2,2′-biphenol [26]. In this system, the total catalytic turnover (mol nitrile per mol Ni) for butadiene hydrocyanation was four times greater than that of the commercial Ni(P(Otol)$_3$)$_4$ catalyst. Bidentate phosph*ine* ligands with large bite angles, such as **3**, are reasonably active and selective catalysts for the Markovnikov hydrocyanation of styrene [27]. Phosphine nickel complexes are typically poor hydrocyanation catalysts; however, the authors attribute the nearly tetrahedral bite angle of the ligand (106°) to the enhanced catalytic activity.

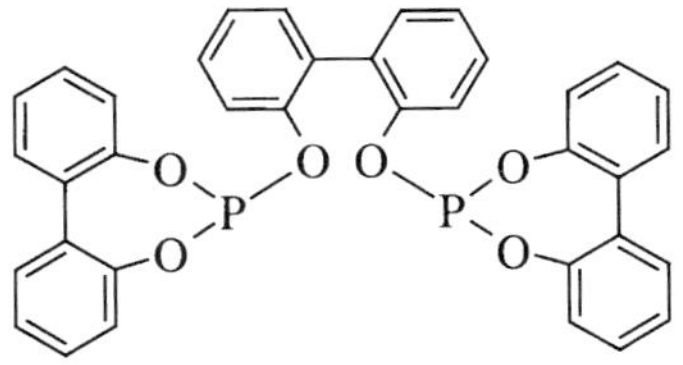

2.6.4.2
Catalytic Asymmetric Hydrocyanation

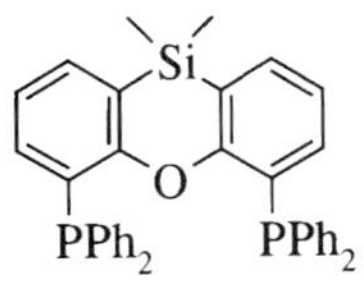

Most of the early studies on the transition metal-catalyzed asymmetric hydrocyanation of alkenes focused on the asymmetric hydrocyanation of norbornene or its derivatives. Although high facial selectivity was observed, only modest enantioselectivities and yields were obtained [28]. To date, the highest *ee* (enantiomeric excess) reported for this substrate class is 48% obtained using a BINAPHOS palladium complex (Eq. 12) [29].

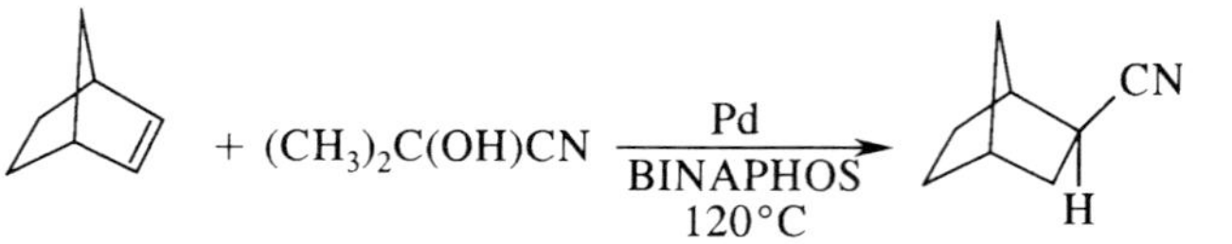

$$\text{norbornene} + (CH_3)_2C(OH)CN \xrightarrow[120°C]{Pd \atop BINAPHOS} \text{norbornyl-CN} \quad (12)$$

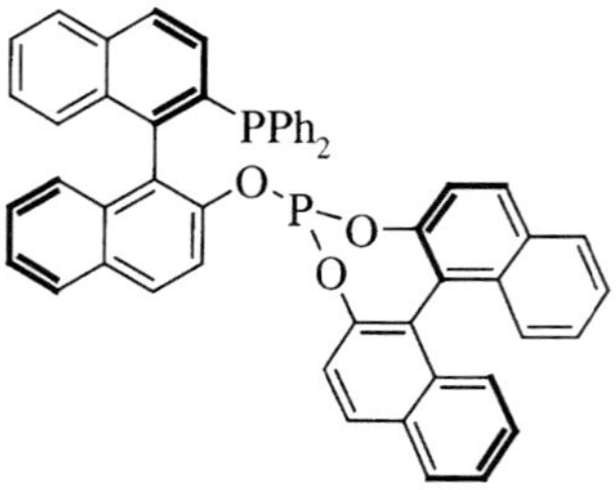

R,S-BINAPHOS

The highest enantioselectivities have been obtained for the asymmetric hydrocyanation of vinylarenes using carbohydrate-based phosphinite nickel catalysts [30]. The asymmetric, Markovnikov hydrocyanation of these substrates gives rise to useful precursors for optically active profen drugs (Eq. 13) [16, 30]. Using the 2,3-disubstituted glucodiarylphosphinite ligand **4**, for example, the Naproxen precursor MVN (6-methoxy-2-vinylnaphthalene) was hydrocyanated with complete regioselectivity in 85–91% *ee* (*S*) (Eq. 14). This catalyst system was remarkably active, giving maximum reaction rates of 2000 turnovers h^{-1} (turnover = mol nitrile per mol Ni) and 700–800 total turnovers at room temperature.

$$ArCH=CH_2 \xrightarrow{HCN} ArCH(CN)CH_3 \xrightarrow{H_2O} ArCH(CO_2H)CH_3 \quad (13)$$

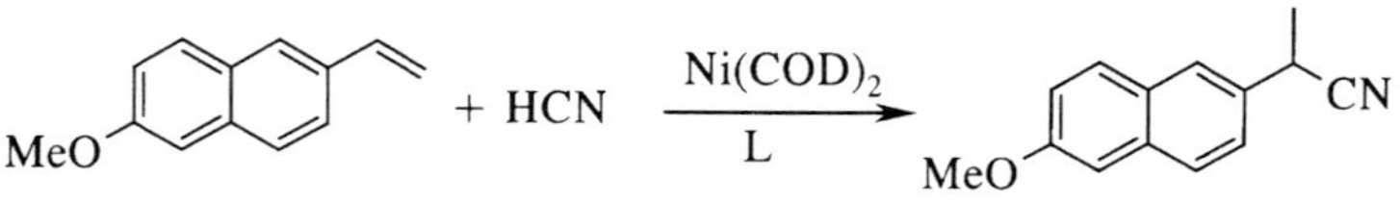

MVN + HCN $\xrightarrow[L]{Ni(COD)_2}$ MeO-naphthyl-CH(CN)CH₃

4

A key finding of this work was the importance of ligand electronic effects. A study of the effect of the phosphorus–aryl substituents showed a pronounced increase in the enantioselectivity as the electron-withdrawing power of the substituent increased. For example, the MVN hydrocyanation *ee* increased from 16 to 78% as the *meta* substituents in **4** were varied in the series Me, H, F, CF$_3$ ($\sigma_{m\bullet}$ = –0.07, 0, 0.34, 0.43, respectively). On the other hand, ligand *electronic asymmetry* proved to be important in the fructose-based phosphinite ligand system **5** [30e,f]. In this case, the electronic differentiation of the two phosphorus sites was used to maximize the enantioselectivity. Thus, the incorporation of a more electron-withdrawing phosphorus-aryl group in the 4-hydroxy position than in the 3-hydroxy position was crucial to obtaining the highest *ee*'s (e.g. X=3,5-$(CF_3)_2$, Y=H, 89% *ee*; X=3,5-$(CF_3)_2$, Y=3,5-$(CF_3)_2$, 56% *ee*; X=H, Y=3,5-$(CF_3)_2$, 58% *ee*).

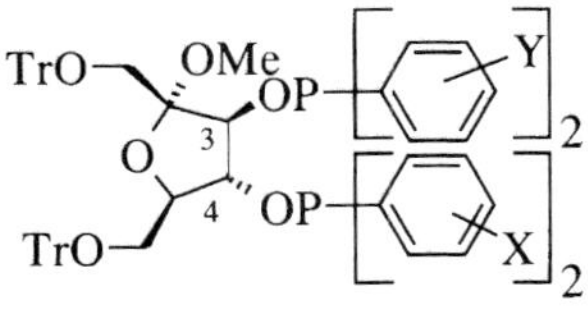

5

Another intriguing feature of these carbohydrate-based hydrocyanation catalysts is that the site of phosphorus substitution on the carbohydrate ring controls the sense of product chirality (i.e. *R* vs. *S*). In the glucophosphinite ligands **6** and **7**, phosphorus substitution at the 2,3-hydroxy positions leads to a predominance of the *S* nitrile whereas substitution at the 3,4-hydroxy positions leads to the *R* nitrile. The 3,4-substituted fructophosphinites **5** have the same local diol chirality as the 3,4-glucophosphinites and thus produce an excess of the *R* nitrile. Thus a judicious choice of the appropriate carbohydrate, the site of phosphorus substitution, and the ligand electronic elements gave either the *R* or *S* nitrile from MVN in 85% *ee*. Notably, these ligand control elements appear to have validity in other asymmetric reactions, such as the rhodium-catalyzed hydrogenation of alkenes [31]. In this case, electron-*donating* phosphorus aryl substituents dramatically enhanced the enantioselectivity of the hydrogenation reaction.

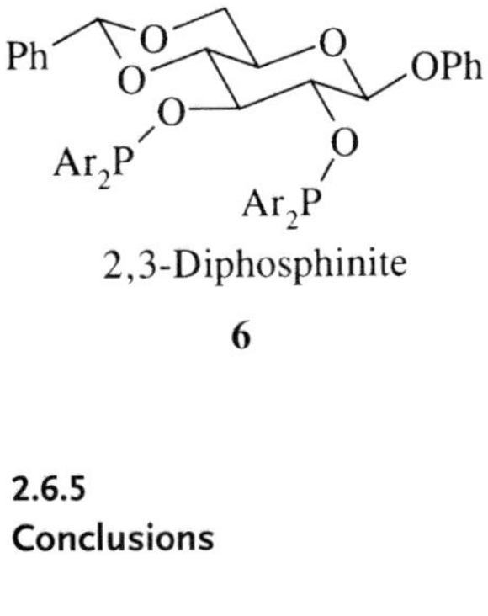

2,3-Diphosphinite
6

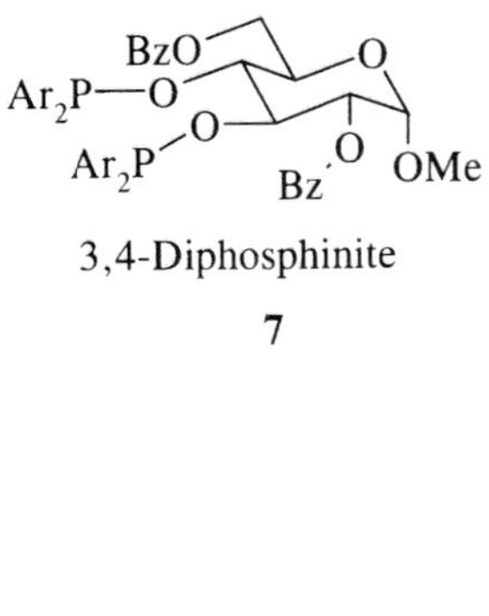

3,4-Diphosphinite
7

2.6.5
Conclusions

Transition metal-catalyzed hydrocyanations provide a direct route from alkenes or alkynes to a number of functional groups including nitriles, amines, aldehydes, and carboxylic acids. Further developments in ligand and catalyst design will undoubtedly lead to new, more active and selective catalysts. In particular, there is tremendous potential for the development of new hydrocyanation catalysts for asymmetric catalysis. Although the inherent danger of working with hydrogen cyanide has probably discouraged many chemists from using this reaction [32], transition metal-catalyzed hydrocyanations should continue to be a useful tool for the practicing synthetic chemist.

2.6.6
References

1. W. C. Drinkard, R. V. J. Lindsey, US Pat., 3,496,215, **1970**.
2. W. C. Drinkard, US Pat., 3,496,217, **1970**.
3. W. C. Drinkard, R. J. Kassal, US Pat., 3,496,217, **1970**.
4. C. A. Tolman, *J. Am. Chem. Soc.* **1970**, *92*, 2953–2956.
5. C. A. Tolman, R. J. McKinney, W. C. Seidel, J. D. Druliner, W. R. Stevens, *Advances in Catalysis*, **1985**, *33*, 1–46.
6. O. S. Andell, J. E. Bäckvall, *Organometallics* **1986**, *5*, 2350–2355.
7. C. A. Tolman, W. C. Seidel, J. D. Druliner, P. J. Domaille, *Organometallics* **1984**, *3*, 33–38.
8. C. A. Tolman, *J. Am. Chem. Soc.* **1970**, *92*, 4217–4222.
9. C. A. Tolman, *J. Am. Chem. Soc.* **1970**, *92*, 2956–2965.
10. J. D. Druliner, A. D. English, J. P. Jesson, P. Meakin, C. A. Tolman, *J. Am. Chem. Soc.* **1976**, *98*, 2156–2160.
11. C. A. Tolman, W. C. Seidel, *J. Am. Chem. Soc.* **1974**, *96*, 2774–2780.
12. C. A. Tolman, *J. Am. Chem. Soc.* **1974**, *96*, 2780–2789.
13. P. J. Arthur, D. C. England, B. C. Pratt, G. M. Whitman, *J. Am. Chem. Soc.* **1954**, *76*, 5364–5367.
14. W. A. Schulze, J. A. Mahan, US Pat., 2,422,859, **1947**.
15. A. L. Casalnuovo, R. J. McKinney, C. A. Tolman in *The Encyclopedia of Inorganic Chemistry*, John Wiley, New York, **1994**.
16. W. A. Nugent, R. J. McKinney, *J. Org. Chem.* **1985**, *50*, 5370–5372.
17. W. R. Jackson, P. Perlmutter, A. J. Smallridge, *J. Chem. Soc., Chem. Commun.* **1985**, 1509–1510.
18. W. R. Jackson, P. Perlmutter, A. J. Smallridge, *Tetrahedron Lett.* **1988**, *29*, 1983–1984.
19. W. R. Jackson, C. G. Lovel, *Aust. J. Chem.* **1983**, *36*, 1975–1982.
20. G. D. Fallon, N. J. Fitzmaurice, W. R. Jackson, P. Perlmutter, *J. Chem. Soc., Chem. Commun.* **1985**, 4–5.
21. W. R. Jackson, C. G. Lovel, *J. Chem. Soc., Chem. Commun.* **1982**, 1231–1232.
22. T. Funabiki, H. Sato, N. Tanaka, Y. Yamazaki, S. Yoshida, *J. Mol. Catal.* **1990**, *62*, 157–169.

2.7
Cyclopropanation

Andreas Pfaltz

2.7.1
Introduction

Cyclopropanes are versatile intermediates in organic synthesis that can be converted to a variety of useful products by cleavage of the strained three-membered ring [1]. There are also numerous natural and synthetic cyclopropane derivatives with interesting physiological activities [2]. Therefore, great efforts have been made to develop efficient stereoselective methods for the synthesis of cyclopropanes [3]. In particular, the cyclopropanation of olefins with diazo compounds has received considerable attention and during the last two decades. Efficient homogeneous metal catalysts have been found which have strongly enhanced the scope of this reaction [4]. By careful selection of a specific metal–ligand combination, the catalyst properties can be adjusted to the specific requirements of a particular application and, moreover, the use of chiral ligands makes it possible to carry out such transformations enantioselectively [5, 6]. In addition to cyclopropanation, the metal-catalyzed decomposition of diazo compounds can also result in other synthetically useful processes such as insertion into C–H and other X–H bonds, or ylide formation [4]. While metal-catalyzed reactions of olefins with diazo compounds are now well established and widely used, promising developments have also become apparent in other areas, such as catalytic Simmons–Smith reactions. This chapter reviews the principal catalytic methods for cyclopropanation with special emphasis on enantioselective transformations.

2.7.2
Metal-catalyzed Decomposition of Diazo Compounds

Among the many different transition metal compounds that are known to catalyze the extrusion of dinitrogen from diazo compounds, the most general and most widely used catalysts are copper and dinuclear rhodium(II) complexes. Although the catalytic cycle and the structures of the intermediates are not known in detail, there is ample evidence that metal carbene complexes are involved as short-lived intermediates [4, 7]. Attempts to detect these elusive carbenoid species have not been successful. However, ruthenium and osmium carbene complexes have recently been isolated from stoichiometric reactions with diazoacetates and shown to be active catalysts for cyclopropanation and olefin formation [8, 9].

Palladium complexes are also efficient catalysts. They differ considerably in mechanism and scope from Cu and Rh complexes [4, 10, 11]. Whereas Cu and Rh catalysts are best suited to the cyclopropanation of electron-rich olefins with α-diazocarbonyl compounds, Pd complexes are the catalysts of choice for reactions with diazoalkanes and electron-deficient C–C double bonds or strained alkenes [11].

The most versatile reagents are α-diazocarbonyl compounds such as diazoacetates because they are readily prepared, stable, and easily handled. Besides diazo compounds, other carbene precursors such as iodonium ylids [12], sulfonium ylids [13], or lithiated alkylsulfones [14] have been also used but, at present, the scope of these reagents is still limited [4c].

The impressive selection of new transition metal catalysts available today allows the chemist to solve many problems of chemo-, regio-, and stereoselectivity [4–6]. By proper choice of the catalyst, it is often possible to differentiate between competing reaction pathways based on steric or electronic properties of the reactants or by making use of neighboring group effects. Most notable advances have been made in the enantiocontrol of cyclopropanation and C–H insertion with chiral Cu and Rh catalysts (see Sections 2.7.3 and 2.7.4). More recently, promising results have also been obtained with chiral Ru(II) complexes [8], whereas attempts to develop chiral Pd catalysts for enantioselective cyclopropanation have not been successful so far.

2.7.3
Enantioselective Cyclopropanation with Copper Catalysts

The development of chiral Cu catalysts for enantioselective cyclopropanation was initiated by the pioneering work of Nozaki et al. in the mid 1960s [15]. Subsequent systematic optimization of the ligand structure in the research group of Aratani at Sumitomo Co. [16] resulted in a dramatic improvement of enantioselectivity (Scheme 1). The catalyst **2** gave ee's of >90% in the reaction of diazoacetates with trisubstituted olefins such as **1** and allowed the development of an industrial process for the production of 2,2-dimethylcyclopropanecarboxylic acid from iso-butene. This product, which is formed with high ee, is a precursor of cilastatin, an enzyme inhibitor used as a drug for suppressing the *in vivo* degradation of the antibiotic iminipenem.

A new class of chiral ligands, C_2-symmetric semicorrins, was introduced in 1986 [17]. Of the various derivatives, the semicorrin **3** with two bulky substituents at the stereogenic centers was found to be the most effective ligand. Using the stable crystalline Cu(II) complex **4** as a catalyst precursor, ee's of >90% could be achieved in the cyclopropanation of terminal and disubstituted olefins (Scheme 2) which exceeded the enantioselectivities reported for catalyst **2** (Scheme 1). With trisubstituted olefins, on the other hand, Aratani's catalyst **2** is more effective [16]. There is ample evidence that the active catalyst is a mono(semicorrinato)copper(I)

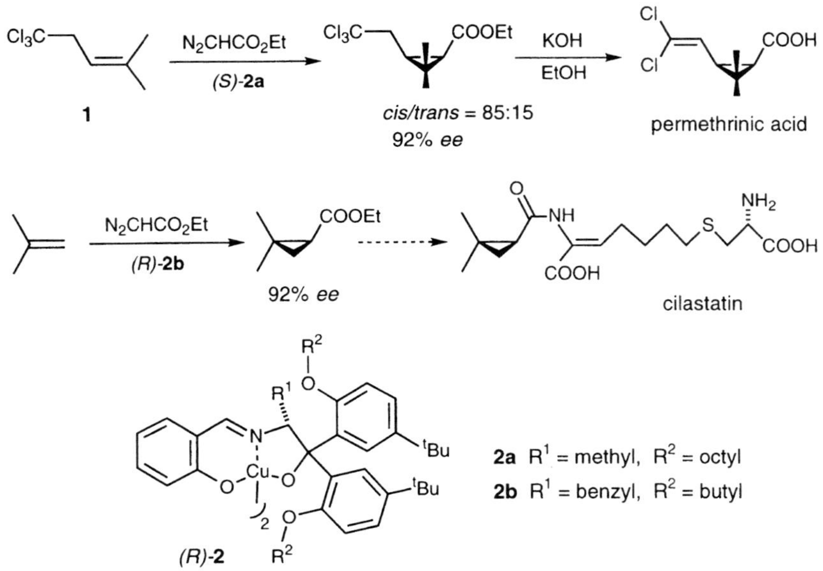

2.7.3 Enantioselective Cyclopropanation with Copper Catalysts

Scheme 1 The Sumitomo process for enantioselective cyclopropanation developed by Aratani et al. [16].

complex which is generated from complex 4 either by heating in the presence of the diazo compound or by treatment with phenylhydrazine at ambient temperature. Cyclopropanation with diazomethane was also briefly investigated [17b]. Using catalyst 4 and (E)-1-phenylpropene or methyl cinnamate as substrates, selectivities of 70–75% ee were obtained.

The development of structurally related ligands such as 5, 6, and 7 led to even more selective catalysts [18–20]. The cationic Cu(I) complex prepared from the bisoxazoline 7 and CuOTf is the most efficient catalyst available today for the cyclopropanation of terminal olefins with diazoacetates (Scheme 2) [19]. Evans et al., who developed this catalyst, achieved >99% ee in the reaction of ethyl diazoacetate with iso-butene using substrate/catalyst ratios as high as 1000:1. For some trisubstituted and 1,2-disubstituted (Z)-olefins, Lowenthal and Masamune found the bisoxazoline 6 to be superior [20]. This is illustrated by the reaction of 2,5-dimethyl-2,4-hexadiene leading to chrysanthemates. The enantioselectivities in this case were comparable to the ee's obtained with Aratani's catalyst.

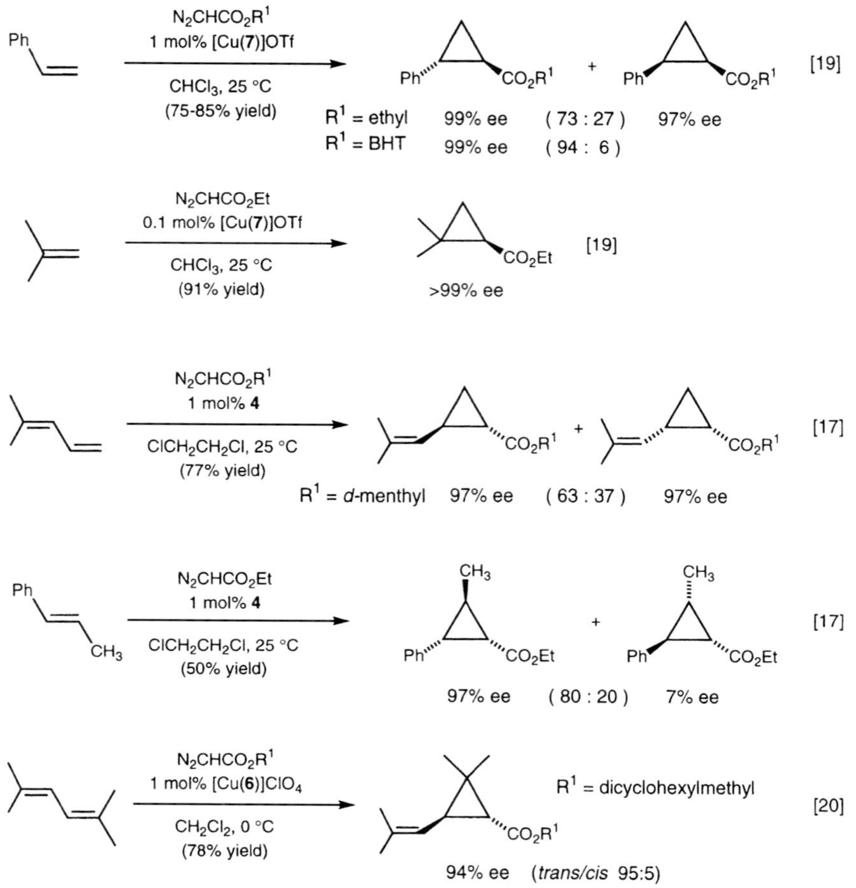

2.7 Cyclopropanation

Scheme 2 Enantioselective cyclopropanation with copper complexes of semicorrins and bisoxazolines.

With all Cu catalysts, the *trans/cis* selectivities in the cyclopropanation of monosubstituted olefins are only moderate. The *trans/cis* ratio depends mainly on the structure of the diazo ester rather than the chiral ligand. As shown in the reaction of styrene (Scheme 2), more bulky ester groups favor the formation of *trans* product. With 1,2-disubstituted olefins, on the other hand, the chiral ligand also influences the *trans/cis* ratio. The enantio- and *trans/cis*-selectivities observed with (semicorrinato)copper complexes and related catalysts can be rationalized by a structural model of the postulated copper–carbene intermediate [17, 18c].

Catalyst **4** was also employed in intramolecular cyclopropanation reactions of alkenyl diazoketones (Scheme 3) [21]. The enantioselectivities were strongly dependent on the substitution pattern of the C=C bond and varied between 94% ee for **9a** and 14% ee for the corresponding dimethyl-substituted analog **9b**. Interestingly, analogous cyclizations of allyl and homoallyl diazoacetates gave disappointingly low ee's. For this class of substrates, Doyle's dinuclear rhodium complexes are, in general, more efficient catalysts (see Section 2.7.4). Recently, Shibasaki and co-workers reported an example of a highly selective intramolecular cyclopropanation reaction of a silyl enol ether [22]. The most effective ligand in this case was a bisoxazoline of type **7** in which the two *tert*-butyl substituents were replaced by

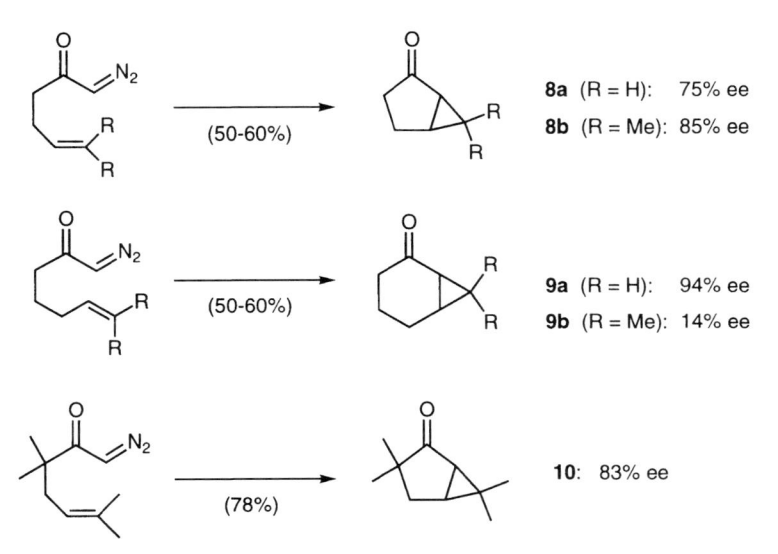

8a (R = H): 75% ee
8b (R = Me): 85% ee

9a (R = H): 94% ee
9b (R = Me): 14% ee

10: 83% ee

Scheme 3 Intramolecular cyclopropanation with catalyst **4**. Conditions: 3 mol% of catalyst, activation with phenylhydrazine, 1,2-dichloroethane, 23 °C [21].

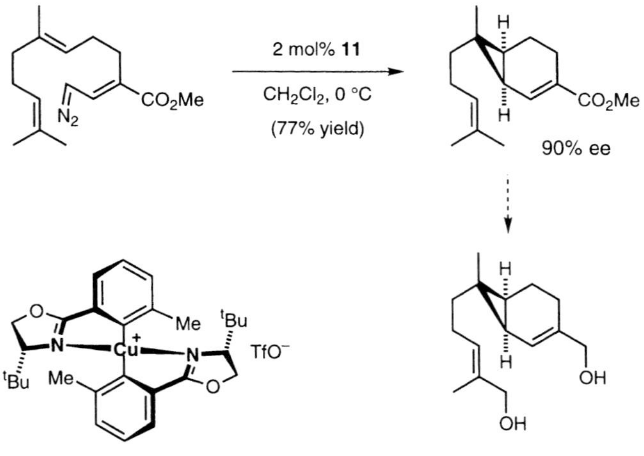

2.7 Cyclopropanation

Scheme 4 Synthesis of sirenin **11** by intramolecular cyclopropanation [23].

(Me$_3$SiO)Me$_2$C groups. Corey and co-workers [23] have developed an interesting new bisoxazoline ligand with a biphenyl backbone which was successfully applied in the key-step of the synthesis of the chemotactic factor sirenin (Scheme 4). The crystalline copper complex **11** emerged as the most effective catalyst for this reaction after extensive screening of a series of different Cu and Rh complexes. A variety of other ligands have been tested in copper-catalyzed cyclopropanations [5b], however, none of them offer real advantages over semicorrin and bisoxazoline derivatives.

The selected examples in Schemes 1–4 illustrate that a universal ligand, which gives optimum results with all substrates, does not exist. Often, extensive screening of different catalysts is necessary to obtain useful enantioselectivities. Therefore, it is extremely important that the ligand synthesis is flexible and allows variation of the structure over a wide range. In this respect, many of the ligands discussed in this section are ideal because they are modular and readily assembled from a large selection of different building blocks.

Good enantioselectivities have also been achieved with chiral cobalt(II) complexes in certain cases [24]. However, the *trans/cis* selectivities are generally low and, therefore, the scope of these catalysts is limited. As mentioned before, a promising new class of chiral catalysts are the (pybox)Ru(II) complexes such as **12** developed by Nishiyama and co-workers [8, 25]. A remarkable feature of these catalysts is the significantly higher *trans* selectivity in the cyclopropanation of monosubstituted olefins compared to copper catalysts. Nishiyama's work has also led to important mechanistic insights. Using bulky diazo reagents such as 2,4,6-tri(*tert*-butyl)phenyl diazoacetate, it was possible to isolate the postulated ruthenium–carbene intermediates, to characterize their structure by NMR spectroscopy, and to study the transfer of the ruthenium-bound carboalkoxycarbene to styrene.

12   **13**   Rh$_2$(5S-MEPY)$_4$

2.7.4
Dinuclear Rhodium Catalysts

Dinuclear rhodium(II) complexes are highly efficient and remarkably versatile catalysts for the extrusion of dinitrogen from diazo compounds [4–6, 10]. Dirhodium(II) carboxylates, especially Rh$_2$(OAc)$_4$, are widely used as catalysts for cyclopropanation and other useful transformations of diazo compounds such as C–H insertion, ylide formation, and cycloadditions. As exemplified by structure **13**, the two rhodium centers are held together by a Rh–Rh single bond and four bridging carboxylate ligands. Each rhodium atom has one vacant coordination site. It is assumed that the catalytic process is initiated by coordination of the diazo compound at one of these sites, followed by elimination of dinitrogen leading to a rhodium-carbene complex. The reactivity and selectivity can be modulated by varying the steric and electronic properties of the ligands, e.g. by replacing acetate by trifluorobutyrate or a more electrondonating carboxamide ligand [26]. This often allows effective switching between competing reaction pathways such as cyclopropanation and C–H insertion (Scheme 5). In the reaction of the diazo ketone **15**, rhodium perfluorobutyrate Rh$_2$(pfb)$_4$ catalyzes exclusively the insertion into the tertiary C–H bond [26]. With the corresponding caprolactam complex Rh$_2$(cap)$_4$ only the cyclopropane **16** is formed, whereas dirhodium(II) acetate produces a mixture.

First attempts to develop enantioselective dirhodium catalysts derived from chiral carboxylates were unsuccessful. In general, the stereogenic centers in such complexes are too remote from the active site to have a significant effect on the catalytic process. The breakthrough came with the development of chiral carboxamide complexes by Doyle [5a, 27]. One of the most effective catalysts, Rh$_2$(5R-MEPY)$_4$ or Rh$_2$(5S-MEPY)$_4$, is readily prepared from (R)- or (S)-pyroglutamic acid methyl ester. Of the four possible stereoisomers, the one shown which has one pair of nitrogen atoms coordinated in a *cis* arrangement on each rhodium atom, is selectively formed. In contrast to chiral carboxylate complexes, the substituents at stereogenic centers are in close proximity to the vacant coordination site where the catalytic reaction takes place. It has been postulated that the interaction between the polar ester groups and the electrophilic carbene ligand plays an important role in the enantioselection process [5a, 27].

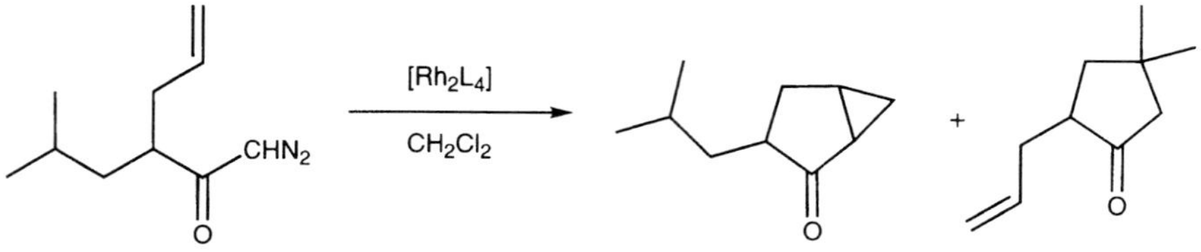

2.7 Cyclopropanation

Catalyst	Yield	16	:	17
[Rh$_2$(OAc)$_4$]	97% yield	44	:	56
[Rh$_2$(pfb)$_4$]	56% yield	0	:	100
[Rh$_2$(cap)$_4$]	76% yield	100	:	0

Scheme 5 Cyclopropanation vs. C,H insertion in the rhodium-catalyzed reaction of diazo ketone 15 [26].

The remarkable potential of these catalysts became apparent in intramolecular cyclopropanation reactions of allyl and homoallyl diazoacetates (Scheme 6) [27]. In collaborative studies, covering a wide range of differently substituted substrates, the groups of Doyle, Martin, and Müller demonstrated that these useful cyclization reactions generally proceed in good yield and with high enantioselectivities. As already mentioned, chiral copper catalysts, in general, give poor results with substrates of this type (for exceptions, see Ref. [27b]), whereas for analogous enantioselective cyclizations of alkenyl diazo ketones (Scheme 3), copper complexes are superior to dirhodium catalysts. The often different selectivity of Cu and Rh catalysts is illustrated in Scheme 7. Whereas the Cu complex derived from bisoxazoline 7 catalyzes formation of a macrocyclic lactone, only cyclization to the γ-lactone is observed with Rh$_2$(MEPY)$_4$ as catalyst [27c]. Quite intriguing results have been obtained with racemic diazoacetates derived from chiral allylic alcohols

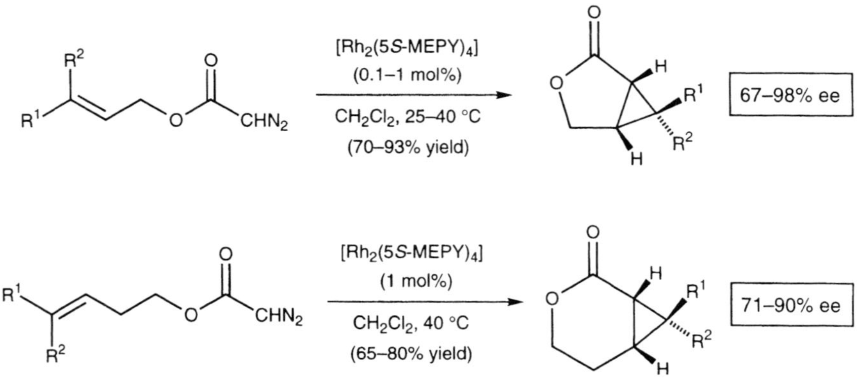

Scheme 6 Intramolecular cyclopropanation with chiral dirhodium carboxamide catalysts [27].

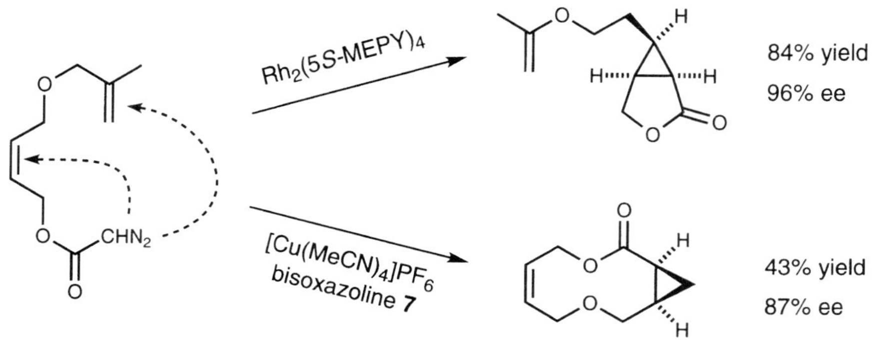

Scheme 7 Rhodium- vs. copper-catalyzed intramolecular cyclopropanation [27c].

(Scheme 8). In certain cases the two enantiomers undergo completely different transformations, one involving fragmentation, the other intramolecular cyclopropane formation [28]. Intramolecular cyclopropanation of allylic and homoallylic carboxamides is more difficult, although high enantioselectivities have been achieved with some substrates [29].

In general, dirhodium catalysts are not as efficient in enantioselective *inter*molecular cyclopropanations as chiral copper catalysts. However, there are exceptions. The conversion of acetylenes to cyclopropenes with diazoacetates proceeds surprisingly well and with remarkable ee's (Scheme 9) [30]. Although, at first sight, the use of chiral carboxylates as ligands does not seem promising (see above), high ee's have been recently reported in intermolecular cyclopropanations of olefins with vinyl- and phenyldiazoacetates using dirhodium complexes with N-(arylsulfonyl)amino acids [31].

Insertion into C–H bonds can be a competing process in rhodium-catalyzed cyclopropanations (see Scheme 5). If the challenging problems of chemo-, regio- and stereoselectivity are overcome, these reactions can be very useful in synthesis. This is the case for intramolecular C–H insertions which show a strong preference for the formation of five-membered rings [4, 6, 32]. The reactivity of C–H

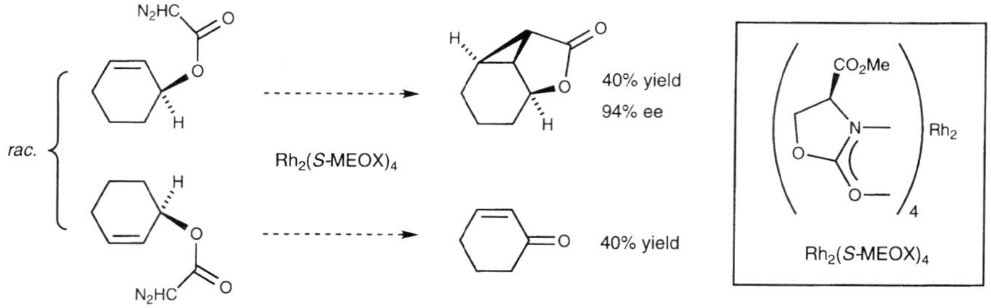

Scheme 8 Enantiomer differentiation in the rhodium-catalyzed cyclopropanation of racemic allylic diazoacetates [28].

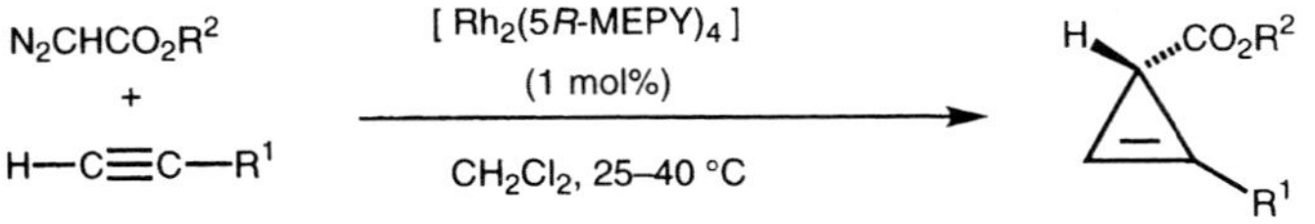

2.7 Cyclopropanation

Scheme 9 Enantioselective cyclopropanation of acetylenes with chiral dirhodium carboxamide catalysts [30].

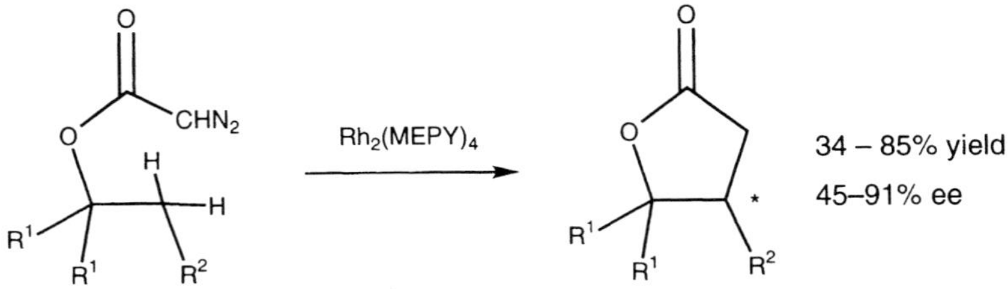

R^1	R^2	% yield	% ee
CH_2OMe	Et	73	69
CH_2OMe	t-Bu	56	78
CH_2OMe	d-Menthyl	43	≥ 94
CH_2OMe	l-Menthyl	45	43
n-Bu	d-Menthyl	46	86
t-Bu	d-Menthyl	51	77

Scheme 10 Enantioselective rhodium-catalyzed C–H insertion [33].

34 – 85% yield
45–91% ee

bonds generally follows the order tertiary > secondary > primary. But there are other factors to be considered as well, and product distribution is often the result of a subtle balance of steric, electronic, and conformational effects. Alkyl diazoacetates which can be converted to γ-lactones are a particularly useful class of substrates. As shown in Scheme 10, $Rh_2(MEPY)_4$ is an efficient enantioselective catalyst for this type of transformation [33]. Various other diazo compounds have been studied as substrates and in many cases useful levels of selectivity have been achieved by screening different chiral dirhodium catalysts. The remarkable variety of transformations that can be catalyzed by dirhodium complexes demonstrates how useful and versatile these catalysts are. Their scope is often complementary to the copper catalysts discussed in Section 2.7.3, and taken together, Cu and Rh catalysts make enantioselective cyclopropanations one of the most efficient and most reliable methods for enantioselective C–C bond formation.

2.7.5
Simmons–Smith Reaction

The Simmons–Smith reaction, involving methylene transfer from an organozinc reagent to a C–C double bond, is one of the standard methods for cyclopropanation [34]. A useful modification of the original procedure has been introduced by Furukawa et al. [35]. Instead of the insoluble Zn–Cu couple, diethylzinc is used to generate the active reagent from a geminal diiodide. Winstein et al.'s observation of a strong directive effect by adjacent hydroxyl groups [36] has initiated the development of useful diastereoselective cyclopropanation reactions [37]. More recently, enantioselective variants which are based on diethylzinc and stoichiometric amounts of a chiral additive have become available [37]. The most effective additive is a dioxaborolane derived from butylboronic acid and tartaric acid bis(dimethylamide) [38]. This useful and practical reagent developed by Charette has been successfully applied in several syntheses of complex natural products [39]. As shown by Kobayashi, Denmark, and Charette, enantioselectivity can also be induced by substoichiometric amounts of a chiral promoter such as a bis(sulfonamide) derived from cyclohexane-1,2-diamine [40] or a TADDOL-titanium complex [41] (Scheme 11). Although the scope of these catalytic methods is still limited, the results are promising and show the direction for further development.

2.7.6
Kulinkovich Hydroxycyclopropane Synthesis

An intriguing new process for the synthesis of hydroxycyclopropanes has been discovered by Kulinkovich et al. (Scheme 12) [42]. When carboxylic esters are treated with an excess of Grignard reagent in the presence of 0.2–1.0 equivalents

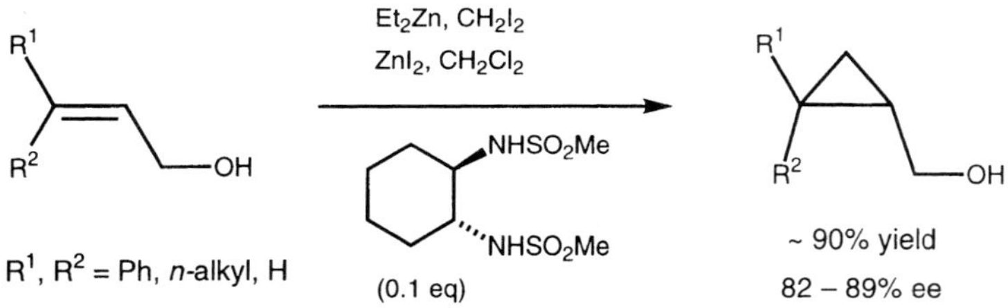

Scheme 11 Enantioselective Simmons–Smith reaction [40b].

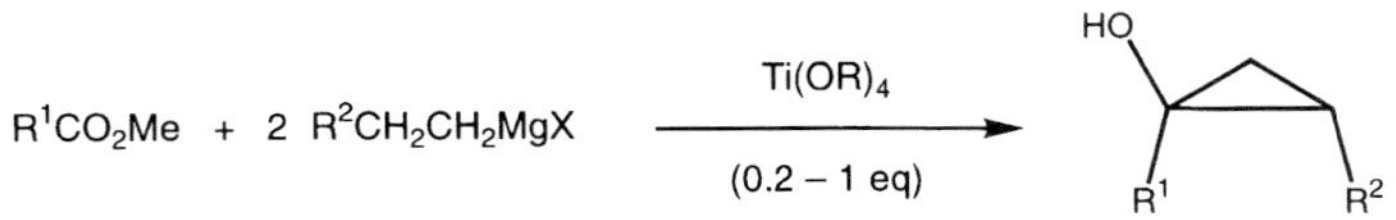

Scheme 12 Kulinkovich hydroxycyclopropane synthesis [42].

of Ti(O*i*-Pr)$_4$, 1,2-*trans*-disubstituted hydroxycyclopropanes are formed stereoselectively. Most likely, the reaction proceeds via a titanacyclopropane intermediate. Recently, the method has been extended to the synthesis of aminocyclopropanes starting from carboxamides [43]. An enantioselective variant based on a chiral bis(TADDOL)-titanium complex has also been reported [44]. All these studies indicate a considerable potential for this promising new method which will certainly be developed further in coming years.

2.7.7 References

1. H. N. C. Wong, M.-Y. Hon, C.-W. Tse, Y.-C. Yip, J. Tanko, T. Hudlicky, *Chem. Rev.* **1989**, *89*, 165.
2. H. W. Liu, C. T. Walsh, *Biochemistry of the cyclopropyl group*, in *The Chemistry of the Cyclopropyl Group* (Eds.: S. Patai, Z. Rappoport), Vol. 2, Wiley, Chichester, **1987**, Chap. 16; C. J. Suckling, *Angew. Chem.* **1988**, *100*, 555; *Angew. Chem., Int. Ed. Engl.* **1988**, *27*, 537.
3. J. Salaün, *Chem. Rev.* **1989**, *89*, 1247.
4. (a) G. Maas, *Topics Curr. Chem.* **1987**, *137*, 75. (b) M. P. Doyle, *Chem. Rev.* **1986**, *86*, 919. (c) T. Ye, A. McKervey in *The Chemistry of the Cyclopropyl Group* (Ed.: Z. Rappoport), Wiley, Chichester, **1995**, Vol. 2, Chap. 11. (d) T. Ye, A. M. McKervey, *Chem. Rev.* **1994**, *94*, 1091.
5. (a) M. P. Doyle in *Catalytic Asymmetric Synthesis* (Ed.: I. Ojima), VCH, New York, **1993**, Chap. 3; M. P. Doyle, *Rec. Trav. Chim. Pays-Bas* **1991**, *110*, 305. (b) V. K. Singh, A. DattaGupta, G. Sekar, *Synthesis* **1997**, 137.
6. A. Padwa, D. J. Austin, *Angew. Chem.* **1994**, *106*, 1881; *Angew. Chem., Int. Ed. Engl.* **1994**, *33*, 1797.
7. For recent mechanistic studies, see: (a) M. C. Pirrung, A. T. Morehead Jr., *J. Am. Chem. Soc.* **1996**, *118*, 8162. (b) D. W. Hartley, T. Kodadek, *J. Am. Chem. Soc.* **1993**, *115*, 1656 and Refs. 8, 9.
8. S.-B. Park, N. Sakata, H. Nishyiama, *Chem. Eur. J.* **1996**, *2*, 303.
9. D. A. Smith, D. N. Reynolds, L. K. Woo, *J. Am. Chem. Soc.* **1993**, *115*, 2511; J. P. Collman, E. Rose, G. D. Venburg, *J. Chem. Soc., Chem. Commun.* **1993**, 934.
10. A. J. Anciaux, A. Demonceau, A. F. Noels, R. Warin, A. J. Hubert, P. Teyssie, *Tetrahedron* **1983**, *39*, 2169; A. J. Anciaux, A. J. Hubert, A. F. Noels, N. Petiniot, P. Teyssie, *J. Org. Chem.* **1980**, *45*, 695; review: J. Adams, D. M. Spero, *Tetrahedron* **1991**, *47*, 1765.
11. U. Mende, B. Raduchel, W. Skuballa, H. Vorbruggen, *Tetrahedron Lett.* **1975**, *9*, 629; J. Kottwitz, H. Vorbruggen, *Synthesis* **1975**, 636.
12. J. N. C. Hood, D. Lloyd, W. A. MacDonald, T. M. Shepherd, *Tetrahedron* **1982**, *38*, 3355; R. M. Moriarty, O. Prakash, R. K. Vaid, L. Zhao, 7. *Am. Chem. Soc.* **1989**, *111*, 6443; L. Hatjiarapoglou, A. Varvoglis, N. W. Alcock, G. A. Pike, *J. Chem. Soc., Perkin Trans. 1* **1988**, 2839; R. M. Moriarty, J. Kim, L. Guo, *Tetrahedron Lett.* **1993**, *34*, 4129; P. Müller, D. Fernandez, *Helv. Chim. Acta* **1995**, *78*, 947.
13. T. Cohen, G. Herman, T. M. Chapman, D. Kuhn, *J. Am. Chem. Soc.* **1974**, *96*, 5627; B. Cimetiere, M. Julia, *Synlett* **1991**, 271.
14. Y. Gai, M. Julia, J.-N. Verpeaux, *Synlett* **1991**, 56; Y. Gai, M. Julia, J.-N. Verpeaux, *Synlett* **1991**, 269.
15. H. Nozaki, S. Moriuti, H. Takaya, R. Noyori, *Tetrahedron Lett.* **1966**, 5239; H. Nozaki, H. Takaya, S. Moriuti, R. Noyori, *Tetrahedron* **1968**, *24*, 3655.
16. T. Aratani, Y. Yoneyoshi, T. Nagase, *Tetrahedron Lett.* **1975**, 1707; *Tetrahedron Lett.* **1977**, 2599; *Tetrahedron Lett.* **1982**, *23*, 685; T. Aratani, *Pure Appl. Chem.* **1985**, *57*, 1839.

17 (a) H. Fritschi, U. Leutenegger, A. Pfaltz, *Angew. Chem.* **1986**, *98*, 1028; *Angew. Chem., Int. Ed. Engl.* **1986**, *25*, 1005; H. Fritschi, U. Leutenegger, A. Pfaltz, *Helv. Chim. Acta* **1988**, *71*, 1553. (b) A. Pfaltz in *Modern Synthetic Methods 1989* (Ed.: R. Scheffold), Springer-Verlag, Berlin, **1989**, pp. 199–248. (c) A. Pfaltz, *Acc. Chem. Res.* **1993**, *26*, 339.

18 (a) D. Müller, G. Umbricht, B. Weber, A. Pfaltz, *Helv. Chim. Acta* **1991**, *74*, 232. (b) U. Leutenegger, G. Umbricht, Ch. Fahrni, P. von Matt, A. Pfaltz, *Tetrahedron* **1992**, *48*, 2143. (c) A. Pfaltz in *Advances in Catalytic Processes* (Ed.: M. P. Doyle), JAI Press, 1995, Vol. 1, pp. 61–94.

19 D. A. Evans, K. A. Woerpel, M. M. Hinman, M. M. Faul, *J. Am. Chem. Soc.* **1991**, *113*, 726; D. A. Evans, K. A. Woerpel, M. J. Scott, *Angew. Chem.* **1992**, *104*, 439; *Angew. Chem., Int. Ed. Engl.* **1992**, *31*, 430.

20 R. E. Lowenthal, A. Abiko, S. Masamune, *Tetrahedron Lett.* **1990**, *31*, 6005; R. E. Lowenthal, S. Masamune, *Tetrahedron Lett.* **1991**, *32*, 7373.

21 C. Piqué, B. Fähndrich, A. Pfaltz, *Synlett* **1995**, 491. For analogous cyclizations using Aratani's catalyst, see: W. G. Dauben, R. T. Hendricks, M. J. Luzzio, H. P. Ng, *Tetrahedron Lett.* **1990**, *31*, 6969.

22 R. Tokunoh, H. Tomiyama, M. Sodeoka, M. Shibasaki, *Tetrahedron Lett.* **1996**, *37*, 2449.

23 T. G. Gant, M. C. Noe, E. J. Corey, *Tetrahedron Lett.* **1995**, *48*, 8745. For closely related ligands, see: Y. Uozumi, H. Kyota, E. Kishi, K. Kitayama, T. Hayashi, *Tetrahedron: Asymmetry* **1996**, *7*, 1603.

24 T. Fukuda, T. Katsuki, *Synlett* **1995**, 825; G. Jommi, R. Pagliarin, G. Rizzi, M. Sisti, *Synlett* **1993**, 833; A. Nakamura, A. Konishi, Y. Tatsuno, S. Otsuka, *J. Am. Chem. Soc.* **1978**, *100*, 3443; A. Nakamura, A. Konishi, R. Tsujitani, M. Kudo, S. Otsuka, *J. Am. Chem. Soc.* **1978**, *100*, 3449.

25 S.-B. Park, K. Murata, H. Matsumoto, H. Nishiyama, *Tetrahedron: Asymmetry* **1995**, *6*, 2487; H. Nishiyama, Y. Itoh, H. Matsumoto, S.-B. Park, K. Itoh, *J. Am. Chem. Soc.* **1994**, *116*, 2223; H. Nishiyama, Y. Itoh, Y. Sugawara, H. Matsumoto, K. Aoki, K. Itoh, *Bull. Chem. Soc. Jpn.* **1995**, *68*, 1247 and Ref. [8].

26 A. Padwa, D. J. Austin, A. T. Price, M. A. Semones, M. P. Doyle, M. N. Protopopova, W. R. Winchester, A. Tran, *J. Am. Chem. Soc.* **1993**, *115*, 8669 and Ref. [6].

27 (a) M. P. Doyle, R. J. Pieters, S. F. Martin, R. E. Austin, C. J. Oalman, P. Müller, *J. Am. Chem. Soc.* **1991**, *113*, 1423; M. P. Doyle, R. E. Austin, A. S. Bailey, M. P. Dwyer, A. B. Dyatkin, A. V. Kalinin, M. M. Y. Kwan, S. Liras, C. J. Oalmann, R. J. Pieters, M. N. Protopopova, C. E. Raab, G. H. P. Roos, Q.-L. Zhou, S. F. Martin, *J. Am. Chem. Soc.* **1995**, *117*, 5763; S. F. Martin, M. R. Spaller, S. Liras, B. Hartmann, *J. Am. Chem. Soc.* **1994**, *116*, 4493. (b) M. P. Doyle, C. S. Peterson, Q.-L. Zhou, H. Nishiyama, *Chem. Commun.* **1997**, 211. (c) M. P. Doyle, C. S. Peterson, D. L. Parker Jr., *Angew. Chem.* **1996**, *108*, 1439; *Angew. Chem., Int. Ed. Engl.* **1996**, *35*, 1334.

28 M. P. Doyle, A. B. Dyatkin, A. V. Kalinin, D. A. Ruppar, S. F. Martin, M. R. Spaller, S. Liras, *J. Am. Chem. Soc.* **1995**, *117*, 11021.

29 M. P. Doyle, A. V. Kalinin, *J. Org. Chem.* **1996**, *61*, 2179; M. P. Doyle, M. Y. Eismont, M. N. Protopopova, M. M. Y. Kwan, *Tetrahedron Lett.* **1994**, *50*, 1665.

30 M. N. Protopopova, M. P. Doyle, P. Müller, D. Ene, *J. Am. Chem. Soc.* **1992**, *114*, 2755; M. P. Doyle, M. N. Protopopova, P. Müller, D. Ene, E. Shapiro, *J. Am. Chem. Soc.* **1994**, *116*, 8492.

31 H. M. L. Davies, P. R. Bruzinski, D. H. Lake, N. Kong, M. J. Fall, *J. Am. Chem. Soc.* **1996**, *118*, 6897; M. P. Doyle, Q.-L. Zhou, C. Charnsangavej, M. A. Longoria, M. A. McKervey, C. F. Garcia, *Tetrahedron Lett.* **1996**, *37*, 4129.

32 D. F. Taber, R. E. Ruckle Jr., *J. Am. Chem. Soc.* **1986**, *108*, 7686.

33 M. P. Doyle, A. van Oeveren, L. J. Westrum, M. N. Protopopova, T. W. Clayton Jr., *J. Am. Chem. Soc.* **1991**, *113*, 8982; M. P. Doyle, Q.-L. Zhou, C. E. Raab, G. H. P. Roos, *Tetrahedron Lett.* **1995**, *36*, 4745; M. P. Doyle, A. V. Kali-

NIN, D. G. ENE, *J. Am. Chem. Soc.* **1996**, *118*, 8837; P. MÜLLER, P. POLLEUX, *Helv. Chim. Acta* **1994**, 77, 645; for Cu- and Ag-catalyzed C–H insertions, see: K. BURGESS, H.-J. LIM, A. M. PORTE, G. A. SULIKOWSKI, *Angew. Chem.* **1996**, *108*, 192; *Angew. Chem., Int. Ed. Engl.* **1996**, *35*, 220.

34 H. E. SIMMONS, R. D. SMITH, *J. Am. Chem. Soc.* **1958**, *80*, 5323; H. E. SIMMONS, T. L. CAIRNS, S. A. VLADUCHNIK, C. M. HOINESS, *Org. React.* **1973**, *20*, 1; K.-P. ZELLER, H. GUGEL, *Houben-Weyl: Methoden der Organischen Chemie*, (Ed.: M. REGITZ), Thieme, Stuttgart, **1989**, Vol. E 19b, pp. 195–211.

35 J. FURUKAWA, N. KAWABATA, J. NISHIMURA, *Tetrahedron Lett.* **1966**, 3353.

36 S. WINSTEIN, J. SONNENBERG, L. DEVRIES, *J. Am. Chem. Soc.* **1959**, *81*, 6523.

37 A. B. CHARETTE, J.-F. MARCOUX, *Synlett* **1995**, 1197.

38 A. B. CHARETTE, H. JUTEAU, *J. Am. Chem. Soc.* **1994**, *116*, 2651; A. B. CHARETTE, S. PRESCOTT, C. BROCHU, *J. Org. Chem.* **1995**, *60*, 1081.

39 See, e.g.: A. B. CHARETTE, H. LEBEL, *J. Am. Chem. Soc.* **1996**, *118*, 10327; A. G. M. BARRETT, D. HAMPRECHT, A. J. P. WHITE, D. J. WILLIAMS, *J. Am. Chem. Soc.* **1996**, *118*, 7863; J. R. FALCK, B. MEKONNEN, J. YU, J.-Y. LAI, *J. Am. Chem. Soc.* **1996**, *108*, 6096; J. D. WHITE, T.-S. KIM, M. NAMBU, *J. Am. Chem. Soc.* **1995**, *117*, 5612.

40 (a) H. TAKAHASHI, M. YOSHIOKA, M. SHIBASAKI, M. OHNO, N. IMAI, S. KOBAYASHI, *Tetrahedron Lett.* **1995**, *57*, 12013; H. TAKAHASHI, M. YOSHIOKA, M. OHNO, S. KOBAYASHI, *Tetrahedron Lett.* **1992**, *33*, 2575. (b) S. E. DENMARK, B. L. CHRISTENSON, D. M. COE, S. P. O'CONNOR, *Tetrahedron Lett.* **1995**, *36*, 2215; S. E. DENMARK, B. L. CHRISTENSON, S. P. O'CONNOR, *Tetrahedron Lett.* **1995**, *36*, 2219; S. E. DENMARK, S. P. O'CONNOR, *J. Org. Chem.* **1997**, *62*, 584 and 3390.

41 A. B. CHARETTE, C. BROCHU, *J. Am. Chem. Soc.* **1995**, *117*, 11367.

42 O. G. KULINKOVICH, S. V. SVIRIDOV, D. A. VASILEVSKI, *Synthesis* **1991**, 234; J. LEE, H. KIM, J. K. CHA, *J. Am. Chem. Soc.* **1996**, *118*, 4198; *J. Am. Chem. Soc.* **1995**, *117*, 9919; A. DE MEIJERE, S. I. KOZHUSHKOV, T. SPATH, N. S. ZEFIROV, *J. Org. Chem.* **1993**, *58*, 502.

43 V. CHAPLINSKI, A. DE MEIJERE, *Angew. Chem.* **1996**, *108*, 491; *Angew. Chem., Int. Ed. Engl.* **1996**, *35*, 413.

44 E. J. COREY, S. ACHYUTHA RAO, M. C. NOE, *J. Am. Chem. Soc.* **1994**, *116*, 9345.

2.8
Cyclomerization of Alkynes

H. Bönnemann and W. Brijoux

2.8.1
Introduction

Acetylene (ethyne) is the simplest hydrocarbon with a triple bond. Because of its strongly unsaturated character and high free energy of formation ($\Delta H = 226.9$ kJ/mol at 298.15 K) [1] it reacts readily with many other organic or inorganic compounds. In the presence of catalysts it can react with itself to form benzene and cyclooctatetraene as well as linear polymers ("cuprene"). The catalyzed cocyclization with hydrogen cyanide or nitriles leads to pyridine and its derivatives.

The next alkyne or the first derivative of acetylene is propyne (methyl acetylene). The alkyl substitution enhances the stability and normally reduces the reactivity of the C≡C triple bond. The higher stability of propyne could be deduced from its lower heat of formation, which is reduced to 185.6 kJ/mol at 298.15 K [1], so propyne often replaces acetylene in special applications. Additionally it allows the catalyzed synthesis of benzene or cyclooctatetraene derivatives and of trisubstituted pyridines in the cocyclization of propyne with nitriles.

Disubstituted acetylene, with an internal C≡C triple bond, leads in the cyclization reactions to hexasubstituted benzene, octasubstituted cyclooctatetraene, and pentasubstituted pyridine.

Acetylene is still used as a C_2 building block for both fine chemicals and industrial applications. Vinylation reactions (Hg-, Zn-, or Cd-catalyzed addition of compounds with active hydrogen atoms like water, alcohols, or acids), ethynylation reactions with alkaline catalysts (addition of carbonyl compounds with conservation of the triple bond), and the metal carbonyl-catalyzed carbonylation reactions (reaction with carbon monoxide and compounds with mobile hydrogen atoms) are the most important technical applications. Of additional industrial interest is the cyclotetramerization of acetylene to cyclooctatetraene discovered by W. Reppe [2] at BASF (Badische Anilin- und Soda-Fabriken, Ludwigshafen, Germany) in 1940.

The transition metal-catalyzed cyclotrimerization of acetylene (Eq. (1)) was discovered by M. Berthelot [3] way back in the last century using heterogeneous systems.

2.8 Cyclomerization of Alkynes

$$3 \text{ HC}\equiv\text{CH} \rightarrow \text{[benzene]} \tag{1}$$

The merits of homogeneous catalysts in this field were demonstrated most convincingly by W. Reppe. As early as 1948, W. Reppe, O. Schlichting, K. Klager, and T. Toepel reported the discovery of the "cyclic polymerization of acetylene" to cyclooctatetraene (Eq. (2)) using nickel catalysts [4]. This discovery represented a true milestone in transition metal catalysis.

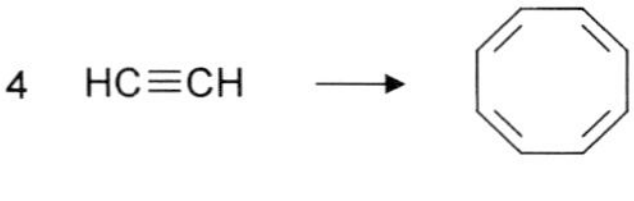

$$4 \text{ HC}\equiv\text{CH} \longrightarrow \text{[cyclooctatetraene]} \tag{2}$$

The mechanism of this remarkable reaction is little understood. Originally it was formulated as a concerted "zipper" process [6]. Recently, re-investigators of this reaction propose a *bis*-(cyclooctatetraene) dinickel complex as the active catalyst for the cyclotetramerization of acetylene [6]. Labelling experiments with mono-^{13}C-ethyne rule out cyclobutadiene, carbyne, and benzene Ni complexes as intermediates of the catalytic process [7].

Monosubstituted alkynes may be included in this cyclization, giving 1,2,4,7-, 1,2,4,6-, and 1,3,5,7-tetrasubstituted cyclooctatetraene derivatives [8]. Special cases are the cyclotetramerization of 1-phenylpropyne, giving the octasubstituted C$_8$-product besides the hexasubstituted benzene derivative [9] (Eq. (3)) and the (PMe$_3$)$_2$ Ni cod catalyzed dimerization of biphenylenes which were accessible from 1,2-dialkynylbenzene and acetylene – as described in Chapter 2.3 – to the corresponding tetrabenzocyclooctatetraene [10] according to Eq. (4).

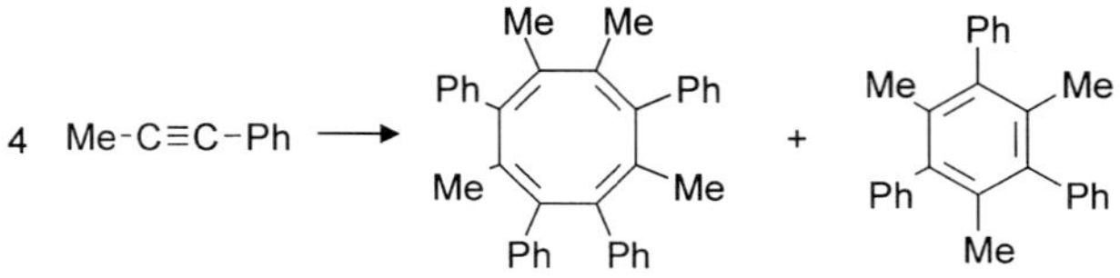

$$4 \text{ Me-C}\equiv\text{C-Ph} \longrightarrow \text{[octasubstituted COT]} + \text{[hexasubstituted benzene]} \tag{3}$$

$$2 \text{ [dialkynylbenzene]} + 2 \text{ HC}\equiv\text{CH} \longrightarrow 2 \text{ [biphenylene]} \longrightarrow \text{[tetrabenzocyclooctatetraene]} \tag{4}$$

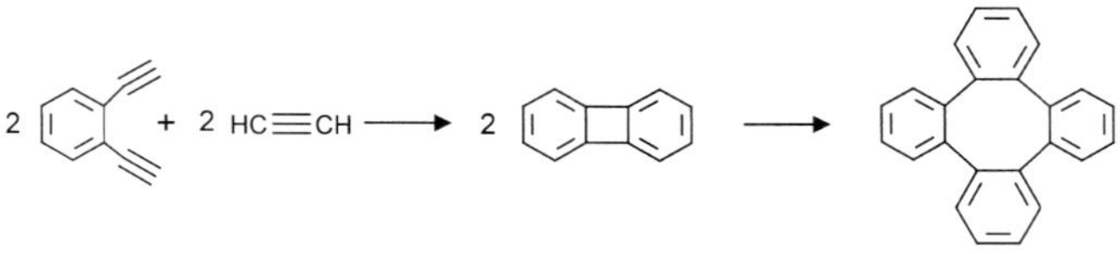

In 1973, H. Yamazaki and Y. Wakatsuki [11] first reported the homogeneous catalytic cycloaddition of alkynes and nitriles to pyridines, and later they investigated the mechanism of the reaction. At the same time K. P. C. Vollhardt [12] developed a number of elegant synthetic applications in organic chemistry, especially for the synthesis of steroids and phenylenes. Since 1974, H. Bönnemann and co-workers [13] have focused their work on the development of a "one-pot synthesis" of pyridine derivatives and of highly reactive organocobalt catalyts for the synthesis of α-substituted pyridines in homogeneous phase according to Eq. (5).

$$2\ HC\equiv CH + R-C\equiv N \rightarrow \underset{N\ \ R}{\bigcirc} \qquad (5)$$

N. E. Schore in 1988 [14] as well as S. Saito and Y. Yamamoto in 2000 [15] published comprehensive reviews on cyclomerization reactions of alkynes mediated by transition metal complexes. A detailed report about the metal complex catalysis in the synthesis of pyridines was given by U. M. Dzhemilev, F. A. Selimov, and G. A. Tolstikov in 2001 [16].

This article focuses on the transition metal-catalyzed formation of ring systems using alkynes in the homogeneous phase. Because of the great number of possible products, this survey had to be restricted to the formation of 6-membered carbo- and heterocycles and their homologs.

2.8.2
Transition Metal-Catalyzed Syntheses of 6-Membered Carbocycles

The cyclotrimerization of acetylene to benzene (see Eq. (1)) is highly exothermic. The free energy of this process was estimated to be 595 kJ per mol of product [17]. Monosubstituted acetylenes give 1,2,4- or 1,3,5-trisubstituted benzene derivatives (Eq. (6)). The regioselectivity of the cyclization may be controlled by the electronic properties and the sterical demand of the catalyst and the substrates as well as by the reaction conditions. Because of the inherent sensitivity of most organometallic catalysts to heteroatoms, this reaction is mainly limited to alkyl-, alkenyl- or arylsubstituted acetylenes. Educts containing polar hetero atoms are only processed by very few homogeneous catalysts, e.g., cobalt catalysts.

$$3\ R-C\equiv CH \rightarrow \text{1,2,4-isomer} + \text{1,3,5-isomer} \qquad (6)$$

(R = alkyl, aryl, CO_2Me, etc.)

The mechanism of acetylene trimerization at the most used catalyst, η^5-cp Co, was investigated by Th. A. Albright and co-workers in a theoretical study at the *ab initio* and the density functional theory levels. They found a stepwise synthesis of the benzene via η^5-cp Co $(\eta^2\text{-}C_2H_2)$ and η^5-cp Co $(\eta^2\text{-}C_2H_2)_2$, then ring closure to η^5-cp-cobalta-cyclopentadiene, and the direct formation of η^5-cp Co $(\eta^4\text{-}C_6H_6)$ by the addition of the third acetylene without the formerly postulated intermediate η^5-cp-cobalta-cycloheptatriene [18].

A good survey of the cobalt-mediated [2+2+2] cycloaddition reactions of alkenes and alkynes to various carbocycles is given by K.P.C. Vollhardt in two articles [19]. The diyne reaction of 1,4-, 1,5-, 1,6-, and 1,7-diynes via rhodium complexes was reviewed by E. Müller [20].

2.8.2.1
Benzenes and Cyclohexadienes

In practice, the alkyne cyclotrimerization to benzene and its derivatives may be performed using both homogeneous and heterogeneous catalysts. Many catalysts give good yields in the cyclotrimerization of unsymmetrically substituted terminal and also internal alkynes. As mentioned above, in the case of terminal alkynes, 1,2,4- and 1,3,5-trisubstituted benzenes are formed (Eq. (6)). For example, a chromium(VI) catalyst trimerizes propyne to give pseudocumene and mesitylene in a 4:1 ratio [21]. The cyclotrimerization of 1-hexyne, 1-octyne, methyl propiolate, and phenylacetylene at organorhodium half-sandwich complexes was investigated by G. Ingrosso et al. [22]. In the case of the alkyl-substituted acetylenes, the regioselectivity of the trimerization was found to be independent of the rhodium catalyst applied. The cyclization of methyl propiolate at the η^5-flu Rh catalyst gave a higher portion of the symmetrically substituted benzene derivatives than found at the η^5-ind Rh complex. The η^5-ind Rh *bis*-(ethene) complex was found to be unusually selective in the cyclotrimerization of 3,3-dimethyl-1-butyne, giving a 76% yield of 1,2,4-tri-*t*-butylbenzene [23]. The 1,3,5-isomer is available from the 3,3-dimethyl-1-butyne in the presence of $PdCl_2$ [24]. This type of catalytic alkyne reaction has been reviewed by P.M. Maitlis [25]. The kinetics of the cyclotrimerization of $CH_3CO_2-C\equiv C-CO_2CH_3$ and hex-3-yne, respectively, catalyzed by diverse η^5-cp Rh L_2 complexes (L=alkene, CO, PF_3), was investigated by B.L. Booth and co-workers [26]. They found that the rates of this rhodium-mediated cyclotrimerization depend on the nature of the ligand L. The cyclization of 1,1,1-d_3-but-2-yne was analyzed by G.M. Whitesides and W.J. Ehmann, with the result that only in the case of the $AlCl_3$-catalyzed reaction does the cyclization proceed via an intermediate of cyclobutadiene-like symmetry [27]. The homogeneous cyclotrimerization of internal alkynes with a heterogeneous Pd catalyst in the presence of $(CH_3)_3SiCl$ was described by W.F. Maier and A.K. Jhingan [28], where the active species of the catalysis was formed by the reaction of the silane with the charcoal-supported Pd. The cyclooligomerization of terminal and internal alkynes under phase transfer conditions by the $RhCl_3$-Aliquat® 336 catalyst was published by J. Blum et al. Whereas mono- and dialkylated acetylenes react exclusively to the corresponding benzene derivatives, the phenyl-substituted alkynes gave, as well as the benzenes, 2,3-disubstituted 1-phenylnaphthalene [29]. In special cases, by using mono-substituted alkynes, the exclusive synthesis of one benzene derivative is possible. Thus, S. Saito et al. described the cyclotrimerization of 1-perfluoroalkylenynes in the presence of $Ni(PPh_3)_4$ to 1,2,4-trisubstituted benzene derivatives in highly regioselective yield [30].

The regiochemical product distribution of the cocyclization of two or three different alkynes occurs statistically. In some cases carefully controlled reaction conditions allow one to isolate a main product from mixed cyclotrimerizations. For example, 1,2,3,4-tetraphenyl-5,6-diethylbenzene can be obtained, cobalt catalyzed, from tolane and hex-3-yne in good yield [31]. The selective synthesis of radiolabeled toluene and *p*-xylene via co-cyclotrimerization of acetylene and propyne was obtained using a heterogeneous chromium catalyst, as described in [23, 32]. M. Mori and co-workers published the synthesis of biaryls using an *in situ* prepared Ni catalyst. They started either from an alkyne bearing a phenyl group and two equivalents of acetylene or from a,ω-diynes having a phenyl group at the a-position and one acetylene [33].

For environmental reasons, water is the preferred reaction medium ("green chemistry"), and therefore the cyclotrimerization of alkynes should be carried out in water or water/alcohol mixtures. B. E. Eaton et al. reported the synthesis of hexasubstituted benzene according to Eq. (7) in water/methanol mixtures catalyzed by water soluble cobalt complexes of type R-cp Co cod (R=ester or keto group) [34].

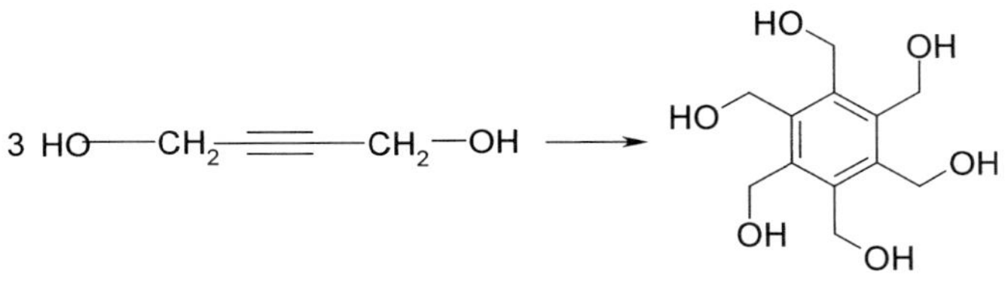

$$3 \ HO-CH_2-\!\!\!=\!\!\!-CH_2-OH \longrightarrow \text{hexakis(hydroxymethyl)benzene} \quad (7)$$

Instead of a second or third alkyne, an alkene C=C double bond may be incorporated into the cyclotrimerization reaction. Iron [35], rhodium [36], nickel [37], palladium [38], or cobalt [39] catalysts have been used to form cyclohexa-1,3-dienes.

$$2 \ R^1\text{-C}\equiv\text{CH} + R^2\text{-CH=CH}_2 \rightarrow \text{trisubstituted cyclohexa-1,3-dienes} \quad (8)$$

(R^1, R^2 = alkyl, aryl)

However, in preparative use this catalytic cocyclization is disturbed by consecutive side reactions of the resulting dienes such as cycloaddition or dehydrogenation. H. Suzuki et al. [40] have reported the straight palladium-catalyzed cocyclization reaction of $C_2(CO_2Me)_2$ and norbornene.

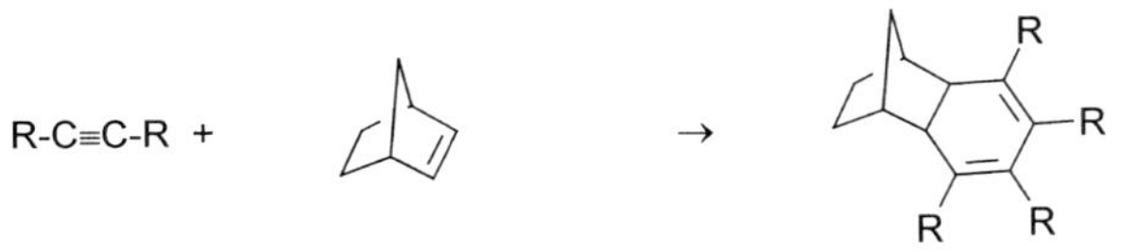

$$R\text{-C}\equiv\text{C-R} + \text{norbornene} \rightarrow \text{product} \quad (9)$$

(R = CO_2Me)

2.8 Cyclomerization of Alkynes

K. Jonas and M.G.J. Tadic [41] have investigated the homogeneous cobalt-catalyzed co-cyclotrimerization of acetylene and olefines. The reaction with η^5-ind Co bis-(ethene) as the catalyst was carried out with ethene, α-olefines, and 2-butene, as well as cyclohexene and cyclooctene (Eqs. (10) and (11)).

$$2\ HC{\equiv}CH + H_2C{=}CH{-}R \rightarrow \text{[cyclohexadiene-R]} \tag{10}$$

(R = H, alkyl, CO_2Me)

$$2\ HC{\equiv}CH + \text{[cyclohexene]} \rightarrow \text{[hexahydronaphthalene]} \tag{11}$$

The reaction according to Eq. (11) occurs exclusively to *cis*-hexahydronaphthalene (*cis*-hexaline), a product which is otherwise accessible only by multistep synthetic pathways [42].

Recently, Korean researchers published the catalyzed cyclization of 1,6-enynes according to Eq. (12), forming bicyclic cyclohexa-1,4-diene derivatives in the presence of the Wilkinson catalyst and $AgBF_4$ [43].

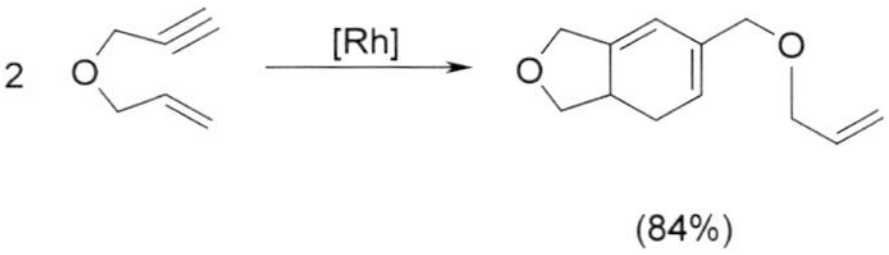

$$2\ \text{[allyl propargyl ether]} \xrightarrow{[Rh]} \text{[bicyclic product]} \tag{12}$$

(84%)

D.W. Macomber [44] reported the [2+2+2]-cycloaddition reaction of diphenylacetylene or $C_2(CO_2Me)_2$ and *endo*-dicyclopentadiene or norbornylene, respectively, in the presence of η^5-cp Co dicarbonyl or η^5-Me-cp Co dicarbonyl in refluxing toluene.

Intramolecular cyclohexa-1,3-diene syntheses have been developed by K.P.C. Vollhardt et al. [45]. Enediynes with a terminal double bond react in *iso*-octane at 100 °C in the presence of η^5-cp Co dicarbonyl, giving the three ring system according to Eq. (13) [46].

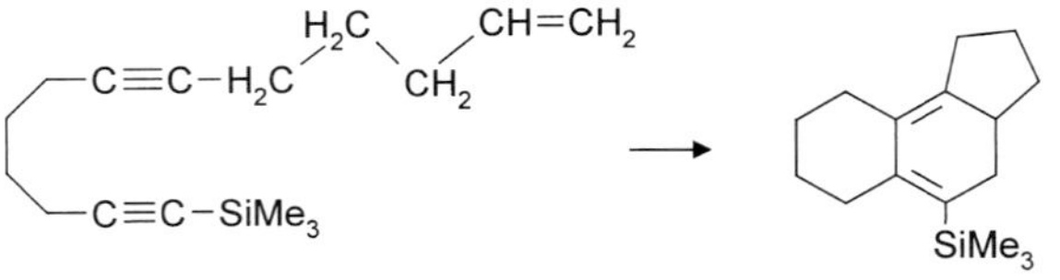

$$\text{[enediyne with }{-}C{\equiv}C{-}H_2C{-}CH_2{-}CH{=}CH_2\text{ and }{-}C{\equiv}C{-}SiMe_3\text{]} \rightarrow \text{[tricyclic product with }SiMe_3\text{]} \tag{13}$$

C-ring dienyl steroids and B-ring aromatic steroids (Eq. (14)) have been made accessible with appropriate precursors in a remarkably high stereoselective process [47]. Intramolecular cycloaddition reactions of enediynes containing terminal alkyne groups have also been observed by K.P.C. Vollhardt [48] (Eq. (15)).

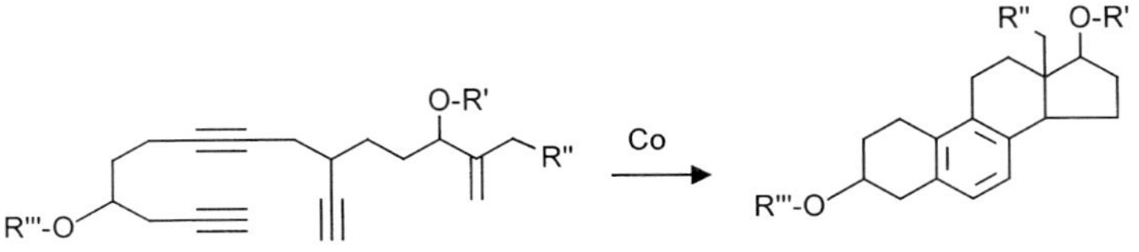

2.8.2 Transition Metal-Catalyzed Syntheses of 6-Membered Carbocycles

(14)

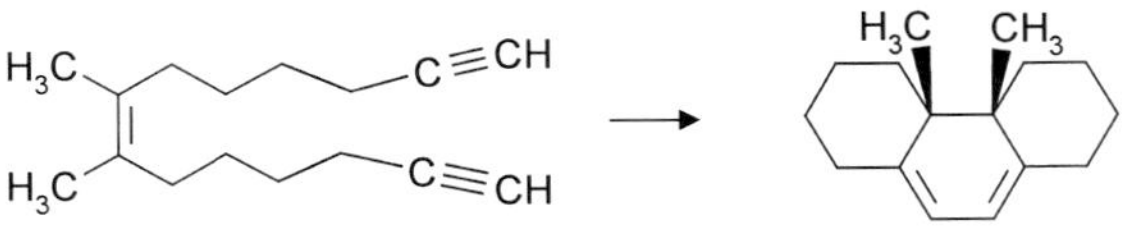

(15)

Cyclohexa-1,4-dienes have been synthesized by the Fe-catalyzed reaction of 1,3-dienes and internal alkynes (Eq. (16)). Generally, the yields are quite good and the reaction conditions are very mild [49].

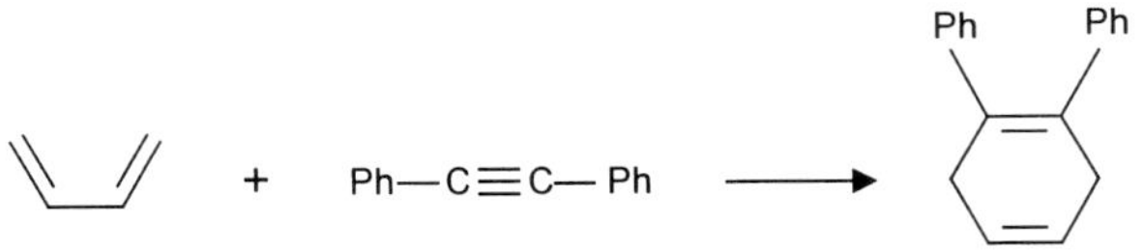

(16)

The high reactivity of Ziegler catalysts may be exploited in this process, but only alkynes not capable of self-trimerization can be used. So *bis*-trimethylsilyl acetylene reacts with numerous substituted 1,3-dienes in the presence of $Et_2AlCl/TiCl_4$ to give cyclohexa-1,4-dienes in ca. 70% yields [50].

Finally the Rh-catalyzed synthesis of cyclo-hexenones via a [4+2] annulation of 4-alkynals with mono- or disubstituted alkynes (Eq. (17)) was published by K. Tanaka and G. C. Fu [51].

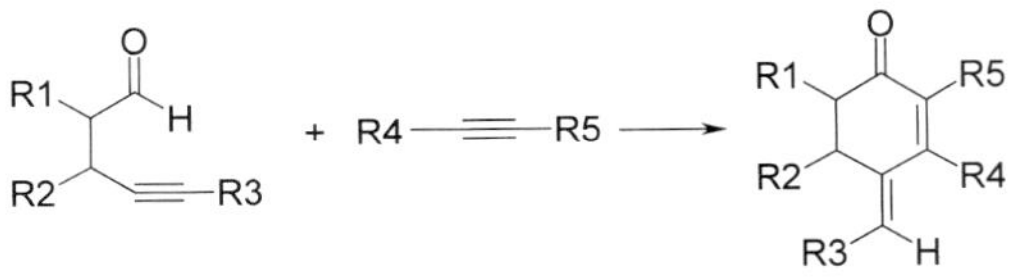

(17)

2.8.2.2
Quinones

The organotransition metal synthesis of quinones has been studied by several authors [52]. Starting from internal alkynes and metal carbonyls, tetrasubstituted quinones were synthesized in good yields.

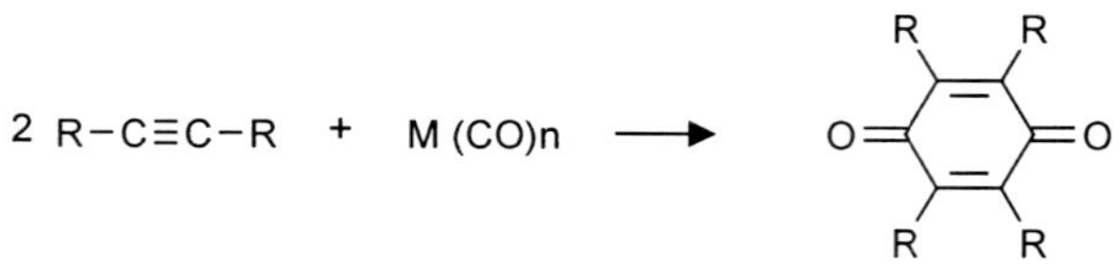

$$2\ R-C\equiv C-R\ +\ M(CO)_n \longrightarrow \underset{R\ \ R}{\overset{R\ \ R}{O=\underset{}{\bigcirc}=O}} \tag{18}$$

L. S. Liebeskind and co-workers extended the reaction to naphthoquinones by reacting phthaloyl metal complexes with disubstituted acetylenes [53]. Anthraquinones are accessible by the cobalt-catalyzed reaction of alkynyl ketones with alkynes [12].

2.8.2.3
Phenylenes

o-Diethynylbenzene, available from o-diiodobenzene, can easily cocyclize with internal alkynes to 2,3-disubstituted diphenylenes [54] at the η^5-cp Co dicarbonyl complex as catalyst precursor.

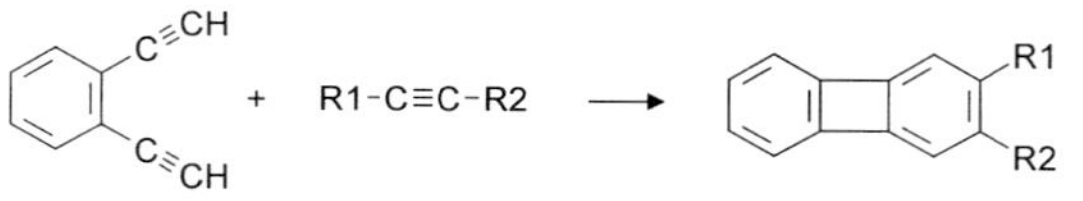

$$\tag{19}$$

(R^1, R^2 = H, alkyl, aryl, CO$_2$Me, SiMe$_3$)

In the case of $R^1 = R^2 = SiMe_3$, the successive synthesis of polyphenylenes has been reported. Subsequent iodination of the trimethylsilyl group generates a new o-diiodoarene as the educt for the subsequent o-diethynylarene, which can react with further bis-(trimethylsilyl)-acetylene forming terphenylene and so on (Eq. (20)). Multiphenylenes synthesized in this way have been claimed to represent a new type of organic semiconductors [55].

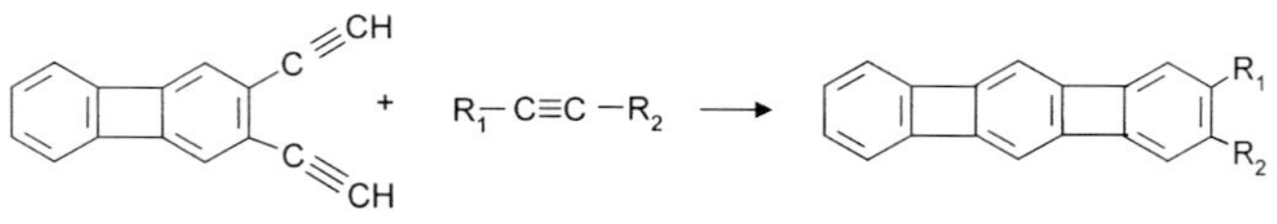

$$\tag{20}$$

(R$_1$, R$_2$ = H, SiMe$_3$)

2.8.2.4
Naphthalenes and Phenanthrenes

Naphthalenes and phenanthrenes have been selectively synthesized by the cocyclization of arynes with internal alkynes in the presence of Pd complexes. Depending on the type of Pd-catalyst, either naphthalene or phenanthrene derivatives are formed. So, for example, in the case of Pd(PPh$_3$)$_4$ phenanthrenes are the major products whereas naphthalenes are formed by using Pd$_2$(dba)$_3$ as the catalyst [56].

M. Catellani and co-workers obtained phenanthrene derivatives by the reaction of ortho-substituted aryl iodides with diphenyl- or alkylphenylacetylenes. For the synthesis according to Eq. (21) they used $Pd(acetate)_2/K_2CO_3/(butyl)_4NBr$/norbonene as catalyst [57].

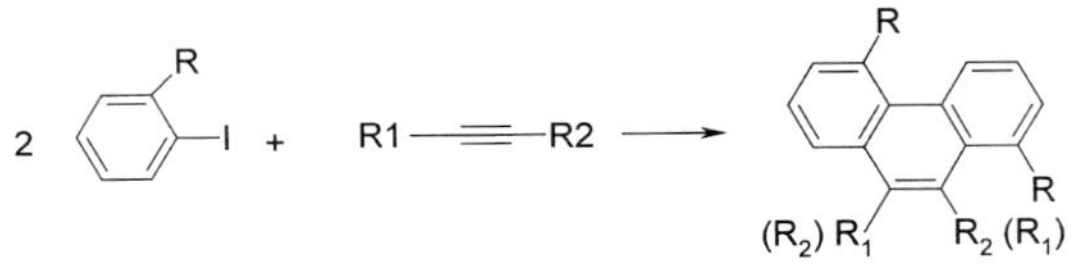

$$2 \text{ Ar-I} + R_1-\equiv-R_2 \longrightarrow \text{phenanthrene} \quad (21)$$

2.8.3
Transition Metal-Catalyzed Syntheses of 6-Membered Heterocycles

Although quite spectacular results have been obtained in the field of transition metal-catalyzed transformations [58] of olefines and alkynes (e.g., see above), reactions which could lead to heterocyclic compounds have been relatively neglected. As mentioned above, an obvious reason for this is that substrates containing heteroatoms like nitrogen, oxygen, or sulfur could coordinate to the metal and suppress catalytic activity. Nevertheless, some interesting early examples of transition metal-catalyzed syntheses of heterocycles have been reported, and have been reviewed by C. W. Bird [59].

2.8.3.1
Pyranes, Pyrones, Pyridones, and Sulfur-Containing Heterocycles

Recently, the incorporation of carbon dioxide, which enables lactones to be synthesized from alkynes, has begun to attract attention [60–64]. Whereas the homogeneous reaction of carbon dioxide with butadiene (Eq. (22)) is catalyzed by Pd complexes [65], the reaction with alkynes (Eq. (23)) proceeds via Ni(0) systems with electron-donating small phosphine ligands [61].

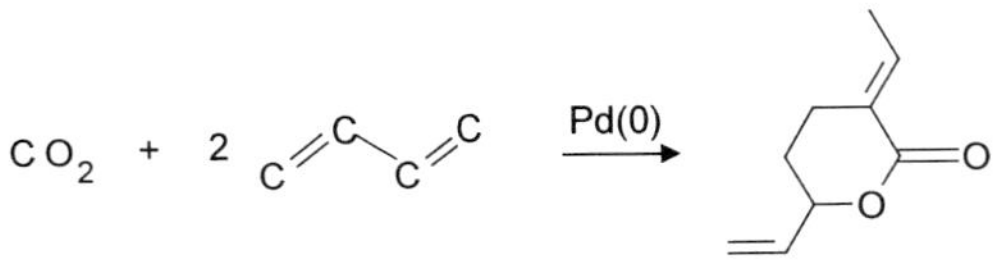

$$CO_2 + 2 \text{ butadiene} \xrightarrow{Pd(0)} \text{lactone} \quad (22)$$

$$CO_2 + 2 \text{ R-C}\equiv\text{C-R} \xrightarrow{Ni(0)} \text{pyrone} \quad (23)$$

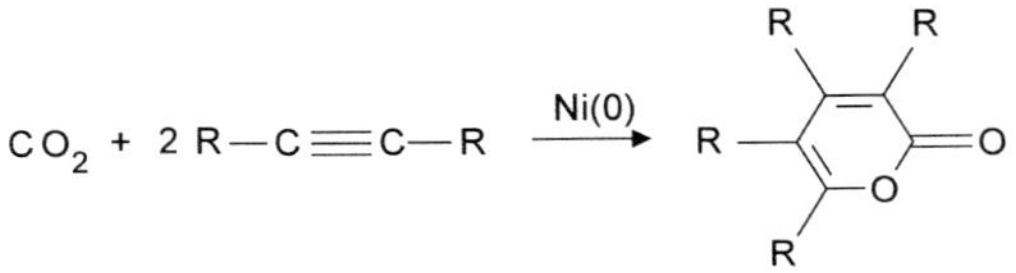

R = H, alkyl, functional group

2.8 Cyclomerization of Alkynes

The substituents R at the acetylene can be widely spread by this type of Ni-catalyzed reaction, so that a diverse range of 2-pyrones can be synthesized. Beside mono- and dialkylated acetylenes, alkynes with functional groups such as -OR and -COOR can also be incorporated in this catalytic reaction. Dialkynes, e.g., butadiyne, lead to poly-2-pyrones. IR investigations of the system tetramethylethylenediamine/Ni(0)/hex-3-yne/CO_2 show that the product tetraethyl-2-pyrone will be formed even at room temperature. The first step of the catalytic cyclooligomerization is the formation of a metallacyclic carboxylate of hex-3-yne and CO_2. This complex could be isolated from the reaction mixture, and its structure was determined by X-ray analysis [63]. Other hetero cumulenes such as carbon disulfide, carbodiimides, and isothiocyanates can also be incorporated in such a reaction. But the cyclization reactions of diphenyl acetylene ("tolane") and the above-mentioned hetero cumulenes are performed via heteroatom free cobalt metallacycles ("coboles") (Eq. (24)) [66, 67]. On the other hand, the analogous cocyclization of substituted alkynes and isocyanates forming 2-pyridones occurs in the presence of a rhodium complex (Eq. (25)) [68].

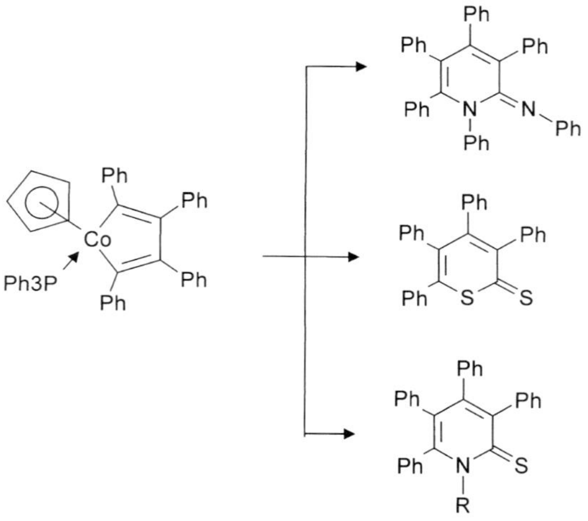

(24)

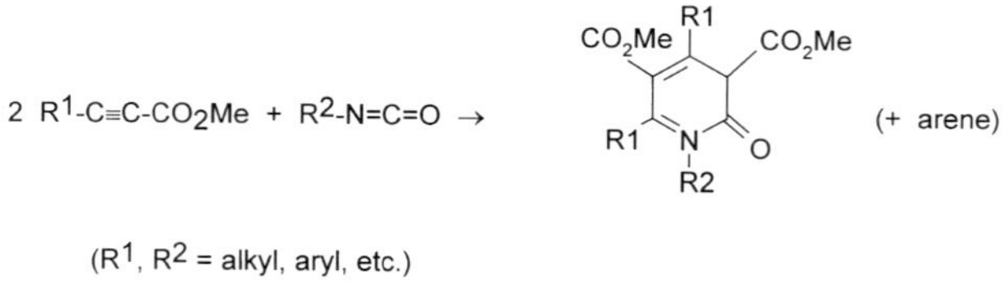

$$2\ R^1\text{-C}{\equiv}\text{C-CO}_2\text{Me} + R^2\text{-N=C=O} \rightarrow \qquad (+ \text{ arene})$$

(R^1, R^2 = alkyl, aryl, etc.)

(25)

The η^5-cp Co-mediated [2+2+2] cycloadditions of alkynes with aldehydes or ketones to fused pyrans was investigated by K.P.C. Vollhardt and co-workers [69]. Starting from diynes and ketones, alkynyl aldehydes and disubstituted acetylenes, or dialkynes with a carbonyl function, they synthesized various bi- and tricyclic 2H-pyrans (Eq. (26)) and/or their $α,β,γ,δ$-unsaturated carbonyl isomers.

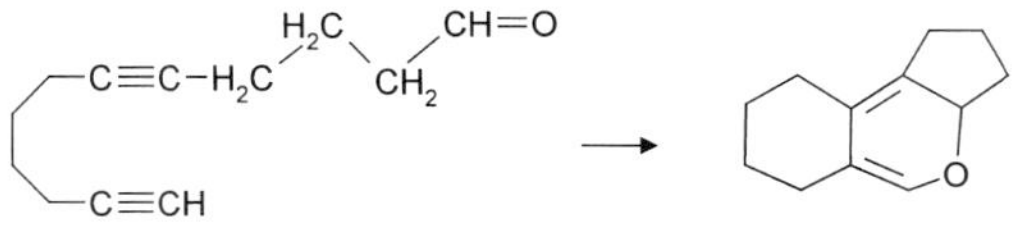

$$\begin{array}{c}\text{-C}\equiv\text{C-H}_2\text{C}\diagdown\overset{\text{H}_2\text{C}}{\text{CH}_2}\diagup\overset{\text{CH}=\text{O}}{}\\ \text{-C}\equiv\text{CH}\end{array} \longrightarrow \text{[bicyclic pyran product]} \qquad (26)$$

A similar reaction but restricted to aldehydes and internal diynes was found by T. Tsuda et al. using Ni(0) as catalyst [70].

Remarkably, the η^5-cp Co-catalyzed cycloaddition of acrolein and diphenyl acetylene in the presence of a small amount of methyl acetate occurs selectively at the carbonyl rather than at the C=C double bond to give vinylpyran (Eq. (27)) [71].

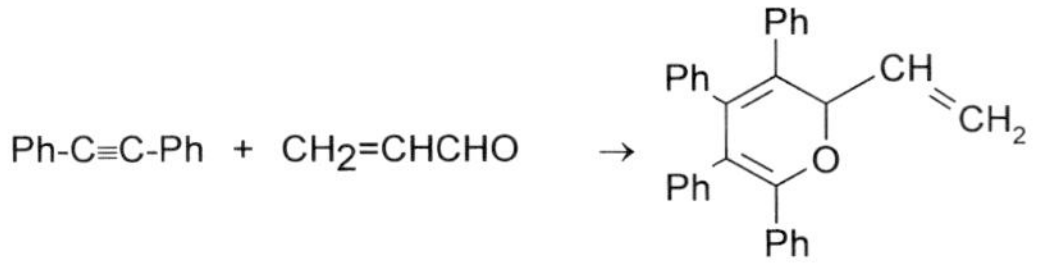

$$\text{Ph-C}\equiv\text{C-Ph} + \text{CH}_2=\text{CHCHO} \longrightarrow \text{[2,3,6-triphenyl-2-vinyl-2H-pyran]} \qquad (27)$$

Imidazole and indole derivatives have also been cocyclized with disubstituted acetylenes at the η^5-cp Co catalyst to give the 6-membered nitrogen-containing heterocycles 4a,9a-dihydro-9H-carbazoles or, after oxidation, precursors for strychnine (Eqs. (28)–(30)) [72, 73].

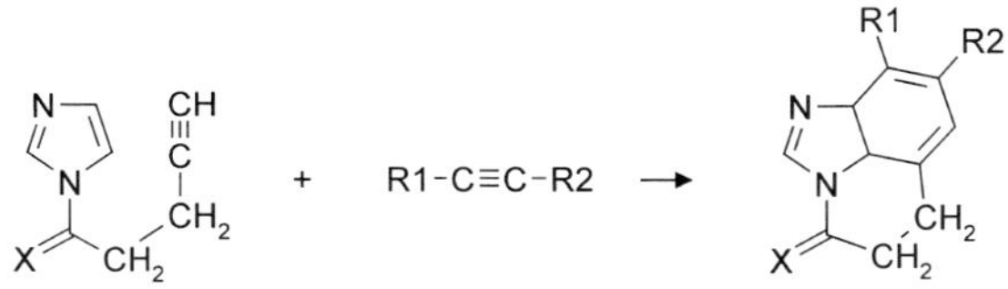

$$\text{[imidazole-tethered enyne]} + \text{R1-C}\equiv\text{C-R2} \longrightarrow \text{[dihydroimidazo-fused product]} \qquad (28)$$

(R^1, R^2 = Si(CH$_3$)$_3$, CH$_2$CH$_3$, CO$_2$CH$_3$, OCH$_3$; X = O, H$_2$)

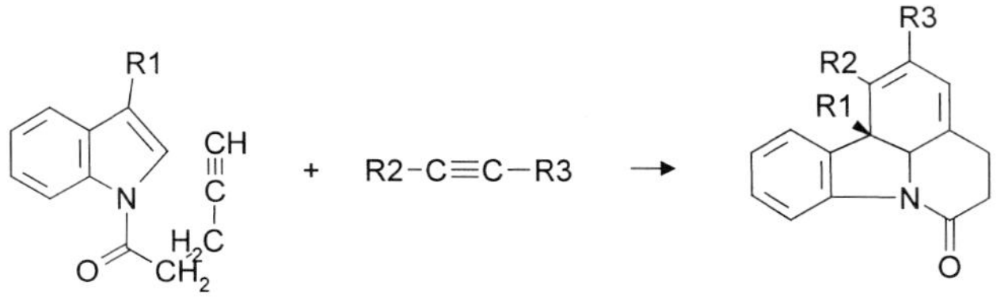

$$\text{[indole-tethered enyne]} + \text{R2-C}\equiv\text{C-R3} \longrightarrow \text{[carbazole-fused product]} \qquad (29)$$

(R^1 = H, CH$_3$, CH$_2$CH$_2$NHR, CH$_2$CH$_2$OR, COCH$_2$C=CH;
R^2, R^3 = Si(CH$_3$)$_3$, OCH$_3$, CO$_2$CH$_3$)

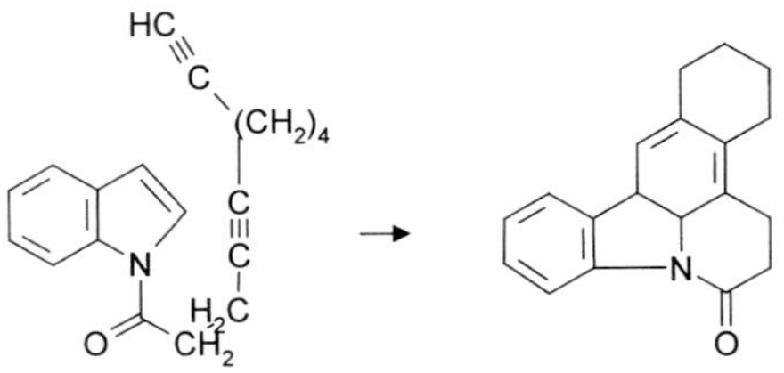

[Structural scheme] (30)

2.8.3.2
Pyridines

Pyridine and its derivatives are industrially important fine chemicals. Their isolation from coal tar is decreasing in volume, whereas synthetic manufacture using selective methods has increased rapidly in the last decades. The classic pyridine syntheses have been extensively reviewed by Abramovitch [74]. Many of them rely on the condensation of aldehydes or ketones with ammonia in the vapor phase. However, these processes suffer from unsatisfactory selectivity. Soluble organocobalt catalysts allow a selective one-step access to pyridine and a wide range of α-substituted derivatives from acetylene and the corresponding cyano compounds (see Eq. (5)).

The basic cocyclotrimerization (Eq. (31)) was first observed in 1876 by Sir William Ramsey [75], who obtained small amounts of pyridine from acetylene and hydrogen cyanide in a red-hot iron tube.

$$2\ HC{\equiv}CH\ +\ HC{\equiv}N\ \rightarrow\ \text{[pyridine]} \qquad (31)$$

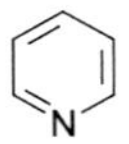

As mentioned above, the homogeneous catalytic [2+2+2]-cycloaddition of alkynes and nitriles to pyridines was first discovered by H. Yamazaki and Y. Wakatsuki [11] using the phosphine-stabilized cobalt(III) complex (Scheme 1). In this complex, two alkyne molecules are already linked together forming a 5-membered metallacycle ("cobole"). The reaction with a nitrile gives the corresponding pyridine derivative.

H. Bönnemann and co-workers [13] observed at the same time the cocyclization (Eq. (5)) at cobalt catalysts prepared *in situ*, as well as by using highly active phosphine-free organocobalt(I) diolefine complexes (Schemes 2–5).

The *in situ* system (Eq. (32)) may be recommended for the quick exploration of new synthetic applications in research laboratories which are not specialized in organometallic techniques, because the cobalt salts can be used in the hydrated form under air and no sophisticated ligands are necessary [76].

$$CoCl_2 \cdot 6\ H_2O\ /\ NaBH_4\ +\ \text{Nitrile / Alkyne} \qquad (32)$$

Cobalt(I) halide complexes of the type [XCoL₃] having a moderate activity in the synthesis of 2-alkylpyridines are also easily accessible (Eq. (33)) [77].

$$CoX_2 + L + Red. \rightarrow XCoL_3 \qquad (33)$$

[X = Cl, Br, I; Red. = NaBH₄, Zn; L = P(C₆H₅)₃, P(OEt)₃, P(OC₃H₇)₃]

Two types of pre-prepared organocobalt complex proved to be most effective catalysts for the cocyclization of alkynes and nitriles: the allyl-cobalt type, where the organic group is η^3-bonded to the metal (Scheme 2), and also the η^5-cp Co and η^5-ind Co half-sandwich compounds (Schemes 3 and 4). During the catalytic cycle in the case of the η^3-allyl-cobalt catalyst a 12-electron system is regenerated, whereas in the case of the η^5-cp Co and η^5-ind Co complexes the catalytic reaction involves a 14-electron moiety. In fact, the cobalt-catalyzed pyridine synthesis was one of the first examples where η^5-cp groups were used as controlling ligands in homogeneous catalysis [13f, 13g]. The modification of the basic η^5-cp ligand systems by additional substituents, R, transferring electron-donating or -withdrawing effects to the η^5-cp group results in strong changes in catalyst activity and selectivity. In addition, η^6-borininato ligands may be used as 6π-electron ligands for cobalt (Scheme 5).

Since the phosphine-stabilized cobalt(III) complex proved to be unsuitable for practical purposes, H. Yamazaki and Y. Wakatsuki later turned to cobaltocene (η^5-cp₂Co) as a catalyst for the pyridine synthesis [78, 79]. With the same system, P. Hardt at Lonza AG developed various procedures for preparing pyridine derivatives [80]. An elegant model reaction by H. Yamazaki and Y. Wakatsuki [78] showed that cobaltocene can be regarded as the precursor for the actual catalyst (Eq. (34)): prior to the catalysis, η^5-cp₂Co is converted by alkyne into an η^5-cp Co diene complex.

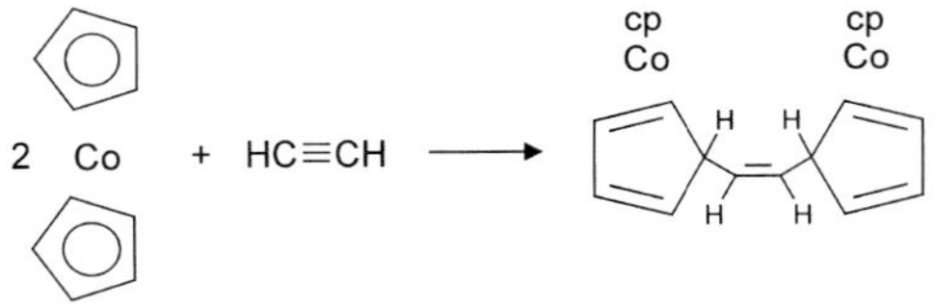

$$2\ Co(cp)_2 + HC{\equiv}CH \longrightarrow \text{(cp)Co–CH=CH–CH=CH–Co(cp)} \qquad (34)$$

Arene-solvated cobalt atoms, obtained by reacting cobalt vapor and arenes, have been used by Italian workers to promote the conversion of α,ω-dialkynes and ni-

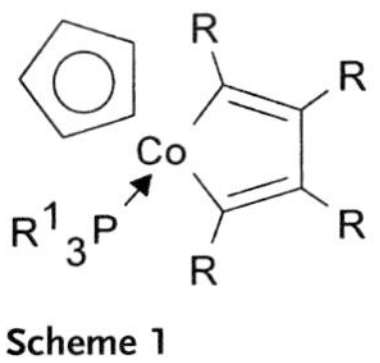

Scheme 1

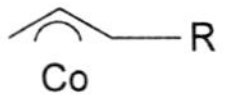

2.8 Cyclomerization of Alkynes

Scheme 2 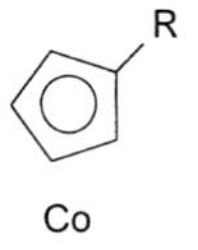   **Scheme 3** 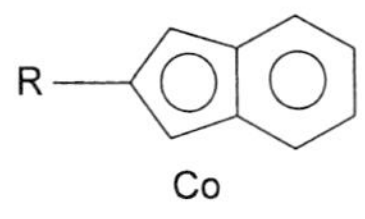   **Scheme 4** 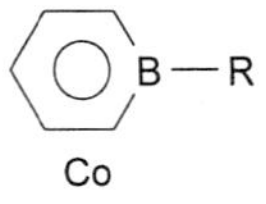   **Scheme 5**

triles, giving alkynyl-substituted pyridines [81]. η^6-toluene iron(0) complexes as well as η^6-phosphinine Fe cod have also been utilized for the cocyclotrimerization of acetylene and alkylcyanides or benzonitrile, giving α-substituted pyridine derivatives. However, the catalytic activity cannot compete successfully with the η^5-cp Co systems, and the transformation to the industrially important 2-vinylpyridine fails: acrylonitrile cannot be cocyclotrimerized with acetylene at the iron catalyst [82].

In 1989, G. Oehme, H. Pracejus, and W. Schulz reported a photo-assisted synthesis of α-substituted pyridines under mild conditions using η^5-cp Co complexes as the catalyst. Mixtures of alkylcyanides and acetylene were irradiated with light of 360–500 nm at room temperature [83]. To overcome the side reaction forming benzene from three ethyne molecules, B. Heller and G. Oehme extended the photo-assisted reaction by irradiation of the reaction mixture with sunlight, using water as solvent with the inclusion of surfactants. The low solubility of acetylene in water results in a very high pyridine-to-benzene ratio (chemoselectivity) in the catalysis [84]. A. W. Fatland and B. E. Eaton used water-soluble Co complexes for the synthesis of pyridines starting with 2-butyne-1-ol or 2-butyne-1,4-diol, respectively, and diverse nitriles. In all reactions they found no benzene side products from the competing alkyne cyclotrimerization [85]. An overview of the pyridine synthesis in water was given by B. Heller [86].

Whereas alkynes undergo cyclotrimerization in superheated and supercritical water, the cyclization of alkynes with acetonitrile fails because of the hydrolysis of the nitrile under such conditions [87].

H. Bönnemann and co-workers [88] as well as others [89] have tried acetylacetonato- and η^5-cp Rh complexes as catalysts in the pyridine synthesis. Resin-attached η^5-cp-rhodium complexes are active in the cocyclization of alkynes and nitriles, and, similarly to the cobalt case, the activity was found to depend on the nature of the η^5-ligands bonded to rhodium [89]. However, rhodium catalysts are generally less effective than the analogous cobalt systems.

The substituent on the alkyne R^2 and the cyano group R^1 can be widely varied (Eq. (35)).

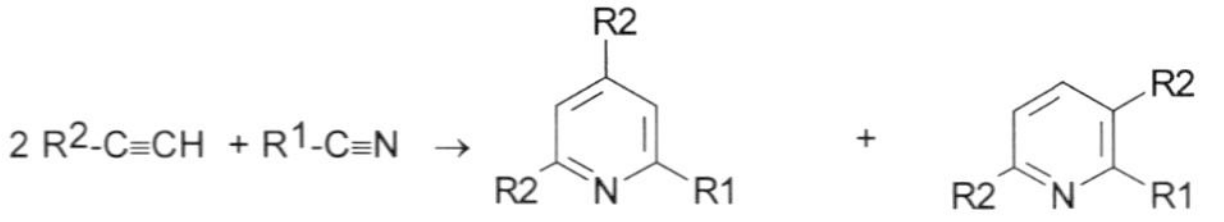

$$2\ R^2\text{-C}{\equiv}\text{CH} + R^1\text{-C}{\equiv}\text{N} \rightarrow \quad\quad\quad\quad\quad\quad\quad\quad\quad\quad (35)$$

(R^1, R^2 = H, alkyl, aryl, CO_2Me, etc.)

H. Bönnemann and co-workers developed the basic catalytic reaction (see Eq. (27)) into a general synthetic method for the selective preparation of pyridines. Only small amounts of benzene derivatives are formed as the byproduct.

2.8.3.2.1 Pyridine

The parent compound has been prepared under mild conditions using the homogeneous η^6-1-phenylborininato Co cod catalyst (Eq. (27)) [90]. However, the turnover number was very limited (about 100). The strong incentive for further developments lies in the fact that both HCN and acetylene are cheap bulk chemicals in industry.

The introduction of boron into the carbocyclic ligand attached to the cobalt enhances the catalyst lifetime considerably, probably via the suppression of the protolytic 1,4-addition of HCN to the olefinic cobaltacycle; the resulting cyano-substituted 1,3-dienes cannot be displaced from the cobalt center by acetylene, and the catalytic cycle is stopped (Eq. (36)).

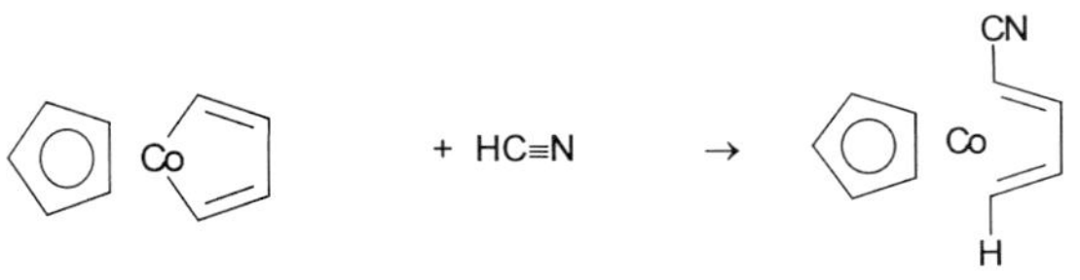

$$\text{[Co complex]} + HC\equiv N \rightarrow \text{[Co-CN complex]} \tag{36}$$

2.8.3.2.2 Alkyl-, Alkenyl- and Arylpyridines

A two-step process for the production of α-picoline has been commercialized by DSM in the Netherlands. Acrylonitrile is first reacted with a large excess of acetone [91] (Eq. (37)). In the liquid phase at 180 °C and 2.1 MPa, a monocyanoethylation product is formed, initially catalyzed by a primary amine and a weak acid. The ring closure in the vapor phase giving α-picoline is catalyzed by a palladium contact.

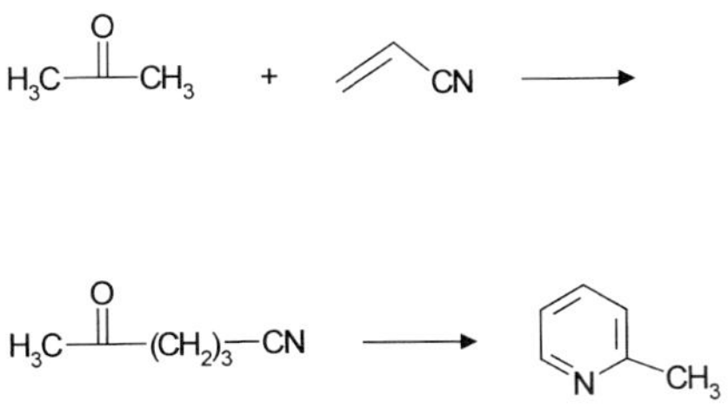

$$H_3C-CO-CH_3 + CH_2=CH-CN \longrightarrow$$

$$H_3C-CO-(CH_2)_3-CN \longrightarrow \text{2-methylpyridine} \tag{37}$$

Nippon Steel has developed an interesting liquid-phase process for α-picoline from ethylene and ammonia [92]. The catalyst is reminiscent of the well-known Wacker process, viz. the Pd^{2+}/Cu^{2+} redox system (Eq. (38)).

$$4\ H_2C=CH_2 + 4\ Pd(NH_3)^{2+} \rightarrow \text{[3,4-dimethyl-2-methylpyridine]} \qquad (38)$$

The preferred catalysts for the one-step cocyclization of acetylene and acetonitrile (or alkylcyanides in general) to give α-picoline (or 2-alkyl-pyridines), are η^5-cp Co cod or η^5-Me$_3$Si-cp Co cod (see Eq. (5)). The α-picoline synthesis is best performed in pure nitrile without any additional solvent. The acetonitrile is saturated at 20–25 °C with acetylene at 1.7 MPa. This allows the acetylene to be added in one batch at the start of the reaction. At the reaction temperature (130–150 °C), a maximum of 6 MPa may be reached, which slowly drops as the acetylene is consumed. Alternatively, a constant acetylene pressure of 2 MPa is maintained with the help of a compressor connected to the autoclave. The yields can be as high as 80% based on a 25% nitrile conversion, and the product may be easily separated from the reaction mixture. The pyridine/benzene selectivity reaches 21:1. For further experimental data see [13 d].

A significant outlet for α-picoline is the production of 2-chloro-6-(trichloromethyl)-pyridine, which is used as a nitrification inhibitor in agriculture chemistry and the manufacture of the defoliant 4-amino-2,5,6-trichloropicolinic acid. However, the major commercial outlet for α-picoline is still its use as a starting material for the two-step production of 2-vinylpyridine. The total yield of 2-vinylpyridine formed via Eq. (39) can be as high as 90%.

$$\text{2-methylpyridine} + H_2C=O \rightarrow \text{2-vinylpyridine} \qquad (39)$$

2-Vinylpyridine may also be obtained in almost quantitative yields from 2-alkylaminopyridine derivatives (directly available through cobalt catalysis) using a supported (e.g., Al$_2$O$_3$) alkali metal hydroxide (Eq. (40)) [93].

$$\text{[2-aminomethylpyridine deriv.]} \rightarrow \text{[2-vinylpyridine]} + HN\begin{smallmatrix}R1\\R2\end{smallmatrix} \qquad (40)$$

(R^1 = R^2 = alkyl, cycloalkyl, etc.)

α-Ethylpyridine, α-undecylpyridine, and other α-alkylpyridines can be prepared in an analogous way from acetylene and the alkylcyanides. The preferred catalyst is the η^5-Me$_3$Si-cp Co system. 2-Undecylpyridine is formed similarly (94% yield) and can be easily separated from the reaction mixture. The yields of conventional alkylation reactions [94] lie between 22 and 54%, suggesting that cobalt catalysis might be an attractive alternative for large-scale productions. The hydrochlorides and methiodides of a number of 2-alkylpyridines (Eq. (41)) have an effect on the aqueous surface tension and show antibacterial properties. The salts of 2-pentade-cylpyridine show the best results [95].

2-alkylpyridine + CH$_3$I → N-methyl-2-alkylpyridinium iodide (41)

Starting from optically active nitriles, C. Botteghi and co-workers [96] have applied the cobalt-catalyzed reaction for the preparation of optically active 2-substituted pyridines (Eq. (42)). The chiral centre is maintained during the alkyne nitrile cocyclization reaction. This reaction has recently been extended to the synthesis of bipyridyl compounds having optically active substituents [97] and provides an access to chiral ligands of potential interest in transition metal-catalyzed asymmetric synthesis.

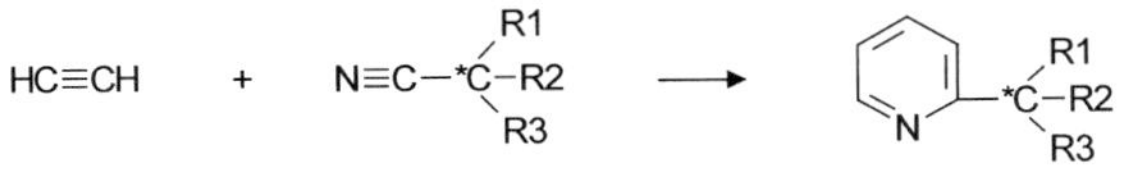

$$HC\equiv CH + N\equiv C-\overset{R1}{\underset{R3}{\overset{|}{C}}}-R2 \longrightarrow \underset{N}{\bigcirc}-\overset{R1}{\underset{R3}{\overset{|}{C}}}-R2 \quad (42)$$

(R^i = H, alkyl, aryl, CO$_2$Me, etc., R^1 ≠ R^2 ≠ R^3)

The reaction of monosubstituted alkynes with nitriles (see Eq. (35)) gives a mixture of isomeric trialkylpyridines (collidines). Collidines have been prepared using η^5-cp-Co cod at 130 °C with high turnover numbers [13d]. Especially the reaction of hex-1-yne with acetonitrile with η^5-cp Co cod as catalyst was investigated by J. S. Viljoen and J. A. K. Plessis. They found that the rate of the photochemically activated reaction can be accelerated by increasing the reaction temperature [98].

The cobalt-catalyzed cocyclization of benzonitrile and acetylene at η^5-cp Co cod gives 2-phenylpyridine in high yield (Eq. (43)) [13g].

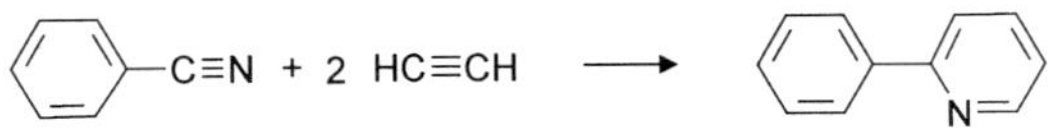

$$\bigcirc-C\equiv N + 2\ HC\equiv CH \longrightarrow \bigcirc-\underset{N}{\bigcirc} \quad (43)$$

The catalytic reaction forming pyridine derivatives may also be carried out using two different alkynes. For example, the cocyclization of acetylene and propyne with acetonitrile yields a mixture of dimethylpyridines (lutidines) in addition to α-picoline and the isomeric collidines. The cocyclization (Eq. (44)), however, turned out to be non-selective. For experimental details see [13h].

HC≡CH + CH$_3$-C≡CH + CH$_3$-C≡N →

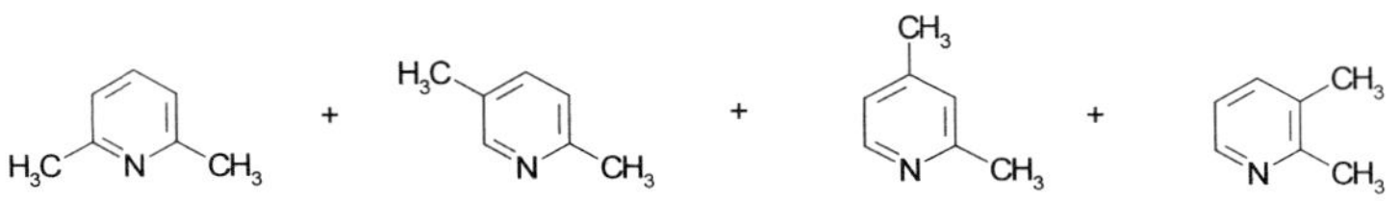

$$\underset{H_3C\ \ N\ \ CH_3}{\bigcirc} + \underset{\ \ \ \ N\ \ CH_3}{\overset{H_3C}{\bigcirc}} + \underset{\ \ \ N\ \ CH_3}{\overset{CH_3}{\bigcirc}} + \underset{\ \ \ N\ \ CH_3}{\overset{CH_3}{\bigcirc}} \quad (44)$$

Mixtures of phenylacetylene, hex-1-yne, and acetonitrile may also be cocyclotrimerized at η^5-cp Co cod. This reaction gives the trisubstituted pyridine derivatives in statistical distribution [98c].

The most interesting application from an industrial point of view is the cobalt-catalyzed one-step synthesis of 2-vinylpyridine (Eq. (45)).

$$2\ HC{\equiv}CH\ +\ \text{CH}_2{=}CH{-}CN\ \longrightarrow\ \text{2-vinylpyridine} \tag{45}$$

This way, the fine chemical can be manufactured using equal amounts by weight of the comparatively inexpensive components acetylene and acrylonitrile. The 2-vinylpyridine synthesis must be carried out in pure acrylonitrile below 130–140 °C, otherwise acrylonitrile and the product 2-vinylpyridine undergo thermal polymerization [99]. Therefore only very active catalysts can be applied in the reaction of Eq. (41). The best results were obtained using η^6-1-phenylborininato Co cod as the catalyst (productivity: 2.78 kg 2-vinylpyridine per g cobalt). A solution of the catalyst in acrylonitrile is saturated with acetylene at 2 MPa and then heated up to 130 °C (for experimental details see [13e]). The catalytic turnover number exceeds 2000 (2 h). Remarkably, no pseudo-Diels-Alder reaction at the C=C double bond of the acrylonitrile was observed. (Eq. (46)).

$$2\ HC{\equiv}CH\ +\ \text{CH}_2{=}CH{-}CN\ \not\longrightarrow\ \text{product} \tag{46}$$

This heterocyclization reaction may also be performed under normal pressure in toluene solution by irradiation with light, but the turnover numbers are less than by thermal catalysis. Fumaronitrile, maleonitrile, or allyl cyanide could not be reacted with acetylene by irradiation [100].

The outlet for 2-vinylpyridine is the manufacture of copolymers for the use in tire cord binders. The tire cord is treated first with a resorcinol-formaldehyde polymer and then with a terpolymer made from 15% 2-vinylpyridine, styrene, and butadiene. This treatment gives the close bonding of tire cord to rubber essential in the production of tires [101]. Consequently, the cord-tire markets dictate the demand. 2-Vinylpyridine is also an additive in a dying processes for acrylic fibers: 1–5% of copolymerized 2-vinylpyridine provide the reactive sites for the dye.

2.8.3.2.3 2-Amino- and 2-Alkylthiopyridines

A wide variety of substituents at the cyano group are tolerated by the cobalt catalyst. For example, monomeric cyanamide reacts with acetylene in the presence of η^6-borininato Co half-sandwich complexes to give 2-aminopyridine [13e] (Eq. (47)).

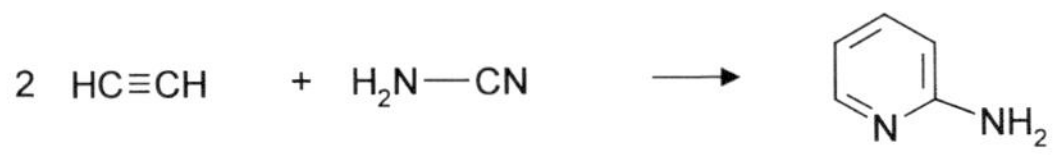

$$2\ HC{\equiv}CH\ +\ H_2N{-}CN\ \longrightarrow\ \underset{N}{\bigcirc}{-}NH_2 \tag{47}$$

2-Aminopyridine, which is of practical interest, is conventionally prepared by the substitution of the pyridine ring via the so-called Chichibabin reaction using sodium amide in dimethylaniline (Eq. (48)).

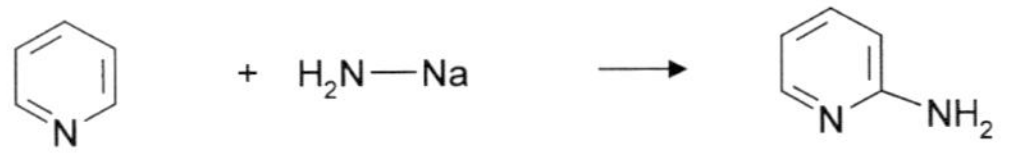

$$\underset{N}{\bigcirc}\ +\ H_2N{-}Na\ \longrightarrow\ \underset{N}{\bigcirc}{-}NH_2 \tag{48}$$

The product is obtained in 85% yield by treating with aqueous NaOH followed by distillation [102]. 2-Aminopyridine is used in the manufacture of several chemotherapeutics, dyes for acrylic fibers, and as an additive for lubricants [103]. Alkylthiocyanates can also be used as the cyano component, and react [104] to give 2-alkylthiopyridines (see Eq. (19)), which are otherwise accessible only by multistep synthetic pathways [105]. The catalytic reaction (Eq. (49)) seems to offer an easy entry into the pyrithione systems.

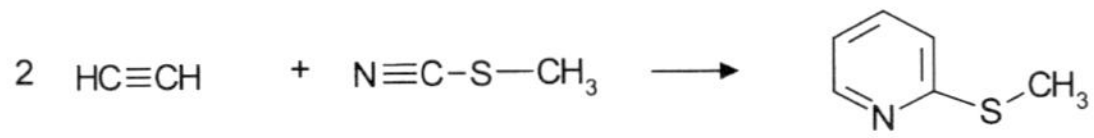

$$2\ HC{\equiv}CH\ +\ N{\equiv}C{-}S{-}CH_3\ \longrightarrow\ \underset{N}{\bigcirc}{-}S{-}CH_3 \tag{49}$$

The classical access to this is given in Eq. (50). 2-Chloropyridine-N-oxide reacts with sodium hydrogen sulphide to give pyrithione, which, in the form of its zinc salt, is added to hair cosmetics as a general antifungal agent [106].

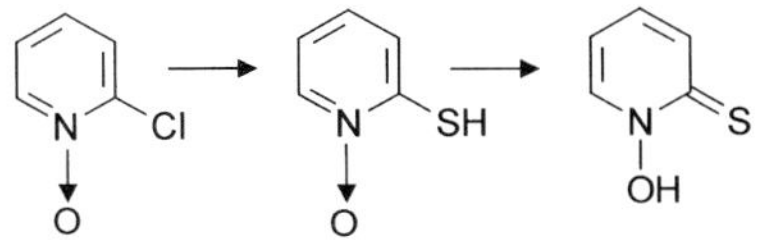

$$\underset{O^-}{\underset{N^+}{\bigcirc}}{-}Cl\ \longrightarrow\ \underset{O^-}{\underset{N^+}{\bigcirc}}{-}SH\ \longrightarrow\ \underset{OH}{\underset{N}{\bigcirc}}{=}S \tag{50}$$

2.8.3.3
Bipyridyls

The industrial route for 2,2'-bipyridyl consists in the dehydro-dimerization of pyridine on Raney nickel using a process developed by the Imperial Chemical Industries [107]. 2,2'-Bipyridyl reacts with ethylene bromide to give 1,1'-ethylene-2,2'-bipyridilium dibromide (diquat). The production of one ton of diquat (which is widely used as a herbicide) requires 1.2 tons of pyridine (Eq. (51)).

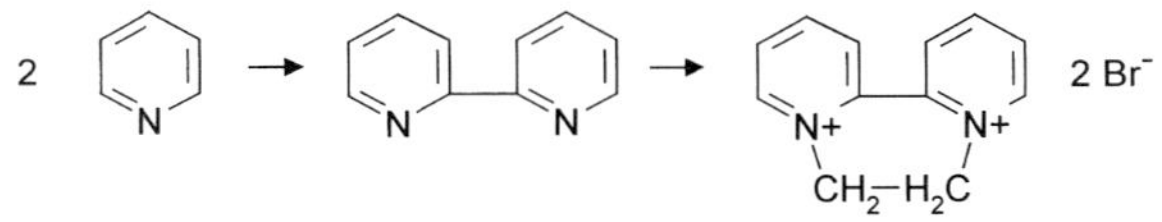

$$2\ \underset{N}{\bigcirc}\ \longrightarrow\ \underset{N}{\bigcirc}{-}\underset{N}{\bigcirc}\ \longrightarrow\ \underset{\underset{CH_2{-}H_2C}{N^+}}{\bigcirc}{-}\underset{N^+}{\bigcirc}\quad 2\ Br^- \tag{51}$$

2.8 Cyclomerization of Alkynes

The cobalt-catalyzed synthesis enables 2,2'-bipyridyl to be prepared directly from 2-cyanopyridine and acetylene in a 72% yield with a 2-cyanopyridine conversion of 21% (Eq. (52)).

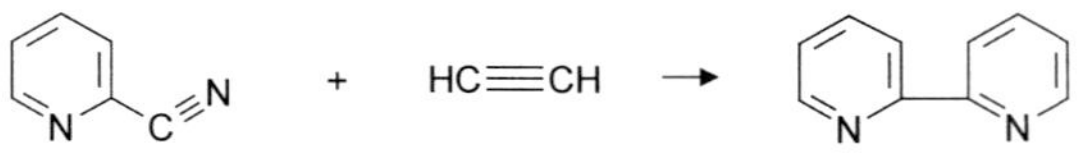

$$\text{2-pyridyl-C}\equiv\text{N} + \text{HC}\equiv\text{CH} \rightarrow \text{2,2'-bipyridyl} \qquad (52)$$

This reaction has to be carried out in benzene or toluene, and a comparatively high acetylene pressure has to be maintained in order to achieve a sufficiently high, stationary alkyne concentration in the solution (for experimental details see 13d). Starting from readily available cyanopyridines, reaction with alkynes leads to substituted bipyridyls (Eq. (53)).

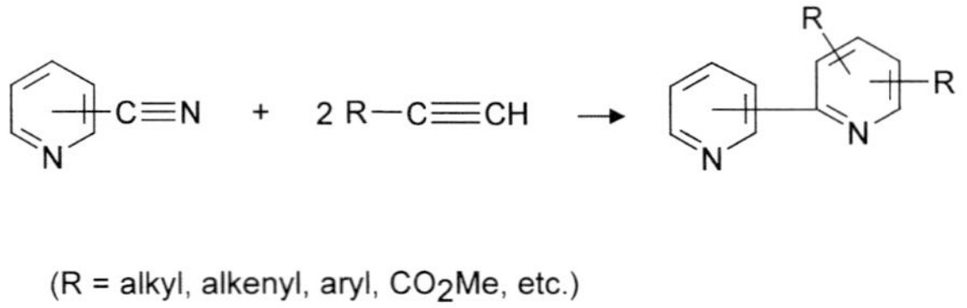

$$\text{pyridyl-C}\equiv\text{N} + 2\,\text{R}-\text{C}\equiv\text{CH} \rightarrow \text{substituted bipyridyl} \qquad (53)$$

(R = alkyl, alkenyl, aryl, CO_2Me, etc.)

Polynuclear pyridine derivatives can also be synthesized [108]. Use of cyanoalkylpyridine and acetylene as the substrates gives the respective parent bipyridyl. Substituted alkynes give two positional isomers (Eq. (54)).

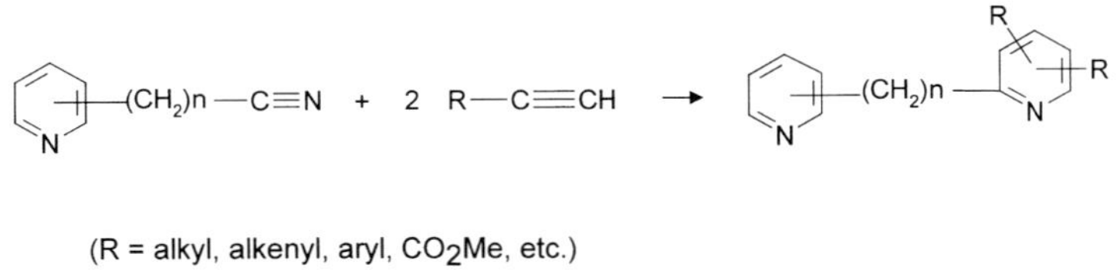

$$\text{pyridyl}-(\text{CH}_2)_n-\text{C}\equiv\text{N} + 2\,\text{R}-\text{C}\equiv\text{CH} \rightarrow \text{product} \qquad (54)$$

(R = alkyl, alkenyl, aryl, CO_2Me, etc.)

2.8.3.4
Isoquinolines

A reaction pathway to 3-substituted isoquinolines via coupling of aryl- and alkenyl-substituted terminal acetylenes with the *t*-butylimines of *ortho*-iodobenzaldehydes in the presence of a Pd catalyst was found by R.C. Larock and co-workers (Eq. (55)). In addition, they reported the synthesis of isoquinoline heterocycles by the Cu-catalyzed cyclization of iminoalkynes. In both cases the isoquinolines were prepared in good to excellent yields. The total synthesis of the natural product decumbenine B has also been accomplished by employing this new reaction route [109, 110].

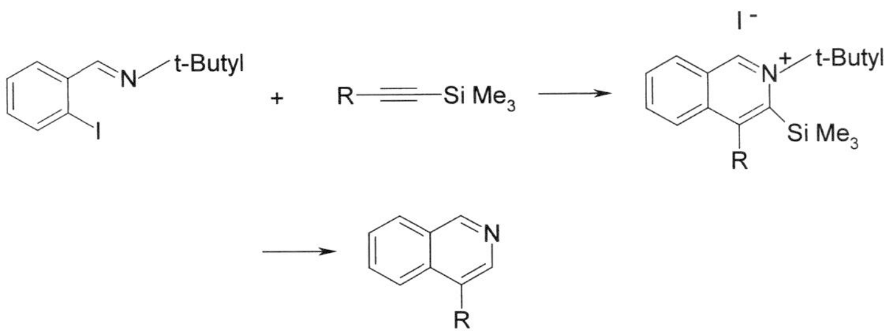

2.8.3 Transition Metal-Catalyzed Syntheses of 6-Membered Heterocycles

(55)

2.8.3.5
Miscellaneous

An interesting variation is the reaction of α,ω-diynes on η^5-cp Co diene complexes. 1,7-Octadiyne initially undergoes an intramolecular process to give, in the presence of excess nitrile, derivatives of tetrahydroisoquinoline in ca. 60% yield (Eq. (56)).

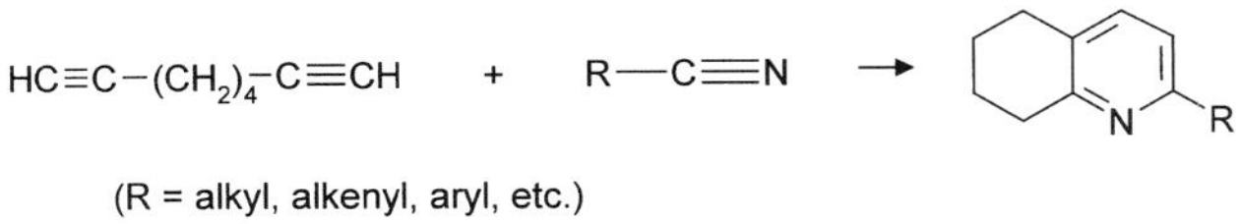

$$HC\equiv C-(CH_2)_4-C\equiv CH \quad + \quad R-C\equiv N \longrightarrow$$

(R = alkyl, alkenyl, aryl, etc.)

(56)

The annelated pyridine is also obtained with η^5-cp Co dicarbonyl as catalyst [12b]. Using this variant of the cobalt-catalyzed cycloaddition, K. Schleich et al. [111] opened up a new route to pyridoxine (vitamin B_6) as its hydrochloride (Eq. (57)).

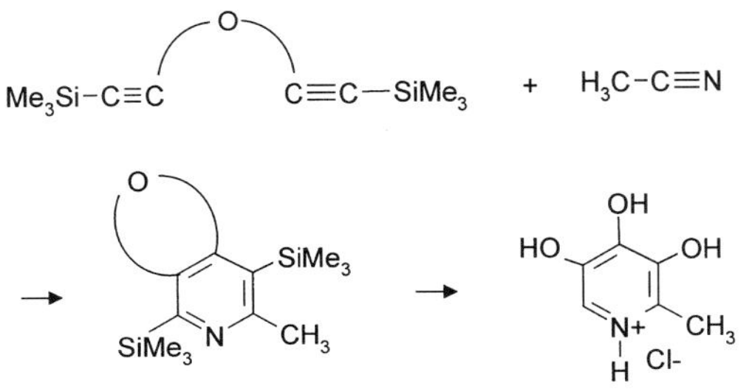

(57)

Applying the versatility of cobalt-catalyzed pyridine formation (see Eq. (5)), K.P.C. Vollhardt [112] has extensively varied the basic reaction. Using rather sophisticated alkyne and nitrile precursors with η^5-cp Co dicarbonyl as the catalyst, a number of polyheterocyclic systems having physiological interest were prepared. Using Eq. (51), a synthetic route to the isoquino-[2,1-5]-2,6-naphthyridine nucleus (Eq. (58)) was developed [113].

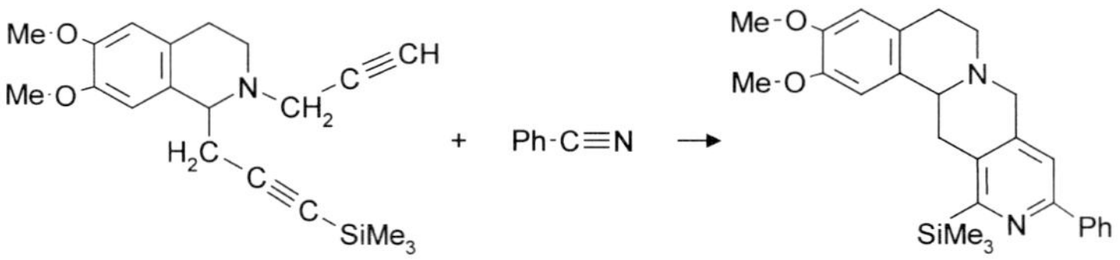

2.8 Cyclomerization of Alkynes

(58)

6-Heptynenitrile was incorporated into the indole system, giving a pyridine derivative (Eq. (59)) related to the ergot alkaloids [114].

(59)

(R^1, R^2, R^3 = alkyl, aryl, Si(CH$_3$)$_3$, etc.)

C. Saa et al. prepared 7,7′- and 8,8′-spiropyridines by means of a Co-catalyzed double cyclization of bis-alkynenitriles and alkynes. These spiropyridines will be used as novel C_2-symmetric ligands [115].

Polypyridines may be obtained by the cobaltocene-catalyzed cycloaddition copolymerization of diynes with nitriles. T. Tsuda and H. Maehara found this remarkable synthesis by the reaction of 1,11-dodecadiyne with acetonitrile in toluene at 150 °C (Eq. (60)). The molecular weight of the polymer was up to 18 000. The length of the methylene chain connecting the two C≡C triple bonds controls the synthesis: shorter chains than eight CH_2 groups give only the intramolecular process and no polymer product [116].

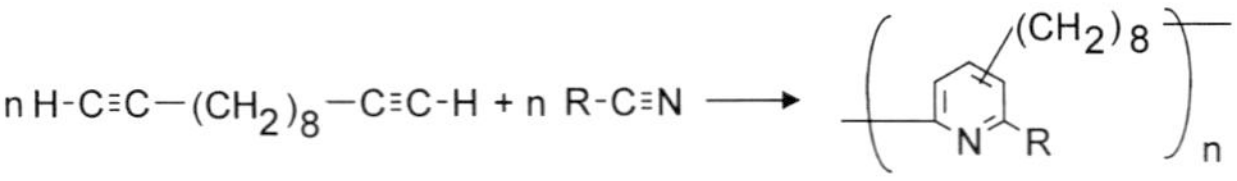

(60)

2.8.4 List of Abbreviations

cp = cyclopentadienyl
ind = indenyl
flu = fluorenyl
Ph = phenyl
Me = methyl
Et = ethyl
cod = cycloocta-1,5-diene
dba = dibenzylideneacetone

2.8.5 References

1. Ullmann's Encyclopedia of Industrial Chemistry, Volume A 1, **1985**, 97
2. (a) Reppe, W., *Neue Entwicklungen auf dem Gebiet der Chemie des Acetylens und Kohlenoxids*, Springer, Berlin Göttingen – Heidelberg, **1949**; (b) Reppe, W., *Chemie und Technik der Acetylen-Druckreaktionen*, 2nd edn., Verlag Chemie, Weinheim **1952**
3. (a) M. Berthelot, *Liebigs Ann. Chem.* **1866**, *141*, 173; (b) M. Berthelot, *Hebd. Seances Acad. Sci.*, **1866**, 905
4. W. Reppe, O. Schlichting, K. Klager, T. Toepel, *Justus Liebigs Ann. Chem.* **1948**, 560, 1
5. (a) G. N. Schrauzer, S. Eichler, *Chem. Ber.* **1962**, *95*, 550; (b) G. N. Schrauzer, P. Glockner, S. Eichler, *Angew. Chem.* **1964**, *76*, 28; *Angew. Chem., Int. Ed. Engl.* **1964**, *3*, 185
6. (a) W. Geibel, G. Wilke, R. Goddard, C. Krüger, R. Mynott, *J. Organomet. Chem.* **1978**, *160*, 139; (b) G. Wilke, *Angew. Chem.* **1988**, *100*, 189; *Angew. Chem., Int. Ed. Engl.* **1988**, *27*, 185
7. (a) R. E. Colborn, K. P. C. Vollhardt, *J. Am. Chem. Soc.* **1981**, *103*, 6259; (b) R. E. Colborn, K. P. C. Vollhardt, *J. Am. Chem. Soc.* **1986**, *108*, 5470
8. (a) P. Cini, N. Palladino, A. Santambrogio, *J. Chem. Soc. C* **1967**, 836; (b) J. R. Leto, M. F. Leto, *J. Am. Chem. Soc.* **1961**, *83*, 2944
9. L. H. Simons, J. J. Lagowski, *Fund. Res. Homogeneous Catal.* **1978**, *2*, 73
10. H. Schwager, S. Spyroudis, K. P. C. Vollhardt, *J. Organomet. Chem.*, **1990**, *382*, 191
11. (a) H. Yamazaki, Y. Wakatsuki, *Tetrahedron Lett.* **1973**, 3383; (b) H. Yamazaki, Y. Wakatsuki, *J. Organomet. Chem.* **1977**, *139*, 157; (c) Y. Wakatsuki, H. Yamazaki, *J. Organomet. Chem.* **1977**, *139*, 169; (d) Y. Wakatsuki, H. Yamazaki, *J. Chem. Soc. Dalton Trans.* **1978**, 1278
12. (a) K. P. C. Vollhardt, *Acc. Chem. Res.* **1977**, *10*, 1; (b) A. Naiman, K. P. C. Vollhardt, *Angew. Chem.* **1977**, *89*, 758; *Angew. Chem., Int. Ed. Engl.* **1977**, *16*, 708; (c) J. R. Fritch, K. P. C. Vollhardt, *Angew. Chem.* **1980**, *92*, 570; *Angew. Chem., Int. Ed. Engl.* **1980**, 559; (d) G. Ville, K. P. C. Vollhardt, M. J. Winter, *J. Am. Chem. Soc.* **1981**, *103*, 5267; (e) J. P. Tane, K. P. C. Vollhardt, *Angew. Chem.* **1982**, *94*, 642; (f) J. R. Fritch, K. P. C. Vollhardt, *Organometallics* **1982**, *1*, 590; (g) J. S. Drage, K. P. C. Vollhardt, *Organometallics* **1982**, *1*, 1545; (h) D. J. Brien, A. Naiman, K. P. C. Vollhardt, *J. Am. Chem. Soc.* **1982**, *104*, 133
13. (a) H. Bönnemann, R. Brinkmann, H. Schenkluhn, *Synthesis* **1974**, 575; (b) Studiengesellschaft Kohle m. b. H. (H. Bönnemann, H. Schenkluhn) US Pat. 4006149 (**1975**); (c) H. Bönnemann, *Angew. Chem.* **1978**, *90*, 517, *Angew. Chem.*,

Int. Ed. Engl. **1978**, *17*, 505; (d) H. Bönnemann, W. Brijoux, *Aspects Homogeneous Catal.* **1984**, *5*, 75; (e) H. Bönnemann, W. Brijoux, R. Brinkmann, W. Meurers, R. Mynott, W. von Philipsborn, T. Egolf, *J. Organomet. Chem.* **1984**, *272*, 231; (f) H. Bönnemann, *Angew. Chem.* **1985**, *97*, 264; *Angew. Chem., Int. Ed. Engl.* **1985**, *24*, 248; (g) H. Bönnemann, W. Brijoux, *Aspects Homogeneous Catal.* **1984**, *5*, 165; (h) H. Bönnemann, W. Brijoux, *Adv. Heterocycl. Chem.* **1990**, *48*, 177

14 N. E. Schore, *Chem. Rev.* **1988**, *88*, 1081
15 S. Saito, Y. Yamamoto, *Chem. Rev.* **2000**, *100*, 2901
16 U. M. Dzhemilev, F. A. Selimov, G. A. Tolstikov; *Arkivoc* **2001** part IX 85
17 S. W. Benson, *Thermochemical Kinetics* **1968**, Wiley, New York
18 J. H. Hardesty, J. B. Koerner, Th. A. Albright, G.-Y. Lee, *J. Am. Chem. Soc.* **1999**, *121*, 6055
19 (a) K. P. C. Vollhardt, *Acc. Chem. Res.* **1977**, *10*, 1; (b) K. P. C. Vollhardt, *Angew. Chem.* **1984**, *96*, 525
20 E. Müller, *Synthesis* **1974**, 761
21 R. A. Ferrieri, A. P. Wolf, *J. Phys. Chem.* **1984**, *88*, 2256
22 A. Borrini, P. Diversi, G. Ingrosso, A. Lucherini, G. Serra, *J. Mol. Catal.* **1985**, *30*, 181
23 P. Caddy, M. Green, E. O'Brien, L. E. Smart, P. Woodward, *J. Chem. Soc., Dalton Trans.* **1980**, 962
24 (a) P. M. Maitlis, *Acc. Chem. Res.* **1976**, *9*, 93; (b) P. M. Maitlis, E. A. Kelly, *J. Chem. Soc., Dalton Trans.* **1979**, 167; (c) F. Canziani, C. Allevi, L. Garlaschelli, M. C. Malatesta, A. Albinati, F. Ganazzoli, *J. Chem. Soc., Dalton Trans.* **1984**, 2637
25 P. M. Maitlis, *J. Organomet. Chem.* **1980**, *200*, 161
26 K. Abdulla, B. L. Booth, C. Stacey, *J. Organomet. Chem.* **1985**, *293*, 103
27 G. M. Whitesides, W. J. Ehmann, *J. Am. Chem. Soc.* **1969**, *91*, 3800
28 A. K. Jhingan, W. F. Maier, *J. Org. Chem.* **1987**, *52*, 1161
29 (a) I. Amer, T. Bernstein, M. Eisen, J. Blum, K. P. C. Vollhardt, *J. Mol. Catal.* **1990**, *60*, 313; (b) I. Amer, J. Blum, K. P. C. Vollhardt, *J. Mol. Catal.* **1990**, *60*, 323
30 S. Saito, T. Kawasaki, N. Tsuboya, Y. Yamamoto, *J. Org. Chem.* **2001**, *66*, 796
31 (a) W. Hübel, C. Hoogsand, *Chem. Ber.* **1960**, *93*, 103; (b) O. S. Mills, G. Robinson, *Proc. Chem. Soc.* **1964**, 187
32 M. Speranza, R. A. Ferrieri, A. P. Wolf, F. Cacace, *J. Labeled Compd. Radiopharm.* **1982**, *19*, 61
33 Y. Sato, K. Ohashi, M. Mori, *Tetrahedron Lett.* **1999**, *40*, 5231
34 M. S. Sigman, A. W. Fatland, B. E. Eaton; *J. Am. Chem. Soc.* **1998**, *120*, 5130
35 A. Carbonaro, A. Greco, G. Dall'Asta, *Tetrahedron Lett.* **1968**, 5129
36 D. M. Singleton, *Tetrahedron Lett.* **1973**, 1245
37 A. Chalk, *J. Am. Chem. Soc.* **1972**, *94*, 5928
38 L. D. Brown, K. Itoh, H. Suzuki, K. Hirai, J. A. Ibers, *J. Am. Chem. Soc.* **1978**, *100*, 8232
39 E. Dunach, R. L. Halterman, K. P. C. Vollhardt, *J. Am. Chem. Soc.* **1985**, *107*, 1664
40 H. Suzuki, K. Itoh, Y. Ishii, K. Simon, J. A. Ibers, *J. Am. Chem. Soc.* **1976**, *98*, 8494
41 M. G. J. Tadic, Ph. D. Thesis **1990**, Ruhr-Universität Bochum, FRG
42 (a) W. G. Dauben, M. S. Kellog, *J. Am. Chem. Soc.* **1980**, *102*, 4456; (b) W. G. Dauben, E. G. Olson, *J. Org. Chem.* **1980**, *45*, 3377
43 Ch. H. Oh, H. R. Sung, S. H. Jung, Y. M. Lim; *Tetrahedron Lett.* **2001**, *42*, 5493
44 D. W. Macomber, A. G. Verma, *Organometallics* **1988**, *7*, 1241
45 K. P. C. Vollhardt, *Pure Appl. Chem.* **1985**, *57*, 1819
46 (a) E. D. Sternberg, K. P. C. Vollhardt, *J. Am. Chem. Soc.* **1980**, *102*, 4841; (b) E. D. Sternberg, K. P. C. Vollhardt, *J. Org. Chem.* **1984**, *49*, 1564
47 (a) E. D. Sternberg, K. P. C. Vollhardt, *J. Org. Chem.* **1984**, *49*, 1574; (b) H. Butenschön, M. Winkler, K. P. C. Vollhardt, *J. Chem. Soc., Chem. Commun.* **1986**, 388; (c) S. H. Lecker, N. H. Nguyen, K. P. C. Vollhardt, *J. Am. Chem. Soc.* **1986**, *108*, 856

48 T. R. Gadek, K. P. C. Vollhardt, *Angew. Chem., Int. Ed. Engl.* **1981**, *20*, 802
49 (a) A. Carbonaro, A. Greco, G. Dall'Asta, *J. Org. Chem.* **1968**, *33*, 3948; (b) A. Carbonaro, A. Greco, G. Dall'Asta, *J. Organomet. Chem.* **1969**, *20*, 177; (c) H. tom Dieck, R. Diercks, *Angew. Chem., Int. Ed. Engl.* **1983**, *22*, 778
50 K. Mach, H. Antropiusova, L. Petrusova, F. Turecek, V. Hanus, P. Sedmera, J. Schraml, *J. Org. Chem.* **1985**, *289*, 331
51 K. Tanaka, G. C. Fu, *Org. Lett.* **2002**, *4* (6), 933
52 (a) W. Reppe, H. Vetter, *Justus Liebigs Ann. Chem.* **1953**, *582*, 133; (b) H. W. Sternberg, R. Markby, I. Wender, *J. Am. Chem. Soc.* **1985**, *107*, 1009; (c) R. S. Dickson, H. P. Kirsch, *Aust. J. Chem.* **1974**, *27*, 61; (d) J. L. Davidson, M. Green, F. G. A. Stone, A. J. Welch, *J. Chem. Soc., Dalton Trans.* **1976**, 738; (e) S. McVey, P. M. Maitlis, *J. Organomet. Chem.* **1969**, *19*, 169; (f) F. Canziani, M. C. Malatesta, *J. Organomet. Chem.* **1975**, *90*, 235
53 L. S. Liebeskind, S. L. Baysdon, M. S. South, S. Iyer, J. P. Leeds, *Tetrahedron* **1985**, *41*, 5839.
54 B. C. Berris, Y.-H. Lai, K. P. C. Vollhardt, *J. Chem. Soc., Chem. Commun.* **1982**, 953.
55 B. C. Berris, G. H. Hovakeemian, Y.-H. Lai, H. Mestdagh, K. P. C. Vollhardt, *J. Am. Chem. Soc.* **1985**, *107*, 5670
56 D. Peña, D. Pérez, E. Guitián, L. Castedo, *J. Am. Chem. Soc.* **1999**, *121*, 5827
57 M. Catellani, E. Motti, S. Baratta, *Org. Lett.* **2001**, *3* (23), 3611
58 G. Wilkinson, F. G. A. Stone, E. W. Abel (Eds.) *Comprehensive Organometallic Chemistry*, Vols. 7 and 8, Pergamon Press **1982**
59 C. W. Bird, *J. Organomet. Chem.* **1973**, *47*, 281
60 (a) Y. Inoue, Y. Itoh, H. Hashimoto, *Chem. Lett.* **1978**, 633; (b) Y. Inoue, Y. Itoh, H. Kazama, H. Hashimoto, *Bull. Chem. Soc. Jpn.* **1980**, *53*, 3329
61 (a) D. Walther, H. Schönberg, E. Dinjus, J. Sieler, *J. Organomet. Chem.* **1987**, *334*, 377; (b) D. Walther, *Coord. Chem. Rev.* **1987**, *79*, 135
62 H. Hoberg, Y. Peres, A. Milchereit, S. Gross, *J. Organomet. Chem.* **1987**, *345*, C17
63 D. Walther, G. Bräunlich, R. Kempe, J. Sieler, *J. Organomet. Chem.* **1992**, *436*, 109
64 D. Walther, *Nachr. Chem. Tech. Lab.* **1992**, *40*, 1220
65 (a) A. Behr, *Chem. Ing. Tech.* **1985**, *57*, 893; (b) A. Behr, *Angew. Chem.,* **1988**, *100*, 681; (c) P. Braunstein, D. Matt, D. Nobel, *Chem. Rev.* **1988**, *88*, 747
66 (a) H. Yamazaki, Y. Wakatsuki, *Kagaku Sosetsu* **1981**, *32*, 161; (b) H. Yamazaki, *J. Synth. Org. Chem.* **1987**, *45*, 244
67 (a) R. A. Earl, K. P. C. Vollhardt, *J. Am. Chem. Soc.* **1983**, *105*, 6991; (b) P. Diversi, G. Ingrosso, A. Lucherini, S. Malquori, *J. Mol. Catal.* **1987**, *40*, 267
68 S. T. Flynn, S. E. Hasso-Henderson, A. W. Parkins, *J. Mol. Catal.* **1985**, *32*, 101
69 D. F. Harvey, B. M. Johnson, Ch. S. Ung, K. P. C. Vollhardt, *Synlett*, **1989**, 15
70 T. Tsuda, T. Kiyoi, T. Miyane, T. Saegusa, *J. Am. Chem. Soc.* **1988**, *110*, 8570
71 H. Bönnemann, X. Chen, *Proc. Swiss Chem. Soc. Autumn Meet.* Bern, **1987**, 39
72 (a) G. S. Sheppard, K. P. C. Vollhardt, *J. Org. Chem.* **1986**, *51*, 5496; (b) R. Boese, H.-J. Knölker, K. P. C. Vollhardt, *Angew. Chem.* **1987**, *99*, 1067
73 (a) R. B. Woodward, M. P. Cava, W. D. Ollis, A. Hunger, H. U. Daeniker, K. Schenker, *Tetrahedron* **1963**, *19*, 247; (b) D. B. Grotjahn, K. P. C. Vollhardt, *J. Am. Chem. Soc.* **1986**, *108*, 2091; (c) K. P. C. Vollhardt, *Lect. Heterocycl. Chem.* **1987**, *9*, 61
74 R. A. Abramovitch, *Chem. Heterocycl. Compd.* **1974/1975**, *14*, Suppl. Parts 1–4,
75 (a) W. Ramsay, *Philos. Mag.* 5, **1876**, *4*, 269; (b) W. Ramsay, *Philos. Mag.* 5. **1877**, *5*, 24; (c) N. Ljubawin, *J. Russ. Phys. Chem. Ges.* **1885**, 250; (d) R. Meyer, A. Tanzen, *Ber. Dtsch. Chem. Ges.* **1913**, *46*, 3186
76 H. Bönnemann, H. Schenkluhn, Studiengesellschaft Kohle mbH, Ger. Pat. 2416295 (**1974**)

77 (a) M. Aresta, M. Rossi, A. Sacco, Inorg. Chim. Acta **1969**, *3*, 227; (b) P. Diversi, A. Guisti, G. Ingrosso, A. Lucherini, *J. Organomet. Chem.* **1981**, *205*, 239

78 Y. Wakatsuki, H. Yamazaki, *Synthesis* **1976**, 26

79 H. Yamazaki, Y. Wakatsuki, Jap. OS 7,725,780 (1975); *Chem. Abstr.* **1977**, *87*, 68168

80 P. Hardt, (a) Swiss Pat. Appl. 12,139-75 (1975), DOS 2,615,309 (1976), (b) Swiss Pat. Appl. 13079-76 (**1976**), DOS 2742541 (**1978**), (c) Swiss Pat. Appl. 9471-77 (**1977**), DOS 2742542 (**1979**), US Pat. 4196387 (**1980**)

81 G. Vitulli, S. Bertozzi, M. Vignali, R. Lazzaroni, P. Salvadori, *J. Organomet. Chem.* **1987**, *326*, C33

82 (a) U. Schmidt, U. Zenneck, *J. Organomet. Chem.* **1992**, *440*, 187; (b) D. Böhm, F. Koch, S. Kummer, U. Schmidt, U. Zenneck, *Angew. Chem.* **1995**, *107*, 251; (c) F. Knoch, F. Kremer, U. Schmidt, U. Zenneck, P. Le Floch, F. Mathey, *Organometallics* **1996**, *15*, 2713

83 (a) W. Schulz, H. Pracejus, G. Oehme, *Tetrahedron Lett.* **1989**, *30*, 1229; (b) B. Heller, J. Reihsig, W. Schulz, G. Oehme, *Appl. Organomet. Chem.* **1993**, *7*, 641; (c) B. Heller, D. Heller, G. Oehme, *J. Mol. Catal. A* **1996**, *110*, 211

84 (a) B. Heller, G. Oehme, *J. Chem. Soc., Chem. Commun.* **1995**, 179; (b) P. Wagler, B. Heller, J. Ortner, K.-H. Funken, G. Oehme, *Chem. Ing. Tech.* **1996**, *68*, 823; (c) B. Heller, XXX. Jahrestreffen deutscher Katalytiker, Eisenach (**1997**)

85 A. W. Fatland, B. E. Eaton, *Organic Lett.* **2000** *2* (20), 3131

86 B. Heller, *Nachr. Chem. Tech. Lab.* **1999** *47* (1), 9

87 E. J. Parsons, *Chemtech* **1996**, July, 30

88 (a) D. M. M. Rohe, Ph. D. Thesis **1979**, RWTH Aachen, FRG; (b) H. Bönnemann, Studiengesellschaft Kohle mbH Ger. Pat. DE 3117363.2 (**1981**); (c) H. Bönnemann, Studiengesellschaft Kohle mbH US Pat. 4588815 (**1984**)

89 P. Diversi, L. Ermini, G. Ingrosso, A. Lucherini, *J. Organomet. Chem.* **1993**, *447*, 291

90 (a) G. Herberich, W. Koch, H. Leuken, *J. Organomet. Chem.* **1978**, *160*, 17; (b) H. Bönnemann, B. Bogdanovic, Studiengesellschaft Kohle mbH Ger. Pat. Appl. 310550.1 (**1982**); (c) H. Bönnemann, B. Bogdanovic, Studiengesellschaft Kohle mbH Eur. Pat. Appl. 83/101,246.3 (1983)

91 (a) N.V. Stamicarbon (J.M. Deumens, S. H. Green) Br. Pat. 1,304,155 (1973); (b) N.V. Stamicarbon (J.M. Deumens, S.H. Green) US Pat. 3780082 (**1973**); (c) *Chem. Mark. Rep.* **1977**

92 (a) Y. Kusunoki, H. Okazeku, *Hydrocarbon Process.* **1974**, *53* (11), 129, 131; (b) Y. Kusunoki, H. Okazaki, *Nippon Kagaku Kaishi* **1981**, *12*, 1969; (c) Y. Kusunoki, H. Okazaki, *Nippon Kagaku Kaishi* **1981**, *12*, 1971

93 (a) P. Hardt, Lonza AG Swiss Pat. Appl. 76/14399 (**1976**); (b) P. Hardt, Lonza AG DOS 2751072 (**1978**)

94 (a) S. Goldschmidt, M. Minsinger, Ger. Pat. 952807 (**1956**); (b) J.P. Wibaut, C. Hoogzand, *Chem. Weekblad* **1956**, *52*, 357

95 M.J. Birchenough, *J. Chem. Soc.* **1951**, 1263

96 (a) D. Tatone, Trane Cong Dich, R. Nacco, C. Botteghi, *J. Org. Chem.* **1975**, *40*

97 C. Botteghi, private communication (**1975**)

98 (a) J.S. Viljoen, J.A.K. Plessis, *J. Mol. Catal.* **1993**, *79*, 75; (b) J.A.K. du Plessis, J.S. Viljoen, *J. Mol. Catal. A: Chemical* **1995**, *99*, 71; (c) J.A.K. du Plessis, J.S. Viljoen, *9th Int. Symposium on Homogeneous Catalysis* **1994**, *A24*, 132

99 R. Brinkmann, private communication (**1982**)

100 F. Kabaret, B. Heller, K. Kortus, G. Oehme, *Appl. Organomet. Chem.* **1995**, *9*, 651

101 D.B. Wootton, *Dev. Adhes.* **1977**, *1*, 181

102 Schering AG, DT 663891 (**1938**)

103 P. Arnall, N.R. Clark, *Chem. Process. (London)* **1971**, *17* (10), 9, 11–13, 15

104 H. Bönnemann, G.S. Natarajan, *Erdöl, Kohle, Erdgas, Petrochemie* **1980**, *33*, 328

105 R.A. Abramovitch, *Chem. Heterocycl. Compd.* **1975**, *14*, Suppl. Part 4, Chap. 15, 189

106 (a) E. Shaw, J. Bernstein, K. Losse, W. A. Lott, *J. Am. Chem. Soc.* **1950**,

107 (a) G. M. Badger, W. H. F. Sasse, *Adv. Heterocycl. Chem.* **1963**, *2*, 179; (b) M. A. E. Hodgson, *Chem. Ind. (London)* **1968**, 49; (c) L. A. Summers, *The Bipyridinium Herbicides*, Academic Press, New York, **1980**

108 H. Bönnemann, R. Brinkmann, *Synthesis* **1975**, 600

109 Q. Huang, J. A. Hunter, R. C. Larock, *J. Org. Chem.* **2002**, *67* 3437

110 (a) K. R. Roesch, R. C. Larock, *Org. Lett.* **1999** *1* (4), 553; (b) K. R. Roesch, H. Zhang, R. C. Larock; *J. Org. Chem.* **2001** 668042; (c) K. R. Roesch, R. C. Larock; *J. Org. Chem.* **2002**, *67,* 86

111 R. E. Geiger, M. Lalonde, H. Stoller, K. Schleich, *Helv. Chim. Acta* **1984**, *67*, 1274

112 (a) K. P. C. Vollhardt, J. E. Bercaw, R. G. Bergman, *J. Am. Chem. Soc.* **1974**, *96*, 4996; (b) C. A. Parnell, K. P. C. Vollhardt, *Tetrahedron* **1985**, *41*, 5791; (c) K. P. C. Vollhardt, *Lect. Heterocycl. Chem.* **1987**, *9*, 59

113 (a) D. J. Brien, A. Naiman, K. P. C. Vollhardt, *J. Chem. Soc., Chem. Commun.* **1982**, 133; (b) K. P. C. Vollhardt, *Lect. Heterocycl. Chem.* **1987**, *9*, 60

114 K. P. C. Vollhardt, *Int. Congr. Heterocycl. Chem.* **1987**, 11th, Heidelberg, FRG

115 J. A. Varela, L. Castedo, C. Saá, *Organic Lett.* **1999**, *1* (13), 2141

116 T. Tsuda, H. Maehara, *Macromolecules* **1996**, *29*, 4544.

2.9
Isomerization of Olefin and the Related Reactions

Sei Otsuka and Kazuhide Tani

2.9.1
Introduction

This chapter deals with isomerization of olefins catalyzed by transition metal complexes emphasizing its asymmetric application. We will briefly mention asymmetric skeletal rearrangement producing optically active olefins. However, intramolecular carbon–carbon bond forming reactions, e.g. asymmetric intramolecular hydroacylation of 4-pentenals, involving some kind of isomerization of olefin, are not included here because they are treated elsewhere in this book see Chapter 2.8 (first edition). Also excluded are related reactions like intramolecular cyclization of 1,5-dienes.

2.9.2
Allylamines

In the late 1970s, studies aiming to find asymmetric isomerization catalysis started in Takasago. There were industrial incentives in choosing allylamines such as **1** and **2** as the subtrates for the enantioselective isomerization to obtain enamines **3** or **4** which can be hydrolyzed to give citronellal **7**. Enantiomerically pure citronellal **7** is a rather expensive starting material to produce chiral terpenoids, as the maximum optical purity of commercial citronellal is less than 80% ee [1]. For the asymmetric isomerization of substrate **1** or **2**, Takasago first tried low-valent cobalt complexes, generally prepared by reducing the metal ion with organoaluminum in the presence of a chiral phosphine [2]. The catalyst gave only modest enantioselectivity for enamine **3** or **4**, but subsequent work by Otsuka's group produced spectacular results [3–7]. Thus, the asymmetric isomerization of geranyl-2 and neryldiethylamine **1** with a Rh(I)–(S)-BINAP (BINAP=2,2'-bis(diphenylphosphino)-1,1'- binaphthyl **5**) catalyst (*vide infra*) produced (R,E)- **3** and (S,E)-enamine **4**, respectively. The geometry of the double bond is 100% E. Rh(I)–(R)-BINAP catalyst produced also respective enantiomers in the opposite sense (Scheme 1). Every route in this scheme is achieved with comparable enantioselectivity ($\sim$ 98% ee). The high optical yields as well as the high chemical yields (>96%) met the industrial requirements for production of terpenoids [6d,e].

Transition Metals for Organic Synthesis, Vol. 1, 2nd Edition.
Edited by M. Beller and C. Bolm
Copyright © 2004 WILEY-VCH Verlag GmbH & Co. KGaA, Weinheim
ISBN: 3-527-30613-7

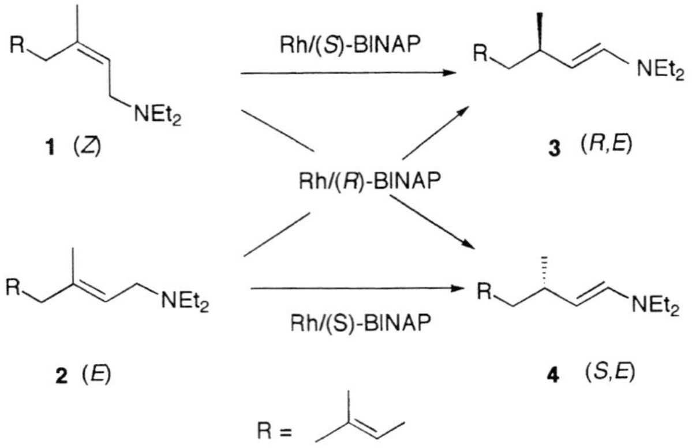

Scheme 1 Correlation diagram between stereochemistry of substrates, catalysts, and products for the Rh(I)–BINAP-catalyzed asymmetric isomerization of allylamines.

2.9.2.1
Characteristics of the Catalysis

The catalytic performance deserves a few comments. (1) Among a variety of Rh(I) complex structures, cationic tetra-coordinate complexes [Rh(diphosphine)L$_2$]$^+$ (L=solvent molecule or L$_2$=diene) were found to be quite active. The cationic nature coupled with the aryl substituents of BINAP induces an appropriate Lewis acidity on the metal center assisting substrate coordination through the amino group. In the absence of this effective amine coordination, the catalysis does not take place smoothly. This is shown [3c, 6a,b,d, 7] by slow reaction observed for a fully alkylated diphosphine Rh(I) complex catalyst and for such a substrate as N-acylallylamines. (2) The seven-membered chelation is pliable. This flexibility was interpreted to contribute to reduce the activation free energy [4, 6a]. Consistently, the slow catalytic rate was observed for [Rh(diphos)(cod)]$^+$ which forms a stable and rigid five-membered chelate ring. [Rh(diphos)$_2$]$^+$ is completely inert even above 100 °C [5]. (3) The C_2 disposition of four phenyl groups of BINAP, especially the two equatorially oriented phenyl rings, are the main gears which govern the spacial disposition of substrate substituents. Takasago has found appropriate catalytic activity at high temperature with fairly stable [Rh(binap)$_2$]$^+$. Steric congestion in this complex apparently assists the ligand dissociation observed at high temperature (>80 °C). The active species must be the mono-BINAP complex since the stereochemistry and degree of chiral recognition were exactly the same as those obtained with the mono-BINAP complex (Scheme 1) [5]. The marvelous chemical stability of [Rh(binap)$_2$]$^+$ enabled Takasago to achieve a turnover of c. 400 000 per mol catalyst. Apparently the presence of an extra BINAP molecule prevents deteriorative reactions caused by impurities. For example, [Rh(binap)(MeOH)$_2$]$^+$ reacts with water resulting in an inert trinuclear Rh(I) cluster [8]. The long life of [Rh(binap)$_2$]$^+$ invokes a term '*lid-on-off*' mechanism [6f].

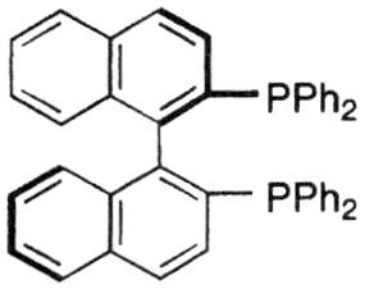

(R)-(+)-BINAP
(R)-5

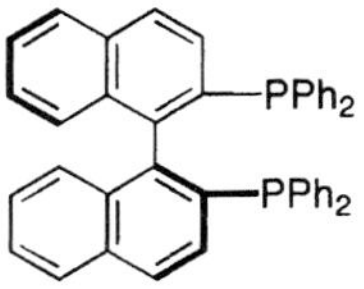

(S)-(−)-BINAP
(S)-5

2.9.2.2
Mechanisms

The mechanism of this catalysis has been extensively studied [4, 7]. The reaction kinetics are characterized with product-inhibiting catalysis. The catalytic process is initiated with coordination of the substrate nitrogen atom, followed by a stereospecific β-hydrogen elimination through a transient iminium complex **6a** or **6b** resulting in an overall enantioselective 1,3-hydrogen shift, a result verified by enantiotopically deuterated substrates.

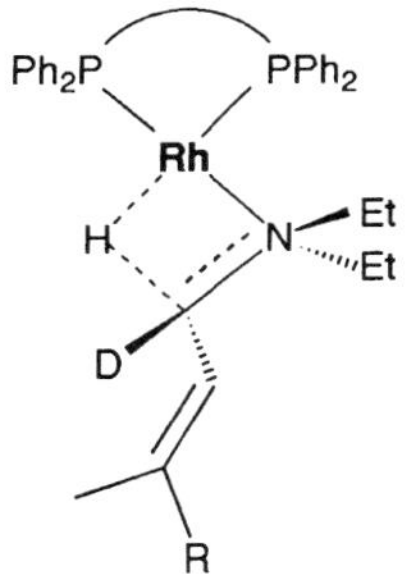

6a

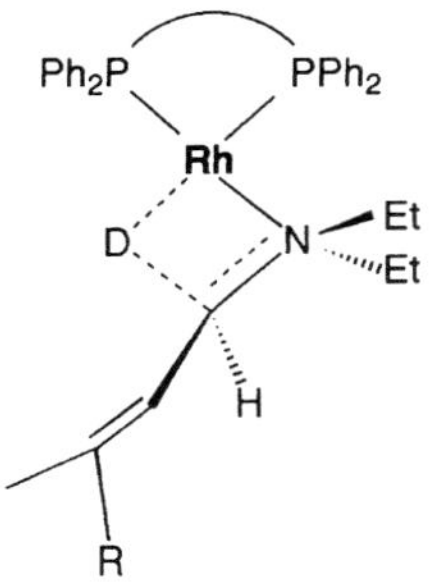

6b

2.9.2.3
Synthetic Applications

Based on the aforementioned isomerization catalysis, Takasago has been producing (−)-menthol (c. 2000 t year^{-1}) since 1983 (Scheme 2). The Lewis-acid-catalyzed ring closure of citronellal **7** to form (−)-isopulegol involves formation of two more chiral centers. The present commercial production of one particular chiral form out of eight possible optical isomers is rather remarkable. For some more details of this synthesis see Refs. [3c, 6b,d,e].

Methoprene is an insect growth regulator, which exhibits remarkable activity about 100 times that of natural juvenile hormone for pest management of yellow fever mosquito. Also it is used as an insecticide for cockroaches. The synthetic in-

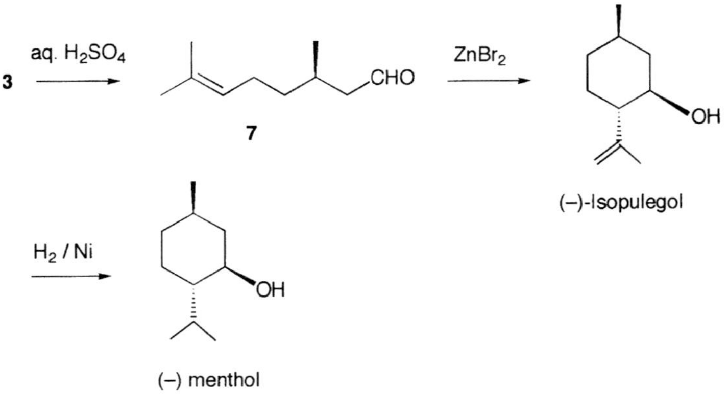

Scheme 2 Takasago's menthol synthesis.

termediate (S)-(−)-7-methoxy-3,7-dimethyloctanal **9** must be optically pure. Therefore, synthesis based on natural citronellal would not be feasible. The [Rh{(+)-binap}$_2$]$^+$-catalyzed isomerization of 7-methoxygeranylamine followed by acid hydrolysis provides the intermediate, (S)-7-methoxycitronellal **9** with satisfactory optical purity (98% ee) and chemical yield (97). The acid-catalyzed methoxylation of the (S)(E)-citronellalenamine **4** (98% ee) provides (S)(E)-7-methoxycitronellalenamine **8** which gave **9** in 79% yield without racemization (Scheme 3) [6b, 9].

In addition to the large-scale production of menthol Takasago is producing many chiral fragrance chemicals based on the Rh-catalyzed isomerization reac-

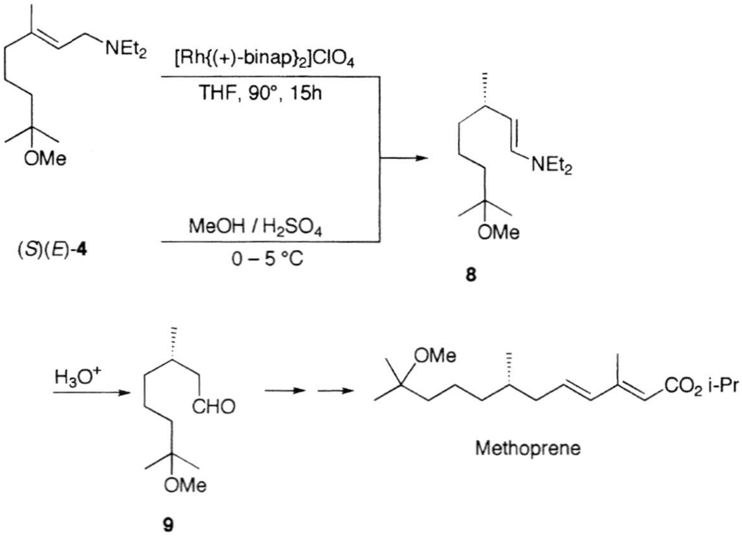

Scheme 3 Synthetic route of methoprene.

tion. Representative examples are: (+)-citronellol (**10**), (−)-7-hydroxycitronellal (**11**), (S)-7-methoxycitronellal (**9**), (S)-3,7-dimethyl-l-octanol (**12**).

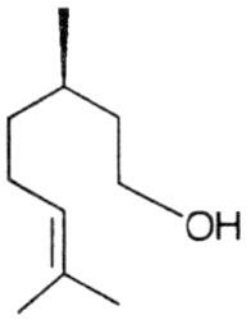

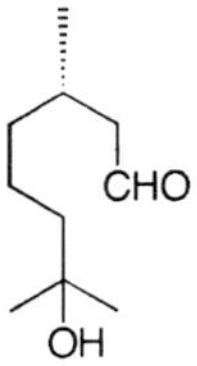

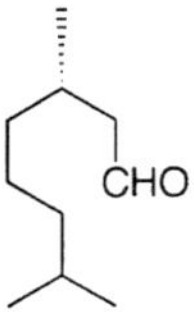

2.9.3
Allyl Alcohols

Allyl alcohols, **13–17**, can be isomerized with [Rh(binap)(sol)$_2$]$^+$ (THF, 60 °C, 24 h) to the corresponding saturated aldehydes or ketones as expected. The chemical yield are generally moderate [7]. The enantioselectivity and chemical yield for the isomerizations of prochiral allyl alcohols **18** or **19** to produce the corresponding chiral aldehyde **20** or **21** were much lower compared to the Rh(I)-catalyzed isomerization of allylamines (*vide supra*). However, kinetic resolution of the racemic allylic alcohol **22** [10] was achieved using the [Rh{(R)-binap}(MeOH)$_2$]$^+$-catalyzed isomerization. The R enantiomer of 4-hydroxy-2-cyclopentenone (R)-**22** which is a key chiral building block for prostaglandin synthesis was obtained in 91% ee and 27% yield.

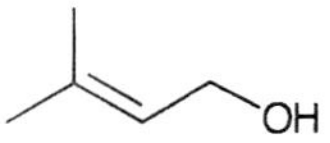

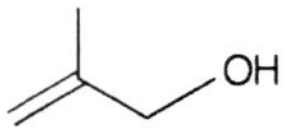

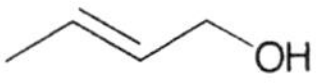

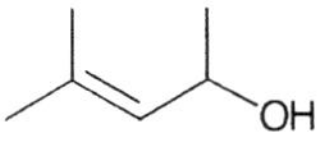

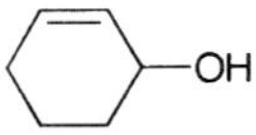

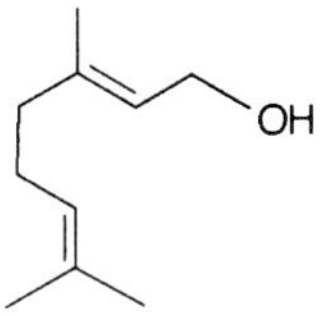

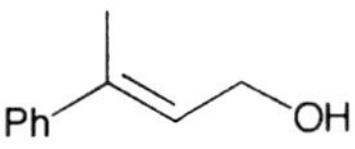

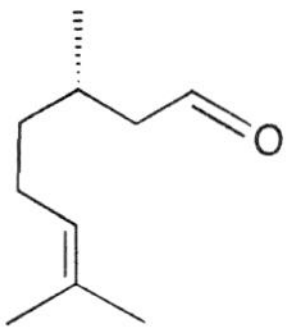

Due to the facile tautomerism simple aliphatic enols were believed to be elusive intermediates in the isomerization of allyl alcohols to ketonic compounds [11]. In 1988, Chin et al. [12] have found that even a simple aliphatic enol like 2-methylprop-1-en-1-ol **23** is quite stable in the absence of solvents as well as in aprotic solvents. Thus, the catalytic isomerization of **14** with [Rh(CO)(PPh$_3$)$_3$]ClO$_4$ gave the enol **23**. Similarly, several enols were isolated from the catalytic enolization of allyl alcohols using [Rh(CO)(PPh$_3$)$_3$]ClO$_4$ [12, 13], [Ir(COD)(PhCN)(PPh$_3$)$_3$]ClO$_4$ [13], or [Rh(diphosphine)(solvent)$_2$]$^+$ [14] as a catalyst. In contrast to the Rh(I)-catalyzed isomerization of allylamines, the isomerization of aliphatic allyl alcohols with Rh(I) catalysts generally produced a mixture of (E)- and (Z)-enols. For example, the isomerization of 2-ethylprop-2-en-l-ol **24** with [Rh(CO)(PPh$_3$)$_3$]ClO$_4$ at room temperature produced (E)- and (Z)-2-methylbut-1-en-1-ol **25**, in c. 1:6~9 (Eq. 1). Both the E and Z isomers undergo slow ketonization with the same Rh(I) catalyst. The enol production appears to proceed via a 1,3-hydrogen shift mechanism involving hydrido-π-allyl intermediates and the catalytic ketonization proceeds via a hydrido-π-oxoallyl intermediate. Isomerization of **26** with the [Rh{(S)-binap}]$^+$ catalyst provided the optically active aldehyde **28** in 18% ee (Eq. 2) [17], indicating an enantioselective tautomerization of enol **27** to the chiral aldehyde **28**.

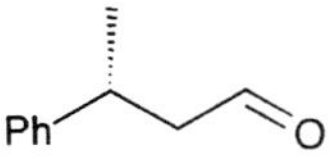

21 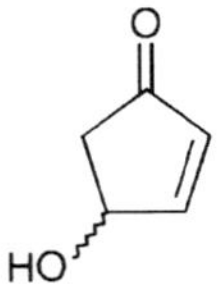 **22** 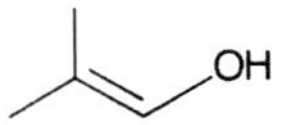 **23**

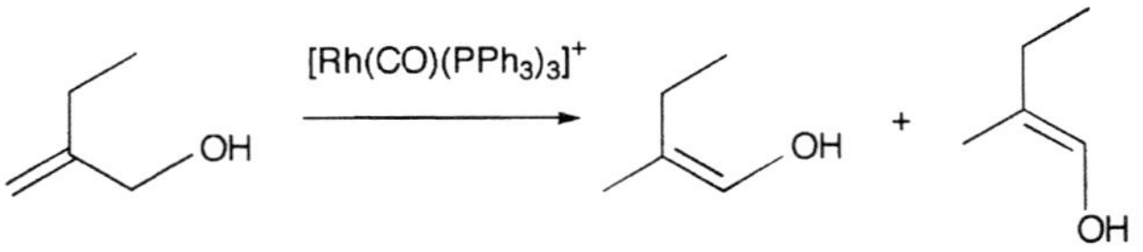

$$\text{(1)}$$

24 **Z-25** **E-25**

$$\text{(2)}$$

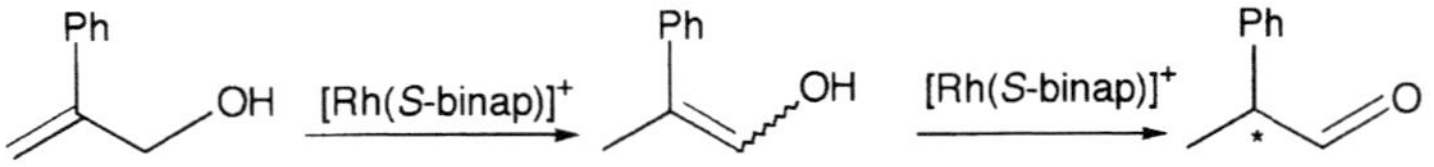

26 **27** **28**

Recently, catalytic stereospecific isomerization of racemic allyl alcohol was reported [15]. A catalytic amount of [Ru{(R)-binap}(H)(MeCN)(THF)$_2$]BF$_4$ isomerized rac-but-3-en-2-ol **29** stereoselectively to give a simple enol (Z)-but-2-en-2-ol **30**. A partial kinetic resolution of the R and S enantiomers of rac-**29** is involved. The ee was 42% (S) at 50% conversion. The isomerization of **29** using rhodium(I)-bis(phosphine) catalysts was substantially less stereoselective [14].

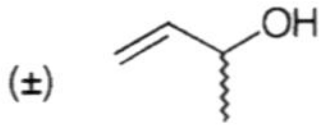

29 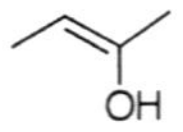 **30**

2.9.4
Allyl Ethers

Rh(I)-catalyzed isomerization of *meso*-ene-1,4-diol **31** in the presence of *S*-BINAP, gave the optically active hydroxyketone **32** in 43% ee with quantitative yield (Eq. 3) [16]. The corresponding bis-ethers **33**, especially bis(silyl) ethers, gave excellent selectivities; desilylation of the enol ether intermediate **34** with *n*-Bu_4NF gave the hydroxyketone **35a** (R=H; from **33a**) or silyloxyketone **35b** (R=TBS; TBS=Si(*t*-Bu)Me_2, from **33b**) with 93–98% ee (Eq. 4). The chemical and enantiomeric efficiencies of the method are comparable to those of the enzymatic procedures [17]. Interestingly, the directions of the enantioselection for the isomerization of the bis-ethers **33** were opposite to that of the free diol **31** with the same catalyst, although the reason being unknown. A mechanism involving a suprafacial 1,3-hydrogen migration has been proposed based on a deuterium labeling experiment [16]. The product obtained can be transformed easily into versatile chiral synthons such as a chiral 2,5-cyclohexadienone synthon **36** [17] as well as 4-hydroxy-2-cyclohexanone **37** (R=H) and its TBS ether **37** (R=TBS) [18], which serve as starting materials for the synthesis of the medicinally important compounds ML-263A and FK-506, respectively.

31 **32** (3)

33 **34** **35**

a: R = TMS (n = 1, 2)
b: R = TBS (n = 1, 2)
TMS = $SiMe_3$; TBS = $SiBu^tMe_2$

a: R = H (n = 1, 2)
b: R = TBS (n = 1, 2)
ca. 95 % yield
94-98 %ee (4)

(–)-**36**

37 (R = H, TBS)

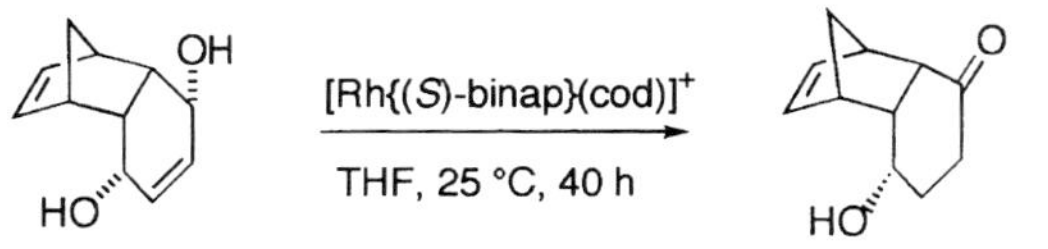

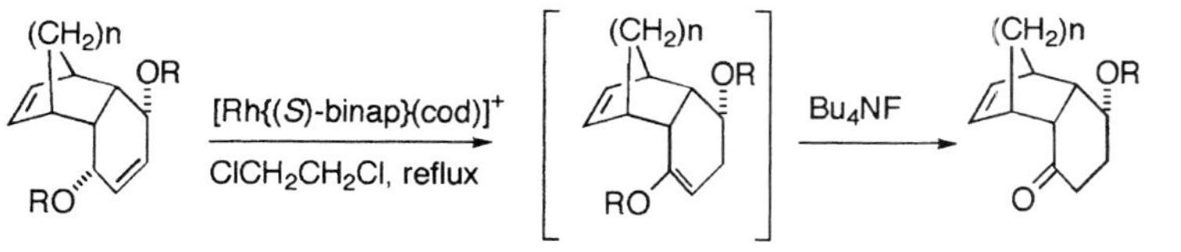

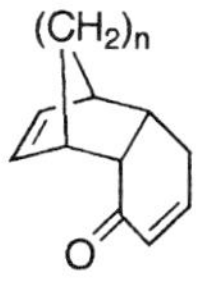

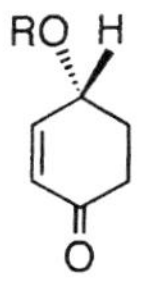

The asymmetric isomerization of allyl silyl ethers has been extended to the monocyclic *meso*-substrate, *cis*-3,7-bis(*tert*-butyldimethylsiloxy)cycloheptene, **38** [19]. The isomerization with a catalytic amount of (S)-binap-Rh(I) catalyst gave a mixture of the optically active silyl enol ether **39** and siloxy ketone (**40**, R=TBS) which can be further converted without isolation to (S)-4-hydroxycycloheptenone (**40**, R=H) in 71% ee quantitatively. The (R)-enantiomer has been used for the synthesis of (–)-(S)-physoperuvine **41**, a major alkaloid of *Physalis peruviana*.

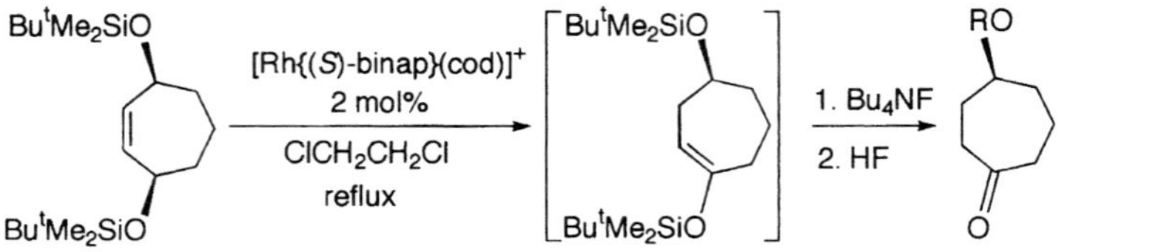

$$R = TBS; 96\,\%$$
$$R = H; 95\%; 71\,\%ee \tag{5}$$

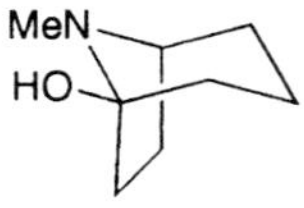

(–)-(S)-physoperuvine **41**

Asymmetric isomerizations of prochiral 4,7-dihydro-1,3-dioxepins **42** to 4,5-dihydro-1,3-dioxepins **43** with $Ru_2Cl_4(diop)_3$ or $[Rh(cod)Cl]_2/diop$ activated by $NaBH_4$ (Eq. 6) gave only low enantioselectivities (25% ee, $R = C_2H_4CN$) [20].

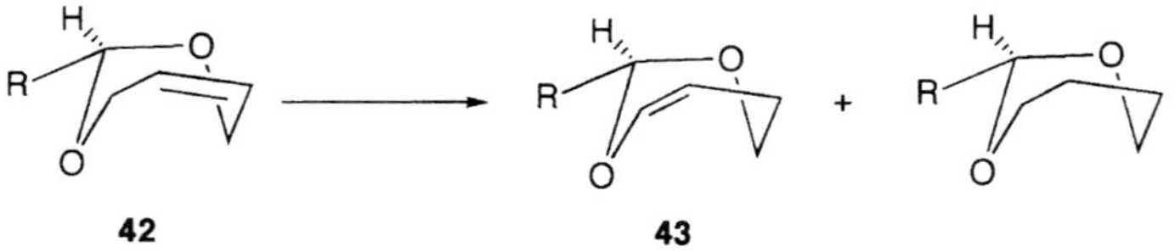

$$R = n\text{-}Bu, C_2H_4Cl, R = C_2H_4CN \tag{6}$$

2.9.5
Unfunctionalized Olefins

Despite extensive studies on the transition-metal-catalyzed isomerization of simple olefins in connection with hydrogenation or hydroformylation reactions [21], only very low asymmetric inductions [22] have been observed. However, a chiral *ansa*-bis(indenyl)titanium complex (1R,2R,4R,5R)-**47** activated by $LiAlH_4$ isomerized *meso,trans*-4-*tert*-butyl-1-vinylcyclohexane **44** to the alkene *S*-**45** in 80% ee (23 °C, 24 h) [23].

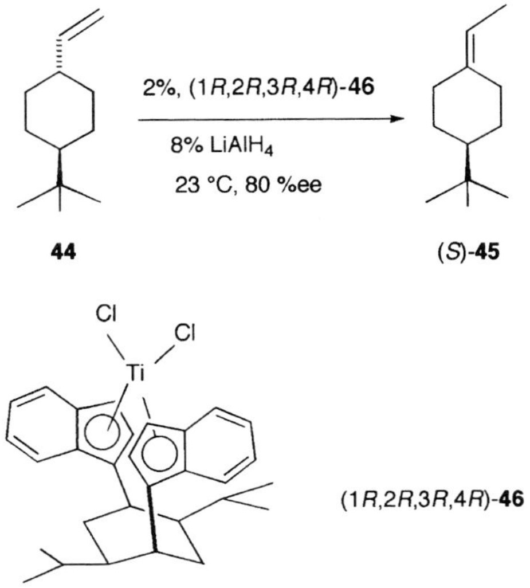

2.9.6
Asymmetric Skeletal Rearrangements

2.9.6.1
Epoxides

Achiral Pd(0)–phosphine complex catalyzed transformation of epoxide to allylic alcohols has been reported [24]. Asymmetric catalytic skeletal rearrangements were also achieved. In addition, achiral epoxides **47** are isomerized to optically active allylic alcohols **48** with catalytic amounts of cob(I)alamin in protic solvents (Eq. 8) [25]. The catalyst precursor is vitamin B_{12s} (cob(I)alamin) obtained from vitamin B_{12} (hydroxocobalamin hydrochloride or cyanocobalamin) by *in situ* two-electron reduction with, e.g. Zn/NH_4Cl. Several achiral epoxides have been isomerized with 1–3 mol% of cob(I)alamin to the corresponding (R)-allyl alcohols in moderate optical yields; cyclopentene oxide (65% ee), cyclohexene oxide (c. 40% ee), cis-2-butene oxide (26% ee). For the mechanistic studies see Ref. [26].

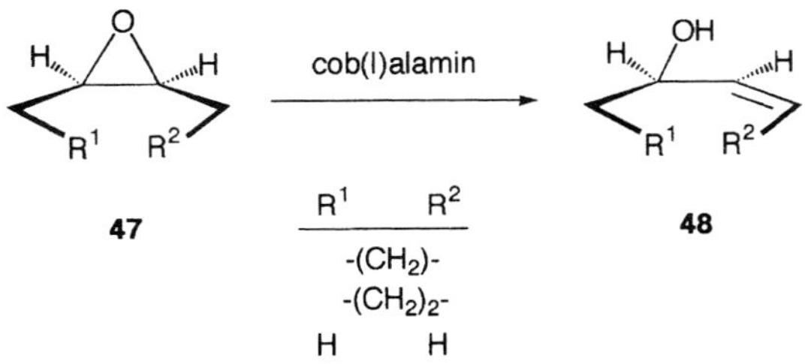

2.9.6.2
Aziridines

As an extension of asymmetric isomerization of epoxides to allyl alcohols, the enantioselective isomerization of aziridines to optically active allylamines has been reported [27]. Achiral N-acylaziridines **49** isomerized with a catalytic amount of cob(I)alamin in MeOH at room temperature to optically active (R)-N-acyl(cycloalk-2-en-1-yl)amines **50** (Eq. 9) with much higher enantioselectivities than the case of the isomerization of epoxides, though the yield and enantiomeric excess depend on the structure of the aziridine. In the case of the 7- and 8-ring aziridines, no isomerization was observed.

$$(CH_2)_n \diagdown NCOR \xrightarrow[\text{MeOH}]{\text{cob(I)alamin}} (CH_2)_n \diagdown NHCOR$$

49   **50**   (9)

R=Ph, n=2: 90%, 91 %ee
R=t-BuO, n=1: 56%, 87 %ee
R=t-BuO, n=2: 64%, 95 %ee

2.9.7
References

1. P. Werkhoff, *Proceeding 12th Int. Cong. Frav. Frag & Ess. Oils,* Vienna, Austria, **1992**; B.D. Sully, P.L. Williams, *Perfum. Essent. Oil Rec.* **1968**, *59*, 365.
2. H. Kumobayashi, S. Akutagawa, S. Otsuka, *J. Am. Chem. Soc.* **1978**, *100*, 3949–3950.
3. (a) K. Tani, T. Yamagata, S. Otsuka, S. Akutagawa, H. Kumobayashi, T. Taketomi, H. Takaya, A. Miyashita, R. Noyori in *Asymmetric Reactions and Processes in Chemistry (ACS Symposium Series 185)* (Eds.: E.L. Eliel, S. Otsuka), American Chemical Society, Washington, DC, **1982**, Chap. 13. (b) K. Tani, T. Yamagata, S. Otsuka, S. Akutagawa, H. Kumobayashi, T. Taketomi, H. Takaya, A. Miyashita, R. Noyori, *J. Chem. Soc., Chem. Commun.* **1982**, 600–601. (c) K. Tani, T. Yamagata, S. Akutagawa, H. Kumobayashi, T. Taektomi, H. Takaya, A. Miyashita, R. Noyori, S. Otsuka, *J. Am. Chem. Soc.* **1984**, *106*, 5208–5217.
4. S.-I. Inoue, H. Takaya, K. Tani, S. Otsuka, T. Sato, R. Noyori, *J. Am. Chem. Soc.* **1990**, *112*, 4897-4905.
5. K. Tani, T. Yamagata, Y. Tatsuno, K. Tomita, S. Akutagawa, H. Kumobayashi, S. Otsuka, *Angew. Chem., Int. Ed. Engl.* **1985**, *24*, 217–219.
6. For reviews see: (a) S. Otsuka, K. Tani in *Asymmetric Synthesis* (Ed.: J. D. Morrison), Academic Press, New York, **1985**, Vol. V, Chap. 6. (b) S. Otsuka, K. Tani, *Synthesis* **1991**, 665–680. (c) R. Noyori, *Asymmetric Catalysis in Organic Synthesis,* Wiley, New York, **1994**, Chap. 3. (d) S. Akutagawa, K. Tani in *Catalytic Asymmetric Synthesis* (Ed.: I. Ojima), VCH, **1993**, Chap. 2. (e) S. Akutagawa, *Appl. Catal. A: General* **1995**, *128*, 171–207. (f) S. Otsuka, *Acta Chem. Scand.* **1996**, *50*, 353–360.
7. K. Tani, *Pure Appl. Chem.* **1985**, *57*, 1845–1854.

8 T. Yamagata, K. Tani, Y. Tatsuno, T. Saito, *J. Chem. Soc., Chem. Commun.* **1988**, 466–468.
9 H. Kumobayashi, PhD Thesis, Osaka University, **1986**.
10 M. Kitamura, K. Manabe, R. Noyori, H. Takaya, *Tetrahedron Lett.* **1987**, *28*, 4719.
11 H. Hart, *Chem. Rev.* **1979**, *79*, 515.
12 C. S. Chin, S. Y. Lee, J. Park, S. Kim, *J. Am. Chem. Soc.* **1988**, *110*, 8244–8245.
13 C. S. Chin, B. Lee, S. Kim, J. Chun, *J. Chem. Soc., Dalton Trans.* **1991**, 443–448.
14 S. H. Bergens, B. Bosnich, *J. Am. Chem. Soc.* **1991**, *113*, 958–967.
15 J. A. Wiles, C. E. Lee, R. McDonald, S. H. Bergens, *Organometallics* **1996**, *15*, 3782–3784.
16 K. Hiroya, Y. Kurihara, K. Ogasawara, *Angew. Chem., Int. Ed. Engl.* **1995**, *34*, 2287–2289.
17 S. Takano, Y. Higashi, T. Kamikubo, M. Moriya, K. Ogasawara, *Synthesis* **1993**, 948–950; M. Moriya, T. Kamikubo, K. Ogasawara, *Synthesis* **1995**, 187–190.
18 S. J. Danishefsky, B. Simomura, *J. Am. Chem. Soc.* **1989**, *111*, 958–967; A. B. Jones, M. Yamaguchi, A. Patten, S. J. Danishefsky, J. A. Ragan, D. B. Smith, S. L. Schreiber, *J. Org. Chem.* **1989**, *54*, 17–19.
19 K. Hiroya, K. Ogasawara, *J. Chem. Soc., Chem. Commun.* **1995**, 2205–2206.
20 H. Frauenrath, T. Philips, *Angew. Chem., Int. Ed. Engl.* **1986**, *25*, 274.
21 For review see: G. W. Parshall, S. D. Ittel, *Homogeneous Catalysis. The Applications and Chemistry of Catalysis by Soluble Transition Metal Complexes,* John Wiley, New York, **1992**, Chap. 2.
22 U. Matteoli, M. Bianchi, P. Frediani, G. Menchi, C. Botteghi, M. Marchetti, *J. Organomet. Chem.* **1984**, *263*, 243–246.
23 Z. Chen, R. Halterman, *J. Am. Chem. Soc.* **1992**, *114*, 2276–2277.
24 M. Suzuki, Y. Oda, R. Noyori, *J. Am. Chem. Soc.* **1979**, *707*, 1623–1625.
25 H. Su, L. Walder, Z.-d. Zhang, R. Scheffold, *Helv. Chim. Acta* **1988**, *71*, 1073–1078.
26 P. Bonhote, R. Scheffold, *Helv. Chim. Acta* **1991**, *74*, 1425–1444.
27 Z.-d. Zhang, R. Scheifold, *Helv. Chim. Acta* **1993**, *76*, 2602–2615.

2.10
Coupling of Aryl and Alkyl Halides with Organoboron Reagents (Suzuki Reaction)

Alexander Zapf

2.10.1
Introduction

Palladium-catalyzed cross-coupling reactions of aryl halides, triflates, and related compounds with nucleophilic organometallic reagents have attracted great attention in synthetic organic chemistry during the last three decades. One of the most important reactions in this field is the Suzuki coupling, which utilizes organoboron reagents as nucleophiles (Scheme 1) [1]. Although these can be tertiary boranes, arylboronic acids are most often used in these reactions. They can easily be prepared from trialkylborates with Grignard [2] or organolithium reagents [3], from aryl halides or triflates and (di)boranes under palladium catalysis [4], or by direct transition metal-catalyzed borylation of arenes via CH-activation [5]. Because of their stability toward air and moisture, their tolerance toward many functional groups, and their low toxicity, the coupling of arylboronic acids with aryl halides has probably become the most important method for the synthesis of unsymmetric biaryls. Biaryls [6] are ubiquitous substructures in natural products [7], pharmaceuticals and agrochemicals [8], ligands [9], and new materials [10], rendering the Suzuki methodology useful not only for academic research but also for industrial production of fine chemicals [11].

A wide variety of catalyst systems has been developed for Suzuki reactions. Not only palladium but also especially nickel has been applied as the active metal center [12]. The advantage of nickel catalysts is their inherent high reactivity towards aryl chlorides, so that the use of specially designed ligands is not required. Unfortunately, nickel tends to undergo side reactions, and thus only low catalyst productivities can be obtained. Although simple palladium catalysts are less active in many cases, extremely high turnover numbers can be obtained with more sophisticated ones. As ligands not only phosphines but also N-heterocyclic carbenes and N- or S-donor ligands can be used. The most important examples will be de-

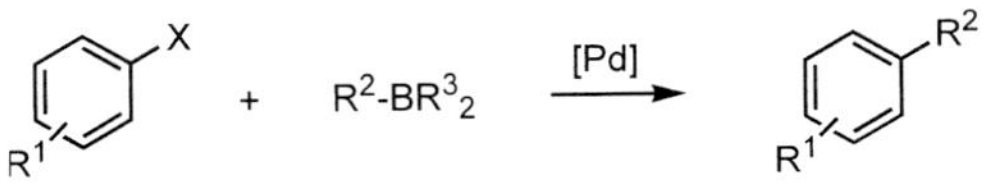

Scheme 1 The Suzuki reaction.
X=Cl, Br, I, OTf etc.; R^2=aryl, alkenyl, alkynyl, alkyl; R^3=OH, OR, alkyl.

Transition Metals for Organic Synthesis, Vol. 1, 2nd Edition.
Edited by M. Beller and C. Bolm
Copyright © 2004 WILEY-VCH Verlag GmbH & Co. KGaA, Weinheim
ISBN: 3-527-30613-7

scribed in this chapter. Additionally, heterogeneous [13] or immobilized catalysts [14] and biphasic systems [15] have been introduced with varying success. Water and ionic liquids have also been used as reaction media for these transformations [16]. Some examples of ligand-free palladium catalysts have been described [17], and even one example of a transition metal-free system for the coupling of aryl bromides under microwave heating has recently been reported [18].

Because of the tremendous number of publications dealing with Suzuki type chemistry that have appeared in the last years [19], this overview must be restricted to some aspects of this wide field. Thus, it covers, after some short general remarks, only reactions of aryl halides, with special emphasis on *chloro*arenes and some recent developments in the area of *alkyl*-X coupling reactions.

2.10.2
Mechanism

The principle of the catalytic cycle of Suzuki reactions is depicted in Scheme 2. It starts with a coordinatively unsaturated palladium(0) species with usually one or two donor ligands such as phosphines or carbenes, for example. This palladium complex inserts into the C-X bond of the electrophile (oxidative addition), resulting in an organopalladium(II) complex with the leaving group X⁻ or another anionic or neutral ligand (base, solvent, etc.) coordinated to the metal center. Then, a transmetalation reaction with the organoboron reagent takes place, leading to a diorganopalladium(II) complex and X-BR$_2$. After isomerization from *trans* to *cis*, the desired coupling product is reductively eliminated from the complex, regenerating the catalytically active palladium(0) species [19a, 20].

In case of electrophiles with low reactivity, especially aryl chlorides, the oxidative addition is considered to be the rate-determining step. Electron-withdrawing substituents were found to increase the reaction rate, whereas electron-donating substituents decrease it. When the oxidative addition proceeds fast (often with aryl

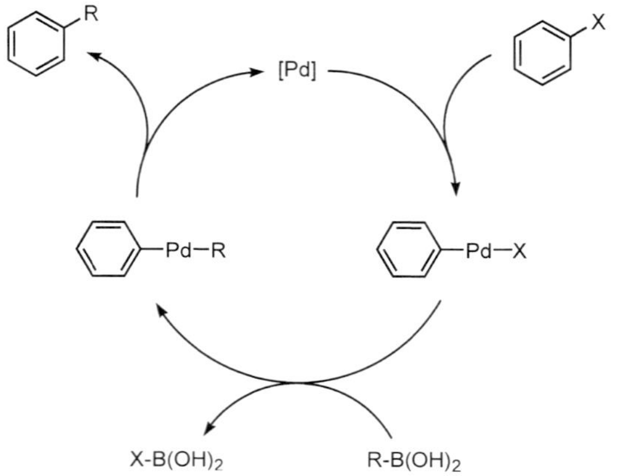

Scheme 2 Simplified mechanism of the Suzuki reaction. [Pd]=12 or 14 electron Pd(0) complex.

iodides, bromides, or triflates), the transmetalation generally becomes the limiting factor. To facilitate the transfer of the organic group from the boron to the palladium center, its nucleophilicity needs to be increased. Most often this is achieved by the addition of Brønsted or Lewis bases with high affinity to boron (alkoxides, hydroxide, fluoride [21], etc.). These additives can be only partially soluble in the reaction medium (suspension or biphasic conditions). There are several models for the way the transmetalation proceeds, depending on the particular reaction conditions [19a]. For example, borate complexes are formed, which transfer their most labile group to the palladium catalyst. Alternatively, X$^-$ can be substituted by an alkoxide on palladium, and the resulting R-Pd-OR' complex directly reacts with the neutral organoboron reagent because of the high oxophilicity of boron. As in the case of the electrophilic coupling partners, electron-poor arylboronates react faster than electron-rich ones.

2.10.3
Coupling of Aryl Halides

2.10.3.1
Phosphine Ligands

Initially, tetrakis(triphenylphosphine)palladium(0) was used as the catalyst for Suzuki reactions [19a, 22]. Then, *in situ* mixtures of palladium(0) or palladium(II) sources and triphenylphosphine were used, leading to similar results. These catalysts were suitable for the coupling of many aryl bromides, iodides, and triflates, but also for π-deficient heteroaryl chlorides. For example, chloropyridines, -pyrimidines, and -pyrazines can be arylated employing simple triphenylphosphine [23] or 1,4-bis(diphenylphosphino)butane as the ligand [24]. But a number of substrates require more active catalysts. Special phosphines have to be used, in many cases most conveniently as *in situ* mixtures with palladium(II) acetate, which is reduced to the catalytically active palladium(0) species under reaction conditions. The phosphine ligand can act as the reductant, but boronic acid, base, or solvent are also potentially reducing agents.

Before the mid 1990s there was no general protocol for the Suzuki coupling of aryl *chlorides*. Although chloroarenes are among the substrates which can be activated by palladium catalysts only with difficulties [25] they are especially interesting for coupling reactions on a large scale because of their easy availability and lower price compared to the corresponding bromide, iodide or triflate derivatives [26]. In 1995, Beller and Herrmann described the use of palladacycle **1** for Heck [27] and Suzuki [28] reactions of aryl halides. This was the signal for a rapid development of new, highly active catalysts for the general coupling of chloroarenes. Compound **1** is simply obtained by reacting tri(*o*-tolyl)phosphine with palladium(II) acetate and is stable against air and moisture even at high temperatures [29]. It is believed that **1** is slowly reduced to a tri(*o*-tolyl)phosphine-palladium(0) complex, which is the true catalyst. Although a mechanistic cycle via Pd(II)/

Pd(IV) without previous reduction of the catalyst cannot be excluded, it has become very unlikely [30]. With palladacycle **1**, high turnover numbers can be obtained for non-activated aryl bromides and electron-deficient aryl chlorides. Non-activated aryl chlorides require higher catalyst amounts.

Further investigations have shown that not only does tri(o-tolyl)phosphine lead to an active catalyst, but so also do many other P-donor ligands, provided that they are applied in significant excess to palladium and the reaction temperature is high enough. Under these conditions the reactivity of the catalyst towards aryl chlorides is sufficient, and catalyst deactivation (palladium black precipitation) can be reduced by the high concentration of stabilizing ligands. Even electron-poor phosphites can be utilized as ligands in *in situ* mixtures (TON 820 000 for 2-bromo-6-methoxynaphthalene) [31] or as preformed cyclometalated palladium complexes (TON 30 000 for 4-bromoanisole) [32], although phosphites were previously believed to be unsuitable for oxidative addition reactions with unreactive aryl halides because of their ability to withdraw electron density from the palladium center. Bedford and co-workers demonstrated that these high catalyst productivities do not result from an increase in reaction rate, but rather from greater catalyst longevity. When cyclometalated palladium phosphite complexes such as **2** or **3** are applied in combination with tricyclohexylphosphine, even electron-rich aryl chlorides can be coupled with exceedingly high turnover numbers (TON almost 2 million for 4-chlorotoluene) [33]. Catalytic systems with just a phosphite or tricyclohexylphosphine as the ligand show significantly lower productivities.

With sterically congested triarylphosphine **4**, Fu and co-workers were able to couple electron-rich aryl chlorides at temperatures as low as 70 °C [34]. Good yields require 1.5 mol% $Pd_2(dba)_3$, and at least one *ortho*-substituent either on the aryl chloride or on the boronic acid seems to be necessary. Interestingly, the coupling of strongly activated aryl chlorides can be performed at room temperature, albeit with low turnover numbers (5 mol% $Pd(OAc)_2$). The trimethylsilyl group in **4** has been found to be crucial for good catalytic activity as well as the methyl groups on the second cyclopentadienyl ring. Similarly, the substituted tris(ferrocenyl) phosphine **5** was described for the coupling of different types of aryl chlor-

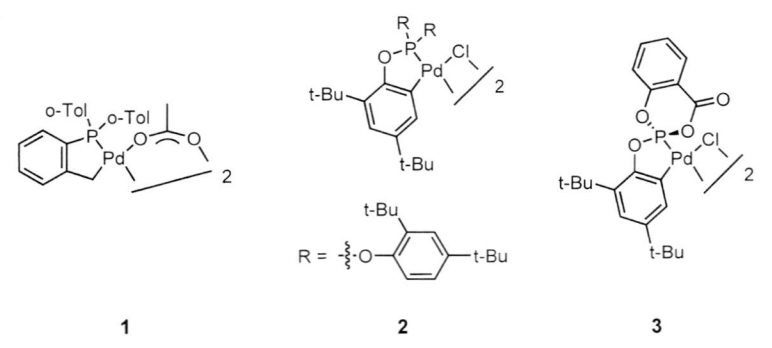

Fig. 1 Palladacycles.

ides [35]. Moderate to good yields can be obtained with only 0.03 mol% palladium at 60 °C with cesium carbonate as the base.

Even milder reaction conditions can be applied when electron-rich and sterically demanding phosphines are used as ligands. The group of Buchwald has developed a number of such ligands with one biaryl and two bulky alkyl substituents on the phosphorus atom. In the presence of **6**, activated and deactivated aryl chlorides and bromides can be coupled at room temperature with just 0.5–1 mol% palladium. At elevated temperatures (100 °C), high turnover numbers (ca. 2000 for 4-chlorotoluene, ca. 100 000 for 4-*tert*-butylbromobenzene) can be obtained with ligand **7**. For all these reactions, potassium phosphate and potassium fluoride have been found to be the best bases in THF, dioxane, or toluene as solvent [36].

Sterically hindered substrates such as 2-chloro-*m*-xylene and *o*-tolylboronic acid can also be coupled successfully to give tri-*ortho*-methyl-substituted biphenyl, but require somewhat harsher reaction conditions (100 °C, 1 mol% Pd) [36 b]. The synthesis of tetra-*ortho*-methyl-substituted biaryls can only be accomplished with high yield if the phenanthrene-based ligand **8** is used (2 mol% $Pd_2(dba)_3$, 8 mol% **4**, 110 °C). It is believed that interactions of ligated palladium and the π-electrons of the C^9–C^{10} bond of the phenanthrene substituent are decisive for the improved performance of **8** compared to simple biaryl phosphino ligands [37].

Ligands with similar stereoelectronic properties have been used by Guram and Beller (Fig. 4). Instead of a biaryl backbone, Guram introduced an acetophenone moiety which was protected as its ethyleneglycol ketal (**9**) [38]. This phosphine was found to act as a P,O-chelating ligand and therefore leads to more stable palladium complexes than similar non-chelating phosphines. The Suzuki reaction of

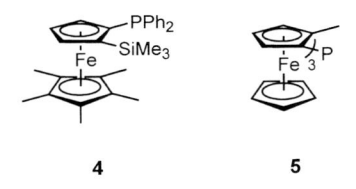

Fig. 2 Ferrocenylphosphines.

Fig. 3 Buchwald's ligands.

aryl iodides, bromides, and chlorides, even electron-rich and *ortho*-substituted ones, affords the desired coupling products in high yields.

A different ligand backbone was introduced by Beller and co-workers. They applied 2-(dialkylphosphino)-*N*-arylpyrroles (PAP) as ligands for the Suzuki coupling of various aryl chlorides [39]. These ligands are easily prepared by lithiation of *N*-arylpyrroles and reaction with chlorophosphines. The *tert*-butyl derivative PAP-*t*Bu (**10**) has been shown to lead to the most active catalyst for *meta*- and *para*-substituted chloroarenes. In the case of bulky *ortho*-substituted aryl chlorides, the sterically less demanding ligand PAP-Cy (**11**) sometimes gives better results. Turnover numbers up to ca. 20000 have been described at only 60 °C, and even at room temperature 800 turnovers can be obtained with deactivated 4-chlorotoluene (0.1 mol% Pd(OAc)$_2$).

Earlier on, Littke and Fu introduced tri-*tert*-butylphosphine as a ligand for the Suzuki coupling of chloroarenes [40]. The optimum palladium/ligand ratio has been found to be ca. 1:1, suggesting a monophosphine-palladium complex as the catalytically active species. At room temperature, very good yields are obtained for aryl iodides, bromides, and electron-poor aryl chlorides in the presence of potassium fluoride (0.5 mol% Pd$_2$(dba)$_3$). However, electron-rich aryl chlorides require elevated reaction temperatures and a higher catalyst loading (70–100 °C, 1.5 mol% Pd$_2$(dba)$_3$). A remarkable reactivity has been found for 4-chlorophenyl triflate. Although generally an Ar-OTf bond can be activated more easily than an Ar-Cl bond, it is the chloride that reacts first with high selectivity. Subsequently the triflate group can be substituted by using a different catalyst system based on tricyclohexylphosphine, which on the other hand is not able to activate Ar-Cl bonds under the applied conditions. A significant drawback of the Pd/P(*t*-Bu)$_3$ catalyst compared to other active systems is the extreme sensitivity of the ligand towards oxygen, whereas the Buchwald and PAP ligands are reasonable stable in air.

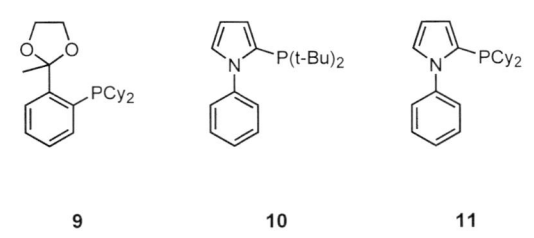

Fig. 4 The ligands of Guram and Beller.

$$R_3P-Pd\underset{Br}{\overset{Br}{<>}}Pd-PR_3$$

12a PR$_3$ = P(1-Ad)(*t*-Bu)$_2$
12b PR$_3$ = P(*t*-Bu)$_3$

Fig. 5 Monophosphine palladium(I) complexes.

Hartwig and co-workers used 1-adamantyl-di-*tert*-butylphosphine- or tri-*tert*-butylphosphine-coordinated palladium complexes (**12a** and **b**) for room temperature couplings of aryl bromides and chlorides [41]. Palladium is formally in the oxidation state +1 in these catalysts, but they readily disproportionate into monophosphine-palladium(0) and monophosphine-palladium(II) bromide or are reduced to two equivalents of R_3P-Pd(0) under reaction conditions. The palladium(0) complex is then responsible for exceedingly high turnover frequencies in Suzuki couplings. Reactions go to completion within several minutes at room temperature (0.5 mol% **12a**)! Here, simple potassium hydroxide is used as the base and THF as the solvent.

A ligand with steric and electronic properties similar to Fu's and Hartwig's ligands is di-1-adamantyl-*n*-butylphosphine (**13**), which was developed for palladium-catalyzed coupling reactions in the group of Beller [42]. With this ligand, turnover numbers of ca. 20 000 can be realized for non-activated aryl chlorides at 100 °C with potassium phosphate as the base [43].

Hartwig and co-workers have found that the sterically hindered ferrocenyl dialkylphosphine **14** is suitable for the Suzuki reaction of aryl bromides and chlorides [44]. Using $Pd_2(dba)_3$ and **14**, aryl bromides can be coupled at room temperature in many cases (0.5–4 mol% Pd). Activated aryl chlorides require at least 45 °C and deactivated ones at least 80 °C for good yields (0.5–5 mol% Pd).

As mentioned above, the catalytically active species in palladium-catalyzed coupling reactions is believed to be a 12- or 14-electron palladium(0) complex. Thus, a catalyst precursor consisting of a Pd-L fragment with only weakly coordinated co-ligands which are easily removed under reaction conditions should be advantageous. The *in situ* complex formation with possible side reactions is not necessary in this case. Accordingly, Beller and co-workers employed monophosphine-palladium(0)-diene complexes for the Suzuki coupling of aryl chlorides [45]. Among the tested phosphines (PPh_3, PCy_3, and **7**) Buchwald's ligand turned out to be the best. But the diene part is also important for catalytic performance. Diallylether complexes (e.g., **15**) give higher yields of the desired biaryls than the more stable dvds complexes (dvds = 1,1,3,3-tetramethyl-1,3-divinyldisiloxane).

Palladium-monophosphine fragments have also been developed by Bedford and co-workers [46]. For example, the tricyclohexylphosphine-containing palladacycle **16** with palladium in the oxidation state +2 displays high activity in the coupling reaction of 4-chloroanisole with phenylboronic acid at 100 °C (dioxane, Cs_2CO_3)

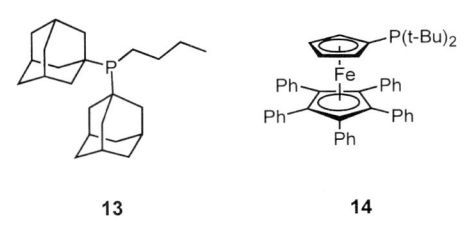

13   **14**

Fig. 6 Bulky alkylphosphines.

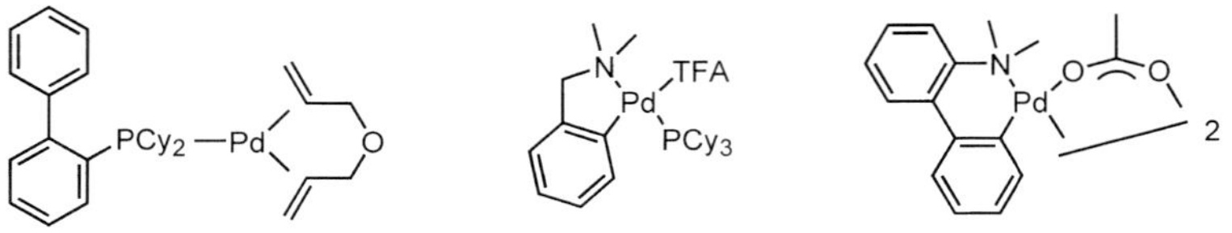

Fig. 7 Palladium complexes as pre-catalysts.

[47]. Buchwald's ligand **7** instead of tricyclohexylphosphine gives similar results, but surprisingly tri-*tert*-butylphosphine leads to significantly lower yields.

Secondary phosphine ligands for Suzuki reactions have been described by Schnyder and Indolese [48]. These are applied in combination with phosphorus-free palladacycles like **17**, and the true ligand is probably formed under basic reaction conditions by reductive elimination from the phosphido-coordinated palladacycle with P-C bond formation. Di-2-norbornylphosphine and **17** gave the best result in the coupling of 4-chlorotoluene with phenylboronic acid at 100 °C (100% yield, 1 mol% Pd). Di-1-adamantylphosphine and di-*tert*-butylphosphine are also suitable for this reaction.

A novel ligand concept for palladium-catalyzed coupling reactions was introduced by Li (DuPont). He applied a secondary phosphine oxide (namely di-*tert*-butylphosphine oxide) which tautomerizes to the corresponding phosphinous acid and then coordinates to the metal catalyst. Under basic reaction conditions the phosphinous acid can be deprotonated, generating an electron-rich tertiary phosphino ligand which leads to highly active catalysts for various coupling reactions of aryl halides [49]. Di-*tert*-butylphosphine oxide is air-stable, and, as well as cesium fluoride, cheaper bases like potassium carbonate or sodium *tert*-butoxide can be used for coupling chloroarenes with arylboronic acids.

Chelating ligands are not suited for Suzuki reactions in many cases. Often they block free coordination sites on the catalyst, leading to highly stable but non-productive palladium complexes. One important exception of this rule is the Tedicyp ligand **18** developed by Feuerstein et al. [50, 51]. Using this tetradentate ligand, activated heteroaryl bromides can be coupled with arylboronic acids with exceedingly high turnover numbers of more than 10 million [50a]. Substituted bromobenzenes and boronic acids without steric congestion react similarly efficiently [50b–e]. For the synthesis of 2,2'-disubstituted biaryls, only slightly higher catalyst concentrations are required (ca. 10^{-4} mol% Pd), but 2,2,2'-trisubstituted biaryls need significantly more catalyst (ca. 1 mol% Pd) to obtain useful yields [50b]. Electron-deficient aryl chlorides can be coupled with high turnover numbers (TON ca. 1000 for 4-chloroacetophenone, TON 6.8 million for 2-chloro-5-(trifluoromethyl)-nitrobenzene), whereas non-activated chloroarenes are not converted to an appreciable extent [51].

Other catalysts with chelating ligands are orthometalated bisphosphinite complexes ("pincer complexes") (**19**). These have been applied in the coupling of aryl bromides and 4-chloronitrobenzene [52], but the turnover numbers observed are in the range of those for ligand-free catalyst systems.

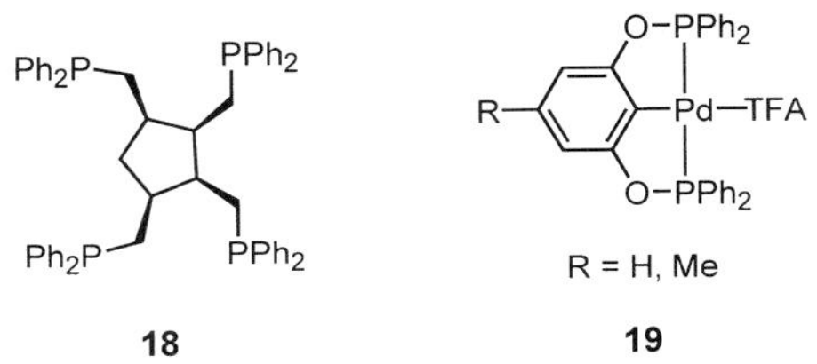

Fig. 8 Chelating P-donor ligands.

A library of easily accessible chelating P,N-ligands has been developed by Kempe and co-workers for various catalytic reactions. P-Functionalized aminopyridines are suitable both for palladium- and nickel-catalyzed coupling reactions of electron-rich and electron-poor aryl chlorides with phenylboronic acid (1 mol% catalyst, 60 °C) [53].

The first industrial applications of the Suzuki methodology utilizing phosphine ligands have appeared in recent years. Accordingly, o-(p-tolyl)benzonitrile (OTBN) is produced on a multiton scale by Clariant AG (Frankfurt am Main) by reaction of o-chlorobenzonitrile with p-tolylboronic acid in the presence of a palladium/TPPTS catalyst (TPPTS = P(m-NaOSO$_2$C$_6$H$_4$)$_3$; Scheme 3) [54]. Ethylene glycol is used as the solvent, and a small amount of a sulfoxide or sulfone is added to increase the catalyst stability. At the end of the reaction, two phases form with the catalyst and salts in the polar phase and the product in the non-polar phase. This biphasic procedure allows for recycling of the homogeneous catalyst.

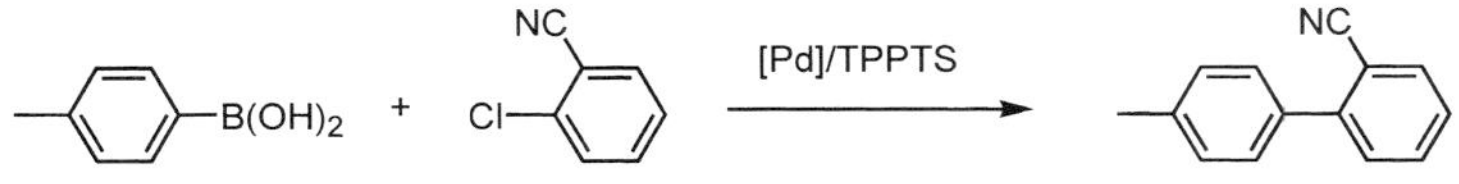

Scheme 3 Synthesis of OTBN (Clariant AG).

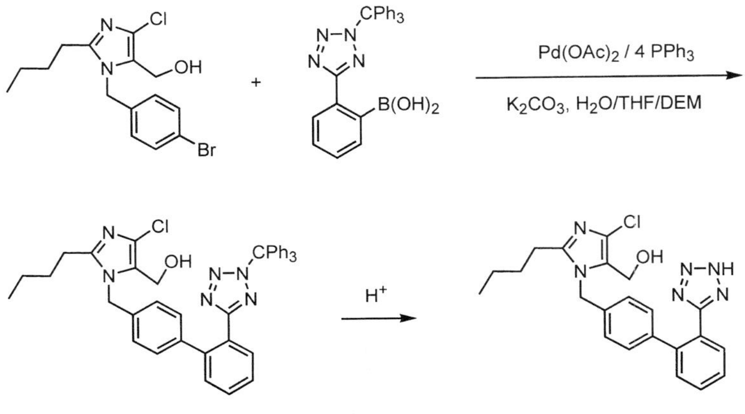

Scheme 4 Synthesis of Losartan (Merck).

OTBN is functionalized further to give a family of pharmaceutically active AT II antagonists. A more convergent route to these bioactive molecules has been developed at Merck [55]. In the synthesis of Losartan, the biaryl coupling is performed in the final stage of the protocol (Scheme 4). A simple mixture of palladium(II) acetate and triphenylphosphine is used as the catalyst and a mixture of THF and diethoxymethane as the solvent. A defined amount of water is necessary to ensure high reaction rates and yields.

2.10.3.2
Carbene Ligands

N-Heterocyclic carbenes (NHC) have emerged as a useful alternative to phosphines as ligands in many transition metal-catalyzed reactions [56]. The first example of their application in a Suzuki reaction was reported by Herrmann et al. [57]. Palladium complex **20** with a chelating biscarbene was used for the coupling of aryl bromides and 4-chloroacetophenone, respectively, with phenylboronic acid. At 120 °C good to excellent yields were obtained with 1 mol% catalyst and potassium carbonate as the base.

A more active catalyst for the coupling of (non-activated) aryl chlorides was described by Nolan and co-workers shortly after this finding [58]. They applied the monodentate IMes ligand **21** in an *in situ* mixture with $Pd_2(dba)_3$. Dioxane is used as the solvent, and a temperature of 80 °C is sufficient to obtain excellent yields with electron-poor and electron-rich chloroarenes. Unfortunately, these good results are only achieved when two equivalents of cesium carbonate are used as the base, and a comparatively high catalyst loading is required (≥ 2 mol% Pd). On the other hand, the synthesis of a defined catalyst is not necessary here, and also the free carbene need not be liberated prior to the coupling reaction. The corresponding imidazolium salt, which is stable toward air and moisture, can be applied directly, giving even better yields than utilizing the preformed free carbene.

Fürstner et al. have published a general and simple protocol for the coupling of aryl chlorides with 9-R-9-BBN (Scheme 5) [59]. They used *in situ* mixtures of imidazolium salts and palladium(II) acetate in the presence of potassium methoxide or aqueous sodium hydroxide and found that the "unsaturated" IPr (**22**) is superior to its "saturated" counterpart SIPr (**23**).

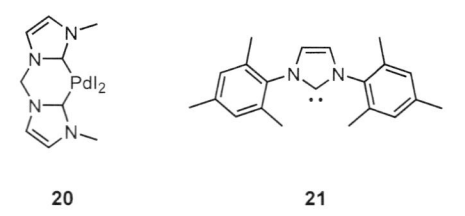

Fig. 9 NHC as ligands for Suzuki reactions.

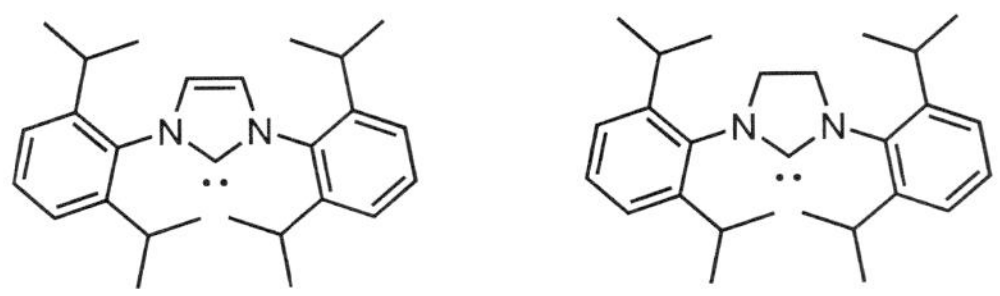

22 **23**

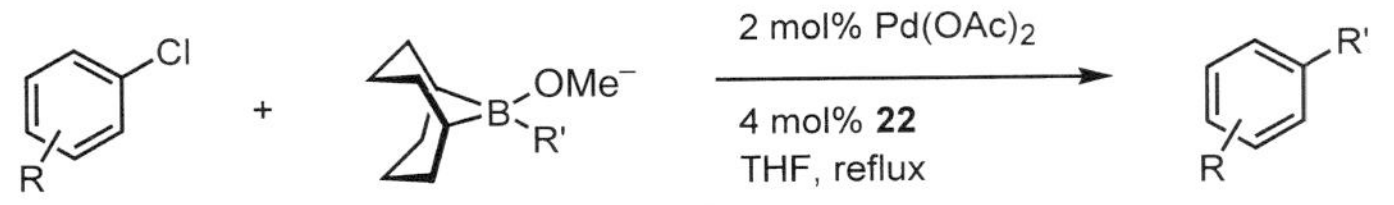

Scheme 5 Coupling of chloroarenes with 9-R'-9-BBN. R' = alkyl, allyl, alkynyl.

The first monocarbene-palladium(0) complexes have been synthesized and introduced as catalysts for C-C coupling reactions by Beller and co-workers. As an additional stabilizing ligand for the palladium center, either 1,6-dienes (especially 1,1,3,3-tetramethyl-1,3-divinyldisiloxane – dvds) [60] or quinones [61] can be used. Naphthoquinone complexes are good catalysts for the coupling of aryldiazonium tetrafluoroborates with arylboronic acids. With 0.1 mol% of **24**, good to excellent yields of the desired biaryls are obtained in methanol at 50 °C [61a].

Recently, Nolan and co-workers synthesized a series of NHC-palladium(II) (allyl)chloride complexes for palladium-catalyzed coupling reactions [62]. For the Suzuki reaction of (hetero)aryl chlorides, IPr and IMes turned out to be the best ligands, yielding the desired products in high yields at 80 °C (0.5–2 mol% Pd).

In most cases an excess of carbene ligand with respect to palladium is not required, as carbenes hardly dissociate from the metal center and thus prevent catalyst deactivation more efficiently than phosphines do. Moreover, a higher carbene/palladium ratio is often, but not always, detrimental. The (IMes)$_2$Pd complex is totally inactive for the coupling of 4-chlorotoluene with phenylboronic acid under conditions where a 1:1 mixture of Pd(0) and IMes gives 93% of 4-methylbiphenyl [63a]. The catalytic activity of these biscarbene complexes is strongly dependent on the steric bulk of the ligands. With *tert*-butyl groups on both nitrogen atoms of the imidazol-2-ylidene, for instance, some reactivity can be observed (68% yield), whereas *iso*-propyl or cyclohexyl groups again lead to almost inactive complexes [63a]. Surprisingly, 1-adamantyl groups give a very active catalyst, which enables the Suzuki coupling of *meta*- and *para*-substituted aryl chlorides even at room temperature [63b]. In addition to its high activity, this complex displays a comparatively good catalytic productivity. A maximum TON of 570 is reported for the reaction of 4-chlorotoluene at room temperature.

Chelating bis-carbene ligands with both a suitable linker and bulky groups on the nitrogen atoms, e.g. **25**, are also effective in the arylation of chloroarenes (80 °C) [64]. Additionally, a carbene ligand (**26**) with a hemilabile pyridine donor function has been employed in the coupling of highly reactive 4-bromoacetophenone with phenylboronic acid and exhibited a moderate TON of ca. 100 000 [65]. The pincer bis-carbene palladium complex **27** has been described for the same reaction, but was found to

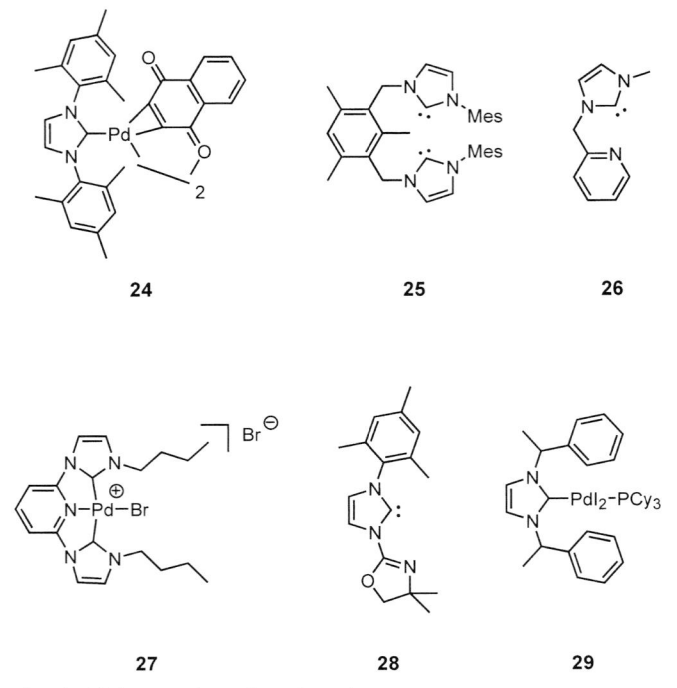

Fig. 10 N-Heterocyclic carbene ligands.

be less active (TON 350) [66]. For solubility reasons, DMF had to be used as the solvent at 110 °C. Ligand **28** with an additional oxazoline moiety catalyzes the Suzuki coupling of aryl bromides and electron-deficient aryl chlorides (3 mol% Pd, 80 °C) [67]. Moreover, palladium complexes bearing one NHC and one phosphine donor ligand have been tested in the Suzuki coupling of aryl bromides and chlorides [68]. The steric bulk of the substituents both on the carbene and the phosphine ligand have been found to be decisive for the catalytic activity. Compared to the corresponding biscarbene and bisphosphine complexes the "mixed ligand catalyst" **29** gives slightly higher yields, probably because of its higher thermal stability, but the homoleptic phosphine complex exhibits a significantly higher initial activity.

2.10.3.3
Other Ligands

As well as catalyst systems based on phosphines or carbenes, a number of other types of ligands have appeared in literature during the last few years. Mostly N-donor ligands are used to stabilize the palladium catalyst, but few examples of S-donor ligands have also emerged. For example, the sulfur-containing palladacycle **30** has been described for the coupling of aryl halides with arylboronic acids [69]. While bromoarenes and electron-poor chloroarenes can be activated at room temperature to give very good yields of the corresponding substituted biphenyls (0.5 mol% **30**), non-activated aryl chlorides give low yields (< 50%) even under drastic reaction conditions (130 °C) and at high catalyst loadings (0.5–2 mol% **30**).

Nolan and co-workers have synthesized a number of 1,4-diazabutadienes within their two-step procedure for the preparation of imidazolium salts. These diimines have been employed successfully in the Suzuki reaction of aryl bromides and activated aryl chlorides [58c, 70]. With 3 mol% palladium(II) acetate and 3 mol% **31**, excellent yields are obtained even for *ortho*-substituted aryl bromides (Cs_2CO_3, dioxane, 80 °C).

Oxime-palladacycles like **32** have been described by Nájera and co-workers for the coupling of chloroarenes and phenylboronic acid in aqueous media (neat water or methanol/water) [71]. Electron-deficient aryl chlorides can be arylated with low catalyst loadings (0.01–1 mol%) at 100 °C or even at room temperature (1 mol% Pd). Here, the addition of at least 50 mol% of tetra-*n*-butylammonium bromide is necessary to enhance the reaction rate. Similarly, the bispyridine-palladium complex **33** has been employed for coupling reactions of aryl bromides and chlorides in neat water or mixtures of water and DMF or methanol [72].

2.10.4
Coupling of Alkyl Halides and Tosylates

The palladium-catalyzed cross-coupling methodology using *alkyl* halides and other *alkyl* electrophiles is significantly underdeveloped compared to reactions at $C(sp^2)$ centers [73]. This is mainly because of two problems which arise during the activation of alkyl-X compounds via an oxidative addition pathway: The oxidative addition itself proceeds quite sluggishly at sp^3 carbon atoms, and the resulting alkyl-palladium(II)-X complex is prone to undergo fast β-hydride elimination, yielding the corresponding olefin and HX [74]. Most of the examples, known some years ago, of palladium-catalyzed functionalization reactions of alkyl halides covered the coupling of benzylic or other derivatives which lack β-hydrogen atoms or the combination of two allyl moieties following a slightly different reaction mechanism [75]. Only alkyl *iodides* were described to react with any *generality* in the Suzuki re-

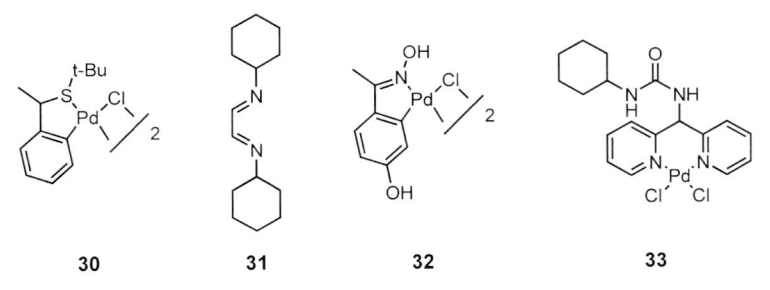

Fig. 11 S- and N-donor ligands for Suzuki reactions.

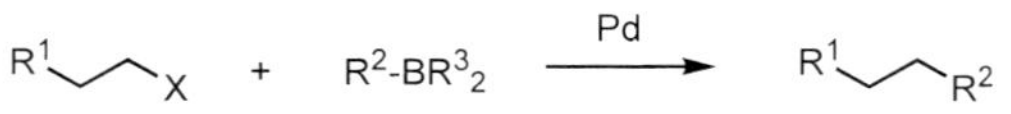

Scheme 6 The Suzuki coupling of alkyl halides and tosylates. X = Br, Cl, OTs; R^3_2 = (OH)$_2$, 9-BBN.

action using tetrakis(triphenylphosphine)palladium(0) as the catalyst [76]. Another early protocol for the coupling of alkyl halides with organometallic reagents was developed by Knochel and co-workers for nickel-catalyzed Negishi reactions, which benefit from an acceleration of the reductive elimination step [77]. The first palladium-catalyzed Suzuki reactions with alkyl bromides, chlorides, or tosylates were published by Fu and co-workers in 2001 (Scheme 6). Surprisingly, they found that alkyl bromides react with 9-alkyl-9-BBN in the presence of 4 mol% palladium(II) acetate and 8 mol% tricyclohexylphosphine at room temperature with potassium phosphate monohydrate as the base in THF [78a]. Tri-*iso*-propylphosphine also leads to an active catalyst, but other electron-rich phosphines with significantly different cone angles (tri-*n*-butyl-, tri-*tert*-butylphosphine, etc.) give, as the main product, the corresponding olefin by elimination of hydrogen bromide. Functional groups like ethers, esters, nitriles, or chlorides are tolerated as in Suzuki reactions with aryl or vinyl halides. When boronic acids are used instead of 9-BBN derivatives, di-*tert*-butylmethylphosphine gives a more productive catalyst than tricyclohexylphosphine, and potassium *tert*-butoxide has to be used as the base in *tert*-amyl alcohol [78b]. Amazingly, the oxidative addition product of 3-phenylpropyl-1-bromide and Pd[MeP(*t*-Bu)$_2$]$_2$ could be isolated and characterized crystallographically, demonstrating that alkyl-palladium complexes are not necessarily unstable.

Alkyl *chlorides* can be coupled with 9-alkyl-9-BBN under similar conditions, albeit with lower turnover numbers than alkyl bromides [79]: 5 mol% Pd$_2$(dba)$_3$ and 20 mol% tricyclohexylphosphine are required for good yields. Cesium hydroxide hydrate has to be used as the base in dioxane at 90 °C. As for the Suzuki reaction of alkyl bromides the presence of stoichiometric amounts of water, introduced as crystal water of the base, are necessary. Furthermore, the coupling of alkyl *tosylates* and 9-alkyl-9-BBN is possible using 4 mol% palladium(II) acetate and 16 mol% di-*tert*-butylmethylphosphine [80]. Here, simple sodium hydroxide serves as the base. It was found that the oxidative addition proceeds with inversion of the stereochemistry at the substituted carbon atom. During the reductive elimination the stereochemistry is not affected, so that the overall reaction gives the coupling product with inversion of stereochemistry.

Despite this progress, the palladium-catalyzed Suzuki reaction of alkyl-X derivatives is restricted to primary alkyl substituents at the moment. The coupling of secondary or tertiary derivatives constitutes an unsolved problem so far.

2.10.6
Summary and Outlook

Three major problems in Suzuki chemistry remain to be solved: (1) the coupling of secondary and tertiary alkyl halides, (2) the efficient synthesis of tetra-*ortho*-substituted biaryls, and (3) asymmetric biaryl coupling reactions.

Initial results on asymmetric Suzuki reactions have already been published: Cammidge and Crépy synthesized (*R*)-(–)2,2′-dimethyl-1,1′-binaphthyl with 50% yield and 85% *ee* using 3 mol% palladium(II) chloride, 6 mol% (*S*)-(*R*)-PFNMe, and cesium fluoride as the base in refluxing DME (6 days, PFNMe = 1-diphenylphosphino-2-[1-(dimethylamino)ethyl]ferrocene) [81]. Yin and Buchwald reported on a catalyst system comprising monodentate BINAP-type ligands. Enantioselectivities up to 92% could be obtained when a weakly coordinating functional group (NO_2 or $P(O)(OEt)_2$) was present in the *ortho*-position of the aryl bromide (Scheme 7) [82]. These first results demonstrate that an asymmetric Suzuki reaction is possible in principle, but significant improvement is still required here.

An academically interesting variant of the Suzuki protocol has recently been described by Kim and Yu [83]. Aryl fluorides, which are hardly ever used for transition metal-catalyzed coupling reactions because of their low reactivity, were coupled with arylboronic acids. However, a nitro group *ortho* to the fluorine substituent is required to facilitate insertion of the palladium catalyst into the C-F bond. Then, simple tetrakis(triphenylphosphine)palladium(0) is suited as the catalyst precursor at temperatures as low as 65 °C.

During the last decade, the Suzuki coupling methodology has been developed to a versatile instrument for C-C bond-forming reactions. Significant progress was realized, especially in the functionalization of aryl chlorides. High turnover numbers with simple catalyst systems will facilitate further technical applications in the near future – not only for the synthesis of precious biologically active agents. Furthermore, some interesting new types of ligands have emerged for coupling reactions which will be the basis for further developments.

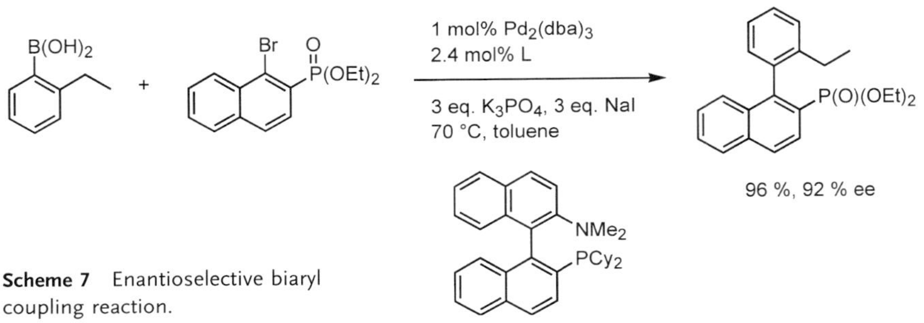

Scheme 7 Enantioselective biaryl coupling reaction.

2.10.7
References

1 (a) N. MIYAURA, T. YANAGI, A. SUZUKI, *Synth. Commun.* **1981**, *11*, 513; (b) A. SUZUKI, *Acc. Chem. Res.* **1982**, *15*, 178.
2 (a) F.R. BEAN, J.R. JOHNSON, *J. Am. Chem. Soc.* **1932**, *54*, 4415; (b) R.M. WASHBURN, E. LEVENS, C.F. ALBRIGHT, F.A. BILLIG, *Org. Synth. CV* **1963**, *4*, 68.
3 D.J. CUNDY, S.A. FORSYTH, *Tetrahedron Lett.* **1998**, *39*, 7979.
4 (a) M. MURATA, S. WATANABE, Y. MASUDA, *J. Org. Chem.* **1997**, *62*, 6458; (b) L. ZHU, J. DUQUETTE, M. ZHANG, *J. Org. Chem.* **2003**, *68*, 3729; (c) Review: T. ISHIYAMA, N. MIYAURA, *J. Organomet. Chem.* **2000**, *611*, 392.
5 (a) K. MERTINS, A. ZAPF, M. BELLER, *J. Mol. Catal.* **2004**, *207*, 21 (b) T. ISHIYAMA, J. TAKAGI, J.F. HARTWIG, N. MIYAURA, *Angew. Chem.* **2002**, *114*, 3182; *Angew. Chem. Int. Ed.* **2002**, *41*, 3056; (c) J. TAKAGI, K. SATO, J.F. HARTWIG, T. ISHIYAMA, N. MIYAURA, *Tetrahedron Lett.* **2002**, *43*, 5649; (d) M.K. TSE, J.Y. CHO, M.R. SMITH III, *Org. Lett.* **2001**, *3*, 2831; (e) J.-Y. CHO, C.N. IVERSON, M.R. SMITH III, *J. Am. Chem. Soc.* **2000**, *122*, 12868; (f) C.N. IVERSON, M.R. SMITH III, *J. Am. Chem. Soc.* **1999**, *121*, 7696.
6 For cross-coupling reactions in biaryl syntheses see: S.P. STANFORTH, *Tetrahedron* **1998**, *54*, 263.
7 (a) E.J.-G. ANCTIL, V. SNIECKUS, *J. Organomet. Chem.* **2002**, *653*, 150; (b) K. DURAIRAJ, *Curr. Science* **1994**, *66*, 833.
8 (a) N. YASUDA, *J. Organomet. Chem.* **2002**, *653*, 279; (b) H.-U. BLASER, A.F. INDOLESE, A. SCHNYDER, *Curr. Science* **2000**, *78*, 1336.
9 (a) J. CLEWS, A.D.M. CURTIS, H. MALKIN, *Tetrahedron* **2000**, *56*, 8735; (b) F.Y. KWONG, K.S. CHAN, *Organometallics* **2000**, *19*, 2058; (c) T. HAYASHI, *Acc. Chem. Res.* **2000**, *33*, 354; (d) S.L. BUCHWALD, D.W. OLD, J.P. WOLFE, M. PALUCKI, K. KAMIKAWA, US 113478 (2001).
10 (a) R. ZENK, S. PARTZSCH, *Chim. Oggi* **2003**, *21*, 72; (b) X. ZHAN, S. WANG, Y. LIU, X. WU, D. ZHU, *Chem. Mat.* **2003**, *15*, 1963; (c) S. MERLET, M. BIRAU, Z.Y. WANG, *Org. Lett.* **2002**, *4*, 2157; (d) M. HIRD, G.W. GRAY, K.J. TOYNE, *Mol. Cryst. Liq. Cryst.* **1991**, *206*, 187; (e) E. POETSCH, *Kontakte* **1988**, 15.
11 (a) M. BELLER, A. ZAPF, in Handbook of organopalladium chemistry for organic synthesis (ed. E.-I. NEGISHI), Wiley-Interscience, New York, **2002**, 1209; (b) M. BELLER, A. ZAPF, *Top. Catal.* **2002**, *19*, 101; (c) J.G. DE VRIES, *Can. J. Chem.* **2001**, *79*, 1086.
12 (a) D. ZIM, V.R. LANDO, J. DUPONT, A.L. MONTEIRO, *Org. Lett.* **2001**, *3*, 3049; (b) J.-C. GALLAND, M. SAVIGNAC, J.-P. GENÊT, *Tetrahedron Lett.* **1999**, *40*, 2323; (c) S. SAITO, S. OHTANI, N. MIYAURA, *J. Org. Chem.* **1997**, *62*, 8024; (d) A. INDOLESE, *Tetrahedron Lett.* **1997**, *38*, 3513; (e) S. SAITO, M. SAKAI, N. MIYAURA, *Tetrahedron Lett.* **1996**, *37*, 2993.
13 (a) B.M. CHOUDARY, S. MADHI, N.S. CHOWDARI, M.L. KANTAM, B. SREEDHAR, *J. Am. Chem. Soc.* **2002**, *124*, 14127; (b) C.R. LEBLOND, A.T. ANDREWS, Y. SUN, J.R. SOWA, *Org. Lett.* **2001**, *3*, 1555; (c) M.T. REETZ, E. WESTERMANN, *Angew. Chem.* **2000**, *112*, 170; *Angew. Chem. Int. Ed.* **2000**, *39*, 165; (d) B.H. LIPSHUTZ, J.A. SCALFANI, P.A. BLOMGREN, *Tetrahedron* **2000**, *56*, 2139; (e) M.T. REETZ, R. BREINBAUER, K. WANNINGER, *Tetrahedron Lett.* **1996**, *37*, 4499; (f) G. MARCK, A. VILLINGER, R. BUCHECKER, *Tetrahedron Lett.* **1994**, *35*, 3277.
14 (a) Q.-S. HU, Y. LU, Z.-Y. TANG, H.-B. YU, *J. Am. Chem. Soc.* **2003**, *125*, 2856; (b) R.S. VARMA, K.P. NAICKER, *Tetrahedron Lett.* **1999**, *40*, 439; (c) I. FENGER, C. LE DRIAN, *Tetrahedron Lett.* **1998**, *39*, 4287; (d) S.-B. JANG, *Tetrahedron Lett.* **1997**, *38*, 1793.
15 (a) M. BELLER, J.G.E. KRAUTER, A. ZAPF, *Angew. Chem.* **1997**, *109*, 793; *Angew. Chem. Int. Ed. Engl.* **1997**, *36*, 772; (b) C. BIANCHINI, G. GIAMBASTIANI, *Chemtracts* **2003**, *16*, 485.
16 (a) R.B. BEDFORD, M.E. BLAKE, C.P. BUTTS, D. HOLDER, *Chem. Commun.* **2003**, 466; (b) N.E. LEADBEATER, M. MARCO, *J. Org. Chem.* **2003**, *68*, 888; (c) L. BOTELLA, C. NÁJERA, *J. Organomet. Chem.*

2002, 663, 46; (d) C. Dupuis, K. Adiey, L. Charruault, V. Michelet, M. Savignac J.-P. Genêt, *Tetrahedron Lett.* 2001, 42, 6523; (e) C. J. Mathews, P. J. Smith, T. Welton, *Chem. Commun.* 2000, 1249; (f) P. Machnitzki, M. Tepper, K. Wenz, O. Stelzer, E. Herdtweck, *J. Organomet. Chem.* 2000, 602, 158; (g) C. J. Mathews, P. J. Smith, T. Welton, *Chem. Commun.* 2000, 1249; (h) E. Paetzold, G. Oehme, *J. Mol. Catal.* 2000, 152, 69.

17 (a) Y. Deng, L. Gong, A. Mi, H. Liu, Y. Jiang, *Synthesis* 2003, 337; (b) L. M. Klingensmith, N. E. Leadbeater, *Tetrahedron Lett.* 2003, 44, 765; (c) G. A. Molander, B. Biolatto, *Org. Lett.* 2002, 4, 1867; (d) N. E. Leadbeater, M. Marco, *Org. Lett.* 2002, 4, 2973; (e) D. Zim, A. L. Monteiro, J. Dupont, *Tetrahedron Lett.* 2000, 41, 8199; (f) N. A. Bumagin, D. A. Tsarev, *Tetrahedron Lett.* 1998, 39, 8155; (g) D. Badone, M. Baroni, R. Cardamone, A. Ielmini, U. Guzzi, *J. Org. Chem.* 1997, 62, 7170; (h) M. Moreno-Mañas, F. Pajuelo, R. Pleixats, *J. Org. Chem.* 1995, 60, 2397; (i) T. I. Wallow, B. M. Novak, *J. Org. Chem.* 1994, 59, 5034.

18 (a) N. E. Leadbeater, M. Marco, *J. Org. Chem.* 2003, 68, 5660; (b) N. E. Leadbeater, M. Marco, *Angew. Chem.* 2003, 115, 1445; *Angew. Chem. Int. Ed.* 2003, 42, 1407.

19 (a) N. Miyaura, A. Suzuki, *Chem. Rev.* 1995, 95, 2457; (b) A. Suzuki, in Metal-catalyzed Cross-coupling Reactions (Eds.: F. Diederich, P. J. Stang), Wiley-VCH, Weinheim, 1998, 49; (c) A. Suzuki, *J. Organomet. Chem.* 1999, 576, 147.

20 (a) K. Matos, J. A. Soderquist, *J. Org. Chem.* 1998, 63, 461; (b) G. B. Smith, G. C. Dezeny, D. L. Hughes, A. O. King, T. R. Verhoeven, *J. Org. Chem.* 1994, 59, 8151.

21 S. W. Wright, D. L. Hagemann, L. D. McClure, *J. Org. Chem.* 1994, 59, 6095.

22 For phosphine scrambling with PPh$_3$ see D. F. O'Keefe, M. C. Dannock, S. M. Marcuccio, *Tetrahedron Lett.* 1992, 33, 6679.

23 (a) O. Lohse, P. Thevenin, E. Waldvogel, *Synlett* 1999, 45; (b) G. Cooke, H. A. de Cremiers, V. M. Rotello, B. Tarbit, P. E. Vanderstraeten, *Tetrahedron* 2001, 57, 2787.

24 (a) M. B. Mitchell, P. J. Wallbank, *Tetrahedron Lett.* 1991, 32, 2273; (b) N. M. Ali, A. McKillop, M. B. Mitchell, R. A. Rebelo, P. J. Wallbank, *Tetrahedron Lett.* 1992, 48, 8117.

25 V. V. Grushin, H. Alper, *Chem. Rev.* 1994, 94, 1047.

26 (a) A. F. Littke, G. C. Fu, *Angew. Chem.* 2002, 114, 4350; *Angew. Chem. Int. Ed.* 2002, 41, 4176; (b) H. Gröger, *J. Prakt. Chem.* 2000, 342, 334; (c) R. Stürmer, *Angew. Chem.* 1999, 111, 3509; *Angew. Chem. Int. Ed.* 1999, 38, 3307.

27 W. A. Herrmann, C. Brossmer, K. Öfele, C.-P. Reisinger, T. Riermeier, M. Beller, H. Fischer, *Angew. Chem.* 1995, 107, 1989; *Angew. Chem. Int. Ed. Engl.* 1995, 34, 1845.

28 M. Beller, H. Fischer, W. A. Herrmann, K. Öfele, C. Brossmer, *Angew. Chem.* 1995, 107, 1992; *Angew. Chem. Int. Ed. Engl.* 1995, 34, 1848.

29 V. P. W. Böhm, W. A. Herrmann, *Chem. Eur. J.* 2001, 7, 4191.

30 (a) M. Beller, T. H. Riermeier, *Eur. J. Inorg. Chem.* 1998, 29; (b) W. A. Herrmann, W. P. W. Böhm, C.-P. Reisinger, *J. Organomet. Chem.* 1999, 576, 23.

31 A. Zapf, M. Beller, *Chem. Eur. J.* 2000, 6, 1830.

32 D. A. Albisson, R. B. Bedford, S. E. Lawrence, P. N. Scully, *Chem. Commun.* 1998, 2095.

33 (a) R. B. Bedford, C. S. J. Cazin, S. L. Hazelwood, *Angew. Chem.* 2002, 114, 4294; *Angew. Chem. Int. Ed.* 2002, 41, 4120; (b) R. B. Bedford, S. L. Hazelwood, M. E. Limmert, *Chem. Commun.* 2002, 2610; (c) R. B. Bedford, S. L. Hazelwood, P. N. Horton, M. B. Hursthouse, *J. Chem. Soc., Dalton Trans.* 2003, 4164.

34 S.-Y. Liu, M. J. Choi, G. C. Fu, *Chem. Commun.* 2001, 2408.

35 T. E. Pickett, C. J. Richards, *Tetrahedron Lett.* 2001, 42, 3767.

36 (a) D. W. Old, J. P. Wolfe, S. L. Buchwald, *J. Am. Chem. Soc.* 1998, 120, 9722; (b) J. P. Wolfe, R. A. Singer, B. H. Yang, S. L. Buchwald, *J. Am. Chem. Soc.* 1999, 121, 9550; (c) J. P. Wolfe, S. L. Buch-

WALD, *Angew. Chem.* **1999**, *111*, 2570; *Angew. Chem. Int. Ed.* **1999**, *38*, 2413.

37 J. YIN, M. P. RAINKA, X.-X. ZHANG, S. L. BUCHWALD, *J. Am. Chem. Soc.* **2002**, *124*, 1162.

38 (a) X. BEI, T. CREVIER, A. S. GURAM, B. JANDELEIT, T. S. POWERS, H. W. TURNER, T. UNO, W. H. WEINBERG, *Tetrahedron Lett.* **1999**, *40*, 3855; (b) X. BEI, H. W. TURNER, W. H. WEINBERG, A. S. GURAM, J. L. PETERSEN, *J. Org. Chem.* **1999**, *64*, 6797.

39 A. ZAPF, R. JACKSTELL, F. RATABOUL, T. RIERMEIER, A. MONSEES, C. FUHRMANN, N. SHAIKH, U. DINGERDISSEN, M. BELLER, *Chem. Commun.* **2004**, *4*, 38.

40 (a) A. F. LITTKE, G. C. FU, *Angew. Chem.* **1998**, *110*, 3586; *Angew. Chem. Int. Ed.* **1998**, *37*, 3387; (b) A. F. LITTKE, C. DAI, G. C. FU, *J. Am. Chem. Soc.* **2000**, *122*, 4020.

41 J. P. STAMBULI, R. KUWANO, J. F. HARTWIG, *Angew. Chem.* **2002**, *114*, 4940; *Angew. Chem. Int. Ed.* **2002**, *41*, 4746.

42 For the synthesis of BuPAd$_2$ see: (a) A. TEWARI, M. HEIN, A. ZAPF, M. BELLER, *Synthesis* **2004**, 935; (b) A. EHRENTRAUT, A. ZAPF, M. BELLER, *Synlett* **2000**, 1589.

43 A. ZAPF, A. EHRENTRAUT, M. BELLER, *Angew. Chem.* **2000**, *112*, 4315; *Angew. Chem. Int. Ed.* **2000**, *39*, 4153.

44 N. KATAOKA, Q. SHELBY, J. P. STAMBULI, J. F. HARTWIG, *J. Org. Chem.* **2002**, *67*, 5553.

45 M. GÓMEZ ANDREU, A. ZAPF, M. BELLER, *Chem. Commun.* **2000**, 2475.

46 R. B. BEDFORD, C. S. J. CAZIN, S. J. COLES, T. GELBRICH, M. B. HURSTHOUSE, V. J. M. SCORDIA, *J. Chem. Soc., Dalton Trans.* **2003**, 3350.

47 (a) R. B. BEDFORD, C. S. J. CAZIN, S. J. COLES, T. GELBRICH, P. N. HORTON, M. B. HURSTHOUSE, M. E. LIGHT, *Organometallics* **2003**, *22*, 987; (b) R. B. BEDFORD, C. S. J. CAZIN, *Chem. Commun.* **2001**, 1540.

48 A. SCHNYDER, A. F. INDOLESE, M. STUDER, H.-U. BLASER, *Angew. Chem.* **2002**, *114*, 3820; *Angew. Chem. Int. Ed.* **2002**, *41*, 3668.

49 (a) G. Y. LI, *J. Org. Chem.* **2002**, *67*, 3643; (b) G. Y. LI, *Angew. Chem.* **2001**, *113*, 1561; *Angew. Chem. Int. Ed.* **2001**, *40*, 1513.

50 (a) M. FEUERSTEIN, H. DOUCET, M. SANTELLI, *Tetrahedron Lett.* **2001**, *42*, 5659; (b) M. FEUERSTEIN, H. DOUCET, M. SANTELLI, *Tetrahedron Lett.* **2001**, *42*, 6667; (c) M. FEUERSTEIN, D. LAURENTI, C. BOUGEANT, H. DOUCET, M. SANTELLI, *Chem. Commun.* **2001**, 325; (d) M. FEUERSTEIN, D. LAURENTI, H. DOUCET, M. SANTELLI, *Synthesis* **2001**, 2320; (e) M. FEUERSTEIN, F. BERTHIOL, H. DOUCET, M. SANTELLI, *Synlett* **2002**, 1807.

51 M. FEUERSTEIN, H. DOUCET, M. SANTELLI, *Synlett* **2001**, 1458.

52 R. B. BEDFORD, S. M. DRAPER, P. N. SCULLY, S. L. WELCH, *New J. Chem.* **2000**, *24*, 745.

53 T. SCHAREINA, R. KEMPE, *Angew. Chem.* **2002**, *114*, 1591; *Angew. Chem. Int. Ed.* **2002**, *41*, 1521.

54 S. HABER, in Aqueous-Phase Organometallic Catalysis (eds. B. CORNILS, W. A. HERRMANN), Wiley-VCH, Weinheim, **1998**, 444.

55 R. D. LARSEN, A. O. KING, C. Y. CHEN, E. G. CORLEY, B. S. FOSTER, F. E. ROBERTS, C. YANG, D. R. LIEBERMAN, R. A. REAMER, D. M. TSCHAEN, T. R. VERHOEVEN, P. J. REIDER, Y. S. LO, L. T. ROSSANO, A. S. BROOKES, D. MELONI, J. R. MOORE, J. F. ARNETT, *J. Org. Chem.* **1994**, *59*, 6391.

56 W. A. HERRMANN, *Angew. Chem.* **2002**, *114*, 1342; *Angew. Chem. Int. Ed.* **2002**, *41*, 1290.

57 W. A. HERRMANN, C.-P. REISINGER, M. SPIEGLER, *J. Organomet. Chem.* **1998**, *557*, 93.

58 (a) C. ZHANG, J. HUANG, M. L. TRUDELL, S. P. NOLAN, *J. Org. Chem.* **1999**, *64*, 3804; (b) G. A. GRASA, M. S. VICIU, J. HUANG, C. ZHANG, M. L. TRUDELL, S. P. NOLAN, *Organometallics* **2002**, *21*, 2866; (c) A. C. HILLIER, G. A. GRASA, M. S. VICIU, H. M. LEE, C. YANG, S. P. NOLAN, *J. Organomet. Chem.* **2002**, *653*, 69.

59 A. FÜRSTNER, A. LEITNER, *Synlett* **2001**, 290.

60 R. JACKSTELL, M. GÓMEZ ANDREU, A. FRISCH, K. SELVAKUMAR, A. ZAPF, H. KLEIN, A. SPANNENBERG, D. RÖTTGER, O. BRIEL, R. KARCH, M. BELLER, *Angew.*

Chem. **2002**, *114*, 1028; *Angew. Chem. Int. Ed.* **2002**, *41*, 986.

61 (a) K. Selvakumar, A. Zapf, A. Spannenberg, M. Beller, *Chem. Eur. J.* **2002**, *8*, 3901; (b) K. Selvakumar, A. Zapf, M. Beller, *Org. Lett.* **2002**, *4*, 3031.

62 M. S. Viciu, R. F. Germaneau, O. Navarro-Fernandez, E. D. Stevens, S. P. Nolan, *Organometallics* **2002**, *21*, 5470.

63 (a) V. P. W. Böhm, C. W. K. Gstöttmayr, T. Weskamp, W. A. Herrmann, *J. Organomet. Chem.* **2000**, *595*, 186; (b) C. W. K. Gstöttmayr, V. P. W. Böhm, E. Herdtweck, M. Grosche, W. A. Herrmann, *Angew. Chem.* **2002**, *114*, 1421; *Angew. Chem. Int. Ed.* **2002**, *41*, 1363.

64 C. Zhang, M. L. Trudell, *Tetrahedron Lett.* **2000**, *41*, 595.

65 D. S. McGuinness, K. J. Cavell, *Organometallics* **2000**, *19*, 741.

66 J. A. Loch, M. Albrecht, E. Peris, J. Mata, L. W. Faller, R. H. Crabtree, *Organometallics* **2002**, *21*, 700.

67 V. César, S. Bellemin-Laponnaz, L. H. Gade, *Organometallics* **2002**, *21*, 5204.

68 (a) T. Weskamp, V. P. W. Böhm, W. A. Herrmann, *J. Organomet. Chem.* **1999**, *585*, 348; (b) W. A. Herrmann, V. P. W. Böhm, C. W. K. Gstöttmayr, M. Grosche, C.-P. Reisinger, T. Weskamp, *J. Organomet. Chem.* **2001**, *617/618*, 616.

69 D. Zim, A. S. Gruber, G. Ebeling, J. Dupont, A. L. Monteiro, *Org. Lett.* **2000**, *2*, 2881.

70 G. A. Grasa, A. C. Hillier, S. P. Nolan, *Org. Lett.* **2001**, *3*, 1077.

71 L. Botella, C. Nájera, *Angew. Chem.* **2002**, *114*, 187; *Angew. Chem. Int. Ed.* **2002**, *41*, 179.

72 C. Nájera, J. Gil-Moltó, S. Karlström, L. R. Falvello, *Org. Lett.* **2003**, *5*, 1451.

73 T.-Y. Luh, M.-k. Leung, K.-T. Wong, *Chem. Rev.* **2000**, *100*, 3187.

74 D. J. Cárdenas, *Angew. Chem.* **2003**, *115*, 398; *Angew. Chem. Int. Ed.* **2003**, *42*, 384.

75 M. Méndez, J. M. Cuerva, E. Gómez-Bengoa, D. J. Cárdenas, A. M. Echavarren, *Chem. Eur. J.* **2002**, *8*, 3620.

76 T. Ishiyama, S. Abe, N. Miyaura, A. Suzuki, *Chem. Lett.* **1992**, 691.

77 (a) R. Giovannini, T. Stüdemann, G. Dussin, P. Knochel, *Angew. Chem.* **1998**, *110*, 2512; *Angew. Chem. Int. Ed.* **1998**, *37*, 2387; (b) D. J. Cárdenas, *Angew. Chem.* **1999**, *111*, 3201; *Angew. Chem. Int. Ed.* **1999**, *38*, 3018.

78 (a) M. R. Netherton, C. Dai, K. Neuschütz, G. C. Fu, *J. Am. Chem. Soc.* **2001**, *123*, 10099; (b) J. H. Kirchhoff, M. R. Netherton, I. D. Hills, G. C. Fu, *J. Am. Chem. Soc.* **2002**, *124*, 13662.

79 J. H. Kirchhoff, C. Dai, G. C. Fu, *Angew. Chem.* **2002**, *114*, 2025; *Angew. Chem. Int. Ed.* **2002**, *41*, 1945.

80 M. R. Netherton, G. C. Fu, *Angew. Chem.* **2002**, *114*, 4066; *Angew. Chem. Int. Ed.* **2002**, *41*, 3910.

81 A. N. Cammidge, K. V. L. Crépy, *Chem. Commun.* **2000**, 1723.

82 J. Yin, S. L. Buchwald, *J. Am. Chem. Soc.* **2000**, *122*, 12051.

83 Y. M. Kim, S. Yu, *J. Am. Chem. Soc.* **2003**, *125*, 1696.

2.11
Transition Metal-Catalyzed Arylation of Amines and Alcohols

Alexander Zapf, Matthias Beller, and Thomas H. Riermeier

2.11.1
Introduction

Aromatic amines and aryl ethers constitute important structural units in natural products as well as pharmaceuticals and agrochemicals. An ingenious solution to the synthesis of this class of compounds is the use of metal-catalyzed processes in which aryl halides are coupled with amines or alcohols. While related palladium-catalyzed C–C coupling reactions have been extensively studied in the last three decades (see Chapters 2.10 and 2.13), there has been a growing interest in catalytic C–N and C–O coupling reactions since the mid-1990s [1].

2.11.2
Catalytic Amination Reactions

2.11.2.1
Palladium-Catalyzed Arylation of Aromatic and Aliphatic Amines

First attempts to use palladium catalysts for C–N bond-forming reactions were reported by Migita and co-workers [2]. They showed that aryl bromides reacted with tributyl-*N,N*-diethylaminostannane as transamination reagent in the presence of palladium catalysts to yield the corresponding aromatic amines (Scheme 1).

According to investigations by Hartwig et al. [3], the actual catalytically active species in this amination reaction is believed to be a bis(tri-*o*-tolylphosphine)palla-

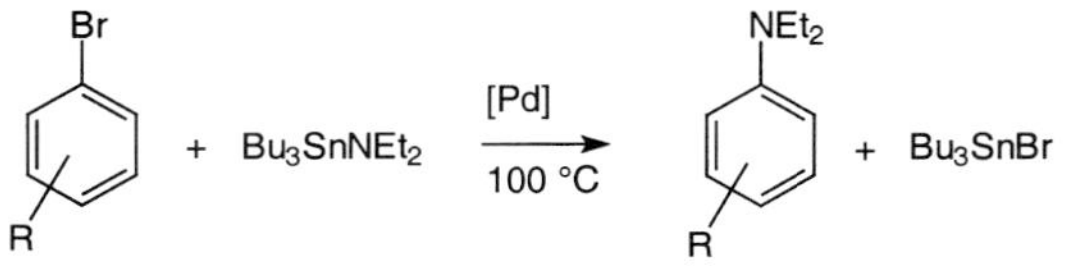

Scheme 1 First palladium-catalyzed amination of aryl halides. [Pd] = $PdCl_2[P(o\text{-tol})_3]_2$. R = H, 2-$CH_3$, 3-$CH_3$, 4-$CH_3$, 4-$OCH_3$, 4-$COCH_3$, 4-$NO_2$, 4-$N(CH_3)_2$.

Transition Metals for Organic Synthesis, Vol. 1, 2nd Edition.
Edited by M. Beller and C. Bolm

ISBN: 3-527-30613-7

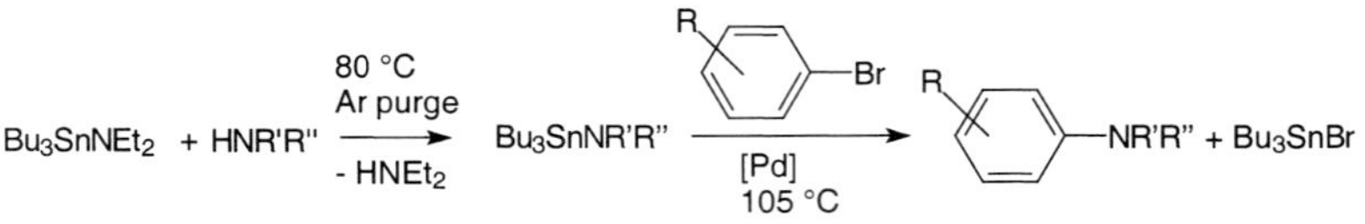

Scheme 2 Palladium-catalyzed amination of aryl halides using tin amides. [Pd] = PdCl$_2$[P(o-tol)$_3$]$_2$. R = 4-CH$_3$, 3-CH$_3$, 4-CF$_3$, 3-OCH$_3$, 4-CO$_2$Et; R' = CH$_2$C$_6$H$_5$, C$_6$H$_5$; R'' = H, CH$_3$.

dium(0) complex. The catalytic cycle starts with an oxidative addition of the palladium(0) complex into the aryl-halogen bond. The resulting arylpalladium(II) complex reacts with the tin amide with transmetalation; this step is postulated to be rate determining. Subsequent reductive elimination of the aminated arene liberates the active palladium(0) species again. By clever combination of a transamination reaction of tributyl-N,N-diethylaminostannane with higher boiling amines and palladium catalysis, Guram and Buchwald succeeded in extending the amination method (Scheme 2) [4]. It was reported that secondary aliphatic and aromatic amines react with substituted aryl bromides to afford the corresponding arylamines in good yields.

The use of stoichiometric amounts of tin amides is the main disadvantage of this type of C–N coupling reaction both for ecological reasons and with regard to practicability. Thus, it was an important improvement when Buchwald et al. [5] and Hartwig et al. [6] reported the first catalytic aminations of aryl bromides with free amines in 1995. In general this palladium-catalyzed coupling reaction, nowadays called the Buchwald-Hartwig reaction, occurs in the presence of a stoichiometric amount of a sterically hindered base such as NaOtBu in toluene or tetrahydrofurane at temperatures of 65–100 °C (Scheme 3).

Tertiary arylamines are generally formed in good to excellent yields (Tab. 1). As a by-product, the hydrodehalogenated benzene derivative is sometimes observed. This arises from β-hydride elimination at the amido arylpalladium complex, giving an unstable hydrido arylpalladium complex, which subsequently undergoes reductive elimination to give the corresponding arene. Interestingly, the base employed has a decisive influence on the course of the reaction. Whereas in the presence of silyl amides the rate-determining step in the catalytic cycle is the oxidative addition of the aryl halide to bis(tri-o-tolylphosphine)palladium(0), the formation

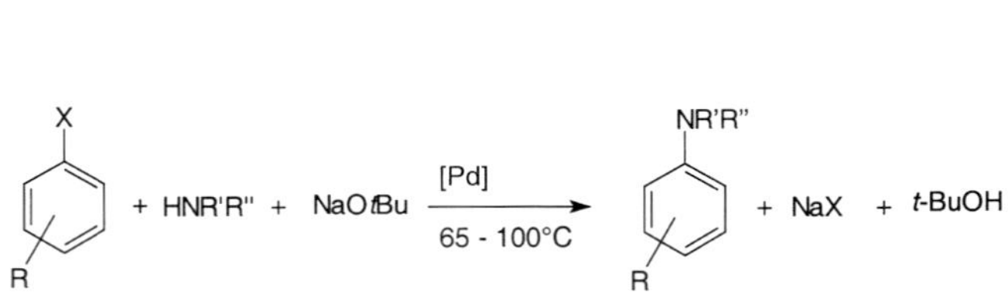

Scheme 3 Catalytic amination of aryl halides using amines (Buchwald-Hartwig reaction). [Pd] = PdCl$_2$[P(o-tol)$_3$]$_2$. X = Br, I; R = 4-C$_4$H$_9$, 4-CF$_3$, 4-OCH$_3$, 4-C$_6$H$_5$, 4-N(CH$_3$)$_2$; R' = C$_6$H$_5$, C$_6$H$_{13}$; R'' = H, CH$_3$; R'-R'' = (CH$_2$)$_5$, CH$_2$CH$_2$N(CH$_3$)CH$_2$CH$_2$.

Tab. 1 Amination of aryl halides using bis(tri-o-tolylphosphine)palladium(0) as catalyst.

Entry Ref.	Aryl halide	Amine	Base	Arylamine	mol% Pd	Yield (%)
1 [9]	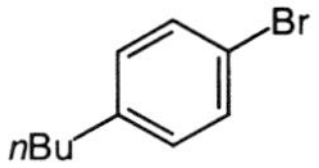 4-nBu-C6H4-Br	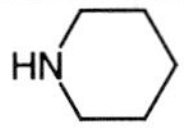 HN(piperidine)	LiN(SiMe3)2	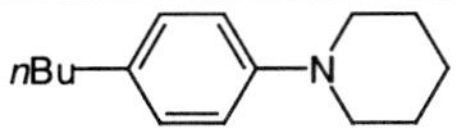 nBu-C6H4-N(piperidine)	5	89
2 [9]	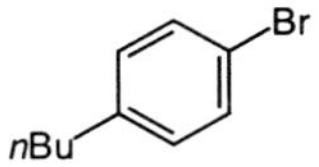 4-nBu-C6H4-Br	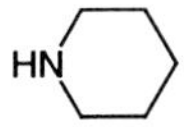 HN(piperidine)	NaOtBu	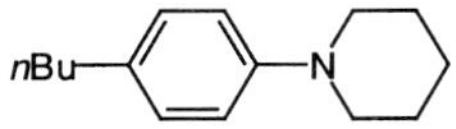 nBu-C6H4-N(piperidine)	5	89
3 [9]	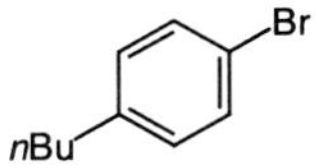 4-nBu-C6H4-Br	HNEt2	LiN(SiMe3)2	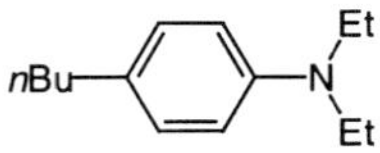 nBu-C6H4-NEt2	5	40
4 [9]	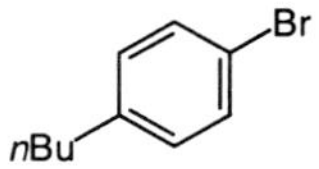 4-nBu-C6H4-Br	HNEt2	LiOtBu	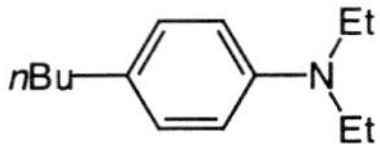 nBu-C6H4-NEt2	5	<2
5 [8]	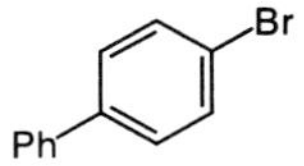 4-Ph-C6H4-Br	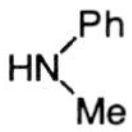 HN(Ph)Me	NaOtBu	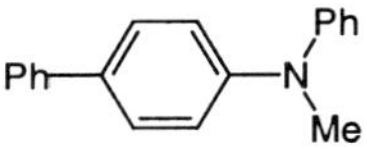 Ph-C6H4-N(Ph)Me	2	88
6 [8]	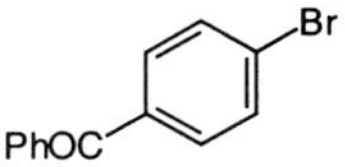 4-PhOC-C6H4-Br	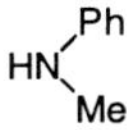 HN(Ph)Me	NaOtBu	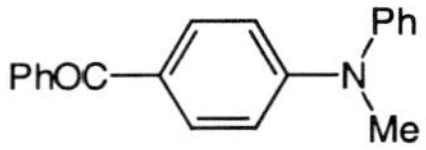 PhOC-C6H4-N(Ph)Me	2	89
7 [8]	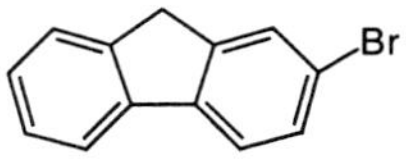 2-Br-fluorene	 HN(CH2CH2)2NMe	NaOtBu	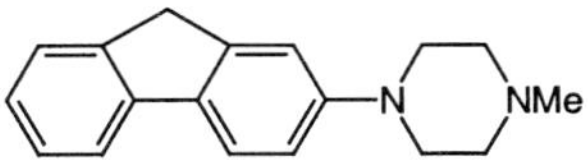 fluorene-N(CH2CH2)2NMe	2	79
8 [8]	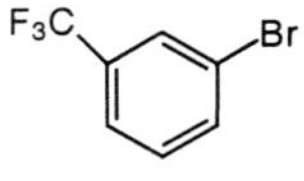 3-F3C-C6H4-Br	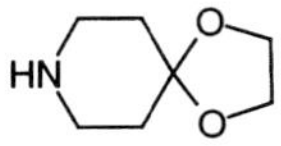 HN(spiro-dioxolane-piperidine)	NaOtBu	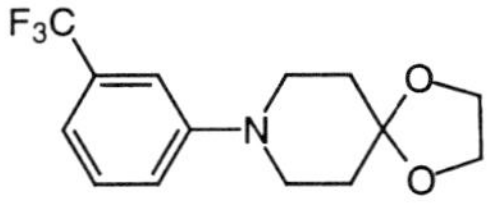 F3C-C6H4-N(spiro-dioxolane-piperidine)	2	67
9 [10]	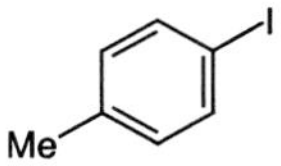 4-Me-C6H4-I	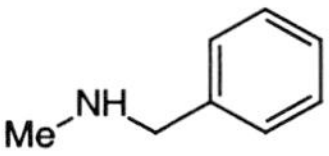 Me-NH-CH2Ph	NaOtBu	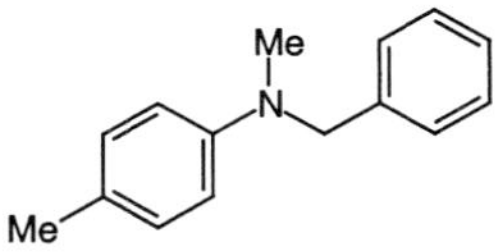 4-Me-C6H4-N(Me)CH2Ph	1	79
10 [10]	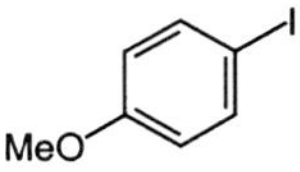 4-MeO-C6H4-I	 HN(morpholine)	NaOtBu	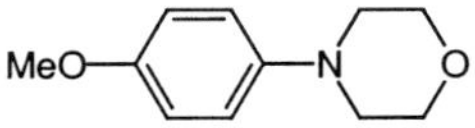 MeO-C6H4-N(morpholine)	1	66

and reductive elimination of the amido arylpalladium complex is decisive for the rate of the reaction in the presence of LiOtBu [6].

As shown in Tab. 1, results using sodium tert-butoxide as base are similar to those using lithium amides as base (entries 1 and 2). Both are superior to lithium tert-butoxide (entries 3 and 4). Since NaOtBu is easier to handle, it seems to be the base of choice for this reaction. In the presence of tri-o-tolylphosphine as li-

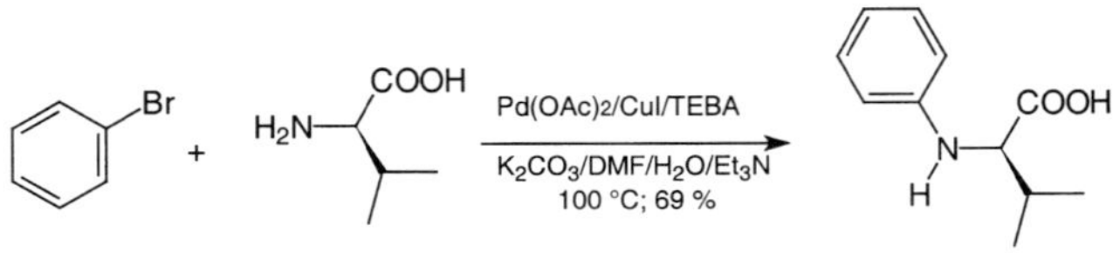

Scheme 4 Catalytic amination of amino acids.

gand, the methodology is limited mainly to secondary amines as starting materials. Nevertheless, an interesting application of this new method appeared shortly afterwards in the literature: Zhao and co-workers from Roche Bioscience demonstrated elegantly the synthetic potential of the amination reaction for the synthesis of various N-arylpiperazines [7]. In the case of piperazine itself, the appropriate choice of reaction conditions leads to either a symmetrical N,N'-bisarylpiperazine or N-monoarylpiperazine in good yield. Amination reactions with C-substituted unsymmetrical piperazines proceeded with high regioselectivity, allowing facile preparation of several novel arylpiperazines. Two other research groups, Ward et al. from Boehringer Ingelheim Pharmaceuticals and Willoughby et al. from Merck Research Laboratories, used a solid phase variant of the Buchwald-Hartwig amination protocol as a tool for combinatorial synthesis [8]. Kanbara et al. succeeded in the synthesis of poly(aryleneamine)s by polycondensation of aryl dibromides with diamines [9]. Based on the findings of Beletskaya et al. [10], who showed that the amination reaction is possible in aqueous-organic emulsions under the co-catalysis of copper(I) salts, Ma and Yao were able to arylate chiral α-amino acids by this method without significant racemization (Scheme 4) [11].

Extending the methodology, dihydroindoles, dihydroquinolines, and other N-heterocycles were synthesized by simple intramolecular trapping reactions, starting from aminoalkyl [12] or aminoaryl [13] substituted aryl bromides (Scheme 5). The intramolecular amination has been achieved using tetrakis(triphenylphosphine) palladium or $Pd_2(dba)_3/P(t\text{-Bu})_3$ as catalyst. Here, best results are obtained with mixtures of NaOtBu and potassium carbonate as base.

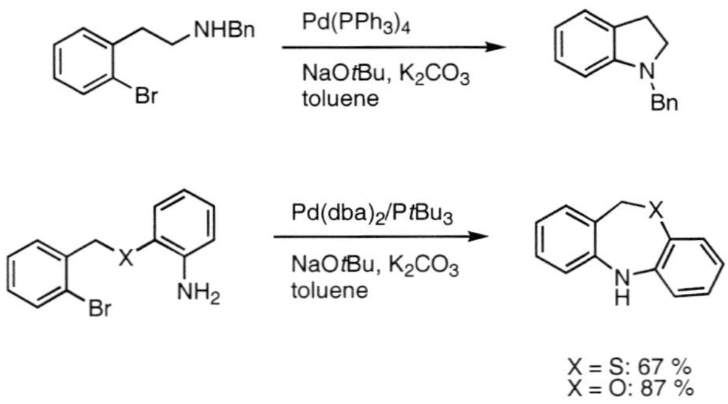

Scheme 5 Intramolecular palladium-catalyzed amination of aryl halides.

The usefulness of palladium-catalyzed aryl aminations was demonstrated also in natural product synthesis relatively shortly after the discovery of the reaction [14]. Here, the key step of the total synthesis of the toad poison dehydrobufotenine was the intramolecular amination of the corresponding aryl iodide (Scheme 6).

Based on detailed studies on the reactivity of certain potential intermediates of the catalytic cycle, Buchwald et al. and Hartwig et al. proposed a well-considered mechanism for the palladium-catalyzed amination of aryl halides (Scheme 7) [15].

The catalytic cycle starts with the oxidative addition of the aryl halide on a low ligated palladium(0) complex. After coordination of the amine to the metal center, deprotonation occurs to give an anionic amido complex which is stable only at low temperatures. At higher temperatures a tri-coordinated neutral amido compound is formed. Subsequent reductive elimination leads to the product and a palladium(0) complex, which starts a new catalytic cycle.

In the years 1995–98, typically 1–5 mol% of palladium complexes containing monodentate triarylphosphines were used as catalysts for the amination reactions. In the case of tri-o-tolylphosphine, its steric bulk leads to superior reactivity by favoring low coordination numbers at the metal center. Shortly after developing the Pd/P(o-tol)$_3$ catalyst system, both the groups of Buchwald [16] and Hartwig [17] described "second-generation amination catalysts" based on chelating bisphosphine ligands. While Hartwig and co-workers used 1,1'-bis(diphenylphosphino)ferrocene as ligand, Buchwald et al. employed racemic BINAP. As shown in Tab. 2, these newer catalyst systems work for the cross coupling of a variety of primary and secondary amines with both aryl bromides and aryl iodides. Next, the aryl amination methodology was extended to the coupling of aryl triflates, also using chelating ligands [18].

Noteworthy are the superior yields for the coupling of amines with *ortho*-substituted halides and halopyridines [19]. In the case of the synthesis of aminopyri-

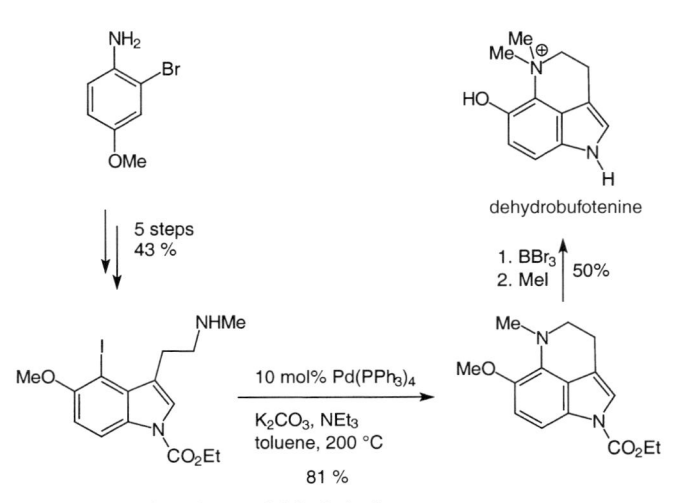

Scheme 6 Total synthesis of dehydrobufotenine.

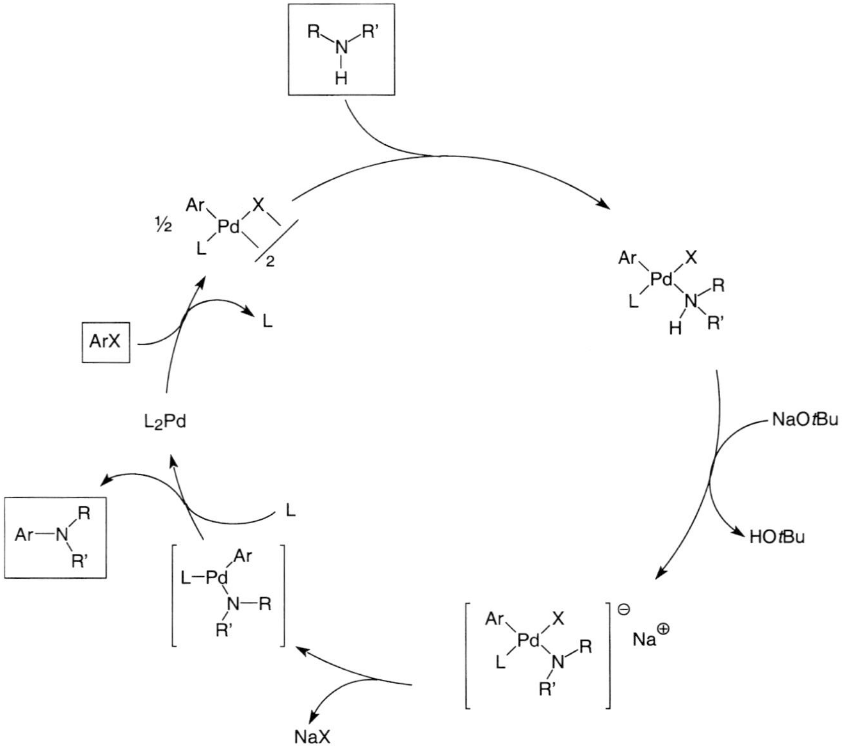

Scheme 7 Proposed mechanism of the Buchwald-Hartwig amination reaction.

dines, the improved catalyst activity is explained by the ability of chelating ligands to prevent the formation of aminopyridyl palladium complexes that terminate the catalytic cycle. Interestingly, electron-rich aryl bromides (entry 2) gave similarly high yields to those with electron-poor aryl halides (entry 1). The sterically hindered 1-bromo-2,5-dimethylbenzene can be coupled with N-methylpiperazine even in the presence of 0.05 mol% palladium (entry 4). Thus, catalyst turnover numbers as high as 2000 were realized for the first time. When primary amines were used, only small amounts of double arylated products were detected. In general, coupling reactions of acyclic aliphatic amines led to the formation of significant amounts of dehalogenated aromatic derivatives as by-products because of the higher ability of these amines to undergo β-hydride elimination during the catalytic cycle. This problem was solved by using the ferrocenyl-derived phosphinoether ligand 1 [20]. Thus, the ratio of desired product to reduced aryl halide in the reaction of 4-bromo-*tert*-butylbenzene with dibutylamine is increased to 39:1 from 13:1 using the P(*o*-tol)$_3$ or 1:5 using the BINAP ligand. Similar improvements in the selectivity in arylations of primary alkylamines were achieved by using chelating [21] or monodentate [22] phosphine ligands based on the ferrocene core with

Tab. 2 Aminations of aryl bromides and iodides catalyzed by palladium complexes with chelating ligands ("second-generation catalysts").

Entry Ref.	Aryl halide	Amine	Arylamine	mol% Pd	Yield (%)
1 [14]	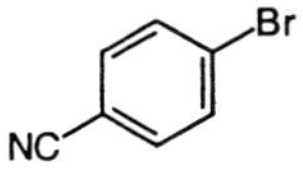 4-Br-C6H4-CN	nHexNH2	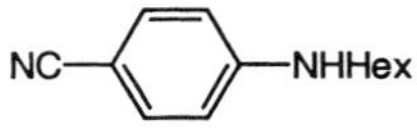 4-NC-C6H4-NHHex	0.5	98
2 [14]	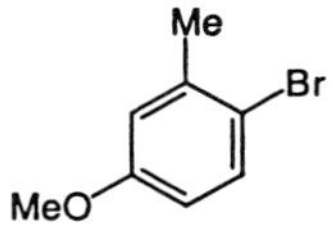 2-Me-4-MeO-C6H3-Br	nHexNH2	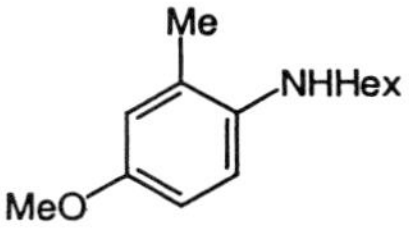 2-Me-4-MeO-C6H3-NHHex	0.5	95
3 [14]	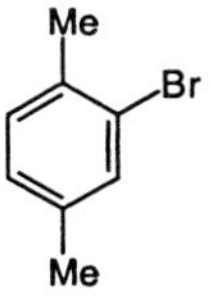 2,5-Me2-C6H3-Br	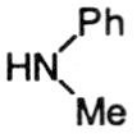 HN(Ph)Me	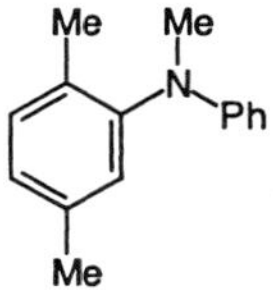 2,5-Me2-C6H3-N(Ph)Me	0.5	94
4 [14]	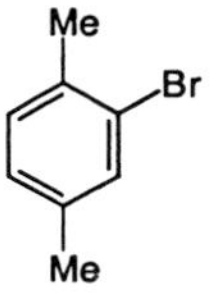 2,5-Me2-C6H3-Br	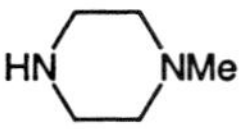 HN(CH2CH2)2NMe	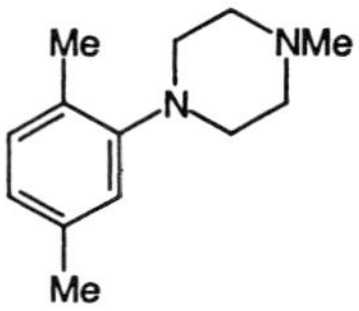 2,5-Me2-C6H3-N(CH2CH2)2NMe	0.05	94
5 [15]	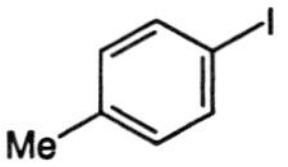 4-Me-C6H4-I	H2N-Ph	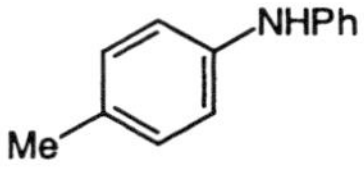 4-Me-C6H4-NHPh	5	92
6 [15]	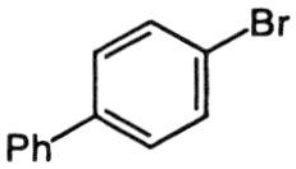 4-Ph-C6H4-Br	H2N-Ph	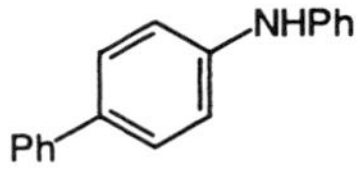 4-Ph-C6H4-NHPh	5	94
7 [15]	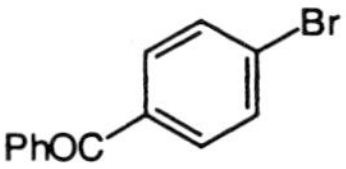 4-PhOC-C6H4-Br	nBuNH2	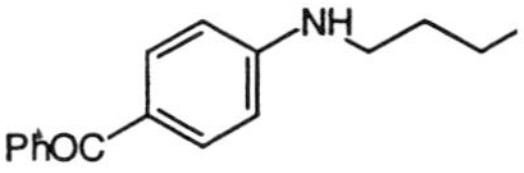 4-PhOC-C6H4-NHBu	5	96
8 [15]	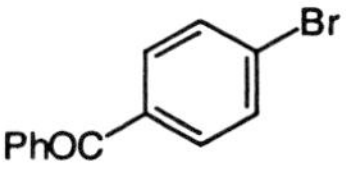 4-PhOC-C6H4-Br	iBuNH2	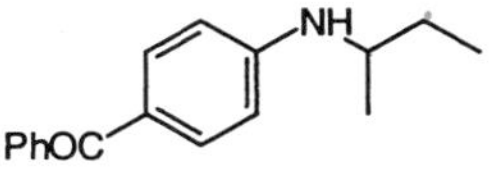 4-PhOC-C6H4-NH-iBu	5	84
9 [16]	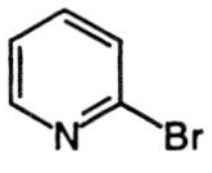 2-Br-pyridine	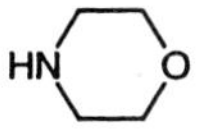 HN(CH2CH2)2O	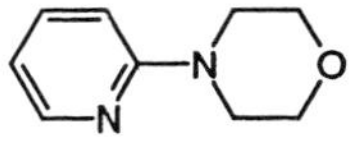 2-morpholino-pyridine	1	87
10 [16]	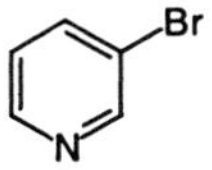 3-Br-pyridine	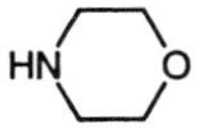 HN(CH2CH2)2O	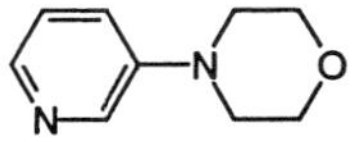 3-morpholino-pyridine	1	75

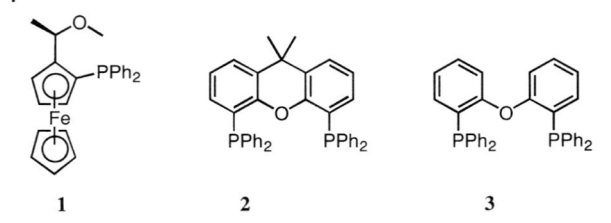

Fig. 1 Ligands for amination reactions.

sterically demanding alkyl groups (*t*Bu or Cy) on the phosphorus. Other useful ligands for the coupling of difficult amines are the Xantphos ligand (**2**), which was introduced for the amination of aryl bromides and triflates by van Leeuwen [23], and DPEphos (**3**) for the arylation of anilines [24] (Fig. 1).

Concerning the mechanism in the presence of "second-generation catalysts", the reductive elimination seems to be rate determining. Thus, the advantage of chelating ligands is to promote this step in the catalytic cycle. Detailed mechanistic studies of the reductive elimination reveal that the C–N bond-forming step is accelerated by electron-withdrawing groups on the aryl group and by electron-donating groups on the amido ligand [25].

Synthetic applications using catalysts with "second generation" ligands include the synthesis of aryl-substituted polyamine compounds. For example, 1,2-diaminoethylene reacts selectively with 4-bromobiphenyl (64% yield) to give the monoarylated product [26], and oligoanilines can be built up by repeated amination and protecting group chemistry [27]. Substituted indoles have been synthesized by arylation of hydrazones, followed by Fischer indole synthesis (Scheme 8) [28].

Substituted 2-aryl-2*H*-indazoles were synthesized by cyclization of the corresponding 1-[(*ortho*-bromophenyl)methyl]-1-arylhydrazines [29]. During a study of the enantioselective rearrangement of allylic imidates to allylic amides, Overman et al. used the palladium-catalyzed amination for the determination of the absolute configuration of certain products [30]. In addition, the coupling of optically active amines with a stereogenic center a to the nitrogen was described without racemization [31].

By using cesium carbonate as base, the catalytic amination of aryl-X derivatives with sensitive functional groups was achieved [32]. Thus, aminations are possible in the presence of esters, aldehydes, enolizable ketones, or nitro groups. Further-

Scheme 8 Synthesis of indoles from aryl bromides and hydrazones.

more, the amination of aryl triflates is improved using cesium carbonate because there are fewer side reactions [33].

Apart from aryl bromides and aryl iodides, chloroarenes constitute an interesting class of starting materials because of their availability and low cost. Thus, the coupling reaction of 4-trifluoromethyl-1-chlorobenzene with piperidine in the presence of so-called palladacycles, e.g., trans-di(μ-acetato)-bis[o-(di-o-tolylphosphino)benzyl]dipalladium(II) [34] was studied as a model reaction (Scheme 9) [35]. Crucial for the success of this C–N bond-forming reaction was the use of potassium tert-butoxide as base and reaction temperatures >120 °C. Turnover numbers up to 900 and yields up to 80% have been obtained in this reaction. Small amounts of the meta-substituted aniline were observed. This is explained by aryne intermediates which can be formed under these reaction conditions.

Other early palladium-based catalyst systems for the amination of aryl chlorides used tricyclohexylphosphine [36] or tri-tert-butylphosphine as ligands [37]. Pd catalysts for chloroarene amination have been improved dramatically during recent years, mainly by the groups of Buchwald and Hartwig. Hartwig's group focused on the use of commercially available tri-tert-butylphosphine as ligand [38], whereas Buchwald and co-workers developed a new class of ligands characterized by a o-biaryl and two sterically demanding alkyl groups on the phosphorus ("Buchwald ligands", Fig. 2), which turned out to be stable in air in contrast to PtBu$_3$ [39]. Alkyl bromides and chlorides can be coupled with anilines or alkylamines at room temperature or at 70 °C with P(t-Bu)$_3$ or (o-biphenyl)P(t-Bu)$_2$ as ligand and sodium tert-butoxide or cesium carbonate as base [38, 39]. Turnover numbers of almost 2000 were obtained in the amination of non-activated aryl chlorides using the latter catalyst systems [39]. The highest turnover frequencies have been reported for [P(t-Bu)$_3$PdBr]$_2$, which is supposed to decompose to mono-ligated Pd(0) and Pd(II)Br$_2$. This catalyst is capable of coupling 4-chloroanisole with di-n-butylamine within 15 min at room temperature to give 87% of the desired product (1 mol% Pd) [38b]. Sterically encumbered N-alkylanilines can also be arylated easily with bromoarenes by applying this catalyst system. A simple 1:1 mixture of Pd(OAc)$_2$ and P(t-Bu)$_3$ gave lower yields as well as chelating "second generation catalysts" [40]. Buchwald also immobilized his ligands by linking them with Merrifield Resin and showed that they are suitable for the amination of a variety of aryl iodides, bromides, and chlorides, although with lower TON compared to the homogeneous systems [41]. Further work in this group revealed that the palladacyclic compound obtained by simple stirring of 2-(di-tert-butylphos-

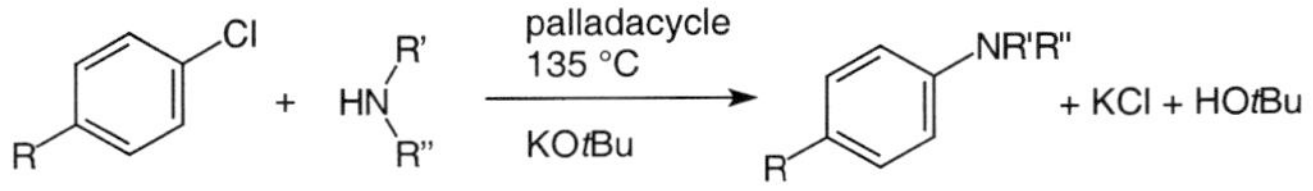

Scheme 9 Palladium-catalyzed amination of activated aryl chlorides. Palladacycle: trans-di(μ-acetato)-bis[o-(di-o-tolylphosphino)benzyl]dipalladium(II); R=CF$_3$, COPh; R'=C$_4$H$_9$, C$_6$H$_5$; R''=C$_4$H$_9$, CH$_3$; R'-R''=(CH$_2$)$_5$, CH$_2$CH$_2$OCH$_2$CH$_2$.

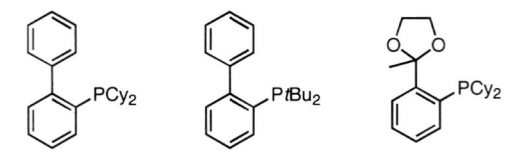

Fig. 2 Buchwald's and Guram's ligands for aryl chloride amination.

phino)biphenyl with palladium(II) acetate in toluene at room temperature is also an active pre-catalyst for the coupling of all kinds of aryl chlorides with primary and secondary amines [42]. The advantage of this new complex catalyst is its higher stability toward air compared to the corresponding *in situ* system, but in general the catalyst productivity is lower for the cyclometallated catalyst (TON < 200).

More recently, Bedford found that changing the palladium source from Pd(OAc)$_2$ to the cyclopalladated complex 4 (Fig. 3) led to an increase in the catalyst activity by a factor of up to 6.5, depending on the phosphine ligand applied [43]. Buchwald's 2-(dicyclohexylphosphino)biphenyl ligand again formed the most efficient catalyst.

At Symyx Technologies, Guram et al. developed a class of ligands similar to those of Buchwald for amination of aryl chlorides (Fig. 2) [44].

Here, a cyclic ketal *ortho* to the phosphorus builds up a substructure closely related to the biaryl ligands. Anilines and secondary and primary alkylamines were arylated in high yields in the presence of these ligands and 2 mol% Pd(dba)$_2$. Guram supposes that the formation of a P,O-chelate is responsible for the high catalytic activity [44b], but a comparison with Buchwald's results points to a pure steric effect.

Detailed studies on the amination of five-membered heterocyclic halides have been published very recently [45]. Tri-*tert*-butylphosphine is the ligand of choice for these substrates, but relatively high concentrations of catalyst (2–5 mol%) are required to obtain satisfactory yields with oxygen, nitrogen (with protected N–H), and even sulfur-containing heterocycles. Different carbazoles were obtained by sequential amination of aryl bromides and cyclization under C–H activation (Scheme 10) [46].

Hartwig described the use of aqueous hydroxide as an inexpensive and air-stable base for the amination of aryl bromides and chlorides in a liquid-liquid or liquid-solid two-phase system [47]. When cetyltrimethylammonium bromide was

Fig. 3 Pre-catalyst 4 for the amination of aryl halides.

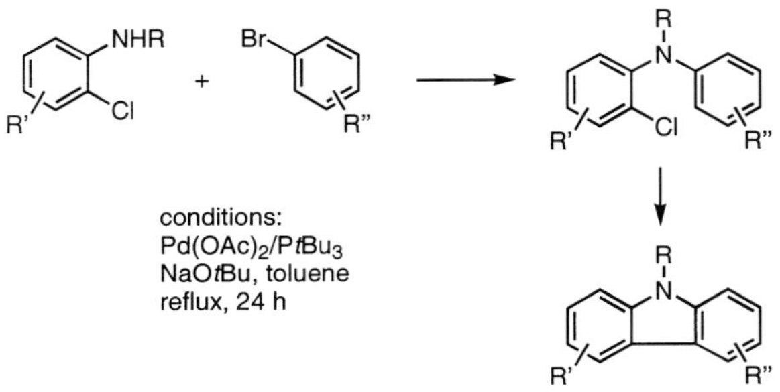

Scheme 10 Synthesis of carbazoles by palladium-catalyzed tandem reaction of 2-chloro anilines and bromoarenes [46a].

applied as a phase transfer catalyst, similar results were obtained compared to those obtained with the more expensive base sodium *tert*-butoxide. Amazingly, functional groups, such as esters, enolizable ketones, nitriles, and nitro groups, which were not tolerated by *tert*-butoxide, caused no problems under the new reaction conditions. An improved functional group tolerance was also observed by Buchwald when he employed lithium bis(trimethylsilyl)amide as a base in the coupling of aryl halides containing hydroxyl, amide, or enolizable keto groups [48].

A novel type of ligand for coupling reactions was developed by Li of DuPont. He found that secondary phosphine oxides, which are in equilibrium with the corresponding phosphinous acid, lead to active catalysts for the amination of non-activated aryl chlorides (Scheme 11) [49]. The ligands are air-stable and strongly electron-donating in the presence of base, but no TON higher than 20 has been described for different amination reactions.

We applied adamantylphosphines successfully in the amination of different aryl chlorides [50]. Diarylamines with *o*-substituents on both aryl groups are accessible in high yields with only 0.5 mol% Pd(OAc)$_2$ and 1 mol% di(1-adamantyl)-*n*-butylphosphine (Scheme 12).

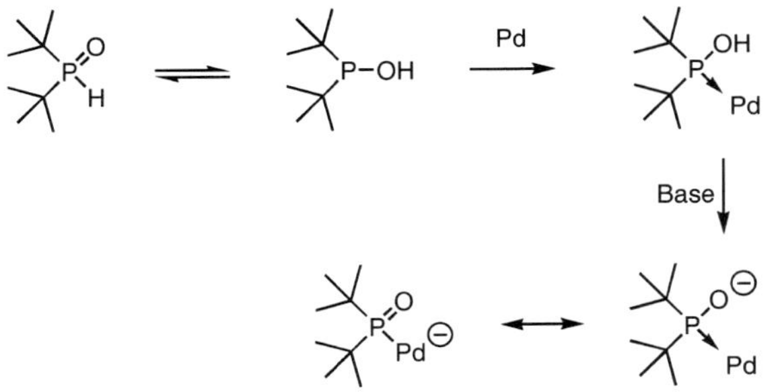

Scheme 11 Secondary phosphine oxides as new ligands.

Scheme 12 Synthesis of highly sterically congested diarylamines.

Solvias developed catalyst systems comprising palladacycles and secondary phosphines with bulky alkyl substituents for the amination of aryl chlorides [51]. The fate of the catalyst ingredients is not clear, but most likely an initial C–P coupling reaction between the palladacycle and the phosphine generates a sterically demanding and electron-rich tertiary phosphine ligand, which leads to the highly active catalysts.

Regarding substrate scope, the amination methodology has been extended in recent years to sulfoximines [52], indoles [38, 53, 54], carbamates [38, 55], amides [55, 56], sulfonamides [55, 56], pyrrole [54], carbazole [54], oxazolidinones [56], and ureas [56] using the different types of phosphine ligands described above.

During recent years N-heterocyclic carbenes have attracted much attention as ligands for a number of transition metal-catalyzed reactions involving Heck and Suzuki reactions, hydrosilylation, and metathesis, for example [57]. The first amination of aryl halides using Pd-carbene catalysts was described by Nolan et al. in 1999 [58]. They found that aryl chlorides as well as bromides and iodides can be coupled with primary and secondary alkyl and aryl amines in the presence of 2 mol% palladium with potassium *tert*-butoxide as base at 100 °C. The ligand of choice was 1,3-bis(2,6-di-*iso*-propylphenyl)imidazol-2-ylidene (IPr, Fig. 4), which was generated *in situ* from its hydrochloride salt. A ligand/palladium ratio of 2:1 was utilized, but a ratio of 1:1 has also been mentioned leading to a higher activity (TOF) of the catalyst. Less bulky carbene ligands have been found to be almost inactive in these reactions. Shortly after these findings, Hartwig published a similar protocol for the amination of aryl chlorides with the "saturated carbene" 1,3-bis(2,6-di-*iso*-propylphenyl)-4,5-dihydroimidazol-2-ylidene as ligand (SIPr, Fig. 4) [59]. Anilines and secondary alkylamines can be coupled even at room temperature in the presence of 1 mol% Pd. Under harsher reaction conditions (100 °C) turnover numbers up to 5000 were obtained.

Isolated biscarbene palladium complexes, which have been prepared from allylpalladium(II) chloride dimer and carbene or by ligand exchange from Pd[P(*o*-tol)$_3$]$_2$, have been described to catalyze the reaction of 4-chlorotoluene with some

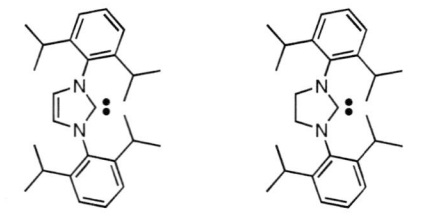

Fig. 4 1,3-Bis(2,6-di-*i*-propylphenyl)imidazol-2-ylidene (IPr) and 1,3-bis(2,6-di-*iso*-propylphenyl)-4,5-dihydroimidazol-2-ylidene (SIPr).

amines, albeit with low turnover numbers (<50) [60]. In contrast, the isolated monocarbene palladium complex [(IPr)PdCl$_2$]$_2$ is more efficient, giving complete conversion applying 1 mol% palladium [60a]. These results demonstrate that one carbene per palladium is sufficient for stabilizing and activating the palladium catalyst. More equivalents of carbene ligand are not required as they retard or even suppress the catalyst's activity. Interestingly, the use of Pd-carbene catalysts has been extended to the arylation of imines with aryl bromides and chlorides using IPrHCl as ligand precursor [61]. The coupling of less nucleophilic indoles has been realized using SIPrHCl [61]. Investigations into the oxidative addition step demonstrate the influence of the carbene ligand [62]: the reaction of 4-chlorotoluene with Pd(SIPr)$_2$ leads to decomposition of the complex and formation of the 2-(p-tolyl)-substituted imidazolium salt, whereas the corresponding oxidative addition complex with the N,N'-di-tert-butyl-substituted carbene ligand is stable.

2.11.2.2
Palladium-Catalyzed Synthesis of Primary Anilines

The amination of aryl halides with simple ammonia would lead to primary anilines – important compounds per se and starting materials for further derivatizations. Unfortunately, two severe problems arise with these reactions: first, reductive elimination from an Ar-Pd-NH$_2$ complex is considered to be much more difficult than that from an Ar-Pd-NHR complex and has not been observed so far, and second, the resulting amine is more reactive than simple ammonia, thus leading to double arylated products. Therefore, suitable N-protecting groups have to be chosen, which are easily cleaved off after the coupling reaction. The first example of this strategy was the amination of benzophenone imine. The protecting benzophenone can be removed by simple hydrolysis of the reaction product (Scheme 13). Phosphine ligands [63] are suitable for this reaction, but so are also N-heterocyclic carbene ligands [61].

Allylamine and diallylamine have also been used as coupling partners [64], as well as benzylamine and diphenylmethylamine [65]. Subsequent cleavage of the allyl or benzyl protecting group is straightforward and well known for many amines. Also, the reaction of aryl bromides or chlorides with lithium bis(trimethylsilyl)amide or triphenylsilylamine in the presence of base leads to the desired anilines after acidic cleavage of the corresponding silyl protecting group [66].

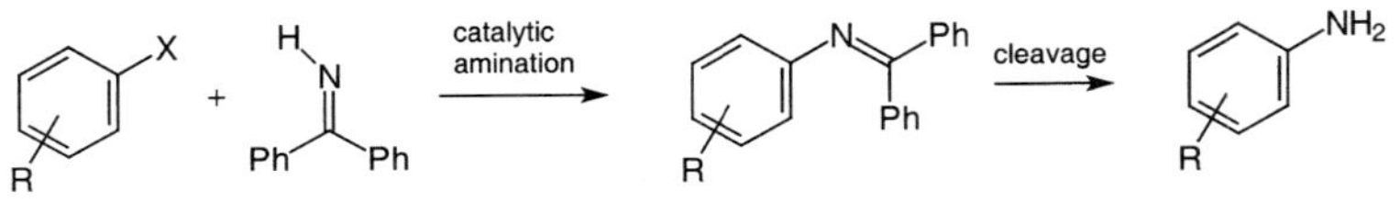

Scheme 13 Synthesis of substituted anilines.

2.11.2.3
Nickel-Catalyzed Arylation of Primary and Secondary Amines

The first nickel-catalyzed amination of aryl chlorides was described by Wolfe and Buchwald in 1997 [67]. Because Ni(0) complexes are more nucleophilic than their Pd(0) counterparts, the oxidative addition of aryl chlorides to these compounds is relatively easy. Therefore, reactions of all kinds of chloroarenes and also some chloropyridines with a variety of amines can be effected by combination of a Ni(0) or Ni(II) (which is reduced by MeMgBr *in situ*) precatalyst with simple dppf or 1,10-phenanthroline. Turnover numbers (TON) of up to 50 have been obtained [67]. Regarding the mechanism, it is likely that the coupling involves an electron transfer from an Ni(0) species to the aryl chloride and does not proceed via a classical oxidative addition (Ni(0) → Ni(II)) as in the case of palladium catalysts. Later on, somewhat milder reaction conditions have been described by Brenner and Fort, who used sodium hydride and sodium *tert*-amyloxide in the presence of Ni(OAc)$_2$, 2,2'-bipyridine, and styrene to improve catalyst lifetime by minimizing the competitive reductive dehalogenation of the aryl halide [68a, b]. However, still 10–20 mol% Ni have to be applied under these conditions and only secondary alkylamines (in most cases cyclic derivatives) have been arylated. Under similar conditions, (hetero)aryl di- and trichlorides could be mono-aminated selectively or aminated repeatedly with different amines [68c–e]. Also, piperazine has been arylated selectively once or twice [68f].

The first Ni-carbene catalyst system has also been described by the group of Fort [69]. 1,3-Bis(2,6-di-*iso*-propylphenyl)imidazolium chloride (IPrHCl) or the 4,5-dihydro analog (SIPrHCl) in combination with Ni(acac)$_2$ gives a highly active catalyst for the coupling of aryl chlorides with secondary alkylamines and anilines. Here, 2 mol% of catalyst are necessary for the arylation of cyclic amines, whereas 5 mol% are required for the coupling of acyclic amines and anilines, respectively. The presence of styrene or another hydrogen scavenger is not required in the carbene-based catalyst system. Cyclization of *ortho*-aminoalkyl-substituted chloroarenes could also be realized under these conditions with SIPrHCl as ligand [70]. Five- to seven-membered *N*-heterocycles with an annelated benzene moiety were synthesized in moderate to excellent yields.

2.11.2.4
Copper-Catalyzed Arylation of Primary and Secondary Amines

The well-known copper-mediated Ullmann [71] and Goldberg reactions [72] for the synthesis of aryl amines and amides both suffer from the required drastic reaction conditions: high temperatures, highly polar solvents, and large amounts of copper are needed for successful transformations. In 1999 Buchwald et al. described an improved variant for the arylation of imidazoles in the presence of copper catalysts. In the presence of 10 mol% (CuOTf)$_2 \cdot$PhH, stabilized by 1,10-phenanthroline and dba, both electron-poor and electron-rich aryl iodides were coupled with substituted imidazoles [73]. Later, a new protocol for the arylation of amides was also developed by

Buchwald and co-workers. Here, aryl iodides, bromides, and even aryl chlorides were coupled with amides in the presence of 1–10 mol% copper(I) iodide and *trans*-1,2-diaminocyclohexane as catalyst in dioxane at 110 °C [74]. In addition, substituted indoles and related *N*-heterocycles were reacted with aryl iodides in the presence of 1 mol% of Cu catalyst. Ethylenediamine has been also used as a ligand [75]. Improved results were obtained when chelating *secondary* amine ligands were applied [76]. Functional groups such as sulfur (thiophene), hydroxyl, or amino groups are tolerated under these conditions. Also amido-substituted furans, which are intermediates for the synthesis of *N*-heterocycles, for example, have been prepared following this protocol [77]. Researchers of Merck extended this method for the preparation of primary anilines [78]. The use of N-protecting groups for the coupling step is not required here, as for similar palladium-catalyzed reactions. The use of 0.5 mol% copper(I) oxide was sufficient to effect the reaction under mild conditions (80 °C, 50 psi) in ethylene glycol or *iso*-propanol [79].

A soluble, defined copper(I) complex with a 1,10-phenanthroline and a triphenylphosphine ligand has been used for the coupling of diphenylamine with aryl halides in toluene at 110 °C [80]. The arylation of primary alkylamines has been realized by the use of *N,N*-diethylsalicylamide as ligand in DMF or under solvent-free conditions at 90 °C [81]. Under the same reaction conditions, the cyclization of *o*-aminoalkyl-substituted aryl bromides and chlorides to form annelated five- and six-membered heterocycles is possible. A general protocol for the coupling of aryl iodides with primary or secondary alkylamines or with anilines relies on *α*-amino acids as ligands (e.g., L-proline) [82]. Here, relatively low reaction temperatures (60–90 °C) are sufficient, but 10 mol% CuI has to be applied as the catalyst and DMSO as the solvent. This finding is based on earlier work concerning the arylation of *α*-amino acids under relatively mild conditions (90 °C) [83]. It was found that amino acids, especially those with larger hydrophobic groups, undergo the arylation reaction quite smoothly, probably by building up a template around the copper catalyst. The amino acid should be coordinated via the carboxylate function, whereas the aryl iodide may form a π-complex with the copper, bringing nitrogen and arene close together.

Not only aryl halides, but also arylboronic acids have been subjected to copper-catalyzed amination reactions. Amines, amides, imides, ureas, sulfonamides, carbamates, and *N*-heterocycles can be arylated with moderate to very good yields at room temperature in the presence of stoichiometric amounts of copper (Scheme 14) [84].

Regarding the mechanism of this coupling procedure, it is believed that one acetate ligand of Cu(OAc)$_2$ is substituted by the deprotonated *N*-heterocycle, re-

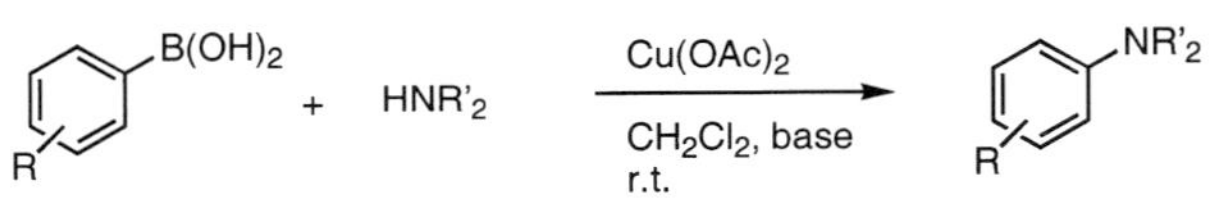

Scheme 14 Copper-mediated amination of arylboronic acids.

sulting in a copper amide complex. The second acetate ion should be replaced by the aryl residue of the boronic acid in a transmetalation reaction. Subsequent reductive elimination of the arylated amine occurs directly from this complex or after oxidation of the Cu(II) to Cu(III) by air [85].

The first N-arylation of arylboronic acids that can be performed in a catalytic manner was described by Collmann et al. in 2000 [86]. According to this procedure, imidazoles were arylated in the presence of 10 mol% [Cu(OH)(TMEDA)]$_2$Cl$_2$ under an atmosphere of oxygen. Water can also be used as solvent instead of CH$_2$Cl$_2$ [86b]. The speculated mechanism is somewhat different from the one formerly postulated: the first step would be substitution of the hydroxide ligand by the aryl group, followed by coordination of imidazole. Deprotonation of the heterocycle occurs together with oxidation of the Cu(II) center by oxygen. The resulting Cu(III) complex reductively eliminates the product. The liberated Cu(I) fragment is re-oxidized to Cu(II) by oxygen, which again reacts with the arylboronic acid.

The arylation of anilines and primary and secondary alkylamines applying arylboronic acids was also realized by the addition of catalytic amounts of myristic acid to enhance the catalyst solubility [87]. Vigorous stirring is necessary to ensure a sufficient rate of oxygen uptake for re-oxidation of the copper catalyst. The scope of the protocol has been significantly extended by the use of a stoichiometric amount of an oxidant, e.g., pyridine N-oxide, TEMPO or NMO [88]. Here, not only N-heterocycles and amines can be arylated, but also amides, imides, and sulfonamides.

2.11.3
C–O Coupling Reactions

Until recently, the nucleophilic substitution of aliphatic iodides with phenolates, the so-called Ullmann ether synthesis [71], and the direct nucleophilic substitution of activated aryl halides constituted the most important tools for aryl ether synthesis. However, these methods generally require harsh reaction conditions, a large excess of the alcohol, or undesirable solvents, and often give unsatisfactory results. Thus, there is a need for new practical catalytic methods. After the first report of palladium-catalyzed C–O bond-forming reactions in 1996 [89], this new type of ether synthesis has been elaborated to a relatively general method [1d, 1e, 1h, 1i]. The development started with the intramolecular cyclization of tertiary alcohols shown in Scheme 15. This reaction proceeds smoothly in the presence of palladium, a chelating phosphine like Tol-BINAP or dppf, and NaOtBu or K$_2$CO$_3$ as the base.

Shortly after the invention of the palladium-catalyzed *intra*molecular alkoxylations, the first *inter*molecular aryl ether syntheses were reported by Hartwig et al. Studies of the reductive elimination of aryl *tert*-butoxy palladium complexes to give aryl *tert*-butyl ethers led to a catalytic coupling process of activated aryl bromides and sodium *tert*-butoxide (Scheme 16) [90]. In the presence of 10 mol% of

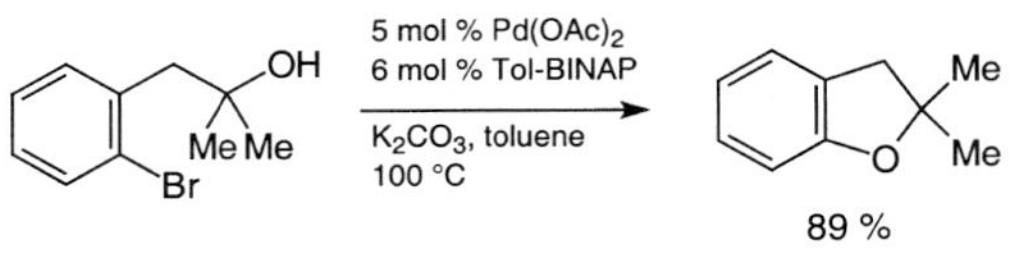

Scheme 15 Intramolecular C–O coupling reaction.

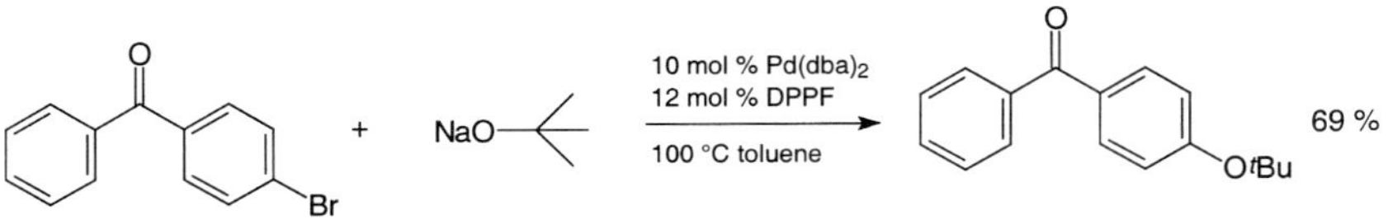

Scheme 16 Palladium-catalyzed synthesis of aryl *tert*-butyl ethers.

a Pd(0) precursor and the chelating ligand dppf, yields of up to 69% have been obtained for various aryl *tert*-butyl ethers.

Also based on mechanistic studies of the C–O reductive elimination step from palladium aryl alkoxide complexes [91], Buchwald et al. extended the methodology to palladium-catalyzed C–O coupling reactions of aryl bromides and primary and secondary aliphatic alcohols as starting materials (Scheme 17) [92].

The performances of palladium and nickel catalyst systems for the synthesis of alkyl-protected phenols have been compared by Hartwig et al. [93]. They found that Pd-dppf catalysts are superior for the coupling of *tert*-butoxide with electron-deficient aryl bromides and chlorides, whereas nickel systems are preferred for the synthesis of the corresponding methyl and *tert*-butyldimethyl silyl ethers. Kinetic investigation of C–O coupling reactions by Buchwald [91] and Hartwig [90] have shown that the C–O bond-forming step is the rate-determining step in this cross-coupling reaction. It is believed that a nucleophilic attack of an alkoxide on the *ipso*-carbon of the Pd-aryl group leads to the formation of the desired ether. As phenoxides are less nucleophilic than alkoxides, the formation of diaryl ethers was not described until Hartwig et al. reported on the coupling of electron-rich sodium phenoxides with electron-poor bromoarenes [94]. Reaction rate and yield could be improved for the arylation of sodium phenoxide by applying an electron-poor, CF_3-substituted dppf ligand (Scheme 18).

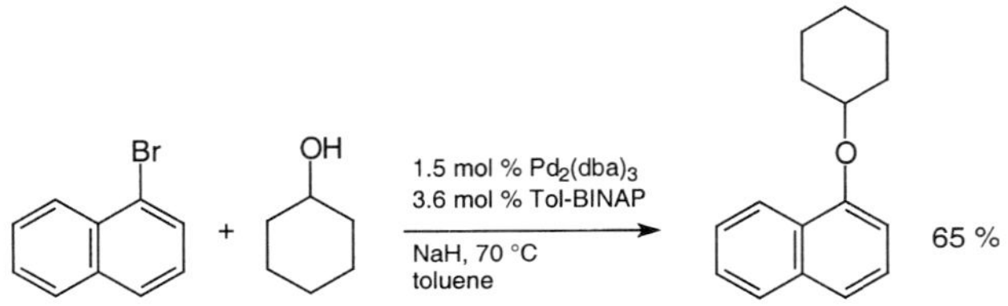

Scheme 17 Palladium-catalyzed synthesis of alkyl aryl ethers by using NaH as the base.

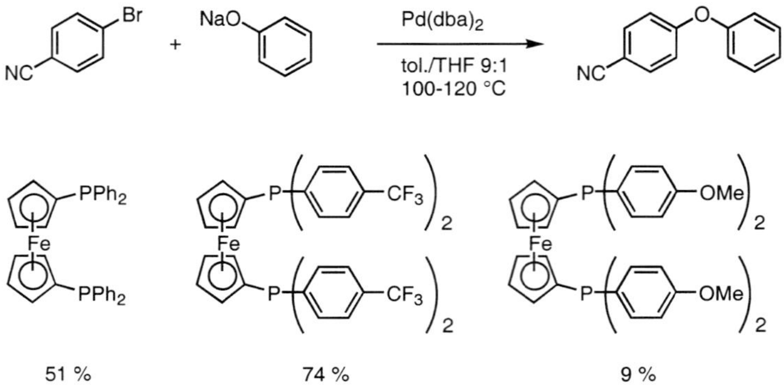

Scheme 18 Electronic influence of the ligand on the palladium-catalyzed coupling of 4-bromobenzonitrile and sodium phenoxide.

The Pd-BINAP and dppf catalyst systems were utilized for the first intramolecular diaryl ether synthesis in 2000 [95], when a variety of substituted dibenzoxepino[4,5-*d*]pyrazoles were prepared with yields of up to 69% (Fig. 5).

Further investigations on the dppf system revealed that this ligand is partly converted to di-*tert*-butylphenylphosphine and di-*tert*-butylphosphino ferrocene under coupling conditions [96]. The latter phosphine turned out to be responsible for the high catalytic activity, demonstrating that no chelating ligand is required for efficient C–O coupling reactions. This new catalyst system was applied successfully for the reaction of non-activated aryl bromides and chlorides with electron-rich sodium phenoxides or sodium *tert*-butoxide (Tab. 3).

Although an active catalyst is obtained by employing di-*tert*-butylphosphino ferrocene, it is not this phosphine itself, which is part of the active species, but a pentaarylated derivative, which results from perarylation of the unsubstituted cyclopentadienyl ring of the ferrocenyl moiety (Scheme 19) [97].

Ph$_5$FcP*t*Bu$_2$ forms an extremely active catalyst, enabling the coupling of aryl bromides and electron-deficient aryl chlorides with different alkoxides even at room temperature (5 mol% Pd(dba)$_2$) [97]. After Hartwig's finding that reduction of the electron donor capability of the dppf ligand increases the yields in coupling reactions with simple sodium phenoxide, the successful application of Buchwald's electron-rich bulky dialkylarylphosphines for the coupling of a plethora of aryl bromides, chlorides, and triflates with phenols was quite amazing [98]. Apart from the generality of these catalysts, another advantage is the possibility to use simply

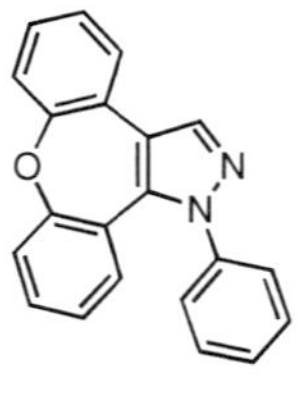

Fig. 5 Dibenzoxepino[4,5-*d*]pyrazoles.

2.11.3 C–O Coupling Reactions

Tab. 3 Arylation of sodium alkoxides (Pd(dba)$_2$, FcPtBu$_2$-hydrolysis by CF$_3$CO$_2$H + CF$_3$SO$_3$H).

Entry	ArX	NaOR	Product	Temp. (°C)	Yield (%)
1	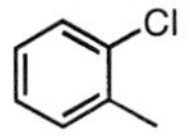 2-chlorotoluene	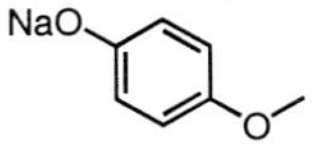 NaO-C$_6$H$_4$-OMe (4-)	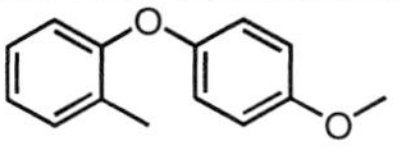 2-methylphenyl 4-methoxyphenyl ether	80	82
2	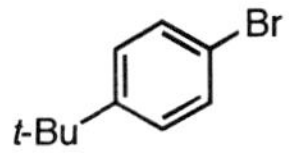 4-bromo-t-butylbenzene	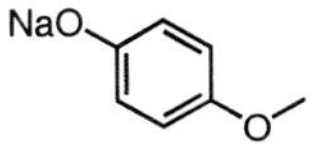 NaO-C$_6$H$_4$-OMe (4-)	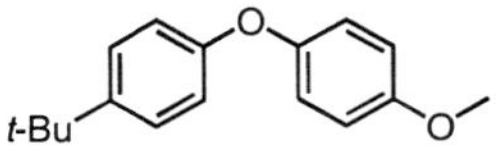 4-t-butylphenyl 4-methoxyphenyl ether	110	74
3	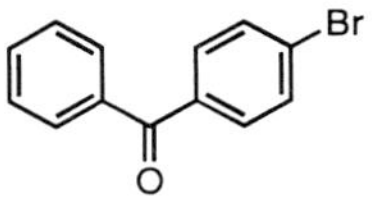 4-bromobenzophenone	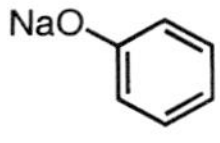 NaO-C$_6$H$_5$	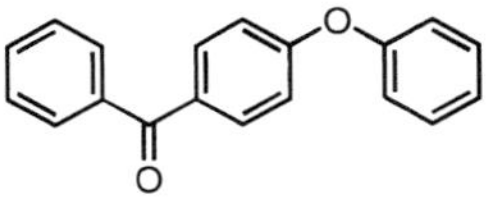 4-(phenoxy)benzophenone	110	63
4	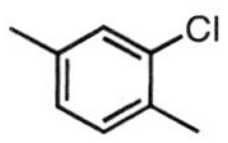 2-chloro-4-methyl... (Cl-arene with methyl)	NaOtBu	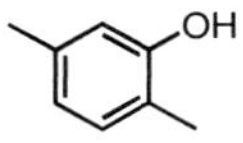 corresponding phenol (OH)	85	71

in situ mixtures of the desired phenols in combination with K$_3$PO$_4$ or NaH instead of pre-forming the corresponding sodium phenoxides (Tab. 4).

Buchwald's ligands were also applied successfully for the coupling of non-activated and deactivated aryl bromides and chlorides with sodium *tert*-butoxide. These reactions proceeded smoothly with 1–2.5 mol% palladium at 100 °C [99]. Furthermore, the intramolecular C–O coupling methodology has been applied for *o*-hydroxyalkyl-substituted bromo- and chloro-arenes to give five- to seven-membered oxygen heterocycles [100]. Interestingly, enantiomerically enriched alcohols were coupled under these conditions without loss of optical purity (Scheme 20) [100b]. Notably, the intermolecular coupling of a variety of aryl bromides and chlorides with primary aliphatic alcohols, e.g., *n*-butanol, which often led to hydrodehalogenation of the haloarene, was achieved in the presence of these catalyst systems [101].

With regard to the ligand, it is interesting to note that Watanabe et al. described the use of simple PtBu$_3$ for the coupling of activated and unactivated aryl bromides and chlorides with sodium *tert*-butoxide [102]. In addition to palladium-based C–O

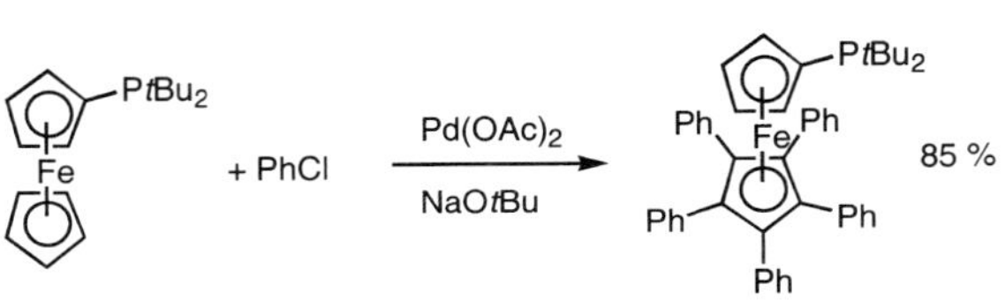

FcPtBu$_2$ + PhCl $\xrightarrow{\text{Pd(OAc)}_2,\ \text{NaO}t\text{Bu}}$ Ph$_5$FcPtBu$_2$ (85%)

Scheme 19 *In situ* formation of Ph$_5$FcPtBu$_2$.

Tab. 4 Diarylether synthesis with Buchwald ligands.

Entry	ArX	ROH	Product	Base	Ligand	Yield (%)
1	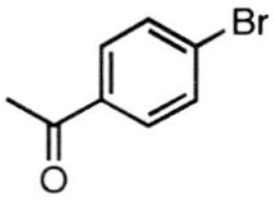 4-Br-C6H4-C(O)CH3	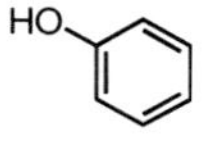 PhOH	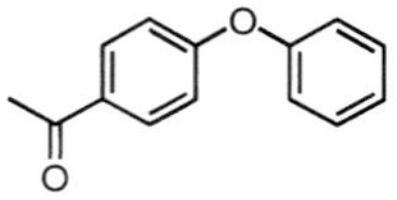 4-CH3C(O)-C6H4-O-Ph	K_3PO_4	L1	94
2	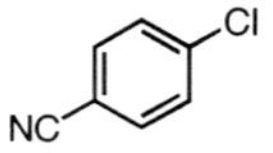 4-Cl-C6H4-CN	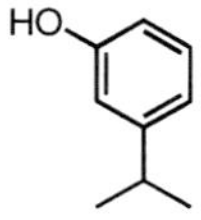 3-iPr-C6H4-OH	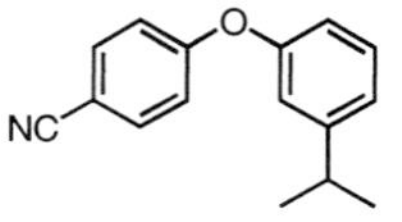 4-NC-C6H4-O-C6H4-3-iPr	K_3PO_4	L1	91
3	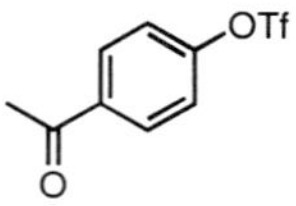 4-OTf-C6H4-C(O)CH3	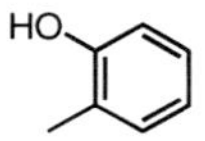 2-Me-C6H4-OH	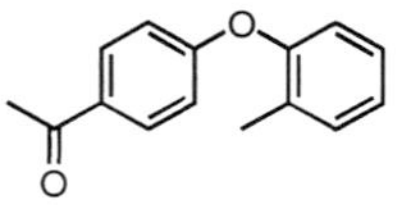 4-CH3C(O)-C6H4-O-C6H4-2-Me	K_3PO_4	L1	84
4	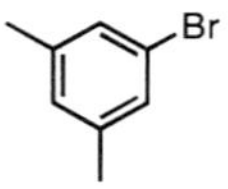 3,5-diMe-C6H3-Br	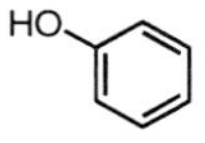 PhOH	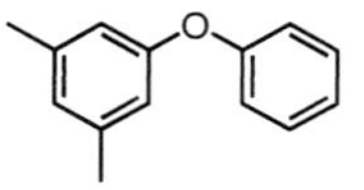 3,5-diMe-C6H3-O-Ph	K_3PO_4	L2	83
5	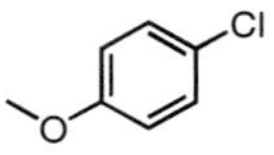 4-Cl-C6H4-OMe	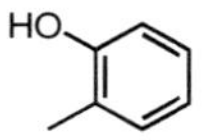 2-Me-C6H4-OH	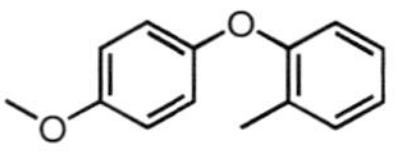 4-MeO-C6H4-O-C6H4-2-Me	K_3PO_4	L3	73
6	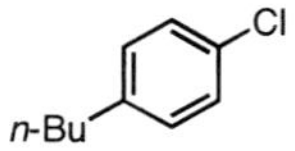 4-Cl-C6H4-n-Bu	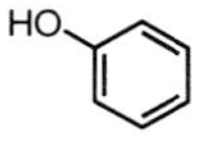 PhOH	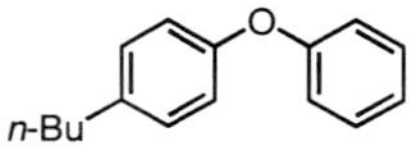 4-n-Bu-C6H4-O-Ph	NaH	L2	61

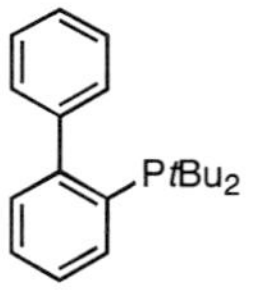

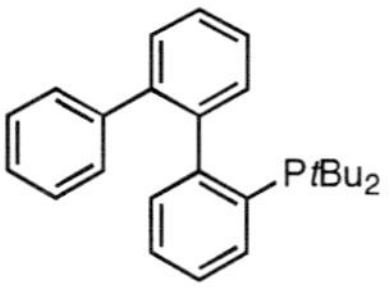

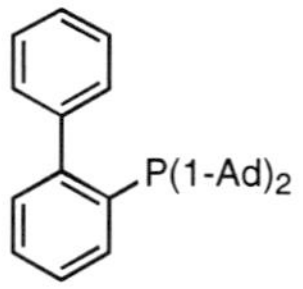

L1: 2-biphenyl-P^{t}Bu$_2$ L2: 2',6'-biphenyl-P^{t}Bu$_2$ L3: 2-biphenyl-P(1-Ad)$_2$

coupling reactions, the copper-catalyzed diaryl ether synthesis from aryl bromides or iodides and phenols was reported in 1997 (Scheme 21) [103]. The catalyst system employed is characterized by a catalytic amount of a copper salt in combination with cesium carbonate as the base. A stoichiometric amount of a carboxylic acid should be added if less soluble phenols are used. In contrast to the previously known procedures, these conditions are compatible with a wide variety of functionalities, including ethers, ketones, carboxylic acids, esters, dialkylamines, nitriles, and nitro groups. Interestingly, there is no significant influence of the electronic nature either

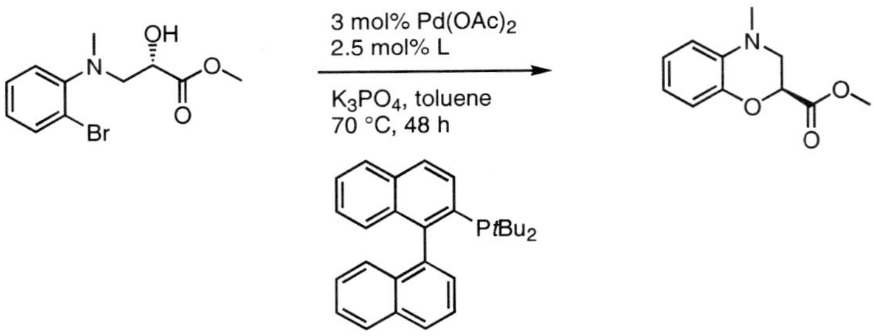

Scheme 20 Palladium-catalyzed cyclization of an optically active alcohol.

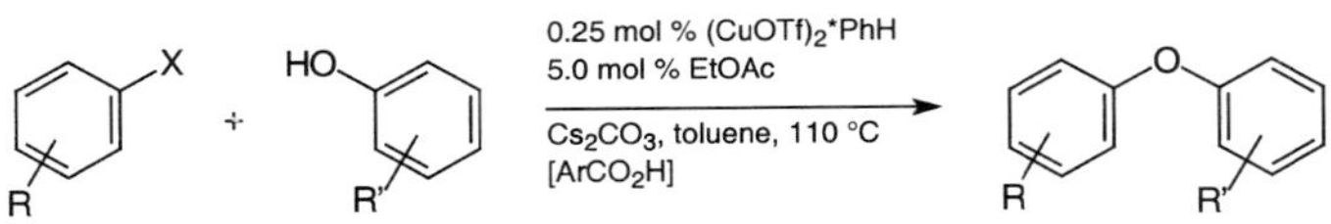

Scheme 21 Copper-catalyzed diaryl ether synthesis.

of the aryl halide or the phenol on the product yield. Some examples of this new development from the Buchwald group are given in Tab. 5.

Recently, a beneficial effect of microwave heating on the reaction rate and yield in the arylation of phenols has been reported by He and Wu [104]. The coupling of aryl iodides and bromides proceeds relatively fast (1–3 h), but high temperatures (195 °C) and a large amount of the CuI catalyst (10 mol%) are required.

The influence of chelating and non-chelating N-donor ligands on the copper-catalyzed reaction of bromoarenes with some aliphatic or aromatic alcohols has been investigated by Hauptmann and co-workers [105]. The results obtained from an automated parallel screening of reaction conditions using a 96-member library of pyridine derivatives as co-catalysts suggest that bidentate ligands with relatively small bite angles, e.g., 2-aminopyridines, are good additives for the investigated coupling reactions. A more general protocol for the coupling of primary and secondary alkanols with aryl iodides in the presence of pyridine-type ligands has been described by Buchwald and co-workers [106]. Here, 1,10-phenanthroline is required as a ligand for the Cu-catalyst, and the reaction is run in neat alcohol or with toluene as the solvent. Apart from simple copper salts, triphenylphosphine copper(I) complexes have also been used as a soluble catalyst source for the coupling of aryl bromides with electron-rich phenols [107]. For aryl bromides with strongly electron-withdrawing substituents in the *p*-position, the alkoxylation proceeds even in the absence of a catalyst in NMP at relatively low temperatures (70 °C) by simple nucleophilic substitution [107 a].

Although somewhat out of the scope of this review, it is interesting to note that in addition to aryl halides, arylboronic acids can be coupled with alcohols in copper-mediated reactions [108]. Best yields in the coupling with substituted phenols are obtained in the presence of stoichiometric amounts of copper(II). Neverthe-

Tab. 5 Copper(I) triflate-catalyzed synthesis of diaryl ethers.

Entry	Aryl halide	Phenol	Aryl ether	Yield (%)
1	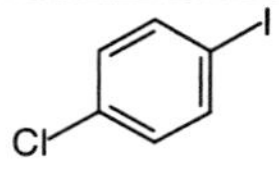 4-Cl-C6H4-I	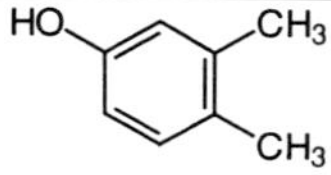 HO-C6H3(CH3)2 (3,4-dimethyl)	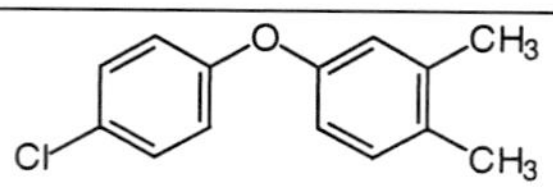 4-Cl-C6H4-O-C6H3(CH3)2	89
2	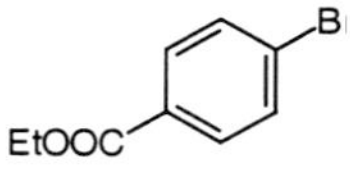 4-EtOOC-C6H4-Br	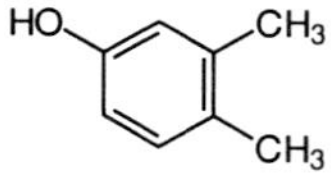 HO-C6H3(CH3)2 (3,4-dimethyl)	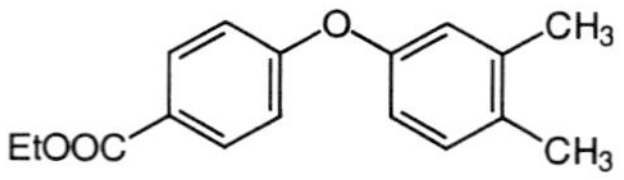 4-EtOOC-C6H4-O-C6H3(CH3)2	80
3	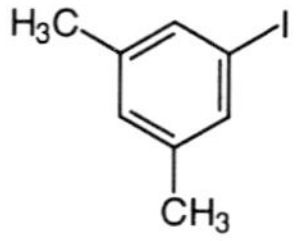 3,5-(CH3)2-C6H3-I	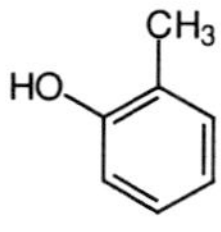 2-CH3-C6H4-OH	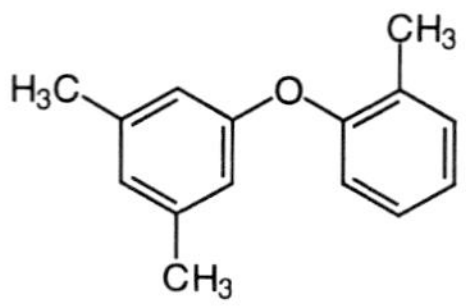 3,5-(CH3)2-C6H3-O-C6H4-2-CH3	87
4	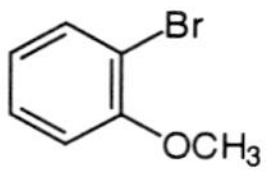 2-OCH3-C6H4-Br	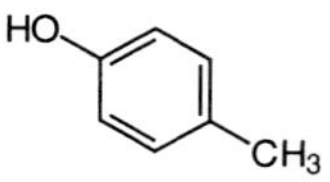 4-CH3-C6H4-OH	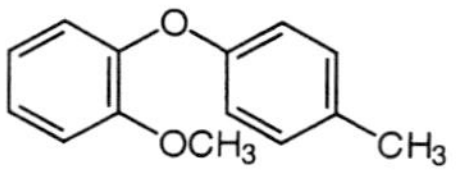 2-OCH3-C6H4-O-C6H4-4-CH3	79 [a]
5	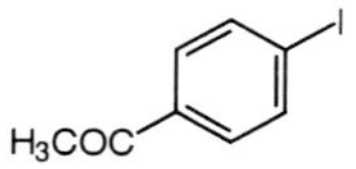 4-H3COC-C6H4-I	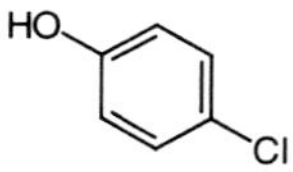 4-Cl-C6H4-OH	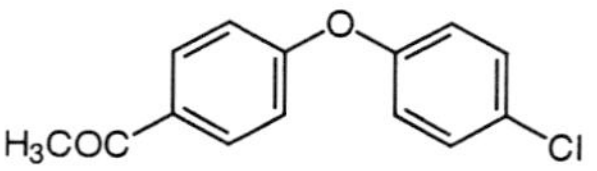 4-H3COC-C6H4-O-C6H4-4-Cl	93 [a]

a) Reaction performed in the presence of 2.0 eq. 1-naphthoic acid and 5 Å molecular sieves.

less, a catalytic version of this reaction has also been described, albeit with a low turnover number (<8) under an atmosphere of oxygen [108b,c]. Here, a tertiary amine is required as an additive to promote the reaction. Noteworthy are the high tolerance of functional groups (e.g., iodo, nitro, or ester functions) and the mild reaction conditions (room temperature up to 50 °C) that allow, for instance, the coupling of protected amino acid derivatives without any racemization or N-arylation (Scheme 22) [108b]. Despite its synthetic usefulness, the mechanism of this coupling reaction has so far remained unclear.

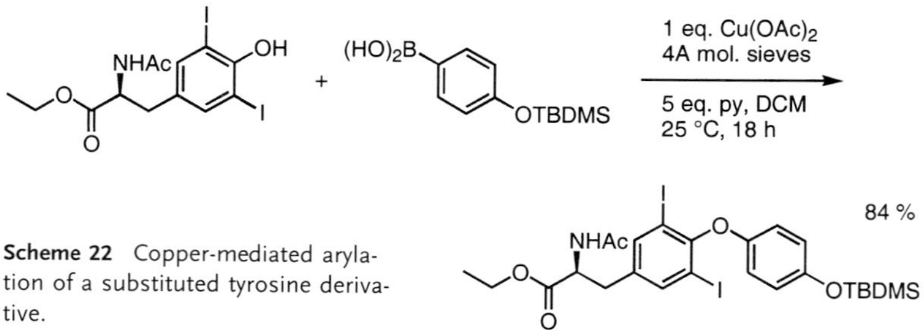

Scheme 22 Copper-mediated arylation of a substituted tyrosine derivative.

Reagents: 1 eq. Cu(OAc)$_2$, 4 Å mol. sieves, 5 eq. py, DCM, 25 °C, 18 h. Yield: 84%.

In conclusion, palladium- and copper-catalyzed C–N and C–O bond-forming reactions have become reliable methods for the synthesis of a variety of anilines and aryl ethers. The development of these new catalytic methods has taken place at a rapid pace. It is evident that these methods provide exciting opportunities for future organic synthesis.

2.11.4 References

1. (a) M. Beller, *Angew. Chem.* **1995**, *107*, 1436; *Angew. Chem. Int. Ed. Engl.* **1995**, *34*, 1316; (b) M. Beller, T. H. Riermeier in Organic Synthesis Highlights III (Eds.: J. Mulzer, H. Waldmann), Wiley-VCH, Weinheim, **1998**, p. 126; (c) J. F. Hartwig, *Synlett* **1997**, 329; (d) J. F. Hartwig, *Angew. Chem.* **1998**, *110*, 2154; *Angew. Chem. Int. Ed.* **1998**, *37*, 2047; (e) F. Theil, *Angew. Chem.* **1999**, *111*, 2493; *Angew. Chem. Int. Ed.* **1999**, *38*, 2345; (f) B. H. Yang, S. L. Buchwald, *J. Organomet. Chem.* **1999**, *576*, 125; (g) P. W. N. M. van Leeuwen, P. C. J. Kamer, J. N. H. Reek, R. Dierkes, *Chem. Rev.* **2000**, *100*, 2741; (h) S. Höger, *Chem. unserer Zeit* **2001**, *35*, 102; (i) J. F. Hartwig in Handbook of Organopalladium Chemistry for Organic Synthesis (Ed.: E. Negishi), Wiley-Interscience, New York, **2002**, Vol. 1, p. 1051.
2. M. Kosugi, M. Kameyama, T. Migita, *Chem. Lett.* **1983**, 927.
3. F. Paul, J. Patt, J. F. Hartwig, *J. Am. Chem. Soc.* **1994**, *116*, 5969.
4. A. S. Guram, S. L. Buchwald, *J. Am. Chem. Soc.* **1994**, *116*, 7901.
5. (a) A. S. Guram, R. A. Rennels, S. L. Buchwald, *Angew. Chem.* **1995**, *107*, 1456; *Angew. Chem. Int. Ed. Engl.* **1995**, *34*, 1348; (b) J. P. Wolfe, S. L. Buchwald, *J. Org. Chem.* **1996**, *61*, 1133.
6. J. Louie, J. F. Hartwig, *Tetrahedron Lett.* **1995**, *36*, 3609.
7. S. Zhao, A. K. Miller, J. Berger, L. A. Flippin, *Tetrahedron Lett.* **1996**, *37*, 4463.
8. (a) Y. D. Ward, V. Farina, *Tetrahedron Lett.* **1996**, *37*, 6993; (b) C. A. Willoughby, K. T. Chapman, *Tetrahedron Lett.* **1996**, *37*, 7181.
9. T. Kanbara, A. Honma, K. Hasegawa, *Chem. Lett.* **1996**, 1135.
10. D. V. Davydov, I. P. Beletskaya, *Russ. Chem. Bull.* **1995**, *44*, 1141.
11. D. Ma, J. Yao, *Tetrahedron Asymm.* **1996**, *7*, 3075.
12. J. P. Wolfe, R. A. Rennels, S. L. Buchwald, *Tetrahedron* **1996**, *52*, 7525.
13. B. J. Margolis, J. J. Swidorski, B. N. Rogers, *J. Org. Chem.* **2003**, *68*, 644.
14. A. J. Peat, S. L. Buchwald, *J. Am. Chem. Soc.* **1996**, *118*, 1028.
15. (a) R. A. Wiedenhoefer, H. A. Zhong, S. L. Buchwald, *Organometallics* **1996**, *15*, 2745; (b) J. Louie, F. Paul, J. F. Hartwig, *Organometallics* **1996**, *15*, 2794; (c) J. Louie, J. F. Hartwig, *Angew. Chem.* **1996**, *108*, 2531; *Angew. Chem. Int. Ed. Engl.* **1996**, *35*, 2359; (d) L. M. Alcazar-Roman, J. F. Hartwig, A. L. Rheingold, L. M. Liable-Sands, I. A. Guzei, *J. Am. Chem. Soc.* **2000**, *122*, 4618; (e) L. M. Alcazar-Roman, J. F. Hartwig, *J. Am. Chem. Soc.* **2001**, *123*, 12905; (f) J. P. Stambuli, M. Bühl, J. F. Hartwig, *J. Am. Chem. Soc.* **2002**, *124*, 9346; (g) U. K. Singh, E. R. Strieter, D. G. Blackmond, S. L. Buchwald, *J. Am. Chem. Soc.* **2002**, *124*, 14104.
16. (a) J. P. Wolfe, S. Wagaw, S. L. Buchwald, *J. Am. Chem. Soc.* **1996**, *118*, 7215; (b) J. P. Wolfe, S. L. Buchwald, *J. Org. Chem.* **2000**, *65*, 1144; (c) X.-X. Zhang, M. C. Harris, J. P. Sadighi, S. L. Buchwald, *Can. J. Chem.* **2001**, *79*, 1799.
17. (a) M. S. Driver, J. F. Hartwig, *J. Am. Chem. Soc.* **1996**, *118*, 7217; (b) B. C. Hamann, J. F. Hartwig, *J. Am. Chem. Soc.* **1998**, *120*, 3694.
18. (a) J. P. Wolfe, S. L. Buchwald, *J. Org. Chem.* **1997**, *62*, 1264; (b) J. Louie, M. S. Driver, B. C. Hamann, J. F. Hartwig, *J. Org. Chem.* **1997**, *62*, 1268.

19 S. Wagaw, S. L. Buchwald, *J. Org. Chem.* **1996**, *61*, 7240.

21 J.-F. Marcoux, S. Wagaw, S. L. Buchwald, *J. Org. Chem.* **1997**, *62*, 1568.

21 B. C. Hamann, J. F. Hartwig, *J. Am. Chem. Soc.* **1998**, *120*, 7369.

22 N. Kataoka, Q. Shelby, J. P. Stambuli, J. F. Hartwig, *J. Org. Chem.* **2002**, *67*, 5553.

23 (a) Y. Guari, D. S. van Es, J. N. H. Reek, P. C. J. Kramer, P. W. N. M. van Leeuwen, *Tetrahedron Lett.* **1999**, *40*, 3789; (b) Y. Guari, G. P. F. van Strijdonck, M. D. K. Boele, J. N. H. Reek, P. C. J. Kramer, P. W. N. M. van Leeuwen, *Chem. Eur. J.* **2001**, *7*, 475.

24 J. P. Sadighi, M. C. Harris, S. L. Buchwald, *Tetrahedron Lett.* **1998**, *39*, 5327.

25 M. S. Driver, J. F. Hartwig, *J. Am. Chem. Soc.* **1997**, *119*, 8232.

26 I. P. Beletskaya, A. G. Bessmertnykh, R. A. Guilard, *Tetrahedron Lett.* **1997**, *38*, 2287.

27 R. A. Singer, J. P. Sadighi, S. L. Buchwald, *J. Am. Chem. Soc.* **1998**, *120*, 213.

28 S. Wagaw, B. H. Yang, S. L. Buchwald, *J. Am. Chem. Soc.* **1998**, *120*, 6621.

29 J. J. Song, N. K. Yee, *Org. Lett.* **2000**, *2*, 519.

30 M. Calter, T. K. Hollis, L. E. Overman, J. Ziller, G. G. Zipp, *J. Org. Chem.* **1997**, *62*, 1449.

31 S. Wagaw, R. A. Rennels, S. L. Buchwald, *J. Am. Chem. Soc.* **1997**, *119*, 8451.

32 J. P. Wolfe, S. L. Buchwald, *Tetrahedron Lett.* **1997**, *38*, 6359.

33 J. Åhman, S. L. Buchwald, *Tetrahedron Lett.* **1997**, *38*, 6363.

34 (a) W. A. Herrmann, C. Brossmer, K. Öfele, C.-P. Reisinger, T. Riermeier, M. Beller, H. Fischer, *Angew. Chem.* **1995**, *107*, 1989; *Angew. Chem. Int. Ed. Engl.* **1995**, *34*, 1844; (b) W. A. Herrmann, C. Brossmer, C.-P. Reisinger, T. H. Riermeier, K. Öfele, M. Beller, *Chem. Eur. J.* **1997**, *3*, 1357.

35 (a) M. Beller, T. H. Riermeier, C.-P. Reisinger, W. A. Herrmann, *Tetrahedron Lett.* **1997**, *38*, 2073; (b) T. H. Riermeier, A. Zapf, M. Beller, *Top. Catal.* **1997**, *4*, 301.

36 N. P. Reddy, M. Tanaka, *Tetrahedron Lett.* **1997**, *38*, 4807.

37 (a) M. Nishiyama, T. Yamamoto, Y. Koie, *Tetrahedron Lett.* **1998**, *39*, 617; (b) T. Yamamoto, M. Nishiyama, Y. Koie, *Tetrahedron Lett.* **1998**, *39*, 2367; (c) M. Watanabe, M. Nishiyama, T. Yamamoto, Y. Koie, *J. TOSOH Res.* **1999**, *43*, 37.

38 (a) J. F. Hartwig, M. Kawatsura, S. I. Hauck, K. H. Shaughnessy, L. M. Alcazar-Roman, *J. Org. Chem.* **1999**, *64*, 5575; (b) J. P. Stambuli, R. Kuwano, J. F. Hartwig, *Angew. Chem.* **2002**, *114*, 4940; *Angew. Chem. Int. Ed.* **2002**, *41*, 4746.

39 (a) D. W. Old, J. P. Wolfe, S. L. Buchwald, *J. Am. Chem. Soc.* **1998**, *120*, 9722; (b) J. P. Wolfe, S. L. Buchwald, *Angew. Chem.* **1999**, *111*, 2570; *Angew. Chem. Int. Ed.* **1999**, *38*, 2413; (c) J. P. Wolfe, H. Tomori, J. P. Sadighi, J. Yin, S. L. Buchwald, *J. Org. Chem.* **2000**, *65*, 1158.

40 M. Prashad, X. Y. Mak, Y. Liu, O. Repič, *J. Org. Chem.* **2003**, *68*, 1163.

41 C. A. Parrish, S. L. Buchwald, *J. Org. Chem.* **2001**, *66*, 3820.

42 D. Zim, S. L. Buchwald, *Org. Lett.* **2003**, *5*, 2413.

43 R. B. Bedford, C. S. J. Cazin, *Organometallics* **2003**, *22*, 987.

44 (a) X. Bei, A. S. Guram, H. W. Turner, W. H. Weinberg, *Tetrahedron Lett.* **1999**, *40*, 1237; (b) X. Bei, T. Uno, J. Norris, H. W. Turner, W. H. Weinberg, A. S. Guram, *Organometallics* **1999**, *18*, 1840.

45 M. W. Hooper, M. Utsunomiya, J. F. Hartwig, *J. Org. Chem.* **2003**, *68*, 2861.

46 (a) R. B. Bedford, C. S. J. Cazin, *Chem. Commun.* **2002**, 2310; (b) I. C. F. R. Ferreira, M.-J. R. P. Queiroz, G. Kirsch, *Tetrahedron* **2002**, *58*, 7943.

47 R. Kuwano, M. Utsunomiya, J. F. Hartwig, *J. Org. Chem.* **2002**, *67*, 6479.

48 M. C. Harris, X. Huang, S. L. Buchwald, *Org. Lett.* **2002**, *4*, 2885.

49 (a) G. Y. Li, *Angew. Chem.* **2001**, *113*, 1561; *Angew. Chem. Int. Ed.* **2001**, *40*, 1513; (b) G. Y. Li, G. Zheng, A. F. Noonan, *J. Org. Chem.* **2001**, *66*, 8677.

50 A. Ehrentraut, A. Zapf, M. Beller, *J. Mol. Catal.* **2002**, *182/183*, 515.

51 A. Schnyder, A. F. Indolese, M. Studer, H.-U. Blaser, *Angew. Chem.* **2002**, *114*, 3820; *Angew. Chem. Int. Ed.* **2002**, *41*, 3668.

52 (a) C. Bolm, J. P. Hildebrand, *Tetrahedron Lett.* **1998**, *39*, 5731; (b) C. Bolm, J. P. Hildebrand, *J. Org. Chem.* **2000**, *65*, 169; (c) C. Bolm, J. P. Hildebrand, J. Rudolph, *Synthesis* **2000**, 911.
53 D. W. Old, M. C. Harris, S. L. Buchwald, *Org. Lett.* **2000**, *2*, 1403.
54 G. Mann, J. F. Hartwig, M. S. Driver, C. Fernández-Rivas, *J. Am. Chem. Soc.* **1998**, *120*, 827.
55 J. Yin, S. L. Buchwald, *Org. Lett.* **2000**, *2*, 1101.
56 J. Yin, S. L. Buchwald, *J. Am. Chem. Soc.* **2002**, *124*, 6043.
57 (a) W. A. Herrmann, *Angew. Chem.* **2002**, *114*, 1342; *Angew. Chem. Int. Ed.* **2002**, *41*, 1290; (b) A. C. Hillier, G. A. Grasa, M. S. Viciu, H. M. Lee, C. Yang, S. P. Nolan, *J. Organomet. Chem.* **2002**, *653*, 69.
58 J. Huang, G. Grasa, S. P. Nolan, *Org. Lett.* **1999**, *1*, 1307.
59 S. R. Stauffer, S. Lee, J. P. Stambuli, S. I. Hauck, J. F. Hartwig, *Org. Lett.* **2000**, *2*, 1423.
60 (a) S. Caddick, F. G. N. Cloke, G. K. B. Clentsmith, P. B. Hitchcock, D. McKerrecher, L. R. Titcomb, M. R. V. Williams, *J. Organomet. Chem.* **2001**, *617/618*, 635; (b) L. R. Titcomb, S. Caddick, F. G. N. Cloke, D. J. Wilson, D. McKerrecher, *Chem. Commun.* **2001**, 1388.
61 G. A. Grasa, M. S. Viciu, J. Huang, S. P. Nolan, *J. Org. Chem.* **2001**, *66*, 7729.
62 (a) S. Caddick, F. G. N. Cloke, P. B. Hitchcock, J. Leonard, A. K. de K. Lewis, D. McKerrecher, L. R. Titcomb, *Organometallics* **2002**, *21*, 4318; (b) D. S. McGuiness, K. J. Cavell, B. W. Skelton, A. H. White, *Organometallics* **1999**, *18*, 1596.
63 J. P. Wolfe, J. Åhman, J. P. Sadighi, R. A. Singer, S. L. Buchwald, *Tetrahedron Lett.* **1997**, *38*, 6367.
64 S. Jaime-Figueroa, Y. Liu, J. M. Muchowski, D. G. Putman, *Tetrahedron Lett.* **1998**, *39*, 1313.
65 G. Mann, M. S. Driver, J. F. Hartwig, *J. Am. Chem. Soc.* **1998**, *120*, 827.
66 (a) S. Lee, M. Jørgensen, J. F. Hartwig, *Org. Lett.* **2001**, *3*, 2729; (b) X. Huang, S. L. Buchwald, *Org. Lett.* **2001**, *3*, 3417.
67 J. P. Wolfe, S. L. Buchwald, *J. Am. Chem. Soc.* **1997**, *119*, 6054.
68 (a) E. Brenner, Y. Fort, *Tetrahedron Lett.* **1998**, *39*, 5359; (b) E. Brenner, R. Schneider, Y. Fort, *Tetrahedron* **1999**, *55*, 12829; (c) C. Desmarets, R. Schneider, Y. Fort, *Tetrahedron Lett.* **2000**, *41*, 2875; (d) C. Desmarets, R. Schneider, Y. Fort, *Tetrahedron* **2001**, *57*, 7657; (e) C. Desmarets, R. Schneider, Y. Fort, *Tetrahedron Lett.* **2001**, *42*, 247; (f) E. Brenner, R. Schneider, Y. Fort, *Tetrahedron Lett.* **2000**, *41*, 2881.
69 (a) B. Gradel, E. Brenner, R. Schneider, Y. Fort, *Tetrahedron Lett.* **2001**, *42*, 5689; (b) C. Desmarets, R. Schneider, Y. Fort, *J. Org. Chem.* **2002**, *67*, 3029.
70 R. Omar-Amrani, A. Thomas, E. Brenner, R. Schneider, Y. Fort, *Org. Lett.* **2003**, *5*, 2311.
71 General review: J. Lindley, *Tetrahedron* **1984**, *40*, 1433.
72 B. Renger, *Synthesis* **1985**, 856.
73 A. Kiyomori, J.-F. Marcoux, S. L. Buchwald, *Tetrahedron Lett.* **1999**, *40*, 2657.
74 A. Klapars, J. C. Antilla, X. Huang, S. L. Buchwald, *J. Am. Chem. Soc.* **2001**, *123*, 7727.
75 S.-K. Kang, D.-H. Kim, J.-N. Park, *Synlett* **2002**, 427.
76 A. Klapars, X. Huang, S. L. Buchwald, *J. Am. Chem. Soc.* **2002**, *124*, 7421.
77 A. Padwa, K. R. Crawford, P. Rashatasakhon, M. Rose, *J. Org. Chem.* **2003**, *68*, 2609.
78 F. Lang, D. Zewge, I. N. Houpis, R. P. Volante, *Tetrahedron Lett.* **2001**, *42*, 3251.
79 F. Y. Kwong, A. Klapars, S. L. Buchwald, *Org. Lett.* **2002**, *4*, 581.
80 R. K. Gujadhur, C. G. Bates, D. Venkataraman, *Org. Lett.* **2001**, *3*, 4315.
81 F. Y. Kwong, S. L. Buchwald, *Org. Lett.* **2003**, *5*, 793.
82 D. Ma, Q. Cai, H. Zhang, *Org. Lett.* **2003**, *5*, 2453.
83 D. Ma, Y. Zhang, J. Yao, S. Wu, F. Tao, *J. Am. Chem. Soc.* **1998**, *120*, 12459.
84 (a) D. M. T. Chan, K. L. Monaco, R.-P. Wang, M. P. Winters, *Tetrahedron Lett.* **1998**, *39*, 2933; (b) P. Y. S. Lam, C. G. Clark, S. Saubern, J. Adams, M. P. Winters, D. M. T. Chan, A. Combs, *Tetrahedron Lett.* **1998**, *39*, 2941; (c) D. J.

Cundy, S. A. Forsyth, *Tetrahedron Lett.* **1998**, *39*, 7979.

85 P. Y. S. Lam, C. G. Clark, S. Saubern, J. Adams, K. M. Averill, D. M. T. Chan, A. Combs, *Synlett* **2000**, 674.

86 (a) J. P. Collman, M. Zhong, *Org. Lett.* **2000**, *2*, 1233; (b) J. P. Collman, M. Zhong, L. Zeng, S. Costanzo, *J. Org. Chem.* **2001**, *66*, 1528.

87 J. C. Antilla, S. L. Buchwald, *Org. Lett.* **2001**, *3*, 2077.

88 P. Y. S. Lam, G. Vincent, C. G. Clark, S. Deudon, P. K. Jadhav, *Tetrahedron Lett.* **2001**, *42*, 3415.

89 M. Palucki, J. P. Wolfe, S. L. Buchwald, *J. Am. Chem. Soc.* **1996**, *118*, 10333.

90 G. Mann, J. F. Hartwig, *J. Am. Chem. Soc.* **1996**, *118*, 13109.

91 (a) R. A. Widenhoefer, H. A. Zhong, S. L. Buchwald, *J. Am. Chem. Soc.* **1997**, *119*, 6787; (b) R. A. Widenhoefer, S. L. Buchwald, *J. Am. Chem. Soc.* **1998**, *120*, 6504.

92 M. Palucki, J. P. Wolfe, S. L. Buchwald, *J. Am. Chem. Soc.* **1997**, *119*, 3395.

93 G. Mann, J. F. Hartwig, *J. Org. Chem.* **1997**, *62*, 5413.

94 G. Mann, J. F. Hartwig, *Tetrahedron Lett.* **1997**, *38*, 8005.

95 R. Olivera, R. SanMartin, E. Domínguez, *Tetrahedron Lett.* **2000**, *41*, 4357.

96 (a) G. Mann, C. Incarvito, A. L. Rheingold, J. F. Hartwig, *J. Am. Chem. Soc.* **1999**, *121*, 3224; (b) J. F. Hartwig, *Pure Appl. Chem.* **1999**, *71*, 1417.

97 Q. Shelby, N. Kataoka, G. Mann, J. F. Hartwig, *J. Am. Chem. Soc.* **2000**, *122*, 10718.

98 A. Aranyos, D. W. Old, A. Kiyomori, J. P. Wolfe, J. P. Sadighi, S. L. Buchwald, *J. Am. Chem. Soc.* **1999**, *121*, 4369.

99 C. A. Parrish, S. L. Buchwald, *J. Org. Chem.* **2001**, *66*, 2498.

100 (a) K. E. Torraca, S.-I. Kuwabe, S. L. Buchwald, *J. Am. Chem. Soc.* **2000**, *122*, 12907; (b) S.-I. Kuwabe, K. E. Torraca, S. L. Buchwald, *J. Am. Chem. Soc.* **2001**, *123*, 12202.

101 K. E. Torraca, X. Huang, C. A. Parrish, S. L. Buchwald, *J. Am. Chem. Soc.* **2001**, *123*, 10770.

102 M. Watanabe, M. Nishiyama, Y. Koie, *Tetrahedron Lett.* **1999**, *40*, 8837.

103 J.-F. Marcoux, S. Doye, S. L. Buchwald, *J. Am. Chem. Soc.* **1997**, *119*, 10539.

104 H. He, Y.-J. Wu, *Tetrahedron Lett.* **2003**, *44*, 3445.

105 P. J. Fagan, E. Hauptmann, R. Shapiro, A. Casalnuovo, *J. Am. Chem. Soc.* **2000**, *122*, 5043.

106 M. Wolter, G. Nordmann, G. E. Job, S. L. Buchwald, *Org. Lett.* **2002**, *4*, 973.

107 (a) R. Gujadhur, D. Venkataraman, *Synth. Commun.* **2001**, *31*, 2865; (b) R. K. Gujadhur, C. G. Bates, D. Venkataraman, *Org. Lett.* **2001**, *3*, 4315.

108 (a) D. M. T. Chan, K. L. Monaco, R.-P. Wang, M. P. Winters, *Tetrahedron Lett.* **1998**, *39*, 2933; (b) D. A. Evans, J. L. Katz, T. R. West, *Tetrahedron Lett.* **1998**, *39*, 2937; (c) P. Y. S. Lam, G. Vincent, C. G. Clark, S. Deudon, P. K. Jadhav, *Tetrahedron Lett.* **2001**, *42*, 3415.

2.12
Catalytic Enantioselective Alkylation of Alkenes by Chiral Metallocenes

Amir H. Hoveyda

2.12.1
Introduction

The development of catalytic C–C bond forming reactions that proceed under mild conditions in an enantioselective fashion (>90% *ee*) is an important and challenging task in chemical synthesis [1]. Within this context, chiral C2-symmetric ansa-metallocenes, also referred to as bridged metallocenes, have found extensive use as catalysts that effect bond-forming processes in an enantioselective manner [2]. In general, bridged ethylene(bis-tetrahydroindenyl)-metallocene dichlorides (**1–3**, Scheme 1) put forth attractive options for the design of asymmetric reactions because of their geometrically-constrained structure and relative ease of preparation. This chapter is a brief review of the ability of these transition metal complexes to effect the *catalytic and enantioselective alkylation of olefins*.

2.12.2
Zr-Catalyzed Enantioselective Carbomagnesation Reactions

2.12.2.1
Catalytic Enantioselective Addition Reactions

The zirconocene-catalyzed addition of Grignard reagents to alkenes (carbomagnesation) has been developed as a method for enantioselective C-C bond formation. As illustrated in Tab. 1, in the presence of 2.5–10 mol% nonracemic (EBTHI)ZrCl$_2$ (or (EBTHI)Zr-binol) and EtMgCl as the alkylating agent, five-, six-, and seven-membered unsaturated heterocycles undergo facile asymmetric ethyl-

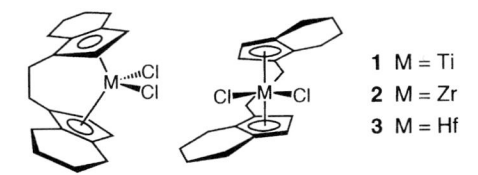

1 M = Ti
2 M = Zr
3 M = Hf

Scheme 1 Group IV ethylene-bridged bis(tetrahydroindenyl) systems.

Transition Metals for Organic Synthesis, Vol. 1, 2nd Edition.
Edited by M. Beller and C. Bolm
Copyright © 2004 WILEY-VCH Verlag GmbH & Co. KGaA, Weinheim
ISBN: 3-527-30613-7

Tab. 1 (EBTHI)Zr-catalyzed enantioselective ethylmagnesation of unsaturated heterocycles[a]

Entry	Substrate	Product	ee [%]	Yield [%]
1		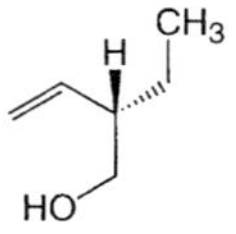	>97	65
2		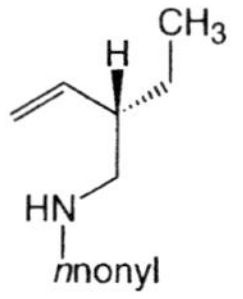	>95	75
3	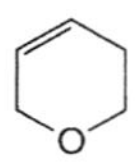	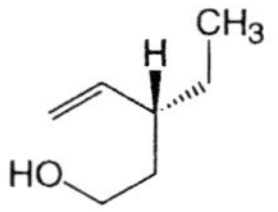	95	73
4	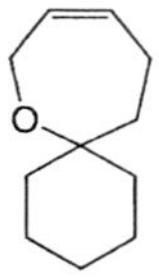	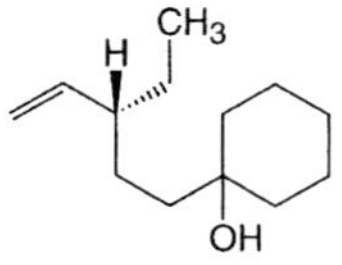	92	75

a) Reaction conditions: 10 mol% (R)-2, 5.0 equivalents EtMgCl, THF, 22 °C for 6–12 h. Entry 1 with 2.5 mol% (R)-2.

magnesation [3]. The rate of carbomagnesation in the terminal alkenes of the reaction products is sufficiently slower, so that unsaturated alcohols and amines can be isolated in high yield (the second alkylation is not generally stereoselective).

The stereoselective ethylmagnesation shown in entry 1 of Tab. 1 has been utilized as a key step in the first enantioselective total synthesis of the antifungal agent Sch 38516 [4]. As illustrated in Scheme 2, further functionalization of the Zr-catalyzed ethylmagnesation product through three subsequent catalytic procedures yields the requisite carboxylic acid synthon in >99% ee.

The catalytic cycle that we have proposed to account for the enantioselective ethylmagnesations is illustrated in Scheme 3. Asymmetric carbomagnesation is initiated by the chiral zirconocene-ethylene complex (R)-3, formed upon reaction of dichloride (R)-2 with EtMgCl [Eq. (a); the dichloride salt or the binol complex may be used with equal efficiency] [5]. Coupling of the alkene substrate with (R)-3 leads to the formation of the metallacyclopentane intermediate i. In the proposed catalytic cycle, reaction of i with EtMgCl affords zirconate ii, which undergoes Zr–Mg ligand exchange to yield iii. Subsequent β-hydride abstraction, accom-

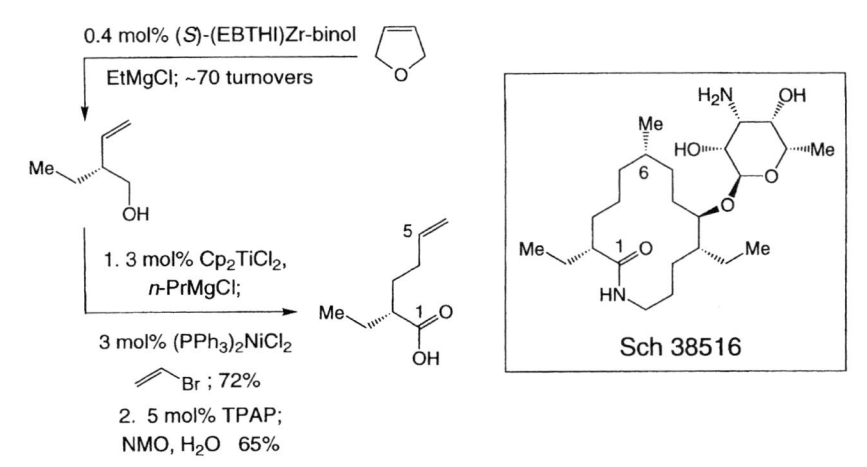

2.12.2 Zr-Catalyzed Enantioselective Carbomagnesation Reactions

Scheme 2 Demonstration of the utility of (EBTHI)Zr-catalyzed ethylmagnesation in the enantioselective synthesis of the macrolactam aglycon of Sch 38516.

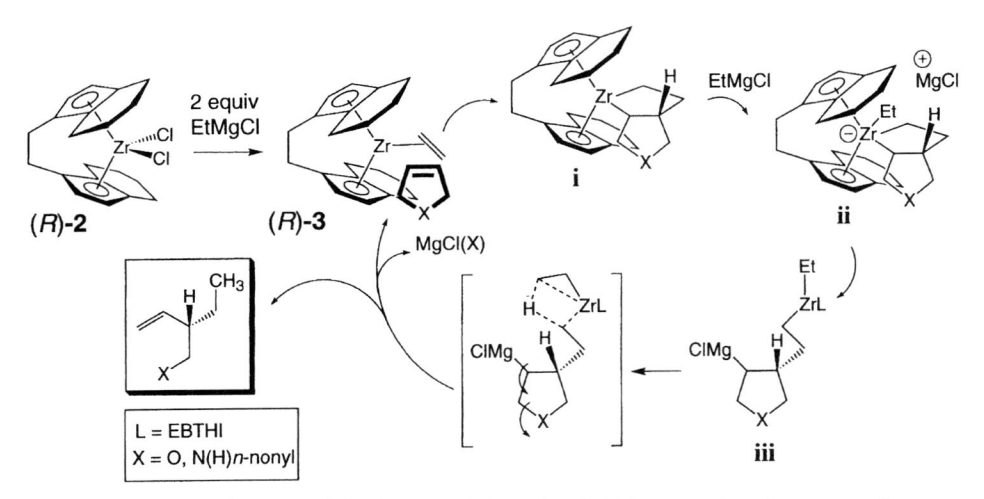

Scheme 3 Catalytic cycle proposed for the (EBTHI)Zr-catalyzed ethylmagnesation of unsaturated heterocycles.

panied by intramolecular magnesium-alkoxide elimination, leads to the release of the carbomagnesation product and regeneration of 3 [6].

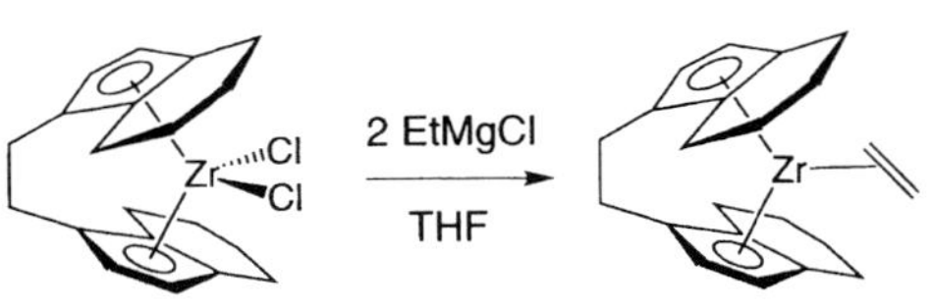

(a)

An important aspect in the carbomagnesation of six-membered and larger heterocycles is the exclusive intermediacy of metallacyclopentanes where the C–Zr bond is formed a to the heterocycle C–O bond. Whether the regioselectivity in the zirconacycle formation is kinetically nonselective and rapidly reversible, or whether formation of the metallacycle is kinetically selective (stabilization of electron density upon formation of the C–Zr bond by the adjacent C–O) [7], has not been rigorously determined. However, as will be discussed below, the regioselectivity with which the intermediate zirconacyclopentane is formed is critical in the (EBTHI)Zr-catalyzed kinetic resolution of heterocyclic alkenes.

Why does the (EBTHI)Zr system induce such high levels of enantioselectivity in the C–C bond formation process? It is plausible that the observed levels of enantioselection arise from minimization of unfavorable steric and torsional interactions in the complex that is formed between 3 and the heterocycle substrates (Scheme 3). The alternative mode of addition, illustrated in Fig. 1, would lead to costly steric repulsions between the olefin substituents and the cyclohexyl group of the chiral ligand [6]. Thus, reactions of simple terminal olefins under identical conditions results in little or no enantioselectivity. This is presumably because in the absence of the alkenyl substituent (of the carbon that bonds with Zr in **i**) the aforementioned steric interactions are ameliorated and the olefin substrate reacts indiscriminately through the two modes of substrate-catalyst binding represented in Fig. 1.

These alkylation processes become particularly attractive when used in conjunction with the powerful catalytic ring-closing metathesis protocols. The requisite starting materials can be readily prepared in high yield and catalytically [8]. The examples shown in Scheme 4 demonstrate that synthesis of the heterocyclic alkene and subsequent alkylation can be carried out in a single vessel to afford unsaturated alcohols and amides in good yield and >99% *ee* (judged by GLC analysis) [9].

Catalytic alkylations where higher alkyls of magnesium are used (Tab. 2) proceed less efficiently (35–40% isolated yield) but with similarly high levels of enantioselection (>90% *ee*). As illustrated in Eqs. (b)–(d), in the chemistry of zirconocene-alkene complexes that are derived from the longer chain alkylmagnesium halides several additional selectivity issues present themselves: the derived transition metal-alkene complex can exist in two diastereomeric forms, exemplified in Eqs. (b)–(d) with (R)-**8** *anti* and *syn*; reaction through these stereoisomeric complexes can lead to the formation of different product diastereomers. The data in Tab. 2 indicate that the mode of addition shown in Eq. (c) is preferred.

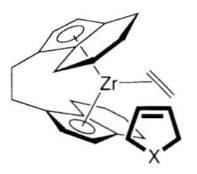

Favored   Disfavored

Fig. 1 Substrate-catalyst (3) interactions favor a specific mode of alkene insertion into the zirconocene-alkene complex.

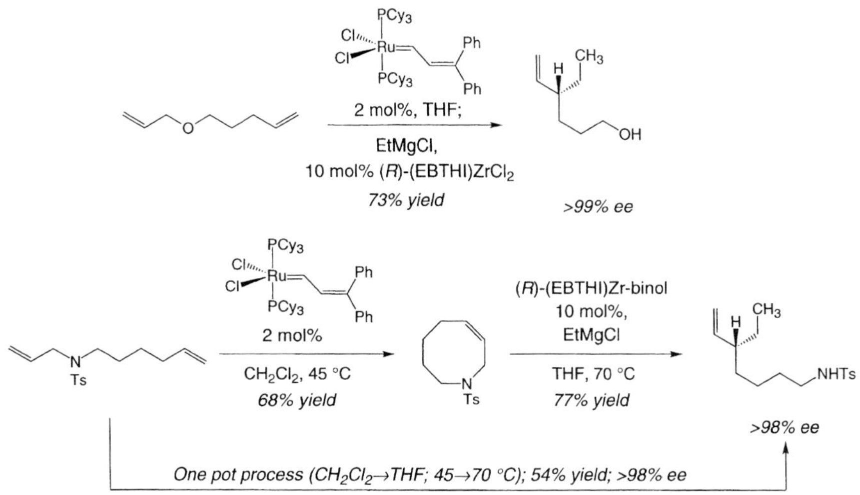

2.12.2 Zr-Catalyzed Enantioselective Carbomagnesation Reactions

Scheme 4 Ru-catalyzed ring closing metathesis processes, in conjunction with Zr-catalyzed enantioselective alkylation reactions provide a convenient protocol for efficient synthesis of optically pure materials.

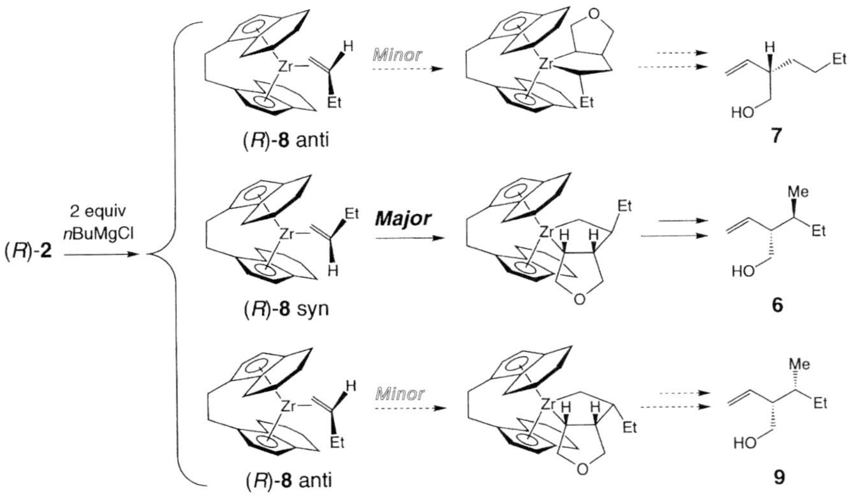

(b – d)

As illustrated in Eqs. (b) and (d), the carbomagnesation process can afford either the *n*-alkyl or the branched product. Alkene substrate insertion from the more substituted front of the zirconocene–alkene system affords the branched isomer [Eq. (d)], whereas reaction from the less substituted end of the (EBTHI)Zr–olefin system leads to the formation of the straight chain product [Eq. (b)]. The results

Tab. 2 (EBTHI)Zr-catalyzed enantioselective carbomagnesation of unsaturated heterocycles with longer chain alkylmagnesium halides[a]

Entry	Substrate	Major product(s)	Temp. [°C]	RMgCl	Regio-selectivity	ee [%]	Diastereo-selectivity
1	 (2,3-dihydrofuran)	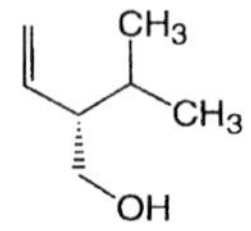 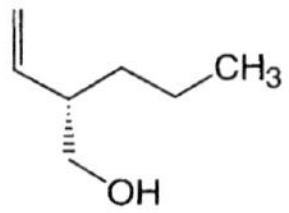 **4**	22	n-PrMgCl	2:1	99 (4), 99 (5)	–
2	(3,4-dihydro-2H-pyran)	**5**	70	n-PrMgCl	20:1	94 (4)	–
3	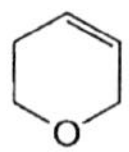 (2,3-dihydrofuran)	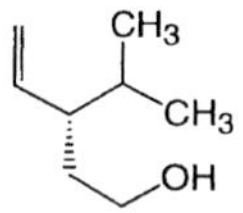 —	22	n-PrMgCl	>25:1	98	–
4	 (2,3-dihydrofuran)	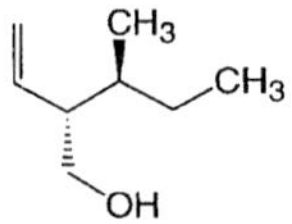 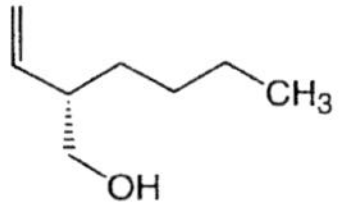 **6**	22	n-BuMgCl	2:1	>99 (6), >99 (7)	15:1
5	(3,4-dihydro-2H-pyran)	**7**	70	n-BuMgCl	15:1	90 (6)	13:1
6	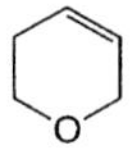 (3,4-dihydro-2H-pyran)	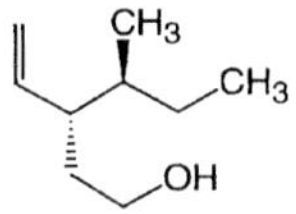 —	22	n-BuMgCl	>25:1	>95	>25:1

a) Reaction conditions: 5 equivalents alkylMgCl, 10 mol% (R)-**2**, 16 h; all yields: 35–40% after silica gel chromatography.

shown in Tab. 2 indicate that, depending on the reaction conditions, products derived from the two isomeric metallacyclopentane formations can be competitive.

Detailed studies shed light on the mechanistic intricacies of asymmetric catalytic carbomagnesations, allowing for an understanding of the above trends in regio- and stereoselectivity [6]. Importantly, the mechanistic studies indicate that there is no preference for the formation of either the *anti* or the *syn* (EBTHI)Zr–olefin isomers (e.g. **8** *anti* vs. **8** *syn):* it is only that one metallocene-alkene diastereomer *(syn)* is more reactive. Moreover, it has been shown that zirconacyclopentane intermediates (i in Scheme 3) do not spontaneously eliminate to the derived zirconocene-alkoxide; Zr–Mg ligand exchange is likely a prerequisite for the alkoxide elimination and formation of the terminal alkene.

2.12.2.2
Zr-Catalyzed Kinetic Resolution of Unsaturated Heterocycles

As the data in Tab. 3 indicate, in the presence of catalytic amounts of nonracemic (EBTHI)ZrCl$_2$, a variety of unsaturated pyrans can be resolved effectively to deliver these synthetically useful heterocycles in excellent enantiomeric purity [10]. A number of important issues in connection to the catalytic kinetic resolution of pyrans are noteworthy: (1) Reactions performed at elevated temperatures (70 °C) afford recovered starting materials with significantly higher levels of enantiomeric purity, compared to processes carried out at 22 °C. For example, the 2-substituted pyran shown in entry 1 of Tab. 3, when subjected to the same reaction conditions but at room temperature, is recovered after 60% conversion in 88% *ee* (vs. 96% *ee* at 70 °C). (2) Consistent with molecular models illustrated in Fig. 2, 6-substituted pyrans (Tab. 3, entry 2) are not resolved effectively. (3) Pyrans that bear a C5 group are resolved with high selectivity as well (entry 4). In this class of substrates, one enantiomer reacts more slowly, presumably because its association with the zirconocene-alkene complex leads to sterically unfavorable interactions between the C5 alkyl unit and the coordinated ethylene ligand.

As the representative data in Tab. 4 indicate, the Zr-catalyzed resolution technology may be applied to medium ring heterocycles as well; in certain instances (e.g. entries 1 and 2) the recovered starting material can be obtained with outstanding enantiomeric purity.

Availability of oxepins that carry a side chain containing a Lewis basic oxygen atom (entry 2, Tab. 4) has further implications in enantioselective synthesis: the derived alcohol, benzyl ether, or MEM-ethers, where resident Lewis basic heteroatoms are less sterically hindered, undergo diastereoselective uncatalyzed alkylation reactions readily when treated with a variety of Grignard reagents [11]. The examples shown below (Scheme 5) serve to demonstrate the synthetic potential of these stereoselective alkylation technologies. Thus, resolution of the TBS-protected oxepin **10**, conversion to the derived alcohol and diastereoselective alkylation with *n*-BuMgBr affords **11** with >96% *ee* in 93% yield. As shown in Scheme 5, alkylation of (*S*)-**12** with an alkyne-bearing Grignard agent (→ (*S*)-**13**), allows for a subsequent Pauson-Khand cyclization to provide the corresponding bicycle **14** in the optically pure form.

Tab. 3 (EBTHI)Zr-catalyzed kinetic resolution of unsaturated pyrans[a)]

Entry	Substrate		Conversion [%]	Cat. [mol%]	Unreacted subs. confi., ee [%]
1	(pyran with CH₂CHMe₂ substituent)		60	10	R, 96
2	(pyran with cyclohexyl substituent)		60	10	S, 41
3	(pyran with Me and CH₂CH₂OR substituents)	a R=MgCl b R=TBS	56 60	10 10	R, >99 R, >99
4	(pyran with Me-alkyl chain)		58	20	R, 99
5	(pyran with Me and OR substituents)	a R=MgCl b R=TBS	63 61	10 10	R, >99 R, 94

a) Reaction conditions: indicated mol% (R)-**2**, 5.0 equivalents of EtMgCl, 70 °C, THF. Mass recovery in all reactions is >85%.

Zirconocene-catalyzed kinetic resolution of dihydrofurans is also possible, as illustrated in Scheme 6 [12]. Unlike their six-membered ring counterparts, both of the heterocycle enantiomers react readily, but through distinctly different reaction pathways, to afford – in high diastereomeric and enantiomeric purity – constitutional isomers that are readily separable. A plausible reason for the difference in

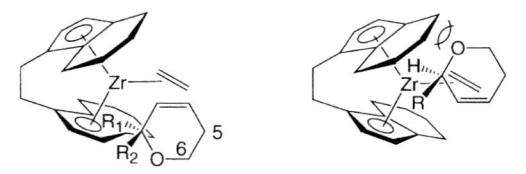

I R_1 = H, R_2 = alkyl
II R_1 = alkyl, R_2 = H

III R = alkyl

Fig. 2 Preferential association of one pyran enantiomer with (R)-(EBTHI)Zr-ethylene complex.

2.12.2 Zr-Catalyzed Enantioselective Carbomagnesation Reactions

Tab. 4 (EBTHI)Zr-catalyzed kinetic resolution of 2-substituted medium ring heterocycles [a]

Entry	Substrate	Conversion [%]	Time	Unreacted subs. config., ee [%]
1	(oxepine with CH₂CH(Me)Me substituent)	58	30 min	R, >99
2	(oxepine with CH₂OTBS substituent)	63	100 min	R, 96
3	(benzoxepine with CH₂CH₂Me substituent)	60	8 h	R, 60
4	(benzoxocine with CH₂CH₂Me substituent)	63	11 h	R, 79

a) Reaction conditions: 10 mol% (R)-2, 5.0 equivalents of EtMgCl, 70 °C, THF. Mass recovery in all reactions is >85%.

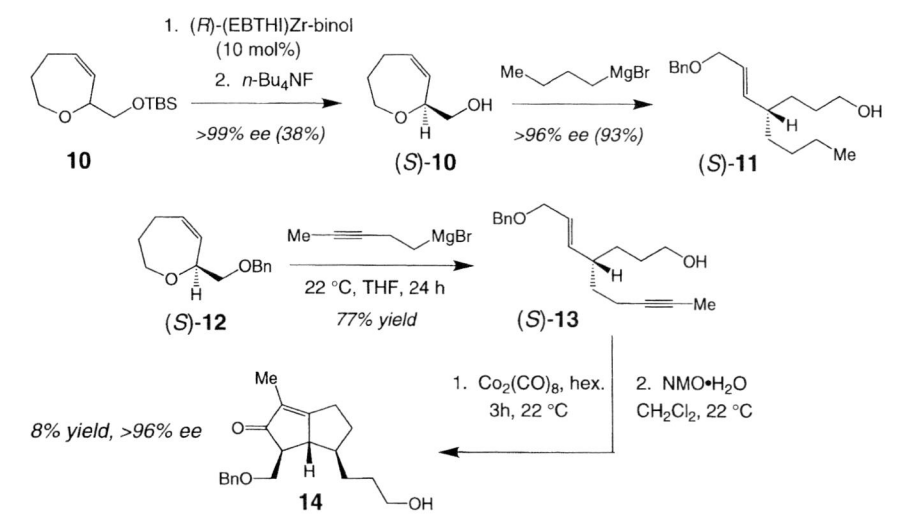

Scheme 5 Chiral medium-ring heterocycles that have been resolved by the Zr-catalyzed kinetic resolution are subject to highly diastereoselective alkylations that afford synthetically useful materials in the optically pure form.

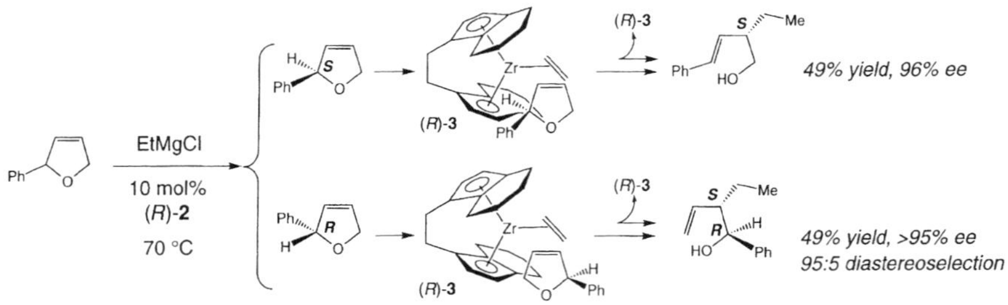

Scheme 6 (EBTHI)Zr-catalyzed kinetic resolution of dihydrofurans.

the reactivity pattern of pyrans and furans is that, in the latter group of compounds, both olefinic carbons are adjacent to a C–O bond: C–Zr bond formation can take place at either end of the C–C -system.

2.12.2.3
Zr-Catalyzed Kinetic Resolution of Cyclic Allylic Ethers

As depicted in Eqs. (e)–(g), kinetic resolution of a variety of cyclic allylic ethers is effected by asymmetric Zr-catalyzed carbomagnesation. Importantly, in addition to six-membered ethers, seven- and eight-membered ring systems can be readily resolved by the Zr-catalyzed protocol.

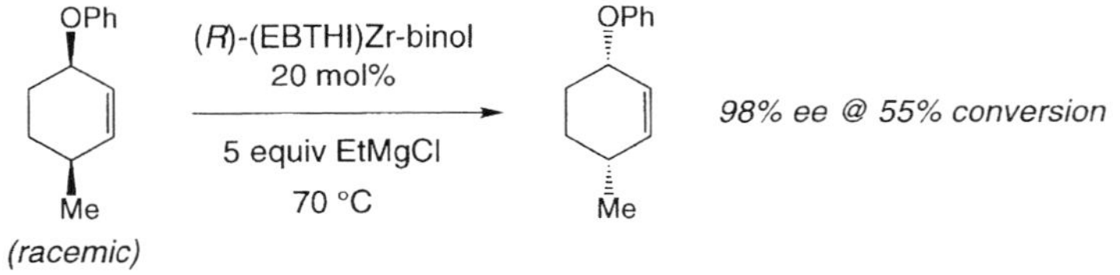

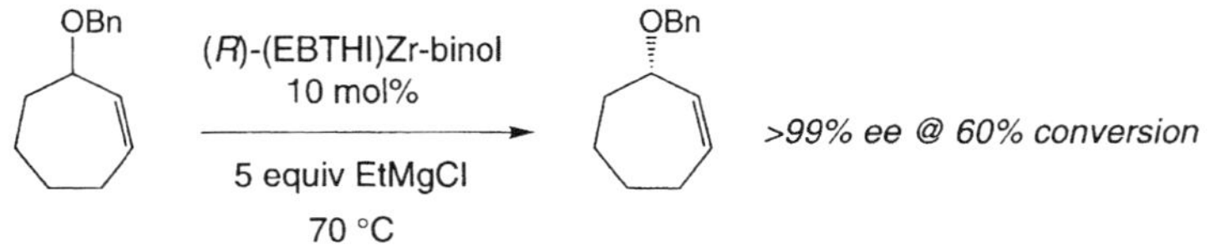

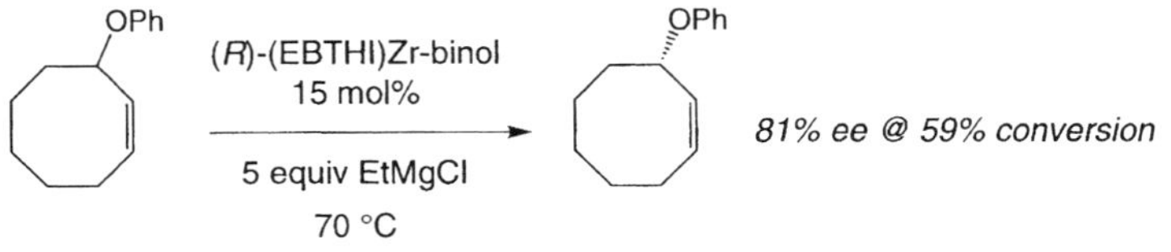

The synthetic versatility and significance of the Zr-catalyzed kinetic resolution of cyclic allylic ether is readily demonstrated in the example provided in Scheme 7. Optically pure starting allylic ether, obtained by the above mentioned catalytic ki-

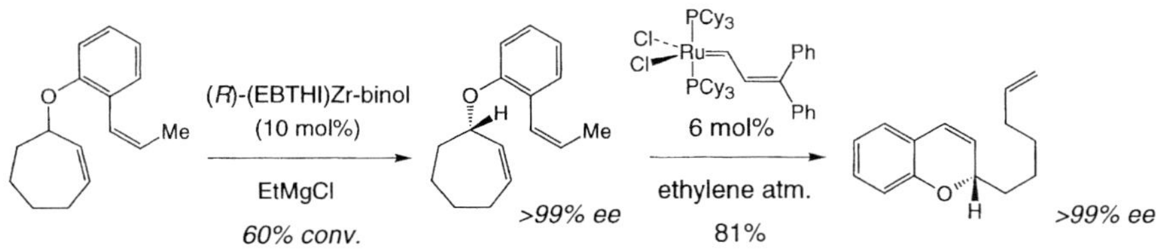

Scheme 7 Tandem Zr-catalyzed kinetic resolution and Ru-catalyzed rearrangement affords chiral chromenes in high enantiomeric purity.

netic resolution, undergoes a facile Ru-catalyzed rearrangement to afford chromene derivatives in >90% ee. [13].

2.12.2.4
Other Related Catalytic Enatioselective Olefin Alkylations

The zirconocene-catalyzed enantioselective carbomagnesation accomplishes the addition of an alkylmagnesium halide to an alkene, where the resulting carbo-metallation product is suitable for a variety of additional functionalization reactions (see Schema 2). Excellent enantioselectivity is obtained in reactions with Et-, n-Pr-, and n-BuMgCl, and the catalytic resolution processes allow for preparation of a variety of nonracemic heterocycles. Nonetheless, the development of reaction processes where a larger variety of olefinic substrates and alkylmetals (e.g. Me-, vinyl-, phenylmagnesium halides, etc.) can be added to unfunctionalized alkenes efficiently and enantioselectively stands as a challenging goal in enantioselective reaction design. As illustrated below (Eqs. (h) and (i)), recent reports by Negishi and co-workers, where Erker's nonbridged chiral zirconocene [14] is used as a catalyst, is an important and impressive step towards this end [15].

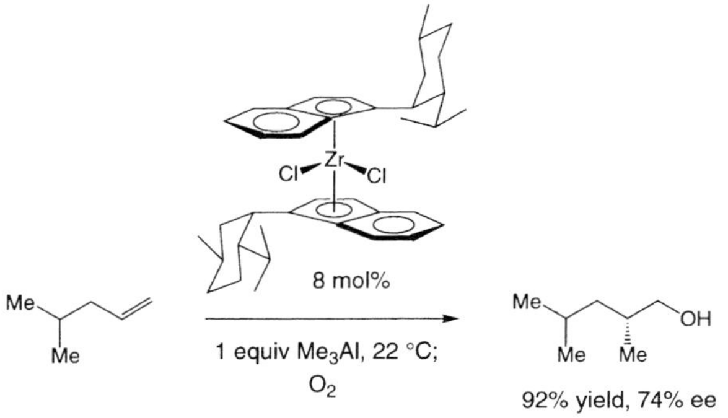

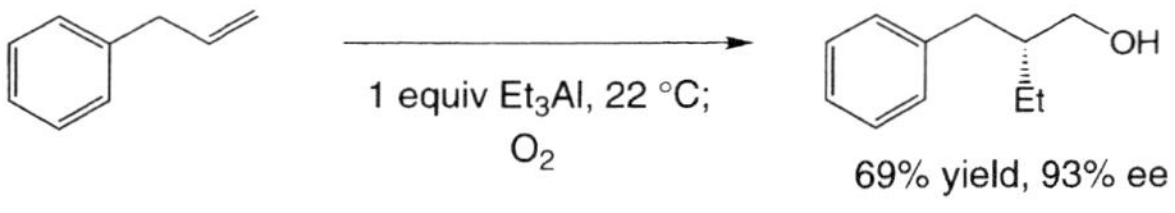

2.12.3
Summary and Outlook

The chemistry described in this chapter demonstrates that chiral *ansa*-metallocene complexes can be used to effect an important reaction that is largely unprecedented in classical organic chemistry: addition of alkylmagnesium halides to unactivated olefins. Although EBTHI metallocenes have proven to be effective at promoting the above enantioselective transformations, the equipment required to prepare such catalysts (glovebox and high pressure hydrogenation apparatus), as well as costs associated with the required metallocene resolution (nonracemic binaphtol = $ 45 per g), suggests more attractive catalyst alternatives may be desired [16]. Promising advances toward more facile syntheses of inexpensive and chiral (EBTHI)MX$_2$ equivalents may eventually provide more practical alternatives to this powerful class of transition metal catalysts. There is little doubt that future exciting discoveries in the area of design and development of useful asymmetric catalytic C–C bond forming transformations are in the making.

2.12.4
References

1 For a general overview of recent advances in this ares, see: I. OJIMA (Ed.), *Catalytic Asymmetric Synthesis*, VDH, New York, **1993**.

2 For a recent review, see: A. H. HOVEYDA, J. P. MORKEN, *Angew. Chem.* **1996**, *108*, 1378–1401; *Angew. Chem., Int. Ed. Engl.* **1996**, *35*, 1262–1284.

3 J. P. MORKEN, M. T. DIDIUK, A. H. HOVEYDA, *J. Am. Chem. Soc.* **1993**, *115*, 6997–6998. See also: H.-G. SCHMALZ, *Nachr. Chem. Lab.* **1994**, *42*, 724–729.

4 (a) A. F. HOURI, Z. XU, D. A. COGAN, A. H. HOVEYDA, *J. Am. Chem. Soc.* **1995**, *117*, 2943–2944. (b) Z. XU, C. W. JOHANNES, S. S. SALMAN, A. H. HOVEYDA, *J. Am. Chem. Soc.* **1996**, *118*, 10926–10927; see also: H.-G. SCHMALZ, *Angew. Chem.* **1995**, *107*, 1981–1984; *Angew. Chem., Int. Ed. Engl.* **1995**, *34*, 1833–1836.

5 A. H. HOVEYDA, J. P. MORKEN, *J. Org. Chem.* **1993**, *58*, 4237–4244.

6 M. T. DIDIUK, C. W. JOHANNES, J. P. MORKEN, A. H. HOVEYDA, *J. Am. Chem. Soc.* **1995**, *117*, 7097–7104.

7 (a) A. S. GURAM, R. F. JORDAN, *Organometallics* **1990**, *9*, 2190–2192. (b) *Organometallics* **1991**, *10*, 3470–3479.

8 (a) G. C. FU, R. H. GRUBBS, *J. Am. Chem. Soc.* **1992**, *114*, 7324–7325. (b) G. c. FU, R. H. GRUBBS, *J. Am. Chem. Soc.* **1993**, *115*, 3800–3801. (c) R. H. GRUBBS, S. J. MILLER, G. C. FU, *Acc. Chem. Res.* **1995**, *107*, 1981–1984; *Angew. Chem., Int. Ed. Engl.* **1995**, *34*, 1833–1836 and references therein. (e) R. R. SCHROCK, J. S. MURDZEK, G. C. BAZAN, J. ROBBINS, M. DIMARE, M. O'REGAN, *J. Am. Chem. Soc.* **1990**, *112*, 3875–3886. (f) G. C. BAZAN, R. R. SCHROCK, H.-N. CHO, V. C. GIBSON, *Marcomolecules* **1991**, *24*, 4495–4502.

9 M. S. VISSER, N. M. HERON, M. T. DIDIUK, J. F. SAGAL, A. H. HOVEYDA, *J. Am. Chem. Soc.* **1996**, *118*, 4291–4298.

10 J. P. MORKEN, M. T. DIDIUK, M. S. VISSER, A. H. HOVEYDA, *J. Am. Chem. Soc.* **1994**, *116*, 3123–3124.

11 N. M. HERON, J. A. ADAMS, A. H. HOVEYDA, *J. Am Chem. Soc.* **1997**, *119*, 6205–6206.

12 M. S. VISSER, A. H. HOVEYDA, *Tetrahedron* **1995**, 4383–4394.

13 J. P. A. HARRITY, M. S. VISSER, J. S. GLEASON, A. H. HOVEYDA, *J. Am Chem. Soc.* **1997**, *119*, 1488–1489.

14 G. ERKER, M. AULBACH, M. KNICKMEIER, D. WINGBERMUHLE, C. KRUGER, M. NOLTE, S. WERNER, *J. Am Chem. Soc.* **1993**, *115*, 4590–4601.

15 (a) D. Y. KONDAKOV, E. NEGISHI, *J. Am Chem. Soc.* **1995**, *117*, 1071–1072. (b) D. Y. KONDAKOV, E. NEGISHI, *J. Am. Chem. Soc.* **1996**, *118*, 1577–1578; see also: D. Y. KONDAKOV, S. WANG, E. NEGISHI, *Tetrahedron Lett.* **1996**, *37*, 3803–3806.

16 For a recent report along these lines, see: L. BELL, R. J. WHITBY, R. V. H. JONES, M. C. H. STANDEN, *Tetrahedron Lett.* **1996**, *37*, 7139–7142.

2.13
Palladium-Catalyzed Olefinations of Aryl Halides (Heck Reaction) and Related Transformations

Matthias Beller, Alexander Zapf, and Thomas H. Riermeier

2.13.1
Introduction

The synthesis of arylated and vinylated olefins is of fundamental importance in organic chemistry. The palladium-catalyzed carbon–carbon coupling of haloalkenes and haloarenes with olefins, generally known as the Heck reaction, provides an efficient gateway into such compounds [1]. As shown in Scheme 1, styrenes and dienes can be prepared directly from the corresponding alkene and aryl or vinyl compounds substituted with a leaving group X, where X=Cl, Br, I, N_2BF_4, OTf, etc. Owing to the possibility of preparing not only simple terminal or 1,2-disubstituted olefins but also numerous complex molecular frameworks, e.g., tertiary and quaternary stereogenic centers, via Heck reactions, this methodology has become one of the most important transition metal-catalyzed transformations in organic synthesis. In addition to olefins and dienes, alkynes can also be used as unsaturated compounds [2]. Often these alkynylations proceed under similar conditions – although the addition of copper co-catalysts is usually required – compared to the classical Heck reaction.

The synthetic usefulness of the Heck reaction is explained by the fact that the methodology is amenable to a variety of easily available starting materials. In general, R must be an aryl or vinyl group for reasons outlined below, although similar reactions of benzyl–X or allyl–X are known. The Heck reaction is remarkably chemoselective. Hence, educts containing most of the known functional groups may be used. The palladium catalysts typically employed are stable to water and air. In addition, palladium intermediates of the catalytic cycle of the Heck reaction can, in principle, undergo various domino reactions. Thus, polycyclizations, olefination-carbonylation or olefination-alkynylation, and other domino sequences are possible.

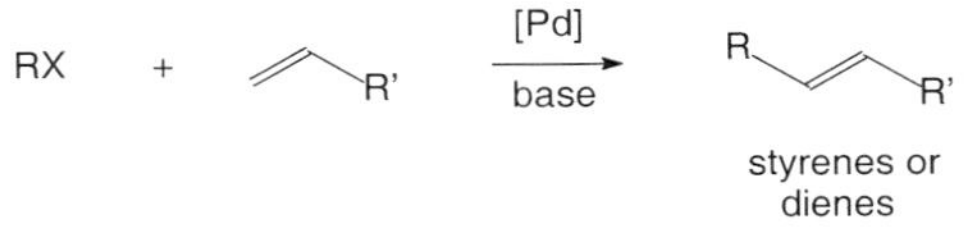

$$RX + \underset{R'}{\diagup\!\!\!\diagup} \xrightarrow[\text{base}]{[Pd]} R\diagup\!\!\!\diagup R'$$

styrenes or dienes

Scheme 1 The Heck olefination reaction: X=Cl, Br, I, N_2BF_4, OTf, etc.; R=aryl, vinyl.

Much progress has been made over the past 10 years and several excellent reviews have appeared in this area [3]. Hence, this chapter will cover selected highlights of the Heck reaction since its discovery in the late 1960s with a particular emphasis on work of the last few years where several exciting breakthroughs have been made including (i) development of more reactive and thermally stable catalytic systems, (ii) new more efficient enantioselective variants, and (iii) expanded applications in organic synthesis.

2.13.2
Mechanism

The generally accepted mechanism of the Heck reaction is shown in Scheme 2. A coordinatively unsaturated palladium(0) species, usually coordinated with weak donor ligands such as triarylphosphines, is assumed to be the real active species. However, Amatore et al. have shown that anionic complexes of the form L_2PdX^- (X = Cl, Br, I, OAc) may be the catalytically active species when strong donor anions are present [4]. There is also some speculation about a Pd(II)/Pd(IV) mechanism, especially in the case of palladacyclic catalysts [5], but these seem to be restricted to special cases if they are valid at all. Hence, the first step of the catalytic cycle is oxidative addition of a haloarene or haloalkene to a palladium(0) compound to form a palladium(II) species (step A). In the next step, an alkene molecule is coordinated concomitantly with the dissociation of L or X^-. In the latter case, a cationic species is formed. The need for free coordination sites for the oxidative addition step and the olefin complexation is consistent with the observation that high concentrations of phosphines retard the reaction rate [6]. In step C, the alkene inserts into the Pd–Ar bond. The insertion process occurs via a four-centered transition state, which requires a planar assembly of the alkene and Pd–Ar bond. Hence, insertion proceeds in a *syn* manner to generate a σ-alkylpalladium complex. Subsequent *syn* β-hydride elimination generates the product olefin and a palladium(II) hydride complex (step D). In the last step, E, the Pd(0) catalyst is reformed after formal reductive elimination of HX in the presence of base.

The rate-determining step of the catalytic cycle is dependent on the starting materials employed. Clearly, the nature of the leaving group X (Scheme 1) effects the rate of oxidative addition (ArI > ArBr >> ArCl). Hence, oxidative addition is likely to be the rate-determining step when aryl chlorides are used. Until recently, inexpensive and readily available chloroarenes and chloroalkenes have not been reactive enough toward Heck reactions to allow efficient catalysis. However, remarkable progress has been made in the area of chloroarene activation in the last 5 years, as outlined below [7].

When unsymmetrically substituted alkenes are employed as substrates, several regioisomeric products are possible. As shown in step C of Scheme 2, the arylpalladium(II) intermediate can, in principle, insert into the α or β position of the coordinated alkene to give the corresponding 1,1- or 1,2-disubstituted alkenes, respectively. Since insertion occurs in a *syn* manner, the 1,2-disubstituted alkene

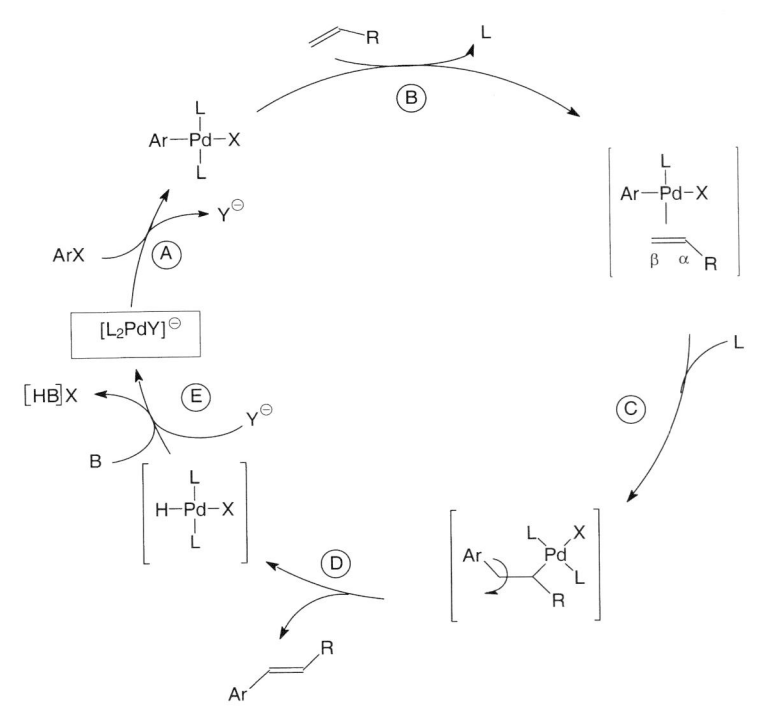

Scheme 2 Textbook mechanism of the Heck reaction.

should be favored based on steric effects. However, poor α or β regioselectivities have been observed for enol ethers and enamines. Valuable contributions have been made by Cabri and Candiani to the solution of this problem. They showed that, with the appropriate conditions, either 1,1- or 1,2-disubstituted enol ethers may be formed preferentially [3b].

As shown in Scheme 3, the coupling of enol ethers with RX compounds (R=aryl, vinyl) in the presence of bidentate phosphine-coordinated palladium(II) species may follow two reaction pathways. If X=OTf, then the 1,1-disubstituted product **7**, corresponding to path 1, is favored. If X=I, Br, or Cl, then the linear 1,2-disubstituted product **4**, corresponding to path 2, is favored. In path 2, coordination of the olefin occurs with concomitant dissociation of one of the two phosphorus atoms of the bidentate ligand. Here, steric factors dominate, and the R group migrates to the least substituted carbon. In path 1, coordination of the olefin is accompanied by the dissociation of the weakly bound counterion to generate cationic complex **5**. The lability of the Pd-OTf bond is well known [8]. The coordination of the π-system in **5** induces an increase in the polarization of the cationic intermediate. Hence, the Ar group, which is formally an anion, preferentially migrates to the carbon atom with the lowest charge density. This leads to the 1,1-disubstituted product **7**.

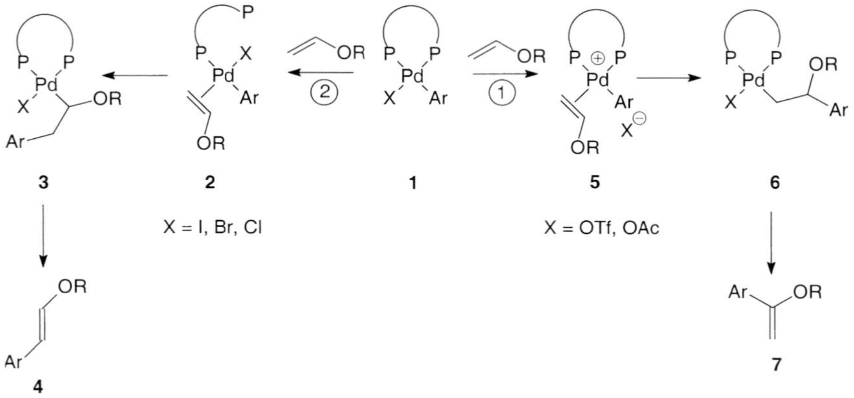

Scheme 3 The two possible coordination insertion pathways.

The electronic nature of the alkene substrate also affects the regioselectivity. Electron-rich olefins react faster via the cationic pathway, while electron-poor olefins react faster via the neutral pathway [3b]. It is also possible to control the regioselectivity of a given Heck reaction using electron-rich olefins, no matter which ArX precursor is employed. For example, addition of halide-sequestering agents such as Ag^+ to the reaction of an aryl/vinyl halide will direct the reaction to proceed via the cationic pathway. Conversely, addition of halide anions to the reaction of aryl/vinyl triflates will result in the reaction following the neutral pathway.

Problems with regio- and stereoselectivity also arise when disubstituted olefins are coupled [9, 10]. For example, in the case of the arylation of 1,2-disubstituted olefins, the origin of E/Z selectivity is of a thermodynamic nature [9c]. On the other hand, functional groups might influence the position of β-hydride elimination. Thus, for α-methacrylic esters the direction of β-hydride elimination is greatly influenced by the kind of base applied, switching from thermodynamic to kinetic control [10].

2.13.3
Catalysts

Palladium(0)-phosphine complexes, such as $Pd(PPh_3)_4$, are often used as Heck reaction catalysts. However, Pd(II) salts in the presence of ligands are clearly more important as *in situ* catalysts. Since the original work of Heck [1], many modifications and improvements of the catalyst system have been reported. For example, Jeffrey has shown that the addition of tetraalkylammonium salts can dramatically improve the reactivity and selectivity of the reaction [11]. Employment of high pressure conditions has also been demonstrated to increase catalyst efficiency [12]. Very recently, the promotion of Heck reactions by microwave irradiation has been reported [13].

One general problem for industrial applications of the Heck reaction is the separation of the products from the expensive palladium catalysts. One solution is to simply anchor the palladium onto a solid support. In principle, after completion of the heterogeneous Heck reaction, the products can be distilled or decanted off, so the catalyst can be used again. Also, heterogeneous catalysts such as palladium dispersed on carbon, silica, magnesia, or zeolites have been described successfully for Heck reactions [14]. In addition, palladium deposited on porous glass [15], palladium clusters [16a–d], and palladium/nickel clusters [16e] are also effective as heterogeneous Heck catalysts. However, most of the described examples of Heck reactions using heterogeneous palladium catalysts have to be performed at high temperatures, and activated bromo- or iodoaromatics or diazonium salts are required for complete conversion. Moreover, so far it has not been unequivocally demonstrated that heterogeneous catalysts can be efficiently reused without loss of activity.

Polymer-bound palladium complexes are a connecting link between homogeneous and heterogeneous catalysts. The active sites are mono-atomic and molecularly defined, but, through the polymeric backbone, a separation from the reaction mixture by simple filtration is possible. Nevertheless, the problems of leaching and deactivation processes have to be solved, as in the case of "purely heterogeneous" catalysts. For example, the palladium complex of a simple diphenylphosphino substituted polymer could be used for the coupling of phenyl iodide with acrylates [17a]. A chelating bisphosphine ligand tethered to polystyrene [17b] has been described as well as a chelating biscarbene ligand, that has been linked to Wang resin [17c].

The use of dendritic diphosphine metal complexes is a similar concept for catalysis at the interface of homogeneous and heterogeneous catalysis [18]. These structurally defined molecules are efficient Heck catalysts and are easily recycled from the reaction mixture. However, similarly to supported palladium catalysts, the "real" reuse (up to ten times and more) without loss of activity has not been demonstrated until today.

Another solution to the separation of the palladium catalyst from the products is two-phase catalysis [19]. Here, the catalyst is sequestered in a hydrophilic phase in which the organic products are insoluble. A water-soluble catalytic system utilizing a sulfonated phosphine (TPPTS 8) has been successfully employed in the Heck reaction [19a–e]. Also, other classes of hydrophilic triarylphosphines for two-phase catalysis have been developed [19f]. For example a carbohydrate-substituted moiety is

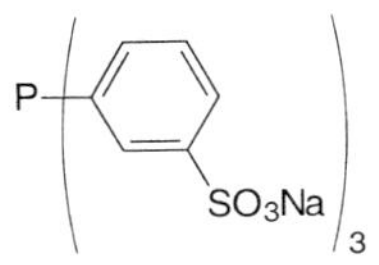

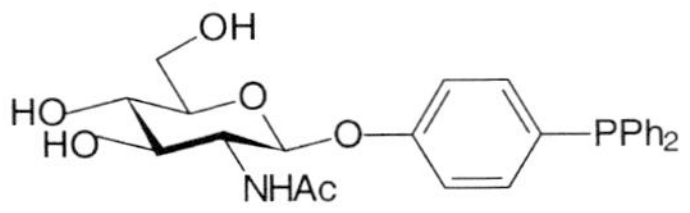

Fig. 1 Selected hydrophilic phosphine ligands for two-phase catalysis.

used to render the catalyst soluble in a hydrophilic phase (**9**). The Heck coupling in aqueous media is also possible in the presence of phase-transfer catalysts [20].

Researchers from DSM have elegantly demonstrated that the recovery of a homogeneous palladium catalyst can be accomplished by its deposition on silica or celite at the end of the reaction. After filtration, oxidation by iodine or bromine solubilizes and re-activates the catalyst for the next run. As oxidative conditions are not compatible with phosphine ligands, only aryl iodides and activated aryl bromides are suitable substrates for this protocol [21].

Intense efforts have been undertaken in the last few years to develop more active and productive catalysts in order to use the synthetic power of Heck and related reactions for large-scale industrial syntheses. To this end, the groups of Herrmann and Beller discovered, based on the work of Spencer [22], that cyclometalated palladium complexes are efficient pre-catalysts for Heck and related reactions (Fig. 2) [23, 24]. These air- and moisture-stable phosphapalladacycles catalyze the carbon–carbon coupling between aryl halides and olefins (Heck reaction) [23], arylboronic acids (Suzuki coupling) [24], and alkynes [2e]. Turnover numbers (TON) and turnover frequencies (TOF) as high as 1 000 000 and 20 000 h^{-1}, respectively, which exceeded all previously published data at that time, have been achieved. Most importantly, an effective activation of electron-deficient chloroarenes has been realized with palladacycles for the first time. As an example, the Heck coupling of 4-chloroacetophenone with n-butyl acrylate is catalyzed with only 0.2 mol% of **10** to produce n-butyl 4-acetylcinnamate in 81% yield [23c].

The increased catalyst productivity of palladacycles is due, in part, to their pronounced thermal stability in solution. Conventional adducts such as Pd(PPh$_3$)$_2$(OAc)$_2$ and Pd$_2$(PPh$_3$)$_2$(μ-OAc)$_2$(OAc)$_2$ are deactivated at temperatures above 120 °C. The depletion of catalyst-stabilizing phosphines causes the precipitation of elemental palladium, resulting in a breakdown of the catalytic cycle. With **10**, palladium deposits and phosphorus–carbon bond cleavage products are not observed at temperatures up to 130 °C [23a]. Hence, palladacycles have an increased lifetime as catalysts and are particularly advantageous for less reactive substrates.

Other new types of very active Heck catalysts have been developed in subsequent years. Milstein and co-workers used Pd(II) species with a tridentate PCP ligand system which are oxygen and moisture stable (**11**) (Fig. 3) [25a]. Furthermore, no catalyst degradation occurs upon prolonged heating at 140 °C under typi-

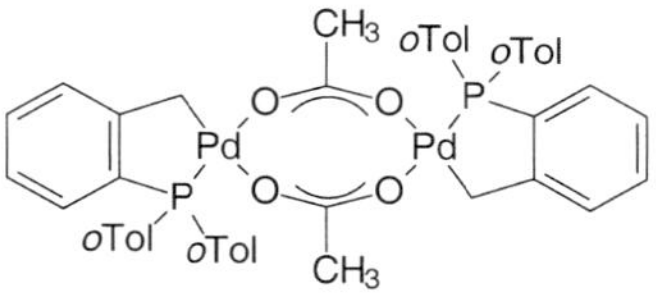

10

Fig. 2 Palladacycle catalyst for use in coupling reactions.

cal Heck reaction conditions. The utility of **11** was demonstrated in the Heck coupling of bromobenzene with methyl acrylate, which is catalyzed with 3.5×10^{-5} mol% of **11** to produce methyl cinnamate in 93% yield. This type of ligand has also been applied for an intramolecular Heck-cyclization to form a five-membered ring [25b]. Unfortunately, no coupling could be observed if the olefinic part was a 1,4-diene, whereas the desired cyclization occurred in the absence of the second double bond. This was found to be in contrast to the behavior of a "classical" palladium(II) acetate/triphenylphosphine catalyst system.

The concept of these so-called "pincer" ligands was extended to bisphosphinito (**12**) and bis(thioether) systems (**13**). The former type is easier to prepare compared to the analog bisphosphine ligands and is an efficient catalyst for the olefination of all kinds of chloroarenes, including electron-rich ones [25c], and for the synthesis of trisubstituted olefins [25d]. The thioether ligands require an activating substituent in 5-position for efficient insertion of Pd into the C–H bond [25e]. Via this substituent, the pincer ligand can be linked to a polymer and thus be recycled several times. Nevertheless, this type of ligand is only useful for the coupling of aryl iodides, but the catalytic reactions can be performed in air as no phosphorus is present in the ligand.

Reetz and co-workers have shown that even non-activated chlorobenzene can be olefinated using a tetraphenylphosphonium salt as co-catalyst [26]. Interestingly, it was found that also electron-poor phosphite ligands are capable of activating aryl bromides and even electron-deficient aryl chlorides when they are applied in some excess to palladium (P/Pd = 10:1 to 100:1) or as a cyclometalated complex [27].

Highly basic and sterically demanding phosphines have been utilized by several groups [28]. In a high-throughput screening of catalysts for the coupling of aryl bromides and chlorides with *n*-butyl acrylate, Hartwig et al. found that di-*tert*-butylphosphinoferrocenes (**14**) led to a highly active catalyst system (Fig. 4) [28a,b]. Comparable results can be obtained in the presence of tri-*tert*-butylphosphine (**15**) [28c,d]. By using this simple (but pyrophoric) phosphine together with dicyclohexylmethylamine as the base, Littke and Fu have been able to couple aryl bromides and activated aryl chlorides even at room temperature. Non-activated chloroarenes require elevated reaction temperatures, but turnover numbers up to 400 can be realized with this catalyst system. Beller and co-workers have described di(1-adamantyl)-*n*-butylphosphine (**16**) to be an equally efficient ligand for the olefination of aryl chlorides [28e]. Secondary dialkylphosphines, e.g., di-*tert*-butylphosphine, are also suitable ligands for Heck reactions, but it is likely that they are arylated *in situ* to give the real catalytically active species [28f,g].

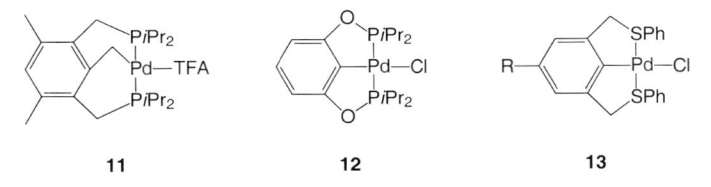

Fig. 3 New efficient catalysts for use in the Heck reaction.

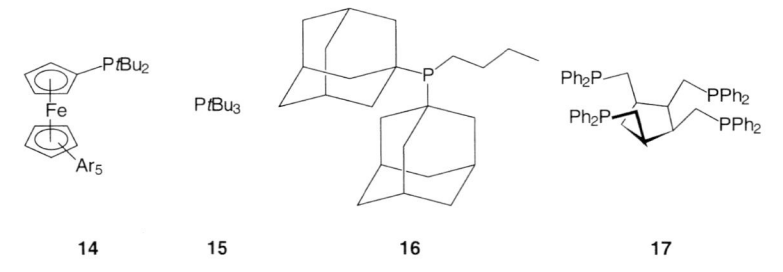

Fig. 4 Phosphine ligands for the activation of aryl chlorides.

An interesting tetrapodal phosphine ligand, Tedicyp (**17**), has been introduced by Feuerstein et al. for palladium-catalyzed coupling reactions [29]. Although it is not very electron rich and it probably does not form highly coordinatively unsaturated palladium complexes, exceedingly high turnover numbers and frequencies have been observed. For example, 3,5-bis(trifluoromethyl)bromobenzene is coupled with *n*-butyl acrylate with TONs of more than 200 million and TOFs up to five million at 130 °C. A similar bidentate ligand leads to TONs and TOFs that are significantly lower [30].

The use of nonaqueous ionic liquids (NAIL) instead of ordinary organic solvents results in an improvement in the catalyst performance in some cases [31]. Here, good results can also be obtained applying ligand-free palladium(II) chloride as catalyst. After distilling off the reactants and products, the catalyst and NAIL can be reused several times without significant loss of catalytic activity.

Carbene-palladium complexes constitute another class of stable and active catalysts [32]. Herrmann et al. have shown that *N*-heterocyclic carbenes derived from imidazole and pyrazole (**18**) exhibit high thermal stability and resist oxidation under conditions that normally destroy traditional palladium-phosphine complexes (Fig. 5). Thus, palladium dicarbene complexes catalyze, after *in situ* reduction, the Heck coupling of activated aryl chlorides with *n*-butyl acrylate [32a]. *In situ* reduc-

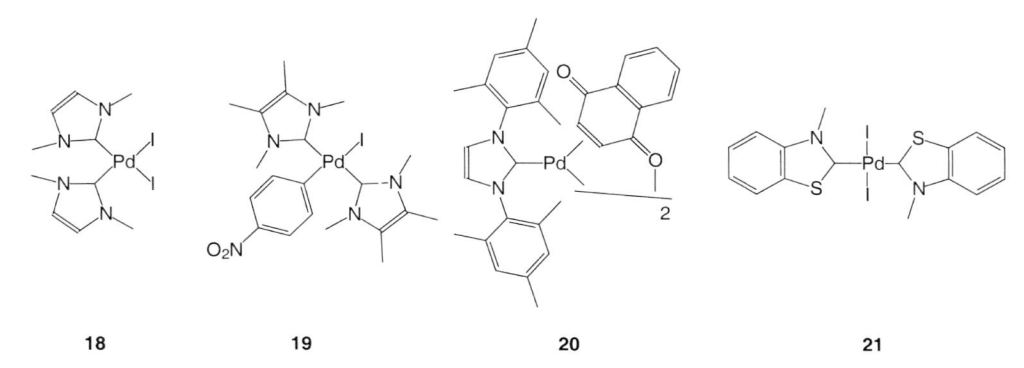

Fig. 5 Heterocyclic carbene catalysts for Heck reactions.

tion is not required if the oxidative addition product, i.e. the corresponding biscarbene arylpalladium halide complex **19**, is used as catalyst [32b]. Monocarbene palladium complexes (**20**) have been introduced as coupling catalysts by Beller and co-workers. They catalyze the Heck reaction of aryldiazonium salts [32c] and aryl chlorides in ionic liquids [32d], respectively.

When simple palladium salts are used as catalysts in 1-butyl-3-methylimidazolium salts as NAILs, the formation of biscarbene complexes after deprotonation of the imidazolium salt by the base, which is required for the olefination reaction, is observed. These complexes catalyze the Heck reaction of iodobenzene and electron-poor aryl bromides with acrylate or styrene [33]. *In situ* mixtures of palladium(II) acetate and imidazolium salts have also been used in organic solvents for the coupling of aryl bromides and aryldiazonium salts, respectively [34].

Other types of heterocyclic carbenes are also suitable as ligands for Heck reactions, as has been demonstrated by the group of Caló. They used the palladium-benzothiazole carbene complex **21** as the catalyst for the olefination of iodo- and bromoarenes [36a,b]. Even α- and β-substituted acrylic esters can be arylated using this catalyst [36c,d], and allylic alcohols are converted to the corresponding β-aryl ketones by isomerization after arylation [36e].

Different types of chelating carbene ligands have been investigated for Heck couplings, but in most cases without a significant improvement in the catalyst performance (Fig. 6). A chelating biscarbene ligand with a binaphthyl core (**22**) has been described for the olefination of phenyl iodide and bromide [35a]. The mixed carbene-phosphine ligand precursor **23** has been used for the coupling of a number of bromoarenes [35b], and the oxazoline-carbene complex **24** catalyzes the formation of stilbenes from aryl bromides and activated aryl chlorides [35c]. Even tridentate "pincer" biscarbene ligands (**25**) can be used for coupling of aryl

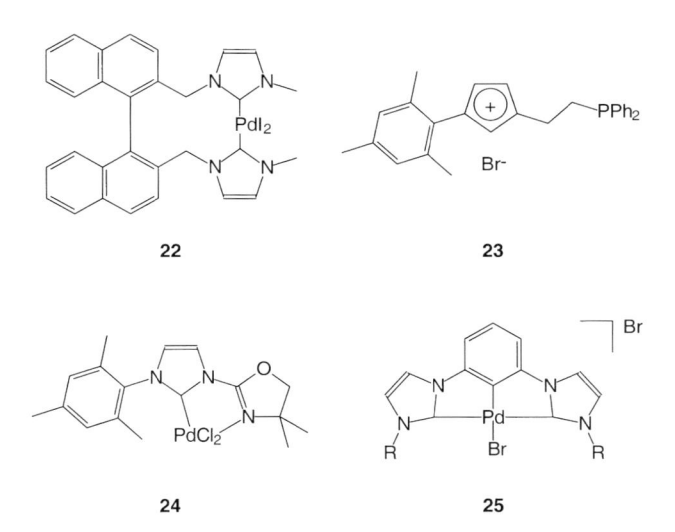

Fig. 6 Chelating carbene ligands.

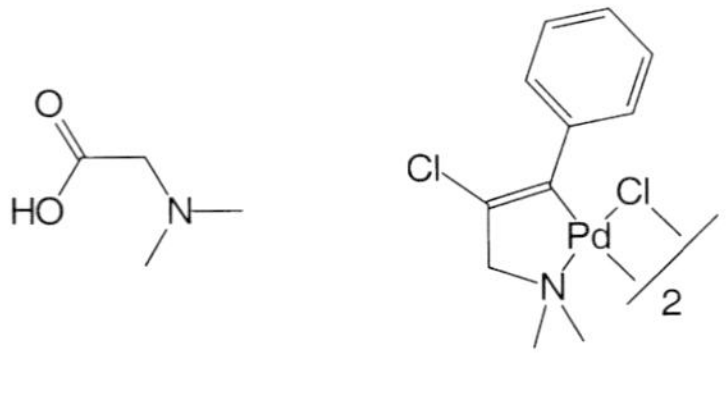

Fig. 7 Unusual catalyst systems for Heck reactions.

iodides, bromides, and activated chlorides [35 d, e]. These ligands are among the best ones known so far for olefination of chloroarenes, giving TONs in the region of 50 000 for 4-chlorobenzaldehyde, for example.

Two unusual homogeneous catalyst systems are worth mentioning (Fig. 7): Reetz et al. have shown that N,N-dimethylglycine (**26**) can prevent palladium precipitation and allows for turnover numbers up to ca. 100 000 for the coupling of bromobenzene and styrene [37 a]. On the other hand the cyclopalladated allylamine **27**, which is easily obtained by reacting palladium(II) chloride with the corresponding propargylamine, has been found to be active for the olefination of aryl iodides, aryl bromides, and electron-deficient aryl chlorides [37 b].

Interestingly, transition metals other than palladium can also be used as catalysts for selected Heck reactions. For example, Iyer et al. have shown that the vinylation of aryl iodides is catalyzed by $Cu(I)X$, $CoCl(PPh_3)_3$, $RhCl(PPh_3)_3$, or $IrCl(CO)(PPh_3)_2$ to give cinnamates and stilbenes in good yields [38]. Unfortunately, bromobenzene and chlorobenzene do not react with methacrylate in the presence of the Co, Rh, or Ir complexes, even at elevated temperatures. The first platinum-catalyzed Heck reaction was recently described by Kelkar et al., who showed that $Pt(COD)Cl_2$ (COD = 1,5-cyclooctadiene) catalyzes the vinylation of aryl iodides [39]. Nickel salts have also been employed for this reaction, although with limited success [40].

Besides Ar–X compounds, wherein X is a "simple" leaving group like halide, triflate or dinitrogen, arylcarbonic acid derivatives have been used as arene sources in Heck-type reactions (Scheme 4). Blaser and Spencer were the first to apply this novel concept in the palladium-catalyzed coupling of acid chlorides with olefins under CO elimination (Blaser-Heck reaction) [41]. The major advantage of this method is that a new class of substrates, i.e. carbonic acids, can be converted to olefins (after activation). This finding has broadened the scope of Heck-type reactions significantly.

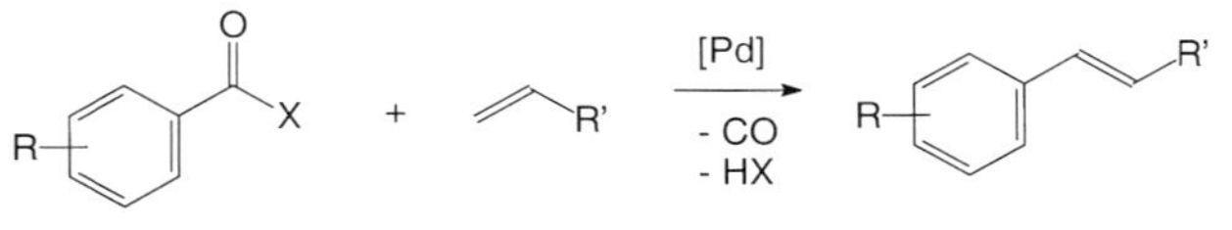

Scheme 4 Heck-type olefination with arylcarbonic acid derivatives.

It took more than 15 years before other acid derivatives were utilized for olefination reactions. Anhydrides of aryl- and heteroarylcarbonic acids have been applied in Heck-type couplings [42] as well as free carbonic acids, which are transferred to suitable mixed anhydrides *in situ*, e.g., by reaction with Boc anhydride [43]. Similarly, esters of arylcarbonic acids with electron-poor phenols (*p*-nitrophenol or pentafluorophenol) can be subjected to this decarbonylative olefination [44]. The supposed mechanism is, in principle, the same for all three reaction variants: after insertion of Pd(0) into the carbonyl–Cl/OR bond, forming an acylpalladium(II) complex, elimination of CO leads to an arylpalladium(II) species, which reacts with olefin in the usual manner. Obviously, on using anhydrides or esters as coupling partners, no stoichiometric amount of salt by-product is formed resulting from the leaving group and base. The "organic leaving groups" carboxylic acid and phenol, respectively, can be re-cycled or burned.

Another way to subject arylcarbonic acids to olefination reactions is to apply the free acids in the presence of (over) stoichiometric amounts of silver carbonate to re-oxidize Pd(0), which is obtained at the end of the coupling sequence consisting of the following steps: (1) decarboxylation of the initially formed Pd(II) carboxylate complex, (2) insertion of olefin into the arylpalladium(II) complex, (3) β-hydride elimination with concomitant product liberation, and (4) reductive elimination of HX [45]. The drawbacks of this protocol are the necessity for large amounts of silver(I) as oxidant and the low catalytic productivity (TON < 5).

2.13.4
Asymmetric Heck Reactions using Chiral Palladium Catalysts

The asymmetric Heck reaction has recently emerged as a powerful tool for the enantioselective synthesis of chiral compounds. The intramolecular asymmetric Heck reaction has received particular attention, although intermolecular variants are also known. Reaction conditions which promote the formation of cationic intermediates (Scheme 3, path 1) are typically employed. However, as discussed below, some substrates give higher enantioselectivities under conditions where the neutral pathway is favored.

Most of the early work on asymmetric Heck reactions is treated in several reviews [3, 46]. Hence, the following chapters will cover the highlights of the asymmetric Heck reaction with a particular focus on work since 1995.

In general, no stereogenic center is generated in the Heck reaction of simple monosubstituted olefins (Scheme 1). However, the strict *syn* carbopalladation of disubstituted olefins generates an intermediate stereogenic center (29). This stereogenic center is conserved only when no β-hydride elimination toward this new chiral center is possible. This is the case if (a) rotation around the carbon–carbon bond to generate the necessary *syn* orientation for dehydropalladation (β-hydride elimination) is not possible, (b) there are no hydrogen atoms on the new stereogenic sp^3 carbon atom (quaternary carbon), (c) the elimination of a β'-group (i.e. SiMe$_3$) is favored [47], (d) the elimination of a β'-hydrogen atom generates a

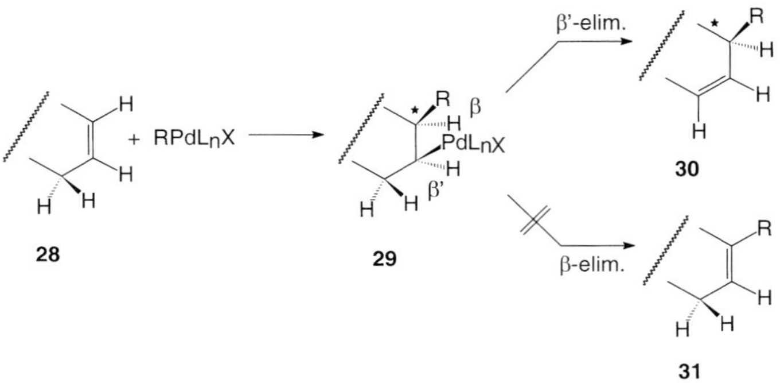

Scheme 5 Formation of a new stereogenic center in Heck reactions of cyclic disubstituted olefins.

thermodynamically more stable product or, (e) the palladium complex undergoes reactions other than β-hydride elimination. In general, requirement (a) is fulfilled by cyclic olefins, which were the first substrates for asymmetric Heck reactions (Scheme 5).

The first example of an asymmetric Heck reaction was reported by Shibasaki and co-workers in 1989 [48]. Based on the fundamental work of Larock and co-workers [49] on the Heck reaction with cyclic olefins, an intramolecular Pd-catalyzed C–C coupling reaction for the synthesis of *cis*-decalin derivatives was developed (Scheme 6). Optimization of the reaction conditions led to the realization of up to 90% *ee* for this process. Similar procedures were used to prepare hydrindanes and indolizidines [50]. In general, *in situ*-generated "Pd(BINAP)" (BINAP: 2,2′-bis(diphenylphosphino)-1,1′-binaphthyl) complexes were used for the coupling of aryl triflates [51]. Low *ee*s were observed when aryl iodides were utilized, presumably because of the dissociation of one of the phosphorus atoms of the chelating phosphine in the catalytically active species via path 2 in Scheme 3.

The related intermolecular reactions of aryl triflates **34** and 2,3-dihydrofuran (**35**) was discovered by Ozawa, Hayashi, and co-workers (Scheme 7) [52]. The 2-aryl-2,3-dihydrofuran product **36** was obtained in high *ee* along with the 2,5-dihydrofuran isomer **37** as the minor product. The reaction could also be extended to cyclic enamides [52b]. In addition, the alkenylation of cyclic olefins is possible [52e]. Migration of the double bond is not possible in the case of 2,2-disubstituted 2,3-dihydrofurans, which can be coupled to one single regioisomer with high *ee*s [53].

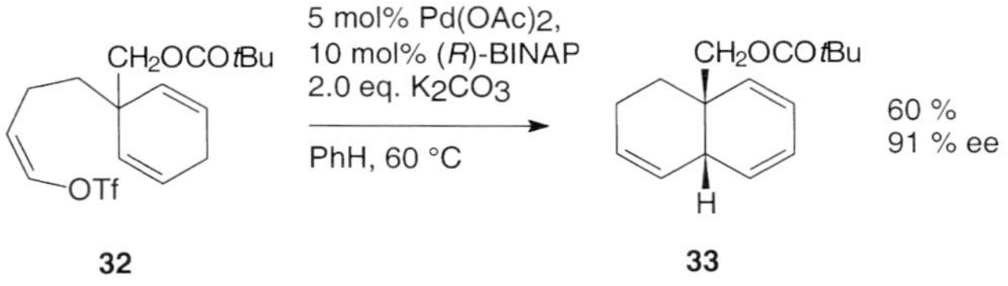

Scheme 6 Shibasaki's synthesis of *cis*-decalin derivatives.

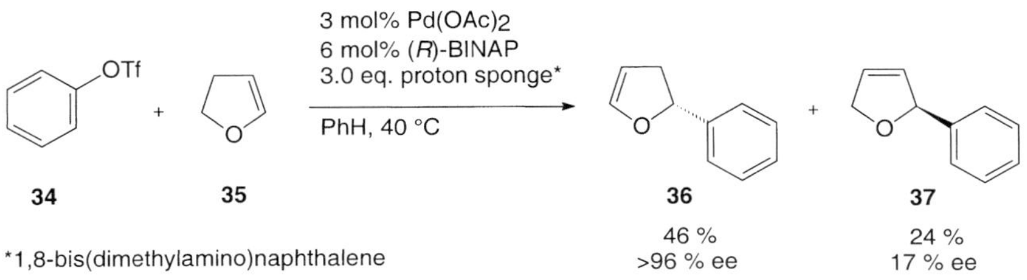

2.13.4 Asymmetric Heck Reactions using Chiral Palladium Catalysts

Scheme 7 Intermolecular asymmetric Heck reaction.

34 + 35 → 36 (46%, >96% ee) + 37 (24%, 17% ee)

Conditions: 3 mol% Pd(OAc)$_2$, 6 mol% (R)-BINAP, 3.0 eq. proton sponge*, PhH, 40 °C

*1,8-bis(dimethylamino)naphthalene

2.13.4.1
Mechanistic Features of Asymmetric Heck Reactions

Overman and co-workers used the Heck reaction with chiral palladium catalysts for the stereoselective synthesis of quaternary carbon atoms (Scheme 8) [54]. They reported high *ee*s for the asymmetric cyclization of aryl iodides in the absence of a silver salt. This finding was difficult to rationalize since cationic intermediates were believed to be necessary for high *ee*s. Otherwise the halide ion remains attached to the metal center and partial dissociation of the bidentate chiral ligand occurs, lowering the enantioselectivity of the process.

In an excellent mechanistic study on the cyclization of the triflate **41** and iodide **42**, Overman and co-workers were able to demonstrate significantly higher *ee*s for the cyclization of the aryl iodide **43** compared to the triflate **41** (Scheme 9) [55]. The enantioselection of cyclization of aryl triflate **41** paralleled that of iodide **42** when halide ions were added to the reaction.

To gain a better mechanistic understanding, the cyclization of iodide **42** with three monodentate analogs of the (R)-BINAP ligand were tested. Under otherwise identical conditions, low *ee*s (<27%) were obtained. This indicates that (R)-BINAP

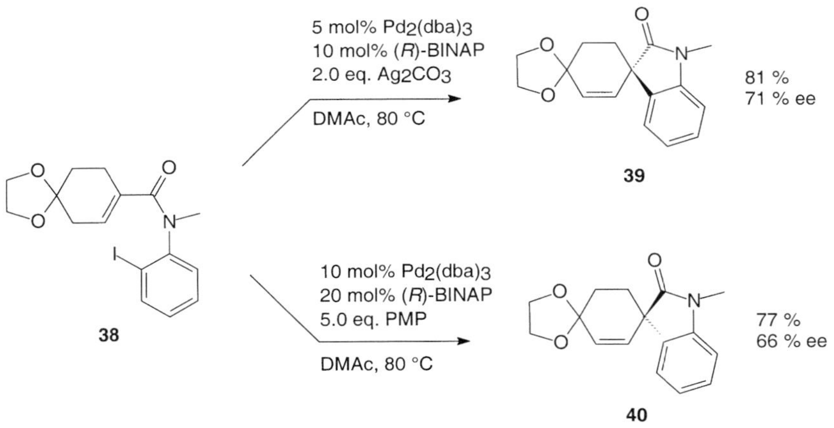

Scheme 8 Overman's Heck coupling for the synthesis of quaternary carbon atoms.

38 → 39 (81%, 71% ee) with 5 mol% Pd$_2$(dba)$_3$, 10 mol% (R)-BINAP, 2.0 eq. Ag$_2$CO$_3$, DMAc, 80 °C

38 → 40 (77%, 66% ee) with 10 mol% Pd$_2$(dba)$_3$, 20 mol% (R)-BINAP, 5.0 eq. PMP, DMAc, 80 °C

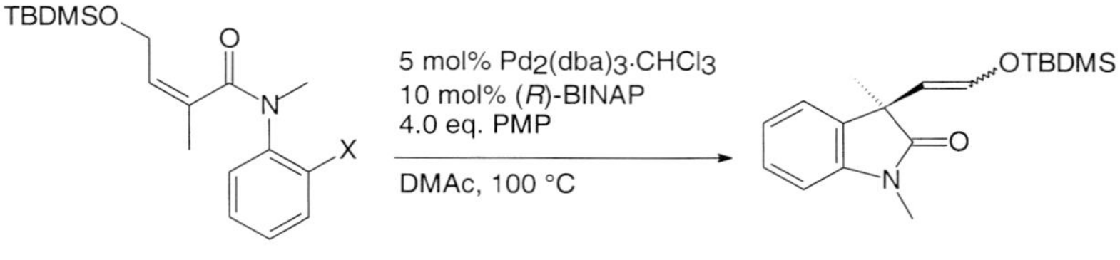

41 X = OTf
42 X = I

43
72 %, 43 % ee from **41**
76 %, 91 % ee from **42**

Scheme 9 Mechanistic studies on the asymmetric Heck reaction.

must remain coordinated via both phosphorus atoms to achieve high enantioselective induction. The enantioselection determining step in this reaction is likely to be the binding of the enantiotopic C=C face to form intermediate **46**. Subsequent carbopalladation generates the new chiral center. To explain the higher enantioselection of iodides compared to triflates, the authors propose a higher C=C-enantioface binding selectivity when aryl iodides are used. As shown in Scheme 10, the new five-coordinate intermediate **45**, which is not observed when triflates are employed, is proposed. The dissociation of halide ions from **45** should be an associative process, while dissociation of triflate is likely to be a dissociative process. Hence, higher enantioface binding selectivity and higher *ee*s should occur with halides. The involution of a fifth coordination site in some intramolecular asymmetric Heck reactions should open an interesting field for the design of new chiral ligands.

The mechanism of the intermolecular Heck reaction of 2,3-dihydrofuran with phenyl triflate (Scheme 7) was the subject of a detailed NMR study by Brown and co-workers [56]. The reaction of the (*S*)-BINAP palladium aryl triflate complex **48** was studied in THF from –70 °C to –40 °C (Scheme 11). The only detectable intermediate is compound **51**, which slowly forms the Heck products at –30 °C. The authors postulate an intramolecular dyotropic shift, since there is no exchange with an excess of 2,3-dihydrofuran. Compound **49** is stabilized by a coordination of the oxygen to the cationic metal center forming an oxapalladacyclopropane structure.

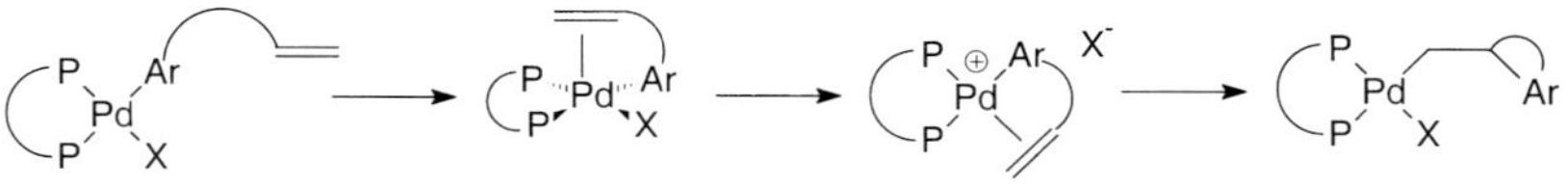

44 **45** **46** **47**

Scheme 10 Mechanistic explanation of the halide influence in asymmetric Heck reactions.

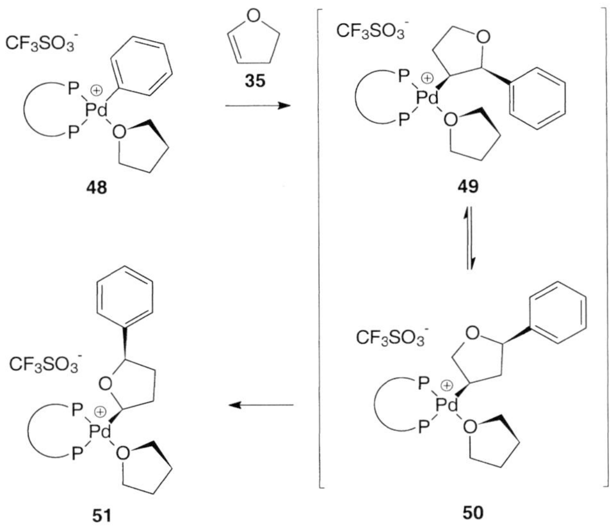

Scheme 11 Mechanistic studies on the Heck reaction of 2,3-dihydrofuran.

2.13.4.2
New Catalyst Systems for Asymmetric Heck Reactions

Despite interesting recent developments, most synthetic chemists still use *in situ* mixtures of a palladium source (Pd(OAc)$_2$, Pd$_2$dba$_3$, etc.) and BINAP as chiral Heck catalysts. Shibasaki's group invented (R)-2,2'-bis(diphenylarsino)-1,1'-binaphthyl ((R)-BINAs), the arsine analog of BINAP, as a chiral ligand and tested it in asymmetric Heck reactions [57]. In the intramolecular asymmetric synthesis of decalin derivatives, BINAs was found to be superior with regard to yield and *ee* compared to BINAP when vinyl iodides were used. On the other hand, BINAs was found to be less effective than BINAP for the cyclization of alkenyl triflates.

Another ligand of the 1,1'-biaryl type, which has found interest in asymmetric Heck reactions, is bis(3,5-di-*tert*-butyl)-MeO-BIPHEP (**52**), originally prepared by Schmid and co-workers [58a]. Pregosin and co-workers tested this ligand in the intermolecular reaction of 2,3-dihydrofuran with aryl triflates (Scheme 7) [58b,c]. By using 3 mol% Pd(OAc)$_2$ and 6 mol% of the ligand, the authors isolated the arylated 2,3-dihydrofuran with a chemical yield of 65% and an *ee* of >98%. The other double bond isomer is formed in low yields only (3%). Thus, this ligand is superior to the classical Pd/BINAP system for this model reaction. Furthermore, the analog 7,7'-dimethoxy-2,2'-bis(diphenylphosphino)-1,1'-binaphthalene [59a] or bis(diphenylphosphino) substituted 3,3'-bithiophene derivatives such as **53** [47b] sometimes give improved *ee*s compared to BINAP in intra- and intermolecular asymmetric Heck reactions.

Pfaltz and co-workers demonstrated that phosphinooxazolines are effective ligands for enantioselective Heck reactions [46c] and enantioselective alkylations of

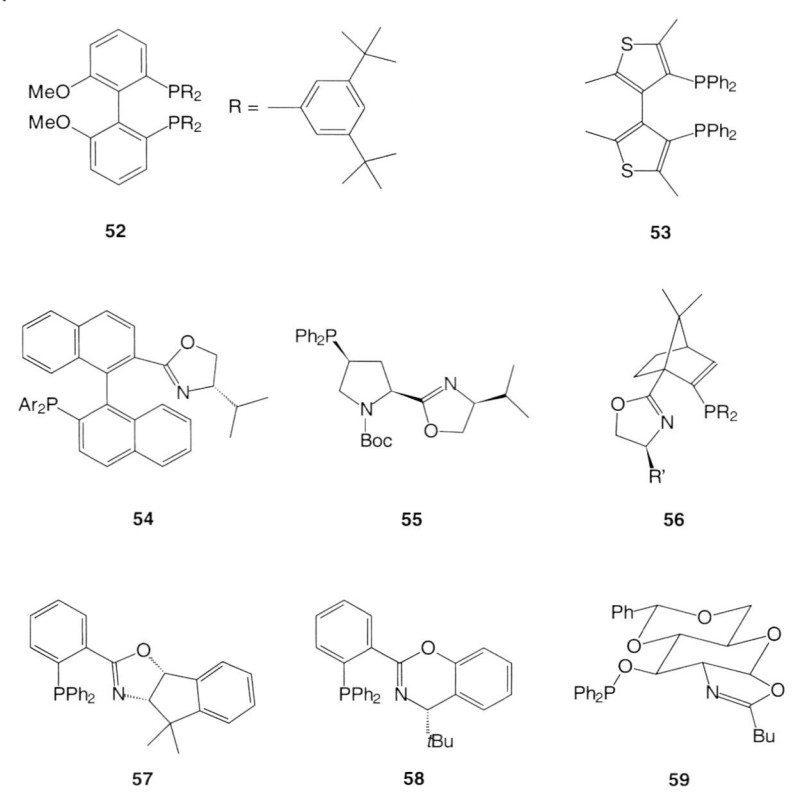

Fig. 8 Chelating ligands for asymmetric Heck reactions.

malonates and analogous systems [60]. With the *tert*-butyloxazoline derivative **62**, a yield of 98% and an *ee* of 99% was obtained for the model reaction of 2,3-dihydrofuran and cyclohex-1-enyl triflate (**60**) in the presence of N,N-di-*iso*-propylethylamine (Scheme 12) [61].

In contrast to the Pd/BINAP system, the Pfaltz system exhibited a low tendency to promote the isomerization of the double bond. Thus, distribution of regioisomers can differ greatly using these catalyst systems, making the P,N-ligands often superior. For example, Pfaltz and co-workers have been able to arylate cyclopentene with high selectivity (80% yield, 86% *ee*, 99:1 regioselectivity).

Interestingly, with ligand **62**, high enantioselectivities (up to 92%) could be achieved at high temperature (120–160 °C) with microwave heating [62]. The reactions proceed comparatively fast (<8 h), and the ratios for the double bond isomers are generally >90:10, depending on the aryl triflate applied. If the phenylene backbone in **62** is substituted by a ferrocenyl moiety, thus incorporating a second stereogenic element into the ligand, high selectivities for the 2-arylated 2,5-dihydrofuran isomer **61** are observed, but *ee*s are somewhat lower than with "standard chiral li-

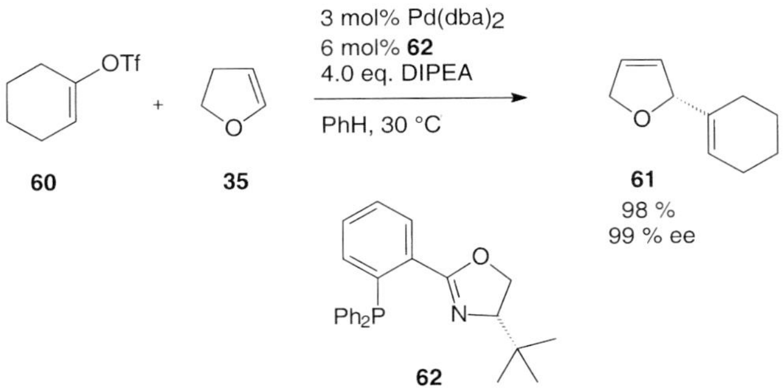

Scheme 12 Phosphinooxazoline ligand for asymmetric Heck reactions.

gands" (<93%) [63]. Similarly, oxazolinyl-substituted binaphthyl ligands **54** result in *ee*s up to 96% [64], showing significant "meta effects" of the Ar groups. A comparison of ligands with opposite axial chirality revealed that the enantioselectivity is mainly determined by the binaphthyl backbone and only to a minor extent by the oxazoline subunit. The proline-based ligand **55** has been introduced by Gilbertson et al. for asymmetric Heck reactions [65 a, b]. Enantioselectivities up to 86% have been described for the coupling of 2,3-dihydrofuran and 1-cyclohexenyl triflate (Scheme 12). Here again, not the stereochemistry within the oxazoline ring but that within the proline ring is decisive for enantioselection in the Heck coupling. Ligands of type **56**, which can be prepared from ketopinic acid, show moderate to very good enantioselectivities in the same test reaction (56–94% *ee*) [65 c].

A Pfaltz-analog ligand **57** with an indane fragment on the oxazoline moiety has also been synthesized and tested in the asymmetric Heck reaction depicted in Scheme 7 [66]. Enantioselectivities are excellent, being similar to the original ligand (up to 98%). Another modification of the N-donor part of the ligand was described by Kündig and co-workers. They used a benzoxazine instead of an oxazoline moiety (**58**), and found that this new ligand gives 91% *ee* in the standard test reaction (Scheme 7, 79% yield) [67]. The chiral environment at the metal center of the catalyst has been shown to be very similar to that of the corresponding oxazoline ligands, which explains the reaction outcome. A phosphinite-oxazoline ligand derived from D-glucosamine (**59**) also gives excellent enantioselectivities (up to 96%) at high yields [68].

A novel ligand concept for asymmetric arylations has been introduced by Zhou and co-workers in 2000 [69]. They used quinolinyloxazolines **63** for the hydroarylation of norbornene with phenyl iodide. The *exo*-product is formed selectively with reasonable yields and enantioselectivities (<75% *ee*) when formic acid/triethylamine is used as hydride source and DMSO as solvent.

Finally, there is a first example of a monodentate ligand for effective asymmetric Heck reaction: the phosphoramidite **64** gives excellent *ee*s of 93–96% in the intramolecular coupling depicted in Scheme 13 [70]. When sterically demanding amines are used as the base (e.g., *N*,*N*-dicyclohexylmethylamine) 70–75% of the desired product has been isolated after two days.

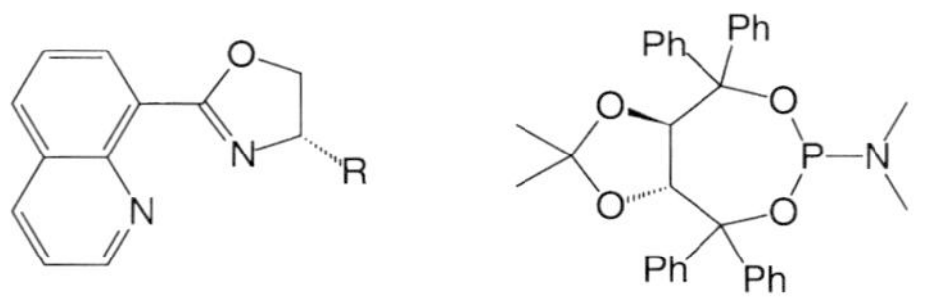

63

64

Fig. 9 Novel ligands for asymmetric arylation reactions.

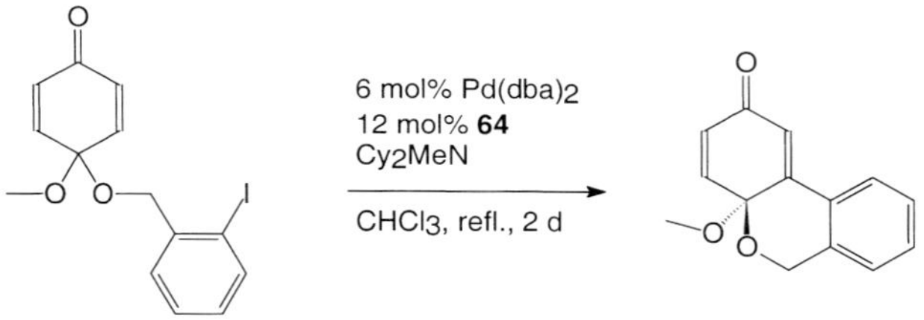

65 → **66**

6 mol% Pd(dba)₂
12 mol% **64**
Cy₂MeN

CHCl₃, refl., 2 d

Scheme 13 Intramolecular asymmetric Heck reaction with a monodentate ligand.

2.13.5
Recent Applications of Heck Reactions for the Synthesis of Natural Products, Complex Organic Building Blocks and Pharmaceuticals

The last decade has seen an extraordinary growth in the use of stereoselective palladium-catalyzed olefinations of aryl and vinyl derivatives, both for complex total synthesis and for valuable organic building blocks. In most cases, intramolecular variants offer shorter ways to assemble structurally simple precursor molecules to congested polycyclic frameworks. Because of the availability of excellent reviews up to 1995 [3a,e,f, 71] mainly work after 1995, but also including some earlier highlights, will be described in the following section.

The synthetic potential of intramolecular Heck reactions is elegantly shown by the work of Overman and his group. As an example, the synthesis of morphine (**69**) deserves special attention [71, 72]. The pivotal cyclization step of the bicyclic precursor **67** to form the tetracyclic morphinan skeleton **68** was accomplished in 60% yield using a catalyst system consisting of 10 mol% Pd(OCOCF₃)₂(PPh₃)₂ in refluxing toluene in the presence of 1,2,2,6,6-pentamethylpiperidine. The resulting intermediate **68** was then transformed into (−)-morphine (**69**) (Scheme 14).

An even more efficient access to pentacyclic opiates utilizes the domino intramolecular Heck insertion-heterocyclization of a suitable trisubstituted 1,3-diene (**70**) [73]. The reaction sequence starts with the Heck reaction to form the tetracyclic π-allyl palladium intermediate **71**, which then undergoes a nucleophilic attack of the tethered alcohol on the π-allyl group to generate the key intermediate **72** (Scheme 15).

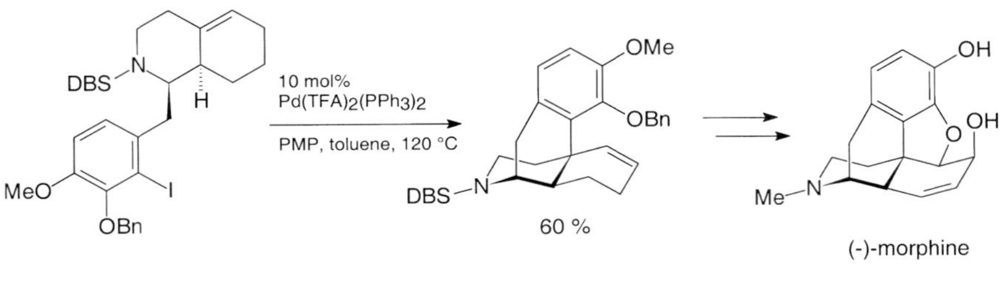

Scheme 14 Synthesis of morphine via intramolecular Heck reaction.

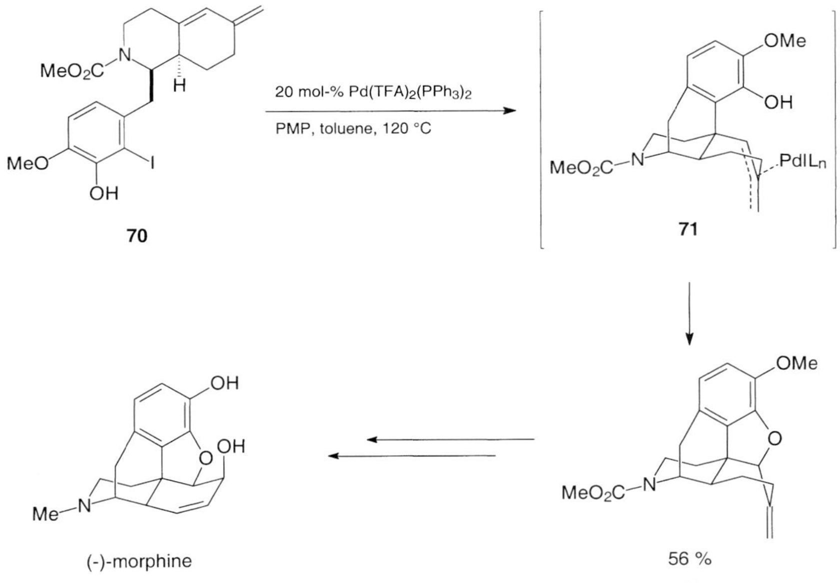

Scheme 15 Synthesis of morphine via a domino process.

Other groups also used intramolecular Heck reactions as a key step for the synthesis of morphine fragments. In this respect, Cheng et al. developed a facile construction of tricyclic and tetracyclic morphine fragments [74]. As an example, the palladium-catalyzed cyclization of **73** to afford the tricyclic ANO fragment **74** is shown in Scheme 16.

Other examples of the use of palladium-catalyzed cyclizations of aryl and vinyl halides with tethered alkenes to form quaternary carbon centers were developed by the Overman group and include the total syntheses of 6a-epipretazettine, tazettine, and scopadulcic acid A and B [75]. As shown in Scheme 17, the key step of the first total synthesis of a scopadulan diterpene is the double Heck cyclization

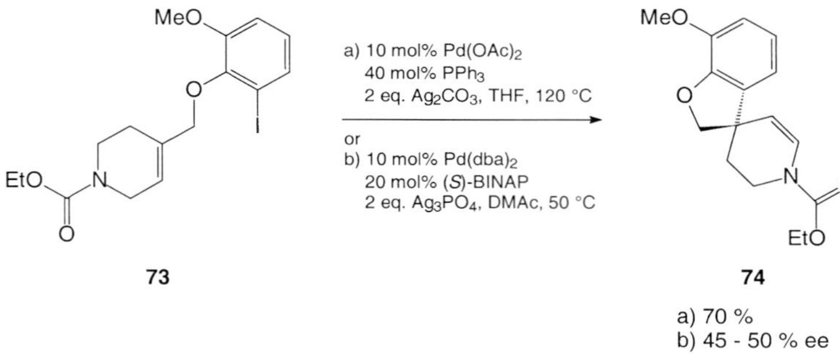

Scheme 16 Synthesis of morphine fragments.

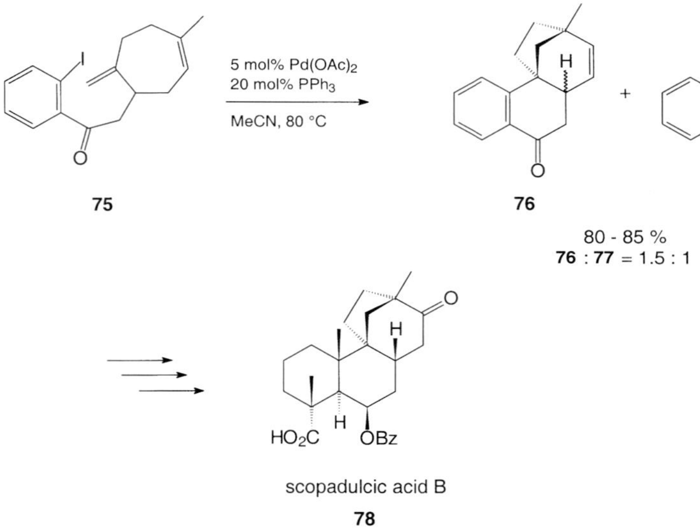

Scheme 17 Total synthesis of racemic scopadulcic acid B.

of dienyl aryl iodide (**75**) to form the tetracyclic enones **76** and **77** in 80–85% combined yield. This critical cyclization was accomplished with a variety of Pd(0) catalysts. Using a catalyst system that had previously been found to be effective in minimizing double bond migration (10 mol% Pd(OAc)$_2$, 2–4 equiv. PPh$_3$, 2 equiv. Ag$_2$CO$_3$ in refluxing acetonitrile), the products were obtained in variable yields of 30–60% as a 3:1 mixture of enones **76** and **77**. Higher yields (80–85%) were obtained using 5 mol% Pd(OAc)$_2$, 20 mol% PPh$_3$, and an excess of triethylamine in refluxing acetonitrile.

Moreover, the power of the palladium-catalyzed selective construction of C–C bonds has been demonstrated in the natural product synthesis of (+)-vernolepin (**79**) [50e,f], (–)-physostigmine (**80**) [54c], dehydrotubifoline (**81**) [76], FR-900482

2.13.5 Recent Applications of Heck Reactions for the Synthesis of Natural Products, Complex Organic

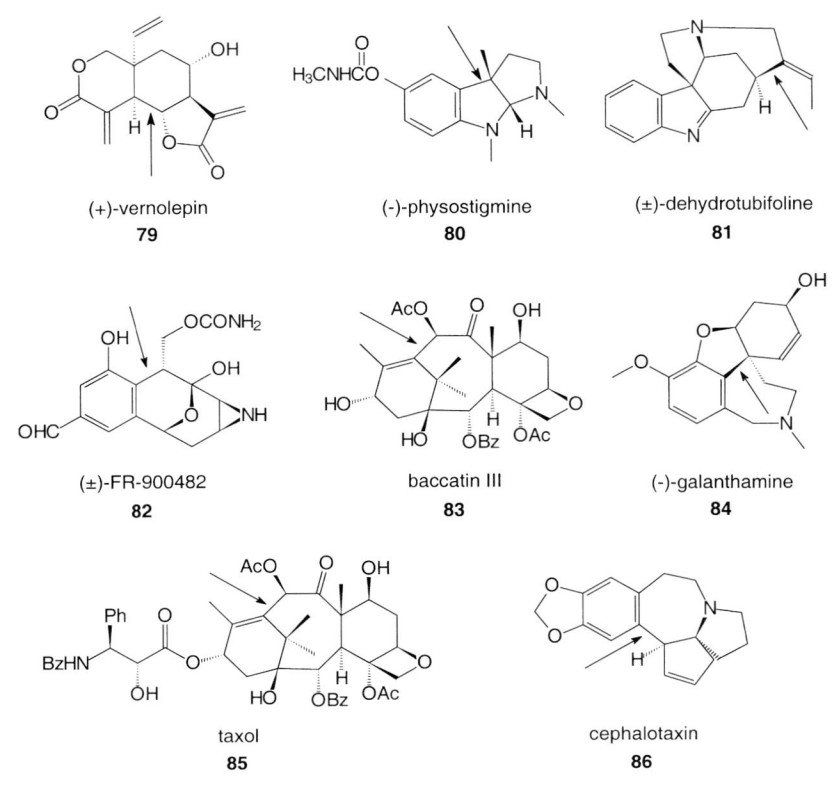

Fig. 10 Selected natural products synthesized using an intramolecular Heck reaction as a key step (arrows indicate the C–C-bond built by the Pd-catalyzed reaction).

(+)-vernolepin **79**, (−)-physostigmine **80**, (±)-dehydrotubifoline **81**, (±)-FR-900482 **82**, baccatin III **83**, (−)-galanthamine **84**, taxol **85**, cephalotaxin **86**

(**82**) [77a,b], baccatin III (**83**) [77c], taxol (**85**) [77c], and recently (−)-galanthamine (**84**) [78], and cephalotaxin (**86**) [79] (Fig. 10).

As mentioned earlier, a disadvantage of a number of palladium-catalyzed C–C bond-forming reactions is the low selectivity with regard to the position of the double bond in the product. Thus, the selective formation of tertiary stereogenic centers was difficult in the past. A solution to this serious problem was developed by Tietze and co-workers, who showed that intramolecular asymmetric Heck reactions can be carried out with high regioselectivity of the double bond formation through the use of an allylsilane as alkene component [47]. In such cases, the trimethylsilyl group is preferentially eliminated and directs the regiochemistry of the double bond formation. Without the terminating trimethylsilyl group, a mixture of several double bond isomers is formed under the reaction conditions. The use of this strategy is shown in an elegant enantioselective total synthesis of the natural product norsesquiterpene 7-demethyl-2-methoxycalamene (**90**) (Scheme 18) [47a,b]. The intramolecular Heck reaction of the allylsilane **88** was accomplished with a catalyst system containing 2.5 mol% of Pd$_2$(dba)$_3$ · CHCl$_3$, 7.0 mol% (*R*)-BI-

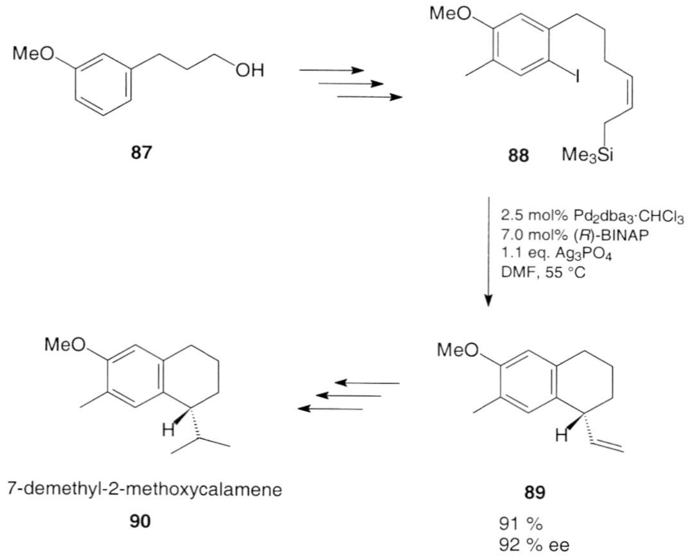

Scheme 18 Synthesis of 7-demethyl-2-methoxycalamene.

NAP and 1.1 equiv. Ag$_3$PO$_4$ in DMF at 80 °C to give the vinyl-substituted tetralin **89** in 91% yield with 92% ee. Three further steps gave pure **90**.

Another question concerning selectivity is the regioselectivity in the cyclization of substrates containing two different double bonds allowing the formation of a five- or a six-membered ring, respectively. Shibasaki and co-workers investigated the cyclization of **91** to **92** and **93**, which are versatile synthetic intermediates for a number of diterpenes such as kaurene (**94**), abietic acid (**95**), and brucetin analog (**96**) (Scheme 19) [80]. The cyclization of diene **91** can follow either a 5-*exo* or a 6-*exo* mode. The 7-*endo* and 6-*endo* insertion modes are not favored because of the severe torsional strain that would result from "in-plane" coordination of the olefin with the cationic intermediate. Assuming a square planar cationic Pd(II) intermediate and an "in-plane" coordination of the olefin, molecular models predicted that the 6-*exo* cyclization path was favored. Indeed, cyclization of **91** gave the 6-*exo* product **92** and the double bond isomer **93** in 62% yield and 95% ee.

Regarding regioselectivity, the regiochemical outcome of palladium-catalyzed cyclizations of starting materials containing 1-halo-5-ene subunits nearly always favors the 5-*exo-trig* pathway in those cases where competition between 6-*endo-* and 5-*exo-trig* closures is possible. However, indoline-based substrates favor ring closure through a 6-*endo* pathway to afford quinolones (Scheme 20) [81].

Other examples of the change from 5-*exo* to 6-*endo* preference were recently reported for the intramolecular carbopalladation of 1,6-enynes by modifying the catalyst [82] as well as the cyclization of vinylic halides using water-soluble Pd/TPPTS catalysts [19c].

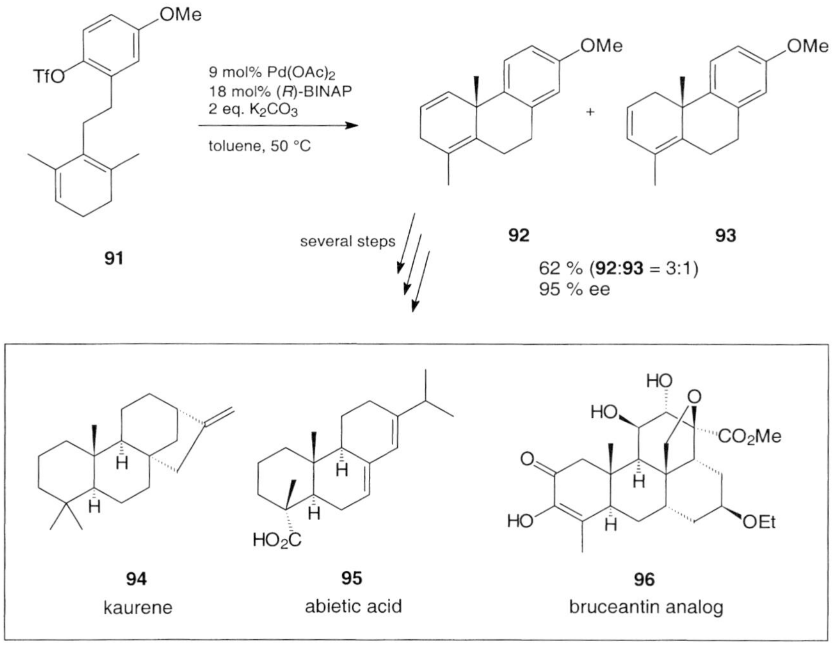

2.13.5 Recent Applications of Heck Reactions for the Synthesis of Natural Products, Complex Organic

Scheme 19 Intramolecular Heck reaction as key step in the synthesis of several natural products.

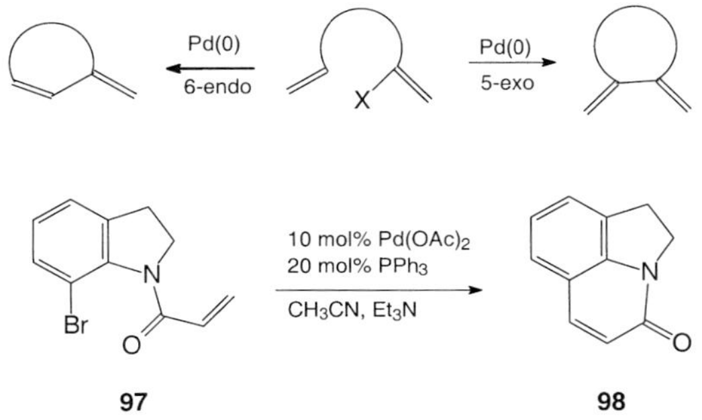

Scheme 20 Palladium-catalyzed cyclization. Example of a 6-*endo-trig* intramolecular cyclization.

Apart from complex natural product synthesis, the Heck reaction has become increasingly important as a general approach to new classes of pharmaceuticals. A recent example from an industrial group demonstrates the usefulness of this strategy. Larsen et al. used the Heck coupling for the synthesis of a new LTD$_4$ receptor antagonist L-699-392 (**102**), which is a possible drug for the control of asthma (Scheme 21) [83]. The palladium-catalyzed coupling of the styrene derivative **99**

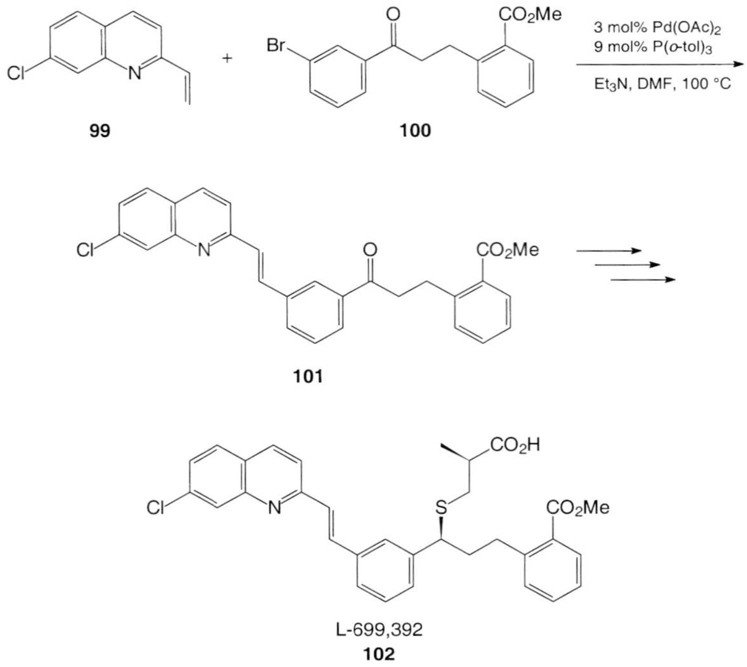

Scheme 21 Synthesis of L-699,392.

was performed with the aryl bromide **100** in the presence of 3 mol% Pd(OAc)$_2$ in DMF. To couple the bromide substrate, the addition of a phosphine ligand was necessary. Here, tri-o-tolylphosphine gave better yields than those with triphenylphosphine.

As discussed before, the Heck reaction terminates the catalytic cycle by a β-hydride elimination. However, if the β-hydride elimination pathway becomes unfavorable or even impossible, the resulting alkylpalladium species can undergo further reactions. Interestingly, it has been reported that solvent effects can to some extent control the competing pathways of further alkene insertion versus β-hydride elimination. For example, aqueous media enable double Heck reactions [84]. The enormous popularity of the Heck methodology in organic synthesis is significantly due to the manifold possibilities of the resulting domino sequences. In this respect, Heck olefination–olefin reduction, olefination–olefination, olefination–cross coupling, olefination–amination, alkynylation–olefination, alkynylation–cross coupling, alkynylation–reduction and many more cascade reactions have been developed. A special case of a domino sequence, although β-hydride elimination takes place, is the Heck olefination–olefin isomerization reaction. Important developments to cascade reactions involving the Heck reaction as one step were made by a number of groups including those of Dyker [85], Grigg [86], Larock [87], de Meijere [88], Negishi [3e, 89], Tietze [79, 90] and many more.

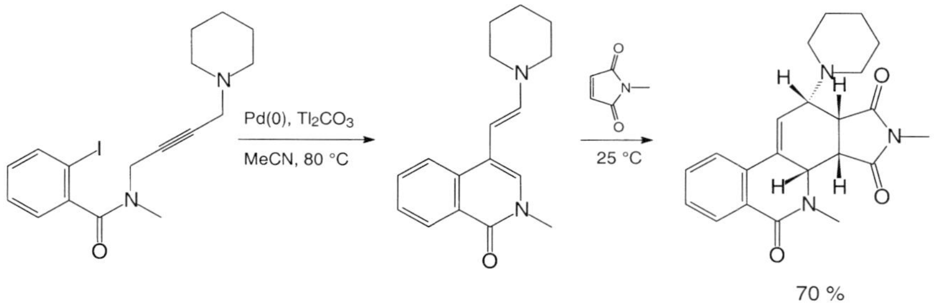

2.13.5 Recent Applications of Heck Reactions for the Synthesis of Natural Products, Complex Organic

Scheme 22 Palladium-catalyzed cyclization–anion capture process.

Grigg and co-workers developed very useful and efficient palladium-catalyzed cyclization–anion capture processes [86]. Although a number of elegant contributions of this group are known, only one example will be illustrated here. The alkyne **103** undergoes a palladium-catalyzed cyclization and Tl$_2$CO$_3$-mediated reaction furnishing the diene **104**, which could be trapped by intermolecular Diels-Alder reaction to give the tetracyclic product **105** in 70% yield (Scheme 22).

In 1993 Shibasaki and co-workers reported an asymmetric domino reaction [91 a, b]. In the initial reports the reaction was limited to acetate and benzylamine as nucleophiles but in 1996 the authors reported the total synthesis of (–)-$\Delta^{9(12)}$-capnellene (**109**) utilizing an asymmetric Heck reaction–carbanion capture process as the key step [91 c]. In this report, the vinyl triflate **106** was cyclized in the presence of various carbanions. For example, the reaction of **106** with malonate-derived carbanion **107** as nucleophile gave **108**, the key intermediate in the total synthesis of **109**, with 75% yield and an *ee* of 66% (Scheme 23).

Cascade processes of vinyl or aryl halides with multiple double or triple bonds achieve dramatic increases in molecular complexity. The asymmetric carbopalladation followed by further reactions of the formed alkylpalladium species was used by Keay and co-workers in the total synthesis of (+)-xestoquinone (**113**, Scheme 24) [92]. The key step in this synthesis is an asymmetric palladium-catalyzed polyene cyclization of the aryl triflate **110**. After the insertion of the first olefin, the palladium intermediate inserts a second olefin intramolecularly and then undergoes a β-hydride elimination to form **113**.

The indole nucleus is an important structural element of a number of natural products and medicinal agents. Chen et al. developed an interesting domino process for the synthesis of substituted indoles based on the palladium-catalyzed coupling of *o*-iodoanilines with ketones (Scheme 25) [93]. The reaction proceeds by initial enamine formation followed by an intramolecular Heck reaction. It was found that an excess of an amine base – in general DABCO was used – is critical to the successful coupling.

Steglich and co-workers applied elegantly simple Heck reactions as well as more complicated domino processes for the synthesis of the slime mold alkaloid arcyriacyanin A [94]. One approach mimics the possible biosynthesis of arcyriacya-

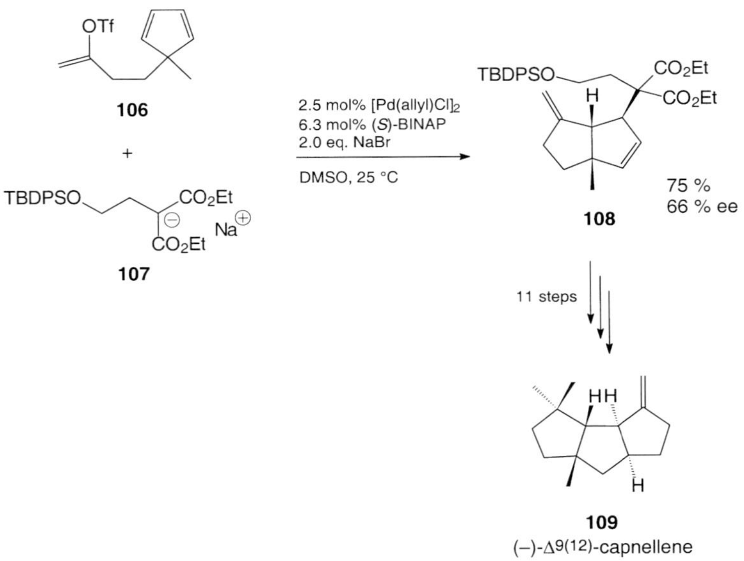

Scheme 23 Total synthesis of (−)-Δ$^{9(12)}$-capnellene.

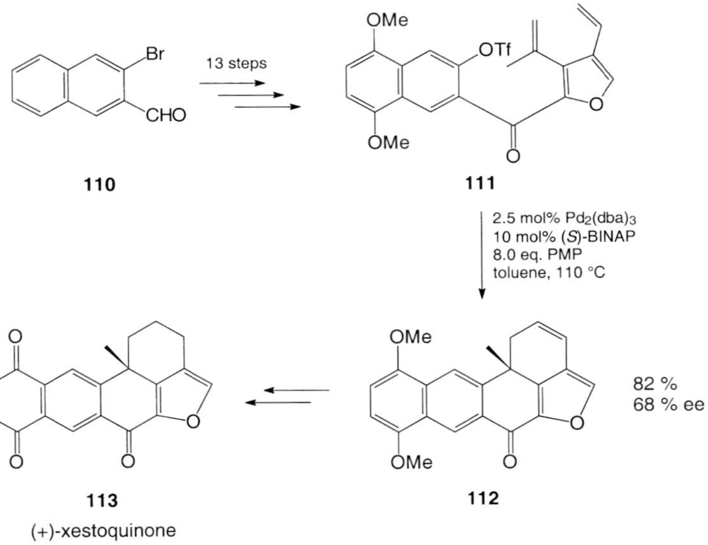

Scheme 24 Total synthesis of (+)-xestoquinone.

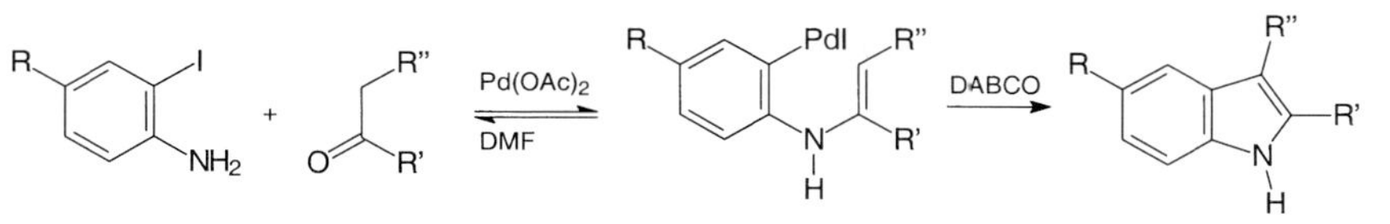

2.13.5 Recent Applications of Heck Reactions for the Synthesis of Natural Products, Complex Organic

Scheme 25 Synthesis of indoles via a domino process.

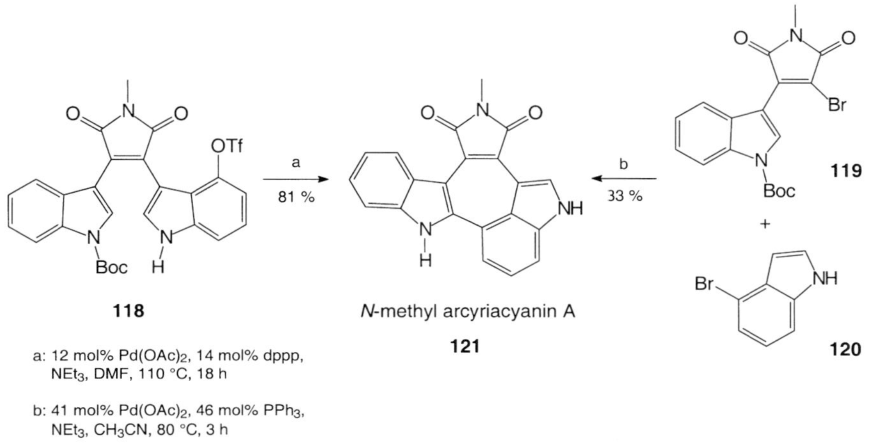

Scheme 26 Synthesis of arcyriacyanin A.

a: 12 mol% Pd(OAc)₂, 14 mol% dppp, NEt₃, DMF, 110 °C, 18 h

b: 41 mol% Pd(OAc)₂, 46 mol% PPh₃, NEt₃, CH₃CN, 80 °C, 3 h

nin A through an intramolecular Heck reaction of the bridged bisindole triflate **118**. Here, the Heck cyclization took place in DMF in the presence of Pd(OAc)$_2$ and 1,3-bis(diphenylphosphino)propane as pre-catalyst. Interestingly, the coupling reaction proceeded with excellent yield (81%) despite unfavorable steric conditions required for the *syn* elimination of "H–Pd–X". It is proposed that the *syn* β-hydride elimination occurs by a base-catalyzed fragmentation. As indicated in Scheme 26, the key step in another approach to arcyriacyanin A is a domino Heck reaction between bromo(indolyl)maleimide **119** and 4-bromoindole **120**. By means of this chemoselective double Heck reaction *N*-methylarcyriacyanin A (**121**) is obtained in 33% yield.

Significant contributions on the use of domino Heck processes for the synthesis of steroids and related molecules were disclosed by Tietze and co-workers [79, 90]. The synthesis of estrone derivatives was easily achieved by double Heck reaction of **122** with a substituted hexahydro-1H-indene (**123**) (Scheme 27) [90a]. Best yields for the first step were obtained by using palladacycle **10**.

Domino processes using the Heck reaction as the second step are also known. For example, in their study on the total synthesis of the marine natural product halenaquinol **130**, Shibasaki and co-workers introduced a cascade reaction of a Su-

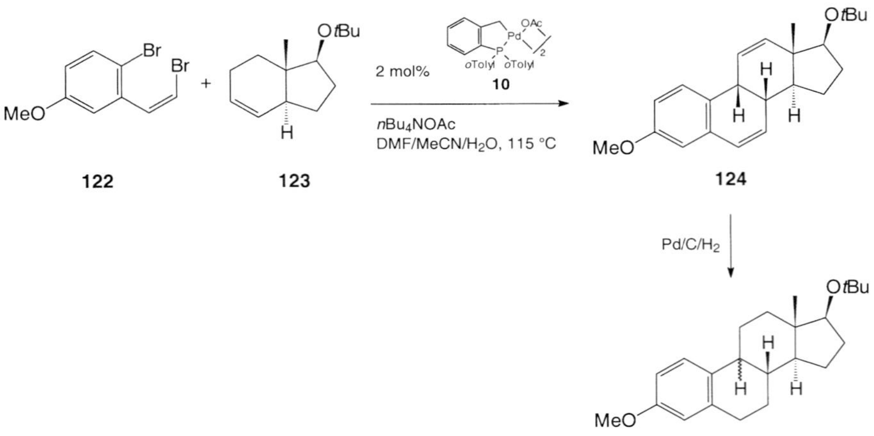

122 123 124

125

Scheme 27 Synthesis of estrone derivatives by a double Heck reaction.

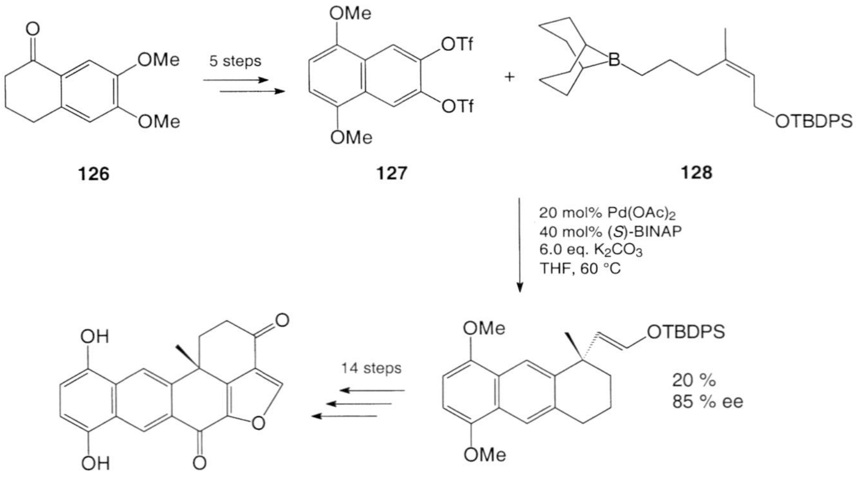

126 127 128

halenaquinol

130

129

Scheme 28 Total synthesis of halenaquinol.

zuki cross-coupling and an asymmetric Heck reaction [95]. The bistriflate **127** was coupled with the alkylborane **128** and gave the non-racemic **129** in a single step (20% yield, 85% *ee*) in the presence of a $Pd(OAc)_2/(S)$-BINAP catalyst system (Scheme 28).

2.13.6
Miscellaneous

Although numerous examples of Heck reactions utilizing ArX and alkenes are known, few examples where the alkene-coupling partner is part of an aromatic ring have been developed. On the one hand, electron-rich aromatics such as indole or thiophene react even intermolecularly with aryl or vinyl halides [96]. On the other hand, intramolecular couplings of phenolates with aryl or vinyl bromides proceed smoothly under basic conditions with palladium catalysts (Scheme 29) [97]. The recently developed palladacycle **10** is particularly suited as catalyst [23]. The reaction is expected to proceed via oxidative addition followed by nucleophilic attack of the ambient phenolate anion to afford a diaryl palladium species. Subsequent reductive elimination gives the observed product in good to excellent yield.

Other recent examples of Heck reactions formally involving CH-activation steps include the arylation of 2-hydroxybiphenyl derivatives [98] and the well-known coupling of aryl halides with norbornene [99].

A recent application of an unusual Heck reaction for natural product synthesis is reported by Steglich and co-workers (Scheme 30). The key step of the synthesis of lamellarine G is the intramolecular cyclization of the pentasubstituted pyrrole

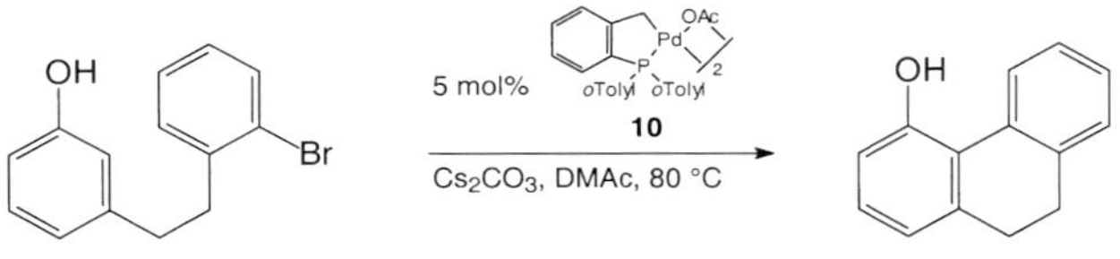

Scheme 29 Intramolecular coupling of phenols with aryl halides.

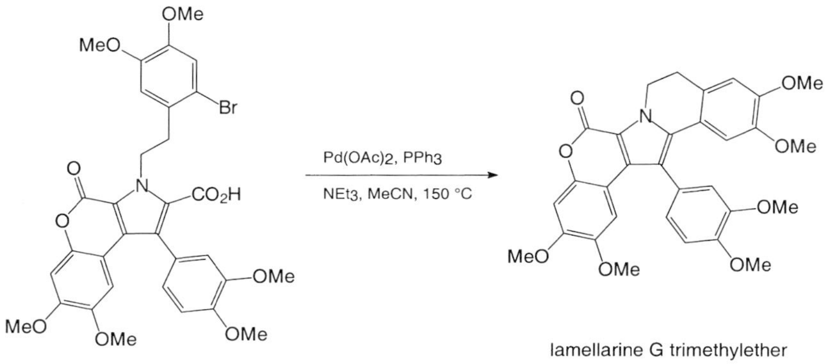

Scheme 30 Synthesis of lamellarine G.

carboxylic acid derivative **133** [100]. This is the first example of a Heck reaction where the coupling of the arylpalladium(II) derivative with the olefin takes place via extrusion of CO_2. Interestingly, the coupling step using the carboxylic acid proceeds with better yields than those achieved in the cyclization of the decarboxylated pyrrole derivative. However, stoichiometric amounts of palladium seem to be necessary for the successful coupling.

2.13.7
Concluding Remarks

From its discovery in the late 1960s, the Heck reaction has become one of the most versatile and powerful tools for the selective construction of carbon–carbon bonds starting from olefins. The method has found numerous applications on the laboratory scale in the synthesis of natural products, fine chemicals, and pharmacologically interesting compounds. The almost unique ability of palladium to catalyze various reactions has been an especially fruitful principle for the design of appropriate domino reactions.

Especially in the last decade, there has been tremendous progress in catalyst efficiency for Heck reactions of simple terminal olefins. However, most Heck reactions described in natural product synthesis still need 5–20 mol% catalyst. In addition to more active catalysts, more selective palladium catalysts which prevent side reactions, e.g., olefin isomerization, are needed. The selective Heck reaction of terminal aliphatic olefins such as hexene or octene is still an unsolved problem. Finally, more applications in natural product synthesis using aryl chlorides, bromides, mesylates, and diazonium salts instead of the more expensive and reactive aryl iodides and triflates are needed to make the Heck reaction even more useful and economical in the years to come.

2.13.8
References

1 (a) R. F. Heck, *J. Am. Chem. Soc.* **1968**, *90*, 5518; (b) R. F. Heck, *Palladium Reagents in Organic Synthesis*, Academic Press, New York, **1985**.

2 (a) L. Cassar, *J. Organomet. Chem.* **1975**, *93*, 253; (b) H. A. Dieck, R. F. Heck, *J. Organomet. Chem.* **1975**, *93*, 259; (c) M. Alami, F. Ferri, G. Linstrumelle, *Tetrahedron Lett.* **1993**, *34*, 6403; (d) M. Pinault, Y. Frangin, J. P. Gent, H. Zamarlik, *Synthesis* **1992**, 746; (e) W. A. Herrmann, C.-P. Reisinger, K. Öfele, C. Brossmer, M. Beller, H. Fischer, *J. Mol. Catal.* **1996**, *108*, 51.

3 Excellent reviews of Heck and related reactions: (a) A. de Meijere, F. Meyer, *Angew. Chem.* **1994**, *106*, 2437; *Angew. Chem. Int. Ed. Engl.* **1994**, *33*, 2379; (b) W. Cabri, I. Candiani, *Acc. Chem. Res.* **1995**, *28*, 2; (c) T. Jeffery, *Adv. Metal-Org. Chem.* **1995**, *5*, 153; (d) J. Tsuji, *Palladium Reagents and Catalysts – Innovations in Organic Synthesis*, Wiley, Chichester, **1995**; (e) E. Negishi, C. Copéret, S. Ma, S.-Y. Liou, F. Liu, *Chem. Rev.* **1996**, *96*, 365; (f) W. A. Herrmann in *Applied Homogeneous Catalysis* (Eds.: B. Cornils, W. A. Herrmann), Wiley-VCH, Weinheim, **2002**; (g) M. Shibasaki, C. D. J. Boden,

A. KOJIMA, *Tetrahedron* **1997**, *53*, 7371; (h) I. P. BELETSKAYA, A. V. CHEPRAKOV, *Chem. Rev.* **2000**, *100*, 3009; (i) N. J. WHITCOMBE, K. K. HII, S. E. GIBSON, *Tetrahedron* **2001**, *57*, 7449; (j) E. NEGISHI, *Organopalladium Chemistry for Organic Synthesis*, Wiley-Interscience, New York, **2002**, p. 1133.

4 (a) C. AMATORE, M. AZZABI, A. JUTAND, *J. Am. Chem. Soc.* **1991**, *113*, 8375; (b) C. AMATORE, G. BROEKER, A. JUTAND, F. KHALIL, *J. Am. Chem. Soc.* **1997**, *119*, 5176; (c) C. AMATORE, A. JUTAND, *Acc. Chem. Res.* **2000**, *33*, 314.

5 (a) B. L. SHAW, *New. J. Chem.* **1998**, 77; (b) B. L. SHAW, S. D. PERERA, E. A. STALEY, *Chem. Commun.* **1998**, 1361; (c) G. T. CRISP, *Chem. Soc. Rev.* **1998**, *27*, 427.

6 (a) W. A. HERRMANN, C. BROSSMER, K. ÖFELE, M. BELLER, H. FISCHER, *J. Mol. Catal.* **1995**, *103*, 133; (b) F. ZHAO, B. M. BHANAGE, M. SHIRAI, M. ARAI, *J. Mol. Catal.* **1999**, *142*, 383.

7 (a) V. V. GRUSHIN, H. ALPER, *Chem. Rev.* **1994**, *94*, 1047; (b) T. H. RIERMEIER, A. ZAPF, M. BELLER, *Top. Catal.* **1998**, *4*, 301; (c) A. F. LITTKE, G. C. FU, *Angew. Chem.* **2002**, *114*, 4350; *Angew. Chem. Int. Ed.* **2002**, *41*, 4176.

8 A. JUTAND, A. MOSLEH, *Organometallics* **1995**, *14*, 1810.

9 (a) M. LUDWIG, S. STRÖMBERG, M. SVENSSON, B. ÅKERMARK, *Organometallics* **1999**, *18*, 970; (b) H. VON SCHENCK, B. ÅKERMARK, M. SVENSSON, *J. Am. Chem. Soc.* **2003**, *125*, 3503; (c) C. GÜRTLER, S. L. BUCHWALD, *Chem. Eur. J.* **1999**, *5*, 3107.

10 (a) M. BELLER, T. H. RIERMEIER, *Eur. J. Inorg. Chem.* **1998**, 29; (b) W. A. HERRMANN, C. BROSSMER, C.-P. REISINGER, T. H. RIERMEIER, K. ÖFELE, M. BELLER, *Chem. Eur. J.* **1997**, *3*, 1357; (c) M. BELLER, T. H. RIERMEIER, *Tetrahedron Lett.* **1996**, *37*, 6535.

11 T. JEFFREY, *Tetrahedron* **1996**, *52*, 10113 and references therein.

12 (a) S. HILLERS, S. SARTORI, O. REISER, *J. Am. Chem. Soc.* **1996**, *118*, 2087; (b) M. BUBACK, T. PERKOVIĆ, S. REDLICH, A. DE MEIJERE, *Eur. J. Org. Chem.* **2003**, 2375.

13 M. LARHED, A. HALLBERG, *J. Org. Chem.* **1996**, *61*, 9582.

14 (a) R. L. AUGUSTINE, S. T. O'LEARY, *J. Mol. Catal.* **1995**, *95*, 277; (b) M. BELLER, K. KÜHLEIN, *Synlett* **1995**, 441; (c) A. WALI, S. M. PILLAI, V. K. KAUSHIK, S. SATISH, *Appl. Catal.* **1996**, *135*, 83; (d) C. P. MEHNERT, D. W. WEAVER, J. Y. YING, *J. Am. Chem. Soc.* **1998**, *120*, 12289; (e) L. DJAKOVITCH, K. KÖHLER, *J. Mol. Catal.* **1999**, *142*, 275; (f) B. M. CHOUDARY, S. MADHI, N. S. CHOWDARI, M. L. KANTAM, B. SREEDHAR, *J. Am. Chem. Soc.* **2002**, *124*, 14127.

15 (a) J. LI, A. W.-H. MAU, C. R. STRAUSS, *Chem. Commun.* **1997**, 1275; (b) L. TONKS, M. S. ANSON, K. HELLGARDT, A. R. MIRZA, D. F. THOMPSON, J. M. J. WILLIAMS, *Tetrahedron Lett.* **1997**, *38*, 4319.

16 (a) M. T. REETZ, G. LOHMER, *Chem. Commun.* **1996**, 1921; (b) M. BELLER, H. FISCHER, K. KÜHLEIN, C.-P. REISINGER, W. A. HERRMANN, *J. Organomet. Chem.* **1996**, *520*, 257; (c) S. KLINGELHÖFER, W. HEITZ, A. GREINER, S. OESTREICH, S. FÖRSTER, M. ANTONIETTI, *J. Am. Chem. Soc.* **1997**, *119*, 10116; (d) V. CALÓ, A. NACCI, A. MONOPOLI, S. LAERA, N. CIOFFI, *J. Org. Chem.* **2003**, *68*, 2929; (e) M. T. REETZ, R. BREINBAUER, K. WANNINGER, *Tetrahedron Lett.* **1996**, *37*, 4499.

17 (a) L. HONG, E. RUCKENSTEIN, *J. Mol. Catal.* **1992**, *77*, 273; (b) P.-W. WANG, M. A. FOX, *J. Org. Chem.* **1994**, *59*, 5358; (c) J. SCHWARZ, V. P. W. BÖHM, M. G. GARDINER, M. GROSCHE, W. A. HERRMANN, W. HIERINGER, G. RAUDASCHL-SIEBER, *Chem. Eur. J.* **2000**, *6*, 1773.

18 M. T. REETZ, G. LOHMER, R. SCHWICKARDI, *Angew. Chem.* **1997**, *109*, 1559; *Angew. Chem. Int. Ed. Engl.* **1997**, *36*, 1526.

19 (a) S. SENGUPTA, S. BHATTACHARYA, *J. Chem. Soc. Perkin Trans. I* **1993**, 1943; (b) N. A. BUMAGIN, V. V. BYKOV, L. I. SUKHOMLINOVA, T. P. TOLSTAYA, I. P. BELETSKAYA, *J. Organomet. Chem.* **1995**, *486*, 259; (c) S. LEMAIRE-AUDOIRE, M. SAVIGNAC, C. DUPUIS, J.-P. GENÊT, *Tetrahedron Lett.* **1996**, *37*, 2003; (d) B. M. BHANAGE, F.-G. ZHAO, M. SHIRAI, M. ARAI, *Tetrahedron Lett.* **1998**, *39*, 9509; (e) B. M. BHANAGE, M. SHIRAI, M. ASAI, *J. Mol. Catal.* **1999**, *145*, 69; (f) M. BELLER, J. G. E. KRAUTER, A. ZAPF, *Angew. Chem.* **1997**, *109*, 793; *Angew. Chem. Int. Ed. Engl.* **1997**, *36*, 772.

20 I. Basnak, S. Takatori, R.T. Walker, Tetrahedron Lett. **1997**, *38*, 4869.
21 A.H.M. de Vries, F.J. Parlevliet, L. Schmieder-van de Vondervoort, J.H.M. Mommers, H.J.W. Henderickx, M.A.M. Walet, J.G. de Vries, Adv. Synth. Catal. **2002**, *344*, 996.
22 A. Spencer, J. Organomet. Chem. **1984**, *270*, 115.
23 (a) W.A. Herrmann, C. Brossmer, K. Öfele, C.-P. Reisinger, T. Riermeier, M. Beller, H. Fischer, Angew. Chem. **1995**, *107*, 1989; Angew. Chem. Int. Ed. Engl. **1995**, *34*, 1844; (b) M. Beller, T.H. Riermeier, Tetrahedron Lett. **1996**, *37*, 6535; (c) W.A. Herrmann, C. Brossmer, C.-P Reisinger, T.H. Riermeier, K. Öfele, M. Beller, Chem. Eur. J. **1997**, *3*, 1357; (d) M. Beller, T.H. Riermeier, Eur. J. Inorg. Chem. **1998**, 29; (e) W.A. Herrmann, V.P.W. Böhm, C.-P. Reisinger, J. Organomet. Chem. **1999**, *576*, 23.
24 M. Beller, H. Fischer, W.A. Herrmann, K. Öfele, C. Brossmer, Angew. Chem. **1995**, *107*, 1992; Angew. Chem. Int. Ed. Engl. **1995**, *34*, 1848.
25 (a) M. Ohff, A. Ohff, M.E. van der Boom, D. Milstein, J. Am. Chem. Soc. **1997**, *119*, 11687; (b) K. Kiewel, Y. Liu, D.E. Bergbreiter, G.A. Sulikowski, Tetrahedron Lett. **1999**, *40*, 8945; (c) D. Morales-Morales, R. Redón, C. Yung, C.M. Jensen, Chem. Commun. **2000**, 1619; (d) D. Morales-Morales, C. Grause, K. Kasaoka, R. Redón, R.E. Cramer, C.M. Jensen, Inorg. Chim. Acta **2000**, *300-302*, 958; (e) D.E. Bergbreiter, P.L. Osburn, Y.-S. Liu, J. Am. Chem. Soc. **1999**, *121*, 9531.
26 M.T. Reetz, G. Lohmer, R. Schwickardi, Angew. Chem. **1998**, *110*, 492; Angew. Chem. Int. Ed. **1998**, *37*, 481.
27 (a) M. Beller, A. Zapf, Synlett **1998**, 792; (b) D.A. Albisson, R.B. Bedford, P.N. Scully, Tetrahedron Lett. **1998**, *39*, 9793; (c) D.A. Albisson, R.B. Bedford, S.E. Lawrence, P.N. Scully, Chem. Commun. **1998**, 2095.
28 (a) K.H. Shaughnessy, P. Kim, J.F. Hartwig, J. Am. Chem. Soc. **1999**, *121*, 2123; (b) J.P. Stambuli, S.R. Stauffer, K.H. Shaughnessy, J.F. Hartwig, J. Am. Chem. Soc. **2001**, *123*, 2677; (c) A.F. Littke, G.C. Fu, J. Org. Chem. **1999**, *64*, 10; (d) A.F. Littke, G.C. Fu, J. Am. Chem. Soc. **2001**, *123*, 6989; (e) A. Ehrentraut, A. Zapf, M. Beller, Synlett **2000**, 1589; (f) A. Schnyder, A.F. Indolese, M. Studer, H.-U. Blaser, Angew. Chem. **2002**, *114*, 3820; Angew. Chem. Int. Ed. **2002**, *41*, 3668; (g) A. Schnyder, T. Aemmer, A.F. Indolese, U. Pittelkow, M. Studer, Adv. Synth. Catal. **2002**, *344*, 495.
29 M. Feuerstein, H. Doucet, M. Santelli, J. Org. Chem. **2001**, *66*, 5923.
30 S. Sjövall, M.H. Johansson, C. Andersson, Eur. J. Inorg. Chem. **2001**, 2907.
31 V.P.W. Böhm, W.A. Herrmann, Chem. Eur. J. **2000**, *6*, 1017.
32 (a) W.A. Herrmann, M. Elison, J. Fischer, C. Köcher, G.R.J. Artus, Angew. Chem. **1995**, *107*, 2602; Angew. Chem. Int. Ed. Engl. **1995**, *34*, 2371; (b) D.S. McGuinness, K.J. Cavell, B.W. Skelton, A.H. White, Organometallics **1999**, *18*, 1596; (c) K. Selvakumar, A. Zapf, A. Spannenberg, M. Beller, Chem. Eur. J. **2002**, *8*, 3901; (d) K. Selvakumar, A. Zapf, M. Beller, Org. Lett. **2002**, *4*, 3031.
33 L. Xu, W. Chen, J. Xiao, Organometallics **2000**, *19*, 1123.
34 (a) C. Yang, S.P. Nolan, Synlett **2001**, 1539; (b) M.B. Andrus, C. Song, J. Zhang, Org. Lett. **2002**, *4*, 2079.
35 (a) D.S. Clyne, J. Jin, E. Genest, J.C. Gallucci, T.V. RajanBabu, Org. Lett. **2000**, *2*, 1125; (b) C. Yang, H.M. Lee, S.P. Nolan, Org. Lett. **2001**, *3*, 1511; (c) V. César, S. Bellemin-Laponnaz, L.H. Gade, Organometallics **2002**, *21*, 5204; (d) E. Peris, J.A. Loch, J. Mata, R.H. Crabtree, Chem. Commun. **2001**, 201; (e) J.A. Loch, M. Albrecht, E. Peris, J. Mata, J.W. Faller, R.H. Crabtree, Organometallics **2002**, *21*, 700.
36 (a) V. Caló, A. Nacci, L. Lopez, N. Mannarini, Tetrahedron Lett. **2000**, *41*, 8973; (b) V. Caló. R. Del Sol, A. Nacci, E. Schingaro, F. Scordari, Eur. J. Org. Chem. **2000**, 869; (c) V. Caló, A. Nacci, L. Lopez, A. Napola, Tetrahedron Lett. **2001**, *42*, 4701; (d) V. Caló, A. Nacci, A. Monopoli, L. Lopez, A. di Cosmo, Tetrahedron **2001**, *57*, 6071; (e) V. Caló, A. Nacci, A. Monopoli, M. Spinelli, Eur. J. Org. Chem. **2003**, 1382.

37 (a) M. T. Reetz, E. Westermann, R. Lohmer, G. Lohmer, *Tetrahedron Lett.* **1998**, *39*, 8449; (b) C. S. Consorti, M. L. Zanini, S. Leal, G. Ebeling, J. Dupont, *Org. Lett.* **2003**, *5*, 983.

38 (a) S. Iyer, *J. Organomet. Chem.* **1995**, *490*, C27; (b) S. Iyer, C. Ramesh, A. Sarkar, P. P. Wadgaonkar, *Tetrahedron Lett.* **1997**, *38*, 8113.

39 A. A. Kelkar, *Tetrahedron Lett.* **1996**, *37*, 8917.

40 (a) G. P. Boldrini, D. Savoia, E. Tagliavani, C. Trombini, A. Umani Ronchi, *J. Organomet. Chem.* **1986**, *301*, C62; (b) S. A. Lebedev, V. S. Lopatina, E. S. Petrov, I. P. Beletskaya, *J. Organomet. Chem.* **1988**, *344*, 253; (c) R. Sustmann, P. Hopp, P. Holl, *Tetrahedron Lett.* **1989**, *30*, 689.

41 H.-U. Blaser, A. Spencer, *J. Organomet. Chem.* **1982**, *233*, 267.

42 (a) M. S. Stephan, A. J. J. M. Teunissen, G. K. M. Verzijl, J. G. de Vries, *Angew. Chem.* **1998**, *110*, 688; *Angew. Chem. Int. Ed.* **1998**, *37*, 662; (b) M. S. Stephan, J. G. de Vries, *Chemical Industries* **2001**, *82* (Catalysis of Organic Reactions), 379; (c) A. F. Shmidt, V. V. Smirnov, *Kinet. Catal.* **2000**, *41*, 743; (d) A. F. Shmidt, V. V. Smirnov, *Kinet. Catal.* **2002**, *43*, 195; (e) A. J. Carmichael, M. J. Earle, J. D. Holbrey, P. B. McCormac, K. R. Seddon, *Org. Lett.* **1999**, *1*, 997.

43 L. J. Goossen, J. Paetzold, L. Winkel, *Synlett* **2002**, 1721.

44 L. J. Goossen, J. Paetzold, *Angew. Chem.* **2002**, *114*, 1285; *Angew. Chem. Int. Ed.* **2002**, *41*, 1237.

45 A. G. Myers, D. Tanaka, M. R. Mannion, *J. Am. Chem. Soc.* **2002**, *124*, 11250.

46 (a) P. J. Guiry, A. J. Hennessy, J. P. Cahill, *Top. Catal.* **1997**, *4*, 311; (b) M. Shibasaki, E. M. Vogl in *Perspectives in Organopalladium Chemistry for the XXI. Century* (Ed. J. Tsuji), Elsevier, Amsterdam, **1999**, p. 1; (c) O. Loiseleur, M. Hayashi, M. Keenan, N. Schmees, A. Pfaltz in *Perspectives in Organopalladium Chemistry for the XXI. Century* (Ed. J. Tsuji), Elsevier, Amsterdam, **1999**, p. 16.

47 (a) L. F. Tietze, R. Schimpf, *Angew. Chem.* **1994**, *106*, 1138; *Angew. Chem. Int. Ed. Engl.* **1994**, *33*, 1089; (b) L. F. Tietze, T. Raschke, *Liebigs Ann.* **1996**, 1981; (c) L. F. Tietze, K. Thede, R. Schimpf, F. Sannicolò, *Chem. Commun.* **2000**, 583.

48 Y. Sato, M. Sodeoka, M. Shibasaki, *J. Org. Chem.* **1989**, *54*, 4738.

49 (a) R. C. Larock, S. Babu, *Tetrahedron Lett.* **1987**, *28*, 5291; (b) R. C. Larock, B. E. Baker, *Tetrahedron Lett.* **1988**, *29*, 905; (c) R. C. Larock, H. Song, B. E. Baker, W. H. Gong, *Tetrahedron Lett.* **1988**, *29*, 2919; (d) R. C. Larock, D. E. Stinn, *Tetrahedron Lett.* **1988**, *29*, 4687; (e) R. C. Larock, P. L. Johnson, *J. Chem. Soc. Chem. Commun.* **1989**, 1368; (f) R. C. Larock, W. H. Gong, B. E. Baker, *Tetrahedron Lett.* **1989**, *30*, 2603; (g) R. C. Larock, W. H. Gong, *J. Org. Chem.* **1989**, *54*, 2047; (h) R. C. Larock, W. H. Gong, *J. Org. Chem.* **1990**, *55*, 407.

50 (a) A. Sato, M. Sodeoka, M. Shibasaki, *Chem. Lett.* **1990**, 1953; (b) K. Kagechika, M. Shibasaki, *J. Org. Chem.* **1991**, *56*, 4093; (c) Y. Sato, S. Watanabe, M. Shibasaki, *Tetrahedron Lett.* **1992**, *33*, 2589; (d) Y. Sato, T. Honda, M. Shibasaki, *Tetrahedron Lett.* **1992**, *33*, 2593; (e) K. Kondo, M. Sodeoka, M. Mori, M. Shibasaki, *Tetrahedron Lett.* **1993**, *34*, 4219; (f) K. Ohrai, K. Kondo, M. Sodeoka, M. Shibasaki, *J. Am. Chem. Soc.* **1994**, *116*, 11737; (g) S. Nukui, M. Sodeoka, M. Shibasaki, *Tetrahedron Lett.* **1993**, *34*, 4965; (h) T. Takemoto, M. Sodeoka, H. Sasai, M. Shibasaki, *J. Am. Chem. Soc.* **1993**, *115*, 8477; (i) K. Kagechika, T. Ohshima, M. Shibasaki, *Tetrahedron* **1993**, *49*, 1773.

51 For BPPFOH (R)-α-(S)-1',2-bis(diphenylphosphino)ferrocenyl ethyl alcohol as efficient ligand in the asymmetric Heck reaction see: (a) S. Nukui, M. Sodeoka, M. Shibasaki, *Tetrahedron Lett.* **1993**, *34*, 4965; (b) Y. Sato, S. Nukui, M. Sodeoka, M. Shibasaki, *Tetrahedron* **1994**, *50*, 371.

52 (a) F. Ozawa, A. Kubo, T. Hayashi, *J. Am. Chem. Soc.* **1991**, *113*, 1417; (b) F. Ozawa, T. Hayashi, *J. Organomet. Chem.* **1992**, *428*, 268; (c) F. Ozawa, A. Kubo, T. Hayashi, *Tetrahedron Lett.* **1992**, *33*, 1485; (d) F. Ozawa, A. Kubo, T. Hayashi, *Chem. Lett.* **1992**, 2177; (e) F. Ozawa, Y. Kobatake, T. Hayashi, *Tetrahedron Lett.* **1993**, *34*, 2505; (f) T. Hayashi, A. Kubo, F. Ozawa, *Pure Appl. Chem.* **1992**, *64*, 421.

53 (a) A. J. Hennessy, Y. M. Malone, P. J. Guiry, *Tetrahedron Lett.* **1999**, *40*, 9163; (b) A. J. Hennessy, Y. M. Malone, P. J. Guiry, *Tetrahedron Lett.* **2000**, *41*, 2261; (c) A. J. Hennessy, D. J. Connolly, Y. M. Malone, P. J. Guiry, *Tetrahedron Lett.* **2000**, *41*, 7757; (d) T. G. Kilroy, A. J. Hennessy, D. J. Connolly, Y. M. Malone, A. Farrell, P. J. Guiry, *J. Mol. Catal.* **2003**, *196*, 65.

54 (a) N. E. Carpenter, D. J. Kucera, L. E. Overman, *J. Org. Chem.* **1989**, *54*, 5846; (b) A. Ashimory, L. E. Overman, *J. Org. Chem.* **1992**, *57*, 4571; (c) A. Ashimori, T. Matsuura, L. E. Overman, D. J. Poon, *J. Org. Chem.* **1993**, *58*, 6949; (d) A. Ashimori, B. Bachand, L. E. Overman, D. J. Poon, *J. Am. Chem. Soc.* **1998**, *120*, 6477.

55 L. E. Overman, D. J. Poon, *Angew. Chem.* **1997**, *109*, 536; *Angew. Chem. Int. Ed. Engl.* **1997**, *36*, 518.

56 (a) K. K. Hii, T. D. W. Claridge, J. M. Brown, *Angew. Chem.* **1997**, *109*, 1033; *Angew. Chem. Int. Ed. Engl.* **1997**, *36*, 984; (b) K. K. Hii, T. D. W. Claridge, J. M. Brown, A. Smith, R. J. Deeth, *Helv. Chim. Acta* **2001**, *84*, 3043.

57 A. Kojima, C. D. J. Boden, M. Shibasaki, *Tetrahedron Lett.* **1997**, *38*, 3459.

58 (a) R. Schmid, E. A. Broger, M. Cereghetti, Y. Crameri, J. Foricher, M. Lalonde, R. K. Mueller, M. Scalone, G. Schoettel, U. Zutter, *Pure Appl. Chem.* **1996**, *68*, 131; (b) G. Trabesinger, A. Albinati, N. Feiken, R. W. Kunz, P. S. Pregosin, M. Tschoerner *J. Am. Chem. Soc.* **1997**, *119*, 6315; (c) M. Tschoerner, P. S. Pregosin, A. Albinati, *Organometallics* **1999**, *18*, 670.

59 D. Che, N. G. Andersen, S. Y. W. Lau, M. Parvez, B. A. Keay, *Tetrahedron Asymm.* **2000**, *11*, 1919.

60 (a) P. von Matt, O. Loiseleur, G. Koch, A. Pfalz, C. Lefeber, T. Feucht, G. Helmchen, *Tetrahedron Asymm.* **1994**, *5*, 573; (b) G. Koch, G. C. Lloyd-Jones, O. Loiseleur, A. Pfalz, R. Préfôt, S. Schaffner, P. Schinder, P. von Matt, *Recl. Tranv. Chim. Pays-Bas* **1995**, *114*, 206.

61 (a) O. Loiseleur, P. Meier, A. Pfaltz, *Angew. Chem.* **1996**, *108*, 218; *Angew. Chem. Int. Ed. Engl.* **1996**, *35*, 200; (b) A. Pfaltz, *Acta Chem. Scand.* **1996**, *50*, 189.

62 P. Nilsson, H. Gold, M. Larhed, A. Hallberg, *Synthesis* **2002**, 1611.

63 W.-P. Deng, X.-L. Hou, L.-X. Dai, X.-W. Dong, *Chem. Commun.* **2000**, 1483.

64 (a) M. Ogasawara, K. Yoshida, T. Hayashi, *Heterocycles* **2000**, *52*, 195; (b) K. Selvakumar, M. Valentini, P. S. Pregosin, A. Albinati, F. Eisenträger, *Organometallics* **2000**, *19*, 1299; (c) P. Dotta, A. Magistrato, U. Rothlisberger, P. S. Pregosin, A. Albinati, *Organometallics* **2002**, *21*, 3033.

65 (a) S. R. Gilbertson, Z. Fu, D. Xie, *Tetrahedron Lett.* **2001**, *42*, 365; (b) S. R. Gilbertson, D. Xie, Z. Fu, *J. Org. Chem.* **2001**, *66*, 7240; (c) S. R. Gilbertson, Z. Fu, *Org. Lett.* **2001**, *3*, 161.

66 Y. Hashimoto, Y. Horie, M. Hayashi, K. Saigo, *Tetrahedron Asymm.* **2000**, *11*, 2205.

67 (a) E. P. Kündig, P. Meier, *Helv. Chim. Acta* **1999**, *82*, 1360; (b) G. H. Bernardinelli, E. P. Kündig, P. Meier, A. Pfaltz, K. Radkowski, N. Zimmermann, M. Neuburger-Zehnder, *Helv. Chim. Acta* **2001**, *84*, 3233.

68 K. Yonehara, K. Mori, T. Hashizume, K.-G. Chung, K. Ohe, S. Uemura, *J. Organomet. Chem.* **2000**, *603*, 40.

69 (a) X.-Y. Wu, H.-D. Xu, Q.-L. Zhou, A. S. C. Chan, *Tetrahedron Asymm.* **2000**, *11*, 1255; (b) X.-Y. Wu, H.-D. Xu, F.-Y. Tang, Q.-L. Zhou, *Tetrahedron Asymm.* **2001**, *12*, 2565.

70 R. Imbos, A. J. Minnaard, B. L. Feringa, *J. Am. Chem. Soc.* **2001**, *124*, 184.

71 (a) L. E. Overman, *Pure Appl. Chem.* **1994**, *66*, 1423; (b) K. C. Nicolaou, E. J. Sorensen, *Classics in Total Synthesis*, VCH, Weinheim, **1996**, p. 566.

72 C. Y. Hong, N. Kado, L. E. Overman, *J. Am. Chem. Soc.* **1993**, *115*, 11028.

73 C. Y. Hong, L. E. Overman, *Tetrahedron Lett.* **1994**, *35*, 3453.

74 C.-Y. Cheng, J.-P. Liu, M.-J. Lee, *Tetrahedron Lett.* **1997**, *38*, 4571.

75 (a) L. E. Overman, D. J. Ricca, V. D. Tran, *J. Am. Chem. Soc.* **1993**, *115*, 2042; (b) D. J. Kucera, S. J. O'Conner, L. E. Overman, D. J. Ricca, V. D. Tran, *J. Org. Chem.* **1993**, *58*, 5304; (c) L. E. Overman, D. J. Ricca, V. D. Tran, *J. Am. Chem. Soc.* **1997**, *119*, 12031.

76 (a) V. H. Rawal, C. Michoud, *Tetrahedron Lett.* **1991**, *32*, 1695; (b) V. H. Rawal, C. Michoud, R. Monestel, *J. Am. Chem. Soc.*

1993, *115*, 3030; (c) V. H. Rawal, C. Michoud, *J. Org. Chem.* **1993**, *58*, 5583.

77 (a) K. F. McClure, S. J. Danishefsky, *J. Am. Chem. Soc.* **1993**, *115*, 6094; (b) J. M. Schkeryantz, S. J. Danishefsky, *J. Am. Chem. Soc.* **1995**, *117*, 4722; (c) S. J. Danishefsky, J. J. Masters, W. B. Young, J. T. Link, L. B. Snyder, T. V. Magee, D. K. Jung, R. C. A. Isaacs, W. G. Bornmann, C. A. Alaimo, C. A. Coburn, M. J. Di Grandi, *J. Am. Chem. Soc.* **1996**, *118*, 2843.

78 (a) B. M. Trost, F. D. Toste, *J. Am. Chem. Soc.* **2000**, *122*, 11262; (b) B. M. Trost, W. Tang, *Angew. Chem.* **2002**, *114*, 2919; *Angew. Chem. Int. Ed.* **2002**, *41*, 2795.

79 (a) L. F. Tietze, H. Schirok, *Angew. Chem.* **1997**, *109*, 1159; *Angew. Chem. Int. Ed. Engl.* **1997**, *36*, 1124; (b) L. F. Tietze, H. Schirok, *J. Am. Chem. Soc.* **1999**, *121*, 10264.

80 (a) K. Kondo, M. Sodeoka, M. Shibasaki, *J. Org. Chem.* **1995**, *60*, 4322; (b) K. Kondo, M. Sodeoka, M. Shibasaki, *Tetrahedron Asymm.* **1995**, *6*, 2453.

81 J. W. Dankwardt, L. A. Flippin, *J. Org. Chem.* **1995**, *60*, 2312.

82 B. M. Trost, J. Dumas, *Tetrahedron Lett.* **1993**, *34*, 19.

83 R. D. Larsen, E. G. Corley, A. O. King, J. D. Carrol, P. Davis, T. R. Verhoeven, P. J. Reider, M. Labelle, J. Y. Gauthier, Y. B. Xiang, R. J. Zamboni, *J. Org. Chem.* **1996**, *61*, 3398.

84 D. B. Grotjahn, X. W. Zhang, *J. Mol. Catal.* **1997**, *116*, 99.

85 (a) G. Dyker, P. Grundt, *Tetrahedron Lett.* **1996**, *37*, 619; (b) G. Dyker, F. Nerenz, P. Siemsen, P. Bubenitschek, P. G. Jones, *Chem. Ber.* **1996**, *129*, 1264; (c) G. Dyker, J. Körning, F. Nerenz, P. Siemsen, S. Sostmann, A. Wiegand, *Pure Appl. Chem.* **1996**, *68*, 323.

86 For a review see: (a) R. Grigg, *J. Heterocycl. Chem.* **1994**, *31*, 631; (b) S. Brown, S. Clarkson, R. Grigg, V. Sridharan, *J. Chem. Soc. Chem. Commun.* **1995**, 1135; (c) R. Grigg, V. Loganathan, V. Sridharan, *Tetrahedron Lett.* **1996**, *37*, 3399.

87 R. C. Larock, M. A. Mitchell, *J. Am. Chem. Soc.* **1978**, *100*, 180.

88 (a) A. de Meijere, S. Bräse in *Perspectives in Organopalladium Chemistry for the XXI. Century* (Ed. J. Tsuji), Elsevier, Amsterdam, **1999**, p. 88; (b) S. Bräse, A. de Meijere, *Angew. Chem.* **1995**, *107*, 2741; *Angew. Chem. Int. Ed. Engl.* **1995**, *34*, 2545; (c) K. Albrecht, A. de Meijere, *Chem. Ber.* **1994**, *127*, 2539; (d) O. Reiser, B. König, K. Meerholz, J. Heinze, T. Wellauer, F. Gerson, R. Frim, M. Rabinovitz, A. de Meijere, *J. Am. Chem. Soc.* **1993**, *115*, 3511.

89 (a) C. Coperet, S. Ma, E.-i. Negishi, *Angew. Chem.* **1996**, *108*, 2255; *Angew. Chem. Int. Ed. Engl.* **1996**, *35*, 2125; (b) E.-i. Negishi, C. Coperet, S. Ma, T. Mita, T. Sugihara, J. M. Tour, *J. Am. Chem. Soc.* **1996**, *118*, 5904.

90 (a) L. F. Tietze, T. Nöbel, M. Spescha, *Angew. Chem.* **1996**, *108*, 2385; *Angew. Chem. Int. Ed. Engl.* **1996**, *35*, 2259; (b) L. F. Tietze, R. Ferraccioli, *Synlett* **1998**, 145; (c) L. F. Tietze, S. Petersen, *Eur. J. Org. Chem.* **2001**, 1619; (d) L. F. Tietze, W.-R. Krahnert, *Chem. Eur. J.* **2002**, *8*, 2116.

91 (a) K. Kagechika, T. Ohshima, M. Shibasaki, *Tetrahedron* **1993**, *49*, 1773; (b) K. Kagechika, M. Shibasaki, *J. Org. Chem.* **1991**, *56*, 4093; (c) T. Ohshima, K. Kagechika, M. Adachi, M. Sodeoka, M. Shibasaki, *J. Am. Chem. Soc.* **1996**, *118*, 7108.

92 S. P. Maddaford, N. G. Andersen, W. A. Cristofoli, B. A. Keay, *J. Am. Chem. Soc.* **1996**, *118*, 10766.

93 C.-Y. Chen, D. R. Liebermann, R. D. Larsen, T. R. Verhoeven, P. J. Reider, *J. Org. Chem.* **1997**, *62*, 2676.

94 M. Brenner, G. Mayer, A. Terpin, W. Steglich, *Chem. Eur. J.* **1997**, *3*, 70.

95 A. Kojima, T. Takemoto, M. Sodeoka, M. Shibasaki, *J. Org. Chem.* **1996**, *61*, 4876.

96 I. Basnak, S. Takatori, R. T. Walker, *Tetrahedron Lett.* **1997**, *38*, 4869.

97 (a) D. D. Hennings, S. Iwasa, V. H. Rawal, *J. Org. Chem.* **1997**, *62*, 2; (b) D. D. Hennings, S. Iwasa, V. H. Rawal, *Tetrahedron Lett.* **1997**, *38*, 6379.

98 T. Satoh, Y. Kawamura, M. Miura, M. Nomura, *Angew. Chem.* **1997**, *109*, 1820; *Angew. Chem. Int. Ed. Engl.* **1997**, *37*, 1740.

99 M. Catellani, L. Ferioli, *Synthesis* **1996**, 769.

100 A. Heim, A. Terpin, W. Steglich, *Angew. Chem.* **1997**, *109*, 158; *Angew. Chem. Int. Ed. Engl.* **1997**, *37*, 115.

2.14
Palladium-Catalyzed Allylic Substitutions

Andreas Heumann

2.14.1
Introductory Remarks and Historical Background [1]

Allylic substrates are important compounds in nature [2], and industrially produced allyl halides or allyl alcohol [3] are starting materials for fine chemicals and polymeric allyl resins. Bringing a *p*-orbital into interaction with the π-orbital in reactive allyl anions, radicals, or cations results in electron delocalization and the overlap of three adjacent *p*-orbitals. The resulting ease of replacement of the allylic (leaving) group by an exchange with carbon and heteroatoms makes allylic substitution a versatile process [4] in organic synthesis. The intramolecular version is an especially powerful method for cyclization to unsaturated cyclic substrates.

An important feature of the allylic system is the formation of π-allyl transition metal complexes, with the allyl group acting as a η^3-ligand. Here the reactivity of the allyl unit is no longer governed by the allylic bond but by the electronic properties of the ligand-metal system. If the metal is a strong electron acceptor (e.g. Pd^{2+}), the η^3-allyl group acts as cationic C_3H_5 and palladium-mediated allylic substitution [5] will become nucleophilic in nature.

Actually, at the end of the 1950s, the first (*stoichiometric*) allylic substitution involving palladium was the (thermally induced) reductive elimination of a π-allyl-palladium chloride complex (1) (Scheme 1) [6, 7].

Later, the carbonylation (500 atm) of allylic chlorides catalyzed by π-allylpalladium chloride complexes was reported [8]: this was the first metal-catalyzed *catalytic* allylic substitution [9]. An important step to synthetic applications was the observation that π-allylpalladium chloride reacts with external nucleophiles such as sodium ethyl malonate or 1-morpholino-1-cyclohexene with C–C bond formation [10]. The catalytic nucleophilic substitution reaction was found a couple of years later [11–14].

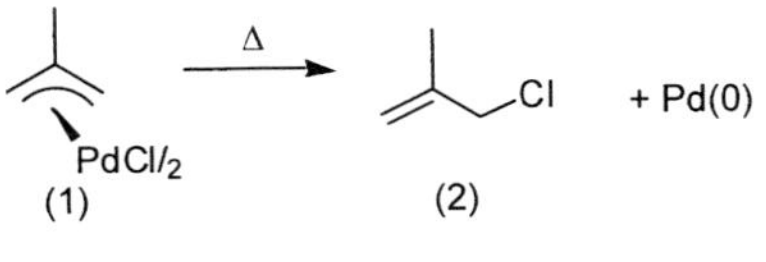

Scheme 1 Thermal decomposition of $(C_4H_7PdCl)_2$ [6].

2.14.2
Reactions of π-Allyl Palladium Complexes [15]

π-Allyl palladium complexes [16] are efficient catalysts for several important reactions, such as the dimerization of olefins [17], dimerization-oligomerization of butadiene [18], or hydrovinylation of alkenes [19]. Palladium-catalyzed allylic oxidations with (moderate) chiral induction are mediated by the chiral bis[acetoxy(3,2,10-η-pinene) palladium(II) [20]. Meanwhile, better results (<97% ee) are obtained in these reactions with cationic binaphthylbisoxazoline palladium complexes [21].

π-Allylpalladium complexes are electrophilic in nature. They react with any kind of nucleophilic reagent, in general, at the less substituted of the two termini of the allyl complex [22], together with formation of palladium(0) (Scheme 2).

It is of major importance for synthetic applications that a catalytic process should be possible, since the π-allylpalladium complex is both readily formed from organic precursors and is subsequently substituted by multiple nucleophilic agents (Scheme 2). In addition, oxidation of π-allylpalladium complexes leads to the incorporation of oxygen functionalities, and thermal decomposition leads to dienes.

Under certain conditions, cyclopropanation takes place via attack on the central atom of a π-allylpalladium complex stabilized by N,N-donor ligands (7) (Scheme 3), using a weakly stabilized nucleophile (pK_a range of protonated carbanions: 20–30) [23].

The collapse of the intermediate palladacyclobutane and elimination of Pd takes place in the presence of CO [24]. The ligand control [25], the nature of nucleophiles such as phenoxide ion [26] or PhO⁻/carbon nucleophile combination [27] can orient the nucleophilic attack to the central or terminal carbon of the π-allylpalladium complex.

Scheme 2 Nucleophilic substitution of π-allylpalladium complexes.

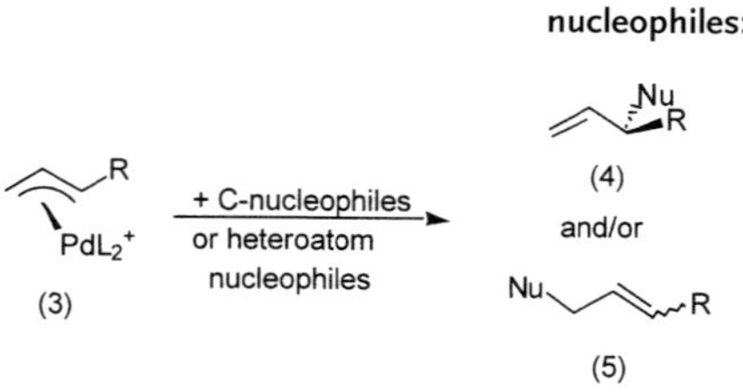

nucleophiles: XYCH⁻ (X,Y=CO_2R, CN, SO_2Ph)
⁻CH_2NO_2, 1,3-diketone, β-ketoester, enamines, Cp⁻, enolates, allyl, aryl stannanes, organometals (Al, Mg, Tl, Zn, Zr), CO, alkenes, dienes

COOR, OH, $OSiR_3$
β-ketoester, 1,3-diketone ('O-alkylation')
NR_2 (R=H, alkyl, aryl, Ts) isocyanates, carbamates

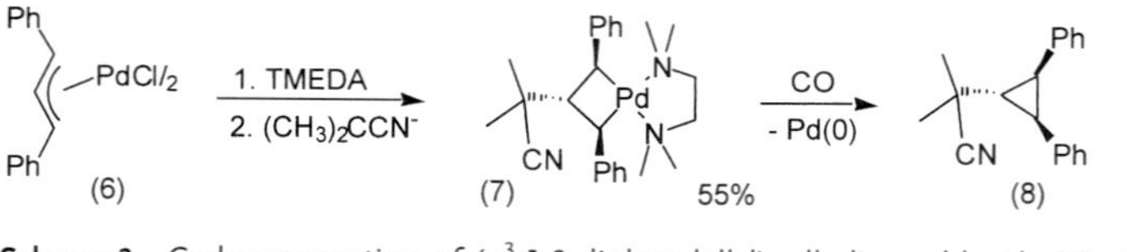

Scheme 3 Cyclopropanation of ($η^3$-1,3-diphenylallyl)palladium chloride [23, 24].

2.14.3
Catalytic Introduction of Nucleophiles

The reaction of any suitable organic precursor (allylic oxygen, nitrogen, sulfur, silicon, tin, phosphorus or halogen derivatives, vinyl epoxides, vinyl cyclopropanes, and 1,3-dienes) [28] with a suitable palladium catalyst permits a catalytic allylic substitution reaction. The catalyst may be a preformed π-allylpalladium complex or Pd(0), generally stabilized by a P- or N-donor ligand. Typical palladium(0) sources are Pd(PPh$_3$)$_4$ [29] or Pd$_m$(dba)$_n$ catalysts. More often the resulting active Pd(0) catalyst is conveniently formed *in situ* from palladium(II) acetate and PPh$_3$ [30] or via Wacker-type olefin oxidation. In the case when $[Pd(\eta^3\text{-}C_3H_5)(\mu\text{-}Cl)]_2$ is the catalyst precursor, it has been shown [31] that, because of the presence of chloride ion, the active species is (σ-allyl)palladium complex. The oxidative addition [32] of the allylic (and related) organic substrate to Pd(0) is often reversible, and a base is required. Most popular organic substrates are allylic esters and allylic carbonates [33]. In the latter compound, the driving force is the irreversible formation of CO_2 together with the base (RO$^-$) under neutral conditions [34]. This base is necessary to generate the anion of certain carbon nucleophiles and has to be added when allylic acetates are used. The more simple direct allylation using allyl alcohols is still less common, but interesting results have been published with aldehydes [35] and amines [36]. From a synthetic point of view, the intramolecular substitution is a powerful tool for the synthesis of any kind of mono- and polycyclic systems [37]. Cross-coupling [38], allylic carbonylation [9c], and phase transfer reactions [39] are treated elsewhere in these series.

Regarding carbon nucleophiles, the most significant reactions concern malonate-type soft carbon anions (CRYZ$^-$, Y, and Z being electron-withdrawing groups such as ketone, carboxylate, nitrile, sulfonate, sulfoxide, nitro, imine) as nucleophiles, permitting smooth and controlled carbon–carbon bond formations. This so called Trost-Tsuji reaction [40] has a broad scope, and the numerous synthetic applications have been reviewed extensively [5, 15]. Other soft carbon nucleophiles are anions from nitromethane and cyclopentadiene, enamines, and enolates (Scheme 2). Hard and nonstabilized alkyl anions or organometallic reagents such as aryl stannanes, Grignard reagents, organo-Al, -Tl, -Zn, and -Zr react also but differ in their stereochemical behavior. π-Allylpalladium complexes react too, after *Umpolung* with diethyl zinc, for example, as nucleophilic species [41].

Nitrogen compounds and especially secondary amines are good nucleophiles for allylic substitution; with primary amines or ammonia the bisallylation may be competitive with monoallylation, resulting in reduced chemoselectivity. In general, O-nucleophiles are less active than N-nucleophiles. Especially alcohols are poor nucleophiles, although pyranoside primary alcohols have been allylated. The O-alkylation, sometimes after activation as O-silylated alcohols, is especially interesting in cyclization reactions. Reaction with sulfur (thiol, alkylsilyl sulfide, sulfinate), phosphorus (phosphide, phosphites [42]) and selenium (diselenides) lead to the respective allylic heteroatomic compounds.

Allylic substitution polymerization (catalyst: Pd(0)-dppb, polymer: $M_n = 22000$) was developed with (Z)-1,4-diacetoxybut-2-ene and employing substrates with various functional groups such as carbonyl and ether functions [43].

Hetero-π-allyl systems are a special case of allylic substrates worth mentioning. Here, oxa-π-allyl palladium complexes [44] have been postulated in the oxidation of silyl enol ethers to cyclic ketones [45], but were formulated as Pd(II) enolates some time later [46]. $β,γ$-Unsaturated amino acid derivates are formed via 2-aza-π-allyl palladium complexes [47].

2.14.4
Mechanism – Stereochemistry

The stereochemistry of palladium-catalyzed substitution has been a subject of extensive studies [48–51]. The reaction of allylic esters proceeds with a high degree of stereospecificity (Scheme 4). In the first reaction step, the leaving group is replaced by the metal with inversion of configuration via coordination of palladium to the allylic double bond (*anti* route). An example of a *syn* route has been designed with phosphine-containing allylic leaving groups [52] and the reaction of allylic chlorides under particular reaction conditions [53]. The subsequent nucleophilic substitution reaction of the intermediate palladium-π-allyl complex is dependent on the nature of the nucleophile [54]: soft nucleophiles (stabilized carbanions and many heteroatom nucleophiles) react with a second inversion of the stereochemistry. Thus, the reaction of the allylic precursor (**9**) to give substrates (**9**) → (**12r**) (Nu: amine, ether, acetate, or malonate-type anions [55]) proceeds with overall retention of configuration.

If the incoming substituent is a hard nucleophile (such as hydride or nonstabilized carbon nucleophile), the attack at the carbon skeleton is preceded by coordination of the nucleophile to the palladium, and the final substitution proceeds via *cis* migration. The result is an overall inversion of configuration (**9**) → (**12i**).

There are only a few reactions known which do not follow this general rule [56]; however, by carefully controlling the kinetics in these transformations [57], side reactions, such as those between the palladium(0) catalyst and the palladium π-allyl intermediate [58], can be diminished. Unselective alkylations with malonate-type nucleophiles can be improved to give 100% selectivity.

New ways to improve regiocontrol are also explored. The effect of ring strain [59] or coordination with removable heteroatom groups [60] has been used to drive reaction selectivity efficiently.

Ionic intermediates in allylic alkylation may not be rigorously symmetric [61], and the nature of ion pairs definitely affects the selectivity of the final substitution. This *memory effect* has important implications for the outcome of enantioselective allylations. This effect was also used in isotopic desymmetrization with "minimal effort and with great effect" [62].

Many nucleophiles in these reactions are also good leaving groups, and the reaction conditions determine whether kinetic or thermodynamic factors control the

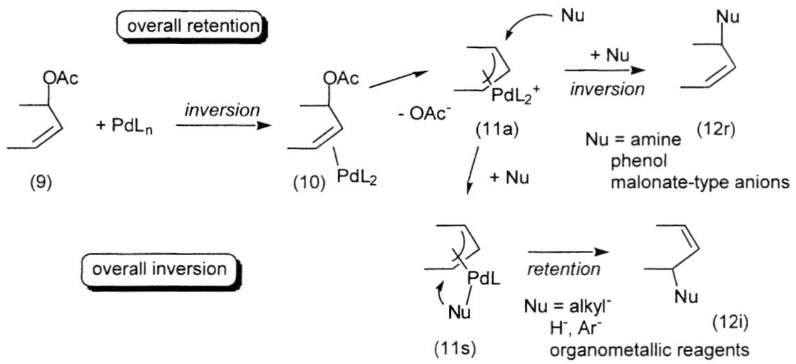

Scheme 4 Stereochemistry of palladium-catalyzed allylic substitution.

regiochemistry. Even for carbon nucleophiles it has recently been shown, with dienyl acetates, that palladium-catalyzed allylic C-C bond cleavage can take place under standard conditions and that mechanistic interpretation might be less simple, even in allylic alkylations [63]. Interestingly, allylic carbonate also may react without loss of CO_2 [64], or even be formed from CO_2, alcohols and allylic chloride under pressure [65].

2.14.5
Allylic Reductions – Hydrogenolysis – Eliminations

Allylic hydrogenolysis leads to olefins [66, 67]. This is after all an elegant method to introduce regio- and stereoselectively deuterium atoms with $NaBD_4$ [68] or to form chiral alkenes [69]. Commonly used reducing agents are alumino- and borohydrides, hydrosilanes, tin hydride, formic acid, or ammonium formate [70]. Electrolysis is also possible. A particular field of current application is the ion capture of cyclized π-allyl Pd species [71]. Allylic elimination is a particular reaction in Pd-allyl chemistry that might be interesting for obtaining chiral 1,3-dienes [72, 73].

2.14.6
Protective Groups

The Pd- (but also Rh- and Ir-) catalyzed cleavage of allyl esters is used as a key step in efficient strategies for protecting molecules with acid- and base-labile functional groups [74]. Carboxylic acids, amines, and alcohols can be blocked as allyl carbonates, allyl esters, allyl ethers, and allyl urethanes.

2.14.7
Trimethylenemethane (TMM) Cycloadditions

Palladium-catalyzed 2-acetoxymethyl-3-allyl trimethylsilane addition to electron-deficient alkenes is an elegant and general method for the obtention of five-membered rings (Scheme 5) [75]. The intermediate is a metal-stabilized π-allyl 1,3-dipole (15).

The [3+2] type TMM reactions have been extended to [4+3] cycloadditions in heterocyclic series [76] and to fullerene substrates [77]. Enantioselective versions with chiral enones have appeared [78].

2.14.8
Allylic Rearrangements

Palladium(0) and palladium(II) catalyze electrocyclic rearrangements [79] of allylic substrates such as oxy-Cope and aza-Cope transpositions [80, 81] or more general heteroatom to heteroatom rearrangements (e.g., O to S, S to N, etc.). A bis(η^3-allyl)palladium(II) intermediate has been proposed for the transposition of 1,5-dienes [82]. The allyl ester transposition has been successfully used for complete chirality transfer [83], asymmetric rearrangement of allylic imidates [84], and deracemization of cyclic allyl esters [85]. The Pd(II) catalysis is also called "cyclization-induced rearrangement", since a probable mechanism consists in the nucleophilic attack of the heteroatom to the metal-coordinated alkene part.

2.14.9
Enantioselective Reactions

The allylic substitution, via the addition to (intermediate) π-allyl complexes with concomitant displacement of the metal, may be considered as a displacement at sp^3 carbon (σ-allyl Pd bond), and is quite different from most other highly enantiocontrolled transition metal-catalyzed addition to π-systems [86]. From 1977, the search for an efficient enantioselective system parallels the general development of palladium-catalyzed allylic substitutions with malonate-type [87] and other soft carbon [88] nucleophiles. A comprehensive discussion of the rapidly developing field has recently appeared from the pioneer in the area [89, 90]. In general, rather disappointing results with chiral diphosphines, so successful in asymmetric

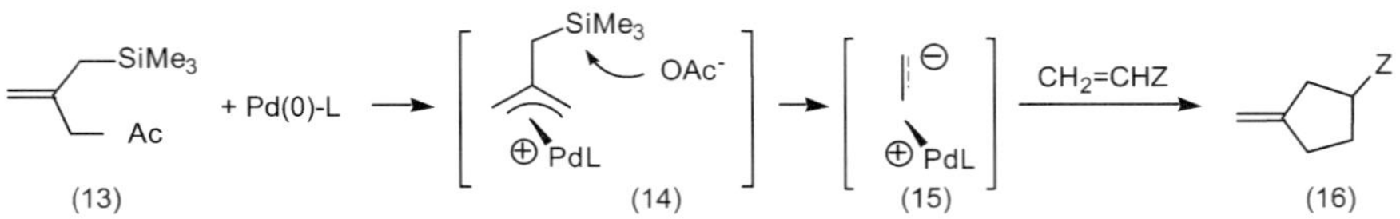

Scheme 5 [3+2] TMM cycloaddition reaction.

hydrogenation in this period [91], revealed the complexity [92] of enantiocontrol in allylic substitutions. Only quite recently, this trend was completely reversed with the profusion of more selective ligands [93] and the development of new concepts [94]. According to B.M. Trost, four different mechanistic conditions determine enantioselectivity: (1) enantiotopic alkene complexation, (2) enantiotopic leaving groups, (3) enantiotopic termini, and (4) enantiotopic nucleophilic additions. The occurrence of the stereodiscriminating bond-breaking and bond-forming steps outside of the coordination sphere as well as fluxional π-allyl Pd complexes furthermore complicate the reaction pathway. After modest evolution during nearly 20 years, the mid 1990s saw the start of a literal explosion, and it seems difficult today to have an exact overview of all available new chiral ligands. One essential feature for high chiral induction seems to be the chelating character of the ligand, but good results have also been published with monodentate species [95, 96]. Recently, the following chiral ligands have been employed quite successfully to induce a high degree of stereoselectivity. The first ligand type is the ferrocenylphosphines, with lateral, heteroatom-containing side arms, able to coordinate from the front and the back [97]. A second type includes the impressive number of recently developed systems that coordinate via two (identical or different) heteroatoms and where five- or six-membered heterometallacycles generate the efficient π-allyl palladium system (Fig. 1). A (nonexhaustive) list of these comprises diphosphines such as BINAP [98] or related phosphites BINAPO [99], bisphosphinites [100], pinane diphosphine [101], diamines such as bisaziridines [102], bispyrrolidines [103], aza-semicorrins [104], bisoxazolines [105], sparteine [106], bisdihydrooxazoles [107], phenanthrolines [108], bisheteroatomic ligands with P/N [109], N/Se [110], N/S [111], P/COOH [112], tris and higher heteroatoms P/P/N (BPPFA) [113], and P/P/COOH [114]. However, most general applicability is related to Trost's DPPA ligands diphosphinebisamides (18) [89, 94, 115], where the front is efficiently oriented by two diphenylphosphine linkers and the back locked by a chiral *scaffold*. A third type of ligands with multiple stereogenic elements is just emerging as a tool for improving the efficiency of catalytic processes [116].

Usually, the allylic alkylation of 1,3-diphenylallyl esters (23, R=Ph) serves as a test reaction, and many of the above-mentioned ligands give inductions higher than 90% ee. Other allylic alkylations, but also highly enantioselective allylic amination [97, 107], allylic silylation [117], or allylic sulfonylation [118], were of interest.

A special case of enantioselective allylations, where the question of the symmetry of π-allyl Pd intermediates is addressed [61b], are the reactions of symmetrical (*meso*) π-allyl complexes. With the same substrates, extensive NMR studies [119]

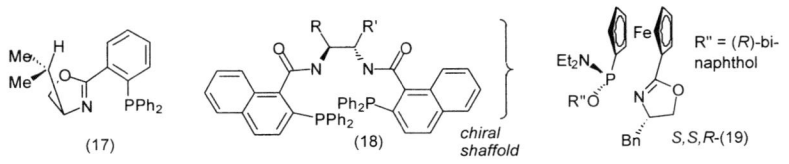

Fig. 1 Types of high-induction chiral ligands [94, 120, 121].

Scheme 6 Alkylation of 1,3-allyl systems.

(21) R = Ph, Nu = CH$_2$NO$_2$
*L = (17), > 96%, > 99%ee
(22) R = Me, Nu = CH$_3$C(CO$_2$CH$_3$)$_2$
*L = (18), > 80%, > 86%ee
*L = S,S,R-(19), 83%, 94%ee

of isolated (η^3-1,3-dialkylallyl) (phosphanyloxazoline)palladium complexes confirmed the proposition of a transition state (23) in enantioselective allylic alkylations [120].

In allylic alkylations of acyclic substrates, e.g., the 1,3-dimethylallyl system (20, R = Me), the high flexible π-allyl system can be *mastered* by ferrocene P,N-ligands (S,S,R)-19 [121] or by the chiral *invertoner* amide propeller ligands (18, R, R' = cyclohexane) [122] (induction of 94 and 86% ee, respectively) (Scheme 6).

Here a certain P–Pd–P "bite angle" θ [123] in Pd-alkene coordination compound (24) is required for good induction, depending on the substrate (cyclic or acyclic) (Scheme 7). This is caused by a subtle accommodation of rigidity of the amide-ester part and the geometry of the *trans* ethane in the chiral *scaffold*. Together with the two phosphine *linkers*, this generates chiral pockets which envelop and direct the reactants in the enantioselective reaction step.

The construction of quaternary carbon atoms is a difficult problem in synthesis. For a couple of years the asymmetric allylic alkylation (AAA) of carbonyl compounds mediated with chiral pocket ligands has offered elegant solutions for this purpose, giving rise to highly enantioselective reactions [124, 125]. It is remarkable that the creation of a chiral quaternary carbon is, under certain conditions, accompanied by the formation of a second chiral carbon with high diastereoselectivity [126]. This is exemplified by the synthesis of α-alkylated amino acids (Scheme 8).

One can conclude that good results and good enantioselectivities of 70–95% ee (best > 99%) [120a] with turnover numbers (TON) of 100 are routinely attained today, but the challenge will now be to search for catalysts with much higher TONs and to improve the levels to broad(er) (technical) applicability. One step in this direction is reported from Takemoto in Japan. No chiral ligand is required in the asymmetric allylation (94% ee) of glycine imino esters when a combination of tri-

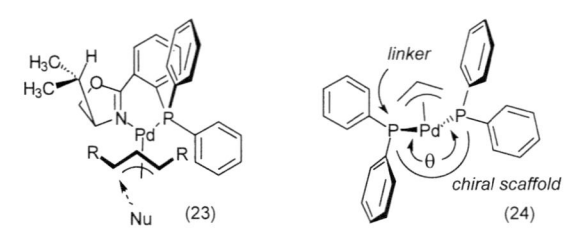

Scheme 7 Pd π-allyl intermediates of oxazoline [120] and DPPA [89, 94] ligands.

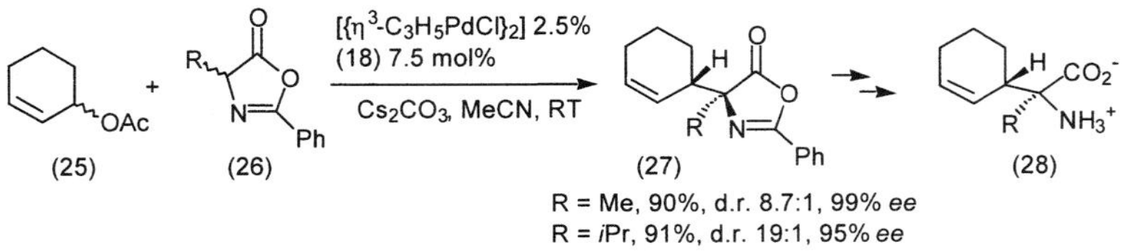

Scheme 8 Asymmetric synthesis of α-alkylated amino acids [126a].

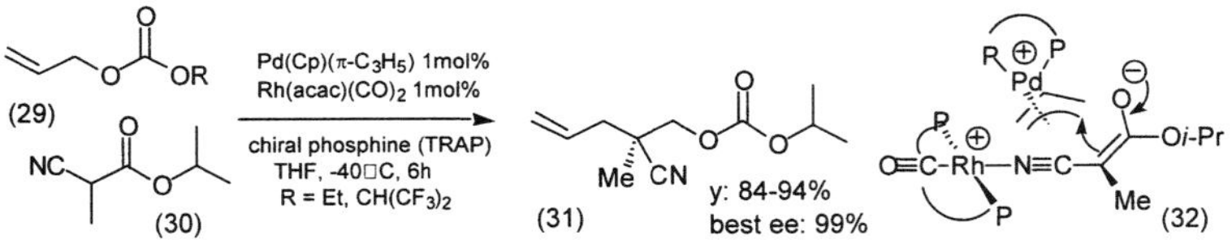

Scheme 9 Rhodium-palladium-catalyzed combined alkylation of activated nitriles [128].

phenyl phosphite and *O*-methyl-*N*-anthracenyl cinchonidinium iodide is used under phase-transfer (50% KOH, toluene) conditions [127].

A further way to proceed is demonstrated by the *combined alkylation of activated nitriles* [128], where the combination of two metal systems assures high yields and the highest enantioselectivity.

An electrophilic π-allyl palladium complex is formed from allylic carbonates generating simultaneously the base (OR⁻) that deprotonates the activated nitrile (**30**) (Scheme 9). In the Rh(I)-coordinated enolate (**32**), the orientation of enolate is controlled by the C_2 symmetrically and *trans*-bound TRAP ligand, a stereoorientation that reduces the C–C bond-forming step to one accessible enantioface. The rhodium complex induces chirality but only "bimetallic" Rh–Pd catalyst ensures both high yield and selectivity.

2.14.10
Preparative Glossary

Organic synthesis preparations: alkylation with malonate [129], amination [130] reaction of epoxyalkene [131], and allylic cross-coupling [132]; also Chapter 5 and a part of Chapter 6 in Heck's useful book [133].

2.14.11
References and Notes

1 R. W. FRIESEN in *Science of Synthesis. Houben-Weyl Methods of Molecular Transformations*, Vol. 1, (Ed.: M. LAUTENS), Georg Thieme, Stuttgart – New York, 2001, pp. 113–264.

2 For example, allylisothiocyanate in allyl mustard oil; the name *allyl* was intro-

duced by Wertheim in 1844 from *allium* (garlic). See (a) O.-A. NEUMÜLLER, *Römpps Chemie Lexikon*, Franck'sche Verlagshandlung, Stuttgart, **1979**, 8th edn., p. 140, and (b) A. NICKON, E. F. SILVERSMITH, *Organic Chemistry: The Name Game. Modern Coined Terms and their Origin*, Pergamon, New York, **1987**, p. 285.

3   KIRK-OTHMER, ENCYCLOPEDIA OF CHEMICAL TECHNOLOGY, Wiley, New York.

4   R. H. DEWOLFE, W. G. YOUNG, *Chem. Rev.* **1956**, *56*, 753–901.

5   Books: (a) J. TSUJI, *Palladium Reagents and Catalysts, Innovations in Organic Synthesis*, Wiley, Chichester, **1995**; (b) E. NEGISHI, in *Handbook of Organopalladium Chemistry for Organic Synthesis* (Ed.: A. DE MEIJERE), 2-Volume Set, Wiley, Chichester, **2002**.

6   (a) R. HÜTTEL, J. KRATZER, *Angew. Chem.* **1959**, *71*, 456. (b) R. HÜTTEL, *Synthesis* **1970**, 225–255.

7   It is interesting to note that this realization parallels very closely the other history-making discovery in organic palladium chemistry, the Wacker reaction. The Wacker team also isolated the first π-allyl Pd complex: J. SMIDT, W. HAFNER, *Angew. Chem.* **1959**, *71*, 284.

8   Preparation: W. T. DENT, R. LONG, A. J. WILKINSON, *J. Chem. Soc.* **1964**, 1585–1588.

9   (a) J. TSUJI, J. KIJI, S. IMAMURA, M. MORIKAWA, *J. Am. Chem. Soc.* **1964**, *86*, 4350–4353; (b) W. T. DENT, R. LONG, G. H. WHITFIELD, *J. Chem. Soc.* **1964**, 1588–1594.

10 J. TSUJI, H. TAKAHASHI, M. MORIKAWA, *Tetrahedron Lett.* **1965**, 4387–4388.

11 T. M. SHRYNE, E. J. SMUTNY, D. P. STEVENSON, US patent 3493617 **1970**; *Chem. Abstr.* **1970**, *72*, 78373e.

12 W. E. WALKER, R. M. MANYIK, K. E. ATKINS, M. L. FARMER, *Tetrahedron Lett.* **1970**, 3817–3820.

13 K. E. ATKINS, W. E. WALKER, R. M. MANYIK, *Tetrahedron Lett.* **1970**, 3821–3824.

14 Ref. [5a], Chapter 4.2. The book of Professor Tsuji contains exhaustive literature citations.

15 Reviews: (a) S. A. GODLESKI in *Comprehensive Organic Synthesis* (Eds.: B. M. TROST, I. FLEMING), Pergamon Press, Oxford, **1991**, p. 585–661; (b) B. M. TROST, T. R. VERHOEVEN in *Comprehensive Organometallic Chemistry* (Ed.: G. WILKINSON), Pergamon, Oxford, **1982**, pp. 799–938.

16 Reviews: (a) π-Allylmetal derivatives in organic synthesis: R. BAKER, *Chem. Rev.* **1973**, *73*, 487–530. (b) R. BAKER, *Chem. Ind.* **1980**, 816–824.

17 G. M. DIRENZO, P. S. WHITE, M. BROOKHART, *J. Am. Chem. Soc.* **1996**, *118*, 6225–6234.

18 J. TSUJI, *Adv. Organomet. Chem.* **1979**, *17*, 141–193.

19 P. W. JOLLY, G. WILKE in *Applied Homogeneous Catalysis by Organometallic Complexes* Vol. 2 (Eds.: B. CORNILS, W. A. HERRMANN), VCH, Weinheim, **1996**, pp. 1024–1048.

20 (a) T. HOSOKAWA, S.-I. MURAHASHI, *Acc. Chem. Res.* **1990**, *23*, 49–54; (b) T. HOSOKAWA, T. UNO, S. INUI, S.-I. MURAHASHI, *J. Am. Chem. Soc.* **1981**, *103*, 2318–2323.

21 Y. UOZUMI, K. KATO, T. HAYASHI, *J. Am. Chem. Soc.* **1997**, *119*, 5063–5064. (b) Y. UOZUMI, H. KYOTA, K. KATO, M. OGASAWARA, T. HAYASHI, *J. Org. Chem.* **1999**, *64*, 1620–1625.

22 For a theoretical study, see K. J. SZABO, *J. Am. Chem. Soc.* **1996**, *118*, 7818–7826.

23 (a) A. WILDE, A. R. OTTE, H. M. R. HOFFMANN, *J. Chem. Soc., Chem. Commun.* **1993**, 615–616; (b) A. R. OTTE, A. WILDE, H. M. R. HOFFMANN, *Angew. Chem. Int. Ed. Engl.* **1994**, *33*, 1280–1282.

24 H. M. R. HOFFMANN, A. R. OTTE, A. WILDE, S. MENZER, D. J. WILLIAMS, *Angew. Chem. Int. Ed. Engl.* **1995**, *34*, 100–102.

25 (a) A. M. CASTANO, A. ARANYOS, K. J. SZABO, J.-E. BÄCKVALL, *Angew. Chem. Int. Ed. Engl.* **1995**, *34*, 2551–2553. (b) A. ARANYOS, K. J. SZABO, A. M. CASTANO, J.-E. BÄCKVALL, *Organometallics* **1997**, *16*, 1058–1064.

26 (a) M. G. ORGAN, M. MILLER, *Tetrahedron Lett.* **1997**, *38*, 8181–8184. (b) M. G. ORGAN, M. MILLER, Z. KONSTANTINOU, *J. Am. Chem. Soc.* **1998**, *120*, 9283–9290.

27 (a) J. KADOTA, H. KATSURAGI, Y. FUKUMOTO, S. MURAI, *Organometallics* **2000**, *19*, 979–983. (b) J. KADOTA, S. KOMORI, Y.

Fukumoto, S. Murai, *J. Org. Chem.* **1999**, *64*, 7523–7527.
28 Reviews: (a) Ref. [4], (b) Ref. [14]
29 C. F. J. Barnard, M. J. H. Russell in *Comprehensive Coordination Chemistry* (Eds.: G. Wilkinson, R. D. Gillard, J. A. McCleverty), Pergamon, Oxford, **1987**, Vol. 5, pp. 1099–1130.
30 C. Amatore, A. Jutand, M. A. M'Barki, *Organometallics* **1992**, *11*, 3009–3013.
31 C. Amatore, A. Jutand, M. A. M'Barki, G. Meyer, L. Mottier, *Eur. J. Inorg. Chem.* **2001**, 873–880.
32 C. Amatore, A. Jutand, A. Suarez, *J. Am. Chem. Soc.* **1993**, *115*, 9531–9541.
33 F. Guibe, Y. Saint M'Leux, *Tetrahedron Lett.* **1981**, *22*, 3591–3594.
34 Review: J. Tsuji, I. Minami, *Acc. Chem. Res.* **1987**, *20*, 140–145.
35 M. Kimura, Y. Horino, R. Mukai, S. Tanaka, Y. Tamaru, *J. Am. Chem. Soc.* **2001**, *123*, 10401–10402.
36 Y.-J. Shue, S.-C. Yang, H.-C. Lain, *Tetrahedron Lett.* **2003**, *44*, 1481–1485.
37 Reviews: (a) metallocene: W. Oppolzer, *Angew. Chem. Int. Ed. Engl.* **1989**, *28*, 38–52; (b) A. Heumann, M. Réglier, *Tetrahedron* **1995**, *51*, 975–1015 corr. 9509; (c) cascade reactions: A. Heumann, M. Réglier, *Tetrahedron* **1996**, *52*, 2989–3046; (d) B. M. Trost, *Angew. Chem., Int. Ed. Engl.* **1989**, *28*, 1173–1192.
38 Ref. [5b], Section 2.10 (E. Negishi).
39 Section 3.1, Volume 2 (D. Sinou).
40 (a) J. Tsuji, *Tetrahedron* **1986**, *42*, 4361–4401; (b) B. M. Trost, *Tetrahedron* **1977**, *33*, 2615–2649.
41 (a) Y. Tamaru, A. Tanaka, K. Yasui, S. Goto, S. Tanaka, *Angew. Chem. Int. Ed. Engl.* **1995**, *34*, 787–789. (b) Review: Y. Tamaru, *J. Organomet. Chem.* **1999**, *576*, 215–231. (c) Cf. also K. J. Szabo, *Chem. Eur. J.* **2000**, *6*, 4413–4421.
42 Phosphines provoke elimination to dienes.
43 (a) N. Nomura, K. Tsurugi, M. Okada, *J. Am. Chem. Soc.* **1999**, *121*, 7268–7269. (b) N. Nomura, K. Tsurugi, M. Okada, *Angew. Chem. Int. Ed. Engl.* **2001**, *40*, 1932–1935.
44 A. Ohsuka, T. W. Wardhana, H. Kurosawa, I. Ikeda, *Organometallics* **1997**, *16*, 3038–3043.
45 Y. Ito, H. Aoyama, T. Hirao, A. Mochizuki, T. Saegusa, *J. Am. Chem. Soc.* **1979**, *101*, 494–496.
46 Y. Ito, M. Nakatsuka, N. Kise, T. Saegusa, *Tetrahedron Lett.* **1980**, 2873–2876.
47 M. J. O'Donnell, M. Li, W. D. Bennett, T. Grote, *Tetrahedron Lett.* **1994**, *35*, 9383–9386.
48 Review: B. M. Trost, *Acc. Chem. Res.* **1980**, *13*, 385–393.
49 (a) J.-E. Bäckvall, R. E. Nordberg, K. Zetterberg, B. Åkermark, *Organometallics* **1983**, *2*, 1625–1629. (b) Review: B. Åkermark, J. E. Bäckvall, K. Zetterberg, *Acta Chem. Scand.* **1982**, *B36*, 577–585.
50 Review: H. Kurosawa, *J. Organomet. Chem.* **1987**, *334*, 243–253.
51 Review: J. C. Fiaud in *Metal Promoted Selectivity in Organic Synthesis* (Eds.: M. Graziani, A. J. Hubert, A. F. Noels), Kluwer Academic, Dordrecht, **1991**, pp. 107–131.
52 (a) I. Stary, P. Kocovsky, *J. Am. Chem. Soc.* **1989**, *111*, 4981–4982. (b) C. N. Farthing, P. Kocovsky, *J. Am. Chem. Soc.* **1998**, *120*, 6661–6672. c) M. E. Krafft, A. M. Wilson, Z. Fu, M. J. Procter, O. A. Dasse, *J. Org. Chem.* **1998**, *63*, 1748–1749.
53 (a) H. Kurosawa, S. Ogoshi, Y. Kawasaki, S. Murai, M. Miyoshi, I. Ikeda, *J. Am. Chem. Soc.* **1990**, *112*, 2813–2814; (b) H. Kurosawa, H. Kajimura, S. Ogoshi, H. Yoneda, K. Miki, N. Kasai, S. Murai, I. Ikeda, *J. Am. Chem. Soc.* **1992**, *114*, 8417–8424.
54 E. Keinan, Z. Roth, *J. Org. Chem.* **1983**, *48*, 1769–1772.
55 Acetate may react with overall retention or inversion.
56 Examples: (a) B. M. Trost, E. Keinan, *J. Am. Chem. Soc.* **1978**, *100*, 7779–7781; (b) B. M. Trost, T. R. Verhoeven, *J. Am. Chem. Soc.* **1980**, *102*, 4730–4743; (c) R.-E. Nordberg, J.-E. Bäckvall, *J. Organomet. Chem.* **1985**, *385*, C24–C26; (d) E. Keinan, M. Sahai, Z. Roth, A. Nudelman, J. Herzig, *J. Org. Chem.* **1985**, *50*, 3558–3566; (e) M. Moreno-Manas, J. Ribas, A. Virgili, *J. Org. Chem.* **1988**, *53*, 5328–5335.

57 J.-E. BÄCKVALL, K. L. GRANBERG, A. HEUMANN, Isr. J. Chem. **1991**, *31*, 17–24.
58 K. L. GRANBERG, J.-E. BÄCKVALL, J. Am. Chem. Soc. **1992**, *114*, 6858–6863.
59 M. E. KRAFFT, M. SUGIURA, K. A. ABBOUD, J. Am. Chem. Soc. **2001**, *123*, 9174–9175.
60 K. ITAMI, T. KOIKE, J. YOSHIDA, J. Am. Chem. Soc. **2001**, *123*, 6957–6958.
61 (a) J. C. FIAUD, J. L. MALLERON, Tetrahedron Lett. **1981**, *22*, 1399–1402. (b) B. M. TROST, R. C. BUNT, J. Am. Chem. Soc. **1996**, *118*, 235–236. (c) T. HAYASHI, M. KAWATSURA, Y. UOZUMI, J. Am. Chem. Soc. **1998**, *120*, 1681–1687.
62 (a) G. C. LLOYD-JONES, S. C. STEPHEN, Chem. Eur.J. **1998**, *4*, 2539–2549. (b) G. C. LLOYD-JONES, Synlett **2001**, 161–182.
63 Y. I. M. NILSSON, P. G. ANDERSSON, J. E. BÄCKVALL, J. Am. Chem. Soc. **1993**, *115*, 6609–6613.
64 (a) A. P. DAVIS, B. J. DORGAN, E. R. MAGEEAN, J. Chem. Soc., Chem. Comm. **1993**, 492–494. (b) C. AMATORE, S. GAMEZ, A. JUTAND, G. MEYER, M. MORENO-MANAS, L. MORRAL, R. PLEIXATS, Chem. Eur. J. **2000**, *6*, 3372–3376.
65 W. D. MCGHEE, D. P. RILEY, M. E. CHRIST, K. M. CHRIST, Organometallics **1993**, *12*, 1429–1433.
66 H. HEY, H.-J. ARPE, Angew. Chem. Int. Ed. Engl. **1973**, *12*, 928–929.
67 Review: J. TSUJI, T. MANDAI, Synthesis **1996**, 1–24.
68 M. H. RABINOWITZ, Tetrahedron Lett. **1991**, *32*, 6081–6084.
69 T. HAYASHI, M. KAWATSURA, H. IWAMURA, Y. YAMAURA, Y. UOZUMI, J. Chem. Soc., Chem. Commun. **1996**, 1767–1768.
70 M. OSHIMA, T. SAKAMOTO, Y. MARUYAMA, F. OZAWA, I. SHIMIZU, A. YAMAMOTO, Bull. Chem. Soc. Jpn. **2000**, *73*, 453–464.
71 B. BURNS, R. GRIGG, V. SANTHAKUMAR, V. SRIDHARAN, P. STEVENSON, T. WORAKUN, Tetrahedron **1992**, *48*, 7297–7320.
72 (a) T. HAYASHI, K. KISHI, Y. UOZUMI, Tetrahedron: Asymmetry **1991**, *2*, 195–198. (b) E. B. KOROLEVA, P. G. ANDERSSON, Tetrahedron: Asymmetry **1996**, *7*, 2467–2470.
73 J. M. TAKACS, E. C. LAWSON, F. CLEMENT, J. Am. Chem. Soc. **1997**, *119*, 5956–5957.
74 Reviews: (a) M. SCHELHAAS, H. WALDMANN, Angew. Chem. Int. Ed. Engl. **1996**, *35*, 2056–2083. (b) P. J. KOCIENSKI, Protecting Groups, Georg Thieme, Stuttgart, **2000**, 2nd edn. c) F. GUIBÉ, Tetrahedron **1997**, *53*, 13509–13556 and **1998**, *54*, 2967–3042.
75 Reviews: (a) B. M. TROST, Angew. Chem. Int. Ed. Engl. **1986**, *25*, 1–20. (b) D. M. T. CHAN in Comprehensive Organic Synthesis (Eds.: B. M. TROST, I. FLEMING), Pergamon Press, Oxford, **1991**, pp. 271–314.
76 B. M. TROST, C. M. MARRS, J. Am. Chem. Soc. **1993**, *115*, 6636–6645.
77 C. K. F. SHEN, K.-M. CHIEN, T.-Y. LIU, T.-I. LIN, G.-R. HER, T.-Y. LUH, Tetrahedron Lett. **1995**, *36*, 5383–5384.
78 B. M. TROST, B. YANG, M. L. MILLER, J. Am. Chem. Soc. **1989**, *111*, 6482–6484.
79 Reviews: (a) L. E. OVERMAN, Angew. Chem. Int. Ed. Engl. **1984**, *23*, 579–586. (b) L. S. HEGEDUS in Comprehensive Organic Synthesis (Eds.: B. M. TROST, I. FLEMING), Pergamon Press, Oxford, **1991**, Vol. 4, pp. 551–569.
80 T. G. SCHENCK, B. BOSNICH, J. Am. Chem. Soc. **1985**, *107*, 2058–2066.
81 More recent examples, (a) M. SUGIURA, T. NAKAI, Tetrahedron Lett. **1996**, *37*, 7991–7994. (b) K. ITAMI, D. YAMAZAKI, J. YOSHIDA, Org. Lett. **2003**, *5*, 2161–2164.
82 H. NAKAMURA, H. IWAMA, M. ITO, Y. YAMAMOTO, J. Am. Chem. Soc. **1999**, *121*, 10850–10851.
83 P. A. GRIECO, T. TAKIGAWA, S. L. BONGERS, H. TANAKA, J. Am. Chem. Soc. **1980**, *102*, 7587–7588.
84 (a) M. CALTER, T. K. HOLLIS, L. E. OVERMAN, J. ZILLER, G. G. ZIPP, J. Org. Chem. **1997**, *62*, 1449–1456. (b) Review: T. K. HOLLIS, L. E. OVERMAN, J. Organomet. Chem. **1999**, *576*, 290–299. (c) Y. DONDE, L. E. OVERMAN, J. Am. Chem. Soc. **1999**, *121*, 2933–2934.
85 B. M. TROST, M. G. ORGAN, J. Am. Chem. Soc. **1994**, *116*, 10320–10321.
86 (a) R. NOYORI, Asymmetric Catalysis in Organic Synthesis, Wiley, New York, **1994**. (b) OJIMA, Catalytic Asymmetric Synthesis, VCH, New York, **1993**.
87 B. M. TROST, P. E. STREGE, J. Am. Chem. Soc. **1977**, *99*, 1649–1651.
88 J. C. FIAUD, A. HIBON DE GOURNAY, M. LARCHEVEQUE, H. B. KAGAN, J. Organomet. Chem. **1978**, *154*, 175–185.

89 Review: B. M. Trost, D. L. van Vranken, *Chem. Rev.* **1996**, *96*, 395–422.
90 Other reviews: (a) A. Pfaltz, *Chimia* **2001**, *55*, 708–714. b) G. Helmchen, *J. Organomet. Chem.* **1999**, *576*, 203–214. c) O. Reiser, *Angew. Chem. Int. Ed. Engl.* **1993**, *32*, 547–549. (d) C. G. Frost, J. Howarth, J. M. J. Williams, *Tetrahedron: Asymmetry* **1992**, *3*, 1089–1122.
91 Review: D. Arntz, A. Schafer in *Metal Promoted Selectivity in Organic Synthesis* (Eds.: A. F. Noels, M. Graziani, A. J. Hubert), Kluwer, Dordrecht, **1991**, pp. 161–189.
92 P. B. Mackenzie, J. Whelan, B. Bosnich, *J. Am. Chem. Soc.* **1985**, *107*, 2046–2054.
93 Personal accounts of this period: (a) J. M. J. Williams, *Synlett* **1996**, 705–710. (b) A. Pfaltz, *Synlett* **1999**, 835–842.
94 Review: B. M. Trost, *Acc. Chem. Res.* **1996**, *29*, 355–364.
95 G. Brenchley, M. Fedouloff, M. F. Mahon, K. C. Molloy, M. Wills, *Tetrahedron* **1995**, *51*, 10581–10592.
96 Reviews: (a) T. Hayashi, *Acc. Chem. Res.* **2000**, *33*, 354–362. (b) T. Hayashi, *J. Organomet. Chem.* **1999**, *576*, 195–202.
97 (a) T. Hayashi, A. Yamamoto, Y. Ito, E. Nishioka, H. Miura, K. Yanagi, *J. Am. Chem. Soc.* **1989**, *111*, 6301–6311. (b) *Ferrocenes – Homogeneous Catalysis. Organic Synthesis. Material Science*, (Eds. A. Togni, T. Hayashi), VCH, Weinheim, **1995**.
98 M. Yamaguchi, T. Shima, T. Yamagishi, M. Hida, *Tetrahedron Lett.* **1990**, *31*, 5049–5052.
99 H. Yoshizaki, H. Satoh, Y. Sato, S. Nukui, M. Shibasaki, M. Mori, *J. Org. Chem.* **1995**, *60*, 2016–2021.
100 D. S. Clyne, Y. C. Mermet-Bouvier, N. Nomura, T. V. RajanBabu, *J. Org. Chem.* **1999**, *64*, 7601–7611.
101 R. L. Halterman, H. L. Nimmons, *Organometallics* **1990**, *9*, 273–275.
102 P. G. Andersson, A. Harden, D. Tanner, P.-O. Norrby, *Chem. Eur. J.* **1995**, *1*, 12–16.
103 H. Kubota, M. Nakajima, K. Koga, *Tetrahedron Lett.* **1993**, *34*, 8135–8138.
104 A. Pfaltz, *Acc. Chem. Res.* **1993**, *26*, 339–345.
105 A. K. Ghosh, P. Mathivanan, J. Cappiello, *Tetrahedron: Asymmetry* **1998**, *9*, 1–45.
106 A. Togni, *Tetrahedron: Asymmetry* **1991**, *2*, 683–690.
107 P. von Matt, G. C. Lloyd-Jones, A. B. E. Minidis, A. Pfaltz, L. Macko, M. Neuburger, M. Zehnder, H. Rüegger, P. S. Pregosin, *Helv. Chim. Acta* **1995**, *78*, 265–284.
108 E. Pena-Cabrera, P.-O. Norrby, M. Sjögren, A. Vitagliano, V. De Felice, J. Oslob, S. Ishii, D. O'Neill, B. Åkermark, P. Helquist, *J. Am. Chem. Soc.* **1996**, *118*, 4299–4313.
109 (a) P. von Matt, O. Loiseleur, G. Koch, A. Pfaltz, C. Lefeber, T. Feucht, G. Helmchen, *Tetrahedron: Asymmetry* **1994**, *5*, 573–584. (b) A. Togni, U. Burckhardt, V. Gramlich, P. S. Pregosin, R. Salzmann, *J. Am. Chem. Soc.* **1996**, *118*, 1031–1037. (c) D.-R. Hou, J. H. Reibenspies, K. Burgess, *J. Org. Chem.* **2001**, *66*, 206–215. Reviews: (d) G. Helmchen, A. Pfaltz, *Acc. Chem. Res.* **2000**, *33*, 336–345. (e) F. Agbossou, J.-F. Carpentier, F. Hapiot, I. Suisse, A. Mortreux, *Coord. Chem. Rev.* **1998**, *178–180*, 1615–1645.
110 J. Sprinz, M. Kiefer, G. Helmchen, M. Reggelin, G. Huttner, O. Walter, L. Zsolnai, *Tetrahedron Lett.* **1994**, *35*, 1523–1526.
111 (a) J. V. Allen, S. J. Coote, G. J. Dawson, C. G. Frost, C. J. Martin, J. M. J. Williams, *J. Chem. Soc. Perkin Trans. 1* **1994**, 2065–2072. (b) G. Chelucci, M. A. Cabras, *Tetrahedron: Asymmetry* **1996**, *7*, 965–966.
112 (a) G. Knühl, P. Sennhenn, G. Helmchen, *J. Chem. Soc., Chem. Comm.* **1995**, 1845–1846. (b) E. J. Bergner, G. Helmchen, *Eur. J. Org. Chem.* **2000**, 419–423.
113 (a) G. Zhu, M. Terry, X. Zhang, *Tetrahedron Lett.* **1996**, *37*, 4475–4478. (b) W. Zhang, T. Hirao, I. Ikeda, *Tetrahedron Lett.* **1996**, *37*, 4545–4548.
114 A. Yamazaki, K. Achiwa, *Tetrahedron: Asymmetry* **1995**, *6*, 51–54.
115 B. M. Trost, B. Breit, S. Peukert, J. Zambrano, J. W. Ziller, *Angew. Chem. Int. Ed. Engl.* **1995**, *34*, 2386–2388.

116 K. Muniz, C. Bolm, *Chem. Eur. J.* **2000**, *6*, 2309–2316.

117 T. Hayashi, A. Ohno, S. Lu, Y. Matsumoto, E. Fukuyo, K. Yanagi, *J. Am. Chem. Soc.* **1994**, *116*, 4221–4226.

118 (a) H. Eichelmann, H.-J. Gais, *Tetrahedron: Asymmetry* **1995**, *6*, 643–646. (b) B.M. Trost, M.G. Organ, G.A. O'Doherty, *J. Am. Chem. Soc.* **1995**, *117*, 9662–9670.

119 Supported by X-ray crystallographic studies and quantum-chemical calculations.

120 (a) H. Rieck, G. Helmchen, *Angew. Chem. Int. Ed. Engl.* **1995**, *34*, 2687–2689. (b) H. Steinhagen, M. Reggelin, G. Helmchen, *Angew. Chem. Int. Ed. Engl.* **1997**, *36*, 2108–2110. (c) J. Junker, B. Reif, H. Steinhagen, B. Junker, I.C. Felli, M. Reggelin, C. Griesinger, *Chem. Eur. J.* **2000**, *6*, 3281–3286. d) M. Kollmar, B. Goldfuss, M. Reggelin, F. Rominger, G. Helmchen, *Chem. Eur. J.* **2001**, *7*, 4913–4927.

121 S.-L. You, X.-Z. Zhu, Y.-M. Luo, X.-L. Hou, L.-X. Dai, *J. Am. Chem. Soc.* **2001**, *123*, 7471–7472.

122 (a) B.M. Trost, R.C. Bunt, *Angew. Chem. Int. Ed. Engl.* **1996**, *35*, 99–102. (b) B.M. Trost, A.C. Krueger, R.C. Bunt, J. Zambrano, *J. Am. Chem. Soc.* **1996**, *118*, 6520–6521.

123 P. Dierkes, P.W.N.M. van Leeuwen, *J. Chem. Soc. Dalton Trans.* **1999**, 1519–1529.

124 β-Ketoesters: (a) B.M. Trost, R. Radinov, E.M. Grenzer, *J. Am. Chem. Soc.* **1997**, *119*, 7879–7880. (b) B.M. Trost, C. Jiang, *J. Am. Chem. Soc.* **2001**, *123*, 12907–12908. (c) α-Aryl ketones: B.M. Trost, G.M. Schroeder, J. Kristensen, *Angew. Chem. Int. Ed. Engl.* **2002**, *41*, 3492–3495. (d) Alcohol pronucleophiles: B.M. Trost, E.J. McEachern, F.D. Toste, *J. Am. Chem. Soc.* **1998**, *120*, 12702–12703.

125 (a) α-Acetamido-β-ketoesters: R. Kuwano, Y. Ito, *J. Am. Chem. Soc.* **1999**, *121*, 3236–3237. (b) Ketones: S.-L. You, X.-L. Hou, L.-X. Dai, X.-Z. Zhu, *Org. Lett.* **2001**, *3*, 149–151.

126 (a) B.M. Trost, X. Ariza, *Angew. Chem. Int. Ed. Engl.* **1997**, *36*, 2635–2637. (b) B.M. Trost, J.-P. Surivet, *J. Am. Chem. Soc.* **2000**, *122*, 6291–6292.

127 M. Nakoji, T. Kanayama, T. Okino, Y. Takemoto, *J. Org. Chem.* **2002**, *67*, 7418–7423.

128 M. Sawamura, M. Sudoh, Y. Ito, *J. Am. Chem. Soc.* **1996**, *118*, 3309–3310.

129 J.E. Bäckvall, J.O. Vagberg, *Organic Syntheses Coll. Vol. VIII* **1993**, 5–8.

130 J.E. Nyström, T. Rein, J.E. Bäckvall, *Organic Syntheses Coll. Vol. VIII* **1993**, 9–13.

131 D.R. Deardorff, D.C. Myles, *Organic Syntheses Coll. Vol. VIII* **1993**, 13–16.

132 E. Negishi, H. Matsushita, *Organic Syntheses Coll. Vol. VII* **1990**, 245–248.

133 R.F. Heck, *Palladium Reagents in Organic Synthesis*, Academic Press, London, **1985**.

2.15
Alkene and Alkyne Metathesis in Organic Synthesis

Oliver R. Thiel

2.15.1
Introduction

Only slightly more than a decade has passed since the discovery of the first highly active and well-defined homogeneous catalysts for olefin metathesis and the first reports of the use of ring-closing metathesis (RCM) in organic synthesis. Nonetheless, the metathesis reaction has already made an enormous impact on the field of modern organic synthesis. Since this field has been extensively reviewed in the last couple of years, only a brief overview and a summary of the most current developments are given in this review [1, 2].

2.15.2
Alkene Metathesis

The term olefin metathesis defines the mutual exchange of alkylidene fragments between two alkenes. The intramolecular version of the reaction leads to cyclic products and is therefore described as ring-closing metathesis (RCM). The intermolecular reaction between two olefins is described as cross metathesis (CM). In the early days of metathesis chemistry, the reaction was catalyzed by heterogeneous early transition metal catalysts that were activated with alkylating reagents. These reaction conditions are not well tolerated by most functional groups, and therefore application of the reaction was limited to very simple alkenes. Nonetheless, the reaction found industrial application in the Shell Higher Olefin Process (SHOP) and in the ring-opening metathesis polymerization (ROMP) of strained olefins [3]. The mechanistic proposal of Chauvin (Scheme 1) [4], that involved metal carbenes as the active catalytic species and a series of [2+2] cycloadditions and cycloreversion as elemental steps, spurred fundamental research in various organometallic laboratories [5].

This research eventually led to the discovery of well-defined transition metal carbene complexes as catalysts for metathesis, thereby enabling the success of this reaction in organic synthesis in the last decade (Scheme 2). The first broadly used catalyst was the molybdenum complex **1** introduced by Schrock in 1990 [6]. This

Transition Metals for Organic Synthesis, Vol. 1, 2nd Edition.
Edited by M. Beller and C. Bolm

ISBN: 3-527-30613-7

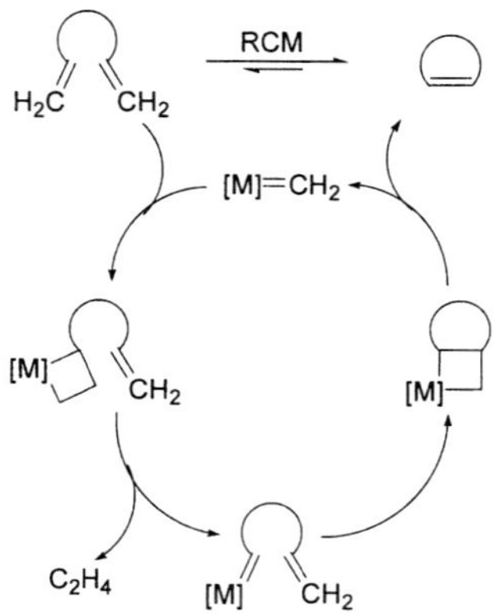

Scheme 1

complex displays remarkably high catalytic activity, but because of the high sensitivity toward air and moisture it has recently been replaced by second-generation ruthenium complexes in many applications. Grubbs introduced ruthenium carbene complexes of type **2** in 1992 [7]. The excellent functional group compatibility and the robustness of these catalysts made them the most popular catalysts until the late nineties. The major drawback compared to the Mo-complex **1** was their diminished catalytic activity; many substrates could not be converted with the first-generation ruthenium complexes. Herrmann introduced N-heterocyclic carbenes (NHC) as ligands for the ruthenium complexes [8]. Based on this work, Grubbs, Nolan and Herrmann independently described the ruthenium complexes of type **3**, where one of the phosphine ligands of complex **2** is replaced by an unsaturated NHC-ligand [9]. This catalyst shows greatly enhanced reactivity in olefin metathesis. Catalyst **4** bearing a saturated NHC-ligand instead of the phosphine was described shortly thereafter by Grubbs, and this catalyst leads to even higher catalytic activity [10]. The recyclable catalyst **5**, introduced by Hoveyda [11], yields better results than the parent complex **4** in selected applications [12]. Subsequently, modified versions of catalyst **5** have appeared in the literature showing even faster initiation rates [13]. Currently, catalysts **1**, **2**, **4** and **5** are commercially available.

The enhanced catalytic activity of the NHC-substituted catalysts **3** and **4** was initially attributed to facilitated phosphine dissociation due to the large *trans*-effect of

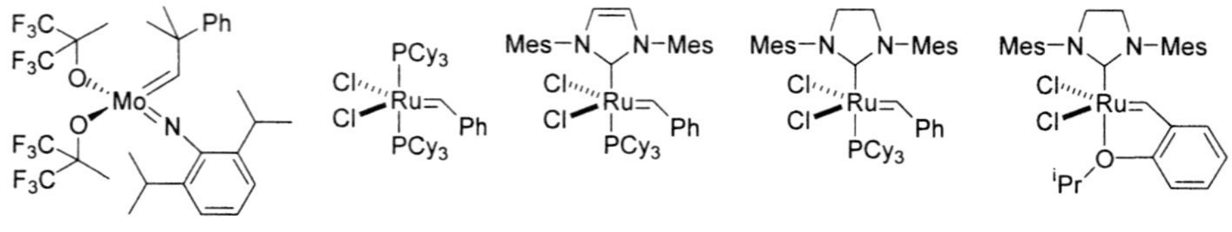

Scheme 2

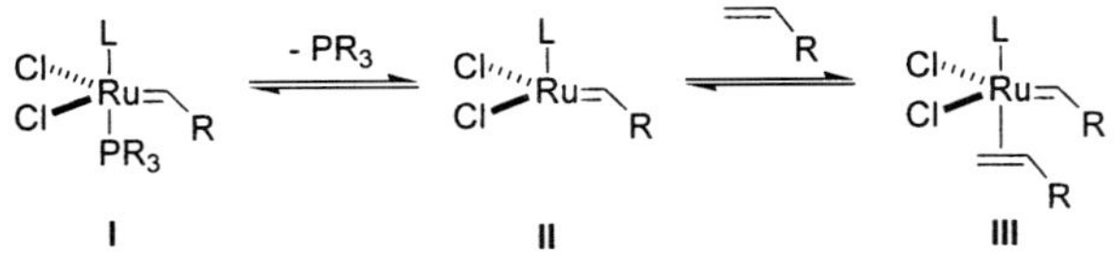

Scheme 3

the NHC-ligands. Elegant kinetic work by Grubbs revealed another explanation for the increase in reactivity (Scheme 3) [14]. Dissociation of the phosphine from **I** leads to the 14-electron intermediate **II**. The phosphine dissociation constant is lower for the NHC-complex **4** than for the parent complex **2**. The increase of reactivity is mainly due to a higher propensity of complex **4** to coordinate π-donors (olefins) in the presence of σ-donors (phosphines). Subsequent theoretical work is in good agreement with the experimental results and underscored the importance of steric factors for the enhanced reactivity of the NHC-substituted complexes [15].

One general problem associated with the use of transition metal catalysts is the removal of metal impurities from the products. In the case of the ruthenium catalysts this can be of great importance, since there have been reports of product decomposition in the presence of catalyst residues. Treatment with tris(hydroxymethyl)phosphine [16a], Pb(OAc)$_4$ [16b], DMSO [16c] or activated carbon [16d] have been described as effective methods for the removal of ruthenium from the reaction mixtures.

Substantial efforts have been devoted to the immobilization of the ruthenium catalysts; an efficient catalyst recovery would facilitate the workup and make the metathesis reaction in organic synthesis more attractive for industry. Hoveyda developed a variant of catalyst **5** which is immobilized on a dendrimer [11]. Several groups described the immobilization of the ruthenium catalysts on soluble polymers [17] and on monolithic materials [18]. Most of the supported complexes show lower catalytic activity than their parent systems, but usually the catalyst can be reused for several cycles.

Although dichloromethane and toluene are used as solvents in most metathesis reactions, the use of alternative solvents might improve product isolation and catalyst recovery. Supercritical carbon dioxide was employed successfully as solvent for ROMP and RCM [19]. Ionic liquids have also been employed as solvents for the olefin metathesis. Extraction of the ionic liquid with an organic solvent yields the product; the ruthenium catalyst remains in the ionic liquid and can be reused over several cycles [20].

The ruthenium-based catalysts tolerate most polar functional groups, the only notable exceptions being amines, thioethers, and phosphines. The classical ruthenium catalyst had only limited activity toward sterically encumbered double bonds. With the second-generation catalysts, tri- and even tetrasubstituted double bonds can be assembled efficiently (Tab. 1) [9a, 21]. The formation of trisubstituted double bonds is even feasible in the formation of macrocycles [22]. Another major advancement over the first-generation catalyst is the enhanced reactivity to-

Tab. 1

Product	Catalyst (mol%)	Yield (%)	Ref.
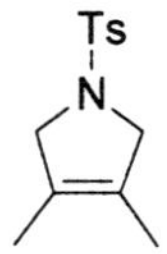 (pyrroline with Ts on N, dimethyl)	3 [2.5]	96	21
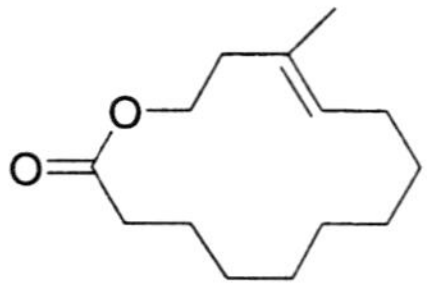 (macrolactone with trisubstituted alkene)	2 [10]	65	22
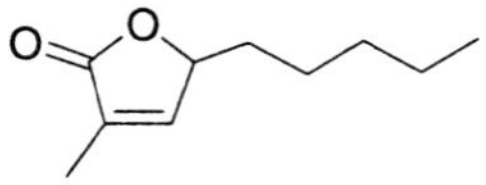 (butenolide with pentyl and methyl substituents)	3 [5]	92	21
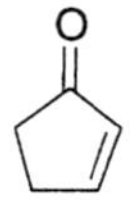 (cyclopentenone)	4 [5]	93	23
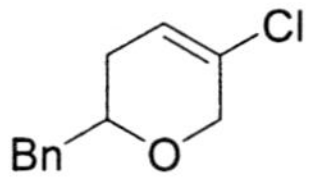 (Bn, O, Cl-substituted dihydropyran)	4 [10]	84	24
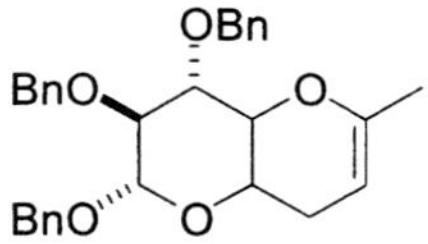 (bis-BnO, OBn bicyclic pyranopyran with methyl)	4 [20]	89	26
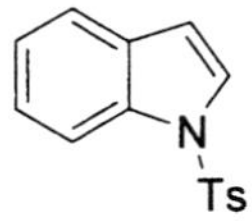 (N-Ts indole derivative)	4 [5]	94	27
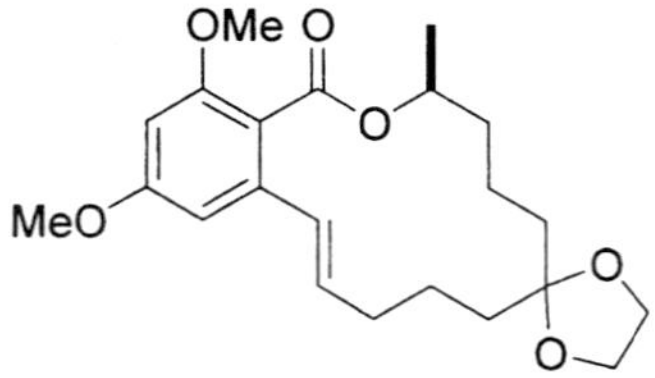 (macrocycle with OMe, MeO, dioxolane)	3 [5]	91	28

ward electron-poor olefins. Acrylates, methacrylates, and α,β-unsaturated ketones cyclize smoothly with catalyst **3** or **4** [21, 23]. The same holds true for vinyl chlorides [24] and vinyl fluorides [25]. Electron-rich olefins cyclize very well with the molybdenum-based catalysts, but with the ruthenium-based catalysts relatively high catalyst loadings are necessary [26]. Styrene derivatives are also attractive substrates for the RCM now, and successful examples range from indoles [27] to macrolides [28].

The olefin metathesis is a reversible reaction, but only the reactivity of the NHC-substituted catalysts toward 1,2-disubstituted double bonds allows the formation of the most thermodynamically favored products. This feature can enhance the utility of metathesis greatly. Smith exploited this reversibility in the synthesis of cylindrocyclophane F (Scheme 4) [29]. Out of the seven possible dimeric products, the thermodynamically favored compound is formed in good selectivity. Undesired isomers can be transformed to the desired product by resubjecting them to the metathesis reaction.

The difference between kinetic and thermodynamic product control also becomes apparent in the (*E*/*Z*)-selectivity of the metathesis reaction. The use of catalyst **3** in a synthetic study toward herbarumin leads to the thermodynamically favored (*Z*)-isomer (Scheme 5) [30]. The desired kinetically favored (*E*)-isomer can be obtained by using the less active catalyst **2**. In this case the (*E*/*Z*)-selectivity could be controlled by the choice of the catalyst; in other cases the selectivity is highly dependent on small changes in the substrate. This became apparent in the synthesis of Coleophomones B and C by Nicolaou (Scheme 6) [31].

In addition to the above-mentioned examples, numerous other applications of RCM in the synthesis of natural products have been reported. These range from the formation of simple 5- or 6-membered rings to the formation of macrolides. A whole compilation of these examples is not possible in this chapter, and therefore only three particularly noteworthy structures are discussed here (Scheme 7). The formation of a triene by RCM was explored in the synthesis of Oximidine II [32]. The synthesis of the proposed structure of Amphidinolide A showcased that several olefins can survive the RCM as innocent bystanders [33]. The synthesis of Nakadomarin A utilized RCM for the closure of both the 8- and 15-membered azacycles [34]. Here again a weakness of current RCM methodology came into play. The undesired (*E*)-isomer was obtained as the major product when the macrocyclic ring was closed.

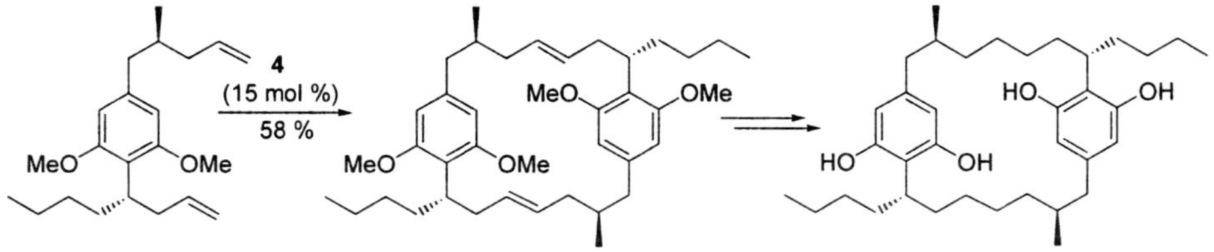

Scheme 4

Cylindrocyclophane F

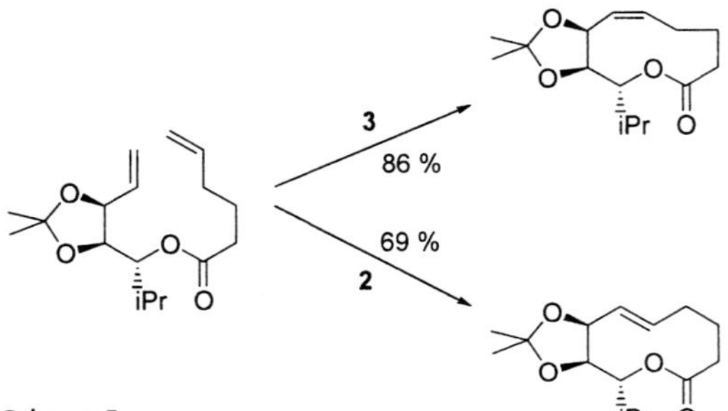

Scheme 5

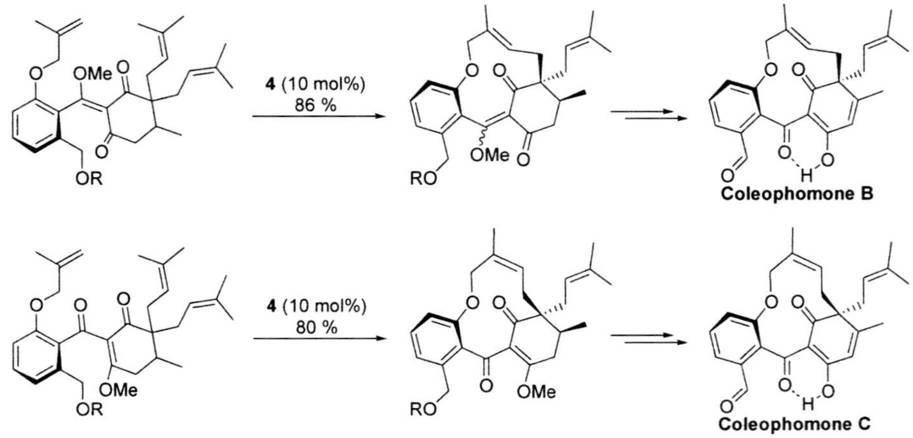

Scheme 6

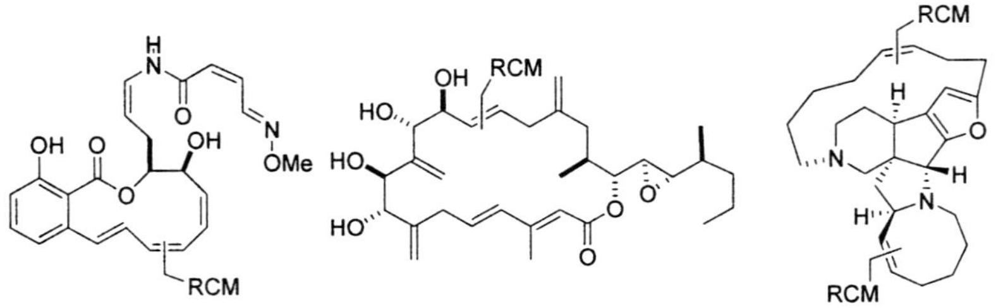

Scheme 7

Oximidine II

Proposed structure of Amphidinolide A

Nakadomarin A

Compared to RCM, cross metathesis (CM) has played only a minor role in organic synthesis thus far [2d]. This is mainly due to the inherent selectivity problem. A cross metathesis between two unfunctionalized olefins will lead to a statistical mixture of all possible products (Scheme 8). The second selectivity issue con-

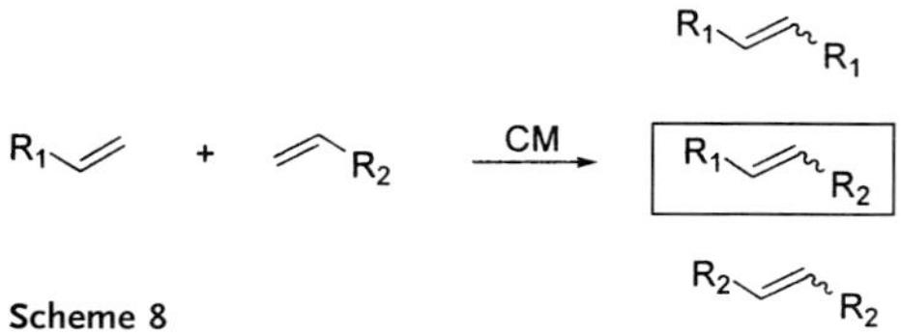

2.15.2 Alkene Metathesis

Scheme 8

$R_1{\diagup} + {\diagup}R_2 \xrightarrow{CM} \boxed{\begin{array}{c} R_1{\diagup}{\diagdown}R_1 \\ R_1{\diagup}{\diagdown}R_2 \\ R_2{\diagup}{\diagdown}R_2 \end{array}}$

cerns the stereochemistry. For unfunctionalized double bonds, mixtures are usually formed, the (E)-olefin being the major component in most cases. Some of these selectivity issues have been solved in the past years. Depending on the catalyst system, kinetically different reactivity towards functionalized olefins is observed, thereby allowing selective CM reactions (Tab. 2) [23, 35]. When one of the coupling partners is electron-poor or sterically very hindered, good yields of the desired CM product can be obtained, even when equimolar amounts of the two partners are utilized. The coupling with acrylates and methacrylates can be an efficient alternative to the Wittig reaction in organic synthesis.

Asymmetric olefin metathesis is a field that has received broad attention in recent years. Hoveyda and Schrock developed several molybdenum-based catalyst systems that give very high yields and selectivities in asymmetric RCM and asym-

Tab. 2

Olefin	CM Partner	Product	Yield (%)	E/Z
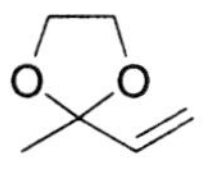 dioxolane-vinyl	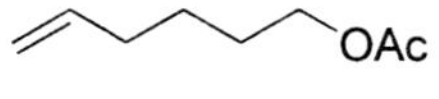 ⟶⟶⟶OAc	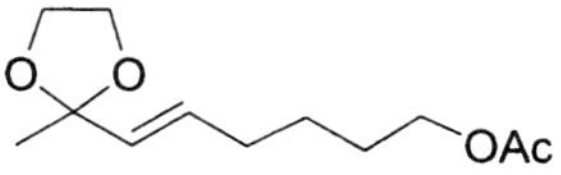 dioxolane-CH=CH-(CH₂)₃-OAc	91	>20:1
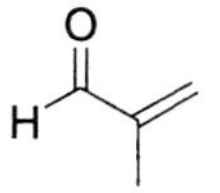 methacrolein	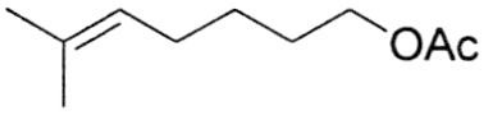 (CH₃)₂C=CH-(CH₂)₃-OAc	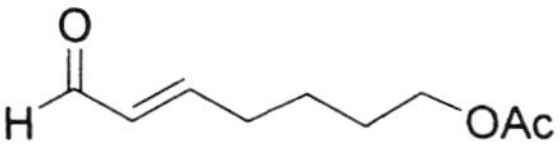 OHC-C(CH₃)=CH-(CH₂)₃-OAc	97	>20:1
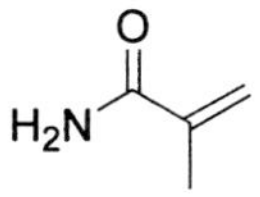 methacrylamide	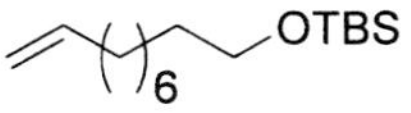 CH₂=CH-(CH₂)₆-OTBS	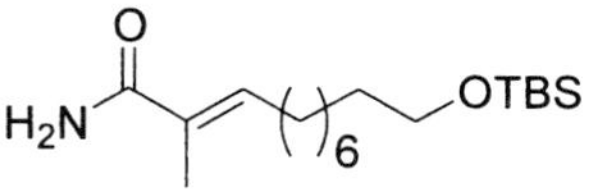 H₂N-CO-C(CH₃)=CH-(CH₂)₆-OTBS	71	>20:1
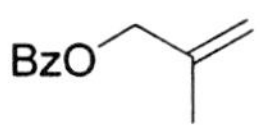 BzO-C(CH₃)=CH₂	 CH₂=CH-(CH₂)₃-OAc	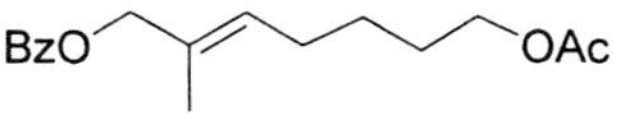 BzO-C(CH₃)=CH-(CH₂)₃-OAc	80	4:1
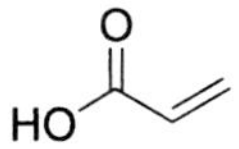 acrylic acid	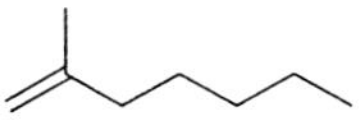 (CH₃)₂C=CH-(CH₂)₂-CH₃	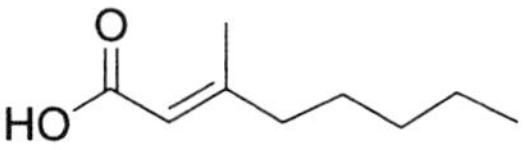 HOOC-C(CH₃)=CH-(CH₂)₂-CH₃	83	2:1
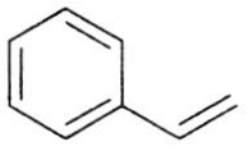 styrene	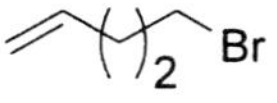 CH₂=CH-(CH₂)₂-Br	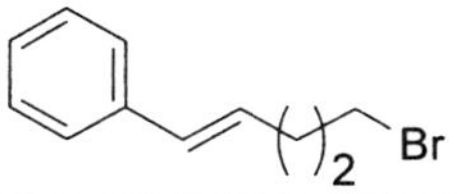 Ph-CH=CH-(CH₂)₂-Br	90	>20:1

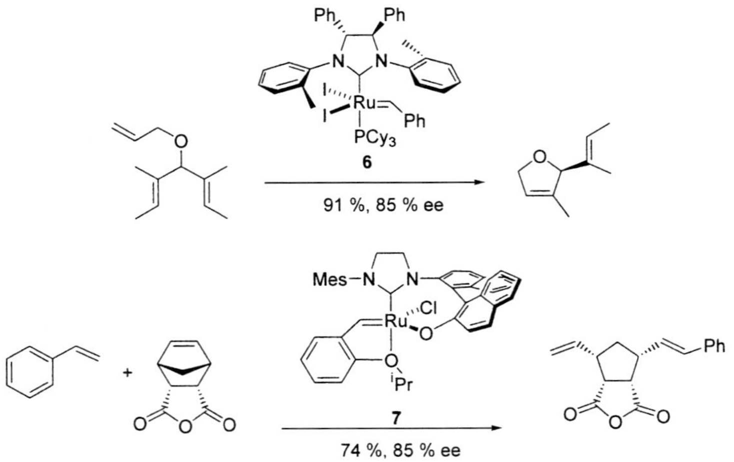

Scheme 9

91 %, 85 % ee

74 %, 85 % ee

metric ring-opening metathesis (ROM) [36]. So far, these catalyst systems have found only limited applications in organic synthesis of more complex products. Recently, the first chiral ruthenium-based catalysts appeared in the literature (Scheme 9). Grubbs introduced catalyst **6**, in which the NHC-ligand is derived from a chiral diamine [37]. Hoveyda reported catalyst **7**, a BINOL-modified version of catalyst **5** [38]. Both systems give moderate to good enantioselectivities in the asymmetric RCM, and **7** gives good enantioselectivities in the asymmetric ROM.

2.15.3
Enyne Metathesis

The first enyne metathesis with a ruthenium carbene catalyst was described by Mori. In this transformation an enyne is transformed into a 1,3-diene [39, 40]. The reaction can be performed in an inter- or intramolecular fashion, it is completely atom economical, and the obtained products are useful templates for further transformations (Tab. 3). The advent of the new Ru-NHC catalyst expanded the substrate scope, so that sterically more encumbered systems are now accessible [41]. The reaction is very general in the formation of five- and six-membered rings, and in some cases the reaction yields are improved when the reaction is performed under an atmosphere of ethylene [42]. An interesting application is the enyne metathesis of siloxyalkynes. Treatment of the reaction product with HF yields a,β-unsaturated ketones [43]. Simple alkynes can be utilized in the cross-enyne metathesis with ethylene [44].

Very interesting is the enyne metathesis for the formation of macrocyclic 1,3-dienes. Depending on the ring sizes, the exo- or endo-cyclization product can be

2.15.3 Enyne Metathesis

Tab. 3

Substrate	Product	Yield (%)	Ref.
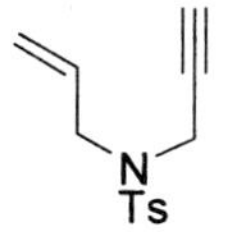 (allyl-N(Ts)-propargyl)	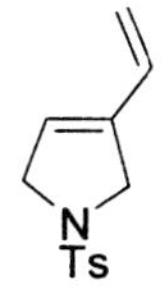 3-vinyl-N-Ts-pyrroline	90	42
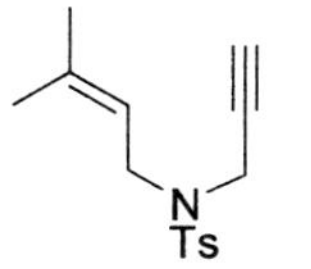 (prenyl-N(Ts)-propargyl)	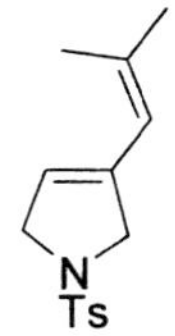 3-isopropenyl-N-Ts-pyrroline	97	41
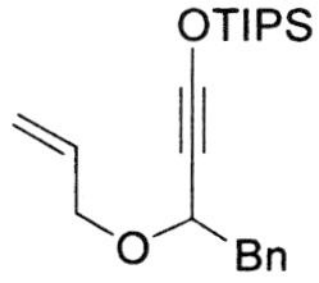 (allyl-O-CH(Bn)-C≡C-OTIPS)	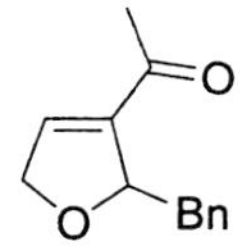 acetyl dihydrofuran with Bn	86	43
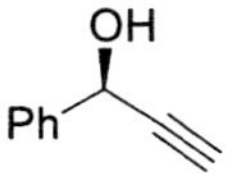 Ph-CH(OH)-C≡CH	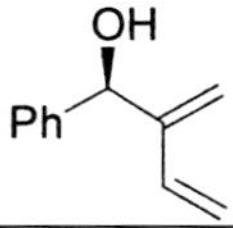 Ph-CH(OH)-C(=CH$_2$)-CH=CH$_2$	68	44

obtained [45]. For ring sizes >11 the endo-cyclization mode is observed. The biomimetic synthesis of longithorone used intramolecular enyne metathesis to access both macrocyclic 1,3-dienes (Scheme 10) [46]. Subsequent intra- and transannular Diels-Alder reactions lead to the natural product.

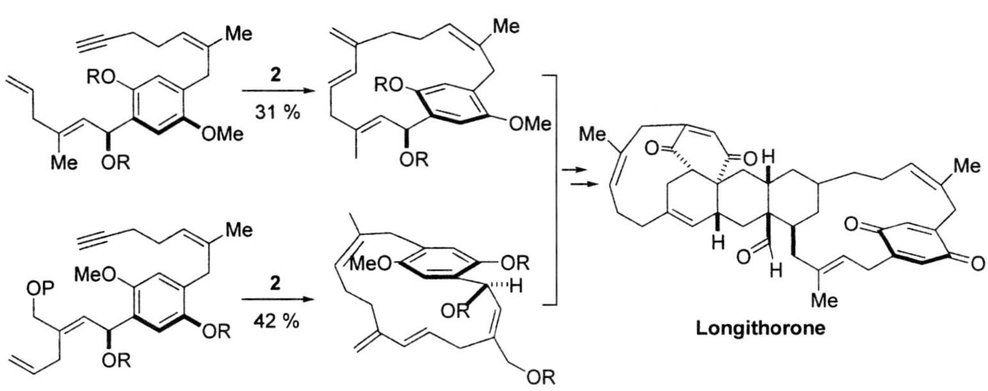

Scheme 10

Longithorone

2.15.4
Alkyne Metathesis

The major current drawback of alkene metathesis remains the lack of (E/Z)-selectivity concerning the formed double bond. Major advancements have been made in this area since the introduction of the NHC-substituted Ru-catalysts, but the stereochemical outcome of the RCM reaction is still very substrate-dependent. Subtle changes in the starting material can have huge influences on the product ratio, and currently there is no predictive tool available.

Fürstner introduced an alternative indirect approach for the stereoselective formation of macrocyclic alkenes [47]. Despite being a relatively old and mechanistically well-understood reaction, alkyne metathesis was until recently mainly used in macromolecular chemistry. Use in organic synthesis was limited to alkyne cross-metathesis reactions of phenylacetylenes. Ring-closing alkyne metathesis (RCAM) offers a very straightforward approach to cycloalkynes. The reaction is fairly general for ring sizes ≥12, and most functional groups are tolerated by the available catalysts. Three different systems have been used successfully as catalysts for RCAM (Scheme 11). Initially, the well-defined tungsten-alkylidene complex **8** was used as catalyst for the reaction, and the molybdenum complex **9** was used subsequently [48]. The third alternative is the use of an *in situ* system, where the active catalyst is generated under the reaction conditions from $Mo(CO)_6$ and phenol additives [49]. The latter system requires relatively harsh conditions, which limits its use for the synthesis of compounds with sensitive functional groups. The mechanism of formation of the presumed active catalyst species, the metal-alkylidene complex, is currently not well understood for the Mo systems.

Initial applications of the RCAM were toward the stereoselective formation of macrocyclic (Z)-alkenes (Scheme 12). A simple Lindlar hydrogenation of the cycloalkyne leads selectively to the (Z)-alkene.

Among the natural products that were obtained with this methodology were macrocyclic alkaloids, cyclophanes, and macrocyclic glycosides [50]. The synthesis of prostaglandin E_2-1,15-lactone demonstrated the compatibility of very sensitive structural moieties with the reaction conditions (Scheme 13) [51]. The synthesis of epothilone C clearly showcased the advantage of RCAM over RCM in certain cases. The early application of RCM in the synthesis of the epothilones [52] showed the potential of this reaction for advanced organic synthesis, but it also struggled with the selectivity problem. All reported examples led to mixtures of stereoisomers, in

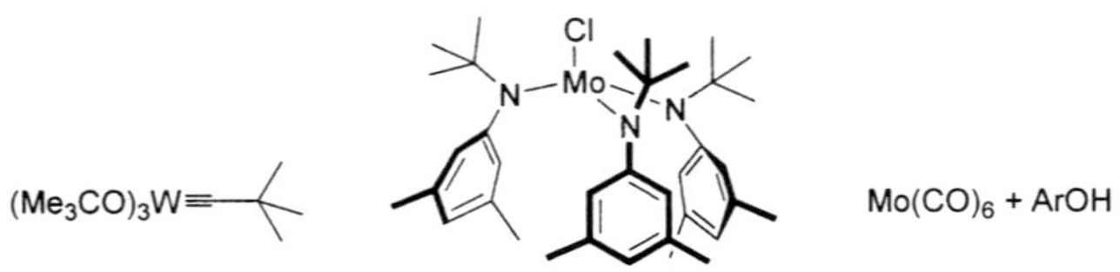

$(Me_3CO)_3W\equiv$ —

8

9

$Mo(CO)_6 + ArOH$

10

Scheme 11

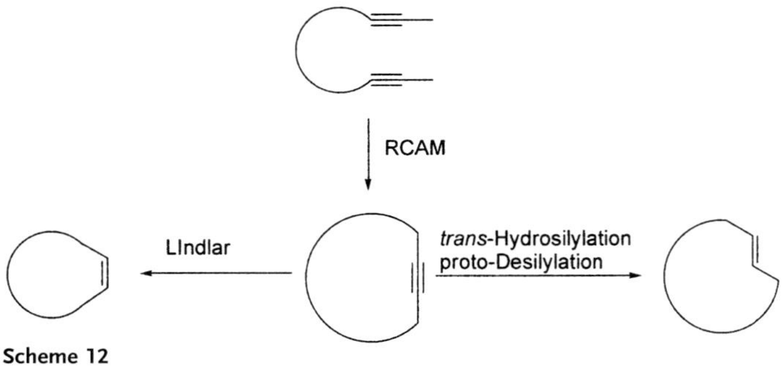

Scheme 12

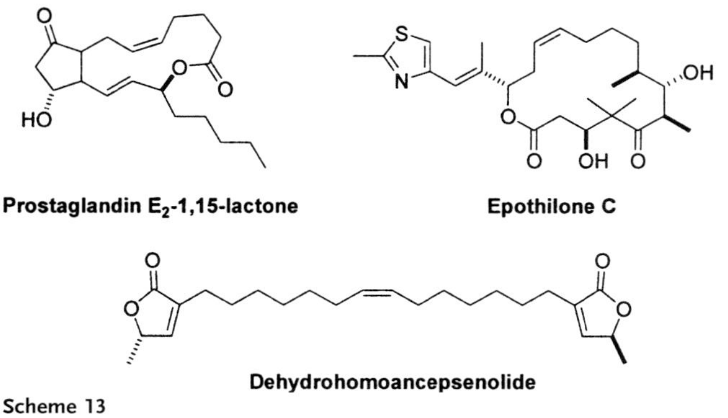

Prostaglandin E₂-1,15-lactone **Epothilone C**

Dehydrohomoancepsenolide

Scheme 13

some cases the undesired (*E*)-isomer being obtained as major product. Using the RCAM/Lindlar-hydrogenation approach, epothilone C could be synthesized as a single stereoisomer [48b]. The scope of the reaction expands to the alkyne cross-metathesis, and this was demonstrated in a synthesis of dehydrohomoancepsenolide [53].

The complementary approach to (*E*)-alkenes through RCAM was hampered by the fact that until recently there was no general method for the reduction of alkynes to (*E*)-alkenes in the presence of sensitive groups. A recent discovery by Trost expands the possibilities of RCAM. Alkynes can be *trans*-hydrosilylated with high selectivity in the presence of [Cp*Ru(MeCN)$_3$]PF$_6$ [54]. The corresponding vinylsilanes can be protodesilylated with CuI/TBAF or AgF. This new reaction will expand the scope of the RCAM metathesis.

2.15.5
Outlook

The emergence of the metathesis reaction into the field of organic synthesis has changed the way most organic chemists plan and execute their synthesis of complex targets like no other reaction in the last decade. Several elegant applications of the reaction have appeared in the literature in the last years. While RCM already has become a standard tool in most organic laboratories, the application of cross metathesis and the alkyne metathesis still has growth potential.

2.15.6
References

1 For a detailed overview: *Handbook of Metathesis* (Ed.: R.H. GRUBBS), Wiley-VCH, Weinheim, **2003**.
2 For leading reviews see: (a) A.H. HOVEYDA, R.R. SCHROCK, *Angew. Chem. Int. Ed.* **2003**, *42*, 4592. (b) S.J. GANNON, S. BLECHERT, *Angew. Chem. Int. Ed.* **2003**, *42*, 1900. (c) T.M. TRNKA, R.H. GRUBBS, *Acc. Chem. Res.* **2001**, *34*, 18. (d) A. FÜRSTNER, *Angew. Chem. Int. Ed.* **2000**, *39*, 3012. (e) S.K. ARMSTRONG, *J. Chem. Soc. Perkin Trans. 1* **1998**, 371. (f) R.H. GRUBBS, S. CHANG, *Tetrahedron* **1998**, *54*, 4413. (g) *Alkene Metathesis in Organic Synthesis* (Ed.: A. FÜRSTNER), Springer, Berlin, **1998**.
3 K.J. IVIN, J.C. MOL, *Olefin Metathesis and Metathesis Polymerisation*, Academic Press, San Diego, **1997**.
4 J.-L. HÉRISSON, Y. CHAUVIN, *Makromol. Chem.* **1971**, *141*, 161.
5 For an interesting account on the early days of metathesis see: A.M. ROUHI, *Chem. Eng. News* **2002**, *80* (51), 34.
6 R.R. SCHROCK, J.S. MURDZEK, G.C. BAZAN, J. ROBBINS, M. DIMARE, M. O'REGAN, *J. Am. Chem. Soc.* **1990**, *112*, 3875.
7 (a) S.T. NGUYEN, L.K. JOHNSON, R.H. GRUBBS, J.W. ZILLER, *J. Am. Chem. Soc.* **1992**, *114*, 3974. (b) P. SCHWAB, M.B. FRANCE, J.W. ZILLER, R.H. GRUBBS, *Angew. Chem. Int. Ed.* **1995**, *34*, 2039.
8 T. WESKAMP, W.C. SCHATTENMANN, M. SPIEGLER, W.A. HERRMANN, *Angew. Chem. Int. Ed.* **1998**, *37*, 2490; Corrigendum: *Angew. Chem. Int. Ed.* **1999**, *38*, 262.
9 (a) M. SCHOLL, T.M. TRNKA, J.P. MORGAN, R.H. GRUBBS, *Tetrahedron Lett.* **1999**, *40*, 2247. (b) J. HUANG, E.D. STEVENS, S.P. NOLAN, J.L. PETERSON, *J. Am. Chem. Soc.* **1999**, *121*, 2674. (c) T. WESKAMP, F.J. KOHL, W. HIERINGER, D. GLEICH, W.A. HERRMANN, *Angew. Chem. Int. Ed.* **1999**, *38*, 2416.
10 M. SCHOLL, S. DING, C.W. LEE, R.H. GRUBBS, *Org. Lett.* **1999**, *1*, 953.
11 S.B. GARBER, J.S. KINGSBURY, B.L. GRAY, A.H. HOVEYDA, *J. Am. Chem. Soc.* **2000**, *122*, 8168.
12 S. RANDL, S. GESSLER, H. WAKAMATSU, S. BLECHERT, *Synlett* **2001**, 430.
13 (a) H. WAKAMATSU, S. BLECHERT, *Angew. Chem. Int. Ed.* **2002**, *41*, 2403. (b) K. GRELA, S. HARUTYUNYAN, A. MICHROWSKA, *Angew. Chem. Int. Ed.* **2002**, *41*, 4038.
14 M.S. SANFORD, J.A. LOVE, R.H. GRUBBS, *J. Am. Chem. Soc.* **2001**, *123*, 6543.
15 (a) L. CAVALLO, *J. Am. Chem. Soc.* **2002**, *124*, 8965. (b) S.F. VYBOISHCHIKOV, M. BÜHL, W. THIEL, *Chem. Eur. J.* **2002**, *8*, 3962. (c) C. ADLHART, P. CHEN, *Angew. Chem. Int. Ed.* **2002**, *41*, 4485.
16 (a) H.D. MAYNARD, R.H. GRUBBS, *Tetrahedron Lett.* **1999**, *40*, 4137. (b) L.A. PAQUETTE, J.D. SCHLOSS, I. EFREMOV, F. FABRIS, F. GALLOU, J. MENDEZ-ANDINO, J. YANG, *Org. Lett.* **2000**, *2*, 1259. (c) Y.M. AHN, K.L. YANG, G.I. GEORG, *Org. Lett.* **2001**, *3*, 1411. (d) J.H. CHO, B.M. KIM, *Org. Lett.* **2003**, *5*, 531.
17 (a) S.C. SCHÜRER, S. GESSLER, N. BUSCHMANN, S. BLECHERT, *Angew. Chem. Int. Ed.* **2000**, *39*, 3898. (b) S.J. CONNON, A.M. DUNNE, S. BLECHERT, *Angew. Chem. Int. Ed.* **2002**, *41*, 3835. (c) M. AHMED, A.G.M. BARRETT, D.C. BRADDOCK, S.M. CRAMP, P.A. PROCOPIOU, *Tetrahedron Lett.* **1999**, *40*, 8657. (d) Q. YAO, *Angew. Chem. Int. Ed.* **2000**, *39*, 4060.
18 M. MAYR, B. MAYR, M.R. BUCHMEISTER, *Angew. Chem. Int. Ed.* **2001**, *40*, 3839. (b) J.S. KINGSBURY, S.B. GARBER, J.L. GIFTOS, B.L. GRAY, M.M. OKAMOTO, R.A. FARRER, J.T. FOURKAS, A.H. HOVEYDA, *Angew. Chem. Int. Ed.* **2001**, *40*, 3898.
19 A. FÜRSTNER, L. ACKERMANN, K. BECK, H. HORI, D. KOCH, K. LANGEMANN, M. LIEBL, C. SIX, W. LEITNER, *J. Am. Chem. Soc.* **2001**, *123*, 9000.

20 (a) N. Audic, H. Clavier, M. Mauduit, J.-C. Guillemin, *J. Am. Chem. Soc.* **2003**, *125*, 9248. (b) Q. Yao, Y. Zhang, *Angew. Chem. Int. Ed.* **2003**, *42*, 3395.

21 A. Fürstner, O. R. Thiel, L. Ackermann, H.-J. Schanz, S. P. Nolan, *J. Org. Chem.* **2000**, *65*, 2204.

22 A. Fürstner, O. R. Thiel, L. Ackermann, *Org. Lett.* **2001**, *3*, 449.

23 A. K. Chatterjee, J. P. Morgan, M. Scholl, R. H. Grubbs, *J. Am. Chem. Soc.* **2000**, *122*, 3783.

24 W. Chao, S. M. Weinreb, *Org. Lett.* **2003**, *5*, 2505.

25 S. S. Salim, R. K. Bellingham, V. Satcharoen, R. C. D. Brown, *Org. Lett.* **2003**, *5*, 3403.

26 J. D. Rainier, J. M. Cox, S. P. Allwein, *Tetrahedron Lett.* **2001**, *42*, 179.

27 M. Arisawa, Y. Terada, M. Nakagawa, A. Nishida, *Angew. Chem. Int. Ed.* **2002**, *41*, 4733.

28 A. Fürstner, O. R. Thiel, N. Kindler, B. Bartkowska, *J. Org. Chem.* **2000**, *65*, 7990.

29 A. B. Smith III, C. M. Adams, S. A. Kozmin, D. V. Paone, *J. Am. Chem. Soc.* **2001**, *123*, 5925.

30 A. Fürstner, K. Radkowski, C. Wirtz, R. Goddard, C. W. Lehmann, R. Mynott, *J. Am. Chem. Soc.* **2002**, *124*, 7061.

31 K. C. Nicolaou, G. Vassilikogiannakis, T. Montagnon, *Angew. Chem. Int. Ed.* **2002**, *41*, 3276.

32 X. Wang, J. A. Porco, Jr. *J. Am. Chem. Soc.* **2003**, *125*, 6040.

33 R. E. Maleczka, Jr., L. R. Terrell, F. Geng, J. S. Ward III, *Org. Lett.* **2002**, *4*, 2841.

34 T. Nagata, M. Nakagawa, A. Nishida, *J. Am. Chem. Soc.* **2003**, *125*, 7484.

35 (a) A. K. Chatterjee, T.-L. Choi, D. L. Sanders, R. H. Grubbs, *J. Am. Chem. Soc.* **2003**, *125*, 11360. (b) A. K. Chatterjee, R. H. Grubbs, *Angew. Chem. Int. Ed.* **2002**, *41*, 3171. (c) T.-L. Choi, A. K. Chatterjee, R. H. Grubbs, *Angew. Chem. Int. Ed.* **2000**, *39*, 1277. (d) A. K. Chatterjee, R. H. Grubbs, *Org. Lett.* **1999**, *1*, 1751.

36 A. H. Hoveyda, R. R. Schrock, *Chem. Eur. J.* **2001**, *7*, 945.

37 T. J. Seiders, D. W. Ward, R. H. Grubbs, *Org. Lett.* **2001**, *3*, 3225.

38 J. H. Van Veldhuizen, S. B. Garber, J. S. Kingsbury, A. H. Hoveyda, *J. Am. Chem. Soc.* **2002**, *124*, 4954.

39 A. Kinoshita, M. Mori, *Synlett* **1994**, 1020.

40 For a recent review see: C. S. Poulsen, R. Madsen, *Synthesis* **2003**, 1.

41 A. Fürstner, L. Ackermann, B. Gabor, R. Goddard, C. W. Lehmann, R. Mynott, F. Stelzer, O. R. Thiel, *Chem. Eur. J.* **2001**, *7*, 3236.

42 M. Mori, N. Sakakibara, A. Kinoshita, *J. Org. Chem.* **1998**, *63*, 6082.

43 M. P. Schramm, D. S. Reddy, S. A. Kozmin, *Angew. Chem. Int. Ed.* **2001**, *40*, 4274.

44 J. A. Smulik, S. T. Diver, *Org. Lett.* **2000**, *2*, 2271.

45 E. C. Hansen, D. Lee, *J. Am. Chem. Soc.* **2003**, *125*, 9582.

46 M. Layton, C. A. Morales, M. D. Shair, *J. Am. Chem. Soc.* **2002**, *124*, 773.

47 (a) A. Fürstner, G. Seidel, *Angew. Chem. Int. Ed.* **1998**, *37*, 1734. (b) A. Fürstner, O. Guth, A. Rumbo, G. Seidel, *J. Am. Chem. Soc.* **1999**, *121*, 11108.

48 (a) A. Fürstner, C. Mathes, C. W. Lehmann, *J. Am. Chem. Soc.* **1999**, *121*, 9453; (b) A. Fürstner, C. Mathes, C. W. Lehmann, *Chem. Eur. J.* **2001**, *7*, 5299.

49 (a) A. Mortreux, M. Blanchard, *Chem. Commun.* **1974**, 786. (b) K. Grela, J. Ignatowska, *Org. Lett.* **2002**, *4*, 3747.

50 (a) A. Fürstner, A. Rumbo, *J. Org. Chem.* **2000**, *65*, 2608. (b) A. Fürstner, F. Stelzer, A. Rumbo, H. Krause, *Chem. Eur. J.* **2002**, *8*, 1856. (c) A. Fürstner, K. Radkowski, J. Grabowski, C. Wirtz, R. Mynott, *J. Org. Chem.* **2000**, *65*, 8758.

51 A. Fürstner, K. Grela, C. Mathes, C. W. Lehmann, *J. Am. Chem. Soc.* **2000**, *122*, 11799.

52 K. C. Nicolaou, F. Roschangar, D. Vourloumis, *Angew. Chem. Int. Ed.* **1998**, *37*, 2014.

53 A. Fürstner, T. Dierkes, *Org. Lett.* **2000**, *2*, 2463.

54 (a) B. M. Trost, Z. T. Ball, T. Jöge, *J. Am. Chem. Soc.* **2002**, *124*, 7922. (b) A. Fürstner, K. Radkowski, *Chem. Commun.* **2002**, 2182.

2.16
Homometallic Lanthanoids in Synthesis: Lanthanide Triflate-catalyzed Synthetic Reactions

Shū Kobayashi

2.16.1
Introduction

Lanthanides have larger radius and specific coordination numbers compared with typical transition metals. They are known to act as strong Lewis acids because of their hard character and their strong affinity toward carbonyl oxygens [1]. Among these compounds, lanthanide trifluoromethanesulfonates (lanthanide triflates) were expected to be one of the strongest Lewis acids because of the strongly electron-withdrawing trifluoromethanesulfonyl group. On the other hand, their hydrolysis was postulated to be slow on the basis of their hydration energies and hydrolysis constants [2]. In fact, while most metal triflates are prepared under strict anhydrous conditions, lanthanide triflates were reported to be prepared in aqueous solution [3, 4]. After finding that lanthanide triflates are stable and act as Lewis acids in water [5], many synthetic reactions using these triflates as catalysts have been developed [6]. This chapter surveys them, especially focusing on carbon–carbon bond-forming reactions.

2.16.2
Lewis Acid Catalysis in Aqueous Media

Lewis acid-catalyzed carbon–carbon bond-forming reactions have been of great interest in organic synthesis because of their unique reactivities, selectivities, and for the mild conditions used [7]. While various kinds of Lewis acid-promoted reactions have been developed, many of which have been applied in industry, these reactions must be carried out under strict anhydrous conditions. The presence of even a small amount of water stops the reaction, because most Lewis acids immediately react with water rather than the substrates and decompose or deactivate. This fact has restricted the use of Lewis acids in organic synthesis. On the other hand, the utility of aqueous reactions is now generally recognized [8]. It is desirable to perform the reactions of compounds containing water of crystallization or other water-soluble compounds in aqueous media, because tedious procedures to remove water are necessary when the reactions are carried out in organic solvents. Moreover, aqueous reac-

Transition Metals for Organic Synthesis, Vol. 1, 2nd Edition.
Edited by M. Beller and C. Bolm
Copyright © 2004 WILEY-VCH Verlag GmbH & Co. KGaA, Weinheim
ISBN: 3-527-30613-7

tions of organic compounds avoid the use of harmful organic solvents. Lanthanide triflates were found to be stable Lewis acids in water, and many useful aqueous reactions using lanthanide triflates as catalysts have been reported.

2.16.2.1 Aldol Reactions

It was found that the hydroxymethylation reaction of silyl enol ethers with commercial formaldehyde solution proceeded smoothly by using lanthanide triflates as Lewis acid catalysts [5, 9]. The reactions were most effectively carried out in commercial formaldehyde solution–THF media. The amount of the catalyst necessary for efficient transformations was examined by taking the reaction of the silyl enol ether derived from propiophenone with commercial formaldehyde solution as a model. The reaction was found to be catalyzed even by 1 mol% ytterbium triflate (Yb(OTf)$_3$): 1 mol% (90% yield); 5 mol% (90% yield); 10 mol% (94% yield); 20 mol% (94% yield); 100 mol% (94% yield).

$$\text{HCHO aq.} + \underset{R^1}{\overset{OSiMe_3}{\diagup\!\!\!\diagdown}} R^2 \xrightarrow[\text{H}_2\text{O/THF, rt}]{\text{cat. Ln(OTf)}_3} \underset{R^2}{R^1\overset{O}{\diagup\!\!\!\diagdown}} OH \qquad (1)$$

77–94%

Formaldehyde is a versatile reagent as it is one of the most highly reactive Cl electrophiles in organic synthesis [10], however dry gaseous formaldehyde has been required for many reactions. Clearly the use of commercial formaldehyde solution, which is an aqueous solution containing 37% formaldehyde and 8–10% methanol, is advantageous because it is cheap, easy to handle, and does not self-polymerize easily [11]. However, the use of this reagent has been strongly restricted due to the existence of a large amount of water. For example, the titanium tetrachloride (TiCl$_4$)-promoted hydroxymethylation reaction of a silyl enol ether was carried out by using trioxane as a HCHO source under strict anhydrous conditions [12, 13].

It was also found that lanthanide triflates were effective for the activation of aldehydes other than formaldehyde [9, 14]. The aldol reaction of silyl enol ethers with aldehydes proceeded smoothly to afford the aldol adducts in high yields in the presence of a catalytic amount OfYb(OTf)$_3$, gadolinium triflate (Gd(OTf)$_3$), or lutetium triflate (Lu(OTf)$_3$) in aqueous media (water–THF). Diastereoselectivities were generally good to moderate. One feature in the present reaction is that water-soluble aldehydes, for instance, acetaldehyde, acrolein, and chloroacetaldehyde can be reacted with silyl enol ethers to afford the corresponding cross aldol adducts in high yields. Some of these aldehydes are commercially supplied as water solutions and are appropriate for direct use. Phenylglyoxal monohydrate also works well. Furthermore, salicylaldehyde and 2-pyridinecarboxaldehyde can be successfully employed. The former has a free hydroxy group which is incompatible with metal enolates or Lewis acids, and the latter is generally difficult to use under the influence of Lewis acids because of the coordination of the nitrogen atom to the Lewis acids resulting in the deactivation of the acids.

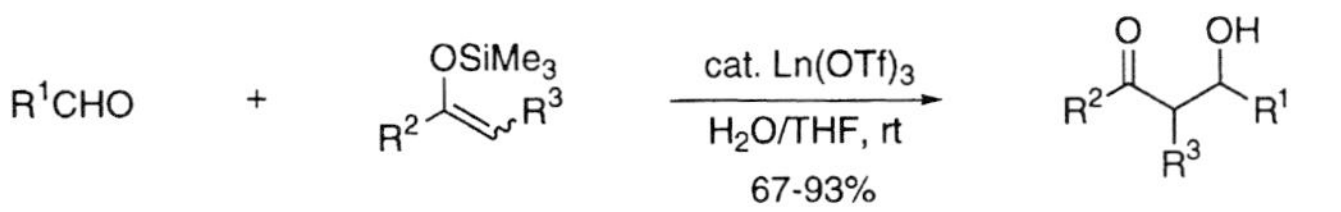

R^1CHO + $R^2\text{-CH=CH-}R^3$ (OSiMe₃) →[cat. Ln(OTf)₃][H₂O/THF, rt, 67-93%] $R^2\text{-CO-CHR}^3\text{-CH(OH)-}R^1$ (2)

The aldol reactions of silyl enol ethers with aldehydes also proceed smoothly in water–ethanol–toluene [15]. The reactions proceed much faster than in water–THF. Furthermore, the new solvent system realizes continuous use of the catalyst by a very simple procedure. Although the water–ethanol–toluene (1:7:4) system is one phase, it easily becomes two phases by adding toluene after the reaction is completed. The product is isolated from the organic layer by a usual work-up. On the other hand, the catalyst remains in the aqueous layer, which is used directly in the next reaction without removing water. It is noteworthy that the yields of the second, third, and fourth runs are comparable to that of the first run (Eq. 3).

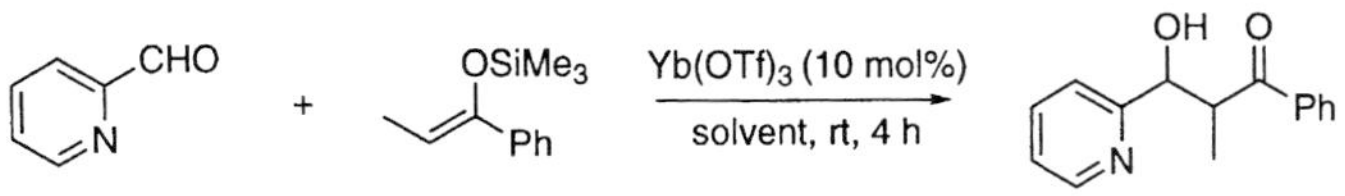

pyridyl-CHO + CH₂=CH-Ph (OSiMe₃) →[Yb(OTf)₃ (10 mol%)][solvent, rt, 4 h] pyridyl-CH(OH)-CH(Me)-CO-Ph (3)

Solvent		Yield / %
H₂O / EtOH / Toluene	(1:7:4)	70
	(1:10:4)	96
H₂O / THF	(1:4)	12

2.16.2.2
Allylation Reactions

Synthesis of homoallylic alcohols by the reaction of allyl organometallics with carbonyl compounds is an important process in organic synthesis [16]. The allylation reactions of carbonyl compounds were found to proceed smoothly under the influence of 5 mol% Sc(OTf)₃ by using tetraallyltin as an allylating reagent [17]. The corresponding homoallylic alcohols were obtained in high yields. Ketones could also be used in the reaction. In most cases, the reactions were successfully carried out in aqueous media. It is noteworthy that unprotected sugars reacted directly to give the adducts in high yields. The allylated products are intermediates for the synthesis of higher sugars [18]. Moreover, aldehydes containing water of crystallization such as phenylglyoxal monohydrate reacted with tetraallyltin to give the di-allylated adduct in high yield.

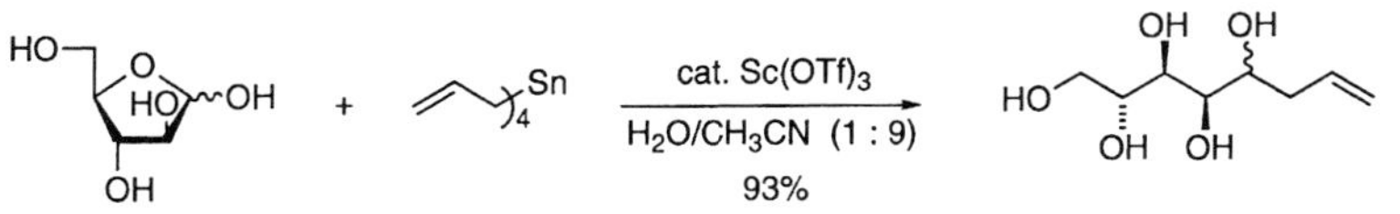

sugar(HO-CH₂-[CHOH]-O-CH-OH) + (CH₂=CH-CH₂)₄Sn →[cat. Sc(OTf)₃][H₂O/CH₃CN (1:9), 93%] HO-CH₂-CH(OH)-CH(OH)-CH(OH)-CH(OH)-CH₂-CH=CH₂ (4)

As a catalyst Yb(OTf)₃ is also effective in the present allylation reactions. For example, 3-phenylpropionealdehyde reacted with tetraallyltin in the presence of

5 mol% Yb(OTf)$_3$ to afford the adduct in a 90% yield [15]. In addition, the water–ethanol–toluene system could be successfully applied to the present allylation reactions, and continuous use of the catalyst was realized.

2.16.2.3
Diels–Alder Reactions

Although many Diels–Alder reactions have been carried out at higher reaction temperatures without catalysts, heat sensitive compounds in complex and multistep syntheses cannot be employed. While Lewis acid catalysts allow the reactions to proceed at room temperature or below with satisfactory yields in organic solvents, they are often accompanied by diene polymerization and excess amounts of the catalyst are often needed to catalyze carbonyl-containing dienophiles [19].

It was found that the Diels–Alder reaction of naphthoquinone with cyclopentadiene proceeded in the presence of a catalytic amount of a lanthanide triflate in H$_2$O–THF (1:9) at room temperature to give the corresponding adduct in a 93% yield (endo/exo = 100/0) [20].

$$\text{naphthoquinone} + \text{cyclopentadiene} \xrightarrow[\text{H}_2\text{O/THF (1:9)}]{\text{Sc(OTf)}_3 \text{ (10 mol\%)}} \text{adduct} \quad (5)$$

93% yield, endo/exo = 100/0

2.16.2.4
Micellar Systems

Quite recently, scandium triflate (Sc(OTf)$_3$)-catalyzed aldol reactions of silyl enol ethers with aldehydes were successfully carried out in micellar systems [21]. While the reactions proceeded sluggishly in pure water (without organic solvents), remarkable enhancement of the reactivity was observed in the presence of a small amount of a surfactant. In these systems, versatile carbon–carbon bond-forming reactions proceeded in water without using any organic solvents.

Lewis acid catalysis in micellar systems was first found in the model reaction of the silyl enol ether of propiophenone with benzaldehyde. While the reaction proceeded sluggishly in the presence of 0.2 equivalents Yb(OTf)$_3$ in water, remarkable enhancement of the reactivity was observed when the reaction was carried out in the presence of 0.2 equivalents Yb(OTf)$_3$ in an aqueous solution of sodium dodecylsulfate (SDS, 0.2 equivalents, 35 mM), and the corresponding aldol adduct was obtained in a 50% yield. In the absence of the Lewis acid and the surfactant (water-promoted conditions) [12], only 20% yield of the aldol adduct was isolated after 48 h, while a 33% yield of the aldol adduct was obtained after 48 h in the absence of the Lewis acid in an aqueous solution of SDS. Judging from the critical micelle concentration, micelles would be formed in these reactions [22], Although several organic reactions in micelles were reported, there was no report on Lewis acid catalysis in micelles.

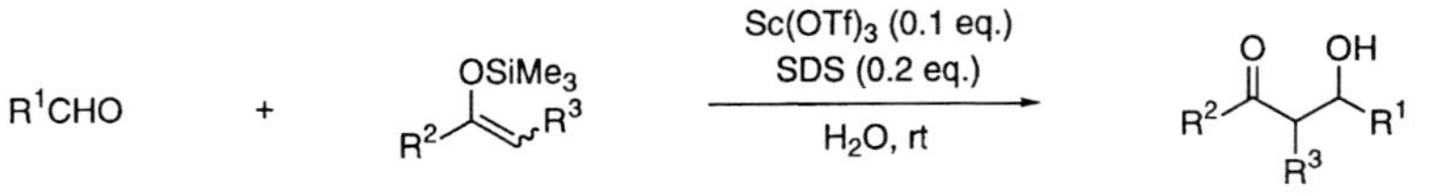

2.16.2 Lewis Acid Catalysis in Aqueous Media

$$R^1CHO \ + \ \underset{R^2}{\overset{OSiMe_3}{\diagup}}\!\!R^3 \quad \xrightarrow[H_2O, \ rt]{Sc(OTf)_3 \ (0.1 \ eq.) \\ SDS \ (0.2 \ eq.)} \quad R^2\underset{R^3}{\overset{O \quad OH}{\diagdown\!\diagup}}R^1$$

Tab. 1 Sc(OTf)$_3$-Catalyzed aldol reactions in micellar systems.

Aldehyde	Silyl enol ether	Yield [%]
PhCHO	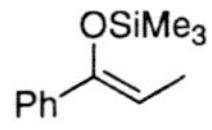 OSiMe$_3$ / Ph	88 a)
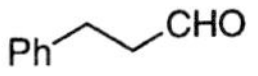 Ph⁀CHO	1	86 b)
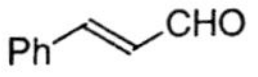	1	88 c)
Ph⁀CHO		
HCHO	1	82 d)
PhCHO	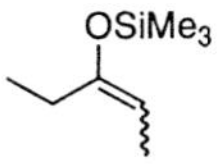 OSiMe$_3$ (trisubstituted)	88 e)
PhCHO	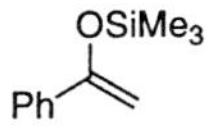 OSiMe$_3$ / Ph	75 f, g)
PhCHO	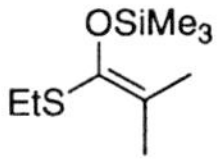 OSiMe$_3$ / EtS	94
PhCHO	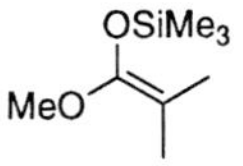 OSiMe$_3$ / MeO	84 g)

a) Syn/anti = 50/50.
b) Syn/anti = 45/55
c) Syn/anti = 41/59.
d) Comercially available HCHO aq. (3 ml), **1** (0.5 mmol), Sc(OTf)$_3$ (0.1 mmol), and SDS (0.1 mmol) were combined.
e) Syn/anti = 57/43.
f) Sc(OTf)$_3$ (0.2 eq.) was used.
g) Additional silyl enolate (1.5 eq.) was charged after 6 h.

Several examples of the Sc(OTf)$_3$-catalyzed aldol reactions in micellar systems are shown in Tab. 1. Not only aromatic, but also aliphatic and α,β-unsaturated aldehydes react with silyl enol ethers to afford the corresponding aldol adducts in high yields. Formaldehyde–water solution also works well. Even the ketene silyl acetal **2**, which is known to hydrolyze very easily in the presence of a small amount of water, reacts with an aldehyde in the present micellar system to afford the corresponding aldol adduct in a high yield.

In addition the allylation reactions of aldehydes with tetraallyltin proceeded smoothly in micellar systems using Sc(OTf)$_3$ as a catalyst [23]. Again the reactions were successfully carried out in the presence of a small amount of a surfactant in water without using any organic solvents, to afford the corresponding homoallylic alcohols in high yields.

It is conceivable that Lewis acid catalysis in micellar sytems will lead to clean and environmentally friendly processes, and it will become a more important topic in the near future.

2.16.2.5
Recovery and Reuse of the Catalyst

Lanthanide inflates are more soluble in water than in organic solvents such as dichloromethane. Very interestingly, almost 100% of lanthanide triflates is easily recovered from the aqueous layer after the reaction is completed and it can be reused without significant loss of catalyst activity. For example, the first use (20 mol% of $Yb(OTf)_3$) in the reaction of the silyl enol ether of propiophenone with formaldehyde–water solution leads to 94% yield and the second use to 91% yield and the third use to 93% yield. The reactions are usually quenched with water and the products are extracted with an organic solvent (for example, dichloromethane). The lanthanide triflates stay in the aqueous layer and after removal of water the catalyst can be used in the next reaction (Scheme 1). Thus, lanthanide triflates are expected to solve some severe environmental problems induced by mineral acid- or Lewis acid-promoted reactions in industry chemistry [24].

2.16.3
Activation of Nitrogen-containing Compounds

As for activation of nitrogen-containing compounds by Lewis acids, many Lewis acids are deactivated or sometimes decomposed by the nitrogen atoms of starting materials or products, and even when the desired reactions proceed, more than

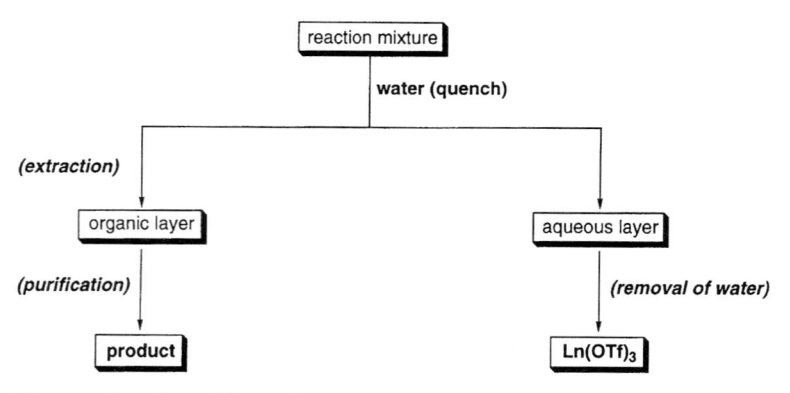

```
                    reaction mixture
                           │
                    water (quench)
          ┌────────────────┴────────────────┐
    (extraction)                              │
          │                                   │
     organic layer                       aqueous layer
          │                                   │
   (purification)                     (removal of water)
          │                                   │
        product                           Ln(OTf)_3
```

Scheme 1 Recycling of the catalyst.

stoichiometric amounts of the Lewis acids are needed because the acids are trapped by the nitrogen atoms. Clearly, it is desirable to activate nitrogen-containing compounds catalytically. Lanthanide triflates have been demonstrated to be effective for such purpose.

2.16.3.1
Mannich-type Reaction

The Mannich and related reactions provide one of the most fundamental and useful methods for the synthesis of β-amino ketones or esters. Although the classical protocols include some severe side reactions, new modifications using preformed iminium salts and imines have been developed [25]. Among them, reactions of imines with enolate components, especially silyl enolates, provide useful and promising methods leading to β-amino ketones or esters. The first report using a stoichiometric amount of $TiCl_4$ as a promoter appeared in 1977 [26], and since then, some efficient catalysts have been developed [27].

Regarding lanthanide triflates as catalyst, it was found that the reactions of imines with silyl enolates proceeded smoothly in the presence of 5 mol% $Yb(OTf)_3$ to afford the corresponding β-amino ester derivatives in good to high yields. Yttrium triflate $(Y(OTf)_3)$ was also effective, and the yield was improved when $Sc(OTf)_3$ was used instead of $Yb(OTf)_3$ as a catalyst. Not only silyl enolates derived from esters, but also that derived from a thioester worked well to give the desired β-amino esters and thioester in high yields. In the reactions of the silyl enolate derived from benzyl propionate, *anti* adducts were obtained in good selectivities. In addition, the catalyst could be recovered after the reaction was completed and could be reused.

$$\underset{R^1}{\overset{R^2}{N}}\!\!\!\diagdown\!\!\!\underset{H}{} + \underset{R^4}{\overset{OSiMe_3}{R^3\diagup\diagdown R^5}} \xrightarrow[CH_2Cl_2, 0\,°C]{cat.\ Ln(OTf)_3} \underset{R^3\ R^4}{\overset{R^2\diagdown NH\ O}{R^1\diagup\diagdown R^5}} \qquad (6)$$
47-97%

While the catalytic reactions of imines with silyl enolates are successfully carried out using lanthanide triflates, many imines are hygroscopic, unstable at high temperatures, and difficult to purify by distillation or column chromatography. Thus, it is desirable from a synthetic point of view that imines, generated *in situ* from aldehydes and amines, immediately react with silyl enolates and provide β-amino esters in a one-pot reaction. However, most Lewis acids cannot be used in this reaction because they decompose or deactivate in the presence of the amines and water that exist during imine formation. Judging from the unique properties of lanthanide triflates, they are expected to be used as catalysts for the above one-pot preparation of β-amino esters from aldehydes.

Indeed, the one-pot synthesis of β-amino esters from aldehydes has been successfully achieved by using a catalytic amount of $Yb(OTf)_3$ [28]. After testing various combinations (Tab. 2), it was found that β-amino esters were obtained in

$R^1CHO + R^2NH_2 + $ [silyl enolate with R^3, R^4, R^5, OSiMe$_3$] $\xrightarrow{\text{Yb(OTf)}_3 \text{ (5-10 mol\%)}}_{\text{Additive, CH}_2\text{Cl}_2\text{, rt}}$ R^1-CH(NHR^2)-C(R^3)(R^4)-C(O)R^5

Tab. 2 One-pot synthesis of β-amino esters from aldehydes.

Entry	R^1	R^2	Silyl enolate	Additive[a]	Yield%
1	Ph	Ph	$\underset{\mathbf{2}}{\overset{\text{OSiMe}_3}{\diagdown}}\text{OMe}$	MS4A MgSO$_4$	90 89
2	Ph	Bn	2	MS4A	85
3	Ph	p-MeOPh	2	MgSO$_4$	91[b]
4	Ph	o-MeOPh	2	MS4A	96
5	Ph	Ph	OSiMe$_3$ / SEt	MS4A	90
6	Ph	Bn	OSiMe$_3$ / SEt	MS4A	62[b]
7	Ph	p-MeOPh	OSiMe$_3$ / SEt	MS4A	79 84[b], 87[b,c]
8	Ph	C$_4$H$_9$	2	MS4A	89
9	PhCO[d]	Ph	2	MgSO$_4$	82
10	PhCO[d]	Ph	OSiMe$_3$ / Ph	MgSO$_4$	87
11	PhCH=CH	p-MeOH	2	MgSO$_4$	92[e]
12	Ph(CH$_2$)$_2$	Bn	2	MgSO$_4$	83[f]
13	C$_4$H$_9$	Bn	2	MgSO$_4$	77[f]
14	C$_8$H$_{17}$	Bn	2	MgSO$_4$	81[f]
15	C$_8$H$_{17}$	Ph$_2$CH	2	MgSO$_4$	89[g]

a) MS4A or MgSO$_4$ was used. Almost comparable yields were obtained in each case.
b) CH$_3$CN was used as a solvent.
c) Sc(OTf)$_3$ was used instead of Yb(OTf)$_3$.
d) Monohydrate.
e) C$_2$H$_5$CN, −78 °C.
f) −78 °C to 0 °C
g) 0 °C.

high yields and that no adducts between aldehydes and the silyl enolates were observed in any reaction. As an example, aliphatic aldehydes reacted with amines and silyl enolates to give the corresponding β-amino esters in high yields. Phenylglyoxal monohydrate also worked well in this reaction, although the imine derived from phenylglyoxal is unstable [29]. As for the diastereoselectivity of this reaction, good results were obtained after examination of the reaction conditions (Tab. 3). While *anti* adducts were produced preferentially in the reactions of benzaldehyde,

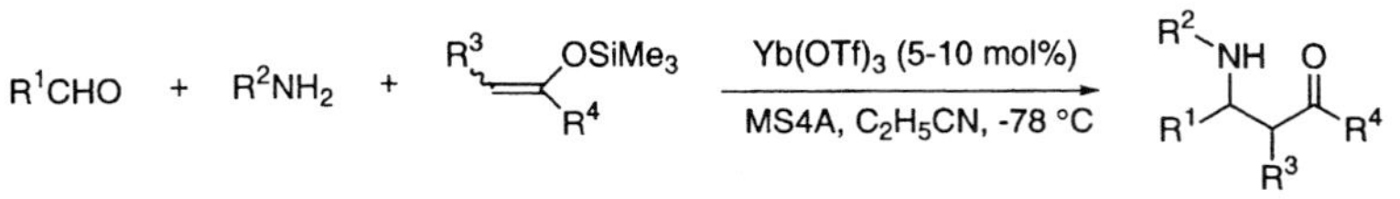

2.16.3 Activation of Nitrogen-containing Compounds

$$R^1CHO + R^2NH_2 + \underset{R^4}{\overset{R^3}{\diagup}}\!\!\!-\!\!\!OSiMe_3 \xrightarrow[MS4A, C_2H_5CN, -78\ °C]{Yb(OTf)_3\ (5\text{-}10\ mol\%)} \underset{R^3}{R^1}\!\!\!\diagup\!\!\!\overset{R^2\text{-}NH}{\underset{}{}}\!\!\!\overset{O}{\diagup}\!\!\!R^4$$

Tab. 3 Diastereoselective one-pot synthesis of β-amino esters from aldehydes.

Entry	R^1	R^2	Silyl enolate	Yield %	syn/anti[a]
1	Ph	Bn	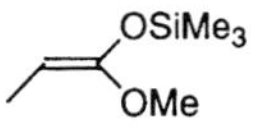 OSiMe$_3$ / OMe	90	1/13.3
2	Ph	Bn	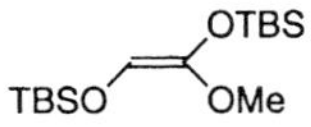 TBSO / OTBS / OMe	78	1/9.0
3	Ph(CH$_2$)$_2$	Ph$_2$CH	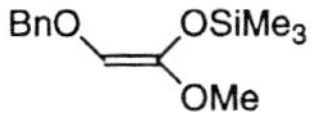 BnO / OSiMe$_3$ / OMe	88	8.1/1
4	C$_4$H$_9$	Ph$_2$CH	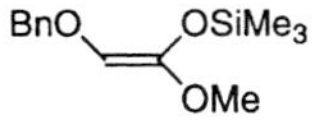 BnO / OSiMe$_3$ / OMe	90	8.1/1
5	(CH$_3$)$_2$CHCH$_2$	Ph$_2$CH	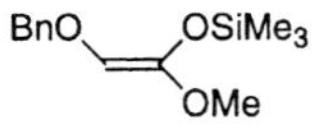 BnO / OSiMe$_3$ / OMe	86	7.3/1

a) Determined by ^{1}H NMR analysis.

syn-adducts were obtained with high selectivities in the reactions of aliphatic aldehydes.

The high yields of the present one-pot reactions depend on the unique properties of lanthanide triflates as the Lewis acid catalysts. Although TiCl$_4$ and TMSOTf are known to be effective for the activation of imines [26, 30], the use of even stoichiometric amounts of TiCl$_4$ and TMSOTf instead of lanthanide triflate in the present one-pot reactions gives only trace amounts of the product in both cases.

One-pot preparation of a β-lactam from an aldehyde, an amine, and a silyl enolate has been achieved on the basis of the present reaction [28].

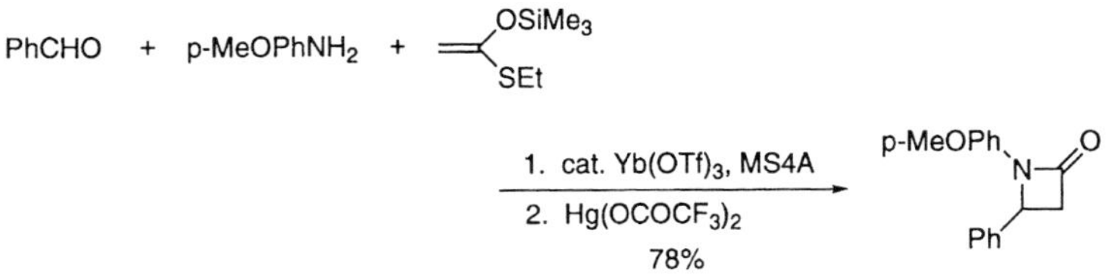

$$\text{PhCHO} + p\text{-MeOPhNH}_2 + \underset{SEt}{\overset{OSiMe_3}{=\!\!\!=}} \xrightarrow[\substack{1.\ cat.\ Yb(OTf)_3,\ MS4A \\ 2.\ Hg(OCOCF_3)_2 \\ 78\%}]{} \underset{Ph}{\overset{p\text{-MeOPh}}{\underset{}{\diagdown}N\!\!\!-\!\!\!\square\!\!\!=\!\!\!O}} \qquad (7)$$

Vinyl ethers reacted with imines in the presence of a catalytic amount of Ln(OTf)$_3$, to afford the corresponding β-amino ketones [31]. In addition, the reactions proceeded smoothly by the combination of aldehydes, amines, and vinyl ethers in aqueous media [32]. The procedure is very simple: in the presence of

10 mol% of Yb(OTf)$_3$, an aldehyde, an amine, and a vinyl ether are combined in a solution of THF–water (9:1) at room temperature to afford a β-amino ketone (Eq. 8). Commercially available formaldehyde and chloroacetaldehyde water solutions are used directly and the corresponding β-amino ketones were obtained in good yields. Phenylglyoxal monohydrate, methyl glyoxylate, an aliphatic aldehyde, and an α,β-unsaturated ketone also work well to give the corresponding β-amino esters in high yields. Other lanthanide triflates could also be used as catalysts. For example, the reaction of phenylglyoxal monohydrate, p-chloroaniline, and 2-methoxypropene, proceeds with 90% (Sm(OTf)$_3$), 94% (Tm(OTf)$_3$), and 91% (Sc(OTf)$_3$) yields although iminium salts and imines are quite unstable [33].

$$R^1CHO + R^2NH_2 + \underset{R^3}{\overset{OMe}{\diagup\!\!\diagdown}} \xrightarrow[\text{H}_2\text{O/THF (1:9)}]{\text{cat. Yb(OTf)}_3} \underset{R^1}{\overset{R^2\text{NH} \quad O}{\diagdown\!\!\diagup\!\!\diagdown\!\!\diagup R^3}} \quad (8)$$

55%-quant.

2.16.3.2
Aza Diels–Alder Reactions

The aza Diels–Alder reaction is among the most powerful synthetic tools for constructing nitrogen-containing six-membered heterocycles [34]. Although Lewis acids often promote these reactions, more than stoichiometric amounts of the acids are required due to the strong coordination of the acids to nitrogen atoms [34]. Again lanthanide triflates proved to be efficient catalysts in these reactions.

Hence, in the presence of 10 mol% Yb(OTf)$_3$, N-benzylideneaniline reacts with 2-trimethylsiloxy-4-methoxy-1,3-butadiene (Danishefsky's diene, 3) [35] in acetonitrile at room temperature to afford the corresponding imino Diels–Alder adduct, a tetrahydropyridine derivative, in 93% yield (Eq. 9) [36].

$$\underset{\underset{Ph}{\diagup}\overset{N}{\diagdown}H}{\overset{R^1}{\bigcirc}} + \underset{3}{\overset{OSiMe_3}{\diagup\!\!\diagdown\!\!\diagup OMe}} \xrightarrow[\text{CH}_3\text{CN, rt}]{\text{cat. Ln(OTf)}_3} \underset{Ph}{\overset{R^1}{\bigcirc\!\!-\!\!N\diagdown\!\!\diagup\!\!\diagdown O}} \quad (9)$$

The adduct was obtained quantitatively when Sc(OTf)$_3$ was used as a catalyst. On the other hand, the reaction of N-benzylideneaniline with cyclopentadiene was performed under the same reaction conditions. Surprisingly, the reaction course changed in this case and a tetrahydroquinoline derivative was obtained in 69% yield (Eq. 10) [37]. In this reaction, the imine worked as an azadiene toward one of the double bonds of cyclopentadiene as a dienophile [29, 38]. Using 2,3-dimethylbutadiene, mixtures of tetrahydropyridine and tetrahydroquinoline derivatives were obtained.

2.16.3 Activation of Nitrogen-containing Compounds

[Scheme: PhCH=N-C6H4-R1 + cyclopentadiene, cat. Ln(OTf)3, CH3CN, rt → tetrahydroquinoline product] (10)

A vinyl sulfide, a vinyl ether, and a silyl enol ether worked equally well as dienophiles to afford tetrahydroquinoline derivatives in high yields [39, 40]. As for the lanthanide triflates, heavy lanthanides such as Er, Tm, and Yb gave better results.

One synthetic problem in the imino Diels–Alder reactions is the stability of imines under the influence of Lewis acids. It is desirable that the imines activated by Lewis acids are immediately trapped by dienes or dienophiles. In 1989, Sisko and Weinreb reported a convenient procedure for the imino Diels–Alder reaction of an aldehyde and a 1,3-diene with N-sulfinyl p-toluenesulfonamide via N-sulfonyl imine produced *in situ*, by using a stoichiometric amount of $BF_3 \cdot OEt_2$ as a promoter [41].

Because of the usefulness and efficiency of one-pot procedures, three-component coupling reactions between aldehydes, amines, and alkenes via imine formation and imino Diels–Alder reactions were examined by using lanthanide triflate as a catalyst. In the presence of 10 mol% $Yb(OTf)_3$ and magnesium sulfate, benzaldehyde was treated with aniline and **3** successively in acetonitrile at room temperature. The three-component coupling reaction proceeded smoothly to afford the corresponding tetrahydropyridine derivative in an 80% yield (Eq. 11) [37]. It is noteworthy that $Yb(OTf)_3$ kept its activity and effectively catalyzed the reaction even in the presence of water and the amine. When typical Lewis acids such as $BF_3 \cdot OEt_2$ and $ZnCl_2$ (100 mol%) were used instead of the $Yb(OTf)_3$ under the same reaction conditions, lower yields were observed (23 and 12%, respectively). Use of $Sc(OTf)_3$ slightly improved the yield. In the reaction between benzaldehyde, anisidine, and cyclopentadiene under the same reaction conditions, the reaction course changed and the tetrahydroquinoline derivative was obtained in a 56% yield. A vinyl sulfide and a vinyl ether, and a silyl enol ether worked well as dienophiles to afford tetrahydroquinoline derivatives in high yields. Phenylglyoxal monohydrate reacted with amines and **3** or cyclopentadiene to give the corresponding tetrahydropyridine or quinoline

[Scheme: PhCHO + PhNH2 + CH2=C(OSiMe3)-CH=CH-OMe (3), cat. Yb(OTf)3, CH3CN, rt, 80% → tetrahydropyridinone product] (11)

derivatives in high yields. The three-component coupling reactions proceeded even in aqueous solution, and commercial formaldehyde–water solution could be used directly (Eq. 12). Most lanthanide triflates tested were effective in the three-component coupling reactions. These reactions provide very useful routes for the synthesis of pyridine and quinoline derivatives.

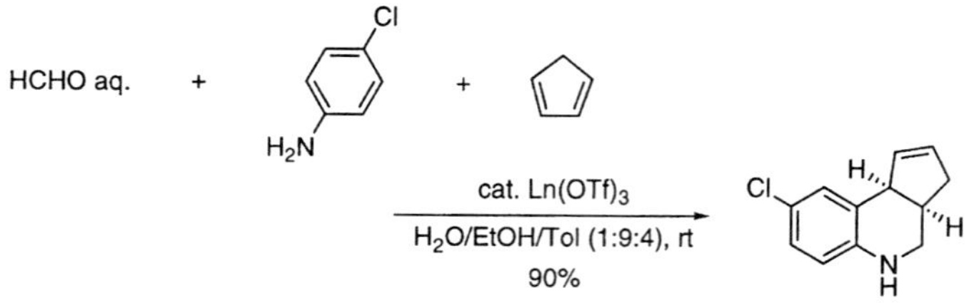

$$\text{HCHO aq.} + \text{4-Cl-C}_6\text{H}_4\text{NH}_2 + \text{cyclopentadiene} \xrightarrow[\text{H}_2\text{O/EtOH/Tol (1:9:4), rt}]{\text{cat. Ln(OTf)}_3} \text{product} \quad 90\% \tag{12}$$

In the reactions of **4a–c** with cyclopentadiene, a vinyl sulfide, or a vinyl ether (**4a–c** work as azadienes), **4c** gave the best yields, while the yields using **4b** were lowest. The HOMO and LUMO energies and coefficients of **4a–c** and protonated **4a–c** were calculated, however, these data did not correspond to the differences in reactivity between **4a–c**, if the reactions were postulated to proceed via concerted [4+2]-cycloaddition. On the other hand, the high reactivity of **4c** toward electrophiles compared to **4a** and **b** may be accepted by assuming a stepwise mechanism.

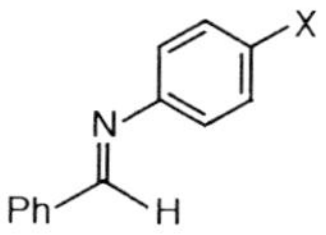

4a : X = H
4b : X = OMe
4c : X = Cl

The reaction of **4a** with 2-methoxypropene was carried out in the presence of Yb(OTf)$_3$ (10 mol%). The main product was tetrahydroquinoline derivative **5a**, and small amounts of quinoline **6a** and β-amino ketone dimethylacetal **7a** were also obtained (Eq. 13).

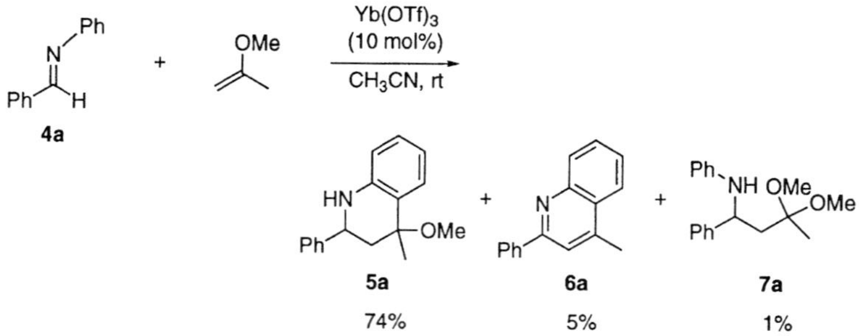

$$\mathbf{4a} + \text{2-methoxypropene} \xrightarrow[\text{CH}_3\text{CN, rt}]{\text{Yb(OTf)}_3 \text{ (10 mol\%)}} \mathbf{5a} + \mathbf{6a} + \mathbf{7a} \tag{13}$$

5a 74%   6a 5%   7a 1%

On the other hand, the three-component coupling reaction between benzaldehyde, aniline, and 2-methoxypropene gave only a small amount of tetrahydroquinoline derivative **5a**, and the main products in this case were β-amino ketone **8a** and its dimethylacetal **7a** (Eq. 14). Similar results were obtained in the reaction of **4b** with 2-methoxypropene and the three-component coupling reaction between benzaldehyde, anisidine, and 2-methoxypropene.

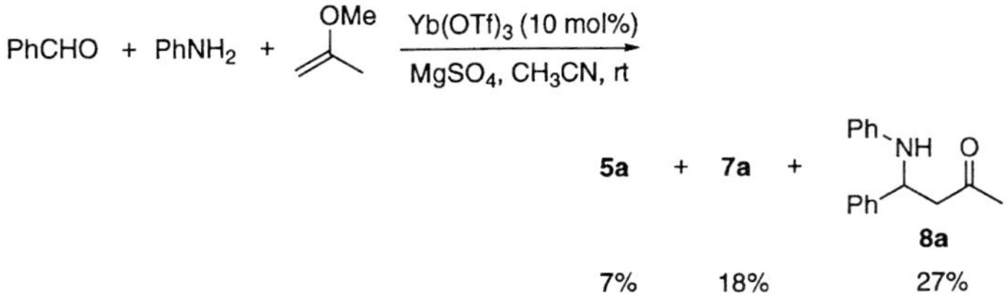

$$\text{PhCHO} + \text{PhNH}_2 + \underset{\text{OMe}}{\diagup\!\!\!\diagdown} \xrightarrow[\text{MgSO}_4, \text{CH}_3\text{CN, rt}]{\text{Yb(OTf)}_3 \text{ (10 mol\%)}} \textbf{5a} + \textbf{7a} + \underset{\textbf{8a}}{\text{Ph-NH-CH(Ph)-CH}_2\text{-C(=O)}} \quad (14)$$

7%   18%   27%

A possible mechanism of these reactions is shown in Scheme 2. Intermediate **9** is quenched by water and methanol generated *in situ* to afford **7** and **8**, respectively. While **5** is predominantly formed from **9** under anhydrous conditions, formation of **7** and **8** predominated in the presence of even a small amount of water. It is noted that these results suggest a stepwise mechanism in these types of imino Diels–Alder reactions [42].

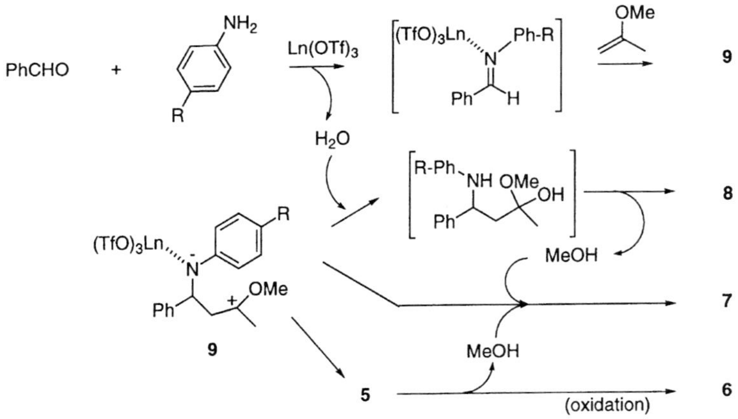

Scheme 2 A possible mechanism of the three-component coupling reaction.

2.16.3.3
1,3-Dipolar Cycloaddition

It has been reported that lanthanide triflates were excellent catalysts for the reactions of nitrones with dipolarophiles leading to isoxazolidine derivatives [43], whose reductive cleavage gave a range of compounds such as β-hydroxy ketones and β-amino alcohols, etc. [44]. Three-component coupling reactions of aldehydes, hydroxylamines, and alkenes also proceeded smoothly in the presence of a catalytic amount of a lanthanide triflate, to afford the isoxazolidine derivatives in high yields with high diastereoselectivities [45].

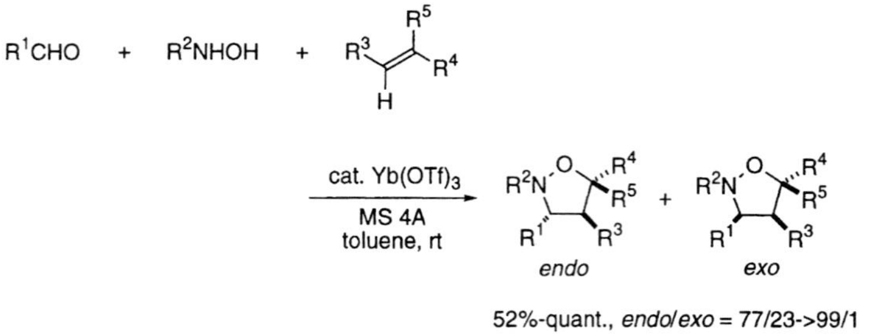

R¹CHO + R²NHOH + [alkene with R³, R⁴, R⁵, H]

$\xrightarrow[\text{toluene, rt}]{\substack{\text{cat. Yb(OTf)}_3 \\ \text{MS 4A}}}$ endo isoxazolidine + exo isoxazolidine (15)

52%-quant., endo/exo = 77/23->99/1

2.16.3.4
Reactions of Imines with Alkynyl Sulfides

Recently, a new reaction of imines with alkynyl sulfides has been developed [46]. The reaction was effectively catalyzed by a lanthanide triflate to afford α,β-unsaturated thioimidates. [2+2]-Cycloaddition and successive fragmentation mechanisms were suggested and the reaction was successfully applied to the intramolecular version for the synthesis of cycloalkane derivatives.

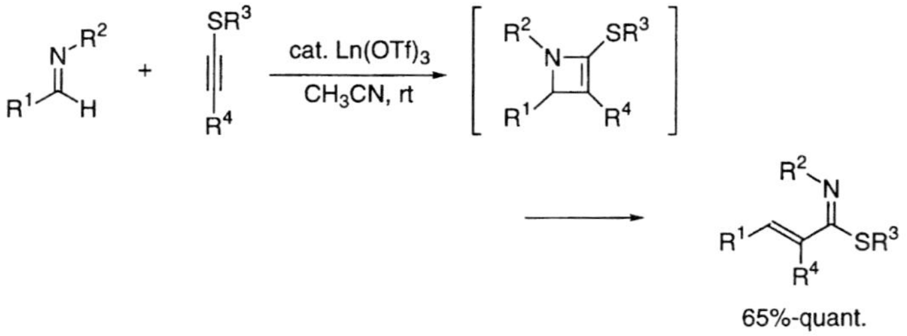

imine + alkynyl sulfide $\xrightarrow[\text{CH}_3\text{CN, rt}]{\text{cat. Ln(OTf)}_3}$ [2+2 cycloadduct] → α,β-unsaturated thioimidate (16)

65%-quant.

2.16.4
Asymmetric Catalysis

2.16.4.1
Asymmetric Diels–Alder Reaction

Recently, some efficient asymmetric Diels–Alder reactions catalyzed by chiral Lewis acids have been reported [47]. The chiral Lewis acids employed in these reactions are generally based on traditional acids such as titanium, boron, or aluminum reagents, and they are well modified to realize high enantioselectivities. Although lanthanide compounds were expected to be Lewis acid reagents, only a few asymmetric reactions catalyzed by chiral lanthanide Lewis acids were reported. Danishefsky's pioneering work demonstrated that Eu(hfc)$_3$ (a NMR shift reagent) catalyzed hetero-Diels–Alder reactions of aldehydes with siloxydienes, but enantiomeric excesses were moderate [48].

Later on it was demonstrated that chiral Yb triflates can be prepared *in situ* from Yb(OTf)$_3$, (*R*)-(+)-binaphthol ((*R*)-(+)-BINOL), and a tertiary amine at 0 °C for 0.5 h in dichloromethane.

$$\text{Yb(OTf)}_3 \text{ (20 mol\%)} + \text{1.2 eq (R)-BINOL} \xrightarrow[\text{MS 4A}]{} \xrightarrow[\text{0 °C, 30 min}]{\text{2.4 eq. amine}} \text{"chiral Yb triflate"} \quad (17)$$

In the presence of the chiral Yb triflate, 3-(2-butenoyl)-1,3-oxazolidin-2-one (**10**) reacted with cyclopentadiene at room temperature to afford the Diels–Alder adduct in an 87% yield *(endo/exo=76/24)* and the enantiomeric excess of the *endo* adduct was shown to be 33%. The amine employed at the stage of the preparation of the chiral catalyst strongly influenced the diastereo and enantioselectivities. In general, bulky amines gave better results and 70, 75, and 71% ee's were observed when diisopropylethylamine, *cis*-2,6-dimethylpiperidine, and *cis*-1,2,6-trimethylpiperidine were used, respectively. In addition, a better result was obtained when the amine was combined with 4 Å molecular sieves (*cis*-1,2,6-trimethylpiperidine, 91% yield, *endo/exo*=86/14, *endo*=90% ee), and the enantiomeric excess was further improved to 95% when the reaction was carried out at 0 °C [49].

Interestingly, high selectivities were obtained when the diene and the dienophile were added after stirring Yb(OTf)$_3$, (*R*)-(+)-BINOL, and a tertiary amine at 0 °C for 0.5 h in dichloromethane (the original catalyst system). These results seemed to be ascribed to the aging of the catalyst, but the best result (77% yield, *endo/exo*=89/11, *endo*=95% ee) was obtained when the mixture (the substrates and 20 mol% of the catalyst) was stirred at 0 °C for 20 h. It was suggested that the dienophile (**10**) is effective in preventing the catalyst from aging.

After screening several additives other than **10**, it was discovered that some additives were effective not only in stabilizing the catalyst but also *in controlling the enantiofacial selectivities in the Diels–Alder reaction*. When 3-acetyl-1,3-oxazolidin-2-one (**11**) was combined with the original catalyst system (to form *catalyst A)*, the *endo* adduct was obtained in 93% ee and the absolute configuration of the product was 2*S*, 3*R* (Tab. 4). On the other hand, when acetyl acetone derivatives were mixed with the catalyst, *reverse enantiofacial selectivities were observed*. The *endo* adduct with an absolute configuration of 2*R*, 3*S* was obtained in 81% ee when 3-phenylacetylacetone (PAA) was used as an additive *(catalyst B)*. In these cases, the chiral source was the same (*R*)-(+)-BINOL. Therefore, *the enantioselectivities were controlled by the achiral ligands, 3-acetyl-1,3-oxazolidin-2-one and PAA* [50].

The same selectivities were observed in the reactions of other 3-acyl-1,3-oxazolidin-2-ones. Thus, by using the same chiral source ((*R*)-(+)-BINOL), both enantiomers of the Diels–Alder products could be prepared (Tab. 5). Traditional methods required both enantiomers of chiral sources in order to prepare both enantiomers stereoselectively [51], but the counterparts of some chiral sources are of poor quality or are hard to obtain (for example, sugars, amino acids, alkaloids, etc.).

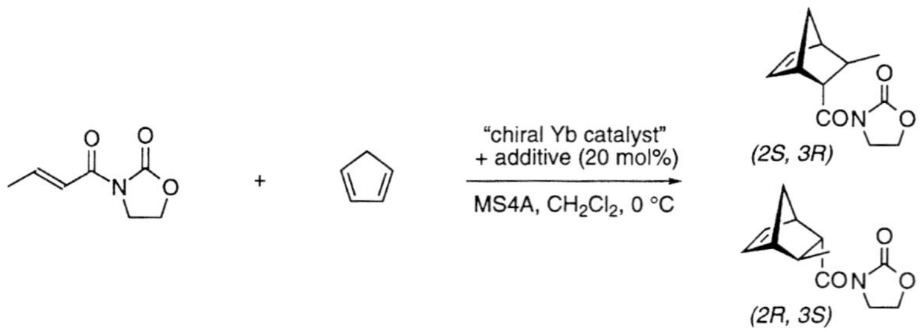

Tab. 4 Effect of additives.

Additive	Yield [%]	endo/exo	2S,3R/2R,3S	ee [%][a]
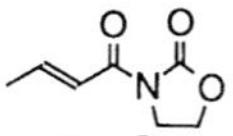 3-crotonoyl-1,3-oxazolidin-2-one (**10**)	66	87/13	94.0/6.0	88
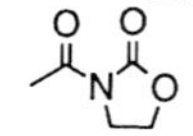 3-pivaloyl-1,3-oxazolidin-2-one (**11**)	77	89/11	96.5/3.5	93
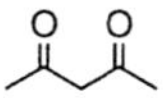 2,4-pentanedione	80	88/12	22.5/77.5	55
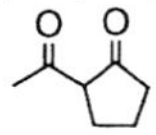 2-acetylcyclopentanone	36	81/19	19.0/81.0	62
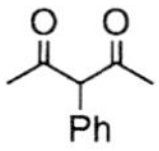 3-methyl-2,4-pentanedione	69	88/12	15.5/84.5	69
1-phenyl-1,3-butanedione (Ph)	83	93/7	9.5/90.5	81[b]

a) Enantiomer ratios of endo adducts.
b) 1,2,2,6,6-Pentamethylpiperidine was used instead of cis-1,2,6-trimethylpiperidine. Yb(OTf)$_3$, MS4A, and the additive were stirred in dichloromethane at 40 °C for 3 h.

This exciting effect is believed to be strongly dependent on the specific coordination number of Yb(III) [52] (Scheme 3).

Although Sc(OTf)$_3$ has slightly different properties compared with lanthanide triflates, the chiral Sc catalyst could be prepared similarly from Sc(OTf)$_3$, (R)-(+)-BINOL, and a tertiary amine in dichloromethane [53]. The catalyst was also found to be effective in the Diels–Alder reactions of acyl-1,3-oxazolidin-2-ones with dienes. Again, the amines employed in the preparation of the catalyst influenced the enantioselectivities strongly. For example, in the Diels–Alder reaction of 3-(2-butenoyl)-1,3-oxazolidin-2-one (**10**) with cyclopentadiene (CH$_2$Cl$_2$, 0 °C), the enantiomeric excesses of the *endo* adduct depended crucially on the amines employed; aniline, 14% ee; lutidine, 46% ee; triethylamine, 51% ee; 2,2,6,6-tetramethylpiperidine, 51% ee; diisopropylethylamine, 69% ee; 2,6-dimethylpiperidine, 69% ee; 1,2,2,6,6-pentamethylpiperidine, 72% ee; and cis-1,2,6-trimethylpiperidine, 84% ee. The highest enantioselectivities were observed when cis-1,2,6-trimethylpiperidine was employed as an amine. 3-(2-Butenoyl)-3-cinnamoyl-, and 3-(2-hexenoyl)-1,3-ox-

Tab. 5 Synthesis of both enantiomers of the Diels–Alder products of cyclopentadiene and dienophiles by use of catalysts A and B.

Dienophile	Yield [%]	Catalyst A endo/exo	2S,3R/2R,3S	ee [%][b]
(acryloyl-oxazolidinone)	77	89/11	96.5/3.5	93
	77	89/11	97.5/2.5	95[c]
Ph-(cinnamoyl-oxazolidinone)	40	81/19	91.5/8.5[d]	83
	34	80/20	93.0/7.0	86
Pr-(crotonyl-type-oxazolidinone)	81	80/20	91.5/8.5	83[c]

Dienophile	Yield [%]	Catalyst B[a] endo/exo	2S,3R/2R,3S	ee [%][b]
(acryloyl-oxazolidinone)	83	93/7	9.5/90.5	81
Ph-(cinnamoyl-oxazolidinone)	60	89/11	10.5/89.5[d]	79
	51	89/11	8.5/91.5[d]	83
	51	89/11	5.5/94.5[d]	89[e]
Pr-(crotonyl-type-oxazolidinone)	81	91/9	10.0/90.0	80
	85	91/9	9.0/91.0	82[e]
	60	91/9	7.5/92.5	85[f]

Catalyst A: Yb(OTf)$_3$ + (R)-(+)-binaphthol + cis-1,2,6-trimethylpiperidine + MS4A + 3-acetyl-1,3-oxazolidin-2-on3 (7)
Catalyst B: YB(OTf)$_3$ + (R)-(+)-binaphthol + cis-1,2,6-trimethylpiperidine + MS4A + 3-acetyl-1,3-phenylacetylacetone (PAA)
a) 1,2,2,6,6-Pentamethylpiperidine was used instead of 1,2,6-trimethylpiperidine
b) Enantiomer ratios of endo adducts
c) Without additive.
d) 2R,3R/2R,3S.
e) Tm(OTf)$_3$ was used instead of Yb(OTf)$_3$.
f) Er(OTf)$_3$ was used instead of Yb(OTf)$_3$.

azolidin-2-ones reacted with cyclopentadiene smoothly in the presence of the chiral Sc catalyst to afford the corresponding Diels–Alder adducts in high yields and high selectivities (Tab. 6). It should be noted that even 3 mol% of the catalyst was enough to complete the reaction yielding the *endo* adduct in 92% ee.

The chiral Sc catalyst was also found to be effective for the Diels–Alder reactions of an acrylic acid derivative [54]. 3-Acryloyl-1,3-oxazolidin-2-one reacted with 2,3-dimethylbutadiene to afford the corresponding Diels–Alder adduct in 78% yield and 73% ee, whereas the reaction of 3-acryloyl-1,3-oxazolidin-2-one with cyclohexadiene gave a 72% ee for the *endo* adduct (88% yield, *endo/exo* = 100/0). Similar to the chiral Yb catalyst, aging was observed in the chiral Sc catalyst. It was found that **11** or 3-benzoyl-1,3-oxazolidin-2-one was a good additive for stabiliza-

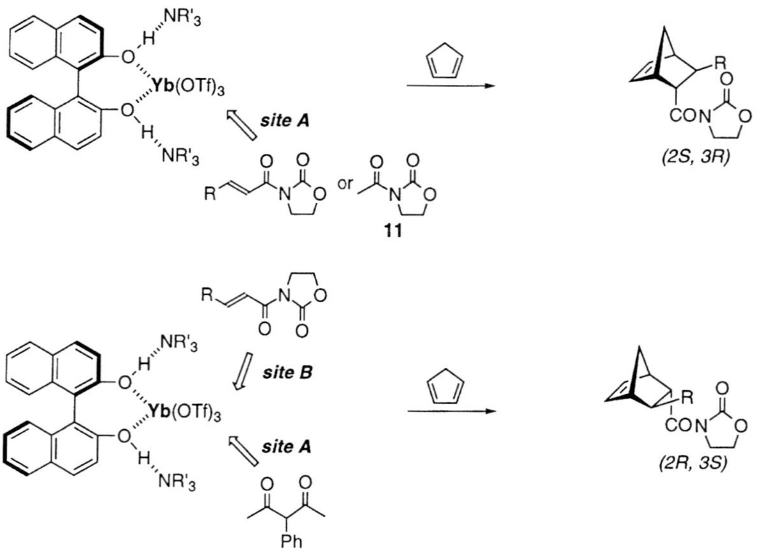

Scheme 3 Synthesis of both enantiomers using the same chiral source.

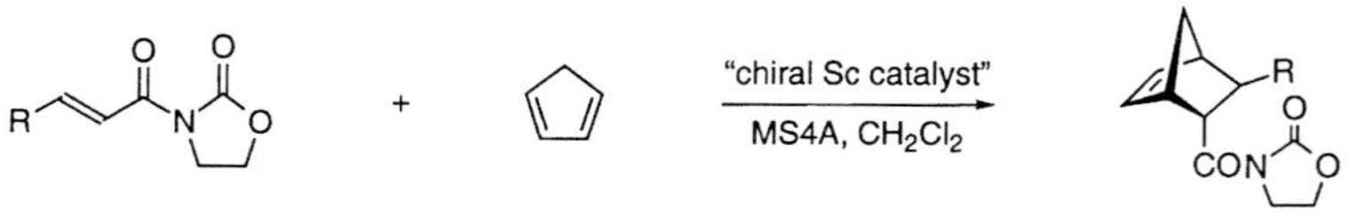

Tab. 6 Enantioselective Diels–Alder reactions using a chiral scandium catalyst.

R	Catalyst/mol%	Yield [%]	endo/exo	ee [%] (endo)
Me	20	94	89/11	92 (2S,3R)
	10	84	86/14	96 (2S,3R)
	5	84	87/13	93 (2S,3R)
	3	83	87/13	92 (2S,3R)
Ph	20	99	89/11	93 (2S,3R)
	10	96	90/10	97 (2S,3R)
Pr	20	95	78/22	74 (2S,3R)
	10	86	78/22	75 (2S,3R)

tion of the catalyst, but that reverse enantioselectivities by additives were not observed. This can be explained by the coordination numbers of Yb(III) and Sc(III); while Sc(III) is known to coordinate up to seven ligands, the specific coordination numbers of Yb(III) allow up to 12 ligands [52, 55].

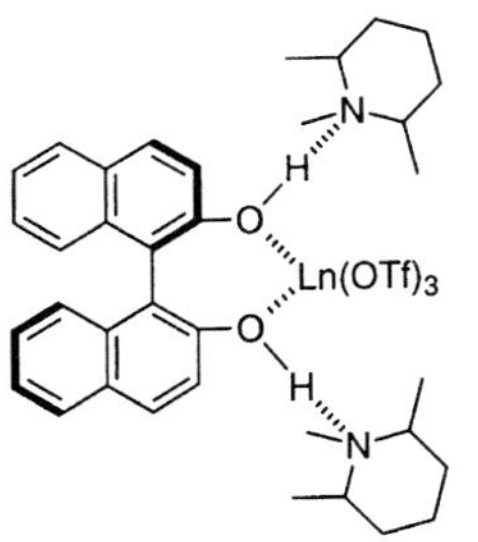

Scheme 4 Chiral lanthanide catalysts.

As for the chiral lanthanide and scandium catalysts, the following structures have been postulated (Scheme 4) [56]. The most characteristic point of the catalyst structure is the existence of hydrogen bonds between the phenolic hydrogens of (R)-(+)-BINOL and the nitrogens of the tertiary amines. The ^{13}C NMR spectra of the complexes indicate these interactions, and the existence of the hydrogen bonds was confirmed by the IR spectra [57]. The coordination mode of these catalysts may be similar to that of the lanthanide(III)–water-or–alcohol complex [55]. Clearly, the structure is quite different from those of conventional chiral Lewis acids based on aluminum [58], boron [59], or titanium [60]. In the present chiral catalysts, the axial chirality of (R)-(+)-BINOL is transferred via the hydrogen bonds to the amine parts, which shield one side of the dienophile effectively. This is consistent with the experimental results showing that amines employed in the preparation of the chiral catalysts strongly influenced the selectivities and that bulky amines gave better selectivities. Moreover, since the amine part can be freely chosen, the design of efficient catalyst systems is easier compared to other catalysts on the basis of (R)-(+)-BINOL. Although some 'modified' binaphthols were reported to be effective as chiral sources, their preparations often require long steps [58a, 61].

Inverse electron-demand asymmetric Diels–Alder reactions of 2-pyrone derivatives using the above chiral ytterbium catalyst were reported [62]. Using vinyl ethers as dienophiles, bulky ethers gave higher selectivities. When phenylvinylsulfide was used, the corresponding Diels–Alder product was obtained with more than 95% enantiomeric excess (Eq. 18). It was reported that the reactions did not proceed in the presence of Yb(OTf)$_3$ alone (without the ligand and the amine).

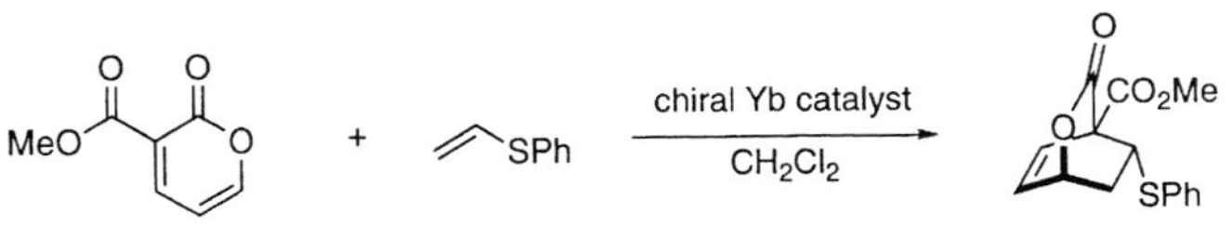

$$\text{MeO} \underset{O}{\overset{O}{\bigcirc}} + \diagup\!\!\!\diagdown\text{SPh} \xrightarrow[\text{CH}_2\text{Cl}_2]{\text{chiral Yb catalyst}} \text{product with CO}_2\text{Me and SPh} \qquad (18)$$

2.16.4.2
Asymmetric [2+2]-Cycloaddition

Enantioselective [2+2]-cycloaddition of a vinyl sulfide with an oxazolidone derivative proceeds smoothly in the presence of a catalytic amount of the chiral lantha-

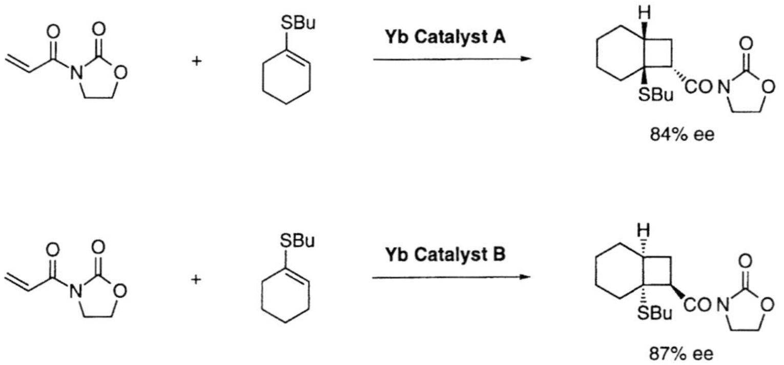

Scheme 5 Asymmetric [2+2]-cycloaddition. Synthesis of both enantiomers using the same chiral source and a choice of achiral ligands.

nide catalyst [63]. Both enantiomers can be prepared by using a single chiral source with an appropriate achiral ligand (Scheme 5).

2.16.4.3
Asymmetric Aza Diels–Alder Reaction

Aza Diels–Alder reactions are one of the most basic and versatile reactions for the synthesis of nitrogen-containing heterocyclic compounds [34, 64]. Although asymmetric versions using chiral auxiliaries or a stoichiometric amount of a chiral Lewis acid have been reported [27j, 65], examples using a catalytic amount of a chiral source were unprecedented.

The first example has been reported by using a chiral lanthanide catalyst [66]. The reaction of N-benzylidene-2-hydroxyaniline (**12a**) with cyclopentadiene proceeded under the influence of 20 mol% of a chiral ytterbium Lewis acid prepared from Yb(OTf)$_3$, (R)-(+)-1,1'-bi-naphthol (BINOL), and diazabicyclo-[5,4,0]-undec-7-ene (DBU), to afford the corresponding 8-hydroxyquinoline derivative [67] in a high yield. The enantiomeric excess of the *cis* adduct was 40%. It was indicated that the phenolic hydrogen of **12a** would interact with DBU, which should interact with the hydrogen of (R)-(+)-BINOL [56], to decrease the selectivity. Additives which interact with the phenolic hydrogen of **12a** were examined. When 20 mol% N-methylimidazole (MID) was used, 91% ee of the *cis* adduct was obtained, however, the chemical yield was low. Other additives were screened and it was found that the desired tetrahydroquinoline derivative was obtained in a 92% yield with high selectivities (*cis/trans*=>99/1, 71% ee), when 2,6-di-*tert*-butyl-4-methylpyridine (DTBMP) was used.

Other substrates were tested, and the results are summarized in Tab. 7. Vinyl ethers also worked well to afford the corresponding tetrahydroquinoline derivatives in good to high yields with good to excellent diastereo- and enantioselectivities (entries 1–9). Use of 10 mol% of the chiral catalyst also gave the adduct in high yields and selectivities (entries 2 and 6). As for additives, 2,6-di-*tert*-butylpyri-

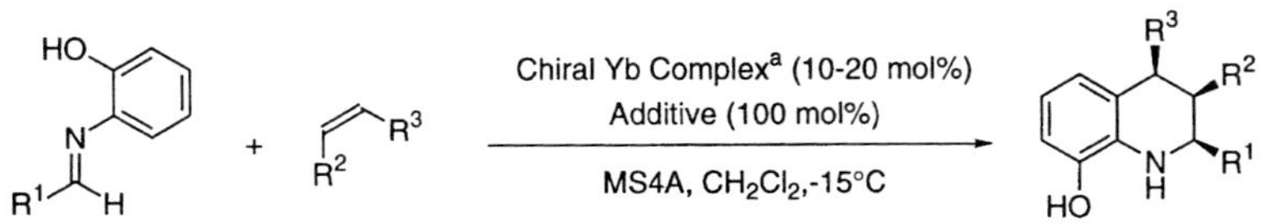

Tab. 7 Asymmetric synthesis of tetrahydroquinoline derivatives.

Entry	R¹	Alkene		Additive[b]	Yield [%]	cis/tasns	ee [%] (cis)
1[c]	Ph (12a)	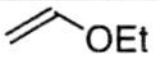 OEt	(13a)	DTBP	58	94/6	61
2[c]	Ph (12a)	13a		DTBP	52	94/6	77
3	α-Naph (12b)	13a		DTBP	69	>99/1	86
4	α-Naph (12b)	13a		DPP	65	99/1	91
5	α-Naph (12b)	13a		DTBMP	74	>99/1	91
6	α-Naph (12b)	13a		DTBMP	62	98/2	82
7	α-Naph (12b)	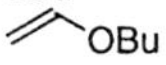 OBu		DTBMP	80	66/34	70
8	α-Naph (12b)	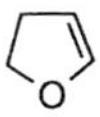 furan	(13b)	DTBMP	90	91/9	78
9	α-Naph (12b)	13b		DPP	67	93/7	86
10	α-Naph (12b)	 cyclopentene	(13c)	DTBMP	69	>99/1	68
11[d]	c-C₆H₁₁ (12c)	13c		DTBMP	58	>99/1	73

a) Prepared from Yb(OTf)₃, (R)-(+)-BINOL, and DBU.
b) DTBP 2,6-Di-*tert*-butylpyridine. DTBMP: 2,6-Di-*tert*-butyl-4-methylpyridine. DPP: 2,6-Diphenylpyridine.
c) –45 °C   d) Sc(OTf)₃ was used. See text

dine (DTBP) gave the best result in the reaction of imine **12a** with ethyl vinyl ether (**13a**), while higher selectivities were obtained when DTBMP or 2,6-diphenylpyridine (DPP) were used in the reaction of imine **12b** with **13a**. While use of butyl vinyl ether decreased the selectivities (entry 7), dihydrofuran reacted smoothly to achieve high levels of selectivity (entries 8, 9).

2.16.4.4
Asymmetric 1,3-Dipolar Cycloaddition

A catalytic asymmetric 1,3-dipolar cycloaddition of a nitrone with a dipolarophile was carried out using chiral lanthanide catalysts [45, 68]. The chiral catalyst, which was effective in asymmetric Diels–Alder reactions, was readily prepared from Yb(OTf)₃, (R)-(+)-BINOL, and *cis*-1,2,6-trimethylpiperidine. The reaction of benzylbenzylideneamine N-oxide with 3-(2-butenoyl)-1,3-oxazolidin-2-one was performed in the presence of the chiral catalyst (20 mol%) to yield the desired isoxazolidine in 75% yield with perfect diastereoselectivity (*endo/exo*=>99/1) (Scheme 6). The enantiomeric excess of the *endo* adduct was 73% ee determined by HPLC

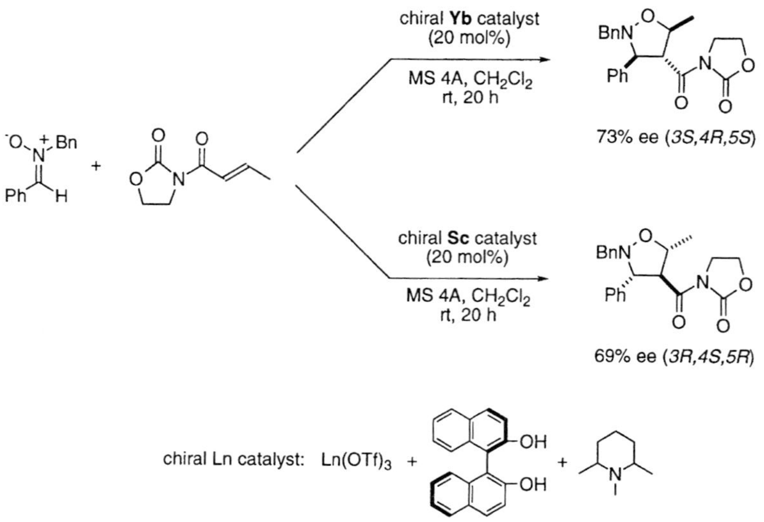

Scheme 6 Asymmetric 1,3-dipolar cycloaddition. Synthesis of both enantiomers using the same chiral source and a choice of lanthanides.

analysis. On the other hand, it was found that reverse enantioselectivity was observed when a chiral scandium catalyst was used instead of the chiral ytterbium catalyst.

2.16.5
Miscellaneous

Many useful synthetic reactions using lanthanide triflates as catalysts in organic solvents have been developed. Although this chapter does not cover them all, the characteristic points are that the reactions are completed by using only a catalytic amount of the triflate in most cases, and that the catalyst can be easily recovered after the reactions are completed and can be reused. There are many kinds of Lewis acid-promoted reactions in the chemical industry which use large amounts of acids or Lewis acids, thus creating important and severe environmental problems. From the standpoints of their catalytic use and reusability, lanthanide triflates are expected to be new types of catalysts providing solutions for these problems.

One example is the Friedel–Crafts acylation. Friedel–Crafts alkylation and acylation reactions are fundamental and important processes in organic synthesis as well as in industrial chemistry [69]. While the alkylation reaction proceeds in the presence of a catalytic amount of a Lewis acid such as $AlCl_3$ or BF_3, the acylation reaction requires more than a stoichiometric amount of catalyst due to the consumption of the Lewis acid by coordination to produced aromatic ketones. In addition, rather drastic reaction conditions, tedious work-up procedures, etc., remain as severe problems to overcome. Although some catalysts such as activated iron sulfates [70], iron oxides [71], heteropoly acid [72], trifluoromethanesulfonic acid [73], diphenylboryl hexachloroantimonate [74], or more recently, hafnium triflate (Hf(OTf)$_4$) [75] for the Friedel–Crafts acylation reaction have been reported, development of more efficient and economical catalysts is still required.

Friedel–Crafts acylation of substituted benzenes proceeded in the presence of a catalytic amount of a lanthanide triflates [76] in acetic anhydride, acetonitrile, or nitromethane as solvent. Using nitromethane gave the highest yield of the desired product. On the other hand, in carbon disulfide, dichloromethane, or nitrobenzene, the reaction mixture was heterogeneous and the yields of the acylation product were very poor. A quantitative acylation product was obtained when 20 mol% Yb(OTf)$_3$ was used. In the presence of 5 mol% of the catalyst the desired acylation product was obtained in 79% yield.

Several substituted benzenes were subjected to Yb(OTf)$_3$-catalyzed acetylation. Although the acetylation of benzene did not occur, activated benzenes gave the products in moderate to high yields.

One of the features of the present Friedel–Crafts reaction is the recovery and re-use of the catalyst. The yields of 4-methoxyacetophenone in the second and third uses of the catalyst were almost the same as that in the first use. Furthermore, almost 90% of the Yb(OTf)$_3$ catalyst was easily recovered from the aqueous layer by simple extraction.

Sc(OTf)$_3$ was found to be more efficient than Yb(OTf)$_3$ and Y(OTf)$_3$, e.g. in the case of the reaction of mesitylene, Sc(OTf)$_3$ afforded a much higher yield in a shorter reaction time (73%, 1 h) than Yb(OTf)$_3$ did (16%, 18 h) [77].

$$\text{MeO-C}_6\text{H}_5 \xrightarrow[\text{(MeCO)}_2\text{O}]{\text{Sc(OTf)}_3 \text{ (1 mol\%)}} \text{MeO-C}_6\text{H}_4\text{-COMe} \quad 62\% \tag{19}$$

Several aromatic compounds were subjected to the Sc(OTf)$_3$-catalyzed acylation reactions. For example, the acetylation of thioanisole, o- or m-dimethoxybenzene gave a single acetylated product in an excellent yield. Benzoylation of anisole also proceeded smoothly in the presence of a catalytic amount of Sc(OTf)$_3$. Although both benzoic anhydride and benzoyl chloride were effective in the reactions, benzoic anhydride gave a slightly higher yield of 4-methoxybenzophenone. In each reaction, formation of the other isomer was not detected by GLC. In addition, it was found that addition of LiClO$_4$ as a co-catalyst improved the yields and that acylation of toluene was acheived [78].

The catalytic activities of the recovered catalyst were also examined in the acetylation of anisole. The yields of 4-methoxyacetophenone in the second and third uses of the catalyst were almost the same as that in the first use.

In conclusion, the characteristic features of lanthanide triflate-catalyzed Friedel–Crafts acylation are as follows: (1) A catalytic amount of the triflate is enough to complete the reaction (TON = 62). (2) Sc(OTf)$_3$ is the most active among lanthanide triflates in the catalytic Friedel–Crafts acylation. (3) Lanthanide triflates can be recovered and reused.

Another example is the acylation of alcohols, which is among the most fundamental and important organic reactions [79]. Sc(OTf)$_3$ was found to be a useful Lewis acid catalyst for the acylation of alcohols with acid anhydrides. The catalytic activity of Sc(OTf)$_3$ has been attained in the acylation of primary alcohols as well as sterically-hindered secondary and tertiary alcohols.

It was also found that lanthanide triflates were effective in the direct acylation reactions [80]. The direct acylation of alcohols with carboxylic acids, the Fischer esterification, can be brought about with mineral acids or sulfonic acids conventionally. Although the only byproduct is water in this process, the reaction is reversible and typically large excesses of either alcohols or carboxylic acids are required. In addition, the use of strong mineral acids leads to highly acidic waste streams posing an environmental problem for industrial processes. Lanthanide triflates have been found to be efficient catalysts for the direct acetylation of primary, secondary, and tertiary alcohols. The acetylation was successfully carried out in acetic acid in the presence of a catalytic amount of a lanthanide triflate. The catalyst could be recovered quantitatively and reused with no loss of activity.

Lanthanide triflates also catalyze the nitration of a range of simple aromatic compounds in good to excellent yield using stoichiometric quantities of 69% nitric acid [81]. The only byproduct is water and the catalysts could be readily recycled by simple extraction.

2.16.6
References

1 Review: G. A. MOLANDER, *Chem. Rev.* **1992**, *92*, 29.
2 C. F. BAES Jr., R. E. MESMER, *The Hydrolysis of Cations*, John Wiley, New York, **1976**, p. 129.
3 K. F. THORN, US PATENT 3615169, **1971**; *Chem. Abstr.* **1972**, *76*, 5436a.
4 (a) J. H. FORSBERG, V. T. SPAZIANO, T. M. BALASUBRAMANIAN, G. K. LIU, S. A. KINSLEY, C. A. DUCKWORTH, J. J. POTERUCA, P. S. BROWN, J. L. MILLER, *J. Org. Chem.* **1987**, *52*, 1017. See also: (b) S. COLLINS, Y. HONG, *Tetrahedron Lett.* **1987**, *28*, 4391. (c) M.-C. ALMASIO, F. ARNAUD-NEU, M.-J. SCHWING-WEILL, *Helv. Chim. Acta* **1983**, *66*, 1296. Cf. (d) J. M. HARROWFIELD, D. L. KEPERT, J. M. PATRICK, A. H. WHITE, *Aust. J. Chem.* **1983**, *36*, 483.
5 S. KOBAYASHI, *Chem. Lett.* **1991**, 2187.
6 S. KOBAYASHI, *Synlett* **1994**, 689.
7 (a) D. SCHINZER (Ed.), *Selectivities in Lewis Acid Promoted Reactions*, Kluwer, Dordrecht, **1989**. (b) M. SANTELLI, J.-M. PONS, *Lewis Acids and Selectivity in Organic Synthesis*, CRC, Boca Raton, **1995**.
8 REVIEW: (a) C.-J. LI, *Chem. Rev.* **1993**, *93*, 2023. (b) A. LUBINEAU, J. ANGE, Y. QUENEAU, *Synthesis* **1994**, 741.
9 S. KOBAYASHI, I. HACHIYA, *J. Org. Chem.* **1994**, *59*, 3590.
10 (a) B. B. SNIDER, D. J. RODINI, T. C. KIRK, R. CORDOVA, *J. Am. Chem. Soc.* **1982**, *104*, 555. (b) B. B. SNIDER, In: *Selectivities in Lewis Acid Promoted Reactions* (Ed.: D. Schinzer), Kluwer, London, **1989**, pp. 147–167. (c) K. MARUOKA, A. B. CONCEPCION, N. HIRAYAMA, H. YAMAMOTO, *J. Am. Chem. Soc.* **1990**, *112*, 7422. (d) K. MARUOKA, A. B. CONCEPCION, N. MUR-

ase, M. Oishi, H. Yamamoto, *J. Am. Chem. Soc.* **1993**, *115*, 3943.

11 TMSOTf-mediated aldol-type reaction of silyl enol ethers with dialkoxymethanes was also reported. S. Murata, M. Suzuki, R. Noyori, *Tetrahedron Lett.* **1980**, *21*, 2527.

12 Lubineau reported the water-promoted aldol reaction of silyl enol ethers with aldehydes, however the yields and the substrate scope are not yet satisfactory, (a) A. Lubineau, *J. Org. Chem.* **1986**, *51*, 2142. (b) A. Lubineau, E. Meyer, *Tetrahedron* **1988**, *44*, 6065.

13 (a) T. Mukaiyama, K. Narasaka, T. Banno, *Chem. Lett.* **1973**, 1011. (b) T. Mukaiyama, K. Banno, K. Narasaka, *J. Am. Chem. Soc.* **1974**, *96*, 7503.

14 S. Kobayashi, I. Hachiya, *Tetrahedron Lett.* **1992**, 1625.

15 S. Kobayashi, I. Hachiya, Y. Yamanoi, *Bull. Chem. Soc. Jpn.* **1994**, *67*, 2342.

16 Review: Y. Yamamoto, N. Asao, *Chem. Rev.* **1993**, *93*, 2207.

17 I. Hachiya, S. Kobayashi, *J. Org. Chem.* **1993**, *58*, 6958.

18 (a) W. Schmid, G. M. Whitesides, *J. Am. Chem. Soc.* **1991**, *113*, 6674. (b) E. Kim, D. M. Gordon, W. Schmid, G. M. Whitesides, *J. Org. Chem.* **1993**, *58*, 5500.

19 (a) D. Yates, P. E. Eaton, *J. Am. Chem. Soc.* **1960**, *82, 4436*. (b) T. K. Hollis, N. P. Robinson, B. Bosnich, *J. Am. Chem. Soc.* **1992**, *114*, 5464. Review: (c) W. Carruthers, *Cycloaddition Reactions in Organic Synthesis,* Pergamon Press, Oxford, **1990**.

20 S. Kobayashi, I. Hachiya, M. Araki, H. Ishitani, *Tetrahedron Lett.* **1993**, *34*, 3755.

21 S. Kobayashi, T. Wakabayashi, S. Nagayama, H. Oyamada, *Tetrahedron Lett.* **1997**, *38*, 4559.

22 (a) J. H. Fendler, E. J. Fendler, *Catalysis in Micellar and Macromolecular Systems,* Academic Press, London, **1975**. (b) P. M. Holland, D. N. Rubingh (Eds.), *Mixed Surfactant Systems,* ACS, Washington, DC, **1992**. (c) C. J. Cramer, D. G. Truhlar (Eds.), *Structure and Reactivity in Aqueous Solution,* ACS, Washington, DC, **1994**. (d) D. A. Sabatini, R. C. Knox, J. H. Harwell (Eds.), *Surfactant-enhanced Subsurface Remediation,* ACS, Washington, DC, **1994**.

23 S. Kobayashi, T. Wakabayashi, H. Oyamada, *Chem. Lett.* **1997**, 831.

24 J. Haggin, *Chem. Eng. News* **1994**, April 18, 22.

25 E. F. Kleinman, *Comprehensive Organic Synthesis* (Ed.: B. M. Trost), Pergamon Press, Oxford, **1991**, Vol. 2, p. 893.

26 I. Ojima, S. Inaba, K. Yoshida, *Tetrahedron Lett.* **1977**, 3643.

27 (a) K. Ikeda, K. Achiwa, M. Sekiya, *Tetrahedron Lett.* **1983**, *24*, 4707. (b) T. Mukaiyama, K. Kashiwagi, S. Matsui, *Chem. Lett.* **1989**, *1397*. (c) T. Mukaiyama, H. Akamatsu, J. S. Han, *Chem. Lett.* **1990**, 889. (d) M. Onaka, R. Ohno, N. Yanagiya, Y. Izumi, *Synlett* **1993**, 141. For a stoichiometric use: (e) J.-E. Dubois, G. Axiotis, *Tetrahedron Lett.* **1984**, *25*, 2143. (f) E. W. Colvin, D. G. McGarry, *J. Chem. Soc., Chem. Commun.* **1985**, 539. (g) S. Shimada, K. Saigo, M. Abe, A. Sudo, M. Hasegawa, *Chem. Lett.* **1992**, 1445. (h) S. Kobayashi, S. Iwamoto, S. Nagayama, *Synlett* **1997**, 1099. For an enantioselective version: (i) K. Hattori, H. Yamamoto, *Tetrahedron* **1994**, *50*, 2785. (j) K. Ishihara, M. Miyata, K. Hattori, T. Tada, H. Yamamoto, *J. Am. Chem. Soc.* **1994**, *116*, 10520. (k) H. Ishitani, M. Ueno, S. Kobayashi, *J. Am. Chem. Soc.* **1997**, *119*, 7153.

28 S. Kobayashi, M. Araki, M. Yasuda, *Tetrahedron Lett.* **1995**, *36*, 5773.

29 V. Lucchini, M. Prato, G. Scorrano, P. Tecilla, *J. Org. Chem.* **1988**, *53*, 2251.

30 G. Guanti, E. Narisano, L. Banfi, *Tetrahedron Lett.* **1987**, *28*, 4331.

31 S. Kobayashi, H. Ishitani, unpublished.

32 S. Kobayashi, H. Ishitani, *J. Chem. Soc., Chem. Commun.* **1995**, 1379.

33 Grieco *et al.* reported *in situ* generation and trapping of imminium salts under Mannich-like conditions, (a) S. D. Larsen, P. A. Grieco, *J. Am. Chem. Soc.* **1985**, *107*, 1768. (b) P. A. Grieco, D. T. Parker, *J. Org. Chem.* **1988**, *53*, 3325 and references therein.

34 (a) S. M. Weinreb, *Comprehensive Organic Synthesis* (Ed.: B. M. Trost), Pergamon Press, Oxford, **1991**, Vol. 5, p. 401. (b) D. L. Boger, S. M. Weinreb, *Hetero Diels–Alder Methodology in Organic Syn-*

thesis, Academic Press, San Diego, **1987**, Chaps. 2 and 9.
35 S. DANISHEFSKY, T. KITAHARA, *J. Am. Chem. Soc.* **1974**, *96*, 7807.
36 S. KOBAYASHI, M. ARAKI, H. ISHITANI, S. NAGAYAMA, I. HACHIYA, *Synlett* **1995**, 233.
37 (a) S. KOBAYASHI, H. ISHITANI, S. NAGAYAMA, *Chem. Lett.* **1995**, 423. (b) S. KOBAYASHI, H. ISHITANI, S. NAGAYAMA, *Synthesis* **1995**, 1195.
38 (a) D. L. BOGER, *Tetrahedron* **1983**, *39*, 2869. (b) P. A. GRIECO, A. BAHSAS, *Tetrahedron Lett.* **1988**, *29*, 5855 and references therein.
39 Y. MAKIOKA, T. SHINDO, Y. TANIGUCHI, K. TAKAKI, Y. FUJIWARA, *Synthesis* **1995**, 801.
40 As for the reactions of vinyl ethers, see: (a) T. JOH, N. HAGIHARA, *Tetrahedron Lett.* **1967**, 4199. (b) L. S. POVAROV, RUSSIAN CHEM. REV. **1967**, *36*, 656. (c) D. F. WORTH, S. C. PERRICINE, E. F. ELSAGER, *J. HETEROCYCLIC CHEM.* **1970**, *7*, 1353. (d) T. KAMETANI, H. TAKEDA, Y. SUZUKI, H. KASAI, T. HONDA, *HETEROCYCLES* **1986**, *24*, 3385. (e) Y. S. CHENG, E. HO, P. S. MARIANO, H. L. AMMON, *J. Org. Chem.* **1985**, *56*, 5678 and references therein. As for the reactions of vinyl sulfides, see K. NARASAKA, T. SHIBATA, *Heterocycles* **1993**, *35*, 1039.
41 J. SISKO, S. M. WEINREB, *Tetrahedron Lett.* **1989**, *30*, 3037.
42 Y. NOMURA, M. KIMURA, Y. TAKEUCHI, S. TOMODA, *Chem. Lett.* **1978**, 267.
43 S. KOBAYASHI, R. AKIYAMA, M. KAWAMURA, H. ISHITANI, UNPUBLISHED.
44 (a) J. J. TUFARIELLO, *1,3-Dipolar Cycloaddition Chemistry* (Ed.: A. PADWA), John Wiley, Chichester, **1984**, Vol. 2, p. 83. (b) K. B. G. TORSSELL, *Nitrile Oxides, Nitrones and Nitronates in Organic Synthesis*, VCH, Weinheim, **1988**.
45 S. KOBAYASHI, R. AKIYAMA, M. KAWAMURA, H. ISHITANI, *Chem. Lett.* **1997**, 1039.
46 H. ISHITANI, S. NAGAYAMA, S. KOBAYASHI, *J. Org. Chem.* **1996**, *61*, 1902.
47 Review: (a) K. NARASAKA, *Synthesis* **1991**, 1. (b) H. B. KAGAN, O. RIANT, *Chem. Rev.* 92, 1007. See also: (c) S. HASHIMOTO, N. KOMESHITA, K. KOGA, *J. Chem. Soc., Chem. Commun.* **1979**, 437. (d) K. NARASAKA, N. IWASAWA, M. INOUE, T. YAMADA, M. NAKASHIMA, J. SUGIMORI, *J. Am. Chem. Soc.* **1989**, *111*, 5340. (e) C. CHAPUIS, J. JURCZAK, *Helv. Chim. Acta* **1987**, *70*, 436. (f) K. MARUOKA, T. ITOH, T. SHIRASAKI, H. YAMAMOTO, *J. Am. Chem. Soc.* **1988**, *110*, 310. (g) E. J. COREY, R. IMWINKELRIED, S. PIKUL, Y. B. XIANG, *J. Am. Chem. Soc.* **1989**, *111*, 5493. (h) D. KAUFMANN, R. BOESE, *Angew. Chem., Int. Ed. Engl.* **1990**, *29*, 545. (i) E. J. COREY, N. IMAI, H. Y. ZHANG, *J. Am. Chem. Soc.* **1991**, *113*, 728. (j) D. SARTOR, J. SAFFRICH, G. HELMCHEN, C. J. RICHARD, H. LAMBERT, *Tetrahedron: Asymmetry* **1991**, *2*, 639. (k) K. MIKAMI, M. TERADA, Y. MOTOYAMA, T. NAKAI, *Tetrahedron: Asymmetry* **1991**, *2*, 643. (l) D. A. EVANS, S. J. MILLER, T. LECTKA, *J. Am. Chem. Soc.* **1993**, *775*, 6460. (m) K. ISHIHARA, Q. GAO, H. YAMAMOTO, *J. Am. Chem. Soc.* **1993**, *775*, 10412. (n) K. MARUOKA, N. MURASE, H. YAMAMOTO, *J. Org. Chem.* **1993**, *58*, 2938. (o) J. G. SEERDEN, H. W. SCHEEREN, *Tetrahedron Lett.* **1993**, *34*, 2669.
48 (a) M. BEDNARSKI, C. MARING, S. DANISHEFSKY, *Tetrahedron Lett.* **1983**, *24*, 3451. See also: (b) M. QUIMPERE, K. JANKOWSKI, *J. Chem. Soc., Chem. Commun.* **1987**, 676. Quite recently, Shibasaki et al. reported catalytic asymmetric nitro aldol reactions using a chiral lanthanum complex *as a base*, (c) H. SASAI, T. SUZUKI, N. ITOH, K. TANAKA, T. DATE, K. OKUMURA, M. SHIBASAKI, *J. Am. Chem. Soc.* **1993**, *115*, 10372 and references therein.
49 S. KOBAYASHI, I. HACHIYA, H. ISHITANI, M. ARAKI, *Tetrahedron Lett.* **1993**, *34*, 4535.
50 (a) S. KOBAYASHI, H. ISHITANI, *J Am. Chem. Soc.* **1994**, *116*, 4083. (b) S. KOBAYASHI, H. ISHITANI, I. HACHIYA, M. ARAKI, *Tetrahedron* **1994**, *50*, 1623.
51 S. C. STINSON, *Chem. Eng. News* **1993**, Sept. 27, 38.
52 F. A. COTTON, G. WILKINSON, *Advanced Inorganic Chemistry; Fifth Edition*, John Wiley, New York, **1988**, p. 973.
53 S. KOBAYASHI, M. ARAKI, I. HACHIYA, *J. Org. Chem.* **1994**, *59*, 3758.
54 K. NARASAKA, H. TANAKA, F. KANAI, *Bull. Chem. Soc. Jpn.* **1991**, *64*, 387.
55 REVIEW: F. A. HART IN *Comprehensive Coordination Chemistry* (Ed.: G. WILKIN-

son), Pergamon Press, New York, **1987**, Vol. 3, p. 1059.

56 S. Kobayashi, H. Ishitani, M. Araki, I. Hachiya, *Tetrahedron Lett.* **1994**, *35*, 6325.

57 J. Fritsch, G. Zundel, *J. Phys. Chem.* **1981**, *85*, 556.

58 (a) K. Maruoka, H. Yamamoto, *J. Am. Chem. Soc.* **1989**, *111*, 789. (b) J. Bao, W.D. Wulff, A.L. Rheingold, *J. Am. Chem. Soc.* **1993**, *115*, 3814.

59 K. Hattori, H. Yamamoto, *J. Org. Chem.* **1992**, *57*, 3264. See also Ref. **56f]**.

60 (a) M.T. Reetz, S.-H. Kyung, C. Bolm, T. Zierke, *Chem. Ind.* **1986**, *824*. (b) K. Mikami, M. Terada, T. Nakai, *J. Am. Chem. Soc.* **1990**, *112*, 3949.

61 J. Bao, W.D. Wulff, A.L. Rheingold, *J. Am. Chem. Soc.* **1993**, *115*, 3814.

62 I.E. Marko, G.R. Evans, *Tetrahedron Lett.* **1994**, *45*, 2771.

63 H. Ishitani, S. Kobayashi, *The 68th Annual Meeting of the Chemical Society of Japan.*

64 (a) T. Kametani, H. Kasai, *Studies in Natural Product Chem.* **1989**, *3*, 385. (b) V.I. Grigos, L.S. Povarov, B.M. Mikhailov, *Izv. Akad. Nauk SSSR, Ser. Khim.* **1965**, 2163; *Chem. Abstr.* **1966**, *64*, 9680.

65 (a) H. Waldmann, *Synthesis* **1994**, 535. (b) E. Borrione, M. Prato, G. Scorrano, M. Stiranello, *J. Chem. Soc., Perkin Trans. 1* **1989**, 2245.

66 H. Ishitani, S. Kobayashi, *Tetrahedron Lett.* **1996**, *37*, 7357.

67 Some interesting biological activities have been reported in 8-hydroxyquinoline derivatives. For example: (a) B.S. Rauckman, M.Y. Tidwell, J.V. Johnson, B. Roth, *J. Med. Chem.* **1989**, *32*, 1927. (b) J.V. Johnson, B.S. Rauckman, D.P. Baccanari, B. Roth, *J. Med. Chem.* **1989**, *32*, 1942. (c) R.J. Ife, T.H. Brown, D.J. Keeling, C.A. Leach, M.L. Meeson, M.E. Parsons, D.R. Reavill, C.J. Theobald, K.J. Wiggall, *J. Med. Chem.* **1992**, *35*, 3413. (d) R. Sarges, A. Gallagher, T.J. Chambers, L.-A. Yeh, *J. Med. Chem.* **1993**, *36*, 2828. (e) F. Mongin, J.-M. Fourquez, S. Rault, V. Levacher, A. Godard, F. Trecourt, G. Queguiner, *Tetrahedron Lett.* **1995**, *36*, 8415.

68 S. Kobayashi, M. Kawamura, *J. Am. Chem. Soc.*, submitted.

69 (a) G.A. Olah, *Friedel–Crafts Chemistry*, Wiley-Interscience, New York, **1973**. (b) H. Heaney, *Comprehensive Organic Synthesis* (Ed.: B.M. Trost), Pergamon Press, Oxford, **1991**, Vol. 2, p. 733. (c) G.A. Olah, R. Krishnamurti, G.K.S. Prakash, *Comprehensive Organic Synthesis* (Ed.: B.M. Trost), Pergamon Press, Oxford, **1991**, Vol. 3, p. 293.

70 (a) M. Hino, K. Arata, *Chem. Lett.* **1978**, 325. (b) K. Arata, M. Hino, *Bull. Chem. Soc. Jpn.* **1980**, *53*, 446.

71 K. Arata, M. Hino, *Chem. Lett.* **1980**, 1479.

72 (a) K. Nomita, Y. Sugaya, S. Sasa, M. Miwa, *Bull. Chem. Soc. Jpn.* **1980**, *53*, 2089. (b) T. Yamaguchi, A. Mitoh, K. Tanabe, *Chem. Lett.* **1982**, 1229.

73 F. Effenberger, G. Epple, *Angew. Chem., Int. Ed. Engl.* **1972**, *11*, 300.

74 T. Mukaiyama, H. Nagaoka, M. Ohshima, M. Murakami, *Chem. Lett.* **1986**, 165.

75 (a) I. Hachiya, M. Moriwaki, S. Kobayashi, *Tetrahedron Lett.* **1995**, *36*, 409. (b) I. Hachiya, M. Moriwaki, S. Kobayashi, *Bull. Chem. Soc. Jpn.* **1995**, *68*, 2053.

76 A. Kawada, S. Mitamura, S. Kobayashi, *J. Chem. Soc., Chem. Commun.* **1993**,1157.

77 A. Kawada, S. Mitamura, S. Kobayashi, *Synlett* **1994**, 545.

78 A. Kawada, S. Mitamura, S. Kobayashi, *J. Chem. Soc., Chem. Commun.* **1996**, 183.

79 (a) K: Ishihara, M. Kubota, H. Kurihara, H. Yamamoto, *J. Am. Chem. Soc.* **1995**, *777*, 4413. (b) K. Ishihara, M. Kubota, H. Kurihara, H. Yamamoto, *J. Org. Chem.* **1996**, *61*, 4560.

80 A.G.M. Barrett, D.C. Braddock, *J. Chem. Soc., Chem. Commun.* **1997**, 351.

81 F.J. Waller, A.G.M. Barrett, D.C. Braddock, D. Ramprasad, *J. Chem. Soc., Chem. Commun.* **1997**, 613.

2.17
Lanthanide Complexes in Asymmetric Two-Center Catalysis

Masakatsu Shibasaki, Hiroaki Sasai, and Naoki Yoshikawa

2.17.1
Heterobimetallic Lanthanide Complexes in Asymmetric Two-Center Catalysis

2.17.1.1
Introduction

Optically active heterobimetallic lanthanide complexes $M_3[Ln(binol)_3]$ (M = alkali metal, Ln = rare earth, H_2binol = 1,1'-bi-2-naphthol), abbreviated as LnMB (B = binaphthoxide), are readily prepared from corresponding rare earth trichlorides (hydrate) or rare earth *iso*-propoxides (Scheme 1). Because these complexes are stable in the presence of either oxygen or moisture, the inexpensive hydrate of rare earth trichlorides can be employed as a starting material [1, 2]. Scheme 1 shows the structures of the resulting complexes of LnMB, which were unequivocally determined by LDI-TOF mass spectroscopy and X-ray crystallography [3–6]. These complexes function as a Brønsted base and as a Lewis acid [7], just like an enzyme, making possible a variety of efficient catalytic enantioselective reactions [8–12]. Spectral analyses and computational simulations of the LnMB-catalyzed enantioselective reactions support the synergistic cooperation of the metals in LnMB [13].

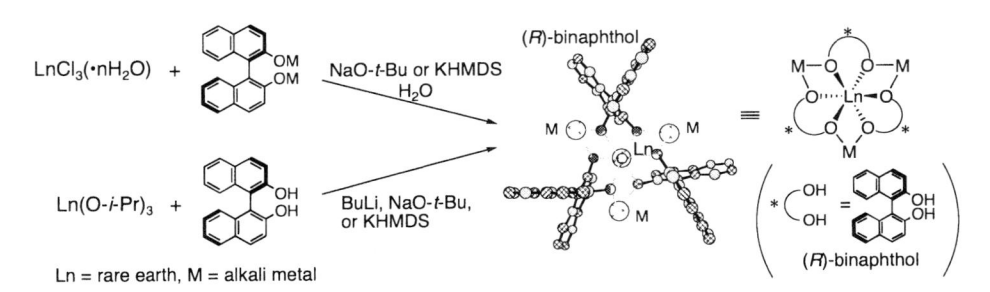

Scheme 1 Preparation and structure of rare earth-alkali metal-binaphthoxide complexes.

Transition Metals for Organic Synthesis, Vol. 1, 2nd Edition.
Edited by M. Beller and C. Bolm
Copyright © 2004 WILEY-VCH Verlag GmbH & Co. KGaA, Weinheim
ISBN: 3-527-30613-7

2.17.1.2
Catalytic Enantioselective Nitroaldol Reactions Promoted by LnLB Catalysts

The nitroaldol (Henry) reaction is a powerful synthetic tool utilized in the construction of numerous natural products and other useful compounds. In 1992, we reported the first general and effective catalytic enantioselective nitroaldol reaction, which proceeds efficiently in the presence of catalytic amounts of LnLB ($Li_3[Ln(binol)_3]$) [14]. The resulting nitroaldols are readily converted to β-amino alcohols or α-hydroxy carbonyl compounds. Effective application of the LnLB-catalyzed nitroaldol reaction results in convenient syntheses of several kinds of biologically important compounds [4, 15–18]. For example, 10 mol equivalents of nitromethane (**2**) and **8** at –50 °C in the presence of 3.3 mol% of (*R*)-LLB (**9**, $Li_3[La(R\text{-}binol)_3]$) produced a 76% yield of nitroaldol **10** in 92% *ee* (Scheme 2). After further conversions, (*S*)-(–)-pindolol (**11**) was synthesized from 4-hydroxyindole in only four steps [18].

We also developed the diastereo- and enantioselective nitroaldol reactions using nitroethane, nitropropane, or nitroethanol as substrates (Tab. 1) [19]. Although

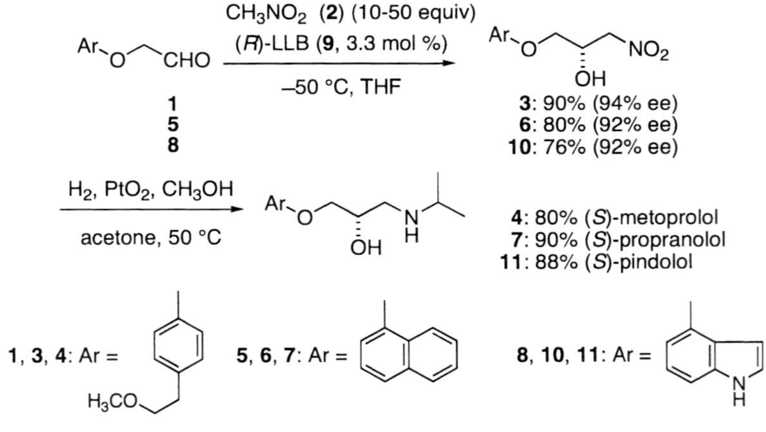

Scheme 2 Catalytic asymmetric syntheses of β-blockers using (*R*)-LLB as a catalyst.

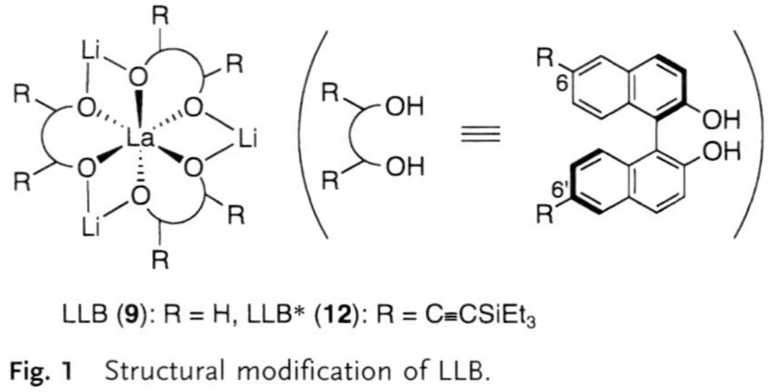

LLB (**9**): R = H, LLB* (**12**): R = C≡CSiEt₃

Fig. 1 Structural modification of LLB.

Tab. 1 Diastereo- and enantioselective nitroaldol reactions

$$RCHO + R'CH_2NO_2 \xrightarrow[\text{THF}]{\text{catalyst (3.3 mol \%)}} R\overset{OH}{\underset{syn}{\wedge}}\overset{R'}{\underset{NO_2}{}} + R\overset{OH}{\underset{anti}{\wedge}}\overset{R'}{\underset{NO_2}{}}$$

13: R=PhCH$_2$CH$_2$
23: R=CH$_3$(CH$_2$)$_4$

14: R'=CH$_3$
17: R'=Et
20: R'=CH$_2$OH

15 (syn), 16 (anti): R=PhCH$_2$CH$_2$, R'=CH$_3$
18 (syn), 19 (anti): R=PhCH$_2$CH$_2$, R'=Et
21 (syn), 22 (anti): R=PhCH$_2$CH$_2$, R'=CH$_2$OH
24 (syn), 25 (anti): R=CH$_3$(CH$_2$)$_4$, R'=CH$_2$OH

Entry	Substrates	Catalyst	Time (h)	Temp. (°C)	Products	Yield (%)	syn:anti	ee (%) (syn)
1	13 + 14	LLB (9)	75	−20	15 + 16	79	74:26	66
2	13 + 14	LLB* (12)	75	−20	15 + 16	70	89:11	93
3	13 + 14	LLB* (12)	115	−40	15 + 16	21	94:6	97
4	13 + 17	LLB (9)	138	−40	18 + 19	89	85:15	87
5	13 + 17	LLB* (12)	138	−40	18 + 19	85	93:7	95
6	13 + 20	LLB (9)	111	−40	21 + 22	62	84:16	66
7	13 + 20	LLB* (12)	111	−40	21 + 22	97	92:8	97
8	23 + 20	LLB (9)	93	−40	24 + 25	79	87:13	78
9	23 + 20	LLB* (12)	93	−40	24 + 25	96	92:8	95

LLB (9) gave unsatisfactory results in terms of stereoselectivity, substitution at the 6,6'-position of the binaphthol led to the formation of superior catalysts (e.g., 12) (Fig. 1). These modified catalysts produced much higher *syn*-selectivity and enantioselectivity in all cases (Tab. 1).

2.17.1.3
Second-Generation LLB Catalyst

Enantioselective nitroaldol reactions promoted by LLB or its derivatives required at least 3.3 mol% of asymmetric catalysts for efficient conversion. Thus, it was desirable to reduce the required amount of asymmetric catalysts and to accelerate the reactions. As depicted in Scheme 3, generation of the lithium nitronate is a slow reaction. To accelerate this step, we added a base to the LLB catalyst. After many attempts, second-generation LLB (LLB-II), prepared from LLB, H$_2$O (1 mol equiv. to LLB), and butyllithium (0.9 mol equiv. to LLB), accelerated the catalytic enantioselective nitroaldol reactions even with a reduced amount (1 mol%) (Scheme 4) [20]. The structure of LLB-II has not been unequivocally determined. We propose, however, that it is a complex of LLB and LiOH. A proposed reaction course for an improved catalytic enantioselective nitroaldol reaction is shown in Scheme 3.

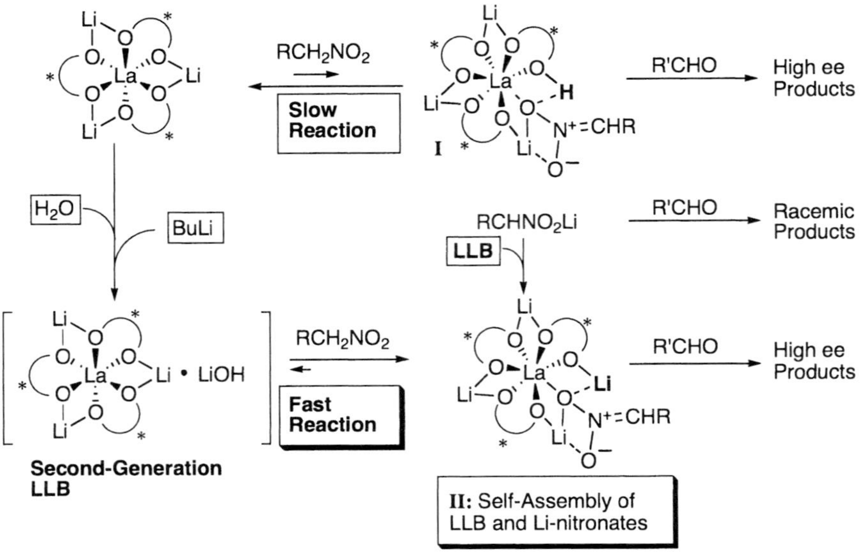

Scheme 3 Proposal mechanism of catalytic asymmetric nitroaldol reaction promoted by LLB or LLB-II.

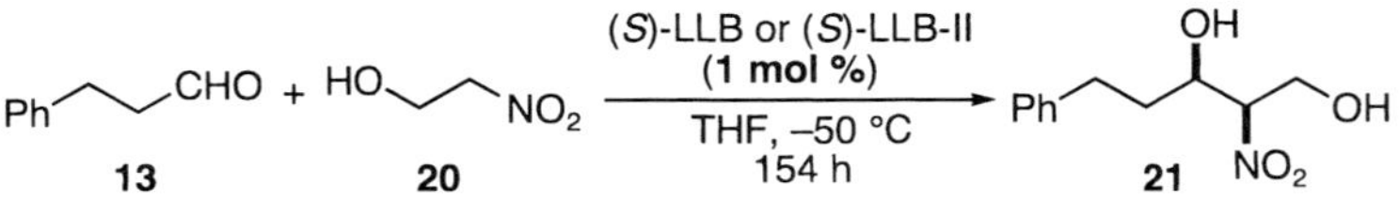

(S)-LLB: yield trace
(S)-LLB + H$_2$O + BuLi: yield 76% (syn/anti = 94/6), 96% ee (syn)
(LLB-II)

Scheme 4 Acceleration of nitroaldol reaction by second-generation LLB.

2.17.1.4
Catalytic Asymmetric Conjugate Additions by LnSB

The catalytic asymmetric Michael reaction is one of the most important synthetic methods for obtaining chiral products. 1,3-Dicarbonyl compounds in particular are highly promising Michael donors. Although LSB (Na$_3$[La(binol)$_3$]) was ineffective as an asymmetric catalyst for nitroaldol reactions, it was effective in the catalytic asymmetric Michael reaction of various enones with either malonates or β-keto esters, affording Michael adducts in up to 92% ee and almost quantitative yield [13, 21]. Some typical results are summarized in Tab. 2.

Heterobimetallic complexes containing sodium (LnSB) (Na$_3$[Ln(binol)$_3$]) also promoted the conjugate additions of thiols to α,β-unsaturated carbonyl compounds [22, 23]. In particular, LSB (Na$_3$[La(binol)$_3$]) and SmSB (Na$_3$[Sm(binol)$_3$])

Tab. 2 Catalytic asymmetric Michael reactions promoted by (R)-LMB (10 mol%)

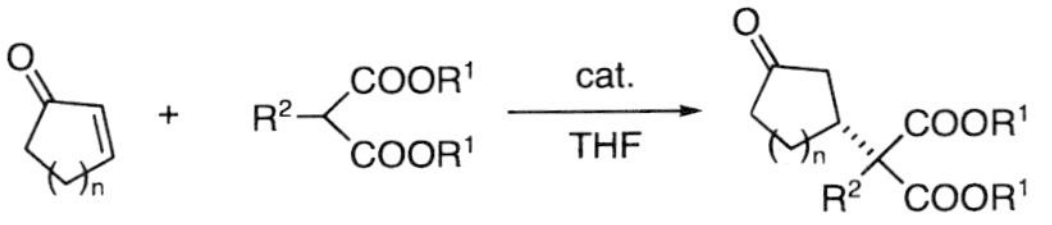

26: n=2
27: n=1

28: R^1=Bn, R^2=H
29: R^1=Bn, R^2=CH_3
30: R^1=CH_3, R^2=H
31: R^1=Et, R^2=H

32: n=2, R^1=Bn, R^2=H
33: n=2, R^1=Bn, R^2=CH_3
34: n=2, R^1=CH_3, R^2=H
35: n=2, R^1=Et, R^2=H
36: n=1, R^1=Bn, R^2=CH_3

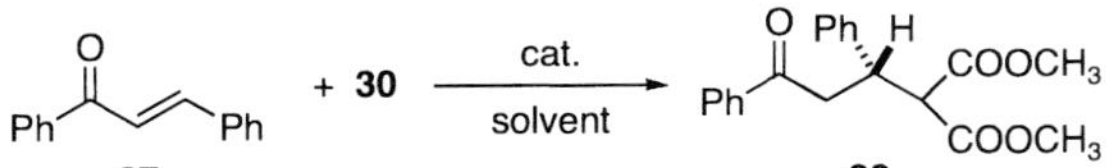

Entry	Enone	Michael donor	Catalyst	Temp. (°C)	Time (h)	Product	Yield (%)	ee (%)
1	26	28	LSB	0	24	32	97	88
2	26	28	LSB	rt	12	32	98	85
3	26	28	LLB	rt	12	32	78	2
4	26	28	LPB [b]	rt	12	32	99	48
5	26	29	LSB	0	24	33	91	92
6	26	29	LSB	rt	12	33	96	90
7	26	30	LSB	rt	12	34	98	83
8	26	31	LSB	rt	12	35	97	81
9	27	29	LSB	−40	36	36	89	72
10	37	30	LSB	−50	36	38	62	0
11 [a]	37	30	LSB	−50	24	38	93	77

a) Toluene was used as solvent.
b) $K_3[La(binol)_3]$. See Section 2.17.1.5.

had excellent performance, giving the products with up to 93% *ee* (Tabs. 3 and 4). The chirality of products 55–59 (Tab. 4) should be generated at the protonation step. Catalytic asymmetric protonation in combination with the conjugate addition of thiols was achieved.

2.17.1.5
Catalytic Enantioselective Hydrophosphonylations

α-Aminophosphonic acids are interesting compounds that can be used in the design of enzyme inhibitors. The absolute configuration of the α-carbon strongly influences the biological properties. We succeeded in developing the first catalytic enantioselective hydrophosphonylation of imines using LnPB ($K_3[Ln(binol)_3]$)

Tab. 3 Catalytic asymmetric conjugate addition of thiols to enones

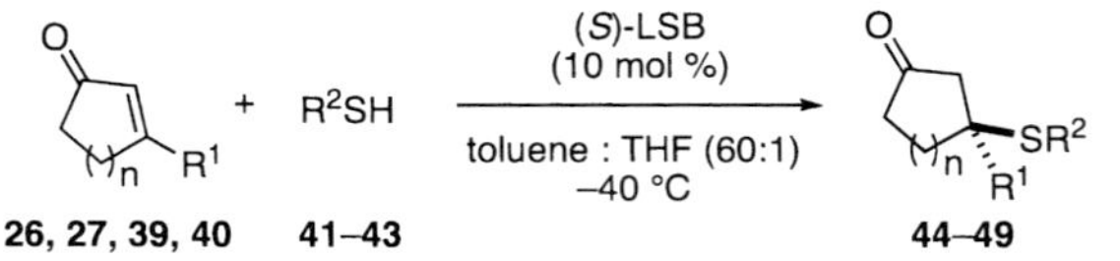

Enone	R^2		Product	Time	Yield (%)	ee (%)
n=2, R^1=H (26)	4-t-BuPh (41)		44	20 min	93	84
26	Ph (42)		45	20 min	87	68
26	PhCH$_2$ (43)		46	14 h	86	90
n=1, R^1=H (27)	43		47	4 h	94	56
n=3, R^1=H (39)	43		48 [a]	41 h	87	83
n=2, R^1=Me (40)	43		49 [a,b]	43 h	56	85

a) 20 mol% of catalyst was used, and toluene was used as solvent.
b) Reaction at –20 °C.

Tab. 4 Catalytic asymmetric protonations in conjugate addition of thiols

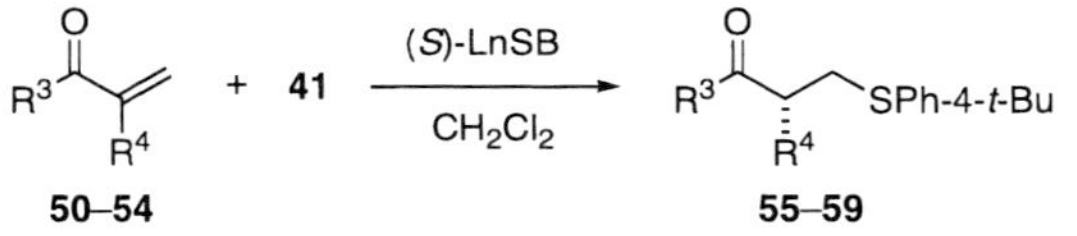

Enone			Product	Ln	Catalyst (mol%)	Temp. (°C)	Time (h)	Yield (%)	ee (%)
R^3	R^4								
EtO	Me	50	55	La	20	–20	48	44	75
EtO	Me	50	55	La	20	–20	48	50	82
EtS	Me	51	56	La	20	–78	2	93	90
EtS	Me	51	56	La	10	–78	8	90	88
EtS	Me	51	56	Sm	10	–78	7	86	93
EtS	Me	51	56	Sm	2	–78	6	89	88
EtS	i-Pr	52	57	Sm	10	–78	7	78	90
EtS	PhCH$_2$	53	58	Sm	10	–78	7	89	87
EtS	Ph	54	59	Sm	10	–93	1	98	84

a) Toluene was used as solvent.

complexes, which gives optically active α-aminophosphonates in modest to high enantiomeric excess (Scheme 5) [24–28].

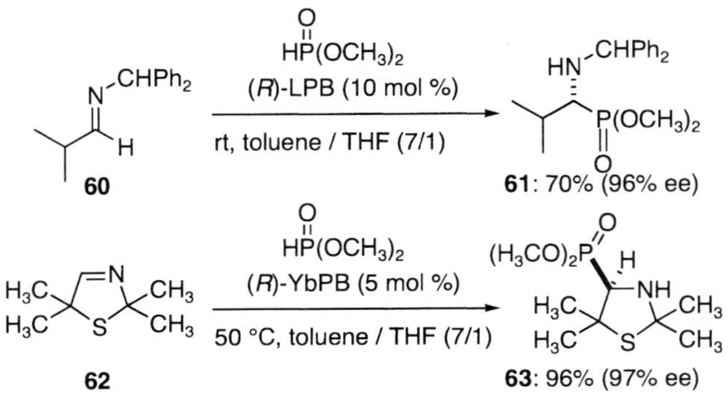

Scheme 5 Catalytic asymmetric hydrophosphonylation of imines promoted by LnPB.

2.17.1.6
Enantioselective Direct Aldol Reactions

The aldol reaction is well established in organic chemistry as a remarkably useful synthetic tool. Although highly enantioselective processes have been achieved using only catalytic amounts of chiral promoters, most of the methodologies require the preformation of latent enolates **65**, such as ketene silyl acetals, using stoichiometric amounts of silylating agents (Scheme 6, top). Because of the growing demand for an atom-economic process, the development of a *direct* catalytic asymmetric aldol reaction (Scheme 6, bottom), which should eliminate the need to preform latent enolates, is an exciting and challenging subject [29, 30].

In 1997, we discovered the promotion of enantioselective aldol reactions of unmodified ketones **64** with a broad range of applicable substrates [31]. The reactions were catalyzed by the LLB complex to afford the desired aldol products **69** with up to 94% *ee*. In 1999, a large acceleration of this reaction was achieved using a heteropolymetallic catalyst that was prepared from LLB, KOH, and H$_2$O,

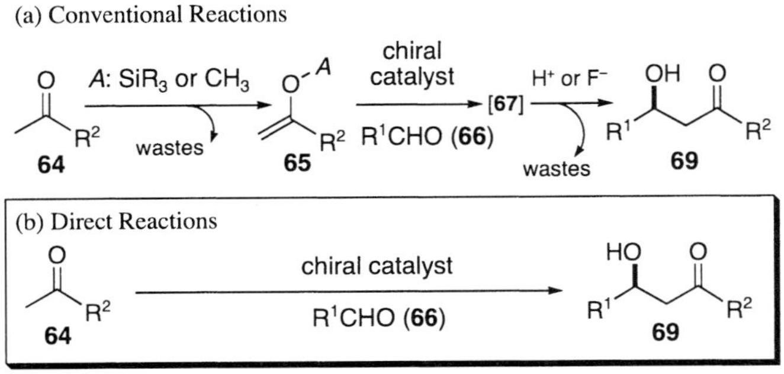

Scheme 6 Aldol-type addition of latent enolates and direct aldol reaction of unmodified ketones.

Tab. 5 Direct catalytic asymmetric aldol reactions promoted by a heteropolymetallic asymmetric catalyst (LLB-KOH-H$_2$O)

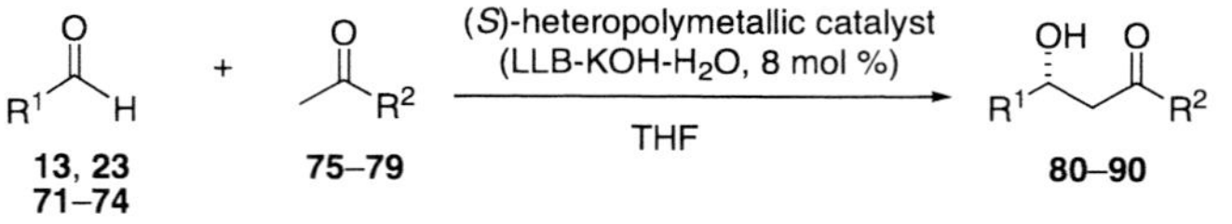

$$R^1CHO + R^2COCH_3 \xrightarrow[\text{THF}]{\text{(S)-heteropolymetallic catalyst (LLB-KOH-H}_2\text{O, 8 mol \%)}} R^1CH(OH)CH_2COR^2$$

13, 23, 71–74   75–79   80–90

70: R^1 = t-Bu
71: R^1 = PhCH$_2$C(CH$_3$)$_2$
72: R^1 = BnOCH$_2$C(CH$_3$)$_2$
73: R^1 = i-Pr

74: R^1 = Et$_2$CH
13: R^1 = PhCH$_2$CH$_2$
23: R^1 = n-C$_5$H$_{11}$

75: R^2 = Ph
76: R^2 = CH$_3$
77: R^2 = Et
78: R^2 = 3-NO$_2$-C$_6$H$_4$
79: R^2 = cyclopentanone

Entry	Aldehyde (R^1)	Ketone[a] (R^2) (eq)	Catalyst (mol%)	Temp. (°C)	Aldol	Time (h)	Yield (%)	ee (%)
1	70	75 (5)	8	–20	80	15	75	88
2	71	75 (5)	8	–20	81	28	85	89
3	71	76 (10)	8	–20	82	20	62	76
4[b]	71	77 (15)	8	–20	83	95	72	88
5	72	75 (5)	8	–20	84	36	91	90
6[c]	72	75 (5)	8	–20	84	24	70	93
7	73	75 (5)	8	–30	85	15	90	33
8	73	78 (3)	8	–50	86	70	68	70
9	74	78 (3)	15	–45	87	96	60	80
10	13	78 (3)	8	–40	88	31	50	30
11	23	78 (5)	30	–50	89	96	55	42
12	71	79 (5)	8	–20	90	99	95	76/88 (syn:anti=93:7) (syn/anti)

a) Excess ketone was recovered after the reaction.
b) H$_2$O: 8 mol%.
c) 5.7 mmol (**72**) scale.

allowing for a reduction in the amount of catalyst from 20 mol% to 3–8 mol% with a significantly shorter reaction time (Tab. 5) [32].

The heteropolymetallic catalyst (LLB + KOH + H$_2$O) was successfully applied to the direct aldol reaction of 2-hydroxyacetophenones (e.g., **94** and **95**) without the need to protect the hydroxyl group, providing a valuable method for enantioselective synthesis of *anti*-1,2-diols (Tab. 6) [33, 34].

A catalyst that was prepared from LLB, LiOH, and H$_2$O promoted the direct aldol reaction of glycinate Schiff bases **104** with aldehydes, providing access to β-hydroxy-α-amino acid esters **105** (Eq. 1) [35].

Tab. 6 Diastereo- and enantioselective direct catalytic aldol reaction of 2-hydroxyacetophenones with aldehydes: catalytic asymmetric synthesis of *anti*-1,2-diols

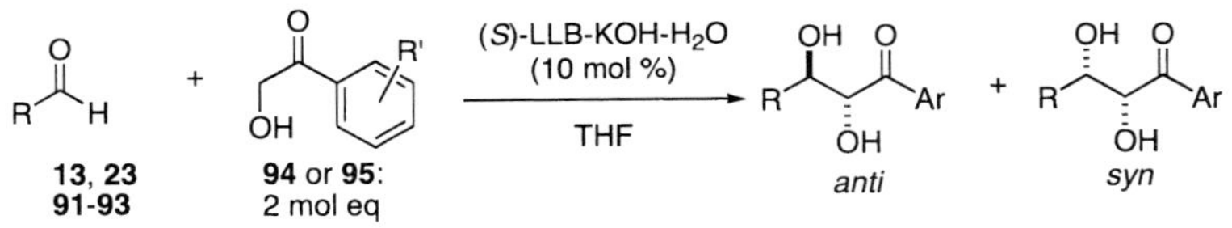

13, 23
91-93

94 or 95:
2 mol eq

91: R = $C_6H_5(CH_2)_3$
23: R = n-C_5H_{11}
92: R = *trans*-3-nonenyl
93: R = 2-methylpropyl
13: R = $C_6H_5(CH_2)_2$

94: R' = H
95: R' = 4-Me

96: R = $C_6H_5(CH_2)_3$, R' = H
97: R = n-C_5H_{11}, R' = H
98: R = *trans*-3-nonenyl, R' = H
99: R = 2-methylpropyl, R' = H
100: R = $C_6H_5(CH_2)_2$, R' = H
101: R = $C_6H_5(CH_2)_3$, R' = 4-Me
102: R = n-C_5H_{11}, R' = 4-Me

Entry	Aldehyde	Ketone	Products	Temp. (°C)	Time (h)	Yield (%)	dr (anti:syn)	ee (%) (anti/syn)
1	91	94	96	−50	24	84	84:16	95/74
2	91	94	96	−50	40	78	78:22	92/70
3	23	94	97	−50	24	84	74:26	94/84
4	92	94	98	−50	28	90	72:28	94/83
5	93	94	99	−50	24	86	65:35	90/83
6	13	94	100	−50	24	89	69:31	95/87
7	91	95	101	−40	35	90	83:17	97/85
8	23	95	102	−40	13	96	82:18	96/83

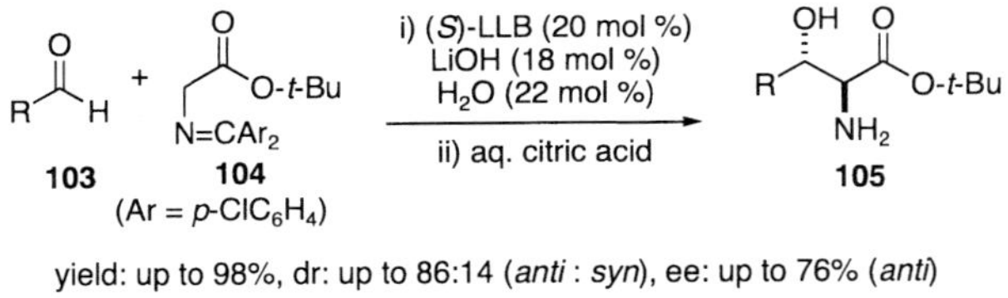

103
104 (Ar = p-ClC_6H_4)

i) (S)-LLB (20 mol %)
LiOH (18 mol %)
H_2O (22 mol %)
ii) aq. citric acid

105

(1)

yield: up to 98%, dr: up to 86:14 (*anti* : *syn*), ee: up to 76% (*anti*)

2.17.2
Alkali Metal-Free Lanthanide Complexes in Asymmetric Two-Center Catalysis

2.17.2.1
Catalytic Enantioselective Epoxidations

As a different type of catalyst, alkali metal-free lanthanide complexes were developed for conjugate additions of nucleophiles to a,β-unsaturated carbonyl compounds. Catalysts that were formed upon mixing Ln(O-i-Pr)$_3$ and BINOL (Eq. 2) possessed catalytic activity toward a Michael reaction [36].

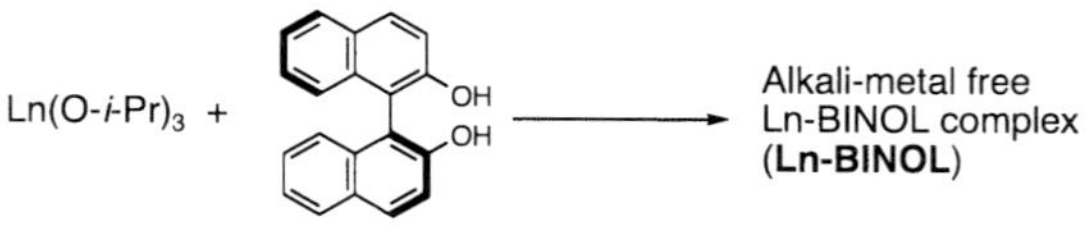

$$\text{Ln(O-}i\text{-Pr)}_3 + \text{[BINOL]} \longrightarrow \begin{array}{c}\text{Alkali-metal free}\\\text{Ln-BINOL complex}\\\textbf{(Ln-BINOL)}\end{array} \qquad (2)$$

This alkali metal-free complex (Ln-BINOL) was further examined in asymmetric epoxidations. α,β-Epoxy ketones were produced in a highly enantioselective manner by the reaction of *trans*-enones with an oxidant in the presence of the alkali metal-free lanthanide complex (Tab. 7, left) [37]. Epoxidation of *cis*-enones, which might suffer from isomerization of the enones, was also achieved using the same catalyst (Eq. 3) [38]. The epoxidation of *trans*-enones was intensively investigated to accelerate the reaction. In 1998, Inanaga et al. reported that the catalytic activity was enhanced by the addition of Ph$_3$P=O [39]. We further improved this procedure by employing Ph$_3$As=O as an additive (Tab. 7, right) [40].

Tab. 7 Catalytic enantioselective epoxidations of enones promoted by alkali metal-free lanthanide complexes

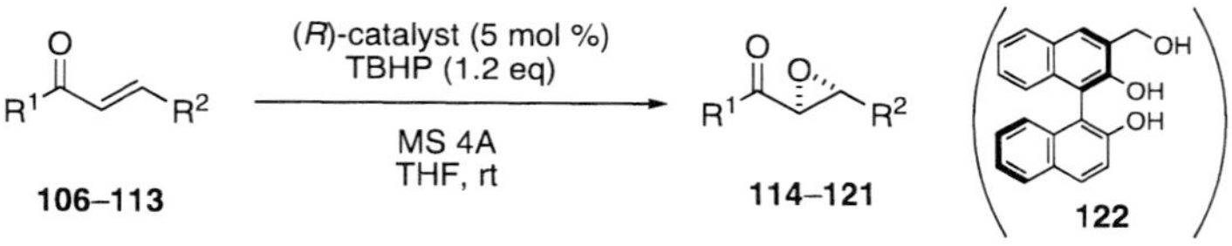

106, 114: R^1 = Ph, R^2 = Ph
107, 115: R^1 = o-MOMO-C$_6$H$_4$, R^2 = Ph
108, 116: R^1 = Ph, R^2 = i-Pr
109, 117: R^1 = t-Bu, R^2 = Ph
110, 118: R^1 = i-Pr, R^2 = Ph
111, 119: R^1 = CH$_3$, R^2 = Ph
112, 120: R^1 = CH$_3$, R^2 = CH$_2$CH$_2$Ph
113, 121: R^1 = CH$_3$, R^2 = C$_5$H$_{11}$

Entry	Enone	Epoxide	Ln-BINOL			La-BINOL-Ph$_3$As=O		
			Time (h)	Yield (%)	ee (%)	Time (h)	Yield (%)	ee (%)
1	106	114	7	93	91 [a), b)]	0.25	99	96
2	106	114	44	95	89 [a), b)]	3	97	89
3	107	115	20	85	85 [a)]	4	91	95
4	108	116	7	95	94 [a), b)]	1.5	95	94
5	109	117				7	94	98
5	110	118	159	55	88 [b), c), d)]	8	72	95
6	111	119	96	83	94 [b), c)]	6	92	>99
7	112	120	118	91	88 [b), c), d)]	1.5	98	92
8	113	121	67	71	91 [b), c), d)]	1.5	89	95

a) Ln = La.
b) 3-(Hydroxymethyl)-BINOL (**122**, 1.25 mol eq to Ln) was used as a ligand.
c) Ln = Yb.
d) Catalyst: 8 mol%.

2.17.2 Alkali Metal-Free Lanthanide Complexes in Asymmetric Two-Center Catalysis

$$\text{C}_5\text{H}_{11}\overset{\displaystyle\text{O}}{\diagdown\!\!\!\diagup}\text{C}_3\text{H}_7 \quad \xrightarrow[\text{MS 4A, THF, rt, 127 h}]{\substack{(R)\text{-Yb(O-}i\text{-Pr)}_3 \text{ (10 mol \%)} \\ \mathbf{122} \text{ (14 mol \%)} \\ \text{TBHP (3 mol eq)}}}$$

(3)

$$\text{C}_5\text{H}_{11}\overset{\text{O}}{\triangle}\overset{\text{O}}{\diagup}\text{C}_3\text{H}_7 \quad + \quad \text{C}_5\text{H}_{11}\overset{\text{O}}{\triangle}\overset{\displaystyle\text{O}}{\diagdown}\text{C}_3\text{H}_7$$

80%, 96% ee   trace

The La-BINOL-Ph$_3$As=O catalyst had excellent performance compared to other catalysts, so that the epoxidation of chalcone was completed in 3 min with 10 mol% of catalyst loading. The scope of this process is also very broad: a series of enones could be used as substrates to afford the corresponding epoxides with excellent chemical yield and excellent enantiomeric excess (Tab. 7, right). Moreover, α,β-unsaturated carboxylic acid imidazolides underwent a very efficient epoxidation in the presence of the La-BINOL-Ph$_3$As=O catalyst (Tab. 8) [41]. This method should

Tab. 8 Catalytic asymmetric epoxidation of cinnamic acid imidazolides promoted by La-BINOL-Ph$_3$As=O complex

123, 133: R = Ph
124, 134: R = 4-Cl-C$_6$H$_4$
125, 135: R = 4-Br-C$_6$H$_4$
126, 136: R = 4-MeO-C$_6$H$_4$
127, 137: R = PhCH$_2$CH$_2$
128, 138: R = (pentenyl)
129, 139: R = (propenyl)
130, 140: R = (Ph-pentenyl)
131, 141: R = (ketone chain)
132, 142: R = cyclohexyl

Entry	Imidazolide	Epoxide	Time (h)	Yield (%)	ee (%)
1	123	133	3.5	86	92
2 [a]	123	133	12	73	85
3	124	134	5	91	93
4 [b]	125	135	4	86	89
5	126	136	6	80	91
6	127	137	1	86	83
7	128	138	2	93	86
8	129	139	1.5	92	79
9	130	140	2	85	82
10	131	141	4	81	81
11	132	142	4	72	88

a) Catalyst: 5 mol%.
b) The 4-methylimidazolide was used.

be highly useful, because the products serve as versatile intermediates for a wide range of building blocks.

2.17.3
La-Linked-BINOL Complex

Catalysts that are easily handled and reusable are preferable from a practical point of view. Linked-BINOL **143** offers a highly effective framework for the development of a stable and efficient catalyst. La-linked-BINOL catalyst (**144**), prepared from La(O-i-Pr)$_3$ and linked-BINOL (**143**) (Eq. 4) [42], promotes the enantioselective Michael reaction of a broad range of substrates (Tab. 9) [43].

Tab. 9 Catalytic asymmetric Michael reactions promoted by La-linked-BINOL (**144**)

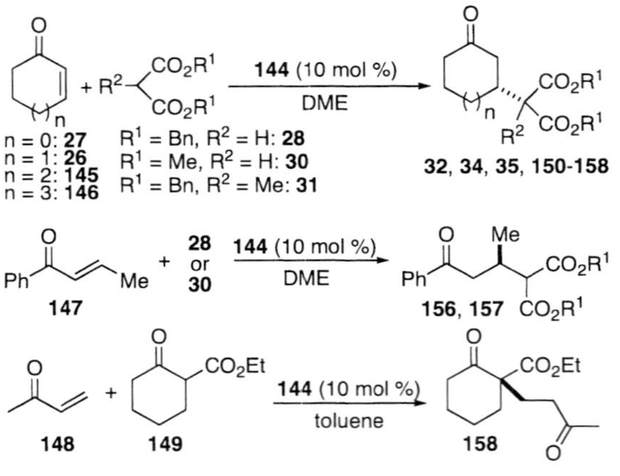

Entry	Acceptor	Donor	Temp. (°C)	Time (h)	Product	Yield (%)	ee (%)
1	27	28	4	85	150	85	>99
2	27	30	4	85	151	96	>99
3	26	28	4	85	32	98	>99
4	26	30	rt	72	34	95	>99
5[a]	26	31	rt	84	35	84	98
6	145	28	4	85	152	96	>99
7	145	30	4	85	153	97	>99
8[a]	146	30	rt	96	154	82	99
9	146	28	4	120	155	61	82
10	147	28	−40	56	156	97	78
11	147	30	−40	56	157	95	74
12	148	149	−30	36	158	97	75

a) The reaction was carried out in DME/THF (9/1).

2.17.4 Enantioselective Cyanosilylation of Aldehydes Catalyzed by Ln-Ln Homobimetallic Complexes

Tab. 10 Asymmetric Michael reaction using recycled La-linked-BINOL (**144**)

26 + 28 $\xrightarrow[\text{DME, 4 °C, 110 h}]{(R,R)\text{-La-linked-BINOL }\mathbf{144}\text{ (10 mol \%)}}$ 32

Cycle	1	2	3	4
Yield (%) [a]	82	94	68	50
ee (%)	>99	>99	99	98

a) Isolated yield.

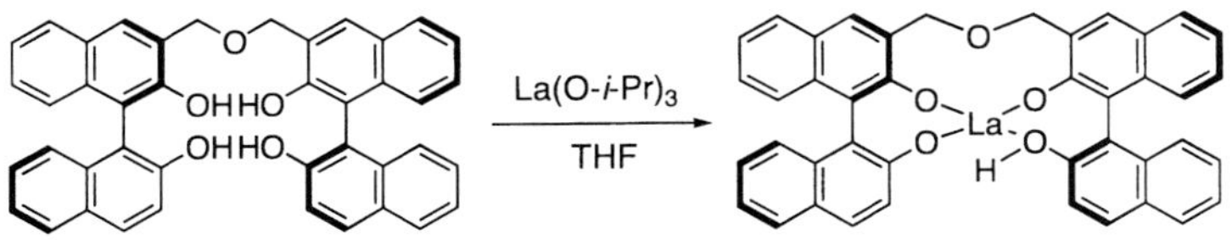

(R,R)-linked-BINOL **143** $\xrightarrow[\text{THF}]{\text{La(O-i-Pr)}_3}$ **144** (4)

The stability of this catalyst was demonstrated in the reaction of **26** and **28** (Tab. 10). The catalyst was precipitated by the addition of pentane to the reaction mixture, and powdered catalyst was obtained by filtration. The recovered catalyst was reused several times, giving the Michael adduct with almost constant enantiomeric excess [43].

2.17.4
Enantioselective Cyanosilylation of Aldehydes Catalyzed by Ln-Ln Homobimetallic Complexes

Whereas the aforementioned organic transformations are catalyzed by the synergistic function of Lewis acidity and Brønsted basicity, a cyanide addition using silylated reagents requires a different strategy for activation of the cyanide source. Transmetalation of the silyl group by an electronically more positive metal is a method used to generate a reactive cyanide species. This mode of catalysis was demonstrated by lanthanide homobimetallic complexes in the reaction of ketones with TMSCN. The catalyst (Gd-**159**) that was prepared from a sugar-based ligand (**160**) [44] and Gd(O-i-Pr)$_3$ promoted the cyanosilylation of various ketones in a highly enantioselective manner (Tab. 11) [45]. ESI-MS analysis suggested the formation of a bimetallic species **178** (Scheme 7), wherein Ln-CN bonds were generated from TMSCN. The resulting cyanide should be more nucleophilic than that in TMSCN and reacts smoothly with ketones, which are activated and fixed in near proximity by the other Ln center.

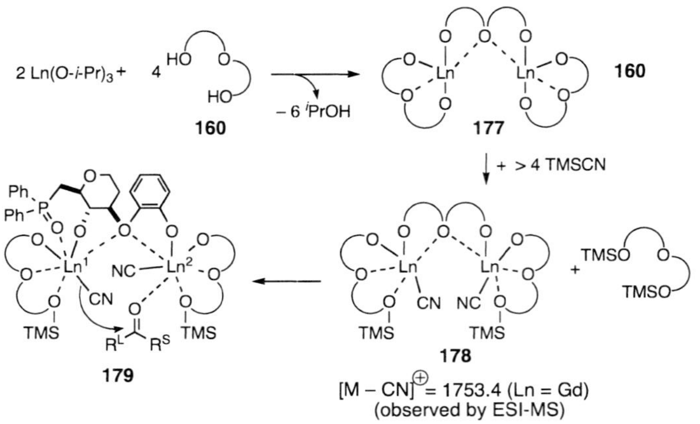

Scheme 7 Working model for the catalyst structure and the reaction mechanism for enantioselective cyanosilylation of ketones.

Tab. 11 Enantioselective cyanosilylation of ketones catalyzed by Gd-159

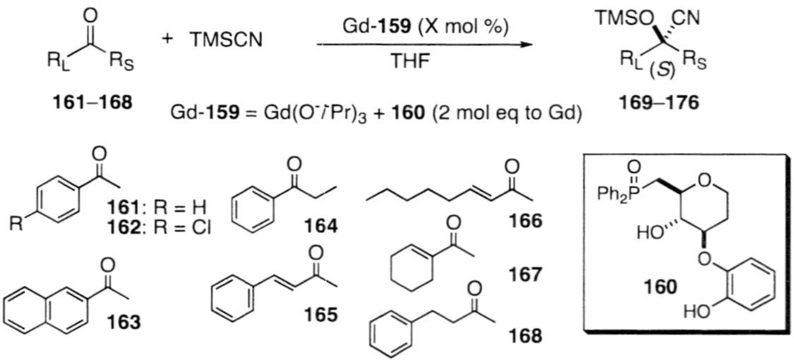

Gd-159 = Gd(O^{i}Pr)$_3$ + 160 (2 mol eq to Gd)

Entry	Ketone	Product	Gd (mol%)	Temp. (°C)	Time (h)	Yield (%)	ee (%)
1	161	169	5	−40	2	92	92
2	162	170	5	−60	55	89	89
3	163	171	5	−60	24	95	87
4	164	172	5	−60	14	93	97
5	165	173	10	−60	14	97	86
6	166	174	15	−60	18	87	80
7	167	175	15	−60	4	95	89
8	168	176	5	−60	1	90	62

2.17.5
Conclusions

The lanthanide complexes that contain multiple sites for the activation of substrates enable various efficient enantioselective reactions in a manner analogous to enzyme chemistry. Several such processes are now being investigated for potential industrial applications. We hope that the findings discussed in this section will be recognized as a significant landmark in the development of the field of catalytic asymmetric synthesis.

2.17.6
References

1. H. Sasai, T. Suzuki, N. Itoh, M. Shibasaki, *Tetrahedron Lett.* **1993**, *34*, 851–854.
2. H. Sasai, S. Watanabe, M. Shibasaki, *Enantiomer* **1997**, *2*, 267–271.
3. H. Sasai, T. Suzuki, N. Itoh, K. Tanaka, T. Date, K. Okamura, M. Shibasaki, *J. Am. Chem. Soc.* **1993**, *115*, 10372–10373.
4. E. Takaoka, N. Yoshikawa, Y. M. A. Yamada, H. Sasai, M. Shibasaki, *Heterocycles* **1997**, *46*, 157–163.
5. H. C. Aspinall, J. L. M. Dwyer, N. Greeves, A. Steiner, *Organometallics* **1999**, *18*, 1366–1368.
6. H. C. Aspinall, J. F. Bickley, J. L. M. Dwyer, N. Greeves, R. V. Kelly, A. Steiner, *Organometallics* **2000**, *19*, 5416–5423.
7. T. Morita, T. Arai, H. Sasai, M. Shibasaki, *Tetrahedron: Asymmetry* **1998**, *9*, 1445–1450.
8. M. Shibasaki, H. Sasai, T. Arai, *Angew. Chem., Int. Ed. Engl.* **1997**, *36*, 1236–1256.
9. M. Shibasaki, H. Sasai, T. Arai, T. Iida, *Pure Appl. Chem.* **1998**, *70*, 1027–1034.
10. M. Shibasaki, T. Iida, Y. M. A. Yamada, *J. Synth. Org. Chem. Jpn.* **1998**, *56*, 344–356.
11. M. Shibasaki, *Chemtracts: Org. Chem.* **1999**, *12*, 979–998.
12. M. Shibasaki, N. Yoshikawa, *Chem. Rev.* **2002**, *102*, 2187–2209.
13. H. Sasai, T. Arai, Y. Satow, K. N. Houk, M. Shibasaki, *J. Am. Chem. Soc.* **1995**, *117*, 6194–6198.
14. H. Sasai, T. Suzuki, S. Arai, T. Arai, M. Shibasaki, *J. Am. Chem. Soc.* **1992**, *114*, 4418–4420.
15. H. Sasai, N. Itoh, T. Suzuki, M. Shibasaki, *Tetrahedron Lett.* **1993**, *34*, 855–858.
16. H. Sasai, W.-S. Kim, T. Suzuki, M. Shibasaki, *Tetrahedron Lett.* **1994**, *35*, 6123–6126.
17. H. Sasai, T. Suzuki, N. Itoh, M. Shibasaki, *Appl. Organometal. Chem.* **1995**, *9*, 421–426.
18. H. Sasai, Y. M. A. Yamada, T. Suzuki, M. Shibasaki, *Tetrahedron* **1994**, *50*, 12313–12318.
19. H. Sasai, T. Tokunaga, S. Watanabe, T. Suzuki, N. Itoh, M. Shibasaki, *J. Org. Chem.* **1995**, *60*, 7388–7389.
20. T. Arai, Y. M. A. Yamada, N. Yamamoto, H. Sasai, M. Shibasaki, *Chem. Eur. J.* **1996**, *2*, 1368–1372.
21. H. Sasai, E. Emori, T. Arai, M. Shibasaki, *Tetrahedron Lett.* **1996**, *37*, 5561–5564.
22. E. Emori, T. Arai, H. Sasai, M. Shibasaki, *J. Am. Chem. Soc.* **1998**, *120*, 4043–4044.
23. E. Emori, T. Iida, M. Shibasaki, *J. Org. Chem.* **1999**, *64*, 5318–5320.
24. H. Sasai, S. Arai, Y. Tahara, M. Shibasaki, *J. Org. Chem.* **1995**, *60*, 6656–6657.
25. H. Gröger, Y. Saida, S. Arai, J. Martens, H. Sasai, M. Shibasaki, *Tetrahedron Lett.* **1996**, *37*, 9291–9292.
26. H. Sasai, M. Bougauchi, T. Arai, M. Shibasaki, *Tetrahedron Lett.* **1997**, *38*, 2717–2720.

27 H. Gröger, Y. Saida, H. Sasai, K. Yamaguchi, J. Martens, M. Shibasaki, *J. Am. Chem. Soc.* **1998**, *120*, 3089–3103.
28 K. Yamakoshi, S. J. Harwood, M. Kanai, M. Shibasaki, *Tetrahedron Lett.* **1999**, *40*, 2565–2568.
29 B. Alcaide, P. Almendros, *Eur. J. Org. Chem.* **2002**, 1595–1601.
30 M. Shibasaki, N. Yoshikawa, S. Matsunaga, in *Comprehensive Asymmetric Catalysis*, E. N. Jacobsen, A. Pfaltz, H. Yamamoto (Eds.), Springer, New York, in press.
31 Y. M. A. Yamada, N. Yoshikawa, H. Sasai, M. Shibasaki, *Angew. Chem., Int. Ed. Engl.* **1997**, *36*, 1871–1873.
32 N. Yoshikawa, Y. M. A. Yamada, J. Das, H. Sasai, M. Shibasaki, *J. Am. Chem. Soc.* **1999**, *121*, 4168–4178.
33 N. Yoshikawa, N. Kumagai, S. Matsunaga, G. Moll, T. Ohshima, T. Suzuki, M. Shibasaki, *J. Am. Chem. Soc.* **2001**, *123*, 2466–2467.
34 N. Yoshikawa, T. Suzuki, M. Shibasaki, *J. Org. Chem.* **2002**, *67*, 2556–2565.
35 N. Yoshikawa, M. Shibasaki, *Tetrahedron* **2002**, *58*, 8289–8298.
36 H. Sasai, T. Arai, M. Shibasaki, *J. Am. Chem. Soc.* **1994**, *116*, 1571–1572.
37 M. Bougauchi, S. Watanabe, T. Arai, H. Sasai, M. Shibasaki, *J. Am. Chem. Soc.* **1997**, *119*, 2329–2330.
38 S. Watanabe, T. Arai, H. Sasai, M. Bougauchi, M. Shibasaki, *J. Org. Chem.* **1998**, *63*, 8090–8091.
39 K. Daikai, M. Kamaura, J. Inanaga, *Tetrahedron Lett.* **1998**, *39*, 7321–7322.
40 T. Nemoto, T. Ohshima, K. Yamaguchi, M. Shibasaki, *J. Am. Chem. Soc.* **2001**, *123*, 2725–2732.
41 T. Nemoto, T. Ohshima, M. Shibasaki, *J. Am. Chem. Soc.* **2001**, *123*, 9474–9475.
42 S. Matsunaga, T. Ohshima, M. Shibasaki, *Adv. Synth. Catal.* **2002**, *344*, 3–15.
43 Y.-S. Kim, S. Matsunaga, J. Das, A. Sekine, T. Ohshima, M. Shibasaki, *J. Am. Chem. Soc.* **2000**, *122*, 6506–6507.
44 S. Masumoto, K. Yabu, M. Kanai, M. Shibasaki, *Tetrahedron Lett.* **2002**, *43*, 2919–2922.
45 K. Yabu, S. Masumoto, S. Yamasaki, Y. Hamashima, M. Kanai, W. Du, D. P. Curran, M. Shibasaki, *J. Am. Chem. Soc.* **2001**, *123*, 9908–9909.

2.18
Bismuth Reagents and Catalysts in Organic Synthesis

Axel Jacobi von Wangelin

2.18.1
Introduction

Although bismuth-mediated organic transformations have been known for 70 years, bismuth is still viewed as a stranger among metals in organic synthesis. This is particularly astonishing in view of the striking properties that predestine bismuth for a more general utilization. Bismuth metal and inorganic bismuth compounds are important starting materials for industrial purposes (cosmetics, alloys, ceramics, pigments, etc.) and thus are commercially available at cheap prices. The uniqueness of bismuth as a heavy metal is further characterized by its low toxicity. In spite of long-standing investigations of their structural properties [1], easy-to-handle bismuth salts have hardly found any general synthetic application in recent years. However, there has been a resurgence of interest in the properties of bismuth owing to its application in disparate fields such as pharmaceuticals [2], superconductors, catalysts, and reagents for organic synthesis.

Bismuth is almost exclusively present in oxidation states (III) and (V), with Bi(V) being strongly oxidizing because of the inert pair and the relativistic effect. The first application of bismuth for organic transformations was reported by Challenger in 1934 and involved the oxidation of alcohols with Bi(V) [3]. Apart from extensive applications as a promoter (in mixed phases with Mo, Fe, Al, or Zn) in propylene oxidations, subsequent years witnessed the extensive use of pentavalent organobismuth compounds in oxidation and acylation reactions. It was not until the mid-1980s that the potential of bismuth for carbon-carbon bond-forming reactions started to be exploited. Considerable efforts have been made toward the use of bismuth reagents and catalysts for oxidation, allylation, aldol, acylation, Diels-Alder, and arylation reactions. The most significant developments in bismuth chemistry for organic synthesis are highlighted below. The interested reader is referred to some recent general reviews, which also cover the use of bismuth oxides and halides in oxidation [4] and reduction [5] reactions. These are not within the scope of this chapter and so are not dealt with here [6, 7].

Transition Metals for Organic Synthesis, Vol. 1, 2nd Edition.
Edited by M. Beller and C. Bolm
Copyright © 2004 WILEY-VCH Verlag GmbH & Co. KGaA, Weinheim
ISBN: 3-527-30613-7

2.18.2
Carbon-Carbon Bond-Forming Reactions

2.18.2.1
Bismuth(0)

Bismuth(0)-mediated Barbier-type allylation of aldehydes with high chemoselectivity over ketones was first disclosed by Wada in 1985 [8]. Similar reactions have been realized with *in situ*-generated Bi(0) by using combinations of BiCl$_3$ and reducing agents (Zn, Al, Mg, NaBH$_4$). The general protocol tolerates nitrile, ester, halide, and alcohol functionalities. With crotyl halides, reaction selectively occurs at the γ-position to give the *erythro* α-substituted allylic alcohol with high diastereoselectivity [9] (Scheme 1).

Improvements in the procedure include the use of water as solvent and the realization of a catalytic procedure [10]. The proposed mechanism of the catalytic variant is shown in Scheme 2. The intermediate allylic metal species discriminates aldehydes from ketones and the carbonyl group from conjugated double bonds.

Bismuth(0)-catalyzed allylations of aldehydes in aqueous solution have stimulated further studies, which have resulted in the development of allylations of imines, enamines, and sulfonyl chlorides [11]. Furthermore, extensions have been made to include Reformatsky-type reactions of α-bromoketones [12], reductive coupling of imines [13], reductive cross-couplings of aldehydes with α-diketones [14], and allenylation-propargylations of aldehydes. Li reported on regio- and diastereoselective metal-mediated allenylations of aldehydes and disclosed superior activity

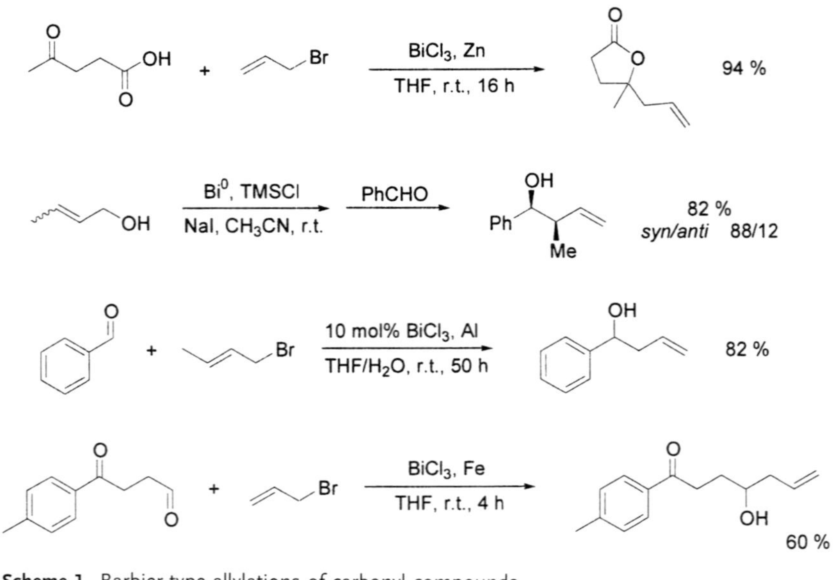

Scheme 1 Barbier-type allylations of carbonyl compounds.

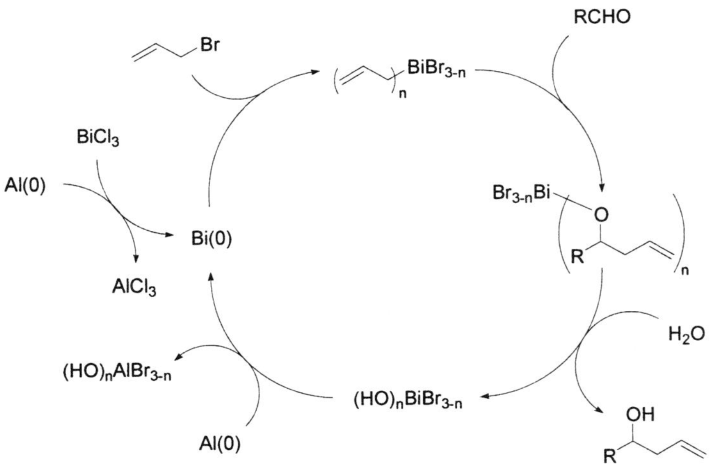

Scheme 2 Mechanism for the Bi(0)-catalyzed allylation of aldehydes.

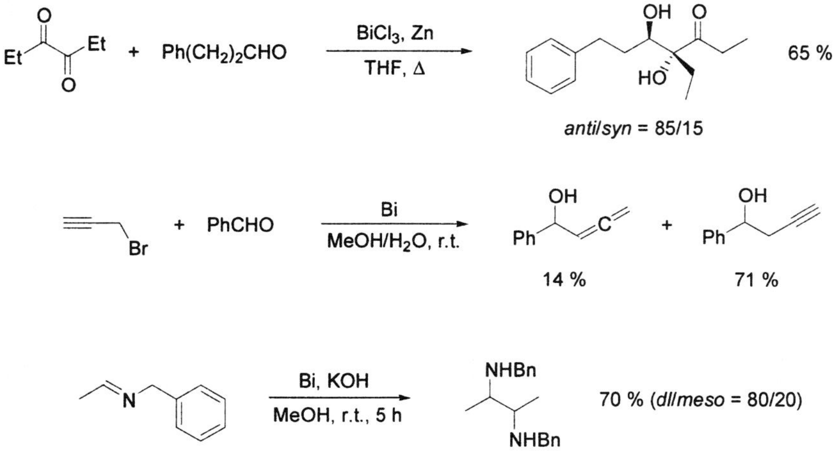

Scheme 3 Further examples of Bi(0)-mediated C-C bond-forming reactions.

of bismuth and indium, with preferential formation of the homopropargylic alcohol [15] (Scheme 3).

2.18.2.2
Bismuth(III)

Bismuth chloride is an effective catalyst for Mukaiyama-aldol reactions. Aldehydes, acetals, ketones, and α,β-unsaturated ketones were shown to cleanly react with TMS-enol ethers. Addition of metal iodides such as ZnI_2 can result in re-

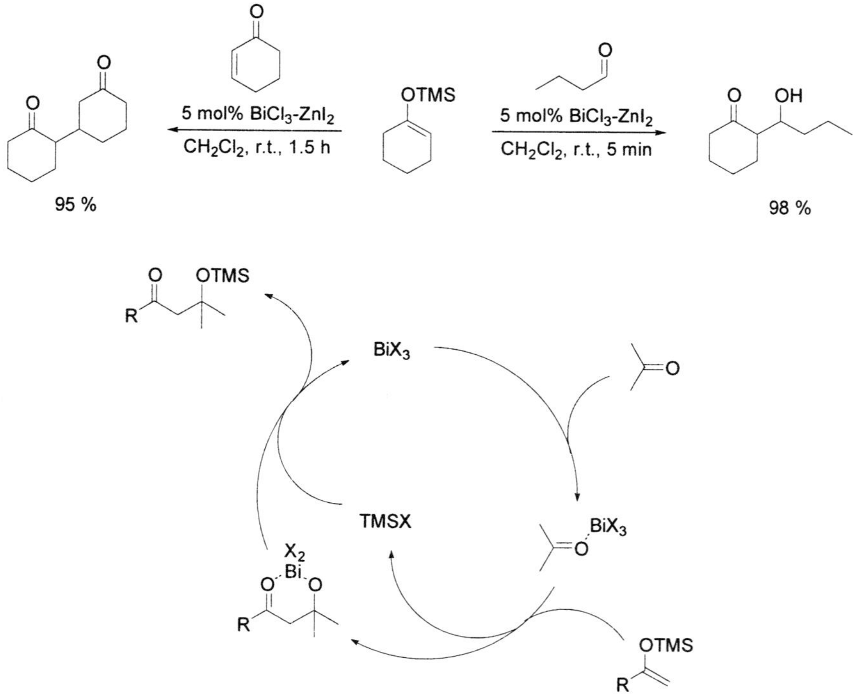

Scheme 4 Bi(III)-catalyzed Michael and aldol reactions (top) and mechanism of the Bi(III)-catalyzed Mukaiyama aldol reaction (bottom).

markably enhanced activity [16]. The formation of BiI_3 was confirmed by X-ray powder diffraction analysis, and the mechanism is proposed to involve carbonyl activation by Lewis-acidic Bi(III) and formation of an unstable chelate, which is subject to decomposition by *in situ*-generated TMS halide (Scheme 4).

The first selective C-acylation of silyl enolates was accomplished by Dubac in 1996 with binary $BiCl_3/MI_x$ catalyst systems [17]. Upon addition of 1.5 equivalents of ZnI_2 per bismuth, the selectivity (C/O-acylation) is greatly increased in favor of the C-acylated product. Similar reactions with allylsilanes exclusively afford the γ-acylated products without accompanying decarbonylation or double-bond isomerization [18] (Scheme 5). The same binary BiX_3/ZnI_2 catalyst has also been employed in Michael-type additions of furan derivatives and tandem aldol-halogenation reactions [19]. $Bi(OTf)_3$ was also reported to be an efficient catalyst precursor for Mukaiyama aldol reactions, though *in situ*-formed TMSOTf was suggested to act as the real catalyst. Arylketones can be synthesized in high yields via Friedel-Crafts acylation in the presence of catalytic $Bi(OTf)_3$ with activities superior to most of the known metal triflate catalysts (Al, Ln, Sc, etc.) [20]. Mechanistic studies by Dubac revealed the intermediacy of mixed anhydride RCO_2Tf as the true acylating agent. Activation of Bi(III) catalysts in Friedel-Crafts acylations of arenes in ionic liquid solvent systems has recently been disclosed [21].

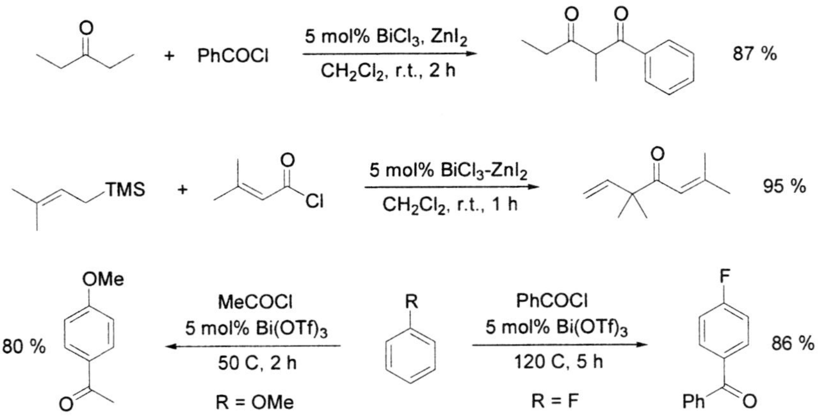

Scheme 5 Bi(III)-catalyzed acylation reactions.

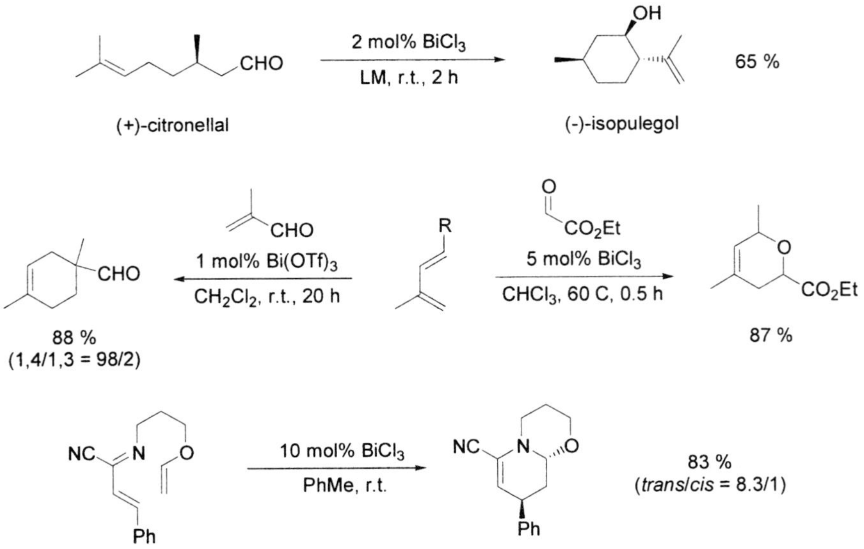

Scheme 6 Bi(III)-catalyzed ene and Diels-Alder reactions.

In the 1990s, Dubac and Suzuki studied bismuth(III) salts as catalysts for Diels-Alder reactions under mild conditions. Both Bi(OTf)$_3$ and BiCl$_3$ were shown to exhibit excellent catalytic activity for reactions of all-carbon dienes with activated dienophiles. In hetero Diels-Alder reactions with aldehyde dienophiles, high Diels-Alder/ene product ratios were observed [22]. Even in the presence of water, Bi(OTf)$_3$ catalyzes hetero Diels-Alder reactions with glyoxylic acid [23]. Ene reactions have been performed in the presence of 2 mol% BiCl$_3$. For example, Dubac reported on the selective intramolecular ene cyclizations of citronellal [24] (Scheme 6).

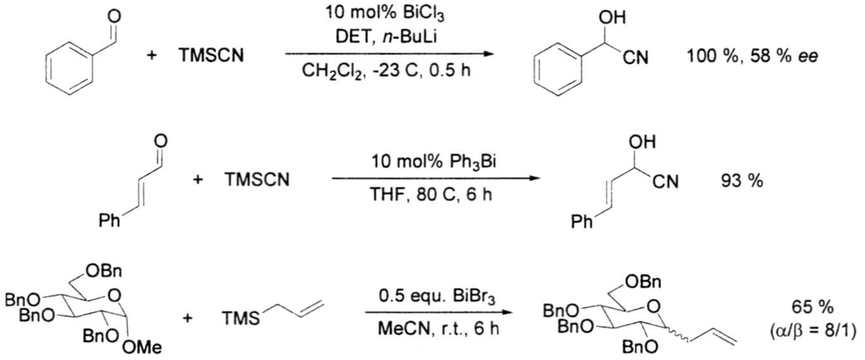

Scheme 7 Selected examples of cyanation and allylation reactions of aldehydes.

Bismuth chloride was demonstrated to catalyze the olefination of aldehydes with active methylene compounds [25]. With α,β-unsaturated carbonyl compounds, Michael addition is observed in high yields under microwave irradiation. The allylation of aldehydes and ketones with allyltin or silicon nucleophiles has been shown to give high yields of the corresponding homoallylic alcohols in the presence of catalytic $BiCl_3$ [26]. The cyanation of carbonyl compounds with TMSCN cleanly gave the cyanhydrins in quantitative yields. An asymmetric version of this reaction has been studied with chiral tartrate-based bismuth(III) complexes, but engendered only moderate enantioselectivities ($ee < 58\%$) [27]. Bismuth bromide was also found to be a useful Lewis acid catalyst for allylations and cyanations of carbonyl compounds with silicon compounds under mild conditions [28]. Suzuki and Wada further extended the allylation methodology to the C-allylation of glycopyranosides (Scheme 7).

Recently, the highly efficient $Bi(OTf)_3$-catalyzed conjugate addition of indoles to α,β-unsaturated enones has been realized in high yields at room temperature [29]. With substituted quinones, reaction occurs at the more hindered quaternary carbon (Scheme 8).

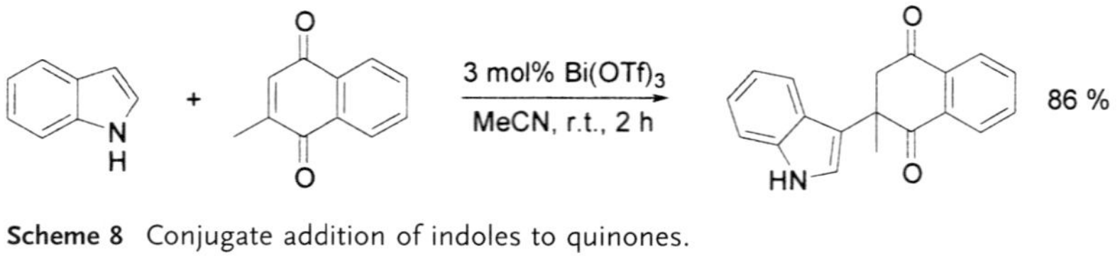

Scheme 8 Conjugate addition of indoles to quinones.

2.18.2.3
Organobismuth Compounds

Most of the early research activities concerning organobismuth reagents for carbon-carbon bond-forming reactions were devoted to C-arylations of arenes [30]. However, most procedures require activation of the arenes, and, hence, phenols, naphthols, 1,3-dicarbonyl compounds, and related compounds have been the substrates of choice. C-Arylations of phenols and naphthols can be conducted under basic conditions with tri- and tetraarylbismuth(V) compounds, while pentaphenylbismuth(V) allows for neutral conditions (see also Scheme 10) [31]. Compared to arylbismuth(V) compounds, tetraarylbismuthonium salts Ar_4BiX allow for far milder reaction conditions and shorter reaction times, and were shown to be less prone to oxidative side reactions. High chemoselectivity (C- over O-alkylation) can be effected with phenols bearing electron-donating groups (Scheme 9) [32].

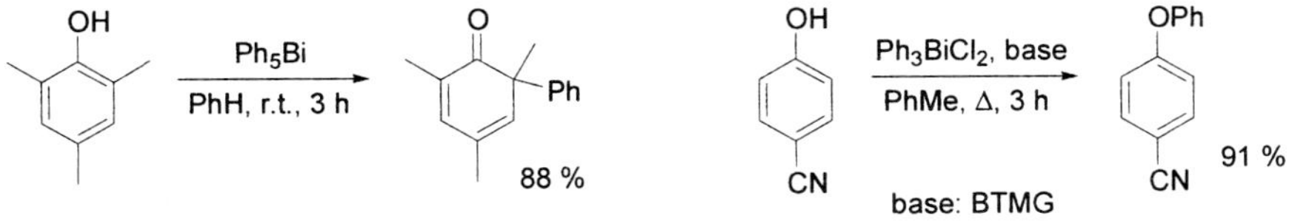

Scheme 9 Bismuth(V)-mediated C- and O-arylations of phenols.

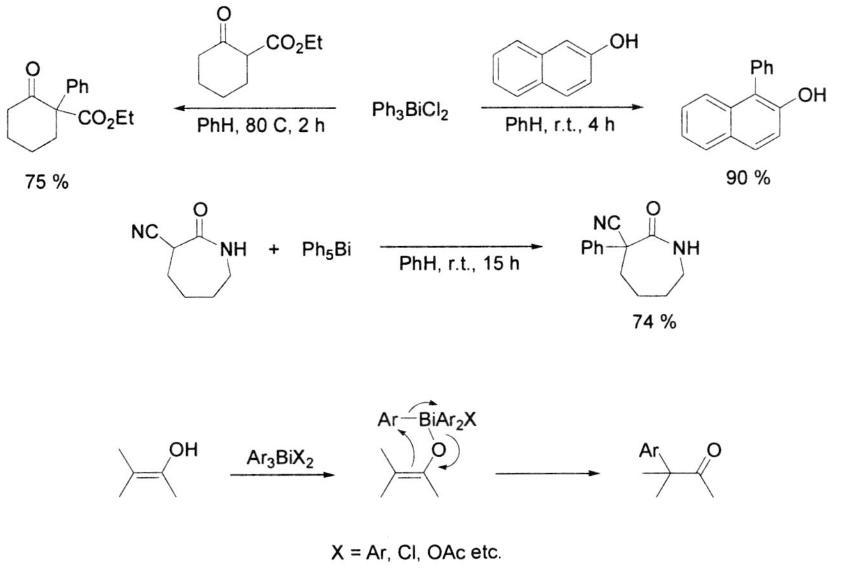

Scheme 10 Organobismuth(V)-mediated C-arylations.

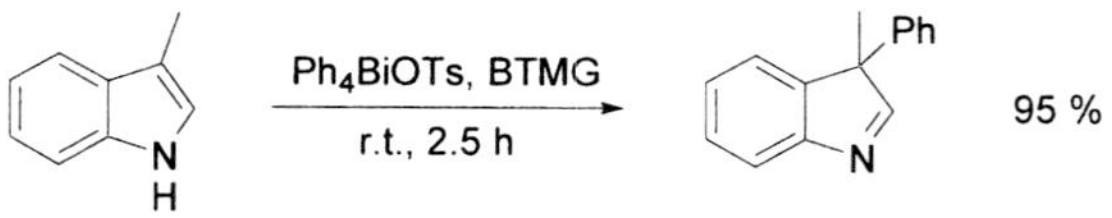

Scheme 11 Bismuth(V)-mediated β-arylation of indoles.

The C-arylation of 1,3-dicarbonyl compounds such as β-diketones, β-ketoesters, and malonic esters is readily achieved with numerous organobismuth(V) compounds, although selective monoarylation is difficult to control in most cases [31, 33]. Under basic conditions, dimedone was converted to its *a,a*-diphenyl derivative in the presence of tetraphenylbismuthonium salt, whereas the similar reaction with triphenylbismuthine gives an ylide [34]. Triarylbismuth carbonates have been used for the C-arylation toward the synthesis of isoflavanones and hydroxycoumarins [35]. The mechanistic pathway of the C-arylation involves the intermediacy of a pentavalent bismuth alkoxide, which has been isolated in some cases. The subsequent reductive elimination of the bismuth(III) compound is believed to proceed in concerted fashion [31] (Scheme 10).

Non-enolizable substrates are generally not arylated by arylbismuth(V) compounds under neutral conditions, but their enolate anions have been demonstrated to undergo facile *a*-arylation with Ph_3BiCl_2, Ph_3BiCO_3, and tetraphenylbismuth(V) compounds [34, 35]. Similar reactions have been performed with anions generated from nitroalkanes and indoles [36] (Scheme 11).

The thermal decomposition of pentaphenylbismuth gives rise to the formation of benzyne, which can be reacted with conjugated dienes to afford the corresponding Diels-Alder adducts [4f].

Transition metal-catalyzed reactions have recently been added to the arsenal of reactions with organobismuth(V) substrates. Early work by Barton established arylbismuth(V) compounds to be useful substrates for palladium-catalyzed aryl-aryl homocoupling and arylation reactions [37]. Quantitative aryl-aryl couplings have been effected with a wide variety of tri- and pentavalent organobismuth compounds in the presence of triethylamine and catalytic $Pd(OAc)_2$. Arylations of acyl chlorides, indoles [38], alkynes [39], allyl and propargyl halides have been accomplished by copper or palladium catalysts. Uemura reported on the carbonylative homocoupling of various Ar_3Bi derivatives in the presence of $RhCl(H_2O)_3$ [40] (Scheme 12).

The third important class of organobismuth(V) compounds for organic synthesis are bismuthonium salts and ylides. Upon treatment of dimedone with triphenylbismuth carbonate, a stabilized bismuthonium ylide [34] can be prepared which undergoes ready reaction with various nucleophiles. In the presence of copper(I) catalysts, 1,3-oxazoles [34], sulfonium ylides [41], and furan [42] derivatives can be obtained from isothiocyanates, dimethyl sulfide, and terminal acetylides, respectively. However, the yields generally range below 50%. Heterocyclic sulfones can be obtained from reaction with a sulfene [43], whereas aldehydes afford cyclopropane derivatives via a sequential Wittig-type olefination and Corey-type cyclopropanation reactions [44] (Scheme 13).

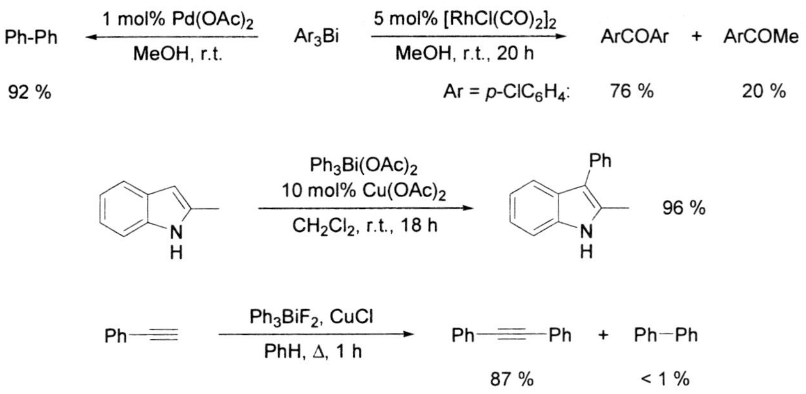

Scheme 12 Transition metal-catalyzed arylation reactions.

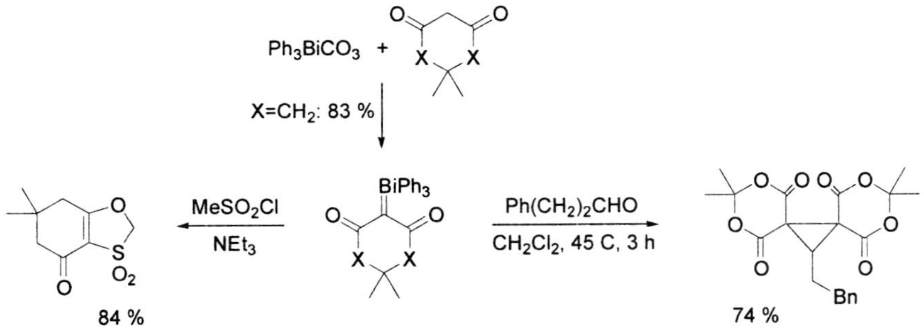

Scheme 13 Typical reactions of bismuthonium ylides derived from dimedone.

Triphenylbismuthonium 2-oxoalkylides have been exploited for the synthesis of diverse building blocks. Scheme 14 illustrates the synthesis of tropolones [45] and α-substituted ketones [46]. Interestingly, the reaction mode leading to oxiranes [47] and aziridines [48] is in marked contrast to those observed for the lighter homologs P, As, and Sb, which undergo Wittig-type olefination with imines and aldehydes.

Alkenylbismuthonium salts have been utilized by Suzuki in efficient preparations of substituted olefins [49]. With styrenes, cyclopropylidenes can be obtained in good yields (Scheme 15).

Triorganylbismuthines, such as Bu_3Bi and Ph_3Bi, are also capable of mediating C-C bond-forming reactions such as the allylation of ketones [50] and Diels-Alder reactions with α,β-unsaturated esters [51].

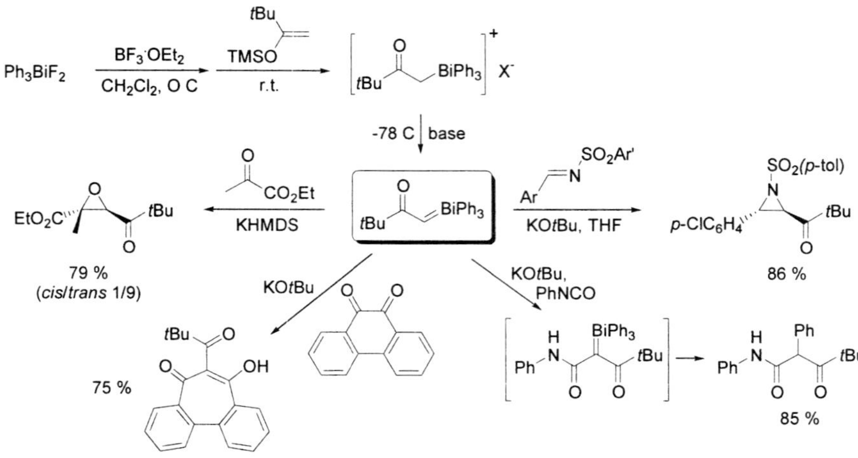

Scheme 14 Selected reactions of (2-oxoalkyl)triphenylbismuthonium salts.

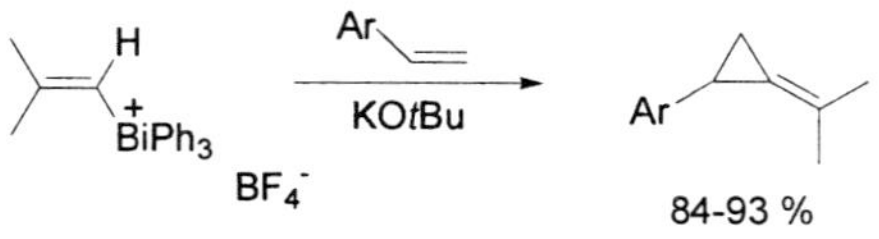

Scheme 15 Cyclopropanation of styrenes with alkenyltriphenylbismuthonium salts.

2.18.3
Carbon-Heteroatom Bond-Forming Reactions

Important C-O bond-forming processes are summarized in Schemes 16 and 17. Tandem cyclopropane ring cleavage and ester functionalization of thujopsene has been catalyzed by $Bi_2(SO_4)_3$ in moderate yield [52]. Bismuth(III) halides are effective catalysts for acetal deprotections [53] and etherifications. Catalytic amounts of $BiBr_3$ have been used for the synthesis of symmetrical and unsymmetrical ethers via reductive etherification of aldehydes in the presence of triethylsilane. Ketones proved unreactive under these reaction conditions [54]. Suzuki utilized the reductive heterocoupling for the single-step preparation of crownophanes with olefinic and acetylenic linkages [54b].

Mohan reported on the utilization of catalytic $Bi(OTf)_3$ for the efficient formation and deprotection of tetrahydropyranyl ethers under solvent-free conditions [55].

Recently, Evans elucidated the role of catalytic $BiBr_3$ in etherification reactions [56]. A tandem two-component etherification protocol was used for the synthesis of 2,6-disubstituted tetrahydropyrans, in which excellent *cis/trans* selectivity could be obtained for either stereoisomer through the judicious choice of the nucleo-

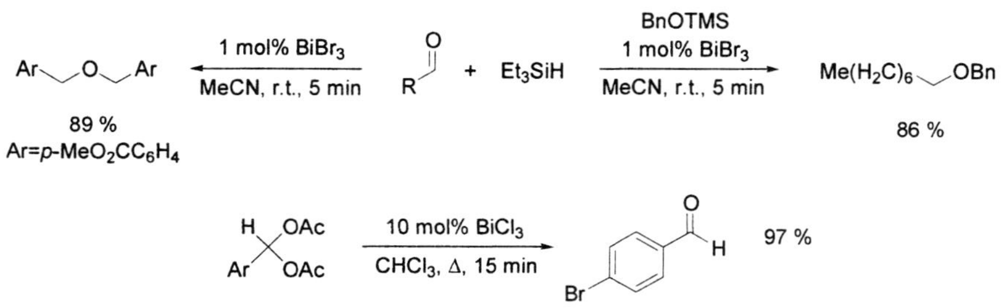

Scheme 16 Bismuth(III)-catalyzed reductive etherification of aldehydes (top) and deprotection of acetals.

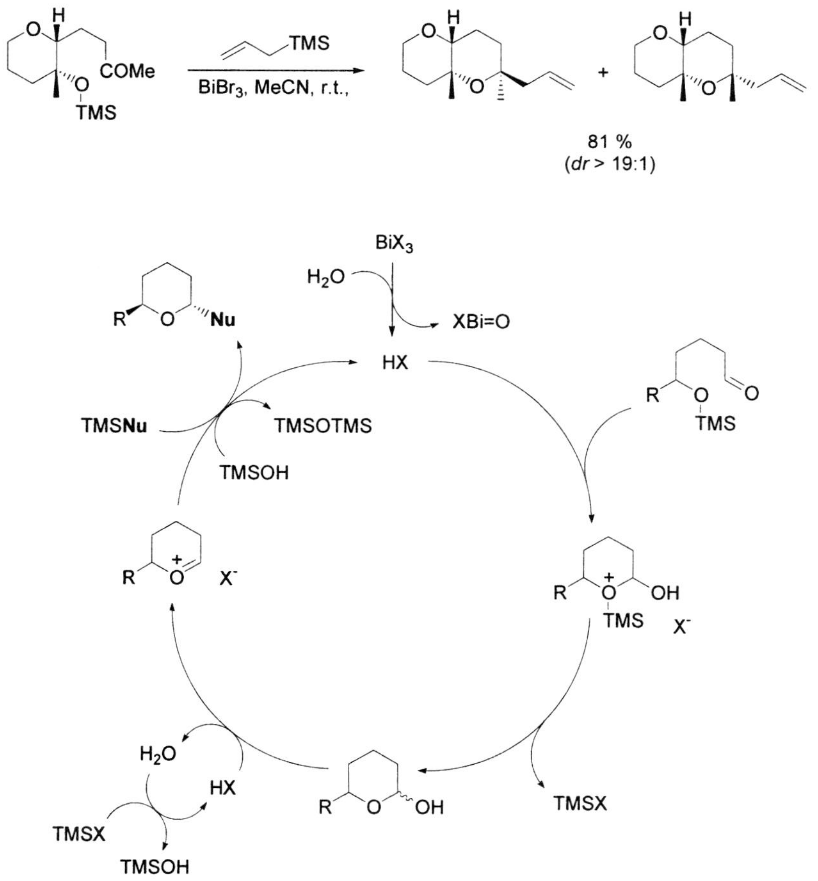

Scheme 17 Evans' etherification protocol for the synthesis of tetrahydropyrans.

phile and substrate. Furthermore, this work provided evidence for HBr and BrBi=O to be responsible for the catalysis (Scheme 17).

Torii expanded the scope of the bismuth-catalyzed allylation reactions to include C-S bond-forming processes for the synthesis of hydroxycephem and methylenepanem derivatives [57]. Remarkably high turnover numbers have been achieved in bismuth halide- or sulfate-catalyzed sulfenylations [20c] such as thioacetalizations of aldehydes and ketones and conjugate additions of thiophenol to α,β-unsaturated carbonyl compounds [58]. Recently, $BiCl_3$ was shown to catalyze the conversion of oxiranes to thiiranes and the sulfonation of arenes [59]. The mechanism of the latter is believed to involve ligand exchange with $Bi(OTf)_3$ to form $ArSO_3Tf$, which acts as the real catalyst. Thus, $Bi(OTf)_3$ can be viewed as a practical alternative to hygroscopic and corrosive TfOH. The sulfonation of substituted arenes usually gives regioisomers, though steric hindrance favors *para*-substitution in most cases (Scheme 18).

Alcohol, silyl ether, and acetate functions can be efficiently substituted with halogen atoms under mild conditions in the presence of catalytic bismuth(III) halide and TMSX as the actual halide source [19b, 60]. With stoichiometric bismuth(III) halide, highly chemoselective substitution reactions with haloalkanes have been performed. When adjacent halides are present, the mechanism involves chelation of the halides to form a bismuth(V) species [61] (Scheme 19).

O-Arylation [30] of alcohols, phenols, enols, and 1,3-dicarbonyl compounds can be effected with Ar_5Bi, Ar_4BiX, and Ar_3BiX_2 (X=halides, acetates) [31]. Moderate enantioselectivities have been obtained from selective mono O-arylation of *meso* 1,2-diols in the presence of triphenylbismuth acetate and copper pyridyloxazoline

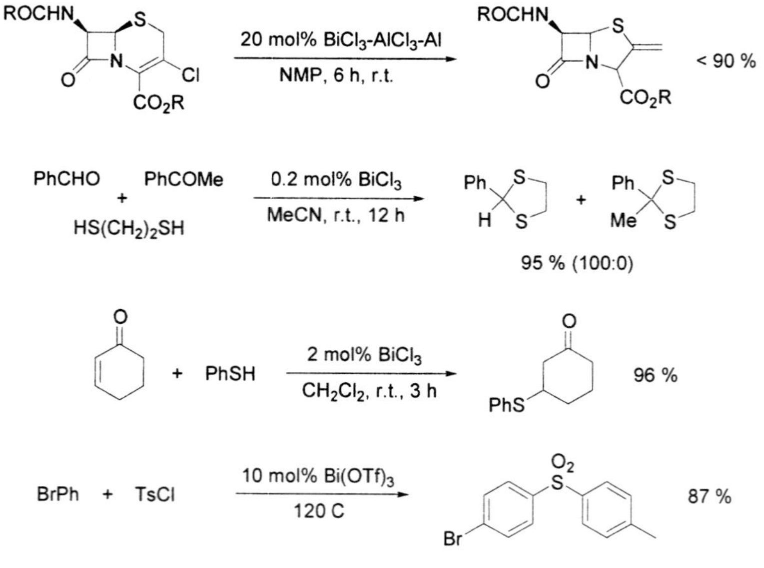

Scheme 18 Selected bismuth-catalyzed C-S bond-forming processes.

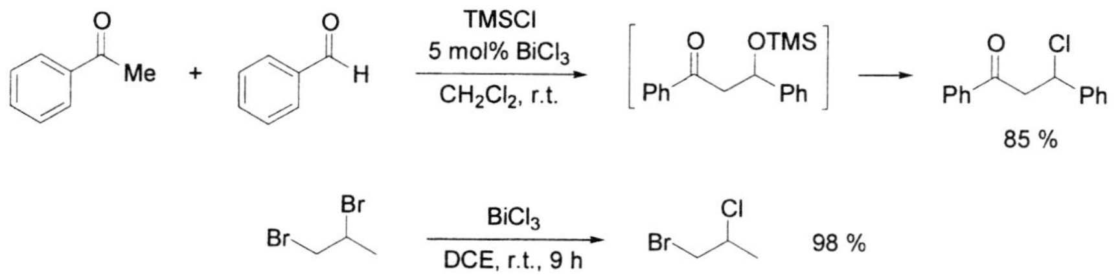

Scheme 19 Bismuth-mediated halogenation reactions.

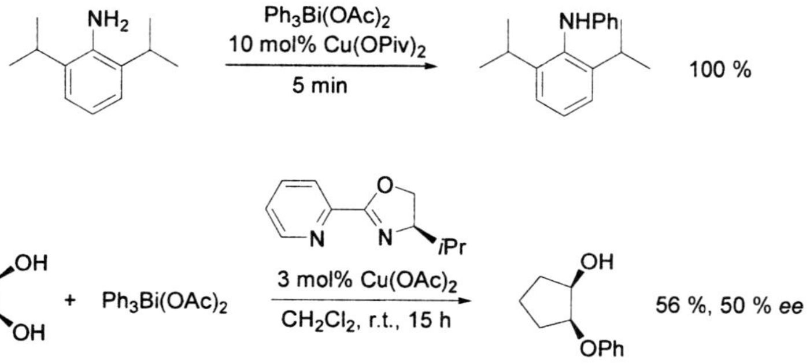

Scheme 20 Copper-catalyzed arylation of anilines and phenols.

catalysts [62]. Several examples of efficient N-arylations [30] of amines, amides, imides, indoles, lactams, and anilines have been reported [63]. Generally, the addition of catalytic amounts of copper, copper(II) acetate, or pivalate [64] significantly improves the arylating ability of most arylbismuth sources [65] (Scheme 20).

Bismuthonium salts have also been successfully employed as synthetic intermediates for the construction of carbon-heteroatom bonds. For example, Suzuki demonstrated the utilization of allylbismuthonium tetrafluoroborates for the synthesis of heteroatom-substituted allyl species (Scheme 21) [66].

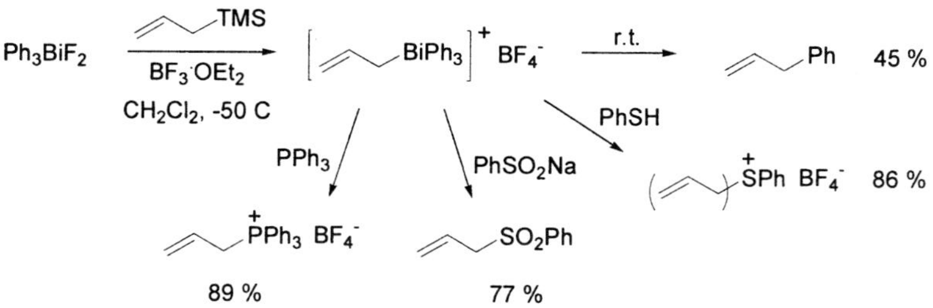

Scheme 21 Carbon-heteroatom bond-forming reactions with allyltriphenylbismuthonium salts.

2.18.4
Outlook

Bismuth is generally considered to be the least toxic heavy metal, and for many years its compounds have entertained a rich inorganic chemistry. In spite of these beneficial properties, applications of bismuth derivatives to organic synthesis were rare. However, in recent decades there has been revived interest in bismuth reagents and catalysts in organic synthesis. While most of the early investigations have focused on the structural aspects of organobismuth compounds and their stoichiometric use as oxidizing and arylating agents, more recent years have witnessed profound studies of their catalytic performance. Important applications of bismuth-based catalysts range from oxidations of alcohols and Barbier-type allylations of carbonyl compounds to Lewis acid catalysis of cyanation, Diels-Alder, aldol, and Michael reactions. The combination of low toxicity and cost with their proven catalytic activity in various reactions will undoubtedly prompt further research efforts toward the development of new and chiral bismuth catalysts and will result in a wider variety of applications to organic synthesis.

2.18.5
References

1 C. Silvestru, H. J. Breunig, H. Althaus, Chem. Rev. **1999**, 99, 3277.
2 For a recent review on biological and medicinal aspects of bismuth compounds, see: G. G. Briand, N. Burford, Chem. Rev. **1999**, 99, 2601.
3 F. Challenger, O. V. Richards, J. Chem. Soc. **1934**, 405.
4 For selected examples, see: (a) C. Coin, V. Le Boisselier, I. Favier, M. Postel, E. Duach, Eur. J. Org. Chem. **2001**, 735; (b) N. Komatsu, A. Taniguchi, M. Uda, H. Suzuki, Chem. Commun. **1996**, 1847; (c) N. Komatsu, M. Uda, H. Suzuki, Chem. Lett. **1997**, 1229; (d) D. H. R. Barton, D. J. Lester, W. B. Motherwell, M. T. B. Papoula, J. Chem. Soc., Chem. Commun. **1980**, 246; (e) D. H. R. Barton, J. P. Kitchin, D. J. Lester, W. B. Motherwell, M. T. B. Papoula, Tetrahedron **1981**, 37 (Suppl. 1), 73; (f) D. H. R. Barton, J.-P. Finet, W. B. Motherwell, C. Pichon, J. Chem. Soc., Perkin Trans. 1 **1987**, 251.
5 For selected examples, see: (a) P.-D. Ren, S.-F. Pan, T.-W. Dong, S.-H. Wu, Synth. Commun. **1996**, 26, 763; (b) P.-D. Ren, Q.-H. Jin, Z.-P. Yao, Synth. Commun. **1997**, 27, 2577; (c) H. Suzuki, Tetrahedron Lett. **1997**, 38, 7219. For an application of catalytic BiX_3 in asymmetric imine hydrogenation, see: (d) K. Satoh, M. Inenaga, K. Kanai, Tetrahedron Asymm. **1998**, 9, 2657.
6 Organobismuth Chemistry (H. Suzuki, Y. Matano, Eds.), Elsevier, Amsterdam **2001**.
7 (a) M. Wada, H. Ohki, Yuki Gosei Kagaku Kyokaishi **1989**, 47, 425; (b) H. Suzuki, T. Ikegami, Y. Matano, Synthesis **1997**, 249.
8 M. Wada, K.-y. Akiba, Tetrahedron Lett. **1985**, 26, 4211.
9 For some examples, see: (a) M. Wada, H. Ohki, K.-y. Akiba, Tetrahedron Lett. **1986**, 27, 4771; (b) M. Wada, H. Ohki, K.-y. Akiba, J. Chem. Soc., Chem. Commun. **1987**, 708; (c) M. Wada, T. Fukuma, M. Morioka, T. Takahashi, N. Miyoshi, Tetrahedron Lett. **1997**, 38, 8045; (d) M. Wada, M. Honma, Y. Kuramoto, N. Miyoshi, Bull. Chem. Soc. Jpn. **1997**, 70, 2265; (e) N. Miyoshi, M. Nishio, S. Murakami, T. Fukuma, M. Wada, Bull. Chem. Soc. Jpn. **2000**, 73, 689.

10 M. Wada, H. Ohki, K.-y. Akiba, *Bull. Chem. Soc. Jpn.* **1990**, *63*, 1738.

11 (a) P. J. Bhuyan, D. Prajapati, J. S. Sandhu, *Tetrahedron Lett.* **1993**, *34*, 7975; (b) L. Tussa, C. Lebreton, P. Mosset, *Chem. Eur. J.* **1997**, *3*, 1064; (c) M. Baruah, A. Baruah, D. Prajapati, J. S. Sandhu, *Synlett* **1998**, 1083.

12 Z. Shen, J. Zhang, H. Zou, M. Yang, *Tetrahedron Lett.* **1997**, *38*, 2733.

13 B. Baruah, D. Prajapati, J. S. Sandhu, *Tetrahedron Lett.* **1995**, *36*, 6747.

14 N. Miyoshi, T. Fukuma, M. Wada, *Chem. Lett.* **1995**, 999.

15 X.-H. Yi, Y. Meng, X.-G. Hua, C.-J. Li, *J. Org. Chem.* **1998**, *63*, 7472.

16 (a) H. Ohki, M. Wada, K.-y. Akiba, *Tetrahedron Lett.* **1988**, *29*, 4719; (b) M. Wada, E. Takeichi, T. Matsumoto, *Bull. Chem. Soc.* **1991**, *64*, 990; (c) C. Le Roux, H. Gaspard Iloughmane, J. Dubac, *J. Org. Chem.* **1993**, *58*, 1835.

17 C. Le Roux, S. Mandrou, J. Dubac, *J. Org. Chem.* **1996**, *61*, 3885.

18 C. Le Roux, J. Dubac, *Organometallics* **1996**, *15*, 4646.

19 (a) C. Le Roux, M. Maraval, M. E. Borredon, H. Gaspard Iloughmane, J. Dubac, *Tetrahedron Lett.* **1992**, *33*, 1053; (b) C. Le Roux, H. Gaspard Iloughmane, J. Dubac, *J. Org. Chem.* **1994**, *59*, 2238.

20 (a) J. R. Desmurs, M. Labrouillere, C. Le Roux, H. Gaspard, A. Laporterie, J. Dubac, *Tetrahedron Lett.* **1997**, *38*, 8871; (b) C. Le Roux, L. Ciliberti, H. L. Robert, A. Laporterie, J. Dubac, *Synlett* **1998**, 1249; (c) For a review, see: C. Le Roux, J. Dubac, *Synlett* **2002**, 181.

21 S. Gmouh, H. Yang, M. Vaultier, *Org. Lett.* **2003**, *5*, 2219.

22 (a) B. Garrigues, F. Gonzaga, H. Robert, J. Dubac, *J. Org. Chem.* **1997**, *62*, 4880; (b) H. Robert, B. Garrigues, *Tetrahedron Lett.* **1998**, *39*, 1161; (c) M. Labrouillere, C. Le Roux, H. Gaspard, A. Laporterie, J. Dubac, J. R. Desmurs, *Tetrahedron Lett.* **1999**, *40*, 285; I. A. Motorina, D. S. Grierson, *Tetrahedron Lett.* **1999**, *40*, 7215.

23 H. L. Robert, C. Le Roux, J. Dubac, *Synlett* **1998**, 1138.

24 L. Peidro, C. Le Roux, A. Laporterie, J. Dubac, *J. Organomet. Chem.* **1996**, *521*, 397.

25 B. Baruah, A. Baruah, D. Prajapati, J. S. Sandhu, *Tetrahedron Lett.* **1997**, *38*, 1449.

26 S. L. Serre, J.-C. Guillemin, T. Karapti, L. Soos, L. Nyulkgzi, T. Veszpremi, *J. Org. Chem.* **1998**, *63*, 59.

27 M. Wada, T. Takahashi, T. Domae, T. Fukuma, N. Miyoshi, K. Smith, *Tetrahedron Asymm.* **1997**, *8*, 3939.

28 N. Komatsu, M. Uda, H. Suzuki, T. Takahashi, T. Domae, M. Wada, *Tetrahedron Lett.* **1997**, *38*, 7215.

29 A. Vijender Reddy, K. Ravinder, T. Venkateshwar Goud, P. Krishnaiah, T. V. Raju, Y. Venkateshwarlu, *Tetrahedron Lett.* **2003**, *44*, 6257.

30 For a recent review on arylations with organobismuth compounds, see: G. I. Elliott, J. P. Konopelski, *Tetrahedron* **2001**, *57*, 5683.

31 (a) D. H. R. Barton, N. Y. Bhatnagar, J.-C. Blazejewski, B. Charpiot, J.-P. Finet, D. J. Lester, W. B. Motherwell, M. T. B. Papoula, S. P. Stanforth, *J. Chem. Soc., Perkin Trans. 1* **1985**, 2657; (b) D. H. R. Barton, N. Y. Bhatnagar, J.-P. Finet, W. B. Motherwell, *Tetrahedron* **1986**, *42*, 3111; (c) A. Fedorov, S. Combes, J.-P. Finet, *Tetrahedron* **1999**, *55*, 1341; (d) S. Combes, J.-P. Finet, *Tetrahedron* **1999**, *55*, 3377.

32 D. H. R. Barton, N. Y. Bhatnagar, J.-P. Finet, J. Khamsi, W. B. Motherwell, S. P. Stanforth, *Tetrahedron* **1987**, *43*, 323.

33 M. S. Akhtar, W. J. Brouillette, D. V. Waterhouse, *J. Org. Chem.* **1990**, *55*, 5222.

34 (a) D. H. R. Barton, J.-C. Blazejewski, B. Charpiot, J.-P. Finet, W. B. Motherwell, M. T. B. Papoula, S. P. Stanforth, *J. Chem. Soc., Perkin Trans. 1* **1985**, 2667; (b) H. Suzuki, T. Murafuji, T. Ogawa, *Chem. Lett.* **1988**, 847.

35 D. H. R. Barton, D. M. Donnelly, J.-P. Finet, P. H. Stenson, *Tetrahedron* **1988**, *44*, 6387.

36 (a) J. J. Lalonde, D. E. Bergbreiter, C.-H. Wong, *J. Org. Chem.* **1988**, *53*, 2323; (b) D. H. R. Barton, J.-P. Finet, C. Gian-

NOTTI, F. HALLEY, *J. Chem. Soc., Perkin Trans. 1* **1987**, 241.

37 D. H. R. BARTON, N. OZBALIK, J. C. SARMA, *Tetrahedron* **1988**, *44*, 5661.

38 D. H. R. BARTON, J.-P. FINET, J. KHAMSI, *Tetrahedron Lett.* **1988**, *29*, 1115.

39 S. A. LERMONTOV, I. M. RAKOV, N. S. ZEFIROV, P. J. STANG, *Tetrahedron Lett.* **1996**, *37*, 4051.

40 T. OHE, T. TANAKA, M. KURODA, C. S. CHO, K. OHE, S. UEMURA, *Bull. Chem. Soc. Jpn.* **1999**, *72*, 1851.

41 H. SUZUKI, T. MURAFUJI, *Bull. Chem. Soc. Jpn.* **1990**, *63*, 950.

42 T. OGAWA, T. MURAFUJI, K. IWATA, H. SUZUKI, *Chem. Lett.* **1989**, 325.

43 T. OGAWA, T. MURAFUJI, H. SUZUKI, *J. Chem. Soc., Chem. Commun.* **1989**, 1749.

44 T. OGAWA, T. MURAFUJI, H. SUZUKI, *Chem. Lett.* **1988**, 849.

45 (a) Y. MATANO, H. SUZUKI, *Chem. Commun.* **1996**, 2697; (b) M. M. RAHMAN, Y. MATANO, H. SUZUKI, *Synthesis* **1999**, 395; (c) M. M. RAHMAN, Y. MATANO, H. SUZUKI, *J. Chem. Soc., Perkin Trans. 1* **1999**, 1533.

46 Y. MATANO, *J. Chem. Soc., Perkin Trans. 1* **1994**, 2703.

47 T. OGAWA, T. IKEGAMI, T. HIKASA, N. ONO, H. SUZUKI, *J. Chem. Soc., Perkin Trans. 1* **1994**, 3479.

48 Y. MATANO, M. YOSHIMUNE, H. SUZUKI, *J. Org. Chem.* **1995**, *60*, 4663.

49 Y. MATANO, M. YOSHIMUNE, N. AZUMA, H. SUZUKI, *J. Chem. Soc., Perkin Trans. 1* **1996**, 1971.

50 Y.-Z. HUANG, Y. LIAO, *Heteroatom Chem.* **1991**, *2*, 297.

51 I. SUZUKI, Y. YAMAMOTO, *J. Org. Chem.* **1993**, *58*, 4783.

52 K. ABE, M. ITO, *Bull. Chem. Soc.* **1978**, *51*, 319.

53 I. MOHAMMADPOOR-BALTORK, H. ALIYAN, *Synth. Commun.* **1999**, *29*, 2741.

54 (a) Y. WATANABE, C. NAKAMOTO, S. OZAKI, *Synlett* **1993**, 115; (b) N. KOMATSU, J. ISHIDA, H. SUZUKI, *Tetrahedron Lett.* **1997**, *38*, 7219.

55 J. R. STEPHENS, P. L. BUTLER, C. H. CLOW, M. C. OSTWALD, R. C. SMITH, R. S. MOHAN, *Eur. J. Org. Chem.* **2003**, 3827.

56 P. A. EVANS, J. CUI, S. J. GHARPURE, R. J. HINKLE, *J. Am. Chem. Soc.* **2003**, *125*, 11456.

57 (a) H. TANAKA, S. SUMIDA, Y. NISHIOKA, N. KOBAYASHI, Y. TOKUMARU, Y. KAMEYAMA, S. TORII, *J. Org. Chem.* **1997**, *62*, 3610; (b) H. TANAKA, Y. TOKUMARU, Y. KAMEYAMA, S. TORII, *Chem. Lett.* **1997**, 1221.

58 N. KOMATSU, M. UDA, H. SUZUKI, *Synlett* **1995**, 984.

59 (a) S. REPICHET, C. LE ROUX, P. HERNANDEZ, J. DUBAC, *J. Org. Chem.* **1999**, *64*, 6479; (b) S. REPICHET, C. LE ROUX, J. DUBAC, *Tetrahedron Lett.* **1999**, *40*, 9233.

60 M. LABROUILLERE, C. LE ROUX, H. GASPARD ILOUGHMANE, J. DUBAC, *Synlett* **1994**, 723.

61 B. BOYER, E. M. KERAMANE, S. ARPIN, J.-L. MONTERO, J.-P. ROQUE, *Tetrahedron* **1999**, *55*, 1971.

62 H. BRUNNER, U. OBERMANN, P. WIMMER, *Organometallics* **1989**, *8*, 821.

63 (a) D. H. R. BARTON, J.-P. FINET, C. PICHON, *J. Chem. Soc., Chem. Commun.* **1986**, 65; 54; (b) D. H. R. BARTON, J.-P. FINET, J. KHAMSI, *Tetrahedron Lett.* **1986**, *27*, 3615; (c) S. OMBES, J.-P. FINET, *Tetrahedron* **1998**, *54*, 4313.

64 T. ARNAULD, D. H. R. BARTON, E. DORIS, *Tetrahedron* **1997**, *53*, 4167.

65 (a) T. HARADA, S. UEDA, T. YOSHIDA, A. INOUE, M. TAKEUCHI, N. OGAWA, A. OKU, *J. Org. Chem.* **1994**, *59*, 7575; (b) K. M. J. BRANDS, U.-H. DOLLING, R. B. JOBSON, G. MARCHESINI, R. A. REAMER, J. M. WILLIAMS, *J. Org. Chem.* **1998**, *63*, 6721.

66 Y. MATANO, M. YOSHIMUNE, H. SUZUKI, *Tetrahedron Lett.* **1995**, *36*, 7475.

3
Transition Metal-Mediated Reactions

3.1
Fischer-Type Carbene Complexes

Karl Heinz Dötz and Ana Minatti

3.1.1
Synthesis and Reactivity

Carbene complexes [1] are characterized by a divalent carbon species coordinated to a transition metal. The properties and the reactivity of carbene complexes can be tuned by the choice of the metal and its oxidation state as well as by the design of its coligand sphere. Two types of carbene complexes have to be distinguished: Fischer-type carbene complexes are characterized by a low-valent (middle to late) transition metal and strong acceptor coligands such as carbon monoxide. The carbene carbon atom behaves as an electrophilic center and is usually stabilized through π-donation. Therefore their charge distribution is best explained by canonical forms (a) and (c) in Fig. 1.

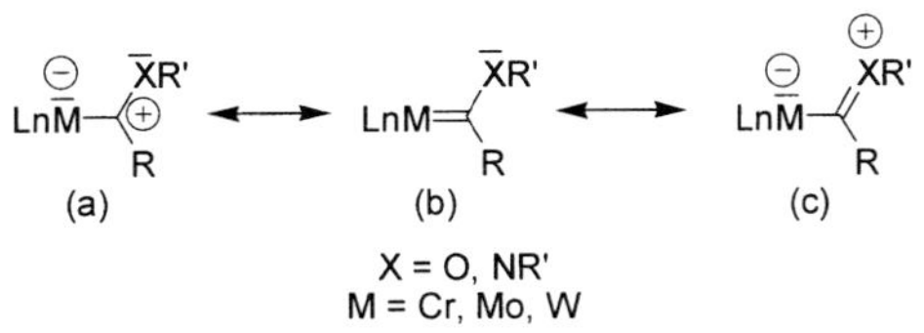

Fig. 1 Canonical forms of Fischer-type carbene complexes.

In contrast, a Schrock-type metal carbene complex bears a higher valent (early to middle) transition metal attached to efficient donor coligands. As a consequence, the polarity of the metal-carbon bond is reversed and imposes considerable nucleophilicity onto the carbene carbon atom.

Up to the present time, in most cases the synthesis of Fischer carbene complexes has followed the classical route developed by Fischer, which is based on the consecutive addition of an organolithium nucleophile and a carbon electrophile across a carbonyl ligand [2]. Important extensions of the synthetic strategy involve the trapping of highly reactive acyloxycarbene complexes with enantiopure alcohols leading to optically active alkoxycarbene complexes and the addition of pentacarbonylmetalate dianions to carboxylic acid chlorides or amides followed by alky-

Transition Metals for Organic Synthesis, Vol. 1, 2nd Edition.
Edited by M. Beller and C. Bolm
Copyright © 2004 WILEY-VCH Verlag GmbH & Co. KGaA, Weinheim
ISBN: 3-527-30613-7

lation or deoxygenation ("Semmelhack-Hegedus route") [3]. More specialized approaches have been developed for difluoroboroxycarbene complexes [4, 5] as well as for cycloalkylidene and tricyclic diaryl carbene complexes, which are based on the cycloisomerization of alkynols assisted by pentacarbonylchromium compounds and on the transformation of diazo precursors, respectively [6, 7].

Since the discovery by Fischer and Maasböl, Fischer carbene complexes have been developed to valuable building blocks or key intermediates for the construction of organic molecules. Their reactivity pattern reflects both ligand-centered and metal-centered reactions.

The pronounced electrophilicity of the carbene carbon atom favors the attack of carbon and heteroatom nucleophiles [8]. The aminolysis reaction allows the synthesis of optically active aminocarbene complexes and has recently been applied to label amino acid derivatives [9]. Electrophiles may add to the heteroatom carbene substituent [10]. Metal-stabilized carbanions may be generated by deprotonation of the acidic α-CH groups. Finally, carbonyl ligands undergo a thermal or photochemical substitution for other type of ligands, which allows modification of the coordination sphere.

3.1.2
Carbene-Ligand Centered Reactions

3.1.2.1
Carbon-Carbon Bond Formation *via* Metal Carbene Anions

The pentacarbonyl metal moiety of Fischer carbene complexes serves as a strong electron-withdrawing group, thus rendering the hydrogen atoms α to the carbene carbon highly acidic [11]. Deprotonation with organolithium bases generates carbene complex-stabilized conjugated carbanions, which react with various electrophiles to give α-substituted carbene complexes and may be exploited in carbon-carbon bond formation. For example, aldol condensation reactions starting from the lithiation of α-silylated tungsten amino carbene complexes followed by addition of 1,1'-diformylferrocene give rise to ferrocene-bridged conjugated Fischer carbene complexes [12].

When treated with lithium bases, aminocarbenes generally form a mixture of two rotamers, which may reduce their usefulness in stereoselective synthesis.

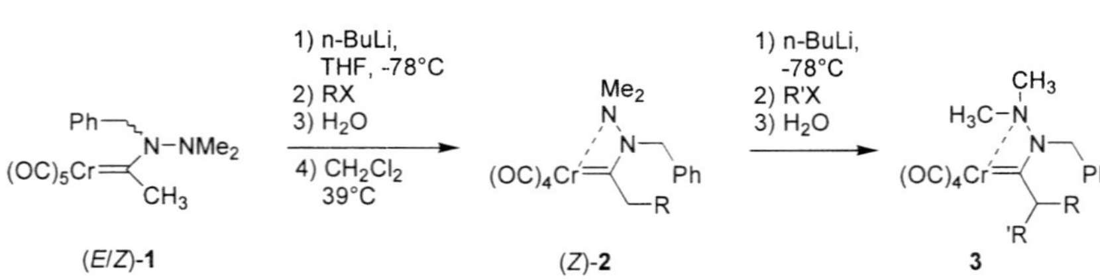

Scheme 1 α-Alkylation of aminocarbene complexes.

This problem has been overcome for the alkyl(hydrazino)carbene complex **1** (Scheme 1) [13]. Equilibration at the stage of the deprotonated (*E*)- and (*Z*)-mixture, thermal decarbonylation and subsequent alkylation lead exclusively to the (*Z*)-rotamer **2**.

An interesting feature of carbene chelate (*Z*)-**2** is the fact that – after treatment with *n*-butyllithium – it can add further electrophiles and thus generate a stereogenic center at the *α*-carbon position to yield **3**, a reaction that is mostly hampered for related aminocarbenes bearing an alkyl group on the carbene carbon. A chiral morpholinyl carbene anion derived from methylcarbene complex **4** allows for a diastereoselective Michael-type addition to nitroolefins to give **5** (Scheme 2) [14].

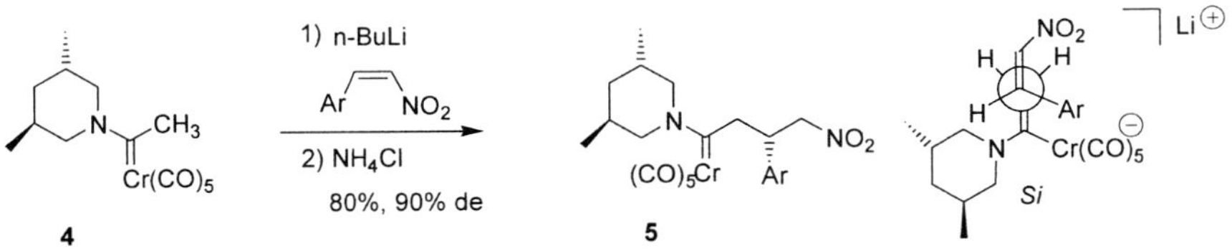

Scheme 2 Diastereoselective Michael addition of nitroolefins (Ar = 4-Cl-Ph).

An improved diastereoselectivity and shorter reaction times are observed when an electron-withdrawing substituent is present on the aromatic ring of the (*Z*)-nitroalkene and the reaction is carried out in the presence of crown ethers. These results and the stereochemical outcome of the reaction can be rationalized by the assumption that the *Si* face of the nitrostyrene reacts with the anion with the electron-poor aryl group facing the negatively charged $Cr(CO)_5^-$ moiety because of favorable electronic interaction. An enantioselective version of this chemistry has already been applied in the total synthesis of (*R*)-(−)-baclofen [15].

A different synthetic strategy is employed for the formation of the carbene complex-stabilized carbanion **7**, which is prepared *via* a coupling of the 1-lithioglucal **6** with a chromiumpentacarbonyl source (Scheme 3) [16]. This type of organometallic sugars undergo a highly diastereoselective alkylation in favor of the axial diastereomer **8**. Among further interesting reactions of carbene complex stabilized anions is an iodine oxidation [17].

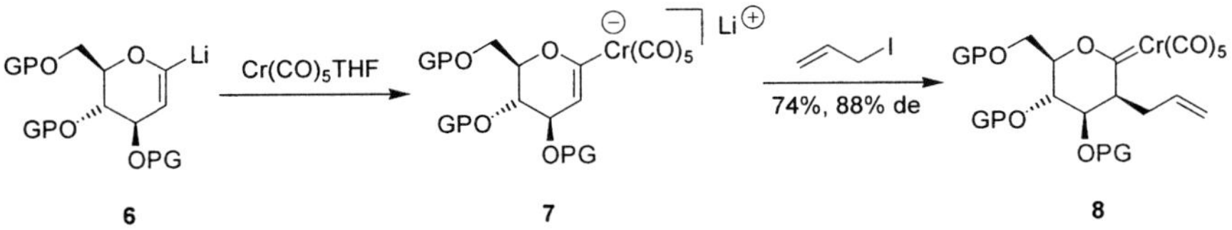

Scheme 3 Stereoselective *α*-allylation of vinyl chromates (PG = TIPS).

3.1.2.2
Carbon Nucleophile Addition to α,β-Unsaturated Carbene Complexes

Usually, organolithium reagents add to the carbene carbon atom of Fischer aryl(alkoxy) carbene complexes in a simple 1,2-addition reaction [18]. However, this regioselectivity is hampered by bulky alkoxy groups such as (–)-menthyloxy or (–)-8-phenylmenthyloxy, which favor a regioselective nucleophilic conjugate addition of either secondary or tertiary alkyllithium or aryllithium compounds to the aromatic nucleus [19]. The phenylcarbene complex **9** undergoes a 1,6-addition, affording the deconjugated cyclohexadienyl carbene complex **10** (Scheme 4). A more electron-rich aromatic ring such as that in the 4-methoxy-substituted carbene complex analog **9'** leads to a 1,4-addition of the organolithium compound to yield **11**. All intermediates resulting from the initial addition step were trapped with methyl triflate. These regio- and stereoselective nucleophilic aromatic 1,6- or 1,4-addition reactions represent a novel dearomatization procedure. Complementarily, a 1,2-addition to the sterically hindered carbene carbon atom takes place with lithium acetylides [20]. The resulting propargylic carbanion-type intermediate could be trapped regioselectively with electrophiles.

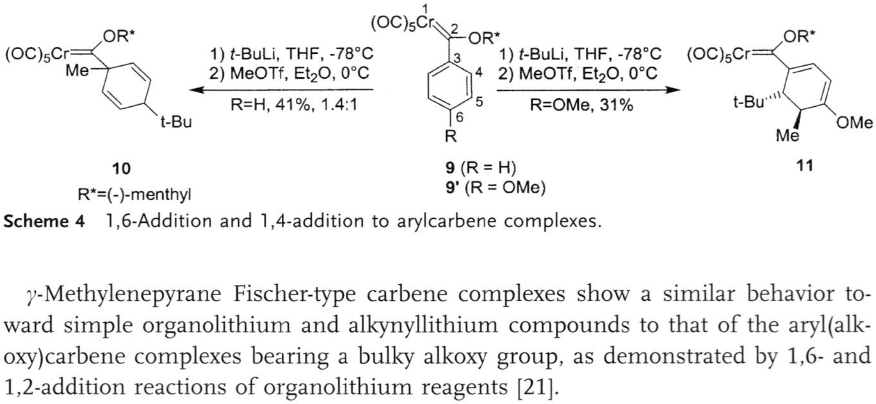

Scheme 4 1,6-Addition and 1,4-addition to arylcarbene complexes.

γ-Methylenepyrane Fischer-type carbene complexes show a similar behavior toward simple organolithium and alkynyllithium compounds to that of the aryl(alkoxy)carbene complexes bearing a bulky alkoxy group, as demonstrated by 1,6- and 1,2-addition reactions of organolithium reagents [21].

The first example of an asymmetric Michael addition of organolithium compounds to chiral Fischer alkenylcarbene complexes was achieved by reaction of alkyllithium reagents with the optically active Fischer vinylcarbene complex **12** derived from (–)-8-phenylmenthol (Scheme 5) [22].

The observed high regioselectivity can be attributed to the bulky alkoxy group bound to the carbene carbon atom; the excellent diastereoselectivity reflecting the sense of facial discrimination, however, is best explained in terms of the model shown in Scheme 5. The most stable conformation of [(–)-8-phenylmenthyloxy](alkenyl)carbene complexes, which is favored by an alkene-arene π-stacking effect, shields the (*Re,Re*) side of the double bond and forces a nucleophilic attack on the Michael acceptor from the side opposite to the phenyl group [22].

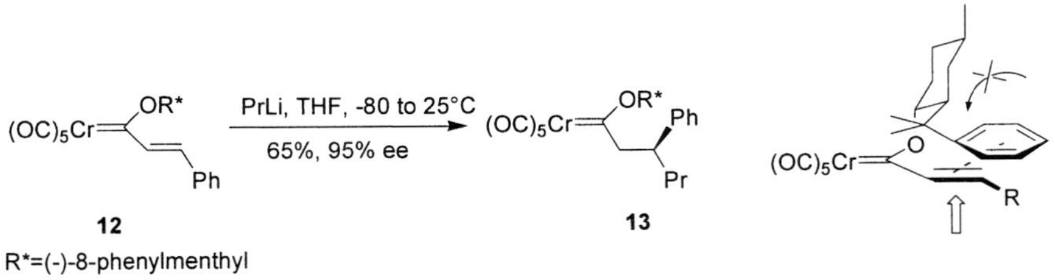

Scheme 5 Asymmetric 1,4-addition of alkyllithium compounds.

R*=(–)-8-phenylmenthyl

As lithium enolates are softer nucleophiles than alkyllithium compounds, they are more prone to undergo conjugate addition to Michael acceptors. The course of the nucleophilic addition to alkenylcarbene complexes can be influenced by steric requirements of the enolate. The treatment of cyclic alkenylcarbene complexes **14** with lithium enolates of either methyl ketones, methyl enones, or methyl ynones in the presence of an excess of PMDTA leads – after hydrolysis and oxidative decomplexation – to bicyclic cyclopentenol derivatives **15**, which were isolated as single diastereomers (Scheme 6) [23].

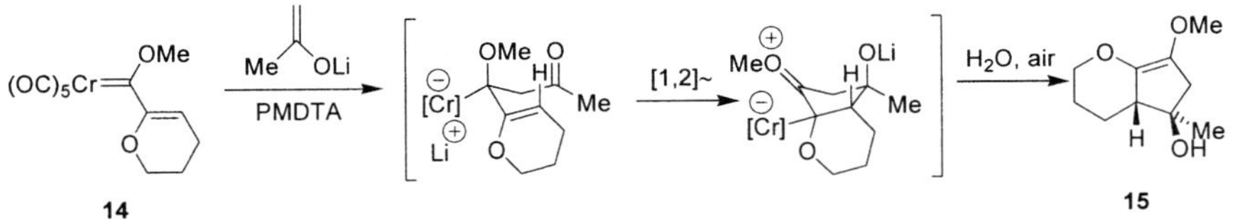

Scheme 6 1,2-Addition of lithium enolates (PMDTA = N,N,N',N',N''-Pentamethyldiethylenetriamine).

Alkenylcarbene complexes bearing an acyclic C=C bond afford diastereomerically pure cycloheptenone derivatives. The formation of these products involves two key steps: 1,2-addition of the lithium enolate to the carbene complex is followed by cyclization initiated by a [1, 2] shift of the pentacarbonyl chromium fragment.

Substituted enolate anions add to the alkenylcarbene complexes in a 1,4-fashion. The Michael addition reactions of ketone and ester enolates to enantiomerically pure [(–)-8-phenylmenthyloxy](alkenyl)carbene complexes of chromium proceed with high levels of asymmetric induction and high *syn* diastereoselectivity when substituted lithium enolates are involved [22, 24]. Complementarily, because of a rigid substrate conformation, a high *syn* selectivity has been achieved in the addition of related ketone (*E*)-lithium enolates to carbonyl-chelated imidazolidinone chromium vinylcarbene complexes [25]. The formation of Michael adducts with either high *anti* or *syn* selectivity in the reaction of lithium enolates of achiral *N*-protected glycine esters with the chiral alkoxy(alkenyl)carbene chromium complex **12** can be controlled by the choice of the nitrogen-protecting group to yield **16** and **17**, respectively (Scheme 7) [26].

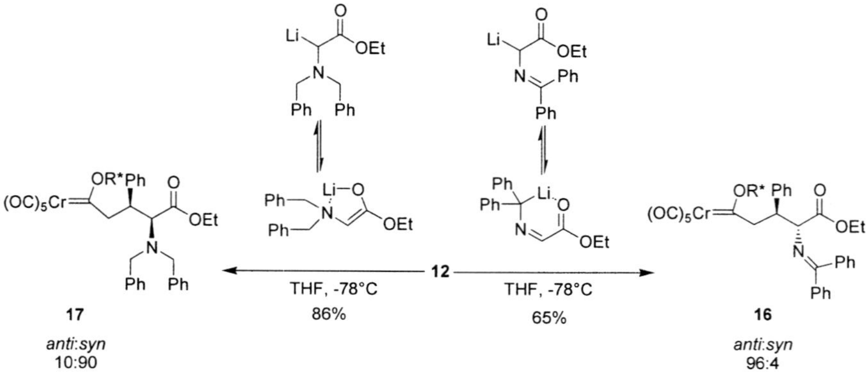

Scheme 7 1,4-Addition of lithium enolates.

This type of reaction represents a novel synthetic approach to racemic and optically enriched β-substituted glutamic or pyroglutamic acids, which are formed after sequential deprotection, oxidation, and hydrolysis of the initial 1,4-adducts.

The reaction of alkenylcarbene complexes with weaker nucleophiles like enamines is described in Section 3.1.4. A novel synthetic route to functionalized bicyclo[3.2.1]octane and bicyclo[3.3.1]nonane skeletons with high diastereo- and enantioselectivity comprises an α,β,β'-annulation reaction of cyclopentanone enamine and alkenylcarbene complexes that is unprecedented in organometallic chemistry [27].

3.1.3
Metal-Centered Reactions

3.1.3.1
[3+2+1] Benzannulation

The [3+2+1] benzannulation of α,β-unsaturated Fischer carbene complexes with alkynes provides one of the most powerful tools to generate densely substituted benzenoid compounds [28]. Within the synthesis of benchrotrenes, the concept of atom economy is convincingly preserved, as this type of reaction represents a highly efficient one-pot procedure. The formal [3+2+1]-cycloaddition involves an α,β-unsaturated carbene ligand (C_3-synthon) 18, an alkyne (C_2-synthon) and a carbonyl ligand (C_1-synthon) and takes place within the coordination sphere of the chromium(0), which acts as a metal template (Scheme 8). The usually obtained 4-methoxyphenol derivative 19 remains coordinated to the $Cr(CO)_3$ fragment, which may be exploited in an activation of the arene toward subsequent transformations. The generally accepted mechanism for this reaction as based on experimental and kinetic studies is consistent with more recent theoretical calculations and is depicted in Scheme 8 [29]. The first and rate-determining step involves a thermal

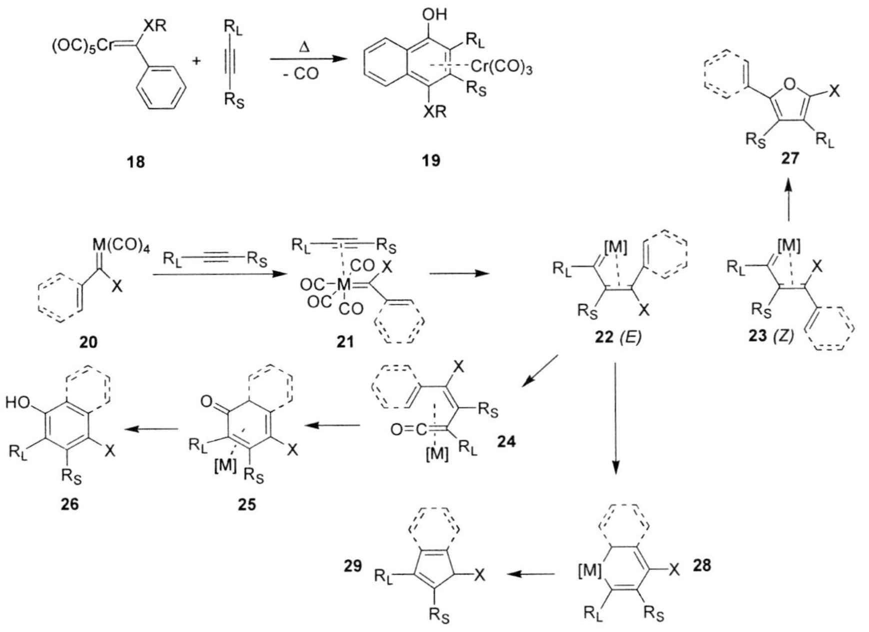

Scheme 8 Mechanism of the benzannulation reaction.

decarbonylation of the starting pentacarbonyl carbene complex to yield the coordinatively unsaturated tetracarbonyl carbene complex **20** [30]. A $\eta^1:\eta^3$-vinylcarbene complex analog **30** corresponding to this reaction intermediate has been isolated (Scheme 9) [31]. This coordinatively unsaturated intermediate is trapped by the alkyne to yield **21**. A structural analog **31** displaying an intramolecular alkyne coordination has been characterized by X-ray analysis [32]. The subsequent insertion of the alkyne into the metal-carbene bond occurs with high regioselectivity and affords the η^3-vinylcarbene complex **22/23** [33]. A related species **32** has been isolated from the reaction of an aminocarbene complex [31b]. Two different isomers of the η^3-metallatriene may be formed, of which only the (E)-isomer **22** is able to undergo an insertion of a carbonyl ligand to generate the s-cis η^4-vinylketene **24**. The enaminoketene complex **33** and the silyl vinylketene **34** have been synthe-

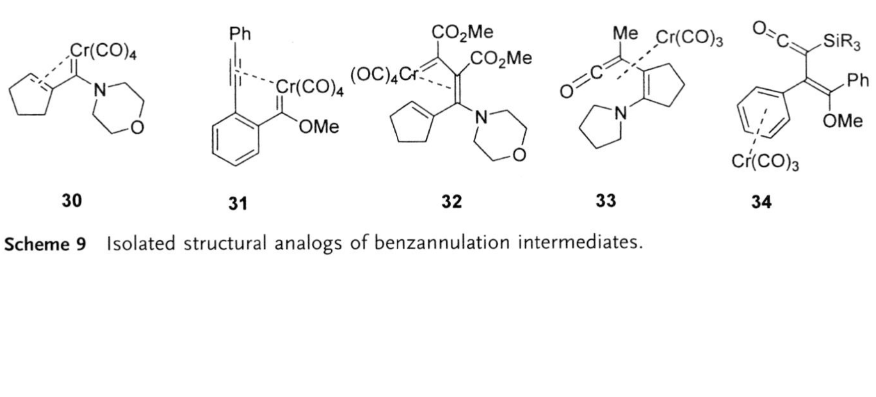

Scheme 9 Isolated structural analogs of benzannulation intermediates.

sized as structural analogs of this type of intermediate [34, 35]. The final electrocyclic ring closure affords a cyclohexadienone complex **25** which tautomerizes to give the naphthol complex **26**. Cyclohexadienone complex **36** missing a hydrogen atom for tautomerization has been obtained in the annulation of the highly functionalized carbene complex **35** with 1-pentyne (Scheme 10) [36]. If the reaction time is prolonged to 48 h, aromatization *via* a [1,5] sigmatropic rearrangement has been observed to yield **37** [35].

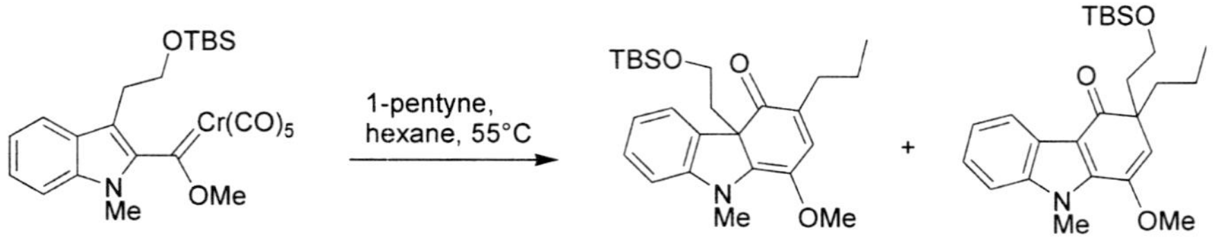

35	**36**	**37**
	t=1.5h 53%	0%
	t=48h 0%	38%

Scheme 10 [1,5] Sigmatropic rearrangement of cyclohexadienone.

The stability of the annulation product is enhanced by *in situ* protection of the phenolic group (e.g., silylation) or by subsequent oxidative demetalation of the chromium(0) fragment [37]. Because of further oxidation, this route often generates the corresponding quinones.

The observed chemoselectivity depends on the nature of the metal template [38], the carbene substitution pattern, and the reaction conditions (e.g., solvent, alkyne concentration, temperature) [39]. The highest selectivity for benzannulation was found for chromium alkoxy(aryl or alkenyl)carbene complexes reacted in donor solvents like ethers or benzene. Carbene complexes containing molybdenum and tungsten favor competing pathways that lead to indenes **29** and furanes **27**. The (Z)-isomer of η^3-vinylcarbene complex intermediate **23** is responsible for the formation of the furan skeleton [40]. Substitution of the alkoxycarbene substituent for a better electron-releasing amino group results in exclusive cyclopentannulation to yield **29** because of the thermal stability of the metal-carbonyl bond, which hampers the primary decarbonylation and impedes the CO incorporation into the final product [41]. Strongly coordinating solvents like acetonitrile may even lead to the formation of cyclobutenones, albeit in moderate yields [39]. This competing pathway is also realized when the η^4-vinylketene complex **24** undergoes a 4π-electrocyclization. The benzannulation affords η^6-arene Cr(CO)$_3$ complexes, which contain a plane of chirality due to the unsymmetric arene substitution pattern and therefore can serve as powerful reagents in stereoselective synthesis and asymmetric catalysis [42].

Since the resolution of planar chiral arene chromium complexes can be rather tedious, diastereoselective approaches toward optically pure planar chiral products appear highly promising. Different general strategies in order to achieve a diastereoselective benzannulation have been devised and differ only in the respective attachment of the chiral auxiliary [43]. For the use of achiral Fischer carbene complexes, enantiopure alkynes have been employed. For example, propargylic alcohols allow for diastereoselective benzannulation [44]. On the other hand, a low-temperature benzannulation could be achieved for the reaction of chiral pool-derived 2-alkynylglucose derivative **38** with the non-stabilized diphenyl carbene chromium complex to yield **39**, although the diastereoselectivity remained rather low (Scheme 11) [45]. The most general approach to diastereoselective benzannulation relies on chiral alkoxy auxiliaries, which are readily available from the terpene or carbohydrate pool and can be incorporated into the carbene ligand *via* the common acylation-alcoholysis sequence. The best performance (80% *de*) within a series of terpenoid alcohols was observed for (–)- and (+)-menthol (Scheme 11) [46, 47].

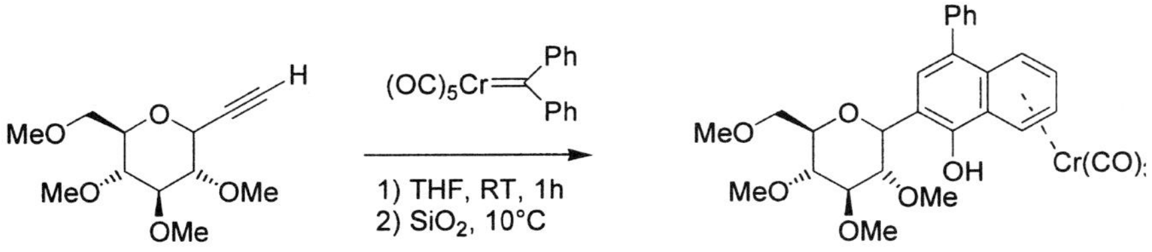

38 **39** (30%, 1.25:1 d.r.)

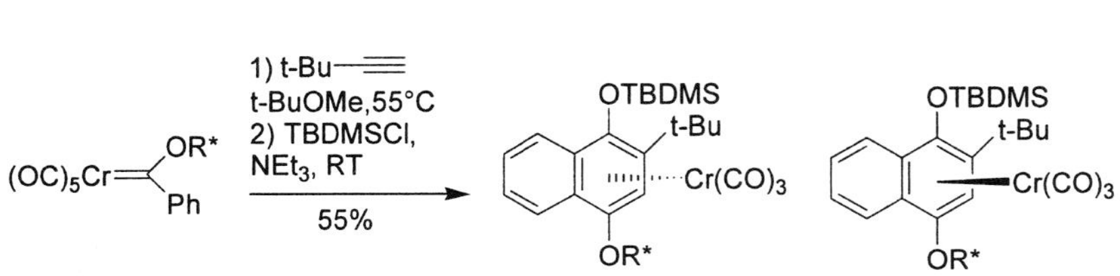

40 **41** **42**

	41		42	
R*=(-)-menthyl	10	:	1	82% de
R*=(+)-menthyl	1	:	9.2	80% de

Scheme 11 Diastereoselective benzannulation.

When the benzannulation is carried out with unsymmetrical alkynes, the major regioisomer generally bears the larger alkyne substituent next to the phenolic group, suggesting that the regioselectivity is mainly governed by the difference in steric demands of the two alkyne substituents [48]. A reversal of this regioselectivity may be achieved either by an intramolecular version of the benzannulation, where the alkyne functionality is incorporated in the alkoxy chain [49], or by the use of stannyl acetylenes [50] and alkynylboranes [51]. The benzannulation of the Fischer carbene complex **43** with alkynylboronates of varying size results in the

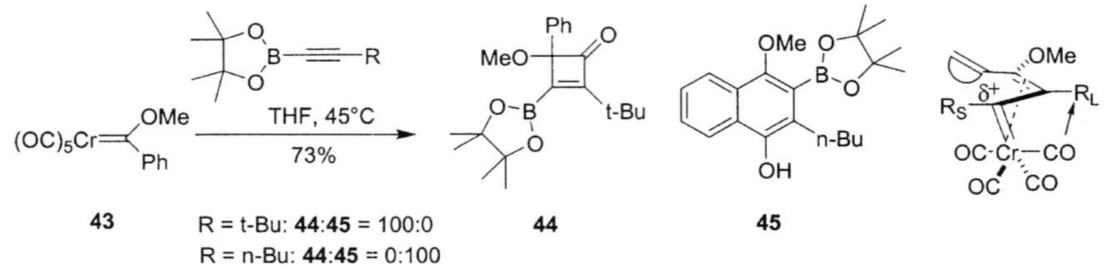

Scheme 12 Benzannulation with inverse regioselectivity.

formation of naphthylboronic esters **45** of inverse regiochemistry. However, sterically hindered alkynylboranes afforded only cyclobutenones **44** (Scheme 12). The installation of the electron-withdrawing boronate moiety far away from the electrophilic carbene center and a Lewis acid/base interaction $[CO \rightarrow B(OR)_2]$ in the η^3-metallatriene intermediate offer two possible explanations based on electronic criteria for this inversion in regioselectivity.

An unexpected varying regiochemistry in intramolecular benzannulation has also been observed in the synthesis of cyclophanes [52].

With regard to the construction of diaryl ethers in synthetic endeavors toward complex natural products bearing sensitive functional groups, the benzannulation offers a potentially attractive method. The formation of diaryl ethers from the coupling of alkynes with aryloxy α,β-unsaturated carbene complexes has been convincingly demonstrated [53]. An unconventional strategy for biaryl synthesis also relies upon the benzannulation of Fischer carbene complexes with aryl acetylenes [54]. Here, the main idea consists in forming the biaryl bond before constructing the final arene ring. A series of substituted aryl acetylenes were examined to determine possible steric and electronic effects. Methyl, methoxy, chloro and N-amide substituents in 2-position give moderate to good yields of products, whereas carbonyl derivatives and the nitro group are deleterious. The chromium tricarbonyl complexes stabilized by protective silylation of the phenol function contain two different stereogenic elements – planar and axial chirality – as demonstrated for **47** and **48** (Scheme 13) [55]. The diastereomeric ratio of both atropisomers can be controlled by the reaction conditions. An *in situ* silylation (method **A**) affords an 89:11 mixture of both isomers. On the other hand, if the silylation is performed

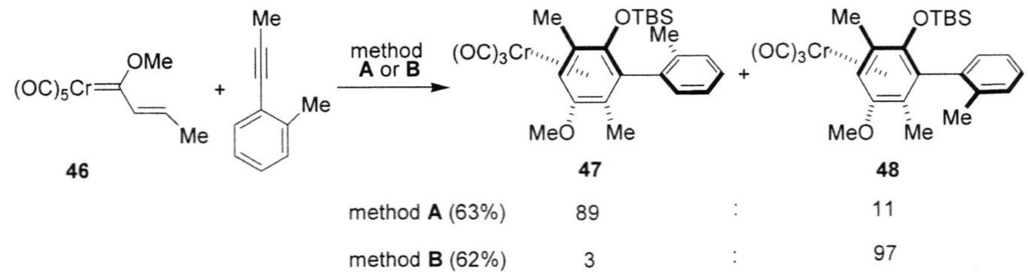

Scheme 13 Concomitant generation of planar and axial chirality.

in a separate step after the benzannulation has been completed (method **B**), a 97:3 ratio is obtained in favor of the other diastereomer.

The use of a stereogenic carbon center allowed for an efficient asymmetric induction in the benzannulation reaction toward axially chiral intermediates in the synthesis of configurationally stable ring C functionalized derivatives of allocolchicinoids [56]. The benzannulation of carbene complex **49** with 1-pentyne followed by oxidative demetalation led to the isolation of the single diastereomer **50** (Scheme 14).

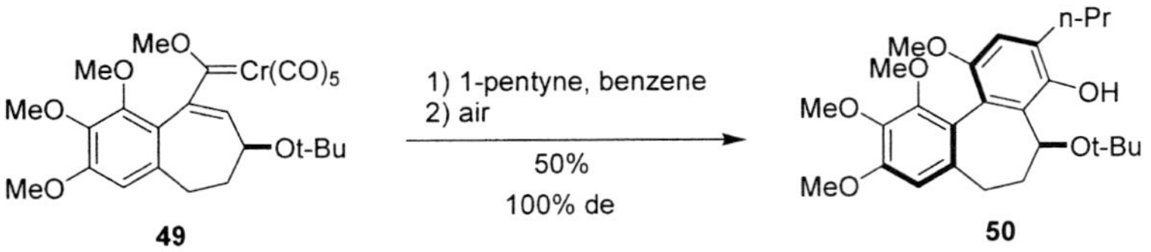

Scheme 14 Diastereoselective benzannulation.

A bidirectional benzannulation of the axial chiral biscarbene complex **51** affords a bis-$Cr(CO)_3$-coordinated biphenanthrene derivative **52**, which combines elements of axial and planar chirality (Scheme 15) [57]. In moderate diastereoselectivity, four diastereomers are formed, which can be converted to a single C_2-symmetric bisquinone **53** upon oxidative workup. In order to introduce a distinct chiral element, a novel type of C_2-symmetric biscarbene complex bearing a silyl-bridged binaphthol moiety was synthesized [58]. A double benzannulation reaction affords upon oxidation a bisquinone as a major product characterized by a combination of axial and helical chirality.

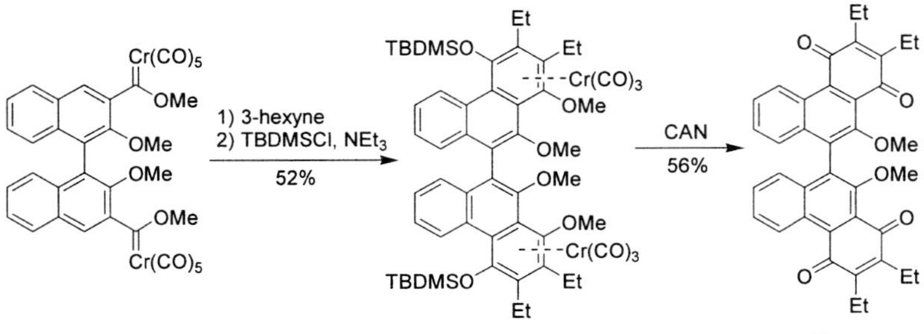

Scheme 15 Bidirectional benzannulation of an axial chiral biscarbene complex.

While the benzannulation reaction normally creates an angular annulation pattern, even in cases where an *ortho*-substitution has been applied to force the annulation into a linear pathway, a surprising linear benzannulation has been observed for the dibenzofurylcarbene complex **54** (Scheme 16) [59]. Chromatographic work-

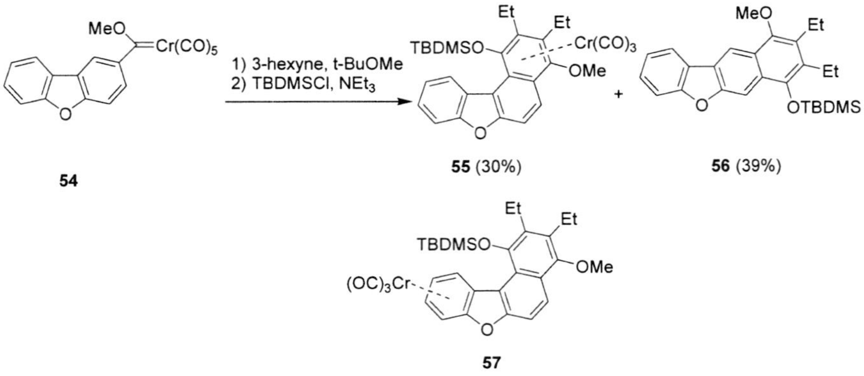

Scheme 16 Synthesis of benzonaphthofurans *via* benzannulation.

up of the reaction products afforded the uncoordinated benzo[b]naphtho[2,3-d]furan **56** as an unprecedented linear benzannulation product along with the expected angular benzonaphthofuran Cr(CO)$_3$ complex **55**. The kinetic benzo[b]-naphtho[1,2-d]furan benzannulation product undergoes a thermally induced haptotropic metal migration along which the metal is shifted to the opposite terminal benzene ring to yield **57** [59]. The molecular structures of **55**, **56** and the haptotropic rearrangement product **57** have been established by X-ray analysis.

In a more ambitious work, a tetradirectional benzannulation of a tetrakis(alkoxy)carbene complex with 3-hexyne led, after demetalation, to the corresponding tetranaphthoxy derivative in high yield [60]. This type of reaction is diastereounselective and leads to a mixture of seven Cr(CO)$_3$ complex isomers.

The benzannulation reaction serves as an important synthetic tool in the preparation of complex natural products, for example, in the synthesis of highly oxygenated polycyclic aromatic cores. Earlier applications to natural product synthesis concentrated on vitamins [61], steroids [62], and antibiotics [63]. More recently, the successful completion of the convergent synthetic steps in the total synthesis of deoxyfrenolicin [64], menogaril [65] and olivin [66] were reported.

In an extension to the usual reaction pathway for alkenylcarbene complexes that react with alkynes chemoselectively to the six-membered benzannulation product, dienylcarbene complexes **58** undergo a higher order reaction. Their reaction with alkynes leads exclusively to the formation of eight-membered carbocycles **59** (Scheme 17) [67]. Nevertheless, the reaction can be regarded as a Dötz benzannulation analog, as the major key steps sequentially involve the insertion of the alkyne and of a carbon monoxide molecule. The main difference is the subsequent eight-electron-cyclization of the trienyl-ketene complex intermediate to yield a Cr(CO)$_3$ complex, which, after loss of the metal moiety, releases the trienylone product **59**.

If dienylcarbene complexes are reacted with isocyanides, *o*-alkoxyanilines are obtained from the electrocyclic ring closure of the initially formed ketenimine intermediate [68]. This reaction has been successfully applied to the preparation of analogs of indolocarbazole natural products [69] and to the total synthesis of calphostins [70].

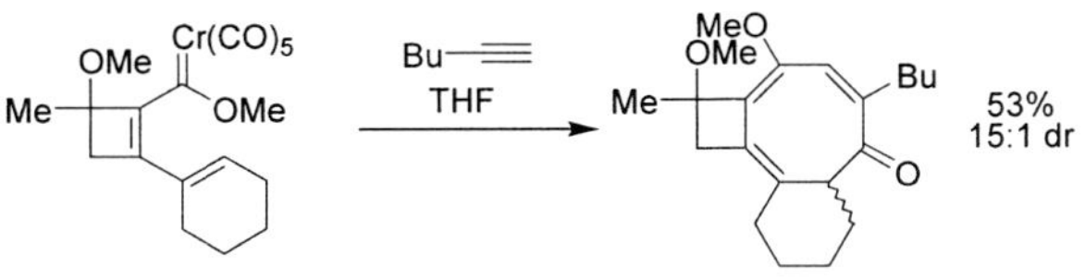

Scheme 17 Synthesis of cyclooctatrienones.

3.1.3.2
Cyclopropanation Reactions

The transfer of the carbene ligand to polarized alkenes affords cyclopropanes. Thus, alkenes bearing either electron-withdrawing [71, 72] or electron-donating [73, 74] groups undergo a formal [2+1] cycloaddition with heteroatom-stabilized metal carbene complexes under thermal conditions in an intermolecular fashion. The reaction with electron-rich olefins such as enol ethers must be carried out under high CO pressure in order to avoid a competing olefin metathesis process. On the other hand, a rapid decarbonylation pre-equilibrium followed by coordination of the alkene to the unsaturated tetracarbonyl intermediate is characteristic of the cyclopropanation reaction with electron-deficient olefins (Scheme 18). The olefin-tetracarbonyl-carbene complex leads, either *via* a discrete metallacyclobutane or *via* direct carbene transfer, to the corresponding cyclopropane products with different diastereoisomeric excesses. An interesting example is the [2+1] cycloaddition of the glycosylidene carbene complex **60** with the acrylate, which yields the spirocyclopropane **61** as a single diastereoisomer (Scheme 18) [75].

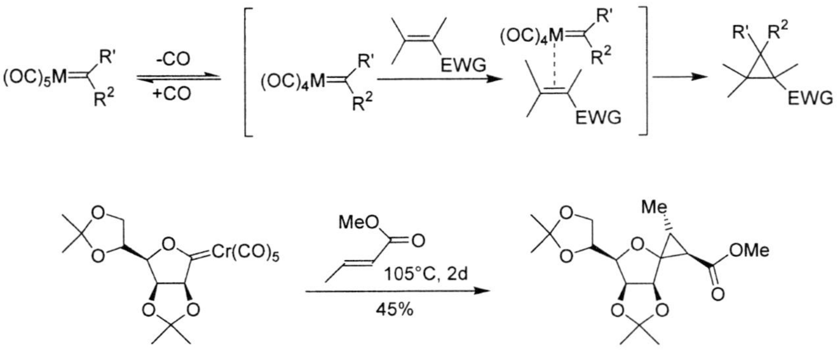

Scheme 18 Cyclopropanation of electron-deficient alkenes.

The cyclopropanation of non-functionalized and therefore electronically neutral alkenes normally requires a strongly electrophilic metal carbene complex such as non-heteroatom-stabilized metal carbene complexes or cationic iron carbene complexes [76]. However, a heteroatom-stabilized 2-ferrocenyl-alkenyl Fischer carbene complex **62** has recently been reacted with 1-hexene to afford the corresponding vinylcyclopropane **63** in high yield and with 97% *de* (Scheme 19) [77]. A series of electronically neutral alkenes were consequently examined in the cyclopropanation reaction with (2-phenyl)- and (2-ferrocenyl)alkenyl carbene complexes with regard to scope and limitation [78]. A high degree of diastereoselectivity and moderate to good yields were achieved with either terminal or acyclic and cyclic 1,2-disubstituted simple olefins. The mechanism of this thermal intermolecular cyclopropanation reaction with electronically neutral alkenes is assumed to proceed *via* the reaction mechanism established for the related cyclopropanation of electron-deficient olefins. The major diastereomer in each case was that in which the alkenyl group is *cis* with respect to the vicinal cyclopropane methine proton.

Different results were obtained in the cyclopropanation reaction of 2-haloalkoxy(alkenyl)carbene chromium complexes **64** with simple alkenes (Scheme 19) [79]. The reaction of **64** with cyclopentene gave the single diastereoisomer **65**, which, upon reaction with *n*-butyllithium, released the free hydroxyl derivative **66**. In case of the chloro compound, a subsequent lithiation resulted in a 5-*exo*-trig-formation of a three-/five-membered spiroheterocycle.

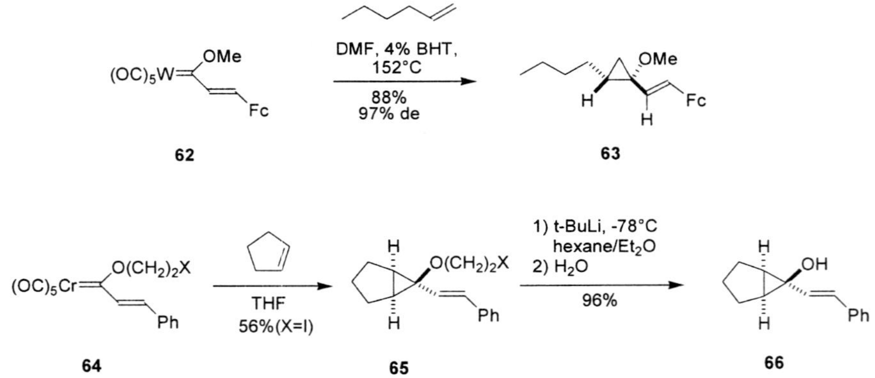

Scheme 19 Cyclopropanation of electronically neutral alkenes (BHT = 2,6-di-*tert*-butyl-4-methylphenol).

The idea of performing the cyclopropanation reaction with Fischer carbene complexes not only in a diastereoselective but also in an enantioselective fashion was first envisaged in 1973 [80]. A chiral monophosphine attached to the chromium center in such a way led only to a very low enantiomeric excess of the cyclopropane. A conceptually novel approach was recently devised by employing Fischer carbene complexes bearing chiral metal centers, which resulted from the exchange of the two CO ligands for an enantiopure, CO-emulating bidentate phosphite ligand [81]. The diastereoselectivities achieved in the cyclopropanation reaction of the racemic chiral-at-metal molybdenum and chromium complexes with acceptor-substituted and non-activated alkenes were in all cases significantly higher than those reported from the series with the respective pentacarbonyl complexes. Unfortunately, enantioselective studies reveal a rather modest selectivity regarding face distinction. A much more successful strategy is based on alkenyl oxazolines used as chiral auxiliaries in electron acceptor-substituted alkenes (Scheme 20) [82]. Preliminary studies involving the phenyl carbene complex **67** and the achiral alkenyl oxazoline **69** revealed not only an excellent yield (89%) but also a superb diastereomeric ratio (*trans*:*cis*=>97:<3) for the cyclopropane **71**. The extension of the reaction to the chiral non-racemic alkenyl oxazoline **70** was developed to the first diastereo- and enantioselective cyclopropanation with a heteroatom-stabilized group 6 Fischer carbene complex. The asymmetric induction for the major *trans* diastereomer **72** is extraordinarily high (98%), and, moreover, a convincing *cis*/*trans* ratio as high as 94:6 was observed.

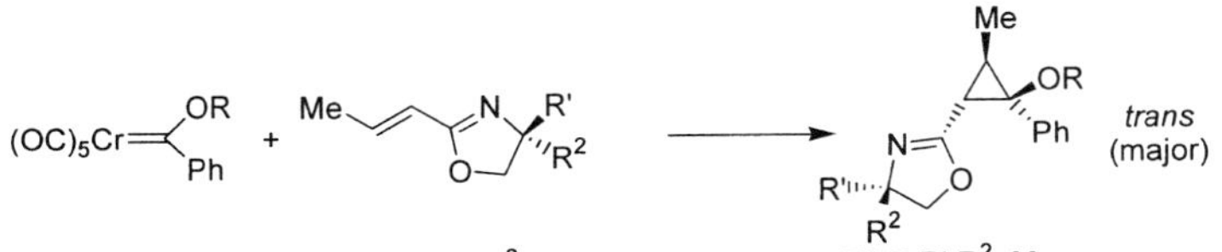

67 R = Me
68 R = i-Pr

69 R',R^2=Me
70 R' = t-Bu, R^2=H

71 R,R',R^2=Me
72 R=i-Pr, R'=t-Bu, R^2=H

Scheme 20 Diastereoselective cyclopropanation.

The cyclopropanation protocol can be applied to 1,3-dienes as well [83]. Electron-poor 1,3-dienes generally fail to react with α,β-unsaturated carbene complexes, except with the methoxy(phenyl)carbene complex, which has been applied to the highly regio- and stereoselective synthesis of vinylcyclopropane derivatives [84]. To this end, thermal reactions of these α,β-unsaturated carbene complexes with 1,3-dienes such as silyloxydiene derivatives reveal a competition between direct cyclopropanation and subsequent Cope-type [3,3] sigmatropic rearrangement, which results in either formation of cycloheptadiene or in direct 5-membered ring formation [85]. Of particular interest is the cyclopropanation reaction of transition metal carbene complexes with 1,3-diene pentafulvenes **73** (Scheme 21) [86]. Not only was a virtually total control of diastereoselectivity in favor of the *endo* cycload-

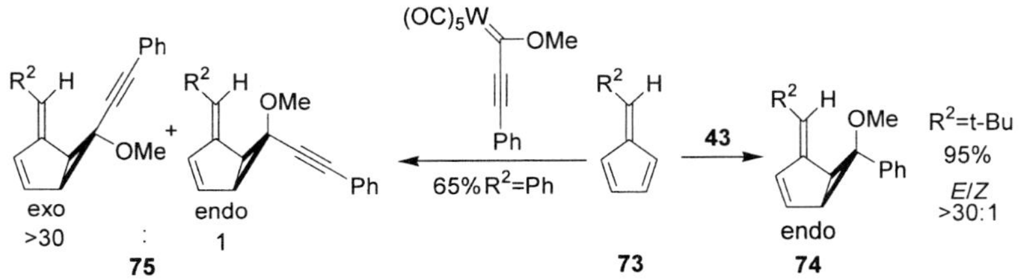

75 **73** **74**

Scheme 21 Cyclopropanation of fulvenes.

duct **74** achieved with chromium alkoxycarbene **43**, but it also was an extension of the cyclopropanation reaction to Fischer alkynyl carbene complexes for the first time. In contrast to simple carbene complexes, the alkynyl complexes led preferentially to the formation of the *exo* diastereoisomer **75**. The control of the *exo/endo* diastereoselectivity is dictated by the metal-OMe or metal-alkynyl coordination, respectively, in the relevant intermediate before reductive elimination of the metal fragment.

A versatile and diastereoselective method for the synthesis of cyclopropyl fused γ-lactones has recently been developed [87].

3.1.3.3
Photoinduced Reactions of Carbene Complexes

Irradiation into the metal-to-ligand charge transfer absorption band of Fischer carbene complexes (350–450 nm) results in the insertion of a *cis*-CO ligand into the Cr=C bond and subsequent formation of a tetracarbonylchromium(0)-coordinated ketene (Scheme 22) [88]. This sequence has been the basis for the design of numerous transformations.

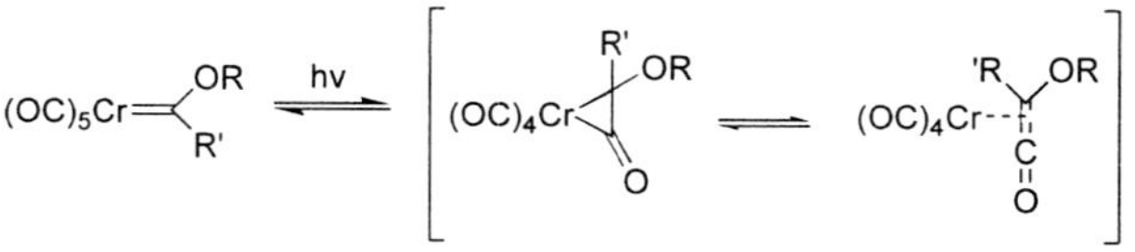

Scheme 22 Photoinduced carbene-CO coupling.

One synthetic aspect of metal carbene photochemistry refers to [2+2] cycloaddition reactions of the ketene intermediate with imines or alkenes to give β-lactams or cyclobutanones, respectively. Bis-carbene complex **76** reacts with an achiral N-protected imidazoline to yield the carbon-linked β-lactam bis(azapenam) as an unselective 1:1 mixture of diastereomers (Scheme 23) [89]. Irradiation of optically active imidazoline **77** in the presence of biscarbene **76** followed by deprotection of the Cbz group gave the optically active bis(azapenam) **78** as a single diastereomer. This approach has been extended to the corresponding tri- and tetra(ethylene glycol) linked bis-carbene complexes [90].

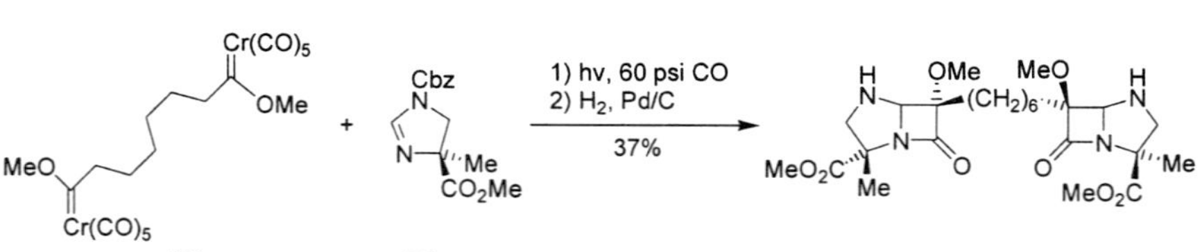

Scheme 23 Asymmetric synthesis of bis(azapenam).

A photochemical [2+2]-cycloaddition reaction between a chiral secondary aminocarbene complex and an amino acid-derived imine is the key step in the synthesis of the biologically active β-lactam 1-carbacephalothin [91]. A detailed review on synthetic aspects of chromium carbene photochemistry has been published recently [1i].

The behavior of chromium carbene complexes bearing additional ligands was studied in context with their photocarbonylation in the presence of imines (Scheme 24) [92]. The pentacarbonyl alkoxycarbene complex 79 reacts with imine 82 in a [2+2] cycloaddition to yield β-lactam 83 as a *cis/trans*-mixture. If one CO ligand is replaced by a strong σ-donor such as Bu$_3$P in 80 the photocarbonylation is inhibited. On the other hand, a soft σ-donor such as Ph$_3$P in 81 is compatible with the lactam formation. Additional theoretical calculations suggest that the HOMO-LUMO gap for complexes featuring an electron-rich phosphine is widened, which suppresses conversion to the vinylketene complex.

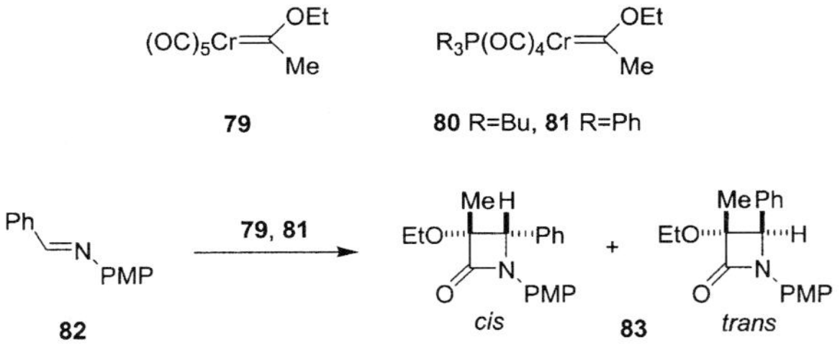

Scheme 24 Synthesis of β-lactams.

A broad range of substituted cyclobutanones have been synthesized *via* photolysis of alkoxycarbene complex 84 with a variety of electron-rich alkenes (Scheme 25) [93]. As electron-deficient alkenes are reluctant to add to ketenes, β-acceptor-substituted cyclobutanones may be generated from β-donor-substituted cyclobutanones by subsequent transformation. The presence of a chiral oxazolidinone auxiliary in the alkene 85 allows for a highly diastereoselective reaction [94]. These functionalized, optically active cyclobutenones 86 have been utilized, among others, as key intermediates in the synthesis of carbocyclic nucleoside analogs and aminocyclitols [95].

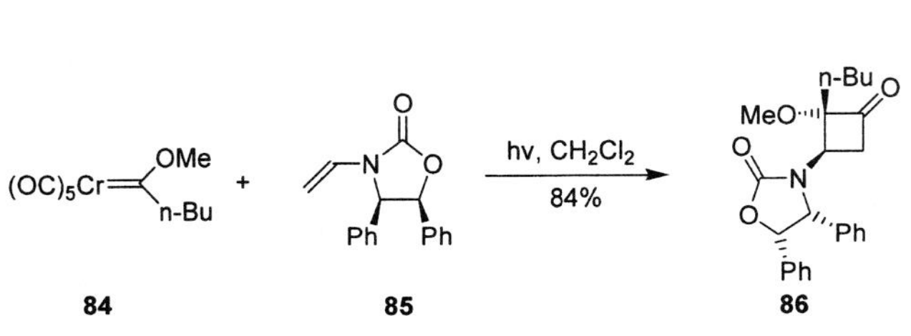

Scheme 25 Synthesis of cyclobutanones.

Unexpected results were obtained from the photochemical reaction of chromium imino-carbene complexes: upon irradiation, 3-aza-1-chroma-1,3-butadiene undergoes a [3+2] cyclopentannulation with alkenes and alkynes affording 1-pyrroline and 2H-pyrrole derivatives [96]. The photochemical reaction in the presence of a nitrogen-containing double bond leads exclusively to the formation of azadienes, suggesting a metathesis process [97].

Another application of photogenerated metal-coordinated ketenes is based on the addition of protic nucleophiles and has been exploited in the synthesis of amino acids and peptides [98].

The chiral oxazolidine auxiliary in aminocarbene complex **88**, successfully applied to asymmetric β-lactam formation, also facilitates an enantioselective synthesis of amino acids **89** (Scheme 26). Since both enantiomers of the auxiliary may be obtained from the corresponding phenyl glycine enantiomers, both natural (S) and non-natural (R) amino acid esters are accessible *via* this route.

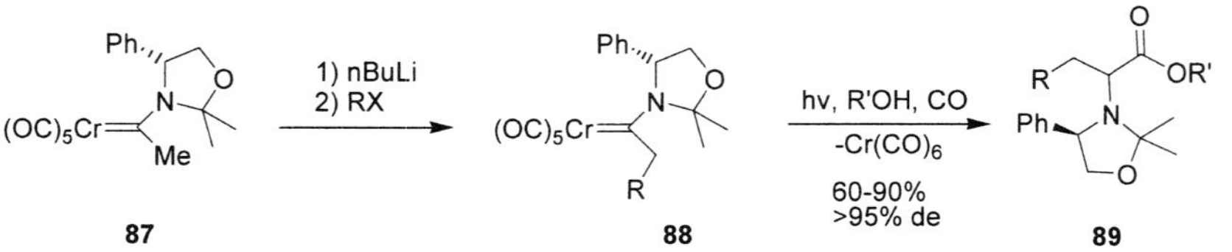

Scheme 26 Asymmetric synthesis of amino acid derivatives.

Appropriate photoconditions generate the analog ketene equivalents from iminopyranosylidene complexes [99]. The intermediate generated from the imino-D-*ribo*-pyranosylidene complex **90** was trapped with methanol with complete β-stereoselectivity to give the homologous imino aldonic ester **91** in high yield (70%) (Scheme 27).

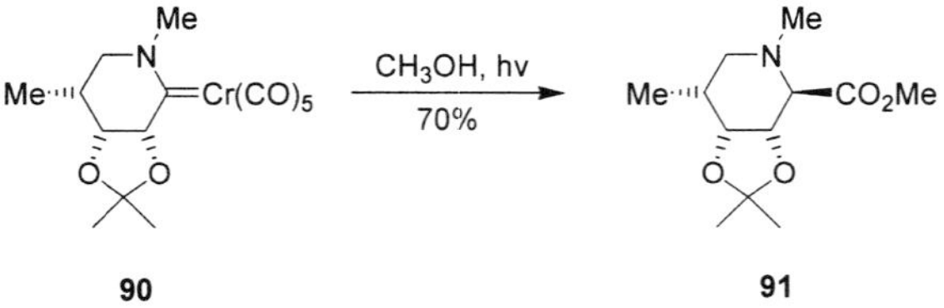

Scheme 27 Photoinduced C-glycosidation.

3.1.4
Synthesis of Five-Membered Carbocycles

α,β-Unsaturated Fischer carbene complexes turned out to be potent dipolarophiles in 1,3-dipolar cycloadditions. They have been exploited in the synthesis of enantiomerically pure Δ²-pyrazolines, which are of interest because of their biological activity, their physical applications, and also their use as starting materials for further transformations. The [3+2] cycloaddition of the chiral non-racemic (–)-8-phenylmenthol-derived alkenyl carbene complex **12** with the *in situ* generated 1,3-dipole nitrilimine followed by one-pot oxidation with PNO afforded the enantiomerically pure Δ²-pyrazoline **92** with high regio- (>95:5) and diastereoselectivity (92:8 *dr*) (Scheme 28) [100]. The absolute stereochemistry of the product was suggested to be (4R,5S) on the basis that in the reactive conformation of **12** the phenyl group on the chiral auxiliary shields the top face of the double bond upper face by π,π-orbital overlap, inducing the dipole to attack selectively from the (*Si,Si*)-bottom face. Enantiomerically pure 3-alkoxycarbonyl-Δ²-pyrazolines **93** have been prepared in a one-pot procedure including a [3+2] cycloaddition using complex **12** and trimethylsilyldiazomethane in the key step [101].

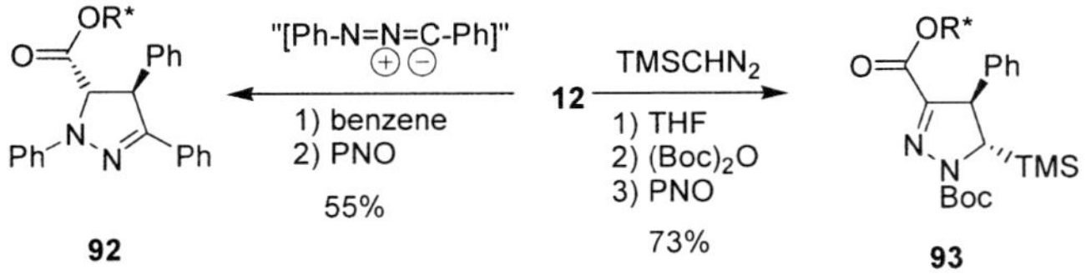

Scheme 28 Stereoselective synthesis of Δ²-pyrazoline derivatives (PNO = pyridine *N*-oxide)

The 1,3-dipolar cycloaddition of complex **12** with the *in situ*-generated functionalized azomethine ylide **94** gives the tetracarbonyl cycloadduct **95** as a single regioisomer in which the metal fragment is stabilized by coordination of sulfur to the metal center which controls the regioselectivity of the reaction (Scheme 29) [102]. The [3+2] cycloaddition is highly diastereoselective because of the crucial influence of the chiral auxiliary on the incoming dipole. The usefulness of this methodology for the synthesis of pharmaceutically interesting pyrrolidinone derivatives has been demonstrated by the synthesis of (+)-rolipram. Another application of the chiral non-racemic carbene complex **12** as a dipolarophile has been reported for the reaction with the formal azomethine ylide dipole **96** (Scheme 29) [103]. The aldimine glycine

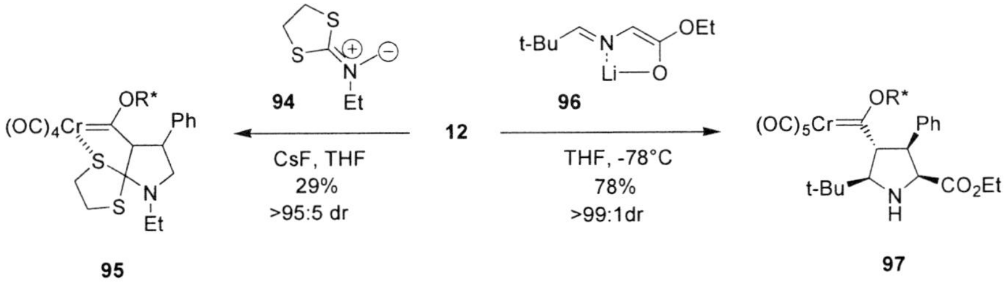

Scheme 29 Stereoselective synthesis of pyrrolidine derivatives.

ester enolate is added in a highly regio-, stereo-, and enantioselective manner, giving the *syn,exo*-cycloadduct **97** as a single regio- and diastereoisomer. The tetrasubstituted pyrrolidine cycloadduct can be converted by successive oxidation and hydrolysis into trisubstituted 4-carboxy prolines with either natural or non-natural configuration. Remarkably, the cycloadduct is not formed by a concerted [3+2] cycloaddition but rather through a stepwise mechanism comprising the addition of the enolate to the *α,β*-unsaturated carbene complex to give the corresponding *syn* Michael adduct followed by a 5-*endo-trig* ring closure.

Cyclopentenones may arise from a [3+2] carbocyclization of alkenylcarbene complexes with enamines [104]. Another example of an asymmetric [3+2] cycloaddition reaction of a Fischer alkenyl carbene complex with an imine is the synthesis of 3-pyrroline derivatives **98** (Scheme 30) [105]. In the course of the experiments, the effect of a Lewis-acid additive was studied. Finally, the [3+2] cycloaddition between the (−)-8-phenylmenthyloxy carbene complex **12** and the imine in the presence of catalytic amounts of Sn(OTf)$_2$ exhibited the highest *trans/cis* selectivity (84:16), while maintaining excellent diastereofacial selectivity (95:5). Subsequent acid hydrolysis afforded the optically pure 2,5-disubstituted-7-pyrrolidinone.

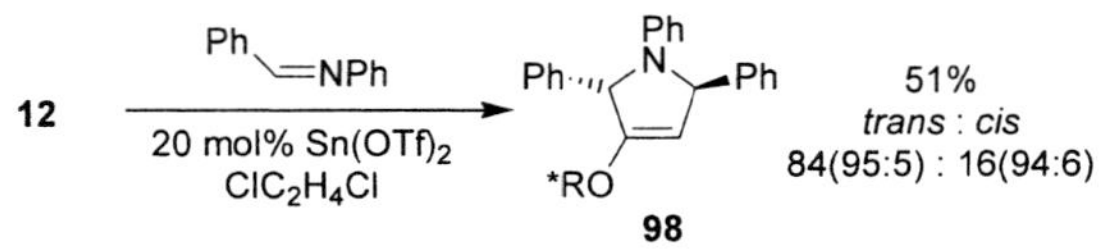

Scheme 30 Asymmetric synthesis of pyrroline derivatives.

Alkylidenecyclopentenones and dialkylidenecyclopentenones are accessible through a coupling of propargylic alcohols and 2-alkyne-1,4-diol derivatives, respectively, with cyclopropylcarbene chromium complexes [106]. The synthesis of 4-alkylidenecyclopentenones can be achieved in either of two ways: (1) intramolecular alkyne-carbene complex coupling when the propargyl oxygen is within the tether or (2) intermolecular coupling involving an internal alkyne when the propargyl oxygen is part of the smaller alkyne substituent. The successful formation of the desired product depends on the effectiveness of the *β*-elimination process at the cyclopentadienide stage and therefore on the leaving-group ability of the propargyl substituent.

Amino substituents provide a strong driving force for the 6π-cyclization of 1-metalla-1,3,5-hexatrienes to cyclopentadienes [107]. *β*-Amino-*α,β*-unsaturated alkoxycarbene complexes **99** react with alkynes to yield 6-amino-1-metalla-1,3,5-hexatrienes **100** (Scheme 31) [108]. These intermediates may give either 5-(1′-dialkylaminoalkylidene)-4-alkoxycyclopent-2-enones **101** via formal [2+2+1] cycloaddition due to a carbonyl insertion, subsequent 1,5-cyclization, and loss of Cr(CO)$_3$, or amino-alkoxy-substituted cyclopentadiene derivatives **102** via formal 6π-cyclization followed by reductive elimination.

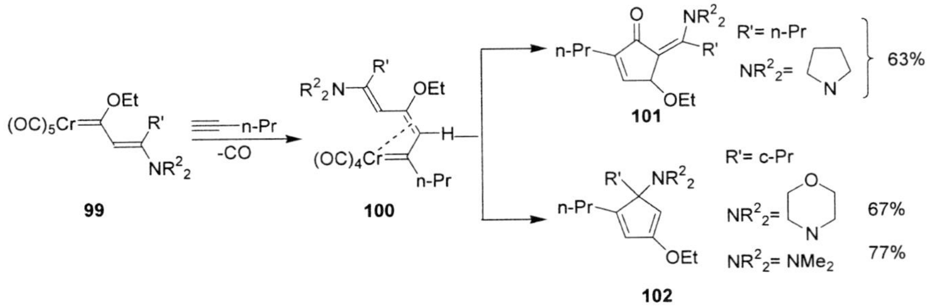

Scheme 31 Coupling of β-amino-α,β-unsaturated alkoxycarbene complexes with alkynes.

The reaction of a silyl-β-aminovinylcarbene complex with an excess of phenylacetylene results in an unexpected spirocyclization to give a mixture of spiro[4.4]nonatriene isomers which differ in the position of one double bond [109].

Another reaction mode of β-amino-α,β-unsaturated alkoxy-carbene complexes in the presence of alkynes and pyridine is inherent to β-cycloalkenyl-substituted β-dialkylaminopropenylidene metal complexes **103**. These 4-amino-1-chroma-1,3,5-hexatrienes apparently undergo a 6π-cyclization more rapidly than alkyne insertion (Scheme 32) [110]. A subsequent reductive elimination forms the cyclohexane-annulated cyclopentadiene, which equilibrates by [1,5]-hydrogen shift, and the resulting intermediate preferentially reacts with the alkyne to afford a single [4+2] cycloadduct **104** in a highly regio- and diastereoselective manner. Trapping the intermediate with different pyran-2-ylidene complexes affords angularly fused tricyclic or steroid-related angularly fused ring systems, respectively, in high chemical yields [111]. If neither an alkyne nor pyridine is present, a spontaneous 1,5-cyclization to the ring-annulated pentacarbonyl-η^1-cyclopentadienyl metal complex **106** is observed [112]. Subsequent heating in pyridine did not liberate the cyclopentadiene ligand of type **104**.

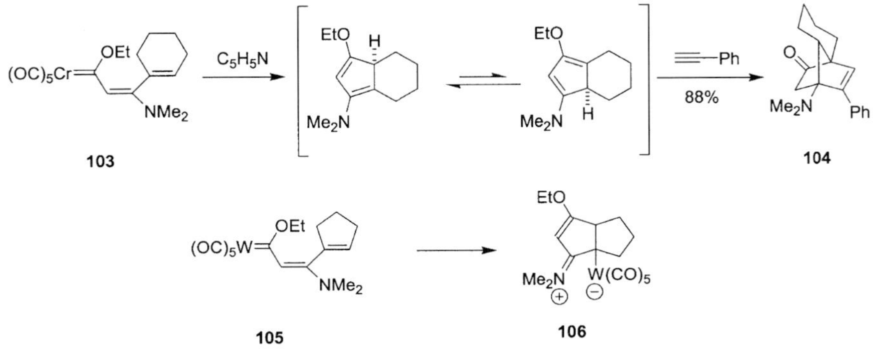

Scheme 32 Coupling of [β-(cycloalkenyl)dialkylaminopropenylidene] metal complexes with alkynes.

3.1 Fischer-Type Carbene Complexes

Isobenzofuran derivatives fused to furans, thiophenes, or benzene can be generated *via* coupling of a methoxy carbene complex with 3-alkynyl-2-heteroaromatic carboxaldehydes or 2-alkynylbenzoyl derivatives, respectively [113]. A subsequent intra- or intermolecular Diels-Alder reaction, leading to oxanorbornene derivatives, occurs if the carbene complex features a remote alkene substituent or if an external dienophile is offered.

Diels-Alder reactions have also been effectuated directly with vinylic or α,β-acetylenic carbene complexes as dienophiles [114]. In order to achieve diastereo- and enantioselective control, metal carbene complexes bearing a chiral metal center or difluoroboroxy carbene complexes have been used [115, 116].

3.1.5
Group 6 Metal Carbenes in Catalytic Carbene Transfer Reactions

Only very recently, group 6 metal carbene complexes have been used as the carbene source in transmetalation reactions. Until then, the transfer of a carbene ligand from a Fischer-type group 6 carbene complex to another metal center had been a rare process [117]. As, for example, the dimerization of chromium-coordinated carbene ligands typically requires temperatures above 130 °C [118], the main idea was to generate a more reactive metal-carbenoid complex intermediate to effect the carbene transfer at temperatures lower than those typically required for group 6 carbene complexes. Therefore, the viability of the inter- and intramolecular dimerization of group 6 alkoxycarbene complexes in the presence of a Pd-catalyst as a promoter was studied, and the experimental results proved the formation of linear and cyclic derivatives at room temperature (Scheme 33) [119]. Whether the catalyst has any influence on the nature of the reaction products depends strongly on the substituent at the carbene complex. For arylcarbene complexes 107, a series of Pd(II)- or Pd(0)-catalysts are equally efficient in promoting the carbene ligand dimerization. In contrast, the nature of the catalyst has a dramatic in-

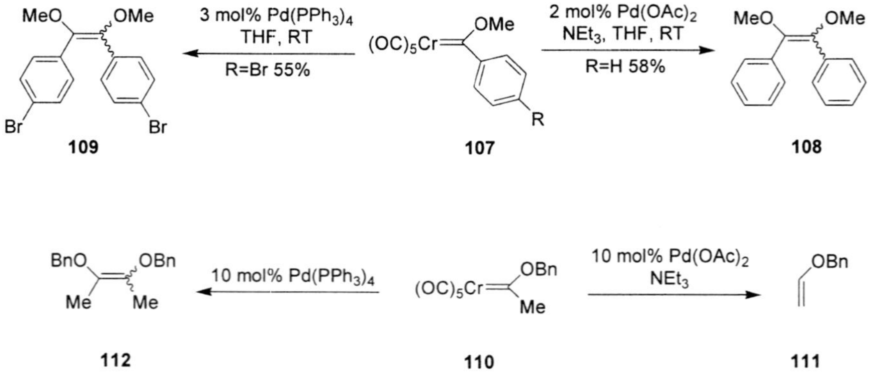

Scheme 33 Chemoselectivity of palladium catalysts in carbene dimerization.

fluence on the dimerization of methylcarbene ligands such as in **110**. A remarkable difference in the chemoselectivity results from the presence of catalytic amounts of Pd(OAc)$_2$/Et$_3$N or Pd(PPh$_3$)$_4$, respectively. Whereas the latter induces specifically the carbene dimerization to give an E/Z-mixture of ene diether **112**, the Pd(OAc)$_2$/Et$_3$N system effects a base-induced hydrogen migration to give enol ether **111**. This transmetalation strategy was successfully applied to the synthesis of conjugated polyenes, endiyne derivatives, and conjugated polyene systems with metal moieties at the terminus of the conjugated system and with cyclic dimerization products. A mechanistic explanation of the results obtained in these reactions is based on the transmetalation of the initial chromium carbene complex **113** onto Pd to form a new Pd-carbene complex **115**, probably through a heterobimetallic cyclopropane intermediate **114**. The nature of the reaction products is defined from the evolution of this Pd carbene intermediate (Scheme 34).

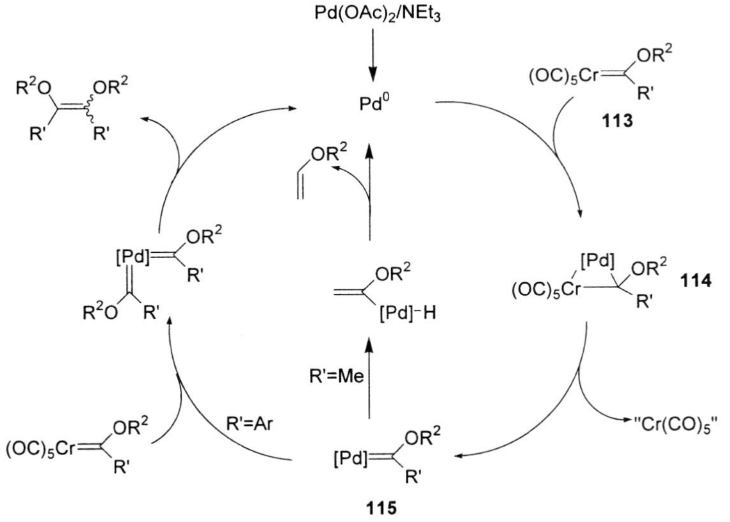

Scheme 34 Catalytic cycle for the transmetalation.

Another type of reaction involving Fischer carbene complexes realized at milder conditions and with higher selectivity due to transmetalation is the cyclization of aminometallahexatrienes to yield vinylcyclopentadienes [120]. Within this context, morpholino derivatives of tungsten and chromium were shown to react smoothly at 20 °C with a variety of terminal alkynes in the presence of 2.5 mol% [(COD)RhCl]$_2$ to give the corresponding vinylcyclopentadiene as a single isomer. On the other hand, vinylcyclopentadienes were obtained under mild conditions by a condensation of 1-alken-3-ynes with 4-amino-1-metalla-1,3-butadienes under catalysis with 2 mol% RhCl$_3 \cdot$ 3H$_2$O in methanol [121].

3.1 Fischer-Type Carbene Complexes

Copper(I) compounds efficiently catalyze reactions of Fischer carbene complexes as well. The formation of a spirocyclic vinylcyclopentadiene from a tungsten carbene complex was found to be promoted by catalytic amounts of CuI (5 mol%) in the presence of NEt$_3$ (8 mol%) [122]. Unfortunately, the replacement of NEt$_3$ by chiral diamines did not lead to any enantioselectivity.

A cross-coupling reaction between Fischer carbene complexes and ethyl diazoacetate providing push-pull alkenes such as **117** was effectuated using 15 mol% CuBr as a catalyst (Scheme 35) [123]. Whereas no stereocontrol of the E/Z-configuration was observed with the achiral methoxycarbene complex, its chiral menthyloxycarbene analog **116b** afforded exclusively the (E)-isomer **117**. When the catalyst was substituted for [Cu(MeCN)$_4$][PF$_6$], an unexpected result was obtained: in the presence of 15 mol% catalyst and EDA only a dimerization of the methoxycarbene ligand took place, with the (E)-isomer **118** favored in a 10:1 ratio. Exposing the chromium (−)-menthyloxycarbene **116b** to 50 mol% [Cu(MeCN)$_4$][PF$_6$] led to the formation of a copper bis-carbene complex **119** and acetonitrile(pentacarbonyl)chromium. The dissolution of this crude product in dichloromethane and diethyl ether was accompanied by a color change and the precipitation of a crystalline substance. These crystals turned out to be the copper(I) carbene complex **120**, and the original chromium carbene complex **116b** was reisolated from the solution. The isolated copper(I) carbene complex is of singular character, since neither the exchange of a carbene ligand for a diethyl ether ligand nor the retransfer of a carbene ligand to the chromium atom has ever been detected. The X-ray structure of **120** provides a rare example of a tricoordinated metal center. Additionally, all copper(I) complexes were characterized by ^{1}H-, ^{13}C-, ^{31}P- and ^{19}F-NMR spectroscopy, demonstrating the existence of a discrete copper(I) carbene complex in solution [124].

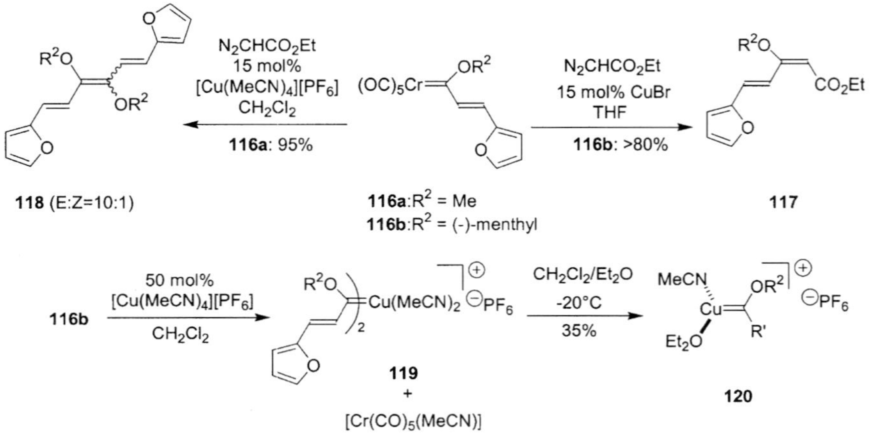

Scheme 35 Synthesis of copper(I) carbene complexes.

3.1.6
References

1. Reviews: (a) K. H. Dötz, H. Fischer, P. Hofmann, F. R. Kreissl, U. Schubert, K. Weiss, *Transition Metal Carbene Complexes*, Verlag Chemie, Weinheim, **1983**. (b) K. H. Dötz, *Angew. Chem.* **1984**, *96*, 573; *Angew. Chem., Int. Ed. Engl.* **1984**, *23*, 587. (c) W. D. Wulff in *Comprehensive Organic Synthesis* (Eds.: B. M. Trost, I. Flemming), Pergamon Press, Oxford, **1991**, Vol. 5, p. 1065. (d) W. D. Wulff in *Comprehensive Organic Synthesis II* (Eds.: E. W. Abel, F. G. A. Stone, G. Wilkinson), Pergamon Press, Oxford, **1995**, Vol. 12, p. 549. (e) L. S. Hegedus in *Comprehensive Organic Synthesis II* (Eds.: E. W. Abel, F. G. A. Stone, G. Wilkinson), Pergamon Press, Oxford, **1995**, Vol. 12, p. 549. (f) D. Harvey, D. M. Sigano, *Chem. Rev.* **1996**, *96*, 271. (g) Y.-T. Wu, A. de Meijere, in *Topics in Organometallic Chemistry* (Ed.: K. H. Dötz), Springer, Heidelberg, **2004**, in press. (h) J. Barluenga, *Pure Appl. Chem.* **1996**, *68*, 543. (i) L. S. Hegedus, in *Topics in Organometallic Chemistry* (Ed.: K. H. Dötz), Springer, Heidelberg, **2004**, in press. (j) R. Aumann, H. Nienhaber, *Adv. Organomet. Chem.* **1997**, *41*, 163. (k) J. Barluenga, *Pure Appl. Chem.* **1999**, *71*, 1385. (l) J. W. Herndon, *Tetrahedron* **2000**, *56*, 1257. (m) A. de Meijere, H. Schirmer, M. Duetsch, *Angew. Chem.* **2000**, *112*, 4124; *Angew. Chem., Int. Ed. Engl.* **2000**, *39*, 3964. (n) J. Barluenga, F. Rodríguez, F. J. Fañanas, J. Florez, in *Topics in Organometallic Chemistry* (Ed.: K. H. Dötz), Springer, Heidelberg, **2004**, in press (o) M. A. Sierra, *Chem. Rev.* **2000**, *100*, 3591. (p) K. H. Dötz, C. Jäkel, W.-H. Haase, *J. Organomet. Chem.* **2001**, *617/-618*, 119. (q) J. W. Herndon, *Coord. Chem. Rev.* **2001**, *214*, 215. (r) J. W. Herndon, *Coord. Chem. Rev.* **2002**, *227*, 1. (s) K. H. Dötz, H. C. Jahr in *Carbene Chemistry* (Ed.: G. Bertrand), Fontis Media S. A., Lausanne, Marcel Dekker Inc., New York, **2002**, p. 231. (t) F. Zaragoza Dörwald, *Metal Carbenes in Organic Synthesis*, Wiley-VCH, Weinheim, 1999. (u) A. Minatti, K. H. Dötz, in *Topics in Organometallic Chemistry* (Ed.: K. H. Dötz), Springer, Heidelberg, **2004**, in press.

2. (a) E. O. Fischer, A. Maasböl, *Angew. Chem.* **1964**, *76*, 645; *Angew. Chem., Int. Ed. Engl.* **1964**, *3*, 580. (b) E. O. Fischer, T. Selmayr, F. R. Kreissl, *Chem. Ber.* **1977**, *110*, 2974.

3. (a) C. W. Rees, E. von Angerer, *J. Chem. Soc., Chem. Commun.* **1972**, 420. (b) M. F. Semmelhack, G. R. Lee, *Organometallics* **1987**, *6*, 1839. (c) R. Imwinkelried, L. S. Hegedus, *Organometallics* **1988**, *7*, 702.

4. J. Barluenga, J. M. Monserrat, J. Flórez, S. García-Granda, E. Martín, *Angew. Chem.* **1994**, *106*, 1451; *Angew. Chem., Int. Ed. Engl.* **1994**, *33*, 1392.

5. J. Barluenga, F. J. Fañanas, *Tetrahedron* **2000**, *56*, 4597.

6. B. Weyershausen, K. H. Dötz, *Eur. J. Inorg. Chem.* **1999**, 1057.

7. J. Pfeiffer, K. H. Dötz, *Organometallics* **1998**, *17*, 4353.

8. (a) E. O. Fischer, M. Leupold, C. G. Kreiter, J. Müller, *Chem. Ber.* **1972**, *105*, 150. (b) E. O. Fischer, G. Kreis, F. R. Kreissl, C. G. Kreiter, J. Müller, *Chem. Ber.* **1973**, *106*, 3910. (c) C. F. Bernasconi, G. S. Perez, *J. Am. Chem. Soc.* **2000**, *122*, 12441.

9. M. Salmain, E. Licandro, C. Baldoli, S. Maiorana, H. Tran-Huy, G. Jaouen, *J. Organomet. Chem.* **2001**, *617/618*, 376.

10. E. O. Fischer, G. Kreis, C. G. Kreiter, J. Müller, G. Huttner, H. Lorenz, *Angew. Chem.* **1973**, *85*, 618; *Angew. Chem., Int. Ed. Engl.* **1973**, *12*, 564.

11. (a) C. P. Casey, R. L. Anderson, *J. Am. Chem. Soc.* **1974**, *99*, 1651. (b) W. D. Wulff, B. A. Anderson, J. Toole, Y.-C. Xu, *Inorg. Chim. Acta* **1994**, *220*, 215. (c) C. F. Bernasconi, *Chem. Soc. Rev.* **1997**, *26*, 299.

12. O. Briel, A. Fehn, W. Beck, *J. Organomet. Chem.* **1999**, *578*, 247.

13. E. Licandro, S. Maiorana, D. Perdicchia, C. Baldoli, C. Graiff, A. Tiripicchio, *J. Organomet. Chem.* **2001**, *617/618*, 399.

14. E. Licandro, S. Maiorana, L. Capella, R. Manzotti, A. Papagni, B. Vandoni,

A. Albinati, S. H. Chuang, J.-R. Hwu, *Organometallics* **2001**, *20*, 4885.

15 E. Licandro, S. Maiorana, C. Baldoli, L. Capella, D. Perdicchia, *Tetrahedron: Asymmetry* **2000**, *11*, 975.

16 C. Jäkel, K. H. Dötz, *Tetrahedron* **2000**, *56*, 2167.

17 K. Fuchibe, N. Iwasawa, *Tetrahedron* **2000**, *56*, 4907.

18 Review: J. Barluenga, J. Flórez, F. J. Fañanás, *J. Organomet. Chem.* **2001**, *624*, 5.

19 J. Barluenga, A. A. Trabanco, J. Flórez, S. García-Granda, E. Martín, *J. Am. Chem. Soc.* **1996**, *118*, 13099.

20 J. Barluenga, A. A. Trabanco, J. Flórez, S. García-Granda, M. A. Llorca, *J. Am. Chem. Soc.* **1998**, *120*, 12129.

21 The nearly quantitative 1,4-addition of alkylcerium reagents to (methoxy)(alkenyl)carbene tungsten complexes has been reported: B. Caro, P. Le Poul, F. Robin-Le Guen, M.-C. Sénéchal-Tocquer, J.-Y. Saillard, S. Kahlal, L. Ouahab, S. Golhen, *Eur. J. Org. Chem.* **2000**, 577.

22 J. Barluenga, J. M. Monserrat, J. Flórez, S. García-Granda, E. Martín, *Chem. Eur. J.* **1995**, *1*, 236.

23 J. Barluenga, J. Alonso, F. Rodríguez, F. J. Fañanás, *Angew. Chem.* **2000**, *112*, 2556; *Angew. Chem., Int. Ed. Engl.* **2000**, *39*, 2459.

24 Y. Shi, W. D. Wulff, *J. Org. Chem.* **1994**, *59*, 5122.

25 W. D. Wulff, *Organometallics* **1998**, *17*, 3116.

26 J. Ezquerra, C. Pedregal, I. Merino, J. Flórez, J. Barluenga, S. Gárcia-Granda, M.-A. Llorca, *J. Org. Chem.* **1999**, *64*, 6554.

27 J. Barluenga, A. Ballesteros, J. Santamaría, R. B. de la Rúa, E. Rubio, M. Tomás, *J. Am. Chem. Soc.* **2000**, *122*, 12874.

28 (a) K. H. Dötz, *Angew. Chem.* **1975**, *87*, 672; *Angew. Chem., Int. Ed. Engl.* **1975**, *14*, 644. (b) K. H. Dötz, P. Tomuschat, *Chem. Soc. Rev.* **1999**, *28*, 187.

29 (a) M. M. Gleichmann, K. H. Dötz, B. A. Hess, *J. Am. Chem. Soc.* **1996**, *118*, 10551. (b) M. Torrent, M. Duran, M. Solá, *J. Am. Chem. Soc.* **1999**, *121*, 1309.

30 H. Fischer, J. Mühlemeier, R. Märkl, K. H. Dötz, *Chem. Ber.* **1982**, *115*, 1355.

31 (a) J. Barluenga, F. Aznar, A. Martín, S. García-Granda, E. Pérez-Carreño, *J. Am. Chem. Soc.* **1994**, *116*, 11191. (b) J. Barluenga, F. Aznar, I. Gutiérrez, A. Martín, S. García-Granda, M. A. Llorca-Baragaño, *J. Am. Chem. Soc.* **2000**, *122*, 1314.

32 K. H. Dötz, T. Schäfer, F. Kroll, K. Harms, *Angew. Chem.* **1992**, *104*, 1257; *Angew. Chem., Int. Ed. Engl.* **1992**, *31*, 1236.

33 P. Hoffmann, M. Hämmerle, G. Unfried, *New J. Chem.* **1991**, *15*, 769.

34 (a) B. A. Anderson, W. D. Wulff, *J. Am. Chem. Soc.* **1990**, *112*, 8615. (b) E. Chelain, A. Parlier, H. Rudler, J. C. Daran, J. Vaissermann, *J. Organomet. Chem.* **1991**, *416*, C5.

35 W. H. Moser, L. Sun, J. C. Huffman, *Org. Lett.* **2001**, *3*, 3389.

36 J. F. Quinn, M. E. Bos, W. D. Wulff, *Org. Lett.* **1999**, *1*, 161.

37 S. Chamberlin, B. Bax, W. D. Wulff, *Tetrahedron* **1993**, *49*, 5531.

38 W. D. Wulff, B. M. Bax, T. A. Brandvold, K. S. Chan, A. M. Gilbert, R. P. Hsung, J. Mitchell, J. Clardy, *Organometallics* **1994**, *13*, 102.

39 K. S. Chan, G. A. Peterson, T. A. Brandvold, K. L. Faron, C. A. Challener, C. Hyldahl, W. D. Wulff, *J. Organomet. Chem.* **1987**, *334*, 9.

40 J. S. McCallum, F. A. Kunng, S. R. Gilbertson, W. D. Wulff, *Organometallics* **1988**, *7*, 2346.

41 (a) A. Yamashita, *Tetrahedron Lett.* **1986**, *27*, 5915. (b) K. H. Dötz, D. B. Grotjahn, *Synlett* **1991**, *6*, 381. (c) K. H. Dötz, T. Leese, *Bull. Soc. Chim. Fr.* **1997**, *134*, 503.

42 (a) C. Bolm, K. Muñiz, *Chem. Soc. Rev.* **1999**, *28*, 51. (b) K. Muñiz in *Topics in Organometallic Chemistry* (Ed.: E. P. Kündig), Springer, Berlin, in press.

43 For a chiral carbene carbon side chain see: R. L. Beddoes, J. D. King, P. Quayle, *Tetrahedron Lett.* **1995**, *17*, 3027.

44 R. P. Hsung, W. D. Wulff, C. A. Challener, *Synthesis* **1996**, 773.

45 D. Paetsch, K. H. Dötz, *Tetrahedron Lett.* **1999**, *40*, 487.

46 K. H. Dötz, C. Stinner, *Tetrahedron: Asymmetry* **1997**, *8*, 1715.

47 R. P. Hsung, W. D. Wulff, S. Chamberlin, Y. Liu, R.-Y. Liu, H. Wang, J. F. Quinn, S. L. B. Wang, A. L. Rheingold, *Synthesis* **2001**, 200.
48 K. H. Dötz, J. Mühlemeier, U. Schubert, O. Orama, *J. Organomet. Chem.* **1983**, *247*, 187.
49 M. F. Gross, M. G. Finn, *J. Am. Chem. Soc.* **1994**, *116*, 10921.
50 S. Chamberlin, M. L. Waters, W. D. Wulff, *J. Am. Chem. Soc.* **1994**, *116*, 3113.
51 M. W. Davies, C. N. Johnson, J. P. A. Harrity, *J. Org. Chem.* **2001**, *66*, 3525.
52 (a) K. H. Dötz, A. Gerhardt, *J. Organomet. Chem.* **1999**, *578*, 223. (b) H. Wang, W. D. Wulff, *J. Am. Chem. Soc.* **2000**, *122*, 9862.
53 S. R. Pulley, S. Sen, A. Vorogushin, E. Swanson, *Org. Lett.* **1999**, *1*, 1721.
54 J. C. Anderson, J. W. Cram, N. P. King, *Tetrahedron Lett.* **2002**, *43*, 3849.
55 L. Fogel, R. P. Hsung, W. D. Wulff, *J. Am. Chem. Soc.* **2001**, *123*, 5580.
56 A. V. Vorogushin, W. D. Wulff, H.-J. Hansen, *J. Am. Chem. Soc.* **2002**, *124*, 6512.
57 P. Tomuschat, E. Kröner, E. Steckhan, M. Nieger, K. H. Dötz, *Chem. Eur. J.* **1999**, *5*, 700.
58 (a) J. Schneider, K. H. Dötz, unpublished results. (b) A. Minatti, K. H. Dötz, unpublished results.
59 H. C. Jahr, M. Nieger, K. H. Dötz, *J. Organomet. Chem.* **2002**, *641*, 185.
60 L. Quast, M. Nieger, K. H. Dötz, *Organometallics* **2000**, *19*, 2179.
61 (a) K. H. Dötz, I. Pruskil, *J. Organomet. Chem.* **1981**, *209*, C4. (b) K. H. Dötz, I. Pruskil, L. Mühlemeier, *Chem. Ber.* **1982**, *115*, 128. (c) K. H. Dötz, W. Kuhn, *Angew. Chem.* **1983**, *95*, 750; *Angew. Chem., Int. Ed. Engl.* **1983**, *22*, 732.
62 J. Bao, W. D. Wulff, V. Dragisch, S. Wenglowsky, R. G. Ball, *J. Am. Chem. Soc.* **1994**, *116*, 7616.
63 (a) M. F. Semmelhack, J. J. Bozell, T. Sato, W. Wulff, E. Spiess, A. Zask, *J. Am. Chem. Soc.* **1982**, *104*, 5850. (b) M. F. Semmelhack, J. J. Bozell, L. Keller, T. Sato, E. Spiess, W. Wulff, A. Zask, *Tetrahedron* **1985**, *41*, 5803. (c) D. L. Boger, O. Hüter, K. Mbiya, M. Zhang, *J. Am. Chem. Soc.* **1995**, *117*, 11839.

64 Y.-C. Xu, D. T. Kohlman, S. X. Liang, C. Errikson, *Org. Lett.* **1999**, *1*, 1599.
65 W. D. Wulff, J. Su, P.-C. Tang, Y.-C. Xu, *Synthesis* **1999**, 415.
66 V. P. Liptak, W. D. Wulff, *Tetrahedron* **2000**, *56*, 10229.
67 J. Barluenga, F. Aznar, M. A. Palomero, *Angew. Chem.* **2000**, *112*, 4514; *Angew. Chem., Int. Ed. Engl.* **2000**, *39*, 4346.
68 J. Barluenga, F. Aznar, M. A. Palomero, *Chem. Eur. J.* **2001**, *7*, 5318.
69 C. A. Merlic, D. McInnes, Y. You, *Tetrahedron Lett.* **1997**, *39*, 6787.
70 C. A. Merlic, C. C. Aldrich, J. Albaneze-Walker, A. Saghatelian, J. Mammen, *J. Org. Chem.* **2001**, *66*, 1297.
71 (a) K. H. Dötz, E. O. Fischer, *Chem. Ber.* **1972**, *105*, 1356. (b) C. P. Casey, M. C. Cesa, *Organometallics* **1982**, *1*, 87. (c) J. W. Herndon, S. U. Turmer, *J. Org. Chem.* **1991**, *56*, 286.
72 Reviews: (a) M. Brookhart, W. B. Studabaker, *Chem. Rev.* **1987**, *87*, 411. (b) D. F. Harvey, D. M. Sigano, *Chem. Rev.* **1996**, *96*, 271.
73 (a) E. O. Fischer, K. H. Dötz, *Chem. Ber.* **1972**, *105*, 3966. (b) B. Dorrer, E. O. Fischer, W. Kalbfus, *J. Organomet. Chem.* **1974**, *81*, C20. (c) W. D. Wulff, D. C. Yang, C. K. Murray, *Pure Appl. Chem.* **1988**, *60*, 137. (d) C. K. Murray, D. C. Yang, W. D. Wulff, *J. Am. Chem. Soc.* **1990**, *112*, 5660. (e) M. Jaeger, M.-H. Prosenc, C. Sontag, H. Fischer, *New. J. Chem.* **1995**, *19*, 911.
74 For an *in situ*-generated nonheteroatom-stabilized chromium carbene complex involved in a catalytic synthesis of spirocyclopropanes from diaryl diazo compounds and electron-rich alkenes, see: (a) J. Pfeiffer, K. H. Dötz, *Angew. Chem.* **1997**, *109*, 2948; *Angew. Chem., Int. Ed. Engl.* **1997**, *36*, 2828. (b) J. Pfeiffer, M. Nieger, K. H. Dötz, *Eur. J. Org. Chem.* **1998**, *1011*, 1.
75 W. C. Haase, K. H. Dötz, M. Nieger, *J. Organomet. Chem.* **2003**, *684*, 153.
76 M. Brookhart, D. Timmers, J. R. Tucker, G. D. Williams, G. R. Husk, H. Brunner, B. Hammer, *J. Am. Chem. Soc.* **1983**, *105*, 6721.

77 J. Barluenga, A. Fernández-Acebes, A. A. Trabanco, J. Flórez, *J. Am. Chem. Soc.* **1997**, *119*, 7591.
78 J. Barluenga, S. López, A. A. Trabanco, A. Fernández-Acebes, J. Flórez, *J. Am. Chem. Soc.* **2000**, *122*, 8145.
79 J. Barluenga, S. López, A. A. Trabanco, A. Fernández-Acebes, J. Flórez, *Chem. Eur. J.* **2001**, *7*, 4723.
80 M. D. Cooke, E. O. Fischer, *J. Organomet. Chem.* **1973**, *56*, 279.
81 J. Barluenga, K. Muñiz, A. Ballesteros, S. Martínez, M. Tomás, *ARKIVOC* **2002**, *(V)*, 110.
82 J. Barluenga, A. L. Suárez-Sobrino, M. Tomás, S. García-Granda, R. Santiago-García, *J. Am. Chem. Soc.* **2001**, *123*, 10494.
83 (a) D. F. Harvey, K. P. Lund, *J. Am. Chem. Soc.* **1991**, *113*, 8916. (b) C. A Merlic, H. D. Bendorf, *Tetrahedron Lett.* **1994**, *35*, 9529.
84 (a) M. Buchert, H.-U. Reissig, *Tetrahedron Lett.* **1988**, *29*, 2319. (b) M. Buchert, H.-U. Reissig, *Chem. Ber.* **1992**, *125*, 2723. (c) M. Buchert, M. Hoffmann, H.-U. Reissig, *Chem. Ber.* **1995**, *128*, 605.
85 M. Hoffmann, M. Buchert, H.-U. Reissig, *Chem. Eur. J.* **1999**, *5*, 876.
86 J. Barluenga, S. Martínez, A. L. Suárez-Sobrino, M. Tomás, *J. Am. Chem. Soc.* **2002**, *124*, 5948.
87 J. Barluenga, F. Aznar, I. Gutiérrez, J. A. Martín, *Org. Lett.* **2002**, *4*, 2719.
88 (a) A. M. McGuire, L. S. Hegedus, *J. Am. Chem. Soc.* **1982**, *104*, 5538. (b) L. S. Hegedus, G. de Weck, S. D'Andrea, *J. Am. Chem. Soc.* **1992**, *114*, 5010.
89 E. Kuester, L. S. Hegedus, *Organometallics* **1999**, *18*, 5318.
90 K. Puntener, M. D. Hellman, E. Kuester, L. S. Hegedus, *J. Org. Chem.* **2000**, *65*, 8301.
91 L. S. Hegedus, R. Imwinkelried, M. Alarid-Sergant, D. Dvorak, Y. Satoh, *J. Am. Chem. Soc.* **1990**, *112*, 1109.
92 A. Arrieta, F. P. Cossío, I. Fernández, M. Gómez-Gallego, B. Lecea, M. J. Mancheño, M. A. Sierra, *J. Am. Chem. Soc.* **2000**, *122*, 11509.
93 L. M. Reeder, L. S. Hegedus, *J. Org. Chem.* **1999**, *64*, 3306.

94 X. Wen, H. Norling, L. S. Hegedus, *J. Org. Chem.* **2000**, *65*, 2096.
95 B. Brown, L. S. Hegedus, *J. Org. Chem.* **1998**, *63*, 8012.
96 P. J. Campos, D. Sampedro, M. A. Rodríguez, *Organometallics* **2000**, *19*, 3082.
97 P. J. Campos, D. Sampedro, M. A. Rodríguez, *Tetrahedron Lett.* **2002**, *43*, 73.
98 L. S. Hegedus, M. A. Schwindt, S. De Lombart, R. Imwinkelried, *J. Am. Chem. Soc.* **1990**, *112*, 2264.
99 K. H. Dötz, M. Klumpe, M. Nieger, *Chem. Eur. J.* **1999**, *5*, 691.
100 J. Barluenga, F. Fernández-Marí, R. González, E. Aguilar, G. A. Revelli, A. L. Viado, F. J. Fañanas, B. Olano, *Eur. J. Org. Chem.* **2000**, 1773.
101 J. Barluenga, F. Fernández-Marí, A. L. Viado, E. Aguilar, B. Olano, S. García-Granda, C. Moya-Rubiera, *Chem. Eur. J.* **1999**, *5*, 883.
102 J. Barluenga, M. A. Fernández-Rodríguez, E. Aguilar, F. Fernández-Marí, A. Salinas, B. Olano, *Chem. Eur. J.* **2001**, *16*, 3533.
103 I. Merino, S. Laxmi Y. R., J. Flórez, J. Barluenga, *J. Org. Chem.* **2002**, *67*, 648.
104 J. Barluenga, M. Tomás, A. Ballesteros, J. Santamaría, C. Brillet, S. García-Granda, A. Piñera-Nicolás, J. T. Vázquez, *J. Am. Chem. Soc.* **1999**, *121*, 4516.
105 H. Kagoshima, T. Okamura, T. Akiyama, *J. Am. Chem. Soc.* **2001**, *123*, 7182.
106 J. W. Herndon, J. Zhu, D. Sampedro, *Tetrahedron* **2000**, *56*, 4985.
107 Review: R. Aumann, *Eur. J. Org. Chem.* **2000**, 17.
108 B. L. Flynn, H. Schirmer, M. Duetsch, A. de Meijere, *J. Org. Chem.* **2001**, *66*, 1747.
109 H. Schirmer, B. L. Flynn, A. de Meijere, *Tetrahedron* **2000**, *56*, 4977.
110 Y.-T. Wu, H. Schirmer, M. Noltemeyer, A. de Meijere, *Eur. J. Org. Chem.* **2001**, 2501.
111 H.-P. Wu, R. Aumann, R. Fröhlich, B. Wibbeling, *Chem. Eur. J.* **2002**, *8*, 910.
112 H.-P. Wu, R. Aumann, R. Fröhlich, B. Wibbeling, O. Kataeva, *Chem. Eur. J.* **2001**, *7*, 5084.
113 (a) Y. Zhang, J. W. Herndon, *J. Org. Chem.* **2002**, *67*, 4177. (b) B. K. Ghorai, S. Menon, D. L. Johnson, J. W. Herndon, *Org. Lett.* **2002**, *4*, 2121.

114 J. Barluenga, F. Aznar, S. Barluenga, M. Fernández, A. Martín, S. García-Granda, A. Piñera-Nicolás, *Chem. Eur. J.* **1998**, *4*, 2280.

115 D. Böttcher, PhD Thesis, University Bonn, **1996**.

116 J. Barluenga, R.-M. Canteli, J. Flórez, S. García-Granda, A. Gutiérrez-Rodríguez, E. Martín, *J. Am. Chem. Soc.* **1998**, *120*, 2514.

117 (a) E. O. Fischer, H.-J. Beck, *Chem. Ber.* **1971**, *104*, 3101. (b) C. P. Casey, L. Anderson, *J. Chem. Soc., Chem. Commun.* **1975**, 895. (c) E. O. Fischer, M. Böck, R. Aumann, *Chem. Ber.* **1981**, *114*, 1853. (d) R.-Z. Ku, J.-C. Huang, J.-Y. Cho, F.-M. Kiang, K. R. Reddy, Y.-C. Chen, K.-J. Lee, J.-H. Lee, G.-H. Lee, S.-M. Peng, S.-T. Liu, *Organometallics* **1999**, *18*, 2145.

118 H. Fischer, S. Zeuner, K. Ackermann, J. Schmid, *Chem. Ber.* **1986**, *119*, 1546.

119 (a) M. A. Sierra, J. C. del Amo, M. J. Mancheño, M. Gómez-Gallego, *J. Am. Chem. Soc.* **2001**, *123*, 851. (b) F. Robin-Le Guen, P. Le Poul, B. Caro, N. Faux, N. Le Poul, S. J. Green, *Tetrahedron Lett.* **2002**, *43*, 3967.

120 R. Aumann, I. Göttker-Schnetmann, R. Fröhlich, O. Meyer, *Eur. J. Org. Chem.* **1999**, 2545.

121 I. Göttker-Schnetmann, R. Aumann, *Organometallics* **2001**, *20*, 346.

122 I. Göttker-Schnetmann, R. Aumann, K. Bergander, *Organometallics* **2001**, *20*, 3574.

123 J. Barluenga, L. A. López, O. Löber, M. Tomás, S. García-Granda, C. Alvarez-Rúa, J. Borge, *Angew. Chem.* **2001**, *113*, 3495; *Angew. Chem., Int. Ed. Engl.* **2001**, *40*, 3392.

124 So far, there has been only a single report on the synthesis of a stable copper(I) carbene complex: B. F. Straub, P. Hofmann, *Angew. Chem.* **2001**, *113*, 1328; *Angew. Chem., Int. Ed. Engl.* **2001**, *40*, 1288.

3.2
Titanium–Carbene Mediated Reactions

Nicos A. Petasis

3.2.1
Introduction

For nearly half a century organotitanium compounds have attracted considerable interest in organic and polymer chemistry. Prompted by the high abundance, low cost, low toxicity, and diverse chemical reactivity of titanium, the study of organotitanium derivatives continues to provide a variety of catalytic as well as stoichiometric synthetic applications. A number of recent reviews on the synthesis, structure, reactivity and synthetic utility of organotitanium compounds have appeared [1–4]. This chapter focuses on the chemistry of titanium carbenes of the general type **1** and **2**, as well as the related carbenoids or geminal dimetallic species **3** and **4**. Some aspects of titanium–carbene chemistry have been previously reviewed [4–8], including their use for carbonyl olefination [9], titanacycle formation [10], olefin metathesis [11], and polymerization reactions [12, 13]. Titanium carbenes have a nucleophilic carbene carbon and belong to the general class of Schrock-type metallocarbenes [14, 15]. Their high reactivity is driven predominantly by the electrophilic and oxophilic nature of titanium and can be modulated by the titanium ligands. A number of theoretical studies of titanium–carbenes have appears [16–22].

$L_n Ti = \begin{matrix} R^1 \\ R^2 \end{matrix}$ $L_n Ti = \!\!= \begin{matrix} R^1 \\ R^2 \end{matrix}$ $L_n Ti \begin{matrix} R^1 \\ M \end{matrix} \begin{matrix} \\ R^2 \end{matrix}$ $L_n Ti \begin{matrix} \\ M \end{matrix} = \begin{matrix} R^1 \\ R^2 \end{matrix}$

   1   2   3   4

3.2.1.1
Precursors to Titanium Carbenes

Among the most prominent types of organotitanium compounds are complexes having two cyclopentadienyl ligands (titanocenes) [3, 23] which have unique reactivity resulting from their higher hydrolytic stability and the lower titanium acidity. The parent titanium carbene in this series is titanocene methylidene (**5**) which has been the focus of numerous studies [5]. Although this species has not been

Transition Metals for Organic Synthesis, Vol. 1, 2nd Edition.
Edited by M. Beller and C. Bolm
Copyright © 2004 WILEY-VCH Verlag GmbH & Co. KGaA, Weinheim
ISBN: 3-527-30613-7

observed in its free form, the corresponding phosphine complexes (**6**) [24] and several of its adducts have been prepared.

A major milestone in the chemistry of titanium carbenes was the isolation of the titanium-aluminum complex (**7**), known as the Tebbe reagent [25–29]. This compound, which is essentially the complex of **5** with Me$_2$AlCl, is prepared by the reaction of trimethyl aluminum with the readily available titanocene dichloride (**8**). In the presence of bases, even as mild as tetrahydrofuran, the Tebbe reagent can give new complexes with **5** with a variety of unsaturated functional groups. The pioneering work of Grubbs and co-workers [5, 6] in this area has led to several applications of this chemistry in organic and polymer synthesis. Thus, while **5** reacts with carbonyls to give olefins, it also converts alkenes to titanacyclobutanes (**10**) [30–34], allenes to alkylidene titanacyclobutanes (**11**) [35, 36], alkynes to titanacyclobutenes (**12**) [37–45], and nitriles to 1,3-diazatitanacyclohexadienes (**13**) [46, 47]. Many of these adducts can be isolated or can undergo further transformations *in situ* to form new titanium-free products.

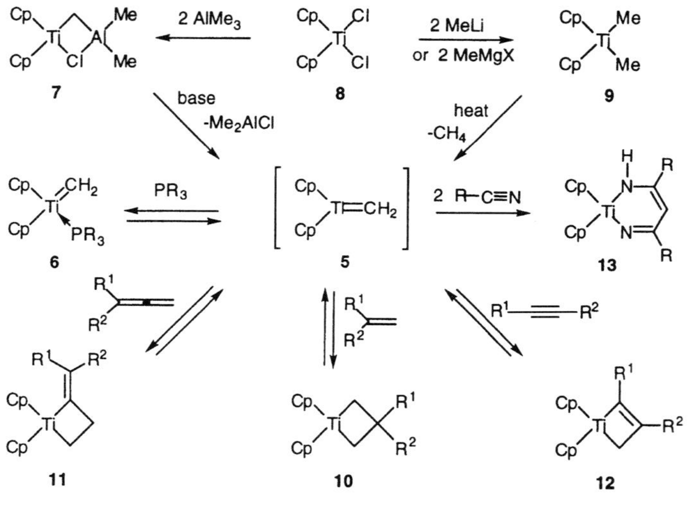

Despite the great synthetic utility of the Tebbe reagent, the presence of aluminum in its structure results in several drawbacks for its use in organic synthesis. In addition to being highly acidic the preparation and handling of this compound is hampered by its extreme sensitivity to air and water.

A superior reagent that exhibits similar reactivity with **7** is dimethyl titanocene (**9**), which is easily prepared from titanocene dichloride (**8**) and methyl lithium or a methyl magnesium halide [48, 49]. This compound tolerates brief exposure to air and water and it is stable at room temperature when kept in solution in the dark. Thus, this reagent has evolved as a mild and practical alternative to the Tebbe reagent for carbonyl methylenations [50–52] and other chemistry, including the formation of **11** [53], **12** [54–57], and **13** [54, 55, 58].

While homologated titanocene-aluminum complexes analogous to **7** are difficult to prepare, the corresponding dialkyl titanocenes (i.e. the homologs of **9**) can be readily obtained from **8** and the appropriate organolithium or Grignard reagent. The thermal stability of these compounds depends heavily on the nature of the C-substituents. Complexes capable of a facile β-hydride elimination are generally unstable at or below room temperature, while others can be stable even at high temperatures. Among the compounds that were shown to exhibit similar reactivity with **9** are the dibenzyl [59], bis(trimethyl-silylmethyl) [53, 60], bis(cyclopropyl) [61], bis(alkenyl) titanocenes [62], and others [63].

3.2.1.2
Geminal Bimetallic Derivatives

A number of reactions carried out by titanium carbenes can also be performed with geminal dimetallic intermediates of the general type **3**. These derivatives are readily formed from the reaction of geminal dihalides (**14**) with zinc or magnesium, followed by reaction with an electrophilic titanium compound. Although their exact structures are not known, these species are believed to have two metal atoms on the same carbon. The geminal dimagnesio intermediate (**15**) was among the first compounds of this type to be reported, and was shown to perform carbonyl methylenations with moderate yields [64]. Reaction of **15** with tianocene dichloride (**8**) gave a Tebbe-type derivative (**16**) [65, 66], which also showed a similar reactivity. The related titanium–zinc compound (**17**) [67] was prepared similarly from the geminal dizinc intermediate (**19**), generated by the reaction of **14** with zinc.

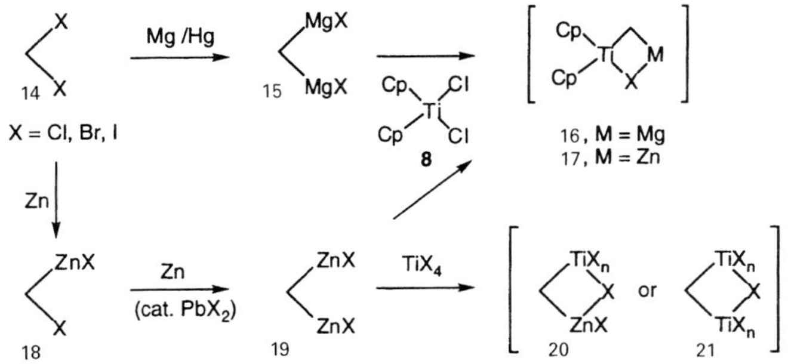

A more widely used procedure involves the reaction of **19** with $TiCl_4$ or a relation TiX_4 derivative to give presumably a geminal dimetallic species such as **20** to **21**. The combination of $CH_2Br_2/Zn/TiCl_4$ and several other similar systems were initially reported by Takai and Oshima [68–71] as effective carbonyl methylenation reagents which have found many applications in synthesis [7, 9]. A modification of this system by Lombardo [72, 73], involving the 'aging' of the reagent of 5 °C prior to use, was found to be more effective for some carbonyl methylenations. More recently it was found [74] that the formation of the geminal dizinc species (**19**) proceeds via a monozinc intermediate (**18**) and is catalyzed by lead, which exists as a minor impurity in some forms of commercially available zinc that are more effective in this process than highly pure zinc.

Homologated geminal dimetallic derivatives are accessible in a similar manner from geminal dibromides, zinc, and titanium tetrachloride in the presence of TMEDA [75–77]. An alternative olefination procedure, using dithioacetals as starting materials which presumably proceeds via geminal dititanium derivatives, was recently reported by Takeda and co-workers [78].

Alkenyl geminal dimetallic derivatives of the general type **23** were first reported by Yoshida and Negishi [79]. They were formed via the carbometallation of alkenyl alanes (**22**). The analogous zinc/zirconium systems (**24**) were recently studied by Knochel and co-workers [80].

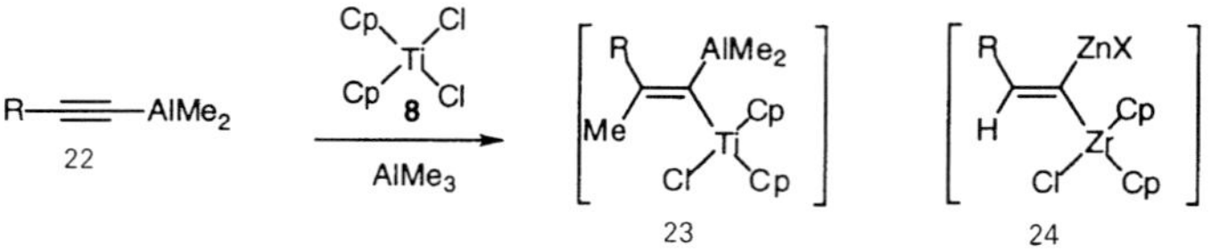

3.2.2
Carbonyl Olefinations

One of the most important applications of titanium carbenes is the conversion of carbonyls (**25**) to olefins (**26**) or allenes (**27**). A number of methods have been developed for this purpose which offer several advantages over Wittig-type processes. Thus, while aldehydes and ketones can be effectively converted to olefins via the Wittig reaction [81–83], the Peterson olefination [84], or other related transformations, these processes involve basic or nucleophilic species and are often not suitable for many types of carbonyl compounds [85], e.g. for the olefination of readily enolizable carbonyls or substrates that undergo facile nucleophilic addition or elimination reactions. Also, sterically hindered substrates often give low yields, while the olefination of esters and lactones is usually not possible.

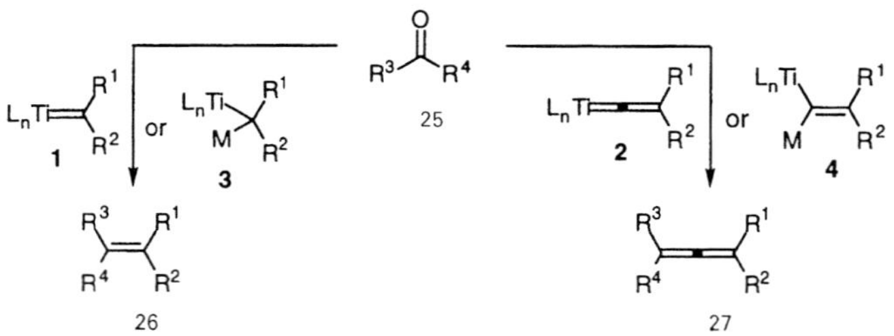

3.2.2.1
Carbonyl Methylenations with the Tebbe Reagent

The Tebbe reagent (**7**) has been used extensively for the methylenation of a variety of carbonyls [9], including aldehydes, ketones, esters, lactones and amides. Aldehydes (e.g. **28** [86]) and ketones can be methylenated in the presence of esters and without epimerization. This reagent is particularly useful for the methylenation of readily enolizable ketones [34] such as γ,δ-unsaturated derivatives (e.g. **29**) as well as ketones that have β-alkoxy or β-halide substituents (e.g. **30** [87]).

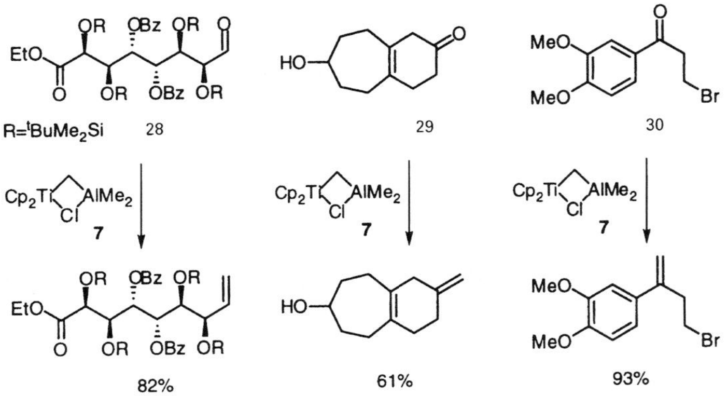

The methylenation of sterically hindered ketones (eg. **31** [88]) was shown to be more efficient with the Tebbe reagent that with the Wittig reagent.

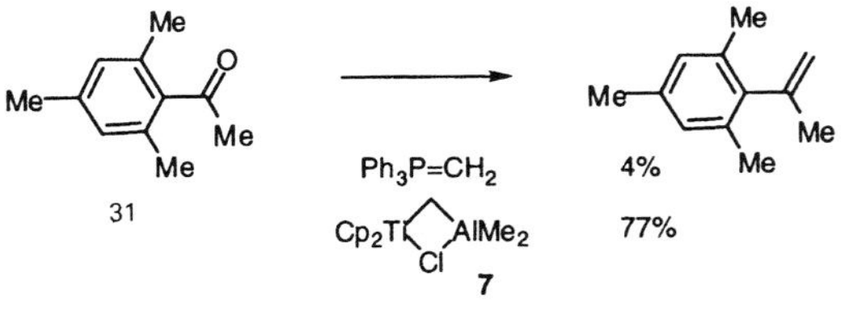

An important feature of the Tebbe reagent is its ability to methylenate esters and lactones. For large scale applications the *in situ* generation of the reagent from titanocene dichloride (**8**) and trimethylaluminum is preferable.

Among the many interesting synthetic applications of this reaction is the methylenation of aldonolactones (e.g. **32**) [89–91].

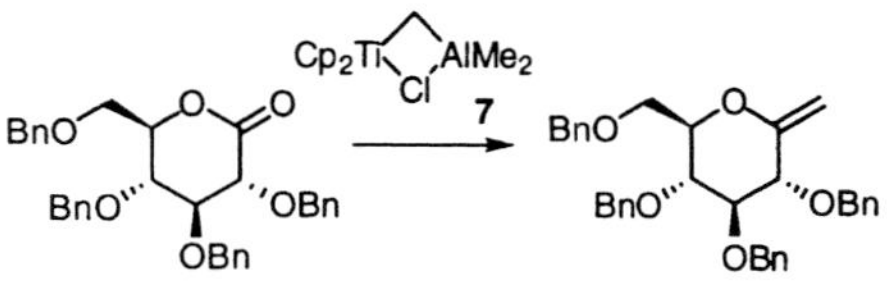

A reaction that can effectively follow the methylenation of esters or lactones is the Claisen rearrangement. Paquette and co-workers [92–97] developed this type of Tebbe–Claisen strategy for the synthesis of eight-membered rings (e.g. **34**) [95, 96] including a variety of complex polycyclic products (e.g. **35**) [97].

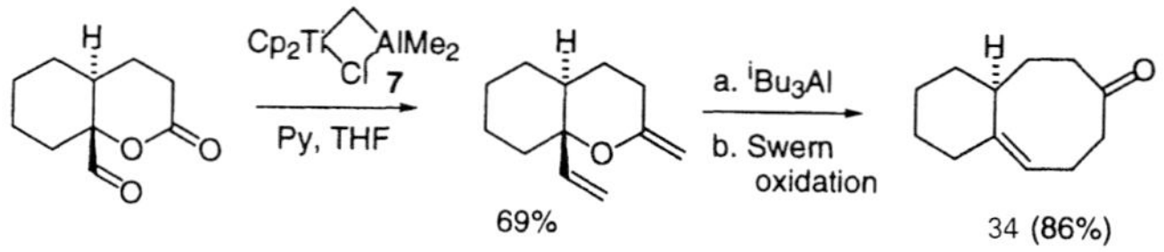

In the presence of pyridine bases the Tebbe reagent (**7**) reacts with olefins at low temperature to form aluminum-free titanacyclobutanes (e.g. **36**), which were extensively studied by Grubbs and co-workers [5, 6, 30, 31, 33, 38]. These thermally labile complexes are also accessible from titanocene dichloride and di-Grignard reagents [65, 98] or from π-allyl titanocene precursors [99]. Upon thermolysis titanacyclobutanes regenerate the titanocene methylidene species (**5**), presumably as its olefin complex [33], which can also be employed for carbonyl methylenations [5, 34].

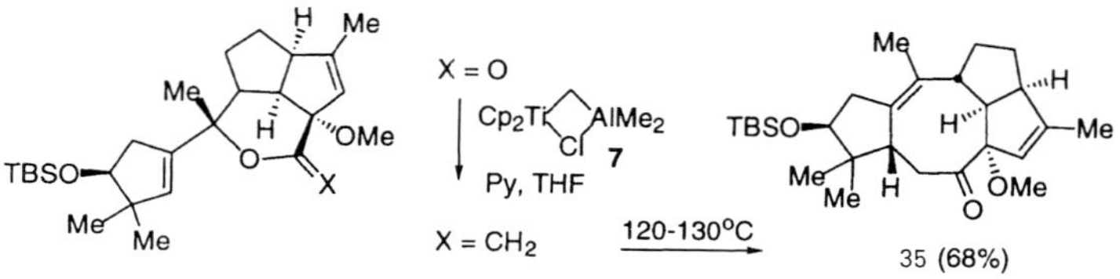

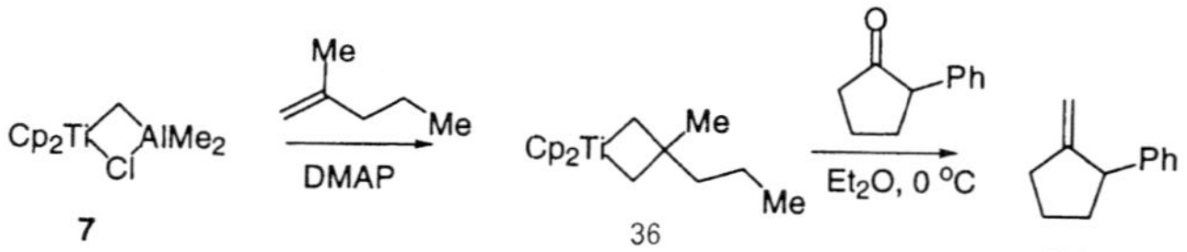

Titanacyclobutanes react with acyl chlorides (e.g. **37**) or anhydrides (e.g. **38**) but instead of methylenation products they form enolates which can participate in subsequent aldol reactions [100].

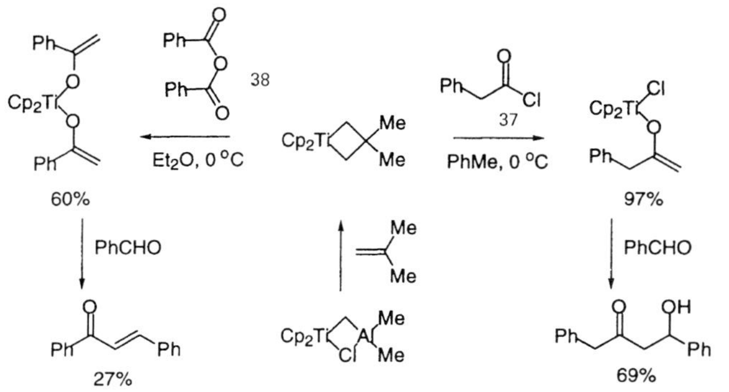

3.2.2.2
Carbonyl Olefinations with Dimethyl Titanocene and Related Derivatives

Although dimethyl titanocene (9) was known for some time, its ability to methylenate carbonyl compounds was discovered only recently [8, 50–52]. Overall, this reagent has reactivity analogous to the Tebbe reagent (7), while it is much easier to prepare and handle. Presumably the methylenation proceeds via the titanocene methylidene species (5) or its derivatives [8, 101].

Upon heating to 60–80 °C dimethyl titanocene (9) can methylenate a variety of carbonyl compounds (39 [50]), including: aldehydes, ketones, esters, lactones, and other heteroatom-substituted carbonyls [51], such as silyl esters, thioesters, selenoesters, acylsilanes, anhydrides, carbonates, amides, and imides.

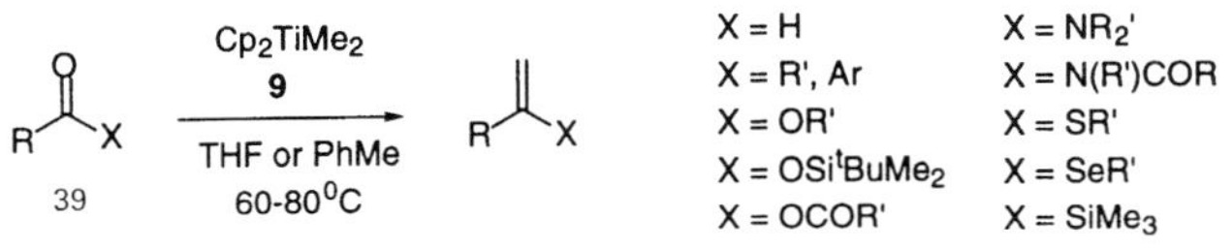

Unlike Wittig-type reagents and similarly to 7, dimethyl titanocene (9) is suitable for the methylenation of base-sensitive substrates such as easily enolizable ketones (40 [50]). Among the sensitive substrates that are preferably methylenated with 9 are the substituted cyclopentanones, such as 41 [102] and 42 [103].

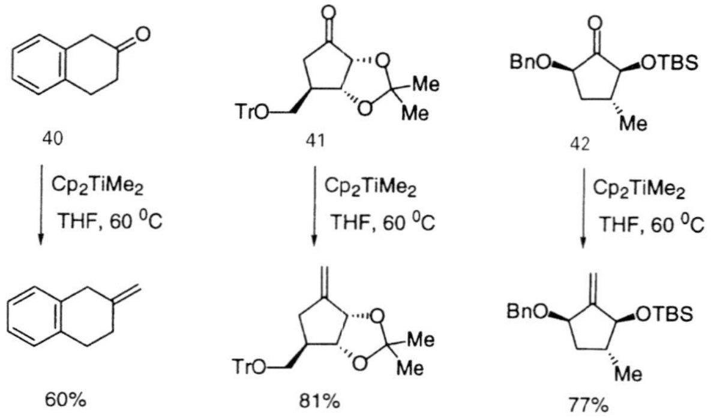

[Structures 40, 41, 42 with Cp₂TiMe₂, THF, 60 °C giving methylenated products in 60%, 81%, 77%]

Although the Tebbe reagent (**7**) is quite acidic due to its aluminum component, **9** is nearly neutral and is suitable for the methylenation of acid-sensitive substrates or the preparation of highly acid-labile products. For example, while the attempted methylenation of spiroketal lactones (**43**) with **7** failed due to undesired fragmentation, the use of **9** was quite effective giving the corresponding enol ethers which could be then hydrogenated or epoxidized to form substituted spiroketals [104]. Similarly, the very labile spiro-bislactone (**44**) could be converted to the corresponding bis-enol ether [51].

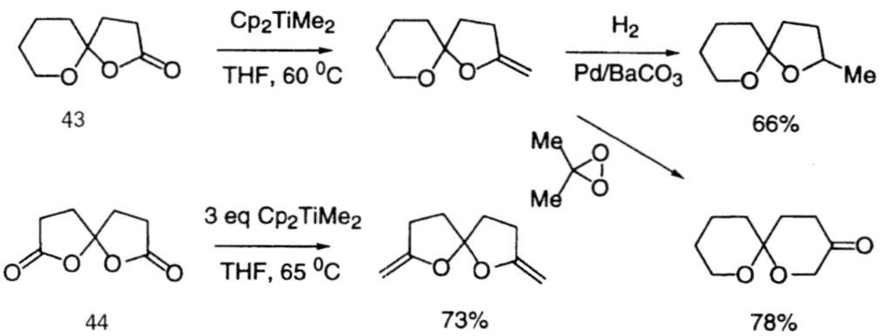

[Structures 43 and 44 with Cp₂TiMe₂ conditions giving enol ethers in 73%, further transformed with H₂/Pd/BaCO₃ to 66% or epoxidized to 78%]

Dimethyl titanocene was also shown to be quite effective for the methylenation of aldonolactones (e.g. **45** [105] and **46** [106]) as well as β-lactones (e.g. **47** [107]), even in the presence of an unprotected secondary hydroxyl group.

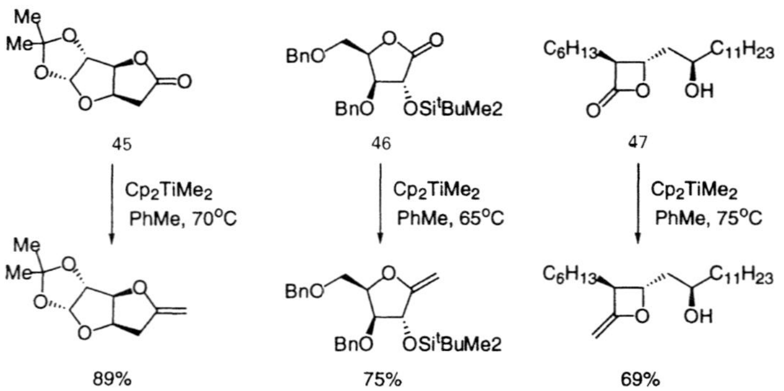

[Structures 45, 46, 47 with Cp₂TiMe₂, PhMe conditions giving methylenated products in 89%, 75%, 69%]

Reaction of **9** with anhydrides or imides (e.g. **48**) proceeds with the methylenation of one or both carbonyl groups, depending on the amount used [51]. With five-membered ring anhydrides and thioanhydrides (e.g. **49**) bis-methylenation can lead to aromatization to form furans or thiophenes [108].

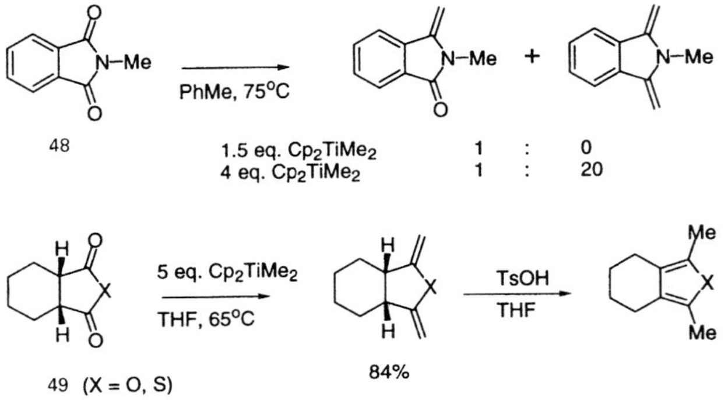

The rates of carbonyl methylenations with dimethyl titanocene are sensitive both to electronic and steric effects. In general, aldehydes are methylenated faster than esters (e.g. **50** [109]) or amides (e.g. **51** [110]). Similarly, ketones are methylenated faster than esters of vinylogous esters (e.g. **52** [52]). Sterically hindered carbonyls undergo a much slower reaction and it is possible to methylenate acetates in the presence of pivaloate esters (e.g. **53** [111, 112]).

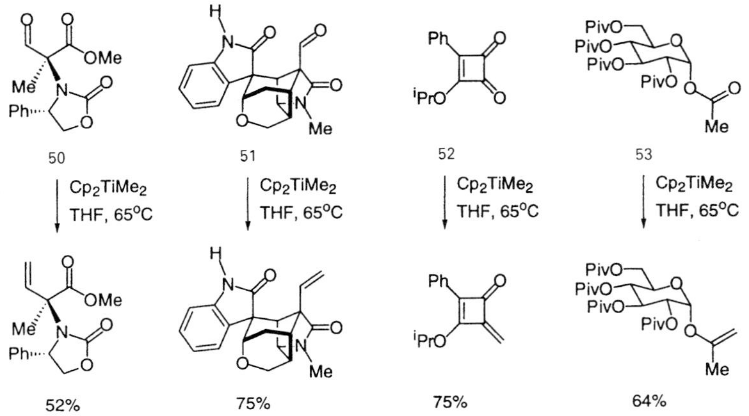

The combination of lactone methylenation with a thermal or aluminum-mediated [3,3]-sigmatropic rearrangement was employed in the synthesis of cyclooctanoids (e.g. **54** [113]) and cembranoids (e.g. **55** [114]).

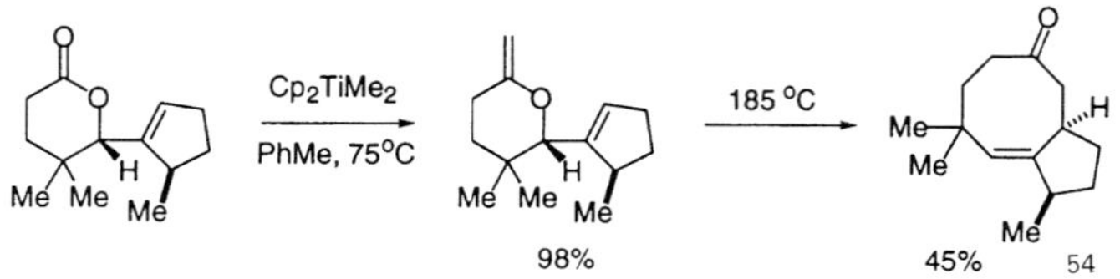

Methylenation of 1,3-dioxolan-4-ones with dimethyl titanocene followed by aluminum-mediated [1,3]-rearrangement gives tetrahydrofurans (e.g. 56 [115]), while the analogous sequence with 1,3-dioxan-4-ones gives tetrahydropyrans (e.g. 57 [116]) in a highly stereocontrolled manner.

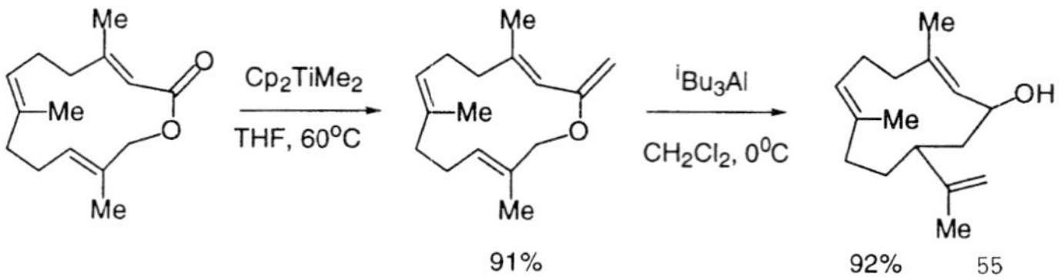

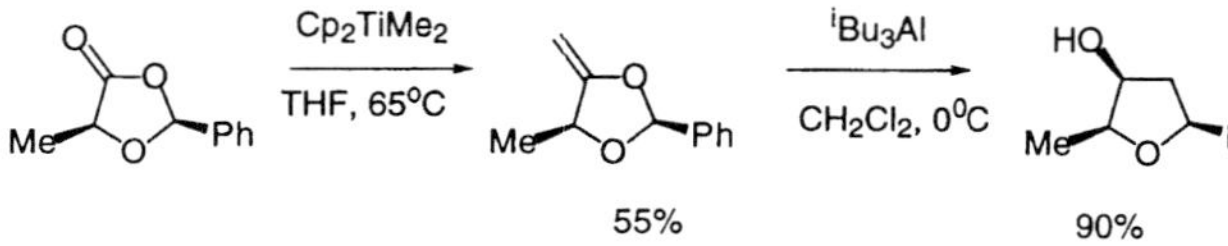

Several other dialkyl titanocenes could be used for the alkylidenation of carbonyl compounds. These include: dibenzyl titanocenes [59] which give phenyl-substituted olefins, bis(trimethylsilylmethyl) titanocene [60] which form vinyl silanes, and bis-cyclopropyl titanocene [61] which affords cyclopropylidenes.

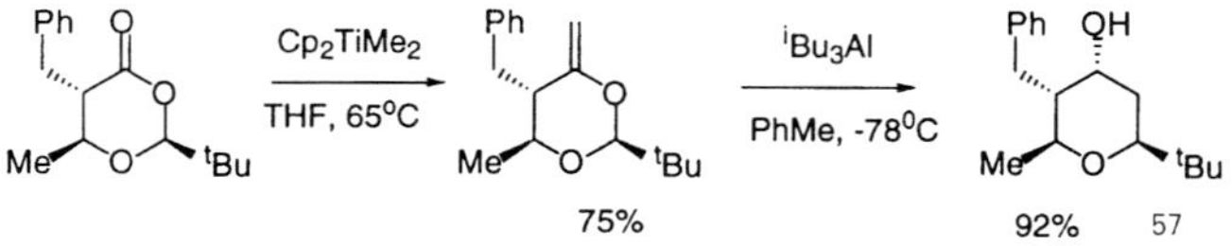

Apart from bis(trimethylsilylmethyl)titanocene mono-cylcopentadienyl tris-(trimethylsilylmethyl) titanium was found to convert carbonyls to alkenyl silanes (58 [60]). This compound can olefinate a variety of carbonyls, including aldeydes, ketones (59 [60]), esters, lactones, and various trifluoromethyl carbonyl compounds (e.g. 60 [117]).

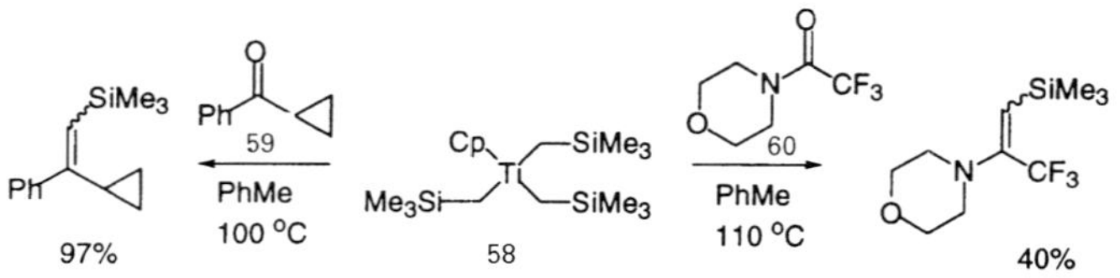

3.2.2.3
Carbonyl Methylenations with CH_2Br_2–Zn–$TiCl_4$ and Related Systems

The initial procedure for the methylenation of carbonyl compounds with the CH_2Br_2–Zn–TiCLt system, reported by Oshima and co-workers [68, 69], involved the brief mixing of a suspension of Zn dust and CH_2Br_2 in THF with a solution of $TiCl_4$ in CH_2Cl_2, followed by subsequent reaction with the carbonyl substrate in CH_2Cl_2 or THF. This *in situ* procedure is most effective for the methylenation of ketones (e.g. **61** [68, 69]) and is less effective with aryl ketones and aldehydes which undergo a competing pinacol-type reductive coupling. However, in many cases the completion of the methylenation requires prolonged stirring at room temperature. The modification introduced by Lombardo [72, 73], involving prior 'low temperature aging' which presumably alters the composition of the reagent, significantly speeds up the methylenation process and limits the exposure of the carbonyl substrate to Zn and $TiCl_4$ leading to increased methylenation yields and suppressed side reactions. This procedure does not epimerize ketones (e.g. **62** [73]) and has been employed extensively in the synthesis of gibberellins and related compounds (e.g. **63** [72, 118]).

The chemoselective methylenation of aldehydes in the presence of ketones can be done with CH_2I_2–Zn–Ti(O*i*-Pr)$_4$ [71]. The reverse selectivity, i.e. the methylenation of a ketone in the presence of an aldehyde, can be accomplished by the *in situ* protection of the aldehyde with Ti(NEt$_2$)$_4$, followed by reaction with CH_2I_2–Zn–$TiCl_4$ and deprotection [71].

Although the reagents of choice for the methylenation of esters and lactones are often the Tebbe reagent (**7**) or dimethyl titanocene (**9**), it is possible to olefinate esters with the CH_2Br_2–Zn–$TiCl_4$ system if TMEDA is mixed with $TiCl_4$ prior to its exposure to Zn and CH_2Br_2 [75]. These conditions were found advantageous over **7** for the olefination of carbohydrate ester inter- mediates (e.g. **64** [119]). A lactone methylenation was similarly accomplished, by using excess of the reagent [120].

64 68%

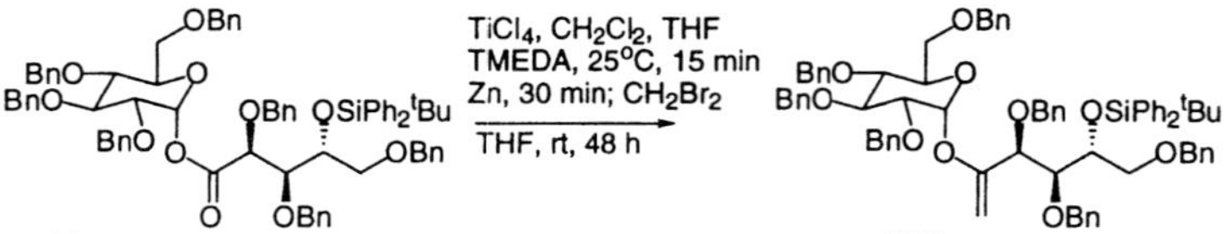

The analogous alkylidenations of carbonyls by the use of 1,1-dihaloalkanes can be accomplished in the presence of TMEDA [75]. Thus, esters and lactones (e.g. **65** [75]) can be converted to the corresponding enol ethers in good yields. Similarly, silyl esters (e.g. **66** [75]) afford silyl enol ethers with predominantly *Z* geometry, which alkylidenation of thioesters (e.g. **67**) and amides (e.g. **68**) gives alkenyl sulfides and enamines, respectively [77].

51% (78%Z) 80% (84%Z) 95% (100%Z) 70% (98% E)

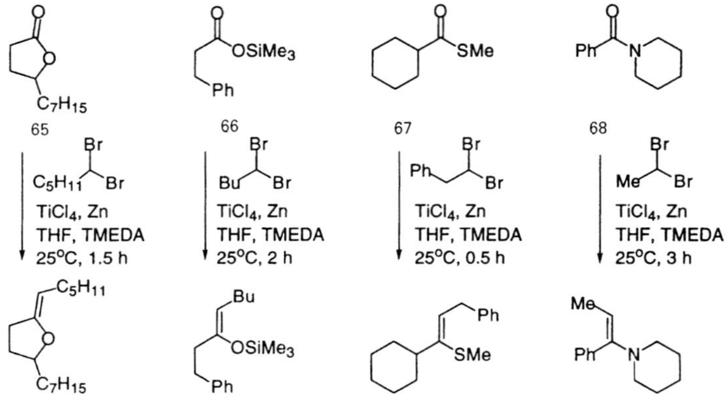

This type of reaction was utilized in a synthesis of spiroketals (e.g. **69** [121]).

69 76 %

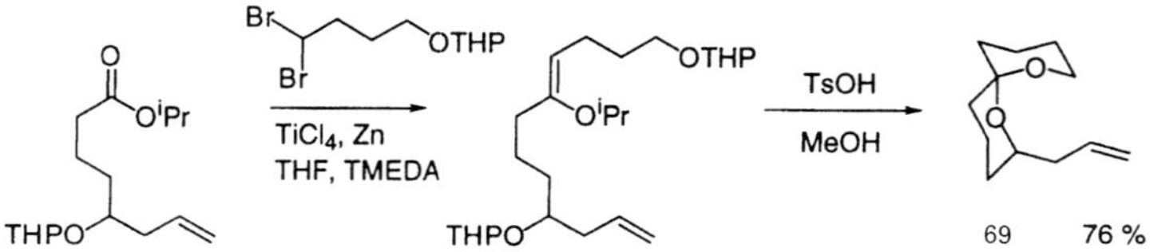

A carbonyl olefination process involving the use of dithioacetals and dithioketals in the presence of the low valent titanium species **72**, was recently reported [78]. This method can be used for the olefination of aldehydes, ketones (e.g. 70), esters, and lactones (e.g. **71**).

77% 70 71 64%

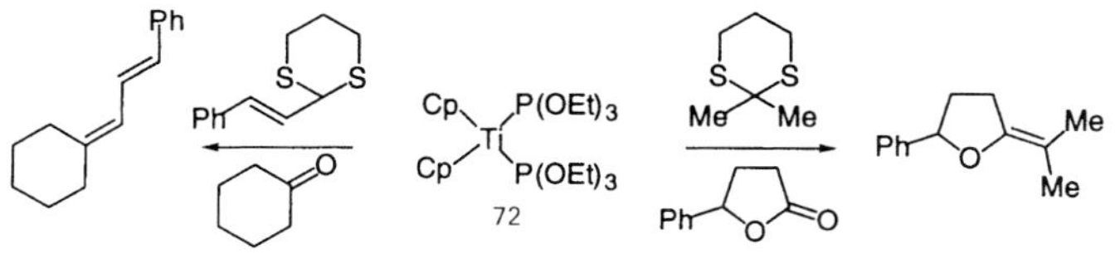

3.2.2.4
Carbonyl Alienations

Aldehydes and ketones can be directly converted to substituted allenes with several titanocene derivatives. Yoshida and Negishi [79] reported the first example of this type of transformation involving the reaction of alkynylalanes (**73**) with titanocene dichloride in the presence of trimethylaluminum to generate a 1,1-dimetalloalkene species (**74**) which was then reacted with carbonyl compounds to give allenes.

73 74 83%

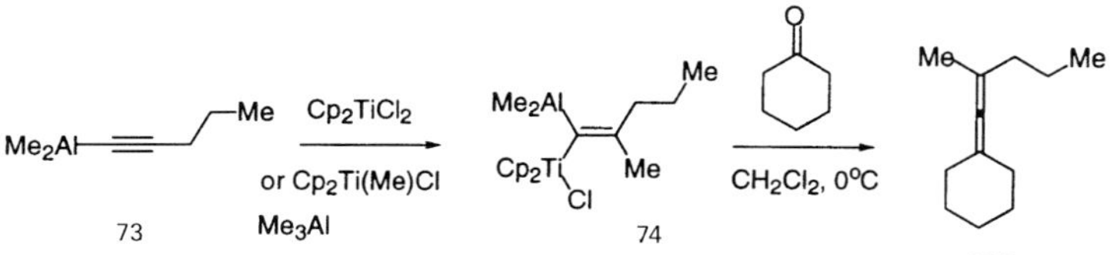

A convenient alienation method [62] involves the formation of alkenyltitanocene derivatives (e.g. **75** or **76**) at low temperature, followed by the *in situ* addition of an aldehyde or ketone and warming to room temperature. The analogous bis(pentamethylcyclopentadienyl) titanium derivatives react with carbonyls to give enolates instead of allenes [122].

75 83%

76 81%

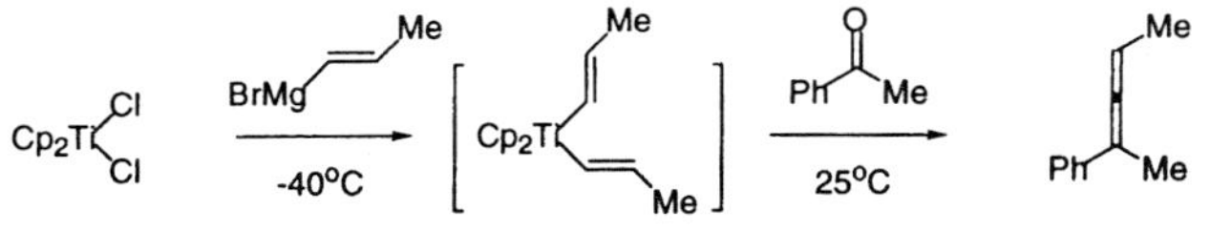

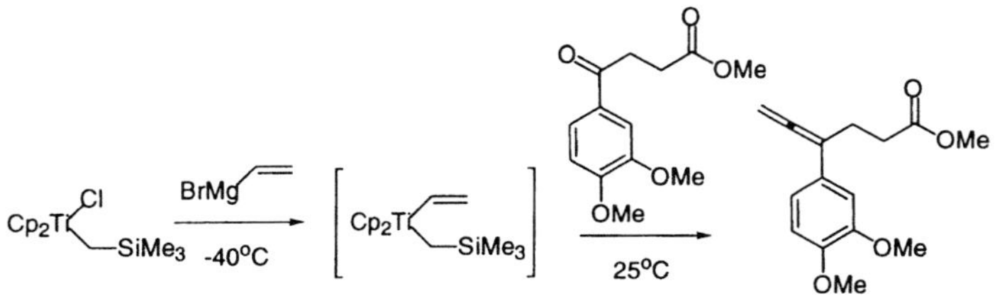

3.2.3
Alkyne Reactions

The reaction of alkynes (**77**) with titanocene methylidene (**5**), generated from a variety of sources including the Tebbe reagent (**7**) [37, 39], titanacyclobutanes (**10**) [38], or dimethyl titanocene (**9**) [54, 55], gives titanacyclobutenes (**12**), which can also be formed from the reaction of cyclopropenes (**78**) with the titanocene bis-phosphine complex (**79**) [123].

Regardless of the formation, titanacyclobutenes (**12**) react with aldehydes or ketones to form homoallylic alcohols after hydrolysis (**80**) [41, 54]), or with nitriles to form β,γ-unsaturated ketones (**81** [41]). They also undergo insertion with carbon monoxide or isonitriles [40], and ring opening to the isomeric vinyl carbenes [56]. Protonolysis of titanacyclobutenes gives methyl-substituted alkenes (**83**), which can also be formed directly via the carbotitanation of alkynes with dimethyl titanocene followed by protonolysis [54, 55]. In some cases, titanacyclobutenes react with aldehydes and ketones to form substituted 1,3-dienes (**84** [45]) and they undergo double nitrile insertion leading to sub-stituted pyridines (**82** [43]). They also react with dichlorophosphines to give phosphacyclobutene derivatives (**85** [42, 44]), while a similar reaction with dichloroarsines gives arsacyclobutenes [44].

3.2.4
Nitrile Reactions

Several reactions of nitriles with titanocene derivatives were studies by Doxsee *et al.* [10]. It was shown that 2 equivalents of nitriles react with *in situ* generated titanocene methylidene (**5**), formed from the Tebbe reagent (**7** [47,124]), or from a titanacyclobutane (**10** [46, 47]). The resulting 1,3-diazatitanacyclohexadienes (e.g. 86) can be hydrolyzed to β-ketoenamines [46] or 4-amino-1-azadienes.

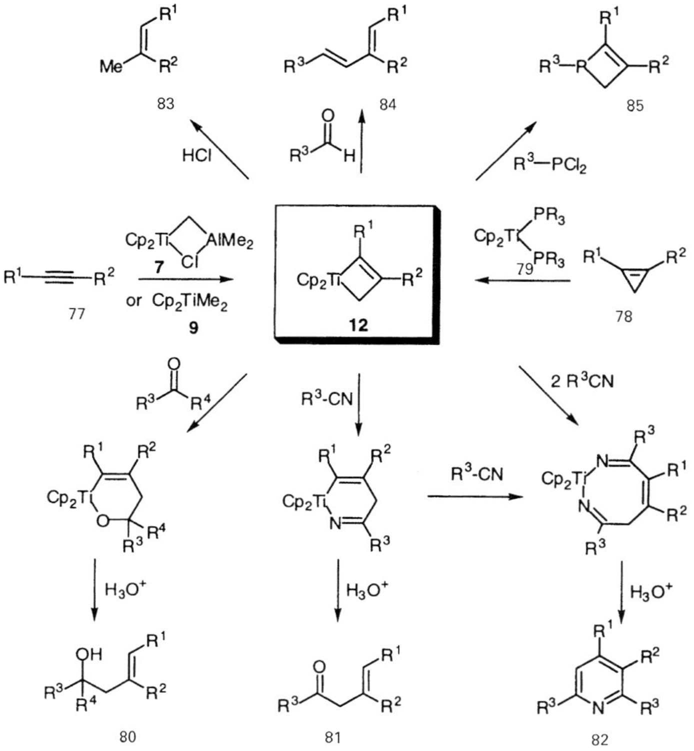

Dimethyl titanocene (**9**) was found to be a more convenient reagent for this process [54, 55, 58].

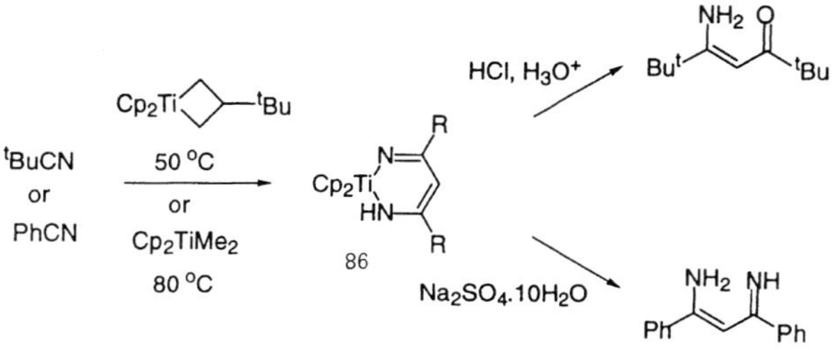

A vinylimido titanocene complex [124, 125] can be generated by the reaction of nitriles with the Tebbe reagent in the presence of DMAP or PMe$_3$. Subsequent addition to ketones, imines, or nitriles gives 3-substituted ketones after hydrolysis [47].

3.2.5
Olefin Metathesis Reactions

Although several other transition metal carbenes are most often used for this purpose, titanocene alkylidenes are capable of performing olefin metathesis reactions. Thus, they add to olefins to form titanacyclobutanes (10) which can then undergo alternative ring opening to generate a new titanocene alkylidene and a new olefin [11]. However, the parent titanocene methylidene species (5) is not very effective for productive metathesis because it is usually more stable than the more substituted alkylidenes and because of the propensity of titanacyclobutanes to undergo ring opening in a manner opposite to their formation to regenerate 5 [126]. This difficulty can be overcome if the starting olefin is highly strained [127] and the productive metathesis pathway gives less strained products.

Grubbs and co-workers reported an elegant application of this concept to the synthesis of polyquinanes based on olefin metathesis followed by intramolecular carbonyl olefination via the resulting titanacyclobutane (87 [126, 128]).

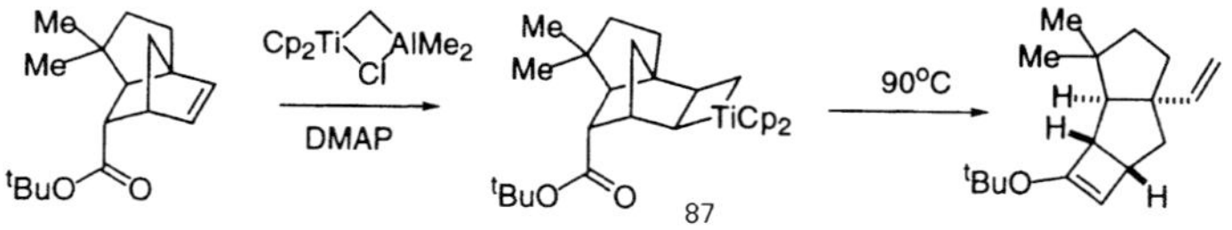

The cyclization of enynes (e.g. 88) via a geminal titanium–aluminum species (89) and a titanocene vinylidene intermediate (90) was reported by Dennehy and Whitby [129]. The resulting titanacyclobutane (91) could then be converted to various titanium-free products.

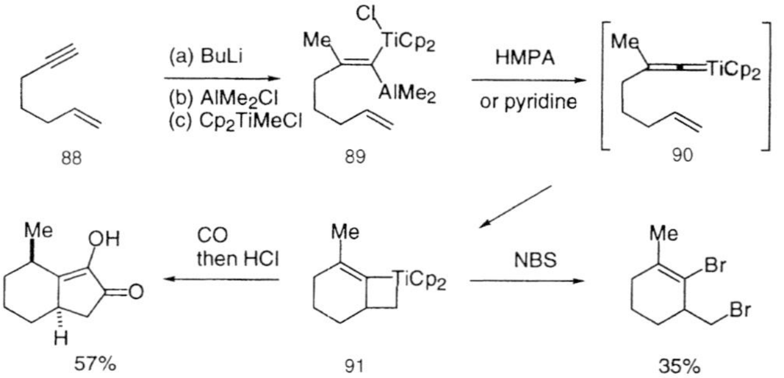

The type of titanium-mediated metathesis chemistry was also adapted for the ring-closing metathesis of dienes, and increasingly popular cyclization process [130]. Nicolaou et al. [131, 132] utilized such a strategy for the synthesis of complex polycyclic ethers via the one-pot olefination of unsaturated esters (e.g. 92 [131] or 93 [132]) with the Tebbe reagent (7) or with dimethyl titanocene (9), followed by olefin metathesis. With acid-labile substrates the use of 9 was preferred over 7.

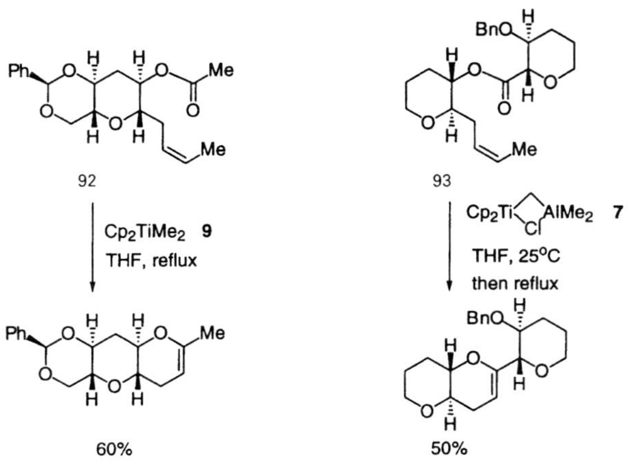

3.2.6
Ring-opening Metathesis Polymerizations (ROMP)

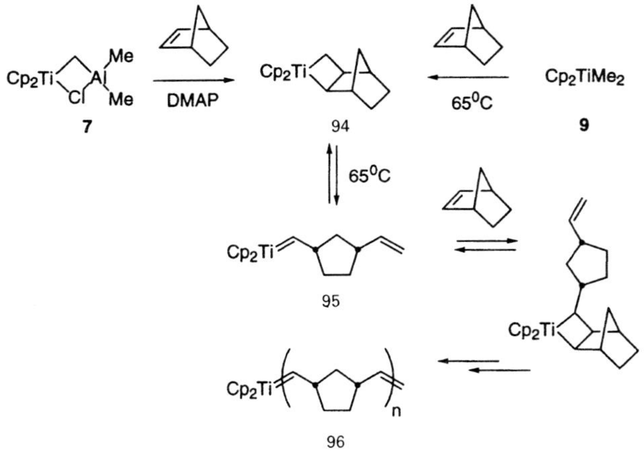

The ring-opening metathesis polymerization (ROMP) of cyclic alkenes [13] has been performed with a variety of transition-metal carbenes. Although certain initiators involving molybdenum, tungsten, or ruthenium are generally more effective for this purpose, a number of titanium-based initiators were identified, which helped elucidate the detailed mechanism of this process. Gilliom and Grubbs [32] reported the first living ROMP of norbornene by using a preformed titanacyclobutane (**94**) prepared with the Tebbe reagent (**7**). A wide variety of other monomers containing strained olefins was also subjected to ROMP [6, 133–135]. The use of dimethyl titanocene (**9**) and other dialkyl titanocenes as initiators of the ROMP of norbornene was also demonstrated [53]. These polymerizations presumably take place via *in situ* generated titanium alkylidene intermediates (e.g. **95**, **96**).

3.2.7
References

1. M.T. REETZ, *Organotitanium Reagents in Organic Synthesis,* Springer-Verlag, Berlin, **1986**.
2. M.T. REETZ in *Organometallics in Synthesis – A Manual* (Ed.: M. SCHLOSSER), John Wiley, New York, **1994**, p. 195.
3. M. BOCHMANN in *Comprehensive Organometallic Chemistry II* (Eds.: E.W. ABEL, F.G.A. STONE, G. WILKINSON), Pergamon Press, Oxford, **1995**, Vol. 12, p. 273.
4. N.A. PETASIS, Y.H. HU, *Curr. Org. Chem.* **1997**, *7*, 249.
5. K.A. BROWN-WENSLEY, S.L. BUCHWALD, L. CANNIZZO, L. CLAWSON, S. HO, D. MEINHARDT, J.R. STILLE, D. STRAUS, R.H. GRUBBS, *Pure Appl. Chem.* **1983**, *55*, 1733.
6. R.H. GRUBBS, W. TUMAS, *Science* **1989**, *243*, 907.
7. J.R. Stille in *Comprehensive Organometallic Chemistry II* (Eds.: E.W. ABEL, F.G.A. STONE, G. WILKINSON), Pergamon Press, Oxford, **1995**, Vol. 12, p. 577.
8. N.A. Petasis, S.P. Lu, E.I. Bzowej, D.K. Fu, J.P. Staszewski, I. Akritopoulou-Zanze, M.A. Patane, Y.H. Hu, *Pure Appl. Chem.* **1996**, *67*, 667.
9. S.H. PINE, *Org. React.* **1993**, *43*, 1.
10. K.M. DOXSEE, J.K.M. MOUSER, J.B. FARAHI, *Synlett* **1992**, 13.
11. R.H. GRUBBS, S.H. PINE IN *Comprehensive Organic Synthesis* (Ed.: B.M. TROST), Pergamon Press, New York, **1991**, Vol. 5, p. 1115.
12. P.D. GAVENS, M. BOTTRILL, J.W. KELLAND, J. MCMEEKING in *Comprehensive Organometallic Chemistry* (Eds.: G. WILKINSON, F.G.A. STONE, E.W. ABEL), Pergamon Press, Oxford, **1982**, Vol. 3, p. 475.
13. J.S. MOORE in *Comphrensive Organometallic Chemistry II* (Eds.: E.W. ABEL, F.G.A. STONE, G. WILKINSON), Pergamon Press, Oxford, **1995**, Vol. 12, p. 1209.
14. R.R. SCHROCK, *Ace. Chem. Res.* **1979**, *12*, 98.
15. W.A. NUGENT, J.M. MAYER, *Metal-Ligand Multiple Bonds,* John Wiley, New York, **1988**.
16. M.M. FRANCL, W.J. HEHRE, *Organometallics* **1983**, *2*, 457.
17. M.M. FRANCL, W.J. PIETRO, R.F. HOUT JR., W.J. HEHRE, *Organometallics* **1983**, *2*, 815.
18. A.R. GREGORY, E.A. MINTZ, *J. Am. Chem. Soc.* **1985**, *107*, 2179.
19. D.S. MARYNICK, C.M. KIRKPATRICK, *J. Am. Chem. Soc.* **1985**, *107*, 1993.
20. T.R. CUNDARI, M.S. GORDON, *J. Am. Chem. Soc.* **1991**, *114*, 539.
21. T.R. CUNDARI, M.S. GORDON, *J. Am. Chem. Soc.* **1991**, *113*, 5231.
22. B. SCHIOTT, K.A. JORGENSEN, *J. Chem. Soc., Dalton Trans.* **1993**, 337.
23. M. BOTTRILL, P.D. GAVENS, J.W. KELLAND, J. MCMEEKING in *Comprehensive Organometallic Chemistry* (Eds.: G. WILKINSON, F.G.A. STONE, E.W. ABEL), Pergamon Press, Oxford, **1982**, Vol. 3, p. 331.
24. J.D. MEINHART, E.V. ANSLYN, R.H. GRUBBS, *Organometallics* **1989**, *8*, 583.
25. F.N. TEBBE, G.W. PARSHALL, G.S. REDDY, *J. Am. Chem. Soc.* **1978**, *100*, 3611.
26. S.H. PINE, R. ZAHLER, D.A. EVANS, R.H. GRUBBS, *J. Am. Chem. Soc.* **1980**, *102*, 3270.
27. S.H. PINE, R.J. PETTIT, G.D. GEIB, S.G. CRUZ, C.H. GALLEGO, T. TIJERINA, R.D. PINE, *J. Org. Chem.* **1985**, *50*, 1212.
28. S.H. PINE, G. KIM, V. LEE, *Org. Synth.* **1990**, *67*, 72.
29. L.F. CANNIZZO, R.H. GRUBBS, *J. Org. Chem.* **1985**, *50*, 2386.
30. J.B. LEE, G.J. GAJDA, W.P. SCHAEFER, T.R. HOWARD, T. IKARIYA, D.A. STRAUS, R.H. GRUBBS, *J. Am. Chem. Soc.* **1981**, *103*, 7358.

31 D. A. Straus, R. H. Grubbs, *Organometallics* **1982**, *7*, 1658.
32 L. R. Gilliom, R. H. Grubbs, *J. Am. Chem. Soc.* **1986**, *108*, 733.
33 E. V. Anslyn, R. H. Grubbs, *J. Am. Chem. Soc.* **1987**, *109*, 4880.
34 L. Clawson, S. L. Buchwald, R. H. Grubbs, *Tetrahedron Lett.* **1984**, *25*, 5733.
35 S. L. Buchwald, R. H. Grubbs, *J. Am. Chem. Soc.* **1983**, *105*, 5490.
36 J. M. Hawkins, R. H. Grubbs, *J. Am. Chem. Soc.* **1988**, *110*, 2821.
37 F. N. Tebbe, R. L. Harlow, *J. Am. Chem. Soc.* **1980**, *102*, 6149.
38 T. R. Howard, J. B. Lee, R. H. Grubbs, *J. Am. Chem. Soc.* **1980**, *102*, 6876.
39 R. J. McKinney, T. H. Tulip, D. L. Thorn, T. S. Coolbaugh, F. N. Tebbe, *J. Am. Chem. Soc.* **1981**, *103*, 5584.
40 J. D. Meinhart, B. D. Santarsiero, R. H. Grubbs, *J. Am. Chem. Soc.* **1986**, *108*, 3318.
41 J. D. Meinhart, R. H. Grubbs, *Bull. Chem. Soc. Jpn.* **1988**, *61*, 111.
42 K. M. Doxsee, G. S. Shen, C. B. Knobler, *J. Am. Chem. Soc.* **1989**, *111*, 9129.
43 K. M. Doxsee, J. K. M. Mouser, *Organometallics* **1990**, *9*, 3012.
44 W. Tumas, J. A. Suriano, R. L. Harlow, *Angew. Chem., Int. Ed. Engl.* **1990**, *29*, 75.
45 K. M. Doxsee, J. K. M. Mouser, *Tetrahedron Lett.* **1991**, *32*, 1687.
46 K. M. Doxsee, J. B. Farahi, *J. Am. Chem. Soc.* **1988**, *110*, 7239.
47 K. M. Doxsee, J. B. Farahi, H. Hope, *J. Am. Chem. Soc.* **1991**, *113*, 8889.
48 v. K. Clauss, H. Bestian, *Justus Liebigs Ann. Chem.* **1962**, 8.
49 J. F. Payack, D. L. Hughes, D. W. Cai, I. F. Cottrell, T. R. Verhoeven, *Org. Prep. Proced. Internat.* **1995**, *27*, 707.
50 N. A. Petasis, E. I. Bzowej, *J. Am. Chem. Soc.* **1990**, *112*, 6392.
51 N. A. Petasis, S. P. Lu, *Tetrahedron Lett.* **1995**, *36*, 2393.
52 N. A. Petasis, Y. H. Hu, D. K. Fu, *Tetrahedron Lett.* **1995**, *36*, 6001.
53 N. A. Petasis, D. K. Fu, *J. Am. Chem. Soc.* **1993**, *115*, 7208.
54 N. A. Petasis, D. K. Fu, *Organometallics* **1993**, *13*, 3776.
55 K. M. Doxsee, J. J. J. Juliette, J. K. M. Mouser, K. Zientara, *Organometallics* **1993**, *72*, 4682.
56 K. M. Doxsee, J. J. J. Juliette, J. K. M. Mouser, K. Zientara, *Organometallics* **1993**, *72*, 4742.
57 K. M. Doxsee, J. J. J. Juliette, K. Zientara, G. Nieckarz, *J. Am. Chem. Soc.* **1994**, *116*, 2147.
58 J. Barluenga, C. D. Losada, B. Olano, *Tetrahedron Lett.* **1992**, *33*, 7579.
59 N. A. Petasis, E. I. Bzowej, *J. Org. Chem.* **1992**, *57*, 1327.
60 N. A. Petasis, I. Akritopoulou, *Synlett* **1992**, 665.
61 N. A. Petasis, E. I. Bzowej, *Tetrahedron Lett.* **1993**, *34*, 943.
62 N. A. Petasis, Y.-H. Hu, *J. Org. Chem.* **1997**, *62*, 782.
63 P. Binger, P. Muller, R. Benn, R. Mynott, *Angew. Chem., Int. Ed. Engl.* **1989**, *28*, 610.
64 C. Cainelli, F. Bertini, P. Grasselli, G. Zubiani, *Tetrahedron Lett.* **1967**, 5153.
65 J. W. Bruin, G. Schat, O. S. Akkerman, F. Bickelhaupt, *Tetrahedron Lett.* **1983**, *24*, 3935.
66 B. J. V. D. Heisteeg, G. Schat, O. S. Akkerman, F. Bickelhaupt, *Tetrahedron Lett.* **1987**, *28*, 6493.
67 J. J. Eisch, A. Piotrowski, *Tetrahedron Lett.* **1983**, *24*, 2043.
68 K. Takai, Y. Hotta, K. Oshima, H. Nozaki, *Tetrahedron Lett.* **1978**, *27*, 2417.
69 K. Takai, Y. Hotta, K. Oshima, H. Nozaki, *Bull. Chem. Soc. Jpn.* **1980**, *53*, 1698.
70 J. Hibino, T. Okazoe, K. Takai, H. Nozaki, *Tetrahedron Lett.* **1985**, *26*, 5579.
71 T. Okazoe, J. Hibino, K. Takai, H. Nozaki, *Tetrahedron Lett.* **1985**, *26*, 5581.
72 L. Lombardo, *Tetrahedron Lett.* **1982**, *23*, 4293.
73 L. Lombardo, *Org. Synth.* **1987**, *65*, 81.
74 K. Takai, T. Kakiuchi, Y. Kataoka, K. Utimoto, *J. Org. Chem.* **1994**, *59*, 2668.
75 T. Okazoe, K. Takai, K. Oshima, K. Utimoto, *J. Org. Chem.* **1987**, *52*, 4410.
76 K. Takai, Y. Kataoka, T. Okazoe, K. Utimoto, *Tetrahedron Lett.* **1988**, *29*, 1065.
77 K. Takai, O. Fujimura, Y. Kataoka, K. Utimoto, *Tetrahedron Lett.* **1989**, *30*, 211.

78 Y. Horikawa, M. Watanabe, T. Fujiwara, T. Takeda, *J. Am. Chem. Soc.* **1997**, *119*, 1127.
79 T. Yoshida, E. Negishi, *J. Am. Chem. Soc.* **1981**, *103*, 1276.
80 C.E. Tucker, B. Greve, W. Klein, P. Knochel, *Organometallics* **1994**, *13*, 94.
81 A. Maercker, *Org. React.* **1965**, *14*, 270.
82 J.I.G. Cadogen, *Organophosphorous Reagents* in Organic Synthesis, Academic Press, London, **1979**.
83 B.E. Maryanoff, A.B. Reitz, *Chem. Rev.* **1989**, *89*, 863.
84 D.J. Ager, *Org. React.* **1990**, *38*, 1.
85 P.J. Murphy, J. Brennan, *Chem. Soc. Rev.* **1988**, 1.
86 N. Ikemoto, L.S. Schreiber, *J. Am. Chem. Soc.* **1992**, *114*, 2524.
87 J.D. Winkler, C.L. Muller, R.D. Scott, *J. Am. Chem. Soc.* **1988**, *110*, 4831.
88 S.H. Pine, G.S. Shen, H. Hong, *Synthesis* **1991**, 165.
89 T.V. RajanBabu, G.S. Reddy, *J. Org. Chem.* **1986**, *51*, 5458.
90 F. Nicotra, L. Panza, G. Russo, *Tetrahedron Lett.* **1991**, *32*, 4035.
91 L. Lay, F. Nicotra, L. Panza, G. Russo, *Synlett* **1995**, 167.
92 W.A. Kinney, M.J. Coghlan, L.A. Paquette, *J. Am. Chem. Soc.* **1985**, *107*, 7352.
93 H.J. Kang, L.A. Paquette, *J. Am. Chem. Soc.* **1990**, *112*, 3252.
94 L.A. Paquette, D. Friedrich, R.D. Rogers, *J. Org. Chem.* **1991**, *56*, 3841.
95 C.M.G. Philippo, N.H. Vo, L.A. Paquette, *J. Am. Chem. Soc.* **1991**, *113*, 2762.
96 L.A. Paquette, C.M.G. Philippo, N.H. Vo, *Can. J. Chem.* **1992**, *70*, 1356.
97 S. Borrelly, L.A. Paquette, *J. Am. Chem. Soc.* **1996**, *118*, 727.
98 B.J.J. van de Heisteeg, G. Schat, O.S. Akkerman, F. Bickelhaupt, *J. Organomet. Chem.* **1986**, *308*, 1.
99 E.B. Tjaden, G.L. Casty, M. Stryker, *J. Am. Chem. Soc.* **1993**, *115*, 9814.
100 J.R. Stille, R.H. Grubbs, *J. Am. Chem. Soc.* **1983**, *105*, 1664.
101 D.L. Hughes, J.F. Payack, D. Cai, T.R. Verhoeven, P.J. Reider, *Organometallics* **1996**, *75*, 663.
102 C. Marschner, G. Penn, H. Griengl, *Tetrahedron* **1993**, *49*, 5067.
103 T. Matsuura, S. Nishiyama, S. Yamamura, *Chem. Lett.* **1993**, 1503.
104 P. DeShong, P.J. Rybczynski, *J. Org. Chem.* **1991**, *56*, 3207.
105 R. Csuk, B.I. Glanzer, *Tetrahedron* **1991**, *47*, 1655.
106 V. Faivre-Buet, I. Eynard, H.N. Nga, G. Descotes, A. Grouiller, *J. Carbohydr. Chem.* **1993**, *72*, 349.
107 L.M. Dollinger, A.R. Howell, *J. Org. Chem.* **1996**, *61*, 7248.
108 M.J. Kates, J.H. Schauble, *J. Org. Chem.* **1994**, *59*, 494.
109 P.J. Colson, L.S. Hegedus, *J. Org. Chem.* **1993**, *58*, 5918.
110 D. Kuzmich, S.C. Wu, D.-C. Ha, C-S. Lee, S. Ramesh, S. Atarashi, J.-K. Choi, D.J. Hart, *J. Am. Chem. Soc.* **1994**, *116*, 6943.
111 H.K. Chenault, L.F. Chafin, *J. Org. Chem.* **1994**, *59*, 6167.
112 H.K. Chenault, A. Castro, L.F. Chafin, J. Yang, *J. Org. Chem.* **1996**, *61*, 5024.
113 N.A. Petasis, M.A. Patane, *Tetrahedron Lett.* **1990**, *31*, 6799.
114 N.A. Petasis, E.I. Bzowej, *Tetrahedron Lett.* **1993**, *34*, 1721.
115 N.A. Petasis, S.P. Lu, *J. Am. Chem. Soc.* **1995**, *117*, 6394.
116 N.A. Petasis, S.P. Lu, *Tetrahedron Lett.* **1996**, *37*, 141
117 J.P. Begue, M.H. Rock, *J. Organomet. Chem.* **1995**, *489*, 1.
118 M. Furber, L.N. Mander, D.L. Patrick, *J. Org. Chem.* **1990**, *55*, 4860.
119 A.G.M. Barrett, L.M. Melcher, B.C.B. Bezuidenhoudt, *Carbohydr. Res.* **1992**, *232*, 259.
120 B.M. Johnson, K.P.C. Vollhardt, *Synlett* **1990**, 209.
121 M. Mortimore, P. Kocienski, *Tetrahedron Lett.* **1988**, *29*, 3357.
122 R. Beckhaus, I. Straub, T. Wagner, *J. Organomet. Chem.* **1994**, *464*, 155.
123 P. Binger, P. Mueller, A.T. Herrmann, P. Philipps, B. Gabor, F. Langhauser, C. Krueger, *Chem. Ber.* **1991**, *124*, 2165.
124 K.M. Doxsee, J.B. Farahi, *J. Chem. Soc., Chem. Commun.* **1990**, 1452.

125 K. M. Doxsee, L. C. Garner, J. J. J. Juliette, J. K. M. Mouser, T. J. R. Weakly, *Tetrahedron* **1995**, *57*, 4321.
126 J. R. Stille, B. D. Santarsiero, R. H. Grubbs, *J. Org. Chem.* **1990**, *55*, 843.
127 L. R. Gilliom, R. H. Grubbs, *Organometallics* **1986**, *5*, 721.
128 J. R. Stille, R. H. Grubbs, *J. Am. Chem. Soc.* **1986**, *108*, 855.
129 R. D. Dennehy, R. J. Whitby, *J. Chem. Soc., Chem. Commun.* **1990**, 1060.
130 R. H. Grubbs, S. J. Miller, G. C. Fu, *Acc. Chem. Res.* **1995**, *28*, 446.
131 K. C. Nicolaou, M. H. D. Postema, C. F. Claiborne, *J. Am. Chem. Soc.* **1996**, *118*, 1565.
132 K. C. Nicolaou, M. H. D. Postema, E. W. Yue, A. Nadin, *J. Am. Chem. Soc.* **1996**, *118*, 10335.
133 F. L. Klavetter, R. H. Grubbs, *J. Am. Chem. Soc.* **1988**, *110*, 7807.
134 T. M. Swager, R. H. Grubbs, *J. Am. Chem. Soc.* **1988**, *110*, 807.
135 T. M. Swager, D. A. Dougherty, R. H. Grubbs, *J. Am. Chem. Soc.* **1988**, *110*, 2973.

3.3
The McMurry Reaction and Related Transformations

Alois Fürstner

3.3.1
Introduction

In the early 1970s Mukaiyama, Tyrlik, and McMurry made the independent and almost simultaneous discovery that low-valent titanium [Ti], prepared by reduction of $TiCl_x$ ($x=3,4$) with an appropriate reducing agent, effects the coupling of aldehydes or ketones to alkenes (Scheme 1) [1–3]. This transformation, which is driven by the high reducing ability and the pronounced oxophilicity of [Ti], has witnessed a considerable scope and has found many applications to advanced organic synthesis [4–6]. The fact that this method provides ready access to strained products and to molecules of theoretical interest which are difficult to prepare otherwise deserves particular mention. Another prominent feature is the template effect exerted by the titanium species which strongly biases intramolecular reactions of dicarbonyl substrates and makes the yields of the cycloalkenes formed essentially independent of the ring size. Moreover, the reaction can be stopped at the intermediate pinacol stage simply by lowering the temperature [4–6].

Despite the vast literature on this reductive C–C bond formation, which is generally referred to as the 'McMurry reaction' in honor of one of its pioneers, con-

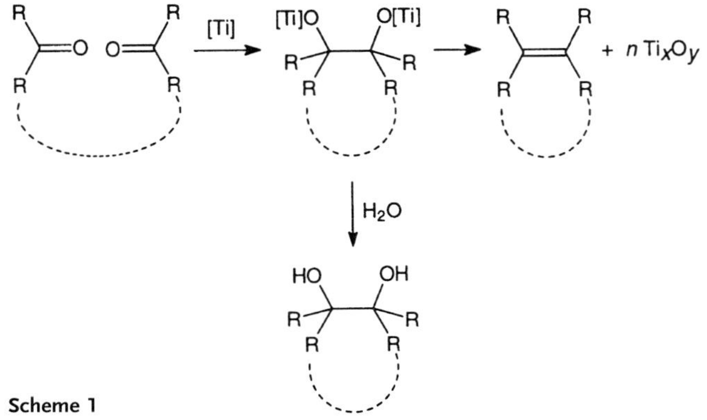

Scheme 1

Transition Metals for Organic Synthesis, Vol. 1, 2nd Edition.
Edited by M. Beller and C. Bolm
Copyright © 2004 WILEY-VCH Verlag GmbH & Co. KGaA, Weinheim
ISBN: 3-527-30613-7

ceptual advancements have been scarce. This situation, however, is likely to change. For example, the actual nature of the 'low-valent' titanium [Ti] formed by one of the standard recipes had been hardly understood for a long period of time; recent in-depth studies, however, shed light on the inorganic part of this reaction and have clearly revised the assumption that metallic Ti particles are necessarily involved [4]. Parallel to this, significantly improved procedures have been developed which make the reaction considerably more convenient and reliable. With regard to the starting materials, a set of functional groups – previously believed to be inert towards [Ti] – were found to undergo efficient intramolecular cross-couplings. This has largely expanded the pool of substrates and opened up a new entry into products other than simple alkenes (e.g. aromatic heterocycles) which were beyond the scope of the conventional McMurry process [4].

Due to the chemical inertness of the titanium oxides formed as the inorganic byproducts, reactions of this type have always been (over)stoichiometric in [Ti]. A very recent achievement concerns a new procedure which allows intramolecular cross-coupling reactions catalytic in titanium to be run for the first time [7]. This new vista may not only fertilize preparative titanium chemistry, but may have impact on other metal-induced C–C bond formations as well [8].

In the following section the present state of the art of low-valent titanium chemistry is briefly summarized. It provides a short overview on the scope and limitations of conventional McMurry olefin syntheses and outlines the recent developments in this field. The reader is also referred to those preparative procedures which – in the author's experience – turned out to be particularly fructuous.

3.3.2
Some Lessons from Inorganic Chemistry: The Family of McMurry Reagents

For a rather long period of time it has been believed that the reduction of $TiCl_3$ with a suitable reducing agent leads to the formation of a slurry of finely dispersed, nonpassivated particles of Ti(0), which were thought to be the actual coupling agent [5, 6a]. Differences in reactivity of differently prepared samples were essentially rationalized in terms of varying particle sizes and textures. Since a variety of reducing agents, including K, Na, Li, Zn, Zn(Cu), Mg, Mg(Hg), C_8K, $LiAlH_4$, etc. has been employed under quite different conditions, a wealth of procedures can be found in the literature.

Only recently, a series of detailed investigations has clearly revised this simplistic picture of the inorganic side of McMurry-type reactions. Thus, Bogdanovic and co-workers [9, 10, 14] have shown that, depending on the mode of preparation, distinctly different low-valent titanium species are formed, each of which may effect reductive C–C bond formations.

Specifically, the reduction of $TiCl_3$ with $LiAlH_4$, as introduced by McMurry and Fleming in their 1974 landmark paper [3] and widely used afterwards, leads to the formation of $[HTiCl(THF)_{\approx 0.5}]$ as the active species [9]. This hydride complex is also formed if $LiAlH_4$ is replaced by activated MgH_2 as the reducing agent.

3.3.2 Some Lessons from Inorganic Chemistry: The Family of McMurry Reagents

The same authors have also studied the TiCl$_3$/Mg reagent combination [10]. In this particular case, the prime product formed is [TiMgCl(THF)$_x$] rather than metallic titanium. Prolonged reaction of TiCl$_3$ with an excess of Mg, however, leads to the formation of [Ti(MgCl)$_2$(THF)$_x$]. The performance of these 'inorganic Grignard reagents' in carbonyl coupling has been probed and quantified in model reactions.

Even more striking are the results obtained in a study of the system TiCl$_3$/Zn or TiCl$_3$(DME)$_{1.5}$/Zn, respectively. Zinc as the reducing agent had been originally proposed by Mukaiyama et al. in 1973 [1], was re-emphasized by McMurry and others [11–13] and is nowadays widely used [4–6]. Therefore it is particularly surprising that Bogdanovic and Bolte have unequivocally shown that Zn (in the form of commercial Zn dust, Zn(Cu) couple, or activated Zn powder) does not reduce TiCl$_3$ as such [14]! *Reduction to a low-valent species, most likely TiCl$_2$, occurs only in the presence of a carbonyl compound.* The Lewis-acidic TiCl$_3$ must first coordinate to the Lewis-basic carbonyl group of the substrate and only then the reduction process mediated by Zn can take place. IR data, a quantitative analysis of the overall process, as well as trapping experiments give a quite detailed picture of the elementary steps along the reaction path [14]. In view of these results it can be safely recommended to skip the 'pre-reduction' step prescribed in the literature [11, 12], if the TiCl$_3$/Zn reagent combination is used to effect carbonyl couplings of any type. Interestingly enough, such a short-cut procedure has already been proposed prior to the publication of Bogdanovic and Bolte: It is the essence of the convenient 'instant method' to mix all ingredients and thus to prepare the active [Ti] species in the presence of the substrate [15].

These and related studies prove that distinctly different low-valent titanium reagents are formed depending on the particular reducing agent chosen and the

Tab. 1 Screening of different titanium species for reductive indole synthesis.

Entry	[Ti]	Formal oxidation state	Isolated yield [%]
1	TiCl$_3$ + 3C$_8$K	0	90
2	TiCl$_3$ + 2C$_8$K	+1	90
3	TiCl$_4$ + 4 K[BEt$_3$H]	0	67
4	Ti(toluene)$_2$	0	75
5	Ti(biphenyl)$_2$	0	70
6	[HTiCl · (THF)$_{0.5}$]	+2	85
7	[TiH$_2$(MgCl$_2$)$_n$(THF)$_2$]	+2	69
8	Cp$_2$Ti(PMe$_3$)$_2$	+2	79

mode of preparation [4]. The notion that carbonyl coupling is a common feature of a quite diverse set of [Ti] species varying in their formal oxidation state, solubility, and ligand sphere was independently confirmed by a model study on the intramolecular oxo-amide cross-coupling reaction of N-benzoyl-2-aminobenzophenone to 2,3-diphenylindole (Tab. 1) [15].

3.3.3
Recommended Procedures

3.3.3.1
Titanium–Graphite and Other Supported Titanium Reagents

Alkali metals have a long tradition as reducing agents for the preparation of activated metals [16] and have been applied to titanium with considerable success. A particularly efficient modification makes use of the well-known potassium–graphite intercalation compound C_8K rather than of molten potassium. C_8K is readily prepared by stirring potassium and the appropriate amount of graphite for ≤10 min at 150 °C under argon. Since the 4s electrons of potassium are delocalized in the π-system of its host, the reduction of $TiCl_3$ added to a suspension of C_8K in an ethereal solvent can occur almost simultaneously at any site of the extended surface area of the laminate, resulting in the formation of nanosized [Ti] particles. Since they become absorbed on the surface of the graphite, which merely acts as an inert support, premature aggregation of the ultrafine [Ti] is retarded and its activity is retained even under quite forcing conditions.

Originally, titanium–graphite obtained by reduction of $TiCl_3$ with $3C_8K$ has been described and successfully applied to various McMurry-type reactions [17]. However, Clive et al. have reported that titanium–graphite obtained from $TiCl_3$ and only $2C_8K$ (thus formally a Ti(+1) species) is preferable and gives excellent results in cases in which other McMurry procedures completely fail [18]. Such a case is the annelation of the A-ring in a total synthesis of (+)-compactin (Scheme 2).

A fairly comprehensive comparative study of these two modifications of titanium–graphite [15] has fully confirmed the conclusions reached by Clive et al., with the system $TiCl_3/2C_8K$ leading in fact to remarkable and well reproducible results. The use of this highly activated form of [Ti] is therefore recommended without reservation. It can be applied to rather acid sensitive substrates and is reactive enough to effect couplings even at low temperature.

However, the use of the pyrophoric C_8K as the reducing agent may intimidate potential users. Although this beautifully bronze-colored intercalation compound is very easy to prepare in a batchwise manner that minimizes the risks [19], the development of a less hazardous but equally efficient alternative is desirable. One step towards this end relies on the use of 'high-surface' sodium on inorganic supports such as Al_2O_3, TiO_2, or NaCl [20]. Reduction of $TiCl_3$ with 2 equivalents of one of these non-pyrophoric reducing agents leads to a low-valent titanium species, which turned out to be highly effective for the coupling of aromatic sub-

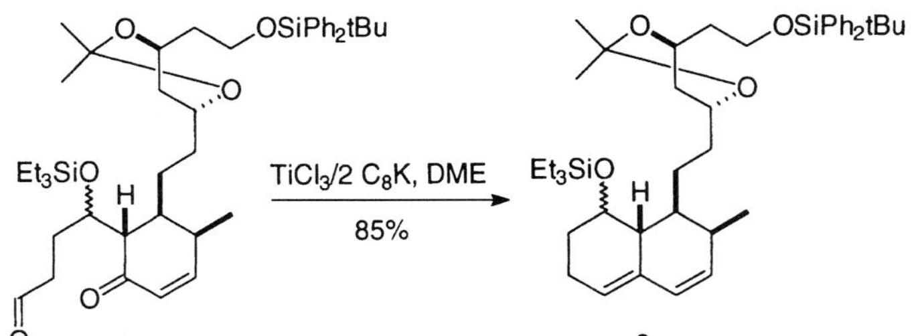

Scheme 2

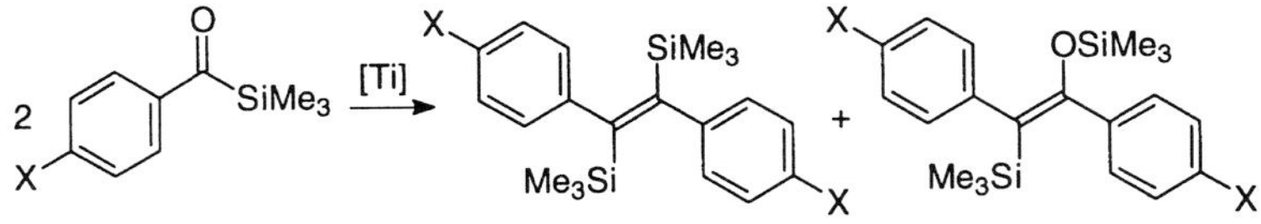

Scheme 3

strates. It also accounts for the first McMurry-type reactions of acylsilanes **3** to 1,2-disilyl-ethene derivatives **4** (X=H, Br, OMe) (Scheme 3) [20b].

3.3.3.2
The TiCl$_3$/Zn Reagent Combinations

Zinc as the reducing agent for TiCl$_x$ (x=3,4) was originally proposed by Mukaiyama et al. [1] and has been widely used ever since [4–6]. An optimized procedure published in 1989 by McMurry et al. recommends to replace commercial Zn dust by Zn(Cu) couple and TiCl$_3$ by TiCl$_3$(DME)$_{1.5}$ [11]. The sky-blue solvate complex can be purified by crystallization and thus helps to avoid problems caused by TiCl$_3$ batches of different quality. This modification has been successfully applied in many cases and led to significant improvements in coupling reactions which are known to be troublesome. For example, the isolated yield of tetra-isopropyl-ethene **7** has been raised to 87%, which nicely contrasts to the 12% originally described using TiCl$_3$/LiAlH$_4$ (Scheme 4) [11].

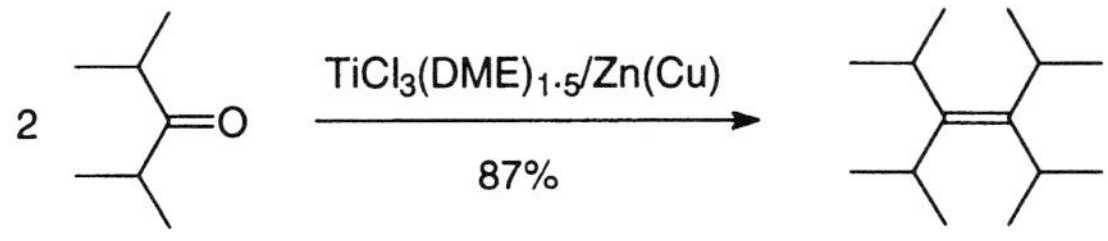

Scheme 4

As outlined in Section 3.3.2, recent investigations have revealed that Zn is unable to reduce $TiCl_3$ or $TiCl_3(DME)_{1.5}$ in the absence of a carbonyl compound [14]. With this result in mind, the 'pre-reduction' of $TiCl_3(DME)_{1.5}$ by Zn(Cu) for 2–5 h prior to the addition of the substrate as recommended in the optimized procedure [11] is unnecessary and can be skipped without loss in performance. In clear contrast to current practice it has been proposed in 1994 to add the carbonyl compound to the other ingredients of a McMurry coupling *prior* to the addition of the $TiCl_x$.

As shown in Scheme 5 this leads to the spontaneous formation of titanium complexes via Lewis-acid/Lewis-base interactions as clearly visible in IR. These complexes can then be reduced by means of Zn or other mild reducing agents that do not affect the substrate on their own. As a consequence the low-valent [Ti] emerges in a site selective manner within the coordination sphere of its substrate and leads to a smooth and efficient conversion (Scheme 5).

Although this 'instant' method [15] has been devised in the context of titanium-induced heterocycle syntheses (cf. Section 3.3.6.3), it soon turned out to be applicable to conventional McMurry reactions as well [15, 21]. It avoids any hazardous reagents, is carried out by simply mixing and heating all components in an inert solvent, and makes the up-scaling of the reactions an easy task. Carbonyl coupling reactions under 'instant' conditions exhibit the same selectivity profile as those

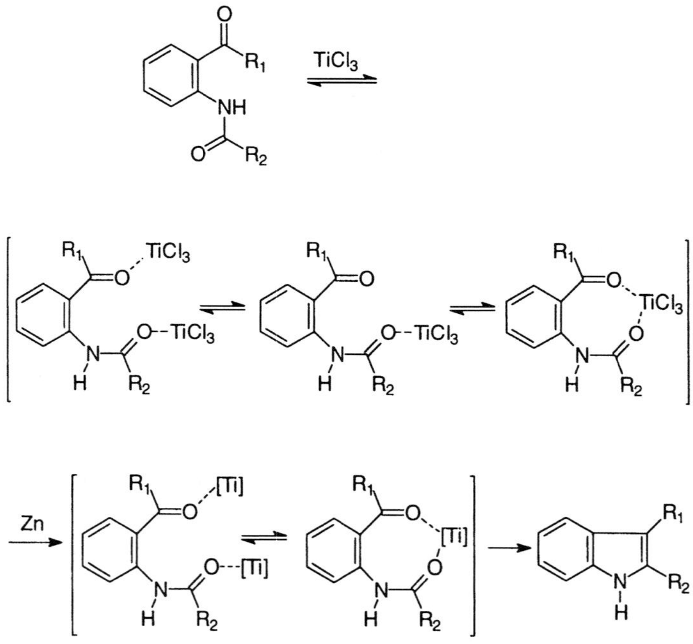

Scheme 5

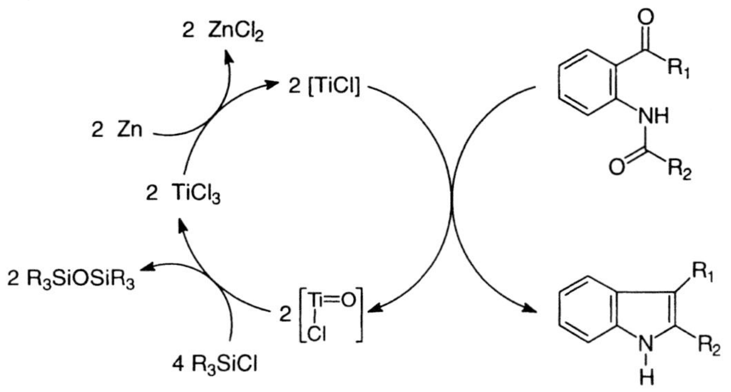

Scheme 6

using pre-formed [Ti]. Limits are set only in the case of very acid sensitive substrates which may suffer from mixing with the Lewis-acidic Ti(3+)-salts.

The formation of the active species in the presence of the substrate is an elementary requirement for catalysis. However, attempted re-reduction of the titanium oxides accumulating in any McMurry-type reaction to the active species by means of Zn is in vain. Therefore all titanium-induced reactions had been notoriously (over)stoichiometric until an indirect method was devised which circumvents this obstacle. It relies on a metathetic ligand exchange of the titanium oxides formed *in situ* with admixed chlorosilanes as formally depicted in Scheme 6. Since this regenerates $TiCl_x$ which will effect the next coupling event under 'instant' conditions, the first *titanium-catalyzed* intramolecular coupling reactions have been possible [7]. The efficiency and the turnover number of this manifold can be properly tuned by the appropriate choice of the chlorosilane.

3.3.3.3
Activation of Commercial Titanium

The ultimate solution for reductive carbonyl coupling might consist of the use of commercial titanium powder as an off-the-shelf reagent. However, this metal of low density and high mechanical strength exhibits an exceptional resistance to almost any kind of chemical attack which renders it ideally suited as a material for applications to industrial plants, marine equipment, and surgical implants. This inertness stems from a thin, nonporous and repairable oxide layer on its surface.

The catalytic scenario displayed in Scheme 6 relies on the reaction of titanium oxides formed *in situ* with admixed chlorosilanes. This suggested to probe whether such additives can also be used to degrade the superficial passivating layer on Ti. In fact, it turned out that a mixture of commercial Ti-powder (<100 mesh) and TMSCl in boiling THF or DME slowly but efficiently induces various inter- as well as intramolecular McMurry reactions of aromatic and *α,β*-unsaturated aldehydes or ketones as well as intramolecular cross-couplings leading to

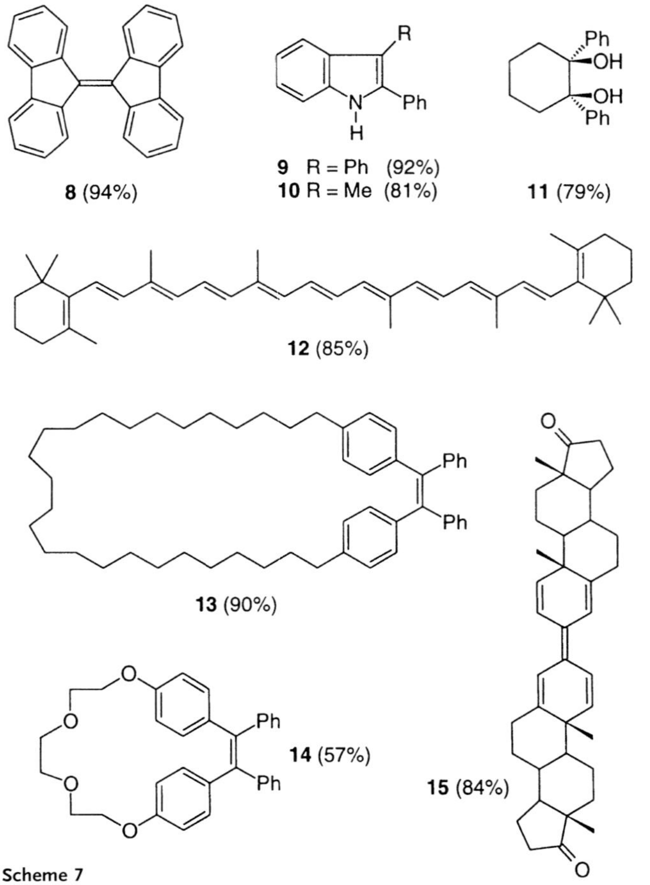

8 (94%)

9 R = Ph (92%)
10 R = Me (81%)

11 (79%)

12 (85%)

13 (90%)

14 (57%)

15 (84%)

Scheme 7

aromatic heterocycles [7, 22]. Purely aliphatic substrates, however, do not react. Some representative examples are shown in Scheme 7. The observed activity stems from altered surface and from altered bulk properties of the titanium particles caused by the treatment with the chlorosilane as can be deduced from a ^{29}Si NMR and electron microscopic study of this reagent combination [7, 23].

3.3.4
McMurry Coupling Reactions in Natural Product Synthesis

Ever since its discovery, the titanium-induced carbonyl coupling process has been applied with considerable success to numerous syntheses of natural products and analogs thereof. Because previous reviews provide a fairly comprehensive overview

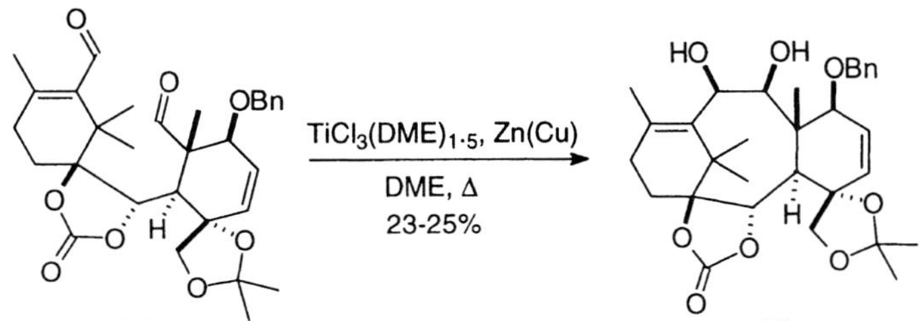

3.3.4 *McMurry Coupling Reactions in Natural Product Synthesis* | 457

Scheme 8

[4–6, 24, 25], the following section is mainly focused on recent highlights which delineate the present state of the art.

A landmark achievement is Nicolaou's total synthesis of Taxol, a promising anti-cancer agent in clinical use, which employs an intramolecular coupling to cyclize the central eight-membered ring of the target [26]. Treatment of the conformation-

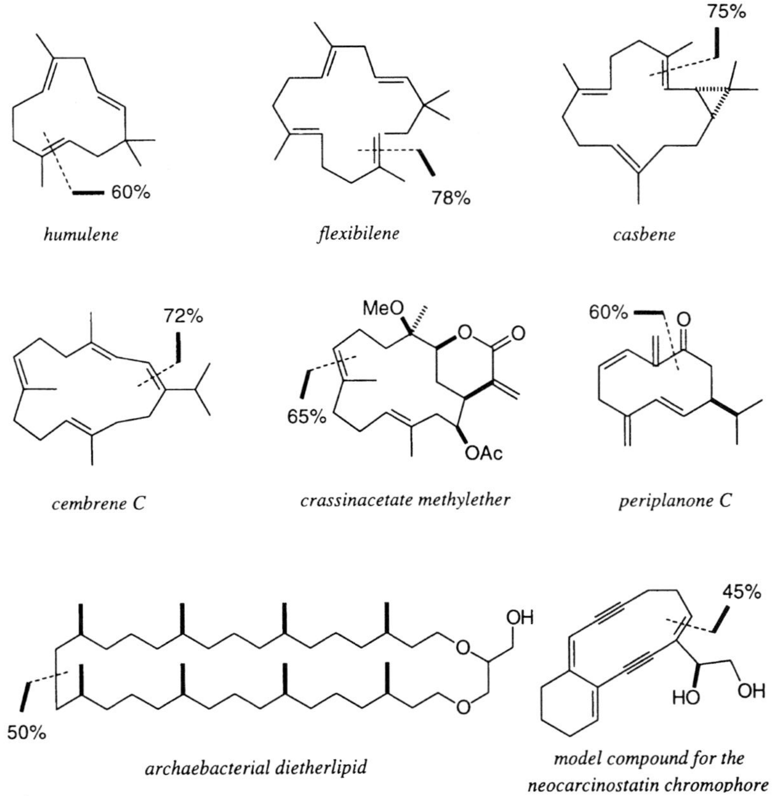

humulene — 60%

flexibilene — 78%

casbene — 75%

cembrene C — 72%

crassinacetate methylether — 65%

periplanone C — 60%

archaebacterial dietherlipid — 50%

model compound for the neocarcinostatin chromophore — 45%

Scheme 9

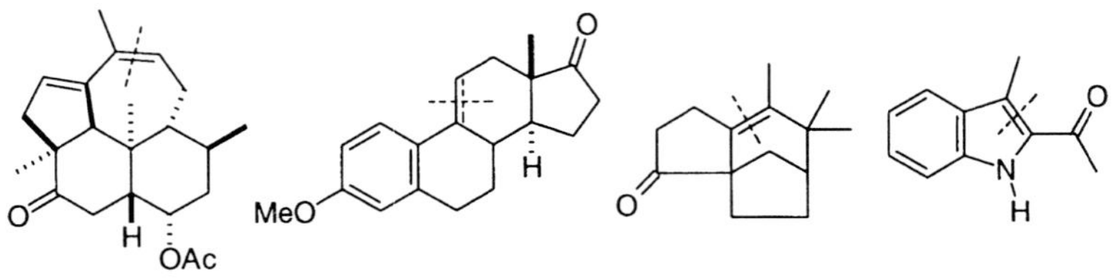

Scheme 10

ally restricted dialdehyde **16** with the TiCl$_3$(DME)$_{1.5}$/Zn(Cu) reagent afforded pinacol **17** in 23–25% yield along with three byproducts (Scheme 8). This seemingly low yield is nevertheless quite respectable if one considers the congestion and the labile functionalities of this particular substrate as well as the fact that titanium-induced reactions of polyoxygenated compounds in general are delicate due to the oxophilicity of [Ti].

Many natural product syntheses highlight the efficiency of McMurry reactions in forming macrocycles. Some representative examples are displayed in Scheme 9 [27]. Among them a recent study is worth mentioning in which a highly strained and thermally very sensitive eleven-membered dienediyne ring serving as a model for the neocarzinostatin chromophore has been obtained. A 36-membered archaebacterial diether lipid constitutes the largest ring to date formed by an intramolecular coupling under high dilution conditions.

Particular emphasis is given to the few examples reported so far in which two carbonyl groups out of three are selectively coupled without affecting the remaining one (Scheme 10). This striking semicompatibility of an unprotected ketone towards [Ti] becomes evident from an efficient approach to the tetracyclic diterpene kempene-2 **18** [28], some syntheses in the estrone series (e.g. **19**) [29], from an elegant preparation of isokhusimone **20** [30], and from the regio- and chemoselective formation of the indolalkaloid salvadoricine **21**, in which even an amide group reacts preferentially over the remaining ketone function [31].

3.3.5
Nonnatural Products

The high driving force of carbonyl coupling reactions stemming from the formation of titanium oxides as the inorganic byproducts can be exploited to build up strained molecules which are difficult to access otherwise [4–6]. Although the long-awaited tetra-*tert*-butylethene could not be reached even by this method, many other spectacular examples highlight this prominent feature of the McMurry reaction (Scheme 11) [32, 33]. It has been estimated that products with a strain energy of ≤ 19 kcal mol^{-1} can be obtained in such a way [5].

Most notable among them is the preparation of bridgehead alkenes such as **23** which upon protonation affords the bicyclic carbocation **24** with a three-center

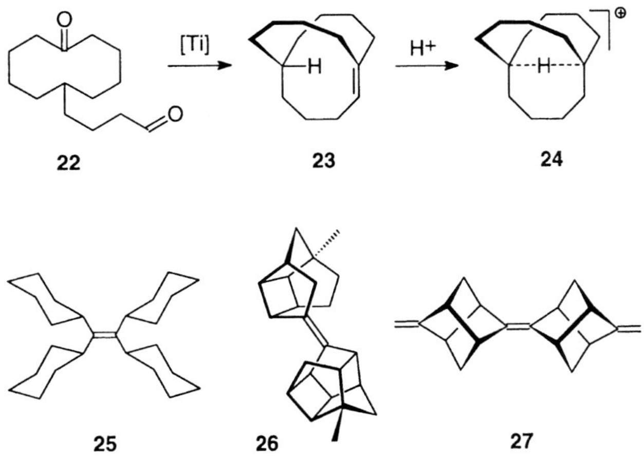

3.3.5 Nonnatural Products | 459

Scheme 11

two-electron C–H–C bond [32]. Likewise, transannular coupling events have been used for tying-off cyclophanes [34] or formylated calix[4]arene derivatives [21], despite the considerable strain that is built up in the products (Scheme 12). In this context the particularly convenient 'instant procedure' [15] was found to be a very suitable means to effect this energetically demanding transformation.

Many applications in the nonnatural product series prominently feature the efficiency of intramolecular McMurry-type reactions in forming cycloalkenes independent of the ring size. These experimental results are interpreted in terms of a strong template effect exerted by the polar surface of the reagent [4–6]. Although this effect is difficult to probe experimentally, a recent study on the synthesis of various crownophanes using a set of different low-valent titanium sources has put forward strong evidence that the preorganization of the dicarbonyl substrate by

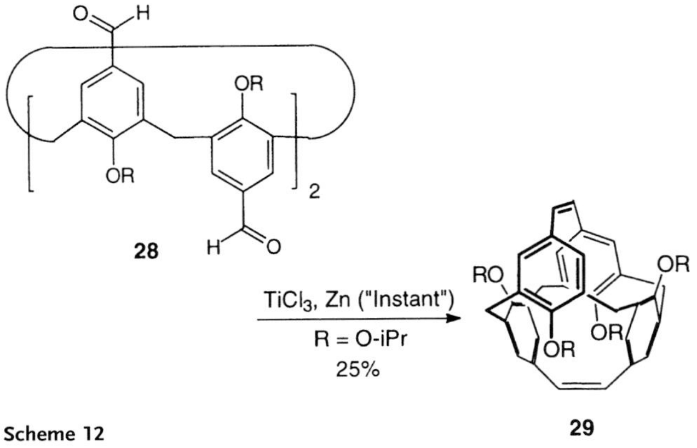

Scheme 12

the [Ti] is by far larger than that exerted by the admixed salts accumulating during its preparation [22]. A particularly challenging example is the preparation of the tetraene **31** which acts as a π-spherand and readily accommodates Ag(+1) in its interior (Scheme 13) [35]. Equally striking are tandem couplings in which an initial intermolecular reaction is followed by an intramolecular one. Although the yields obtained seem to be modest at first sight, it must be taken into account that only the (Z)-isomer formed in the first step can lead to the macrocyclic final product. This strategy has been elegantly applied to the syntheses of polyenes such as **33** (Scheme 14) [36] and to the formation of nonnatural porphyrin analogs [37], which are of relevance for the photodynamic therapy of tumors.

Carbonyl coupling reactions become an increasingly important tool for macromolecular chemistry. Polycondensations of stiff dicarbonyl precursors to polyvinylenes or polypinacols are particularly successful if the substrates carry appropriate

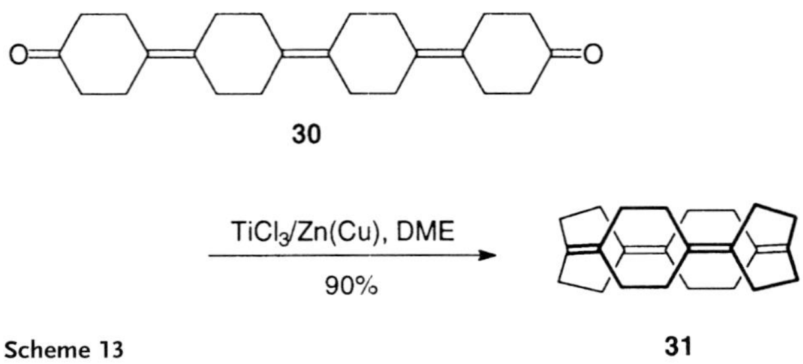

Scheme 13

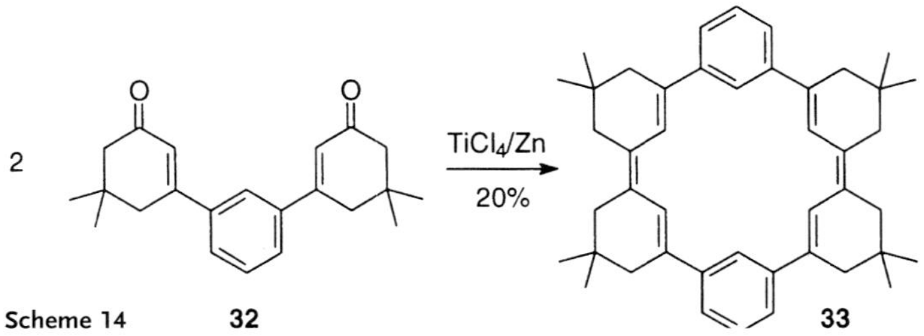

Scheme 14

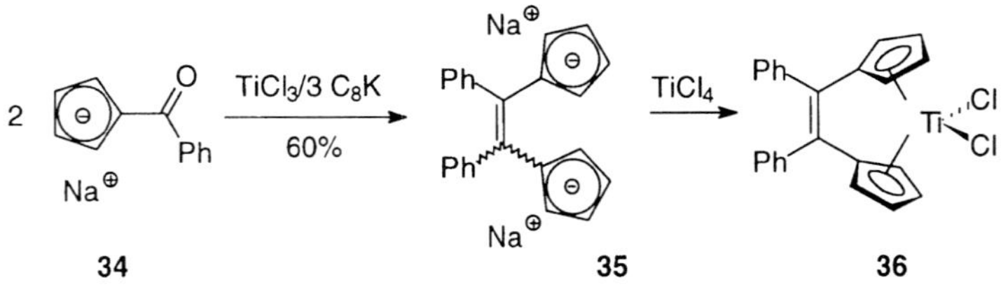

Scheme 15

side chains which solubilize the growing polymer and hence prevent its premature precipitation [38].

Finally, a few applications of McMurry reactions to organometallic compounds deserve mentioning. Most notable among them is the dimerization of acylated cyclopentadienyl salts such as **34** with titanium–graphite (Scheme 15) [39, 40]. Treatment of the products obtained with $TiCl_4$ affords ethene-bridged *ansa*-titanocene derivatives (e.g. **36**), which are of considerable interest as precursors for high-performance Ziegler polymerization catalysts.

3.3.6
Titanium-induced Cross-Coupling Reactions

3.3.6.1
Mixed Couplings of Aldehydes and Ketones

Mixed coupling reactions usually lead to a statistical mixture of all possible alkenes. However, reasonable yields of a single cross-coupling product can be obtained if one of the substrates is used in excess. Several successful isopropylidenations of carbonyl compounds using acetone as the overstoichiometric reaction partner demonstrate this possibility [41].

Preparatively useful results can also be obtained if the substrates in question exhibit sufficiently different redox potentials [41, 42]. Thus, exposure of a 1:1 mixture of a diaryl (or *α,β*-unsaturated) ketone and an aliphatic one results in reasonably high yields of the cross-coupling products (Scheme 16). Mechanistically, this pathway can be accounted for by assuming that the diarylketone undergoes a fast two-electron reduction with formation of a 'carbonyl dianionic' species which then attacks the aliphatic ketone prior to its reduction [41]. This strategy was applied to convenient syntheses of the anticancer agent tamoxifene **42** and analogs thereof [43].

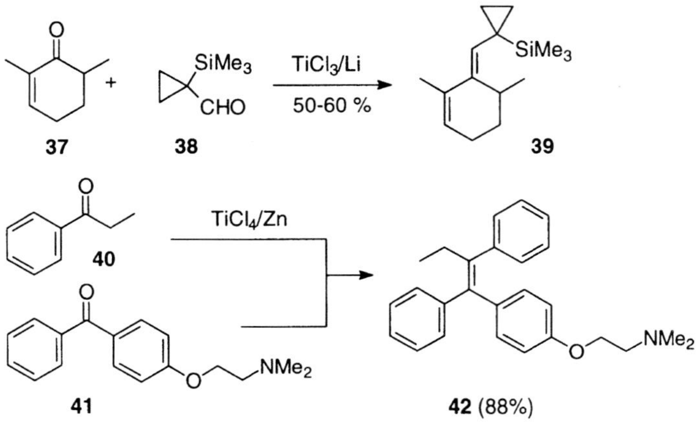

Scheme 16

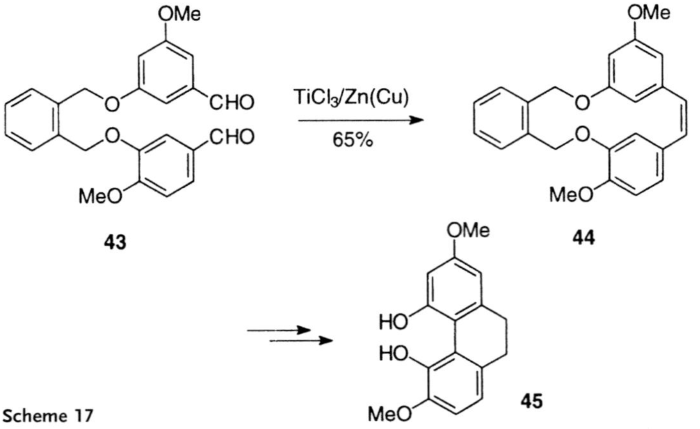

Scheme 17

A very clever way to achieve an efficient cross coupling of two electronically similar aldehydes is described in the context of a total synthesis of cannithrene II **45** (Scheme 17) [44]. The reaction is rendered intramolecular by means of a bis(benzyl ether) bridge attached to the –OH groups of the substrates.

Subsequent ring contraction of the macrocyclic (Z)-stilbene derivative **44** thus obtained to the dihydrophenanthrene core of the target and removal of the ancillary tether complete this elegant synthesis.

3.3.6.2
Keto-Ester Cyclizations

Oxoesters were shown to undergo intramolecular alkylidenation reactions on treatment with low-valent titanium. Aqueous work-up of the reaction mixtures hydrolyzes the enol ethers initially formed and leads to the respective cyclic ketones as the final products (Scheme 18) [45].

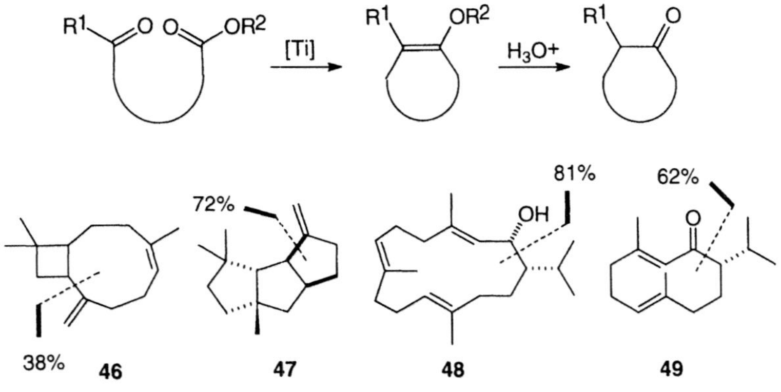

Scheme 18

High yields for this attractive transformation have been obtained with [Ti] formed from TiCl$_3$/LiAlH$_4$ (i.e. [HTiCl] as the active species, cf. Section 3.3.2) in the presence of NEt$_3$ [45] and with titanium–graphite [17, 46]. In contrast to conventional intramolecular couplings of diketones and dialdehydes, the formation of medium-sized rings via this keto-ester cyclization is somewhat less efficient. A strong influence is exerted by the chain length and the steric demand of the OR^2 part of the ester [46]. Applications of this transformation of the syntheses of terpenoid natural products including isocaryophyllene **46** [47], capnellene **47** [48], cembrene **48**, and acoragermacrone **49** [49] feature the potential of this unique cross-coupling strategy (Scheme 18).

3.3.6.3
Synthesis of Aromatic Heterocycles

Treatment of substrates of the general type **50** with low-valent titanium turned out to be an efficient and flexible entry into aromatic heterocycles **51** (Scheme 19) [7, 15, 25, 31, 46, 50–55]. It relies on an intramolecular cross-coupling process of an aldehyde or ketone with functional groups of substantially lower redox potentials which have previously been considered as hardly reactive or even as completely inert towards [Ti]. In addition to esters, this includes amides, carbonates, urethanes and urea derivatives.

As can be seen from Scheme 19, this new reductive cyclization gives ready access to furans, benzo[b]furans, pyrroles, and indoles. It turned out to be quite flexible with regard to the substituents R^2 and R^3 in the enol (X=O) or enamine (X=NR^1, R^1=H, alkyl, aryl, tosyl, etc.) region of the products and is compatible with a wide range of functional groups as well as with pre-existing chiral centers in the substrates. Sterically congested products can also be obtained [15]. Polycarbonyl compounds display a remarkable preference for the formation of the five-

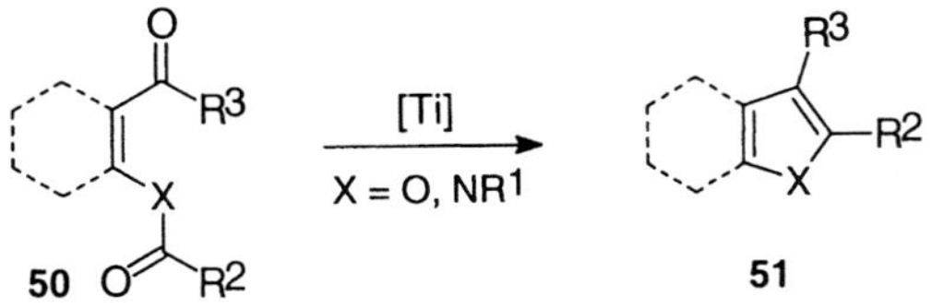

Scheme 19

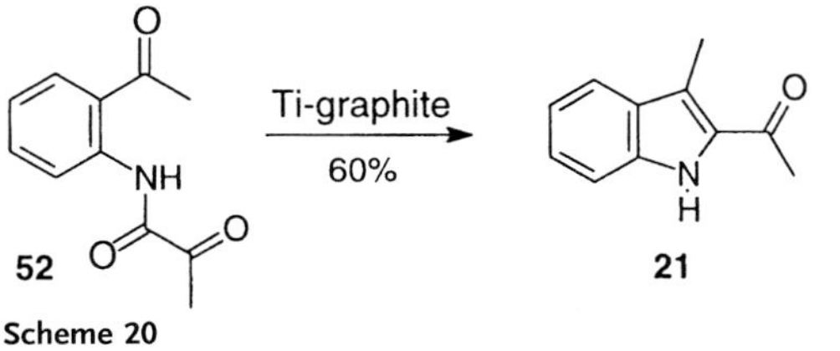

Scheme 20

Tab. 2 Synthesis of indole **58** by means of different titanium samples.

Entry	[Ti]	Solvent	Isolated yield [%]
1	$TiCl_3 + 3 C_8K$	DME	93
2	$TiCl_3 + 2 Na/Al_2O_3$	THF	72
3	$TiCl_3$, Zn ('instant')	THF	87
4	$TiCl_3$ (10 mol%), Zn, TMSCl	MeCN	79
5	Commercial Ti, TMSCl	THF	74

membered heterocyclic rings which accounts for completely chemo- and regio- selective cyclization reactions. This striking bias is evident from the zipper-like closure of poly(oxoamide) precursors to oligoindoles [53]. Even more surprising is the formation of salvadoricine **21** from substrate **52** (Scheme 20) [31]. This specific example shows for the first time that a cross-coupling with an amide may even be favored over a conventional McMurry reaction of two ketone functions.

Transformations of this type can be performed either with titanium–graphite [15, 46, 50–54], by means of the most convenient 'instant' method [15, 51–54] and the catalytic version thereof [7], or with commercial titanium powder/TMSCl [7]. This flexibility allows the reagent and the reaction conditions to be adapted to the specific requirements of the substrates and gives ample room for optimization. For example, Ti-graphite is particularly recommended if acid sensitive substrates are employed that may suffer from mixing with Lewis-acidic $TiCl_x$ ($x = 3, 4$). Tab. 2 summarizes the results obtained with various titanium reagents in the formation of indole **58** which is a known precursor for diazepam.

Because indoles and pyrroles are ubiquitous motives in nature and represent very important pharmacophores, this new approach to heterocycles has also been evaluated in the context of the synthesis of alkaloids and pharmaceutically active compounds. Scheme 21 compiles the targets that have been prepared so far [15, 50–54]. As a representative example, the first total synthesis of secofascaplysin, a bioactive metabolite extracted from the marine sponge *Fascaplysinopsis reticulata*, is outlined in Scheme 22 [52].

Acylation of the easily accessible amine **53** with the acid chloride **54** (prepared from methyl-2-aminobenzoate and oxalyl chloride) gives amide **55**, which is subsequently cross-coupled to indole **56**. Aqueous acidic work-up of the crude reaction mixture results in the cleavage of its acetal and a spontaneous cyclization of the

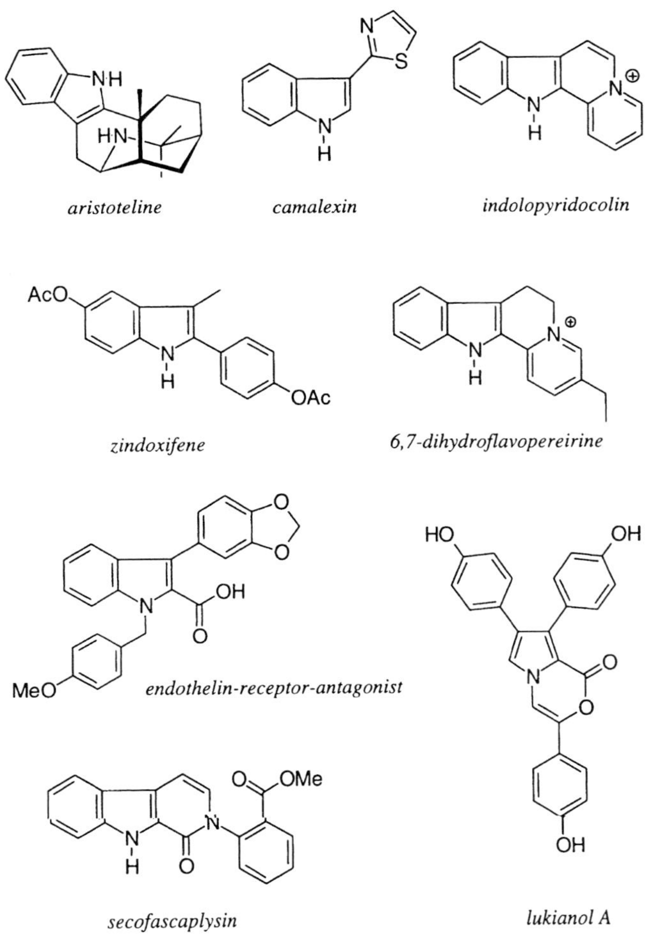

aristoteline

camalexin

indolopyridocolin

zindoxifene

6,7-dihydroflavopereirine

endothelin-receptor-antagonist

lukianol A

secofascaplysin

Scheme 21

C-ring of the alkaloid. Thus, secofascaplysin is available in only two synthetic operations in good yield from very simple precursors [52]. Finally, it should be emphasized that by acylation of a parent aminoketone such as **53** with various acid derivatives and subsequent [Ti]-induced indole formation, a series of analogs of a given target can be prepared in a very straightforward manner. This flexibility of the method in structural terms renders it particularly suitable for establishing structure/activity profiles of physiologically active lead compounds.

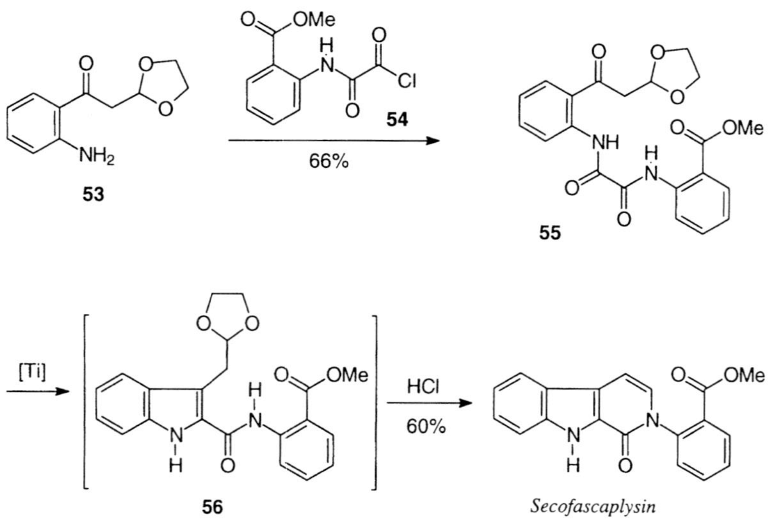

Scheme 22

3.3.7
References

1. T. Mukaiyama, T. Sato, J. Hanna, *Chem. Lett.* **1973**, 1041.
2. S. Tyrlik, I. Wolochowicz, *Bull. Soc. Chim. Fr.* **1973**, 2147.
3. J. E. McMurry, M. P. Fleming, *J. Am. Chem. Soc.* **1974**, *96*, 4708.
4. For a comprehensive treatise see: A. Fürstner, B. Bogdanovic, *Angew. Chem.* **1996**, *108*, 2582; *Angew. Chem., Int. Ed. Engl.* **1996**, *35*, 2442.
5. T. Lectka in *Active Metals. Preparation, Characterization, Applications* (Ed.: A. Fürstner), VCH, Weinheim, **1996**, p. 85.
6. (a) J. E. McMurry, *Chem. Rev.* **1989**, *89*, 1513. (b) D. Lenoir, *Synthesis* **1989**, 883. (c) G. M. Robertson in *Comprehensive Organic Synthesis* (Eds.: B. M. Trost, I. Fleming) Pergamon Press, Oxford, **1991**, Vol. 3, p. 563. (d) R. G. Dushin in *Comprehensive Organometallic Chemistry II* (Ed.: L. S. Hegedus), Pergamon Press, Oxford, **1995**, Vol. 12, p. 1071. (e) C. Betschart, D. Seebach, *Chimia* **1989**, *43*, 39.
7. A. Fürstner, A. Hupperts, *J. Am. Chem. Soc.* **1995**, *117*, 4468.
8. (a) A. Fürstner, N. Shi, *J. Am. Chem. Soc.* **1996**, *118*, 2533. (b) A. Fürstner, N. Shi, *J. Am. Chem. Soc.* **1996**, *118*, 12349 and references therein.
9. L. E. Aleandri, S. Becke, B. Bogdanovic, D. J. Jones, J. Roziere, *J. Organomet. Chem.* **1994**, *472*, 97.
10. L. E. Aleandri, B. Bogdanovic, A. Gaidies, D. J. Jones, S. Liao, A. Michalowicz, J. Rozire, A. Schott, *J. Organomet. Chem.* **1993**, *459*, 87.
11. J. E. McMurry, T. Lectka, J. G. Rico, *J. Org. Chem.* **1989**, *54*, 3748; in the author's experience, the preparation of $TiCl_3(DME)_{1.5}$ in refluxing DME as described by McMurry can be troublesome. Beautifully sky-blue samples (70–80%) of this solvate complex, however, have been reproducibly obtained by stirring $TiCl_3$ with DME at 65 °C for 11 h under Ar.
12. J. E. McMurry, M. P. Fleming, K. L. Kees, L. R. Krepski, *J. Org. Chem.* **1978**, *43*, 3255.
13. D. Lenoir, *Synthesis* **1977**, 553.

14 B. Bogdanovic, A. Bolte, *J. Organomet. Chem.* **1995**, *502*, 109.
15 A. Fürstner, A. Hupperts, A. Ptock, E. Janssen, *J. Org. Chem.* **1994**, *59*, 5215.
16 A. Fürstner, *Angew. Chem.* **1993**, *705*, 171; *Angew. Chem., Int. Ed. Engl.* **1993**, *32*, 164.
17 (a) A. Fürstner, H. Weidmann, *Synthesis* **1987**, 1071. (b) A. Fürstner, R. Csuk, C. Rohrer, H. Weidmann, *J. Chem. Soc., Perkin Trans. 1* **1988**, 1729. (c) G. P. Boldrini, D. Savoia, E. Tagliavini, C. Trombini, A. Umani-Ronchi, *J. Organomet. Chem.* **1985**, *280*, 307. (d) M. A. Araya, F. A. Cotton, J. H. Matonic, C. A. Murillo, *Inorg. Chem.* **1995**, *34*, 5424.
18 (a) D. L. J. Clive, K. S. K. Murthy, A. G. H. Wee, J. S. Prasad, G. V. J. da Suva, M. Majewski, P. C. Anderson, C. F. Evans, R. D. Haugen, L. D. Heerze, J. R. Barrie, *J. Am. Chem. Soc.* **1990**, *112*, 3018. (b) D. L. J. Clive, C. Zhang, K. S. K. Murthy, W. D. Hayward, S. Daigneault, *J. Org. Chem.* **1991**, *56*, 6447.
19 For a review containing detailed procedures for the preparation and handling of C_8K and other metal–graphite reagents see: A. Fürstner (ed.), *Active Metals. Preparation, Characterization, Applications*, VCH, Weinheim, **1996**, p. 381.
20 (a) A. Fürstner, G. Seidel, *Synthesis* **1995**, 63. (b) A. Fürstner, G. Seidel, B. Gabor, C. Kopiske, C. Kruger, R. Mynott, *Tetrahedron* **1995**, *57*, 8875.
21 P. Lhotak, S. Shinkai, *Tetrahedron Lett.* **1996**, *37*, 645.
22 A. Fürstner, G. Seidel, C. Kopiske, C. Kruger, R. Mynott, *Liebigs Ann.* **1996**, 655.
23 A. Fürstner, B. Tesche, *Chem. Mater.* **1988**, *10*, 1968.
24 J. E. McMurry, R. G. Dushin, *Stud. Nat. Prod. Chem.* **1991**, *8*, 15.
25 A. Fürstner in *Organic Synthesis via Organometallics* OSM5 (Ed.: G. Helmchen), Vieweg, Braunschweig, **1997**, p. 309.
26 (a) K. C. Nicolaou, Z. Yang, J. J. Liu, P. G. Nantermet, C. F. Claiborne, J. Renauld, R. K. Guy, K. Shibayama, *J. Am. Chem. Soc.* **1995**, *117*, 645. (b) K. C. Nicolaou, J. J. Liu, Z. Yang, H. Ueno, E. J. Sorensen, C. F. Claiborne, R. K. Guy, C. K. Hwang, M. Nakada, P. G. Nantermet, *J. Am. Chem. Soc.* **1995**, *117*, 634.
27 (a) Humulene and flexibilene: J. E. McMurry, J. R. Matz, K. L. Kees, *Tetrahedron* **1987**, *43*, 5489. (b) Casbene: J. E. McMurry, G. K. Bosch, *J. Org. Chem.* **1987**, *52*, 4885. (c) Cembrene C: Y. Li, W. Li, Y. Li, *Synth. Commun.* **1994**, *24*, 721. (d) Crassin: W. G. Dauben, T. Z. Wang, R. W. Stephens, *Tetrahedron Lett.* **1990**, *31*, 2393. See also: (e) J. E. McMurry, R. G. Dushin, *J. Am. Chem. Soc.* **1990**, *112*, 6942. (f) Periplanone C: J. E. McMurry, N. O. Siemers, *Tetrahedron Lett.* **1994**, *35*, 4505. (g) Archaebacterial lipids: T. Eguchi, T. Terachi, K. Kakinuma, *J. Chem. Soc., Chem. Commun.* **1994**, 137. (h) Neocarcinostatin model: M. Eckhardt, R. Bruckner, *Angew. Chem.* **1996**, *108*, 1185.
28 W. G. Dauben, I. Farkas, D. P. Bridon, C. P. Chuang, K. E. Henegar, *J. Am. Chem. Soc.* **1991**, *113*, 5883.
29 (a) F. E. Ziegler, H. Lim, *J. Org. Chem.* **1982**, *47*, 5229. (b) K. Mikami, K. Takahashi, T. Nakai, T. Uchimaru, *J. Am. Chem. Soc.* **1994**, *116*, 10948.
30 Y. J. Wu, D. J. Burnell, *Tetrahedron Lett.* **1988**, *29*, 4369.
31 A. Fürstner, D. N. Jumbam, *J. Chem. Soc., Chem. Commun.* **1993**, 211.
32 (a) J. E. McMurry, T. Lectka, *J. Am. Chem. Soc.* **1993**, *115*, 10167. (b) J. E. McMurry, T. Lectka, *J. Am. Chem. Soc.* **1990**, *112*, 869. (c) J. E. McMurry, T. Lectka, C. N. Hodge, *J. Am. Chem. Soc.* **1989**, *111*, 8867. (d) J. E. McMurry, C. N. Hodge, *J. Am. Chem. Soc.* **1984**, *106*, 6450. (e) See also: T. S. Sorensen, S. M. Whitworth, *J. Am. Chem. Soc.* **1990**, *112*, 8135.
33 For other recent highlights see: (a) A. P. Marchand, A. Zope, F. Zaragoza, S. G. Bott, H. L. Ammon, Z. Du, *Tetrahedron* **1994**, *50*, 1687. (b) P. R. Brooks, R. Bishop, D. C. Craig, M. L. Scudder, J. A. Counter, *J. Org. Chem.* **1993**, *58*, 5900. (c) R. Gleiter, O. Borzyk, *Angew. Chem.* **1995**, *107*, 1094; *Angew. Chem., Int. Ed. Engl.* **1995**, *34*, 1001. (d) W. von E. Doering, Y. Q. Shi, D. C. Zhao, *J. Am. Chem. Soc.* **1992**, *114*, 10763. (f) U. Grieser, K. Hafner, *Tetrahedron Lett.* **1994**, *35*, 7759.

(g) I. Columbus, S.E. Biali, *J. Org. Chem.* **1994**, *59*, 3402.

34 (a) H. Hopf, C. Mlynek, *J. Org. Chem.* **1990**, *55*, 1361. (b) W.Y. Lee, C.H. Park, H.J. Kim, S. Kim, *J. Org. Chem.* **1994**, *59*, 878.

35 J.E. McMurry, G.J. Haley, J.R. Matz, J.C. Clardy, J. Mitchell, *J. Am. Chem. Soc.* **1986**, *108*, 515.

36 F. Vögtle, C. Thilgen, *Angew. Chem.* **1990**, *102*, 1176; *Angew. Chem., Int. Ed. Engl.* **1990**, *29*, 1162.

37 E. Vogel, *Pure Appl. Chem.* **1996**, *68*, 1355 and reference therein.

38 For leading references see: (a) A.W. Cooke, K.B. Wagener, *Macromolecules* **1991**, *24*, 1404. (b) S. Iwatsuki, M. Kubo, Y. Itoh, *Chem. Lett.* **1993**, 1085. (c) T. Itoh, H. Saitoh, S. Iwatsuki, *J. Polym. Sd: Part A: Polym. Chem.* **1995**, *33*, 1589. (d) M. Rehahn, A.D. Schlüter, *Macromol. Chem. Rapid Commun.* **1990**, *11*, 375. (e) R. Bayer, O. Nuyken, *Kautsch. Gummi Kunstst.* **1996**, *49*, 28.

39 P. Burger, H.H. Brintzinger, *J. Organomet. Chem.* **1991**, *407*, 207.

40 For other examples see: (a) S. Fitter, G. Huttner, O. Walter, L. Zsolnai, *J. Organomet. Chem.* **1993**, *454*, 183. (b) B. Bildstein, P. Denifl, K. Wurst, M. André, M. Baumgarten, J. Friedrich, E. Ellmerer-Müller, *Organometallics* **1995**, *14*, 4334. (c) J. Besançon, J. Szymoniak, C. Moise, *J. Organomet. Chem.* **1992**, *426*, 325.

41 J.E. McMurry, L.R. Krepski, *J. Org. Chem.* **1976**, *41*, 3929.

42 (a) L.A. Paquette, T.-H. Yan, G.J. Wells, *J. Org. Chem.* **1984**, *49*, 3610. (b) S.M. Reddy, M. Duraisamy, H.M. Walborsky, *J. Org. Chem.* **1986**, *57*, 2361. (c) M.M. Cid, J.A. Seijas, M.C. Villaverde, L. Castedo, *Tetrahedron* **1988**, *44*, 6197.

43 (a) P.L. Coe, C.E. Scriven, *J. Chem. Soc., Perkin Trans. 1* **1986**, 475. (b) J. Shani, A. Grazit, T. Livshitz, S. Biran, *J. Med. Chem.* **1985**, *28*, 1504. (c) S. Gauthier, J. Mailhot, F. Labrie, *J. Org. Chem.* **1996**, *61*, 3890. (d) P.C. Ruentiz, C.S. Bourne, K.J. Sullivan, S.A. Moore, *J. Med. Chem.* **1996**, *39*, 4853.

44 I. Ben, L. Castedo, J.M. Saa, J.A. Seijas, R. Suau, G. Tojo, *J. Org. Chem.* **1985**, *50*, 2236.

45 J.E. McMurry, D.D. Miller, *J. Am. Chem. Soc.* **1983**, *105*, 1660.

46 A. Fürstner, D.N. Jumbam, *Tetrahedron* **1992**, *48*, 5991.

47 J.E. McMurry, D.D. Miller, *Tetrahedron Lett.* **1983**, *24*, 1885.

48 M. Iyoda, T. Kushida, S. Kitami, M. Oda, *J. Chem. Soc., Chem. Commun.* **1987**, 1607.

49 (a) W. Li, Y. Li, Y. Li, *Chem. Lett.* **1994**, 741. (b) W. Li, Y. Li, Y. Li, *Synthesis* **1994**, 678. (c) W. Li, Y. Li, Y. Li, *Synthesis* **1994**, 267.

50 A. Fürstner, D.N. Jumbam, G. Seidel, *Chem. Ber.* **1994**, 1125.

51 A. Fürstner, A. Ernst, *Tetrahedron* **1995**, *57*, 773.

52 A. Fürstner, A. Ernst, H. Krause, A. Ptock, *Tetrahedron* **1996**, *52*, 7329.

53 A. Fürstner, A. Ptock, H. Weintritt, R. Goddard, C. Kruger, *Angew. Chem.* **1995**, *107*, 725; *Angew. Chem., Int. Ed. Engl.* **1995**, *34*, 678.

54 A. Fürstner, H. Weintritt, A. Hupperts, *J. Org. Chem.* **1995**, *60*, 6637.

55 A. Fürstner, D.N. Jumbam, N. Shi, Z. *Naturforsch.* **1995**, *50B*, 326.

3.4
Chromium(II)-Mediated and -Catalyzed C-C Coupling Reactions

David M. Hodgson and Paul J. Comina

3.4.1
Introduction

Chromium(II)-based methods have been developed for a wide range of important carbon-carbon bond-forming reactions [1]. The emphasis in this chapter is to provide basic background on the particular types of functional groups which can be reduced to active organochromium reagents, with more detailed information on the more significant methodological developments published from 1998 onwards. As anticipated in the previous chapter on this topic [1a], notable progress in (asymmetric) catalytic processes has occurred since that time, and these advances are now incorporated in the relevant functional group sections below, with some new chromium(II)-mediated transformations involving C=O or C=C reduction in a separate section; some recent significant developments in chromium(III)-mediated or -catalyzed reactions are also briefly discussed.

Many of the carbon-carbon bond-forming reactions developed using chromium(II) chemistry can be broadly summarized as involving the coupling of an organic halide (or equivalent) with a carbonyl compound (Scheme 1).

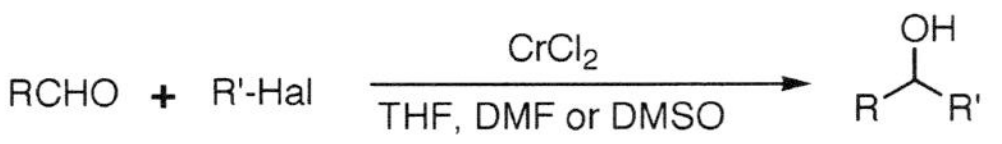

$$\text{RCHO} + \text{R'-Hal} \xrightarrow[\text{THF, DMF or DMSO}]{\text{CrCl}_2} \text{R}\underset{\underset{\text{OH}}{|}}{\text{CH}}\text{R'}$$

Scheme 1

The carbonyl component is normally present *in situ*, because of the presumed instability of the intermediate organometallic species, and so the closest analogy with classical chemistry is found in the Barbier reaction. The reducing power of chromium(II) salts is not as strong as with metals like Li, Mg and Zn, and some exquisite chemoselectivity with chromium(II) salts can be observed between different types of halide and exploited in synthesis. The usual source of chromium(II) is commercially available $CrCl_2$ although, less commonly, *in situ* reduction procedures from $CrCl_3$ have been applied.

Transition Metals for Organic Synthesis, Vol. 1, 2nd Edition.
Edited by M. Beller and C. Bolm
Copyright © 2004 WILEY-VCH Verlag GmbH & Co. KGaA, Weinheim
ISBN: 3-527-30613-7

3.4.2
Allylic Halides

Low-valent chromium salts have traditionally been widely used as reducing agents for a range of functional groups under aqueous conditions. In 1976 Hiyama et al. first had the insight to examine the reduction of organic halides with $CrCl_2$ in aprotic solvents in the presence of a carbonyl compound, on the basis that in the absence of water a transient organochromium intermediate might persist and undergo useful carbon-carbon bond-forming reactions [2]. With allylic halides this has proven to be one of the most useful ways of preparing homoallylic alcohols [1]. A wide variety of substituted allylic halides have since been examined. The reaction shows high chemoselectivity. For example, ester and cyano groups are tolerated in the coupling step, and aldehydes react preferentially in the presence of ketones. This chemoselectivity is a hallmark of most organochromium-mediated transformations. High regio- and stereo-selectivity is also evident in the coupling step.

Excellent regioselectivity is observed using polyhalogenated systems [3]. Insertion of chromium(II) into allylic C-Hal bonds occurs preferentially in the presence of vinylic C-Hal bonds. For example, trihalides **1** react exclusively through reduction of the allylic chloride giving, after reaction with an aldehyde, homoallylic alcohols **2** in good yield and with moderate diastereoselectivity (Scheme 2).

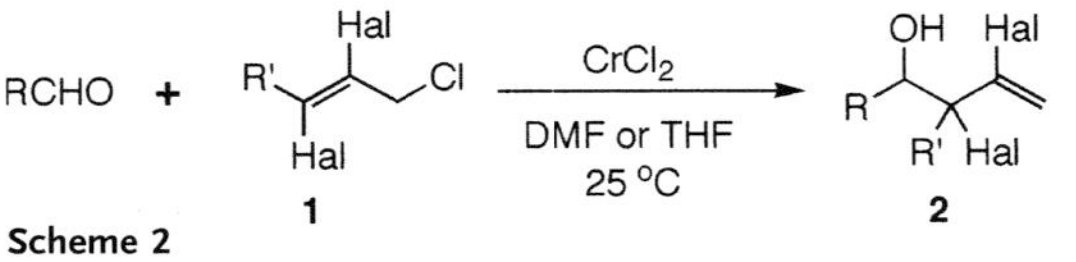

Scheme 2

If an excess of an (aromatic) aldehyde is present in a normal Hiyama-Nozaki allylation reaction, the usual chromium alkoxide products from the addition of an allylchromium to an aldehyde undergo Oppenauer oxidation by excess aldehyde to give β,γ-enones (Scheme 3), although the amount of oxidation is strongly dependent on the substitution pattern of the reaction partners [4]. This oxidation, which is reversible, also poses issues for any asymmetric variants of this reaction, as the newly formed stereocenter is rapidly epimerized.

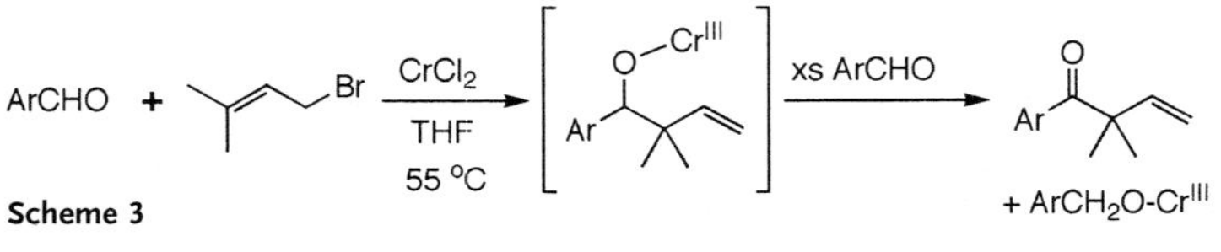

Scheme 3

In an interesting variation of the Hiyama-Nozaki reaction, Hosomi has shown that a Cr(III)-based ate-type reagent, Bu_5CrLi_2, prepared by treating $CrCl_3$ with 5 equivalents of butyllithium, can be efficiently used for the reduction of allyl phosphates to (postulated) allylchromium intermediates, and these then react efficiently with a range of electrophiles, including aldehydes, ketones, imines and isocyanates [5].

Despite the broad utility of chromium(II)-mediated carbon-carbon bond formation in synthesis, further developments in this area were hindered by the toxicity of chromium salts, combined with the fact that the salts are often used in large excess. However, Fürstner reported a significant solution to these concerns by designing a process whereby the chromium(II) [or (III)] salt is used in catalytic amounts [6]. Recycling is achieved by using a combination of Me_3SiCl, for silylation of the chromium alkoxide formed after C-C bond construction, and manganese metal to reduce the chromium(III) salts. The reaction exhibits the usual chemo-, regio- and diastereo-selectivity found in organochromium-mediated transformations and was shown to be applicable to allyl, alkenyl, aryl, and alkynyl halides (and triflates). As little as 1 mol% of Cp_2Cr or $CpCrCl_2 \cdot THF$ was sufficient in the reaction of allylic halides with aldehydes. This work has been the spur for further catalytic developments. It has been shown that the catalytic methodology can be applied to solid-phase chemistry, with the substrate (the aldehyde component) attached to a solid support [7]. Since organochromium compounds react slowly with protic acids it has also been found that collidine hydrochloride, in the presence of bis(di-*iso*-propylphosphino)ethane, could replace Me_3SiCl in the above chemistry, resulting in an acid-mediated chromium-catalyzed allylation of aldehydes [8]. Tanaka has demonstrated the use of tetrakis(dimethylamino)ethylene as an alternative reductant to manganese in the catalytic addition of allylchromiums to aldehydes [9]; the best yields are obtained if a Ni(II) salt (5 mol%) is present and the reactions are carried out in DMF (as opposed to the more usual THF). Boeckman has demonstrated that acrolein acetals are suitable substrates for the generation of alkoxy-substituted allylchromiums in reactions with aldehydes, when they are treated with manganese metal, Me_3SiCl, catalytic NaI, and catalytic $CrCl_2$ [10].

Early efforts to induce asymmetry in chromium(II)-mediated additions to aldehydes used lithium ephedrinate as a chiral ligand, which gave 29% *ee* in the allylation of butyraldehyde [11]. Kishi et al. found significant induction (74% *ee*) in the allylation of benzaldehyde using a dipyridyl chiral ligand; lower *ee*s were observed in the synthesis of allylic alcohols [12]. Kibayashi et al. examined chiral (dialkoxyallyl)chromium complexes, which exerted significant asymmetric induction in reactions with aldehydes [13]. Building on Fürstner's catalytic procedure, Cozzi, Umani-Ronchi and co-workers have further developed the area of chromium(II)-based asymmetric processes [14]: they demonstrated the asymmetric addition of allylic chromiums to aldehydes with high levels of enantioselectivity (up to 90%), using only catalytic quantities of either $CrCl_2$ or $CrCl_3$. The combination of a chiral salen ligand [(R,R)-(N,N'-bis(3,5-di-*tert*-butylsalicylidene)-1,2-cyclohexanediamine], $CrCl_2$ or $CrCl_3$, triethylamine (required to remove the HCl formed during the preparation of the catalyst), manganese, and Me_3SiCl in acetonitrile (a solvent which minimizes the formation of pinacol side-products) proved to be a general method for achieving both diastereoselective and enantioselective addition of allylic halides to a variety of aldehydes (Scheme 4). The amount of chiral ligand used plays a key role in the stereochemical control of the reaction. Thus, by using a 10 mol%

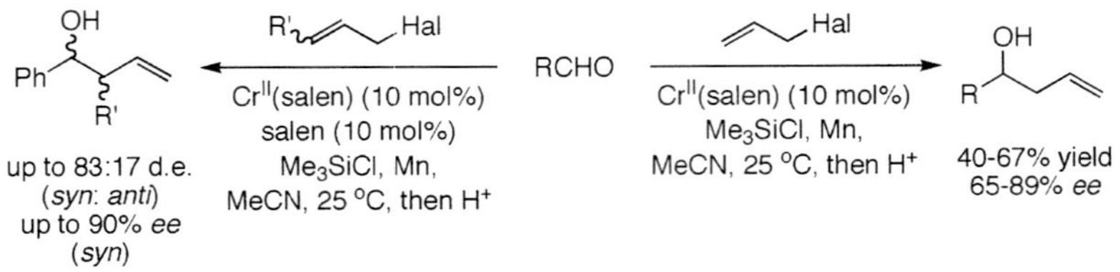

Scheme 4

[Scheme 4: Ph-CH(OH)-CH(R')-CH=CH2 ← R'-CH=CH-Hal + Cr^II(salen) (10 mol%), salen (10 mol%), Me3SiCl, Mn, MeCN, 25 °C, then H+; up to 83:17 d.e. (syn: anti), up to 90% ee (syn). RCHO + CH2=CH-CH2-Hal → R-CH(OH)-CH2-CH=CH2, Cr^II(salen) (10 mol%), Me3SiCl, Mn, MeCN, 25 °C, then H+, 40–67% yield, 65–89% ee]

excess of the Schiff base, complete inversion was observed to *syn* diastereoselection from the *anti* diastereoselection usually observed using chromium salts. Cr(salen)-catalyzed addition of propargyl halides to aromatic aldehydes provides homopropargyl alcohols in satisfactory yields and in up to 56% ee. The excellent chemoselectivity for homopropargyl versus allenyl alcohols is a peculiar characteristic of the process in comparison to other chromium-mediated propargylation reactions.

A novel chiral salen ligand (*endo,endo*-2,5-diamino-norbornane) has recently been shown to lead to improved enantioselectivity in chromium-catalyzed allylations using allylic bromides [15]. New tridentate bis(oxazolinyl)carbazole ligands have also recently been found to be highly effective in asymmetric chromium-catalyzed allylation and methallylation [16].

3.4.3
1,1-Di- and 1,1,1-Trihalides

Whilst the homologation of aldehydes to *E*-1,2-difunctionalized alkenes can often be accomplished using Wittig-type chemistry, certain substitution patterns are impossible to prepare using this strategy. In addition, the stereoselectivity obtained may not be satisfactory, or functional groups elsewhere in the starting aldehyde may not be tolerated in the coupling step. For these cases, *gem*-dichromium reagents derived from *gem*-dihalides have provided some useful solutions [1, 17]. *E*-1,2-Dialkyl-substituted alkenes, alkenyl halides, sulfides, silanes, stannanes [1, 17, 18] and boronic esters [19] can all be prepared using this chemistry, usually with good-to-excellent *E*-stereoselectivity (Scheme 5). Trisubstituted and tetrasubstituted double bonds can also be prepared using similar methodology [20, 21]. 1,2-Addition is observed with α,β-unsaturated aldehydes. However, stereoselectivity for the *E*-isomer is usually slightly lower than that seen with aliphatic and aromatic aldehydes. The reactions have found widespread use in natural product synthesis [1], for example in the construction of an intermediate used to make (–)-periplanone B (Scheme 6) [22].

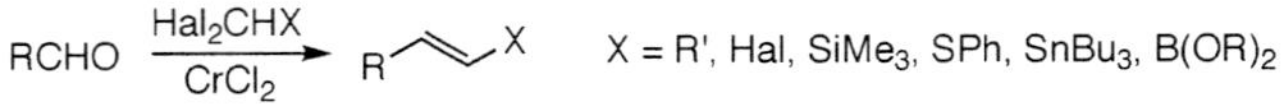

Scheme 5

RCHO + Hal2CHX → (CrCl2) → R-CH=CH-X   X = R', Hal, SiMe3, SPh, SnBu3, B(OR)2

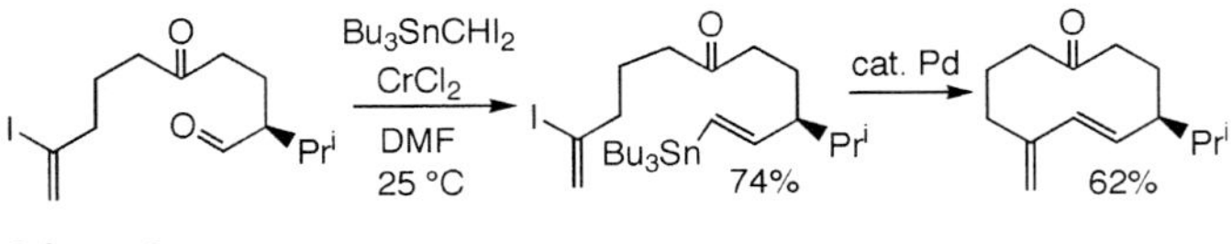

[Scheme 6]

Scheme 6

Takai has demonstrated how relatively subtle differences in reaction conditions can have a dramatic effect on the course of this olefination reaction (Scheme 7). Using a zinc co-reductant in the presence of Me$_3$SiCl, iodoform, and catalytic Cr(II) or Cr(III) results in efficient alkenylation of aldehydes, generating vinyl iodides in good yield and with high E-stereoselectivity [23]. However, with Mn as co-reductant and delaying addition of the aldehyde, Takai has shown that vinylsilanes can also be prepared directly from iodoform rather than from Me$_3$SiCHBr$_2$, which is the usual substrate for this reaction [24]. In this latter process, iodoform is likely converted *in situ* into Me$_3$SiCHI$_2$ through reaction with manganese and Me$_3$SiCl prior to aldehyde addition.

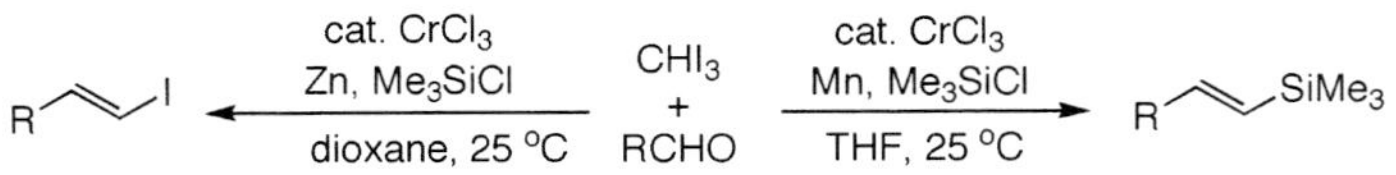

Scheme 7

Falck, Mioskowski, and co-workers have studied the chemistry of 1,1,1-trichloroalkenes **3** (X=H) in the presence of CrCl$_2$, which with aldehydes provides a stereocontrolled synthesis of (Z)-2-chloroalk-2-en-1-ols **5** (Scheme 8) [25]. The experimental evidence supports a reaction pathway via a *gem*-dichromium which eliminates chromium hydride to give a vinyl chromium **4** (in the absence of an aldehyde, addition of water leads to (Z)-1-chloroalk-1-enes). Contemporaneously, Takai and co-workers developed a similar transformation using related substrates **3** (X=OCO$_2$Me) [26]; more recently, Falck and Mioskowski have shown that trichloromethylcarbinols **3** (X=OH) are also effective in this transformation (Scheme 8) [27]. A three-component coupling was found to occur when using carbon tetrachloride and two equivalents of an aldehyde (Scheme 9) [28]; **3** (X = OCrL$_n$) is a likely intermediate. In the absence of an aldehyde, Takai has demonstrated that reaction of carbon tetrachloride with CrCl$_2$ in the presence of a terminal alkene leads to allenes in good yield (Scheme 9) [29]. A 1,1-dichloro-1,1-*gem*-dichromium

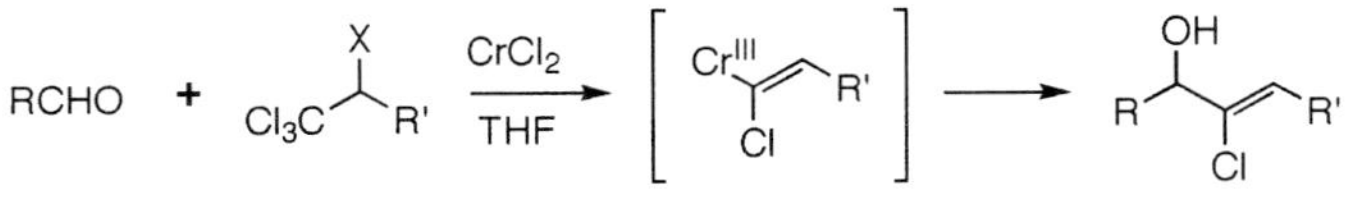

Scheme 8

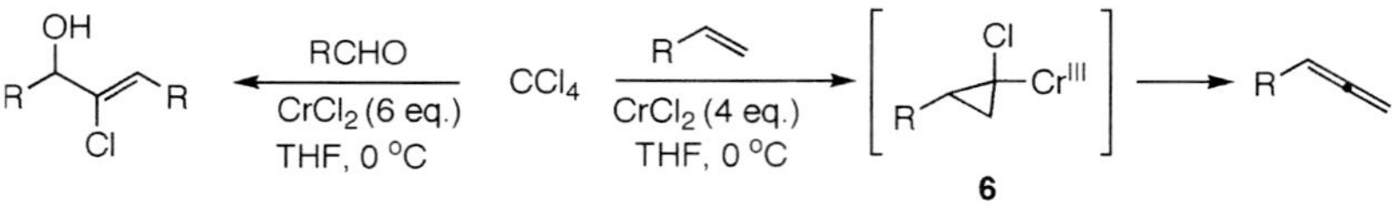

Scheme 9

may effect cyclopropanation, and the resulting α-chloro-α-cyclopropyl chromium (**6**) then rearranges to the allene.

Falck and Mioskowski have reported the preparation of furans via cyclization of 1,1,1-trichloroethyl propargyl ethers using catalytic Cr(II) regenerated by Mn/Me$_3$SiCl [30] (Scheme 10). It was suggested that single-electron transfer leads to a dichloroalkyl radical (**7**), which undergoes cyclization onto the alkyne; a subsequent sequence of reductions and protonation leads to the observed furans. The same researchers have shown that Cr(II)-mediated olefination of aldehydes with trihaloacetates provides a highly stereoselective preparation of (Z)-α-haloacrylates [31].

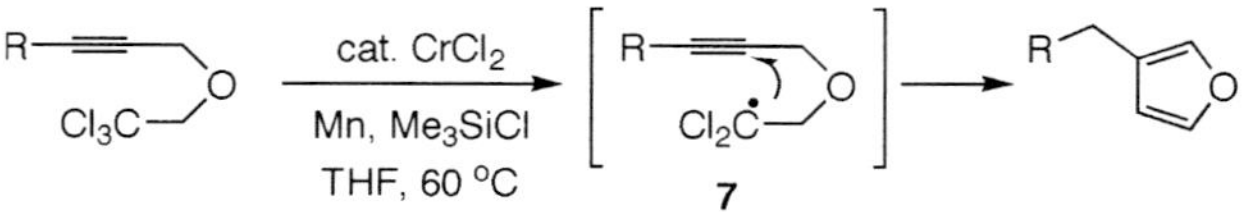

Scheme 10

3.4.4
Alkenyl and Aryl Halides (and Enol Triflates)

Activation of the (sp^2)C-Hal bond and (sp^2)C-O bond of enol triflates for reaction with an aldehyde using CrCl$_2$ usually requires DMF or DMSO as a solvent and trace amounts of a nickel(II) or (much less commonly) palladium(II) catalyst [32]. A variety of different vinyl and aryl halides are useful in this transformation. For example, iodides (**8**) [33], iodoacrylate (**9**) [34], 2-halovinylsulfones (**10**) [35] and 2-bromoallylic acetates (**11**) [36] are all viable substrates under the standard reaction conditions (Fig. 1). The chemistry has also been extended to the addition of vinylchromium species to ketones, either in an intermolecular fashion through the addition of a bipyridyl co-ordinating ligand [37] or in an intramolecular fashion using ketones (**12**) [38]. The products from the intramolecular addition can also undergo subsequent elimination of water, providing a concise route to indenes (Scheme 11). An interesting variation on this reaction makes use of diaryliodonium salts as the "halide" source [39]. These arylate aldehydes in good yield using CrCl$_2$. Addition of functionalized arylzinc iodides to functionalized aldehydes has been achieved by transmetalation with CrCl$_3$ in the presence of Me$_3$SiCl [40]. Chen has shown that aryl iodides can undergo a Wurtz-type homocoupling using a NiCl$_2$/CrCl$_2$/Mn system in the presence of a bipyridyl-type ligand [41].

Chromium(II)-mediated vinylation and arylation of aldehydes have found extensive application in natural product synthesis [1], although occasionally with unex-

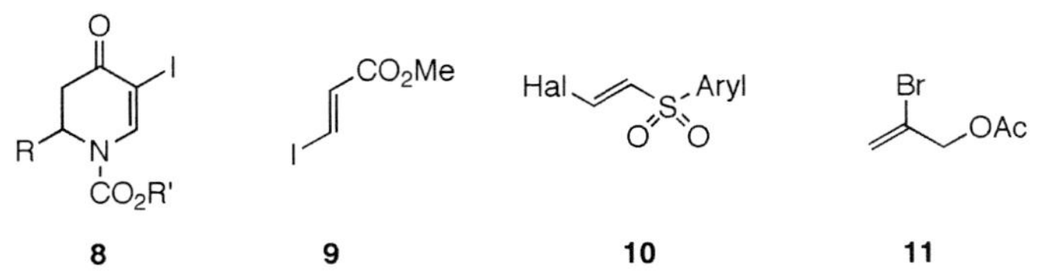

Fig. 1

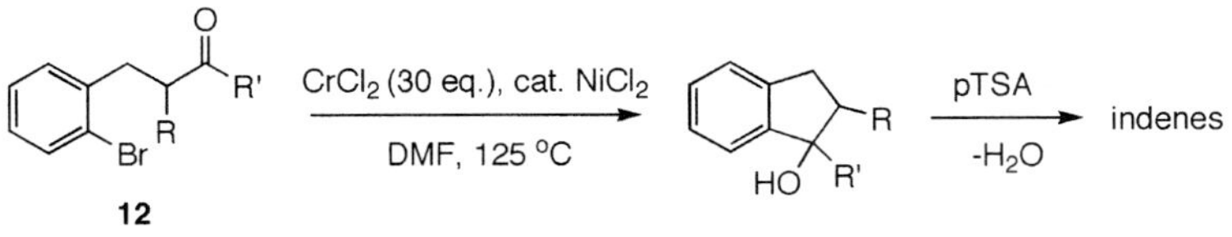

Scheme 11

pected results. In the proposed synthesis of a taxane precursor by intramolecular addition of a vinyl iodide to an aldehyde (Scheme 12), the substrate, in which the iodide was also allylic to a second double bond, cyclized through the "allylic" double bond to give the allene **13** whose structure was confirmed by X-ray analysis [42]. The other diastereoisomer of the starting vinyl iodide (the *cis*-carbonate) gave only a simple reduction product (I replaced by H). Other noteworthy recent examples are found in the syntheses of pinnatoxin by Kishi [43] and of mycalolide by Panek [44], who both made use of two chromium(II)-mediated couplings of vinyl iodides. A number of syntheses have also made use of the increased ease of preparation of medium-ring compounds that organochromium cyclizations lend themselves to. Danishefsky's synthesis of eleutherobin, which also demonstrates the use of 2-bromofurans as suitable halides in this transformation, illustrates the ease with which some ring sizes can be formed using organochromium chemistry [45]; the 9-membered ring containing furanophane **14** was formed in an excellent yield (74%) (Scheme 13). Similarly, Overman has illustrated a 9-membered ring-closure in the preparation of cladiellin diterpenes [46].

Tanaka has described a mixture of catalytic quantities of $CrCl_2$ or $CrCl_3/Zn$ in the presence of catalytic $NiBr_2$, stoichiometric aluminum, and Me_3SiCl for the alkenylation of aldehydes [47]; similarly to the allylation chemistry described earlier,

Scheme 12

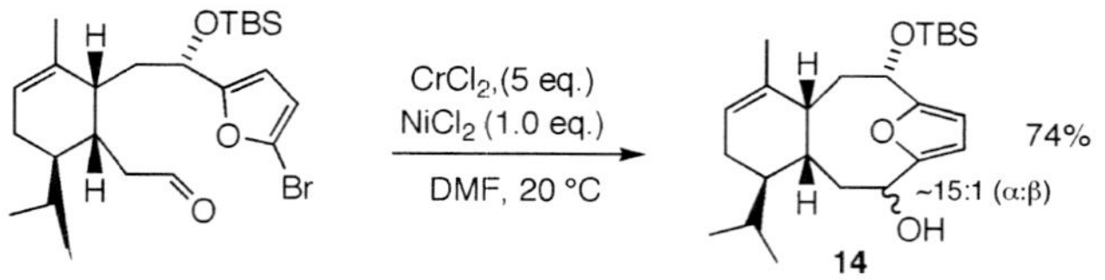

Scheme 13

tetrakis(dimethylamino)ethylene can also function as an alternative reductant to manganese in chromium-catalyzed alkenylations [48]. Independently, Grigg [49], Tanaka [50], and Durandetti [51] have all demonstrated that electrochemical methods can be used to recycle chromium(III) *in situ*, allowing catalytic procedures to take place. Asymmetric induction (up to 75% *ee*) in allylic alcohols using aldehydes with vinylic halide and triflate precursors has been demonstrated using the *endo,endo*-2,5-diamino-norbornane-derived salen ligand mentioned in Section 2 [15] as well as by using tridentate chiral sulfonamide ligands [52].

3.4.5
Alkynyl Halides

Propargylic alcohols can be prepared from alkynyl halides using $CrCl_2$ [53]; the presence of $NiCl_2$ may facilitate the reaction. This reaction has found utility in ring synthesis, as in an elegant approach to *epi*-illudol [54], the synthesis of enediynes [55], and toward eleuthesides [56].

3.4.6
Alkyl Halides

Alkyl halides (and tosylates) can be coupled with aldehydes using $CrCl_2$ in DMF under Co catalysis [57]. From secondary and tertiary alkyl iodides, $CrCl_2$ in DMF is able to generate radicals which add to dienes, prior to further reduction to allylchromiums (e.g., Scheme 14, major regio- and stereoisomer shown) [58].

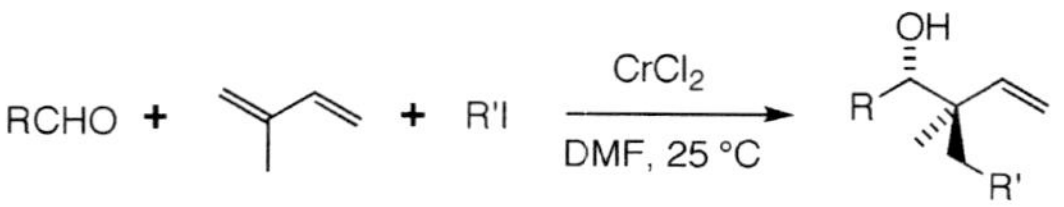

Scheme 14

Chromium(II)-mediated Reformatsky reactions with aldehydes provide access to quaternary centers (which are often prone to retro-aldolization in base-catalyzed processes) and, compared with zinc ester enolates, inversed simple diastereoselectivity [59]. The reaction when applied to ketones allows the synthesis of adjacent quaternary centers [60]. More recently, the addition of benzyl chromium species to aldehydes has been demonstrated in aqueous media in the presence of EDTA [61].

3.4.7
Transformations Involving C=O and C=C Reduction

Takai has shown that 1,3-dienes undergo formation of allylchromiums when treated with $CrCl_2$ in the presence of catalytic vitamin B_{12} (Scheme 15, stereochemistry of major product shown) [62]. The reaction may proceed via chromium(II)-mediated reduction of vitamin B_{12} in the presence of water, forming hydridocobalamin, which undergoes hydrocobaltation of one of the diene double bonds. Homolytic C-Co bond cleavage generates an allylic radical, which is further reduced by chromium(II), generating an allylchromium intermediate.

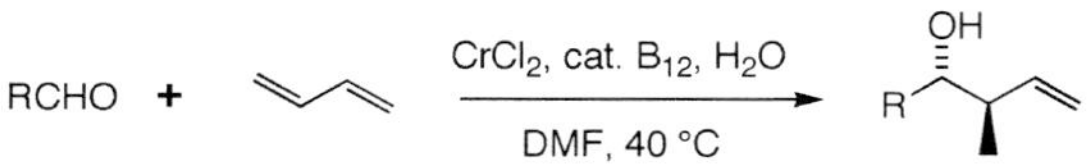

Scheme 15

Takai has also shown that 1,2-disubstituted allylic alcohols are formed regioselectively from reductive coupling of a terminal alkyne and an aldehyde using $CrCl_2$, catalytic $NiCl_2$, and PPh_3 in the presence of a small amount of water in DMF at ambient temperature [63]. The process may proceed via an *in situ*-generated nickel-hydride species which adds across the alkyne, followed by transmetalation of the resulting vinyl nickel to a vinyl chromium which adds to the aldehyde.

Boland has demonstrated that, in the absence of an organo halide or triflate, aldehydes undergo efficient pinacol coupling using the Fürstner catalytic system described earlier [64]. A number of more unusual transformations mediated by chromium(II) have also been reported by Takai. Using $CrCl_2$ in DMF, enones add to aldehydes with concomitant cyclopropanol formation (Scheme 16) [65]. However, if Et_3SiCl is also present, generation of a cross-pinacol-type coupling occurs in which the intermediacy of a silyloxy-substituted allylchromium from the enone is suggested [66]; this process has also been extended to a Cr(II)-catalyzed protocol [67]. A new method for the preparation of 2,5-disubstituted furans has been described using $CrCl_2$–Me_3SiCl–H_2O with conjugated ynones and aldehydes in THF

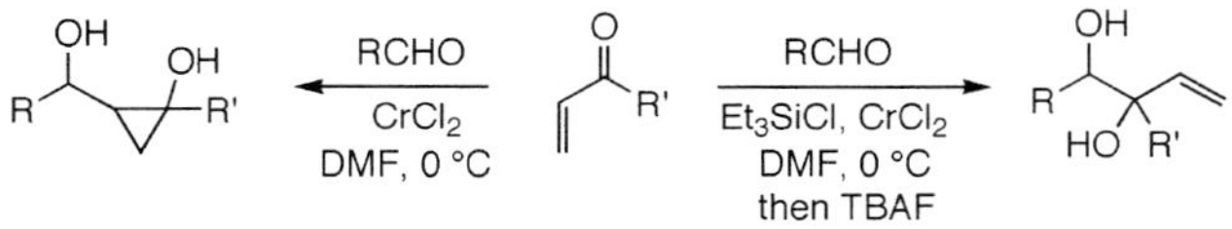

Scheme 16

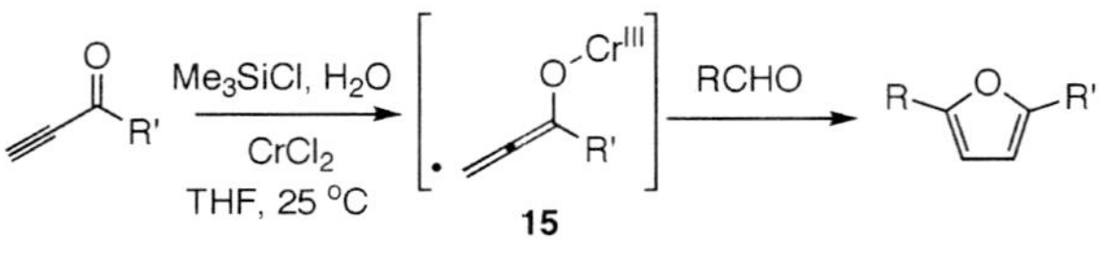

Scheme 17

(Scheme 17) [68]. A plausible reaction sequence commences with generation of an allenyl enolate radical **15**, which is protonated and the resulting vinyl radical reduced prior to aldehyde addition and cyclodehydration. Takai has also shown that imines undergo reaction with CrCl$_2$ in the presence of cat. NiCl$_2$, generating imino-Cr intermediates. The latter are believed to tautomerize to the corresponding Cr-enamines, which react with aldehydes, generating β-(chromiumoxy)imines. Subsequent hydrolysis or reduction leads to the formation of either β-hydroxyketones or β-amino alcohols, respectively [69].

Finally, two related reactions of diynes and enynes, utilising Cr(III) salts, have been described by Oshima. 1,6-Diynes (**16**) undergo [2+2+2] annulation when reacted with methallylchromate (**17**), or methallylmagnesium chloride and catalytic CrCl$_3$ (Scheme 18) [70]. Successive intermolecular-intramolecular alkyne carbometallations followed by an alkene carbometallation ring closure is suggested as a pathway to **18**. In a similar manner, enynes such as **19** lead to dienes **21** (Scheme 19) [71]. Successive intermolecular alkyne- and intramolecular alkene-carbometallations likely generate intermediate **20**, which can be trapped with a range of electrophiles including aldehydes, iodine, acid chlorides, and allylic bromides.

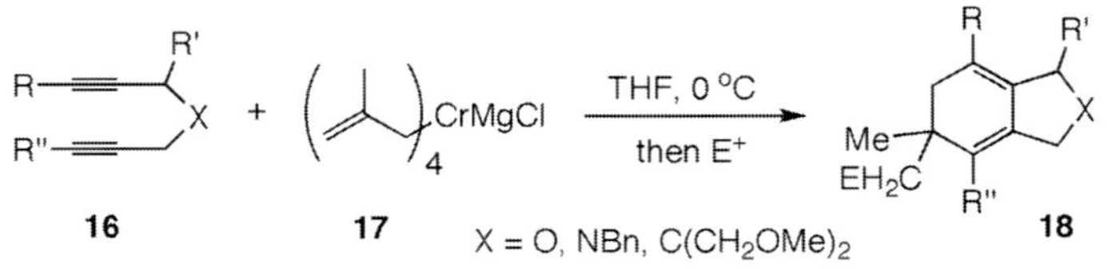

Scheme 18

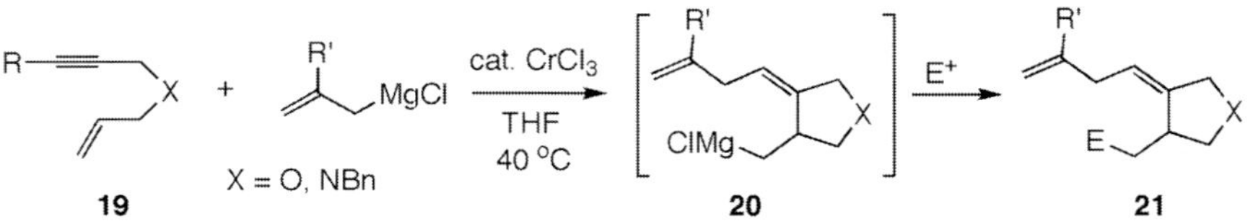

Scheme 19

3.4.8
References

1. Recent reviews: (a) D.M. Hodgson, P.J. Comina in *Transition Metals for Organic Synthesis* (Eds.: M. Beller and C. Bolm), Wiley-VCH, Weinheim, **1998**, Vol. 1, p. 418 and references therein; (b) A. Fürstner, *Chem. Rev.*, **1999**, *99*, 991; (c) M. Avalos, R. Babiano, P. Cintas, J.L. Jiménez, J.C. Palacios, *Chem. Soc. Rev.*, **1999**, *28*, 169; (d) L.A. Wessjohann, G. Scheid, *Synthesis*, **1999**, 1; (e) K. Takai, H. Nozaki, *Proc. Japan. Acad., Ser. B*, **2000**, *76*, 123. (f) R. Baati, *Synlett*, **2001**, 722; (g) M. Semmelhack in *Organometallics in Synthesis: A Manual* (Ed.: M. Schlosser), 2nd edn, John Wiley & Sons, **2002**, p. 1003.
2. Y. Okude, S. Hirano, T. Hiyama, H. Nozaki, *J. Am. Chem. Soc.*, **1977**, *99*, 3179.
3. R. Baati, V. Gouverneur, C. Mioskowski, *J. Org. Chem.*, **2000**, *65*, 1235.
4. H.S. Schrekker, M.W.G. de Bolster, R.V.A. Orru, L.A. Wessjohann, *J. Org. Chem.*, **2002**, *67*, 1975.
5. M. Hojo, R. Sakuragi, S. Okabe, A. Hosomi, *Chem. Commun.*, **2001**, 357.
6. (a) A. Fürstner, N. Shi, *J. Am. Chem. Soc.*, **1996**, *118*, 12349; (b) A. Fürstner, *Pure Appl. Chem.*, **1998**, *70*, 1071.
7. (a) A. Hari, B.J. Miller, *Org. Lett.*, **2000**, *2*, 691; (b) K. Breitenstein, A. Llebaria, A. Delgado, *Tetrahedron Lett.*, **2004**, *45*, 1511.
8. K.H. Shaughnessy, R.C. Huang, *Synth. Commun.*, **2002**, *32*, 1923.
9. M. Kuroboshi, K. Goto, M. Mochizuki, H. Tanaka, *Synlett*, **1999**, 1930.
10. R.K. Boeckman, Jr., R.A. Hudack, Jr., *J. Org. Chem.*, **1998**, *63*, 3524.
11. B. Cazes, C. Vernière, J. Goré, *Synth. Commun.*, **1983**, *13*, 73.
12. C. Chen, K. Tagami, Y. Kishi, *J. Org. Chem.*, **1995**, *60*, 5386.
13. K. Sugimoto, S. Aoyagi, C. Kibayashi, *J. Org. Chem.*, **1997**, *62*, 2322.
14. (a) M. Bandini, P.G. Cozzi, P. Melchiorre, A. Umani-Ronchi, *Angew. Chem. Int. Ed.*, **1999**, *38*, 3357; (b) M. Bandini, P.G. Cozzi, A. Umani-Ronchi, *Angew. Chem. Int. Ed.*, **2000**, *39*, 2327; (c) M. Bandini, P.G. Cozzi, A. Umani-Ronchi, *Polyhedron*, **2000**, *19*, 537; (d) M. Bandini, P.G. Cozzi, P. Melchiorre, S. Morganti, A. Umani-Ronchi, *Org. Lett.*, **2001**, *3*, 1153; (e) M. Bandini, P.G. Cozzi, P. Melchiorre, R. Tino, A. Umani-Ronchi, *Tetrahedron Asymm.*, **2001**, *12*, 1063; (f) M. Bandini, P.G. Cozzi, A. Umani-Ronchi, *Chem. Commun.*, **2002**, 919.
15. A. Berkessel, D. Menche, C.A. Sklorz, M. Schröder, I. Paterson, *Angew. Chem. Int. Ed.*, **2003**, *42*, 1032.
16. (a) M. Inoue, T. Suzuki, M. Nakada, *J. Am. Chem. Soc.* **2003**, *125*, 1140; (b) T. Suzuki, A. Kinoshita, H. Kawada, M. Nakada, *Synlett*, **2003**, 570.
17. (a) D.M. Hodgson, L.T. Boulton in *Preparation of Alkenes: A Practical Approach* (Ed.: J.M.J. Williams), OUP, Oxford, **1996**, p. 81; (b) K. Takai, S. Toshikawa, A. Inoue, R. Kokumai, *J. Am. Chem. Soc.*, **2003**, *125*, 12990.
18. D.M. Hodgson, A.M. Foley, P.J. Lovell, *Tetrahedron Lett.*, **1998**, *39*, 6419.
19. K. Takai, N. Shinomiya, H. Kaihara, N. Yoshida, T. Moriwake, *Synlett*, **1995**, 962.
20. D.M. Hodgson, P.J. Comina, M.G.B. Drew, *J. Chem. Soc., Perkin Trans. 1*, **1997**, 2279.
21. S. Matsubara, M. Horiuchi, K. Takai, K. Utimoto, *Chem. Lett.*, **1995**, 259.
22. (a) D.M. Hodgson, A.M. Foley, P.J. Lovell, *Synlett*, **1999**, 744; (b) D.M. Hodgson, A.M. Foley, L.T. Boulton, P.J. Lovell, G.N. Maw, *J. Chem. Soc., Perkin Trans. 1*, **1999**, 2911.
23. K. Takai, T. Ichiguchi, S. Hikasa, *Synlett*, **1999**, 1268.
24. K. Takai, S. Hikasa, T. Ichiguchi, N. Sumino, *Synlett*, **1999**, 1769.
25. (a) J.R. Falck, D.K. Barma, C. Mioskowski, T. Schlama, *Tetrahedron Lett.*, **1999**, *40*, 2091; (b) R. Baati, D.K. Barma, J.R. Falck, C. Mioskowski, *J. Am. Chem. Soc.*, **2001**, *123*, 9196; (c) D.K. Barma, R. Baati, A. Valleix, C. Mioskowski, J.R. Falck, *Org. Lett.*, **2001**, *3*, 4237; (d) R. Baati, D.K. Barma, U.M.

Krishna, C. Mioskowski, J. R. Falck, *Tetrahedron Lett.*, **2002**, *43*, 959.

26 K. Takai, R. Kokumai, T. Nobunaka, *Chem. Commun.*, **2001**, 1128.

27 R. Baati, D. K. Barma, J. R. Falck, C. Mioskowski, *Tetrahedron Lett.*, **2002**, *43*, 2183.

28 R. Baati, D. K. Barma, J. R. Falck, C. Mioskowski, *Tetrahedron Lett.*, **2002**, *43*, 2179.

29 K. Takai, R. Kokumai, S. Toshikawa, *Synlett*, **2002**, 1164.

30 D. K. Barma, A. Kundu, R. Baati, C. Mioskowski, J. R. Falck, *Org. Lett.*, **2002**, *4*, 1387.

31 D. K. Barma, A. Kundu, H. Zhang, J. R. Falck, C. Mioskowski, *J. Am. Chem. Soc.*, **2003**, *125*, 3218.

32 (a) H. Jin, J. Uenishi, W. J. Christ, Y. Kishi, *J. Am. Chem. Soc.* **1986**, *108*, 5644; (b) K. Takai, M. Tagashira, T. Kuroda, K. Oshima, K. Utimoto, H. Nozaki, *J. Am. Chem. Soc.* **1986**, *108*, 6048; (c) K. Takai, K. Sakogawa, Y. Kataoka, K. Oshima, K. Utimoto, *Org. Synth.* **1993**, *72*, 180.

33 D. L. Comins, A.-C. Hiebel, S. Huang, *Org. Lett.*, **2001**, *3*, 769.

34 H. Yang, X. C. Sheng, E. M. Harrington, K. Ackermann, A. M. Garcia, M. D. Lewis, *J. Org. Chem.*, **1999**, *64*, 242.

35 T. Zoller, D. Uguen, *Eur. J. Org. Chem.*, **1999**, 1545.

36 R. E. Taylor, J. P. Ciavarri, *Org. Lett.*, **1999**, *1*, 467.

37 C. Chen. *Synlett*, **1998**, 1311.

38 R. L. Halterman, C. Zhu, *Tetrahedron Lett.*, **1999**, *40*, 7445.

39 D.-W. Chen, M. Ochiai, *J. Org. Chem.*, **1999**, *64*, 6804.

40 Y. Ogawa, M. Mori, A. Saiga, K. Takagi, *Chem. Lett.* **1996**, 1069.

41 C. Chen, *Synlett*, **2000**, 1491.

42 B. Muller, J.-P. Férézou, J.-Y. Lallemand, A. Pancrazi, J. Prunet, T. Prangé, *Tetrahedron Lett.*, **1998**, *39*, 279.

43 J. A. McCauley, K. Nagasawa, P. A. Lander, S. G. Mischke, M. A. Semones, Y. Kishi, *J. Am. Chem. Soc.*, **1998**, *120*, 7647.

44 J. S. Panek, P. Liu, *J. Am. Chem. Soc.*, **2000**, *122*, 11090.

45 X.-T. Chen, S. K. Battacharya, B. Zhou, C. E. Gutteridge, T. R. R. Pettus, S. J. Danishefsky, *J. Am. Chem. Soc.*, **1999**, *121*, 6563.

46 D. W. C. MacMillan, L. E. Overman, L. D. Pennington, *J. Am. Chem. Soc.*, **2001**, *123*, 9033.

47 M. Kuroboshi, M. Tanaka, S. Kishimoto, K. Goto, H. Tanaka, S. Torii, *Tetrahedron Lett.*, **1999**, *40*, 2785.

48 M. Kuroboshi, M. Tanaka, S. Kishimoto, K. Goto, M. Mochizuki, H. Tanaka, *Tetrahedron Lett.*, **2000**, *41*, 81.

49 R. Grigg, B. Putnikovic, C. J. Urch, *Tetrahedron Lett.*, **1997**, *38*, 6307.

50 M. Kuroboshi, M. Tanaka, S. Hishimoto, H. Tanaka, S. Torii, *Synlett*, **1999**, 69.

51 (a) M. Durandetti, J. Périchon, J.-Y. Nédélec, *Tetrahedron Lett.*, **1999**, *40*, 9009; (b) M. Durandetti, J.-Y. Nédélec, J. Périchon, *Org. Lett.*, **2001**, *3*, 2073.

52 (a) Z.-K. Wan, H. W. Choi, F.-A. Kang, K. Nakajima, D. Demeke, Y. Kishi, *Org. Lett.*, **2002**, *4*, 4431; (b) H. W. Choi, K. Nakajima, D. Demeke, F.-A. Kang, H.-S. Jun, Z.-K. Wan, Y. Kishi, *Org. Lett.*, **2002**, *4*, 4435.

53 K. Takai, T. Kuroda, S. Nakatsukasa, K. Oshima, H. Nozaki, *Tetrahedron Lett.* **1985**, *26*, 5585.

54 M. R. Elliott, A.-L. Dhimane, L. Hamon, M. Malacria, *Eur. J. Org. Chem.*, **2000**, 155.

55 (a) M. Eckhardt, R. Brückner, *Liebigs Ann. Chem.* **1996**, 473; (b) S. Raeppel, D. Toussaint, J. Suffert, *Synlett*, **1998**, 537; (c) G. Rodríguez, D. Rodríguez, M. López, L. Castedo, D. Domínguez, C. Saá, *Synlett*, **1998**, 1282; (d) L. Banfi, G. Guanti, *Tetrahedron Lett.*, **2000**, *41*, 6523; (e) W.-M. Dai, A. Wu, W. Hamaguchi, *Tetrahedron Lett.*, **2001**, *42*, 4211.

56 C. Sandoval, E. Redero, M. A. Mateos-Timoneda, F. A. Bermejo, *Tetrahedron Lett.*, **2002**, *43*, 6521.

57 K. Takai, K. Nitta, O. Fujimara, K. Utimoto, *J. Org. Chem.*, **1989**, *54*, 4732.

58 K. Takai, N. Matsukawa, A. Takahashi, T. Fujii, *Angew. Chem. Int. Ed.*, **1998**, *37*, 152.

59 (a) L. Wessjohann, T. Gabriel, *J. Org. Chem.*, **1997**, *62*, 3772; (b) T. Gabriel, L.

Wessjohann, *Tetrahedron Lett.*, **1997**, *38*, 1363.

60 L. Wessjohann, H. Wild, *Synthesis*, **1997**, 512.

61 K. Micskei, A. Kiss-Szikszai, J. Gyarmati, C. Hajdu, *Tetrahedron Lett.*, **2001**, *42*, 7711.

62 K. Takai, C. Toratsu, *J. Org. Chem.*, **1998**, *63*, 6450.

63 K. Takai, S. Sakamoto, T. Isshiki, *Org. Lett.*, **2003**, *5*, 653.

64 A. Svatos, W. Boland, *Synlett*, **1998**, 549.

65 C. Toratsu, T. Fujii, T. Suzuki, K. Takai, *Angew. Chem. Int. Ed.*, **2000**, *39*, 2725.

66 (a) K. Takai, R. Morita, C. Toratsu, *Angew. Chem. Int. Ed.*, **2001**, *40*, 1116; (b) K. Takai, R. Morita, H. Matsushita, C. Toratsu, *Chirality*, **2003**, *15*, 17.

67 M. Jung, U. Groth, *Synlett*, **2002**, 2015.

68 K. Takai, R. Morita, S. Sakamoto, *Synlett*, **2001**, 1614.

69 K. Takai, N. Katsura, Y. Kunisada, *Chem. Commun.*, **2001**, 1724.

70 T. Nishikawa, H. Kakiya, H. Shinokubo, K. Oshima, *J. Am. Chem. Soc.*, **2001**, *123*, 4629.

71 T. Nishikawa, H. Shinokubo, K. Oshima, *Org. Lett.*, **2002**, *4*, 2795.

3.5
Manganese(III)-Based Oxidative Free-Radical Cyclizations

Barry B. Snider

3.5.1
Introduction

Radical cyclization of alkenes has become a valuable method for the synthesis of cyclic compounds [1–3]. Oxidative free-radical cyclization, in which the initial radical is generated oxidatively and/or the cyclic radical is terminated oxidatively, has considerable synthetic potential, since more highly functionalized products can be prepared from simple precursors. Oxidative formation of an acyclic radical involves the formal loss of a hydrogen atom. In practical terms, this is often accomplished by loss of a proton and oxidation of the resulting anion with a one-electron oxidant such as Mn(III) or Ce(IV) to generate a radical. *The advantage of this method of radical formation is that the precursor is simple and usually readily available.* A potential disadvantage is that the cyclization product may also be susceptible to further deprotonation and oxidation. *Oxidative termination of radical cyclizations is advantageous, since more highly functionalized, versatile products are produced than from reductive terminations that deliver a hydrogen atom.*

The oxidative addition of acetic acid to alkenes reported by Heiba and Dessau [4] and Bush and Finkbeiner [5] in 1968 provides the basis for a general approach to oxidative free-radical cyclization. These oxidative additions have been extensively explored over the past 35 years and have been reviewed recently [6–11]. Mn(III)-based oxidative free-radical cyclizations have been extensively developed since they were first reported in 1984–1985 [12–14]. While the vast majority of the work has used Mn(III) or Mn(III)/Cu(II), other one-electron oxidants, most notably Ce(IV), Fe(III), and Cu(II), have also been employed.

The Mn(III)-based oxidative free-radical cyclization of **1a** and **1b** shown in Scheme 1 serves to introduce the factors that need to be understood to use these reactions in synthesis. Oxidative cyclization of β-keto ester **1a** with Mn(OAc)$_3$ affords a complex mixture of products. Primary and secondary radicals such as **5** are not oxidized by Mn(III). Heiba and Dessau found that Cu(OAc)$_2$ oxidizes secondary radicals 350 times faster than Mn(OAc)$_3$ does and that the two reagents can be used together [15]. Oxidative cyclization of **1a** with 2 equiv. of Mn(OAc)$_3$ and 0.1–1 equiv. of Cu(OAc)$_2$ in acetic acid affords 71% of **6a**. Cu(OAc)$_2$ reacts with radical **5a** to give a Cu(III) intermediate that undergoes oxidative elimination

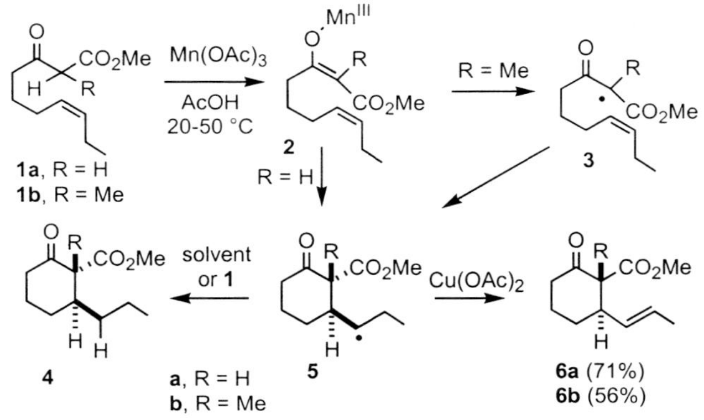

Scheme 1

to give **6a** [13, 16]. A similar oxidative cyclization of **1b** affords 56% of **6b** as the major product.

The first step in the reaction is the loss of a proton to give the Mn(III) enolate **2**. The next step of the reaction with enolates that have a high oxidation potential involves cyclization of the unsaturated Mn(III) enolate **2** to give cyclic radical **5**. This is the operative pathway for R=H. For enolates that have a lower oxidation potential, loss of Mn(II) gives the Mn-free free radical **3**. This is the operative pathway for R=Me. Cyclization of **3b** from the conformation shown gives radical **5b** stereo- and regiospecifically. Finally, Cu(II) oxidation of **5** gives **6** regio- and stereospecifically.

3.5.2
Oxidizable Functionality

β-Keto esters have been used extensively for Mn(III)-based oxidative cyclizations and react with Mn(OAc)$_3$ at room temperature or slightly above [10, 11, 13, 16]. They may be cyclic or acyclic and may be α-unsubstituted or may contain an α-alkyl or chloro substituent. Cycloalkanones are formed if the unsaturated chain is attached to the ketone. γ-Lactones are formed from allylic acetoacetates [10, 11]. Less acidic β-keto amides have seen limited use for the formation of lactams or cycloalkanones. Malonic esters have also been widely used and form radicals at 60–80 °C. Meldrum's acid derivatives react at room temperature [17]. Cycloalkanes are formed if an unsaturated chain is attached to the α-position. γ-Lactones are formed from allylic malonates [10, 11]. β-Diketones have been used with some success for cyclizations to both alkenes and aromatic rings [10, 11]. Other acidic carbonyl compounds such as β-keto acids, β-keto sulfoxides, β-keto sulfones, β-nitro ketones, and β-nitro esters have seen limited use [10, 11, 18]. Oxidative cyclizations of unsaturated ketones can be carried out in high yield in acetic acid at 80 °C if the ketone selectively enolizes to one side and the product cannot enolize [19].

3.5.3
Oxidants and Solvents

Commercially available Mn(OAc)$_3$·2H$_2$O has been used for the majority of oxidative cyclizations. This reagent can also be prepared easily from potassium permanganate and manganous acetate in acetic acid [6]. Anhydrous Mn(OAc)$_3$ is slightly more reactive than the dihydrate. Reaction times with the anhydrous reagent are usually somewhat shorter, but the yields of products are usually comparable. Both trifluoroacetic acid and potassium or sodium acetate have been used with Mn(OAc)$_3$. Use of trifluoroacetic acid as a co-solvent usually increases the rate of the reaction, but often decreases the yield of products. Acetate anion may accelerate enolization and act as a buffer. Addition of lanthanide triflates improves the rate and selectivity of cyclizations of some β-keto esters [20].

Acetic acid is the usual solvent for Mn(OAc)$_3$·2H$_2$O reactions. DMSO, ethanol, methanol, dioxane, and acetonitrile and trifluoroethanol can also be used, although higher reaction temperatures are required and lower yields of products are sometimes obtained [10, 20]. The use of ionic liquids has recently been introduced [21]. The use of ethanol can be advantageous in cyclizations to alkynes [22]. Vinyl radicals formed by cyclization to alkynes are not readily oxidized by Mn(III) and will undergo undesired side reactions unless there is a good hydrogen donor available. Ethanol acts as a hydrogen donor, reducing the vinyl radical to an alkene and giving the α-hydroxyethyl radical, which is oxidized to acetaldehyde by Mn(III).

Mn(OAc)$_3$ is also involved in the termination step. It rapidly oxidizes tertiary radicals to cations, which lose a proton to give an alkene, or react with the solvent, acetic acid, to give acetate esters. Mn(OAc)$_3$ also oxidizes allylic radicals to allylic acetates and oxidizes cyclohexadienyl radicals generated by additions to benzene rings to cations, which lose a proton to regenerate the aromatic system. On the other hand, Mn(OAc)$_3$ oxidizes primary and secondary radicals very slowly, so that hydrogen atom abstraction from solvent or starting material becomes the predominant process.

Cu(OAc)$_2$ is compatible with Mn(OAc)$_3$, and Cu(II) oxidizes primary and secondary radicals to alkenes 350 times faster than Mn(III) does [10, 15]. The Cu(I) that is produced in this oxidation is rapidly oxidized to Cu(II) by Mn(III), so that only a catalytic amount of Cu(OAc)$_2$ is needed, and two equivalents of Mn(OAc)$_3$ are still required. Cu(OAc)$_2$ oxidizes secondary radicals to give primarily E-alkenes and the less-substituted double bond (Hofmann elimination product) [23]. This selectivity is synthetically valuable, since Cu(II) oxidation of primary and secondary radicals formed in oxidative cyclizations often gives primarily or exclusively a single regio- and stereoisomer.

A wide variety of other one-electron oxidants have been used for generating free radicals from β-dicarbonyl compounds, most notably ferric perchlorate in acetonitrile and ceric ammonium nitrate. These oxidants are capable of forming radicals from 1,3-dicarbonyl compounds. However, the oxidant is also necessary for termination of the radical reaction. The nature of the metal, the ligands necessary to obtain the desired oxidation potential, and the solvent needed to achieve solubility of the metal salt all play a crucial role in determining the products formed from oxidation of the cyclized radical [10].

3.5.4
Common Side Reactions

Oxidative cyclization of unsaturated β-dicarbonyl compounds that have two α-hydrogens will give products that still have one α-hydrogen and can be oxidized further. If the product is oxidized at a rate competitive with that of the starting material, mixtures of products will be obtained. For instance, oxidative cyclization of **7** affords 36% of **8** and 10% of dienone **9** formed by further oxidation of **8** [16] as shown in Scheme 2. Competitive oxidation of the product is usually not a problem in intermolecular addition reactions because a vast excess of the oxidizable substrate, such as acetone or acetic acid, is usually used as solvent. Use of excess substrate is not possible in oxidative cyclizations.

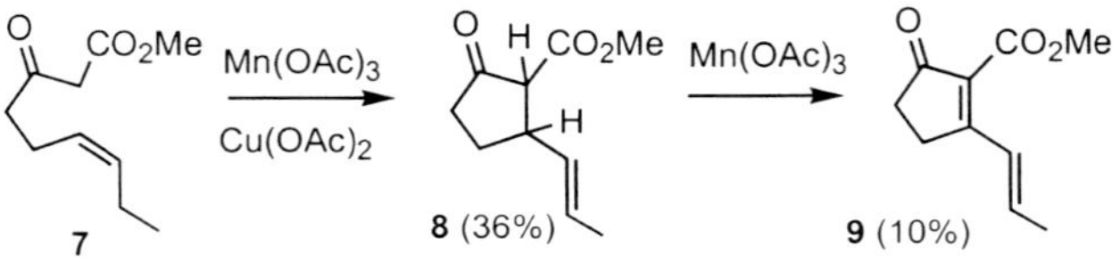

Scheme 2

In some cases, the product is oxidized much more readily than the starting material so that none of the initial product is isolated. These reactions may still be synthetically useful if the products of further oxidation are monomeric. For instance, oxidative cyclization of **10** provides 78% of methyl salicylate (**13**) as shown in Scheme 3 [24–26]. Oxidative cyclization gives radical **11**; oxidation of **11** gives **12**, probably as a mixture of double bond positional isomers. The unsaturated cyclic β-keto ester **12** is more acidic than **11** and is rapidly oxidized further by two

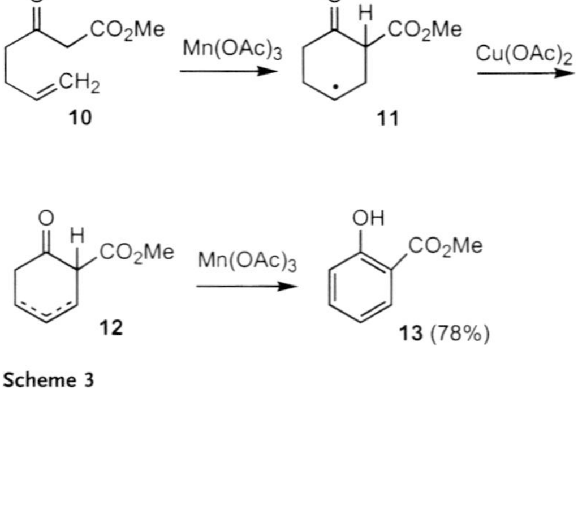

Scheme 3

equivalents of Mn(III) to give a cyclohexadienone that tautomerizes to phenol **13**. The overall reaction consumes 4 equiv. of Mn(OAc)$_3$.

Further oxidation cannot occur if there are no acidic α-hydrogens in the product. α-Chloro substituents serve as protecting groups preventing further oxidation of the product [27–30]. For instance, oxidative cyclization of **14** affords 82% of a 3.1:1 mixture of **15** and **16** [28] as shown in Scheme 4. The other two stereoisomers with the octyl and vinyl groups *cis* are not formed. This mixture was elaborated to avenaciolide (**17**) by a sequence that used an S$_N$2 reaction on the α-chloro lactone to form the second lactone ring.

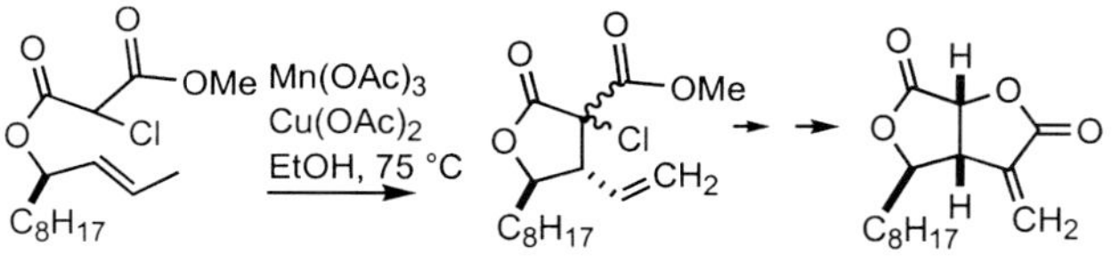

14

15 (62%) β-Cl, α-CO$_2$Me
16 (20%) α-Cl, β-CO$_2$Me

avenaciolide
(**17**)

Scheme 4

3.5.5
Cyclization Substrates

Cyclizations that form a single carbon-carbon bond can be accomplished by oxidative cyclization of unsaturated β-diketones, β-keto esters, or β-keto amides (**18**), which lead to cycloalkanones, unsaturated β-diketones, β-keto esters, or malonate esters (**19**), which lead to cycloalkanes, and unsaturated esters or amides (**20**), which lead to lactams or lactones [10].

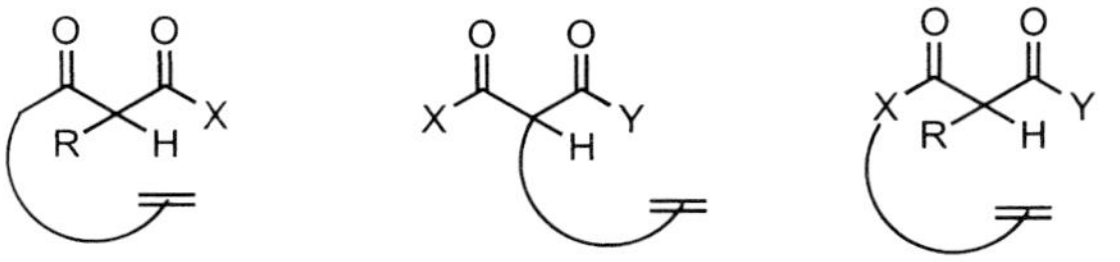

18, X = C, OR, NR$_2$ **19**, X, Y = C, OR, NR$_2$ **20**, X = O, NR
                                                              Y = C, OR, NR$_2$

More complex targets can be made with excellent stereocontrol by tandem oxidative cyclizations. These reactions can be divided into two classes depending on whether the second cyclization is to an aromatic ring or to another double bond. Oxidative cyclization of **21** with 2 equivalents of Mn(OAc)$_3$ in MeOH at 0 °C provides 50–60% of **22** as a single stereoisomer whose structure was established by Clemmensen reduction to give ethyl O-methylpodocarpate (**23**) [13, 24, 31]. Tandem cyclizations can also be terminated by cyclization to an arene conjugated

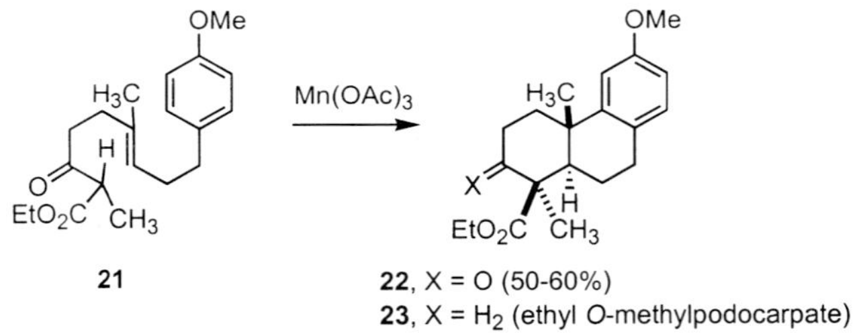

Scheme 5

21 → 22, X = O (50–60%)
23, X = H₂ (ethyl O-methylpodocarpate)

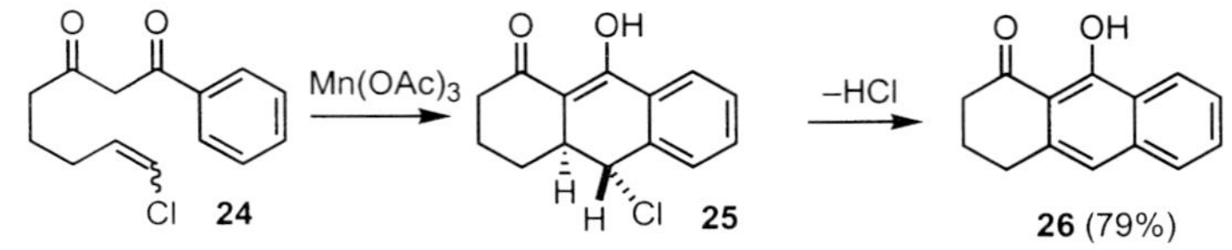

24 → 25 → 26 (79%)

with a carbonyl group. Oxidative cyclization of either the E- or Z-isomer of **24** with Mn(OAc)₃ in acetic acid affords **25**, which undergoes slow loss of hydrogen chloride to afford 79% of the desired naphthol **26** [32–34]. Similar cyclizations were used for the first syntheses of okicenone and aloesaponol III.

The utility of tandem oxidative cyclizations is clearly demonstrated in substrates in which both additions are to double bonds [35, 36]. Oxidative cyclization of **27** with two equivalents of Mn(OAc)₃ and Cu(OAc)₂ in acetic acid at 25 °C affords 86% of bicyclo[3.2.1]octane (**32**). Oxidation affords a keto radical (**28**), which cyclizes exclusively 6-*endo* in the conformation shown to afford tertiary radical **29** with an equatorial allyl group. Chair-chair interconversion provides **30** with an axial allyl group. 5-*exo*-Cyclization of the 5-hexenyl radical of **30** gives **31** as a 2 : 1

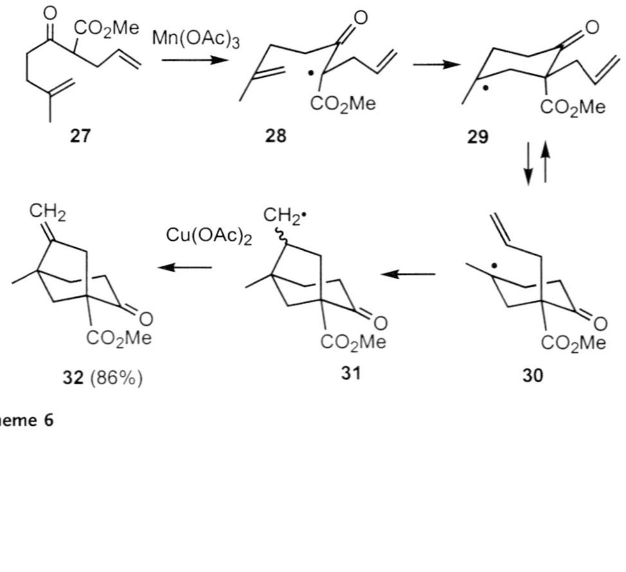

Scheme 6

mixture of *exo*- and *endo*-stereoisomers. Oxidation of both stereoisomers of **31** with Cu(II) provides **32** as shown in Scheme 6.

A wide variety of tandem, triple [36], and even quadruple [37–42] cyclizations can be carried out with multiply unsaturated 1,3-dicarbonyl compounds, as shown in Eqs. 1 [41] and 2 [42], which provide intermediates for steroid and terpene syntheses. High levels of asymmetric induction can be achieved with phenylmenthyl acetoacetate esters and dimethylpyrrolidine acetoacetamides [31]. Recent results indicate that addition of Yb(OTf)$_3$ in trifluoroethanol gives improved asymmetric induction [20].

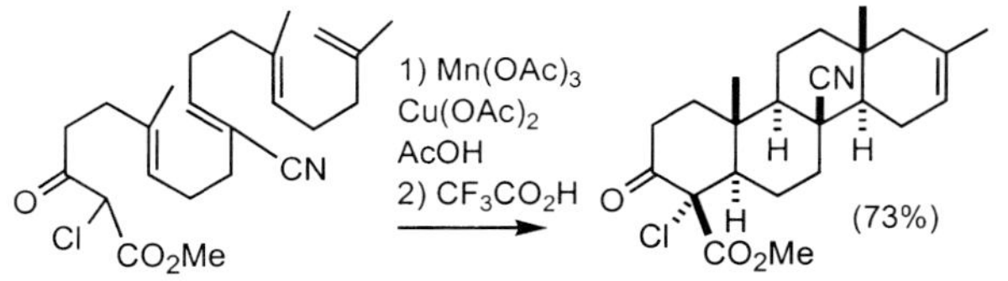

The application of Mn(III)-mediated radical reactions to natural product total synthesis provides an excellent demonstration of the scope and utility of these reactions, since the method must be versatile enough to deal with complex skeletons and diverse functionality. Many examples are presented above, and other recent applications include the vannusal A ring system [43], araliopsine [44], and wentilactone B [45]. The addition of malonate esters to glycals provides a route to modified carbohydrates [46].

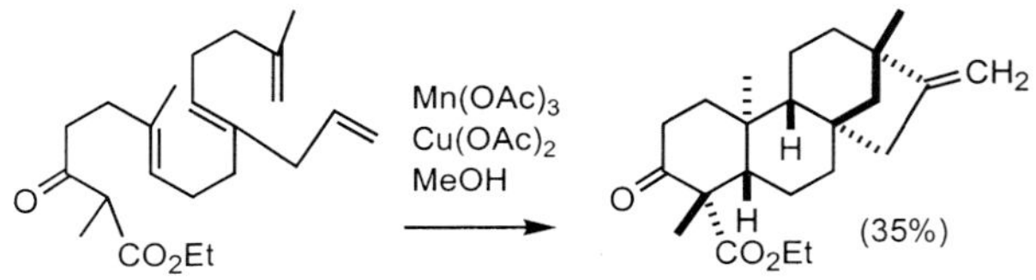

3.5.6
References

1 (a) D. P. Curran, *Synthesis* **1988**, 417 and 489. (b) C. P. Jasperse, D. P. Curran, *Chem. Rev.* **1991**, *91*, 1237.
2 (a) B. Giese, *Radicals in Organic Synthesis: Formation of Carbon-Carbon Bonds*, Pergamon Press, Oxford, New York, **1986**. (b) *C-Radikale*, Vol E 19A in *Houben-Weyl Methoden der Organischen Chimie*, (Eds: M. Regitz, B. Giese), Thieme, Stuttgart, **1989**.
3 P. Rennaud, M. P. Sibi *Radicals in Organic Synthesis, Vols 1 and 2*, Wiley-VCH, Weinheim, 2001
4 E. I. Heiba, R. M. Dessau, W. J. Koehl Jr., *J. Am. Chem. Soc.* **1968**, *90*, 5905.
5 J. B. Bush Jr., H. Finkbeiner, *J. Am. Chem. Soc.* **1968**, *90*, 5903.
6 W. J. de Klein, in *Organic Synthesis by Oxidation with Metal Compounds* (Eds:

W. J. Mijs, C. R. H. de Jonge), Plenum Press, New York, **1986**, pp 261–314.
7. Sh. O. Badanyan, G. G. Melikyan, D. A. Mkrtchyan, *Russ. Chem. Rev.* **1989**, *58*, 286; *Uspekhi Khimii* **1989**, *58*, 475.
8. G. G. Melikyan, *Synthesis*, **1993**, 833.
9. J. Iqbal, B. Bhatia, N. K. Nayyar, *Chem. Rev.* **1994**, *94*, 519.
10. B. B. Snider, *Chem. Rev.* **1996**, *96*, 339.
11. (a) G. G. Melikyan, *Org. React.* **1997**, *49*, 427. (b) G. G. Melikyan, *Aldrichimica Acta* **1998**, *31*, 50.
12. E. J. Corey, M.-C. Kang, *J. Am. Chem. Soc.* **1984**, *106*, 5384.
13. B. B. Snider, R. M. Mohan, S. A. Kates, *J. Org. Chem.* **1985**, *50*, 3659.
14. A. B. Ernst, W. E. Fristad, *Tetrahedron Lett.* **1985**, *26*, 3761.
15. (a) E. I. Heiba, R. M. Dessau, *J. Am. Chem. Soc.* **1971**, *93*, 524. (b) E. I. Heiba, R. M. Dessau, *J. Am. Chem. Soc.* **1972**, *94*, 2888.
16. S. A. Kates, M. A. Dombroski, B. B. Snider, *J. Org. Chem.* **1990**, *55*, 2427.
17. B. B. Snider, R. B. Smith, *Tetrahedron* **2002**, *58*, 25.
18. B. B. Snider, Q. Che, *Tetrahedron* **2002**, *58*, 7821.
19. B. A. M. Cole, L. Han, B. B. Snider, *J. Org. Chem.* **1996**, *51*, 7832.
20. (a) D. Yang, X.-Y. Ye, M. Xu, K.-W. Pang, N. Zou, R. M. Lechter, *J. Org. Chem.* **1998**, *63*, 6446. (b) D. Yang, X.-Y. Ye, S. Gu, M, Xu, *J. Am. Chem. Soc.* **1999**, *121*, 5579. (b) D. Yang, X.-Y. Ye, M. Xu, K.-W. Pang, K.-K. Cheung, *J. Am. Chem. Soc.* **2000**, *122*, 1658. (d) D. Yang, X.-Y. Ye, M. Xu *J. Org. Chem.* **2000**, *65*, 2208. (e) D. Yang, M. Xu, M.-Y. Bian, *Org. Lett.* **2001**, *3*, 111.
21. G. Bar, A. F. Parsons, C. B. Thomas, *Chem. Commun.* **2001**, 1350.
22. B. B. Snider, J. E. Merritt, M. A. Dombroski, B. O. Buckman, *J. Org. Chem.* **1991**, *56*, 5544.
23. B. B. Snider, T. Kwon, *J. Org. Chem.* **1990**, *55*, 1965.
24. R. Mohan, S. A. Kates, M. A. Dombroski, B. B. Snider, *Tetrahedron Lett.* **1987**, *28*, 845.
25. B. B. Snider, J. J. Patricia, *J. Org. Chem.* **1989**, *54*, 38.
26. J. R. Peterson, R. S. Egler, D. B. Horsley, T. J. Winter, *Tetrahedron Lett.* **1987**, *28*, 6109.
27. B. B. Snider, J. J. Patricia, S. A. Kates, *J. Org. Chem.* **1988**, *53*, 2137.
28. B. B. Snider, B. A. McCarthy, *Tetrahedron* **1993**, *49*, 9447.
29. E. J. Corey, A. W. Gross, *Tetrahedron Lett.* **1985**, *26*, 4291.
30. N. Fujimoto, H. Nishino, K. Kurosawa, *Bull Chem. Soc. Jpn.* **1986**, *59*, 3161.
31. Q. Zhang, R. M. Mohan, L. Cook, S. Kazanis, D. Peisach, B. M. Foxman, B. B. Snider, *J. Org. Chem.* **1993**, *58*, 7640.
32. B. B. Snider, R. M. Mohan, S. A. Kates, *Tetrahedron Lett.* **1987**, *28*, 841.
33. B. B. Snider, Q. Zhang, M. A. Dombroski, *J. Org. Chem.* **1992**, *57*, 4195.
34. B. B. Snider, Q. Zhang, *J. Org. Chem.* **1993**, *58*, 3185.
35. B. B. Snider, M. A. Dombroski, *J. Org. Chem.* **1987**, *52*, 5487.
36. M. A. Dombroski, S. A. Kates, B. B. Snider, *J. Am. Chem. Soc.* **1990**, *112*, 2759.
37. P. A. Zoretic, X. Weng, M. L. Caspar, D. G. Davis, *Tetrahedron Lett.* **1991**, *32*, 4819.
38. P. A. Zoretic, M. Wang, Y. Zhang, Z. Shen, A. A. Ribeiro, *J. Org. Chem.* **1996**, *61*, 1806.
39. P. A. Zoretic, Y., Zhang, H. Fang, A. A. Ribeiro, *J. Org. Chem.* **1998**, *63*, 1162.
40. P. A. Zoretic, H. Fang, A. A. Ribeiro, *J. Org. Chem.* **1998**, *63*, 4779.
41. P. A. Zoretic, H. Fang, A. A. Ribeiro, *J. Org. Chem.* **1998**, *63*, 7213.
42. B. B. Snider, J. Y. Kiselgof, B. M. Foxman, *J. Org. Chem.* **1998**, *63*, 7945.
43. K. C. Nicolaou, M. P. Jennings, P. Dagneau, *Chem. Commun.* **2002**, 2480.
44. G. Bar, A. F. Parsons, C. B. Thomas, *Tetrahedron* **2001**, *57*, 4719.
45. A. F. Barrero, M. M. Herrador, J. F. Quílez del Moral, M. V. Valdivia, *Org. Lett.* **2002**, *4*, 1379.
46. T. Linker, *J. Organomet. Chem.* **2002**, *661*, 159.

3.6
Titanium-Mediated Reactions

Rudolf O. Duthaler, Frank Bienewald, and Andreas Hafner

3.6.1
Introduction – Preparation of Titanium Reagents

Titanium, an abundant and non-toxic element, has frequently been used for the modification of organolithium and organomagnesium compounds as well as lithium enolates. The resulting titanium reagents are often well-defined complexes with tetrahedral, trigonal bipyramidal, or octahedral coordination geometry. In its highest oxidation, state titanium is tetravalent; extended penta- or hexacoordination is found in charged "ate-complexes", with neutral ligands, or as a result of aggregate formation. Coordinatively unsaturated titanium centers are *Lewis* acids with preferential binding to oxygen or fluoride nucleophiles. Chemoselectivity was the first advantageous property of titanium reagents discovered, i.e. preferential reaction with aldehydes in the presence of ketones or imines [1]. By virtue of the comparably well-defined and robust coordination geometry conversions involving titanium centers often exhibit excellent stereocontrol [2]. This review is restricted to stoichiometric and catalytic processes, mainly the addition of nucleophiles to aldehydes mediated by complexation of a *Lewis*-acidic Ti(IV)-center with the carbonyl oxygen. The attacking nucleophile is either a ligand of the activating titanium complex or an additional reagent. The stereoselectivity results solely from the bias of the chiral titanium center (Scheme 1). The last section (Section 3.6.5) describes cycloadditions and miscellaneous reactions also mediated by chiral titanium complexes.

For the preparation of individual titanium reagents not covered by this review, the original literature should be consulted. The usual method is ligand exchange on, e.g., titanium tetrachloride, titanium tetra-*iso*-propoxide, or cyclopentadienyltitanium trichloride. Chloride ligands can be displaced by protonated (LH), silylated (LSiMe$_3$), stannylated (LSnR$_3$), or metalated (LM, M: Li, MgX, ZnR, Ag) ligands. Volatiles such as HCl, Me$_3$SiCl, and R$_3$SnCl can be removed by evaporation or co-distillation with solvent, and HCl can also be neutralized with a weak base (e.g., Et$_3$N). Alkoxide ligands can be exchanged through adduct formation with alcohols, whereby the equilibrium is displaced by evaporation of a volatile alcohol (e.g., 2-propanol, Scheme 2). Titanium alkoxides also undergo exchange in a metathetic manner via aggregate formation (Scheme 2). This process is sometimes im-

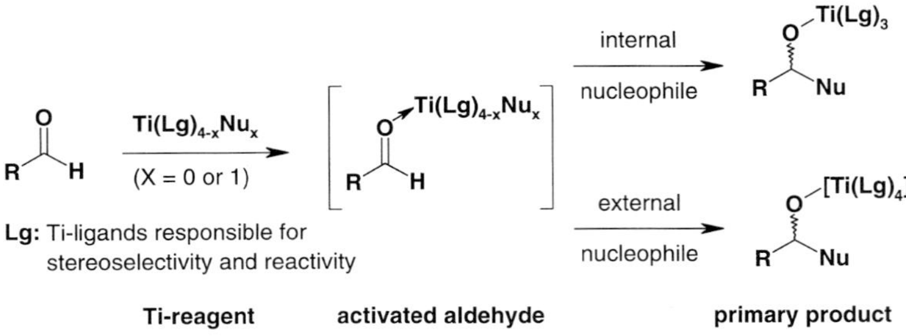

Scheme 1 Main principles of Ti-mediated processes.

portant to regenerate an $L_3Ti(O\text{-}iPr)$ complex from a primary product with an excess of $Ti(O\text{-}iPr)_4$ (Scheme 1). Such a ligand redistribution is of course deleterious if it occurs on chiral ligands during a reaction. This danger is reduced for di- and tridentate chelating ligands or with pentahapto-bound cyclopentadienyl groups. Many titanium compounds are sensitive to humidity. Therefore one has always to be aware that weakly bound ligands such as titanium chlorides or bromides could succumb to hydrolysis, especially in catalytic processes. The resulting hydroxo compounds are strongly acidic and form μ-oxo dimers or trimers readily (Scheme 2). With more water, complete hydrolysis to titanium dioxide proceeds via complex oxo-clusters [3]. An important *caveat* concerns the structural representation of titanium complexes and mechanistic rationalizations deduced thereof. In most cases these formulae merely represent the stoichiometric ratio of titanium and ligands and have little in common with the actual structures, which are often dimeric aggregates or mixtures of higher oligomers. In this chapter, all structures not secured by X-ray diffraction data are represented in brackets.

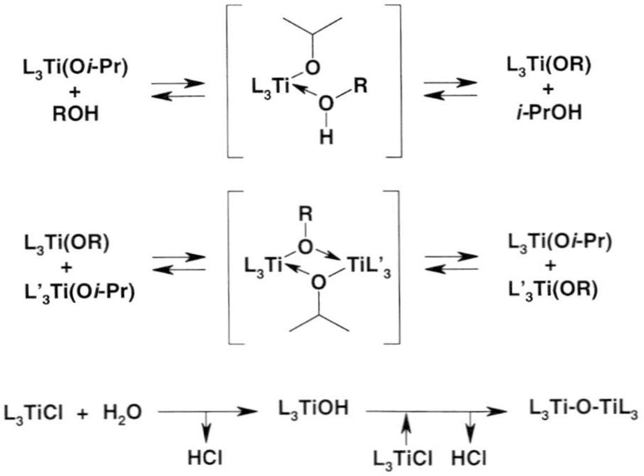

Scheme 2 Alkoxy ligand exchange and hydrolysis of titanates.

3.6.2
Addition of Allyl Nucleophiles to Aldehydes and "Ene" Reactions

Allyltitanium compounds react readily with aldehydes, and a bimolecular mechanism with a cyclic six-membered transition state is postulated [2]. The best stereoselectivity has so far been achieved with monocyclopentadienyl-dialkoxytitanium complexes **a** obtained from the chlorides **1** [4], **2** [5], and *ent*-**2** [5], respectively (Scheme 3). According to ^{1}H- and ^{13}C-NMR studies the allyl substituent is monohapto bound, but the titanium undergoes a very fast 1,3-shift. In the case of terminally monosubstituted allyl groups (R ≠ H) the *trans*-isomer with titanium attached to the unsubstituted terminal carbon always prevails, independently of the nature of the organometallic allyl precursor, which can be an isomeric mixture of allyl-*Grignard* compounds including the secondary isomer **b** or an isomerically pure allyl-Li, e.g., **c** (Scheme 3) [5a]. Besides the parent allyltitanium reagents **a** (R=H), the monosubstituted derivatives **a** with R=CH$_3$, C$_6$H$_5$ [5], vinyl [6], allyl [7], Me$_3$Si, O-alkyl, and O-aryl [5] have been prepared. Reaction with various aldehydes gives the homoallyl alcohols **3** and *ent*-**3** in good yield and excellent stereoselectivity, also in the case of complex chiral structures such as glyceraldehyde acetonide [6]. In the vast majority of cases with R ≠ H, the *anti*-diastereomer of product **3** prevails, implicating chair conformation of the cyclic transition state. The enantioselectivity of allyltitanium reagents **a** derived from chloride **2** or *ent*-**2** is in general superior to that of the corresponding reagents **a** obtained from the diacetone-glucose complex **1** (Scheme 3).

The facile isomerisation of the η^1-allyltitanium compounds precludes fixation of the double bond in the *cis* geometry. The *syn* isomers of the homoallylalcohols **3** can therefore not be obtained by this method. Problems are also encountered with

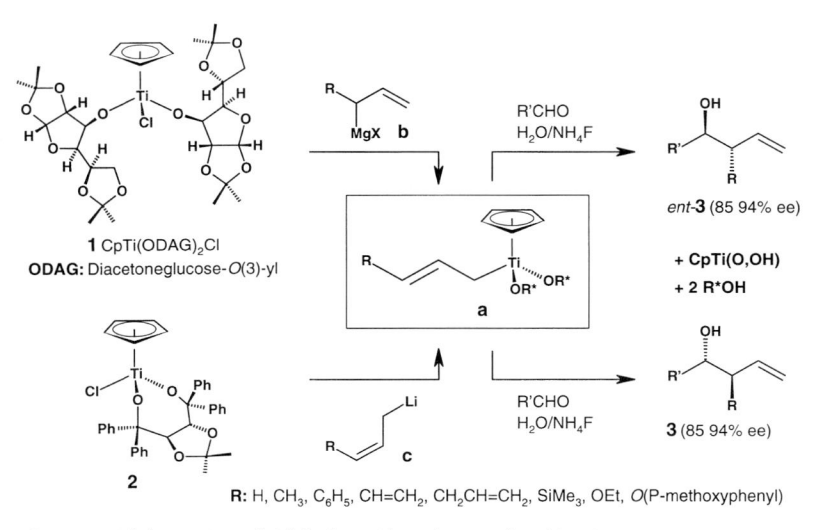

Scheme 3 Allyltitanation of aldehydes with cyclopentadienyltitanium reagents.

substituted allyl groups other than the terminally monosubstituted cases described above. Moderate enantioselectivity (e.g., 73–88% ee for **5**) is thus obtained with methallyltitanium reagents **4** [6, 8] and higher analogs [9] upon addition to aldehydes (Scheme 4). In the case of 1,3-disubstituted allyl groups, the problems are associated with the chiral and fast equilibrating α-carbons. Thus, the position of the thermodynamically controlled equilibrium between **6a** and **6b** is determined by the chiral TADDOL ligand on titanium, which also controls the kinetics of the addition to benzaldehyde, giving **7** from **6a** (matched) and *ent*-**7** from **6b** (mismatched). So far, these competing influences of the chiral titanium center have led to reduced stereoselectivity, 86% ee for **7** and 26% ee for the α-titanated 2-alkenyl-N,N-di-*iso*-propyl-carbamate [6], for which a configuratively stable titanium amide has been described by Hoppe et al. [10].

These alkyltitanium reagents (**a**, R: H, CH$_3$) have recently been applied by Cossy and co-workers for preparing optically active unsaturated alcohols, diols, and polypropionates, in conjunction with synthesizing biologically active compounds [11]. Remarkable is the observation that chiral β-hydroxy-aldehydes (**8**) can be allylated with (R,R)-**9** without protection, giving the products **10** with good yield and excellent stereocontrol, independently of substrate chirality [11b]. Addition of the crotyl reagent (R,R)-**11** to the *meso*-dialdehyde **12** proceeds with excellent Felkin-Anh control, yielding 54% of a single monoadduct **13**, thus establishing five contiguous stereocenters with high selectivity in one step. The enantiomer of **13** is obtained with (S,S)-**11** [11c].

While the allyltitanium reagents are based on stoichiometric amounts of chiral ligands, it has been found that catalytic amounts of titanium complexes derived from 1,1′-binaphthalene-2,2′-diol (BINOL) mediate the enantioselective addition of allyl stannanes **14** [9, 12, 13] to aldehydes, giving the homoallyl alcohols **15** with

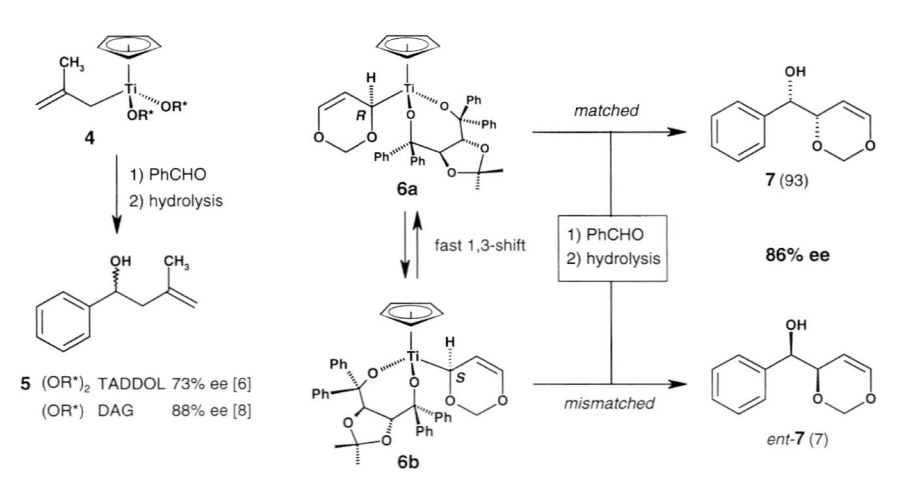

Scheme 4 Cyclopentadienyltitanium allyl reagents with substituents at C(2) and C(1)/C(3), respectively.

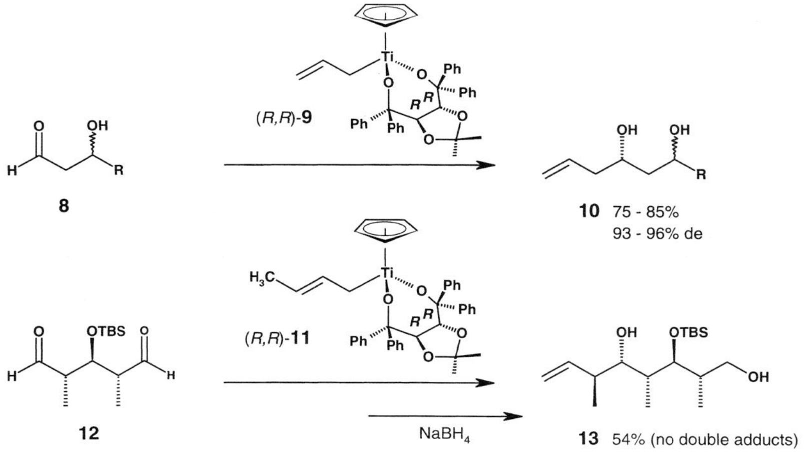

Scheme 5 Synthesis of diols and polypropionate arrays [11].

high enantioselectivity. (S)- or (R)-BINOL is combined in 1:1 or 1:2 ratio with either di-*iso*-propoxy-titanium dichloride [12] or with titanium tetra-*iso*-propoxide [9, 13] (Scheme 6). The dichloride catalyst has recently been shown to catalyze the addition of tetra-allyl tin to aromatic methyl ketones with moderate enantioselec-

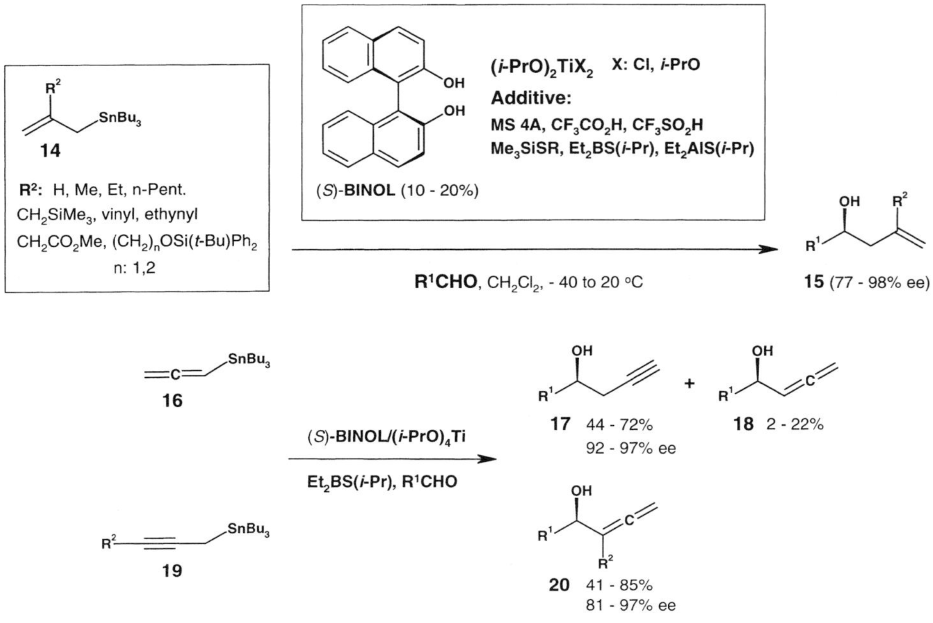

Scheme 6 The BINOL-Ti(IV)-catalyzed addition of allyl stannanes to aldehydes.

tivities [12b], and Keck and co-worker have introduced CH_2CO_2Me as a new C(2)-allyl substituent R^2 of **14** [13d], and applied reagent **14** (R^2:Et) in a total synthesis of *Rhizoxin D* [13e].

A rather confusing variety of recipes, including heating with or without molecular sieves and sometimes also addition of acids (CF_3SO_3H [13a], CF_3CO_2H [13c]), have been applied for optimal results with individual aldehyde/allyl stannane **14** pairs. At least for 2-substituted allyl stannanes, the conditions developed by Brückner and Weigand [9] (mixing of 2 equivalents of BINOL and titanium tetraethoxide for 2 h at room temperature and conducting the reaction at –40 °C) seem to be broadly applicable. A major improvement – mostly related to reaction kinetics – has been brought about by the addition of trimethylsilyl [14a], diethylboryl, and diethylaluminum-sulfides [14b] in stoichiometric quantities. The argument put forward by the authors, Yu and co-workers, is that these additives are trapping the stannyl residue as sulfide and are masking the product alcohols **15** as O-silyl, -boryl, or -aluminyl derivatives, thus preventing inactivation of the titanium catalyst by product coordination. By applying this method, allylstannanes (**14**) with R^2=vinyl or ethynyl [14c], and R^2=CH_2SiMe_3 [14d] have been added to aldehydes, the products **15** of the latter serving as intermediates for tetrahydropyrans. Yet another class of additives are o,o'-tritylamino-benzophenones, which supposedly are bridging two BINOL-titanium complexes, giving more efficient binuclear Lewis acid catalysts [15]. The same catalyst system is also applicable for the addition of allenyl stannane **16** to aldehydes with [14e,f] and without [16] diethylboryl sulfide, affording homopropargylic alcohols (**17**) of high optical purity, contaminated with only minor amounts of allenyl alcohols (**18**). With propargyl stannanes (**19**), on the other hand, enantioselective allenylation of aldehydes has been achieved (→ **20** [14g], Scheme 6). The addition of crotyl stannane to glyoxylate proceeded only with low *syn* selectivity and moderate enantioselectivity (80–86% *ee*) [17]. The chiral amplification for product **15** prepared with a titanium catalyst obtained from partially resolved BINOL is explained by the formation of catalytically inactive aggregates of antipodal titanium centers. Enantiomer-selective inactivation, "chiral poisoning", has been found to be successful with racemic BINOL and an excess of di-*iso*-propyl D-tartrate [18].

The highly desirable catalytic addition of the more versatile allylsilane, exemplified for aldehyde **21** [19b], was achieved only recently with catalyst **22** obtained from BINOL and titanium tetrafluoride [19]. The enantioselectivities vary between 60 and 94% *ee*, with the best values being obtained for substrate aldehydes with a fully substituted α-carbon, e.g., pivalaldehyde (94% *ee*) or **21** (→ **23**, 91% *ee*, Scheme 7). Again, the exact structure of the catalytically active species is unknown, and the non-linear relationship between the optical purity of the BINOL ligand and products points to aggregate formation of **22**. The salient feature of this catalyst is the polar fluorotitanium bonds, which render the titanium center more electrophilic, thereby increasing the Lewis acidity for the aldehyde activation. At the same time, the fluoride might also have an activating influence on the allyl silane by forming a Ti-F-Si bridge [20]. Furthermore, the high Ti-F bond energy affects the efficiency of the catalytic cycle by ensuring regeneration of catalyst **22** and formation of silylated (R=$SiMe_3$) rather than titanated product.

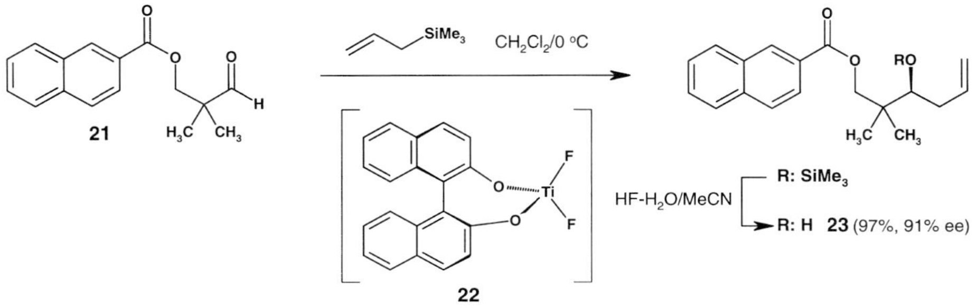

Scheme 7 Fluorotitanate-catalyzed addition of allyl silane to aldehydes.

Mikami, Nakai, and co-workers discovered that the *ene*-reaction of mono-, di-, and trisubstituted olefins with glyoxylic ester **24** can be catalyzed by a titanium catalyst **25** prepared from BINOL and di-*iso*-propoxytitanium dichloride, dibromide, or titanium tetra-*iso*-propoxide [21]. Yield and optical purity of the adducts are usually high, e.g., 98% ee for the product **26** obtained from methylene-cyclohexane **27** (Scheme 8). In addition to simple olefins, this method is also applicable to vinyl selenides and vinyl sulfides. For trisubstituted olefins a more reactive catalyst prepared from 6,6′-dibromo-binaphthol has to be used [21b,c]. Optically active products can also be obtained by desymmetrization of substrates with enantiotopical olefin functions or by kinetic resolution of chiral racemic olefins [21d]. While the intermolecular *ene* reaction is only successful with glyoxylic esters and chloral [21d, 22], less reactive aldehydes have been found to participate in intramolecular reactions [21d]. Interestingly, ketone silyl enolethers (**28**) also react by a prototropic *ene* mechanism with **24** and not by a silatropic Mukaiyama aldol path, as was demonstrated by NMR analysis of the primary products, which turned out to be silyl enolethers (**29**) rather than O-silylated β-hydroxy ketones [23] (Scheme 8).

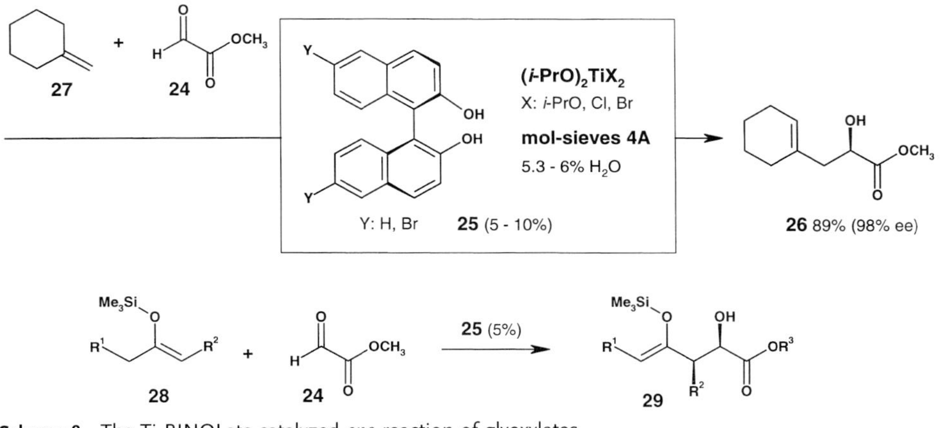

Scheme 8 The Ti BINOLate-catalyzed *ene* reaction of glyoxylates.

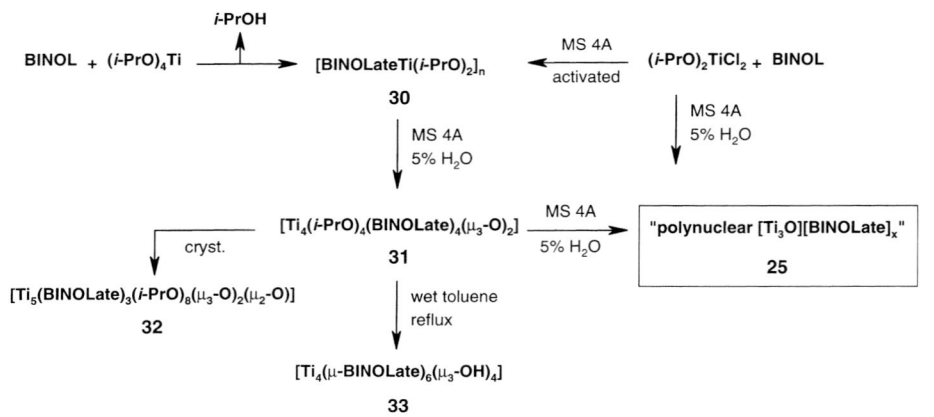

Scheme 9 The BINOL-titanium catalyst system.

The mechanism and the structure of the actual catalyst are still basically unknown. It was evident that molecular sieves (4 Å) are necessary for obtaining an active catalyst, and – quite unexpectedly – it was found that commercially available MS batches, still containing 5.3–6% water, actually serve for exchanging iso-propoxy and chloride ligands by μ_3-oxo bridges (Scheme 9) [24]. Thus, starting from either [BINOLatoTi(O-iPr)$_2$]$_n$ (**30**) or (iPr-O)$_2$TiCl$_2$ and BINOL, partially hydrated molecular sieves lead to the active catalyst **25**, identical according to composition, ^{1}H- and ^{17}O-NMR, and performance. When starting from **30**, the tetranuclear intermediate **31** could be isolated and characterized. Crystallization gave the pentanuclear complex **32**, characterized by X-ray diffraction [24d]. Heating of **31** in wet toluene gives another tetranuclear μ_3-hydroxido complex (**33**), whose structure could also be elucidated by X-ray diffraction [24e]. Furthermore, the presence of equilibrating oligomers – differing for racemic and optically pure complexes – is evident from a positive non-linear relationship between the enantiomeric excess of BINOL and product **26** [25]. This can be exploited by using racemic BINOL for the catalyst preparation and either inactivating one enantiomer by adding optically pure di-iso-propyl tartrate ("chiral poisoning" [22]) or by activating one enantiomer with optically pure BINOL or other axially chiral biphenols [26].

Carreira et al. have recently shown that non-activated aldehydes undergo an ene-type addition with 2-methoxy-propene used as solvent and a catalyst system composed of equimolar amounts of Ti complex **34** and ligand **35** [27]. Acid cleavage of the primary enolether product **36** gives the formal acetone aldol adducts **37** with good yield and up to 98% ee (Scheme 10).

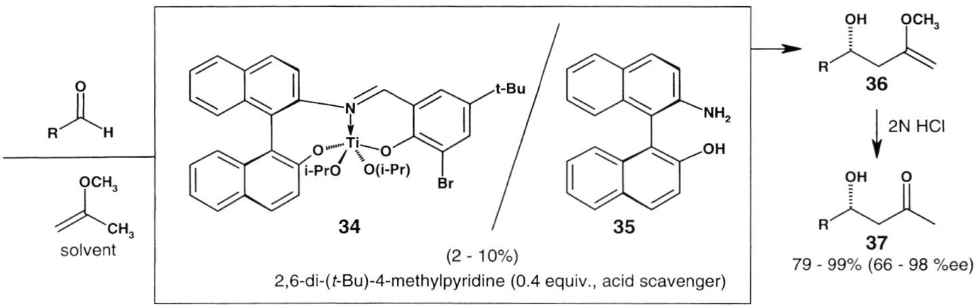

Scheme 10 The Ti-catalyzed *ene* reaction of 2-methoxy-propene.

3.6.3
Aldol-Type Addition of Enolates to Aldehydes

The chiral cyclopentadienyltitanium complexes **1** and **2** obtained from CpTiCl₃ and diacetone-glucose (→ **1**) or TADDOL (→ **2**), respectively, can also be used for highly enantio- and diastereoselective titanium aldol reactions. Thus, when the Li enolate prepared from *tert*-butyl acetate **38** is treated at –78 to –30 °C with a solution of CpTi(ODAG)₂Cl (**1**) in ether or toluene, the titanium enolate **39** is obtained, reaction of which with various aldehydes leads to β-hydroxy-esters (**40**) of over 90% *ee* [2, 4b, 6, 28] (Scheme 11). The stereoselectivity of the actual aldol addition is astonishingly temperature insensitive and can be carried out at 0 °C or even room temperature without much loss of enantioselectivity. The enantiomeric products *ent*-**40** can be obtained by transmetalating the Li enolate with the TADDOL complex **2** (Scheme 3). In this case the optical purity is, however, much lower (78% *ee*) [2]. Unfortunately, no other ligand has so far been found. As L-glu-

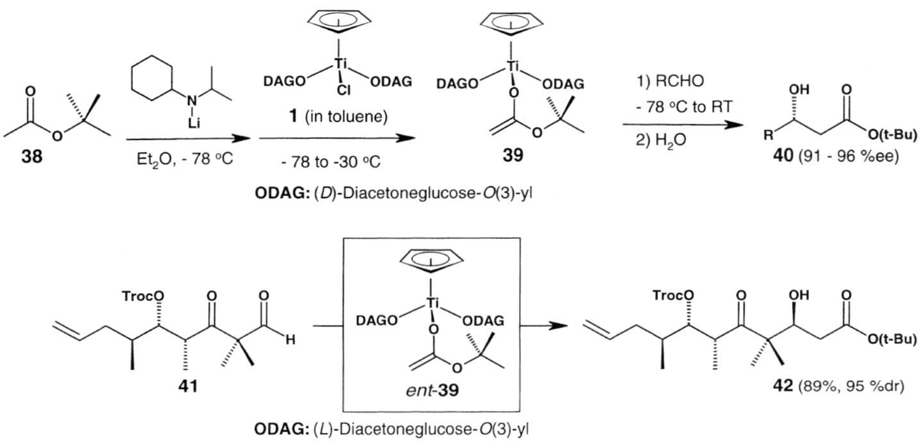

Scheme 11 The aldol reaction of cyclopentadienyltitanium acetate enolates.

cose, needed to prepare *ent*-1, is not readily available and accordingly expensive, this very useful enantioselective acetate aldolization method remains restricted to *Re*-addition. In a recent report on the total synthesis of Desoxyepothilone F [29], Danishefsky and co-workers have still resorted to the L-glucose-derived reagent *ent*-39 to convert the complex and chiral substrate 41 to its acetate aldol 42 with excellent yield and stereocontrol (Scheme 11).

This method is also successful for the propionate aldol addition. The most useful results were obtained with 2,6-dimethylphenyl propionate (43), which according to Heathcock and co-workers [30] cleanly forms the *trans*-enolate (*Z* for Li) 44 upon deprotonation with LDA (Scheme 12). Transmetalation with ethereal CpTi(ODAG)$_2$Cl (1) has to be done at –78 °C, as the Li enolate 44 decomposes at higher temperatures. After 24 h formation of the *E*-titanium enolate, 45 is evident from the reaction with aldehydes, affording *syn*-aldols 46 of high optical purity (91–98% *ee*) in good yield and also with good diastereoselectivity. When, after the initial 24 h of transmetalation at –78 °C, the reaction mixture is warmed for 4 h to –30 °C, the *anti*-aldol 47 becomes the major product except for the conversions of aromatic aldehydes and methacrolein. This can be rationalized by equilibration of 45 to the *Z*-titanium enolate 48. Whereas the optical purity of 47 is also high (94–98% *ee*), the enantiomeric excess of *syn* isomer 46 formed from *Z* enolate 48 can be as low as 47% in cases where substantial amounts are formed (e.g., benzaldehyde, Scheme 12) [2, 4b, 6, 26]. Again this method is restricted to *Re*-addition, as transmetalation with the TADDOL-complex 2 gives poor stereoselectivity [2], and no alternative for L-glucose has been found so far.

Threo-β-hydroxy-α-amino acid esters 49 and 50 can also be prepared by the highly stereoselective Ti-aldol methodology. For this purpose, the *Z*-Li enolate 51 prepared by deprotonation of "stabase"-protected glycine esters 52 with Li cyclohexyl-*iso*-propyl-amide (LICA) in THF is transmetalated either with the diacetoneglucose complex 1 or the TADDOL analog 2. The resulting *E*-titanium enolates 53 and 54, respectively, react smoothly with various aldehydes (Scheme 13). From en-

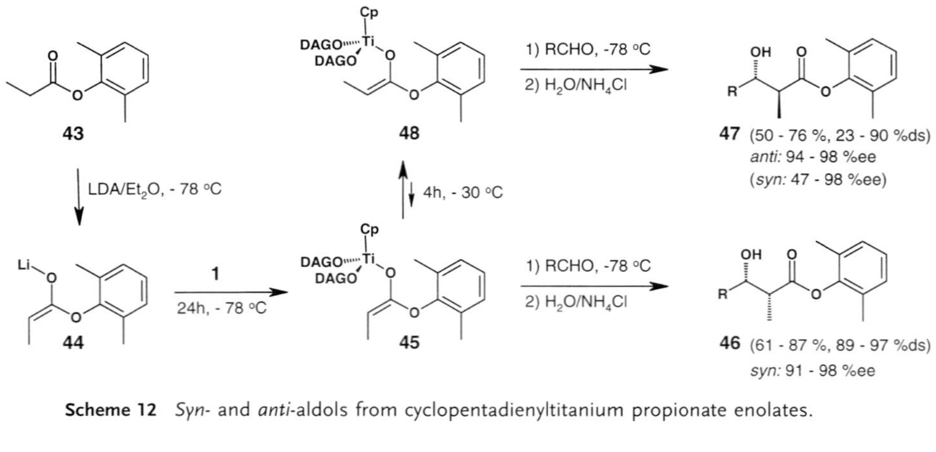

Scheme 12 *Syn*- and *anti*-aldols from cyclopentadienyltitanium propionate enolates.

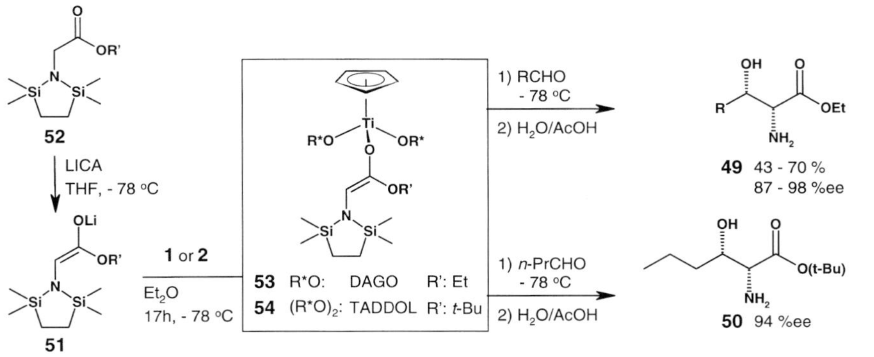

Scheme 13 β-Hydroxy-α-amino acids via aldol reaction with cyclopentadienyltitanium enolates of glycine.

olate **53** the ethyl esters (**49**) of D-configured amino acids are obtained and isolated in fair yield after N-deprotection [2, 4b, 6, 32]. With the exception of the highly reactive glyoxylate, the three other stereoisomers are formed in traces only (≤2%). Again, the reaction with TADDOL as chiral ligand is less enantioselective, and, while reaction of **53** with butanal gives **49** of 98% ee, the enantiomer is obtained with 81% ee only. In this case the situation can, however, be improved considerably by using the corresponding tert-butyl ester **54**, which affords **50** of 94% ee [2, 5b, 6] (Scheme 13).

Since titanium enolates react spontaneously with aldehydes, stoichiometric amounts of chiral ligands are needed for enantiocontrol. Catalytic variants should therefore be based on activation of unreactive systems with chiral Lewis acids. With titanium, a first step in this direction was reported by Oguni and co-workers, who mediated the reaction between aldehydes and diketene **55** with stoichiometric amounts of a titanium complex (**56**) prepared from the tridentate ligand **57** (1.1 equivalents) and Ti(O-iPr)$_4$ [33]. The δ-hydroxy-β-ketoesters **58** were isolated in good yield and with reasonable optical activity (67–84% ee, Scheme 14). While this reaction is most probably still proceeding via a Ti enolate formed in situ from diketene **55** and complex **56**, catalytic versions using 20% of the BINOL-Ti-MS 4 Å system (**25**, cf. Scheme 9) were described shortly afterwards [34a, b]. In this case a Mukaiyama-type aldol reaction between aldehydes and ketene silyl-acetal **59** leads to aldols **60**, isolated with high optical purity (81–98% ee, Scheme 15). More recently, this catalyst was modified by the addition of perfluorophenol and by using methyl-silacyclobutyl ketene acetal [34c], or by adding additional "activating" chiral ligands [34d], and finally by working in supercritical fluids as solvent [34e].

Carreira and co-workers finally combined the two systems by designing a 1,1'-binaphthalene-based chiral Schiff base of salicylaldehyde [35, 36]. But only when 3,5-di-tert-butyl-salicylic acid was introduced as second chelating ligand, the very efficient Ti catalyst **61** was obtained by reaction with Ti(O-iPr)$_4$ (Scheme 16). With only 2% of **61**, O-methyl-O-trimethylsilyl ketene-acetal (**62**) (the O-ethyl analog is

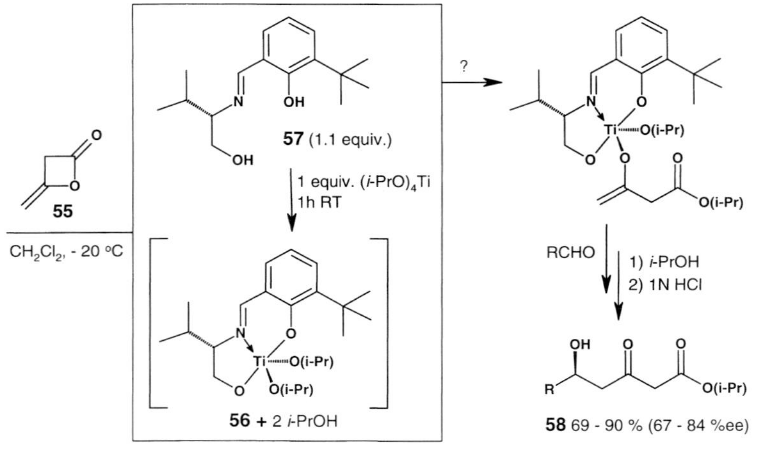

Scheme 14 Ti-mediated addition of diketene to aldehydes.

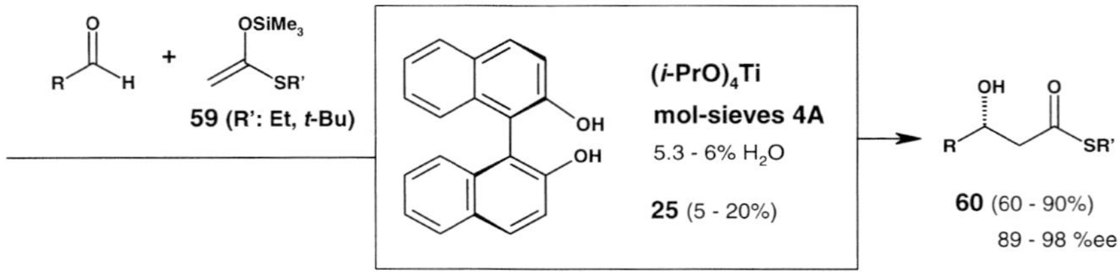

Scheme 15 Mukaiyama aldol reaction catalyzed by Ti BINOLates.

less suited) adds cleanly to aldehydes, affording β-hydroxy-esters **63** after desilylation of precursors **64** – yields and optical purity are excellent [35a]. A more convenient *in situ* preparation of the catalyst **61** has been developed [35b], and this method was successfully applied to α,β-ynals as substrates [35c]. The same catalyst **61** can also be used for the synthesis of aceto-acetate γ-adducts **65**, resulting from the reaction of the silyl enolether derivative **66** [36] (Scheme 16). The addition of **66** and of C(2)- or C(4)-substituted congeners can also be catalyzed with moderate to good enantioselectivity by the BINOL/(iPr-O)$_4$Ti/MS 4 Å-system **25** [37], and was also shown to be successful for Chan's diene-methyl acetoacetate-*bis*-trimethylsilyl enolether [37c].

Mahrwald discovered that 3-pentanone directly adds to aldehydes in the presence of titanium alkoxides, but only when an α-hydroxy acid is added as well. By exchanging (*t*-BuO)$_4$Ti with equimolar amounts of racemic [BINOLatoTi(*t*-BuO)$_2$] preparations (**67**) and using optically pure mandelic acid (**68**) the β-hydroxy-ketones **69** were obtained in excellent stereoselectivity (Scheme 17) [38].

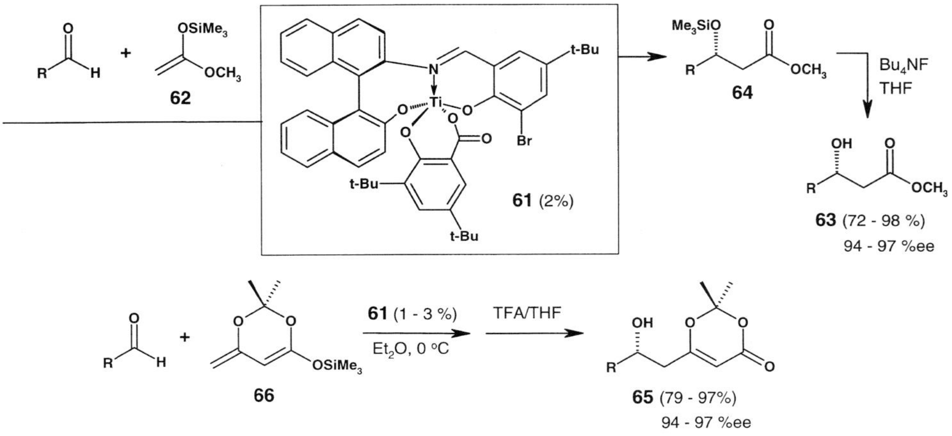

Scheme 16 Mukaiyama aldol reaction catalyzed by the Carreira complex.

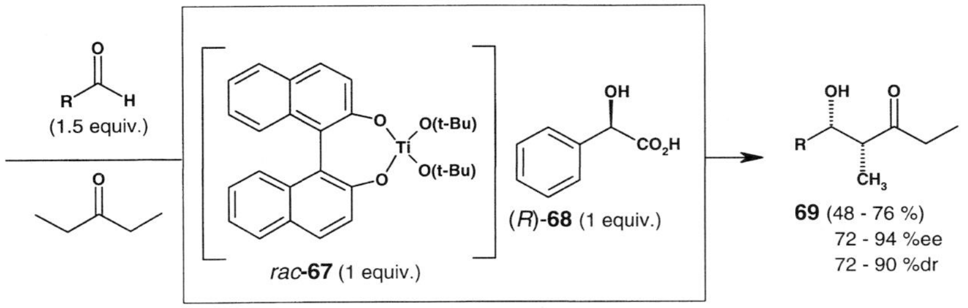

Scheme 17 Direct aldol reaction of diethyl ketone catalyzed by BINOLate-Ti-alkoxides-α-hydroxy acids.

3.6.4
Addition of Alkyl-Nucleophiles to Aldehydes

The use of chiral titanium complexes for the enantioselective addition of nucleophiles attached to the same Ti center also activating the aldehyde carbonyl has been less successful for alkyl groups than for allyl nucleophiles or enolates [2]. On the other hand, chiral titanium complexes have turned out at least as effective catalysts for the enantioselective addition of dialkylzinc compounds to aldehydes [39] as chiral amino alcohols [40], especially in the case of non-aromatic aldehydes and functionalized dialkylzinc compounds. Concerning the catalytic cycle of this process, it is assumed that in a first step the chiral titanium compound Ti(L*)$_4$ is forming a complex **d** with an aldehyde, thereby activating the carbonyl function for the addition of an alkyl residue from (R')$_2$Zn, with possible assistance from the achiral Ti alkoxide Ti(OR*)$_4$ added in stoichiometric amounts to the reaction mixture (Scheme 18). The chiral catalyst T(L*)$_4$ is regenerated by displacement

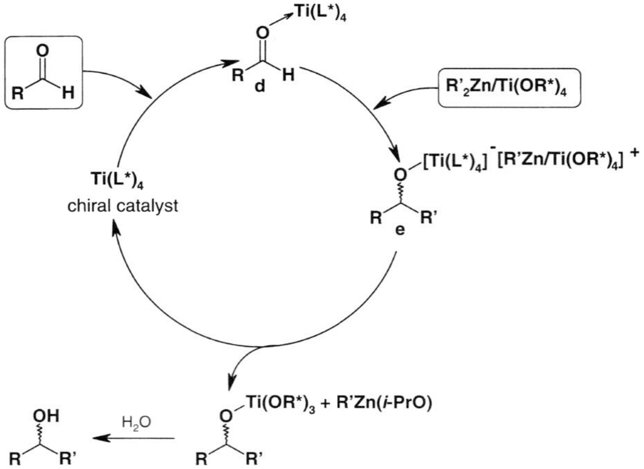

Scheme 18 Catalytic cycle proposed for the Ti-mediated addition of dialkylzinc reagents to aldehydes.

from the primary adduct **e** with achiral Ti(OR*)$_4$. The secondary alcohol **70** is finally obtained from titanate **71** by the hydrolytic workup procedure. Rewarding levels of enantiomeric excess (≥95% *ee*) are obtained for a variety of substrates and dialkylzinc reagents.

The first catalyst, which was introduced by Ohno, Kobayashi, and co-workers [41], is obtained from *trans*-cyclohexane-1,2-diamine *bis*-trifluoromethylsulfonamide (**72**) and Ti(O-*i*Pr)$_4$ in the presence of the diorganozinc reagent. The mechanism of its formation has been elucidated by Walsh and co-workers [42a], and it involves a soluble *bis*-organozinc-sulfonamide intermediate **73**, which is transmetalated to a 5-membered titanium chelate **74**, the structure of which has been secured by X-ray diffraction on crystals of arylsulfonamido analogs [42b] (Scheme 19). This evidence has led to an improved experimental procedure with better reproducibility [42a]. The same catalyst system has also been used extensively and with success by Knochel and co-workers [43], especially for transferring functionalized organozinc residues to aldehydes. More recently, these authors have also introduced trimethylsilyl-methyl and neopentyl as non-transferable organozinc residues, allowing reduction of the amounts of precious functionalized diorganozinc reagents from 1.6–2.4 to 0.8–1.2 equivalents [43f–h]. In addition, new more or less related *bis*-sulfonamido ligands have been introduced [42, 44]. Notable is the derivatization/resolution of *trans*-cyclohexane-1,2-diamine with camphor-sulfonic acid [44f], and the ligand derived from (+)-verbenone [44g].

Very efficient catalysts are furthermore the TADDOLates (**75**) introduced by Seebach and associates [45]. These complexes are prepared from TADDOL ligands and Ti(O-*i*Pr)$_4$ in toluene with azeotropic removal of 2-propanol. Their structure is most probably monomeric, as evidenced by X-ray diffraction on analogous titanates with additional solvent or β-diketone ligands [46] (Scheme 20). To facilitate

3.6.4 Addition of Alkyl-Nucleophiles to Aldehydes | 505

Scheme 19 *Trans*-cyclohexane-1,2-diamine *bis*-sulfonamido-titanium catalysts.

72 R: CF$_3$ 73   74 R: 4-Tol, Mes (X-ray)

catalyst recovery, polymeric linear and branched dendrimeric TADDOL ligands have been prepared by copolymerization. Complexation with titanates then affords polymer-supported catalysts of type **75**, which showed activities and selectivities resembling those of **75** to an astonishingly high degree [47].

In an attempt to extend the excellent stereocontrol of the cyclopentadienyltitanium TADDOLate system from stoichiometric to catalytic reagents, we have prepared the fluoride **76**, which has been characterized by NMR and crystal structure determination [20, 48] (Scheme 20). The polar Ti-F bond of **76** is essential, as the corresponding Ti chloride, Ti trifluoromethylsulfonate, Ti *iso*-propoxide, and methyltitanium compound have no catalytic properties. In the course of these investigations, it also became evident that Et$_2$Zn is not the alkyl-transferring agent. The latter is rather a species, not further characterized, generated from Et$_2$Zn and Ti(O-*i*Pr)$_4$, since good conversions are only obtained if Et$_2$Zn and Ti(O-*i*Pr)$_4$ are mixed at room temperature before conducting the addition to benzaldehyde at –78 °C (70%, 97% *ee*).

Besides the complexes **75** and **76**, other chiral titanium compounds with dioxo ligands related to the TADDOLates [49] and with BINOL ligands [50] have been applied as catalysts for the addition of dialkylzinc reagents to aldehydes. Interestingly, the octahydro-1,1'-binaphthol gave better results than BINOL itself [50c]. The mono-sulfonamido-mono-hydroxy ligand **77** has been applied by Ramón and Yus for titanium-catalyzed additions of diethyl- and dimethylzinc (Scheme 20) [51]. While the stereoselectivity with aldehydes as substrates was rather low [51a], this reagent allowed alkyl addition to aromatic ketones with enantioselectivities up to 89% *ee* for tetralone [51b]. Nugent has reported that a polynuclear Ti-μ-oxo

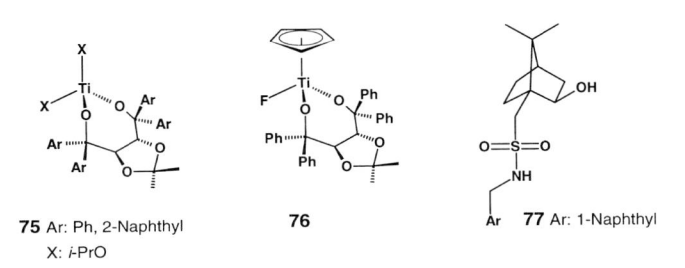

75 Ar: Ph, 2-Naphthyl
X: *i*-PrO

76

77 Ar: 1-Naphthyl

Scheme 20 Titanium catalysts and ligands for the addition of dialkylzinc and other organometallics to aldehydes.

complex with a chiral amino-diol ligand catalyzes the addition of Et$_2$Zn to benzaldehyde without Ti(O-iPr)$_4$ (cf. [2], Scheme 17).

The main advantage of these titanium-based chiral catalysts, when compared to amino alcohols, is good enantiocontrol for a broader variety of aldehydes. While the cyclopentadienyltitanium fluoride **76** shows restrictions in substrate tolerance, especially rewarding results have been obtained with **74** [41–43] and **75** (Ar: 2-naphthyl) [45]. For a long time this method was very much restricted by the availability of salt-free dialkylzinc compounds, which had to be distilled. A much broader variety of suitable reagents including dialkylzinc compounds with functionalized residues are now available by virtue of novel methods for their preparation from Grignard reagents [45 a], from alkyl iodides and Et$_2$Zn [43 c], by Ni-catalyzed metathesis of Et$_2$Zn and olefins [43 d], or by reaction of diethyl-alkylboranes with Et$_2$Zn [43 e].

Remaining disadvantages of the R$_2$Zn/Ti(OR*)$_4$/Ti(L*)$_4$ system are the complexity with three different reagents, the restrictions of solvent, and that only one of the R$_2$Zn groups can be added to the aldehyde carbonyl. It has been discovered that salt-free alkyltitanium tri-iso-propoxides react only sluggishly with aldehydes, especially at low temperature [6, 48, 52]. This reaction is not only catalyzed by Li salts but also by chiral Ti complexes such as **75** and **76** (Scheme 20), giving products of high optical purity [48, 52]. Since salt-free alkyltitanium compounds could until recently only be obtained by distillation, this method was restricted to the relatively volatile and thermally stable CH$_3$Ti(O-iPr)$_3$. Seebach and co-workers have now found new ways for the preparation of such reagents from Grignard compounds with separation of Mg halides precipitated by addition of dioxane or from organolithium compounds and masking of residual Li ions with 12-crown-4. Catalyzed by 20% of TADDOLate (**75**), a wide variety of alkyl residues have been added with good to excellent stereocontrol to aromatic and non-aromatic saturated and unsaturated aldehydes [52]. With benzaldehyde and CH$_3$Ti(O-iPr)$_3$, several cyclopentadienyltitanium TADDOLates also gave 1-phenyl-ethanol (**78**) of high optical purity, but only when used in stoichiometric amounts [48]. Again, the titanium fluoride **76** (Scheme 20) turned out to be a superior catalyst, as **78** of ≥90% *ee* was obtained with good conversion with only 2% of **76** [20, 48] (Scheme 21). With 0.5% of **76**, the optical purity of **78** dropped, however, to 78% *ee* at 60% conversion but was still 87% *ee* at 12% conversion. Similar experiments with 2% and 0.5% of the di-iso-propoxytitanium TADDOLate **75** gave **78** of comparable optical purity, but with low conversions (16% and 6%, respectively), indicating lower catalytic activity, when compared to fluoride **76** (Scheme 21). These results imply that more active catalysts are needed to compete with non-stereoselective background processes. While the stereocontrol induced by catalyst **76** is good for addition of CH$_3$Ti(O-iPr)$_3$ to benzaldehyde and simple aliphatic aldehydes, it failed to catalyze additions to more complex substrates such as glyceraldehyde acetonide [20, 48].

In a recent paper, Mikami and co-workers describe a new system, which is obtained by exchanging the di-iso-propoxy ligands of racemic o,o'-biphenolatotitanium di-iso-propoxides with TADDOL [53]. The reaction of 3,5-bis-trifluoromethylbenzaldehyde with CH$_3$Ti(O-iPr)$_3$ is efficiently catalyzed by 10% of such com-

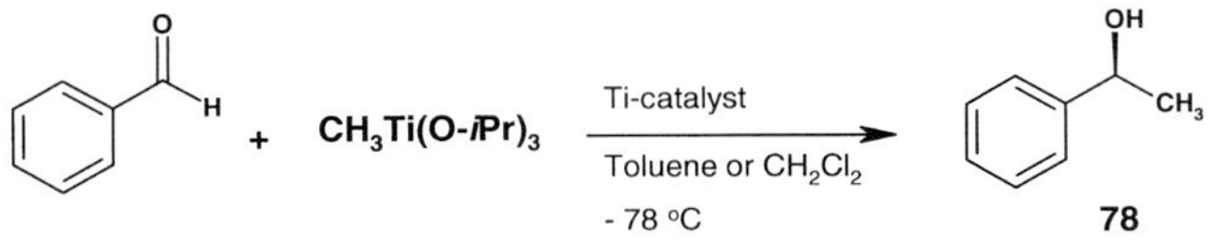

3.6.4 Addition of Alkyl-Nucleophiles to Aldehydes | 507

catalyst	mol %	solvent	time (h)	conv.	% ee
75	20.0	Toluene	19.0	80 %	99
	2.0	CH_2Cl_2	19.0	16 %	92
	0.5	CH_2Cl_2	19.0	6 %	78
76	2.0	CH_2Cl_2	1.5	35 %	95
	2.0	CH_2Cl_2	19.0	77 %	93
	0.5	CH_2Cl_2	1.5	12 %	87
	0.5	CH_2Cl_2	19.0	60 %	78

Scheme 21 Addition of methyltitanium tri-*iso*-propoxide to benzaldehyde catalyzed by Ti TADDOLate (75) and cyclopentadienyltitanium fluoride (76).

plexes, giving 60% of virtually optically pure product in the case of a 3,3′-dimethoxy-BIPOL derivative. Unfortunately, a direct comparison with the TADDOLate complex **75** is not possible for this rather activated substrate.

Trialkylaluminum compounds are available in bulk quantities and would therefore be valuable reagents for the enantioselective addition to aldehydes. Several successful attempts along these lines have already been reported [54]. These conversions take place in THF with 5–20 mol% of a chiral ligand and an achiral Ti compound, in most cases Ti(O-*i*Pr)$_4$, in excess (Scheme 22). It is not unlikely that – in analogy to the bisulfonamide ligand **72** [42] – aluminates are first formed from the ligands and (R^2)$_3$Al, enabling *in situ* formation of chiral titanium catalysts. Suitable ligands are octahydro-BINOL (**79**) [54a], the TADDOL-type diol **80** [54b], TADDOL (**81**) [54c], and the α-sulfonamido-alcohol **82** [54d]. Notable is the

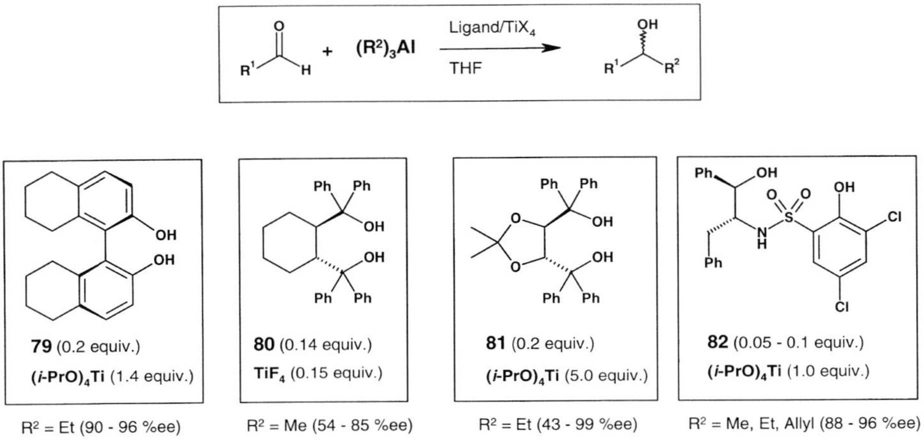

79 (0.2 equiv.)
(*i*-PrO)$_4$Ti (1.4 equiv.)
R^2 = Et (90 - 96 %ee)

80 (0.14 equiv.)
TiF$_4$ (0.15 equiv.)
R^2 = Me (54 - 85 %ee)

81 (0.2 equiv.)
(*i*-PrO)$_4$Ti (5.0 equiv.)
R^2 = Et (43 - 99 %ee)

82 (0.05 - 0.1 equiv.)
(*i*-PrO)$_4$Ti (1.0 equiv.)
R^2 = Me, Et, Allyl (88 - 96 %ee)

Scheme 22 Titanium-catalyzed addition of trialkylaluminum compounds to aldehydes.

ligand **82** for its versatility, also allowing allyl transfer from allyl-diethyl-aluminum [54d], and the titanium fluoride **80**, which operates without excess of achiral titanium compound, giving good results for the more delicate methyl addition [54b].

Various chiral titanium complexes have also been used for enantioselective addition of cyanide to aldehydes. This process and the related Strecker process are not treated in this chapter.

3.6.5
Cycloadditions and Miscellaneous Reactions

Aiming at enantioselective conversions, chiral titanium complexes were also probed as Lewis acid catalysts for cycloadditions. A first success in this direction was reported by Narasaka and co-workers, achieving high enantiocontrol in Diels-Alder reactions of cyclopentadiene and other dienes with oxazolidinone derivatives of various *α,β*-unsaturated carboxylic acids [55]. The catalyst **83** is prepared from $Cl_2Ti(O\text{-}i\text{Pr})_2$ and TADDOL ligand **84** in the presence of molecular sieves (Scheme 23). In analogy to the formation of **75** from $Ti(O\text{-}i\text{Pr})_4$ (Scheme 20), it is assumed that a dichlorotitanium TADDOLate is also generated in this case. Essential for good stereocontrol is the *N*-acyl-oxazolidinone function of the electrophilic olefin, acting as a bidentate Ti ligand. The possibilities of this method are illustrated by the conversion of the borono-acryoyl-oxazolidinone **85** and acetoxy-diene **86** to the highly functionalized cyclohexene-carboxylic acid derivative **87** [55c] (Scheme 23). Subsequently considerable effort has been spent by several groups, mainly to evaluate the influence of the acetal substituents and the aryl residues of the *a,a,a′,a′*-tetraaryl-dioxolane-4,5-dimethanol (TADDOL) ligand on the stereoselectivity and to

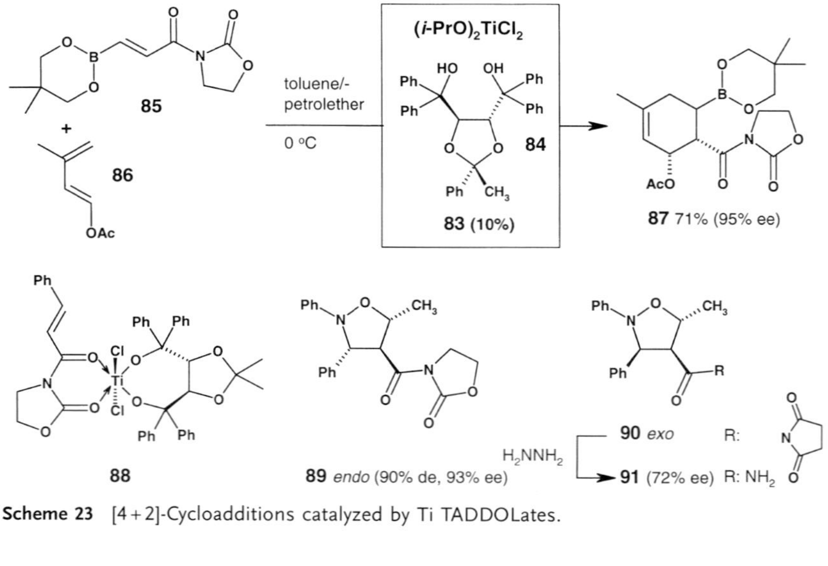

Scheme 23 [4+2]-Cycloadditions catalyzed by Ti TADDOLates.

3.6.5 Cycloadditions and Miscellaneous Reactions | 509

come up with a mechanistic rationalization [56]. From crystal structure determinations of the *N*-cinnamoyl-oxazolidinone adduct **88** [46a] (Scheme 23), it is evident that the transition state involves binding of the dienophile to the Ti catalyst via the *N*-acyl-oxazolidinone function. For a dichlorotitanium TADDOLate there are 5 geometrically different modes for octahedral coordination of an *N*-acyl-oxazolidinone ligand. Careful NMR analysis at −10 °C showed 3 such species in a 70 : 24 : 6 ratio [56b]. The major isomer most probably corresponds to the geometry of the crystal structure **88** with 2 axial chloride ligands. Since the activation for cycloaddition is higher, when one of the oxazolidinone carbonyls has a *trans* relation to one of the weakly bound chlorides, compound (**88**), whose double bond is furthermore in a sterically unbiased environment, most probably does not correspond to the transition state of the cycloaddition. Evidence for proximity to one of the aryl groups and therefore possible steric bias was, on the other hand, observed for the second most abundant species in solution [56b]. However, there are also good arguments for **88** with two axial chlorides being the transition state [56e].

Cycloadditions to *N*-crotonoyl-oxazolidinone were also efficiently catalyzed by polymer-supported Ti TADDOLate catalysts [47]. The addition of nitrones to *N*-crotonoyl-oxazolidinone could also be catalyzed by Ti TADDOLates, affording a mixture of isomeric isoxazolidines [42]. The course of these cycloadditions, giving good results with aryl-substituted nitrones only, is dependent on $(iPr-O)_2TiX_2$ used for the catalyst preparation. With X = Cl the *exo*-adduct of 60% *ee* is the main product, but with the corresponding tosylate the *endo*-isomer **89** is formed in preference and with excellent optical purity (93% *ee*) [57b]. This process could be further improved by replacing the *N*-acyl-oxazolidinone by the more reactive *N*-crotonoyl-succinimide [57c]. With $(iPr-O)_2TiCl_2$ the *exo*-adduct **90** is formed, hydrazinolysis of which affords amide **91** of 72% *ee* (Scheme 23).

The BINOL-derived μ_3-oxo "(Ti)$_3$O" species **25**, generated by controlled hydrolysis with non-activated molecular sieves [24] (Scheme 8), have also been successfully applied as catalysts for Diels-Alder cycloadditions. Again the success is very much dependent on the mode of preparation of the catalyst, and the polycyclic product **92** is obtained from juglone **93** and diene **94** via the unstable primary product **95** in good yield and optical purities varying from 76 to 96% *ee* [21d, 58] (Scheme 24). Such a cycloaddition with benzoquinone was the key step of a recent total synthesis of (−)-ibogamine [58b]. An important drawback of this method is the instability of the primary adducts (e.g., **95**), which have a high tendency to aromatize. This has been circumvented by using quinone monoacetals, rather than free quinones, as dienophiles [58c]. Other examples, obtained by catalysis with **25**, are the methacrolein adduct **96** [58a] and the bicyclus **97**, resulting from addition of a vinyl ether to 2-pyrone [59a] (Scheme 24). Related is the TADDOLato-TiX$_2$-catalyzed inverse-electron-demand-addition of vinyl ethers to enones [59b].

BINOLato-Ti catalysts of type **25** can also be used for hetero-Diels-Alder cycloadditions between reactive dienes such as 1-methoxy-butadiene or the Danishefsky diene **98** (for a general review, see [60]). For the first examples, BINOL was used as ligand, giving already impressive levels of stereoselectivity for many conversions with glyoxylates [21d, 26, 58a, 61a, b], perfluoroalkyl aldehydes [61c], or non-activated alde-

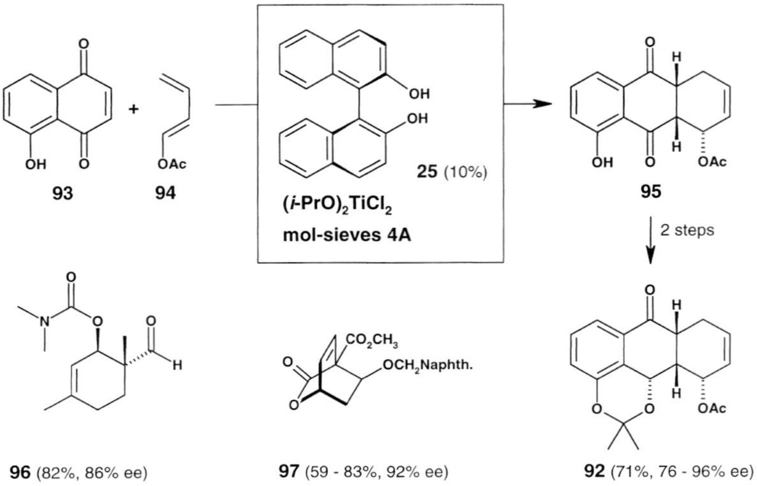

Scheme 24 [4+2]-Cycloadditions catalyzed by Ti BINOLates.

hydes [61 d, e] as substrates. More recently, however, even better and more consistent results were obtained when the catalyst was derived from octahydro-binaphthol (**79**) [62]. With diene **98**, a silyl-enolether, this is most likely a non-concerted process, and the primary product is a Mukaiyama aldol adduct (**99**) [61 d], which upon addition of catalytic amounts of TFA is cyclized to the pyrone **100** (Scheme 25).

Narasaka and co-workers showed that, as well as [4+2]-cycloadditions, the TADDOL complex **83** can also mediate [2+2]-cycloadditions between unsaturated N-acyl-oxazolidinones and electron-rich olefins [55 c, 63]. From the fumaroyl derivative **101** and alkynyl sulfide **102**, cyclobutene **103** is obtained with excellent yield and enantioselectivity, provided that an equimolar amount of titanium complex **83** is used (Scheme 26). In other cases, however, catalytic amounts of **83** (10%) suffice for satisfactory results. Examples include cyclobutanes **104**, adducts of ketene dithioacetal, and **105**, resulting from reaction with a vinyl sulfide. Alkylidene-cyclobutanes such as **106** are in turn obtained from allenyl sulfides (Scheme 26). Related are additions of styrenes to quinones, also catalyzed by complex **83** [64].

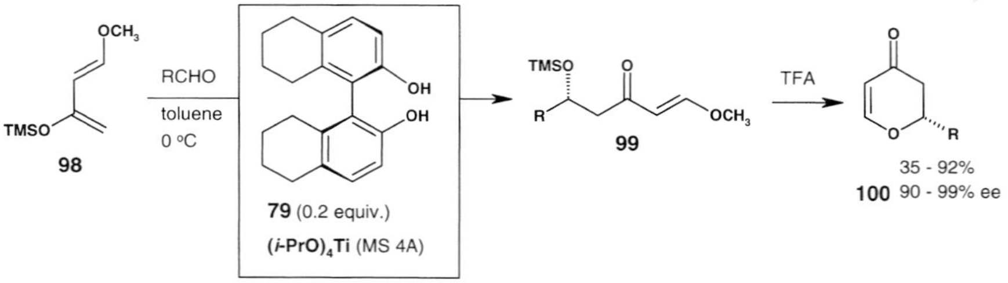

Scheme 25 Hetero-Diels-Alder additions catalyzed by Ti BINOLates.

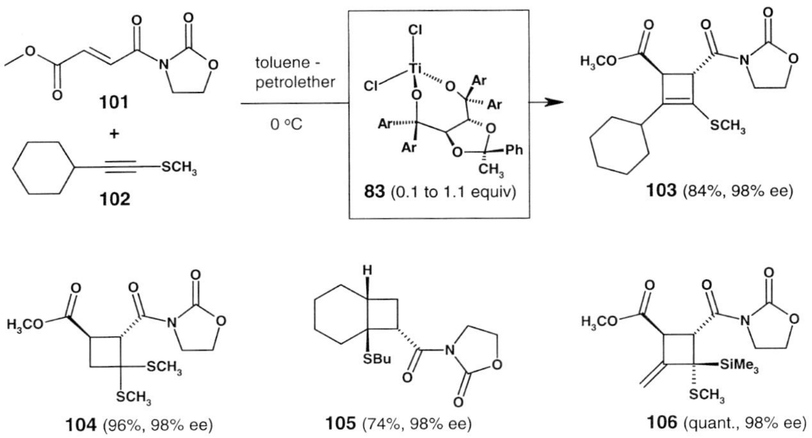

Scheme 26 [2+2]-Cycloadditions catalyzed by Ti TADDOLates.

Another process which is mediated by chiral titanium complexes is the Michael addition [65], an early example being the addition of thioester-derived ketene silyl-acetals to cyclopentenone [65a]. The catalyst is a BINOLato-titanium oxide somewhat related to the Mikami catalyst **25** (cf. Schemes 8 and 9). More recently the dichlorotitanium TADDOLate **107**, related to the di-*iso*-propoxy analog **75** (Scheme 20) and to the Narasaka catalyst **83** (Schemes 23 and 26), has been used to add nucleophiles to β-nitro-styrenes **108** [65b,c] (Scheme 27). While stoichiometric amounts of **107** alone are needed to add dialkylzinc compounds, resulting in the formation of the interesting intermediates **109** [65b], the catalysis of the cycloaddition of trimethylsiloxy-cyclohexene **110** needs 2.5 equivalents of (iPr-O)$_2$TiCl$_2$ in addition [65c]. The primary adducts **111** can either be cleaved to the formal Michael products **112** or are amenable to 1,3-dipolar cycloadditions, e.g., with acetylenes (Scheme 27).

The number of new reactions which can be conducted with high stereocontrol by virtue of chiral titanium complexes is steadily increasing. Reaction of phenethyl-magnesium bromide **113** (2 equivalents) with ethyl acetate leads to optically active cyclopropanol **114** when conducted in the presence of 0.3–1 equivalents of spiro-titanate **115**, first described by Seebach [45c, 66]. A titana-cyclopropane **116**, which can also be viewed as a Ti(II)-styrene complex, has been suggested as intermediate (Scheme 28) [67]. Titanium TADDOLate **117** can also be used for the catalytic enantioselective Simmons-Smith-type cyclopropanation of allylic alcohols **118** [68]. With di-iodomethylzinc a relatively stable Zn alkoxide (**119**) is formed, which subsequently is transformed to cyclopropane **120** by Lewis acid catalysis (Scheme 28). Further enantioselective C-C bond processes, catalyzed by BINOLato titanium preparations, activated by biphenolic additives, are a Friedel-Crafts addition of trifluoroacetaldehyde to phenylethers, giving α-aryl-trifluoroethanols **121** with good *para*-selectivity [69a], and a Mukaiyama-type reaction of ketene-silylacetal and nitrones, giving β-amino acid derivatives of good optical purity, e.g., **122** [69b] (Scheme 28).

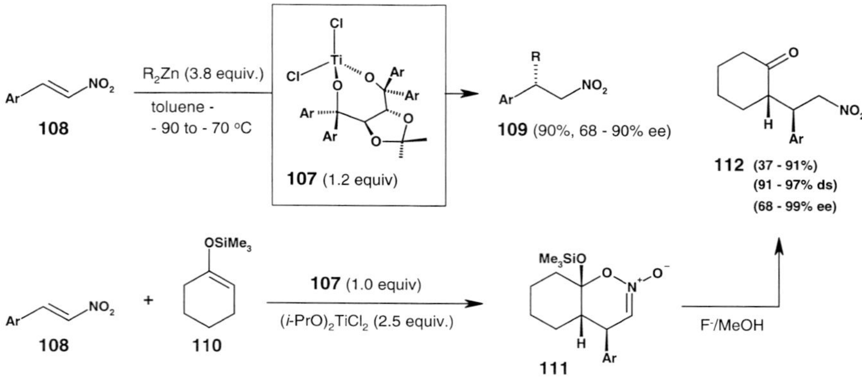

Scheme 27 Titanium-mediated additions to β-nitrostyrenes.

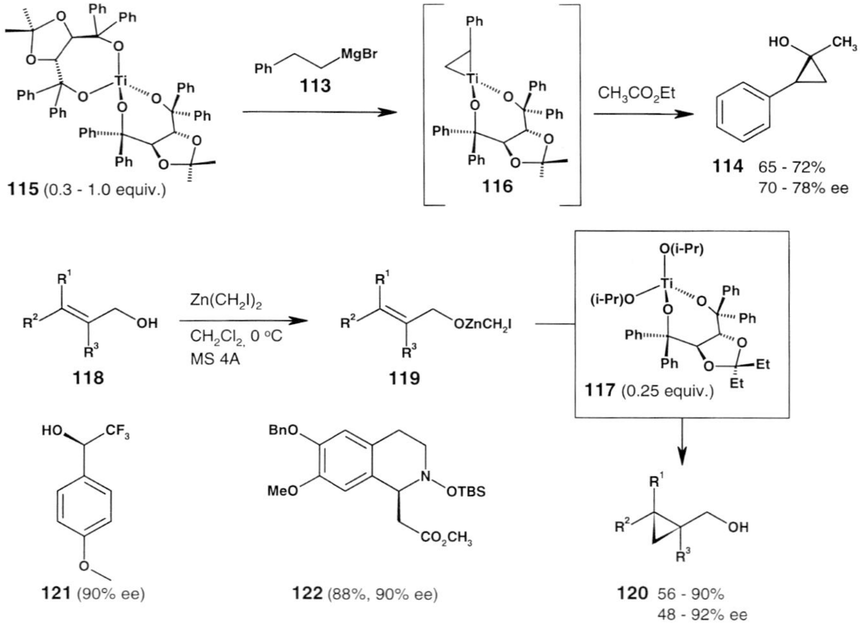

Scheme 28 C-C Bond-forming reactions catalyzed by Ti TADDOLate and Ti BINOLate preparations.

The chiral spiro-titanate **115** (Scheme 28) has been successfully applied for the enantioselective iodocyclization of 2-pentenyl malonates, e.g., **123**, giving cyclopentanes such as **124** with excellent enantioselectivity [46b] (Scheme 29). Togni and co-workers recently reported that Ti TADDOLates catalyze the electrophilic fluorination of β-keto-esters **125** with F-TEDA **126** [70]. The hindered tetra-(1-naphthyl)-

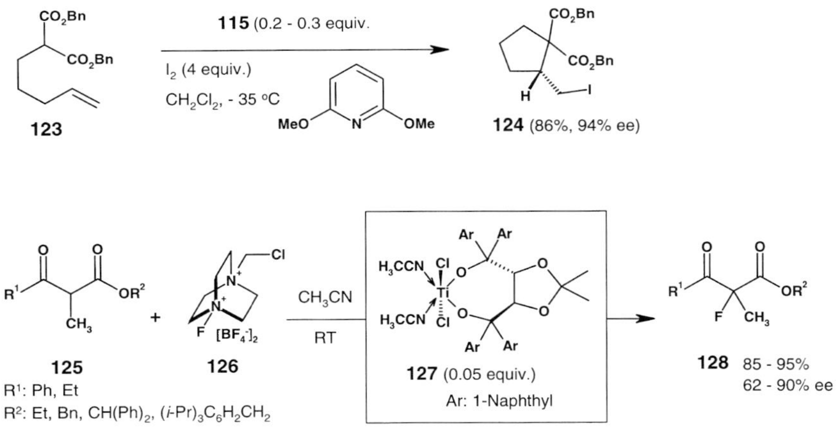

Scheme 29 Ti TADDOLato-mediated iodolactonizations and fluorinations.

TADDOLato-TiCl$_2$ **127** thereby forms chelated enol complexes with **125** by displacement of the acetonitrile ligands. The products **128** are formed with excellent yields and high enantiocontrol, with bulky ester groups R^2 generally giving better selectivities (Scheme 29). This method also works for chlorinations and brominations with N-halogeno-succinimides [70b,c]. Di-*iso*-propoxy-titanium TADDOLates were also used for enantiomer-selective transesterifications of 2-thiopyridyl esters [71a] and alcoholysis of amino acid azolactones [71b]. Equally successful were enantiotopically differentiating alcoholysis reactions of prochiral N-(methylsulfonyl)-dicarboximides and anhydrides [71b,d].

3.6.6
References

1 (a) D. SEEBACH, B. WEIDMANN, L. WIDLER in *Modern Synthetic Methods* (Ed.: SCHEFFOLD, R.); Salle, Frankfurt, **1983**; Vol.3, pp 217–353. (b) M.T. REETZ, *Organotitanium Reagents in Organic Synthesis*; Springer, Berlin, 1986.

2 R.O. DUTHALER, A. HAFNER, *Chem. Rev.* **1992**, *92*, 807–832.

3 (a) N.W. EILERTS, J.A. HEPPERT, M.L. KENNEDY, F. TAKUSAGAWA, *Inorg. Chem.* **1994**, *33*, 4813–4814; (b) V.W. DAY, T.A. EBERSPACHER, Y. CHEN, J. HAO, W.G. KLEMPERER, *Inorg. Chim. Acta* **1995**, *229*, 391–405; (c) J. BLANCHARD, S. BARBOUX-DOEUFF, J. MAQUET, C. SANCHEZ, *New. J. Chem.* **1995**, *19*, 929–941.

4 (a) M. RIEDIKER, R.O. DUTHALER, *Angew. Chem.* **1989**, *101*, 488–490; *Angew. Chem. Int. Ed. Engl.* **1989**, *28*, 494–495; (b) A. HAFNER, R.O. DUTHALER, in *Encyclopedia of Reagents for Organic Synthesis* Vol 2, 1104–1106, J. Wiley, New York, 1995.

5 (a) A. HAFNER, R.O. DUTHALER, R. MARTI, G. RIHS, P. ROTHE-STREIT, F. SCHWARZENBACH, *J. Am. Chem. Soc.* **1992**, *114*, 2321–2336; (b) A. HAFNER, R.O. DUTHALER in *Encyclopedia of Reagents for Organic Synthesis* (Ed.: PAQUETTE, L.A.), Vol. 2, 1106–1108, J. Wiley, New York, **1995**.

6 R.O. DUTHALER, A. HAFNER, P.L. ALSTERS, P. ROTHE-STREIT, G. RIHS, *Pure Appl. Chem.* **1992**, *64*, 1897–1910.

7 R.O. Duthaler, R. Wietzke, unpublished results.
8 R.C. Cambie, J.M. Coddington, J.B.J. Milbank, M.G. Pausler, J.J. Rustenhoven, P.S. Rutledge, G.L. Shaw, P.J. Sinkovich, *Aust. J. Chem.* **1993**, *46*, 583–591.
9 St. Weigand, R. Brückner, *Chem. Eur. J.* **1996**, *2*, 1077–1084.
10 D. Hoppe, Th. Krämer, J.-R. Schwark, O. Zschage, *Pure Appl. Chem.* **1990**, *62*, 1999–2006.
11 (a) J. Cossy, S. BouzBouz, F. Pradaux, C. Willis, V. Bellosta, *Synlett* **2002**, 1595–1606; (b) S. BouzBouz, J. Cossy, *Org. Lett.* **2000**, *2*, 501–504; (c) S. BouzBouz, J. Cossy, *Org. Lett.* **2001**, *3*, 3995–3998.
12 (a) A.L. Costa, M.G. Piazza, E. Tagliavini, C. Trombini, A. Umani-Ronchi, *J. Am. Chem. Soc.* **1993**, *115*, 7001–7002; (b) S. Casolari, D. D'Addario, E. Tagliavini, *Org. Lett.* **1999**, *1*, 1061–1063.
13 (a) G.E. Keck, K.H. Tarbet, L.S. Geraci, *J. Am. Chem. Soc.* **1993**, *115*, 8467–8468; (b) G.E. Keck, L.S. Geraci, *Tetrahedron Lett.* **1993**, *34*, 7827–7828; (c) G.E. Keck, D. Krishnamurthy, M.C. Grier, *J. Org. Chem.* **1993**, *58*, 6543–6544; (d) G.E. Keck, T. Yu, *Org. Lett.* **1999**, *1*, 289–291; (e) G.E. Keck, C.A. Wager, T.T. Wager, K.A. Savin, J.A. Covel, M.D. McLaws, D. Krishnamurthy, V.J. Lee, *Angew. Chem.* **2001**, *113*, 237–240; *Angew. Chem., Int. Ed.* **2001**, *40*, 231–234.
14 (a) Ch.-M. Yu, H.-S. Choi, W.-H. Jung, S.-S. Lee, *Tetrahedron Lett.* **1996**, *37*, 7095–7098; (b) Ch.-M. Yu, H.-S. Choi, W.-H. Jung, H.-J. Kim, J. Shin, *Chem. Commun.* **1997**, 761–763; (c) Ch.-M. Yu, M. Jeon, J.-Y. Lee, J. Seon, *Eur. J. Org. Chem.* **2001**, 1143–1148; (d) Ch.-M. Yu, J.-Y. Lee, B. So, J. Hong, *Angew. Chem.* **2002**, *114*, 169–171; *Angew. Chem. Int. Ed.* **2002**, *41*, 161–163; (e) Ch.-M. Yu, S.-K. Yoon, H.-S. Choi, K. Baek, *Chem. Commun.* **1997**, 763–764; (f) Ch.-M. Yu, H.-S. Choi, S.-K. Yoon, W.-H. Jung, *Synlett* **1997**, 889–890; (g) Ch.-M. Yu, S.-K. Yoon, K. Baek, J.-Y. Lee, *Angew. Chem.* **1998**, *110*, 2504–2506; *Angew. Chem., Int. Ed.* **1998**, *37*, 2392–2395.
15 S. Kii, K. Maruoka, *Tetrahedron Lett.* **2001**, *42*, 1935–1939.
16 G.E. Keck, D. Krishnamurthy, X. Chen, *Tetrahedron Lett.* **1994**, *35*, 8323–8324.
17 S. Aoki, K. Mikami, M. Terada, T. Nakai, *Tetrahedron* **1993**, *49*, 1783–1792.
18 J.W. Faller, D.W.I. Sams, X. Liu, *J. Am. Chem. Soc.* **1996**, *118*, 1217–1218.
19 (a) D.R. Gauthier, Jr., E.M. Carreira, *Angew. Chem.* **1996**, *108*, 2521–2523; *Angew. Chem. Int. Ed. Engl.* **1996**, *35*, 2363–2365; J.W. Bode, D.R. Gautier Jr., E.M. Carreira, *Chem. Commun.* **2001**, 2560–2561.
20 (a) R.O. Duthaler, A. Hafner, *Angew. Chem.* **1997**, *109*, 43–45; *Angew. Chem. Int. Ed. Engl.* **1997**, *36*, 43–45; (b) R.O. Duthaler, A. Hafner, *Fluorotitanium Compounds – Novel Catalysts for the Addition of Nucleophiles to Aldehydes* in *Organic Synthesis Highlights IV* (Ed.: H.-G. Schmalz), Wiley-VCH, Weinheim **2000**, pp 166–171.
21 (a) K. Mikami, M. Terada, T. Nakai, *J. Am. Chem. Soc.* **1990**, *112*, 3949–3954; (b) M. Terada, Y. Motoyama, K. Mikami, *Tetrahedron Lett.* **1994**, *35*, 6693–6696; (c) K. Mikami, Y. Motoyama, M. Terada, *Inorg. Chim. Acta* **1994**, *222*, 71–75; (d) K. Mikami, *Pure Appl. Chem.* **1996**, *68*, 639–644.
22 J.W. Faller, X. Liu, *Tetrahedron Lett.* **1996**, *37*, 3449–3452.
23 K. Mikami, S. Matsukawa, *J. Am. Chem. Soc.* **1993**, *115*, 7039–7040.
24 (a) M. Terada, Y. Matsumoto, Y. Nakamura, K. Mikami, *J. Chem. Soc., Chem. Commun.* **1997**, 281–282; (b) K. Mikami, M. Terada, Y. Matsumoto, M. Tanaka, Y. Nakamura, *Microporous and Mesoporous Materials* **1998**, *21*, 461–466; (c) M. Terada, Y. Matsumoto, Y. Nakamura, K. Mikami, *J. Mol. Catal. A: Chem.* **1998**, *132*, 165–169; (d) M. Terada, Y. Matsumoto, Y. Nakamura, K. Mikami, *Inorg. Chim. Acta* **1999**, *296*, 267–272; (e) K. Mikami, M. Ueki, Y. Matsumoto, M. Terada, *Chirality* **2001**, *13*, 541–544.
25 (a) M. Terada, K. Mikami, *J. Chem. Soc., Chem. Commun.* **1994**, 833–834; (b) K. Mikami, M. Terada, *Tetrahedron* **1992**, *48*, 5671–5680.

26 (a) K. Mikami, S. Matsukawa, *Nature* **1997**, *385*, 613–615; (b) K. Mikami, S. Matsukawa, T. Volk, M. Terada, *Angew. Chem.* **1997**, *109*, 2936–2939; *Angew. Chem., Int. Ed.* **1997**, *36*, 2768–2771.

27 E. M. Carreira, W. Lee, R. A. Singer, *J. Am. Chem. Soc.* **1995**, *117*, 3649–3650.

28 R. O. Duthaler, P. Herold, W. Lottenbach, K. Oertle, M. Riediker, *Angew. Chem.* **1989**, *101*, 490–491; *Angew. Chem. Int. Ed. Engl.* **1989**, *28*, 495–497.

29 Ch. B. Lee, Zh. Wu, F. Zhang, M. D. Chappell, Sh. J. Stachel, T.-Ch. Chou, Y. Guan, S. J. Danishefsky, *J. Am. Chem. Soc.* **2001**, *123*, 5249–5259.

30 St. H. Montgomery, M. C. Pirrung, C. H. Heathcock, *Org. Synth.* **1985**, *63*, 99–108.

31 R. O. Duthaler, P. Herold, S. Wyler-Helfer, M. Riediker, *Helv. Chim. Acta* **1990**, *73*, 659–673.

32 G. Bold, R. O. Duthaler, M. Riediker, *Angew. Chem.* **1989**, *101*, 491–493; *Angew. Chem. Int. Ed. Engl.* **1989**, *28*, 497–498.

33 M. Hayashi, T. Inoue, N. Oguni, *J. Chem. Soc., Chem. Commun.* **1994**, 341–342.

34 (a) K. Mikami, S. Matsukawa, *J. Am. Chem. Soc.* **1994**, *116*, 4077–4078; (b) G. E. Keck, D. Krishnamurthy, *J. Am. Chem. Soc.* **1995**, *117*, 2363–2364; (c) S. Matsukawa, K. Mikami, *Tetrahedron: Asymm.* **1995**, *6*, 2571–2574; (d) S. Matsukawa, K. Mikami, *Enantiomer* **1996**, *1*, 69–73; (e) K. Mikami, S. Matsukawa, Y. Kayaki, T. Ikariya, *Tetrahedron Lett.* **2000**, *41*, 1931–1934.

35 (a) E. M. Carreira, R. A. Singer, W. Lee, *J. Am. Chem. Soc.* **1994**, *116*, 8837–8838; (b) R. A. Singer, E. M. Carreira, *Tetrahedron Lett.* **1997**, *38*, 927–930; (c) R. A. Singer, M. S. Shepard, E. M. Carreira, *Tetrahedron* **1998**, *54*, 7025–7032.

36 (a) R. A. Singer, E. M. Carreira, *J. Am. Chem. Soc.* **1995**, *117*, 12360–12361; (b) Y. Kim, R. A. Singer, E. M. Carreira, *Angew. Chem.* **1998**, *110*, 1321–1323; *Angew. Chem. Int. Ed.* **1998**, *37*, 1261–1263.

37 (a) M. Sato, S. Sunami, Y. Sugita, Ch. Kaneko, *Heterocycles* **1995**, *41*, 1437–1444; (b) M. De Rosa, A. Soriente, A. Scettri, *Tetrahedron: Asymm.* **2000**, *11*, 3187–3195; (c) A. Soriente, M. De Rosa, M. Stanzione, R. Villano, A. Scettri, *Tetrahedron: Asymm.* **2001**, *12*, 959–963.

38 R. Mahrwald, *Org. Lett.* **2000**, *2*, 4011–4012.

39 K. Soai, S. Niwa, *Chem. Rev.* **1992**, *92*, 833–856.

40 R. Noyori, S. Suga, H. Oka, M. Kitamura, *Chem. Rec.* **2001**, *Vol. 1*, 85–100.

41 H. Takahashi, T. Kawakita, M. Ohno, M. Yoshioka, S. Kobayashi, *Tetrahedron* **1992**, *48*, 5691–5700.

42 (a) S. Pritchett, D. H. Woodmansee, T. J. Davis, P. J. Walsh, *Tetrahedron Lett.* **1998**, *39*, 5941–5942: (b) S. Pritchett, D. H. Woodmansee, P. Gantzel, P. J. Walsh, *J. Am. Chem. Soc.* **1998**, *120*, 6423–6424; (c) J. Balsells, P. J. Walsh, *J. Am. Chem. Soc.* **2000**, *122*, 1802–1803; (d) J. Balsells, P. J. Walsh, *J. Am. Chem. Soc.* **2000**, *122*, 3250–3251.

43 (a) R. Ostwald, P.-Y. Chavant, H. Stadtmüller, P. Knochel, *J. Org. Chem.* **1994**, *59*, 4143–4153; (b) St. Nowotny, St. Vettel, P. Knochel, *Tetrahedron Lett.* **1994**, *35*, 4539–4540; (c) P. Knochel, *Chemtracts, Org. Chem.* **1995**, *8*, 205–221; (d) St. Vettel, A. Vaupel, P. Knochel, *Tetrahedron Lett.* **1995**, *36*, 1023–1026; (e) F. Langer, L. Schwink, A. Devasagayaraj, P.-Y Chavant, P. Knochel, *J. Org. Chem.* **1996**, *61*, 8229–8243; (f) St. Berger, F. Langer, Ch. Lutz, P. Knochel, T. A. Mobley, C. K. Reddy, *Angew. Chem.* **1997**, *109*, 1603–1605 *Angew. Chem. Int. Ed.* **1997**, *36*, 1496–1498; (g) Ch. Lutz, P. Knochel, *J. Org. Chem.* **1997**, *62*, 7895–7898; (h) Ch. Lutz, Ph. Jones, P. Knochel, *Synthesis* **1999**, 312–316.

44 (a) K. Ito, Y. Kimura, H. Okamura, T. Katsuki, *Synlett* **1992**, 573–574; (b) K. Soai, Y. Hirose, Y. Ohno, *Tetrahedron: Asymmetry* **1993**, *4*, 1473–1474; (c) X. Zhang, Ch. Guo, *Tetrahedron Lett.* **1995**, *36*, 4947–4950; (d) J. Qiu, Ch. Guo, X. Zhang, *J. Org. Chem* **1997**, *62*, 2665–2668; (e) M. Cernerud, A. Skrinning, I. Bérgère, Ch. Moberg, *Tetrahedron: Asymm.* **1997**, *8*, 3437–3441; (f) Ch.-D. Hwang, B.-J. Uang, *Tetrahedron: Asymm.* **1998**, *9*, 3979–3984; (g) L. A. Paquette, R. Zhou, *J. Org. Chem.* **1999**, *64*, 7929–7934.

45 (a) D. Seebach, L. Behrendt, D. Felix, *Angew. Chem.* **1991**, *103*, 991–992; *Angew. Chem. Int. Ed. Engl.* **1991**, *30*, 1008–1009; (b) B. Schmidt, D. Seebach, *Angew. Chem.* **1991**, *103*, 1383–1385; *Angew. Chem. Int. Ed. Engl.* **1991**, *30*, 1321–1323; (c) D. Seebach, D. A. Plattner, A. K. Beck, Y. M. Wang, D. Hunziker, W. Petter, *Helv. Chim. Acta* **1992**, *75*, 2171–2209; (d) D. Seebach, A. K. Beck, B. Schmidt, Y. M. Wang, *Tetrahedron* **1994**, *50*, 4363–4384.

46 (a) K. V. Gothelf, R. G. Hazell, K. A. Jorgensen, *J. Am. Chem. Soc.* **1995**, *117*, 4435–4436; (b) T. Inoue, O. Kitagawa, O. Ochiai, M. Shiro, T. Taguchi, *Tetrahedron Lett.* **1995**, *36*, 9333–9336; (c) L. Hintermann, D. Broggini, A. Togni, *Helv. Chim. Acta* **2002**, *85*, 1597–1612.

47 (a) D. Seebach, R. E. Marti, T. Hintermann, *Helv. Chim. Acta* **1996**, *79*, 1710–1740; (b) P. B. Rheiner, D. Seebach, *Chem. Eur. J.* **1999**, *5*, 3221–3236; (c) H. Sellner, C. Faber, P. B. Rheiner, D. Seebach, *Chem. Eur. J.* **2000**, *6*, 3692–3705.

48 R. O. Duthaler, A. Hafner, P. L. Alsters, M. Tinkl, G. Rihs, unpublished results, partially presented at the *7th IUPAC Symposium on Organo-Metallic Chemistry directed towards Organic Synthesis*, Sept. 19–23, **1993**, Kobe (Japan).

49 (a) Y. N. Ito, X. Ariza, A. K. Beck, A. Boháč, C. Ganter, R. E. Gawley, F. N. M. Kühnle, J. Tuleja, Y. M. Wang, D. Seebach, *Helv. Chim. Acta* **1994**, *77*, 2071–2110; (b) H. Waldmann, M. Weigerding, C. Dreisbach, Ch. Wandrey, *Helv. Chim. Acta* **1994**, *77*, 2111–2116.

50 (a) M. Mori, T. Nakai, *Tetrahedron Lett.* **1997**, *38*, 6233–6236; (b) F.-Y. Zhang, Ch.-W. Yip, R. Cao, A. S. C. Chan, *Tetrahedron: Asymm.* **1997**, *8*, 585–589; (c) F.-Y. Zhang, A. S. C. Chan, *Tetrahedron: Asymm.* **1997**, *8*, 3651–3655.

51 (a) D. J. Ramón, M. Yus, *Tetrahedron: Asymm.* **1997**, *8*, 2479–2496; (b) D. J. Ramón, M. Yus, *Tetrahedron* **1998**, *54*, 5651–5666.

52 B. Weber, D. Seebach, *Tetrahedron* **1994**, *50*, 7473–7484.

53 M. Ueki, Y. Matsumoto, J. J. Jodry, K. Mikami, *Synlett* **2001**, 1889–1892.

54 (a) A. S. C. Chan, F.-Y. Zhang, Ch.-W. Yip, *J. Am. Chem. Soc.* **1997**, *119*, 4080–4081; (b) B. L. Pagenkopf, E. M. Carreira, *Tetrahedron Lett.* **1998**, *39*, 9593–9596; (c) J.-F. Lu, J.-S. You, H.-M. Gau, *Tetrahedron: Asymm.* **2000**, *11*, 2531–2535; (d) J.-S. You, Sh.-H. Hsieh, H.-M. Gau, *Chem. Commun.* **2001**, 1546–1547.

55 (a) K. Narasaka, N. Iwasawa, M. Inoue, T. Yamada, M. Nakashima, J. Sugimori, *J. Am. Chem. Soc.* **1989**, *111*, 5340–5345; (b) K. Narasaka, H. Tanaka, F. Kanai, *Bull. Chem. Soc. Jpn.* **1991**, *64*, 387–391; (c) K. Narasaka, *Pure Appl. Chem.* **1992**, *64*, 1889–1896.

56 (a) E. J. Corey, Y. Matsumura, *Tetrahedron Lett.* **1991**, *32*, 6289–6292; (b) C. Haase, Ch. R. Sarko, M. DiMare, *J. Org. Chem.* **1995**, *60*, 1777–1787; (c) D. Seebach, R. Dahinden, R. E. Marti, A. K. Beck, D. A. Plattner, F. N. M. Kühnle, *J. Org. Chem.* **1995**, *60*, 1788–1799; (d) E. Wada, W. Pei, Sh. Kanemasa, *Chem. Lett.* **1994**, 2345–2348; (e) K. V. Gothelf, K. A. Jørgensen, *J. Org. Chem.* **1995**, *60*, 6847–6851.

57 (a) K. V. Gothelf, K. A. Jørgensen, *J. Org. Chem.* **1994**, *59*, 5687–5691; (b) K. V. Gothelf, I. Thomsen, K. A. Jørgensen, *J. Am. Chem. Soc.* **1996**, *118*, 59–64; (c) K. B. Jensen, K. V. Gothelf, R. G. Hazell, K. A. Jørgensen, *J. Org. Chem.* **1997**, *62*, 2471–2477.

58 (a) K. Mikami, Y. Motoyama, M. Terada, *J. Am. Chem. Soc.* **1994**, *116*, 2812–2820; (b) J. D. White, Y. Choi, *Org. Lett.* **2000**, *2*, 2373–2376; (c) M. Breuning, E. J. Corey, *Org. Lett.* **2001**, *3*, 1559–1562.

59 (a) G. H. Posner, H. Dai, D. S. Bull, J.-K. Lee, F. Eydoux, Y. Ishihara, W. Welsh, N. Pryor, St. Petr, Jr., *J. Org. Chem.* **1996**, *61*, 671–676; (b) E. Wada, H. Yasuoka, S. Kanemasa, *Chem. Lett.* **1994**, 1637–1640.

60 K. A. Jørgensen, *Angew. Chem.* **2000**, *112*, 3702–3733; *Angew. Chem., Int. Ed.* **2000**, *39*, 3558–3588.

61 (a) M. Terada, K. Mikami, T. Nakai, *Tetrahedron Lett.* **1991**, *32*, 935–938; (b) S. Matsukawa, K. Mikami, *Tetrahedron: Asymm.* **1997**, *8*, 815–816; (c) L. Lévêque, M. Le Blanc, R. Pastor, *Tetrahedron Lett.* **2000**, *41*, 5043–5046; (d) G. E. Keck, X.-Y.

Li, D. Krishnamurthy, *J. Org. Chem.* **1995**, *60*, 5998–5999; (e) S. Kii, T. Hashimoto, K. Maruoka, *Synlett* **2002**, 931–932.

62 (a) B. Wang, X. Feng, X. Cui, Y. Jiang, *Chem. Commun.* **2000**, 1605–1606; (b) B. Wang, X. Feng, Y. Huang, H. Liu, X. Cui, Y. Jiang, *J. Org. Chem.* **2002**, *67*, 2175–2182; (c) J. Long, J. Hu, X. Shen, B. Ji, K. Ding, *J. Am. Chem. Soc.* **2002**, *124*, 10–11.

63 K. Narasaka, Y. Hayashi, H. Shimadzu, Sh. Niikata, *J. Am. Chem. Soc.* **1992**, *114*, 8869–8885.

64 T. A. Engler, M. A. Letavic, J. P. Reddy, *J. Am. Chem. Soc.* **1991**, *113*, 5068–5070.

65 (a) S. Kobayashi, S. Suda, M. Yamada, T. Mukaiyama, *Chem. Lett.* **1994**, 97–100; (b) H. Schäfer, D. Seebach, *Tetrahedron* **1995**, *51*, 2305–2324; (c) D. Seebach, I. M. Lyapkalo, R. Dahinden, *Helv. Chim. Acta* **1999**, *82*, 1829–1840.

66 B. Schmidt, D. Seebach, *Angew. Chem.* **1991**, *103*, 100–101; *Angew. Chem. Int. Ed. Engl.* **1991**, *30*, 99–100.

67 E. J. Corey, S. A. Rao, M. C. Noe, *J. Am. Chem. Soc.* **1994**, *116*, 9345–9346.

68 A. B. Charette, C. Molinaro, Ch. Brochu, *J. Am. Chem. Soc.* **2001**, *123*, 12168–12175.

69 (a) A. Ishii, V. A. Soloshonok, K. Mikami, *J. Org. Chem.* **2000**, *65*, 1597–1599; (b) S.-I. Murahashi, Y. Imada, T. Kuwakami, K. Harada, Y. Yonemushi, N. Tomita, *J. Am. Chem. Soc.* **2002**, *124*, 2888–2889.

70 (a) L. Hintermann, A. Togni, *Angew. Chem.* **2000**, *112*, 4530–4533, *Angew. Chem., Int. Ed.* **2000**, *39*, 4359–4362; (b) L. Hintermann, A. Togni, *Helv. Chim. Acta* **2000**, *83*, 2425–2435; (c) A. Togni, A. Mezzetti, P. Barthazy, C. Becker, I. Devillers, R. Frantz, L. Hintermann, M. Perseghini, M. Sanna, *Chimia* **2001**, *55*, 801–805; (d) St. Piana, I. Devillers, A. Togni, U. Röthlisberger, *Angew. Chem.* **2002**, *114*, 1021–1024, *Angew. Chem., Int. Ed.* **2002**, *41*, 979–982.

71 (a) K. Narasaka, F. Kanai, M. Okuda, N. Miyoshi, *Chem. Lett.* **1989**, 1187–1190; (b) D. Seebach, G. Jaeschke, K. Gottwald, K. Matsuda, R. Formisano, D. A. Chaplin, M. Breuning, G. Bringmann, *Tetrahedron* **1997**, *53*, 7539–7556; (c) D. J. Ramón, G. Guillena, D. Seebach, *Helv. Chim. Acta* **1996**, *79*, 875–894; (d) G. Jaeschke, D. Seebach, *J. Org. Chem.* **1998**, *63*, 1190–1197.

3.7
Zinc-Mediated Reactions

Axel Jacobi von Wangelin and Mathias U. Frederiksen

3.7.1
Introduction

Applications of organozinc chemistry to organic synthesis have gained significant importance over the years. Although organozinc compounds have been known for more than 150 years [1], their advantageous exploitation for selective carbon-carbon bond-forming reactions with high generality and wide scope has been shown only in recent years. Organozinc species constitute versatile nucleophiles, although they show far lower reactivity toward electrophiles – but higher tolerance of functional groups – than their venerable organolithium and organomagnesium relatives. However, the presence of low-lying orbitals facilitates transmetalation to transition metal catalysts (Cu, Ni, Pd, etc.), which provide the basis for a rich synthetic chemistry with organozinc compounds. The combination of their easy preparation, high functional group tolerance, and excellent reactivity and selectivity in the presence of suitable metal catalysts accounts for the numerous applications that organozincs have found in modern organic synthesis [2]. Transition metal-catalyzed cross-coupling reactions with sp^3, sp^2, and sp-electrophiles, addition reactions to C=X bonds and activated olefins, and cyclopropanations are among the most widely used stoichiometric reactions. Important examples of zinc-catalyzed reactions have been reported for carbonyl and aldol addition reactions. This review is intended to familiarize the reader with general aspects of modern organozinc chemistry and emphasize prominent applications to organic synthesis. As a comprehensive treatment of all known methods involving organozinc compounds would go beyond the scope of this chapter, the reader is kindly referred to references given.

3.7.2
Preparations and Coupling Reactions

The various synthetic methodologies for the preparation of organozinc compounds can be categorized as transmetalations and insertions. While early applications of organozinc compounds utilized transmetalation from lithium or magne-

Transition Metals for Organic Synthesis, Vol. 1, 2nd Edition.
Edited by M. Beller and C. Bolm
Copyright © 2004 WILEY-VCH Verlag GmbH & Co. KGaA, Weinheim
ISBN: 3-527-30613-7

sium, these methods were usually accompanied by low functional group tolerance. With the advent of organozinc chemistry in natural product synthesis, functional group tolerance concerns have prompted the utilization of milder methods of preparation. Nowadays, viable preparations of functionalized organozinc compounds include direct insertion into C-X bonds and transmetalations from boron and transition metals.

3.7.2.1
Zinc Insertion into C-X Bonds

Organozinc halides can be prepared by the insertion of zinc (dust) into alkyliodides. This iodine-zinc exchange is the most direct access to organozinc species, whereas the use of bromides and chlorides is largely limited to activated allyl, benzyl, or propargyl compounds. Jackson used ultrasonic activation for the zinc insertion into serine derivative **1** (Scheme 1) [3].

Zinc insertion into $C(sp^2)$-X bonds is usually sluggish, but can be effected at higher temperatures in polar solvents or with activated organylhalides such as hetaryliodides [4]. Where halogen-zinc exchange is slow, the employment of highly activated Rieke® zinc has been shown to facilitate zinc insertions. This highly active powder is prepared by reduction of zinc salts with an alkali metal alone (Na, K) or in combination with an electron acceptor (Li/naphthalene = lithium naphthalenide, $LiC_{10}H_8$) and even allows for the zinc insertion into secondary and tertiary alkylbromides and -iodides. Scheme 2 illustrates the synthetic potential of Rieke® zinc for the chemoselective insertion into a tertiary bromide (over a primary) and a hetaryl iodide for the synthesis of **2** and **3**, respectively [5].

In some cases, the halogen-zinc exchange can also be effected with Et_2Zn under palladium or nickel catalysis [6], in which the initial oxidative addition of the alkyl halide proceeds by a radical mechanism. Scheme 3 shows an application to the stereoconvergent synthesis of cyclopentane **4**.

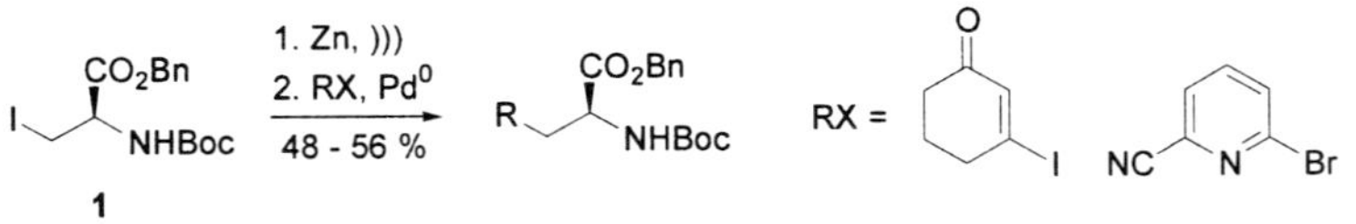

Scheme 1 Amino acid synthesis from iodo-serine (**1**).

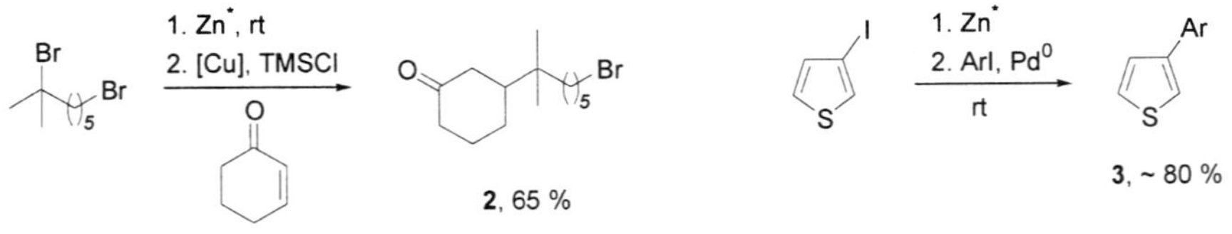

Scheme 2 Organozinc reagents prepared from Rieke zinc.

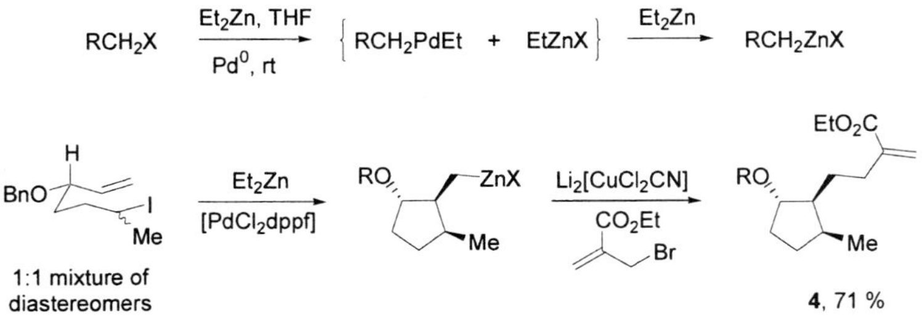

3.7.2 Preparations and Coupling Reactions | 521

Scheme 3 Copper- and palladium-catalyzed halide-zinc exchange.

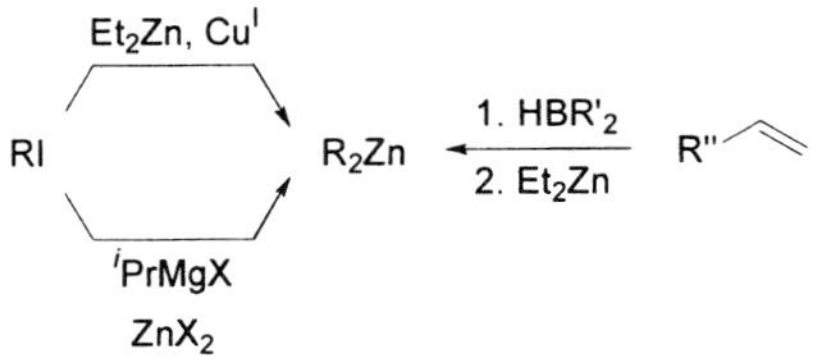

Scheme 4 Synthetic access to diorganozinc species.

Halogen-free diorganozinc species are the reagents of choice in applications to asymmetric reactions, as they undergo facile transmetalation. The preparation of diorganozinc compounds can be effected by (copper-catalyzed) halogen exchange in the presence of diethylzinc or – under milder conditions – di-iso-propylzinc [7]. The latter can be prepared *in situ* from ^{i}PrMgBr and ZnBr$_2$. A more elegant route utilizes olefins as starting materials, which upon hydroboration can be subjected to facile boron-zinc exchange (Scheme 4) [8].

3.7.2.2
Transmetalations

The boron-zinc exchange has been applied to the synthesis of diversely functionalized organozinc species under mild conditions. This strategy proved successful for the preparation of homoallylic zinc such as silyl enolether **5**, which is difficult to access by other methods (Scheme 5) [9].

The mild conditions of the versatile hydroboration–boron-zinc exchange cascade have also entailed applications to the synthesis of chiral secondary organozinc compounds, which can be accessed from the corresponding boranes with retention of configuration [10]. While the chiral information is created by the initial hydroboration, the formed chiral zinc reagents are required to be configurationally stable [11, 12]. Several applications have used such strategy in combination with palladium- or copper-catalyzed coupling reactions. Scheme 6 shows a diastereoselective allylation of substituted styrenes (Z)- and (E)-**6**, which sets up two adjacent

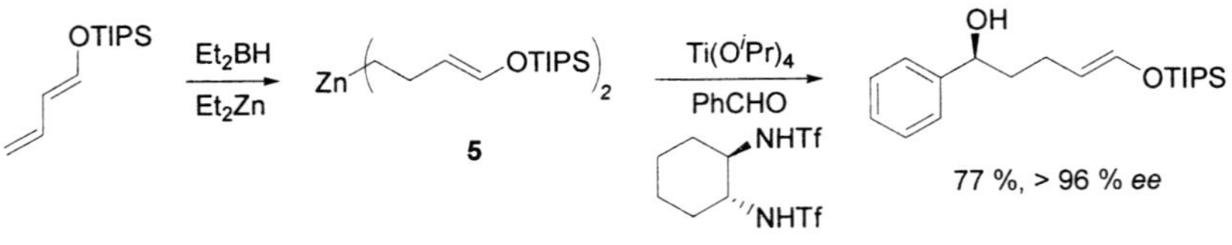

Scheme 5 Sequential hydroboration-B/Zn exchange for the synthesis of homoallylic zinc species.

stereocenters with perfect stereocontrol via sequential asymmetric hydroboration, transmetalation to zinc, and copper-catalyzed coupling with allylbromide [13].

With tetrasubstituted olefins, the intermediate boranes can be subject to thermal rearrangement to the allylic position [14]. Subsequent boron-zinc exchange furnishes the organozinc compound via an overall stereoselective CH activation process. Reactions with substrates bearing diastereotopic allylic hydrogens reveal the potential to generate three adjacent stereocenters. The proposed mechanism of this domino hydroboration-1,2-migration event is shown in Scheme 7 along with a demonstrative application to the highly diastereoselective synthesis of alkynyl cyclopentane **7** [15].

The preparation of alkenylzinc species from alkynes by sequential hydrozirconation and transmetalation has been established as a valuable method, and several elegant applications of this have been reported. The rich addition chemistry of such prepared alkenylzinc reagents to carbonyl compounds will be dealt with in Section 3.7.4.2. An interesting one-pot conversion of alkynes to *trans*-substituted cyclopropanes was reported on by Oshima (Scheme 8A) [16]. The mechanism involves consecutive hydrozirconation, transmetalation to zinc, and cyclopropanation in the presence of CH_2I_2. An extension of this methodology implementing an aldimine addition step with the intermediate zinc species enabled aminomethyl cyclopropanes to be synthesized [17]. When changing the order of addition of reagents (CH_2I_2, then aldimine), homoallylic amines of type **8** can be accessed in good diastereoselectivities (Scheme 8B) [18].

Pioneering studies by Negishi established that zinc(II) salts are able to mediate the Zr-Pd transmetalation via an intermediate zinc species. This realization pro-

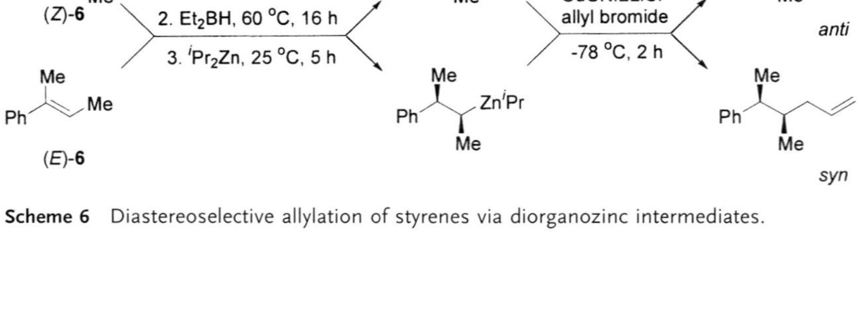

Scheme 6 Diastereoselective allylation of styrenes via diorganozinc intermediates.

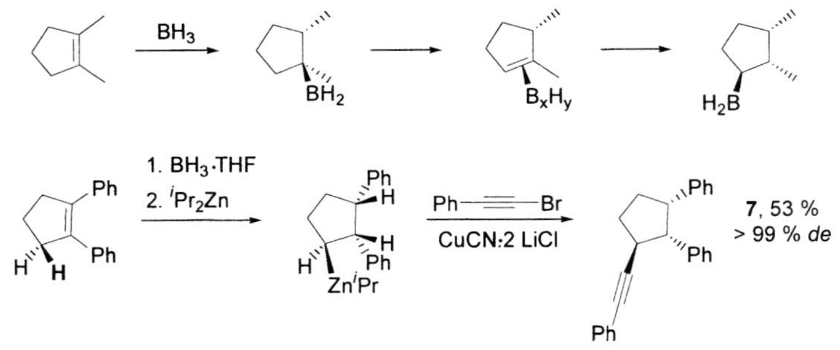

Scheme 7 Allylic functionalization via *pseudo* diastereoselective CH activation.

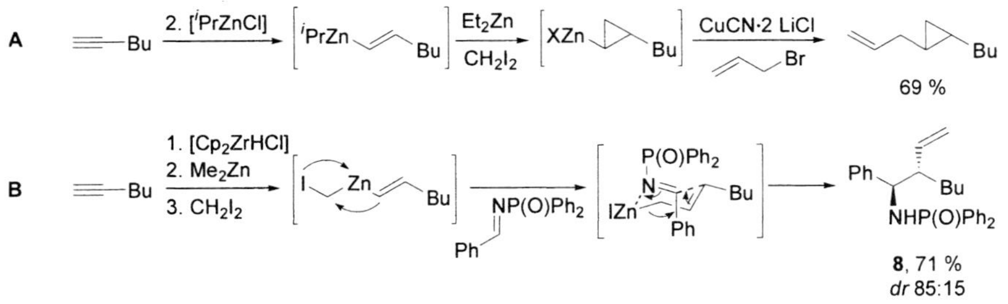

Scheme 8 Hydrozirconation-Zr/Zn exchange in combination with cyclopropanation conditions.

vided the basis for a number of very useful palladium-catalyzed coupling reactions [19]. For example, coupling reactions with C(sp^2) and C(sp) halides constitute an attractive alternative to Suzuki-type reactions and have been applied to the total synthesis of various natural products such as pitiamide A (Scheme 9) [20].

Highly active allylzinc compounds can be prepared by direct zinc insertion or via fragmentation of homoallylzinc alcoholates. Villieras used the former strategy for the diastereoselective synthesis of *α,β*-unsaturated lactam **9** (Scheme 10 A) [21]. The fragmentation of sterically hindered homoallyl alcohol **10**, for which the reduction of steric repulsion by the *tert*-butyl groups is the major driving force, was exploited in the synthesis of allylic compounds **11** and **12** (Scheme 10 B) [22].

Unlike conventionally prepared allylzinc species, allylzinc synthons generated *in situ* from homoallylzinc alcoholates stereoselectively add to aldehydes, which might be attributed to zinc-aldehyde coordination prior to fragmentation, as illustrated by intermediate **13** in Scheme 11. An interesting application is the metalloene reaction for the synthesis of quaternary stereocenters (**14** and **15**, Scheme 11) [23].

Although geminally dimetalated reagents have been known for some time, there was little progress in the development of efficient coupling reactions with electrophiles. Based on Gaudemar's 1971 observation that allyl zinc bromides can add to vinyl Grignard reagents [24], Normant has introduced the allyl zincation of substituted vinyl metals as a particularly powerful method for the preparation of

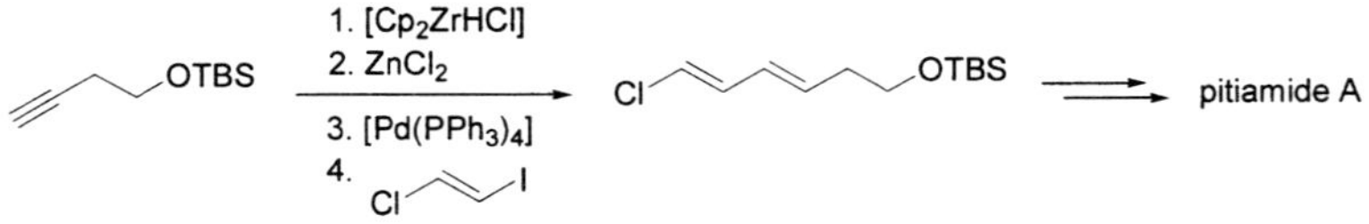

Scheme 9 Palladium-catalyzed vinylzinc-vinylhalide coupling as part of the pitiamide A synthesis.

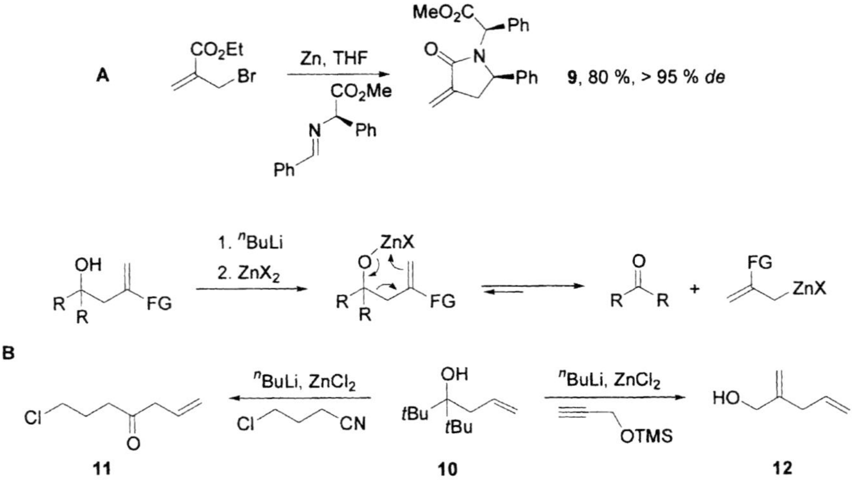

Scheme 10 Allylzinc species from insertion (A) and homoallyl alcoholate fragmentation (B).

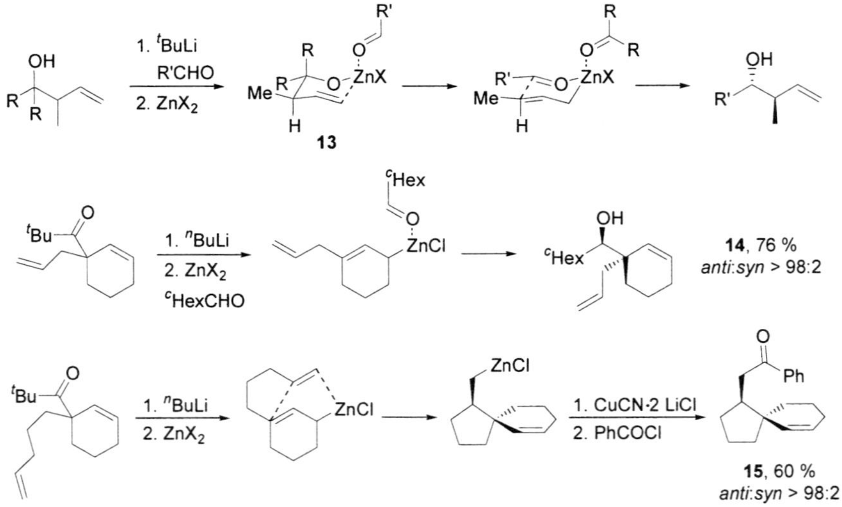

Scheme 11 Homoallylzinc alcoholates as masked allylzinc synthons.

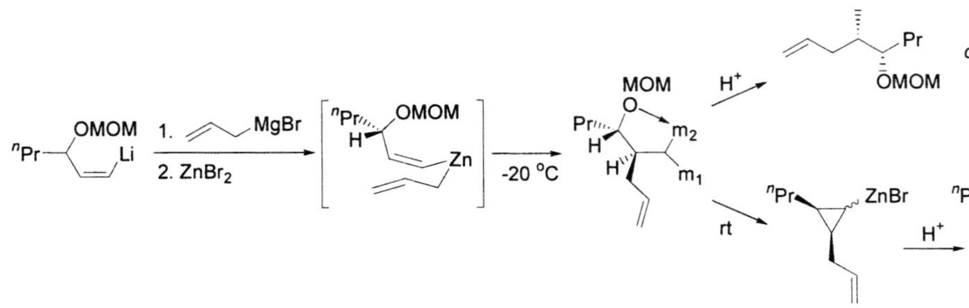

Scheme 12 1,1-Dimetallic reagents for the synthesis of allyl-substituted alcohols and cyclopropanes.

1,1-dimetalated species, which allow for the diastereoselective synthesis of methyl-substituted alkyl chains [25]. 1,1-Dimetallic reagents bearing a methoxymethyl ether function in the γ-position were shown to undergo stereoselective room-temperature cyclization to give substituted cyclopropanes (**16**, Scheme 12) [26].

This zinca-Claisen-type rearrangement has been extended to alkynes and allenes, where the requisite allyl and vinylzinc species are generated via sequential hydrozirconation and transmetalation to zinc. This method allows for the regioselective coupling with γ-substituted allyl species [27]. From a mechanistic standpoint, the dichotomy between a zinco-ene and a zinca-Claisen pathway has been discussed [28].

3.7.3
Cross-Coupling Reactions

3.7.3.1
Unsaturated Coupling Partners

Since the seminal work of Negishi [29], the palladium-catalyzed cross-coupling of aryl and vinyl halides/triflates with organozinc compounds has been developed to a powerful tool for the formation of carbon-carbon bonds with broad scope and high functional group compatibility. Many of the reported examples involve reactions of activated heteroaryl electrophiles, for which a number of catalyst systems give excellent yields under mild conditions. A regioselective Negishi coupling of 5,7-dichloropyrazolo[1,5-a]pyrimidine was used for the synthesis of an angiotensin II receptor antagonist precursor (**17**, Scheme 13) [30]. Other interesting applications of organozinc methodology to hetaryl coupling reactions include the synthesis of 2,2'-bipyridine building blocks, precursors for natural products such as camphothecin [31], and the synthesis of highly functionalized uracil derivatives (**18**, Scheme 13) with Farina's Pd(dba)$_2$/P(2-furyl)$_3$ catalyst system [32]. Reactions of organozinc species with 2,3-dibromoacrylates were demonstrated to provide a synthetically useful basis for highly regioselective domino cross-coupling reactions for the synthesis of enynes

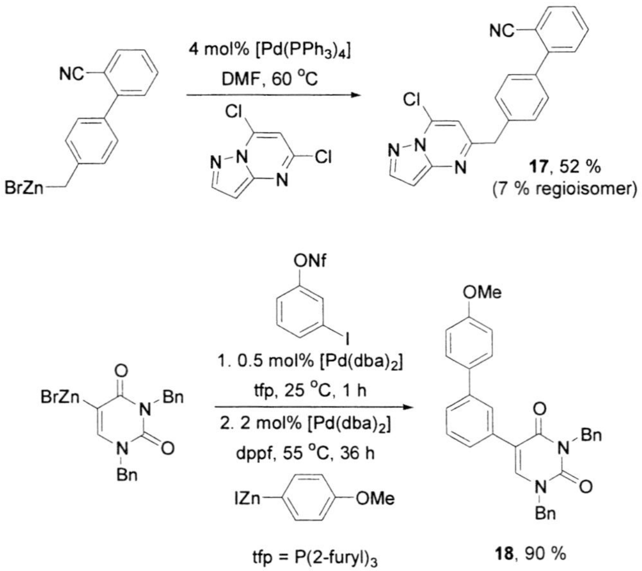

Scheme 13 Selective Negishi cross-coupling toward substituted heterocycles.

[33]. At the other end of the reactivity scale of aryl electrophiles, the quest for efficient palladium-catalyzed couplings of deactivated arylchlorides with aryl and alkylzinc reagents has established [Pd(PtBu$_3$)$_2$] as the most efficient catalyst system [34].

Takei et al. reported on the stereoconvergent insertion of zinc into *syn* and *anti* β-iodoamides **19**, which allowed for the synthesis of *syn* γ-ketoamide **20** in excellent yield (Scheme 14 A) [35]. An elegant example of stepwise coupling reactions with dimetalated reagents has been reported by Utimoto. The double allylation of 1,1-dizincaethane derivative **21** afforded 1,6-hexadiene **22** in a sequence of palladium and copper-catalyzed cross-coupling reactions in good yield (Scheme 14 B) [36].

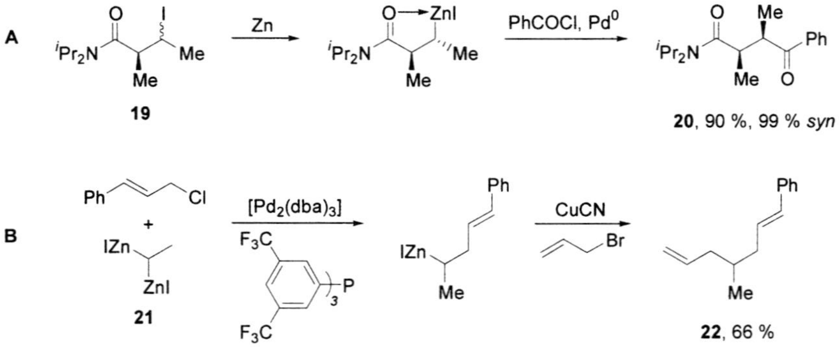

Scheme 14 Palladium- and copper-catalyzed cross-coupling reactions with alkylzinc reagents.

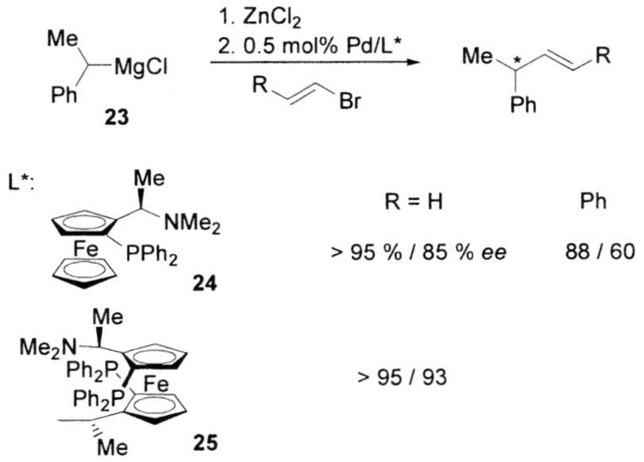

Scheme 15 Asymmetric cross-coupling with 1-phenylethylzinc.

Some effort has also been devoted to the asymmetric cross-coupling of secondary alkylzinc species to vinyl bromides, which provides a powerful tool for the introduction of chirality into allylic positions. The best enantioselectivities in the model reaction of 1-phenylethylzinc reagent **23** with various vinyl bromides were observed in palladium-catalyzed reactions with ferrocenylphosphine ligands **24** and **25**, which were higher than those observed for the reaction with the corresponding Grignard reagent (Scheme 15) [37].

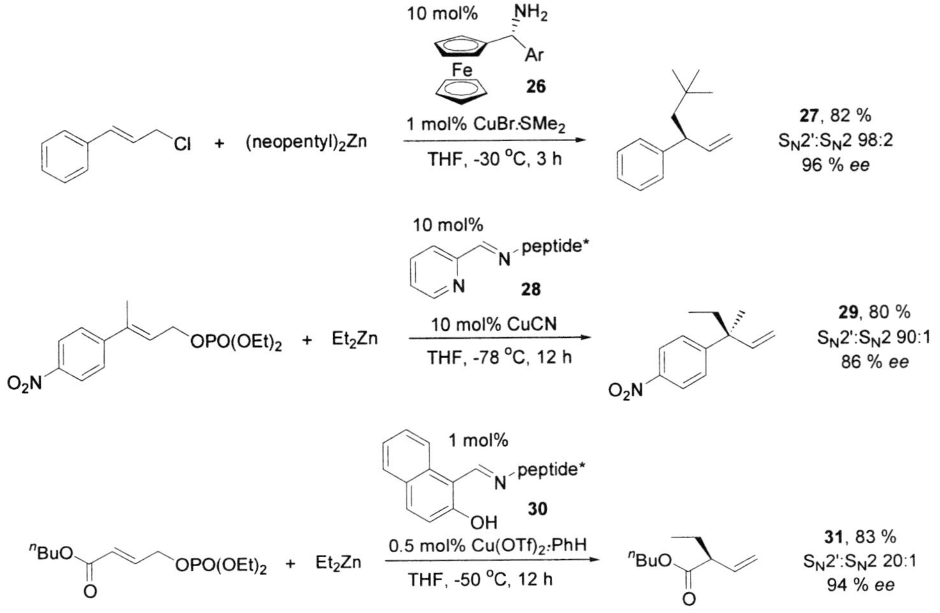

Scheme 16 Enantioselective allylic alkylation with diorganozinc reagents.

The highly enantioselective substitution of allyl chlorides with dialkylzinc compounds has been effected with ferrocenyl amine ligand **26** to give terminal alkenes in a copper(I)-catalyzed S_N2' process (Scheme 16, top), and thus constitutes an alternative entry to chiral allylic systems, such as α-substituted allylbenzene **27** [38]. Reactions with allylic phosphates as electrophiles in the presence of Hoveyda's modular pyridine peptide ligands of type **28** provided quaternary allylic carbon atoms (**29**) with similar site and enantioface control (Scheme 16) [39]. A related phenolic catalyst system (**30**) has been applied to the enantioselective synthesis of α-alkyl-β,γ-unsaturated esters (**31**) and has also been used in a convergent total synthesis of (R)-(–)-elenic acid. Further examples of such enantioselective allylic alkylation reactions with organozinc reagents include Feringa's Cu(I)/phosphoramidite system [40], which gives moderate *ee*s with cinnamyl bromides, and Gennari's Schiff base ligands [41].

3.7.3.2
Saturated Coupling Partners

Although nickel-catalyzed coupling reactions between unsaturated coupling partners are well known, it is only recently that Ni-catalyzed $C(sp^3)$-$C(sp^3)$ bond-forming reactions have been added to the arsenal of organozinc chemistry [42]. Palladium-catalyzed alkyl-alkyl coupling reactions suffer from the slow reductive elimination of the product and hence exhibit β-hydride elimination as the major side reaction; Knochel demonstrated the potential of nickel catalysis for the selective $C(sp^3)$-$C(sp^3)$ coupling with organozinc nucleophiles. Thus, the first efficient alkyl-alkyl cross-coupling reactions with diorganozinc [43] or iodoalkylzinc [44] nucleophiles have been effected, although some substrates require remote unsaturation

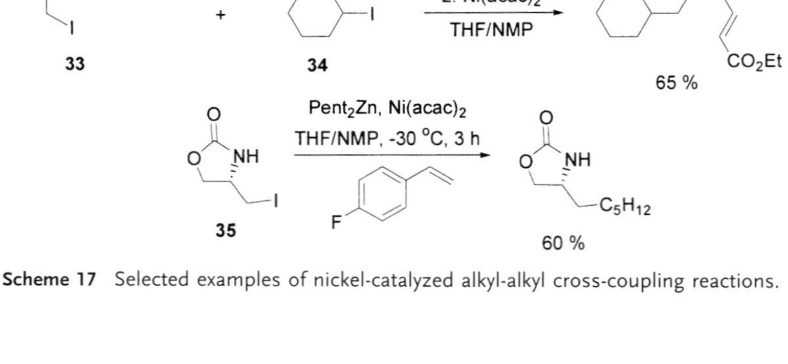

Scheme 17 Selected examples of nickel-catalyzed alkyl-alkyl cross-coupling reactions.

to facilitate the reductive elimination, and the scope of this methodology is largely limited to alkyliodides (**32–35**, Scheme 17). The reductive elimination of the coupling product can also be enhanced by addition of a suitable π-acceptor ligand such as 4-fluorostyrene.

3.7.3.3
Carbometalations

Nickel-catalyzed hydrozincations of olefins in the presence of Et_2Zn have been shown to afford diorganozinc species, which have found application in the synthesis of polyfunctional alcohols (**36**, Scheme 18). The mechanism is believed to involve the intermediacy of a hydridonickel complex, which is generated upon reductive elimination of ethylene [45]. Nickel-catalyzed carbozincations of alkynes have also been used for the *in situ* preparation of organozinc species. Knochel et al. utilized sequential nickel-catalyzed carbozincation and cross-coupling reactions for the synthesis of anti-cancer drug (*Z*)-tamoxifen (**37**) from 1-phenyl-1-butyne [46].

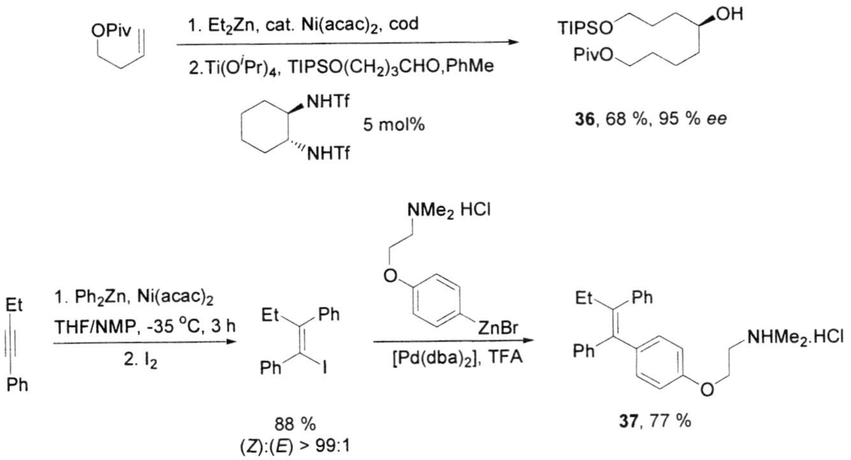

Scheme 18 Nickel-catalyzed carbozincations en route to polyfunctional alcohols and (*Z*)-tamoxifen.

3.7.4
Organozinc Additions to C=X

3.7.4.1
Alkylzinc Nucleophiles

Since the observation of Oguni and Omi in 1984 that diethylzinc can be added to benzaldehyde in the presence of (L)-leucinol with moderate enantioselectivity (49% ee) [47], the catalytic asymmetric diorganozinc addition to aldehydes has matured to a high level of efficiency, and for many years it has served as a test bed for the evaluation of new ligand systems. The subject of organozinc addition to aldehydes has recently been reviewed, and therefore only a condensed overview of the major development will be given here [48]. From the plethora of ligands that were demonstrated to effect highly enantioselective addition, chelating aminoalcohol ligands have emerged as the foremost structural motif. The first ligand that was shown to exhibit high enantioselectivity in the organozinc additions to benzaldehydes was Noyori's 3-*exo*-dimethylaminoisoborneol (DAIB, **38**, Scheme 19) [49].

Ephedrine-based ligands were successfully employed by Soai (**39**) [48a] and Pericas (**40**) [50]. Numerous structurally related aminoalcohol scaffolds with various backbones (e.g., aryl, cycloalkyl, Cp_2Fe, $Cr(CO)_3$, oxazoline, binaphthyl) have also been shown to attain >95% ee in the dialkylzinc addition to aldehydes [48c]. Nugent introduced stable 3-*exo*-morpholinoisoborneol (MIB, **41**) as an advantageous alternative to DAIB, and this gives very high ees even for the Et_2Zn addition to α-branched aldehydes [51]. Both enantiomers of MIB can easily be synthesized from commercially available (R)- and (S)-camphor in only three steps. Diols constitute another family of potent ligands. Seebach carried out extensive studies on titanium-TADDOL(**42**)-catalyzed reactions. Upon structural variation of the diol backbone, very high ees were attained in most cases, although 20 mol% of ligand and superstoichiometric amounts of titanium are used. Highly active BINOL/Ti(O^{i}Pr)$_4$ catalysts were introduced by Nakai and Chan, while Katsuki and Pu (**43**, Scheme 20) reported attractive titanium-free conditions with BINOL-based ligands that contain donor substituents in the 3- and 3'-position. Titanium sulfonamide (**44**) catalysts were first used by Ohno, but despite further improvements of the ligand structure by Walsh, Zhang, and others, the high metal loadings (>1.2 eq.) clearly limit the attractiveness of this class of catalyst systems [48c].

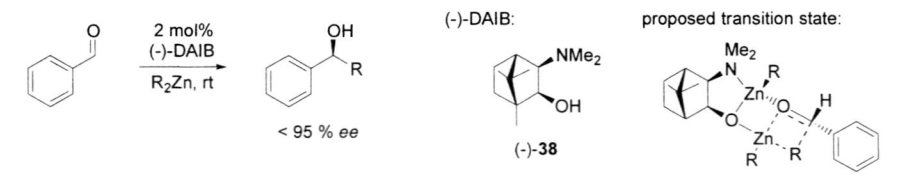

Scheme 19 Noyori's DAIB catalyst for the enantioselective dialkylzinc addition to aldehydes.

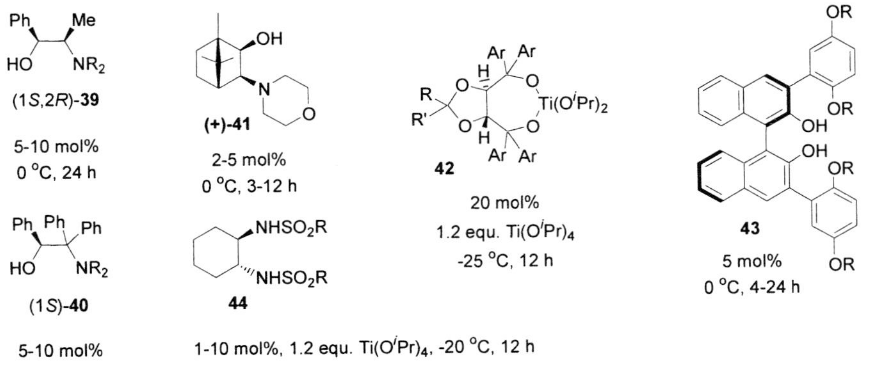

Scheme 20 Representative catalysts for the dialkylzinc addition to aldehydes with >95% ee.

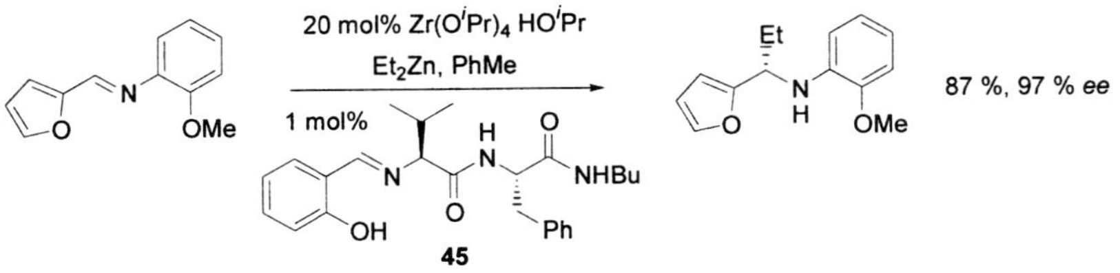

Scheme 21 Zr-catalyzed alkylation of imines with peptide-based Schiff base ligand 45.

As shown in Scheme 21, Hoveyda and Snapper have disclosed efficient Zr-catalyzed alkylations of aryl and alkyl imines promoted by peptide-based ligand 45 in excellent yield and ee (>80%, >92% ee) [52].

3.7.4.2
Arylzinc and Vinylzinc Nucleophiles

While a great deal of research effort has been directed at the development of chiral catalysts for dialkylzinc additions, analogous reactions of aryl-, vinyl- and alkynylzinc compounds are rare. For asymmetric diphenylzinc additions to aldehydes, which suffer the relatively facile non-catalyzed background reaction, Pu reported good enantioselectivities in the presence of binaphthol ligands of type 43. Bolm observed high ees in the same reaction with a 2/1 mixture of diethylzinc/diphenylzinc and a ferrocenyl oxazoline ligand. Reasonable to good ees were obtained by Fu in the diphenylzinc addition to ketones with 15 mol% of DAIB (Scheme 22) [48c].

Enantioselective vinylzinc additions to aldehydes have been described with most of the aforementioned ligands (ephedrines, DAIB, Ti/TADDOL) to afford synthetically useful chiral allylic alcohols [53]. An elegant asymmetric synthesis of allylic alcohols starting from simple alkynes was elaborated by Oppolzer [54]. Monohy-

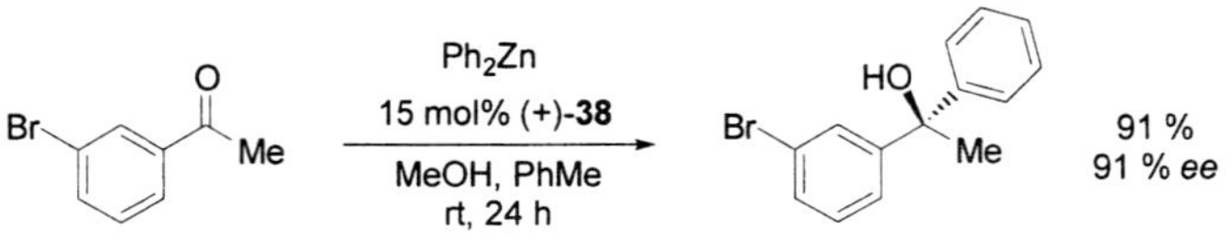

Scheme 22 Enantioselective diphenylzinc addition to ketones with catalytic DAIB.

droboration of alkynes followed by transmetalation to zinc afforded alkenylzinc reagents which, in the presence of DAIB, add to aldehydes to furnish allylic alcohols with >85% ee. The intramolecular variant of this methodology has been applied to the total synthesis of macrocycles such as (R)-(–)-muscone (**46**, Scheme 23) [55] and (+)-aspicilin [56].

Walsh reported on the successfully employed Nugent's MIB ligand (**41**) for the highly enantioselective alkenylzinc addition to aldehydes. Upon sequential combination of Oppolzer's synthesis of allylic alcohols from alkynes and Overman's [3,3] sigmatropic rearrangement of allyl trichloroacetimidates, allylic amines (**47**) were obtained in high enantiopurity [57]. Subsequent oxidative cleavage was demonstrated to provide a general asymmetric synthesis of amino acids (**48**, Scheme 24).

As an extension to the well-established alkenylzinc addition to aldehydes, Walsh also studied the vinylation of ketones. Although diethylzinc shows virtually no reaction with ketones, the combination with ketone in the presence of alkenylborane afforded *cis*-3-hexene-1,6-diols (**49**, Scheme 25) [58]. The mechanism involves an unprecedented reductive dimerization of vinylzinc and insertion of two equivalents of ketone.

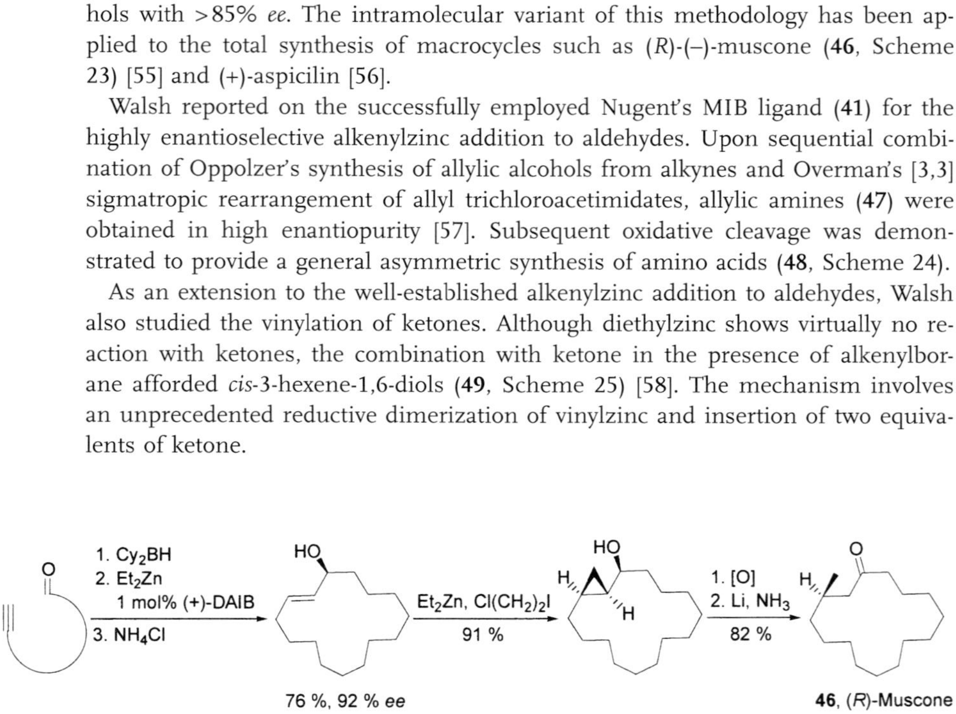

Scheme 23 Oppolzer's muscone synthesis with two zinc-mediated reaction steps.

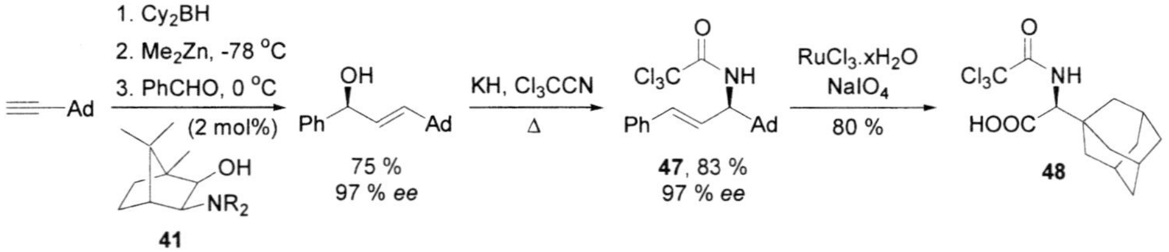

Scheme 24 Asymmetric synthesis of allylic amines and amino acids via alkenylzinc intermediates.

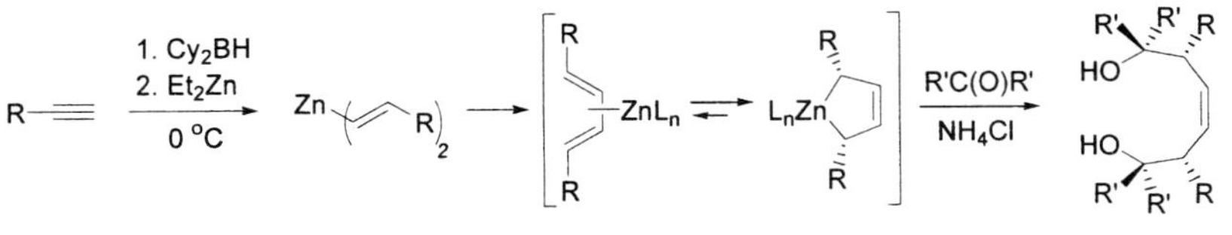

49, 47 - 86 %
dr > 20:1

Scheme 25 Synthesis of hexenediols via reductive coupling of vinylzinc intermediates.

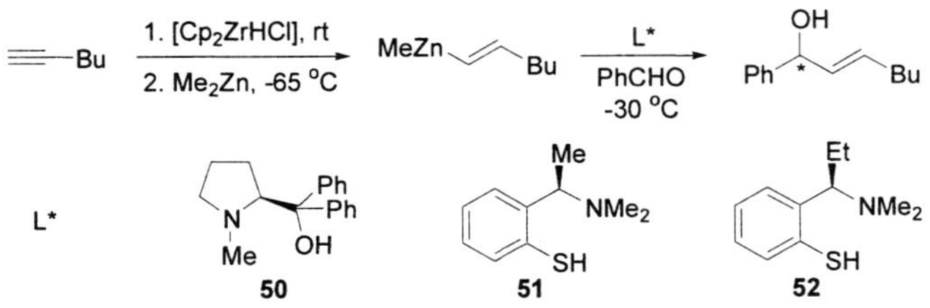

	50		51	52	
	10 mol%	2 mol%	10 mol%	10 mol%	2 mol%
Yield [%]	88	99	76	80	88
ee [%]	81	19	89	95	78

Scheme 26 Vinylzinc species derived from hydrozirconation in stereoselective aldehyde additions.

The sequence of hydrozirconation of alkynes, transmetalation to zinc, and addition to aldehydes has also been demonstrated to provide a practical method for the synthesis of allylic alcohols from alkynes [59]. In the presence of *N,O* or *S,N* ligands (**50–52**), good enantioselectivities have been attained (Scheme 26) [60].

This asymmetric transformation has been used by Danishefsky and Trauner at a late stage in their total synthesis of (+)-halichlorine [61]. Similarly, Jacobsen utilized a diastereoselective variant of the Zr-Zn transmetalation methodology in the presence of an α-chiral ketone for the total synthesis of fostriecin (Scheme 27) [62].

By virtue of simplified product isolation and catalyst recovery, polymeric and dendritic chiral catalysts have become a popular research field in recent years. BINOL- and TADDOL-based systems have emerged as particularly effective polymeric catalysts. However, we would like to refer the interested reader to a recent review on this topic [48c].

3.7.4.3
Alkynylzinc Nucleophiles

Recently, zinc-mediated addition reactions of terminal acetylenes to C=X electrophiles have had a remarkable impact on the development of new carbon-carbon bond-forming methodology, as they allow for a direct access to synthetically useful

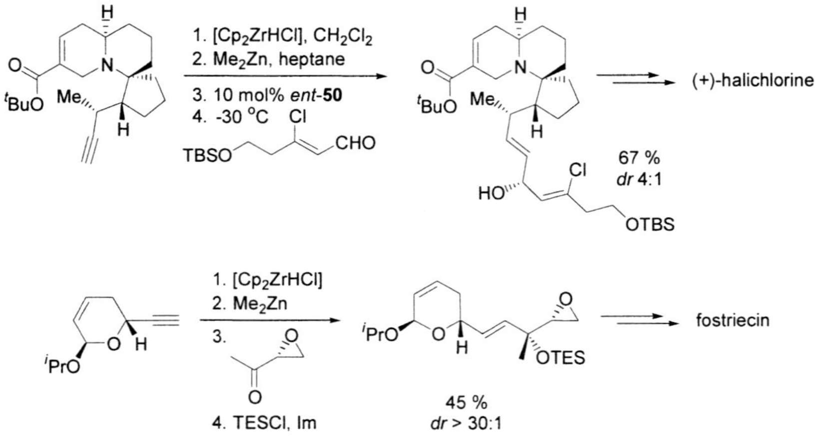

Scheme 27 Zr/Zn exchange in the total syntheses of (+)-halichlorine and fostriecin.

propargylic building blocks. Truly general and operationally simple methods that effect the asymmetric and/or catalytic addition of acetylenes to C=X electrophiles have been difficult to identify. Until very recently, examples of outstanding enantioselection have been restricted to isolated cases exemplified by Merck's synthesis of Efavirenz (**54**) via a highly selective alkynylation of the trifluoroaryl ketone **53** effected by a zinc alkoxide complex (Scheme 28) [63].

A major breakthrough was discovered by Carreira and co-workers, who demonstrated that the combination of Zn(OTf)$_2$ with a tertiary base functioned as a competent reagent system for the generation of metal acetylides *without* resorting to pyrophoric reagents *or* strong bases [64]. Thus, when acetylenes were exposed to nitrones in the presence of catalytic Zn(OTf)$_2$ and Hünig's base, the corresponding propargylic hydroxylamines were obtained in good to excellent yields (Scheme 29) [65].

The resultant nitrones are useful intermediates and can, for example, be further manipulated to give 2,3-dihydroisoxazoles by adding catalytic ZnCl$_2$ and DMAP [66]. Chiral propargylic hydroxylamines can also be obtained in excellent *de* by utilizing a chiral auxiliary in place of the *N*-benzyl group [67]. This methodology has been extended to the catalytic asymmetric addition of acetylenes to aldehydes.

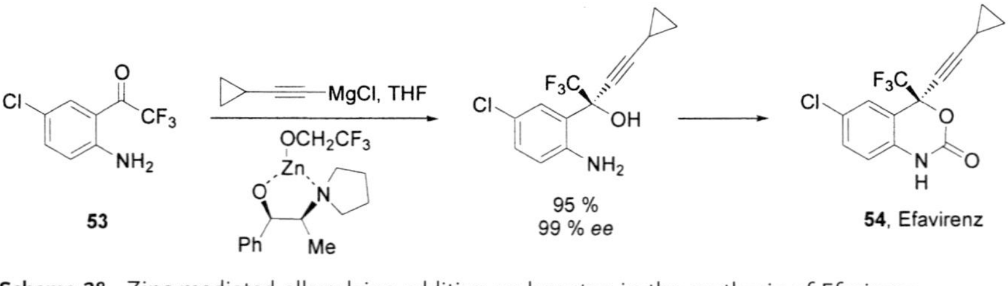

Scheme 28 Zinc-mediated alkynylzinc addition as key step in the synthesis of Efavirenz.

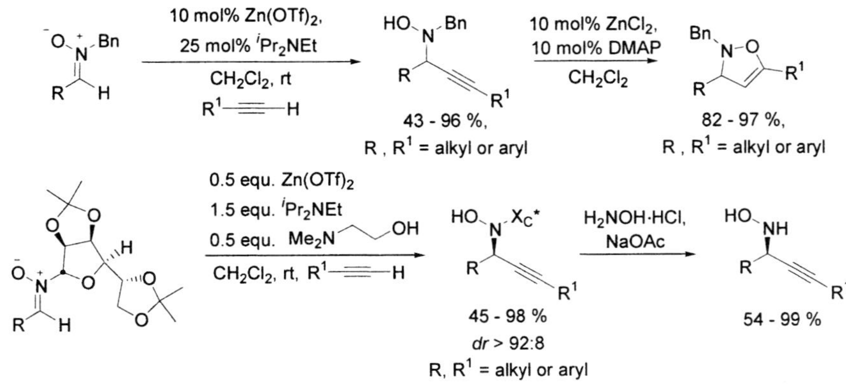

Scheme 29 Carreira's Zn(OTf)$_2$/base system for the synthesis of propargylic hydroxyl amines.

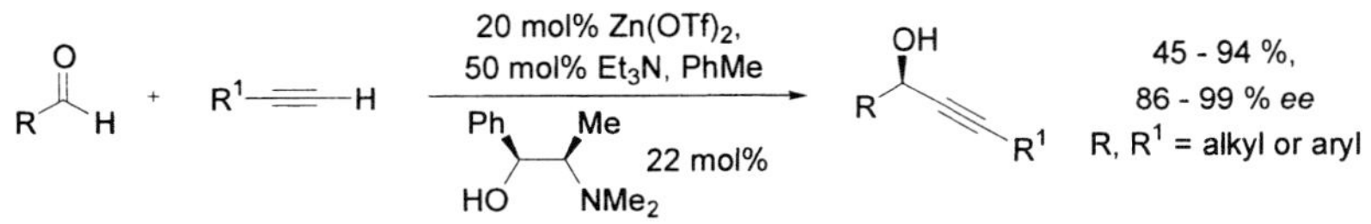

Scheme 30 Catalytic enantioselective zinc acetylide addition to aldehydes.

Thus, chiral propargylic alcohols are obtained by combining aldehydes, acetylenes, and catalytic Zn(OTf)$_2$/Et$_3$N/N-methyl ephedrine (**55**) in up to 99% *ee* (Scheme 30) [68].

Jiang extended the Zn(OTf)$_2$/Et$_3$N-catalyzed generation of zinc acetylides further to the asymmetric alkynylation of α-keto esters using a modified ephedrine ligand [69]. In recent years, a number of efficient methods have been developed for the addition of acetylenes utilizing various dialkylzinc species in combination with a diverse set of catalyst/ligand systems. While these systems do not retain the practicality of Carreira's Zn(OTf)$_2$-catalyzed methods, they nevertheless represent significant additions to the state of the art of organic synthesis. These include self-assembled Ti-catalyzed zinc acetylide addition to aldehydes [70] and Zn(salen)-catalyzed addition to ketones [71]. Hoveyda used Zr-catalyzed alkynylation of arylimines with peptide ligands to afford propargylamines [72], and also reported a related catalyst system for the reciprocal reaction, the enantioselective alkylation of propargylimines [73].

3.7.5
Asymmetric Conjugate Additions

3.7.5.1
Copper Catalysis

In the context of catalytic asymmetric Michael additions, organozinc species in combination with copper-catalysis have provided a particularly fruitful avenue of investigation. Alexakis' seminal discovery that the ephedrine-derived ligand **56**, in combination with various cuprates, catalyzed the addition of diethylzinc to cyclohexenones to give adducts in good yield and *ee* [74] fueled a flurry of activity in the development of asymmetric catalytic additions of dialkylzincs to conjugated electrophiles (Scheme 31) [75].

In spite of these efforts, early methods suffered from lack of scope and variable stereoselectivity. The phosphoramidite-type ligands (e.g., **57**), discovered by Feringa, provided the decisive breakthrough in this regard. The combination of 1 mol% **57** and 0.5 mol% $Cu(OTf)_2$ as the precatalyst and a dialkylzinc species gave the desired products in excellent *ee* (>97%) [76]. Modified phosphoramidite ligands such as **58**, as well as improved experimental procedures, have resulted in essentially quantitative yield and *ee*s of > 99%. An important feature of this methodology is the usefulness of the incipient zinc enolate generated after the conjugate addition. Harnessing this enolate in powerful tandem reaction sequences renders this an extremely useful reaction in synthesis [77]. For example, Feringa demonstrated that the prostaglandin core can be accessed enantioselectively by tandem 1,4-addition aldol reactions of functionalized cyclopentene **59** (Scheme 32) [78].

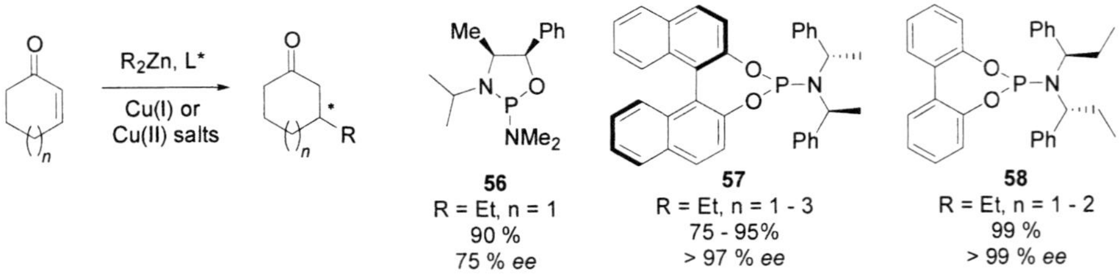

Scheme 31 Powerful ligands in the catalytic asymmetric Michael addition to cycloalkenones.

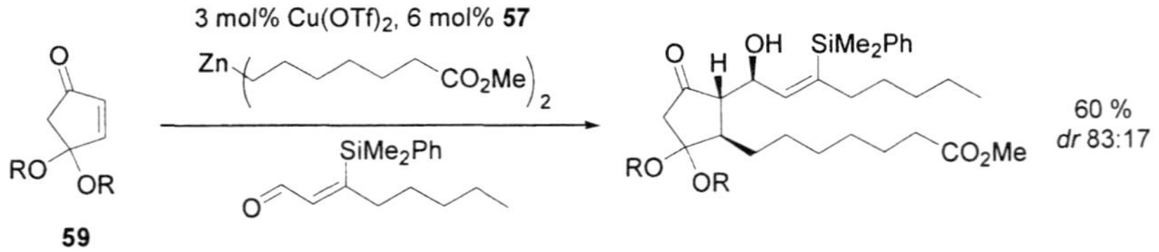

Scheme 32 Formal prostaglandin synthesis via tandem conjugate addition-aldol reactions.

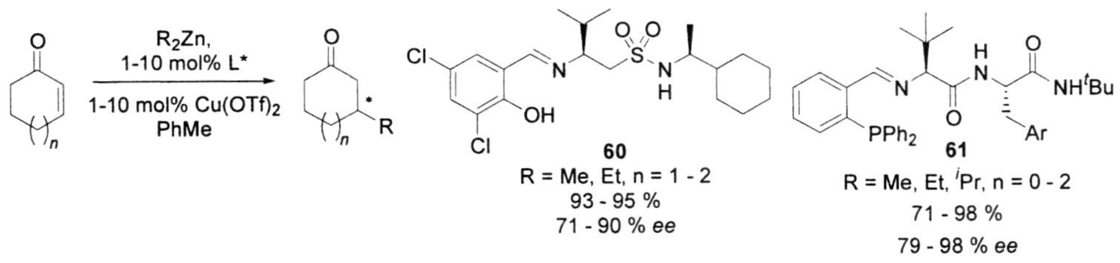

3.7.5 Asymmetric Conjugate Additions | 537

Scheme 33 Modular peptide-based ligands in asymmetric conjugate addition.

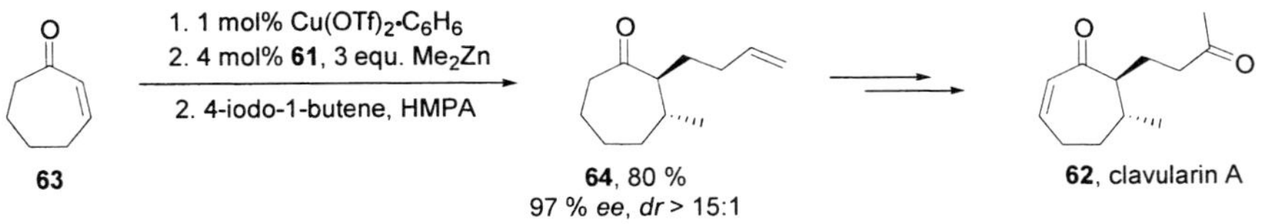

Scheme 34 Enantioselective synthesis of clavularin A.

Other ligand systems based on binaphthalene [79], TADDOL [80], oxazoline derivatives [81] and others [82] have also been utilized in asymmetric conjugate additions, generally proving moderately efficient in terms of both stereoselectivity and scope. It is worth noting that the Michael acceptor is not restricted to cyclic enones; nitrolefins have been used in the Cu-catalyzed conjugate addition of alkylzinc species, albeit generally with lower yields and enantioselectivities [83].

The discovery of novel ligands for asymmetric catalysis through combinatorial synthesis of peptides has gathered momentum in the last few years [84]. In the context of catalytic asymmetric conjugate addition, Gennari and co-workers disclosed the results of a combinatorial ligand discovery effort. Using parallel synthesis of peptide sulfonamide Schiff bases coupled to a high-throughput screen for reactivity and selectivity, ligand **60** was identified as an efficient ligand for the Cu-catalyzed addition of Et$_2$Zn to cycloalkenes (Scheme 33) [85].

This discovery was followed by Hoveyda's disclosure of the highly efficient and versatile peptide-based phosphines **61**. These ligands, when combined with 1 mol% Cu(OTf)$_2$ · C$_6$H$_6$, a dialkylzinc species, and an appropriate cyclic enone, gave the addition product in excellent yield and selectivity (>97% *ee*, R=Et, *n*=0–2) [86]. Upon structural variation, remarkably efficient ligand systems were developed for the catalytic asymmetric conjugate addition of acyclic aliphatic enones [87], trisubstituted cyclic enones [88], nitroalkenes [89], and unsaturated *N*-acyloxazolidinones [90]. This methodology has been employed in a short enantioselective synthesis of clavularin A (**62**, Scheme 34), a potent anti-cancer agent. The synthesis exploits highly selective enantioselective conjugate addition to cycloheptanone (**63**) followed by the diastereoselective alkylation of the incipient zinc enolate to give **64**.

3.7.5.2
Nickel Catalysis

Impressive advances have also been accomplished in this field employing Ni and Co catalysis of the combination with dialkylzincs. Specifically the addition of organozincs to enones, catalyzed by nickel complexes, is one of the most important methods for effecting conjugate additions to sterically encumbered enones [91]. The asymmetric Ni-catalyzed conjugate addition, on the other hand, remains problematic, as the reactions suffer from unsatisfactory enantioselectivities as well as limited substrate scope. Nevertheless, some success has been accomplished, as exemplified by Bolm's work, where the addition product **65** was obtained in good yield and *ee* when exposing 4-chlorochalcone (**66**) to diethylzinc in the presence of catalytic Ni(acac)$_2$ and chiral pyridine ligand **67** (Scheme 35) [92].

Sequential nickel-catalyzed carbozincation of alkynes and Michael additions to enones and carbonyl compounds have been reported on by Montgomery. Applications of such three-component methodology have resulted in efficient syntheses of functionalized olefins and ring systems (Scheme 36) [93]. An extension of the nickel-catalyzed cyclization-transmetalation procedure to allenic substrates has been successfully implemented in a synthesis of (–)-α-kainic acid [94].

3.7.5.3
Oxa- and Phospha-Conjugate Additions

Carbon nucleophiles constitute a small set of possible donors in asymmetric conjugate additions. For example, Enders has described a zinc-mediated asymmetric

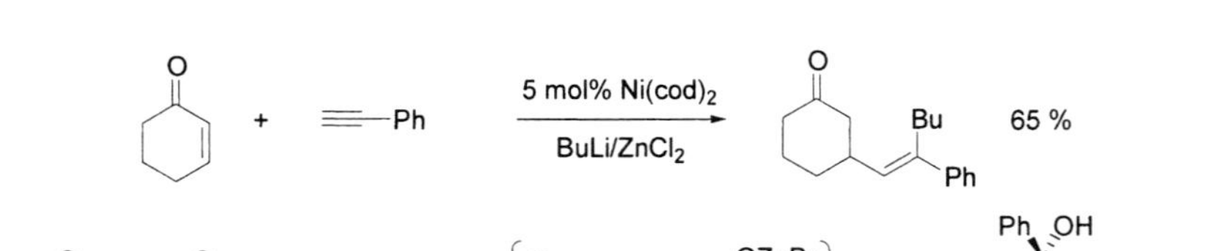

Scheme 35 Ni-catalyzed asymmetric conjugate addition.

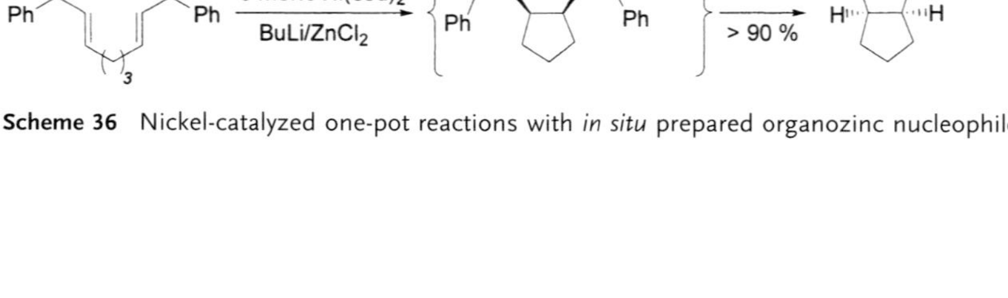

Scheme 36 Nickel-catalyzed one-pot reactions with *in situ* prepared organozinc nucleophiles.

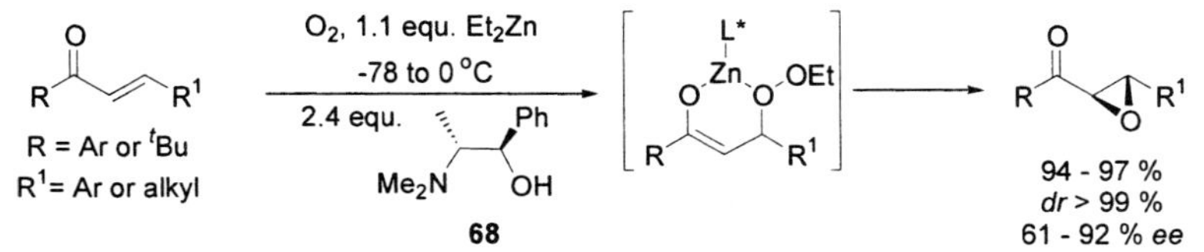

Scheme 37 Asymmetric oxa-Michael additions.

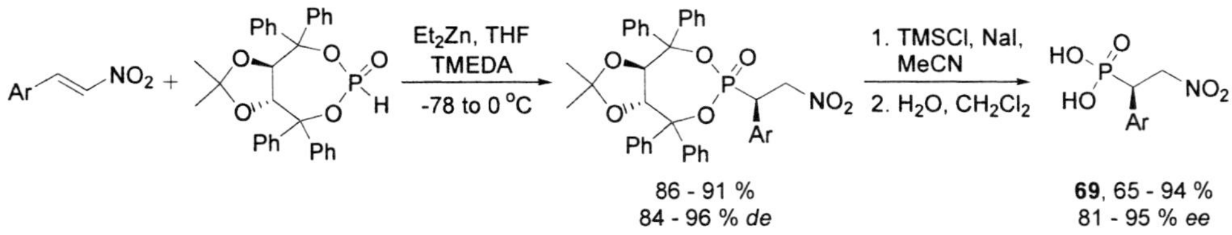

Scheme 38 Phospha-Michael additions to nitroalkenes.

nucleophilic epoxidation of enones (oxa-Michael addition) using N-methyl *pseudo*-ephedrine (**68**) as the chiral ligand (Scheme 37), obtaining epoxyketones in excellent yield and good to excellent *ee* [95].

Enders extended the asymmetric oxa-Michael addition to include nitroalkenes as acceptors, using the same reagent system, obtaining nitrooxiranes in up to 92% *ee* [96].

Few stereoselective methods have been reported that give access to organic phosphonates. Enders reported the first diastereoselective phospha-Michael addition mediated by Et_2Zn and TMEDA. The addition of TADDOL phosphonate to nitroalkenes gives α-substituted β-nitrophosphonic acids such as **69** in high *ee* after cleavage of the auxiliary (Scheme 38) [97].

3.7.6
Aldol Reactions

A family of dinuclear zinc prolinol complexes (**70** and derivatives) were demonstrated by Trost to exhibit excellent activities in direct enantioselective aldol reactions with ketones (Scheme 39) [98]. The reaction utilizes 5 mol% of the catalyst, which is prepared *in situ* by mixing bis(prolinol) ligand **71** with 2 equivalents of Et_2Zn, and is highly atom economical. Extension of the methodology to α-hydroxyketones resulted in an asymmetric synthesis of 1,2-diols as an alternative to the Sharpless AD reaction [99]. A direct Mannich-type reaction in the presence of imines as acceptors afforded *syn* aminoalcohols with high attendant *ees* [100]. With nitromethane as donor, β-hydroxyamines and α-hydroxy carboxylic acids have been obtained via the asymmetric nitroaldol (Henry) reaction [101]. A structurally related ligand system (**72**) has been applied to the desymmetrization of 1,3- and 1,4-

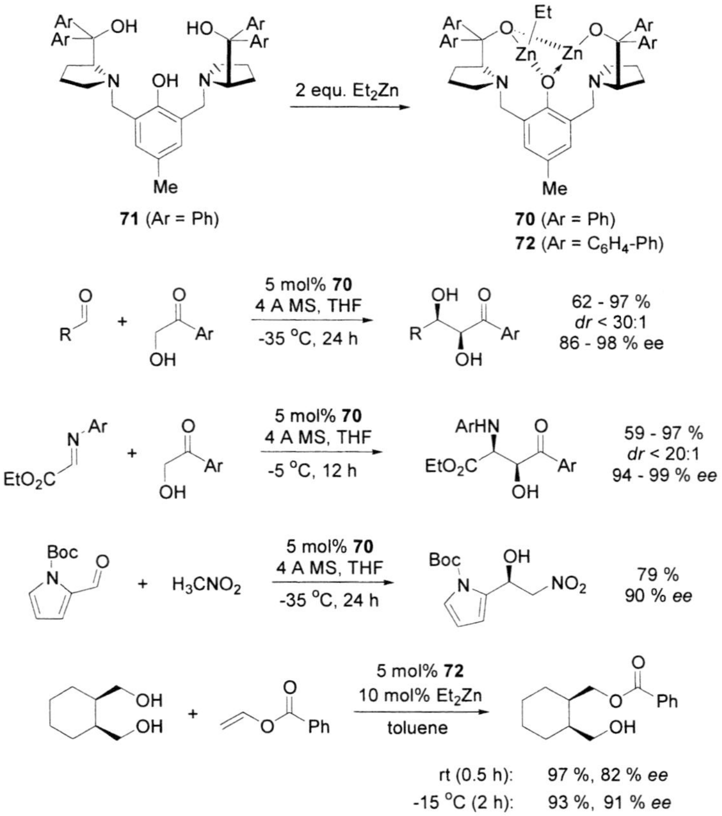

Scheme 39 Dinuclear zinc prolinol catalyst for aldol and desymmetrization reactions.

diols [102]. Structurally related zinc catalysts have been shown to be highly active for the controlled polymerization of lactide [103].

Shibasaki introduced dinuclear zinc catalysts with linked BINOL ligand for direct enantioselective aldol reaction of hydroxyacetophenones with aldehydes (Scheme 40) [104]. In contrast to the related lanthanide system, the reaction affords predominantly the *syn* 1,2-diols (*de* < 94, *ee* < 99%). Optimization studies on the originally reported 2/1 (Et$_2$Zn/L) system resulted in a superior catalyst with a 4/1 stoichiometry that enables reduction of the ligand loading to 0.1 mol% with no loss of selectivity. Furthermore, the reaction is highly atom economical, as only a slight excess (1.1 equ.) of the hydroxyketone donor is required [105].

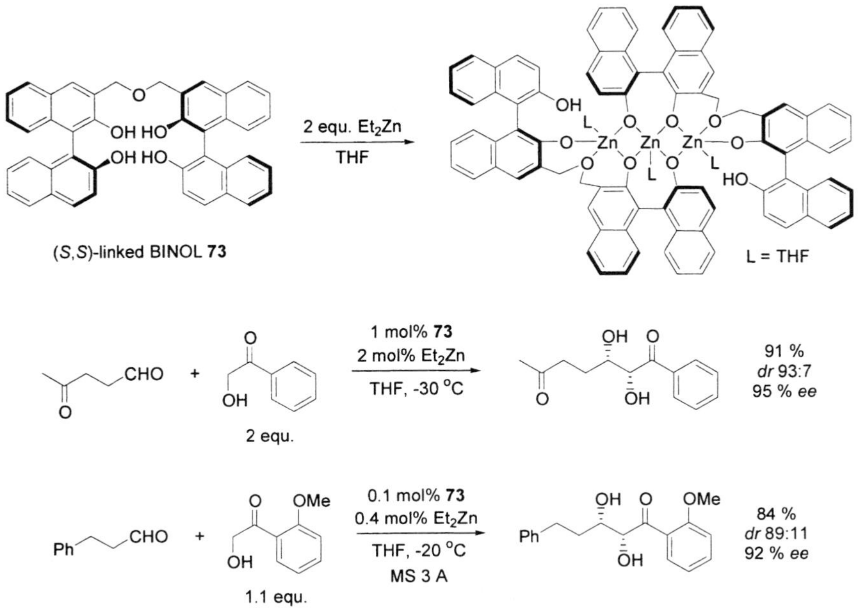

Scheme 40 Linked BINOL ligand for the direct aldol reaction with hydroxyketones.

3.7.7
Cyclopropanation

Since the seminal discovery by Simmons and Smith [106] that cyclopropanations of olefins can be effected with Zn and CH_2I_2, several zinc-mediated protocols have been reported [107]. Halomethylzinc reagents are readily available (prepared from inexpensive materials Zn/XCH_2I; Et_2Zn/CH_2I_2; $CF_3COOH/Et_2Zn/CH_2I_2$; ZnX_2/CH_2N_2), and, because they require only mild reaction conditions, have established zinc-mediated cyclopropanation as a valuable transformation in organic synthesis. Asymmetric variants have also been developed, and this advance has been greatly aided by Winstein's observation that proximal hydroxyl groups effectively direct the cyclopropanation [108], thus opening the door for high diastereoselectivity. Later, auxiliary-controlled stereoselection emerged as a powerful tool for the construction of cyclopropanes. Thus, allylic ethers and acetals equipped with chiral tartrate- or carbohydrate-based auxiliaries have been shown to exhibit high diastereoselectivity. The first practical and general albeit stoichiometric reagent-controlled system was developed by Charette, who described the dioxoborolane **74** as an effective ligand for the enantioselective cyclopropanation of allylic alcohols, giving excellent yields and *ees* (Scheme 41) [109]. The power of this methodology is highlighted by Barrett's exquisite bi-directional approach to the polycyclopropane natural product U-106305 [110].

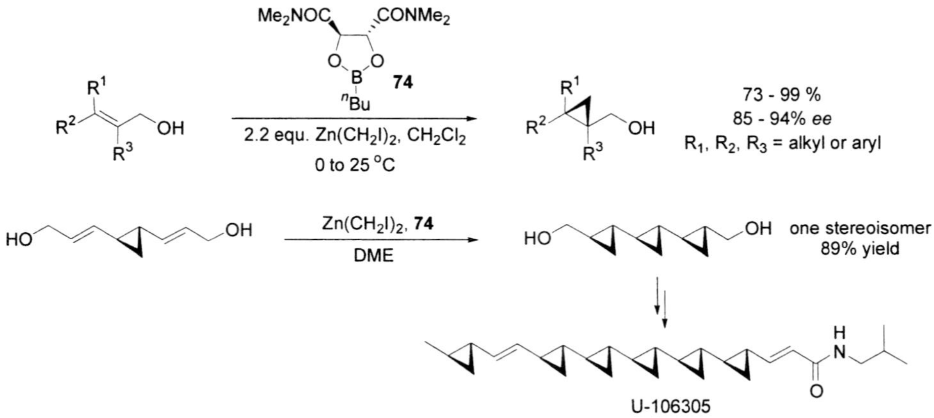

Scheme 41 Asymmetric cyclopropanation of allylic alcohols.

Few catalytic cyclopropanations of allylic alcohols have been described. Nevertheless, Kobayashi reported the first enantioselective cyclopropanation with zinc sulfonamide catalyst (**75** + Et$_2$Zn) in 1992 [111]. This catalyst system proved quite substrate dependent, but very good *ee*s can be obtained. This method is complemented by Charette's Ti TADDOL-catalyzed process [112], which also displays excellent *ee*s with some di- and tri-substituted olefins (Scheme 42).

Innovative applications of the zinc-mediated cyclopropanation in combination with other reactions have been demonstrated. For example, Zerchner disclosed a

Scheme 42 Catalytic asymmetric cyclopropanation of allylic alcohols.

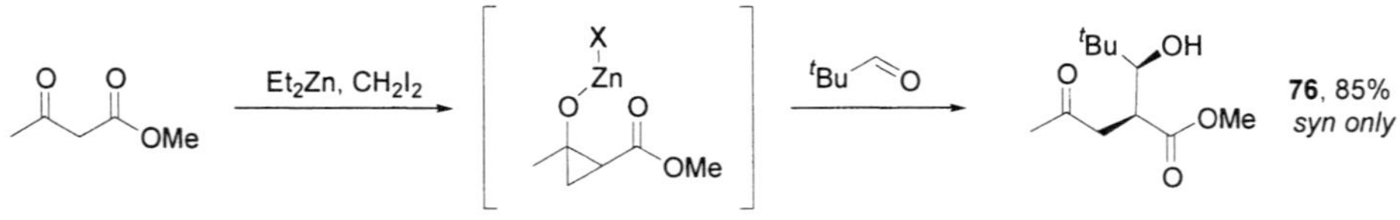

Scheme 43 Tandem cyclopropanation-Reformatsky reaction of β-ketoesters.

tandem chain extension-aldol reaction of β-ketoesters that afforded α-substituted γ-ketoesters such as **76** with good *syn* selectivity (Scheme 43) [113].

The mechanism is believed to involve cyclopropanation of the intermediate zinc ester enolate in the presence of CH_2I_2, followed by rearrangement and Reformatsky-type addition to an aldehyde [114].

3.7.8
Reactions of Zinc Enolates

3.7.8.1
Reformatsky-type Reactions

The Reformatsky reaction, the aldol-type reaction of an enolate formed by oxidative addition of zinc into the C-X bond of an α-halo carbonyl compound, is one of the earliest applications of synthetically useful organozinc chemistry [115]. Although the methodology has been expanded to include enolates of other low-valent metals (Ti, Cr, Sn, Sm, etc.) [116], the insertion of zinc is still the most widely used method. Zinc enolates are relatively stable, and the moderate reactivity as compared to their alkali metal counterparts is responsible for the high chemoselectivity that is observed in reactions with aldehydes and ketones. A particular advantage is the precise site of reaction, which allows for regioselective enolate formation in polycarbonyl compounds. Several elegant applications to total syntheses of complex targets testify to the advantageous use of zinc enolates in aldol-type reactions [115]. Aside from carbonyl species, nitriles play a prominent role as electrophiles, as they lead to β-oxoesters after hydrolytic work-up [117]. Upon preformation of the zinc enolate, the range of electrophiles can be extended to include imines [118], acyl, vinyl, and aryl halides in the presence of palladium or nickel catalysts [119]. Adrian and Snapper developed an efficient, nickel-catalyzed,

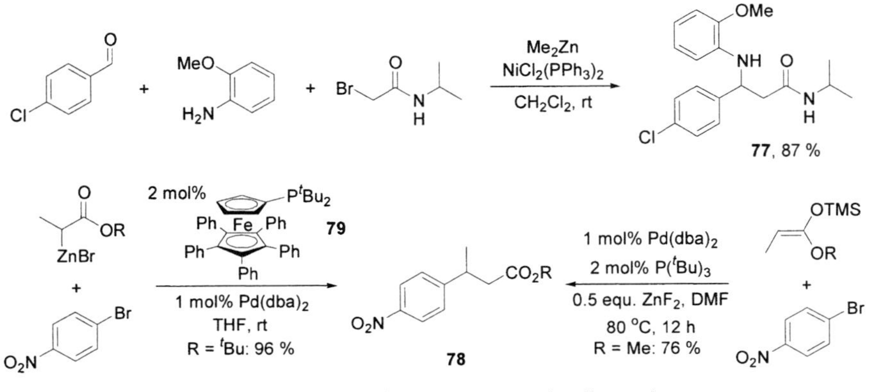

Scheme 44 Reformatsky-type reactions with imines (top) and arylbromides.

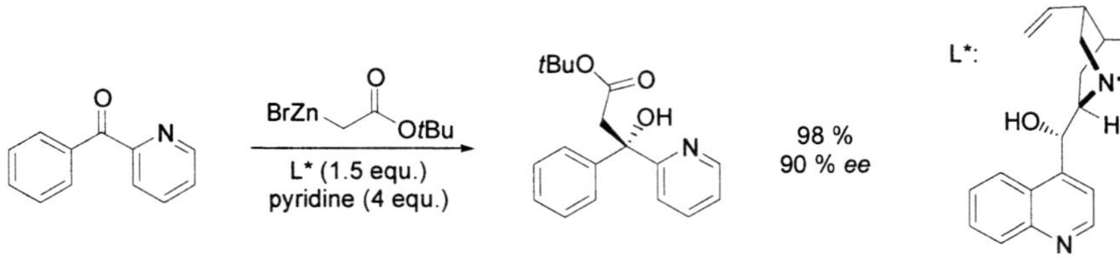

Scheme 45 Chelation-assisted enantioselective Reformatsky reaction with ketones [123 a].

Reformatsky-type three-component condensation (3CC) reaction that affords α-amino carbonyl compounds of type **77** (Scheme 44) [118b]. Hartwig recently reported on the very mild, neutral conditions associated with the palladium-catalyzed α-arylation of esters and amides with functionalized aryl bromides. Generation of the intermediate zinc enolate can be accomplished by Reformatsky-type insertion into an α-halo carbonyl compound or via a silicon-zinc-palladium transmetalation triad (Scheme 44) [120].

Despite considerable efforts, a general and highly stereoselective reaction has not been arrived at so far. Impressive diastereoselectivities have been attained in isolated examples with chiral electrophiles such as oxazolidines [121] or π-complexed aromatic aldehydes [122]. Although chiral amino alcohol (prolinol, ephedrine) and tartrate-based ligands have been shown to give good enantioselectivity in some cases (Scheme 45) [123], the enantioselective Reformatsky reaction still deserves further development with regard to generality and scope.

3.7.8.2
Amino Acid Syntheses

Kazmaier has developed elegant syntheses utilizing zinc enolates as nucleophiles in the synthesis of non-proteogenic amino acids [124]. The power of zinc enolates lies in their inherent ability to form thermally and geometrically stable chelated enolates, thus enabling diastereoselective reactions not accessible with the corresponding lithium enolates [125]. Thus, treating protected amino ester **80** with LHMDS followed by transmetalation with $ZnCl_2$ generates the chelated zinc enolates **81** that undergo highly diastereoselective Pd-catalyzed allylic alkylation with enantiopure allylic carbonates (such as (S)-**82**) to form α-amino acids with high *anti* selectivity (Scheme 46) [126].

Kazmaier and Helmchen have also reported using Pd-catalyzed asymmetric allylic alkylation using racemic allyl carbonates in the presence of chiral phosphine ligands, but with moderate selectivity [127]. Zinc enolates can also be utilized to access *syn* α-amino acids by enlisting a highly diastereoselective Claisen rearrangement of allylic amino esters **83** (Scheme 47) [128, 129].

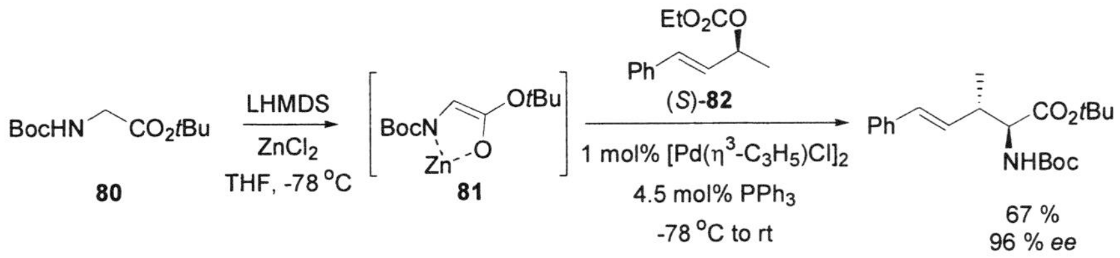

Scheme 46 Zinc enolates in the diastereoselective allylic alkylation toward amino acid derivatives.

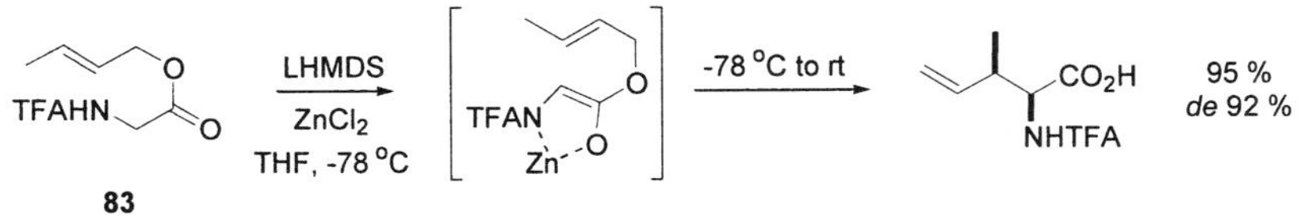

Scheme 47 Zinc-mediated Claisen rearrangement of allylic acetates.

3.7.8.3
Palladium-Catalyzed Reactions

Cossy reported a Pd-catalyzed α-arylation of N-protected piperidones [130, 131]. Interestingly, the success of the reaction was crucially dependent on the use of Zn(II) enolates. Thus deprotonation of N-benzyl piperidone (84) was effected by LHMDS, and the enolate was subsequently transmetalated with $ZnCl_2$ prior to exposure to the palladium-catalyzed reaction (Scheme 48).

Zinc enolates have also been shown to enhance efficiency in the Pd-catalyzed asymmetric allylic alkylation reaction [132]. For example, Fuji documented that the use of Et_2Zn as a base in the reaction between dimethyl malonate and allylic acetate 85 greatly enhanced the ee vis-à-vis the use of other metals and even zinc enolates obtained by deprotonation with a base followed by transmetalation with $ZnCl_2$ (Scheme 49) [133].

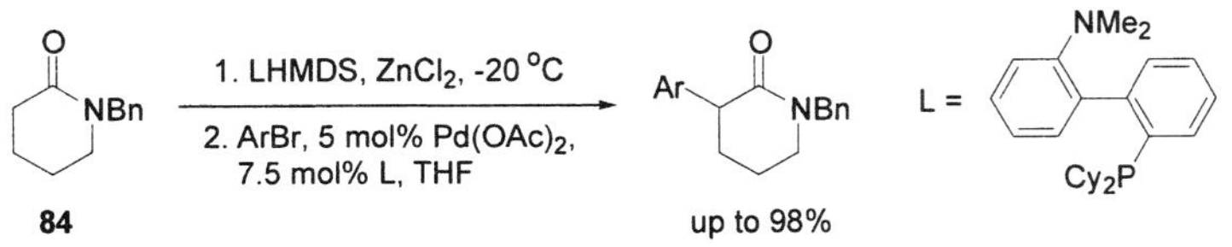

Scheme 48 Pd-catalyzed α-arylation of piperidones.

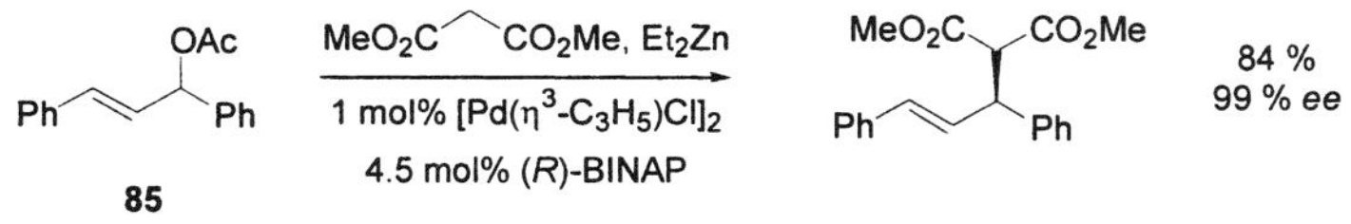

Scheme 49 Pd-catalyzed asymmetric allylic alkylation of Zn-enolates.

3.7.8.4
Miscellaneous Reactions

In an interesting application of zinc enolate chemistry, published by MacMillan, an allenoate variant of the venerable Claisen rearrangement was catalyzed by Zn(OTf)$_2$ (Scheme 50) [134]. Both, the *anti* and the *syn* adduct is obtained in excellent selectivity and yield. Moreover, the reaction displays wide substrate scope, enabling diastereoselective access to 1,2-disubstituted β-amino esters.

Further examples of interesting zinc enolate reactions include the carbocyclization of amino-substituted zinc enolates to give polysubstituted pyrrolidines and piperidines in highly diastereoselective fashion as described by Marek and Normant [135]. The stereochemical outcome of the cyclization toward piperidines is likely governed by the presence of homoallylic substitution (Scheme 51).

Diastereoselective additions of prochiral zinc (*E*)-enolates to chiral 1-acylpyridinium salts have been utilized for the synthesis of 2-substituted 1-acyl-2,3-dihydro-4-pyridones, which constitute versatile building blocks for further synthetic manipulation, as evidenced by the total synthesis of (+)-cannabisativine [136]. Interestingly, the (*E*)-enolate of ketone **86** gives *anti* isomer (*S,S*)-**87** (Scheme 52), whereas open-chain (*Z*)-enolates (because of chelation) lead predominantly to *syn* products. The stereochemistry of the products can thus be set by proper choice of the chiral auxiliary, (+)- or (–)-tcc, and the enolate geometry.

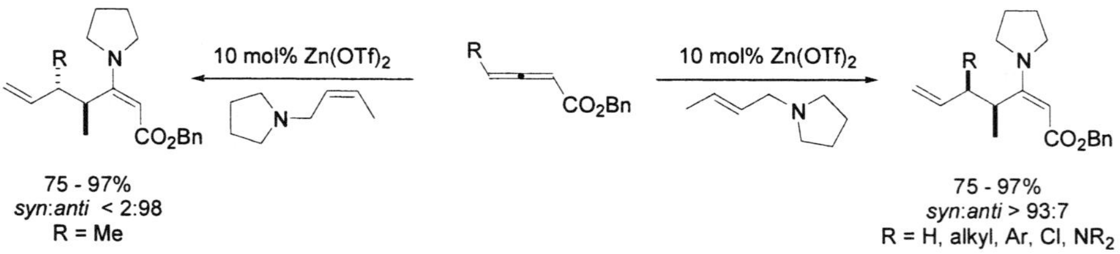

Scheme 50 Zn-catalyzed allenoate-Claisen rearrangement.

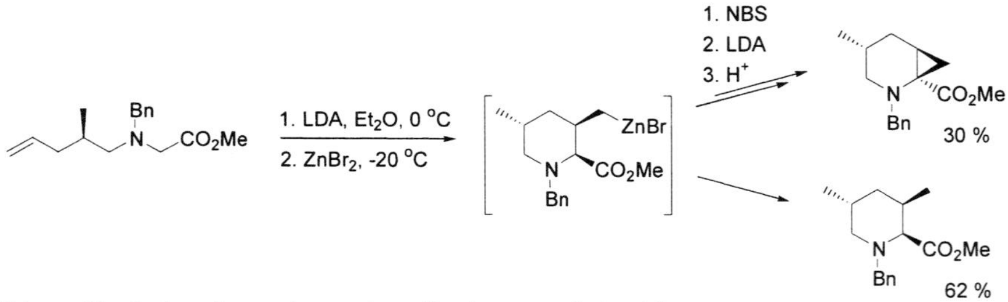

Scheme 51 Amino zinc enolate carbocyclization toward piperidines.

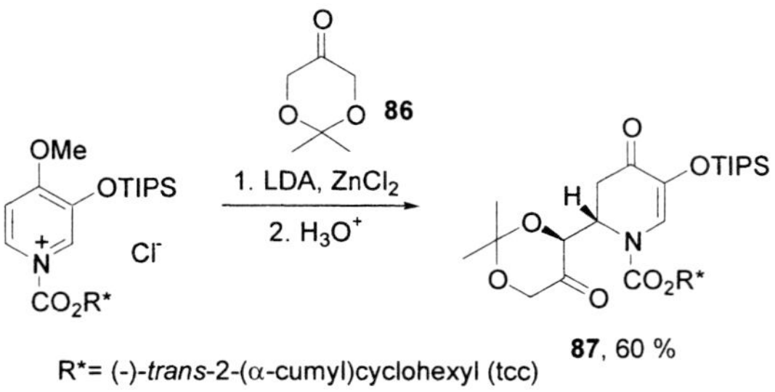

Scheme 52 Diastereoselective zinc enolate addition to pyridinium ions.

3.7.9
Summary and Outlook

Organozinc chemistry has undergone a dramatic development since the early discoveries of Frankland over 150 years ago. In particular, the development of methods that allow for greater functional group tolerance in the preparation of organozinc species has helped to propel organozinc chemistry to the forefront of synthetic methodologies. Furthermore, the compatibility of zinc reagents with many metal-catalyzed processes has greatly enhanced the synthetic toolbox for organic chemists. Particular attention should be drawn to the outstanding contributions made by the catalytic asymmetric addition of zinc organyls to C=X bonds, asymmetric conjugate addition, and direct aldol reactions. On the other hand, more general and atom economic methodologies for catalytic asymmetric cyclopropanation would certainly be welcome. With the rich and varied chemistry of zinc in mind, it is beyond any doubt that chemists will continue to develop innovative and efficient methodologies that go beyond the current catalog of organozinc chemistry.

3.7.10
References

1 E. Frankland, *Liebigs Ann.* **1849**, *71*, 171.
2 (a) A. Boudier, L.O. Bromm, M. Lotz, P. Knochel, *Angew. Chem. Int. Ed.* **2000**, *39*, 4414; (b) P. Knochel, P. Jones (Eds.), *Organozinc Reagents: A Practical Approach*, Oxford University Press, New York, **1999**.
3 M.J. Dunn, R.F. Jackson, J. Pietruszka, D. Turner, *J. Org. Chem.* **1995**, *60*, 2210.
4 T.N. Majid, P. Knochel, *Tetrahedron Lett.* **1990**, *31*, 4413.
5 R.D. Rieke, *Science* **1989**, *246*, 1260; (b) R.D. Rieke, *Aldrich. Acta* **2000**, *33*, 54.
6 (a) H. Stadtmüller, A. Vaupel, C.E. Tucker, T. Stüdemann, P. Knochel, *Chem. Eur.J.* **1996**, *2*, 1204; (b) H. Stadtmüller, R. Lentz, W. Dörner, T. Stüdemann, C.E. Tucker, P. Knochel, *J. Am. Chem. Soc.* **1993**, *115*, 7027; (c) A.V. Kramer, J.A. Labinger, J.S. Bradley, J.A. Osborn, *J. Am. Chem. Soc.* **1974**, *96*, 7145.
7 (a) M.J. Rozema, C. Eisenberg, H. Lütjens, R. Ostwald, K. Belyk, P. Kno-

CHEL, *Tetrahedron Lett.* **1993**, *34*, 3115; (b) M. J. ROZEMA, A. SIDDURI, P. KNOCHEL, *J. Org. Chem.* **1992**, *57*, 1956.

8 (a) L. I. ZAKHARKIN, O. J. OKHLOBYSTIN, *Zh. Obshch. Chim.* **1960**, *30*, 2134; (b) K.-H. THIELE, G. ENGELHARDT, J. KÖHLER, M. ARNSTEDT, *J. Organomet. Chem.* **1967**, *9*, 385.

9 A. DEVASAGAYARAJ, L. SCHWINK, P. KNOCHEL, *J. Org. Chem.* **1995**, *60*, 3311.

10 L. MICOUIN, M. OESTREICH, P. KNOCHEL, *Angew. Chem. Int. Ed.* **1997**, *36*, 245.

11 M. WITANOWSKI, J. D. ROBERTS, *J. Am. Chem. Soc.* **1966**, *88*, 737.

12 For a recent example, see: E. HUPE, I. M. CALAZA, P. KNOCHEL, *Chem. Eur. J.* **2003**, *9*, 2789.

13 A. BOUDIER, F. FLACHSMANN, P. KNOCHEL, *Synlett* **1998**, 1438.

14 (a) S. E. WOOD, B. RICKBORN, *J. Org. Chem.* **1983**, *48*, 555; (b) L. D. FIELD, S. P. GALLAGHER, *Tetrahedron Lett.* **1985**, *26*, 6125; (c) F. LHERMITTE, P. KNOCHEL, *Angew. Chem. Int. Ed.* **1998**, *37*, 2460.

15 H. LAAZIRI, L. O. BROMM, F. LHERMITTE, K. HARMS, P. KNOCHEL, *J. Am. Chem. Soc.* **2000**, *122*, 10218.

16 K. YACHI, H. SHINOKUBO, K. OSHIMA, *Angew. Chem. Int. Ed.* **1998**, *37*, 2515.

17 P. WIPF, C. KENDALL, C. R. J. STEPHENSON, *J. Am. Chem. Soc.* **2003**, *125*, 761.

18 P. WIPF, C. KENDALL, *Org. Lett.* **2001**, *3*, 2773.

19 E. NEGISHI, N. OKUKADO, A. O. KING, D. E. VAN HORN, B. I. SPIEGEL, *J. Am. Chem. Soc.* **1978**, *100*, 2254.

20 S. RIBE, R. K. KONDRU, B. N. BERATAN, P. WIPF, *J. Am. Chem. Soc.* **2000**, *122*, 4608.

21 Y. A. DEMBELE, C. BELAUD, P. HITCHCOCK, J. VILLIERAS, *Tetrahedron: Asymm.* **1992**, *3*, 511.

22 (a) P. JONES, P. KNOCHEL, *J. Org. Chem.* **1999**, *64*, 186; (b) P. JONES, N. MILLOT, P. KNOCHEL, *Chem. Commun.* **1998**, 2405.

23 (a) W. OPPOLZER, H. BIENAYME, A. GENEVOIS-BORELLA, *J. Am. Chem. Soc.* **1991**, *113*, 9660; (b) C. MEYER, I. MAREK, J.-F. NORMANT, *Tetrahedron Lett.* **1996**, *37*, 857.

24 M. GAUDEMAR, *C. R. Acad. Sci. Paris (C)* **1971**, *273*, 1669.

25 P. KNOCHEL, J.-F. NORMANT, *Tetrahedron Lett.* **1986**, *27*, 1039, 1043, 4427, 4431, 5727.

26 D. BERUBEN, I. MAREK, J.-F. NORMANT, N. PLATZER, *J. Org. Chem.* **1995**, *60*, 2488.

27 K. SUZUKI, T. IMAI, S. YAMANOI, M. CHINO, T. MATSUMOTO, *Angew. Chem. Int. Ed.* **1997**, *36*, 2469.

28 A. HIRAI, M. NAKAMURA, E. NAKAMURA, *J. Am. Chem. Soc.* **2000**, *122*, 11791.

29 E. NEGISHI, *Acc. Chem. Res.* **1982**, *15*, 340.

30 T. SHIOTA, T. YAMAMORI, *J. Org. Chem.* **1999**, *64*, 453.

31 N. MURATA, T. SUGIHARA, Y. KONDO, T. SAKAMOTO, *Synlett* **1997**, 298.

32 (a) V. FARINA, B. KRISHNAN, *J. Am. Chem. Soc.* **1991**, *113*, 9585; (b) M. ROTTLÄNDER, P. KNOCHEL, *J. Org. Chem.* **1998**, *63*, 203; (c) J. L. BERILLON, R. WAGNER, P. KNOCHEL, *J. Org. Chem.* **1998**, *63*, 9117.

33 R. ROSSI, F. BELLINA, C. BECHINI, L. MANNINA, P. VERGAMMI, *Tetrahedron* **1998**, *54*, 135.

34 For a review on Negishi coupling reactions with arylchlorides, see: G. C. FU, A. F. LITTKE, *Angew. Chem. Int. Ed.* **2002**, *41*, 4176.

35 M. ASAOKA, M. TANAKA, T. HOUKAWA, T. UEDA, S. SAKAMI, H. TAKEI, *Tetrahedron* **1998**, *54*, 471.

36 K. UTIMOTO, N. TODA, T. MIZUNO, M. KOBATA, S. MATSUBARA, *Angew. Chem. Int. Ed.* **1997**, *36*, 2804.

37 (a) T. HAYASHI, T. HAGIHARA, Y. KATSURO, M. KUMADA, *Bull. Chem. Soc. Jpn.* **1983**, *56*, 363; (b) T. HAYASHI, A. YAMAMOTO, M. HOJO, Y. ITO, *Chem. Commun.* **1989**, 495; (c) T. HAYASHI, A. YAMAMOTO, M. HOJO, K. KISHI, Y. ITO, E. NISHIOKA, H. MIURA, K. YANAGI, *J. Organomet. Chem.* **1989**, *370*, 129.

38 (a) F. DÜBNER, P. KNOCHEL, *Angew. Chem. Int. Ed.* **1999**, *38*, 379; (b) F. DÜBNER, P. KNOCHEL, *Tetrahedron Lett.* **2000**, *41*, 9233.

39 C. A. LUCHACO-CULLIS, H. MIZUTANI, K. E. MURPHY, A. H. HOVEYDA, *Angew. Chem. Int. Ed.* **2001**, *40*, 1456.

40 K. E. MURPHY, A. H. HOVEYDA, *J. Am. Chem. Soc.* **2003**, *125*, 4690.

41 H. MALDA, A. W. VAN ZIJL, L. A. ARNOLD, B. L. FERINGA, *Org. Lett.* **2001**, *3*, 1169.

42 (a) S. Ongeri, U. Piarulli, M. Roux, C. Monti, C. Gennari, *Helv. Chim. Acta* **2002**, 3388; (b) U. Piarulli, P. Daubos, C. Claverie, M. Roux, C. Gennari, *Angew. Chem. Int. Ed.* **2003**, *42*, 234.

43 D. J. Cardenas, *Angew. Chem. Int. Ed.* **2003**, *42*, 384.

44 (a) R. Giovannini, T. Stüdemann, G. Dussin, P. Knochel, *Angew. Chem. Int. Ed.* **1998**, *37*, 2387; (b) R. Giovannini, T. Stüdemann, A. Devasagayaraj, G. Dussin, P. Knochel, *J. Org. Chem.* **1999**, *64*, 3544.

45 A. E. Jensen, P. Knochel, *J. Org. Chem.* **2002**, *67*, 79.

46 P. Knochel in *Transition Metals for Organic Synthesis* (Eds.: M. Beller, C. Bolm), Wiley-VCH, Weinheim, **1998**, p 467.

47 T. Stüdemann, P. Knochel, *Angew. Chem. Int. Ed.* **1997**, *36*, 93.

48 N. Oguni, T. Omi, *Tetrahedron Lett.* **1984**, *25*, 2823.

49 (a) K. Soai, S. Niwa, *Chem. Rev.* **1992**, *92*, 833; (b) R. Noyori in *Asymmetric Catalysis in Organic Synthesis*, Wiley, New York, **1994**, Ch. 5; (c) L. Pu, H.-B. Yu, *Chem. Rev.* **2001**, *101*, 757.

50 (a) M. Kitamura, S. Suga, K. Kawai, R. Noyori, *J. Am. Chem. Soc.* **1986**, *108*, 6071; (b) M. Kitamura, S. Suga, H. Oka, R. Noyori, *J. Am. Chem. Soc.* **1998**, *120*, 9800.

51 K. S. Reddy, L. Sola, A. Moyano, M. Pericas, A. Riera, *J. Org. Chem.* **1999**, *64*, 3969.

52 W. A. Nugent, *Chem. Commun.* **1999**, 1369.

53 (a) J. R. Porter, J. H. Traverse, A. H. Hoveyda, M. L. Snapper, *J. Am. Chem. Soc.* **2001**, *123*, 984; (b) J. R. Porter, J. H. Traverse, A. H. Hoveyda, M. L. Snapper, *J. Am. Chem. Soc.* **2001**, *123*, 10409.

54 (a) W. Oppolzer, R. N. Radinov, *Tetrahedron Lett.* **1988**, *29*, 5645; (b) W. Oppolzer, R. N. Radinov, *Tetrahedron Lett.* **1991**, *32*, 5777; (c) K. Soai, K. Takahashi, *J. Chem. Soc., Perkin Trans. 1* **1994**, 1257; (d) J. L. von dem Bussche-Hünnefeld, D. Seebach, *Tetrahedron* **1992**, *48*, 5719.

55 W. Oppolzer, R. N. Radinov, *Helv. Chim. Acta* **1992**, *75*, 170.

56 W. Oppolzer, R. N. Radinov, *J. Am. Chem. Soc.* **1993**, *115*, 1593.

57 W. Oppolzer, R. N. Radinov, J. De Brabander, *Tetrahedron Lett.* **1995**, *36*, 2607.

58 Y. K. Chen, A. E. Lurain, P. J. Walsh, *J. Am. Chem. Soc.* **2002**, *124*, 12225.

59 C. Garcia, E. R. Libra, P. J. Carroll, P. J. Walsh, *J. Am. Chem. Soc.* **2003**, *125*, 3210.

60 (a) P. Wipf, W. Xu, H. Takahashi, H. Jahn, P. D. G. Coish, *Pure Appl. Chem.* **1997**, *69*, 639; (b) B. Zheng, M. Srebnik, *J. Org. Chem.* **1995**, *60*, 3278.

61 P. Wipf, S. Ribe, *J. Org. Chem.* **1998**, *63*, 6454.

62 D. Trauner, J. B. Schwarz, S. J. Danishefsky, *Angew. Chem. Int. Ed.* **1999**, *38*, 3542.

63 D. E. Chavez, E. N. Jacobsen, *Angew. Chem. Int. Ed.* **2001**, *40*, 3667.

64 L. Tan, C.-Y. Chen, R. D. Tillyer, E. J. J. Grabowski, P. J. Reider, *Angew. Chem. Int. Ed.* **1999**, *38*, 711.

65 D. E. Frantz, R. Fässler, C. S. Tomooka, E. M. Carreira, *Acc. Chem. Res.* **2000**, *33*, 373.

66 D. E. Frantz, R. Fässler, E. M. Carreira, *J. Am. Chem. Soc.* **1999**, *121*, 11245.

67 P. Aschwanden, D. E. Frantz, E. M. Carreira, *Org. Lett.* **2000**, *2*, 2331.

68 R. Fässler, D. E. Frantz, J. Oetiker, E. M. Carreira, *Angew. Chem. Int. Ed.* **2002**, *41*, 3054.

69 N. K. Anand, E. M. Carreira, *J. Am. Chem. Soc.* **2001**, *123*, 9687.

70 B. Jiang, Z. Chen, X. Tang, *Org. Lett.* **2002**, *4*, 3451.

71 X. Li, G. Lu, W. H. Kwon, A. S. C. Chan, *J. Am. Chem. Soc.* **2002**, *124*, 12636, and references therein.

72 P. G. Cozzi, *Angew. Chem. Int. Ed.* **2003**, *42*, 2895, and references therein.

73 J. T. Traverse, A. H. Hoveyda, M. L. Snapper, *Org. Lett.* **2003**, *5*, 3273.

74 L. C. Akullian, M. L. Snapper, A. H. Hoveyda, *Angew. Chem. Int. Ed.* **2003**, *42*, 4244.

75 A. Alexakis, S. Mutti, J. F. Normant, *J. Am. Chem. Soc.* **1991**, *113*, 6332.

76 (a) N. Krause, A. Hoffmann-Röder, *Synthesis* **2001**, 171, (b) A. Alexakis, C. Benhaim, *Eur. J. Org. Chem.* **2002**, 3221.

77 B. L. Feringa, *Acc. Chem. Res.* **2000**, *33*, 346, and references therein.
78 (a) M. Kitamura, T. Miki, K. Nakano, R. Noyori, *Tetrahedron Lett.* **1996**, *37*, 5141; (b) M. Kitamura, T. Miki, K. Nakano, R. Noyori, *Bull. Chem. Soc. Jpn.* **2000**, *73*, 999; (c) A. Alexakis, G. P. Trevitt, G. Bernardinelli, *J. Am. Chem. Soc.* **2001**, *123*, 4358; (d) O. Knopff, A. Alexakis, *Org. Lett.* **2002**, *4*, 3835.
79 L. A. Arnold, R. Naasz, A. J. Minnaard, B. L. Feringa, *J. Am. Chem. Soc.* **2001**, *123*, 5841.
80 For the most recent example, see: O. Huttenloch, J. Spieler, H. Waldmann, *Chem. Eur. J.* **2001**, *7*, 671, and references therein.
81 For the most recent example, see: A. Alexakis, C. Benhaim, *Org. Lett.* **2000**, *2*, 2579, and references therein.
82 R. Shintani, G. C. Fu, *Org. Lett.* **2002**, *4*, 3699.
83 I. J. Krauss, J. L. Leighton, *Org. Lett.* **2003**, *5*, 3201.
84 (a) N. Sewald, V. Wendisch, *Tetrahedron: Asymm.* **1998**, *9*, 1341; (b) J. P. G. Versleijen, A. M. van Leusen, B. L. Feringa, *Tetrahedron Lett.* **1999**, *40*, 5803; (c) A. Alexakis, C. Benhaim, *Org. Lett.* **2000**, *2*, 2579; (d) A. Alexakis, S. Rosset, J. Allamand, S. March, F. Guillen, C. Benhaim, *Synlett* **2001**, 1375.
85 B. Jandeleit, D. J. Schaefer, T. S. Powers, H. W. Turner, W. H. Weinberg, *Angew. Chem. Int. Ed.* **1999**, *38*, 2494.
86 I. Chataigner, C. Gennari, U. Piarulli, S. Ceccarelli, *Angew. Chem. Int. Ed.* **2000**, *39*, 916.
87 S. J. Degrado, H. Mizutani, A. H. Hoveyda, *J. Am. Chem. Soc.* **2001**, *123*, 755.
88 H. Mizutani, S. J. Degrado, A. H. Hoveyda, *J. Am. Chem. Soc.* **2002**, *124*, 779.
89 S. J. Degrado, H. Mizutani, A. H. Hoveyda, *J. Am. Chem. Soc.* **2002**, *124*, 13362.
90 C. A. Luchaco-Cullis, A. H. Hoveyda, *J. Am. Chem. Soc.* **2002**, *124*, 8192.
91 A. W. Hird, A. H. Hoveyda, *Angew. Chem. Int. Ed.* **2003**, *42*, 1276.
92 C. Petrier, J. C. Souza Barbosa, C. Dupuy, J.-L. Luche, *J. Org. Chem.* **1985**, *50*, 5761.
93 C. Bolm, M. Ewald, M. Felder, *Chem. Ber.* **1992**, *125*, 1205.
94 J. Montgomery, *Acc. Chem. Res.* **2000**, *33*, 467.
95 M. V. Chevliakov, J. Montgomery, *J. Am. Chem. Soc.* **1999**, *121*, 11139.
96 D. Enders, J. Zhu, G. Raab, *Angew. Chem. Int. Ed.* **1996**, *35*, 1725.
97 D. Enders, L. Kramps, J. Zhu, *Tetrahedron: Asymm.* **1998**, *9*, 3959.
98 D. Enders, L. Tedeschi, J. W. Bats, *Angew. Chem. Int. Ed.* **2000**, *39*, 4605.
99 B. M. Trost, H. Ito, *J. Am. Chem. Soc.* **2000**, *122*, 12003.
100 B. M. Trost, H. Ito, E. R. Silcoff, *J. Am. Chem. Soc.* **2001**, *123*, 3367.
101 B. M. Trost, L. R. Terrell, *J. Am. Chem. Soc.* **2003**, *125*, 338.
102 B. M. Trost, V. S. C. Veh, *Angew. Chem. Int. Ed.* **2002**, *41*, 861.
103 B. M. Trost, T. Mino, *J. Am. Chem. Soc.* **2003**, *125*, 2410.
104 C. K. Williams, L. E. Breyfogle, S. K. Choi, W. Nam, V. G. Young, Jr., M. A. Hillmyer, W. B. Tolman, *J. Am. Chem. Soc.* **2003**, *125*, 11350.
105 N. Kumagai, S. Matsunaga, N. Yoshikawa, T. Ohshima, M. Shibasaki, *Org. Lett.* **2001**, *3*, 1539.
106 N. Kumagai, S. Matsunaga, T. Kinoshita, S. Harada, S. Okada, S. Sakamoto, K. Yamaguchi, M. Shibasaki, *J. Am. Chem. Soc.* **2003**, *125*, 2169.
107 H. E. Simmons, R. D. Smith, *J. Am. Chem. Soc.* **1958**, *80*, 5323.
108 For a comprehensive review, see: H. Lebel, J.-F. Marcoux, C. Molinaro, A. B. Charette, *Chem. Rev.* **2003**, *103*, 977.
109 S. Winstein, L. De Vries, *J. Am. Chem. Soc.* **1959**, *81*, 6532.
110 A. B. Charette, H. Juteau, *J. Am. Chem. Soc.* **1994**, *116*, 2651.
111 A. G. M. Barrett, D. Hamprecht, A. J. D. White, D. J. Williams, *J. Am. Chem. Soc.* **1997**, *119*, 8608.
112 H. Takahashi, M. Yoshioka, M. Ohno, S. Kobayashi, *Tetrahedron Lett.* **1992**, *33*, 2575.
113 A. B. Charette, C. Molinaro, C. Brochu, *J. Am. Chem. Soc.* **2001**, *123*, 11367.
114 S. Lai, C. K. Zercher, J. P. Jasinski, S. N. Reid, R. J. Staples, *Org. Lett.* **2001**, *3*, 4169.

115 For another example, see: S. Ito, H. Shinokubo, K. Oshima, *Tetrahedron Lett.* **1998**, *39*, 5253.

116 (a) A. Fürstner in *Organozinc Reagents* (Eds.: P. Knochel, P. Jones), Oxford University Press, New York, **1999**, pp 287; (b) A. Fürstner, *Synthesis* **1989**, 571.

117 For some recent contributions, see: Ti(III): (a) J. D. Parrish, D. R. Shelton, R. D. Little, *Org. Lett.* **2003**, *5*, 3615; Cr(II): (b) L. Wessjohann, T. Gabriel, *J. Org. Chem.* **1997**, *62*, 3772; (c) A. Fürstner, *Chem. Rev.* **1999**, *99*, 991; Sn(II): (d) I. Shibata, T. Suwa, H. Sakakibara, A. Baba, *Org. Lett.* **2002**, *4*, 301; Sm(II): (e) G. A. Molander, G. A. Brown, I. Storch de Gracia, *J. Org. Chem.* **2002**, *67*, 3459; (f) S.-I. Fukuzawa, H. Matsuzawa, S.-I. Yoshimitsu, *J. Org. Chem.* **2000**, *65*, 1702.

118 (a) A. S.-Y. Lee, R.-Y. Cheng, O.-G. Pan, *Tetrahedron Lett.* **1997**, *38*, 443; (b) J. J. Duffield, A. C. Regan, *Tetrahedron: Asymmetry* **1996**, *7*, 663.

119 (a) J. C. Adrian, Jr., J. L. Barkin, L. Hassib, *Tetrahedron Lett.* **1999**, *40*, 2457; (b) J. C. Adrian, Jr., M. L. Snapper, *J. Org. Chem.* **2003**, *68*, 2143.

120 See ref. [115a] for more examples.

121 T. Hama, X. Liu, D. A. Culkin, J. F. Hartwig, *J. Am. Chem. Soc.* **2003**, *125*, 11176.

122 C. Andres, A. Gonzalez, R. Pedrosa, A. Perez-Encabo, *Tetrahedron Lett.* **1992**, *33*, 2895.

123 C. Baldoli, P. del Buttero, E. Licandro, A. Papagni, T. Pilati, *Tetrahedron* **1996**, *52*, 4849.

124 For some recent examples, see: (a) A. Ojida, T. Yamano, N. Taya, A. Tasaka, *Org. Lett.* **2002**, *4*, 3051; (b) J. M. Andres, R. Pedrosa, A. Perez-Encabo, *Tetrahedron* **2000**, *56*, 1217; (c) Y. Ukaji, Y. Yoshida, K. Inomata, *Tetrahedron: Asymm.* **2000**, *11*, 733.

125 For general review on the synthesis of non-proteogenic amino acids, see; J.-A. Ma, *Angew. Chem. Int. Ed.* **2003**, *47*, 4290.

126 U. Kazmaier, *Liebigs Ann./Recueil*, **1997**, 285.

127 (a) U. Kazmaier, F. L. Zumpe, *Angew. Chem. Int. Ed.* **2000**, *39*, 802, (b) U. Kazmaier, F. L. Zumpe, *Eur. J. Org. Chem.* **2001**, 4067.

128 T. D. Weiss, G. Helmchen, U. Kazmaier, *Chem. Commun.* **2002**, 1270.

129 For seminal work in this area, see: B. Kuebel, G. Hoefle, W. Steglich, *Angew. Chem. Int. Ed.* **1975**, *14*, 58.

130 U. Kazmaier, H. Mues, A. Krebs, *Chem. Eur. J.* **2002**, *8*, 1850, and references therein.

131 For early work, see: K. H. Shaughnessy, B. C. Hamman, J. F. Hartwig, *J. Org. Chem.* **1996**, *63*, 6546.

132 J. Cossy, A. de Filippis, D. G. Pardo, *Org. Lett.* **2003**, *5*, 3037.

133 For a recent review, see: B. M. Trost, C. B. Lee in *Catalytic Asymmetric Synthesis*, 2nd edn. (Ed.: I. Ojima), Wiley-VCH, New York, **2000**, pp 593.

134 K. Fuji, N. Kinoshita, K. Tanaka, *Chem. Commun.* **1999**, 1895.

135 T. H. Lambert, D. W. C. MacMillan, *J. Am. Chem. Soc.* **2002**, *124*, 13646.

136 (a) E. Lorthiois, I. Marek, J.-F. Normant, *J. Org. Chem.* **1998**, *63*, 566; (b) E. Lorthiois, I. Marek, J.-F. Normant, *J. Org. Chem.* **1998**, *63*, 2442.

137 (a) D. L. Comins, J. T. Kuethe, H. Hong, F. J. Lakner, *J. Am. Chem. Soc.* **1999**, *121*, 2651; (b) J. T. Kuethe, D. L. Comins, *Org. Lett.* **2000**, *2*, 855.

3.8
The Conjugate Addition Reaction

A. Alexakis

3.8.1
Introduction

Conjugate addition is among the basic carbon-carbon bond-forming reactions [1]. An organometallic reagent is required with, most often, a transition metal in stoichiometric or catalytic amount. The substrate is usually a double (or triple) bond to which an electron-withdrawing group (most often a carbonyl group) is attached.

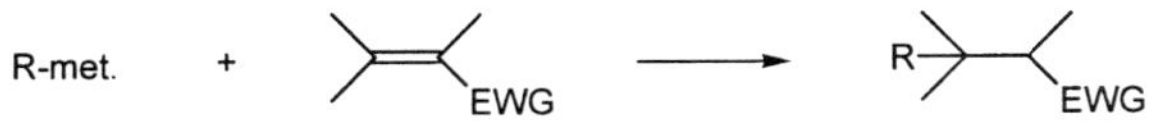

R-met. + >=<EWG ⟶ R->-<EWG

Traditionally, organocopper derivatives were the reagents of choice to undergo this synthetic transformation. Several books and review articles deal with the various aspects of the reaction: reactivity, stereochemistry, and mechanism [2]. This last aspect is still somewhat controversial, as is the exact nature of the reagent itself [3].

This chapter deals with some of the most recent developments placed in the context of reactivity and, particularly, enantioselectivity.

3.8.2
General Aspects of Reactivity

Historically, the conjugate addition reaction was first performed with Grignard reagents and catalytic amounts of copper(I) salts [4]. Later developments used lithium diorganocuprate reagents, which were more reliable and of much wider applicability, since the parent organolithium reagent [5] could be generated in several ways. Nowadays, many other organometallics may also be used, such as organozinc, organotin, oragnoaluminum, organoboron, organozirconium, organosamarium, etc. [6]. By varying the amount of copper(I) salt, several types of reagent can be made available.

3.8 The Conjugate Addition Reaction

$$\text{R-met.} + \text{cat.} \quad \text{CuX} \longrightarrow \text{R-met (cat. CuX)} \quad (1)$$

$$\text{R-met.} + 1 \text{ equiv. CuX} \longrightarrow \text{R-Cu} \quad (2)$$

$$2 \text{ equiv. R-met.} + 1 \text{ equiv. CuX} \longrightarrow \text{R}_2\text{Cu, met} \quad (3)$$

$$3 \text{ equiv. R-met.} + 1 \text{ equiv. CuX} \longrightarrow \text{R}_3\text{Cu, met} \quad (4)$$

Stoichiometric organocopper reagents (Eq. (2)) are usually poorly soluble and of low reactivity. The 2:1 reagent (Eq. (3)), usually called cuprate reagent or Gilman reagent, is the most commonly applied, whereas the 3:1 reagent (Eq.(4)), called higher order cuprate [7], is of increased reactivity but also of increased basicity.

As far the substrate is concerned, there is a notable difference according to the nature of the electron-withdrawing group and the substitution pattern of the double bond. Substitution in the β-position strongly slows down the reaction rate. When the electron-withdrawing group has an ester or amide functionality, the reaction is also severely slowed down. Ketones represent the most common carbonyl group. As for aldehydes, the main problem lies in the competition between the conjugate addition (1,4 addition) and the direct carbonyl attack (1,2 addition).

Among the different classes of R group to be transferred, there are some general trends observed whatever the stoichiometry of the Cu(I) salt. Alkynyl groups react by direct carbonyl attack (1,2 addition). Alkenyl and aryl groups are normally transferred in conjugate fashion, although at a somewhat slower rate than alkyl groups. The main problem is the low thermal stability of the alkenyl copper reagent (in contrast, aryl copper reagents are extremely stable). In the class of alkyl groups, the methyl group (which lacks β hydrogens) is of exceptional stability but very low reactivity.

In general, organometallic reagents based on metals other than copper are of more moderate reactivity. Their use is mainly restricted to chemo- and stereoselective processes and will be discussed below.

In order to increase the reactivity for conjugate addition, Yamamoto [8] studied the compatibility of stoichiometric organocopper reagents (considered as soft) with strong (and hard) Lewis acids. He observed that the combination RCu/BF$_3$ is a suitable alternative to R$_2$CuLi for the transfer of the methyl group to sterically hindered enones.

Me-Cu + [alkene with =O----BF$_3$] ⟶ Me–[C]–CO-R

This RCu/BF$_3$ combination was even more successful in other typical organocopper reactions, such as S$_N$2' substitutions, cleavage of epoxides [9] or asymmetric cleavage of chiral acetals [10]. In all these reactions, it is believed that the activation mainly arises from coordination of the Lewis acid to the oxygen atom of the carbonyl group.

3.8.2 General Aspects of Reactivity | 555

Organomanganese reagents, catalyzed by copper (I) salts, were also reported to be of increased reactivity toward sterically hindered enones, although the exact role of manganese was not clarified [11].

Several groups reported independently the beneficial effect of added Me$_3$SiCl (TMSCl) to the reaction mixture prior to the addition of the enone [12]. Stoichiometric organocopper reagents are totally unreactive toward Me$_3$SiCl, and even Me$_3$SiI! [13]. Organocuprate reagents are also compatible with Me$_3$SiCl, at least below $-50\,°C$ in THF. Even Grignard reagents may be used with catalytic copper (I) salts [12], provided HMPA is added as co-solvent. The dramatic rate enhancement of the conjugate addition and the experimental simplicity have made this method very popular. However, care should be taken in stereoselective processes, as cases of total inversion of selectivity have been reported [12a, 14].

R$_2$CuLi + [alkene with CO-R'] + Me$_3$SiCl ⟶ R-[product with CO-R']

In general, in the presence of Me$_3$SiCl, α,β-ethylenic aldehydes react cleanly in conjugate fashion, with negligible amounts of the competing direct carbonyl attack [15]. Amides and esters [12b], as well as sterically hindered enones, react at a convenient rate. Me$_3$SiCl is also needed for the conjugate addition of organozinc-copper reagents, which may bear additional functionalities [16].

The use of Me$_3$SiI (TMSI) with copper acetylides allows this class of R groups to be transferred in conjugate fashion [17].

R-C≡C-Cu + [alkene with CO-R'] + Me$_3$SiI ⟶ R-≡-[product with CO-R']

Previously, other organometallic combinations such as (R=alkynyl) RAlMe$_2$/Ni cat. [18], RAl(Li)Me$_3$/TMSOTf [19], RZnBr/TMSOTf [20] and RB(9-BBN)/BF$_3$ [21] have been known to undergo alkynylations.

The exact role of TMSCl in accelerating the conjugate addition is still unclear. It was first believed that TMSCl traps the Cu(III) intermediate as a silyl enol ether, thus increasing the rate of the reductive elimination step [12a]. However, recent kinetic studies revealed that conjugate addition sometimes occurs prior to the silylation of the resulting enolate [22]. It was also reported that TMSCl may act as a Lewis acid [23], although it does not accelerate typical Lewis-assisted reactions such as epoxide opening or acetal cleavage. Whatever the exact mechanism, the synthetic result seems wider in scope, since a recent report reveals rate enhancement in the conjugate addition of stabilized organolithium reagents [24].

3.8.3
Enantioselectivity

Most often, the conjugate addition results in the creation of a new stereogenic center. This point had already been noticed in the late 1950s, and there are several reports dealing with this, including a comprehensive review up to 1991 [25a]. More recently, decisive advances have been made in the copper- [26] and rhodium-catalyzed [27] versions of the reaction, resulting in an explosive growth of articles on the subject.

There are several ways to tackle this problem of asymmetry:

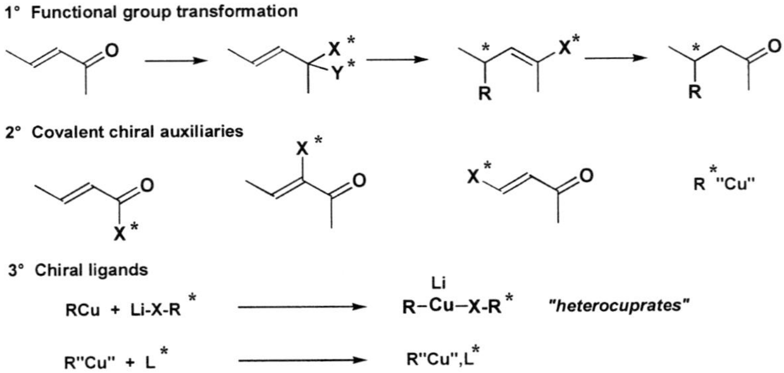

1° Functional group transformation

2° Covalent chiral auxiliaries

3° Chiral ligands

$$RCu + Li\text{-}X\text{-}R^* \longrightarrow R\text{-}Cu\text{-}X\text{-}R^* \quad \text{"heterocuprates"}$$

$$R"Cu" + L^* \longrightarrow R"Cu",L^*$$

One approach is to transform the sp^2 carbon of the carbonyl into another functionality with an sp^3 carbon. The reaction is now a γ allylic substitution. Chiral acetals are the typical example of this class [28], resulting in the formal asymmetric conjugate addition of organolithiums to α,β-ethylenic aldehydes.

Very successful results were obtained by the covalent chiral auxiliary approach, which allows the purification of the major diastereomer, thus affording ultimately a pure enantiomer. The chiral covalent auxiliary may be attached at different positions of the substrate. There are several examples in the literature describing such an approach:

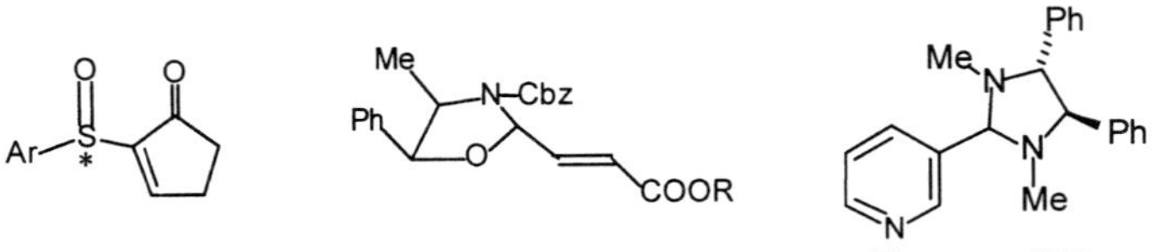

Posner [29]   Scolastico [30]   Mangeney [31]

Chiral esters or amides of various chiral alcohols or amines allow a very large array of chiral auxiliaries [25]. Several of the most popular ones take advantage of the camphor framework. Although the reactivity of such substrates is usually rather low, good chemical yields may be obtained using the various ways to increase the reactivity. Some representative examples are listed below:

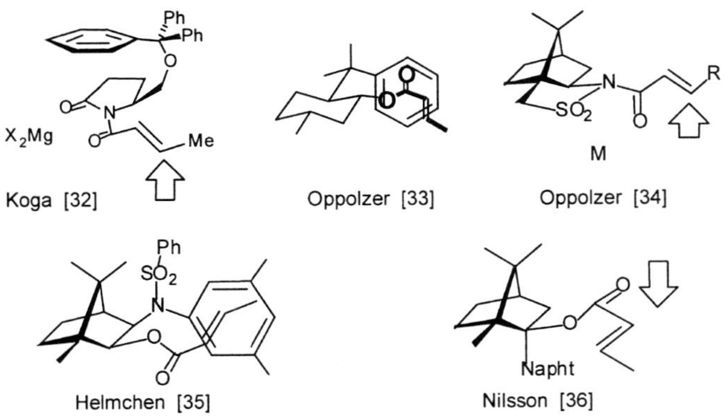

Koga [32] Oppolzer [33] Oppolzer [34]

Helmchen [35] Nilsson [36]

The third approach uses an external chiral moiety. Early studies dealt with heterocuprates, where the chirality is brought about by a chiral alcoholate, thiolate, or amide [25, 37]. However, success came from the external ligand approach, the best one to allow a catalytic use of transition metal and ligand. Cu, Ni, and Rh are the most appropriate metals.

Historically, Cu was the metal of choice. Early reports with (–)-sparteine [38] or chiral solvents [39] and lithium diorganocuprate or triorganozincates gave disappointingly low *ee*s. The first successful example was described by Leyendecker et al. with a ligand derived from hydroxyproline [40]. More recently, Alexakis [41] and Tomioka [42] introduced a new class of ligands based on the concept of chiral trivalent phosphorus derivatives. These ligands were efficient in stoichiometric amounts or with very high catalyst loading.

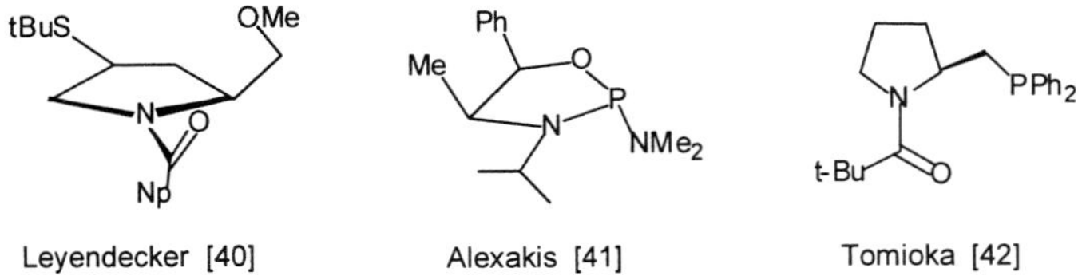

Leyendecker [40] Alexakis [41] Tomioka [42]

The breakthrough in catalysis came with the replacement of organolithium or Grignard reagent, as primary organometallics, by diorganozinc reagents. Although some organozinc species undergo conjugate addition [43], Ni [44] or Cu [45] catalysis is helpful. An even stronger acceleration is found when Cu salts are coordinated by a ligand [46]. The combination of R_2Zn + Cu salt + phosphorus ligand is presently the most widely used. A recent review [26g] covering this precise topic shows that, in just 4–5 years, over 350 chiral ligands have been developed, just for the asymmetric conjugate addition of diethyl zinc to cyclohexenone!

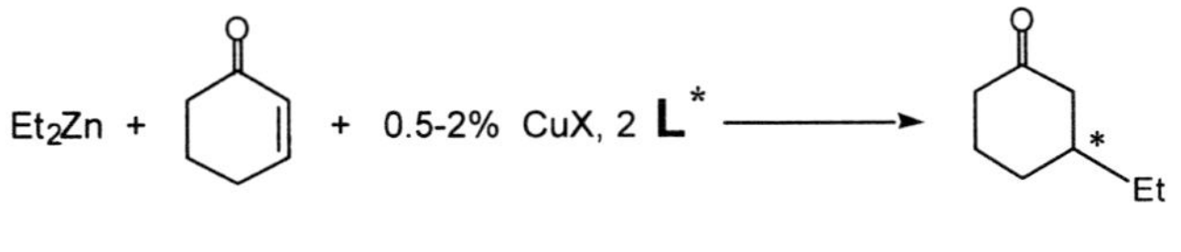

3.8 The Conjugate Addition Reaction

Et$_2$Zn + [cyclohexenone] + 0.5–2% CuX, 2 L* ⟶ [3-ethylcyclohexanone]*

This reaction is wide in scope, allowing very high (>99%) enantioselectivities for a range of substrates and alkyl groups. Even functionalized diorganozincs can be added. Representative Michael acceptors are shown below:

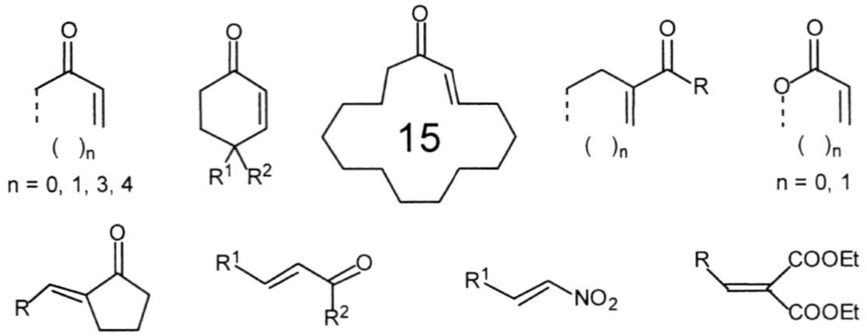

The experimental conditions for all these substrates are rather similar: CuOTf$_2$ or Cu-thiophene carboxylate or CuOAc$_2$ or CuBF$_4$[CH$_3$CN]$_4$ as Cu source; toluene or Et$_2$O as solvent. The Cu:chiral ligand ratio is usually 2:1. Several structural types of chiral ligands are now known, most of them having a phosphorus atom (see [26g] for the complete list of ligands). Representative efficient ligands are shown below:

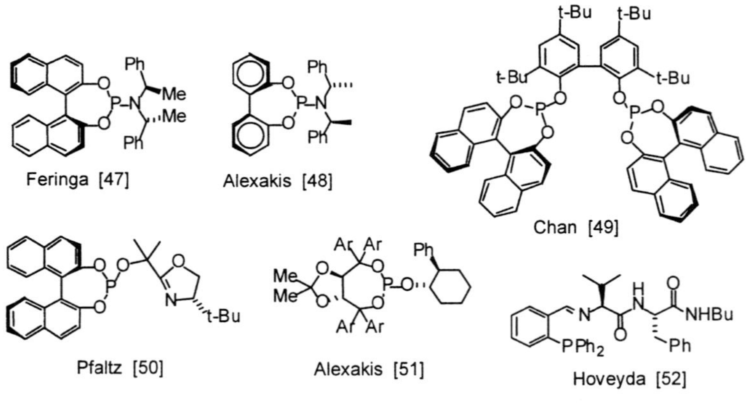

Feringa [47]   Alexakis [48]   Chan [49]

Pfaltz [50]   Alexakis [51]   Hoveyda [52]

More recently, trialkyl aluminum compounds were also realized to be good candidates for the Cu-catalyzed conjugate addition [53]. All these reactions lead ultimately to a zinc (or Al) enolate, which may be trapped by electrophiles other than simple water. Aldehydes react readily [54], but not ketones, to afford the aldol product. Acetals, chiral and achiral, need the assistance of a strong Lewis acid such as $BF_3 \cdot Et_2O$ [55]. Alkylation is feasible with HMPA and excess alkylating agent [52], as well as allylation with Pd catalysis and allyl acetate [47, 54]. Silylation of the enolate allows the very versatile silyl enol ethers to be obtained for further elaboration [56, 57].

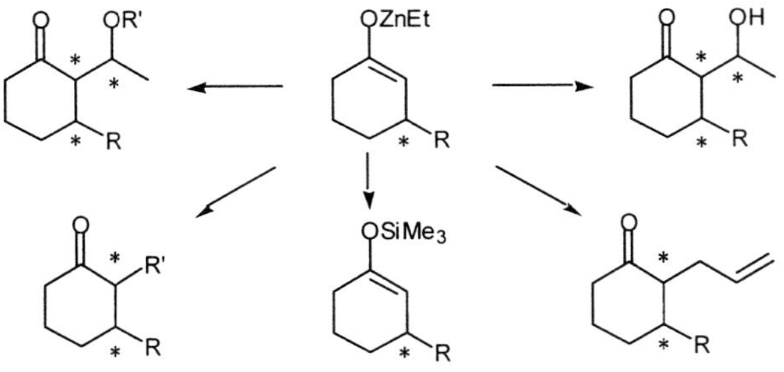

Nickel catalysis with R_2Zn is also known, but limited in scope. Only chalcone-type enones (Ar-CH=CH-CO-Ar′) react, giving some successful enantioselectivities. Representative examples of ligands are shown below:

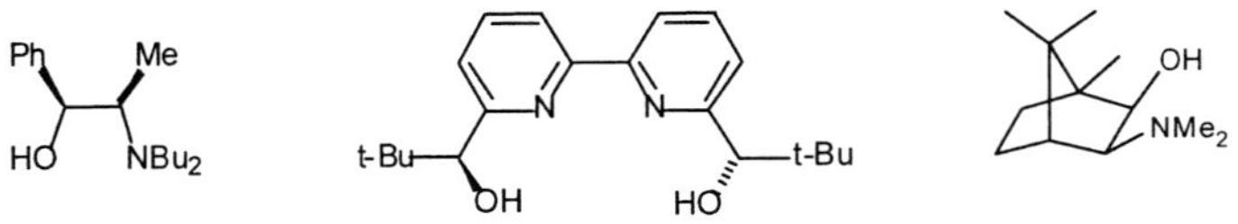

Soai [58a]   Bolm [58b]   Feringa [58c]

In contrast to Cu and Ni, Rh catalysis uses $ArB(OH)_2$ as primary organometallics. Therefore, for the introduction of aryl and alkenyl groups it is more appropriate to shift to Rh catalysis [27, 58]. Since the first report by Miyaura [59], Hayashi et al. have developed the asymmetric version very successfully (ees >95%) both on cyclic and acyclic enones. Other Michael acceptors are also possible, as for copper

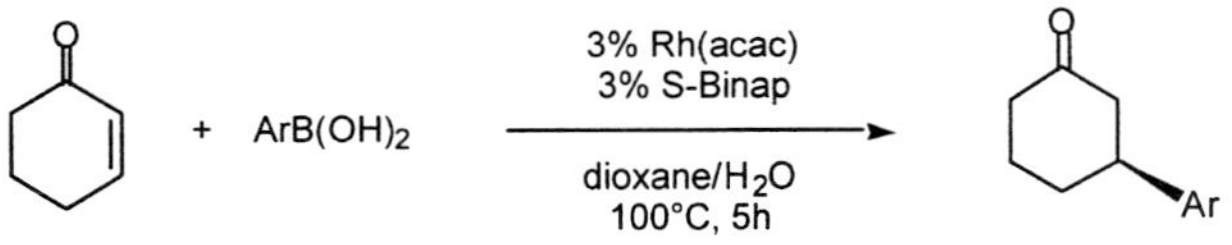

Alkenyl boranes, resulting from hydroboration of alkynes, can also be transferred. An alternative to boronic acids is potassium tetrafluoroborate [60]. The chiral ligand is a diphosphine [27] or a diphosphite [61] or even a monodentate phosphorus ligand [62].

Hayashi [27] Reetz [61] Tomioka [62a] Feringa [62b]

Finally, by analogy to zinc enolates, boron (or titanium) enolates have been trapped by silylation [63], allylation [64], or aldol condensation [65].

3.8.4
References and Notes

1 (a) P. PERLMUTTER, *Conjugate Addition Reaction in Organic Synthesis*, Tetrahedron Organic Chemistry Series, No. 9, Pergamon Press, Oxford, **1992**; (b) Y. YAMAMOTO, *Methods in Organic Chemistry (Houben-Weyl)*, **1995**, vol. 4 (Stereoselective Synthesis), 2041–2057.

2 Books: (a) I. IBUKA *Organocopper Reagents in Organic Synthesis*, Camelia and Rose Press, Osaka, **2000**. (b) N. KRAUSE *Modern Organocopper Chemistry*, Wiley-VCH, Weinheim, **2002**. Reviews: (c) G. H. POSNER *Org. React.* **1972**, *19*, 1–113. (d) R. J. K. TAYLOR *Synthesis* **1985**, 364–392. (e) J. A. Kozlowski in *Comprehensive Organic Synthesis* (Eds.: B. M. TROST, I. FLEMING), Vol. 4 p. 169–198, Pergamon Press, Oxford, **1991**. (f) B. H. LIPSHUTZ, S. SENGUPTA *Org. React.* **1992**, *41*, 135–631. (g) N. KRAUSE *Angew. Chem. Int. Ed. Engl.* **1997**, *36*, 187–204.

3 (a) E. NAKAMURA, S. MORI *Angew. Chem. Int. Ed. Engl.* **2000**, *39*, 3751–3771. (b) S. WOODWARD *Chem. Soc. Rev.* **2000**, *29*, 393–401.

4 M. S. KHARASCH, P. O. TAWNEY *J. Am. Chem. Soc.* **1941**, *63*, 2308.

5 H. GILMAN, R. G. JONES, L. A. WOODS *J. Org. Chem.* **1952**, *17*, 1630.

6 P. WIPF *Synthesis* **1993**, 537–557.

7 B. H. LIPSHUTZ, R. S. WILHELM, J. A. Kozlowski *Tetrahedron* **1984**, *40*, 5005–5038.

8 Y. YAMAMOTO *Angew. Chem. Int. Ed. Engl.* **1986**, *25*, 947–959.

9 A. ALEXAKIS, D. JACHIET, J. F. NORMANT *Tetrahedron* **1986**, *42*, 5607–5619.

10 A. ALEXAKIS, P. MANGENEY *Tetrahedron Asymm.* **1990**, *1*, 477–511.

11 G. CAHIEZ, M. ALAMI *Tetrahedron Lett.* **1989**, *30*, 3541–3544.

12 (a) E. J. COREY, N. W. BOAZ *Tetrahedron Lett.* **1985**, *26*, 6015–6018 and 6019–6022. (b) A. ALEXAKIS, J. BERLAN, Y. BESACE *Tetrahedron Lett.* **1986**, *27*, 1047–1050. (c) E. NAKAMURA, S. MATSUZAWA, Y. HORIGUCHI, I. KUWAJIMA *Tetrahedron Lett.* **1986**, *27*, 4029–4032.

13 I. M. BERGDAHL, E.-L. LINDSTEDT, M. NILSSON, M. OLSSON *Tetrahedron* **1989**, *45*, 535.

14 A. ALEXAKIS, R. SEDRANI, P. MANGENEY *Tetrahedron Lett.* **1990**, *31*, 345–348.

15 C. CHUIT, J. P. FOULON, J. F. NORMANT *Tetrahedron* **1980**, *36*, 2305.

16 P. KNOCHEL, R. SINGER *Chem. Rev.* **1993**, *93*, 2117–2188.

17 M. ERIKSSON, T. ILIEFSKI, M. NILSSON, T. OLSSON *J. Org. Chem.* **1997**, *62*, 182–187.

18 J. SCHWARTZ, D. B. CARR, R. T. HANSEN, F. M. DAYRIT *J. Org. Chem.* **1980**, *45*, 3053.

19 S. KIM, J. H. PARK *Synlett* **1995**, 163.

20 S. KIM, J. M. LEE *Tetrahedron Lett.* **1990**, *31*, 7627.

21 H. Fujishima, E. Tanaka, S. Hara, A. Suzuki *Chem. Lett.* **1992**, 695.

22 (a) S. H. Bertz, G. Miao, B. E. Rossiter, J. P. Snyder *J. Am. Chem. Soc.* **1995**, *117*, 11023–11024. (b) M. Eriksson, A. Johansson, M. Nilsson, T. Olsson *J. Am. Chem. Soc* **1996**, *118*, 10904–10905. (c) D. E. Frantz, D. A. Singleton, J. P. Snyder *J. Am. Chem. Soc.* **1997**, *119*, 3383–3384. (d) *J. Am. Chem. Soc.* **2000**, *122*, 3288–3295.

23 Y. Horiguchi, M. Komatsu, I. Kuwajima *Tetrahedron Lett.* **1989**, *30*, 7087–7090.

24 H. Liu, T. Cohen *Tetrahedron Lett.* **1993**, *36*, 8925–8928.

25 (a) B. E. Rossiter, N. M. Swingle *Chem. Rev.* **1992**, *92*, 771. (b) A. Alexakis in *Organocopper Reagents, a Practical Approach* (Ed.: R. J. K. Taylor), Chapt. 8, pp. 159–183, Oxford University Press, **1994**

26 (a) R. Noyori, *Asymmetric Catalysis in Organic Synthesis*, Wiley, New York, **1994**. (b) I. Ojima *Catalytic Asymmetric Synthesis*, Wiley-VCH, Weinheim, **1993**. (c) A. Alexakis in *Transition Metal Catalysed Reactions* (Eds.: S.-I. Murahashi, S. G. Davies); IUPAC Blackwell Science, Oxford **1999**; p 303. (d) K. Tomioka, Y. Nagaoka in *Comprehensive Asymmetric Catalysis* (Eds.: E. N. Jacobsen, A. Pfaltz, H. Yamamoto), Springer, New York **2000**, p 1105. (e) M. P. Sibi, S. Manyem *Tetrahedron* **2000**, *56*, 8033–8061. (f) N. Krause, A. Hoffmann-Röder *Synthesis* **2001**, 171–196. (g) A. Alexakis, C. Benhaim *Eur. J. Org. Chem.* **2002**, 3221–3236.

27 T. Hayashi *Synlett* **2001**, 879–887.

28 (a) A. Alexakis, P. Mangeney, A. Ghribi, I. Marek, R. Sedrani, C. Guir, J. F. Normant *Pure Appl. Chem.* **1988**, *60*, 49–56. (b) H. Rakotoarisoa, R. Guttierez Perez, P. Mangeney, A. Alexakis *Organometallics* **1996**, *15*,1957–1602.

29 G. H. Posner *Acc. Chem. Res.* **1987**, *20*, 72–78.

30 C. Scolastico *Pure Appl. Chem.* **1988**, *60*, 1689–1698.

31 P. Mangeney, R. Gosmini, S. Raussou, M. Commerçon, A. Alexakis *J. Org. Chem.* **1994**, *59*, 1877–1888.

32 K. Tomioka, T. Suenaga, K. Koga, *Tetrahedron Lett.* **1986**, *27*, 369–372.

33 W. Oppolzer, H. Löher *Helv. Chim. Acta* **1981**, *64*, 2808–2811.

34 (a) W. Oppolzer, G. Poli, A. J. Kingma, C. Starkemann, G. Bernardinelli *Helv. Chim. Acta*, **1987**, *70*, 2201–2214. (b) W. Oppolzer, R. J. Mills, W. Pachinger, T. Stevenson *Helv. Chim. Acta* **1986**, *63*, 1542–1545.

35 G. Helmchen, G. Wegner *Tetrahedron Lett.* **1985**, *26*, 6051–6054.

36 M. Bergdahl, M. Nilsson, T. Olsson, K. Stern *Tetrahedron* **1991**, *47*, 9691–9702.

37 A. Alexakis in *Transition Metals for Organic Synthesis*, ed. M. Beller, C. Bolm, Wiley-VCH, **1998**, Vol. II, pp. 504–513.

38 R. A. Kretchmer *J. Org. Chem.* **1972**, *37*, 2744–2747.

39 W. Langer, D. Seebach *Helv. Chim. Acta* **1979**, *62*, 1710–1722.

40 F. Leyendecker, D. Lancher *New J. Chem.* **1985**, *9*, 13–19.

41 (a) A. Alexakis, S. Mutti, J. F. Normant *J. Am. Chem. Soc.* **1991**, *113*, 6332–6334. (b) A. Alexakis, J. C. Frutos, P. Mangeney *Tetrahedron: Asymmetry* **1993**, *4*, 2427–2430.

42 (a) M. Kanai, K. Koga, K. Tomioka *Tetrahedron Lett.* **1992**, *33*, 7193–7196. (b) M. Kanai, K. Tomioka *Tetrahedron Lett.* **1995**, *36*, 4273–4274. (c) M. Kanai, K. Tomioka *Tetrahedron Lett.* **1992**, *33*, 4275–4278.

43 (a) M. Suzuki, A. Yanagisawa, R. Noyori *J. Am. Chem. Soc.* **1988**, *110*, 4718–4726. (b) J. F. G. A. Jansen, B. L. Feringa *J. Org. Chem.* **1990**, *55*, 4168–4175. (c) C. K. Reddy, A. Devasagayaraj, P. Knochel *Tetrahedron Lett.* **1996**, *37*, 4495–4498.

44 A. E. Greene, J. P. Lansard, J. L. Luche, C. Petrier *J. Org. Chem.* **1984**, *49*, 931–932.

45 E. Nakamura, S. Aoki, K. Sekiya, H. Oshino, I. Kuwajima *J. Am. Chem. Soc.* **1987**, *109*, 8056–8066.

46 (a) M. Kitamura, T. Miki, K. Nakano, R. Noyori *Tetrahedron Lett.* **1996**, *37*, 5141–5144. (b) A. H. M. de Vries, A. Meetsma, B. L. Feringa *Angew. Chem. Int. Ed. Engl.* **1996**, *35*, 2374–2376. (c) A. Alexakis, J. Vastra, P. Mangeney *Tetrahedron Lett.* **1997**, *38*, 7745–7748.

3.8 The Conjugate Addition Reaction

47   B. L. Feringa *Acc. Chem. Res.* **2000**, *33*, 346–353.

48   (a) A. Alexakis, S. Rosset, J. Allamand, S. March, F. Guillen, C. Benham *Synlett* **2001**, 1375-1378. (c) A. Alexakis, C. Benham, S. Rosset, M. Humam *J. Am. Chem. Soc.* **2002**, *124*, 5262–5263.

49   M. Yan, Z. Y. Zhou, A. S. C. Chan *Chem. Commun.* **2000**, *2*, 115–117.

50   I. H. Escher, A. Pfaltz *Tetrahedron* **2000**, *56*, 2879–2888.

51   A. Alexakis, J. Vastra, J. Burton, C. Benhaim, X. Fournioux, A. Van den Heuvel, J.-M. Levoêque, F. Mazé, S. Rosset *Eur. J. Org. Chem.* **2000**, 4011–4027.

52   (a) S. J. Degrado, H. Mizutani, A. H. Hoveyda *J. Am. Chem. Soc.* **2001**, *123*, 755–756. (b) H. Mizutani, S. J. Degrado, A. H. Hoveyda *J. Am. Chem. Soc.* **2002**, *124*, 779–780.

53   (a) Y. Takemoto, S. Kuraoka, N. Hamaue, C. Iwata *Tetrahedron: Asymmetry*, **1996**, *4*, 993–996. (b) L. Liang, A. S. C. Chan *Tetrahedron: Asymmetry* **2002**, *13*, 1393–1396. (c) P. K. Fraser, S. Woodward *Chem. Eur. J.* **2003**, *9*, 776–783.

54   (a) M. Kitamura, T. Miki, K. Nakano, R. Noyori *Tetrahedron Lett.* **1996**, *37*, 5141–5144. (b) *Bull. Chem. Soc. Jpn.* **2000**, *73*, 999–1014.

55   A. Alexakis, G. P. Trevitt, G. Bernardinelli *J. Am. Chem. Soc.* **2001**, *123*, 4358–4359.

56   O. Knopff, A. Alexakis *Org. Lett.* **2002**, *4*, 3835–3837.

57   A. Alexakis, S. March *J. Org. Chem.* **2002**, *67*, 8753–8757.

58   K. Fagnou, M. Lautens *Chem. Rev.* **2003**, *103*, 169–196.

59   (a) K. Soai, T. Hayasaka, S. Ugajin, S. Yokoyama *Chem. Lett.* **1988**, 1571–1572. (b) C. Bolm, M. Ewald *Tetrahedron Lett.* **1990**, *31*, 5011–5012. (c) J. F. G. A. Jansen, B. L. Feringa *Tetrahedron: Asymmetry* **1992**, *3*, 581–582.

60   M. Pucheault, S. Darses, J.-P. Gent *Eur. J. Org. Chem.* **2002**, 3552–3557.

61   M. T. Reetz, D. Moulin, A. Gosberg *Org. Lett.* **2001**, *3*, 4083–4085.

62   (a) M. Kuriyama, K. Nagai, K.-I. Yamada, Y. Miwa, T. Taga, K. Tomioka *J. Am. Chem. Soc.* **2002**, *124*, 8932–8939. (b) J.-G. Boiteau, R. Imbos, A. J. Minaard, B. L. Feringa *Org. Lett.* **2003**, *5*, 681–684.

63   T. Hayashi, N. Tokunaga, K. Yoshida, J. W. Han *J. Am. Chem. Soc.* **2002**, *124*, 12102–12103.

64   K. Yoshida, M. Ogasawara, T. Hayashi *J. Org. Chem.* **2003**, *68*, 1901–1905.

65   D. F. Cauble, J. D. Gipson, M. J. Krische *J. Am. Chem. Soc.* **2003**, *125*, 1110–1111.

3.9
Carbometalation Reactions of Zinc Enolate Derivatives

Daniella Banon-Tenne and Ilan Marek

3.9.1
Introduction

The stereoselective addition of carbon functionalities to unactivated alkenes and alkynes is a significant challenge in organic synthesis since the pioneering work of Bähr and Ziegler in 1927 [1]. Over the last 75 years, but more particulary during the last two decades, an impressive number of reactions have been developed for the intra- and inter-molecular addition of diverse non-stabilized organometallics to a large variety of alkynes, alkenes and allenes. Most of these results were summarized in several reviews and chapters [2]. More recently developed and found to be potentially very interesting is the carbometalation reaction of stabilized zinc organometallics across unactivated unsaturated systems. Therefore, the emphasis of this chapter is placed on this particular type of strategy, but only reactions that possess an alkenyl or alkyl metal species after the carbometalation step, will be reviewed.

3.9.2
Intramolecular Carbometalation

The first intramolecular carbometalation reaction of metalated enolate on unfunctionalized or non-strained double bond was reported only in 1997 [3]. *N*-Methyl-*N*(but-3-enyl)glycinate methyl ester (**1**) was cleanly metalated by treatment with 1.5 equiv. of LDA in Et$_2$O at –40 °C, but after two hours of stirring at room temperature, no cyclization of the corresponding lithium-amino-enolate **1Li** was observed (or of the magnesium-amino-enolate). However, addition of 1.5 equiv. of zinc salt to **1Li** led to the amino-zinc-enolate **1Zn**, resulting in a virtually quantitative 5-*exo-trig* cyclization reaction after 1 h at room temperature to give the cyclic product **2Zn** (Scheme 1).

Hydrolysis of the reaction mixture afforded **2** in 70% isolated yield as a single *cis* diastereomer. The formation of a new functionalized organometallic species was checked by iodinolysis and by reaction with allyl bromide, after transmetalation of the resulting organozinc bromide to an organocopper reagent (Scheme 2) [4].

Transition Metals for Organic Synthesis, Vol. 1, 2nd Edition.
Edited by M. Beller and C. Bolm
Copyright © 2004 WILEY-VCH Verlag GmbH & Co. KGaA, Weinheim
ISBN: 3-527-30613-7

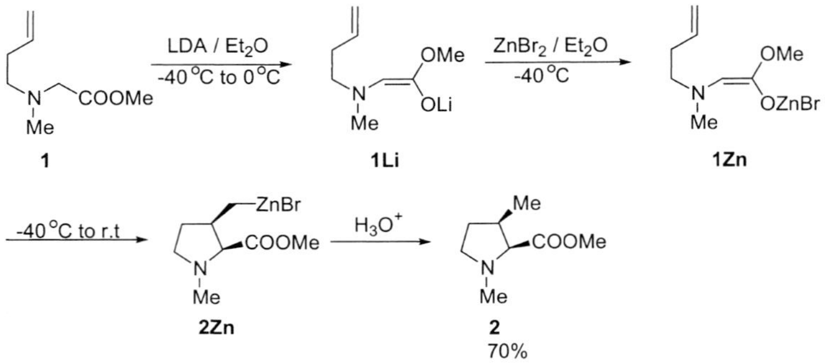

Scheme 1 First carbocyclization of amino-zinc enolate.

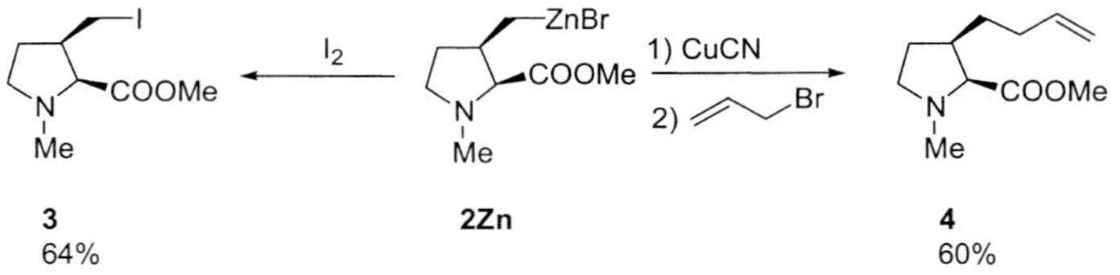

Scheme 2 Functionalization of the organometallic derivatives.

As the Z-configuration of the zinc enolate is imposed by an intramolecular Zn-N chelation [5], the *cis* relative configuration of **2Zn** was attributed to a chair-like transition state, in which the Z-α-amino-zinc-enolate was in the plane parallel to that of the olefinic residue (Scheme 3) [6].

By this simple strategy, several tri- and tetra-substituted pyrrolidines were easily prepared, and the diastereoselectivity of the carbocyclization was studied in detail [7].

When the N-(R)-1-(phenylethyl)-N-(but-3-enyl)glutamate methyl ester **5** was submitted to this metalation-transmetalation cyclization sequence, the chiral cyclic organozinc bromide was diastereoselectively formed, and, after hydrolysis, the chiral β-methyl proline derivative **6** was obtained as a single *cis* diastereomer with a 98:2 diastereomeric ratio and in 93% yield (Scheme 4).

After hydrogenolysis, the secondary amine **7** was obtained in 96% *ee* [3b]. Interestingly, the enantioselectivity of this carbocyclization dropped to 50% when the

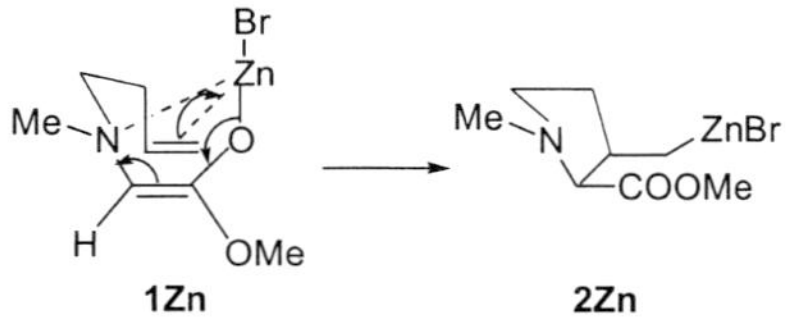

Scheme 3 Mechanistic interpretation for the *cis* stereospecificity.

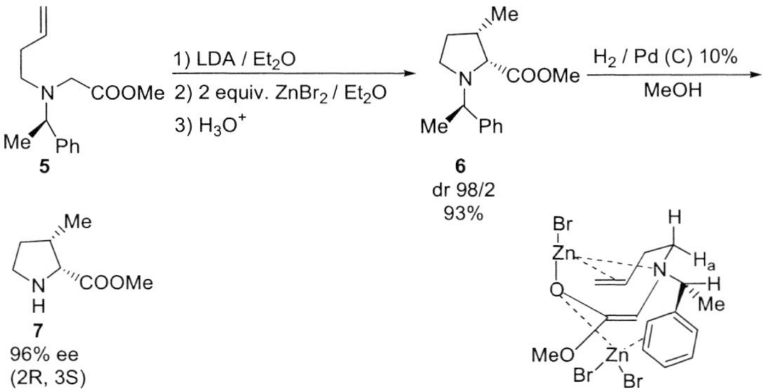

Scheme 4 Carbocyclization of N-(R)-1-phenylethyl)-N-(but-3-enyl)glutamate methyl.

reaction was performed with only 1 equiv. of zinc salt. Moreover, if the aromatic ring of the chiral inductor is replaced by a cyclohexyl ring, no diastereoselection is observed. In view of the above results, the authors have postulated a π chelation between the aromatic ring and the amino-zinc-enolate in the transition state [7]. Knowing that some π chelation between organozinc derivatives and unsaturated systems is well known in the literature [8], the excess of zinc salt, which is necessary for the high diastereoselection, should act as a tether between the aromatic ring and the amino-zinc-enolate, as described in Scheme 4. Therefore, the chiral inductor adopts a position in which the methyl group bound to the chiral center has a lowered eclipsing strain with the two hydrogens in the α-position, when one face of the carbon-carbon double bond is concerned rather than the other.

Several modified chiral amino acids used as probes in structure-activity relationship studies of biologically active peptides, such as 3-prolinomethionines [9], 3-prolinoglutamic acid [10], and 3-alkyl substituted prolines [11] were easily prepared by this methodology, as described in Scheme 5.

The amino-zinc-enolate carbocyclization has also been applied to solid-phase organic synthesis, allowing the preparation of libraries of 3-substituted proline derivatives [12].

When this metalation-transmetalation cyclization was tested on the analogous β-(N-allyl)-aminoester, a reverse addition (dropwise addition of the lithium enolate to an ethereal zinc bromide solution) led to a smooth carbocyclization reaction to give, after hydrolysis or reaction with different electrophiles, the corresponding substituted carbomethoxy-pyrrolidine in good yields (Scheme 6) [13].

Surprisingly, the stereoselectivity of the carbocyclization is now different from the cyclization of α-(N-homoallyl)-amino ester enolate previously described in Scheme 3. A reasonable explanation for the zinc-enolate cyclization of the β-amino-ester could involve a carbon-centered enolate, as for the simple Reformatsky reagent [14], and no longer an oxygen-centered one as described in the case of α-amino-ester. Therefore, the R group should adopt a *pseudo*-equatorial position and the carbomethoxy group a *pseudo*-axial position on the basis of steric hindrance

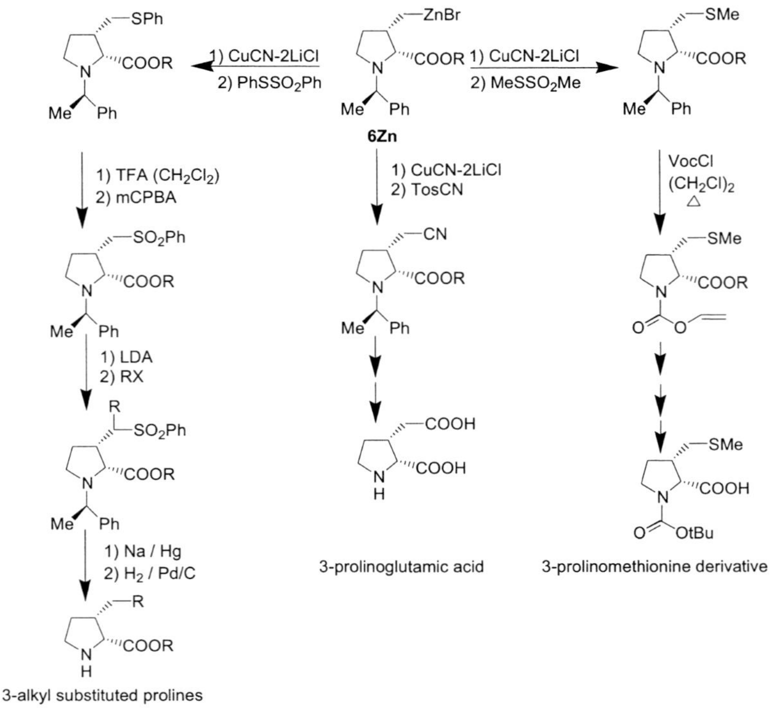

Scheme 5 Synthetic applications of the amino-zinc-enolate carbocyclization.

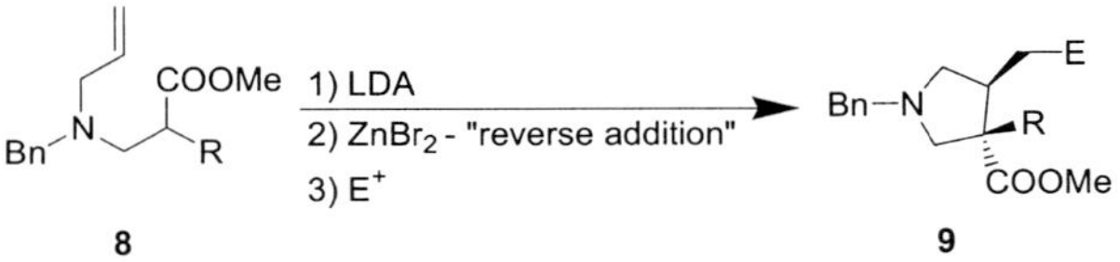

Scheme 6 Substituted carbomethoxy-pyrrolidine.

R = H	E = H	d.r = 87/13	50%
R = Me	E = H	d.r > 99/1	75%
R = Me	E = I$_2$	d.r > 99/1	68%
R = Me	E = PhS	d.r > 99/1	67%

and on the basis of a possible extra chelation with an external zinc species complexed to the nitrogen atom, as described in Scheme 7.

An alternative and elegant method for the preparation of substituted pyrrolidines was recently published and consist in a domino 1,4-addition-carbocyclization-functionalization reaction of different Michael acceptors, such as **10**, with mixed copper-zinc reagent or with a triorganozincate-zinc salt combination [15]. Indeed, the 1,4-addition of triorganozincate reagents to α,β-unsaturated esters **10** leads to the corresponding lithium zincate enolate **11**, which undergoes a subse-

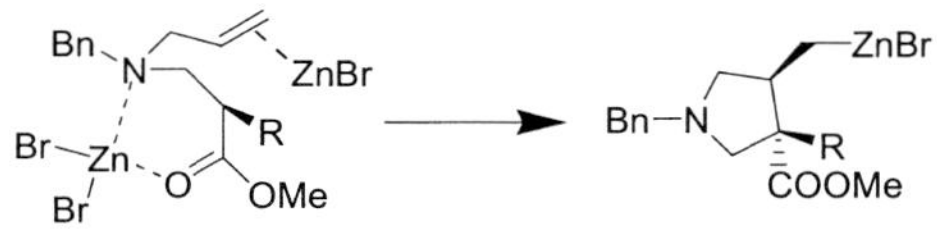

Scheme 7 Mechanistic interpretation for the preparation of carbomethoxy-pyrrolidine.

quent carbocyclization reaction by treatment with ZnBr$_2$ (3 equiv.) to give, after hydrolysis, the carbomethoxy-pyrrolidine **12** in 55% yield as a single diastereomer (Scheme 8).

Even more interesting, the reaction with nBuCu(CN)ZnBr·LiBr (prepared from nBuLi and a mixture of ZnBr$_2$ and CuCN in diethyl ether) gave smoothly the cyclic product in good yield with moderate to excellent stereoselectivities according to the nature of the R group (Scheme 9).

The diastereoselectivity can be improved either by using an excess of zinc salt during the preparation of the organocopper, BuCu(CN)ZnBr·LiBr (d.r 93/7), or by using aryl or vinylic organometallic reagents (such as PhCu(CN)ZnBr·LiBr). In all cases, the carbometalated product can be functionalized with various electrophiles [15]. An N,N-dimethyl hydrazone of an olefinic ketone also undergo the carbocyclization if deprotonated and transmetalated into BuZn(II) cation, as described in Scheme 10 [16].

However, the diastereoselectivity of this carbocyclization (cis/trans = 88/12) is lower than the diastereoselectivity of the intramolecular carbometalation of the zinc-enolate, described in Scheme 3 (d.r > 99/1).

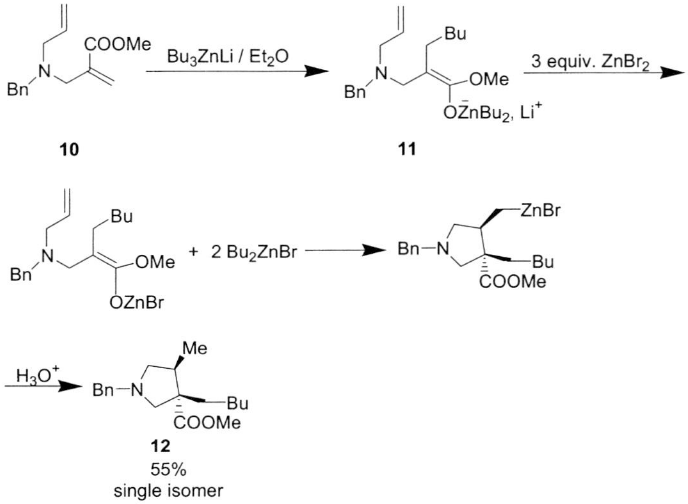

Scheme 8 Domino 1,4-addition-carbocyclization reaction.

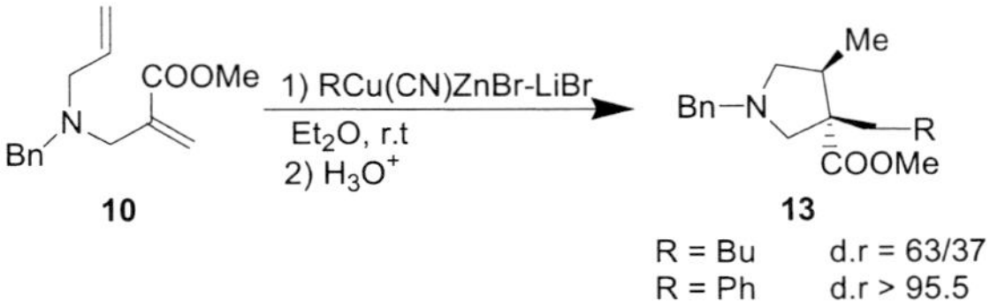

Scheme 9 Domino-1,4-addition-carbocyclization with cuproorganozinc derivatives.

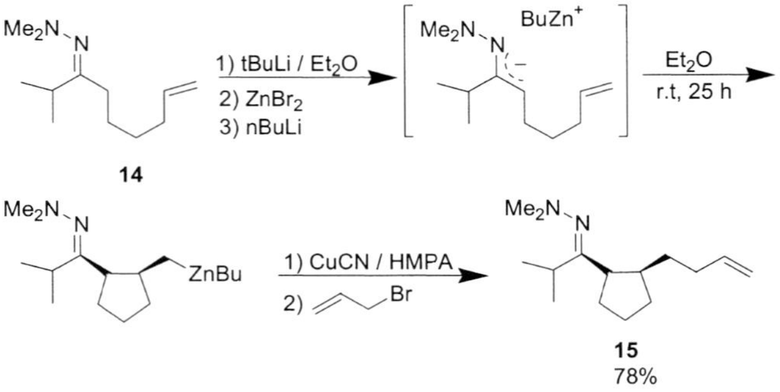

Scheme 10 Intramolecular carbocyclization of zincated hydrazone.

6-*Exo-trig* cyclization of a 6-heptenyl metal to a (cyclohexylmethyl) metal is usually much slower than the analogous 5-*exo-trig* cyclization, and therefore fewer examples of this transformation are known [2]. However, the cyclization reaction of the aminozinc-enolate has been successfully applied for the formation of 6-membered rings, as a new route to polysubstituted piperidine derivatives (Scheme 11) [17].

After metalation in Et$_2$O and transmetalation with zinc bromide, the corresponding Z-amino-zinc-enolate cyclizes at room temperature to give the metalated piperidines **16Zn**. Hydrolysis, iodinolysis, or allylation of the reaction mixture

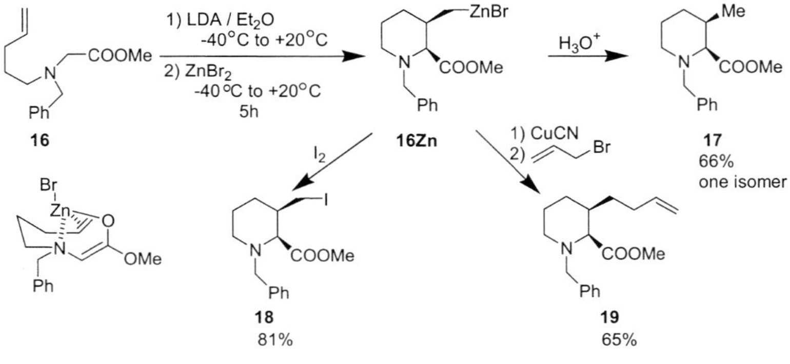

Scheme 11 Formation of functionalized piperidine via 6-*exo-trig* carbocyclization.

after a subsequent transmetalation step afforded the functionalized piperidines **17–19** in 66, 81 and 65% yields, respectively. In all cases, only the *cis* isomer was detected. The stereoselectivity has been explained by a chair-like transition state in which the electrophilic double bond occupies a *pseudo* axial position (the Z-α-amino zinc enolate and the double bond are gauche to each other), as described in Scheme 11 [17].

The stereochemical influence of substituents on the starting linear substrates on the carbocyclization reaction has been studied in detail, and it has been found that the stereochemical outcome of this carbocyclization was mainly due to the presence of a substituent in the homoallylic position [17].

Although much slower, the 6-*exo-trig* cyclization of zincated ketone hydrazone led to the cyclic derivative in good yield with a diastereoselectivity of 91:9 [16].

3.9.3
Intermolecular Carbometalation

The addition reaction of zinc hydrazone to isolated olefins such as ethylene was performed under pressure for four days at 35 °C, leading to the carbometalated product in 88% yield [18].

The zinc hydrazone **21ZnBu** does not isomerize to a more stable isomer such as **22ZnBu**. Here again, the use of the substituted species **20ZnBu** is critical, since, under the same reaction conditions, **20ZnBr** afforded **21** in lower yields

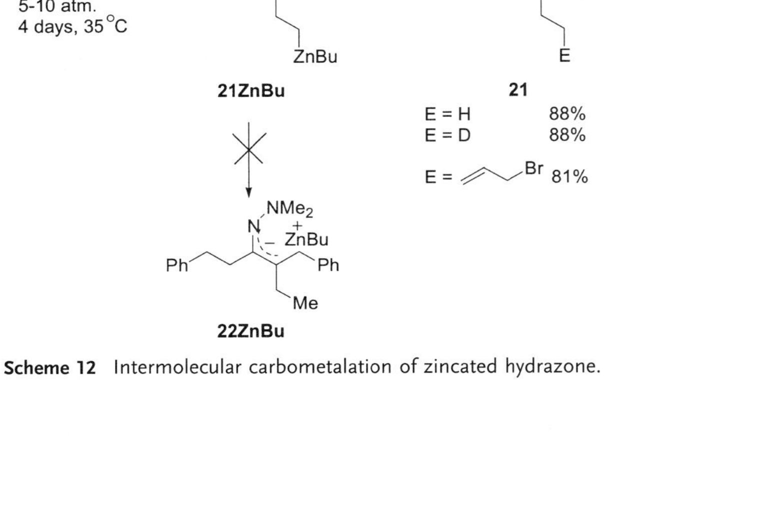

Scheme 12 Intermolecular carbometalation of zincated hydrazone.

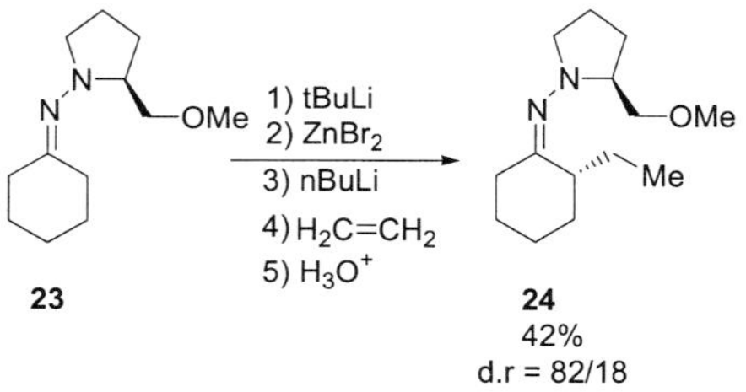

Scheme 13 Diastereoselective carbocyclization of zincated hydrazone.

(22%) [18]. Trapping **21ZnBu** with carbon electrophiles provided a one-pot three-component coupling reaction.

Aliphatic olefins such as 1-octene, as well as aromatic olefins (styrene, *p*-methoxy styrene and *o*-trifluoromethyl styrene), also took part, albeit slowly and in lower yields. On the other hand, vinylsilane [19] and vinylstannane [20] react readily with zincated hydrazones, to give the carbometalated product.

The SAMP hydrazone [21], which was made from cyclohexanone and the optically active SAMP hydrazine, also reacted with excess ethylene to form the allylated product with moderate yield and with a diastereomeric ratio of 82:18 (Scheme 13).

Zinc enolates and zincated hydrazones react with cyclopropenone-acetal in a highly diastereoselective manner to give a β-cyclopropyl-carbonyl derivative (Scheme 14) [19].

This carbometalation reaction takes place in a *cis* manner with a generally high level of 1,2-diastereoselectivity for the newly formed C-C bond. The reaction with an optically active hydrazone is synthetically useful, since the level of selectivity is between 87 and 98%, as described in Scheme 15.

Not unexpectedly for intermolecular carbometalation reactions, unactivated olefins are rather unreactive electrophiles toward enolates, and good yields are obtained only with slightly activated [19, 20] or strained alkenes [21]. On the other hand, the addition of delocalized organometallic moieties, such as allyl or propargyl zinc derivatives, to vinyl magnesium halide proceeds very smoothly at low temperature [22, 23]. A similar pattern was found for the reaction between zincated hydrazone **28Zn** with a vinyl Grignard reagent, which easily takes place at

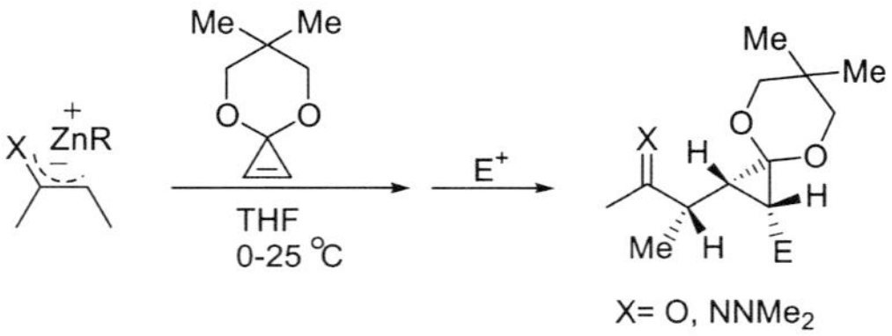

Scheme 14 Preparation of β-cyclopropyl-carbonyl derivatives.

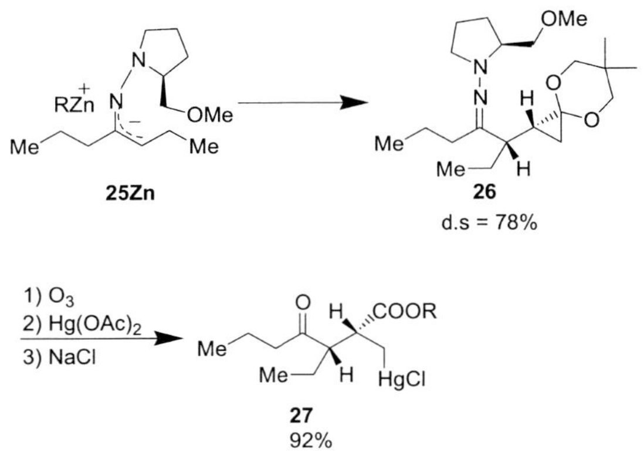

Scheme 15 Enantioselective carbometalation.

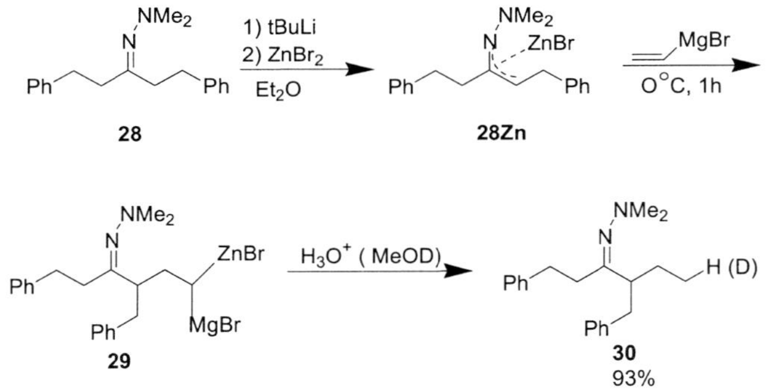

Scheme 16 Carbometalation of zincated hydrazone to vinyl magnesium bromide.

0 °C in less than 1 h, to give the bismetalated hydrazone intermediate **29** in nearly quantitative yield (Scheme 16) [24].

The bismetalated species **29** can also react either with benzaldehyde to give the olefination product **32** or with two different electrophiles such as MeSSMe followed by allyl bromide as described in Scheme 17 for the formation of **31**.

Substituted vinyl Grignard reagents have so far been found to react very sluggishly with the hydrazone anions.

3.9.4
Conclusions

Although this new "olefinic aldol chemistry" is still in its infancy, the addition of zincated enolates or hydrazones to non-activated carbon-carbon double bonds is on the way to becoming a powerful and new method for the preparation of functionalized organometallic derivatives.

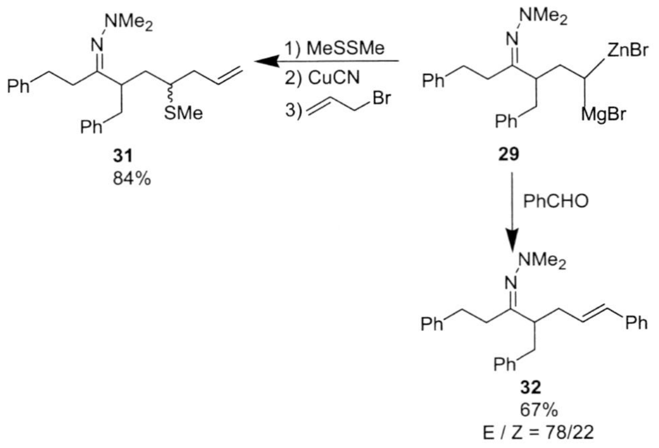

Scheme 17 Reactivity of the bismetalated species.

The simplicity of the protocols, once the right conditions are set up, well warrants the expectation of future developments of this methodology in organic synthesis.

3.9.5
Acknowledgements

This research was supported by Grant No. 2000155 from the United States-Israel Binational Science Foundation (BSF), Jerusalem, Israel and by the Technion V.P.R. Fund-Argentinian Research Fund.

3.9.6
References

1. K. Ziegler, K. Bähr, *Chem. Ber.* **1928**, 253.
2. (a) J. F. Normant, A. Alexakis, *Synthesis* **1981**, 841. (b) W. Oppolzer, *Angew. Chem. Int. Ed. Engl.* **1989**, *28*, 38. (c) E. Negishi, *Pure Appl. Chem.* **1981**, *53*, 2333. (d) P. Knochel in *Comprehensive Organometallic Chemistry II* (Eds.: W. Able, F. G. A. Stone, G. Wilkinson), Pergamon, Oxford **1995**, *11*, 159. (e) P. Knochel in *Comprehensive Organic Synthesis* (Eds.: B. M. Trost, I. Fleming), Pergamon, Oxford **1991**, *4*, 865. (f) Y. Yamamoto, N. Asao, *Chem. Rev.* **1993**, *93*, 2207. (g) E. Negishi, D. Y. Kondakov, *Chem. Rev.* **1996**, *96*, 417. (h) I. Marek, J. F. Normant, in *Carbometalation Reactions in Metal-Catalyzed Cross-Coupling Reactions* (Eds.: F. Diederich, P. Stang), Wiley-VCH, New York **1998**, 271. (i) I. Marek, *J. Chem. Soc. Perkin Trans. 1* **1999**, 535. (j) I. Marek, J. F. Normant in *Carbometalation Reactions* in *Transition Metals for Organic Synthesis*, first edition, Wiley-VCH, Weinheim **1998**, 514. (k) A. G. Fallis, P. Forgione, *Tetrahedron* **2001**, *57*, 5899.
3. (a) E. Lorthiois, I. Marek, J. F. Normant, *Tetrahedron Lett.* **1997**, *38*, 89 (b)

P. Karoyan, G. Chassaing, *Tetrahedron Lett.* **1997**, *38*, 85.

4   J. Knochel, R. D. Singer, *Chem. Rev.* **1993**, *93*, 2117.

5   F. H. Van Der Steen, H. Kleijn, G. J. P. Britovsek, J. T. B. H. Jastizebski, G. Van Koten, *J. Org. Chem.* **1992**, *57*, 3906.

6   (a) C. Meyer, I. Marek, G. Courtemanche, J. F. Normant, *J. Org. Chem.* **1995**, *60*, 863. (b) C. Meyer, I. Marek, J. F. Normant, N. Platzer, *Tetrahedron Lett.* **1994**, *35*, 5645. (c) C. Meyer, I. Marek, J. F. Normant, *Tetrahedron Lett.* **1996**, *37*, 857. (d) E. Lorthiois, I. Marek, C. Meyer, J. F. Normant, *Tetrahedron Lett.* **1995**, *36*, 1263. (e) E. Lorthiois, I. Marek, J. F. Normant, *Bull. Soc. Chem. Fr.* **1997**, *134*, 333.

7   E. Lorthiois, I. Marek, J. F. Normant, *J. Org. Chem.* **1998**, *63*, 2442.

8   (a) D. Beruben, I. Marek, J. F. Normant, N. Platzer, *J. Org. Chem.* **1995**, *60*, 2488 (b) I. Marek, D. Beruben, J. F. Normant, *Tetrahedron Lett.* **1995**, *36*, 3695.

9   P. Karoyan, G. Chassaing, *Tetrahedron; Assymetry* **1997**, *8*, 2025.

10 P. Karoyan, G. Chassaing, *Tetrahedron Lett.* **2002**, *43*, 253.

11 P. Karoyan, G. Chassaing, *Tetrahedron Lett.* **2002**, *43*, 1221.

12 P. Karoyan, A. Triolo, R. Nannicini, D. Giannotti, M. Altanura, G. Chassaing, E. Perrotta, *Tetrahedron Lett.* **1999**, *40*, 71.

13 F. Denes, F. Chemla, J. F. Normant, *Synlett* **2002**, 919.

14 A. Fürstner, *Synthesis* **1989**, 571.

15 F. Denes, F. Chemla, J. F. Normant, *Eur. J. Org. Chem.* **2002**, 3536.

16 E. Nakamura, G. Sakata, K. Kubota, *Tetrahedron Lett.* **1998**, *39*, 2157.

17 E. Lorthiois, I. Marek, J. F. Normant, *J. Org. Chem.* **1998**, *63*, 566.

18 K. Kubota, E. Nakamura, *Angew. Chem. Int. Ed. Engl.* **1997**, *36*, 2491.

19 E. Nakamura, K. Kubota, *Tetrahedron Lett.* **1997**, *38*, 7099.

20 M. Nakamura, K. Hara, G. Sakata, E. Nakamura, *Org. Lett.* **1999**, *1*, 1505.

21 E. Nakamura, K. Kubota, *J. Org. Chem.* **1997**, *62*, 792.

22 I. Marek, J. F. Normant, *Chem. Rev.* **1996**, *96*, 3241.

23 I. Marek, *Chem. Rev.* **2000**, *100*, 2887.

24 E. Nakamura, K. Kubota, G. Sakata, *J. Am. Chem. Soc.* **1997**, *119*, 5457.

3.10
Iron Acyl Complexes

Karola Rück-Braun

3.10.1
Introduction

Transition metal acyl complexes have been found to be useful building blocks in organic synthesis over the past 20 years [1–3]. This especially holds true for various iron acyl compounds [1, 4]. Representative applications and preparations of iron acyl complexes, in particular cyclopentadienyl(dicarbonyl)iron-substituted compounds, are described here, focusing on current developments.

3.10.2
Acyl Complexes Derived from Pentacarbonyl Iron

The chemistry of anionic acyliron(0) intermediates, formed *in situ* from disodium tetracarbonylferrate $Na_2Fe(CO)_4$ (Collman's reagent) and electrophiles such as organo halides or carboxylic acid chlorides 1, has been developed by Collman [5] and Parlman [6] in the past. An example of their application is the synthesis of 1,3-oxazol-5-ones 3 (Scheme 1) starting from pentacarbonyl iron, a carboxylic acid chloride 1 and an imidoyl chloride 2 [7].

More recently, transformations of pentacarbonyl iron via acylferrates to unsymmerial ketones, 1,2-diketones, or carboxylic acid derivatives, e.g., butenolides, have been reported [8].

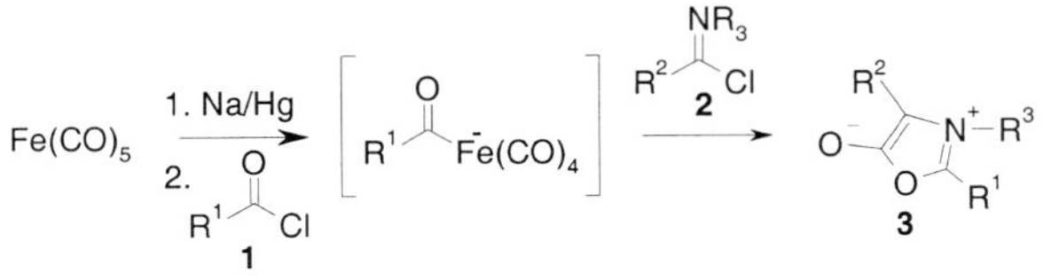

Scheme 1

Transition Metals for Organic Synthesis, Vol. 1, 2nd Edition.
Edited by M. Beller and C. Bolm
Copyright © 2004 WILEY-VCH Verlag GmbH & Co. KGaA, Weinheim
ISBN: 3-527-30613-7

3.10.3
Phosphine-Substituted Chiral-at-Iron Derivatives and Analogs

Iron acyl complexes (5) are commonly prepared by acylation of cyclopentadienyl(dicarbonyl)ferrates with carboxylic acid chlorides (1). Thereby, simple alkyl and aryl as well as α,β-unsaturated acyls have been obtained [2, 9–11]. Sugar acyl iron complexes have been prepared [12]. The synthesis of alkynyl-substituted compounds (6) has been reported by treatment of ferrates with mixed anhydrides (4) derived from carboxylic acids and iso-butylchloroformate in the presence of N-methylmorpholine [13].

Reactions of $[Fe(C_5Me_5)(CO)_4]^+$ with organolithium compounds open up another route to mono- and dinuclear acyl complexes bearing alkyl, aryl, alkynyl, and thienyl moieties [14]. By activation of 2-alkyn-1-ols with $[Fe(C_5Me_5)(CO)_4(H_2O)]^+$, α,β-unsaturated acyl complexes are formed [15]. The formation of acyls starting from iron alkyls by inducing iron to carbon migration of the alkyl residue has been applied in synthesis only in a few cases because of the difficulties in initiating the migration of electron deficient alkyl and aryl residues. However, thermal reactions of alkyl-substituted complexes with triphenylphosphine were used to synthesize racemic phosphine-substituted acyl complexes [1, 2].

α,β-Unsaturated acyl complexes and alkynyl acyl complexes have been applied as dienophiles in Lewis acid-catalyzed Diels-Alder reactions [16, 17]. Michael-type additions of heteronucleophiles, furnishing precursors for the synthesis of β-lactams, have been described [18]. Recently, the asymmetric syntheses of β-lactams, e.g., 10 (Scheme 3) and pseudopeptides via stereoselective conjugate additions of lithium (α-methylbenzyl)-allylamide 8 to α,β-unsaturated iron acyl complexes have been reported by Davies et al. [19].

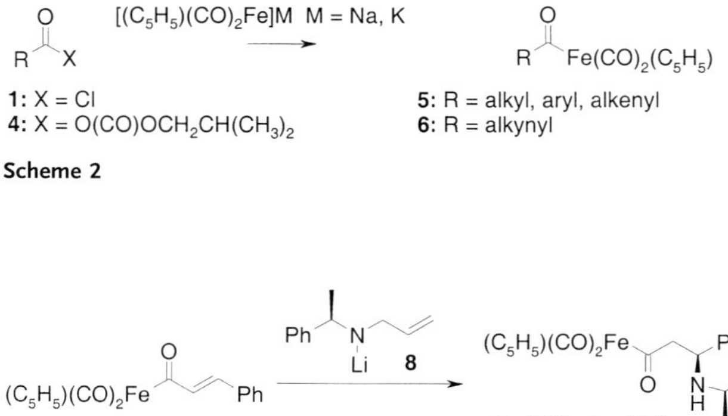

Scheme 2

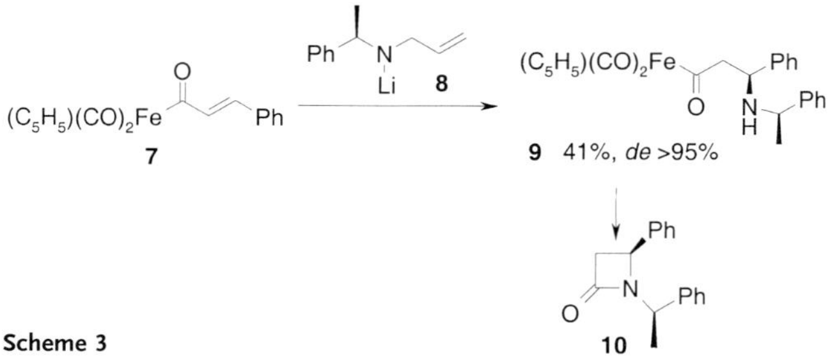

1: X = Cl
4: X = O(CO)OCH$_2$CH(CH$_3$)$_2$

5: R = alkyl, aryl, alkenyl
6: R = alkynyl

Scheme 3

9 41%, de >95%

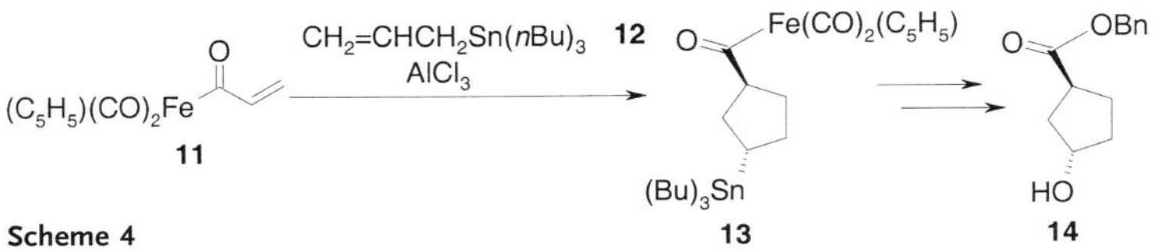

3.10.3 Phosphine-Substituted Chiral-at-Iron Derivatives and Analogs

Scheme 4

Reactions of cyclopentadienyl(dicarbonyl)iron enoyl complexes with allyltributyltin reagents were thoroughly investigated by Herndon et al. and have been found to give five-membered rings, e.g., **13** (8–66%), in the presence of $AlCl_3$ with high stereoselectivity (Scheme 4) [20]. Interestingly, upon replacement of one CO ligand by triphenylphosphine, the resulting complexes proved to be unreactive toward allyltributyltin and $AlCl_3$. Demetalation of the iron fragment was achieved by treatment with NBS in benzyl alcohol. Subsequent reaction of the ester obtained with bromine and oxidation with MCPBA furnished the alcohol **14**.

Achiral and chiral nonheteroatom-stabilized carbene complexes are readily prepared, e.g., from cycloentadienyl(dicarbonyl)iron acyls [21, 22]. Iron carbene complexes have been applied in cyclopropane synthesis [21, 22], intramolecular carbocation alkene cyclizations, and C-H insertion reactions [23, 24]. Iron acyls have been found to be versatile starting materials for heteroatom-stabilized carbenes [25]. (Alkynyl)aminocarbenes derived from alkynyl-substituted acyls [26, 27] have been utilized as dienophiles in Diels-Alder reactions with cyclopentadiene [17].

Asymmetric syntheses employing phosphine-substituted chiral-at-iron acyl complexes were thoroughly examined by Davies [28] and others [29, 30] in the past. Racemic mixtures of the iron acetyl complexes, e.g., **15**, were kinetically resolved via aldol reactions of the derived lithium enolates and camphor (Scheme 5) in the presence of lithium chloride [31]. Chiral enolates derived from the acyl compounds after transmetalation with aluminum-containing Lewis acids (Et_2AlCl) were found to react with aldehydes, yielding predominantly the *anti* diastereomers, whereas by using the copper enolates (CuCN) the *syn* diastereomers were obtained [32]. In addition reactions of the chiral enolates with alkyl halides, ketones, epoxides were found to proceed with excellent stereofacial discrimination [28–30]. In reactions with imines followed by oxidatively induced Fe-C bond β-lactams, e.g., **20**, were obtained [1, 29].

Chiral enoyl complexes **21** are commonly produced by a two-step procedure via aldol addition of aldehydes and the enolate derived from the parent acyl complex **15**, subsequent O-methylation applying NaH/MeI followed by base-catalyzed elimination yielding mixtures of (*E*)- and (*Z*)-isomers, which can be separated by chromatography [28]. a,β-Unsaturated compounds **21** have been applied in Michael additions [28], tandem Michael additions/alkylations [29], cyclopropanations, and Lewis-acid-catalyzed Diels-Alder cycloadditions [28, 33].

Racemic triphenylphosphine-substituted iron acyls have been applied in the synthesis of cationic iron(II) allyloxy carbene complexes, which, after treatment with KH and 18-crown-6 in benzene at $0\,°C$, were found to undergo facile [3,3] sigmatropic Claisen rearrangement to furnish a,β-unsaturated iron acyl complexes [34].

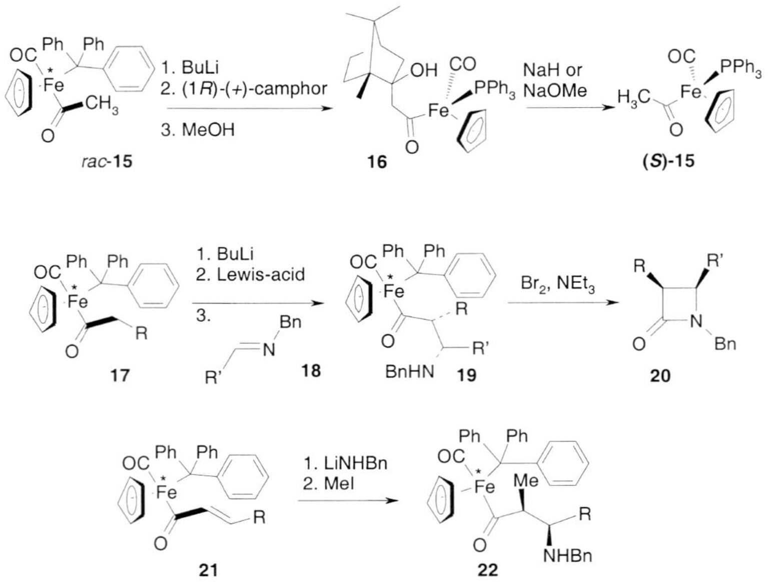

Scheme 5

3.10.4
Diiron Enoyl Acyl Complexes

Hexacarbonyl-diiron bridging α,β-unsaturated acyl complexes have been investigated as dienophiles and dipolarophiles in Diels-Alder reactions with 1,3-dienes and nitrones [35, 36]. The crotonyl compound **23** (Scheme 6) was found to react with Danishefsky's diene **24**, furnishing the adducts **25** in an *exo:endo* ratio of 8:1. With a *tert*-butyl group on the bridging sulfur the Diels-Alder adduct was obtained in 66% yield in an *exo:endo* ratio of 18:1.

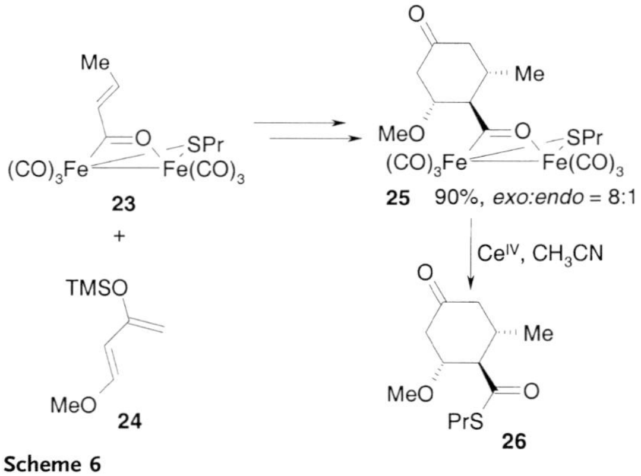

Scheme 6

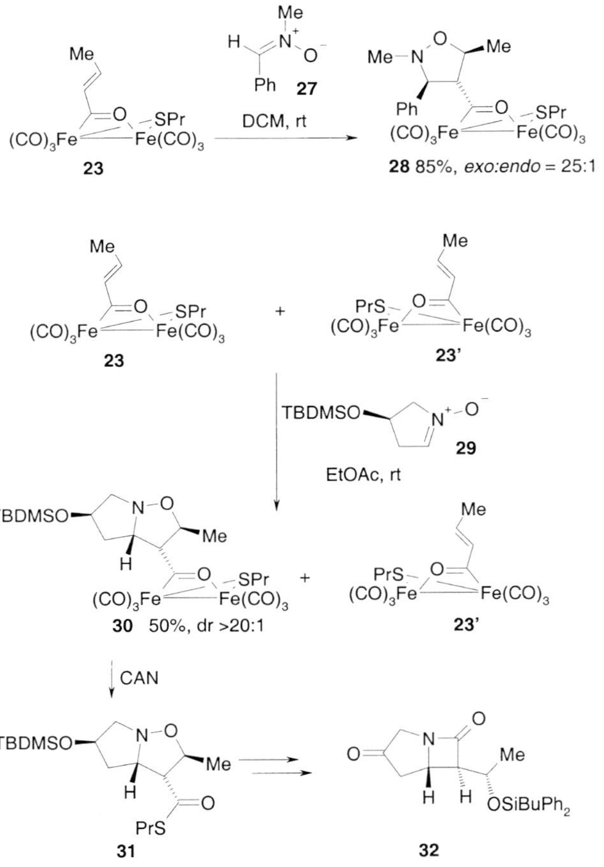

Scheme 7

Air-stable hexacarbonyl-diiron bridging enoyl acyl complexes have been examined in [3+2]-cycloadditions of nitrones [36]. C-Phenyl-N-methylnitrone (**27**) (Scheme 7) as well as other nitrones were found to give 4-substituted isoxazolidines in 20:1 up to 30:1 *endo*:*exo* ratios [36]. Chiral (L)-proline-derived crotonyl derivatives were reported to furnish the 4-isoxazolidines in high diastereoselectivities [37]. With the optically pure chiral nitrone **29** an effective chiral resolution of the enantiomers **23** and **23'** was observed. Based on this chemistry a route to the carbapenem **32** was successfully developed [38].

3.10.5
Iron-Substituted Enones and Enals

In the last decade, iron-substituted enones and enals were investigated as building blocks in organic synthesis. In the past, iron-substituted (*E*)-enones were synthesized from β-chlorovinyl ketones and the ferrate [(C$_5$H$_5$)(CO)$_4$Fe]Na by an addition/elimination process and were applied in organic synthesis as starting materials in carbocation alkene cyclizations [39].

Iron-substituted (Z)-enals **36** (Scheme 8), especially cyclic compounds, have been found to undergo intramolecular cyclocarbonylations furnishing five-membered lactone skeletons upon treatment of the aldehyde functionality with metal hydrides or C-nucleophiles such as organolithiums and Grignard reagents [40–43]. In these studies, β-trifluoromethyl(sulfonyloxy)-substituted cyclic enals derived from β-keto esters were generally employed as starting materials in reactions with the ferrate [(C$_5$H$_5$)(CO)$_4$Fe]Na because of the lability of β-halovinyl aldehydes derived from α-methylene ketones with aliphatic substitution pattern and the formation of mixtures of (*E*)- and (*Z*)-isomers from acyclic precursor molecules **35** [40–45].

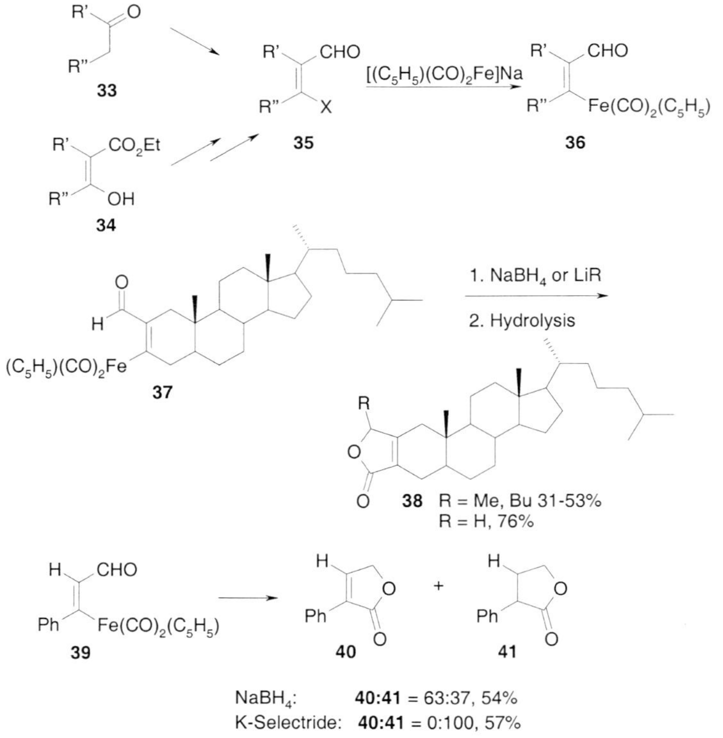

Scheme 8

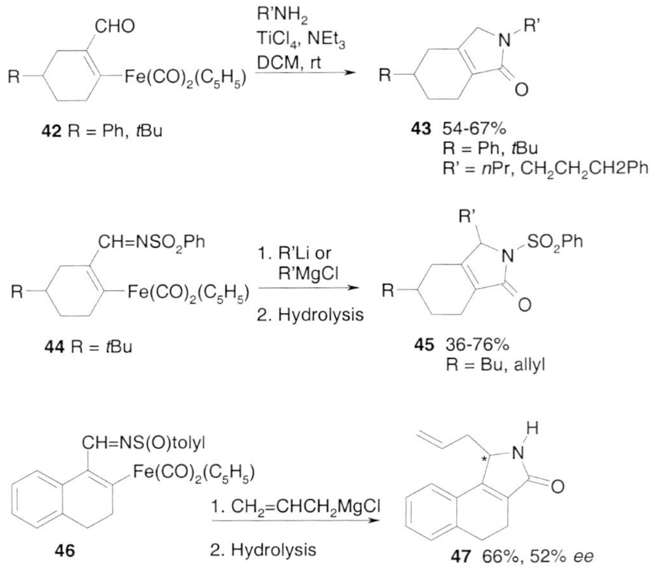

Scheme 9

In reactions with sodium borohydride or K-Selectride, depending upon the reagent and the iron complex employed, α,β-butenolides and saturated γ-lactones were obtained upon hydrolysis (Scheme 8) [40, 41]. According to labeling experiments, the formation of the latter is in accordance with a reduction step involving a π-alkene hydridoiron intermediate [40].

Treatment of iron-substituted (Z)-enals, such as **42**, with electron-rich primary amines in the presence of TiCl$_4$ and triethylamine was found to yield dihydropyrrolones **43** in a one-pot procedure (Scheme 9) [44, 45]. The experimental studies that have been carried out support the key role of the titanium hemiaminal functionality, which is initially formed by attack of the primary amine at the aldehyde group in this cyclocarbonylation reaction cascade. From acceptor-substituted amino compounds, e.g., aniline or benzenesulfonamide, iron-substituted azadienes, e.g., **44**, were synthesized, which proved to be valuable starting materials in cyclocarbonylation reactions with organolithiums and Grignard reagents furnishing 5-substituted α,β-unsaturated γ-lactams **45** (dihydropyrrolones) after hydrolysis (Scheme 9) [46, 47]. A route to optically active compounds was investigated starting from chiral sulfinylimines. Thereby, non-N-protected α,β-unsaturated γ-lactams, e.g., compound **47** (Scheme 9), were obtained exclusively in 27–91% yield starting from p-tolyl-, 2-naphthyl- and tert-butyl-substituted sulfinylimines, but only low to moderate stereoselectivities [47].

3.10.6
References

1. S. G. Davies, *Organotransition Metal Chemistry: Application to Organic Synthesis*, Pergamon Press, Oxford, **1989**.
2. A. J. Pearson, *Iron Compounds in Organic Synthesis*, Academic Press, London, **1994**.
3. L. S. Hegedus, *Organische Synthese mit Übergangsmetallen*, VCH, Weinheim, **1995**.
4. K. Rück-Braun, M. Mikulás, P. Amrhein, *Synthesis* **1999**, 727–744.
5. J. P. Collman, *Acc. Chem. Res.* **1975**, *8*, 342–347.
6. M. P. Cooke, R. M. Parlman, *J. Am. Chem. Soc.* **1975**, *97*, 6863.
7. H. Alper, M. Tanaka, *J. Am. Chem. Soc.* **1979**, *101*, 4245.
8. U. Radhakrishan, M. Periasamy, *Organometallics* **1997**, *16*, 1800–1802 and references cited.
9. R. B. King, M. B. Bisnette, *J. Organomet. Chem.* **1964**, *2*, 15–37.
10. J. A. Gladysz, G. M. Williams, W. Tam, D. L. Johnson, D. W. Parker, J. C. Selover, *Inorg. Chem.* **1979**, *18*, 553–558.
11. J. Kühn, K. Rück-Braun, *J. Prakt. Chem.* **1997**, *339*, 675–678.
12. R. Ehlenz, M. Nieger, K. Airola, K. H. Dötz, *J. Carbohydrate Chem.* **1997**, *16*, 1305–1318.
13. K. Rück-Braun, J. Kühn, *Synlett* **1995**, 1194–1196.
14. J. Kiesewetter, G. Poignant, V. Guerchais, *J. Organomet. Chem.* **2000**, *595*, 81–86.
15. G. Poignant, F. Martin, V. Guerchais, *Synlett* **1997**, 913–914.
16. J. W. Herndon, *J. Org. Chem.* **1986**, *51*, 2853.
17. K. Rück-Braun, J. Kühn, D. Schollmeyer, *Chem. Ber.* **1996**, *129*, 1057–1059.
18. I. Ojima, H. B. Kwon, *Chem. Lett.* **1985**, 1327–1330.
19. S. G. Davies, N. M. Garrido, P. A. McGee, J. P. Shilvock, *J. Chem. Soc., Perkin Trans. 1*, **1999**, 3105–3110.
20. J. W. Herndon, *J. Am. Chem. Soc.* **1987**, *109*, 3165–3166; J. W. Herndon, C. Wu, *Synlett* **1990**, 411; J. W. Herndon, C. Wu, J. J. Harp, *Organometallics* **1990**, *9*, 3157–3171.
21. M. Brookhart, W. B. Studabaker, *Chem. Rev.* **1987**, *87*, 411; M. Brookhart, Y. Liu, E. W. Goldman, D. A. Timmers, G. D. Williams, *J. Am. Chem. Soc.* **1991**, *113*, 927–939.
22. P. Helquist in *Advances in Metal-Organic Chemistry* (Ed.: L. S. Liebeskind), JAI Press, **1991**, pp. 144–194.
23. S. Ishii, S. Zhao, G. Mehta, C. J. Knors, P. Helquist, *J. Org. Chem.* **2001**, *66*, 3449–3458.
24. S. Ishii, S. Zhao, G. Mehta, C. J. Knors, P. Helquist, *J. Am. Chem. Soc.* **2000**, *122*, 5897–5898.
25. W. Petz, *Iron-Carbene Complexes*, Springer, Berlin, **1993**.
26. K. Rück-Braun, J. Kühn, D. Schollmeyer, *Chem. Ber.* **1996**, *129*, 937–944.
27. K. Rück-Braun, J. Kühn, D. Schollmeyer, *Chem. Ber./Recueil* **1997**, *130*, 1647–1654.
28. S. G. Davies, *Aldrichimica Acta* **1990**, *23*, 31; G. Bashiardes, G. J. Bodwell, S. G. Davies, *J. Chem. Soc., Perkin Trans. 1* **1993**, 459–469; S. G. Davies, H. M. Kellie, R. Polywka, *Tetrahedron: Asymmetry* **1994**, *5*, 2563–2570 and references cited.
29. L. S. Liebeskind, M. E. Welker, V. Goedken, *J. Am. Chem. Soc.* **1984**, *106*, 441–443; L. S. Liebeskind, M. E. Welker, *Tetrahedron Lett.* **1985**, *26*, 3079.
30. K. Wisniewski, Z. Pakulski, A. Zamojski, W. S. Sheldrick, *J. Organomet. Chem.* **1996**, *523*, 1–7.
31. S. C. Case-Green, J. F. Costello, S. G. Davies, N. Heaton, C. J. R. Hedgecock, V. M. Humphreys, M. R. Metzler, J. C. Prime, *J. Chem. Soc., Perkin Trans. 1* **1994**, 933–941.
32. S. G. Davies, I. M. Dordor-Hedgecock, P. Warner, *Tetrahedron Lett.* **1985**, *26*, 2125.
33. S. G. Davies, J. C. Walker, *J. Chem. Soc., Chem. Commun.* **1986**, 609; P. W. Ambler, S. G. Davies, *Tetrahedron Lett.* **1988**, *29*, 6979–6982.
34. A. G. M. Barrett, N. E Carpenter, *Organometallics* **1987**, *6*, 2249–2250.

35 S. R. GILBERTSON, X. ZHAO, D. P. DAWSON, K. LEE MARSHALL, *J. Am. Chem. Soc.* **1993**, *115*, 8517–8518.
36 S. R. GILBERTSON, D. P. DAWSON, O. D. LOPEZ, K. LEE MARSHALL, *J. Am. Chem. Soc.* **1995**, *117*, 4431–4432.
37 S. R. GILBERTSON, O. D. LOPEZ, *J. Am. Chem. Soc.* **1997**, *119*, 3399–3400.
38 S. R. GILBERTSON, O. D. LOPEZ, *Angew. Chem. Int. Ed. Engl.* **1999**, *38*, 1116–1119.
39 M. N. MATTSON, P. HELQUIST, *Organometallics* **1992**, *11*, 4.
40 K. RÜCK-BRAUN, C. MÖLLER, *Chem. Eur. J.* **1999**, *5*, 1038–1044.
41 C. MÖLLER, M. MIKULÁS, F. WIERSCHEM, K. RÜCK-BRAUN, *Synlett* **2000**, 182–184.
42 M. MIKULÁS, S. RUST, D. SCHOLLMEYER, K. RÜCK-BRAUN, *Synlett* **2000**, 185–188.
43 M. MIKULÁS, C. MÖLLER, S. RUST, F. WIERSCHEM, P. AMRHEIN, K. RÜCK-BRAUN, *J. Prakt. Chem.* **2000**, *342*, 791–803.
44 K. RÜCK-BRAUN, *Angew. Chem. Int. Ed. Engl.* **1997**, *36*, 509–511.
45 K. RÜCK-BRAUN, T. MARTIN, M. MIKULÁS, *Chem. Eur. J.* **1999**, *5*, 1028–1037.
46 P. AMRHEIN, D. SCHOLLMEYER, K. RÜCK-BRAUN, *Organometallics* **2000**, *19*, 3527–3534.
47 K. RÜCK-BRAUN, P. AMRHEIN, *Eur. J. Org. Chem.* **2000**, 3961–3969.

3.11
Iron–Diene Complexes

Hans-Joachim Knölker

3.11.1
Introduction

The complexation of acyclic and cyclic dienes by the tricarbonyliron fragment is achieved under mild reaction conditions and in high yields. The resulting tricarbonyl[η^4-1,3-diene]iron complexes (iron–diene complexes) offer great potential for synthetic applications, which include use of the tricarbonyliron fragment for acyclic stereocontrol as well as construction of carbocyclic and polyheterocyclic ring systems. This subject was reviewed in the previous edition of this book [1], and therefore only the developments over the last six years are summarized here, covering the literature from the middle of 1997 to the middle of 2003 [2].

3.11.2
Preparation of Iron–Diene Complexes

The transformation of a broad range of cyclic and acyclic 1,3-dienes **1** to the corresponding tricarbonyliron complexes **3** has been achieved using the [η^4-1-azabuta-1,3-diene]tricarbonyliron complex **2** as a tricarbonyliron transfer reagent (Scheme 1, Table 1) [3, 4]. The [η^4-1-azabuta-1,3-diene]tricarbonyliron complex **2** is superior to alternative tricarbonyliron transfer reagents for several reasons:

1. Complex **2** is prepared from the 1-azabuta-1,3-diene **4** under mild reaction conditions and in high yield (nonacarbonyldiiron, THF, ultrasound, room temperature, 88%).

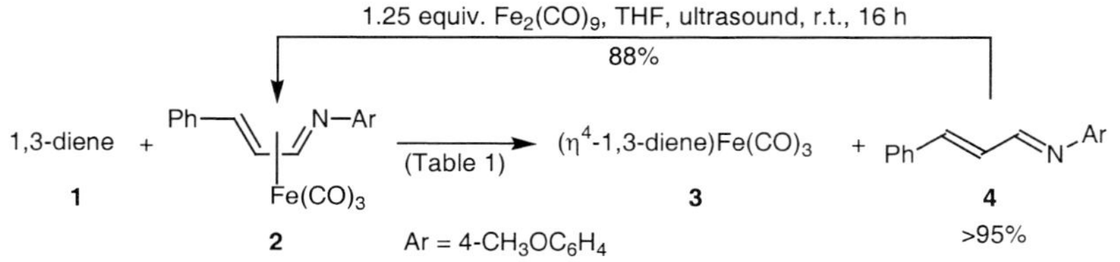

Scheme 1

Transition Metals for Organic Synthesis, Vol. 1, 2nd Edition.
Edited by M. Beller and C. Bolm
Copyright © 2004 WILEY-VCH Verlag GmbH & Co. KGaA, Weinheim
ISBN: 3-527-30613-7

Tab. 1 Complexation of the 1,3-dienes **1** by transfer of the tricabonyliron fragment from the [η^4-1-azabuta-1,3-diene]tricarbonyliron complex **2**

1,3-Diene (1)	Reaction conditions	3, Yield (%)	Ref.
cyclohexa-1,3-diene (**1a**)	THF, 65 °C, 2 h	**3a**, 95	[3]
cyclohexa-1,4-diene	toluene, 110 °C, 24 h	–	[4]
1-methoxycyclohexa-1,3-diene (**1b**)	benzene, 80 °C, 4 h	**3b/3b′**, 64 [a]	[4]
cyclohepta-1,3-diene (**1c**)	benzene, 80 °C, 4.5 h	**3c**, 84	[4]
2,3-dimethylbuta-1,3-diene (**1d**)	benzene, 80 °C, 25 h	**3d**, 71	[4]
hexa-2,4-dienal (**1e**)	toulene, 110 °C, 1 h	**3e**, 69	[4]

[a] 1:1 Mixture of the 1-methoxy- and 2-methoxycyclohexa-1,3-diene–tricarbonyliron complexes (**3b** and **3b′**)

2. The red crystalline compound is stable in the air for months.
3. After transfer of the metal fragment, the free ligand **4** can be recovered in more than 95% yield by crystallization and thus utilized for the regeneration of the transfer reagent **2**.

This last feature provides the basis for the catalytic complexation described below.

A very efficient alternative synthesis of tricarbonyl[η^4-cyclohepta-1,3-diene]iron (**3c**) was developed by the reductive complexation of cycloheptatriene with pentacarbonyliron in the presence of a catalytic amount of sodium borohydride (90% yield) [5]. Also noteworthy is the solid-state complexation by heating the 1,3-diene with a preformed mixture of nonacarbonyldiiron and silica gel (ratio, 1:5). This procedure, which requires no solvent, was applied to a series of acyclic 1,3-dienes, e.g., the complexation of hexa-2,4-dienal **1e** (85 °C, 2 h) afforded complex **3e** in 73% yield [6].

The catalytic complexation of 1,3-dienes was achieved by using pentacarbonyliron in the presence of 12.5 mol% of the 1-azabuta-1,3-diene **4** in dioxane at reflux temperature (Scheme 2). Using 1.5 equivalents of the diene component, the catalytic complexation is very efficient with respect to the tricarbonyliron fragments and was even carried out on a 50 g scale [7]. No excess pentacarbonyliron has to be removed from the reaction mixture and no pyrophoric iron is formed. Therefore, this procedure is non-hazardous on workup in contrast to the classical alternatives. Thus, cyclohexa-1,3-diene **1a** was converted quantitatively to the corresponding complex, tricarbonyl[η^4-cyclohexa-1,3-diene]iron **3a** [7]. The reaction of either 1-methoxycyclohexa-1,3-diene **1b** or the deconjugated isomer, 1-methoxycyclohexa-1,4-diene, under the same conditions at prolonged reaction times both afforded a 1:1 mixture of the two regioisomeric complexes **3b** and **3b′** [4]. This result demonstrates that using appropriate reaction conditions the catalytic complexation of 1,4-dienes provides by concomitant double bond isomerization the corresponding tricarbonyl[η^4-1,3-diene]iron complexes.

The catalytic complexation of the prochiral ligand **1b** with the achiral 1-azabuta-1,3-diene **4** as catalyst afforded the planar-chiral tricarbonyliron complex **3b** as a

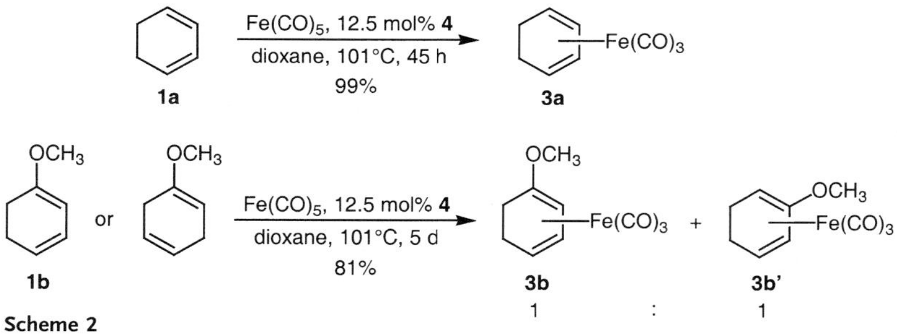

Scheme 2

racemic mixture (Scheme 2). Based on this reaction, a novel, highly efficient asymmetric catalytic complexation was developed [8]. A series of chiral 1-azabuta-1,3-diene catalysts provide complex **3b** enriched in either enantiomer, depending on the chirality of the catalyst [9]. The breakthrough was the finding that additional photolytic induction of this process resulted in a considerably higher yield and an increased enantiomeric excess [10]. Extensive optimization and mechanistic considerations [8] led to a set of standard reaction conditions for the asymmetric catalytic complexation. Using the camphor-derived catalysts (R)-**5** and (S)-**5**, the complexes (S)-**3** and (R)-**3** were obtained quantitatively in 85–86% *ee* (Scheme 3).

The asymmetric catalytic complexation was applied to a wide range of differently substituted prochiral cyclohexa-1,3-dienes and provided the corresponding planar-chiral tricarbonyl[η^4-cyclohexa-1,3-diene]iron complexes in yields over 80% and with *ee*s ranging from 57 to 86% [10]. Thus, nonracemic planar-chiral tricarbonyliron–cyclohexadiene complexes are now becoming easily available starting materials for enantioselective organic synthesis.

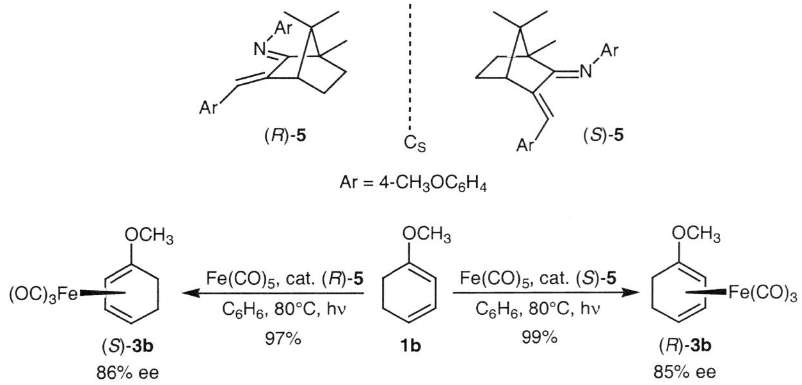

Scheme 3

3.11.3
Iron-Mediated Synthesis of Cyclopentadienones

The iron-mediated [2+2+1] cycloaddition of disilylated terminal diynes **6** with carbon monoxide afforded the tricarbonyl[η^4-cyclopentadienone]iron complexes **7** in yields of about 80% as carbo- and heterobicyclic ring systems, depending on the nature of the linkage (Scheme 4, Table 2) [1, 2c]. Applications of these complexes to organic synthesis require easy access to the corresponding free ligands **8** which are protected from Diels-Alder dimerization because of the steric protection provided by the two trimethylsilyl groups. Demetalation of the complexes **7** with only 4 equivalents of trimethylamine N-oxide in acetone provides the cyclopentadienones **8a–d** in 23–66% yield (method A). The only moderate yields and the failure to obtain compound **8e** are a consequence of double-bond isomerization with concomitant protodesilylation following the initial demetalation to **8** [1, 2c]. Therefore, improved procedures for the demetalation of tricarbonyliron–diene complexes using weakly oxidizing conditions were developed over the past few years. Photolysis of the tricarbonyliron complexes **7** in acetonitrile at low temperature led to a triple ligand exchange of all three carbon monoxide ligands by acetonitrile ligands. Bubbling of air into the cold solution of the intermediate tri(acetonitrile)iron complexes afforded the cyclopentadienones **8a–e** in 56–91% yield (method B) [11]. Moreover, on treatment with sodium hydroxide the tricarbonyl[η^4-cyclopentadienone]iron complexes **7** undergo a transformation related to the classical Hieber reaction. Thus, exchange of a carbon monoxide ligand by a hydrido ligand using NaOH and then further exchange by an iodo ligand using iodopentane led, after addition of phosphoric acid, to the corresponding dicarbonyl[η^5-hydroxycyclopentadienyl]iodoiron complexes, which are readily demetalated to the free ligands **8** on contact with air in the presence of daylight (method C) [12]. Method C generally provides the cyclopentadienones **8** with yields in the range 87–95%, except when base-sensitive substituents (e.g., ester groups) are present (**8c**). It is noteworthy that method B was also successfully applied to the demetalation of tricarbonyliron–butadiene and tricarbonyliron–cyclohexadiene complexes (80–85% yield).

The iron-mediated [2+2+1] cycloaddition of 1,9-bis(trimethylsilyl)nona-1,8-diynes with carbon monoxide to hydroazulene derivatives, previously achieved in only 15% yield [1], has now been improved by application of a three-cycle procedure (43–71% yield) [13]. Easy access to monocyclic tricarbonyl[η^4-cyclopentadie-

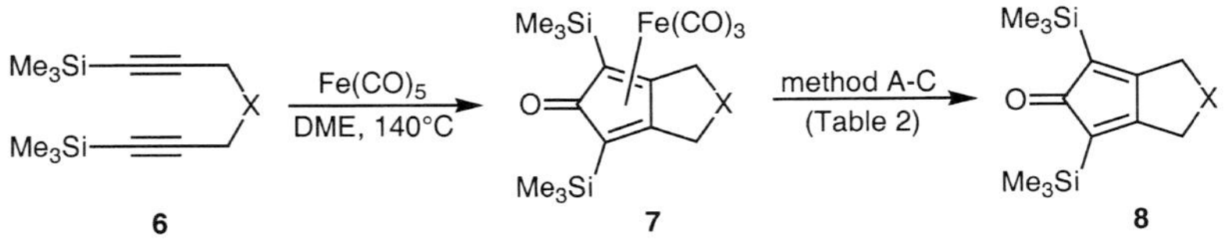

Scheme 4

Tab. 2 Synthesis and demetalations of the tricarbonyl[η^4-cyclopentadienone]iron complexes **7**

6	X	7, Yield (%)	8, Yield (%)		
			Method A[a]	Method B[b]	Method C[c]
a	CH$_2$	78	46	91	93
b	(CH$_2$)$_2$	82	66	89	95
c	C(COOMe)$_2$	84	23	86	12
d	O	85	46	83	89
e	S	76	–	56	87

a) Method A: 4 equiv. Me$_3$NO, acetone, 15–25 °C, 10–30 min [1, 2c]
b) Method B: 1. hv, MeCN, –30 °C; 2. air, –30 °C, 5 min [11]
c) Method C: 1. 1 M NaOH/THF (1:2); 2. C$_5$H$_{11}$I; 3. H$_3$PO$_4$; 4. air, daylight, Et$_2$O/THF, Na$_2$S$_2$O$_3$, Celite, 3 h [12]

none]iron complexes is possible through the use of diyne precursors containing a removable silicon tether [14].

The iron-mediated [2+2+1] cycloaddition of the alkyne-substituted ynamines **9a–c** with carbon monoxide provided the bicyclic tricarbonyl[η^4-3-aminocyclopentadienone]iron complexes **10a–c** with annulated five-, six-, and seven-membered rings, depending on the length of the tether (Scheme 5, Table 3) [15]. Using the alkyne-substituted ynol ethers **9d–f**, the same methodology led to the analogous bicyclic tricarbonyl[η^4-3-alkoxycyclopentadienone]iron complexes **10d–f**, although in this series the yields were lower [16]. Demetalation of the complexes **10a–f** by the application of methods A and C afforded the corresponding free ligands, which were used as dienes in Diels-Alder cycloadditions with alkynes [15, 16].

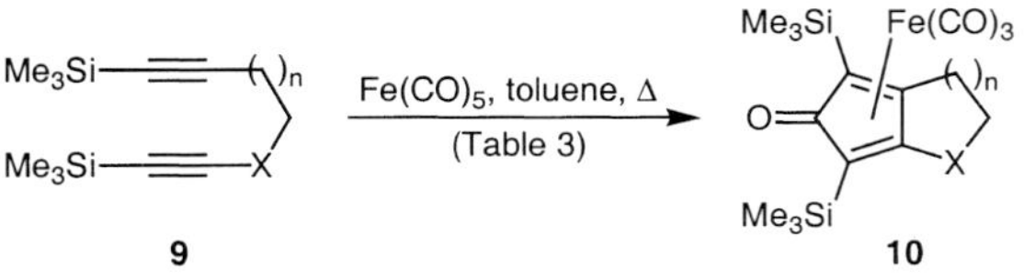

Scheme 5

Tab. 3 Iron-mediated synthesis of the bicyclic 3-amino- and 3-alkoxycyclopentadienones **10**

9	n	X	T (°C)	10, Yield (%)	Ref.
a	1	NTs	110	84	[15]
b	2	NTs	110	65	[15]
c	3	NTs	110	54	[15]
d	1	O	130	37	[16]
e	2	O	130	24	[16]
f	3	O	130	21	[16]

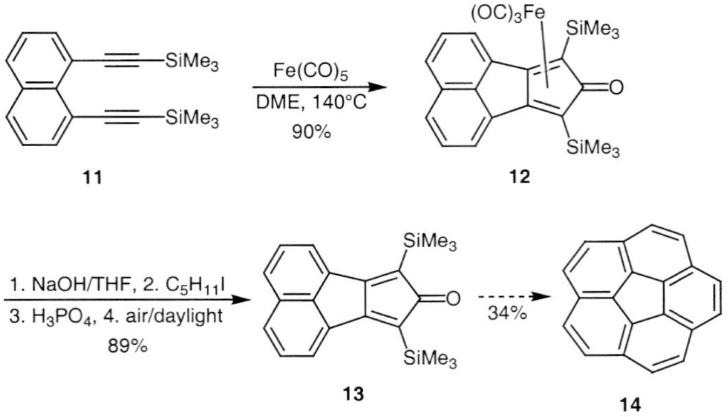

Scheme 6

The utility of this chemistry for organic synthesis was demonstrated by the first applications leading to complex molecules. A simple synthetic route afforded corannulene ($C_{20}H_{10}$), the smallest bowl-shaped polycyclic aromatic hydrocarbon fragment of buckminsterfullerene (C_{60}). The iron-mediated [2+2+1] cycloaddition of 1,8-bis(trimethylsilylethynyl)naphthalene (**11**) with carbon monoxide followed by demetalation of the tricarbonyliron complex **12** using method C provided the cyclopentadienone **13**, which was transformed to corannulene (**14**) in four steps (Scheme 6) [17].

The thermal reaction of 1,2-bis(trimethylsilylpropargyl)-1,2,3,4-tetrahydro-β-carboline (**15**) with pentacarbonyliron and subsequent demetalation of the tricarbonyliron complex **16** using method B afforded the cyclopentadienone **17** (Scheme 7). Diels-Alder cycloaddition of compound **17** with norbornadiene and double protodesilylation led to ($\pm$)-demethoxycarbonyldihydrogambirtannine (**18**), the main alkaloid of the fruit of the African plant *Strychnos usambarensis* (Loganiaceae) [18].

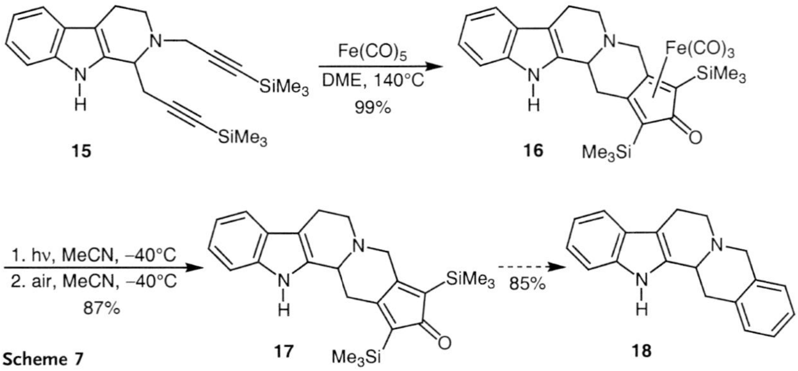

Scheme 7

3.11.4
Synthetic Applications of Iron–Butadiene Complexes

Iron–butadiene complexes have been the subject of extensive structural and theoretical investigations [19], which culminated in the recent discovery of [η^4-s-trans-buta-1,3-diene]tricarbonyliron prepared by photolysis of the classical cis isomer [20]. However, this section will focus again on synthetic aspects.

The Friedel-Crafts acylation of iron–butadiene complexes represents an important tool for stereospecific functionalization [1]. The resulting products can be further utilized for stereoselective aldol reactions. This approach was applied to the synthesis of multiply protected mycosamine (3-amino-3,6-dideoxymannose) [21]. Thus, reaction of the enantiopure α-aminoketone **19** with the protected (R)-(+)-lactaldehyde **20** provided stereoselectively the ketol **21** (Scheme 8). Demetalation of complex **21** with cerium(IV) ammonium nitrate (CAN) led to the diene **22**. Stereoselective reduction using tetramethylammonium triacetoxyhydridoborate, conversion to the corresponding diacetates, and finally ozonolysis afforded the multiply protected mycosamine **23**.

The stereoselective aldol reaction of tricarbonyl[η^4-dienone]iron complexes was further utilized for the enantioselective syntheses of the streptenols C and D [22], 3,6-dideoxyhexoses [23], 3-deoxypentoses [24], and (+)-[6]-gingerdiol [25].

As previously mentioned, complex **24** represents a highly useful building block with a broad potential for organic synthesis [1]. Over the past years several additional interesting applications were elaborated, only two of which are presented here in more detail. Complex **24** is easily transformed into the tricarbonyl[η^5-(1-methoxycarbonyl)pentadienylium]iron salt **25** by reduction and subsequent elimination (Scheme 9) [1]. The nucleophilic attack of malonate anions at the cation of **25** occurs regio- and stereoselectively at an internal position (C2), providing the (pentenediyl)iron complexes **26** (Scheme 9, Table 4) [26]. Based on the reactivity of the complex salt **25** toward stabilized carbon nucleophiles, a remarkable cyclopro-

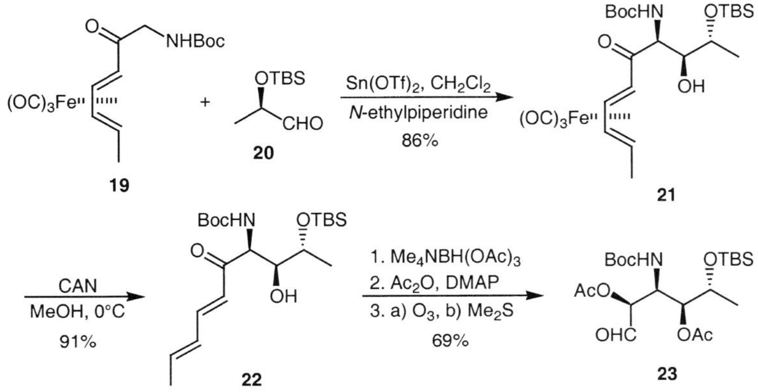

Scheme 8

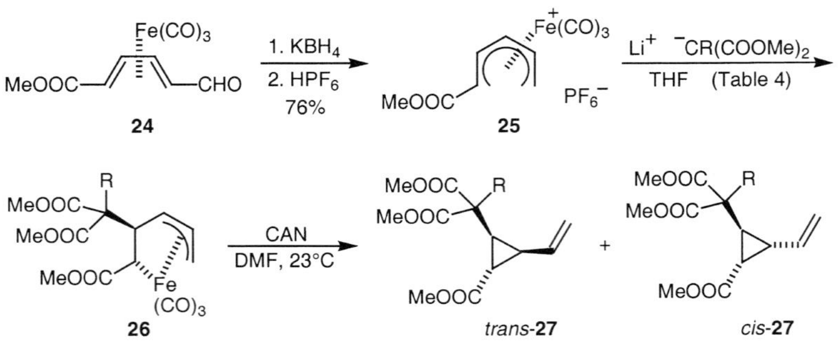

3.11 Iron–Diene Complexes

Scheme 9

Tab. 4 Diastereoselective iron-mediated synthesis of the vinylcyclopropanecarboxylates

	R	26, Yield (%)	27, Yield (%)	Ratio (trans:cis)
a	H	61	70	>10:1
b	Me	66	56	0:1

pane synthesis has been developed [27]. The demetalation of complex **26a**, bearing the dimethyl malonate substituent at C2, provided predominantly the vinylcyclopropanecarboxylate *trans*-**27**. This transformation is the result of an oxidatively induced reductive elimination, which occurs with retention of stereochemistry. However, the oxidative demetalation of the methyl dimethyl malonate derivative **26b** led with inversion of configuration exclusively to the vinylcyclopropanecarboxylate *cis*-**27**. The inversion of configuration is the consequence of a π-σ-π rearrangement. This methodology was applied to the stereoselective synthesis of substituted cyclopropylglycines [28].

In another application complex **24** afforded the methyl ester of α-lipoic acid [29]. The addition of vinylmagnesium bromide provided complex **28** as the major diastereoisomer, which was converted to complex **29** by silyl protection and hydroboration (Scheme 10). A Mitsunobu reaction was used to introduce the sulfur at the primary position, leading to complex **30**. Attack of an appropriate sulfur nucleophile at an intermediate tricarbonyl[η^5-pentadienyl]iron cation followed by demetalation afforded the butadiene **31**. Reduction of the diene with 2,4,6-tri-*iso*-propylbenzenesulfonyl hydrazide (TPSH) and subsequent treatment with potassium carbonate in methanol provided racemic methyl lipoate **32**.

Moreover, complex **24** was used for the stereoselective synthesis of trienes by monohydrogenation of an appended alkyne [30], the asymmetric synthesis of the alkaloid SS20846A [31] (achieved previously using an alternative route [1]), and the preparation of conformationally locked phosphocholines [32].

The asymmetric catalytic alkylation of complex **33** with dipentylzinc in the presence of a chiral aminoalcohol ligand and subsequent O-silylation afforded com-

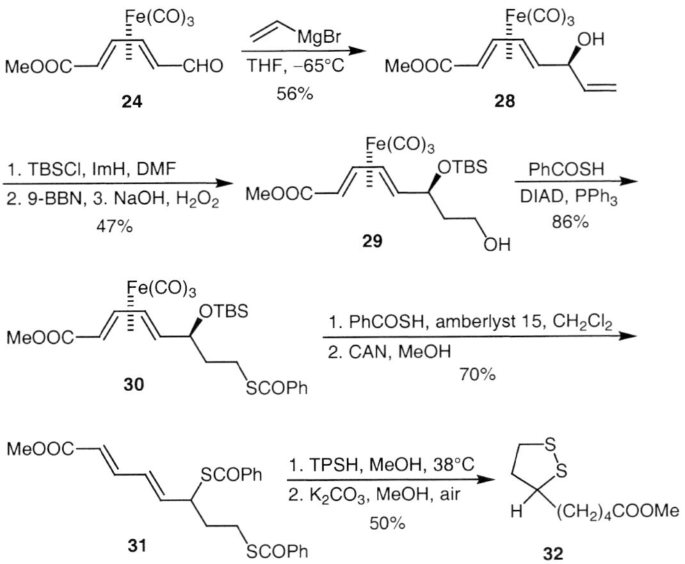

Scheme 10

plex **34** in more than 98% *ee* (Scheme 11) [33]. The preferred diastereoselectivity in this case is the same as that described above for complex **28** (cf. Scheme 10). Wittig-Horner reaction and reduction led to the allylic alcohol **35**. At this stage, a selective dihydroxylation of the free double bond was achieved, because the coordination to the tricarbonyliron fragment protects the residual butadiene fragment. Moreover, the tricarbonyliron fragment once again was used as a stereodirecting group in this transformation. Chemoselective pivaloylation of the primary hydroxy group to complex **36** was followed by chloroacetylation and conversion to the phenyl sulfide **37** with retention of configuration via an intermediate tricarbonyl[η^5-pentadienyl]iron cation. Demetalation and oxidation with *m*-chloroperbenzoic acid (MCPBA) afforded the sulfoxide **38**. A [2,3]-sigmatropic rearrangment of the sulfoxide **38** by treatment with trimethyl phosphite in refluxing methanol, 2-(trimethylsilyl)ethoxymethyl (SEM)-protection, reductive removal of the chloroacetyl group, and repivaloylation gave the skipped diene **39**. The chemo- and diastereoselectivity of the modified Simmons-Smith reaction to the cyclopropane **40** were both controlled by the neighboring hydroxy group. The cyclopropane **40** was transformed to the lipoxygenase inhibitor halicholactone **41** in 10 steps and 6% overall yield [34]. An analysis of the asymmetric total synthesis of this marine natural product shows that the correct absolute configuration of all five stereogenic centers originates from the tricarbonyliron fragment as a stereocontrolling group.

Further interesting synthetic applications of iron–butadiene complexes to organic synthesis include remarkable cases of remote stereocontrol [35], some natural product model studies [36], the enantioselective synthesis of (+)-dienomycin C

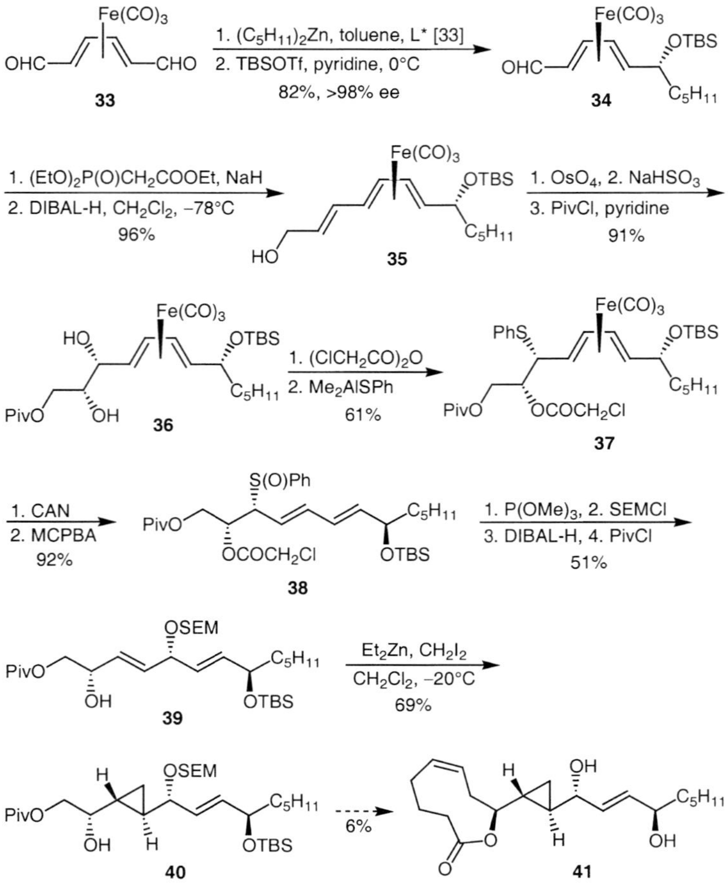

[37], the stereoselective synthesis of 11Z-retinal [38], stereoselective cationic cyclizations of pendant alkenes [39], and the asymmetric synthesis of a building block for amphotericin B [40].

3.11.5
Synthetic Applications of Iron–Cyclohexadiene Complexes

The allylic position of iron–cyclohexadiene complexes is easily functionalized with complete regio- and stereocontrol by a simple procedure (Scheme 12) [1]. Hydride abstraction of complex **3a** affords the iron–cyclohexadienylium salt **42**, which on reaction with appropriate nucleophiles provides the 5-substituted complex **43** by attack *anti* to the tricarbonyliron fragment. Demetalation leads to the correspond-

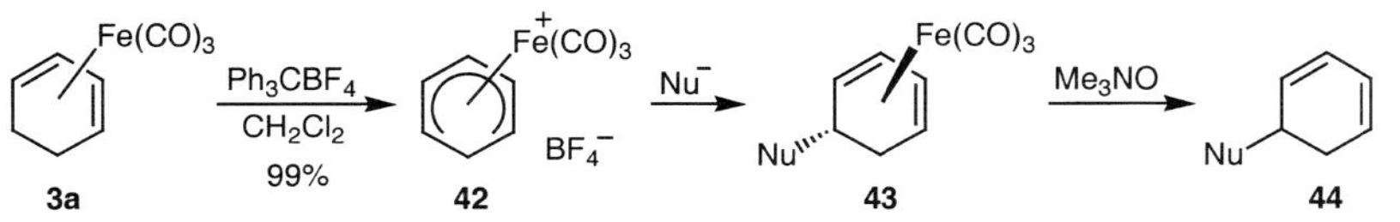

Scheme 12

ing free ligands **44**. This sequence has been used in the synthesis of a range of substituted cyclohexadienes and applied to organic synthesis [41]. The construction of carbazoles and spirocyclic frameworks is described in more detail below.

3.11.5.1
Iron-Mediated Total Synthesis of Carbazole Alkaloids

The iron-mediated construction of the carbazole framework has been applied to the total synthesis of a variety of biologically active carbazole alkaloids [42]. A new approach to the furo[3,2-*a*]carbazole alkaloid furostifoline was developed using the iron-mediated arylamine cyclization (Scheme 13) [43]. Electrophilic substitution of the arylamine **45** by reaction with the complex salt **42** afforded complex **46**. Oxidative cyclization to the carbazole **47** with iodine in pyridine followed by annulation of the furan ring provided furostifoline **48**.

Further recent applications of the iron-mediated arylamine cyclization to the synthesis of carbazole alkaloids have been achieved by using ferricenium hexafluorophosphate/sodium carbonate or very active manganese dioxide as the oxidizing agents and provided hyellazole and isohyellazole [44], carazostatin and *O*-methylcarazostatin [45], carbazomycin G and carbazomycin H [46].

An intriguing example of the synthetic potential of the iron-mediated arylamine cyclization is the bidirectional annulation of two indole units at a central *m*-phenylenediamine (Scheme 14) [47]. Double electrophilic substitution of *m*-phenylenediamine **49** to the dinuclear complex **50** and subsequent double iron-mediated ary-

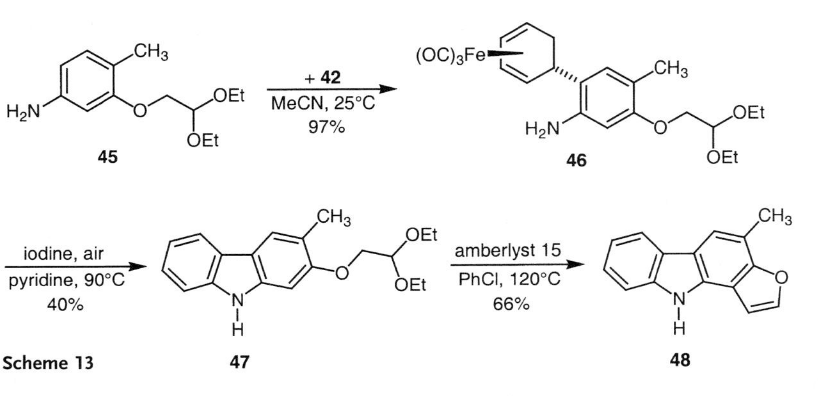

Scheme 13

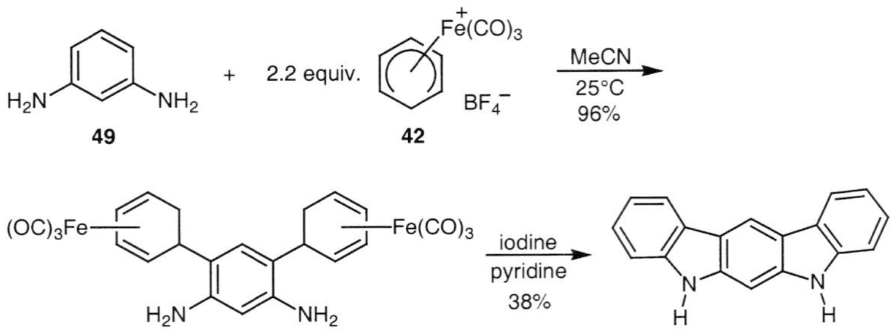

lamine cyclization by iodine in pyridine led in two simple steps to indolo[2,3-b]carbazole **51**.

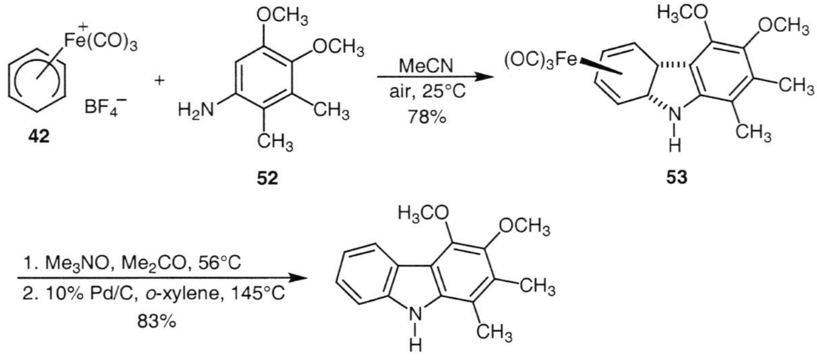

The oxidative cyclization of substituted iron–cyclohexadiene complexes with air as the oxidizing agent provides tricarbonyl[η^4-4a,9a-dihydro-9H-carbazole]iron complexes and represents an alternative procedure for access to carbazoles [1]. This method was applied to a considerably improved route (three steps and 65% overall yield) to the antiobiotic carbazomycin A (Scheme 15) [48]. The electrophilic substitution of the arylamine **52** by reaction with the iron complex salt **42** carried out in air afforded directly the tricarbonyl[η^4-4a,9a-dihydro-9H-carbazole]iron complex **53**. Demetalation followed by dehydrogenation provided carbazomycin A **54**.

The oxidative cyclization using air as oxidizing agent was also used for the total syntheses of carbazomycin B [48], (±)-neocarazostatin B [49], carquinostatin A [50], lavanduquinocin [51], mukonine and mukonidine [52].

3.11.5.2
Iron-Mediated Diastereoselective Spiroannulations

The intramolecular coupling of iron–cyclohexadiene complexes and pendant alkenes was developed into an efficient method for the synthesis of spiroheterocyc-

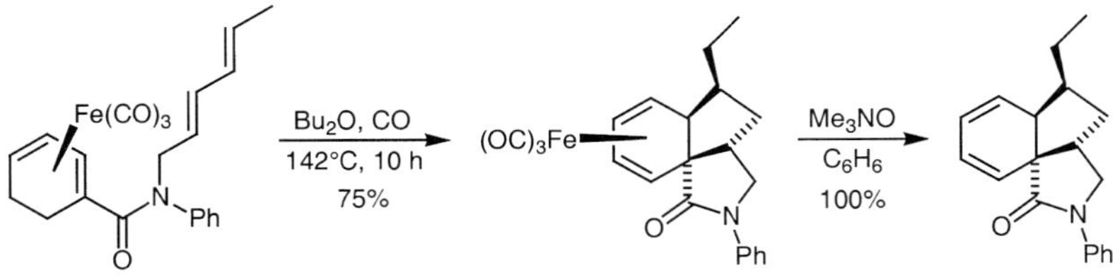

3.11.5 Synthetic Applications of Iron–Cyclohexadiene Complexes | 597

Scheme 16

lic ring systems [53–55]. The most impressive example is the double cyclization of complex **55** with a pendant butadiene, which afforded diastereoselectively the tricyclic complex **56** (Scheme 16) [56]. The diastereoselectivity of the two C–C bond formations, *syn* to the metal, is rationalized by intermediate six-membered metallacycles. The overall transformation appears to be the result of two consecutive ene-type reactions. Demetalation led quantitatively to the tricyclic diene **57**.

An alternative methodology for iron-mediated spiroannulation exploits the reaction of a 1,3-dielectrophile with 1,3-dinucleophiles. Previously it was shown that the addition of arylamines as 1,3-dinucleophiles to the complex salt **58** provides spiroquinolines diastereoselectively [1]. The reaction of the complex salt **58** with the cyclic vinylogous urethane **59** led, depending on the reaction conditions, predominantly either to the spiroindolizidine **60a** or to the spiroindole **60b** (Scheme 17, Table 5) [57]. It is noteworthy that the spiroindole **60b** was obtained as a single diastereoisomer, although it has an additional stereogenic center.

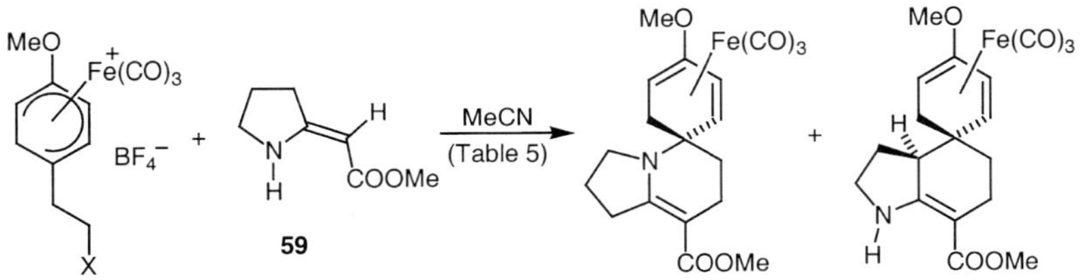

Scheme 17

X = 4-NO$_2$-C$_6$H$_4$COO

Tab. 5 Iron-mediated spiroannulation of the cyclic vinylogous urethane **59**

Reaction conditions	60, Yield (%)	Ratio (60a:60b)
1. rt, 10 h; 2. 82 °C, 13 h	80	6:1
82 °C, 4 d	78	1:5

3.11.6
References

1 H.-J. KNÖLKER, *Transition Metals for Organic Synthesis – Building Blocks and Fine Chemicals, Vol. 1* (Eds.: M. BELLER, C. BOLM), Wiley-VCH, Weinheim, **1998**, Chap. 3.13, p. 534.
2 For further recent reviews on the application of iron–diene complexes to organic synthesis, see: (a) W. A. DONALDSON, *Aldrichimica Acta* **1997**, *30*, 17. (b) L. R. COX, S. V. LEY, *Chem. Soc. Rev.* **1998**, *27*, 301. (c) H.-J. KNÖLKER, *Chem. Soc. Rev.* **1999**, *28*, 151. (d) W. A. DONALDSON, *Curr. Org. Chem.* **2000**, *4*, 837. (e) W. A. DONALDSON, *The Chemistry of Dienes and Polyenes, Vol. 2* (Ed.: Z. RAPPOPORT), Wiley, New York, **2000**, Chap. 11, p. 885. (f) H.-J. KNÖLKER, A. BRAIER, D. J. BRÖCHER, S. CÄMMERER, W. FRÖHNER, P. GONSER, H. HERMANN, D. HERZBERG, K. R. REDDY, G. ROHDE, *Pure Appl. Chem.* **2001**, *73*, 1075.
3 H.-J. KNÖLKER, G. BAUM, N. FOITZIK, H. GOESMANN, P. GONSER, P. G. JONES, H. RÖTTELE, *Eur. J. Inorg. Chem.* **1998**, 993.
4 H.-J. KNÖLKER, B. AHRENS, P. GONSER, M. HEININGER, P. G. JONES, *Tetrahedron* **2000**, *56*, 2259.
5 Y. COQUEREL, J.-P. DEPRÉS, *Chem. Commun.* **2002**, 658.
6 G. F. DOCHERTY, G. R. KNOX, P. L. PAUSON, *J. Organomet. Chem.* **1998**, *568*, 287.
7 H.-J. KNÖLKER, E. BAUM, P. GONSER, G. ROHDE, H. RÖTTELE, *Organometallics* **1998**, *17*, 3916.
8 H.-J. KNÖLKER, *Chem. Rev.* **2000**, *100*, 2941.
9 (a) H.-J. KNÖLKER, H. GOESMANN, H. HERMANN, D. HERZBERG, G. ROHDE, *Synlett* **1999**, 421. (b) H.-J. KNÖLKER, D. HERZBERG, *Tetrahedron Lett.* **1999**, *40*, 3547.
10 H.-J. KNÖLKER, H. HERMANN, D. HERZBERG, *Chem. Commun.* **1999**, 831.
11 H.-J. KNÖLKER, H. GOESMANN, R. KLAUSS, *Angew. Chem.* **1999**, *111*, 727; *Angew. Chem., Int. Ed.* **1999**, *38*, 702.
12 H.-J. KNÖLKER, E. BAUM, H. GOESMANN, R. KLAUSS, *Angew. Chem.* **1999**, *111*, 2196; *Angew. Chem., Int. Ed.* **1999**, *38*, 2064.
13 A. J. PEARSON, X. YAO, *Synlett* **1997**, 1281.
14 A. J. PEARSON, J. B. KIM, *Org. Lett.* **2002**, *4*, 2837.
15 J. D. RAINIER, J. E. IMBRIGLIO, *J. Org. Chem.* **2000**, *65*, 7272.
16 J. E. IMBRIGLIO, J. D. RAINIER, *Tetrahedron Lett.* **2001**, *42*, 6987.
17 H.-J. KNÖLKER, A. BRAIER, D. J. BRÖCHER, P. G. JONES, H. PIOTROWSKI, *Tetrahedron Lett.* **1999**, *40*, 8075.
18 H.-J. KNÖLKER, S. CÄMMERER, *Tetrahedron Lett.* **2000**, *41*, 5035.
19 (a) P. MCARDLE, J. SKELTON, A. R. MANNING, *J. Organomet. Chem.* **1997**, *538*, 9. (b) Ò. GONZÁLEZ-BLANCO, V. BRANCHADELL, *Organometallics* **1997**, *16*, 475. (c) Ò. GONZÁLEZ-BLANCO, V. BRANCHADELL, R. GRÉE, *Chem. Eur. J.* **1999**, *5*, 1722. (d) A. PFLETSCHINGER, H.-G. SCHMALZ, W. KOCH, *Eur. J. Inorg. Chem.* **1999**, 1869. (e) Ò. GONZÁLEZ-BLANCO, V. BRANCHADELL, *Organometallics* **2000**, *19*, 4477.
20 V. BACHLER, F.-W. GREVELS, K. KERPEN, G. OLBRICH, K. SCHAFFNER, *Organometallics* **2003**, *22*, 1696.
21 M. FRANCK-NEUMANN, L. MIESCH-GROSS, C. GATEAU, *Eur. J. Org. Chem.* **2000**, 3693.
22 M. FRANCK-NEUMANN, P. BISSINGER, P. GEOFFROY, *Tetrahedron Lett.* **1997**, *38*, 4469.
23 M. FRANCK-NEUMANN, P. BISSINGER, P. GEOFFROY, *Tetrahedron Lett.* **1997**, *38*, 4473.
24 M. FRANCK-NEUMANN, P. BISSINGER, P. GEOFFROY, *Tetrahedron Lett.* **1997**, *38*, 4477.
25 M. FRANCK-NEUMANN, P. GEOFFROY, P. BISSINGER, S. ADELAIDE, *Tetrahedron Lett.* **2001**, *42*, 6401.
26 W. A. DONALDSON, L. SHANG, C. TAO, Y. K. YUN, M. RAMASWAMY, V. G. YOUNG, *J. Organomet. Chem.* **1997**, *539*, 87.
27 (a) Y. K. YUN, W. A. DONALDSON, *J. Am. Chem. Soc.* **1997**, *119*, 4084. (b) Y. K. YUN, K. GODULA, Y. CAO, W. A. DONALDSON, *J. Org. Chem.* **2003**, *68*, 901.
28 (a) K. GODULA, W. A. DONALDSON, *Tetrahedron Lett.* **2001**, *42*, 153. (b) N. J. WAL-

lock, W. A. Donaldson, *Tetrahedron Lett.* **2002**, *43*, 4541.
29 C. Crévisy, B. Herbage, M.-L. Marrel, L. Toupet, R. Grée, *Eur. J. Org. Chem.* **1998**, 1949.
30 M. Laabassi, P. Mosset, R. Grée, *J. Organomet. Chem.* **1997**, *538*, 91.
31 I. Ripoche, J.-L. Canet, B. Aboab, J. Gelas, Y. Troin, *J. Chem. Soc., Perkin Trans. 1* **1998**, 3485.
32 A. Braun, J.-P. Lellouche, *Tetrahedron Lett.* **2002**, *43*, 727.
33 Y. Takemoto, Y. Baba, A. Honda, S. Nakao, I. Noguchi, C. Iwata, T. Tanaka, T. Ibuka, *Tetrahedron* **1998**, *54*, 15567.
34 (a) Y. Takemoto, Y. Baba, G. Saha, S. Nakao, C. Iwata, T. Tanaka, T. Ibuka, *Tetrahedron Lett.* **2000**, *41*, 3653. (b) Y. Baba, G. Saha, S. Nakao, C. Iwata, T. Tanaka, T. Ibuka, H. Ohishi, Y. Takemoto, *J. Org. Chem.* **2001**, *66*, 81.
35 (a) P. T. Bell, B. Dasgupta, W. A. Donaldson, *J. Organomet. Chem.* **1997**, *538*, 75. (b) Y. Takemoto, N. Yoshikawa, Y. Baba, C. Iwata, T. Tanaka, T. Ibuka, H. Ohishi, *J. Am. Chem. Soc.* **1999**, *121*, 9143. (c) Y. Takemoto, K. Ishii, A. Honda, K. Okamoto, R. Yanada, T. Ibuka, *Chem. Commun.* **2000**, 1445. (d) Y. Cao, A. F. Eweas, W. A. Donaldson *Tetrahedron Lett.* **2002**, *43*, 7831.
36 (a) J. T. Wasicak, R. A. Craig, R. Henry, B. Dasgupta, H. Li, W. A. Donaldson, *Tetrahedron* **1997**, *53*, 4185. (b) B. Dasgupta, W. A. Donaldson, *Tetrahedron: Asymmetry* **1998**, *9*, 3781. (c) A.-A. S. El-Ahl, Y. K. Yun, W. A. Donaldson, *Inorg. Chim. Acta* **1999**, *296*, 261. (d) H. Bärmann, V. Prahlad, C. Tao, Y. K. Yun, Z. Wang, W. A. Donaldson, *Tetrahedron* **2000**, *56*, 2283.
37 I. Ripoche, J.-L. Canet, J. Gelas, Y. Troin, *Eur. J. Org. Chem.* **1999**, 1517.
38 A. Wada, N. Fujioka, Y. Tanaka, M. Ito, *J. Org. Chem.* **2000**, *65*, 2438.
39 (a) M. Franck-Neumann, P. Geoffroy, D. Hanss, *Tetrahedron Lett.* **2002**, *43*, 2277. (b) A. J. Pearson, V. P. Ghidu, *Org. Lett.* **2002**, *4*, 4069.
40 L. Miesch, C. Gateau, F. Morin, M. Franck-Neumann, *Tetrahedron Lett.* **2002**, *43*, 7635.
41 See for example: (a) A. J. Pearson, X. Fang, *J. Org. Chem.* **1997**, *62*, 5284.
(b) E. van den Beuken, S. Samson, E. J. Sandoe, G. R. Stephenson, *J. Organomet. Chem.* **1997**, *530*, 251. (c) H.-J. Knölker, E. Baum, M. Heininger, *Tetrahedron Lett.* **1997**, *38*, 8021. (d) H.-J. Knölker, M. Graf, U. Mangei, *J. Prakt. Chem.* **1998**, *340*, 530. (e) M.-C. P. Yeh, B.-A. Sheu, M.-Y. Wang, *Tetrahedron Lett.* **1998**, *39*, 5987. (f) C. W. Ong, J. N. Wang, T. L. Chien, *Organometallics* **1998**, *17*, 1442. (g) H. Sakurai, T. Ichikawa, K. Narasaka, *Chem. Lett.* **2000**, 508.
42 H.-J. Knölker, K. R. Reddy, *Chem. Rev.* **2002**, *102*, 4303.
43 H.-J. Knölker, W. Fröhner, *Synthesis* **2000**, 2131.
44 H.-J. Knölker, E. Baum, T. Hopfmann, *Tetrahedron* **1999**, *55*, 10391.
45 H.-J. Knölker, T. Hopfmann, *Tetrahedron* **2002**, *58*, 8937.
46 H.-J. Knölker, W. Fröhner, K. R. Reddy, *Eur. J. Org. Chem.* **2003**, 740.
47 (a) H.-J. Knölker, K. R. Reddy, *Tetrahedron Lett.* **1998**, *39*, 4007. (b) H.-J. Knölker, K. R. Reddy, *Tetrahedron* **2000**, *56*, 4733.
48 H.-J. Knölker, W. Fröhner, *Tetrahedron Lett.* **1999**, *40*, 6915.
49 H.-J. Knölker, W. Fröhner, A. Wagner, *Tetrahedron Lett.* **1998**, *39*, 2947.
50 (a) H.-J. Knölker, W. Fröhner, *Synlett* **1997**, 1108. (b) H.-J. Knölker, E. Baum, K. R. Reddy, *Tetrahedron Lett.* **2000**, *41*, 1171.
51 (a) H.-J. Knölker, W. Fröhner, *Tetrahedron* **1998**, *39*, 2537. (b) H.-J. Knölker, E. Baum, K. R. Reddy, *Chirality* **2000**, *12*, 526.
52 H.-J. Knölker, M. Wolpert, *Tetrahedron* **2003**, *59*, 5317.
53 A. J. Pearson, A. Alimardanov, *Organometallics* **1998**, *17*, 3739.
54 A. J. Pearson, I. B. Dorange, *J. Org. Chem.* **2001**, *66*, 3140.
55 A. J. Pearson, X. Wang, *Tetrahedron Lett.* **2002**, *43*, 7513.
56 A. J. Pearson, X. Wang, *J. Am. Chem. Soc.* **2003**, *125*, 638.
57 H.-J. Knölker, E. Baum, H. Goesmann, H. Gössel, K. Hartmann, M. Kosub, U. Locher, T. Sommer *Angew. Chem.* **2000**, *112*, 797; *Angew. Chem., Int. Ed.* **2000**, *39*, 781.

3.12
Chromium-Arene Complexes

Hans-Günther Schmalz and Florian Dehmel

3.12.1
Introduction

Among the numerous types of known transition metal-arene complexes, η^6-arene-$Cr(CO)_3$ complexes have enjoyed particular recognition by synthetic chemists. Since the first report in 1957 [1], a large number of η^6-benzene-$Cr(CO)_3$ derivatives have been prepared, and plenty of information has been collected about their physical properties and chemical reactivity. This chapter intends to highlight some of the synthetically most important aspects of arene-chromium chemistry in a balanced and organized fashion, without any attempt to be comprehensive.

Most η^6-arene-$Cr(CO)_6$ complexes are air-stable, yellow, or red crystalline compounds, but their solutions are usually sensitive toward oxidation, particularly in the presence of light. This sensitivity, however, can be utilized for decomplexation reactions under mild oxidative conditions (e.g., air/sunlight, I_2, or Ce(IV)), allowing an efficient release of the metal-free organic ligand.

From the point of view of organic synthesis, it is most beneficial that the electrophilic $Cr(CO)_3$ group activates the arene ligand in a characteristic fashion. This offers a wide range of useful transformations which cannot be achieved with the free arene itself [2]. Since the bulky $Cr(CO)_3$ tripod effectively shields one π-face of the arene ligand, many reactions proceed with an exceptionally high degree of diastereoselectivity. Another important stereochemical aspect is that (achiral) arene ligands bearing two non-identical substitutents in the 1,2- or 1,3-position give rise to chiral complexes (Scheme 1), which can be viewed as structures possessing a plane of chirality [3].

While such architectures offer interesting opportunities for the design of new chiral ligands and materials, the main application of arene-$Cr(CO)_3$ complexes is

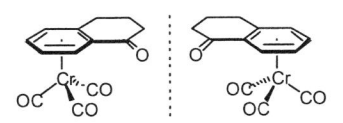

Scheme 1

in the field of total synthesis. In particular, their use as chiral synthetic building blocks opens up novel and competitive strategies for the enantioselective (multi-step) syntheses of complex organic molecules. Of course, the expenditure associated with the introduction of the metal unit pays off, especially in those cases in which the chemical and stereochemical effects of the $Cr(CO)_3$ unit can be exploited in more than one transformation [4].

3.12.2
Preparation

Arene-$Cr(CO)_3$ complexes are usually prepared from the free arenes by thermolysis with $Cr(CO)_6$ in a high-boiling solvent. Special solvent mixtures (e.g., n-Bu_2O/THF 10:1) [5] or sophisticated reaction apparatus [6] have been developed to avoid the troublesome sublimation of $Cr(CO)_6$. The complexation can also be accomplished under much milder conditions using reagents such as $(CH_3CN)_3$-$Cr(CO)_3$, pyrrole-$Cr(CO)_3$ [7], or naphthalene-$Cr(CO)_3$ [8] for the transfer of the $Cr(CO)_3$ fragment.

The complexation of chiral or chirally modified arene ligands often proceeds with high diastereoselectivity [9]. The observed selectivity can be attributed either to repulsive interactions with a sterically demanding substituent [10] or, usually with better selectivity, to a directing pre-coordination of the incoming $Cr(CO)_x$ fragment at a polar functional group (e.g., OH, NR_2) [11]. This was, for instance, utilized for the preparation of non-racemic 1-tetralone-$Cr(CO)_3$ derivatives (Eq. (1)) [12]. Other entries to non-racemic planar-chiral complexes involve either the chemical [13] or enzymatical [14] resolution of racemic mixtures, the diastereoselective transformation of chirally modified complexes [15], or the enantioselective transformation of prochiral complexes [16]. Some arene-$Cr(CO)_3$ complexes are also accessible from Fischer carbene complexes through the Dötz reaction and related processes [17].

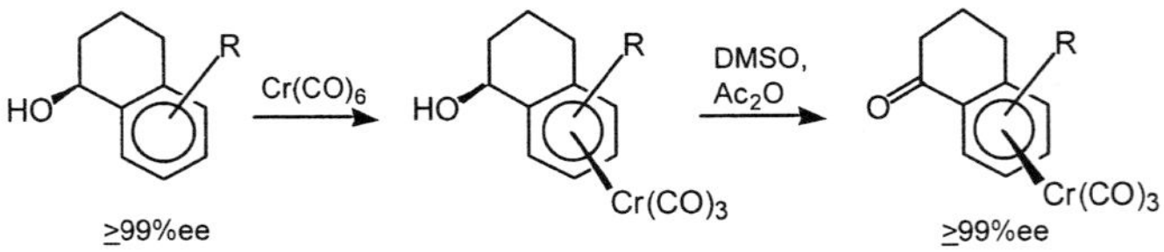

$$\text{(1)}$$

3.12.3
Nucleophilic Addition to the Arene Ring

One of the most prominent and useful effects caused by the (electron-withdrawing) metal fragment is the enhanced reactivity of the arene ring toward nucleophiles [18]. Thus, treatment of chloro- or fluorobenzene-$Cr(CO)_3$ derivatives with a wide range of heteroatom or carbon nucleophiles results in aromatic substitution (S_NAr) (Eq. (2)).

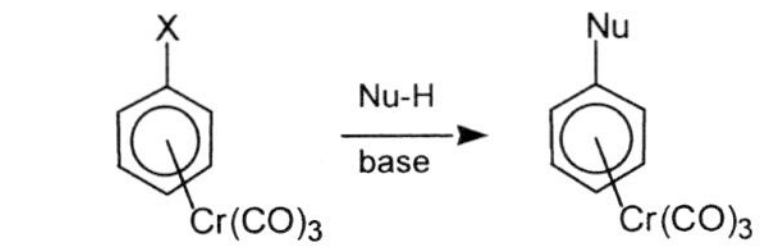

3.12.3 Nucleophilic Addition to the Arene Ring

$$\text{Ar-X} \xrightarrow[\text{base}]{\text{Nu-H}} \text{Ar-Nu} \quad (2)$$

X = F, Cl, (OR);
Nu-H = ROH, RSH, R_2NH, CH-acidic compounds (esters, nitriles, sulfones...)

These reactions may proceed in an *ipso-*, *cine-* or *tele-*mode [19], and even alkoxides can act as a leaving group [20]. For example, the desoxygenation of methoxy-substituted complexes can be accomplished with LiBEt$_3$H (Eq. (3)) [20b].

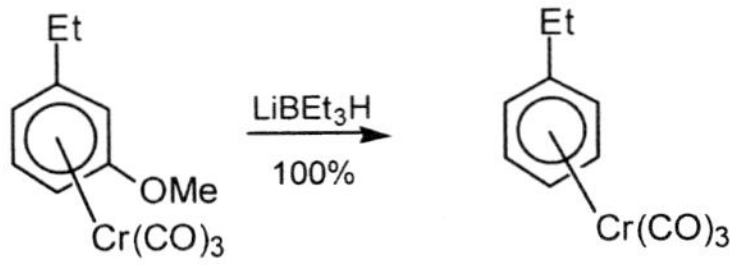

$$\text{(Et-Ar-OMe)Cr(CO)}_3 \xrightarrow[100\%]{\text{LiBEt}_3\text{H}} \text{(Et-Ar)Cr(CO)}_3 \quad (3)$$

Carbon nucleophiles derived from carbon acids with pK_A > 22 even add to benzene-Cr(CO)$_3$ derivatives lacking a good leaving group (Eq. (4)). The resulting η^5-cyclohexadienyl intermediates [21] can either be oxidized with iodine to the decomplexed arenes [22] or converted to dearomatized products, for instance, by protonation with a strong acid to afford cyclohexadienes or by addition of an electrophile (optionally under CO atmosphere), resulting in the formation of *trans*-5,6-disubstituted cyclohexa-1,3-dienes [23]. Treatment of the intermediates with a hydride acceptor gives rise to substituted complexes [24].

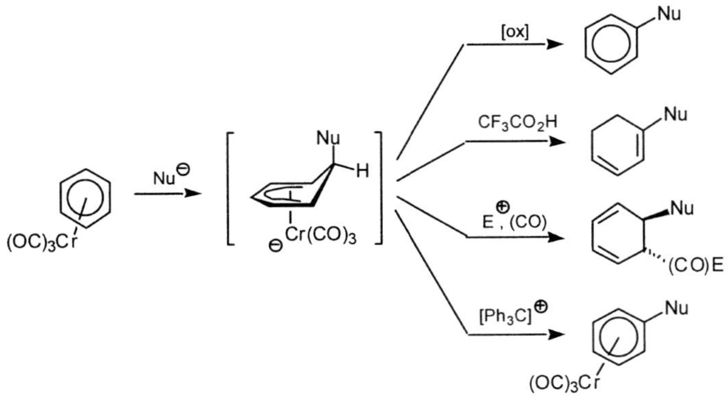

$$(4)$$

In general, the addition of nucleophiles to arenes bearing a donor substituent (OR, NR$_2$) proceeds regioselectively in the meta-position [25], giving access to *meta*-substituted arylethers and anilines. Of particular synthetic value are stereoselective transformations of arenes to yield dearomatized products [23 a]. For example, prochiral benzaldimines can be converted enantioselectively into functionalized cyclohexadienes in the presence of a chiral co-solvent (Eq. (5)) [26], and *ortho*-substituted anisole complexes give access to non-racemic cyclohexenones under appropriate conditions (Eq. (6)) [27].

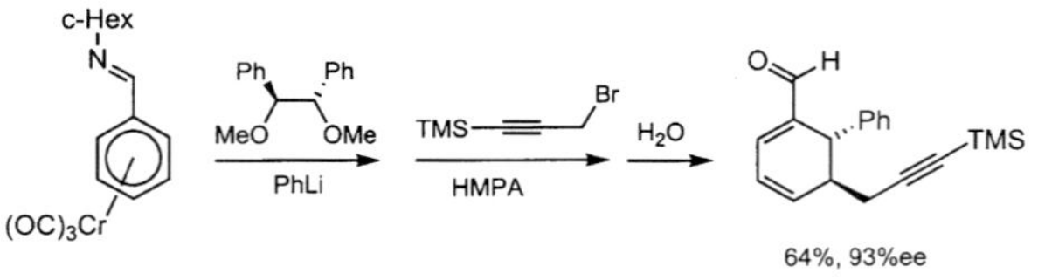

$$(5)$$

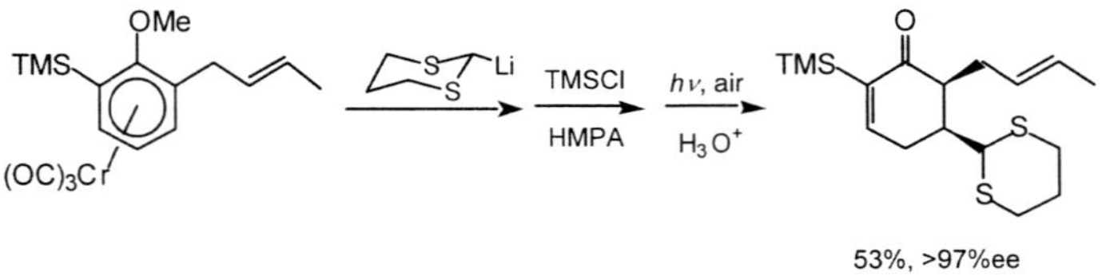

$$(6)$$

It is noteworthy that the addition of nucleophiles to styrene-Cr(CO)$_3$ derivatives usually proceeds in a conjugate fashion, offering interesting applications in total synthesis [28]. As well as anionic nucleophiles, nucleophilic radicals (such as ketyl or ketimine-derived aza-ketyl radicals) may also add to Cr(CO)$_3$-complexed arenes. If the substrate contains a suitable leaving group (OMe, Cl, F), the reactions give rise to (*cine* or *meta-tele*) substituted complexes (Eq. (7)) [29].

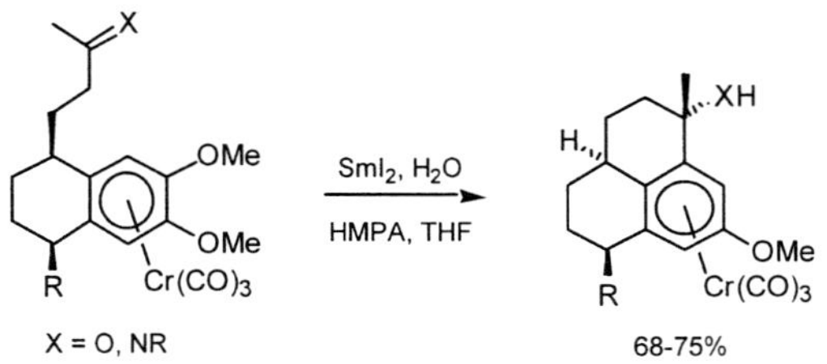

$$(7)$$

3.12.4
Ring Lithiation

Another consequence of the electrophilic nature of the Cr(CO)$_3$ unit is the enhanced acidity of aromatic protons. Therefore, ring deprotonation occurs readily. As in the case of uncomplexed systems, polar substituents exhibit *ortho*-directing effects, allowing the regioselective preparation of numerous alkylated complexes under mild conditions [30] (Eq. (8)).

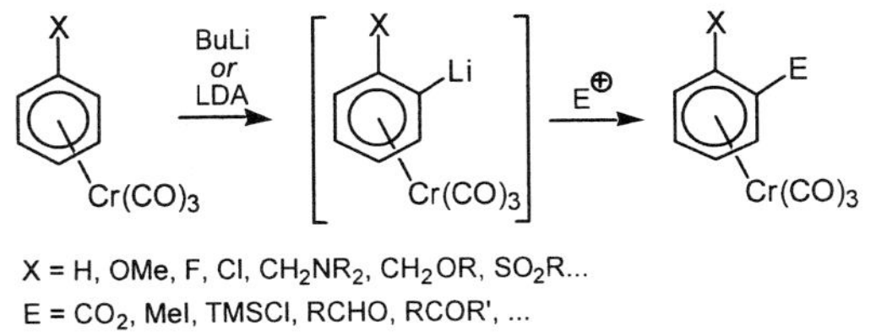

$$X = H, OMe, F, Cl, CH_2NR_2, CH_2OR, SO_2R...$$
$$E = CO_2, MeI, TMSCl, RCHO, RCOR', ...$$

With chiral *ortho*-directing groups, the deprotonation often proceeds with high diastereoselectivity, as in the case shown in Eq. (9) [31].

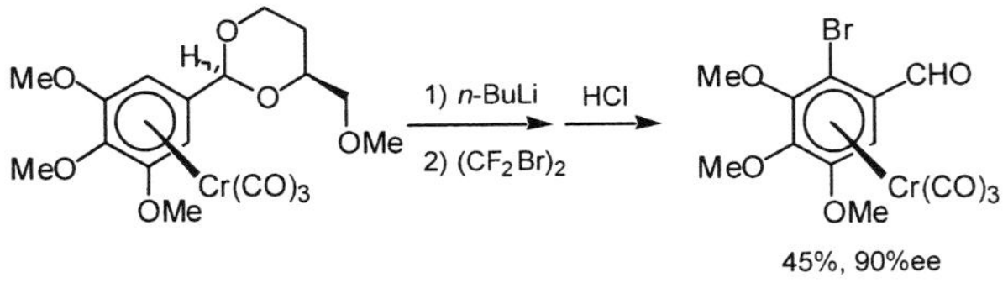

45%, 90%ee   (9)

Moreover, the deprotonation of prochiral substrates can be achieved with high enantioselectivity using chiral bases [32]. An example is shown in Eq. (10) [16e].

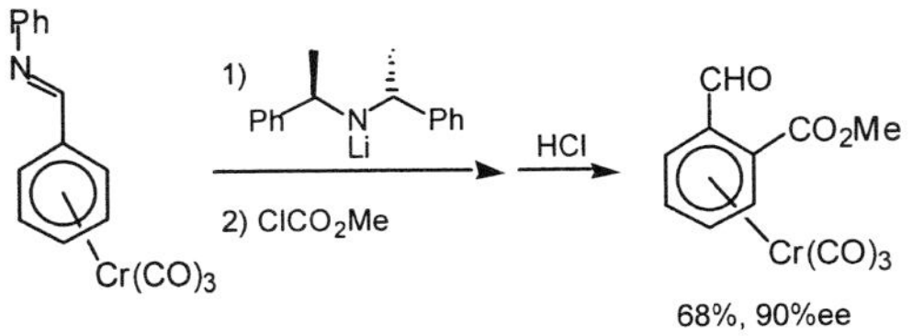

68%, 90%ee   (10)

Meta-lithiations of complexed phenolethers are also possible if bulky triisopropylethers [33] or sterically hindered bases such as LiTMP are used [34].

3.12.5
General Aspects of Side Chain Activation

Toluene-$Cr(CO)_3$ derivatives offer many opportunities for benzylic functionalization because the $Cr(CO)_3$ moiety is able to stabilize either negative or positive charge as well as a radical in the benzylic position. This "chemical hermaphroditism" [35b] can be understood in terms of the resonance structures shown in Scheme 2. Structural and energetical details of such intermediates were calculated using DFT methods [35].

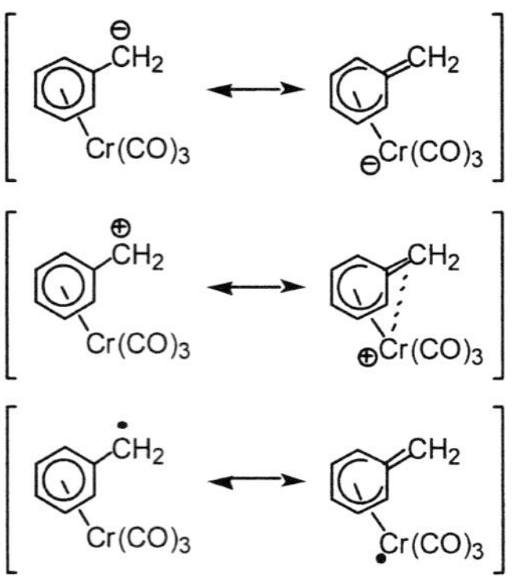

Scheme 2

3.12.6
Side Chain Activation via Stabilization of Negative Charge

The strong stabilization of negative charge in the benzylic position by the $Cr(CO)_3$ fragment allows deprotonations under mild conditions [36]. For example, phthalane-$Cr(CO)_3$ was subsequently methylated in both benzylic positions to selectively form the *cis*-1,3-disubstituted product (Eq. (11)) [37]. However, if a silyl group is introduced first, the second deprotonation occurs at the silylated position to give the 1,1-disubstituted product, which can efficiently be further converted to *trans*-1,3-alkylated products [38] (Eq. (12)).

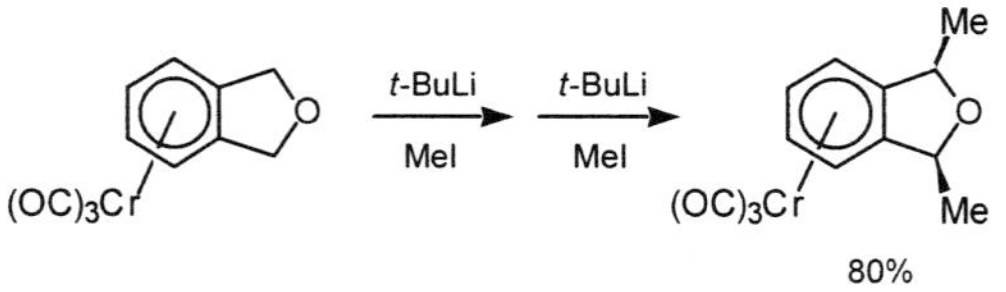

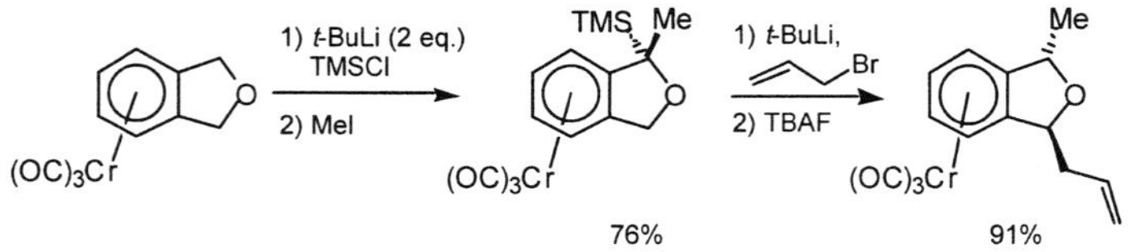

If two competing benzylic positions are present in a substrate, the regioselectivity of the deprotonation is usually controlled by (stereo-) electronic effects [39]. In certain cases, enantiotopic benzylic positions can be discriminated by a chiral base to give rise to optically active alkylation products [40]. The combination of aromatic and benzylic lithiation/alkylation offers efficient strategies for the regio- and stereoselective functionalization of relevant ring systems such as tetrahydroisoquinolins [41] (Eq. (13)) or tetralins. Frequently, more acidic ring positions are temporarily protected by silylation, as in the synthesis of 1-*epi*-helioporin D (Eq. (14)) [42].

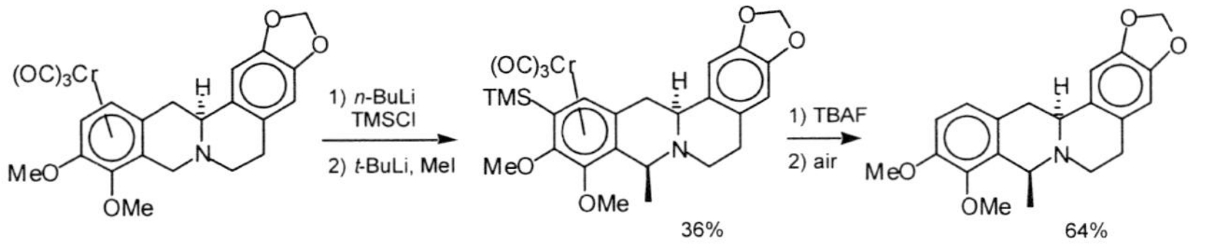

(13)

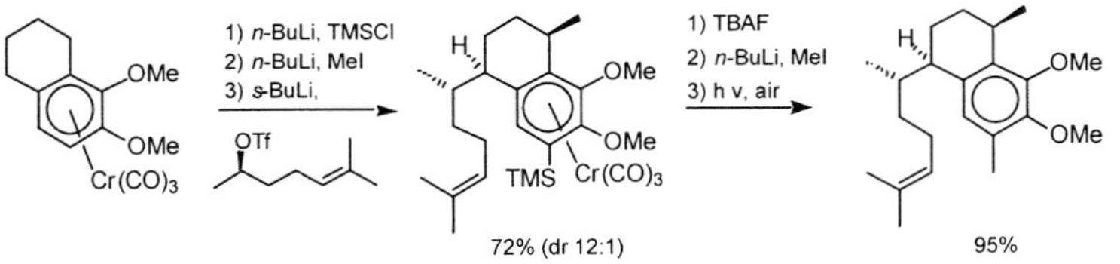

(14)

3.12.7
Side Chain Activation via Stabilization of Positive Charge

The ability of the Cr(CO)$_3$ fragment to efficiently stabilize benzylic carbocations also allows for highly valuable synthetic transformations [43]. For example, benzylic *endo*-alkylated benzocycloalkenes are accessible from the corresponding aryl ketones through nucleophilic addition to the carbonyl group and subsequent ionic hydrogenation of the tertiary alcohol. The hydride attacks the intermediate cation from the less hindered side as shown in Eq. (15) [44].

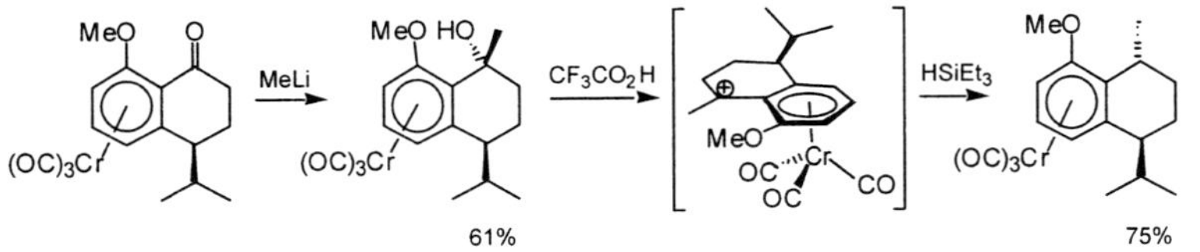

(15)

Because of the coordination of the metal fragment (see Scheme 2), the cationic intermediates have a pronounced configurational stability [35a]. As a consequence, S_N1-type reactions proceed in "acyclic" systems in a highly stereocontrolled fashion (overall retention) [45]. An impressive example is the stereospecific cyclization reaction shown in Eq. (16) leading to a tetrahydrobenzazepine [46].

3.12.8
Stabilization of Radicals in the Benzylic Position

Though much less pronounced, the $Cr(CO)_3$ moiety also causes a significant configurational stability of benzylic radicals [35a]. This can be exploited for highly stereoselective transformations. An example is the electron transfer-mediated umpolung/alkylation of chiral 1-alkyloxy-ethyl-benzene complexes, which proceeds with retention of configuration [47] (Eq. (17)).

The dimerization of ketyl-type radicals derived from planar-chiral benzaldimine or benzaldehyde complexes has been employed for the synthesis of enantiomerically pure 1,2-diols and 1,2-diamines by diastereoselective pinacol coupling [48]. Similarly, the samarium(II)iodide-mediated reaction of complexed *ortho*-substituted benzaldehydes and arylketones with acrylates leads to γ-butyrolactons with virtually complete diastereoselectivity (Eq. (18)) [49].

3.12.9
Additions to Complexed Benzaldehydes and Related Substrates

One of the most frequently exploited concepts in synthetic arene-$Cr(CO)_3$ chemistry is the (nucleophilic) addition to planar-chiral *ortho*-substituted benzaldehyde- or benzimine-$Cr(CO)_3$ to give benzylic chiral products, usually with high diastereoselectivity (Eq. (19)) [50].

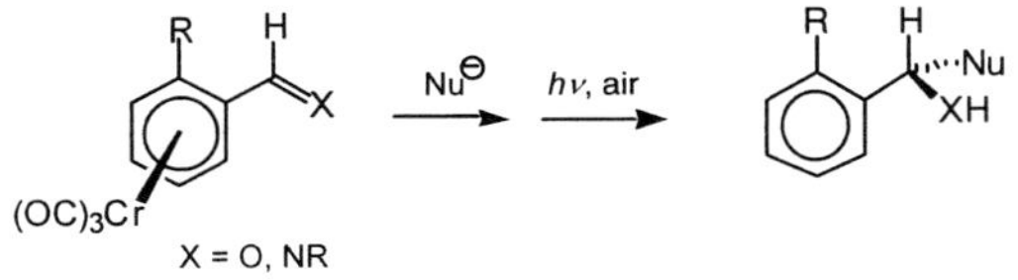

$$\tag{19}$$

As an example, the addition of ester enolates to the *ortho*-TMS-benzaldehyde complex is achieved with high diastereoselectivity. In this case, the addition is followed by a 1,4-silyl-shift and an intramolecular attack of the resulting aryl lithium species to the ester functionality, affording spiro products related to the antibiotic fredericamycin (Eq. (20)) [51].

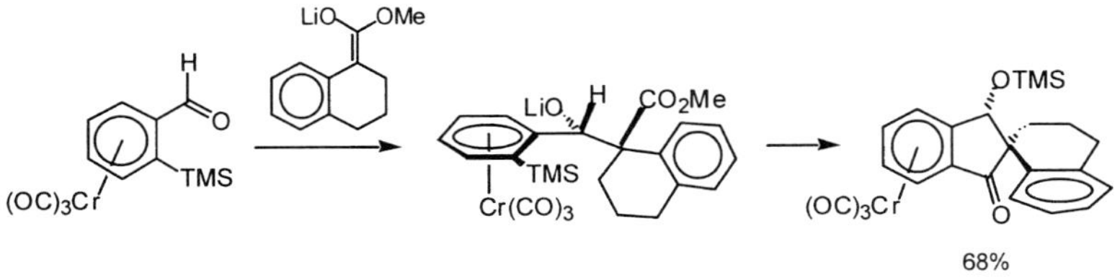

$$\tag{20}$$

68%

Conjugate additions to complexed 3-aryl-enones also proceed with a high degree of diastereoselectivity [52]. Other possible transformations with such substrates involve [4+2] and [2+2]-cycloadditions to deliver interesting building blocks for further elaboration [53]. The addition of vinyl nucleophiles to 1,2-dioxobenzocyclobutene complexes triggers a domino process involving [3.3]-sigmatropic ring expansion and subsequent aldol cyclization to afford interesting products in a highly regio- and diastereoselective fashion (Eq. (21)) [54].

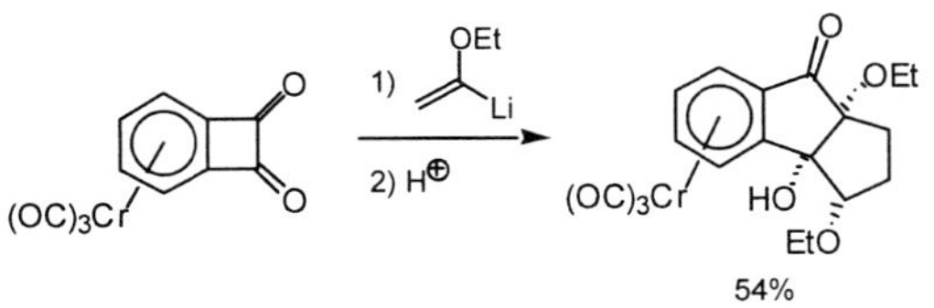

$$\tag{21}$$

54%

3.12.10
Cross-Coupling Reactions

Because of its electron-withdrawing character, the $Cr(CO)_3$-unit strongly activates haloarenes toward the oxidative addition of Pd(0) into the C-X bond. As a consequence, chloro- and even fluoroarene-$Cr(CO)_3$ complexes readily undergo Pd-catalyzed couplings such as Suzuki- [55], Sonogashira- [56], Stille- [57], and Heck-type reactions. An intramolecularly-stabilized organo-indium(III) reagent was used for the introduction of methyl groups [58]. Alkoxycarbonylation reactions of haloarene-$Cr(CO)_3$ complexes also proceed under very mild conditions and allow for the desymmetrization of prochiral dichlorobenzene-$Cr(CO)_3$ complexes, thus affording a catalytic-enantioselective access to planar-chirality [16f] (Eq. (22)).

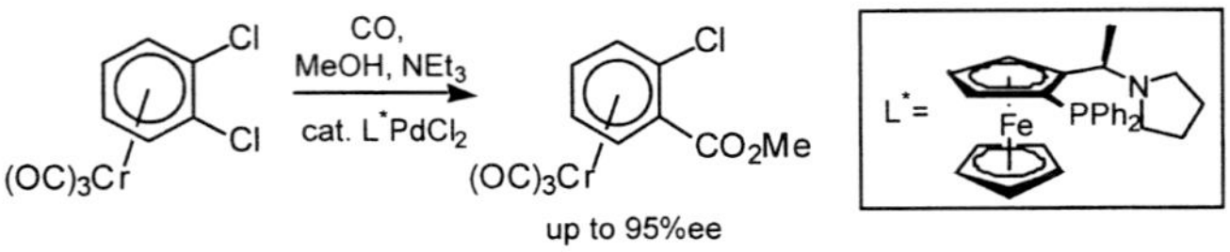

$$\tag{22}$$

Chloroarene complexes with unsaturated side chains undergo intramolecular Heck reactions. If the carbopalladated intermediates are trapped through a subsequent methoxycarbonylation, high diastereoselectivities are observed (Eq. (23)) [59].

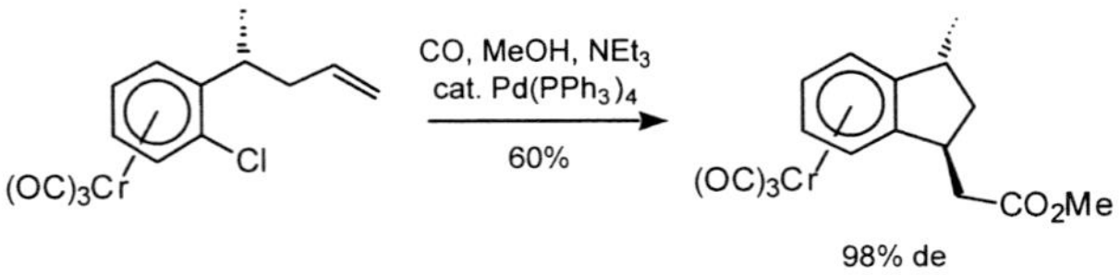

$$\tag{23}$$

A very important feature of cross-coupling reactions between haloarene-$Cr(CO)_3$ complexes and aryl-metals is the fact that the planar chirality of the complex can be transferred into axial chirality of the resulting biaryl system in a highly efficient and controlled manner [60]. For example, the variation of the substitution pattern at the coupling partner allows the diastereoselective preparation of both biarylic atropisomers from one (enantiopure) planar chiral complex (Eq. (24)) [61].

An impressive demonstration of the feasibility of this methodology is found in the enantioselective synthesis of (–)-steganone, in which the biaryl system was diastereoselectively assembled starting from a planar-chiral arene-Cr(CO)$_3$ complex (Eq. (25)) [4 f].

3.12.11
Solid Phase Chemistry

Arene-chromium complexes have also been applied as traceless linkers in solid phase chemistry [62]. By photolysis of an arene-Cr(CO)$_3$ complex in the presence of a phosphane-functionalized polymer, one of the carbonyl ligands of the Cr(CO)$_3$ unit is substituted and the complex attached to the polymer bead. After a desired transformation, such as a nucleophilic substitution, has been performed on the immobilized substrate, the aromatic ligand can be released by oxidative decomplexation (Eq. (26)).

3.12.12
Arene-Cr(CO)₃ Complexes as Catalysts

While most applications of arene-Cr(CO)$_3$ chemistry require stoichiometric amounts of the metal, arene-Cr(CO)$_3$ complexes are also a good source of the "free" Cr(CO)$_3$ unit, which catalyzes a number of highly useful transformations, such as the 1,4-hydrogenation of 1,3-dienes to Z-configured alkenes (Eq. (27)) or 1,5-hydrogen shifts [63]. Only 1,3-dienes that can easily adopt an s-cisoid conformation undergo this reaction; isolated double bonds are not affected.

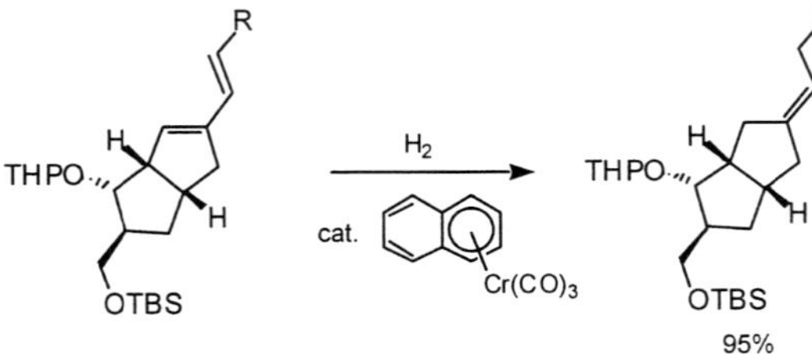

(27)

In recent years, the planar-chiral architecture of *ortho*-disubstituted arene-Cr(CO)$_3$ complexes has become more and more exploited for the construction of chiral ligands for asymmetric catalysis [65]. Successful applications are Rh(I)-mediated hydrogenations of ketones [66], Pd-catalyzed allylic alkylations [67], Diels-Alder cycloadditions [68], addition of diethylzinc to benzaldehyde [69], and Rh(I)-catalyzed isomerization of allylamines [70]. Other convincing examples are the enantioselective hydrosilylation (Eq. (28)) [71] and hydrovinylation (Eq. (29)) [72] of styrene, which both proceed with high enantioselectivity.

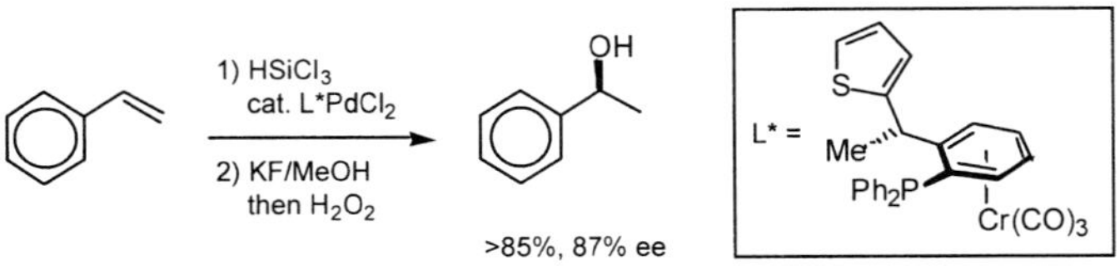

(28)

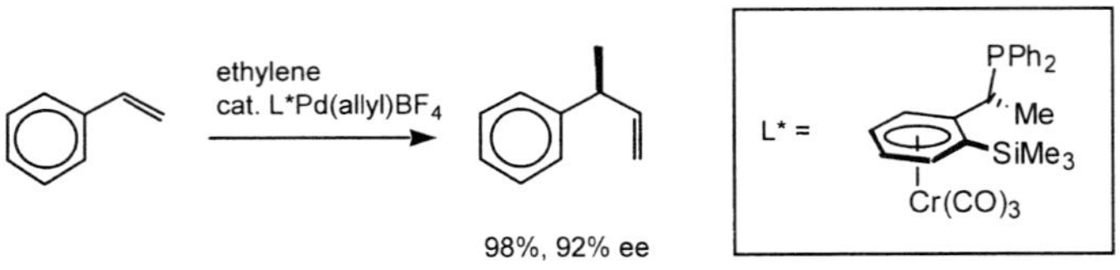

(29)

3.12.13
References

1. (a) E. O. Fischer, K. Öfele, *Chem. Ber.* **1957**, *90*, 2532. (b) E. O. Fischer, K. Öfele, *Z. Naturforsch.* **1958**, *13 B*, 458.
2. For selected reviews, see: (a) L. S. Hegedus, *Transition Metals in the Synthesis of Complex Organic Molecules*, 2nd edn., University Science Books, Sausalito, CA, **1999**, Chap. 10. (b) M. F. Semmelhack in *Comprehensive Organometallic Chemistry II* (Eds.: E. W. Abel, F. G. A. Stone, G. Wilkinson), Pergamon Press, New York, **1995**, Vol. 12, p. 979. (c) M. F. Semmelhack in *Comprehensive Organometallic Chemistry II* (Eds.: E. W. Abel, F. G. A. Stone, G. Wilkinson), Pergamon Press, New York, **1995**, Vol. 12, p. 1017. (d) S. G. Davies, T. D. McCarthy in *Comprehensive Organometallic Chemistry II* (Eds.: E. W. Abel, F. G. A. Stone, G. Wilkinson), Pergamon Press, New York, **1995**, Vol. 12, p. 1039.
3. (a) K. Schlögl, in *Organometallics in Organic Synthesis 2* (Eds.: H. Werner, G. Erker), Springer-Verlag, Berlin, **1989**, p. 63 and ref. cited therein. (b) A. Solladié-Cavallo in *Advances in Metal Organic Chemistry* (Eds.: L. S. Liebeskind), JAI Press, London, **1989**, Vol.2 , p. 99.
4. For selected synthesis, see (a) M. F. Semmelhack, A. Zask, *J. Am. Chem. Soc.* **1983**, *105*, 2034. (b) M. Uemura, H. Nishimura, T. Minami, Y. Hayashi, *J. Am. Chem. Soc.* **1991**, *113*, 5402. (c) A. Majdalani, H.-G. Schmalz, *Synlett* **1997**, 1303. (d) K. Schellhaas, H.-G. Schmalz, J. W. Bats, *Chem. Eur. J.* **1998**, *4*, 57. (e) F. Dehmel, H.-G. Schmalz, *Org. Lett.* **2001**, *3*, 3579. (f) L. G. Monovich, Y. Le Huérou, M. Rönn, G. A. Molander, *J. Am. Chem. Soc.* **2000**, *122*, 52. (g) H. Ratni, E. P. Kündig, *Org. Lett.* **1999**, *1*, 1997.
5. C. A. C. Mahaffy, P. L. Pauson, *Inorg. Synth.* **1979**, *19*, 154.
6. M. Hudecek, S. Toma, *J. Organomet. Chem.* **1990**, *393*, 115.
7. (a) A. Goti, M. F. Semmelhack, *J. Organomet. Chem.* **1994**, *470*, C4–C7. (b) R. Wolfgramm, S. Laschat, *J. Organomet. Chem.* **1999**, *575*, 141.
8. E. P. Kündig, C. Perret, S. Spichiger, G. Bernardinelli, *J. Organomet. Chem.* **1985**, *286*, 183.
9. For a review, see: R. S. Paley, *Chem. Rev.* **2002**, *102*, 1493.
10. See, for instance: (a) G. B. Jones, G. Mustafa, *Tetrahedron: Asymmetry* **1998**, *9*, 2023. (b) K. R. Stewart, S. G. Levine, J. Bordner, *J. Org. Chem.* **1984**, *49*, 4082. (c) T.-L. Ho, K.-Y. Lee, C.-K. Chen, *J. Org. Chem.* **1997**, *62*, 3365.
11. See, for instance: (a) M. Uemura, R. Miyake, M. Shiro, *Tetrahedron Lett.* **1991**, *32*, 4569. (b) E. P. Kündig, J. Leresche, L. Saudan, G. Bernardinelli, *Tetrahedron* **1996**, *52*, 7363. (c) S. G. Davies, C. L. Goodfellow, *J. Organomet. Chem.* **1988**, *340*, 195. (d) M. Uemura, T. Minami, Y. Hayashi, *J. Am. Chem. Soc.* **1987**, *109*, 5277.
12. H.-G. Schmalz, B. Millies, J. W. Bats, G. Dürner, *Angew. Chem.* **1992**, *104*, 640; *Angew. Chem., Int. Ed. Engl.* **1992**, *31*, 631
13. See, for instance: (a) A. Solladié-Cavallo, G. Solladié, E. Tsamo, *J. Org. Chem.* **1979**, *44*, 4189. (b) S. G. Davies, C. L. Goodfellow, *J. Chem. Soc., Perkin Trans. 1* **1990**, 393.
14. For recent work, see: (a) J. A. S. Howell, M. G. Palin, G. Jaouen, B. Malezieux, S. Top, J. M. Ceuse, J. Salaün, P. McArdle, D. Cunningham, M. O'Gara, *Tetrahedron: Asymmetry* **1996**, *7*, 95. (b) M. Uemura, H. Nishimura, S. Yamada, Y. Hayashi, K. Nakamura, K. Ishihara, A. Ohno, *Tetrahedron: Asymmetry* **1994**, *5*, 1673. (c) C. Baldoli, S. Maiorana, G. Carrea, S. Riva, *Tetrahedron: Asymmetry* **1993**, *3*, 767.
15. See, for instance: (a) Y. Kondo, J. R. Green, J. Ho, *J. Org. Chem.* **1993**, *58*, 6182. (b) A. Alexakis, T. Kanger, P. Mangeney, F. Rose-Munch, A. Perrotey, E. Rose, *Tetrahedron: Asymmetry* **1995**, *6*, 2135. (c) M. Uemura, A. Daimon, Y. Hayashi, *J. Chem. Soc., Chem. Commun.* **1995**, 1943.
16. See, for instance: (a) D. A. Price, N. S. Simpkins, A. M. MacLeod, A. P. Watt,

J. Org. Chem. **1994**, *59*, 1961.
(b) M. UEMURA, Y. HAYASHI, Y. HAYASHI, *Tetrahedron: Asymmetry* **1994**, *5*, 1427.
(c) H.-G. SCHMALZ, K. SCHELLHAAS, *Tetrahedron Lett.* **1995**, *36*, 5515. (d) R. WILHELM, I. K. SEBHAT, A. J. P. WHITE, D. J. WILLIAMS, D. A. WIDDOWSON, *Tetrahedron: Asymmetry* **2000**, 11, 5003.
(e) S. PACHE, C. BOTUHU, R. FRANZ, E. P. KÜNDIG, J. EINHORN, *Helv. Chim. Acta* **2000**, *83*, 2436. (f) B. GOTOV, H.-G. SCHMALZ, *Org. Lett.* **2001**, *3*, 1753.

17 For a review, see: (a) K. H. DÖTZ, P. TOMUSCHAT, *Chem. Rev. Soc.* **1999**, *28*, 187; For recent work, see: (b) L. FOGEL, R. P. HSUNG, W. D. WULFF, *J. Am. Chem. Soc.* **2001**, *123*, 5580. (c) K. H. DÖTZ, S. MITTENZWEY, *Eur. J. Org. Chem.* **2002**, 39. (d) R. P. HSUNG, W. D. WULFF, S. CHAMBERLAIN, Y. LIU, R.-Y. LIU, H. WANG, J. F. QUINN, S. L. B. WANG, A. L. RHEINGOLD, *Synthesis*, **2001**, 200.

18 For reviews, see ref. [2c] as well as: M. F. SEMMELHACK in *Comprehensive Organic Synthesis, Vol. 4* (Ed.: B. M. TROST, I. FLEMING), Pergamon, Oxford, **1991**, pp. 517.

19 (a) F. ROSE-MUNCH, E. ROSE, *Eur. J. Inorg. Chem.* **2002**, 1269. (b) F. ROSE-MUNCH, V. GAGLIARDINI, P. RENARD, E. ROSE, *Coord. Chem. Rev.* **1998**, *178–180*, 249. (c) F. ROSE-MUNCH, E. ROSE, *Curr. Org. Chem.* **1999**, *3*, 445.

20 (a) V. GAGLIARDINI, V. ONNIKIAN, F. ROSE-MUNCH, E. ROSE, *Inorg. Chim. Acta* **1997**, *259*, 5029. (b) J.-P. DJUKIC, F. ROSE-MUNCH, E. ROSE, F. SIMON, Y. DROMZEE, *Organometallics* **1995**, *14*, 2027.

21 (a) M. F. SEMMELHACK, H. T. HALL, JR., M. YOSHIFUJI, G. CLARK, *J. Am. Chem. Soc.* **1976**, *98*, 6387. (b) M. F. SEMMELHACK, H. T. HALL, JR., R. FARINA, M. YOSHIFUJI, G. CLARK, T. BARGAR, K. HIROTSU, J. CLARDY, *J. Am. Chem. Soc.* **1979**, *101*, 3535.

22 (a) M. F. SEMMELHACK, G. CLARK, *J. Am. Chem. Soc.* **1977**, *99*, 1675. (b) M. BELLASSOUED, E. CHELAIN, J. COLLOT, H. RUDLER, J. VAISSERMANN, *Chem. Commun.* **1999**, 187. (c) H. RUDLER, V. COMTE, E. GARRIER, M. BELLASSOUED, E. CHELAIN, J. VAISSERMANN, *J. Organomet. Chem.* **2001**, *621*, 284.

23 For a review, see: (a) A. R. PAPE, K. P. KALIAPPAN, E. P. KÜNDIG, *Chem. Rev.* **2000**, *100*, 2917. See also: (b) E. P. KÜNDIG, D. P. SIMMONS, *J. Chem. Soc., Chem. Commun.* **1983**, 1320. (c) E. P. KÜNDIG, A. RIPA, R. LIU, G. BERNARDINELLI, *J. Org. Chem.* **1994**, *59*, 4773. (d) D. BERUBEN, E. P. KÜNDIG, *Helv. Chim. Acta* **1996**, *79*, 1533.

24 A. FRETZEN, A. RIPA, R. LIU, G. BERNARDINELLI, E. P. KÜNDIG, *Chem. Eur. J.* **1998**, *4*, 251.

25 A. PFLETSCHINGER, W. KOCH, H.-G. SCHMALZ, *New J. Chem.* **2001**, *25*, 446.

26 (a) D. AMURRIO, K. KHAN, E. P. KÜNDIG, *J. Org. Chem.* **1996**, *61*, 2258. (b) I. S. MANN, D. A. WIDDOWSON, M. C. CLOUGH, *Tetrahedron* **1991**, *47*, 7991. (c) E. P. KÜNDIG, R. CANNAS, M. LAXMISHA, R. G. LIU, S. TCHERTCHIAN, *J. Am. Chem. Soc.* **2003**, *125*, 5642.

27 (a) M. F. SEMMELHACK, H.-G. SCHMALZ, *Tetrahedron Lett.* **1996**, *37*, 3089.
(b) H.-G. SCHMALZ, K. SCHELLHAAS, *Angew. Chem.* **1996**, *108*, 2277; *Angew. Chem., Int. Ed. Engl.* **1996**, *35*, 2146. See also: ref [4d] as well as (c) A. QUATTROPANI, G. ANDERSON, G. BERNARDINELLI, E. P. KÜNDIG, *J. Am. Chem. Soc.* **1997**, *119*, 4773.

28 (a) M. F. SEMMELHACK, W. SEUFERT, L. KELLER, *J. Am. Chem. Soc.* **1980**, *102*, 6584. (b) M. UEMURA, T. MINAMI, Y. HAYASHI, *J. Chem. Soc., Chem. Commun.* **1982**, 1193. (c) T. J. J. MÜLLER, M. ANSORGE, *Tetrahedron* **1998**, *54*, 1457. (d) M. SAINSBURY, M. F. MAHON, C. S. WILLIAMS, A. NAYLOR, D. I. C. SCOPES, *Tetrahedron* **1991**, *47*, 4195. (e) A. MAJDALANI, H.-G. SCHMALZ, *Tetrahedron Lett.* **1998**, *38*, 4545. (f) F. DEHMEL, J. LEX, H.-G. SCHMALZ, *Org. Lett.* **2002**, *4*, 3915. See also ref. [4e].

29 (a) H.-G. SCHMALZ, S. SIEGEL, J. W. BATS, *Angew. Chem.* **1995**, *107*, 2597; *Angew. Chem., Int. Ed. Engl.* **1995**, *34*, 2383. (b) O. HOFFMANN, H.-G. SCHMALZ, *Synlett* **1998**, 1426. (c) H.-G. SCHMALZ, O. KIEHL, B. GOTOV, *Synlett* **2002**, 1253. (d) O. SCHWARZ, R. BRUN, J. W. BATS, H.-G. SCHMALZ, *Tetrahedron Lett.* **2002**, *43*, 1009.

30 (a) M. F. Semmelhack, J. Bisaha, M. Czarny, *J. Am. Chem. Soc.* **1979**, *101*, 768. (b) R. J. Card, W. S. Trahanovsky, *J. Org. Chem.* **1980**, *45*, 2560. For a review, see ref. [2b]. (c) I. K. Sebhat, Y.-L. Tan, D. A. Widdowson, R. Wilhelm, A. J. P. White, D. J. Williams, *Tetrahedron* **2000**, *56*, 6121.

31 (a) T. Watanabe, M. Shakadou, M. Uemura, *Synlett* **2000**, 1141. (b) K. Kamikawa, T. Watanabe, A. Daimon, M. Uemura, *Tetrahedron* **2000**, *56*, 2325. (c) T. Watanabe, M. Uemura, *J. Chem. Soc., Chem. Commun.* **1998**, 871. (d) K. Kamikawa, A. Tachibana, S. Sugimoto, M. Uemura, *Org. Lett.* **2001**, *3*, 2033. (e) T. Watanabe, M. Shakadou, M. Uemura, *Inorg. Chim. Acta* **1999**, *296*, 80. See also ref. [15].

32 For reviews, see: (a) S. E. Gibson, E. G. Reddington, *Chem. Commun.* **2000**, 989. (b) R. A. Ewin, A. M. MacLeod, D. A. Price, N. S. Simpkins, A. P. Watt, *J. Chem. Soc., Perkin Trans. 1* **1997**, 401. See also: (c) Y.-L. Tan, D. A. Widdowson, R. Wilhelm, *Synlett* **2001**, 1632. (d) Y.-L. Tan, A. J. P. White, D. A. Widdowson, R. Wilhelm, D. J. Williams, *J. Chem. Soc., Perkin Trans. 1* **2001**, 3269. (e) R. Wilhelm, D. A. Widdowson, *Org. Lett.* **2001**, *3*, 3079. See also ref. [16 a–f] and [27 b].

33 (a) I. S. Mann, D. A. Widdowson, M. C. Clough, *Tetrahedron* **1991**, *47*, 7981. (b) N. F. Masters, D. A. Widdowson, *J. Chem. Soc., Chem. Commun.* **1983**, 955.

34 H.-G. Schmalz, T. Volk, D. Bernicke, S. Huneck, *Tetrahedron* **1997**, *53*, 9219.

35 (a) A. Pfletschinger, T. K. Dargel, J. W. Bats, H.-G. Schmalz, W. Koch, *Chem. Eur. J.* **1999**, *5*, 537. (b) C. A. Merlic, J. C. Walsh, D. J. Tantillo, K. N. Houk, *J. Am. Chem. Soc.* **1999**, *121*, 3596. (c) C. A. Merlic, B. N. Hietbrink, K. N. Houk, *J. Org. Chem.* **2001**, *66*, 6738. See also ref. [2 d]

36 For a review, see S. G. Davies, S. J. Coote, C. L. Goodfellow in *Advances in Metal Organic Chemistry* (Eds.: L. S. Liebeskind), JAI Press, London, 1989, Vol. 2, p. 1. For other, more recent examples, see: (b) D. Schinzer, U. Abel, P. G. Jones, *Synlett* **1997**, 632. (c) M. Brisander, P. Caldirola, A. M. Johansson, U. Hacksell, *J. Org. Chem.* **1998**, *63*, 5362.

37 S. J. Coote, S. G. Davies, D. Middlemiss, A. Naylor, *J. Organomet. Chem.* **1989**, *379*, 81.

38 S. Zemolka, H.-G. Schmalz, J. Lex, *Angew. Chem.* **2002**, *114*, 2635; *Angew. Chem., Int. Ed. Engl.* **2002**, *41*, 2525.

39 T. Volk, D. Bernicke, J. W. Bats, H.-G. Schmalz, *Eur. J. Inorg. Chem.* **1998**, 1883.

40 (a) T. Hata, H. Koide, M. Uemura, *Synlett* **2000**, 1145. (b) D. Albanese, S. E. Gibson, E. Rahimian, *J. Chem. Soc., Chem. Commun.* **1998**, 2571. (c) S. E. Gibson, P. O'Brien, E. Rahimian, M. H. Smith, *J. Chem. Soc., Perkin Trans. 1* **1999**, 909. (d) S. E. Gibson, P. Ham, J. R. Jefferson, *J. Chem. Soc., Chem. Commun* **1998**, 123. See also ref. [31a].

41 P. D. Baird, J. Blagg, S. G. Davies, K. H. Sutton, *Tetrahedron* **1988**, *44*, 171.

42 (a) T. Geller, H.-G. Schmalz, J. W. Bats, *Tetrahedron Lett.* **1998**, *39*, 1537. See also: (b) H.-G. Schmalz, M. Arnold, J. Hollander, J. W. Bats, *Angew. Chem.* **1994**, *106*, 77; *Angew. Chem., Int. Ed. Engl.* **1994**, *33*, 109.

43 For a review, see: (a) S. G. Davies, T. J. Donohoe, *Synlett* **1993**, 323. See also: (b) M. Uemura, T. Kobayashi, Y. Hayashi, *Synthesis* **1986**, 386. (b) M. Uemura, T. Minami, Y. Hayashi, *J. Organomet. Chem.* **1986**, *299*, 119.

44 (a) M. Uemura, K. Isobe, K. Take, Y. Hayashi, *J. Org. Chem.* **1983**, *48*, 3855. (b) M. Uemura, K. Isobe, Y. Hayashi, *Chem. Lett.* **1985**, 91. See also ref. [36 a].

45 (a) M. Uemura, T. Kobayashi, K. Isobe, T. Minami, Y. Hayashi, *J. Org. Chem.* **1986**, *51*, 2859. (b) A. Netz, K. Polborn, T. J. J. Müller, *J. Am. Chem. Soc.* **2001**, *123*, 3441. (c) A. Netz, K. Polborn, T. J. J. Müller, *Organometallics* **2001**, *20*, 376. (d) M. Ansorge, K. Polborn, T. J. J. Müller, *Eur. J. Inorg. Chem.* **1999**, 225.

46 S. J. Coote, S. G. Davies, D. Middlemiss, A. Naylor, *Tetrahedron Lett.* **1989**, *30*, 3581.

47 H.-G. Schmalz, C. B. de Koning, D. Bernicke, S. Siegel, A. Pfletschinger, *Angew. Chem.* **1999**, *111*, 1721; *Angew. Chem., Int. Ed. Engl.* **1999**, *38*, 1620.

48 (a) N. Taniguchi, M. Uemura, Tetrahedron 1998, 54, 12775. (b) N. Taniguchi, M. Uemura, Synlett 1997, 51. For a related reaction, see also (c) N. Taniguchi, T. Hata, M. Uemura, Angew. Chem. 1999, 111, 1311; Angew. Chem., Int. Ed. Engl. 1999, 38, 1232.

49 C. A. Merlic, J. C. Walsh, J. Org. Chem. 2001, 66, 2265. (b) C. A. Merlic, J. C. Walsh, Tetrahedron Lett. 1998, 39, 2083.

50 For recent examples, see (a) J. Andrieu, C. Baldoli, S. Maiorana, R. Poli, P. Richard, Eur. J. Org. Chem. 1999, 3095. (b) C. Baldoli, P. Del Buttero, D. Pericchia, T. Pilati, Tetrahedron 1999, 55, 14089. (c) H. Koide, M. Uemura, Tetrahedron Lett. 1999, 40, 3443. (d) S. Maiorana, C. Baldoli, P. Del Buttero, E. Licandro, A. Papagni, M. Lanfranchi, A. Tiripicchio, J. Organomet. Chem. 2000, 593–594, 380. (e) B. C. Maity, V. G. Puranik, A. Sarkar, Synlett 2002, 504. (f) K. Ishimura, T. Kojima, Tetrahedron Lett. 2001, 42, 5037. (g) M. F. Costa, M. R. G. da Costa, M. J. M. Curto, M. Magrinho, A. M. Damas, L. Gales, J. Organomet. Chem. 2001, 632, 27. (h) M. K. McKay, J. R. Green, Can. J. Chem. 2000, 78, 1629.

51 W. H. Moser, J. Zhang, C. S. Lecher, T. L. Frazier, M. Pink, Org. Lett. 2002, 4, 1981.

52 (a) M. Uemura, H. Oda, T. Minami, Y. Hayashi, Tetrahedron Lett. 1991, 32, 4565. (b) V. M. Swamy, S. K. Mandal, Tetrahedron Lett. 1999, 40, 6061. (c) A. Sarkar, S. Ganesh, S. Sur, S. K. Mandal, V. M. Swamy, B. C. Maity, T. S. Kumar, J. Organomet. Chem. 2001, 624, 18. (d) S. K. Mandal, A. Sarkar, J. Org. Chem. 1999, 64, 2454. (e) S. K. Mandal, A. Sarkar, J. Chem. Soc., Perkin Trans. 1 2002, 669.

53 (a) C. Baldoli, S. Maiorana, E. Licandro, G. Zinzalla, M. Lanfranchi, A. Tiripicchio, Tetrahedron: Asymmetry 2001, 12, 2159. (b) P. Del Buttero, C. Baldoli, G. Molteni, T. Pilati, Tetrahedron: Asymmetry 2000, 1927. See also ref. [4g].

54 (a) B. Voigt, M. Brands, R. Goddard, R. Wartchow, H. Butenschön, Eur. J. Org. Chem. 1998, 2719. (b) K. G. Dongol, R. Wartchow, H. Butenschön, Eur. J. Org. Chem. 2002, 1972. For a review, see (c) H. Butenschön, Pure Appl. Chem. 2002, 74, 57.

55 (a) D. A. Widdowson, R. Wilhelm, J. Chem. Soc., Chem. Commun. 1999, 2211. (b) R. Wilhelm, D. A. Widdowson, J. Chem. Soc., Perkin Trans. 1 2000, 3808.

56 T. J. J. Müller, M. Ansorge, K. Polborn, J. Organomet. Chem. 1999, 578, 252.

57 (a) D. Prim, J.-P. Tranchier, F. Rose-Munch, E. Rose, J. Vaissermann, Eur. J. Inorg. Chem. 2000, 901. (b) J.-P. Tranchier, R. Chavignon, D. Prim, A. Auffrant, J. G. Planas, F. Rose-Munch, E. Rose, G. R. Stephenson, Tetrahedron Lett. 2001, 42, 3311.

58 B. Gotov, J. Kaufmann, H. Schumann, H.-G. Schmalz, Synlett 2002, 361 and 1161.

59 E. P. Kündig, H. Ratni, B. Crousse, G. Bernardinelli, J. Org. Chem. 2001, 66, 1852. See also: S. Bräse, Tetrahedron Lett. 1999, 40, 6757.

60 For a review, see: (a) K. Kamikawa, M. Uemura, Synlett 2000, 938. See also ref. [31 a–d] and [32 d].

61 (a) K. Kamikawa, T. Watanabe, M. Uemura, J. Org. Chem. 1996, 61, 1375. (b) M. Uemura, K. Kamikawa, J. Chem. Soc., Chem. Commun. 1994, 2697.

62 (a) M. F. Semmelhack, E. Hilt, J. H. Colley, Tetrahedron Lett. 1998, 39, 7683. (b) S. E. Gibson, N. J. Hales, M. A. Peplow, Tetrahedron Lett. 1999, 40, 1417. (c) A. C. Comely, S. E. Gibson, N. J. Hales, M. A. Peplow, J. Chem. Soc., Perkin Trans. 1 2001, 2526. (d) S. Maiorana, C. Baldoli, E. Licandro, L. Casiraghi, E. Migistris, A. Paio, S. Provers, P. Seneci, Tetrahedron Lett. 2000, 41, 7271. (e) J. H. Rigby, M. A. Kondratenko, Org. Lett. 2001, 3, 3683.

63 M. Sodeoka, M. Shibasaki, Synthesis 1993, 643.

64 K. Kamikawa, S. Sugomoto, M. Uemura, J. Org. Chem. 1998, 63, 8407.

65 For a review, see: C. Bolm, K. Muniz, Chem. Soc. Rev. 1999, 28, 51.

66 C. Pasquier, L. Pélinski, J. Brocard, A. Montreux, F. Agbossou-Niedercorn, Tetrahedron Lett. 2001, 2809.

67 J. W. Han, H.-Y. Jang, Y. K. Chang, Tetrahedron: Asymmetry 1999, 10, 2853.

68 (a) G. B. Jones, M. Guzel, *Tetrahedron: Asymmetry* **2000**, 1267. (b) G. B. Jones, M. Guzel, S. B. Heaton, *Tetrahedron: Asymmetry* **2000**, 4303. (c) G. B. Jones, S. B. Heaton, B. J. Chapman, M. Guzel, *Tetrahedron: Asymmetry* **1997**, *8*, 3625.

69 S. Malfait, L. Pélinski, J. Brocard, *Tetrahedron: Asymmetry* **1998**, *9*, 2595.

70 C. Chapuis, M. Barthe, J.-Y. de Saint Laumer, *Helv. Chim. Acta* **2001**, *84*, 730.

71 I. Weber, G. B. Jones, *Tetrahedron Lett.* **2001**, *42*, 6983.

72 U. Englert, R. Haerter, D. Vasen, A. Salzer, E. B. Eggeling, D. Vogt, *Organometallics* **1999**, *18*, 4390.

3.13
Pauson-Khand Reactions

D. Strübing and M. Beller

3.13.1
Introduction

The facile synthesis of five-membered ring systems constitutes an important topic in organic chemistry. In this context, the Pauson-Khand reaction (PKR) represents an elegant procedure for the preparation of cyclopentenones via transition metal-catalyzed [2+2+1] cycloaddition of an alkyne, an alkene, and carbon monoxide. Since its discovery by Pauson and Khand [1] in 1973, the reaction has received much attention with regard to synthetic elaborations and mechanistic studies [2]. In this review the main focus is on important developments which have taken place in the last decade.

As an early successful example, the reaction of norbornadiene with the phenylacetylene-hexacarbonyldicobalt complex is shown in Eq. (1).

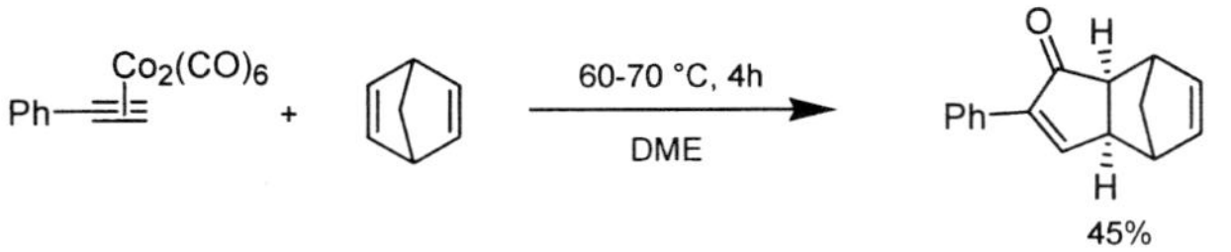

$$Ph-\equiv \quad \xrightarrow{Co_2(CO)_6} \quad + \quad \bigcirc\!\!\!\!\bigcirc \quad \xrightarrow[DME]{60-70\ °C,\ 4h} \quad Ph\text{-cyclopentenone product} \quad 45\% \tag{1}$$

Although the methodology results in a remarkable increase in structural diversity, the Pauson-Khand reaction also had serious disadvantages at that time. For instance, it was necessary to use a stoichiometric amount of dicobaltoctacarbonyl in order to obtain a sufficient amount of the desired products. In general, yields were comparatively low unless strained olefines were used as starting material. Reactions typically gave a mixture of regioisomers if unsymmetrical alkenes and alkynes were used. In many early reported examples it was also essential to use high temperatures over a long period of time. Important progress was reported in 1981 by Schore [3], who showed for the first time an intramolecular PKR using a carbon-tethered enyne precursor. In this case, the PKR gave good yields and led to complete regioselectivity, and it was not necessary to use strained olefins as starting materials.

Even though the PKR has received much attention in the last 30 years, the detailed mechanism still warrants experimental justification. Nevertheless, it is well

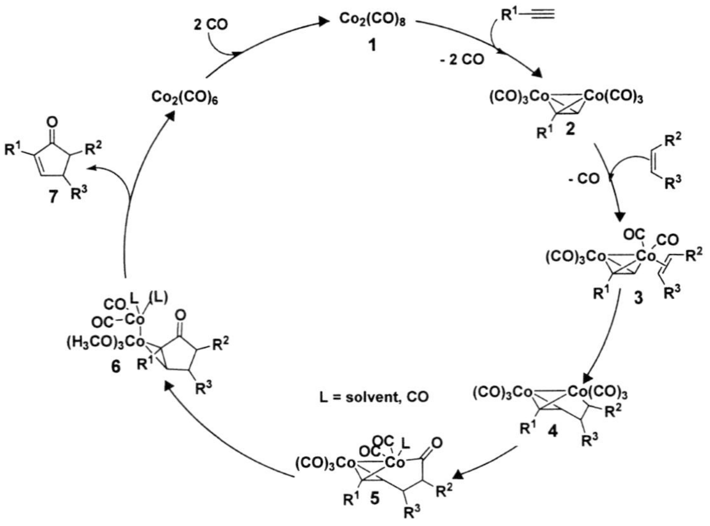

Scheme 1 The proposed mechanism for the Pauson-Khand reaction.

accepted that the reaction starts with the formation of the alkyne-$Co_2(CO)_6$-complex **2**. Apart from this class of complexes, no more detectable intermediates have been described. A proposed mechanism, reported by Magnus [4], is presented in Scheme 1.

After formation of the alkyne-$Co_2(CO)_6$-complex **2**, olefin **3** coordination and insertion at the less hindered end of the alkyne takes place. The metallacycle **4** reacts immediately, with insertion of a CO ligand **5** and reductive elimination of **6**, to liberate the resulting cyclopentenone **7**. It is interesting to note that all the bond-forming steps occur on one cobalt center. The other cobalt atom is acting as an anchor and exerts electronic influences on the bond-forming metal atom via the metal-metal bond [5].

3.13.2
Stoichiometric Pauson-Khand Reactions

To enhance the scope of PKR, various transition metal complexes have been studied for the desired cycloaddition process. For instance $Ni(COD)_2$ [6] and Cp_2TiCl_2/EtMgBr [7] were transformed with various enyne precursors to the resulting metallacycles. Instead of carbon monoxide, isocyanides were used to convert the resulting metallacycles into the corresponding iminocyclopentenes. After an additional acid-catalyzed hydrolysis the desired cyclopentenones were obtained. In addition to dicobaltoctacarbonyl, several other complexes such as iron pentacarbonyl [8], molybdenum hexacarbonyl [9], tungsten pentacarbonyl [10], zirconium [11] or heterobimetallic cobalt/tungsten complexes [12] were reported to furnish cyclopentenones under an atmosphere of carbon monoxide. Despite all these examples, dicobaltoctacarbonyl has become the complex of choice for stoichiometric PKR. This was because of several advantages, for instance, reactivity toward internal and terminal alkynes, a broad tolerance of functional groups, and relatively low costs of the transition metal reagent. On the other hand, the $Co_2(CO)_8$-mediated cyclization was afflicted with problems like the need for high carbon monoxide pressure, high temperatures, and long reaction times. Hence, there was a need to develop methods for the promotion of $Co_2(CO)_8$-mediated PKR. In this connection, Smith and Caple [13] reported that enyne precursors adsorbed on silica gel or alumina could be converted to cyclopentenones with a significant decrease in temperature and reaction time. In 1991 Jeong [14] and Schreiber [15] reported independently about the application of N-methylmorpholine N-oxide and trimethylamine N-oxide as useful reagents for the acceleration of the cobalt-mediated PKR.

In the presence of N-oxides, PKR proceeded smoothly even at room temperature in good yields. It is supposed that the N-oxide acts as an oxidant, which removes one CO ligand of the bond-forming cobalt atom. This results in an oxidative addition of the alkene moiety, which was assumed to be the rate-determining step of the cycloaddition process. Moreover, it was discovered that alkyl-methylsulfides also enable the acceleration of the PKR [16]. Here, best yields were obtained using 3.5 eq. n-butyl methyl sulfide at 83 °C. In addition, Sugihara et al. [17] reported that primary amines also promote the conversion of enyne precursors to the desired cyclopentenones. Quantitative yields were obtained in 1,2-dichloroethane with 3.5 eq. cyclohexylamine at 83 °C in only 5 min. Alternatively, the reaction can be carried out in a 1:3 mixture of dioxane/2 N NH_4OH-solution at 100 °C. Selected examples of promoter-assisted Pauson-Khand reactions are shown in Table 1.

Tab. 1 Selected examples of promoter-assisted Pauson-Khand reactions

Substrates	Products	Condition: yield (%)	Substrates	Products	Conditions: yield (%)
		A: 70 C: – B: 85 D: –	E = CO_2Et		A: 81 C: 92 B: – D: 67
		A: – C: 81 B: 92 D: 85	Ph		A: 80 C: – B: – D: 100

Conditions: **A:** 3 eq. TMANO, CH_2Cl_2, 0.5–3 h, RT; **B:** 6 eq. NMO, CH_2Cl_2, RT, 8–16 h; **C:** 3.5 eq. n-BuSMe, CH_2Cl_2, 2 h, 83 °C; **D:** 3.5 eq. Cyc_{Hex}-NH_2, CH_2Cl_2, 83 °C, 0.5 h or 1,4-dioxane/2M NH_4OH=1/3; 100 °C, 0.5 h.

3.13.3
Catalytic Pauson-Khand Reactions

Compared to the previously discussed cobalt-mediated Pauson-Khand reactions, catalytic variants offer significant benefits because of the avoidance of stoichiometric amounts of waste. It was Rautenstrauch [18] who published the first example of a catalytic PK cycloaddition process. During his investigations of the synthesis of perfume ingredients, he converted heptyne under 40 bar ethylene and 100 bar carbon monoxide to the resulting cyclopentenone in moderate yield (48%). Later, Jeong [19] reported a more convenient catalytic procedure applying 1–3 mol% $Co_2(CO)_8$ in the presence of triphenylphosphite as ligand. Here, the reaction proceeded smoothly under 3 atm carbon monoxide at 110 °C in good yields.

Supercritical fluids have become interesting alternative solvents in industry and scientific research. Because of their easy recovery, they are becoming more and more attractive. In this regard it is interesting to note that Jeong [20] reported that supercritical CO_2 can also promote the PKR. Reactions were typically carried out with $p(CO_2)$ pressures of 112 atm at 37 °C and $p(CO)$ pressures in the range of 15–30 atm at 90 °C. In the presence of 2–5 mol% of $Co_2(CO)_8$, cyclopentenones were obtained in 51–91% yield. More recently, Periasamy et al. reported that a system of 0.4 eq. $CoBr_2$/0.43 eq. Zn is active for the intermolecular cycloaddition at atmospheric carbon monoxide pressure and 110 °C [21]. Isolated yields of cyclopentenones were obtained between 30 and 88%. Chung and co-workers have shown [22] that the $Co(acac)_2/NaBH_4$ system catalyzes both the inter- and intramolecular PKR. Reactions were carried out at 40 atm CO for several days (33–85% yield). Sugihara [23] discovered that methylidynetricobalt nonacarbonyl clusters $Co_3(CO)_9(\mu^3\text{-CH})$ are a good choice of catalyst to enable both the inter- and the intramolecular PKR. Reactions were performed with 1–2 mol% catalyst at a $p(CO)$ pressure of 7 atm and 120 °C. Good to excellent yields (78–91%) were obtained under these conditions.

The addition of Lewis bases like amines and sulfides enables the acceleration of stoichiometric PKRs. Therefore it was not surprising that Hashimoto [24] demonstrated the usefulness of phosphine sulfides to promote the $Co_2(CO)_8$-catalyzed PKR. Here, even at atmospheric pressure of carbon monoxide at 70 °C, cyclopentenones were obtained in excellent yields (ca. 90%). To make a catalytic reaction more applicable for the chemical industry it is often necessary to use a heterogeneous catalyst. Chung [25] and co-workers developed an easy and cheap heterogeneous PK catalyst system. They immobilized 12 wt% metallic cobalt on commercially available charcoal. The resulting catalyst system gave mostly good to excellent yields of different cyclopentenones (61–98%). After the reaction, the heterogenous catalyst is easily filtered off and reused up to ten times without significant loss of activity.

Apart from PKR using cobalt catalysts, other catalytic variants have been developed. For example, $Cp_2Ti(PMe_3)_2$ [26] in combination with $^tBuMe_2SiNC$ provides iminocyclopentenes directly. Instead of $Cp_2Ti(PMe_3)_2$, Buchwald reported on the use of nickel(0) catalysts for the isocyanide cycloaddition reaction [27]. Unfortu-

3.13.3 Catalytic Pauson-Khand Reactions | 623

Tab. 2 Selected examples of catalytic Pauson-Khand reactions

Substrates	Products	Conditions: yield (%)	Substrates	Products	Conditions: yield (%)
E, E, —Me (E = CO₂Et)	E, E, Me (bicyclic enone)	A: 90 F: 65 J: 93 B: 91 I: 78 E: 98 K: 96	O, —Ph (allyl propargyl ether)	O, Ph (bicyclic enone)	A: 51 F: 80 K: 82 B: 70 G: 66 E: 98 H: 57
E, E (enyne)	E, E (bicyclic enone)	A: 82 D: 98 B: 82 K: 55 C: 66	Ts-N (tosyl enyne)	Ts-N (bicyclic enone)	A: 94 C: 85 E: 98
E, E, OAc	E, E, OAc (bicyclic enone)	A: 58 B: 51	Ph-≡, (norbornene)	Ph (tricyclic enone)	C: 100 D: 98 E: 98

Conditions: **A**: 3–5 mol% Co₂(CO)₈, 10–20 mol% P(OPh)₃, DME, 3 atm CO; **B**: 2–5 mol% Co₂(CO)₈, 112 atm CO₂ at 37 °C, 15–30 atm CO at 90 °C; **C**: 0.02–0.05 mol% Co(acac)₂/NaBH₄, 30–40 atm CO, CH₂Cl₂, 100 °C, 48 h; **D**: 2 mol% Co₃(CO)₉(μ³-CH), 120 °C, 7 atm CO, toluene; **E**: 12 wt% Co supported on charcoal, 20–30 atm CO, THF, 7–48 h; **F**: 10 mol% Cp₂Ti(PMe₃), ᵗBuMe₂SiNC, additional hydrolysis; **G**: 5–20 mol% Ni(COD)₂, ligand, 110–120 °C, THF; **H**: 2 mol% Ru₃(CO)₁₂, 10 atm CO, 160 °C, 20 h; **I**: 5 mol% Ru₃(CO)₁₂, DMAc, 15 atm CO, 20 h; **J**: 1–5 mol% [RhCl(CO)₂]₂, 1 atm CO, Bu₂O, 90–100 °C, **K**: 2.5 mol% trans-[RhCl(CO)(dppp)]₂, 1 atm CO, 110 °C, 24 h, toluene.

nately, hydrolysis of the iminocyclopentenes significantly decreases the yield of the desired cyclopentenones. Therefore a new procedure was developed using a Cp₂Ti(CO)₂ catalyst [28]. Here, reactions were carried out with 5–20 mol% of catalyst at 18 psi CO pressure and 90 °C in toluene to give the cycloadducts in yields of 58–95%. In addition to cobalt systems in the last decade, ruthenium and rhodium have been shown to give highly efficient catalysts for PKR. In 1997, Murai [29] and Mitsudo [30] independently described the use of Ru₃(CO)₁₂ as a catalyst for PKR. Only differing in the appropriate solvent, they used almost the same catalyst concentration (2 mol%) at CO pressures of 10–15 atm and temperatures of 140–160 °C. Yields varied in the range of 41–95%.

Rhodium is another transition metal allowing intramolecular PKR. It was shown by Narasaka [31] that [RhCl(CO)₂]₂ allows for cycloaddition reaction (35–91% yield) at atmospheric pressure CO and temperatures of 130–160 °C. Other rhodium complexes like trans-[RhCl(CO)(dppp)]₂ and RhCl(PPh₃)₃ in combination with silver salts were also reported to promote the formation of cyclopentenones starting from enyne precursors in 20–99% yield [32]. Some representative examples of catalytic Pauson-Khand reactions are shown in Table 2.

3.13.4
Stereoselective Pauson-Khand Reactions

During the last decade, various ways of performing stereoselective Pauson-Khand reactions have been developed. For instance, it is possible to use chiral promotors in stoichiometric PKR. Chiral N-oxides are suitable reagents for this purpose. A remarkable example of this type of reaction was reported by Kerr [33]. He used chiral brucine N-oxide as a promoter for the intermolecular PKR of various substituted propargylic alcohols and norbornadiene (Eq. (2)). After optimization, enantioselectivities of up to 78% were achieved in 1,2-dimethoxyethane at −60 °C. The application of chiral sparteine N-oxides was published by Laschat [34], but in general lower enantioselectivities were observed here.

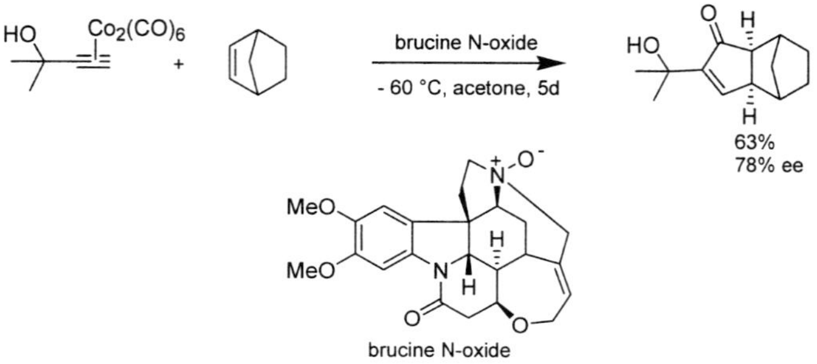

$$(2)$$

In addition to chiral promoters, optically pure enyne precursors have also been applied for the synthesis of enantiomerically pure cylopentenones. For example, Krafft [35] and co-workers reported on the use of chiral cyclopropylidenepropylethynyl dioxolanes for PKR. The desired cyclopentenones were obtained with diastereomeric ratios of up to 1:20 (Eq. (3)).

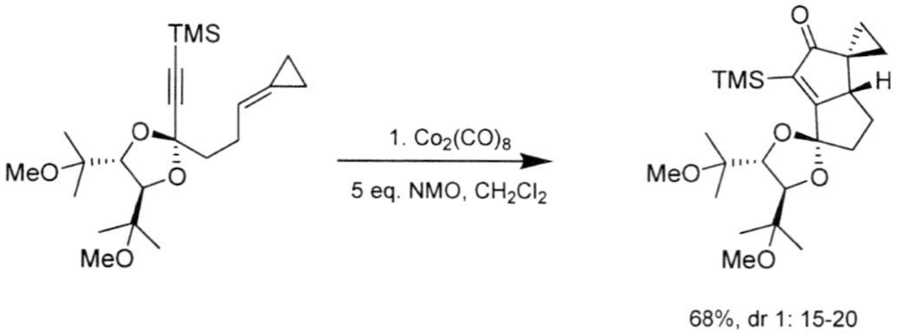

$$(3)$$

Chiral thioethers represent another type of chiral auxiliary for PKR, which have been described by Krafft [36]. A recent example is shown in Eq. (4). Diastereoselectivities of up to 96% were obtained using (1S)-camphor-10-thiol-derived alkyne-dicobaltpentacarbonyl complexes [37].

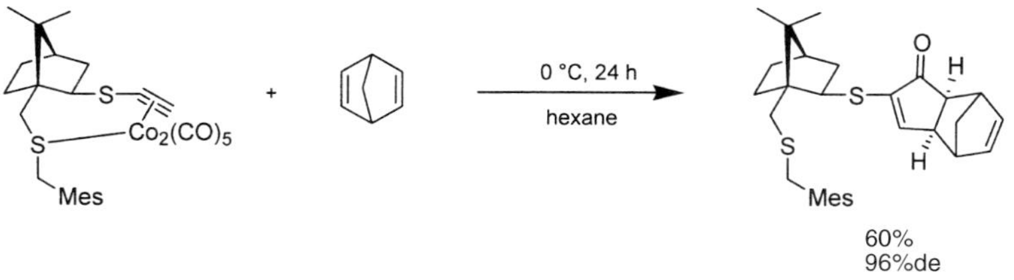

$$\text{(Scheme with Co}_2(\text{CO})_5\text{ complex + norbornene} \xrightarrow[\text{hexane}]{0\,°\text{C, 24 h}} \text{product, 60\%, 96\%de)} \quad (4)$$

The concept of using chiral auxiliaries was successfully established by many other groups in the total synthesis of natural products such as hirsutene [38], β-cupraenone [39] or (+)-15-nor-pentalenene [40]. Even though observed diastereoselectivities are often quite good, the synthesis of chiral auxiliary-based precursors is more or less difficult and costly. Therefore the use of chiral metal complexes in diastereoselective PKR has also been studied. For example, Pericas and Riera reported on the cycloaddition of chiral cobalt complexes with norbornadiene [41]. In the presence of the chiral bidentate P,S-ligand PuPhos-BH$_3$, different substrates reacted smoothly to the desired cyclopentenones in nearly quantitative yield (92–98%) and moderate to excellent enantioselectivities (57–99%) (Eq. (5)).

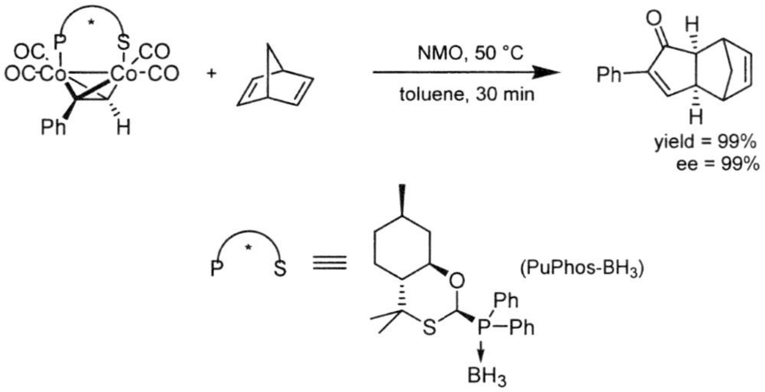

$$\text{(Co complex with PuPhos-BH}_3\text{ + norbornadiene} \xrightarrow[\text{toluene, 30 min}]{\text{NMO, 50\,°C}} \text{product, yield = 99\%, ee = 99\%)} \quad (5)$$

Clearly, a more convenient possibility of creating chiral cyclopentenones would make use of catalytic amounts of a chiral transition metal catalyst. Pioneering work in this area was done by Buchwald and co-workers [42]. They found that the chiral titanocene complex $(S,S)(\text{EBTHI})\text{Ti}(\text{CO})_2$ is a suitable catalyst for the conversion of various enynes to cyclopentenones. Reactions proceeded at 90 °C and 14 psig CO pressure in good yields in the range of 72–96%. The obtained enantioselectivities were also good to excellent, in the range of 70–94%. In the following years, late transition metals were also applied as catalysts for the asymmetric Pauson-Khand reaction. Jeong reported on the combination of [RhCl(CO)$_2$]$_2$, AgOTf, and (S)-BINAP as a catalyst system, which enabled the Pauson-Khand cycloaddition in moderate to excellent yields of 40–99% [43]. The detected *ee*s were between 22 and 96%. In all reported cases, only a moderate carbon monoxide pressure of 1–3 atm was necessary. Shibata reported that, in addition to rhodium complexes, chiral iridium diphosphine complexes catalyze the PKR [44]. Using 10 mol%

Tab. 3 Selected examples of catalytic asymmetric Pauson-Khand reactions

Substrates	Products	Conditions: yield (%), ee (%)	Substrates	Products	Conditions: yield (%), ee (%)
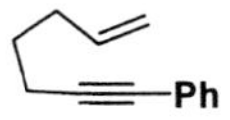 (enyne with Ph)	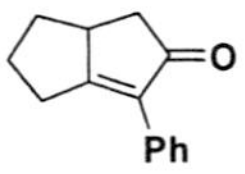 (bicyclic enone with Ph)	A: 70, 87 B: 16, 11 C: –, – D: 61, 51	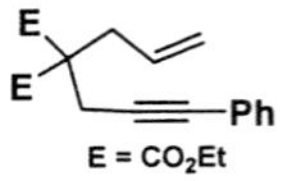 E-CH₂-CH=CH₂ / E-CH₂-C≡C-Ph (E = CO₂Et)	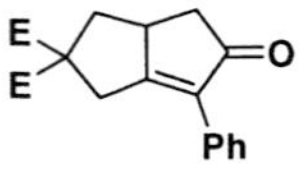 (bicyclic enone with E,E,Ph)	A: 82, 94 B: 75, 75 C: 74, 84 D: –, –
 Ts-N(allyl)(propargyl-Ph)	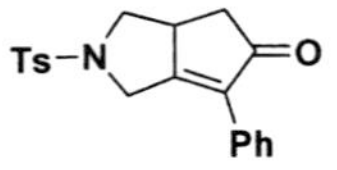 Ts-N bicyclic enone with Ph	A: –, – B: –, – C: 85, 95 D: 99, 71	 E,E-enyne with Me	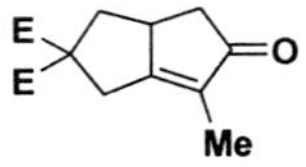 bicyclic enone E,E,Me	A: 87, 90 B: 97, 7 C: –, – D: 93, 71
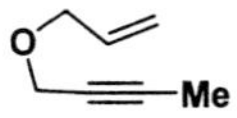 O(allyl)(propargyl-Me)	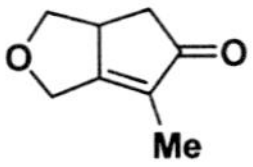 O-bicyclic enone Me	A: –, – B: –, – C: 75, 97 D: 85, 86	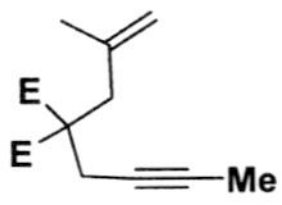 E,E-methallyl-propargyl-Me	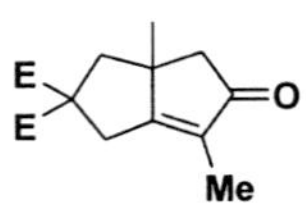 bicyclic enone E,E,Me (with Me branch)	A: 72, 90 B: –, – C: –, – D: –, –
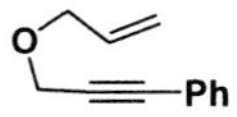 O(allyl)(propargyl-Ph)	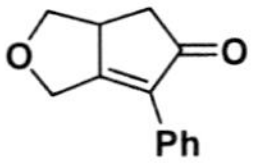 O-bicyclic enone Ph	A: –, – B: –, – C: –, – D: 88, 91	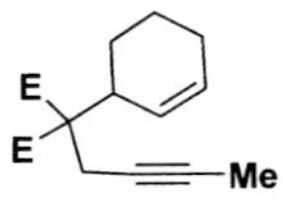 E,E-cyclohexenyl-propargyl-Me	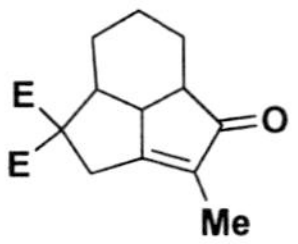 E,E-tricyclic enone Me	A: –, – B: 50, 4 C: –, – D: –, –

Conditions: **A**: 5–20 mol% (*S*,*S*)(EBTHI)Ti(CO)₂, 14 psig CO, toluene, 12 h, 90 °C; **B**: 6 mol% Co₂(CO)₈, 10 mol% ligand, toluene, 1 atm CO, 24 h, 95 °C; **C**: 10 mol% [Ir(COD)Cl]₂ + (*S*)-BINAP, 1 atm CO, refluxing xylene, 20–72 h; **D**: 3 mol% [RhCl(CO)₂]₂, 6 mol% (*S*)-BINAP, 12 mol% AgOTf, THF, 1–3 atm CO, 90–130 °C, 3–20 h.

[Ir(COD)Cl]₂ in combination with (*S*)-tolBINAP as ligand, several cycloaddition products were obtained in yields of 30–85% and *ee*s of 82–98%. Recently, Buchwald has shown that Co₂(CO)₈ in combination with a chiral bisphosphite also enables the formation of chiral PKR products [45]. Cyclization products were obtained in yields of 16–97%. In one case a 75% *ee* was obtained, while in most other reported examples the *ee* was rather low (<20%). Some representative examples of catalytic asymmetric PKR are shown in Table 3.

3.13.5
Synthetic Applications

Because of their extraordinary molecular complexity, PKR became an attractive tool for the synthesis of natural products at an early stage. It is therefore not surprising that many remarkable examples of synthetic applications have been reported during the last thirty years. Thus, PKR was successfully applied in the total synthesis of (+)-epoxydictymene [46], loganin [47], hirsutene [38], and β-cupraenone [39]. Here we only give an update of some interesting examples of recent years. For instance, the total synthesis of ($\pm$)-13-deoxyserratine, a lycopodium alkaloid, was reported by Zard [48]. The reaction sequence included the successful PKR of the corresponding enyne precursor and a Bu$_3$SnH-mediated radical cyclization. The resulting natural product was obtained in an overall yield of 12% (Eq. (6)).

An interesting reaction pathway to the triquinane sesquiterpene ceratopicanol was reported by Mukai [49] in 2002 (Eq. (7)). Although the yield for the Pauson-Khand reaction adduct was high (96%), the observed diastereomeric ratio was comparatively (62:38) low in this case.

The total synthesis of ($\pm$)-magellanine, another example of a lycopodium alkaloid, was reported very recently by Ishizaki and Hoshino [50]. The unique tetracyclic structure was prepared during a reaction sequence including an Ireland-Claisen rearrangement and the required intramolecular PKR. Even though the obtained yields for the cyclization products were moderate to good (35–70%), this example demonstrates the substrate tolerance of the PKR, which was due to the fact that an unprotected hydroxyl group was involved in the reaction sequence. Moreover, a quaternary carbon center was built up in the key PKR step (Eq. (8)).

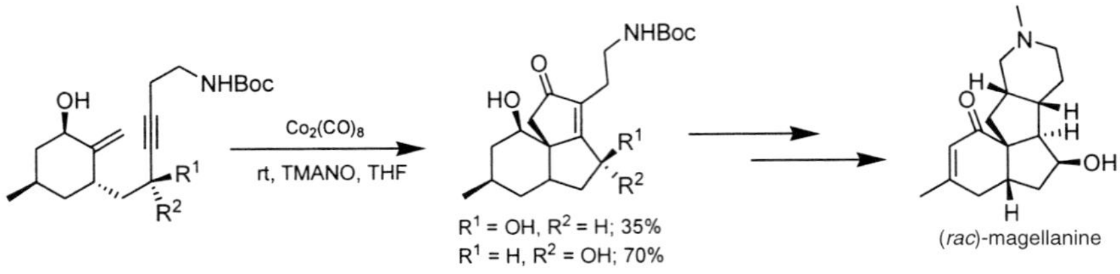

$$\text{(8)}$$

To enlarge the scope of the Pauson-Khand reaction, several groups developed new approaches for related enyne precursors. Very recently, Shibata has shown that catalytic amounts of an [IrCl(CO)(PPh$_3$)$_2$] complex can be used for the efficient conversion of allenynes [51]. Interestingly, reactions were carried out under a very low pressure of CO (0.2 atm) and 120 °C, giving yields of up to 91% (Eq. (9)).

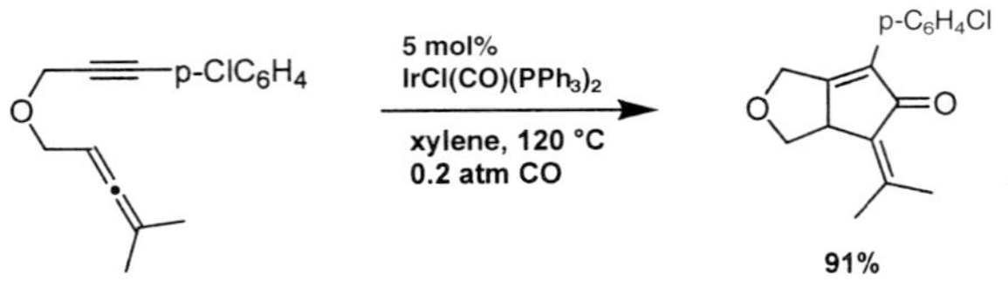

$$\text{(9)}$$

Wender [52] was the first to report on a dienyl-type Pauson-Khand reaction. Best results were obtained using 5 mol% [RhCl(CO)(PPh$_3$)$_2$] as catalyst under atmospheric pressure of CO at room temperature. Various substitutions of the alkyne and dienyl moiety were tolerated, giving the products in yields ranging from 43 to 96% (Eq. (10)).

$$\text{(10)}$$

The first hetero-Pauson-Khand reaction was published by Murai and co-workers [53]. They reported the catalytic conversion of yne-aldehydes to the corresponding bicyclic γ-butenolides (Eq. (11)). They studied different Ru-, Rh-, Co- and Ir-complexes, but Ru$_3$(CO)$_{12}$ was the only catalyst which enabled the cycloaddition process. Reactions were performed with 2 mol% catalyst at 160 °C and a p(CO) of 10 atm. The bicyclic lactones were obtained in yields of 62–93%.

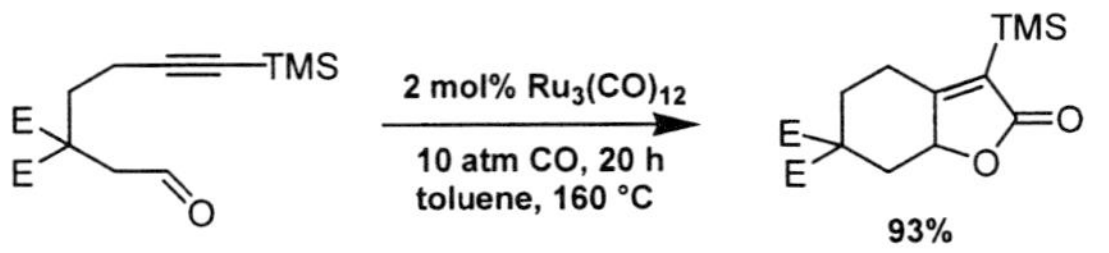

$$\text{(11)}$$

An elegant domino procedure for the synthesis of bicyclopentenones was reported by Jeong [54]. He applied a bimetallic system of [Pd$_2$(dba)$_3$(CHCl$_3$)]/[{RhClCO(dppp)}$_2$] for the sequential construction of an enyne precursor, starting from a malonic acid derivative and allylic acetate, which was converted *in situ* to the cyclopentenone in high yield (Eq. (12)). In the reported case the Pd catalyst is responsible for the allylic substitution reaction, while the Rh complex catalyzes the PKR.

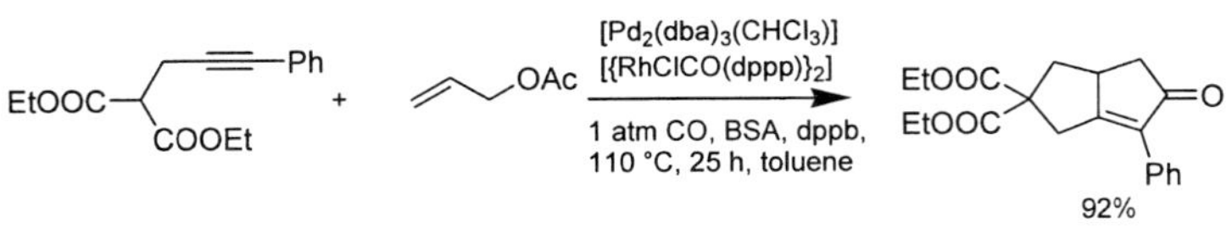

$$\text{EtOOC-C≡C-Ph + AcO-allyl} \xrightarrow[\text{1 atm CO, BSA, dppb, 110 °C, 25 h, toluene}]{[Pd_2(dba)_3(CHCl_3)] \; [\{RhClCO(dppp)\}_2]} \text{bicyclopentenone (92%)} \quad (12)$$

3.13.6
Transfer Carbonylations in Pauson-Khand Reactions

It has been shown, that the PKR is a powerful method for the preparation of highly substituted cyclopentenones. However, the use of the toxic substance carbon monoxide still represents a drawback to many organic chemists. Since the early 1960s, the transition metal-catalyzed decarbonylation of organic oxo compounds has been known as a way to prepare metal carbonyls, which are the key intermediates in Pauson-Khand reactions. Hence, it was suggested that metal carbonyls could be generated without using CO in the presence of enynes.

In 2001 Morimoto and Kakiuchi reported the first catalytic Pauson-Khand reaction using aldehydes as a source of carbon monoxide [55]. It was proposed that the decarbonylation of aldehydes directly leads to the active metal carbonyl catalyst (Scheme 2), which, once formed, enables the desired cycloaddition. ^{13}C-labeling experiments indicate that hardly any free carbon monoxide exists. This implies that CO, generated by decarbonylation of aldehydes, is directly incorporated into the carbonylative coupling. A suitable aldehyde was sought, and it was shown that C$_6$F$_5$CHO and cinnamaldehyde gave the best yields in combination with [RhCl(cod)]$_2$/dppp as catalyst system. In the presence of an excess of aldehyde, en-

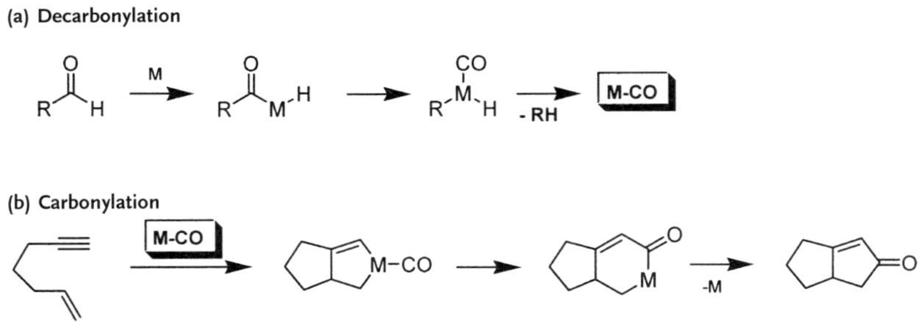

Scheme 2 Proposed reaction mechanism for a PK transfer carbonylation reaction.

Tab. 4 Selected examples of PK-transfer carbonylation reactions

Substrates	Products	Conditions: yield (%)	Substrates	Products	Conditions: yield (%)
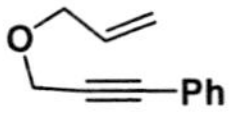 O-allyl-propargyl-Ph	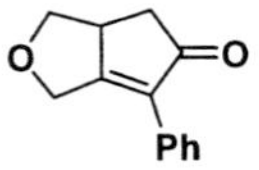 bicyclic enone, Ph	A: – B: 91 C: 87	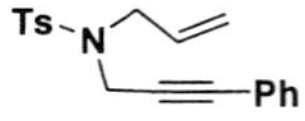 Ts-N-allyl-propargyl-Ph	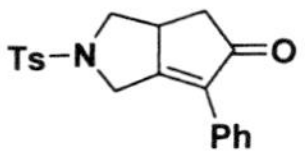 Ts-N bicyclic enone, Ph	A: 98 B: 95 C: 96
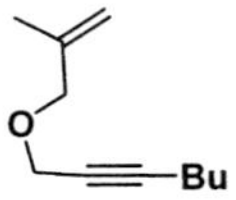 O-methallyl-propargyl-Bu	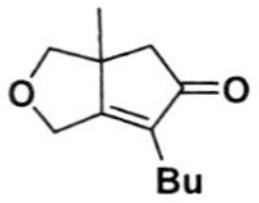 bicyclic enone, Bu	A: 78 B: 92 C: 94	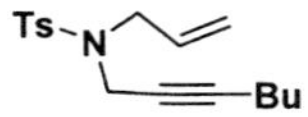 Ts-N-allyl-propargyl-Bu	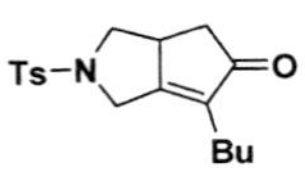 Ts-N bicyclic enone, Bu	A: 89 B: 84 C: –
E,E-diester-allyl-propargyl-Ph (E = CO₂Et)	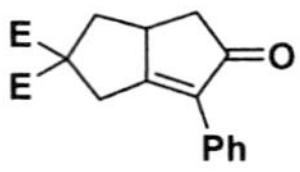 E,E bicyclic enone, Ph	A: 58 B: 60 C: 96	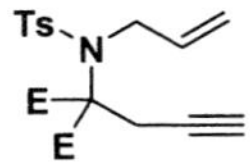 Ts-N, E,E homoallyl-propargyl	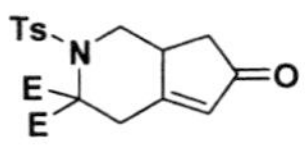 Ts-N, E,E bicyclic enone	A: – B: 36 C: 67

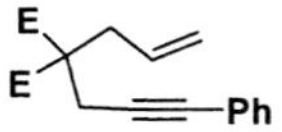

Conditions: **A:** 5 mol% Rh(dppp)$_2$Cl, 20 eq. cinnamaldehyde, 2–24 h, 120 °C; **B:** 5 mol% [RhCl(cod)]$_2$, 11 mol%. dppp, 2 eq. C$_6$F$_5$CHO, xylene, 130 °C; **C:** 5 mol% [RhCl(cod)]$_2$, 10 mol% dppp, 10 mol% TPPTS, 2 eq. SDS, 5–20 eq. HCHO, 2–12 h, refluxing water.

ynes gave the corresponding products in yields ranging from 52 to 97%. A few months later Shibata and co-workers reported that PK-type transfer carbonylation reactions are also possible under solvent-free conditions using a high excess of aldehyde [56]. Reactions went smoothly in yields of 56–98%. It was even shown that cyclopentenones can be obtained enantioselectively with *ee*s of 45–90% when the chiral system [Rh(cod)Cl]$_2$/tolBINAP is used. Very recently, a micelle-containing aqueous PK-type reaction using formaldehyde was reported [57]. In this example decarbonylation and carbonylation reactions are supposed to take place independently in different phases of the reaction system. Formaldehyde as the water-soluble carbon monoxide source is decarbonylated in the aqueous phase enabled by the [RhCl(cod)]$_2$/TPPTS system, while carbonylation takes place in a micelle formed by the surfactant SDS. Isolated yields of cyclopentenones were good to excellent (67–96%). Some representative examples of PK-transfer carbonylation reactions are given in Table 4.

3.13.7
Conclusions and Outlook

During recent years, significant improvements in the area of Pauson-Khand reactions have been achieved. For instance, it is now state of the art to carry out such reactions in a catalytic or even in an enantioselective manner. Despite its many known applications, the PKR still offers unexplored potential for natural product synthesis and more complicated organic building blocks. The recent development of transfer carbonylation is increasing the scope of the methodology and may stimulate organic chemists to use this procedure more often.

3.13.8
References

1 I. U. Khand, G. R. Knox, P. L. Pauson, W. E. Watts, M. I. Foreman, *J. Chem. Soc., Perkin Trans. 1* **1973**, 977–981.
2 (a) N. E. Schore, *Chem. Rev.* **1988**, 88, 1081–1119. (b) N. E. Schore, *Organic Reactions* **1991**, 40, 1–90. (c) N. E. Schore in *Comprehensive Organic Synthesis* (Eds.: B. M. Trost, I. Fleming), Pergamon Press, Oxford, **1991**, Vol. 5, pp. 1037–1064. (d) N. E. Schore in *Comprehensive Organometallic Chemistry II* (Eds.: L. S. Hegedus), Pergamon Press, Oxford, **1995**, Vol. 12, pp. 703–739. (e) O. Geis, H. G. Schmalz, *Angew. Chem.* **1998**, 110, 955–958; *Angew. Chem. Int. Ed.* **1998**, 37, 911–914. (f) S. T. Ingate, J. Marco-Contelles, *Org. Prep. Proc. Int.* **1988**, 30, 121. (g) K. M. Brummond, J. L. Kent, *Tetrahedron* **2000**, 56, 3263–3283. (h) S. E. Gibson (née Thomas), A. Stevenazzi, *Angew. Chem. Int. Ed.* **2003**, 42, 1800–1810.
3 N. E. Schore, M. C. Croudace, *J. Org. Chem.* **1981**, 46, 5436–5438.
4 P. Magnus, L. M. Principle, *Tetrahedron Lett.* **1985**, 26, 4851–4854.
5 M. Yamanaka, E. Nakamura, *J. Am. Chem. Soc.* **2001**, 123, 1703–1708.
6 (a) R. Aumann, H. J. Weidenhaupt, *Chem. Ber.* **1987**, 120, 23–27. (b) K. Tamao, K. Kobayashi, Y. Ito, *J. Am. Chem. Soc.* **1988**, 110, 1286–1288. (c) K. Tamao, K. Kobayashi, Y. Ito, *Synlett* **1992**, 539–546.
7 R. B. Grossmann, S. L. Buchwald, *J. Org. Chem.* **1992**, 57, 5803–5805.
8 (a) A. J. Pearson, R. A. Dubbert, *J. Chem. Soc., Chem. Commun.* **1991**, 202–203. (b) A. J. Pearson, R. A. Dubbert, *Organometallics* **1994**, 13, 1656–1661.
9 N. Jeong, S. J. Lee, B. Y. Lee, Y. K. Chung, *Tetrahedron Lett.* **1993**, 34, 4027–4030.
10 T. R. Hoye, J. A. Suriano, *J. Am. Chem. Soc.* **1993**, 115, 1154–1156.
11 E.-I. Negishi, T. Takahashi, *Acc. Chem. Res.* **1994**, 27, 124–130.
12 R. Rios, M. A. Pericás, A. Moyano, *Tetrahedron Lett.* **2002**, 4903–4906.
13 (a) W. A. Smit, A. S. Gybin, A. S. Shaskov, Y. T. Strychkov, L. G. Kyzmina, G. S. Mikaelian, R. Caple, E. D. Swanson, *Tetrahedron Lett.* **1986**, 27, 1241–1244. (b) W. A. Smit, S. O. Simonyan, G. S. Tarasov, G. S. Mikaelian, A. S. Gybin, I. I. Ibragimov, R. Cable, O. Froen, A. Kraeger, *Synthesis* **1989**, 472–476.
14 N. Jeong, Y. K. Chung, B. Y. Lee, S. H. Lee, S.-E. Yoo, *Synlett* **1991**, 204–206.
15 S. Shambayati, W. E. Crowe, S. L. Schreiber, *Tetrahedron Lett.* **1990**, 31, 5289–5292.
16 T. Sugihara, M. Yamada, M. Yamaguchi, M. Nishizawa, *Synlett* **1999**, 6, 771–773.
17 T. Sugihara, M. Yamada, H. Ban, M. Yamaguchi, C. Kaneko, *Angew. Chem. Int. Ed.* **1997**, 36, 2801–2803.
18 V. Rautenstrauch, P. Megard, J. Conesa, W. Kuster, *Angew. Chem. Int. Ed.* **1990**, 29, 1413–1416.
19 N. Jeong, S. H. Hwang, Y. Lee, Y. K. Chung, *J. Am. Chem. Soc.* **1994**, 116, 3159–3160.
20 S. H. Hwang, Y. W. Lee, J. S. Lim, N. Jeong, *J. Am. Chem. Soc.* **1997**, 119, 10549–10550.
21 T. Rajesh, M. Periasamy, *Tetrahedron Lett.* **1999**, 40, 817–818.
22 N. Y. Lee, Y. K. Chung, *Tetrahedron Lett.* **1996**, 37, 18, 3145–3148.
23 T. Sugihara, M. Yamaguchi, *J. Am. Chem. Soc.* **1998**, 120, 10782–10783.
24 M. Hayashi, Y. Hashimoto, Y. Yamamoto, J. Usuki, K. Saigo, *Angew. Chem.* **2000**, 112, 645–647; *Angew. Chem. Int. Ed.* **2000**, 39, 631–633.

25 S. U. Son, S. I. Lee, Y. K. Chung, *Angew. Chem.* **2000**, *112*, 22, 4318–4320; *Angew. Chem. Int. Ed.* **2000**, *39*, 4158–4160.

26 S. C. Berk, R. B. Grossman, S. L. Buchwald, *J. Am. Chem. Soc.* **1993**, *115*, 4912–4913.

27 M. Zhang, S. L. Buchwald, *J. Org. Chem.* **1996**, *61*, 4498–4499.

28 F. A. Hicks, N. A. Kablaoui, S. L. Buchwald, *J. Am. Chem. Soc.* **1996**, *118*, 9450–9451.

29 T. Morimoto, N. Chatani, Y. Fukomoto, S. Murai, *J. Org. Chem.* **1997**, *62*, 3762–3765.

30 T. Kondo, N. Suzuki, T. Okada, T. Mitsudo, *J. Am. Chem. Soc.* **1997**, *119*, 6187–6188.

31 Y. Koga, T. Kobayashi, K. Narasaka, *Chem. Lett.* **1998**, 249–250.

32 N. Jeong, *Organometallics* **1998**, *17*, 3642–3644.

33 W. J. Kerr, D. M. Lindsay, E. M. Rankin, J. M. Scott, S. P. Watson, *Tetrahedron Lett.* **2000**, *41*, 3229–3233.

34 V. Derdau, S. Laschat, P. G. Jones, *Heterocycles* **1998**, *48*, 1445–1448.

35 L. V. R. Boaga, A. S. Felts, C. Hirosawa, S. Kerrigan, M. E. Krafft, *J. Org. Chem.* **2003**, *68*, 6039–6042.

36 M. E. Krafft, *J. Am. Chem. Soc.* **1988**, *110*, 968–969.

37 I. Marchueta, E. Montenegro, D. Panov, M. Poch, X. Verdaguer, A. Moyano, M. A. Pericás, A. Riera, *J. Org. Chem.* **2001**, *66*, 6400–6409.

38 J. Castro, H. Sörensen, A. Riera, C. Morin, A. Moyano, M. A. Pericás, A. E. Greene, *J. Am. Chem. Soc.* **1990**, *112*, 9388–9391

39 J. Castro, A. Moyano, M. A. Pericás, A. Riera, A. E. Greene, A. Alvarez-Larena, J. F. Piniella, *J. Org. Chem.* **1996**, *61*, 9016.

40 J. Tormo, A. Moyano, M. A. Pericás, A. Riera, *J. Org. Chem.* **1997**, *62*, 4851–4855.

41 (a) X. Verdaguer, A. Moyano, M. A. Pericás, A. Riera, M. A. Maestro, J. Mahía, *J. Am. Chem. Soc.* **2000**, *122*, 10242–10243. (b) X. Verdaguer, M. A. Pericás, A. Riera, M. A. Maestro, J. Mahía, *Organometallics* **2003**, *22*, 1868–1877.

42 (a) F. A. Hicks, S. L. Buchwald, *J. Am. Chem. Soc.* **1996**, *118*, 11688–11689. (b) F. A. Hicks, S. L. Buchwald, *J. Am. Chem. Soc.* **1999**, 121, 7026–7033.

43 N. Jeong, B. K. Sung, Y. K. Choi, *J. Am. Chem. Soc.* **2000**, *122*, 6771–6772.

44 T. Shibata, K. Takagi, *J. Am. Chem. Soc.* **2000**,*122*, 9852–9853.

45 S. J. Sturla, S. L. Buchwald, *J. Org. Chem.* **2002**, *67*, 3398–3403.

46 T. F. Jamison, S. Shambayati, W. E. Crowe, S. L. Schreiber, *J. Am. Chem. Soc.* **1994**, *116*, 5505–5506.

47 N. Jeong, S.-E. Yoo, S. J. Lee, S. H. Lee, Y. K. Chung, *Tetrahedron Lett.* **1991**, *32*, 2137–2140.

48 J. Cassayre, F. Gagosz, S. Z. Zard, *Angew. Chem. Int. Ed.* **2002**, *41*, 1783–1785.

49 C. Mukai, M. Kobayashi, I. J. Kim, M. Hanaoka, *Tetrahedron* **2002**, *58*, 5225–5230.

50 M. Ishizaki, Y. Niimi, O. Hoshino, *Tetrahedron Lett.* **2003**, 44, 6029–6031.

51 T. Shibata, S. Kadowaki, M. Hirase, K. Takagi, *Synlett* **2003**, *4*, 573–575.

52 P. A. Wender, N. M. Deschamps, G. G. Gamber, *Angew. Chem.* **2003**, *115*, 1897–1901; *Angew. Chem. Int. Ed.* **2003**, *42*, 1853–1857.

53 N. Chatani, T. Morimoto, Y. Fukumoto, S. Murai, *J. Am. Chem. Soc.* **1998**, *120*, 5335–5336.

54 N. Jeong, S. D. Seo, J. Y. Shin, *J. Am. Chem. Soc.* **2000**, *122*, 10220–10221.

55 T. Morimoto, K. Fuji, K, Tsutsumi, K. Kakiuchi, *J. Am. Chem. Soc.* **2002**, *124*, 3806–3807.

56 T. Shibata, N. Toshida, K. Takagi, *J. Org. Chem.* **2002**, *67*, 7446–7450.

57 K. Fuji, T. Morimoto, K. Tsutsumi, K. Kakiuchi, *Angew. Chem.* **2003**, *115*, 2511–2513; *Angew. Chem. Int. Ed.* **2003**, *42*, 2409–2411.

Subject Index

Numbers in front of the page numbers refer to Volumes I and II, respectively: e.g., II/254 refers to page 254 in Volume II.

a
1233A II/93
AA reaction II/302
acetal deprotection I/387
acetaldehyde I/336
acetalization I/61
acetals, chiral I/556
acetonitrile I/184, I/186
acetoxylation, allylic II/245, II/246
acetylene I/17, I/171, I/182, I/184, I/186, I/187, I/190, I/511, I/533, I/534
acetylides, terminal I/386
acid
 – anhydride I/358
 – chloride I/280, I/478
 – γ, δ-unsaturated I/122
acoragermacrone I/463
acoustic radiation II/583
acrolein I/181, I/336
 – acetals I/471
acrylates I/325, I/327
acrylic fibers I/189
activation volume II/616
acycloxylation II/258
 – asymmetric II/259–II/263
 – propargylic II/261
acyl chloride I/432
acyl complexes, unsaturated I/576
3-acyl-1,3-oxazolidin-2-ones I/349, I/350
1-acyl-2,3-dihydro-4-pyridones I/546
acylation I/379, I/464
 – *Friedel-Crafts* I/356–I/358, I/382, I/591
acylferrates I/575
acyloxylation reaction, asymmetric II/256
acylperoxomanganese complexes II/350, II/489
acylsilanes I/433, I/453
1-adamantyl-di-*tert*-butylphosphine I/217
adamantylphosphines I/241

adenosine II/327
AD-mix II/285
ADN (adiponitrile) I/149
adriamycin II/70
africanol II/86
agrochemicals I/23, I/41, I/42, I/149
$AlCl_3$ I/356, I/577
alcohol
 – α-allenic I/121
 – acylation I/358
 – allylic I/62, I/63, I/203, I/204, I/307, I/309, I/337, I/470, I/476, I/477, I/511, I/542, II/63
 – – cyclocarbonylation I/127
 – aminoalcohol I/76, II/158, II/326
 – – ligand I/530
 – β-amino I/347, I/364, I/478
 – cyclic allylic II/65
 – fluorinated II/98, II/365
 – halogenated II/46
 – homoallylic I/62, I/384, I/440, I/493
 – homopropargyl I/472
 – oxidation I/379
 – – aerobic, metal-catalyzed II/437–II/473
 – polyfunctional I/529
 – propargylic I/64, I/405, I/416, I/476
 – – chiral I/535
 – unsaturated I/258
aldehyde I/57, I/96, I/103, I/155, I/430, I/435, I/470, I/478, I/491, I/543
 – α,β-unsaturated I/339, I/455, I/472, II/37
 – aliphatic I/342
 – alkyl nucleophiles, addition I/503–508
 – allyl nucleophiles, addition I/493–I/498
 – β-hydroxy I/494
 – carbonyl hydrogenation II/29–II/95
 – mixed couplings I/461, I/462
 – water-soluble I/336
Alder ene reaction I/6

Transition Metals for Organic Synthesis, Vol. 1, 2nd Edition.
Edited by M. Beller and C. Bolm
Copyright © 2004 WILEY-VCH Verlag GmbH & Co. KGaA, Weinheim
ISBN: 3-527-30613-7

Subject Index

aldol   I/93, I/379
- *anti*-aldol   I/500
- *syn*-aldol   I/500
aldol products   I/369
aldol reaction   I/336, I/337, I/398, I/432, I/539, I/540
- addition, enolates to aldehydes   I/499–I/503
- asymmetric   I/370
- enantioselective   I/369–I/371
- hydroformylation/aldol reaction   96, I/97
- intramolecular version   I/94
- *Mukaiyama* type   I/94, I/381, I/382, I/502, I/503
- stereoselective   I/591
aldonolactones   I/431, I/434
Alexakis   I/536, I/557, I/558
Aliquat-336   I/122
alkali metals   I/452
alkaloids   I/37, I/349, I/590
- alkaloid 251F   II/77
- macrocyclic   I/330
- pseudoenantiomeric   II/276
alkane, oxidation   II/221
alkene-arene π-stacking   I/400
alkenes   I/449, I/483
- carbometallation   I/478
- cyclic   I/443
- enantioselective alkylation, by chiral metallocenes   I/257–I/268
- heterogeneous hydrogenation   II/135
- hydrocarboxylation   I/113–I/117
- hydrocyanation   I/149–I/151
- hydroesterification   I/117–I/120
- hydrovinylation   I/308
- internal   I/75
- – oxidation   II/382, II/383
- metathesis   I/321–I/328
- macrocyclic   I/330
- strained   I/570
- terminal   I/58, I/258
alkenol, cyclization   II/384
alkenyl boranes   I/559
alkenyl copper reagent   I/554
alkenyl halides   I/472, I/474–476
alkenylbismuthonium salts   I/387
alkenylborane   I/532
alkenylpyridine   I/185–I/188
alkenylzinc reagents   I/522, I/532
alkoxide ligands   I/491
alkoxycarbene complexes   I/397
- β-amino-α,β-unsaturated   I/416
3-alkoxycarbonyl-Δ^2-pyrazolines   I/415
3-alkoxycyclopentadienones   I/589

alkyl halides   I/211–I/225, I/476
alkyl hydroperoxide   I/20
alkyl peroxides   II/231
alkyl tosylates   I/224
alkylation   I/260
- allylic   I/4, I/6, I/8, I/9, I/311, I/314
- enantioselective, by chiral metallocenes   I/257–I/268
- *O*-alkylation   I/309
alkylidenecyclopentenones   I/416
alkylpyridine   I/185–I/188
- α-alkylpyridine   I/186
alkylrhenium oxide   II/432
alkylsulfones, lithiated   I/158
2-alkylthiopyridine   I/188, I/189
alkylzinc nucleophiles   I/530, 531
alkyne   I/113, I/416, I/417, I/440, I/474, I/485
- cyclomerization   I/171–I/193
- hydrocarboxylation   I/124
- hydrocyanation   I/151–I/153
- hydrosilylation   II/171–II/173
- internal   I/174
- metathesis   I/330, I/331
- oxygenation   II/215
- terminal   I/125, I/174
alkyne reaction   I/440
alkyne-$CO_2(CO)_6$-complex   I/620
alkynols, cycloisomerization   I/398
alkynyl acyl complexes   I/576
alkynyl halides   I/476
alkynylboranes   I/405
alkynylboronates   I/405
alkynylzinc nucleophiles   I/530, I/533–I/535
allene   I/153, I/473
- hydrocarboxylation/hydroesterification   I/120–I/122
allenyl sulfide   I/510
allocolchicinoids   I/407
allyl acetate   I/559
π-allylcomplex   II/243
allyl ethers   I/205, I/206, I/311
- by cyclization of alkenols   II/384
- synthesis from olefins   II/385
allyl halides   I/307, I/470–I/472
allyl organometallics   I/337
allyl resins   I/307
allyl stannanes   I/494
π-allyl transition metal complexes   I/307
allyl urethanes   I/311
allyl zinc bromide   I/523
allylamines   I/199–I/203
- optically active   I/208

2-allylanilines I/129
allylation reaction I/337, I/338, I/379, I/472, I/522
- *Barbier*-type I/380, II/590
- benzaldehyde I/471
- double I/527
- *Hiyama-Nozaki* I/470
- Pd-catalyzed II/566
allylboranes I/88–I/92
allylboration, intramolecular I/91
allylboronates I/92
allylchromium I/471, I/477
allyl-cobalt I/183
allyl-*Grignard* I/493
allylic acetates I/485
allylic acetoacetates I/484
allylic acetoxylation II/245, II/246
allylic alkylation I/4, I/6, I/8, I/311
- asymmetric I/9, I/314
- enantioselective I/527
allylic amines, asymmetric synthesis I/532
allylic bromides I/478
allylic carbonates I/309, I/311
allylic chloride I/307
allylic esters I/309, I/311
allylic ethers, cyclic I/266, 267
allylic imidates I/312
allylic nitriles I/150
allylic oxidation I/308, II/243–263
- copper-catalyzed II/256–II/263
- *Karasch-Sosnovsky* type II/256–II/263
- palladium-catalyzed II/243–II/253
- regioselectivity II/259
allylic radical II/259
allylic silylation I/313
allylic substitution, palladium-catalyzed I/307–I/315
- mechanism I/310, I/311
allylic sulfonylation I/313
1,3-allylic trandposition II/248
π-allylpalladium chloride complex I/307
π-allylpalladium complex I/307, II/244
- 2-aza-π-allyl palladium complex I/310
2-allylphenol I/129
allylsilanes I/88–I/92, I/291, I/382, I/496
allyltitanium I/493, I/494
allyltributyl reagent I/577
aloesaponol III I/488
aluminohydride I/311
5-amino-pyrroles I/152
amides I/244, I/431, I/433, I/435, I/465
amidocarbonylation I/67, I/100–I/103
- aldehydes I/133–I/146

- cobalt-catalyzed I/134–I/140
- domino hydroformylation-amidocarbonylation I/136
- palladium-catalyzed I/141–I/146
amination I/231–I/246, I/315, II/403–II/412
- oxidative II/405, II/406
- reductive I/19, I/97
amine oxidase II/497
amine oxidation II/497–II/505
- dehydrogenative II/497
- metal hydroperoxy species II/499
- metal oxo species II/502
- primary amines II/498, II/501, II/502, II/504, II/505
- secondary amines II/498, II/500, II/501, II/504, II/505
- tertiary amines II/498, II/499, II/502–II/504
amines I/155, I/309
- chiral II/113
- optically active II/189
- primary, arylation I/244
- secondary I/309
- unsaturated I/258
amino acid I/245, I/310, I/349, I/545, I/565
- α- II/14
- asymmetric synthesis I/532
- azolactones I/513
- derivates I/398
- non-proteogenic I/544
- synthesis I/544
amino ketones II/59
4-amino-1,3,5-hexatrienes I/416
4-amino-1-metalla-1,3-butadienes I/419
amino-carbene
- cluster II/499
- aminocarbene complex I/398
aminocyclitols I/413
aminohydroxylation, asymmetric II/275, II/309–II/334
- amide variant II/323–325
- carbamate variant II/320–II/323
- enantioselectivity II/315
- intramolecular II/330, II/331
- nitrogen source II/313, II/314, II/327, II/328
- recent developments II/326–II/334
- regioselectivity II/328–330
- scope II/314, II/315
- secondary-cycle II/331–II/333
- solvent II/314
- sulfonamide variant II/315–320
- three variants, comparison II/312–II/325
- vicinal diamines II/333

aminoketone I/465
aminometallahexatrienes I/419
α-aminophosphonates I/368
α-aminophosphonic acid I/367
2-aminothiopyridines I/188, I/189
amino-zinc-enolate I/565
ammonia I/73
ammonium formate I/311
Amphidinolide A I/326
amphiphilic resin II/444
Amphotericin B I/594
amplification
– asymmetric II/260
– chiral I/496
AM-Ti3 II/338
Andersen method II/479
andrastane-1,4-dione II/6
angiotensin II receptor antagonist I/525
anhydride I/281, I/433
– homogeneous hydrogenation II/98
anilines, primary I/243
anisole I/357
annualtion, (2+2+2) I/478
ansa metallocenes I/257, I/268
anthraquinones I/178
– ligands II/293, II/294
anti-1,2-diols I/370
antibiotics I/408, II/11
antibodies, catalytic II/349
anti-cancer agent I/457, I/537
anti-fungal agent I/258
anti-histaminic agent I/30
anti-inflammatory agents I/31
anti-*Markovnikov* addition I/3, I/149
apomorphine II/346
Ar$_4$BiX I/385
arabitol II/97
araguspongine II/77
araliopsine I/489
Aratani I/158, I/159
ArB(OH)$_2$ I/559
ARCO/HALCON process II/372
arenes I/385
η^6-arene-Cr(CO)$_6$ complexes I/601
aristoteline I/465
aromatic nitro groups, heterogeneous hydrogenation II/132–II/134
aromatic ring, heterogeneous hydrogenation II/136, II/137
aromatic substitution, electrophilic I/8, I/92
arsacyclobutenes I/440
arthrobacillin A II/77
aryl acetylenes I/406

aryl chlorides I/211, I/239
aryl ethers I/231, I/246
aryl fluorides I/225
aryl halides I/211–I/225, I/231, I/474–I/476
– palladium-catalyzed olefination I/271–I/300
aryl iodides I/244
aryl lithium I/609
aryl pincer ligand II/154
aryl propargyl amine I/80
aryl stannanes I/309
arylation I/231–I/253, I/379
– *Heck* arylation I/5
– *N*-arylation I/391
– *O*-arylation I/390
arylboronic acids I/211, I/245, I/251
arylchloride I/526
aryldiazonium salts I/279
aryldiazonium tetrafluoroborates I/221
3-aryl-enones I/609
arylglycins II/323
arylpyridine I/185–I/188
arylzinc
– iodides I/474
– nucleophiles I/531–I/533
aspartame I/134, I/139
aspicillin I/532
asymmetric
– activation II/66–II/68, II/349
– catalysis I/153, I/404
– deactivation II/66–II/68
– isomerization catalysis I/199
– synthesis I/18, I/23, I/57, I/88, I/187
– transformation, second-order II/162
ate-complex I/491
atom economy I/3, I/11, I/12
attractive interaction II/293
autoxidation II/201, II/206, II/208, II/210, II/219, II/349, II/416
– free-radical II/202–II/204
avermectin B$_{12}$ II/5
3-aza-1-chroma-butadiene I/413
1-azabuta-1,3-diene I/585
aza-*Cope* I/312
azadiene I/346
aza-semicorrins I/313
Azinothricins I/38
aziridinations, asymmetric II/389–II/400, II/602, II/603
– copper-catalyzed II/389–II/393
– heterogeneous II/390
– imines as starting materials II/396–II/400

- *Lewis* acids   II/398, II/399
- olefins as starting materials   II/389–II/396
- porphyrin complex   II/395, II/396
- rhodium-catalyzed   II/393, II/394
- salen complexes   II/394, II/395
- ylide reaction   II/399, II/400
aziridines   I/208, II/161, II/389
azirines   II/161

b

Bäckvall   I/23
Baclofen   I/399
Baeyer-Villiger oxidation   II/210, II/267–II/272, II/448, II/555
- asymmetric   II/269, II/271
balanol   II/93
Barbier-type allylation   I/380, II/590
BARF anion, fluorinated   II/547
BASF   I/17, I/19, I/23, I/171
basic chemicals   I/15
bcpm   II/115
BDPP   II/8, II/515
benazepril   II/82
benchrotenes   I/402
benzaldehyde   I/342, I/506, /530, I/571
- allylation   I/471
benzannulation
- (3+2+1) benzannulation   I/402–I/408
- *Dötz*   I/408, II/588, II/605
- intramolecular   I/406
benzene   I/171, I/174–I/177, I/357, I/417
η^6-benzene-Cr(CO)$_3$ I/601
benzo(*b*)furans   I/463
benzo(*b*)naphtol(2,3-*d*)furan   I/407, I/408
benzofused azoles   II/606
benzoic anhydride   I/357
benzoil chloride   I/357
benzonitrile   I/187
benzophenone   II/33
benzoquinone   I/509
benzyl chromium species   I/476
benzyl propionate   I/341
benzylic carbocation   I/607
benzylic lithiation   I/606
benzylic radicals   I/608
BF$_3$ I/356
BF$_3 \cdot$ OEt$_2$ I/345, I/559
Bi(OTf)$_3$ I/383, I/388
Bi(V)   I/379
biaryl synthesis   I/406
biaryls   I/211
BICHEP   II/8
BiCl$_3$ I/383, I/390

bicp   II/115
bicyclo(3.2.1)octane   I/402
bicyclo(3.3.1)nonane   I/402
Biginelli-multicomponent condensation   II/604
bimetallic
- catalyst   II/98
- complex   I/136
- derivatives, geminal   I/429, I/430
- Rh-Pd catalyst   I/315
BINAP   I/23, I/199, I/200, I/235, I/282, I/285, I/313, II/115, II/549
- PEG-bound   II/515
- poly(BINAP)   II/51
- polystyrene-bound   II/51
- (R)-(+)   I/201, II/8
- (S)-(–)   I/201, I/205
- two-phase catalysis   II/521
BINAPHOS   I/31, I/40, I/69, I/72, I/154
binaphthyl diphosphines   I/59
binaphthylbisoxazoline palladium complexes   I/308
binaphtol   I/268, II/535
- ligand   I/531
BINAPO   I/313
BINOL   I/353–I/355, I/374, I/494, I/496, I/533, II/271, II/272, II/369
- BINOL-titanium   I/498, I/530
- ligands   I/505, II/482, II/483
- polymers   II/375
biologically active compounds/substances   II/29, II/92
biomimetic
- reaction   II/299
- systems   II/205
biotechnology   I/17
biphasic aqueous-organic medium   II/513
biphasic conditions   II/349
biphasic system, inverted   II/550
Biphenomycin A   II/93
BIPHEPHOS   I/38, I/39, I/59, I/78, I/94
bipyridines, chiral   II/262, II/263
bipyridyls   I/189, I/190
bis(azapenam)   I/412
bis(aziridines)   II/393
bis(di-*iso*-propylphosphino)ethane   I/471
bis(oxazolinyl)carbazole ligand   I/472
bis(oxazolinyl)pyridines   II/261
bis(pentamethylcyclopentadienyl) titanium   I/439
bis(sulfonamide)   I/167
bis-(trimethylsilyl)acetylene   I/178
bishydrooxazoles   I/313

bismethylenation I/435
bismuth I/379–I/392
– bismuth(0) I/380, I/381
– bismuth(III) I/381–I/384
bismuthonium salt I/386
bisoxazolines I/313, II/260, II/261, II/397, II/533
– ligand I/162, II/389–II/391
bisphosphine-ferrocene II/184, II/185
bisphosphinites I/313
bispyrrolidines I/313
3,5-bis-trifluoromethylbenzaldehyde I/506
bite angle I/153, I/314, II/516, II/521
Blaser-Heck reaction I/280
blood substitutes II/536
BMS 181100 II/70
BNPPA I/114
Bogdanovic I/450
Bolm I/538, I/559
– *Bolm's* complex II/269
Bönnemann I/172, I/182
BoPhoz II/19
borohydrides I/311
boron I/560
boron-zinc exchange I/521
boronic esters I/472
BOX II/390, II/397
BPE I/23
BPPFA I/313
brefeldin A II/77
Breit I/44
2-bromofurans I/475
Bronsted base I/363
Bu$_5$CrLi$_2$ I/470
Buchwald's ligand I/215, I/239, I/249
Buchwald-Hartwig reaction I/21, I/232, II/602
– mechanism I/236
buckminsterfullerene I/590
bulky phosphite I/51
butadiene I/597
– 1,3-butadiene I/121, I/149
– dimers I/121
– monoepoxide I/9
3-buten-1-ols I/128
butenolides I/575
a,β-butenolides I/581
2-butyne-1,4-diol I/184
2-butyne-1-ol I/184
butyrolactones I/62, I/608
– *a*-methylene-γ-butyrolactones I/128
– γ-butyrolactones I/63, I/127, II/271

c

C(sp^3)-C(sp^3) coupling I/528
C$_8$K I/450
C$_9$ telomers I/121
calix(4)arene I/459
calphostin I/408
camalexin I/465
CAMP II/21
camphor I/530, I/577, I/587
camphothecin I/525
cannabisativine I/546
cannithrene II I/462
capnellene I/463
captopril I/134
carazostatin I/595
carbacyclins II/93
carbametallation I/7
carbanion I/88
carbapalladation I/5
carbapenem I/579, II/94
carba-sugar I/37
carbazole alkaloids I/595, I/596
carbazomycin A I/596
carbazomycin B I/596
carbazomycin G I/595
carbene complex II/565
– *a,β*-unsaturated I/400–I/402, I/411
– difluoroboroxy I/418
– *Fischer*-type I/397–I/420
– photoinduced I/412–I/414
carbene ligands I/220–I/222, I/405, I/409, II/154
– electrophilic I/163
carbene precursors I/158
carbene transfer I/409, I/418
carbenoids I/427
carbocyclation, (3+2) I/416
carbocycle
– 5-membered I/414–I/418
– 6-membered I/173–I/179
carbocyclic nucleosides I/36, I/413
carbocyclization I/564
carbodiimide I/180
carbohydrate I/155, I/437, I/489
carbomagnesation reactions I/257–I/263
– asymmetric I/258, I/263
carbometalation I/529
– intramolecular I/563–I/569
carbon dioxide (CO$_2$) I/179
– compressed II/269, II/545–II/556
– dense II/169
– homogeneous hydrogenation II/98–II/102
– hydrogenation II/548

- supercritical (scCO$_2$) I/78, I/323, II/116, II/197, II/198, II/383, II/438, II/539, II/545, II/551, II/567
carbon disulfide I/357
carbon monoxide I/57, I/106, I/113, I/114, I/117, I/308, I/397, I/619
- ^{13}C-labeled I/115
- copolymerization I/108
carbon nucleophiles I/603
carbon tetrachloride I/473
carbonates I/433
carbonyl compounds
- α,β-unsaturated I/371
- α-halo carbonyl compound I/543
- α-hydroxy carbonyl compounds I/364
carbonyl coupling reactions I/454
- reductive I/455
carbonyl ene reactions I/104, I/105
carbonyl
- ligand I/397
- methylenations I/428
- selectivity II/34–II/38
carbonylation I/21, I/113, I/135, I/307, II/29–II/95
- allylic I/309
carboxylate I/309
- chiral I/165
carboxylic acid I/155, II/95–II/102
- α,β-unsaturated I/508
- two-step synthesis II/428, II/429
carquinostatin I/596
Carreira I/501, I/503, I/534, I/535
carvone II/65
cascade reactions I/294
catalyst
- activity II/129
- deactivation II/361
- homogeneous II/29
- recyclable I/322
- selectivity II/130
- suppliers II/128
catechol dioxygenase II/224
cavitational bubble II/587
C-C bond forming I/268, I/307, I/335, I/379–I/392, I/512, I/519, I/533, I/553, II/551
- catalytic I/257
- fluorous catalysts II/532, II/533
- via metal carbene anions I/398, I/399
C-C cleavage products II/278
C-C coupling I/18, I/20–I/22
- chromium(II)-catalyzed reactions I/469–478
Ce(IV) I/483

cembranoids I/435
cembrene I/463
ceramics I/379
cerium(IV) ammonium nitrate I/591
CF$_3$CO$_3$H I/496
CF$_3$SO$_3$H I/496
C-H bond, activation I/9, I/11, I/523
- allylic I/243
C-H compounds, oxidation II/215–II/236
C-H insertion I/163
- intramolecular I/165
CH-π attraction II/148, II/154
CH$_2$I$_2$ I/541
CH$_2$I$_2$-Zn-Ti(O-iPr)$_4$ I/437
CH$_3$Ti(O-iPr)$_3$ I/506
chalcone I/373
Chan I/558
charcoal II/128
Chauvin I/321
chemical hermaphroditism I/605
chemists enzymes I/4
chemoselectivity I/4, I/5, I/18, II/133
chiral
- auxiliary I/3, I/411, I/556, II/83
- ligand I/146, I/187, 313
- poisoning I/498
- promotor I/167, I/624
CHIRAPHOS I/23
chloroacetaldehyde I/336
chloroarenes I/212, I/239, I/276
4-chlorochalcone I/538
chlorophosphites I/44
chlorosilanes I/455
CHP II/484
chromaoxetane intermediate II/278
chromium II/372, II/373
- catalyst I/175
- chromium(0) I/404
- chromium(II)-catalyzed C-C coupling I/469–478
- chromium(III) I/469
- modification II/34
chromium complex I/410
- planar chiral arene I/404
chromium tricarbonyl complex I/406
chromium-arene complexes I/601–I/612
- as catalysts I/612
- nucleophilic addition, arene ring I/602–I/604
- preparation I/602
- ring lithiation I/604, I/605
- side chain activation
- - general aspects I/605

– – via stabilization of negative charge   I/606, I/607
– – via stabilization of positive charge   I/607, I/608
chrysanthemates   I/159
cinca-*Claisen*-type rearrangement   I/525
cinchinoide-modified catalyst   II/79
cinchona alkaloids   II/131, II/299, II/492
– derivates   II/276
– ligands   II/293–II/295, II/300
cinchonine   I/544
cinnamaldehyde   II/38
cis-3-hexene-1,6-diols   I/532
citronellal 7   I/199, I/383
citronellol   I/203, II/10, II/11
cladiellin diterpenes   I/475
Claisen rearrangement   I/432, I/546
– (3,3) sigmatropic   I/577
– diastereoselective   I/544
Clavularin A   I/537
Clemmensen reduction   I/487
C-O coupling reaction   I/246–I/253
Co(Salophen)   II/245
– dicarbonyl   I/191
– diene complex   I/183
$CO_2(CO)_8$ I/102, I/103, I/119
cobalt   I/106, I/113, I/134–I/140, I/173, I/175
– homogeneous catalyst   I/176
cobalt black   II/34
cobalt chloride   I/122
cobalt metallacycles   I/180
cobalt porphyrin   II/537
cobalt vapor   I/183
cobalt(II) complex   I/162
cobaltacycle   I/184
cocyclization   I/171, I/174, I/175
– cobalt catalyzed   I/187
(COD)RhCl$_2$ 419
codaphniphylline   II/93
Coleophomones B/C   I/325, I/326
collidines   I/187
Collman's reagent   I/575
compactin   I/452
computational studies   II/151
computer-aided analysis   II/92
condensation
– aldol   I/398
– *Biginelli*-multicomponent   II/604
– *Knoevenagel* condensation   I/93
conjugate addition   I/368, I/400, I/553–I/560
– asymmetric   I/536–I/539
– enantioselectivity   I/556–I/560
– reactivity, general aspects   I/553–555

continuous flow
– reactor   II/513
– system   II/550
cooperation, metals   I/363
Cope-type (3,3) sigmatropic rearrangement   I/411
copper catalyst   I/158–I/162, I/244–I/246, I/250, I/536, I/537, II/34, II/186, II/187
copper enolates   I/577
copper(I)
– carbene complex   I/420
– hydrazide   II/472
– salts   I/149, I/386, I/419
copper(II)chloride   I/114
copper(III) species   II/258
copper-carbene   I/160
copper-catalyzed coupling reactions   I/522
copper-containing proteins   II/470
copper-nitrene species   II/392
corannulene   I/590
Corey   I/162, I/386
– model   II/291
cosmetics   I/149, I/379
co-solvent, chiral   I/603
Cossy   I/494
CpTiCl$_3$ I/499
Cr(CO)$_3$ I/416
– fragment   I/402
Cr(CO)$_5^-$ I/399
Cr(CO)$_6$ I/602
Cr(salen)   I/472
Cram selectivity   II/39
CrCl$_2$ I/469, I/473, I/474
CrCl$_3$ I/469, I/470, I/474
cross-coupling   I/21, I/24, I/309, I/419, I/519, II/565, II/566
– asymmetric   I/527
– alkyl-alkyl   I/529
– chromium arene complex   I/610, I/611
– intramolecular   I/450, I/463
– saturated coupling partners   I/528, I/529
– titanium-induced   I/461–I/466
– unsaturated coupling partners   I/525–I/528
crown ethers   I/399
crownophane   I/388, I/459
C-S bond-forming process   I/390
CTAB (cetyltrimethylammonium bromide)   I/121, I/122
Cu catalyst   I/160, I/164
Cu chromite   II/95
Cu complex   I/158, I/162
Cu(I) complex, cationic   I/159

Subject Index

Cu(II) I/483
(Cu(MeCN)$_4$)(PF$_6$) I/420
Cu(OAc)$_2$ I/483, I/485, I/488
Cu(OTf)$_2$ I/536
cumulenes I/180
(±)-α-cuparenone I/106
cuprate I/554
cyanation I/392
cyanhydrins I/384
cyanide addition I/375
2-cyanopyridine I/190
cyanosilylation I/375, I/376
- aldehydes I/375–I/377
cyclization I/442
- carbonylative I/108
- 5-*exo* I/488
- keto-ester cyclization I/462, I/463
- manganese(III) based radical I/483–I/489
- - substrates I/487–I/489
- oxidative I/487
- *Pauson-Khand* cyclization I/263, I/619–I/631, II/601
- reductive I/463
cycloaddition I/11, I/163, I/491, I/509
- 1,3-dipolar I/347, I/348, I/414, I/511
- - asymmetric I/355, 356
- (2+1) I/409
- (2+2) I/321, I/348, I/412, I/413, I/510, I/511, I/609
- - asymmetric I/353, I/354
- - photochemical I/412
- (2+2+1) I/588, I/589, I/619
- (2+2+2) I/174, I/176, I/180, I/182
- (3+2) I/415, I/579, II/615
- - asymmetric I/416
- (3+2+1) I/402
- (4+2) I/346, I/508, I/510, I/609
- (4+3) I/312
- cobaltocene-catalyzed I/192
- electrocatalytic II/570
- intramolecular I/176
- miscellaneous reactions I/508–I/513
- palladium-catalyzed II/614–II/620
- trimethylenemethane (TMM) I/312
cycloalkanones I/484
cycloalkene I/449, I/459
cyclobutanones I/413
cyclobutenones I/406, I/413
cyclocarbonylation I/108
- intramolecular I/126–I/130
cyclodextrins II/521
cycloheptadiene I/411

cycloheptanone I/537
cycloheptatriene I/586
cycloheptatrienones I/409
cyclohexa-1,3-dienes I/175
cyclohexadienes I/174–I/177
cyclohexadienone I/404
cyclohexadienyl radicals I/485
cyclohexane-1,2-diamine I/167
cyclohexene oxide I/207
cyclohexenones I/604
cyclohydrocabonylation, dipeptide I/66
cycloisomerization, alkynols I/398
1,5-cyclooctadiene I/11
cyclooctatetraene I/171, I/172
cyclooligomerization I/180
cyclopentadiene I/150, I/338, I/344, I/346, I/349, I/351, I/354, I/416
cyclopentadienones, iron-mediated synthesis I/588–I/590
cyclopentadienyldicarbonyl I/575
cyclopentadienyltitanium I/499
- fluoride I/506
- reagents I/493
- trichloride I/491
cyclopentanes I/512
(3+2) cyclopentannulation I/413
cyclopentanones I/106, I/433, I/619
cyclopentene I/95, I/536
cyclopentene oxide I/207
cyclopentenones I/416
cyclophanes I/330, I/406, I/459
cyclopropanation I/21, I/157–I/168, I/308, I/409–I/412, I/473, I/511, I/519, I/522, II/391
- asymmetric I/542
- copper-catalyzed I/162
- diastereoselective I/411
- enantioselective I/158–I/162, I/166, I/542
- intramolecular I/160, I/165
- ylide-mediated II/399, II/400
- zinc-mediated reaction I/541–543
cyclopropane I/157, I/409, I/525, I/593
cyclopropanol I/477
cyclopropenes I/440
cyclopropylglycins I/592
cyclopropylidenes I/387
cyclotrimerization I/171, I/173, I/174, II/553, II/579
cycphos II/115
cylindrocyclophane F I/325
cytochrome P450 II/226, II/234, II/256, II/497
C-Zr bond I/266

d

DAIB I/530, I/531
α-damascone II/70
Danishefsky I/348, I/475, I/500
Danishefsky's diene I/344, I/509, I/578
Daunorubicin II/70
Davies I/576, I/577
DEAD-H$_2$ II/458
debenzylation, catalytic II/137–II/140
– *N*-benzyl groups, selective removal II/139, II/140
– *O*-benzyl groups, selective removal II/138, II/139
decarbonylation I/382
decursin II/370
DEGUPHOS I/23
dehydrogenation I/18, I/20
dehydrohomoancepsenolide I/331
demethoxycarbonyldihydro-
  gambirtannine I/590
dendrimer I/323
dendritic crystals II/587
denopamine II/59
– hydrochloride II/70
density functional calculations II/362
deoxyfrenolicin I/408
17-deoxyroflamycoin II/86
deracemization, palladium-catalyzed II/247
Desoxyepothilone F I/500
desulfurization, ultrasound-assisted II/593
deuteriobenzaldehydes II/68, II/69
deuterioformylation I/32
deuterium labeling experiment I/205, II/175
deuterohydrogenation II/158
DFT methods I/605
(DHQ)$_2$AQN II/323
(DHQ)$_2$PHAL II/323, II/327, II/330
(DHQ)$_2$PYR ligand II/323
(DHQD)$_2$-PYR II/491
di(1-adamantyl)-*n*-butyl-phosphine I/241
di-1-adamantyl-di-*tert*-butylphosphine I/217, I/277
di-2-norbornylphosphine I/218
diacetone-glucose I/499
diacetoxylation II/250
dialdehyde I/463
dialkoxylation II/252, II/253
dialkyles I/180
dialkylidenecyclopentenones I/416
dialkylphosphines I/277
dialkylzinc I/536, I/538
– compound I/503, I/506

diamination, asymmetric II/333, II/334
diaryl
– ethers I/406
– ketone I/461, II/51
– methanes II/51
– methanols II/51
diastereoselectivity I/8, I/9, II/38–II/42
– *exo/endo* I/412
diastomer isomeric ballast II/7
1,4-diazabutadienes I/223
1,3-diazatitanacyclohexadienes I/428, I/440
diazo compounds I/157, I/163
– decomposition I/157, I/158
– reaction II/529–II/531
diazoacetates I/158
– allyl/homoallyl I/164
α-diazocarbonyl complex I/158
diazomethane 159
1,3-dicarbonyl I/385
dicarbonylation, oxidative I/121
dichloralkyl radical I/474
dicobaltoctacarbonyl I/149, I/621
dicyclohexylborane II/196
Diels-Alder reaction I/6–I/8, I/338, I/344–I/347, I/379, I/576, I/577, I/588, I/589, II/611
– asymmetric I/348–I/355
– hetero I/348, I/383, I/509, II/610
– imino I/345, I/347
– intermolecular I/417
– intramolecular I/417
– transannular I/329
diene I/308
– conjugated I/121, I/150
– hydrocarboxylation/hydroesterifi-
  cation I/120–I/122
1,3-diene I/6, I/7, I/309, I/411, I/440, I/477
– chiral I/311
– complexation I/586
– cyclohexa I/587
– *Danishefsky's* diene I/344, I/509, I/578
– macrocyclic I/328
1,4-diene I/6
1,5-diene I/312, I/328
diethyl-alkylboranes I/506
diethylzinc I/309, I/530
difluoroboroxycarbene I/398
1,1'-diformyl-ferrocene I/398
1,1-dihalides I/472–I/474
1,1-dihaloalkanes I/438
dihydride-based mechanism II/151
4,5-dihydro-1,3-dioxepins I/206
4,7-dihydro-1,3-dioxepins I/206

dihydrofurans I/264
- 2,3-dihydrofuran I/282
2,3-dihydroisoxazoles I/534
dihydropeptides II/515
dihydropyrroles I/68
dihydroquinolines II/160
dihydroxylation, asymmetric II/275–II/305, II/309, II/311
- directed II/301, II/302
- face selectivity II/287–II/290
- homogeneous II/299, II/300
- kinetic resolutions II/300, q301
- ligand optimization II/285, II/286
- osmylation, mechanism II/278–II/282
- polyenes II/299
- process optimization II/283–II/285
- recent developments II/298–II/305
- secondary-cycle catalysis II/302–II/304
1α,25-dihydroxyvitamin D3 II/79
diiron enoyl acyl complexes I/578, I/579
Diisopromine I/33, I/34
di-*iso*-propoxytitanium I/497, I/513
di-*iso*-propylzinc I/520
diketene I/502
diketones II/84–II/86
- β-diketones, unsaturated I/487
dimedone I/386
dimerization
- methyl acrylate II/567
- olefins I/308
- Pd-catalayzed I/418
dimetalated reagents, geminally I/523
dimetallic species I/427
dimethoxybenzene I/357
dimethyl aminoacetone II/58
2,3-dimethylbutadiene I/351
2,2-dimethylcyclopropanecarboxylic acid I/158
dimethyl malonate I/545
dimethylpyridines I/187
dimethylpyrrolidine acetoacetamides I/489
dimethyl succinate I/118
dimethyl sulfide I/386
dimethyltitanocene II/603
dinitrogen I/157
α,ω-diolefine I/80
DIOP I/23, II/21
diorganocuprate I/553
diorganozinc reagent I/504
dioxirane II/210
dioxygen II/201, II/205
dioxygenase II/224
DIPAMP I/23, II/19, II/25

dipentylzinc I/592
diphenyl phtalazine ligands II/294
diphenyl pyrazinopyridazine ligands II/294
1,3-diphenylallyl esters I/313
1,1-diphenylethylene I/72
2,3-diphenylindole I/452
diphosphines I/559, II/184
- chelating II/159, II/516
- chiral I/312, II/42, II/93
- ferrocene-based II/19
- ligand II/20
- *p*-chiral II/19, II/20
diphosphite I/559
diphospholane derivatives II/15
dirhenium heptoxide II/357
dirhodium catalyst I/165, I/166
disiamylborane II/196
di-*tert*-butylphosphine oxide I/218
dithioacetals I/438
dithioketals I/438
diynes I/478
DMAP I/441
DMSO I/323
domino
- procedure I/629
- reaction (*see* hydroformylation reaction)
- sequences I/294
π-donation I/397
Dötz benzannulation I/408, II/588, II/605
double bond migration II/135
Doyle I/161, I/164
DPEphos I/238
DPP (2,6-diphenylpyridine) I/355
dppb (1,4-bis(diphenylphosphino)butane) I/114, I/117, I/129
DSM I/185
DTBMP I/354
DuPHOS I/23, I/69, II/8, II/10, II/19, II/20, II/184
- Me-DUPHOS II/25
DuPont I/149
dyes I/189

e
Eastmann-Kodak I/45
(EBTHI)Zr-binol I/257
(EBTHI)ZrCl$_2$ I/257, I/263
EDTA I/476
Efavirenz I/534
efficiency, synthesis I/3
electrochemistry II/570
electrophilic substitution I/596

electroreduction   II/591
– coupling   II/570
eleutherobin   I/475
eleuthesides   I/476
β-elimination   I/106, I/416, I/528
– β-hydride elimination   I/32, I/108, I/150, I/223, I/272, I/281, I/294, I/429, II/150
enals   II/37
– iron-substituted   I/580, I/581
enamine   I/65, I/67–I/70, I/99, I/380, I/438, II/113
– asymmetric hydrogenation   II/15–II/19
– β,β-disubstituted   II/22
– enantioselective reduction   II/117, II/118
enantiomer-selective deactivation   II/68
enantioselectivity   I/9, I/312–I/315, II/42–II/69
Enders   I/538
endo,endo-2,5-diamino-norbornane   I/472
enediynes   I/476
ene-reaction, allyl nucleophiles addition   I/493
enol ethers   I/434
enolates   I/88, I/96, I/560, I/563
enols, aliphatic   I/203
enones   I/538
– β,γ-enones   I/470
– iron-substituted   I/580, I/581
environmental
– benefit   II/511
– hazard   I/18
enynes   I/442, I/478, I/525
– 1,6-enynes   I/176
– metathesis   I/328, I/329
– precursor   I/619
enzyme-metal-coupled catalysis   II/145
ephedrine   I/536, I/544
– N-methyl ephedrine   I/535
epinephrine hydrochloride   II/56
Epothilone B   II/77
Epothilone C   I/330, I/331
epoxidation   I/371–I/374
– aerobic   II/349–II/351
– asymmetric   II/210, II/341–II/344
– group III elements   II/369, II/370
– group IV elements   II/370
– group V elements   II/371, II/372
– group VI elements   II/372, II/373
– group VII elements   II/373
– group VIII elements   II/373–II/375
– heterogeneous   II/337
– Jacobsen-Katsuki   I/22, II/211, II/346, II/350

– lanthanoids   II/369, II/370
– manganese-catalyzed   II/344–II/353
– POMs-catalyzed   II/419
– rhenium-catalyzed   II/357–II/365
– Sharpless   II/211
– titanium-catalyzed   II/337–343
– ylide-mediated   II/399
epoxides   I/207
epoxyketone   I/539
eprozinol   II/62
Erker   I/267
esomeprazole   II/480
ESPHOS   I/43, I/44
esters   I/431, I/435
– α,β-unsaturated   I/566
– β-amino   I/341–I/343
– β,γ-unsaturated   I/121
– enolates   I/609
– homogeneous hydrogenation   II/98
– unsaturated   I/442
estrone   I/458
Et$_2$Zn   I/506, I/537
ether
– crown ethers   I/399
– diaryl ether   I/406
– enol ether   I/434
etherification   I/387
ethyl diazoacetate   I/419
ethylene   I/569
E-titanium enolates   I/500
EtMgCl   I/258
etoposide   II/10
eutomer   II/7
Evans   I/159
excitation, photochemical   II/578
5-exo-trig cyclization   I/568
6-exo-trig cyclization   I/568, I/569
Eyring plots   II/282

f

Farina   I/525
fatty acid   II/141
f-binaphane   II/115
Fe catalyst   I/177
Fe(C$_5$Me$_5$)(CO)$_4$ I/576
Fe(III)   I/483
Fenvalerate   I/41, I/42
Feringa   I/528, I/536, I/558–I/560
ferricenium hexafluorophosphate   I/595
FERRIPHOS   II/19
ferrocene   I/314
ferrocenyl diphosphine   II/114
– chiral   II/74, II/85

ferrocenyl oxazoline ligands   I/531
ferrocenylphosphine   I/313, II/176
– ligands   I/528
fine chemicals   I/171, I/182, I/188, I/307, II/29, II/437, II/545, II/573
fine chemical synthesis, industrial   II/29, II/145
– catalyst preparation and application   I/23, I/24
– future   I/24, I/25
– general concepts   I/15, I/16
– hydroformylation   I/29–I/51
– use of transition metals   I/17–I/23
Fischer indole synthesis   I/103, I/104
Fischer-type, carbene complexes   I/397–I/420
– chiral   I/400
– γ-methylenepyrane   I/400
FK506   I/205, II/77
FK906   II/93
flavoenzyme   II/497
fluorinated chiral salen ligand   II/349
fluorinated phosphines   II/548
fluorobenzene   II/467
fluorocarbon solvents   II/373
fluorophase principle   II/455
fluorophilicity   II/528
fluorous biphasic separation   II/169
fluorous biphasic systems (FBS)   II/163
fluorous catalysts   II/527–II/541
– C-C bond forming reactions   II/532, II/533
– diazo compounds, reaction   II/529–II/531
– hydroformylation   II/528
– hydrogenation   II/529
– hydrosilation   II/529
fluorous cobalt
– phthalocynine   II/537
– porphyrin   II/536
fluorous cyclic polyamines   II/536
fluorous phase   II/169
fluorous solvent   II/527, II/528, II/541
fluorous thioethers   II/533
fluorous-soluble catalyst   II/259
fluoxetine hydrochloride   II/62, II/70
Fluspirilene   I/33, I/34
formaldehyde   I/135, I/336
formate ester   I/119
formic acid   I/123, I/124, I/311
Fosfomycin   II/93
fostriecin   I/534
fragrance chemicals, chiral   I/202
Frankland   I/547
fredericamycin   I/609

free-radical
– process   I/20
– reaction   II/201
Friedel-Crafts
– acylation   I/356–I/358, I/382, I/591
– addition   I/511
– alkylation   I/356
fructophosphinites   I/155
fruity perfume   II/70
fullerene   I/312
3(2H)-furanones   I/128
furans   I/266, I/382, I/417, I/435, I/463, I/473
– 2,5-disubstituted   I/477
furostifoline   I/595
Fürstner   I/330, I/471, I/477

g
gadolinium triflate   I/336
Gaudemar   I/523
Gd(O-iPr)   I/375
Gennari   I/537
– Schiff base ligands   I/528
geraniol   II/10, II/11, II/341
gibberellins   I/437
Gif-oxidation   II/216, II/256
Gilman reagent   I/554
gloeosporone   II/77
glufosinate   I/102, I/139, I/140
glycals   I/489
glyceraldehyde   I/493
glycine esters   I/401
glycopeptide   I/143
glycopyranosides   I/384
glycosides, macrocyclic   I/330
glycosyl transferases   I/36
glyoxylates   I/509
glyoxylic acid   I/383
graphite   I/118, I/452
green
– chemistry   I/18, I/175, II/162, II/368
– oxidant   II/226–II/230
– reagent   II/235
– solvent   II/545
– synthese   II/415
Grigg   I/475
Grignard compounds   I/506
Grignard reagent   I/167, I/257, I/263, I/309, I/429, I/432, I/525, I/528, I/553, I/555, I/557, I/580
– inorganic   I/451
– vinyl   I/570
Grubbs   I/24, I/322, I/428, I/432, I/442, I/443

h

Haber-Weiss, decomposition of hydroperoxides II/222
hafnium II/370
hair cosmetics I/189
halichlorine I/534
haliclonadiamine II/93
haloalkoxy(alkenyl)carbene chromium complex I/410
Hammond postulate II/292
haptotropic rearrangement I/408
Hartwig I/544
Hayashi I/559, I/560
HCN I/151
heat carrier II/560
Heathcock I/500
Heck reaction I/5, I/21, I/68, I/271–I/300, I/610, II/561, II/563–565, II/586, II/591, II/615
– asymmetric I/281–I/287
– catalysts I/274–I/281
– coupling II/598, II/599
– mechanism I/272–I/274
– two-phase system II/520
helical chirality I/407
Helmchen I/544, I/557
Henry reaction I/364, I/539
1-heptene I/117
herbicides I/41
Herrmann I/322
heteroaryl chlorides I/213
heterobimetallics I/363–I/371
hetero-cuprate I/557
heterocycles I/257, I/260, I/267, I/344, I/450
– aromatic, synthesis I/463–I/466
heterogeneous catalyst I/105, I/118
heteropolyacid II/223, II/244, II/245
heteropolyanions II/416
– salt II/444
heteropolymetal acids II/369
heteropolymetallic catalyst I/370
hexacarbonyl-diiron I/578
1,6-hexadiene I/527
1-hexene I/410
hexyne
– 1-hexyne I/174
– 3-hexyne I/408
high pressure II/609–II/621
– general principles II/609, II/610
– Lewis acid-catalyzed cycloaddition II/610–II/613
Hiyama I/470
Hiyama-Nozaki allylation reaction I/470

Hoechst I/141
Hofmann elimination product I/485
homo aldol reaction sequence I/93
homocoupling, Wurtz I/474
homogeneous catalyst I/105
– microwave-accelerated II/598
Hoppe I/494
Horner-Wadsworth-Emmons reaction II/620
Hoveyda I/322, I/327, I/528, I/531, I/535, I/558
H-transfer, direct II/152
Hünig's base I/534
hydantoin I/142
hydrazido-copper complex II/461
hydrazine I/104
hydrazone I/103, I/104, I/571, II/119
– N-acyl II/120
β-hydride abstraction I/258
hydride complex I/450
hydroacylation, intramolecular I/199
hydroamination II/403
– base-catalyzed II/410–II/412
– intermolecular II/408
– intramolecular II/406, II/407
– transition metal-catalyzed II/406–II/409
hydroaminomethylation I/71–I/81
– intramolecular I/77
– reductive amination I/97
hydroaminovinylation I/70
hydroarylation I/287
hydroazulene I/588
hydroboration I/520, I/521, I/559
– asymmetric I/522, II/193–II/195
– olefins II/193–II/198
– – application in synthesis II/196, II/197
– rhodium-catalyzed II/620, II/621
– supercritical CO_2 II/197, II/198
hydrocarbons
– biodegradation II/216
– functionalization II/217
– oxidation II/218
hydrocarbonylation I/105–I/109
hydrocarboxylation I/113–I/130
– alkenes I/113–I/117
– allene I/20–I/122
– hydroxyalkyles I/122–I/126
– regioselective I/117
hydrocyanation I/21
– alkene I/149–I/151
– alkyne I/151–I/153
– catalytic asymmetric I/153–I/155
– nickel-catalyzed I/153–I/155
hydrocyclopropane I/167, I/168

Subject Index | 647

hydroesterification I/113–I/130
- alkenes I/117–I/120
- allene I/20–I/122
- asymmetric I/118
- hydroxyalkyles I/122–I/126
hydroformylation reaction
  (oxo reaction) I/21, I/57–I/82, I/137,
  II/403, II/521, II/548
- additional carbon-heteroatom bond
  formation I/60, I/61
- aldol reaction I/98
- amidocarbonylation I/100–I/103
- asymmetric I/61
- carbon nucleophiles I/88–I/92
- carbonyl ene reactions I/104, I/105
- chiral homoallylic alcohols I/63
- enamine I/67–I/70
- fine chemicals synthesis,
  applications I/29–/I51
- fluorous catalysts II/528
- hydroaminomethylation I/71–I/81
- imine I/67–I/70
- internal olefine I/59
- isomerization I/58, I/59
- multiple carbon-carbon bond
  formations I/87–I/109
- nitrogen nucleophiles I/64, I/65
- O,N/N,N-acetals I/65–I/67
- oxygen nucleophiles I/61–I/64
- reduction I/59, I/60
- terminal alkenes I/58
- two-phase catalysis II/516, II/517
hydrogen cyanide I/171
hydrogen peroxide I/20, II/201, II/226–
  II/230, II/358, II/415–II/423, II/431, II/434
- oxidation II/417–II/420
hydrogen shift
- 1,3-hydrogen shift I/201, I/204
- 1,5-hydrogen shift I/417
hydrogenation I/18, I/19, I/22, I/152, I/155
- asymmetric II/7–II/12, II/42, II/43, II/47,
  II/88, II/99, II/136
- – enamines II/14–II/26
- – homogeneous II/69, II/76, II/79
- base-catalyzed II/32
- carbonyl
- – ketones/aldehydes II/29–II/95
- – carbonyl selectivity II/34–II/38
- catalytic II/30
- diastereoselective II/6, II/7, II/41, II/83,
  II/88
- enantioselective II/113–II/122
- heterogeneous II/125–II/141

- – active site, accessibility II/126
- – alkenes II/135
- – apparatus and procedure II/131
- – aromatic nitro groups II/132–II/134
- – aromatic rings II/136, II/137
- – catalysts II/127–II/130
- – catalytic debenzylation II/137–II/140
- – diffusion problems II/126
- – ketones II/134, II/135
- – nitriles II/140, II/141
- – process modifiers II/130, II/131
- – reaction conditions II/131
- – reproducibility II/127
- – separation/handling work-up II/126
- – –special features, catalysts II/126
- homogeneous II/29
- Lindlar I/330
- monoolefins II/4–II/6
- olefin II/3–II/12
- polyolefins II/4–II/6
- stereoselective II/14
- transfer I/19, II/120
- two-phase II/37
β-hydrogen elimination I/201
hydrogenolysis, allylic I/311
hydrolysis
- enzymatic II/252
- palladium-catalyzed II/499
hydrometalation I/106, II/168
hydrooxepans I/91
hydroperoxide II/368
- alkyl I/20
- enantiopure II/271
- Haber-Weiss decomposition II/222
- thermolysis II/202
hydroperoxytungstate II/500
hydrophosphonylations I/367–369
hydrosilanes I/311, II/188
hydrosilylation I/18, I/23, I/612, II/167–
  II/180
- alkenes II/168–II/171
- alkynes II/171–II/173
- asymmetric II/173–II/180
- carbonyl compounds II/182–II/188
- cyclization II/178
- enantioselective II/182
- imine compounds II/188, II/189
- olefins II/167, II/168
- platinum(0)-catalyzed II/168
- styrenes, with trichlorosilane II/174, II/175
hydrotalcite II/444
hydrovinylation I/612, II/552
- alkenes I/308

hydroxamic acid II/372
β-hydroxy-α-amino acid esters I/370
α-hydroxy carbonyl compounds I/364
β-hydroxy-α-amino acids I/500
2-hydroxyacetophenones I/370
hydroxyalkyles, hydrocarboxylation/ hydroesterification I/122–I/126
hydroxyamines
– 1,2-hydroxyamines II/309
– β-hydroxyamines I/539
hydroxyapatite II/505
hydroxycarbonylation II/522
hydroxycephem I/390
hydroxycitronellal I/203
β-hydroxy-esters I/499
hydroxylamine accumulation II/133, II/134
hydroxylation II/205
hydroxy-palladation II/385, II/386
hydrozirconation I/522, I/533
hyellazole I/595

i

ibogamine I/8, I/509
Ibuprofen I/114
IFP process II/4
imidazole I/181
imidazolidinone chromium vinylcarbene complexes I/401
imidazolium salt I/220
imides I/433
imido trioxoosmium(VIII) II/327
imine I/67–I/70, I/99, I/309, I/341, I/380, I/470
– cyclic, enatioselective reduction II/118, II/119
– hydrophosphonylation I/369
– metal hydride complex II/498
– N-alkyl imine, enantioselective reduction II/117, II/118
– N-aryl imine, enantioselective reduction II/114–II/117
– phosphinyl II/120
immobilization I/323, II/522, II/549
Indinavir II/349
indium I/381
indoalkylzinc I/529
indole synthesis, *Fischer* I/103, I/104
indoles I/92, I/325, I/384, I/386, I/463, I/464, I/595
indoline ligands II/295
indolizidine 223AB II/77
indolocarbazole I/408
– indolo(2,3-b)carbazole I/596

innocent solvents II/569
inorganic support I/452
insect growth regulator I/201
instant ylide I/34
intermolecular induction, asymmetric II/90, II/92
iodine I/478
iodine-zinc exchange I/520
iodonium ylids I/158
iodosobenzene II/233–235
ion exchangers II/304
ionic liquids I/278, I/323, I/485, II/15, II/26, II/163, II/349, II/511, II/529, II/549, II/559–II/570
– chiral II/560
– imidazolium-based II/560, II/565
β-ionol II/36
β-ionone II/36
Ir catalyst II/50
$Ir(COD)(PhCN)(PPh_3)_3ClO_4$ I/204
iridium I/106, II/186
– complexes II/120
iron I/175
iron acyl complexes I/575–I/581
iron carbonyl I/113
iron(0) complex I/184
iron(III) complex, binuclear II/490
iron(III)-tetraphenyl porphyrin II/489
iron-butadiene complexes I/591–I/594
iron-cyclohexadiene complexes I/594–I/597
iron-diene complexes I/585–I/597
– preparation I/585–I/587
Ishii oxidation system II/223
iso coumarins II/385
isocaryophyllene I/463
isocyanates I/470
isocyanides I/408
isoflavanones I/386
isohyellazole I/595
isokhusimone I/458
isomerization I/18, I/22, I/23, I/58, I/59
– asymmetric I/206
– unimolecular II/618
isopulegol I/201
isoquinolines I/190, I/191, II/11
– isoquinoline-based pharma II/160
isothiocyanate I/180, I/386
isotopic effect, kinetic II/148
isoxazolidine I/347, I/509
– 4-isoxazolidine I/579
ivermectin II/5

j

Jacobsen-Katsuki epoxidation   I/22, II/211,
   II/346, II/350
Jacobsen-type catalyst   II/349
JosiPHOS   II/8, II/10, II/74, II/120
juglone   I/509

k

Kagan   I/9
kainic acid analogs   I/37
Karasch-Sosnovsky type allylic
   oxidation   II/256–II/263
Karstedt's catalyst   II/168
Katsuki   I/530
Katsuki-type salen ligand   II/349
Kazmaier   I/544
KCN   I/151
Keck   I/496
Keggin structure   II/415, II/417
Kemp's triacid   II/262
ketene dithioacetal   I/510
ketene silyl acetals   I/369
β-keto acids   I/484
β-keto amides   I/487
β-ketoenamines   I/440
keto-ester
– cyclization   I/462, I/463
– β-ketoesters   I/483, I/484, I/487, I/543,
   II/72
ketone   I/309, I/387, I/430, I/435, I/449,
   I/458, I/470, I/532
– α,β epoxy   I/372
– α,β-unsaturated   I/325, I/328, I/440
– aromatic   I/356
– alkyl aryl   II/42–II/51
– amino ketones   II/59
– β-amino   I/341
– β-hydroxy   I/347, I/497, I/502
– β,γ-unsaturated   I/94
– carbonyl hydrogenation   II/29–II/95
– cyclic   I/462
– cyclic aromatic   II/50
– dialkyl   II/54–II/56
– fluoro ketones   II/54
– functionalized   II/69–II/95
– hetero-aromatic   II/52–II/54
– hydrogenation   II/134, II/135
– methylenation   I/437
– mixed couplings   I/461
– unsaturated   II/63–II/66
ketopantolactone   II/72, II/81
keto sulfonates   II/87–II/90
keto sulfones   II/87–II/90
– β-keto sulfones   I/484
keto sulfoxides   II/87–II/90
– β-keto sulfoxides   I/484
Kharasch reaction   II/538
kinetic isotope effects   II/301
kinetic resolution   I/23, I/260, I/263–I/266,
   II/61, II/161, II/162, II/270, II/300, II/301,
   II/442, II/538
– cyclic allylic ethers   I/266, 267
– dynamic   II/90–II/95, II/145
– lipase-assisted dynamic   II/149
Kishi   I/471, I/475
Knochel   I/430, I/504, I/528
Knoevenagel condensation   I/93
Knowles   I/9
Koga   I/557
K-selectride   I/580, I/581
Kulinkovich   I/167, I/168

l

(L)-leucinol   I/530
lactams   I/77, I/129, I/484
– β-lactam   I/39, I/343, I/414
– β-lactam 1-carbacephalothin   I/413
lactones   I/63, I/126, I/179, I/431
– α-alkylidene γ-lactones   I/151
– β-lactones   I/434
– bicyclic   I/129
– γ-lactones   I/166, I/412, I/484
– homogeneous hydrogenation   II/98
– optically active   II/270
lanoprazole   II/491
lanthanide   I/335, II/405, II/613
– asymmetric two-center
   catalysis   I/363–I/377
– chiral   I/353, I/355
– triflate (lathanide trifluooromethane-
   sulfonates)   I/335, I/340, I/343, I/485
lanthanocene catalyst   II/171
lanthanoid alkoxides   II/369
lanthanoids, homometallic   I/335–I/358
– reuse, catalyst   I/340
lavanduquinocin   I/596
LDA   I/563
L-dopa synthesis   I/22
L-DOPS   II/93
leaching   I/275
lead   I/430
Leighton   I/45
levamisole   II/62
levofloxacin   II/62, II/63
Lewis acid   I/150, I/335, I/356, I/363, I/384,
   I/392, I/491, I/511, I/554

- aqueous media, catalysis   I/335–I/340
- aziridination, asymmetric   II/398, II/399
- chiral   I/348, I/353, I/354, I/501
- cycloaddition   II/610–II/613

Lewis bases   I/622, II/361, II/362
Leyendecker   I/557
LHMDS   I/544, I/545
LiAlH$_4$ I/450
LiClO$_4$ I/357
Lidoflazine   I/33, I/34
lid-on-off mechanism   I/200
ligands
- acceleration effect   II/154, II/275, II/276, II/279, II/285, II/304
- alkoxide   I/491
- aminoalcohol   I/530
- anthraquinone   II/293, II/294
- aryl pincer   II/154
- bidentate phosphine   I/153
- binaphtol   I/I/531
- BINOL   I/505, II/482, II/483
- bisoxazoline   I/162, II/389–II/391
- bis(oxazolinyl)carbazole   I/472
- Buchwald's ligand   I/215, I/239, I/249
- carbene   I/163, I/220–I/222, I/405, I/409, II/154
- carbonyl   I/397
- chiral   I/146, I/187, 313
- cinchona alkaloids   II/293–II/295, II/300
- cyclopentadienyl   I/427
- (DHQ)2PYR ligand   II/323
- diphenyl phtalazine   II/294
- diphenyl pyrazinopyridazine   II/294
- diphosphine   II/20, II/41
- – chiral, figure   II/40
- electrophilic carbene   I/163
- ferrocenyl oxazoline   I/531
- ferrocenylphosphine   I/528
- Gennari's Schiff base   I/528
- indoline   II/295
- miscellaneous   II/393
- N-donor   I/222
- N,N-donor   I/308
- nitrogen-based   II/183
- oxazoline   II/184
- P,P,N   II/186
- peptide-based   I/531
- phase transfer, thermoregulated   II/517
- phosphine   I/213–I/220
- phosphorus, monodentate   I/559, II/21
- privileged   II/8
- pyrimidine   II/294
- S-donor ligands   I/221

- Schiff base   II/391–393, II/484–II/486
- sulfonamide, chiral   I/476
- TADDOL   I/508
- – dendrimeric   I/505
- TRAP   I/315
- trialkanolamines   II/484

linalool   I/19
Lindlar hydrogenation   I/330
lipid A   II/77
α-lipoic acid   II/77
liquid-liquid biphasic catalysis   II/562, II/563
lithium diorganocuprate   I/557
LiTMP   I/605
LLB catalyst   I/365, I/369
Ln-BINOL   I/364, I/372
LnPB   I/367
longithorone   I/329
Lonza AG   I/183
Losartan   I/220
LSB   I/366
lubricant   I/189, II/560
Lukianol A   I/465
lutetium triflate   I/336
lutidines   I/187

m

MacMillan   I/546
macrocycles   I/323, I/458
macrodiolide   I/12
magnesium   I/260, I/429
Mahrwald   I/502
malonate   I/312, I/315
- dimethyl malonate   I/545
- esters   I/487, I/489
malonic esters   I/484
mandelic acid   I/502
manganese   I/471, II/373
- manganese-catalyzed epoxidation   II/344–II/353
manganese(III) based radical cyclization   I/483–I/489
manganese(III) complexes   II/345
Mangeney   I/556
man-made catalyst, asymmetric   II/349
Mannich reaction   I/341–I/344
Marek   I/546
Markovnikov
- addition   I/3, I/150
- hydrocyanation   I/153
- product   II/404
Mars-van Krevelen mechanism   II/205, II/206, II/423
Masamune   I/159

MCM-41, mesoporous II/223, II/342, II/352, II/363, II/373
McMurry reaction I/449–I/466
– intramolecular I/455
– natural product synthesis I/456–I/458
– nonnatural products I/458–I/461
MCPBA I/577
Me$_2$AlCl I/428
Me$_3$SiCl I/471, I/473, I/491, I/555
Meerwein-Pondorff-Verley-Oppenauer reaction II/460
mefloquine II/62
melatonin I/36
memory effect I/310
menogaril I/408
menthol I/405
mephenoxalone II/62
Merck I/534
MeRe(O)$_3$ II/502
merulidial I/7
mesoporous material II/337
metal carbene complex I/397, II/396
metal centre, chiral I/418
metal peroxo complex II/368
metal-carbon bond I/397
1-metalla-1,3,5-hexatrienes I/416
metallacyclobutane I/409
metallacyclopentane I/258
metallaoxetane II/292, II/347
– mechanism II/278, II/280, II/282
metallic colloids II/586
metal-ligand bifunctional catalysis II/147–II/149
metallonitrene II/392
metalloporphyrins II/229
metathesis I/21, I/62, I/261, I/414, I/506
– alkene I/321–I/328
– alkyne I/330, I/331
– cross-enyne I/328
– cross metathesis (CM) I/321, I/326
– enyne I/328, I/329
– olefin I/427, I/442, II/553, II/601
– – asymmetric I/327
– ring-opening (ROM) I/328
methacrylates I/325, I/327
methallylation I/472
methallylchromate I/478
methallylesters I/44
methallylmagnesium chloride I/478
methallyltitanium reagent I/494
methane monooxygenase II/226
methoprene I/201
4-methoxyacetophenone I/357

methoxycarbonylation I/125
2-methoxypropene I/346
methyl acrylate I/118
methyl formate I/119, I/121
methyl methacrylate I/118
methyl propiolate I/174
methyl pyruvate II/71
methyl salicylate I/486
methyl trioxorhenium II/268
methyl vinyl phosphinate I/102
methyl-2-methoxymethylacrylate I/120
methylenecycloalkane I/4, I/116
methylenepanem I/390
O-methylpodocarpate I/487
methyltrioxorhenium II/357
Metolachlor II/114
MIB I/530
micellar systems I/338–I/340
micelle I/630
Michael reaction I/366, I/371, I/374, I/384, I/392, I/511, I/536, I/538, I/577
– asymmetric I/367
– chiral I/400
– nitroolefin addition I/399
– oxa-*Michael* addition I/539
microwave I/251, I/274, I/384, II/597–II/607
– irradiation II/163
migratory insertion II/147
Mikami I/497, I/506, I/511
MiniPHOS
– *t*-Bu- II/19
Mitsunobu reaction I/592
Miyaura I/559
Mn(III) enolate I/484
Mn(III)/Cu(II) I/483
Mn(OAc)$_3$ I/483, I/486, I/488
Mn(OAc)$_3$·2H$_2$O I/485
mnemonic device II/277, II/289, II/290
MOD-DIOP II/8
molybdenum I/404, I/410, I/443, II/372, II/373
– catalyzation I/6, I/327, I/330
monocarbene-palladium(0) complex I/221
monocyclopentadienyl-dialkoxytitanium I/493
monooxygenase II/224
– enzymes II/205, II/207, II/210
MonoPHOS II/25
monophosphine II/175, II/176, II/182
– chiral I/410
monophosphine palladium(0) diene I/216
monophosphine palladium(I) complex I/216
monophosphonite II/186

monosulfonated catalyst II/118
montmorillonite I/118, II/25
morphine I/288, II/346
MTO II/357–II/359, II/432
– polymer-supported II/363
Mukaiyama I/451
– aldol reaction I/94, I/381, I/382, I/502, I/503
muscone I/532
mycalolide I/475
mycosamine I/591
mycrene II/520

n

N,N-donor ligands I/308
$Na_2Fe(CO)_4$ I/575
$NaBD_4$ I/311
$NaBH_4$ I/151
N-acetyl amino acid derivates I/100
N-acetylcysteine I/134
N-acyl-α-amino acid I/133
N-acyl-oxazolidinone I/509, I/510
N-acylsarcosine I/134
Nakadomarin A I/325, I/326
Nakai I/497, I/530
nanocluster II/585
nanoparticle II/585, II/588
– palladium II/444, II/528, II/533
nanosized material II/585
– amorphous Fe II/586
naphtalenes I/178, I/179
naphthoquinone I/178, I/338
Naproxen I/114, I/154, II/9, II/515, II/521
Narasaka I/508, I/510, I/511
N-arylpiperazines I/234
natural products I/288–I/298, II/253
N-benzylideneaniline I/344
N-demethylation II/503
N-donor ligands I/222
Negishi I/267, I/430, I/523, I/525
neobenodine II/52, II/70
Neocarazostatin B I/596
neomenthol II/94
ngaione II/86
N-H activation mechanism II/409
N-halogeno-succinimides I/513
NHC (N-heterocyclic carbenes) I/220, I/278, I/322
N-heterocyclic carbenes I/242
N-hydroxyphthalimide (NHPI) II/203, II/204, II/223, II/224
Ni catalyst, heterogeneous II/56
Ni complex I/171
Ni(0) I/179

$Ni(CN)_2$ I/122
$Ni(CN)_4^{2-}$ I/151, I/152
$Ni(P(OAr)_3)_x$ I/150
nickel I/106, I/113, I/150, I/175, I/211, I/244
– arylphosphite complex I/151
– catalysis I/538
– – co-dimerization II/552
– complex I/11
– cyanide I/121
– phosphite I/152
– – complexes, zero-valent I/149
$NiCl_2$ I/476
Nicolaou I/325, I/442
Nilsson I/557
niobium II/371, II/372
Nippon Steel I/185
Nishiyama I/162
nitration I/358
nitrene
– complexes II/578
– precursor II/327
– transfer II/389–II/396
– donor II/390
nitric acid I/358
nitrile I/19, I/155, I/172, I/309, I/315, I/440, I/543
– amidocarbonylation I/143
– chemoselective hydrogenation II/140, II/141
– optically active I/187
– reactions I/440, I/441
– unsaturated I/152
nitro I/309
– β-nitro esters I/484
– β-nitro ketones I/484
nitroaldol reaction I/364, I/365
– enantioselective I/364, I/365
nitroalkanes I/386
nitroalkene I/539
nitroarenes II/132
nitroethane I/364
nitroethanol I/364
nitrolefins I/537
nitromethane I/357
nitrones I/534, II/500
nitropropane I/364
β-nitro-styrene I/511, I/512
N-methallylamides I/103
N-methyl ephedrine I/535
N-methylimidazole I/354
N-methylmorpholine N-oxide I/621
Nolan I/322
nonacarbonyldiiron I/585

nonlinear effect II/260, II/480, II/482, II/486
norbornene I/153, I/175, I/443
Normant I/525, I/546
noroxopenlanfuran II/6
Noyori I/530
Nozaki I/158
nucleophilic substitution I/251, II/613
nucleoside I/9
Nugent I/505
nylon-6,6 I/149

O

O,N/N,N-acetals I/65–I/67
1,7-octadiyne I/191
octahydro-1,1'-binaphtol I/505
octahydro-binaphtol I/510
octene
– 1-octene I/570
– 2-octene I/59
– 4-octene I/59
1-octyne I/174
Oehme I/184
Oguni I/501, I/529
Ojima I/81, I/101
Ojima-Crabtree postulation II/173
okicenone I/488
olefination I/450
– carbonyl I/427, I/430–I/439
– – intramolecular I/442
– decarbonylative I/281
– *Heck* reaction (*see there*)
– *Peterson* olefination I/430
– *Wittig* olefination I/3, I/386, I/430
olefins I/113, I/311, I/428
– α-olefin I/176
– aromatic I/570
– dimerization I/308
– 1,2-disubstituted I/160
– epoxidation II/358–II/375
– fluorinated I/116
– hydroboration II/193–II/198
– hydrosilylation II/167, II/168
– hydrozincation I/529
– internal I/59
– isomerization I/199–I/208
– metathesis I/442, II/601
– – asymmetric I/327
– optically active I/199
– osmylation II/278
– rearrengement II/248
– unfunctionalized I/206
oligoindoles I/464

olivin I/408
one-electron
– oxidants I/485
– transfer process II/203
Oppenauer oxidation I/470, II/149
Oppolzer I/531, I/532, I/557
optical fibers I/46
organic synthesis, transition metals
– atom economy I/11, I/12
– basic aspects I/3–I/12
– chemoselectivity I/4, I/5, I/18
– diastereoselectivity I/8, I/9, II/38–II/42
– enantioselectivity I/9
– regioselectivity I/6, I/7, I/18, I/58
organoaluminium I/553
organobismuth I/385–I/387, I/392
– pentavalent I/379
organoboron I/553
organochromium
– compound I/471
– reagents I/469
organocopper
– derivates I/553
– reagent I/563
organolanthanide complex II/8
organolithium I/397, I/398, I/400, I/491, I/519, I/555, I/576, I/580
organomagnesium I/491, I/519
organomanganese reagents I/555
organometallics, allyl I/337
organonitrile I/149
organorhenium oxides II/357
organosamarium I/553
organotin I/553
organotitanium compounds I/427
organozinc I/553
– reagent I/167, I/519
organozinc bromide I/563
organozirconium I/553
orlistat II/83
orphenadrine II/52, II/70
Orsay reagent II/480
ortho-directing effect I/604
oscillation, microwave-induced II/597
Oshima I/522
osmaoxetane mechanism II/292
– stepwise II/279
osmaoxetanes II/282
osmium carbene complex I/158
osmium tetroxide,
  microencapsulated II/304
osmylation, mechanism II/278–II/282, II/284
– (2+2) mechanism II/279, II/301

Subject Index

– (3+2) mechanism II/279, II/301
Otsuka I/199
Overman I/475, I/532
overoxidation II/482
oxa-conjugate addition I/538
oxalic acid I/124
oxanorbornene I/417
oxazolidine I/544
– chiral I/414
oxazolidone I/509
– derivative I/353
oxazolines I/411
– ligands II/184
oxazolinylferrocenylphosphines II/154
oxepins I/263
oxidation I/18, I/20, I/22, I/379
– 1,4-oxidation, palladium-catalyzed II/249–253
– aerobic II/204, II/224, II/345, II/437–II/473, II/504, II/536, II/586
– alkane II/221
– allylic I/308, II/243–263
– *Baeyer-Villiger* II/210, II/267–II/272, II/448
– basics II/201–II/211
– biomimetic II/244
– C-H compounds II/215–II/236
– direct II/205–II/207
– enantioselective II/211, II/252
– enzymatic I/20
– Gif II/216, II/256
– hydrocarbons II/218
– ligand design II/210
– molecular oxygen II/420–II/423
– *Oppenauer* I/470, II/149
– tandem oxidation-reduction II/162
– TBHP II/257
– TEMPO-mediated II/209
– *Wacker* (*see there*)
oxidative addition I/123, I/141, I/212, I/223, I/224, I/232, I/235, I/272, I/309, I/483, II/150
oxidative cleavage II/427–II/434
– acid formation II/433, II/434
– aldehydes, formation II/432, 433
– keto-compounds, formation II/428, II/429
– one-step II/429, II/430
– optimized catalyst systems/reaction conditions II/430, II/431
oxidative cyclization II/275
oxidative decomplexation I/611
oxidative demetalation I/404
Oximidine II I/325, I/326
oxo reaction (*see* hydroformylation reaction)

oxo transfer process II/358
oxomanganese complex II/350
– oxomanganese(V) complex II/346, II/348, II/351
oxone II/430
oxoruthenium species II/503
oxycarbonylation, intramolecular I/129
oxy-*Cope* I/312
oxygen I/117
– catalytic transfer II/207–II/210
– donors II/208
– molecular II/219–II/224 II/420–II/423
– nucleophiles I/61–I/64, II/386, II/387
– rebound mechanism II/207
oxygen-rebound mechanism 227
ozonolysis II/427

p

paclitaxel II/328
Padova reagent II/492
palladacycle I/213, I/239, I/276
palladacyclobutane I/308
palladium I/100, I/106, I/113, I/118, I/119, I/125, I/141–I/146, I/175, I/185, I/231, I/246, I/543
– acetate I/114, I/118
– complex I/158
– palladium(0) complexes I/120, I/308
– palladium(II) acetate I/309
palladium catalyst I/130, I/141
– allylic substitutions I/307–I/315
– cationic I/125
– hydrolysis II/499
palladium chloride I/114
palladium nanoparticles II/444, II/528, II/533
palladium phosphine I/151
palladium phosphite I/151
palladium-catalyzed reaction I/545
p-allylpalladium intermediate I/12
pancreatic lipase inhibitor II/83
Panek I/475
PAP I/216
Paquette I/432
Paracetamol I/144
Parlman I/575
Pateamine A II/77
Pauson-Khand cyclization I/263, I/619–I/631, II/601
– catalytic I/622, I/623
– hetero I/628
– promotor-assisted I/621
– stereoselective I/624–I/626

- stoichiometric   I/620, I/621
- synthetic applications   I/627
- transfer carbonylation   I/629, I/630
Payne reagent   II/210
Pb(OAc)$_4$ I/323
Pd(0)   I/418
- complex   I/4, I/5, I/127, I/418
- phosphine complex   I/207
PdC   I/124
PdCl$_2$ I/121, I/153
PdCl$_2$(PPh$_3$)$_2$ I/107, I/116, I/117
Pd(dba)$_2$ I/120, I/127
Pd$_2$(dba)$_3$ I/127
Pd(OAc)$_2$ I/122, I/129, I/418
Pd(PPh$_3$)$_4$ I/120, I/178, I/418
PEG   II/396, II/549, II/554
PENNPHOS   I/23
pentacarbonylchromium   I/398
pentadecanoic acid   II/96
1,4-pentadienes   I/106
pentafulvenes   I/411
3-pentanone   I/502
pentaphenylbismuth(V)   I/385
1-pentyne   I/404
peptide   II/375
peptidomimetics   I/37
peptoid   I/144
peracetic acid   II/230
perfluoroalkane   II/527
perfluoralkyl aldehydes   I/509
perfluorinated
- liquids   II/216
- solvents   II/562
perfluorohydrocarbons   II/511
perfluorozinc aromatics   II/588
perfumes   II/69, II/70
periplanone   I/472
peroxides
- alkyl   II/231
- sulfur-containing   II/231–II/233
peroxo spezies   II/269
peroxometal pathway   II/209
peroxotitanium species   II/481
peroxy acid, organic   II/230, II/231
peroxy radicals   II/257
peroxytungstophosphate   II/500, II/501
Peterson olefination   I/430
Pfaltz   I/558
PhanePHOS   II/24, II/25
pharmaceuticals   I/18, I/100, I/149, I/288–I/298, I/379
phase transfer   I/315
- catalyst   I/241
- reaction   I/309
- ligands, thermoregulated   II/517
phenanthrenes   I/178, I/179
1,10-phenantroline   I/245
phenanthrolines   I/313
- chiral   II/263
Phenipiprane   I/33, I/34
Pheniramine   I/30, I/32
phenolates   I/299
phenolethers   I/605
phenols   I/385
phenylacetylenes   I/330
phenyl-CAPP   II/8
phenylenes   I/178
1-phenyl-ethanol   I/506
phenylethanolaminotetraline agonist   II/58
1-phenylethylzinc reagent   I/527
phenyl glycine   I/414
phenylglyoxal   I/342, I/345
phenylhydrazine   I/104
phenylmenthol   I/400, I/414
phenylmenthylacetoacetate esters   I/489
3-phenylpropionealdehyde   I/337
phenylvinylsulfide   I/353
phospha-conjugate addition   I/538
phosphacyclobutene   I/440
(phosphanyloxazoline)palladium
    complex   I/314
phosphine   I/136, I/179
- bidendate   II/182
- dissociation   I/322
- monodentate   II/176
- water-soluble   II/514
phosphine ligands   I/213–I/220
- bidentate   I/153
- secondary   I/218
phosphine oxide   I/218, I/241
phosphine-phosphite, bidendate   II/195
phosphine-Ru complex   II/70
phosphinite nickel catalyst,
    carbohydrate-based   I/154
phosphinooxazolines   I/285
phosphinotricine   I/140
phosphinous acid   I/241
phosphite   I/214, I/277
phosphoramidite   I/287, I/536, II/21, II/175
phosphorus   I/309
photoassisted
- reactions   II/573
- synthesis   I/184
photocarbonylation   I/413
photocatalysis   II/573–II/579
- heterogeneous   II/574

photochemistry   I/412, I/413
photo-complex catalysis   II/573
photolysis   I/413, I/611
– catalyzed   II/574
photolytic induction   I/587
phox   II/115
phtalane-Cr(CO)3   I/606
α-picoline   I/186
Pictet-Sprengler   I/92
pigments   I/379
Pimozide   I/34
pinacol   I/449
pinacolone   II/56
pinane diphosphine   I/313
pincer complexes   I/218, I/277, II/538
pindolol   I/364
α-pinene   II/222, II/576
pinnatoxin   I/475
PINO   II/204
pipecolic acid   I/66
pipecolinic acid   I/138
piperidine   I/568, 569
piperidone   I/545
Pitiamide A   I/523
– synthesis   I/524
pivalophenone   II/45
planar-chirality   II184
platinum diphosphine complexes   II/268
PMDTA   I/401
Pme$_3$ I/441
poly(tartrate ester)   II/341
polyamide   I/100
1,2-polybutadiene, hydrocarboxylation   I/116
polycondensation   I/460
polydimethylsiloxane   II/26
– membrane   II/74
polyenes, dihydroxylation   II/299
polyethylene glycol monomethyl
    ether   II/341
polyfluorooxometalates   II/417
– metal-substituted   II/418
polyisoprenoid substrates   II/301
polyketones   II/568
poly-L-leucine   I/126
polymers   I/149
– synthesis   I/428
– water-soluble   II/516
polymer-supported
– complex   II/163
– tartrates   II/342
polymerization   I/310, I/427
– ROMP (ring-opening metathesis
    polymerization)   I/323, I/443. I/444

polyoxo-heterometallates   II/381
polyoxometalates   II/206, II/415–II/423
– metal-substituted   II/420
Polyoxypeptin A   II/93
polyphenylenes   I/178
polyphosphomolybdate, vanadium-
    containing   II/228
polypropionate   I/495
polyquinanes   I/442
POMs-catalyzed epoxidation   II/419
pony tails   II/528, II/540
porphyrins   I/460, II/218
– complex   II/394–II/396
– hindered   II/206
Posner   I/556
potassium cyanide   I/122
pressure wave   II/583
Pringle   I/153
product-inhibitin catalysis   I/201
profene   I/30, I/154
proline   I/564
prolinol   I/544
propargyl
– halides   I/472
– stannanes   I/496
propargylic hydroxylamines   I/534
propionate aldol addition   I/500
propranolol   II/62
prosopinine   I/66
prostaglandin   II/11
– E$_2$ I/330
prostatomegary   II/44
protease inhibitor   II/600
protecting group   I/243
pseudopeptide   I/576
Pt/Al$_2$O$_3$, alkaloid-modified   II/79
Pt/C   I/20
PTC (phase-transfer catalyst)   I/121
PtCl$_2$(dppb)   I/105
pulegone   II/66
pybox   I/162, II/183
pyrans   I/179, I/266
Δ^2-pyrazolines   I/414, I/415
pyrenorphin   II/77
pyrethroid insecticide   I/22, I/41
pyridine   I/171, I/172, I/182–I/185, I/187,
    I/346, I/417, I/440
2-pyridinecarboxaldehyde   I/336
pyridine-imines   II/183
pyridones   I/179
– 2-pyrridones   I/180
pyridoxine   I/191
2-pyridylphosphine   I/125

pyrimidine ligands II/294
pyroglutamic acid I/402
pyrones I/179
– 2-pyrones I/180, I/353, I/509
pyrroles I/92, I/108, I/463, I/464
– 2H-pyrrole derivates I/414
pyrrolidines I/564, I/566, II/606
1-pyrroline I/413
pyruvic aldehyde dimethylacetal II/59, II/60

q

quaternary carbon I/384
– chiral I/314
quaternary stereocenters I/523
quaternization II/139
quinazoline I/67
quinazolinone I/67
quinine II/79
quinoline I/346
quinone I/177, I/178
– monoacetals I/509

r

(R′)$_2$Zn I/503
R$_2$CuLi I/554
radiation-induced reactions II/583–II/594
radical trap II/390
radicals, stabilization, benzylic position I/608
Raney cobalt II/34
Raney Cu II/98
Raney nickel II/34, II/82, II/86, II/127, II/140, II/161, II/603
rare-earth catalyst II/553
RCM 323
rearrangements
– allylic I/312
– Claisen I/432, I/546, I/544, I/577
– cinca-Claisen-type rearrangement I/525
– Cope-type (3,3) sigmatropic I/411
– haptotropic I/408
– olefin II/248
– (3,3)-sigmatropic I/435
recycle, catalyst II/25, II/26
reducing agents I/450
reduction I/59, I/60, I/449
– allylic I/311
– coupling I/437, I/533
– elimination I/106, I/150, I/212, I/232, I/235, 238, I/246, I/307, I/416, I/528, I/529, I/555, I/592
Reetz I/560
refinement reaction II/403

Reformatsky
– reaction I/476, I/543, 544, II/590
– reagent I/565
regioselectivity I/6, I/7, I/18, I/58, I/273
Re-Os bimetallic catalyst II/96
Reppe I/171
Rh catalysis I/559
Rh complex I/158, I/162
Rh(CO)(PPh$_3$)$_3$ClO$_4$ I/204
Rh(cod)Cl$_2$ I/206
Rh(I) I/199
– enolate I/315
Rh(I)-(S)-BINAP I/199
Rh(OAc)$_4$ I/163
Rh$_2$(cap)$_4$ I/163
Rh$_2$(pfb)$_4$ I/163
RhCl$_3$ I/419
rhenaoxetane II/281
rhenium II/373
– catalyzed epoxidation II/357–II/365
Rhizoxin D I/496
rhodium I/62, I/72, I/80, I/94, I/101, I/106, I/136, I/155, I/174, I/175, II/182–II/186, II/188
– carbonyl complex I/29
– complexes II/120
– monohydride II/151
rhodium(I)-bis(phosphine) catalyst I/204
rhodium(II) complex, dinuclear I/157, I/161, I/163–I/166
Rieke zinc I/520
ring closure
– 5-endo-trig I/416
– electrocyclic I/408
rivastatin II/79
Roelen I/29
roflamycoin II/86
ROH addition II/383–II/385
rolipam I/415
ROMP (ring-opening metathesis polymerization) I/323, I/443. I/444
roxaticin II/79
Ru catalyst I/9
Ru(II) complex
– chiral I/158
– pybox I/162
Ru$_2$Cl$_4$(diop)$_3$ I/206
Ru$_3$(CO)$_{12}$ I/119
Ru-catalyzed oxidation II/503
RuCl$_2$(PPh$_3$)$_3$ I/119
ruthenium I/443, II/186, II/189
– carbene complex I/322, I/328
– catalyst I/323

– – chiral   I/328
– complex   I/11, I/72, I/158, II/121
– hydride   II/149
– monohydride complexes   II/146
– polyoxometalates   II/206
– salen complex   II/449
Rychnovsky   I/45

S

(S)-(–)-7-methoxy-3,7-dimethyloctanal   I/202
(S)-4-hydroxycycloheptenone   I/206
salen complexes   II/394, II/395
salen oxovanadium(IV) complexes   II/486
(salen)Mn(III) complexes, chiral   II/488, II/489
salicylaldehyde   I/336
salicylihalamide A   II/77
salvadoricine   I/458, I/464
SAMP   I/570
SAP methodology   II/521
sarcosinate   I/133, I/136, I/139, I/140
Sc(OTf)$_3$ I/337, I/350
scaffold, chiral   I/313, I/314
scandium
– catalysts   I/353
– triflate   I/338, I/339
scCO$_2$ (supercritical carbon dioxide)   I/78, I/323, II/116, II/383, II/438, II/539, II/545, II/551, II/567
– hydroboration   II/197, II/198
Sch 38516   I/258
Schiff base   II/421
– chiral   I/501
– ligand   II/391–II/393, II/484–486
– peptide sulfonamide   I/537
Schrock   I/321, I/327, I/427
– metal carbene complex   I/397
Scolastico   I/556
scopadulciic acid   I/289
S-donor ligands   I/221
secofascaplysin   I/465, I/466
Seebach   I/504, I/506, I/511
selectride reagent   II/41
selenium   I/309
selenoesters   I/433
(SEM)-protection   I/593
semicorrins   I/158
Semmelhack-Hegedus route   I/397
sensors   II/560
Sharpless   I/22
– AD reaction   I/539, II/305
– epoxidation   II/211
– model   II/291–II/293

Sheldon   I/18, I/36
Shibasaki   I/161, I/540
Shilov reaction   II/217
SHOP (*Shell Higher Olefin Process*)   I/24, I/321
Shvo catalyst   II/149, II/152
(3,3)-sigmatropic rearrangement   I/435
silanes   I/472
silica, mesoporous   II/363
silicate   II/500
silyl cyanide   I/152
silyl enol ether   I/97, I/310, I/336, I/338, I/438, I/510, I/555, I/559
silyl enolates   I/342
silyl esters   I/433
silylaldehyds   I/47
silylation
– allylic   I/313
– hydrosilylation   I/18, I/23
silylmetalation   II/168, II/170
Simmons-Smith reaction   I/157, I/167, I/511, I/541, I/593
single-electron transfer   II/584
single-site reaction   II/568
SIPHOS   II/21
Sn(OTf)$_2$ I/416
Soai   I/530, I/559
sodium dodecylsulfate   I/338
solid phase chemistry   I/471, I/611
solvent
– chiral co-solvent   I/603
– environmentally benign   II/15, II/198
– ionic liquid   I/382
solvent-free products   II/545
sonochemistry   II/583, II/584
sonoelectroreduction   II/589
Sonogashira reaction   I/21, I/610, II/518, II/533, II/566, II/600, II/601
sonolysis   II/587
sparteine   I/313, I/557, II/184, II/489
spiroannulation   I/596, I/597
spiro-bislactone   I/434
spiroindole   I/597
spiroketals   I/434, I/438
SpiroNP   II/20
SpirOP   II/20
spiroquinolines   I/597
SR 5861 1A   II/62
SS20846A   I/592
stannanes   I/472
stannyl acetylene   I/405
steganone   I/611
stereochemistry   I/3

stereomutation II/94
stereoselectivity I/18, I/88
sterepolide I/7
steric tuning II/7
steroids I/76, I/172, I/176, I/297, I/408, I/489, II/234, II/349
Stille I/610
– coupling II/599, II/600
Strecker process I/508
Strukul's catalyst II/270
Stryker's reagent II/187
styrene I/30, I/105, I/150, I/570
– hydrosilylation II/174
styrene-Cr(CO)$_3$ I/604
sugars I/337, I/349
– sugar acyl iron I/576
sulfide oxidation II/479–II/493
– asymmetric II/481
– catalyzed by
– – chiral ruthenium/tungsten complexes II/490, II/491
– – chiral salen manganese(III) complexes II/486, II/487
– – chiral salen vanadium complexes II/486, II/487
– – chiral titanium complexes II/479–II/486
– – iron non-porphyric complexes II/490
– kinetic resolution II/491–II/493
sulfides I/472
sulfinylimines I/581
sulfobacin A II/77
sulfonamide ligands, chiral I/476
sulfonate I/309
sulfonium ylids I/158
sulfonyl chloride I/380
sulfonylation, allylic I/313
sulfoxidation II/479
sulfoxide I/309, II/479
sulfur I/309
Sumitomo I/158, I/159
superconductors I/379
supercritical fluids I/622, II/15
supercritical state II/545
super-heating II/601
superparamagnetic material II/586
support II/128
surfactant II/560
sustainable chemistry I/18
Suzuki reaction I/21, I/211–I/225, I/523, I/610, II/533, II/552, II/561, II/591, II/600
– asymmetric I/225
– mechanism I/212, I/213
Suzuki-Stille coupling II/518

symmetric activation /deactivation II/66–II/68
syn 1,2-diols I/540
syngas I/102

t
TACN complex II/352
TADDOL I/494, I/499, I/501, I/506, I/508, I/510, I/533, I/537, I/539, II/272
– dendrimeric ligands I/505
– titanium complex I/167, I/168, I/530
Tagasso menthol process I/22, I/23, I/201
Takai I/472, I/477
Takasago I/199–I/202
Takemoto I/314
Tamao-Fleming oxidation II/168, II/171
tamoxifen I/461
– (Z)-tamoxifen I/529
tandem
– coupling I/460
– reaction
– – cyclization I/66
– – hydroformylation (see there)
TangPHOS II/20
tantalum II/371, II/372
tantalum tartrate II/343
tartaric acid II/82
tautomerization, enantioselective I/204
Taxol I/457, II/328, II/349
– C13 side chain II/317, II/324
TBHP oxidation II/257, II/479, II/485
(t-BuO)$_4$Ti I/502
Tebbe reagent I/428, I/431–I/433
Tebbe-Claisen strategy I/432
technetium II/373
Tedicyp I/218, I/278
telomerization I/21, II/403
– two-phase catalysis II/520
template effect I/449
TEMPO II/209, II/440, II/449, II/451, II/452
teniposide II/10
terpene I/489
terpenoids, chiral I/199
tetraalkylammonium salts I/274
tetraallyltin I/337
tetraarylbismuth(V) I/385
tetrahydrobenzazepine I/607
tetrahydroisoquinoline I/191
tetrahydrolipstatin II/77, II/83
tetrahydropyrans I/388, I/389, I/436
tetrahydropyridine I/68, I/344, I/345
tetrahydroquinolines I/344, II/160
– derivatives I/354, I/355

tetrakis(triphenyl-phosphine)
    palladium(0)   I/224, I/225, I/234
tetramethylammonium
    triacetoxyhydriborate   I/591
tetraphenylphosphonium salt   I/277
TF-505   II/70
TfOH   I/390
theonellamide F   II/79, II/80
thermolysis   II/202
thermomorphic behavior   II/528
thienamycin   II/11
thioanisole   I/357
thioesters   I/433, I/438
thioimidates,α,β-unsaturated   I/348
thiols   I/368
thiopenes   I/417, I/435
Thorpe-Ingold effect   I/106
three component coupling reaction   I/345–I/347, I/473
threo-β-hydroxy-α-amino acids   I/500
thujopsene   I/388
Ti(3$^+$)-salts   I/455
TiCl$_3$ I/450
TiCl$_3$/LiAlH$_4$ I/463
TiCl$_3$/Mg   I/451
TiCl$_4$ I/336, I/341, I/343, I/429, I/437, I/461
Ti-F bond   I/496, I/505
Ti-MCM-41   II/337
Ti(MgCl)$_2$(THF)$_x$ I/451
Ti(Net$_2$)$_4$ I/437
Ti(O-iPr)$_4$ I/437, I/501, I/504, I/505, I/508
Ti(OR*)$_4$ I/503
tin amide   I/232
tin hydride   I/311
Tischenko disproportionation   II/149
titanacycle formation   I/427
titanacyclobutanes   I/428
titanacyclobutenes   I/440
titanacyclopropane intermediate   I/168
titania, mesoporous   II/587
titania-silica aerogel, amorphous
    mesoporous   II/338
titanium   I/449, II/187, II/188, II/189
– complexes   II/121, II/187
– – chiral   I/491, I/511
– powder   I/455, I/464
– reagents, preparation   I/491, I/492
– titanium-mediated reactions   I/491–I/513
titanium carbenes
– precursors   I/427–I/429
– mediated reactions   I/427–I/444
titanium-catalyzed epoxidation   II/337–II/343

titanium dioxide   I/492
titanium enolates   I/501
titanium-graphite   I/452, I/453, I/461, I/463
titanium oxide   I/450
titanium silicate-1   II/337
titanium silsesquinoxanes   II/339
titanium tartrate
– catalyst, heterogeneous   II/343
– complex   II/481
titanium tetrachloride   I/491
titanium tetrafluoride   I/496
titanium tetra-iso-propoxide   I/491, I/494, I/497
titanocene   I/427, I/429
– bis-cyclopropyl   I/436
– bis(trimethylsilylmethyl)   I/436
– dibenzyl   I/436
– dichloride   I/428
– dimethyl   I/428, I/433–I/436
titanycyclobutanes   I/432
Tm   I/345
TMEDA   I/430, I/438
TMSCl   I/455, I/555
TMSCN   I/375
TMSOTf   I/343, I/382, I/555
α-tocopherol   II/11, II/64, II/70
Tolpropamine   I/33
toluene-Cr(CO)$_3$ I/605
Tomioka   I/557, I/560
TPPTS   I/219, I/275
TPSH   I/592
traceless linkers   I/611
transalkylation, amines   II/498
transannular coupling   I/459
trans-cyclohexane-1,2-diamine *bis*-trifluor-methylsulfonamide   I/504
trans-cyclohexane-1,2-diamine *bis*-trifluor-methylsulfonamido-titanium   I/504
trans-effect   I/322
transfer carbonylation   I/629, I/630
transferhydrogenations   II/145–II/163
– catalysts   II/152–II/154
– general background   II/145, II/146
– hydrogen donors/promotors   II/152
– ligands   II/154, II/155
– mechanism   II/146–II/152
– miscellaneous transfer   II/161–II/163
– substrates   II/155–II/161
transition metal complex   I/4
transition metal-arene complexes   I/601
transmetalation   I/212, I/232, I/477, I/519, I/521–525

Subject Index

transmission electron microscop (TEM) II/591
trans-verbenol II/576
TRAP ligand I/315
β,β-trehalose II/515
trehalose dicorynomycolate II/93
tri(acetonitrile)iron complex I/588
tri(*o*-tolyl)phosphine I/213, I/235
trialkanolamine ligands II/484
trialkyl aluminium compound I/507, I/558
triarylbismuth carbonates I/386
1,4,7-triazacyclononane II/227, II/351–II/353
tricarbonyl(η^4-cyclopentadienone)iron complexes I/589
tricarbonyliron fragment I/585
tricarbonyliron-cyclohexadiene complex I/588
tricarbonyliron-diene complex I/588
trichlorosilane II/174
tricyclohexylphosphine I/129, I/214, I/216, I/224, I/239
triene I/325
triethyl orthoformate I/61
trifluoroacetic acid I/485
trifluoroketo ester II/75
trifluoropropene I/137
1,1,1-trihalides I/472–I/474
1,4,7-trimethyl-1,4,7-triazacyclononane
trimethylaluminium I/439
trimethylamine *N*-oxide I/588, I/621
trimethylenemethane (TMM), cycloaddition I/312
triorganozincates I/557
tri-*o*-tolylphosphine I/231
trioxane I/336
trioxoimidoosmium(VIII) complex II/311
triphenylbismuthonium 2-oxoalkylides I/387
triphenylphosphine I/577
– polyether-substituted II/517
tris(cetylpyridinium) 12-tungstophosphate II/434
tris(hydroxymethyl)phosphine I/323
trisoxazolines II/262
tri-*tert*-butylphosphine I/216, I/239, I/277
tropolones I/387
Trost I/313, I/331, I/539
Trost-Tsuji reaction I/309, II/566
TS-1 II/500
tuneable acidity II/569
tungstate-catalyzed reaction II/500
tungsten I/404, I/443, II/372, II/373
turnover frequency (TOF) II/30
twin coronet iron porphyrin II/489

two-center catalysis, asymmetric I/371–I/377
two-electron reduction I/461
two-phase catalysis I/275, II/511–II/522
– alkylation II/517–II/520
– aqueous-organic systems II/512–II/520
– counter phase catalysis II/521
– coupling reaction II/517–II/520
– hydrocarboxylation II/520
– hydroformylation II/516, II/517
– hydrogenation, unsaturated substrates II/512–II/516
– inverse phase catalysis II/521, II/522
– supported aqueous phase catalyst II/520, II/521
– telomerization II/520
tyrosinase II/218

u

U-106305 I/541, I/542
Uemura I/386
UHP II/500
ultrafine powders II/587
ultrasound, ultrasonic II/583–II/594
– activation I/520
– irradiation II/82
uracil derivates I/525
urea/hydrogen peroxide II/363, II/468
ureidocarbonylation, palladium-catalyzed I/142
urethane I/597

v

van Leeuwen I/60
vanadium II/371, II/372
vanadium complex
– asymmetric II/486
– zeolite-encapsulated II/228
Vannusal A I/489
Vasca complex II/4
Venturello compound II/418
Venturello-Ishii catalytic systems II/418
venyl acetate I/43
verbenone I/504, II/222
vicinal diamines II/333
vicinal diols II/427
vigabatrin I/9
vinyl arenes I/154
vinyl bromide I/527
vinyl carbenes I/440
vinyl chromium I/474
vinyl cyclopropanes I/309
vinyl epoxide I/309
vinyl ethers I/343, I/345, I/509

vinyl fluoride I/45, I/325
vinyl iodides I/473
vinyl radicals I/485
vinyl silane I/119, I/436, I/473
vinyl sulfide I/345
vinylation I/5, I/171
vinylcyclopentadienes I/419
vinylcyclopropane I/410
vinylcyclopropanecarboxylate I/592
vinylic halide I/476
2-vinylpyridine I/184, I/186, I/188
vinylstannane I/570
vinylzinc nucleophiles I/531–I/533
vitamin I/408
– B_{12} I/477, II/591
– B_{12} 12
– D I/5
– E II/11, II/520
– K_3 II/230
Vollhardt I/172
VPI-5, microporous II/223

w

Wacker process I/20, I/185, I/309, II/207, II/379–II/387, II/428, II/441, II/536, II/555
Wacker-Hoechst acetaldehyde process II/379, II/380, II/428
Wacker-Tsuji reaction II/381–II/383
Wakamatsu I/100, I/133
Wakatsuki I/172
Walsh I/504 I/530
water I/113, I/184
– supercritical I/184
water gas shift reaction II/517
Weinreb I/345
Wells-Dawson structure II/417
Wentilactone B I/489
Wilkinson complex $(Ph_3P)_3RhCl$ I/19, II/5, II/6
Wilkinson's catalyst I/105, I/176, II/513, II/548, II/621
Wittig I/472
Wittig olefination I/3, I/386, I/430
Wittig reagent I/88, I/89, I/431, I/433
Wittig-Horner reaction I/593
Wurtz homocoupling I/474

x

XANTPHOS I/59, I/72, I/238

y

$Yb(OTf)_3$ I/489
ylide formation I/163
ylides I/386

ynamines I/589
yne-aldehydes I/628
ynol ethers I/589
ytterbium catalyst I/353
ytterbium triflate I/336–I/338, I/353, I/355
yttrium II/179, II/180
Yus I/505

z

Z-α-haloacrylates I/474
Z-2-chloralk-2-en-1-ols I/473
zeolite Y II/363
zeolites II/230, II/259, II/268, II/339, II/390, II/404, II/588
Zhang I/530
Ziegler I/563
– catalyst I/177
– Co-Fe catalyst II/141
Ziegler polymerization catalyst I/461
Ziegler-Natta system II/8
zinc I/429
– insertion in C-X-bonds I/520, 521
– zinc-mediated reactions I/519–I/I/547
zinc I/451, I/453
zinc acetylide I/535
zinc bromide I/568
zinc enolates I/536, I/544, I/545, I/560, I/564
– reactions I/543–547
zinc ester enolates I/476
zinc prolinol complex I/539
zinc-mediated reactions
– aldol reactions I/539, I/540
– asymmetric conjugate addition I/536–I/539
– carbometalation I/529
– cross-coupling reaction I/525–I/529
– cyclopropanation I/541–I/543
– organozinc addition to C=X I/530–I/535
– preparation/coupling reactions I/519, I/520
– transmetalation I/521–I/525
– zinc enolates, reactions I/543–I/547
– zinc insertion into C-X bonds I/520, I/521
zindoxifene I/465
zircanocene I/257–I/267
zirconium II/370
Zn I/151, I/437
$Zn(OTf)_2$ I/534, I/546
$ZnBr_2$ I/567
ZnI_2 I/382
Zr-Mg ligand exchange I/263
ZrO_2, Cr salt-doped II/97